Manufacturing
Engineer's
Reference
Book

Manufacturing Engineer's Reference Book

Edited by
Dal Koshal
with specialist contributors

Butterworth-Heinemann Ltd
Linacre House, Jordan Hill, Oxford OX2 8DP

A member of the Reed Elsevier group

OXFORD   LONDON   BOSTON
MUNICH   NEW DELHI   SINGAPORE   SYDNEY
TOKYO   TORONTO   WELLINGTON

First published 1993

© Butterworth-Heinemann 1993

British Library Cataloguing in Publication Data
A catalogue record for this book is available from the
British Library

ISBN 0 7506 1154 5

Library of Congress Cataloguing in Publication Data
A catalogue record for this book is available from the
Library of Congress

Every effort has been made to trace the holders of copyright
material. However, if any omissions have been made, the
authors will be pleased to rectify them in future editions
of the book.

Printed and bound in Great Britain by Bath Press Ltd, Avon

Contents

Preface

No reference book on manufacturing engineering will ever be an unqualified success. It cannot be a panacea for those seeking cures for the ills of poor management. At best it provides an insight into the multifarious techniques and processes which combine to enable goods to be produced competitively in terms of cost, quality, reliability and delivery.

Within manufacturing organisations it is to be hoped that specialist knowledge exists in all the relevant fields, extending beyond the confines of the chapters in a reference book. If it does not exist, the companies must ensure that they acquire it by providing suitable training or by employing experts, if they hope to survive.

What purpose is served, then, by this sort of reference book? The editor believes that, apart from those copies which are destined to gather dust in executives' bookcases to help provide an ambience of professional respectability, the book will be useful for top and middle managers who feel the need to widen their perspectives. With this in mind it has been written in compartmentalised form, each section being free-standing and capable of being understood as an introductory text. It will also be of use to engineering students as an adjunct to the more specific texts used in support of their lectures.

Some reference books are primarily compendiums of tabulated data, essential to the task of technical quantification. This volume is not such a compendium: it is mainly a qualitative approach to knowledge gathering from which subsequent quantification may be attempted. It covers those aspects of manufacture which are essential for designing new production systems or for managing exisiting factories. These include materials selection, manufacturing and fabrication processes, quality control, and the use of computers for the control of processes and production management.

The editor wishes to thank the specialist authors who contributed their expertise for their forbearance during the various stages of preparation, together with colleagues, past and present, at the University of Brighton for their suggestions. Special thanks are due to Mr Don Richardson for his helpful suggestions on various chapters.

I would particularly like to thank my wife and family for their continued support during this project.

Dr Dalbir Koshal
University of Brighton

List of Contributors

T Z Blazynski PhD, BSc(Eng), CEng, MIMechE
Formerly Reader in Applied Plasticity
Department of Mechanical Engineering
University of Leeds

John Brydson
Former Head of Department of Physical Sciences and
    Technology
Polytechnic of North London

John L Burbidge OBE, DSc, HonFIProdE, FIMechE,
    FBIM
Visiting Professor in Manufacturing Systems
Cranfield Institute of Technology

T C Buttery BSc, PhD, CEng, MIEE, AMIM
Senior Research Co-ordinator at CIMTEX
School of Engineering and Manufacture
De Montfort University, Leicester

Harry L Cather CEng, MIEE, MBIM, MBA, MSc
Senior Lecturer
Department of Mechanical and Manufacturing Engineering
University of Brighton

E N Corlett DSc, PhD, MEng, BSc(Eng), FEng, FIMechE,
    FIEE, HonFErgS, CPsychol
Emeritus Professor
Institute of Occupational Ergonomics
University of Nottingham
Nottingham NG7 2RD

Roy D Cullum FIED
Editor, *Materials and Manufacture*
Publication Services, Worthing

Brad D Etter PhD
Assistant Professor, Bioengineering Program
Texas A&M University, USA

William T File CEng, MIMechE
William T File & Associates
Consultants in Maintenance Engineering
Aylesbury, Bucks

C J Fraser BSc, PhD, CEng, FIMechE, MInstPet
Reader in Mechanical Engineering
Department of Mechanical Engineering
Dundee Institute of Technology

Jan Glownia PhD, Eng
Academy of Mining and Metallurgy
Institute of Foundry Technology and Mechanisation
Cracow, Poland

R Goss BEng, CEng, MIMechE
Senior Application System Engineer (Industrial Product
    Division)
Loctite (UK) Ltd

E N Gregory CEng, FIM, FWeldI
Head of Advisory Section
The Welding Institute
Cambridge

John Hunt CEng, MIEE
Director
Custom Engineering and Networks
Bookham, Surrey

D Koshal MSc, PhD, CEng, FIMechE, FIEE
Principal Lecturer, Manufacturing Engineering
Department of Mechanical and Manufacturing Engineering
University of Brighton

Gordon M Mair BSc(Hons), DMS, CEng, MIEE. MBIM
Lecturer
Department of Design, Manufacture and Engineering
    Management (MEM Division)
University of Strathclyde

Gerald E Miller PhD
Professor and Chairman, Bioengineering Program
Texas A&M University, USA

John S Milne BSc(Hons), CEng, FIMechE
Senior Lecturer in Mechanical Engineering
Department of Mechanical Engineering
Dundee Institute of Technology

D B Richardson MPhil, DIC, CEng, FIMechE, FIEE
Formerly Principal Lecturer in Manufacturing Engineering
Department of Mechanical and Manufacturing Engineering
University of Brighton

Leslie M Wyatt MA(Cantab), FMI, CEng
Independent Consultant and Technical Author

1

Materials Properties and Selection

L Wyatt

Contents

1.1 Engineering properties of materials

1.1.1 Elastic properties

Elastic or Young's modulus, E (units GPa) The stress required to produce unit strain in the same direction, i, in the absence of restraint in the orthogonal directions:

$$E_i = \sigma_i \varepsilon_i^{-1} \tag{1.1}$$

where σ is the stress and ε the strain which it produces. A standard testing method is described in ASTM E231.

Shear modulus, G (units GPa) The shear stress required to produce unit angular rotation of a line perpendicular to the plane of shear:

$$G = T\phi^{-1} \tag{1.2}$$

where T is the shear stress and ϕ is the angular rotation (in radians).

Bulk modulus, K (units GPa) The hydrostatic pressure p required to effect unit change in volume V:

$$K = pV(\Delta V)^{-1} \tag{1.3}$$

Poisson's ratio ν (dimensionless) The ratio of the strain in a direction orthogonal to the direction of stress to the strain in the direction of stress:

$$\nu = \varepsilon_{j,k}\varepsilon_i^{-1} \tag{1.4}$$

These four basic elastic properties apply to homogeneous and isotropic materials and are related by the equations:

$$E = 3K(1 - 2\nu) \tag{1.5}$$
$$= 2G(1 + \nu) \tag{1.6}$$

In the case of a material which has anisotropic elastic properties the terms used may have different meanings, and stresses and strains should be related using tensor analysis.

1.1.2 Tensile testing parameters

When considering the properties obtained from the tensile test it should be realised that the results are always reported as though the load was applied to the initial cross-section A_0 of the test piece. Any reduction of this cross-section is ignored. The test subjects a sample of material of circular or rectangular cross-section, of a specific gauge length and equipped with end pieces of larger section which taper smoothly to the gauge length.

When subjected to uniaxial tension beyond the limit of proportionality the material within the gauge length elongates plastically, contracts uniformly or locally transversely and work hardens. The stress σ in the material increases but, because of the decrease in the cross-sectional area A, the stress S calculated from the load and the original cross-sectional area A_0 increases more slowly, attains a maximum value S_u and (usually) declines before the specimen breaks.

Limit of proportionality The stress at which elastic behaviour of a material is replaced by a combination of elastic and plastic behaviour, normally expressed as either the *yield stress*, S_y (units MPa), or the *proof stress*, $S_{0.5\%}$, $S_{0.2\%}$, $S_{0.1\%}$ (units MPa), where the departure from elastic behaviour is indicated by the suffix and S (or, in some codes P) is the load:

$$S = \frac{P \left(\text{Yield or strain offset}_{\substack{0.005 \\ 0.002 \\ 0.001}} \right)}{A_0}$$

Ultimate tensile stress S_u (units MPa) The maximum load at which a ductile material fractures in the tensile test divided by the original cross-sectional area. S_u is not to be confused with σ_u which is the true stress:

$$\sigma_u = S_u A_0 A_u^{-1} \tag{1.8}$$

where A_u is the cross-sectional area at the time of failure.

S_u depends on the dimensions of the specimen (the gauge length is normally $0.565\sqrt{A_0}$ but it may be 50 mm or some other value) and the rate of application of stress. Both these parameters should be recorded.

Fatigue endurance Related to S_u rather than S_y. The difference between S_u and S_y is a measure of the safety margin against accidental overload.

Most modern design codes base the permissible stress in a material on a factor (say 66%) of S_y. Some other codes use a factor of S_u as a design criterion. This is cost effective and safe when using a ductile material such as mild steel.

Tensile ductility Reported either as *elongation*, e (units %)

$$e = \frac{\delta}{L} = \frac{L - L_0}{L_0} \times 100 \tag{1.9}$$

where δ is the extension to fracture, or as *reduction in area, A_R* (units %)

$$A_R = \frac{A_0 - A_u}{A_0} \times 100 \tag{1.10}$$

Ductility is the property that confers tolerance to flaws to a material and is also an indication of material quality and correct heat treatment. Standards usually specify a minimum ductility.

Standard procedures for tensile testing are given in BS 18, ASTM E8, ASTM E345 and ASTM B557.

Flexural strength, S (units MPa) The calculated maximum stress on the tensile side of a beam which fails when stressed in bending. It is used to measure the strengths of materials such as cast iron and ceramics which are too brittle to be tested using the standard tensile test. A beam stressed in three-point loading has the maximum stress applied only on one line on the surface. Multiple testing is required to produce results which can be used in design and much higher safety factors (see Section 1.6.10) are required than are used for ductile materials tested using the standard tensile test.

A standard testing method is described in ASTM C580.

1.1.3 Hardness

Hardness is the resistance of a material to permanent deformation by indentation or scratching. It is not a simple intrinsic property of a material but a complex response to a test. Vickers, Brinell and Knoop hardnesses compare the load and the area of the impression produced by an indenter, Rockwell hardness compares the load and the depth of the impression, Shore hardness is a measure of the rebound of an indenter, and Moh hardness measures the ability of one material to scratch another.

Vickers hardness, HV (the dimensions are strictly those of force per unit area, but in practice Vickers and Brinell hardnesses are comparative numbers), is the quotient obtained by dividing the load F (kgf) by the sloping area of the indentation left in the surface of the material (in mm^2) by a 136° pyramidal diamond indenter:

$$HV = \frac{2F \sin (136/2)}{d^2} = 1.854Fd^{-2} \tag{1.11}$$

where d is the diagonal of the indentation.

Hardness is a measure of the wear resistance of a material. Used on metals the Brinell hardness value of a medium carbon steel is directly related to the ultimate tensile stress, whilst the Vickers hardness is related to the proof stress. Vickers, Brinell and Rockwell hardnesses can be used to ensure that heat treatment has been carried out correctly.

Hardness testing of ceramics is carried out with very light loads to avoid failure of the material.

Standards for hardness testing are:

Vickers	BS 427, ASTM E92
Brinell	BS 240, ASTM E10
Rockwell	BS 891, ASTM E18
Schlerscope	ASTM 4448

1.1.4 Fracture toughness and impact testing

1.1.4.1 Fracture toughness testing

Plane strain fracture toughness, K_{Ic} (units $N\text{-}m^{-3/2}$) The limiting stress intensity required to cause crack extension in plane strain at the tip of a crack when the stress is transverse to the crack. K_{2c} and K_{3c} are parameters corresponding to stresses in the plane of the crack.

Standard testing methods are given in BS 5447 and ASTM E399.

Fracture toughness is sometimes denoted by K_{1c} or K_{Ic}.

Elastic–plastic fracture toughness, J_{Ic} The limiting value of the J integral (which is a line or surface integral used to characterise the fracture toughness of a material having appreciable plasticity before fracture) required to initiate a crack in tension from a pre-existing crack.

Stress intensity to initiate stress corrosion, K_{Iscc} (units $N\text{-}m^{-3/2}$) The limiting stress intensity required to initiate propagation of a crack in a specific environment at a specific temperature.

1.1.4.2 Impact testing

In contradiction to fracture toughness testing which quantifies a material property, Izod cantilever and Charpy beam type impact test results are a function of the method of testing. In particular, a machined rather than a fatigue-propagated notch is used. Results are expressed as the energy, J (in joules), required to break the cross-sectional area behind the notch.

Testing a number of specimens of body-centred metals, ceramics and polymers over a range of temperature will reveal a transition temperature below which brittle behaviour is observed. This is reported either as the fracture appearance transition temperature (f.a.t.t. in °C) at which half of the fracture surface is fibrous and half is crystalline, or as the fracture energy transition temperature (in °C) at the inflection in the energy curve. This is a criterion of use for assessing material composition, treatment and behaviour.

Standards for impact testing are BS 131, ASTM E23, ASTM E812 and ASTM E602 (sharp notch tension testing).

1.1.5 Fatigue

S–N curve The graphical relationship between the stress S and the number of cycles N required to cause failure of a material in a fatigue test. This depends on the mean stress, frequency and shape of the stress cycle, the temperature and the environment, all of which should be specified. Note this applies to high cycle fatigue.

High strain fatigue is strain, not stress, related and the plastic strain per cycle resulting in failure ε_p is inversely proportional to $N^{1/2}$ for almost all engineering materials.

Fatigue endurance limit, σ_e (units MPa) The maximum stress below which a material is presumed to be able to endure an infinite number of cycles. This applies only to certain specific engineering materials such as, for example, steel and titanium.

Fatigue limit, σ_{10} (units MPa) The maximum stress below which a material is presumed to be able to endure a specific number of cycles; this is usually of the order of 10^7 to 10^8, but may be lower for specific applications.

The fatigue endurance limit and the fatigue limit are both statistical quantities and depend on the same parameters as the *S–N* curve (see above).

Standard methods for fatigue testing are BS 3518, ASTM E513, ASTM E912, ASTM E206, ASTM E742, ASTM E466, ASTM E606, ASTM 4 468 and ASTM E739. Other ASTM standards are given in *ASTM Standards* Vol. 03;01.

Fatigue life for p% survival (units MPa) The maximum stress below which not less than $p\%$ of tested specimens will survive.

Fatigue notch factor, K_f (dimensionless) Ratio of the fatigue strength of a notched to that of an unnotched specimen.

Fatigue notch sensitivity, g (dimensionless)

$$g = \frac{(K_f - 1)}{(K_t - 1)} \tag{1.12}$$

where K_t is the stress concentration factor.

When g approaches 1 a material is fully sensitive; when g approaches 0 a material is insensitive.

1.1.6 Creep and stress rupture

Creep range The temperature range, usually above half the melting-point temperature (in kelvin), at which the design stress computed from creep or stress rupture is lower than that calculated from yield or 0.2% proof.

Stress to rupture, σ_R (units MPa) The tensile stress at which a material will fail if held at a specific temperature for a specific time, depending on the type of application.

Stress to a certain creep strain, σ_ε (units MPa) The tensile stress at which a material will creep to a specific strain ε (ignoring the initial strain on loading) if held at a specific temperature for a specific time. For a specific material σ_ε and σ_R are related.

Creep rupture elongation (units %) The percentage of the original length by which a creep rupture specimen extends before failure.

Larson–Miller parameter, P A parameter used to extrapolate the results of creep rupture tests carried out at relatively short times to longer times. The rate equation is:

$$P = T(\log t_R + C) \tag{1.13}$$

where T is the absolute temperature and C is an empirically determined constant. Other rate equations have been derived by Sherby–Dorn and Manson–Haferd.

Standards for creep and stress rupture testing are BS 5447 and ASTM E1329 for metals and BS 4618 for plastics.

1.1.7 Thermal properties

Specific heat per unit mass, C_p (units $J\ kg^{-1}\ K^{-1}$) The rate of change of heat content of 1 kg of the material with temperature. Specific heats are often quoted in $J\ g^{-1}\ K^{-1}$ or in compilations of thermodynamic data as cal $mol^{-1}\ K^{-1}$. They may also be quoted as mean specific heats over a range of temperature, usually 25°C to a specific elevated temperature.

Specific heat per unit volume, C_v (units $J\ m^{-3}\ K^{-1}$) The specific heat of a gas at constant volume C_v does not include the work required to expand the gas and is therefore lower than C_p.

Thermal expansion The *linear thermal expansion* α (units K^{-1}) is the fractional increase in length l per degree rise in temperature at a specific temperature T:

$$\alpha(T) = l^{-1}\frac{dl}{dT} \tag{1.14}$$

More commonly, a *mean expansion* between two temperatures

$$\bar{\alpha} = l_o^{-1}\frac{\Delta l}{\Delta T} \tag{1.15}$$

where Δl is the change in length from l_o at temperature T_o when the temperature is changed by $\Delta T = T - T_o$, is quoted. In data compilations T_o is often 25°C.

In anisotropic materials (single crystals or materials having a preferred orientation), the thermal expansion coefficient may differ between each of the three orthogonal directions x_i, x_j and x_k.

Thermal conductivity, λ ($W\ m^{-1}\ K^{-1}$) The heat flow per unit area generated by unit temperature gradient:

$$\lambda = A^{-1}\frac{dQ}{dt}\frac{dl}{dT}$$

where dQ/dt is the rate of heat flow across area A and dT/dl is the temperature gradient. λ is normally a function of temperature and, in anisotropic materials, of direction.

Thermal diffusivity, D (units $m^2\ s^{-1}$) A measure of how fast a heat pulse is transmitted through a solid:

$$D = \frac{\lambda}{\rho C_p}$$

where λ is the thermal conductivity, ρ is the density and C_p is the specific heat.

Thermal diffusivity varies with temperature but can be measured more quickly and accurately than thermal conductivity.

1.1.8 Electrical properties

Volume resistivity, ρ (units Ω-m) The resistance (in ohms) of 1 m^3 of a material:

$$R = \rho l A^{-1} \tag{1.16}$$

where R is the resistance of a body, l is its length and A is its uniform cross-section.

Dielectric breakdown (no standard symbol: units $K\ V\ mm^{-1}$ or $K\ V$) Measured according to IEC 672, BS 1598: 1964, ASTM D116 or DIN 40685.

Relative permittivity (dimensionless) The ratio of the charge storage capacity of a material in an electric field which results from realignment of the crystal structure compared with the charge storage capacity of empty space.

Permittivity (units $A\ s\ V^{-1}\ m^{-1}$) Given by:

$(As)/(Vm)$

where A is the current in amperes, s is the time in seconds, V is the electric protential in volts and m is the length in metres.

Dielectric loss, tan δ (dimensionless) The phase angle introduced by the time taken for polarisation to occur on application of a field. Dielectric loss is frequency and (usually) temperature dependent.

1.1.9 Optical properties

Spectral absorption coefficient, K (units mm^{-1}) The log of the ratio of the incident to the transmitted light intensity through unit thickness:

$$K = (\log I_o - \log I)x^{-1} \tag{1.17}$$

where I_o is the incident intensity, I is the transmitted intensity and x is the thickness (in mm). K varies according to the wavelength of the incident light.

Refractive index, μ (dimensionless) The ratio of the velocity of light in vacuo to that in the medium:

$$\begin{aligned}\mu &= V_{vacuo}V^{-1}{}_{medium}\\ &= \sin i\ \sin^{-1} r\end{aligned} \tag{1.18}$$

where i and r are the incident and refracted angles of the beam to the surface.

1.2 The principles underlying materials selection

1.2.1 Introduction

The need to select a material may arise from a number of circumstances including the following.

(1) An entirely new component is to be developed to perform functions not previously visualised.
(2) A component is required to perform an increased duty which renders the performance of the material previously used unsatisfactory.
(3) The incidence of failure in a material previously specified is too high, or occurs at too early a stage in the life of the component.
(4) Some material shortcoming not strictly related to operational performance has become apparent. A material which was acceptable initially may become unsatisfactory because:
   (a) it has become so expensive, relatively or absolutely, that the equipment, of which the component is a part, can no longer fulfil an economic function;
   (b) it is no longer available locally or globally (or might become unavailable in the event of an emergency); or
   (c) it is no longer acceptable on grounds of health, safety, aesthetics or public sentiment.

Examples of materials which have been developed in answer to the above-listed circumstances are as follows.

(1) The 'magnox' can for the first-generation gas-cooled power reactor.
(2) Superalloy blades of progressively increasing creep resistance culminating (so far) in the directionally solidified castings now used.
(3) Notch ductile aluminium killed steels to replace the materials which failed by brittle fracture in the 'liberty' ships.
(4) (a) Steel-cored aluminium instead of copper conductors for overhead power lines;
    (b) nickel-based superalloys for military jet aircraft after it was realised that the source of cobalt in cobalt-based superalloys were situated in doubtful African or Iron Curtain countries; and
    (c) ceramic fibres to replace asbestos as a binder for heat insulation because of the hazard of 'asbestosis'.

All these examples of material choice were developed by means of the techniques described later in this section. The materials selected have performed entirely satisfactorily, and in those cases where operational parameters were not the cause of replacement the substitute material has in fact performed better than the original.

1.2.2   Techniques of materials selection

There are at least three different techniques by which the optimum material for use in a specific component may be selected.

(1) The *classical procedure* using functional analysis and property specification.
(2) The *imitative procedure* which consists of finding out what material has been used for a similar component.
(3) The *comparative procedure* which consists of postulating that the component be made from some cheap and well understood engineering material, assessing in what ways such a material's performance would be inadequate and from this arriving progressively at the right material.

The classical procedure is the only one that is universally applicable and its use is essential, even when procedure (2) or (3) is followed, to check the findings of the functional analysis and property specification. By itself, however, the classical procedure is expensive and time consuming and requires a considerable amount of prototype testing to ensure that no critical requirement or essential property has been overlooked.

The imitative and comparative procedures, where applicable, provide invaluable shortcuts, save a vast amount of time and money and will help to ensure that no essential parameter has been overlooked. The materials engineer will be wise to employ all three techniques in parallel wherever practicable.

1.2.3   Preliminary examination of design

Whichever of the above procedures is employed, it is essential to commence with an analysis of the function of a component, a critical examination of the design and to establish the property requirements of the material under consideration.

Design affects the materials-selection procedure at all stages. A component may fulfil its function in more than one way due to different designs which result in different materials-property requirements and hence different optimum materials and different manufacturing routes. For example, a box with a hinged lid may be made from two pieces of thin metal sheet and a pin or from one piece of polypropylene.

The effect of design on a manufacturing process is particularly important when considering a materials change in an existing product, for example from metal to plastic or ceramic.

Design and materials selection constitute an iterative process: design affects the optimum material, which in turn affects the optimum design.

1.2.4   The classical procedure

1.2.4.1   Functional analysis

Functional analysis is a formal way of specifying material properties, starting from the function of a component. This involves:

(1) specification of the functions of a component;
(2) specification of the requirements of a component; and
(3) specification of the requirements of the material properties.

1.2.4.2   Function

The overall function should be specified as broadly as possible to allow the greatest number of options in design. Where there are several functions all must be specified. This latter requirement is essential even when the choice of material has been necessitated by the failure of a material to perform one specific function, because a change in a material to make it capable of fulfilling one function may make it incapable of fulfilling another. For example, using a higher tensile steel to carry an increased load may result in brittle fracture under shock.

1.2.4.3   Component requirements

When the functions have been established the component requirements can be identified. For example, the one-piece box mentioned above must be capable of being opened and closed an indefinite number of times. In specifying component requirements it is important to remember that it must be possible to produce the article in the required form, and that the component must withstand the environment in which it is operating, at least for its design life.

1.2.4.4   Materials property requirements

From the component requirements the materials-property requirements can be established. The material for the one-piece box must have an almost infinite resistance to high strain fatigue in air at room temperature. This is obtainable from a polypropylene component manufactured in a specific way.

The property requirements established by the functional analysis may be quantitative or qualitative. For example, the material for an automobile exhaust *must* be sufficiently strong and rigid to withstand weight and gas pressure forces. Quantitative requirements must be established by analysis of the design and operating conditions. In comparison, the requirement to resist corrosion and oxidation is qualitative. Property requirements may also be classified as essential and desirable. The strength requirement in the material for the exhaust is essential. Environmental resistance is often sacrificed to minimise initial cost (even when, as in this example, a more resistant material may be economically superior over the total life of an automobile).

1.2.4.5   Materials requirement check-list

The next stage is the formulation of a 'materials requirement check-list'. Some properties which will feature in this check-list are listed in Section 1.1. The reader should not be discouraged by the length and complexity of this list. It will in many cases become evident that whole ranges of properties (and materials) may safely be ignored at first glance. For example, if the component is required to transmit or refract light the choice of material is immediately limited to a glass, mineral or polymer and the design and property specification is thus also restricted. If electrical conductivity is significant, choice is limited to conducting metals, resistive or semiconducting materials or insulators.

1.2.4.6 Important characteristics

The important characteristics requiring consideration for many engineering components are:

(1) mechanical properties—stiffness, strength and ductility;
(2) physical properties—thermal, electrical, magnetic and optical properties;
(3) environmental resistance and wear, including applicability of corrosion protection;
(4) capacity for fabrication; and
(5) cost, which includes material, manufacturing, operating and replacement costs.

1.2.4.7 Mechanical properties

Resistance to manufacturing and in-service loads is a requirement of all products. The material must not buckle or break when the component comes under load. A product must also have an economic life in fatigue or under creep conditions. Where a number of materials meet the minimum strength and stiffness requirements, a preliminary short-list can be made on the basis of cost per unit strength or unit stiffness (or in the case of transport applications, strength per unit weight).

1.2.4.8 Physical properties

Physical properties such as, for example, specific gravity, are important for most applications. As noted above, for some applications optical or electrical properties may be paramount.

1.2.4.9 Environmental resistance (corrosion)

Resistance is a property whose universal importance has been obscured by the circumstance that it has been inherent in the choice of materials for most common applications.

Corrosion-resistance requirements vary from the absolute, where even a trace of contamination in a fine chemical, food or cosmetic is unacceptable, to the barely adequate where the cheapest material whose integrity will survive the minimum economic life should be chosen.

When assessing corrosion resistance attention should be paid not only to the rate of general corrosion but also to the possibility of localised corrosion which, as described in Section 1.8, may destroy component integrity without significant dimensional changes.

Corrosion mechanisms may cause the disintegration or deterioration of metals, polymers, ceramics, glasses and minerals.

1.2.4.10 Wear resistance

Wear is the product of the relative movement between one component and another or its environment. Prevention of wear depends principally on design and operation, but can be minimised or eliminated by the correct choice of material, material pair, or coatings.

1.2.4.11 Manufacture and cost

Manufacturing routes are selected on the basis of lowest total cost to produce the desired performance. In the past, performance requirements have favoured certain processes such as, for example, forging instead of casting, but more recently attention to quality improvement techniques in casting has levelled up in-service properties and cost is emerging as the deciding factor.

It is difficult to assess the relative total costs of different material/manufacturing route combinations at the early stage of a design and, wherever possible, finalising precise geometries should be delayed until possible materials and manufacturing routes have been identified, otherwise there will be an avoidable cost penalty.

The cost of a component includes:

(1) in-position costs, which comprise material cost (influenced by quality and quantity), manufacturing cost, quality-control cost and administration cost; and
(2) lifetime costs, which comprise servicing, maintenance, warranty, outage and replacement costs.

Costs which accrue at different periods must be discounted to a common date. Differences in discounting rates between different countries or organisations can lead to the selection of different materials for applications which are, in all other respects, identical.

Comprehensive knowledge of the application is important in assessing the relative importance of cost and performance. Cost is paramount in the case of a widely marketed consumer item where small differences in reliability and life have little influence on saleability, but performance is paramount for certain sporting or military applications. There may, for example, be no advantage to be gained by incurring additional expense to prolong the life of a car exhaust system from 5 to 7 years when the purchaser intends to replace the car after 2 years. In cases such as this the sales department must always be consulted before the final material choice is made. On the other hand the material from which a racing car spring is made must have the maximum possible specific rigidity, regardless of cost. The potential rewards for employing the optimum material far outweight any cost saving which might be obtained by choosing the second-best material.

1.2.4.12 Material selection

When the properties of candidate materials have been ascertained (by the procedures discussed later) a short-list should be established. If it is immediately obvious that one material is outstandingly superior the choice is straightforward. Often there is one property requirement that outweighs all the others. When this is the case the choice is simplified. There may, however, be a number of possible materials, or none may meet all requirements.

A number of procedures have been proposed for eliminating all but one of a number of possible materials. These include an advantage/limitation table, an elimination grid,[1] and ranking methods for properties (and the number of properties) that meet the requirements.

Local factors—using a material which is familiar locally, using a material which has a margin in one specific property that may be of value in a future marque of component, or using a material that is suitable for a locally available fabricating or machining technique—will often influence the final choice.

When no material meets the necessary requirements, a careful re-examination may reveal that a change in design, environment or operating conditions will enable satisfactory performance at minimum extra cost. As a last resort it may be possible to arrange for easy replacement after a fixed time, and to hold a supply of spares.

1.2.5 Drawbacks of the classical procedure

Application of the procedures outlined above guarantees success if followed logically and completely and if design, operating and material parameters are thoroughly understood. This is seldom the case in practice. Designs cannot always be evaluated precisely, material properties are seldom specified fully, and it is impossible to predict exactly what an operator will do.

In most cases the classical method requires a considerable amount of mechanical-property evaluation, possibly materials and process development and a substantial programme of prototype testing, before satisfactory performance can be guaranteed. Time may not be available to undertake this. The easiest way to reduce the time required is to use the imitative procedure.

1.2.6 The imitative procedure

The imitative procedure involves finding out what material has been used for the same component (or a component as similar as possible) and using this, an improved material, or a material modified for the difference in conditions. Successful implementation of this procedure not only verifies design and reduces the time for materials-property evaluation but also very substantially reduces prototype testing because the most likely causes of failure have already been experienced and cured. The problem is to ensure that the information obtained is accurate, comprehensive and fully understood.

Even within an organisation, operators' reports are not completely reliable. A report of satisfactory performance may merely mean that the operator knows when the component is about to fail so that he can replace it without extra outage. The operator may have found out how to handle this component and a similar component may fail disastrously in the hands of another operator. These difficulties are compounded when information is obtained from an outside source, whether rival or friendly. Informants do not mean to mislead. The information they withhold is usually information that they cannot imagine the recipient does not already possess.

The ability to obtain information when it is required depends on appropriate organisation. Ideally, there should be a materials engineer who combines knowledge of all the materials and requirements of the organisation with an acquaintanceship or, ideally, friendship with all similar persons throughout the world. The right person is, when presented with a problem, able to contact the person who already has experience of the matter wherever she or he may be and obtain the benefit of that experience. His or her knowledge of the other organisation is comprehensive enough to enable him or her to assess the effect of different procedures between the two organisations.

The chemical industry (as described by Dr Edeleanu[2]) operates a world-wide information system with personnel of this type and has found that information on what can be done and how to do it may most efficiently and quickly be obtained in this way.

1.2.7 The comparative procedure

The comparative procedure for materials selection involves selecting a cheap, tolerant and well-understood material and investigating to what extent its properties fall short of what is required for the component to operate satisfactorily.

A typical example, and one for which this procedure is extremely suitable, is the specification of a material for chemical process plant.[3] A scheme design is produced using carbon steel which is cheap, readily produced, easily fabricated, ductile and, therefore, tolerant of flaws and geometrical irregularities and corrodes uniformly at a predictable rate. If carbon steel is shown not to be satisfactory the unsatisfactory property or properties can be modified. The necessary change may impair other properties but will do so in a predictable way. Thus:

(1) Improved corrosion resistance may be obtained by the use of a steel with a higher chromium, and possibly a higher nickel, content. This will increase cost and probably also delivery time, render design and fabrication more sensitive and may enhance sensitivity to localised corrosion.
(2) Improved strength may be obtained by the use of a steel with increased carbon and alloy content with drawbacks similar to those that apply in the case of the improved-corrosion-resistance material.
(3) A higher temperature of operation may require the use of a creep resisting steel, again with similar drawbacks to (1).
(4) Operating at a lower temperature may require a steel with guaranteed low-temperature properties or may, at the limit, require an aluminium alloy.

Evidently this procedure, with the exception of the case where a change is made to a completely different material, involves changes which are progressive, and whose effects can be foreseen. Therefore the chances of encountering some unexpected drawback are minimised and the requirement for component testing is minimised also.

1.2.8 Information sources

It has so far been assumed that staff charged with material selection have at their disposal a complete range of information on material properties. This may be the case when electronic databases[4] now being developed are perfected. In the meantime staff should have available for reference British, American and possibly German materials standards, and volumes such as the *ASM Metals Handbook*, the *Plastic Encyclopaedia* and as up to date a ceramic work as is available. In addition, the *Fulmer Materials Optimizer*, which shows properties of all types of engineering materials in the form of comparative diagrams, will prove an invaluable guide to materials selection.

When the field has been narrowed down to a few materials the material manufacturer should be consulted. Organisations such as steelmakers or polymer manufacturers possess more information on their products than has been published and also have experience in their application. They can provide valuable guidance on final selection, design and manufacture. Furthermore, it should be remembered that a standard steel obtained from one manufacturer may differ in some relevant characteristic from the same steel purchased from a competitor. A reputable manufacturer is aware of this and should, in addition to extolling advantages, warn of problems which will have to be overcome.

1.2.9 Computerisation of materials selection

Much effort is at present being deployed towards producing databases of material properties. Three recent international conferences have been devoted to this subject[4] and a directory of databases for materials is available.[5] These databases are not necessarily material-selection systems and much interest has been directed to providing systems which will undertake material selection using the classical functional-analysis and property-specification procedure.

It is not possible, legitimately, to computerise the imitative procedure because no organisation can be expected knowlingly to provide another, possibly competing, organisation with access to programmes intimately concerned with its own design philosophy and development programme.

There is, however, no difficulty in computerising the comparative procedure of materials selection. A computer program which will select the optimum material for a specific application can easily be produced if (a) the materials involved form a very closely related family with very similar properties, (b) no novel and unforeseen failure mechanism takes over, and (c) the properties of the candidate materials have been determined comprehensively. Two such programs are known to exist: ICI (EPOS), for the selection of polymers; and a Sandwik program for selection of cutting tools. These are knowledge-based systems dealing with families of essentially similar materials.

A computer program for selecting process-plant materials (as described in Section 1.2.7) would be equally straightforward, provided the requirements could be met by a steel and no unforeseen failure mechanism took over.

The requirements for a computer program to undertake selection by the classical procedure are much more general and much less well defined. The starting point is a product design specification (PDS) which is a functional and formal statement of what is required from the product to be designed, not a description of the product. The PDS contains a material design specficiation (MDS) which, like the PDS, is incomplete and ill defined. The computer must match this MDS to descriptions of existing materials and materials specifications (MS) which may be incomplete and reflect various levels of confidence. The result of the analysis may be a requirement to modify the PDS, to develop a new material, or to acquire additional information concerning specific materials.

A computer system capable of selecting materials requires:

(1) the ability to deal with simple and complex data structures;
(2) powerful structures for data acquisition and updating by augmentation and modification;
(3) the ability to manage sparse data;
(4) the ability to compare incomplete descriptions;
(5) the ability to distinguish the relationships, and sometimes lack of relationships, between materials, or parameters nominally in the same classification; and
(6) the ability to be easily extensible.

The system must take into consideration:

(1) the duty or function of the component;
(2) the materials properties;
(3) the manufacturing route;
(4) shape, dimensions and failure mode; and
(5) the relative cost of the materials, manufacturing routes and designs considered.

In addition, the system requires certain user characteristics so that it can be operated by designers and engineers and free the materials engineer for long-term difficult and strategic problems. It should:

(1) be rapid in use;
(2) require a minimum of learning;
(3) be accessible at different levels to suit different levels of user;
(4) have text and graphical output; and
(5) have recording facilities.

Various procedures for optimised decision-making have been proposed, including linear programming methods[6] and numerical algorithms.[7] There is a tendency to rely on ranking methods which allocate a rank of 0 to 3 for each material property. This introduces an imprecision which should not be necessary in the application of a computer which should be capable of relating property variation with overall cost.

At least one system is available which is claimed to be applicable at the innovation stage of design. A brief description of this system is given below, as an example of methods which could be employed.

PERITUS[8] is a knowledge-based system which comprises three main stages.

(1) A *director* stage which directs the non-specialist to data and knowledge modules. The structure of this is shown in *Figure 1.1*.
(2) A *presort* stage which uses ranking lists to produce a short-list of candidates from the materials indicated by the director stage (see *Figure 1.2*).
(3) An *evaluation and optimisation* stage which can either display the short-list together with deviations from ideality or, where the required modules exist, uses failure

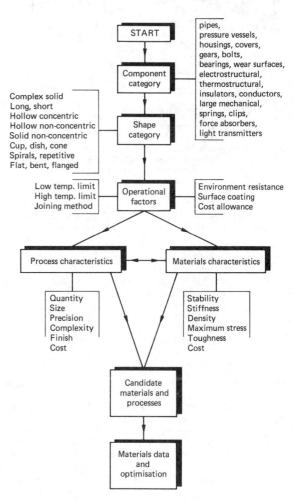

Figure 1.1 The structure and features of the director stage of the PERITUS knowledge-based system for the selection of engineering materials. (Reproduced by permission of *Metals and Materials*)

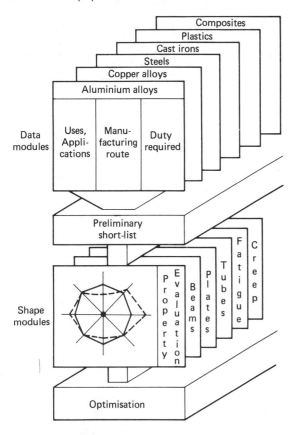

Figure 1.2 A representation of the organisation of the presort, evaluation and optimisation stages of the PERITUS knowledge-based system. (Reproduced by permission of *Metals and Materials*)

modes to optimise component dimensions and properties, presumably including overall cost (see *Figure 1.2*).

This system is modular and additional modules may be added as required or when available.

Further information on the PERITUS system is given by Swindells and Swindells[9] and a conceptual model for materials selections is discussed by Dimaid and Zucker.[10]

1.3 Ferrous metals

1.3.1 Introduction and standards for wrought steels

The development of alloys of iron, which include wrought iron, steel and cast iron, has been essential to a technological society. Iron ore is widespread, cheap and easily mined and from it can be produced alloys with the widest range of properties of any material. Steels may be produced with properties which vary from soft and ductile to strengths and hardnesses which, until the very recent development of the sialons, exceeded those of any relatively inexpensive material.

There are a very large number of wrought steels and numbering systems, German specifications[11] alone include 1400 grades. Because of the rapid evolution of ferrous metallurgy none of these classifications or numbers are ever fully up to date. In engineering, standard steels are commonly referred

to by their AISI/SAE number[12] based on composition (*Table 1.1*). The last two figures indicate the carbon content in the case of non-stainless steels. The corresponding general British specification is BS 970:1983 and earlier. In it steels are identified by a three-digit number denoting the type of steel, followed (in the case of non-stainless steels) by a letter denoting the type of specification (A for analysis, H for hardenability or M for mechanical properties) and two numbers denoting carbon content. For stainless steels the first three figures are identical to the AISI figures, the letter is S, and the final two numbers are coded. In the following accounts of classes of steel the AISI/SAE and BS ranges will, where possible, be given. There are also a number of specifications depending on product or application which do not always follow the BS 970 or AISI/SAE numbering system. Apart from the standard steels there are well-known steels which are recognised by designations used by their orginators and are in no way inferior to standard steels. Many steels of this kind are recognised and used world-wide in aircraft specifications and are made by many different steelmakers, although they have not yet been recognised by the several national bodies that govern steel specifications. Cast irons and cast steels have separate specification numbering systems.

1.3.2 Types of wrought steel

There are at least 11 separate classes of wrought steel, most of which are further subdivided.

1.3.2.1 Carbon steel (AISI 1006-1572; BS 970 000-119)

This is the basic type of steel which far exceeds all other metals in tonnage produced. *Low carbon steels* are subdividied into hot-rolled (see Section 1.4.3.1) and cold-rolled steel.

Hot-rolled steel has low strength (although the higher carbon versions can be heat treated to high hardness in small sections) and low toughness, but is readily available at low cost and is easily formed, welded and machined. *Cold-rolled steel* is harder, has good surface finish and dimensional tolerances, high strength at low cost and good machineability.

1.3.2.2 High strength low alloy steels

These are proprietary steels with low carbon made to SAE 950. They have significantly higher strength and are easily formed and welded.

1.3.2.3 Hardened and tempered steels (AISI/SAE 31–98; BS 970 500–599 including higher carbon steels)

These steels contain sufficient carbon and alloy to enable them to be heat treated to the desired strength and toughness at the design thickness. They may have high toughness and high strength at elevated temperature but they are more expensive than carbon steels and the higher alloy steels have poor weldability and machineability.

1.3.2.4 Case hardening steels

These are steels of relatively low carbon content (final BS specification Nos 12 to 25). They may be surface hardened by carburising, carbonitriding or nitriding when heat treatment will produce a very hard surface and a softer (but, where necessary, strong) ductile core. They are used when wear resistance must be combined with core toughness.

Table 1.1 Basic numbering system for AISI/SAE steels

Numerals and digits	Type of steel and average chemical contents (%)	Numerals and digits	Type of steel and average chemical contents (%)	Numerals and digits	Type of steel and average chemical contents (%)
	Carbon steels		*Nickel–chromium–molybdenum steels*		*Tungsten–chromium steels*
10XX	Plain carbon (1.00% Mn max.)	43XX	Ni 1.82; Cr 0.50 and 0.80; Mo 0.25	71XXX	W 13.50 and 16.50; Cr 3.50
11XX	Resulphurised	43BVXX	Ni 1.82; Cr 0.50; Mo 0.12 and 0.25; V 0.03 min.	72XX	W 1.75; Cr 0.75
12XX	Resulphurised and rephosphorised				
15XX	Plain carbon (max. range >1.00–1.65%Mn)	47XX	Ni 1.05; Cr 0.45; Mo 0.20 and 0.35		*Silicon–manganese steels*
		81XX	Ni 0.30; Cr 0.40; Mo 0.12	92XX	Si 1.40 and 2.00; Mn 0.65, 0.82 and 0.85; Cr 0.00 and 0.65
	Manganese steels	86XX	Ni 0.55; Cr 0.50; Mo 0.20		
13XX	Mn 1.75	87XX	Ni 0.55; Cr 0.50; Mo 0.25		*Low alloy high tensile steels*
		88XX	Ni 0.55; Cr 0.50; Mo 0.35	9XX	Various
	Nickel steels	93XX	Ni 3.25; Cr 1.20; Mo 0.12		
23XX	Ni 3.50	94XX	Ni 0.45; Cr 0.40; Mo 0.12		*Stainless steels AISI (not SAE)*
25XX	Ni 5.00	97XX	Ni 0.55; Cr 0.20; Mo 0.20	2XX	Chromium–manganese–nickel–nitrogen austenitic steels
		98XX	Ni 1.00; Cr 0.80; Mo 0.25	3XX	Chromium–nickel austenitic steels
	Nickel–chromium steels			4XX	Chromium ferritic and martensitic steels
31XX	Ni 1.25; Cr 0.65 and 0.80		*Nickel–molybdenum steels*		
32XX	Ni 1.75; Cr 1.07	46XX	Ni 0.85 and 1.82; Mo 0.20 and 0.25	5XX	Silicon–chromium steels
33XX	Ni 3.50; Cr 1.50 and 1.57	48XX	Ni 3.50; Mo 0.25		
34XX	Ni 3.00; Cr 0.77			XXBXX	*Boron intensified steels* B denotes 'boron steel'
			Chromium steels		
	Molybdenum steels	50XX	Cr 0.27, 0.40, 0.50 and 0.65	XXLXX	*Leaded steels* L denotes 'leaded steel'
40XX	Mo 0.20 and 0.25	51XX	Cr 0.80, 0.87, 0.92, 0.95, 1.00 and 1.05		
44XX	Mo 0.40 and 0.52	501XX	Cr 0.50		
		511XX	Cr 1.02		
	Chromium–molybdenum steels	521XX	Cr 1.45		
41XX	Cr 0.50, 0.80 and 0.95; Mo 0.12, 0.20, 0.25 and 0.30		*Chromium–vanadium steels*		
		61XX	Cr 0.60, 0.80 and 0.95; V 0.10 and 0.15 min.		

1.3.2.5   Stainless steels (AISI 200–499; BS 970 300S–499S)

These contain chromium in amounts above 12% so that the magnetic layer formed on the surface of the iron becomes, at lower chromium levels, a spinel and at higher levels chromic oxide. The introduction of chromium into the oxide layer greatly increases its stability and integrity and provides greatly increased resistance to corrosion and oxidation, but also influences significantly the structure and properties of the underlying metal. The addition of further alloying elements results in five separate classes of stainless steel.

Ferritic stainless steels   These are alloys of iron with up to 18% chromium and relatively small amounts of other alloy. They are ductile, the high chromium versions have good corrosion resistance, and they are inexpensive compared with other stainless steels. However, they have a tendency to grain growth and are therefore difficult to weld.

The recently developed 'low interstitial' ferritic stainless steels have chromium contents between 17 and 30% and very low carbon contents. These are claimed to have outstanding corrosion resistance, but may be difficult to obtain.

Martensitic stainless steels   These are limited in chromium content to about 17% so that they may be hardened by quenching to give high hardness and strength.

Austenitic stainless steels   These avoid the problems which result from the addition of chromium to the ferrite matrix by the addition of nickel and other elements which change the structure to the high-temperature γ form. The resultant alloys may have very high corrosion resistance, good ductility and/or high hot strength. Their very highly alloyed versions merge at iron contents below 50% into nickel alloys (see Section 1.4.5) and the very high creep strength versions are described as 'superalloys of iron'.

Duplex stainless steels   These have compositions which produce a mixed ferrite/austenite structure. They have excellent mechanical strength and corrosion resistance but may be difficult to obtain.

Precipitation hardening stainless steels   These combine very high mechanical strength with excellent corrosion resistance. They require complex treatments.

1.3.2.6   Intermetallic strengthened maraged steels

These steels may be formed or machined in the soft condition and then aged. The maraged steels are very strong and very tough, but are very expensive.

The steel classes so far described are based on composition and structure. Other classifications based mainly on application (which to some extent cut across the classification already described) include the following.

1.3.2.7   Electrical steels

These are very low carbon steels containing about 3% silicon supplied as strip, hot or cold rolled, with the surfaces insulated.

1.3.2.8   Spring steels

These are steels in which a very high hardness can be produced by working, quenching or precipitation hardening. They may be carbon, alloy or stainless.

1.3.2.9   Tool steels

These are steels used for forming or cutting materials. Their essential properties are: high hardness, resistance to wear and abrasion, reasonable toughness and, in the case of high speed steels, high hot hardness.

1.3.2.10   Creep resisting steels

These steels have high creep and creep rupture strengths at high temperatures. They may range from bainitic through martensitic to austenitic steels and superalloys. The higher temperature steels are also oxidation resistant because of chromium additions.

1.3.2.11   Valve steels

These steels have high temperature tensile and creep strengths and good high temperature oxidation and corrosion resistance. In the UK the term is restricted to certain martensitic and austenitic steels, but elsewhere it is used to include all steels which may be used for internal combustion engine poppet valves.

1.3.3   Steelmaking and casting of wrought-steel ingots

The raw materials for steelmaking (iron ore and coke) are converted in the blast furance to molten iron containing about 3% carbon at a rate which varies up to about 8 million ton per annum. (This makes the blast furance by far the most economical process for producing steel, but economies which do not have the capacity to utilise steel production of this magnitude may use the direct reduction of iron ore to sponge and powder instead.) A typical blast furnace is shown in *Figure 1.3*.

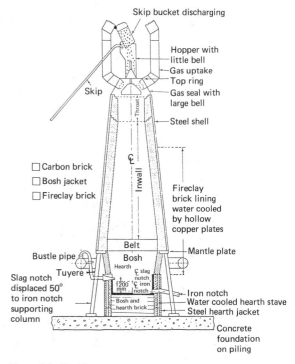

Figure 1.3   Blast furnace

Since World War II the conversion of the iron from the blast furnace to flat products, billet, bar and sections has undergone a revolutionary change due to the development of oxygen argon lancing, vacuum treatment, continuous casting and rapid in-works analysis.

Iron from the blast furnace is conveyed while still molten to the basic oxygen convertor (BOC or BOF) (see *Figure 1.4*) where oxygen gas is passed through it, reducing the carbon content to approximately that of the specified steel. The iron may then proceed direct to the pouring ladle where alloying additions are made and slagging processes undertaken. The molten steel is poured into a lander which conveys it to a water-cooled mould the base of which is formed by the previously poured solid metal. This metal is retracted through the mould which may oscillate about a vertical axis. The whole of the blast-furace output may pass through a single mould of this type (*Figure 1.5(a)*) and the cast metal may be fed directly into a rolling mill which reduces it to plate bar or section, as required (*Figure 1.5(b)*). The process as described is applicable to very large throughputs, but successive ladle charges may differ in composition, the different alloys being separated later in the mill train. Where higher quality or special steels are required the liquid steel from the BOC may be transferred to a vacuum plant for further treatment prior to pouring[13] (see *Figure 1.6*).

The Dortmund Horder (DH) and Ruhr–Stahl Heraeus (RH) processes transfer the metal from a ladle into a superimposed vessel, the DH by sucking it up by vacuum, the RH by driving it up by pressure of argon. A further improvement has been to equip a vacuum argon treatment vessel with electric-arc heating (vacuum arc deoxidation (VAD)[13]) and effectively to produce a secondary steelmaking process (see *Figure 1.7*).

Most special steels, and in particular stainless steels, cannot be decarburised by the BOF because the oxygen which reduces the carbon present also oxidises chromium. Oxidation may be prevented by blowing oxygen in a vacuum, but there are practical problems involved. These may be overcome by the argon oxygen degassing (AOD) process in which an artificial vacuum (insofar as the partial pressume of carbon monoxide is concerned) is produced by diluting the blown oxygen with argon. This process is carried out in a tiltable vessel[14] (see *Figure 1.8*) with base twyers which agitate the vessel contents. With this process the removal of carbon, slag reduction, metal dioxidation, desulphuration and alloying are easily achieved.

Conversion of the molten iron from the blast furnace direct to the finished product in an integrated steelworks not only gives significant economies in energy and labour but can also provide steel of quality equal to or better than was previously available from electric-arc melting.

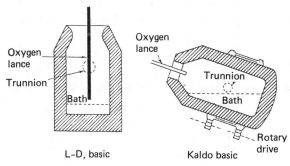

Figure 1.4 Types of basic oxygen convertor (BOC) or basic oxygen furnace (BOF)

In the first place oxygen blown steel is naturally low in nitrogen so that the toughness problems associated with strain ageing are eliminated. However, where, as in austenitic stainless steels, nitrogen has a beneficial effect on tensile strength, controlled amounts may be introduced by replacing some or all of the argon in the AOD process.

The greatest improvement in liquid metal quality is, however, gained by vacuum treatment. Raising and lowering the steel in the DH or RH vessel very significantly reduces the partial pressures of hydrogen and carbon monoxide in the melt and the

$$C + FeO \rightarrow CO + Fe$$

reaction proceeds until equilibrium between carbon and oxygen is established at a lower oxygen level. Line A in *Figure 1.9* represents the oxygen/carbon equilibrum at 1013 mbar (1 atm) over the carbon range 0.03–0.13%; the effect of reducing the pressure (e.g. by vacuum degassing) to 133.3 and 13.33 mbar is shown by lines B and C, respectively. Since both carbon and oxygen are lost to the system, the theoretical effect on a steel initially at 0.05% C is shown by line 1 which connects lines A to C. The actual effect on steels with various vacuum degassing techniques is shown by lines 2–5. Vacuum degassing thus reduces the oxygen content to levels less than half those obtained in the best practice in steels air melted and refined at atmospheric pressure. Consequently, final residual deoxidation can be effected with much smaller amounts of aluminium or silicon and much cleaner steel can be produced with, if required, lower carbon levels. Vacuum carbon deoxidation is extremely beneficial in the production of plate with good through-the-thickness ductile properties because it virtually eliminates planar concentrations of non-metallic inclusions.

Vacuum degassing techniques properly applied to permit carbon deoxidation are the most economical way of upgrading steels and, in particular, low carbon steels. It is further possible to make injections in a stream of inert gas of elements such as calcium or calcium carbide which can reduce sulphur and phosphorus and to replace the sometimes damaging inclusion of silicon and aluminium with other inclusions which improve transverse ductility, fatigue and machineability.

Additional advantages are gained by the substitution of continuous casting for ingot casting. Carbon, sulphur and phosphorus have been shown to segregate enough to give a concentration ratio of 3:1 or 4:1 between the top and bottom of a large conveniently poured ingot. Silicate inclusions segregate to different parts of the ingot. No longitudinal segregation can develop in a continuously cast ingot once the casting process has reached equilibrium. The only longitudinal variation of composition is that between one ladle charge and another; this is revealed by analysis and may be corrected before pouring. There will be some compositional variation across the section, and possibly between dendrites, but at the outside this is unlikely to exceed ±5% of the mean. Continuous casting has the further advantage that the cross-section of the ingot can be much better matched to that of the final rolled product than is possible with a conventionally cast ingot. A large plate requires a large ingot, which in conventional casting must be of large cross-section and may be too large for the available rolls. With continuous casting the size of plate is only limited by handling down the line and the metal quality is improved by the finer structure achieved by casting a smaller section.

However, the quantity of special and stainless steels required seldom warrants production from hot metal in proximity to the blast furnace. Such steels are generally produced from cold metal in the arc furnace. Here the quality of the steel is dependent on careful selection of the charge, so that

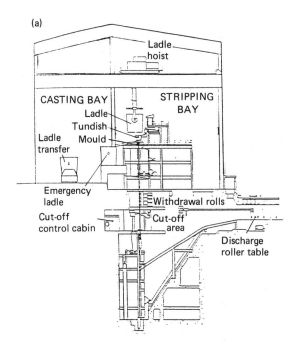

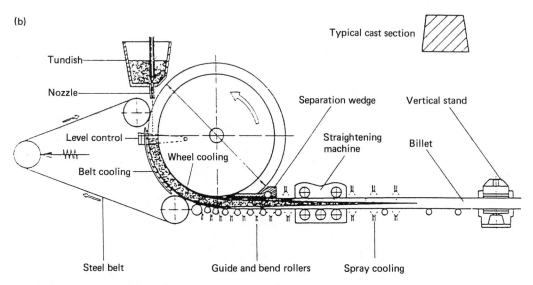

Figure 1.5 Continuous casters: (a) with cut-off; (b) with in-line roll stand. (Reproduced by permission from *Continuous Casting*, Institute of Metals)

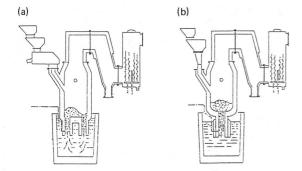

(a) **(b)**

Figure 1.6 Vacuum degassers (deoxidisers, decarburisers): (a) RH; (b) DH. (Reproduced by permission from *Secondary Steel Making for Product Improvement*, Institute of Metals)

quantities of tramp elements (tin, arsenic, antimony, bismuth, copper, etc.), which have serious effects on ductility and cannot be reduced by steelmaking, are minimised. Electric arc furnace steel is usually considered superior to steel made in the BOF, but the application of AOD and VAD can produce equivalent steel from either.

There are other steelmaking processes that will produce very high-quality metal, but all depend on the use of good quality raw material.

The vacuum high frequency furnace will produce metal of low oxygen, nitrogen and hydrogen content, but is not well adapted to a continuous-casting process. It is mainly used to produce small quantities of metal as small ingots (or castings).

The highest quality steel is produced by the consumable vacuum arc or the electroslag remelting (ESR) processes. In both these processes an electrode (or electrodes) is made of steel of the target composition and this electrode is progressively melted by striking an arc between it and a starting pad which is progressively withdrawn (or the electrode is withdrawn) thereby producing a semicontinuous casting (see *Figure 1.10*). The metal is refined on passing through the vacuum in the case of the consumable vaccum arc or through a low-melting-point slag in the case of ESR. The solidifying metal is easily fed from the pool of molten metal and inclusions are dispersed and very substantially reduced in size. The consumable vacuum arc removes all gaseous or gasifiable impurities but does not significantly influence the proportion of non-metallic constituents. ESR can transfer non-metallic inclusions into the slag but does not remove hydrogen. ESR consumables must, therefore, be hydrogen free.

The influence of these melting processes on such important properties as ductility or fatigue strength is very significant. The highest values of properties are not usually improved significantly compared with conventionally melted steel, but the proportion of values falling below a specific standard is very significantly reduced so that a much higher component performance may be guaranteed.

A small proportion of steel is electron beam or plasma melted.

The proportion of steel that is continuous (or semicontinuously) cast has very greatly increased. A large integrated steelworks may have its entire throughput fed to a one-stream continuous casting of 1 m × 2 m section, but batch metal from an electric arc melting shop is often semicontinuously cast in sections down to approximately 10 cm. In such processes mould costs are reduced and throughputs and metal recovery considerably increased compared with the older type of mould. The proportion of wastage at the feeding head and the chances of segregation in the pipe are also reduced.

Ingot is still employed where the quantity of steel required is insufficient to justify the use of continuous casting, or for very large forgings where the cross-section of the ingot required to withstand the reduction needed is too large for it to be continuously cast.

1.3.4 Mechanical working of metals

This topic is included for convenience in this section on ferrous metals, but is more widely applicable.

A metal is worked mechanically either to generate a shape more economically than can otherwise be obtained or to provide improved properties in strength, ductility and/or fatigue (not creep for which a cast structure is superior). The improvement in properties may, overall, be directional to resist a directional stress system, or statistical to ensure that a lower percentage of components fall below a specified level of properties.

Metals can be mechanically worked because they differ from other crystalline solids in that the atoms in a crystal are not linked by valence bonds to adjacent atoms. The free electron metallic bond is non-specific, pulling equally hard in all directions. Metal atoms are therefore bound tightly in certain regular crystalline structures. There are seven crystal systems giving 14 possible space lattices and all engineering metals crystallise in one of three of these lattices (see *Table 1.2* and *Figure 1.11*). Some metals are polymorphic. For example, in the case of iron, austenite is face centred cubic, whilst ferrite is body centred cubic.

These crystalline structures are resistant to tensile stress but can be sheared along certain crystallographic planes. The planes of lowest resistance to shear have the closest packing of atoms in the plane. Each trio of atoms in a close packed layer surrounds a space (or hollow) in which an atom in the next layer can rest. Slip occurs when the atoms in a plane move into the next hollow in the adjacent plane. The shear stress required to move a whole plane of atoms is very high and a perfect metal crystal (such as occurs in 'whiskers') would deform elastically between 3 and 10% before deforming plastically in shear.

The crystals in bulk metal are not perfect. 'Dislocation' rows occur in the lattice where an atom is missing or an additional atom is present. These dislocations allow the planes to shear one row of atoms at a time, thus greatly reducing the critical shearing force. More dislocations come into play as the metal is deformed and, in most cases, two or more dislocations may interact and stop each other. The critical shear stress therefore increases and the metal work hardens, increasing its tensile strength. Finally, no more shear is possible and the metal fractures. If temperature is increased, diffusion and recrystallisation can release stopped dislocations. In hot working diffusion occurs while the metal is worked, but in cold working and annealing the two processes occur separately. In either case the grain of the metal may, if the conditions are chosen correctly, be refined.

Mechanical working has other beneficial effects. The structure of cast metal may contain microsegregates, lumps and planes of constituents which separate out during casting. Mechanical working correctly applied may break up and disperse these to a size at which they can be diffused during heat treatment.

Castings may contain cracks, voids and non-metallic inclusions. Mechanical working correctly applied may close up and remove the cracks and voids (or aggravate them so that the part must obviously be rejected), and break up non-metallic inclusions. Alternatively, defects may be aligned and elongated in the direction of the tensile component of stress so that they do not impair resistance to fracture caused by tension

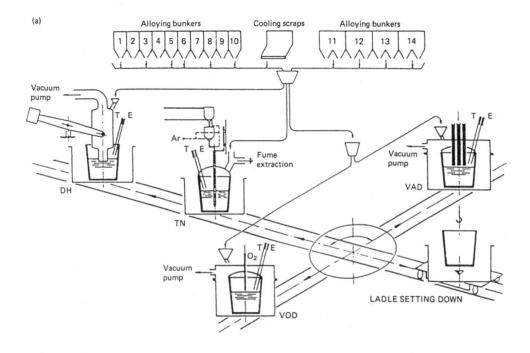

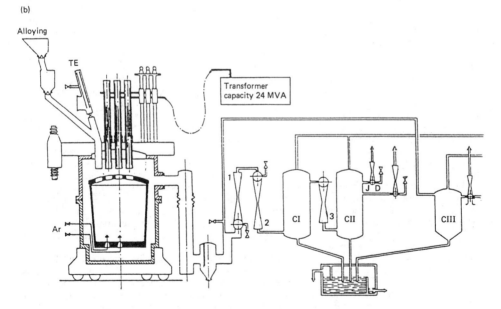

Figure 1.7 Secondary steelmaking plant: (a) general layout; (b) vacuum argon deoxidation (VAD) unit. T, temperature lance; E, sampling lance; DH, Dortmund Horder; TN, Thyssen powder injection unit for desulphurisation; VOD, vacuum oxygen deoxidation. (Reproduced by permission from *Secondary Steel Making for Product Improvement*, Institute of Metals)

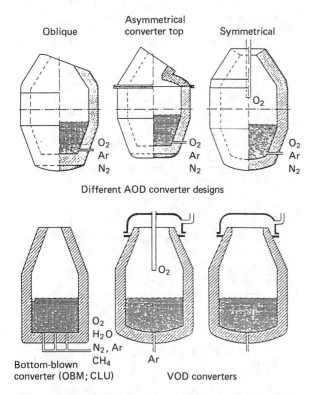

Different AOD converter designs

Bottom-blown
converter (OBM; CLU)   VOD converters

Figure 1.8 Converter processes for oxidising and treating non-alloy
and high chromium alloy premelts. (Reproduced from *Electric Furnace
Steel Production*, by permission of Wiley)

Line 1   Theoretical   Line 4A Fixed ladle degassing
Line 2A ⎫ Stream degassing   Line 4B Fixed ladle degassing
Line 2B ⎭ Deep drawing steels   plus agitation
Line 3A ⎫ Rimming   Line 5A Fixed ladle degassing
Line 3B ⎭   Line 5B Rimming steels

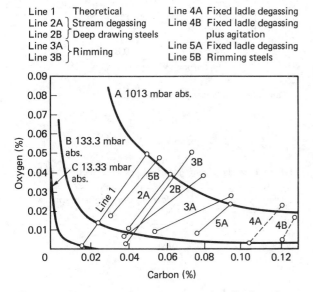

Figure 1.9 Decrease in carbon and oxygen produced by degassing

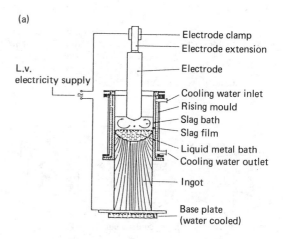

(a)

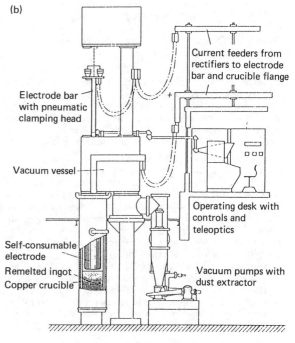

(b)

Figure 1.10 (a) Electroslag remelting. (b) Vacuum arc remelting.
(Reproduced from *Electric Furnace Steel Production*, by permission of
Wiley)

Table 1.2 Space lattices of engineering metals

Lattice	Metals
Face-centred cubic	Aluminium, titanium,* lead, copper, iron,* cobalt,* nickel,* gold, silver
Body-centred cubic	Iron,* vanadium, chromium,* niobium, molybdenum, tungsten, zirconium
Hexagonal close packed	Titanium,* chromium,* cobalt,* nickel,* magnesium, zinc

* Metals having polymorphic habit.

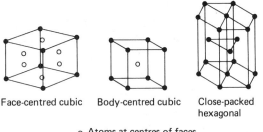

Face-centred cubic   Body-centred cubic   Close-packed hexagonal

o Atoms at centres of faces or body of cube

Figure 1.11 Three simple space-lattice-system unit cells

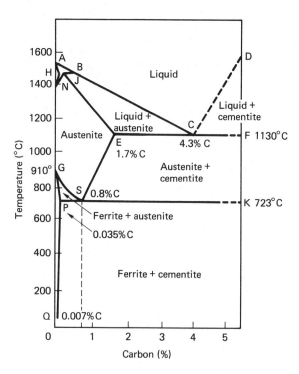

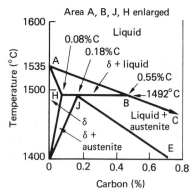

Figure 1.12 Iron–carbon composition diagram

or fatigue. Of the three space lattice systems shown in *Figure 1.11* the face centred cubic has more planes on which slip can occur than the other two (which are themselves greatly superior to other lattice systems). Face centred cubic austenite is, therefore, more ductile than body centred cubic ferrite, and steel is usually hot worked in the austenite (γ) phase.

Hot working Before working, the steel (unless already hot from continuous casting) is reheated into the γ phase. Care must be taken to prevent excessive scaling and to avoid overheating (or burning) the steel. The limiting temperatures for carbon steels can be estimated from *Figure 1.12* and are given in *Table 1.3*. Further forging temperature ranges are given in *Table 1.4*. Grain growth occurs when steel is held in the γ phase, but subsequent working reduces grain size. *Figure 1.13* illustrates these effects. The grain size of mild steel increases from ASTM 10 to ASTM 6 on heating to 900°C, but six rolling passes reduces it to ASTM 10. Heating to 900°C has no effect on the grain size of two other steels, but in both cases heating to higher temperatures increases grain size and rolling reduces it again. The finer austenite resulting from hot working transforms to a finer grain ferrite on cooling.

1.3.4.1 Rolling

Rolling is the most economical way of working steel if sufficient quantities are to be produced. Steel for strip is cast into slabs, scarfed to ensure a defect-free slab and hot rolled using automatic gauge control. Carbon steels may be 'controlled rolled', that is finish rolled at relatively low temperatures thereby inducing a fine ferritic grain structure with improved tensile strength and notch toughness.

Hot rolled strip is cold rolled to reduce thickness and give the required surface finish and forming qualities. Finishing

Table 1.3 Maximum forging temperatures for plain carbon* steels

Carbon (%)	Maximum forging temperature (°C)
0.1	1315
0.3	1286
0.5	1259
0.7	1215
0.9	1176
1.1	1133

* Maximum forging temperatures for alloy steels may be up to 50°C lower.

Table 1.4 Forging temperature ranges, tool and hot work die steels (American* and British practice)

Type of steel	Composition (%)					Temperature—American (°C)		Temperature—British (°C)		Remarks
	C	Cr	Mo	W	V	Start	Finish	Start	Finish	
Carbon tool (water hardening)	0.7 /1.2 0.85/1.1	— —	— —	— —	— —	— 981–1093	— 815	950 —	750 —	
Tungsten (oil hardening)	0.9 /1.15 1.25	0.5 0.5	— —	0.5/1.6 1.5	— —	981–1063 —	871 —	— 950	— 750	General tools and dies Drill and saw steel
Tungsten finishing steel (water hardening)	0.1 /0.3	—	—	3.5/3.75	—	1008–1063	871	950–1010	800	Preheat slowly to 842°C. Slow cool in insulating material. For fine finishing cuts, cold dies, punches, gauges, etc.
Tungsten high speed steel (pil or salt bath hardening)	0.8	4	—	18	1	1120–1176	926	1050–1100	900	Slow cool and anneal after forging
Hot work die steel chromium–molybdenum	0.3	5	1	1.25	—	1093–1149	898	1050–1090	850	Preheat 650–700°C, slow cool, anneal
Hot work steel tungsten	0.35	3	—	10	0.5	1120–1176	898	1100	900	Preheat 800–850°C, slow cool, anneal

* American data from Sub-Committee on Tool Steels, *Metals Handbook*, pp. 991–1031 (1939)

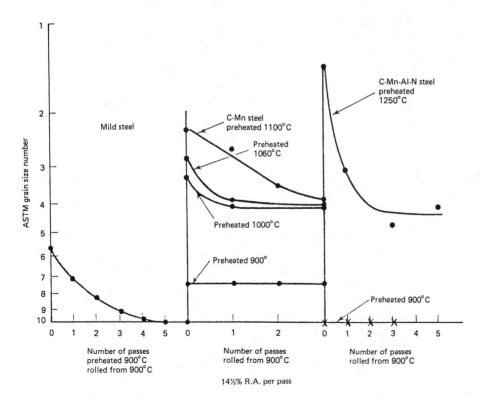

Figure 1.13 Effect of heating temperature and number of roll passes on austenite grain size

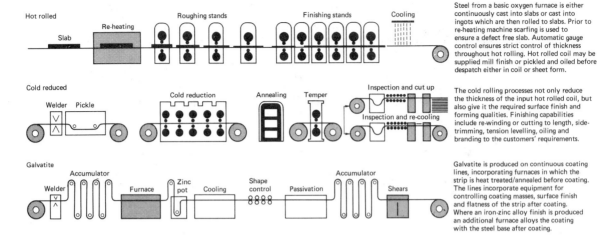

Figure 1.14   Mill trains for producing hot- and cold-rolled strip

Steel from a basic oxygen furnace is either continuously cast into slabs or cast into ingots which are then rolled to slabs. Prior to re-heating machine scarfing is used to ensure a defect free slab. Automatic gauge control ensures strict control of thickness throughout hot rolling. Hot rolled coil may be supplied mill finish or pickled and oiled before despatch either in coil or sheet form.

The cold rolling processes not only reduce the thickness of the input hot rolled coil, but also give it the required surface finish and forming qualities. Finishing capabilities include re-winding or cutting to length, side-trimming, tension levelling, oiling and branding to the customers' requirements.

Galvatite is produced on continuous coating lines, incorporating furnaces in which the strip is heat treated/annealed before coating. The lines incorporate equipment for controlling coating masses, surface finish and flatness of the strip after coating. Where an iron-zinc alloy finish is produced an additional furnace alloys the coating with the steel base after coating.

capabilities include rewinding to coil or cutting to length, side trimming and oiling.

Roll trains for producing hot and cold rolled strip are shown in *Figure 1.14*.

By the use of a cluster mill such as the Senzimir (see *Figure 1.15*), strip down to 0.1 mm thickness can be produced. Grooved rolls are used to produce blooms, billets, bar and sections. Hollow sections can be produced by suitably designed mill trains and strip can be passed through forming rolls which turn the edges towards each other to form a seamed tube which is then welded continuously by inert metal arc or electric resistance. Seamless tube may be formed by cross rolling a billet in a Mannesman piercer in which the rolling action produces a tensile stress at the centre of the work piece.

1.3.4.2 Forging

Forging is the process of working hot metal between dies either under successive blows or by continuous squeezing. It may be used to break down an ingot into a bloom or bar, to work down an ingot or billet to a rough finished shape before finishing, or to make a forging. There are two essential differences between forging and rolling. It is almost always possible (a) to design the dies and to arrange the sequence of forging to impose a higher ratio of compression to tension forces than is possible by rolling, and (b) to ensure that the grain of the metal is in a preferred direction and not purely longitudinal as in rolling. For breaking down an ingot in, for example, tool, high speed and some stainless and heat resisting steels that have a two-phase structure, hot rolling would (at least before the development of electroslag remelting or vacuum arc casting) lead to ruptures due to the strong tension forces induced. Such ingots are usually broken down by hammer cogging. Forging is also used for making very large components such as turbine rotors which are usually 'open die' forged in a press. Large forging ingots have a cross-section the circumference of which comprises a number of arcs meeting at cusps because this shape minimises surface cracking during casting. The first forging operation removes the cusps to form an approximately circular cross-section and the forging is then drawn out through successive shape changes from octagon to square and back to octagon using dies of the shape shown in *Figure 1.16*. If the geometry of forging and press permit, the

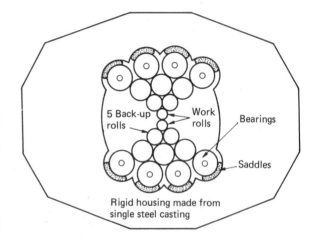

Figure 1.15   Sendzimir cluster reversing mill

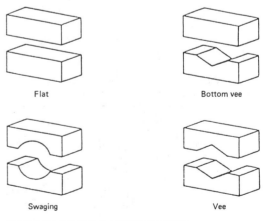

Figure 1.16   Die shapes for open-die forging

forging is upended and upset to produce some radial grain flow and it is then drawn out again.

Hollow forgings are made by punching a hole in the centre of a cylindrical work piece and 'becking', working the die against a stiff bar passing through the forging and supported on V blocks.

Small closed die forgings are made in two dies attached to the hammer ram and bed which have successive cavities to shape the stock progressively into final shape in the last or finish cavity. For larger forgings a number of dies are made to perform one operation each. A wide variety of shapes can be made, depending only on the ability to make and extract the component from two meeting dies with a parting line which may or may not be planar.

Most of the steels listed in BS 970 can be forged and will give properties appropriate to their section. Carbon steel forgings for engineering purposes are listed in BS 24 and BS 29 and forgings for fired and unfired pressure vessels are given in BS 1503.

The most demanding requirements for aircraft and similar requirements are met by ESR or consumable vacuum arc cast ingot, usually made to the manufacturer's own specifications (agreed where appropriate by official inspecting bodies).

There are some 30 hot and cold metal working processes which can be considered to be some version of forging.

1.3.4.3 Drawing and extrusion

A third important group of metal working processes involves shaping metal by pushing ('extruding') or drawing ('drawing') through a die. Extrusion is usually performed hot, whilst drawing is usually done cold.

Extrusion Hot extrusion consists of placing a hot cylindrical billet into a container and either forcing a die with a centrally placed orifice onto one end (indirect extrusion) or applying pressure to a ram at the other end, the die being held stationary (direct extrusion). In either case metal is extruded through the die in the form of an elongated bar having the same cross-section as the orifice.

Extrusion is applied widely to non-ferrous metals which soften at lower temperatures than do steel or hard metal dies. Bar, and very complex sections can be economically produced and cut into a wide variety of shapes. Tubes and hollow sections can be made either by extruding metal through bridge-type dies in which the metal stream separates and rewelds at a later point, or by extruding a hollow billet over a central mandril. Non-ferrous metals are extruded hot or cold. Only the softer steels are extruded cold.

Hot extrusion of steel requires a glass lubricant. Very rigid and powerful presses are required and their high cost plus the need for machined billets limits the process to high cost steel, unusual shapes and tubes (which are mandril extruded). Besides extrusion there are other methods of tube production.

Drawing Usually, after a 'semi' has been produced hot it is cold drawn to reduce the diameter, wall thickness or both. Cold drawing through dies is used to produce wire, tube and light bar. This process requires considerable skill and attention to detail in die design, choice of lubricants, wire rod cleansing and baking to remove hydrogen introduced during cleaning.

1.3.5 Constitution and heat treatment of ferritic steels

The versatility and adaptability of steel depends on the ability to vary and control the constitution, distribution and nature of the microstructure constituents by varying chemical composition, heat treatment and hot and cold working. This provides a wide variation in mechanical properties over a range extending from very low temperatures to around 1000°C.

To indicate how this variation and control is achieved, it is necessary to consider matters such as allotropy, solid solutions, constituents, phases, equilibrium, and the iron–carbon composition diagram.

1.3.5.1 Allotropy of iron

Each of the allotropic modifications of pure iron (see *Figure 1.12*) is stable within certain ranges of temperature as shown in *Table 1.5*. The change from one modification to the other which occurs on heating or cooling through the critical temperatures 910°C and 1400°C is accompanied by recrystallisation.

1.3.5.2 Solid solutions

Solid solutions are formed when a metallic solid dissolves one or more elements or compounds. The solute elements can diffuse as in liquid solutions. If the atomic size of the solvent and solute atoms is similar the solute atoms tend to replace the solvent atoms in an unchanged space lattice but, if they are substantially smaller, they will most probably be situated interstially between the solvent atoms.

Pure iron at room temperature is composed of grains of α iron (ferrite). Ferrite is capable of dissolving in limited quantity elements such as silicon, phosphorus, nickel, copper, arsenic, etc., so that commercial steel of ultra-low carbon content will have ferrite grains with these elements in solid solution. The α iron in commercial steel is referred to as 'ferrite'. γ iron (austenite) also dissolves elements to form solid solutions. If, as is usually the case, γ iron dissolves an element to a greater extent than α iron, when the change to α iron occurs, then on cooling through the critical temperature the α must accommodate more atoms of the element than its solubility allows and becomes supersaturated. Under suitable conditions excess atoms will precipitate to arrive at a stable condition.

The high solubility of iron carbide (Fe_3C), 'cementite' in austenite, and its relatively low solubility in ferrite is the basis for the heat treatment of hardened and tempered steel. Although steels containing cementite are strictly metastable, for all practical purposes they may be considered as stable because the precipitation of graphite which is the stable form of carbon occurs too slowly to be of interest except under special conditions.

1.3.5.3 Phases, equilibrium and the iron–carbon phase diagram

Constituents are the components of a metal alloy, visible under the microscope after suitable etching. Phases are physically and chemically homogenous entities separated from the rest of the alloy by definite bounding surfaces. Phases in solid steel are austenite, ferrite, cementite, graphite, alloy carbides

Table 1.5 Stable temperature ranges of the allotropic modifications of pure iron

Modification	Crystal structure	Temperature range of stability (°C)
Delta, δ	Body centred	1535–1400
Gamma, γ	Face centred	1400–910
Alpha, α	Body centred	⩽910

and other intermetallics. All phases are constituents but some constituents are not necessarily phases, e.g. pearlite is a constituent but it consists of the phases cementite and ferrite in a particular arrangement.

The phase diagram represents the phases present, at given temperatures, which are in equilibrium with each other. Equilibrium is attained by slow heating or cooling to allow constitutional changes to be completed.

The iron–carbide phase diagram (*Figure 1.12*) shows the effect of adding carbon to molten iron on the temperature ranges of stability of the δ, γ and α phases in the solidified alloy and the effect of the varying solubility of iron carbide (cementite) on the phases present at temperatures from solidification downwards. The diagram represents nominally pure alloys; similar diagrams for other iron alloy systems (Fe–Cr, Fe–Mn, etc.) show the effect of other elements such as chromium, manganese, etc., on the range of stability of the phases. Since commercial plain carbon steels contain manganese and other elements in varying degrees *Figure 1.12* does not accurately represent the critical temperatures but is a useful starting point for heat treatment.

1.3.5.4 Equilibrium decomposition of austenite: consideration of the iron–carbon constitution diagram

Point A in *Figure 1.12* is at 1535°C, the highest temperature for the existence of the solid δ phase in carbon free iron; line ABC shows the fall in the temperature, at which the first phase to separate from the solidifying liquid appears, as carbon content is increased. Line AHJEF (the solidus line) shows the temperatures below which all alloys are solid. As is shown in *Figure 1.17* pure iron changes from face-centred (austenite) to body-centred cubic (ferrite) at 910°C; adding carbon lowers this temperature as shown by line GS and, as the solubility of cementite in ferrite is less than in austenite the change results in an increase in the carbon content of the remaining austenite which, on further cooling, then forms more ferrite, again with an increase in carbon content of the remaining austenite. This process continues until the last small amount of austenite contains 0.80%C which changes at the fixed temperature of 723°C to a mixture of cementite and ferrite called 'pearlite' or 'eutectoid', PSK being the 'eutectoid line'.

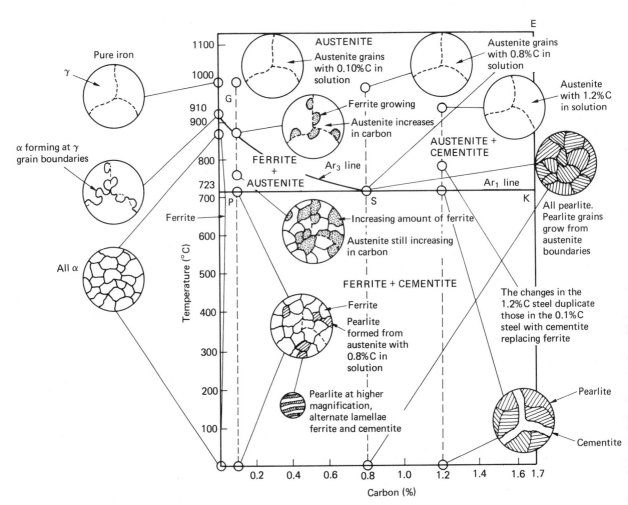

Figure 1.17 Steel portion of the iron–carbon composition diagram. Changes in grain structure and constituents with decreasing temperature. The AR₃,Ar₁ line GS,SK is the line at which austenite transforms on *cooling* during heat treatment. The Ac₃,Ac₁ line (not shown) is the line at which ferrite–pearlite–cementite transforms to austenite on *heating* and is a few degrees, depending on the rates of heating and cooling, higher than the Ar₃,Ar₁ line

Alloys below and above 0.8%C, (hypo- and hyper-eutectoid) both contain ferrite and cementite when cooled to room temperature, but the mode of occurrence (the microstructural constituents of hypoeutectoid steels being ferrite and pearlite, those of hypereutectoid being cementite and pearlite) of the phases is different with markedly different effect on mechanical properties. Descriptive sketches or actual photomicrographs may be used to illustrate the mode of occurrence of the phases. The constituents in the phase fields and the changes in constituents on cooling for pure iron, 0.10%C and 0.80%C steel are shown schematically in *Figure 1.17* when equilibrium conditions exist at extremely slow cooling or heating so that the diffusion of carbon is not inhibited.

The solid solubility of carbon in γ iron is 1.7% (point E in *Figure 1.17*). Hypereutectoid steels between 0.8%C and 1.7%C precipitate cementite (6.8%C) when cooled to line SE which reduces the carbon level of the remaining austenite, thus requiring a further lowering of temperature before more cementite can precipitate. Continuous cooling down to the eutectoid line PSK, therefore, results in continuous precipitation of cementite until, at the eutectoid, the austenite contains 0.8%C and precipitates ferrite and cementite simultaneously to produce pearlite. Steels contain up to 1.7%C and cast irons contain more than 1.7%C.

It is common practice to use terms such as 'primary austenite', to differentiate austenite which freezes out of hypoeutectoid alloys from austenite formed by heating up into the austenite phase field. Also, ferrite formed along line GS (*Figure 1.17*) from steels with less than 0.8%C is called 'hypereutectoid ferrite' and cementite formed along SE is called 'proeutectoid cementite'. The ferrite phase in the eutectoid constituent pearlite is called 'pearlitic ferrite' and the cementite 'pearlitic cementite'.

The microstructural constituents shown schematically in *Figure 1.17* are the basis of most commercial steels, but the proportions in which they occur and the mode of their occurrence (distribution and form) are profoundly influenced by rate of cooling from the austenite phase field. The equilibrium conditions necessary to the behaviour shown in *Figure 1.17* and the approximate temperature indicated, do not obtain in steel heat treatment practice, except possibly when very large forging ingots are annealed.

Moreover, there is a thermal lag in the allotropic change so that the eutectoid line PSK (*Figure 1.17*) lies at a temperature approximately 3°C higher during heating up than on cooling down. With normal rates of heating and cooling of steel the eutectoid change is raised or lowered with faster heating or cooling; consequently, in practice, a plain carbon steel has a temperature range in which the eutectoid temperature can vary according to heating or cooling rates. This also applies to the position of the austenite to ferrite change line GS.

However, *Figure 1.17* illustrates an essential feature of steel heat treatment, namely that austenite in transforming through the lines GS, SE or PSK develops a number of grains of the new constituents in each austenite grain, thereby refining the grain structure. The mechanical properties of steels consisting of ferrite and pearlite are strongly influenced by the average grain size of ferrite as well as the amount and type of pearlite (coarse lamellar, fine lamellar, etc.). The yield stress varies linearly with the reciprocal of the square root of the grain size.

On heating steel through the critical temperatures into the austenitic phase field the behaviour observed on cooling is reversed in the following manner.

Steel with 0.1%C On passing through the lower critical temperature Ac_1, which is higher than Ar_1, the pearlite areas first transform to austenite of 0.8%C content. This austenite grows by dissolving the surrounding ferrite grains as the temperature is raised and its carbon content is reduced. However, the austenite areas developing from the pearlite consists of numerous crystals so that just above line GS, when the structure is wholly austenitic containing 0.1%C, it consists of numerous small austenite grains. Heating to higher temperatures in the austenite phase field causes grain growth, some grains growing by absorbing smaller ones around them.

Eutectoid steel, 0.8%C On heating above the lower critical temperature Ac_1 (which coincides with the upper critical temperature Ac_3 at the eutectoid composition), theoretically the pearlite should transform to austenite of 0.8%C content. In practice it does so over a temperature range, the ferrite lamellae absorbing cementite to form a lower carbon austenite which then dissolves the remaining cementite. Grain growth follows on heating to higher temperatures in the austenite phase field.

Hypereutectoid steel, 1.2%C At the eutectoid line the pearlite starts to transform to austenite of 0.8%C content. As the temperature is raised through the austenite plus cementite phase field, proeutectoid cementite is gradually dissolved by the austenite adjacent to it and eventually, by carbon diffusion, above the upper critical temperature the austenite attains a uniform carbon content of 1.2%. Grain growth follows on heating to higher temperatures in the austenite phase field.

The above simple behaviour of carbon steel relies on adequate time for diffusion of carbon being available. When the time at the required temperature is reduced the diffusion is inhibited in varying degrees, with a pronounced effect on the transformation changes. Thus, in plain carbon steels, increasing the rate of cooling through the critical temperature range Ar_3–Ar_1 lowers this range and alters the proportions of ferrite and pearlite. Steels with less than 0.25%C show refinement of ferrite grains, the growth of individual grains being suppressed, and the pearlitic constituent has finer cementite lamellae. Steels with more than 0.25%C show an increased amount of pearlite and decreased ferrite.

If the steel contains 0.35%C or above, it is possible by a sufficient increase in cooling rate to produce a structure consisting entirely of pearlite. This pearlite will differ from equilibrium pearlite in having very thin cementite lamellae separated by wide ferrite lamellae. Since pearlite (with hard cementite lamellae) is the main contributor to tensile strength in ferrite–pearlite steels, its proportion and morphology in the structure are prime considerations for heat-treatment practice. If the cooling rate through the critical range is increased still further, the austenite transformation may be entirely suppressed, and the steel remains as unstable austenite down to a lower temperature when transformation begins with the formation of lower temperature products, e.g. martensite and bainite. When this happens the steel has been cooled at its 'critical rate'.

Martensite has special characteristics which are of great importance in the heat treatment of steel. The essential difference between the mode of formation of martensite and that of pearlite is that the change from the face-centred cubic austenite lattice to the body-centred cubic ferrite lattice occurs in martensite formation without carbon diffusion, whereas to form pearlite carbon diffusion must take place producing cementite and ferrite. The effect of this is that the carbon atoms strain the α martensite lattice producing microstresses and considerable hardness. The higher the carbon content the greater the hardness of martensite (*Figure 1.18*) and the lower the temperature at which the change to martensite begins.

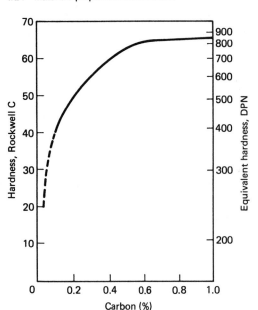

Figure 1.18 Hardness of martensite related to carbon content

period, a start and then a gradual increase in speed of decomposition of the austenite which reaches a maximum at about 50% transformation and then a slow completion. An isothermal transformation diagram which gives a summary of the progress of isothermal decomposition of austenite at all temperatures between A_3 and the start of martensitic transformation can be constructed. This is done by quenching small specimens of a steel (which have been held for the same time at a fixed temperature in the austenite field above Ar_3) to the temperature at which transformation is desired, holding for various times at this temperature and determining the proportion of transformed austenite. Such a diagram provides information on the possibilities of applying isothermal heat treatment to bring about complete decomposition of the austenite just below A_1 (isothermal annealing) or just above M_s (austempering), or of holding the steel at subcritical temperatures for a suitable period to reduce temperature gradients set up in quenching without break down of the austenite as in martempering or stepped quenching. Furthermore, if the steel is air hardening or semi-air-hardening the cooling rate during most welding processes exceeds the 'critical rate' so, by using the isothermal diagram, the preheat temperatures and time necessary to hold a temperature to avoid martensite and obtain a bainitic structure can be assessed.

The principle of the isothermal diagram, also known as the time–temperature transformation (T-T-T) diagram, is illustrated schematically in *Figure 1.19*. The dotted lines showing the estimated start and finish of transformation indicate the uncertainty of determining with accuracy the start and finish. The main feature of isothermal transformation, the considerable difference in time required to complete transformation at different temperatures within the pearlitic and bainitic temperature ranges, should be noted. These diagrams vary in form for different steels. They also vary according to austenitising temperature (coarseness of γ grains) and the extent to which carbides are dissolved in the austenite.

In alloy steels containing chromium, molybdenum or tungsten, segregation and carbide banding (size or carbides) varies and can affect the extent of carbide solution. In applying these diagrams it is usual to allow a considerably longer time for completion of transformation than the time indicated on the diagram, in order to cover the inherent uncertainties in individual consignments of steels.

Furthermore, martensite, which is characterised by an acicular appearance, forms progressively over a temperature range as the temperature falls; if the temperature is held constant after the start no further action takes place. Martensite formation produces an expansion related to the carbon content. The mechanical properties of martensite depend on the carbon content; low carbon martensites (less than 0.08%C) have reasonable ductility and toughness, high carbon martensites have no ductility or toughness and extreme hardness and, because of the state of internal stress, are very liable to spontaneous cracking. Thus low carbon martensite can be used for industrial purposes, e.g. welded 9%Ni steels for low-temperature applications have low carbon martensitic heat-affected zones. High carbon martensite must be tempered before it is allowed to cool to room temperature, e.g. carbon tool steels are water quenched to exceed the critical cooling rate, but the tool is withdrawn from the bath while still hot and immediately tempered.

1.3.5.5 Isothermal decomposition of austenite

Reference was made in the previous section to the fact that if the γ to α transformation is suppressed by fast cooling, the austenite is in an unstable condition. If, before reaching the temperature at which martensite begins to form, the cooling is arrested and the steel is held at a constant temperature, the unstable austenite will transform over a period of time to a product which differs markedly from pearlite and has some visual resemblance to martensite in being acicular. This structure is called bainite; it is formed over a range of temperatures (about 550–250°C) and its properties depend to some degree on the transformation temperature. Bainite formed at a lower temperature is harder than bainite formed at a higher temperature. It is tougher than pearlite and not as hard as martensite. It differs fundamentally from the latter by being diffusion dependent as is pearlite.

This type of transformation, at constant temperature, is important in the heat treatment of steel and is called 'isothermal transformation'. It is characterised by an induction

1.3.5.6 Effect of carbon and alloying elements on austenite decomposition rate

As the carbon content is increased the isothermal diagram is moved to the right which indicates that austenite transformation is rendered more sluggish. Alloying elements increase the induction period thus delaying the start and they also increase the time necessary for completion. Furthermore, the effect of adding alloying elements is cumulative but, because they have different specific effects on transformation in the pearlitic or bainitic ranges, it is not generally possible to predict the behaviour of multialloy steels.

1.3.5.7 Decomposition of austenite under continuous cooling conditions

It will be appreciated that, while the isothermal transformation diagram provides the basic information about the characteristics of isothermal transformation for austenite of given composition, grain size and homogeneity, the common heat treatments used in steel manufacture such as annealing, normalising or quenching are processes which subject the austenite to continuous cooling. This does not necessarily invalidate the use of isothermal diagram data for continuous

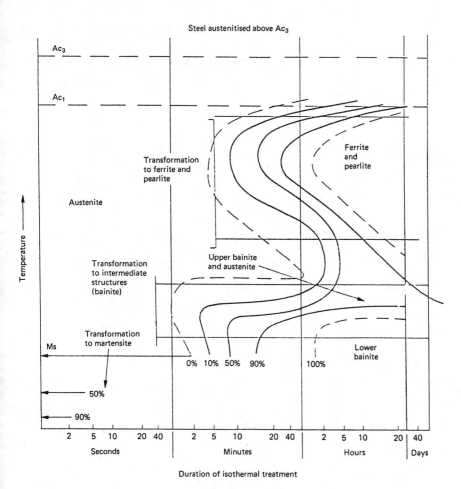

Figure 1.19 Schematic isothermal transformation diagram

cooling conditions because, as the steel passes through successively lower temperatures, the microstructures appropriate to transformation at the different temperatures are formed to a limited extent depending on the time allowed instead of proceeding to completion. The final structure consists of a mixture which is determined by the tendency to form specific structures on the way down, this tendency being indicated by the isothermal diagram.

The time allowed for transformation in the ferrite–pearlite and intermediate (bainite) regions obviously depends on cooling rate. A continuous transformation diagram will therefore have as its essential features means for indicating the amount of ferrite, pearlite, bainite and martensite which is obtained at various defined cooling rates; these are usually appropriate to heat treatment or selected welding cooling rates. Such a diagram is shown schematically in *Figure 1.19*.

The effect of continuous cooling is to lower the start temperatures and increase the incubation period so the transformation time tends to be below and to the right of the isothermal line for the same steel, these effects increasing with increasing cooling rate. As indicated in *Figure 1.19* the time axis may be expressed in any suitable form; e.g. as transformation time (*Figure 1.20(a)*) or as the bar diameter for bars (*Figure 1.20(b)*).

The positions of the lines defining the transformation products obviously vary according to the steel composition and austenitising temperature. Diagrams for welding applications, in which five cooling rates appropriate to the main fusion welding processes are applied to various steel thicknesses, have been produced by the Welding Institute, Cambridge, and by other welding research institutions in connection with the development of weldable high tensile steels.

Manipulation of composition and heat treatment gives rise to the several classes of steel already listed.

1.3.6 Carbon/carbon–manganese steels

Rolled or hollow sections of carbon steels with carbon below about 0.36% constitute by far the greatest tonnage of steels used. Besides the general specification of steels by analysis they are sold by specification depending on product form and BS 970 is applied mainly to bar.

1.3.6.1 Weldable structural steels (specifications BS 4360: 1970 and ISO R630)

These steels have yield strengths depending on section between 210 and 450 MN m^{-2} achieved by carbon additions between 0.16 and 0.22%, manganese up to 1.6% and, for some qualities, niobium and vanadium additions.

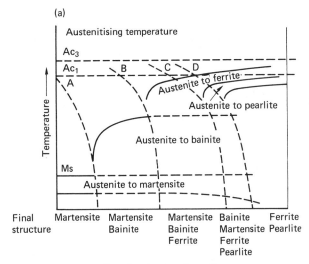

(a)

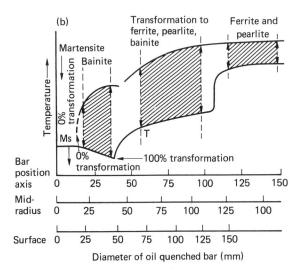

(b)

Figure 1.20 Continuous cooling time–temperature–transformation diagrams. (a) Applicable to forgings, plates and sections. (b) Applicable to heat treatment of bars

1.3.6.2 Structural plates

These products exemplify more than any others the quality improvements that the improvements in steelmaking described in Section 1.3.3 have produced in tonnage steels. Plates can now be obtained with:

(1) lower maximum sulphur levels (as low as 0.008%);
(2) improved deoxidation with low inclusions and controlled morphology;
(3) very low hydrogen levels resulting from vacuum degassing;
(4) greater control of composition resulting from secondary steelmaking units and rapid in plant analysis, low inclusions and controlled morphology;
(5) guaranteed high impact and elongation in the transverse direction; and

(6) high impact values at low temperature in heat affected zones.

There are many private specifications, primarily for material for offshore structures. For example, British Steel Corporation's 'Hized' plate will give reduction in area values through the plate thickness of around 25%.

Plates with superior properties, such as are used for oil pipelines, are made by controlled rolling steels such as BS 4360, grade 50E containing up to 0.1%Nb and/or 0.15%V and, although this is not explicitly specified, small amounts of nitrogen. Controlled rolling produces appreciably higher strength, e.g. yield and tensile values up to 340 and 620 MPa in a very fine grained steel due to precipitation of carbonitrides and the low carbon equivalent promotes weldability.

Besides plates, weldable structural steels are available in the form of flats, sections, round and square bars, blooms and billets for forging, sheet, strip and tubes. The range of flats, sections and bar is slightly restricted compared with plates, and properties show minor variations.

A very wide range of beams guides and columns may be fabricated by automatic welding of plate steels.

Increased use is being made of hollow sections, because they take up less space than angles or I sections, decrease wind resistance and allow increased natural lighting and because, with care in design, they need not be protected on the inside, and are cheaper to paint. Cold forming sections increases strength and improves finish.

Forgings in weldable structural steels are included in BS 970.

Tubes specified in BS 6323 may be hot or cold finished, seamless or welded in various ways. Yield strengths of hot finished carbon steel tubes vary between 195 and 340 MPa and cold finished between 320 and 595 MPa.

Cold finished tubes are available in a variety of heat treatments.

The cheapest available steels to the specifications listed may, if purchased from a reputable steel maker, be used with confidence for most engineering purposes (with the exception of pressure vessels). If service conditions are known to be onerous, more demanding specifications and increased testing may be required.

1.3.6.3 Pressure vessel steels

The range of engineering plates, tubes, forgings (and, included here for convenience, castings) is matched by equivalent specifications for pressure vessel steels.

Pressure vessel plate steels, specified in BS 1501: 1980: Part 1 are similar to structural steels, but differ in the following ways.

(1) Pressure vessel steels are supplied to positive dimensional tolerances, instead of the specified thickness being the mean. A batch of pressure vessel plates will, therefore, weigh more than the equivalent batch of structural plates (and cost more). A tensile test must be carried out on every plate (two for large plates) instead of one test per 40 t batch.
(2) Elevated temperature proof tests are specified for all pressure vessel plates.
(3) All pressure vessel plates have the nitrogen content specified and some the soluble aluminium content.
(4) All pressure vessel plates are supplied normalised.
(5) Pressure vessel tube steels are similar to those used for plates but, to facilitate cold bending, some of the grades are softer. The relevant specifications are: for seamless tube BS 3601: 1974; for electric welded tube BS 3602: 1978; and for submerged arc welded tube BS 3603: 1977.

(6) Yield points lie between 195 and 340 MPa and Charpy V notch impact must exceed 27 J at −50°C.

(7) For lower temperature service steels with up to 9%N, austenitic stainless or even martempered steels should be used.

Carbon–manganese steel forgings for pressure vessels are specified in BS 1503: 1980. Materials are available with yield strengths varying (depending on section) between 215 and 340 MPa.

Carbon–manganese steel castings for pressure vessels are specified in BS 1504: 1976. These castings may contain up to about 0.25% chromium, molybdenum, nickel and copper (total maximum 0.8%) and 0.2% proof stresses range between 230 and 280 MPa.

1.3.6.4 Coil and sheet steel

Basic oxygen furnace (BOF) steel is continuously cast into slabs and rolled hot to coil or cut sheet.

Hot rolled strip is available in thicknesses above 1.6 mm up to 6.5 mm pickled and oiled and 12.7 mm as rolled in widths varying up to 1800 mm in:

(1) forming and drawing quality aluminium killed,
(2) commercial quality, and
(3) tensile qualities to BS 1449: Part 2 and BS 4360,

in a variety of specified minimum yield strengths above 280 MPa. Weathering steel, which develops an adherent coating of oxides and raised pattern floor plate, is also available hot rolled.

Cold reduced strip is available in thicknesses above 0.35 mm up to 3.175 mm and in widths varying up to 1800 mm in:

(1) forming and drawing qualities (typically 180 MPa yield ultimate tensile strength (UTS) 620–790 MPa to BS 1449: Part 1); and
(2) tensile qualities with yield points for low carbon phosphorus containing steels of 125 and 270 MPa and microalloyed with niobium of 300 and 350 MPa.

Cold rolled narrow strip is available to BS 1449 and other more exacting specifications in thicknesses between 0.1 and 4.6 mm and widths up to 600 mm.

Cold rolled strips may be supplied in a variety of finishes, hot dip galvansied to BS 2989, electrogalvanised, electro zinc coated, ternplate (coated with a tin–lead alloy which facilitates forming and soldering) or coated with a zinc–aluminium alloy with exceptional corrosion resistance.

1.3.6.5 Steel wire

Wire with carbon contents ranging from 0.65 to 0.85% is specified in BS 1408. Carbon steel wire in tensile strengths of 1400–12 050 MPa for coiled springs and 1400–1870 MPa for zig-zag and square-form springs are listed in BS 4367: 1970 and BS 4368: 1970, respectively. The heat treatment of wires, including annealing and patenting differs appreciably from other heat-treatment processing.

The increase in tensile strength as the amount of drawing increases is shown for three carbon ranges in *Figure 1.21*. Ductility falls as the tensile strength increases (*Figure 1.22*). When the limit of reduction has been reached the wire must be heat treated to remove the hard drawn structure and replace it by a suitable structure for further reduction. For low carbon steel this treatment is an anneal, just below the lower critical temperature, which recrystallises the ferrite grains to an equiaxed form. Medium and high carbon wires are generally

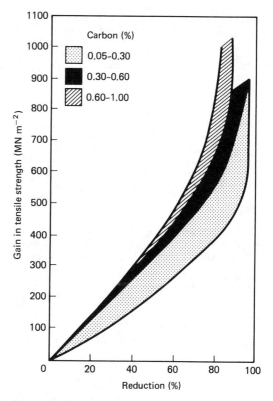

Figure 1.21 Increase in tensile strength related to amount of reduction in wire drawing for three levels of carbon

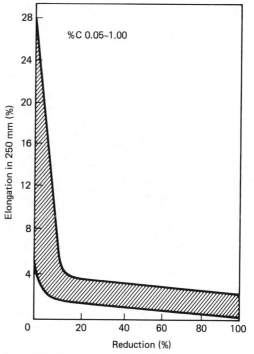

Figure 1.22 Decrease in ductility related to amount of reduction in wire drawing

patented (farily fast cooling from above the upper critical point by air cooling or quenching in lead) to give a coarse pearlitic structure which will draw to very high tensile strengths. Additional to subcritical annealing and patenting, the heat treatments used in wire production include, normalising, annealing, hardening and tempering and austempering, all of which are designed to confer structures and properties which have particular relevance to the requirements of specific wire applications.

The tensile strength obtainable depends on carbon content and an approximate indication of the relationship for annealed, patented and hardened and tempered wire is shown in *Figure 1.23*. Wire has a relatively large surface-to-volume ratio so that any decarburisation due to heat treatment has a proportionately more significant effect than in heavier steel products. Consequently, wire heat treatment is conducted in specialised equipment (i.e. salt baths, atmosphere controlled furnaces, etc.) aimed at minimising any such difficulties.

Cold drawing through dies requires considerable skill and attention to detail in die design, lubricants, wire rod cleansing and baking to remove hydrogen introduced during cleaning.

1.3.7   High strength low alloy steels

High strength low alloy steels (HSLA steels) are proprietary steels manufactured to SAE 950 or ASTM 242 with carbon (0.22% maximum), manganese (1.25% maximum) and such other alloying elements as will give the minimum yield point prescribed for various thicknesses ranging between 12 and 60 mm. Steels are available with yield points ranging from 275 to 400 MPa, and the restriction on carbon and manganese content is intended to ensure weldability. Quenched and tempered welded steels with significantly higher yields are also available.

Reduction in weight of steel gained by utilising the higher yield stress in design is unlikely to reduce the cost of the material compared with that of the greater weight of a standard weldable structural steel purchased from British Steel Corporation.

Cost benefits arise, however, from handling the smaller quantity and welding the reduced thickness of the steel and, in transport applications from increased pay load, decreased fuel costs, freedom from weight restrictions and reduced duty imposed on other components of the vehicle.

Attention must be paid to the following factors.

(1)   The modulus of elasticity of a HSLA steel is the same as that of other ferritic steels. Therefore any design which is buckling critical will require stresses and thus sections identical to those of steels of lower strengths, and there will be no saving in the quantity of steel.

(2)   Stress intensity is proportional to the second power of stress and fatigue growth rate per cycle is proportional to the fourth power of the range of stress per cycle. If brittle or fatigue fracture is a ruling parameter in design, a much more severe standard of non-destructive testing is needed for a component made from steel operating at a higher stress. In the limit the critical defect size may fall below the limit of detection.

(3)   The notch ductility of an HSLA steel varies greatly according to the alloying elements used by the steelmaker.

If there is a risk of brittle fracture, values of Charpy V notch energy and transition temperature should be specified by the designer. Spectacular failures have resulted from ignoring these precepts.

1.3.8   Electrical steels

Electrical steels comprise a class of steel strip which is assembled and bolted together in stacks to form the magnetic cores of alternating current plant, alternators, transformers and rotors. Its essential properties are low losses during the magnetising cycle arising from magnetic hysteresis and eddy currents, high magnetic permeability and saturation value, insulated surfaces, and a low level of noise generation arising from magnetostriction.

These parameters are promoted by maintaining the contents of carbon, sulphur and oxygen to the minimum obtainable and increasing grain size which together minimise hysteresis loss and incorporating a ferrite soluble element (usually silicon) at a level of 3% to increase resistivity and thus reduce eddy current loss. The thickness of the steel must be optimised—reduction in thickness minimises the path available for eddy currents but reduces the packing fraction and hence the proportion of iron available and increases handling problems. The surfaces are coated with a mineral insulant to prevent conduction of eddy currents from one lamination to the next. Accurate control of thickness and flatness minimises stress when the laminations are bolted together and, therefore, reduces magnetostrictive noise which is promoted by stress.

There are two principal grades of electrical steel differing essentially in loss characteristics. Hot-rolled strip is supplied to ASTM 840-85 in gauges of 0.47 and 0.64 mm with guaranteed losses of 13.2 and 16 W kg^{-1} at 15 kG induction and 60 Hz. Cold-rolled strip is supplied to ASTM 843–85 in gauges of

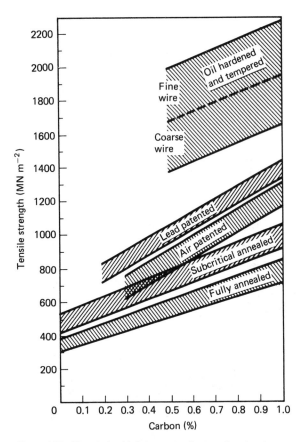

Figure 1.23 The relationship between tensile strength and carbon content for wire

0.27, 0.3 and 0.35 mm with guaranteed losses of 1.10, 1.17 and 1.27 W kg^{-1}, respectively, at 17 kG induction and 50 Hz.

Cold-rolled strip is manufactured by first rolling a sulphurised steel, followed by a programme of rolling and heat treatment which eliminates sulphur and produces a Goss or 'rooftop' texture. In this structure the [1 0 0] crystallographic direction, which is the one most easily magnetised, lies longitudinally in the strip. Cold-rolled strip is normally used for large alternators and transformers where the saving in lost power (and the problems of disposing of heat generated) outweigh the additional cost compared with hot rolled strip.

1.3.9 Hardened and tempered steels

At a carbon content above about 0.35%, or less when alloying elements are present, useful increases in strength may be obtained by transformation. The most important class of steel to which this procedure is applied is the 'hardened and tempered steels'. These will be chosen from AISI/SAE 1035–4310 and BS 970 080A32–945A40.

1.3.9.1 Heat treatment

The steel heat treatments, quenching and tempering, austempering, martempering, annealing and isothermal annealing can be described most simply by means of the isothermal diagram (*Figure 1.20*). (There are other heat treatment procedures, notably ageing and controlled rolling).

1.3.9.2 Quenching and tempering (Figure 1.24(a))

Steel quenched to martensite is hard and brittle due to the carbon being in unstable solid solution in a body-centred tetragonal lattice[15] and has high internal stresses. Heating (tempering) at 100°C causes separation of a transition phase, ε, iron carbide ($Fe_{2.2}C$) from the matrix, this being the first stage of tempering; slight hardening may occur initially. As the temperature is increased, relief of stress and softening occurs due to cementite formation and release of carbon from the matrix. The steel becomes significantly tougher.

Steels of suitable composition quenched fully to martensite and tempered at appropriate temperatures give the best combination of strength and toughness obtainable. There is a tendency, varying with different steels, for a degree of embrittlement to occur when tempering within the range 250–450°C, so steels are either tempered below 250°C for maximum tensile strength, or above about 550°C for a combination of strength, ductility and toughness due to increasing coalescence of carbides.

1.3.9.3 Austempering (Figure 1.24(b))

The purpose of this treatment is to produce bainite from isothermal treatment; lower bainite is generally more ductile than tempered martensite at the same tensile strength but lower in toughness. The main advantage of austempering is that the risk of cracking, present when quenching out to martensite, is eliminated and bainitic steels are therefore used for heavy section pressure vessels.

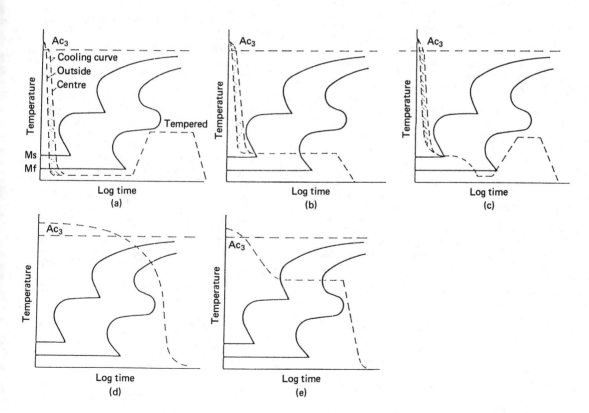

Figure 1.24 Isothermal diagrams showing the heat treatment of steel. (a) Quenching and tempering to give tempered martensite. (b) Austempering to give lower bainite. (c) Martempering to give tempered martensite. (d) Annealing to give ferrite and pearlite. (e) Isothermal annealing to give ferrite and pearlite

1.3.9.4   Martempering (Figure 1.24(c))

The risk of cracking inherent in quenching to martensite can be reduced considerably while retaining transformation to martensite by quenching into a salt bath which is at a temperature slightly above that at which martensite starts to form and then, after soaking, allowing the steel to air cool to room temperature. Distortion in quenching is a problem in pieces of non-uniform section and this is also considerably reduced by martempering.

1.3.9.5   Annealing (Figure 1.24(d))

Maximum softness is attained by annealing, involving slow cooling through the ferrite–pearlite field. The pearlitic structure developed provides optimum machinability in medium carbon steels.

1.3.9.6   Isothermal annealing (Figure 1.24(e))

This treatment is used to produce a soft ferrite–pearlite structure. Its advantage over annealing is that, with appropriate steels and temperatures, it takes less total time because cooling down both to and from the isothermal treatment temperature may be done at any suitable rate, provided the material is not too bulky or being treated in large batches.

1.3.9.7   Hardenability of steel

In this context 'hardenability' refers to the depth of hardening not the intensity. Hardening intensity in a quench is dependent on the carbon content. Plain carbon steels show relatively shallow hardening; they are said to have 'low hardenability'. Alloy steels show deep hardening characteristics, to an extent depending primarily on the alloying elements and the austenitic grain size.

Hardenability is a significant factor in the application of steels for engineering purposes. Most engineering steels for bar or forgings are used in the oil quenched and tempered condition to achieve optimum properties of strength and toughness based on tempered martensite. It is in this connection that hardenability is important; in general, forgings are required to develop the desired mechanical properties through the full section thickness.

Since the cooling rate in a quench must be slower at the centre of a section than at the surface, the alloy content must be such as to induce sluggishness in the austenite transformation sufficient to inhibit the ferrite–pearlite transformation at the cooling rate obtaining at the centre of the section. It follows that, for a given steel composition and quenching medium, there will be a maximum thickness above which the centre of the section will not cool sufficiently quickly except in those steels which have sufficient alloy content to induce transformation to martensite in air cooling (air hardening steels).

The practical usefulness of engineering steels, ignoring differences in toughness, can therefore be compared on the basis of this maximum thickness of ruling section which must be taken into account when considering selection of steel for any specific application.

A method for determining hardenability is to cool a bar of standard diameter and length by water jet applied to one end only. The cooling rate at any position along the bar will progressively decrease as the distance from the water sprayed end increases. The hardness is determined on flats ground at an angle of 180° on the bar surface. The greater the hardenability the further along the bar is a fully martensitic structure developed. This method of assessment is known as the 'Jominy end-quench test': for full details see BS 3337.

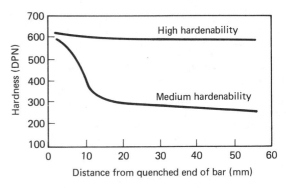

Figure 1.25   End-quench (Jominy) curves for steels of medium and high hardenability

Typical end quench (Jominy) curves for steels of medium and high hardenability are shown in *Figure 1.25*. A relationship between end-quench hardenability curves and the diameter of oil quenched bars is shown in *Figure 1.26*. This can be used to choose a size of bar which will harden fully.

Jominy curves are provided by the SAE/AISI for steels to which the letter 'H' is added to the specification number and to BS 970 steels with the letter 'H' in the specification.

Alternatively, a steel which will through harden to the required yield stress at the design diameter may be selected from *Table 1.6*.

1.3.9.8   The function of alloying elements in engineering alloy steels

Aside from specialised functions—corrosion resistance, abrasion resistance, etc.—alloying elements are most widely used in engineering alloy steels with carbon in the range 0.25–0.55% or less than 0.15% for case hardening. Their function is to improve the mechanical properties compared with carbon steel and, in particular, to make possible the attainment of these properties at section thicknesses which preclude the use of shallow hardening carbon steels, water quenched. They increase hardenability and, thereby, allow a lower carbon content to be used than would be required in a carbon steel and the use of a softer quenching medium, e.g. oil. This substantially reduces quench cracking risks.

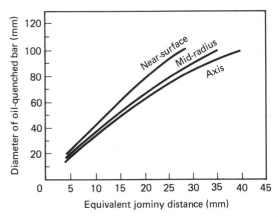

Figure 1.26   Relationship between end-quench hardenability curves and oil-quenched bars

The alloying elements used are manganese, nickel, chromium, vanadium and aluminium (as grain refining element). An important function of alloying elements, by rendering austenite transformations sluggish, is to make possible treatments which depend on an arrested quench followed by a timed hold at somewhat elevated temperature (austempering and martempering) which reduce internal stress and minimise distortion and cracking risks.

For full effectiveness in increasing hardenability, the alloy elements should be completely dissolved in the austenite before quenching; this is no problem with manganese and nickel, but chromium, molybdenum and vanadium form carbides which, in the annealed steel prior to quenching, may be of comparatively large size and, owing to a slower solution rate than cementite, are more difficult to dissolve. Solution temperatures may therefore be increased and/or times increased.

The effect of alloying elements when tempering is important; in general they retard the rate of softening during tempering compared with carbon steel, but in this respect the effect of the carbide formers chromium, molybdenum and vanadium is much greater than that of the other elements. They increase the tempering temperatures required for a given degree of softening, which is beneficial for ductility and toughness. Molybdenum and vanadium, at higher levels, confer an increase in hardness at higher tempering temperatures, due to alloy carbide precipitation; this is 'secondary hardening' and is the basis of hardness in heat treatment alloy tool steels. The effect of individual elements on the properties of steel is given in *Table 1.7*.

1.3.10 Free cutting steels

Most free cutting steels and those with the largest number of, and the most important, applications are carbon/carbon–manganese steels. Some hardened and tempered and a few stainless steels are also free cutting.

AISI/SAE free cutting carbon/carbon–manganese steels have 11 or 12 as the first two digits instead of 10 and the BS 970 designations have as the first digit a '2' while the second and third figures indicate the mean, or the maximum, sulphur content.

Free cutting steels are really composites with additions which form a soft particulate second phase which acts as 'chip breaker' during machining. This reduces tool wear, greatly diminishes the time and cost of machining, and makes it easier to obtain a good finish.

The addition is usually sulphur in amounts of 0.1–0.33%. These steels were formerly manufactured by using a less effective sulphur removing slag, but the present procedure is to resulphurise and the additional processing stage results in a slightly higher price for free cutting steels. There is no systematic nomenclature for direct hardening resulphurised alloy steels.

Additions of lead in amounts of 0.15–0.35% in addition to sulphur make steel even easier to machine.

Specifications indicate leaded steels by inserting an 'L' as an additional third letter in AISI/SAE grade numbers or adding 'Pb' to BS 970 grade designations.

Free cutting austenitic steels are limited to 303 or 303 Se which are standard 18/8 304 steels with sulphur or selenium additions. Free cutting versions of 13%Cr, steels are available to BS 970 416S21, 416S29 and 416S37.

The particulate phase in free cutting steels reduces their resistance to fatigue and may introduce other drawbacks. Free cutting steels may be safely used in low duty applications in non-aggressive environments for components which are not to be welded. It is essential, however, to ensure that components

for severe duties are not made from them. This is of great importance when ordering components from a machining firm which will supply components made from free cutting steel wherever possible to reduce costs. In particular, the designation 18/8 should not be used when ordering a steel as the supplier can supply 303 *or* 304. The AISI number should always be specified.

1.3.11 Case hardening steels

Case hardening produces a very hard wear and fatigue resisting surface on a core which is usually softer but stronger and tougher than that of a hardened and tempered steel. Besides its obvious advantages, case hardening usually improves fatigue endurance, partly because of the compressive stress induced at the surface.

There are at least five different processes:

(1) surface hardening;
(2) carburising;
(3) carbonitriding;
(4) nitriding; and
(5) ion implantation.

1.3.11.1 Surface hardening

Surface hardening is achieved by austenitising only the surface of the steel by applying a high heat flux by electrical induction or by direct flame impingement, and then quenching in moving air, water or oil. Any steel of sufficiently high carbon content may be surface hardened. Those most usually employed are carbon and free cutting steels with 0.45–0.65%C and hardened and tempered steels with 0.35 to 0.55%C.

The properties of the core are those to which the steel has originally been heat treated while hardnesses of from 50 to 65 Rockwell C are produced on the case. These hardnesses are lower than those available from other case hardening processes but surface hardening is very versatile.

The depth of case produced by induction hardening may be varied by varying the frequency from 0.64 mm at 600 kHz to 5 mm at 1 kHz. This is a much thicker case than can be produced by any other method and is very valuable for combating abrasive wear.

In flame hardening the surface is heated by one or more gas burners before quenching. The process can be applied to work pieces whose shape and size preclude other methods of case hardening.

1.3.11.2 Carburising

Any carbon, free cutting or direct hardening alloy steel with 0.23% or less carbon is suitable for carburising. The steel should be chosen according to the properties desired in the core. BS 960 and SAE publish lists of carburising steels with hardenability data. Core strengths between 500 and 1310 MPa are available and Charpy impact toughness up to 55 J (68 with 5%Ni, 0.15%Mo steel). Case hardnesses of 64 Rockwell C for low hardenability steels and 60 Rockwell C for high hardenability steels can be obtained and the case, which contains a proportion of cementite, is hard wearing.

Carburising is achieved by exposing the surface of the steel to a gas or liquid with a high carburising potential at a temperature up to 925°C. Surfaces not required to be carburised should be masked, possibly by copper plating or better; the carburised layer should be machined off before it has been hardened. There are three processes.

In pack carburising the component(s) are placed in a heat-resisting box surrounded by a carburising powder consist-

Table 1.6 BS 970 and BS 4670 steels classified by tensile strength and maximum diameter, hardened and tempered*

Max. diameter (in.)	(mm)	40–50 ton in.$^{-2}$ 630–790 MN m^{-2}			45–55 ton in.$^{-2}$ 690–850 MN m^{-2}			50–60 ton in.$^{-2}$ 750–930 MN m^{-2}			55–65 ton in.$^{-2}$ 85–1000 MN m^{-2}			60–70 ton in.$^{-2}$ 930–1080 MN m^{-2}		
½	12	070M26	0.26C	215–430	080M36 212M36	0.36C 0.36C	245–480 310–495	080M46 150M28 212M44	0.46C 0.28C1.5Mn 0.44C	280–555 325–570 370–540	080M50 150M36 225M44	0.50C 0.36C1.5Mn 1.5Mn	280–595 355–635 415–600			
¾	19	080M30 216M28	0.3C 0.28C	230–450 325–430	080M40 120M19	0.4C 0.19C1.2Mn	345–510 265–510	120M36	0.3C1.2Mn	340–570	070M55	0.55C	310–620			
⅞	22							503M40	1Ni	295–585						
1⅛	28	120M19	1.2Mn	265–510	080M46 150M19 120M28 120M36 216M36	0.46C 1.5Mn 1.2Mn 1.2Mn 1.5Mn	280–555 295–510 310–510 355–570 310–510	080M50 150M36 225M44	0.50C 1.2Mn 1.5Mn	280–585 355–635 415–600	530M40 606M36	1Cr 1.5MnMo	510–680 510–680	606M30 605M36 640M40 708M40 945M38	1.5MnMo 1.5MnMo 1.25NiCr 1CrMo 1.5MnNiCrMo	510–850 480–850 510–755 510–755 480–850
2½	64	216M36 212M36 225M36 150M19	0.36C 0.36C 0.36C 1.5Mn	310–510 310–495 370–480 295–510	503M40 080M50 150M28 150M36 212M44	1Ni 0.5C 1.5Mn 1.5Mn 0.44C	295–585 280–585 325–570 355–635 370–540	070M55 530M40 606M36 905M31	0.55C 1Cr 1.5MnMo 1.25CrAlMo	310–620 510–680 510–680 510–585	605M30 605M36 640M40 708M40 905M39 945M38	1.5MnMo 1.5MnMo 1.25NiCr 1CrMo 1.5CrAlMo 1.5MnNiCrMo	510–850 480–850 510–755 510–755 510–680 480–850	608M38 653M31 709M40 818M40	1.5MnMo 3NiCr 1CrMo 1.5NiCrMo	480–850 570–755 480–850 540–850
4	100	080M46 120M28 120M36 503M40	0.46C 1.2Mn 1.2Mn 1Ni	280–555 310–510 340–570 295–585	070M55 212M44 225M44 530M40 606M36 905M31	0.55C 0.44C 1.5Mn 1Cr 1.5MnMo 1.5CrAlMo	310–620 370–540 415–600 510–680 510–680 510–585	605M30 605M36 640M40 708M40 905M39 945M38	1.5MnMo 1.5MnMo 1.25NiCr 1CrMo 1.5CrAlMo 1.5MnNiCrMo	510–850 480–850 570–755 510–755 510–680 480–850	526M60 680M38 653M31 709M40 816M40	0.75Cr 1.5MnMo 3NiCr 1CrMo 1.5NiCrMo	620–740 480–850 570–755 480–850 540–850	917M40	1.5NiCrMo	635–1240
6	150	150M28 150M36 503M40 785M19	1.5Mn 1.5Mn 1Ni 1.5MnNiMo	325–530 355–635 295–585 430–465	605M30 605M36 608M38 640M40 708M40 905M39 945M38	1.5MnMo 1.5MnMo 1.5MnMo 1.25NiCr 1CrMo 1.5CrAlMo 1.5MnNiCrMo	510–850 480–850 480–850 510–755 510–755 510–680 480–850	608M38 653M31 709M40 816M40	1.5MnMo 3NiCr 1CrMo 1.5NiCrMo	480–850 570–755 480–850 540–850	722M24 817M40 823M30 826M31 830M31	3CrMo 1.5NiCrMo 2NiCrMo 2.5NiCrMo 3NiCrMo	635–755 635–1240 635–1235 635–1235 635–940	722M24 826M31 826M40 830M31	3CrMo 2.5NiCrMo 2.5NiCrMo 3NiCrMo	635–755 635–1235 635–1235 635–940
10	250	503M40 785M19 722M29	1Ni 1.5MnNiMo 3.25CrMo	295–585 430–465 450–780	605M36 608M38 709M40 945M38 711M40 722M29	1.5MnMo 1.5MnMo 1CrMo 1.5MnNiCrMo 1CrMo 3.25CrMo	480–850 480–850 480–850 480–850 480–600 450–780	608M38 709M40 816M40 711M40 818M40 826M31	1.5MnMo 1CrMo 1.5NiCrMo 1CrMo 1.5NiCrMo 2.5NiCrMo	480–850 480–850 540–850 480–600 590–780 635–1235	722M24 817M40 823M30 826M31 830M31 818M40 826M31 826M40 897M39 722M29 976M33	3CrMo 1.5NiCrMo 2NiCrMo 2.5NiCrMo 3NiCrMo 1.5NiCrMo 2.5NiCrMo 2.5NiCrMo 3.25CrMoV 3.25CrMo 3.25NiCrMoV	635–755 635–1240 635–1235 635–1235 635–940 590–780 635–1235 640–900 640–940 450–780 700–980	823M30 826M31 826M40 818M40 826M31 976M33 897M39	2NiCrMo 2.5NiCrMo 2.5NiCrMo 1.5NiCrMo 2.5NiCrMo 3.25NiCrMoV 3.25CrMoV	635–1235 635–1235 635–1235 590–780 635–1235 700–980 640–940
20	500	785M19 722M29	1.5MnNiMo 3.25CrMo	430–465 450–780	711M40 722M29	1CrMo 32.5CrMo	480–600 450–780	818M40 826M31 722M29	1.5NiCrMo 2.5NiCrMo 3.25CrMo	590–780 635–1235 450–780	818M40 826M31 826M40 722M29 976M33 897M39	1.5NiCrMo 2.5NiCrMo 2.5NiCrMo 3.25CrMo 3.25NiCrMoV 3.25CrMoV	590–780 635–1235 640–900 450–780 700–980 640–940	818M40 976M33 897M39	1.5NiCrMo 3.25NiCrMoV 3.25CrMoV	590–780 700–980 640–940
39	1000	722M29	3.25CrMo	450–780	722M29	3.25CrMo	450–780	818M40 826M31 722M29	1.5NiCrMo 2.5NiCrMo 3.25CrMo	590–780 635–1235 450–780	826M40 976M33 897M39	2.5NiCrMo 3.5NiCrMoV 3.25CrMoV	640–900 700–980 640–940	976M33 897M39	3.25NiCrMoV 3.25CrMoV	700–980 640–940

* Reproduced from *The Fulmer 'Optimiser'* by permission of Elsevier Ltd.

1 For each tensile-strength block, the three sub-columns denote:
  1st Present designation of the steel in BS 970. the 'M' in the designation indicates that the steel is to be ordered to specific mechanical property requirements. However, many steels may be ordered to analysis only, when the letter becomes 'A', or to analysis and hardenability (end-quench) specification, when the letter becomes 'H'. The first three figures indicate the broad analysis classification, the last two the carbon content.
  2nd Type of steel by broad analysis. These do not give full analyses, only the medium content of the leading element and a list of other alloying elements. For the straight carbon steels, the carbon content is repeated here, but for the alloy steels the carbon content can be inferred from the designation. Compositions are given as per cent by weight.
  3rd Yield stress range available for the steel at the specified equivalent diameter.
2 Steels marked with an * are free-machining qualities.

Tensile strength														
65–75 ton in.⁻² 1000–1160 MN m⁻²			70–80 ton in.⁻² 1080–1240 MN m⁻²			75–85 ton in.⁻² 1160–1310 MN m⁻²			80–90 ton in.⁻² 1240–1390 MN m⁻²			>100 ton in.⁻² >1540 MN m⁻²		
605M30	*1.5MnMo*	510–850												
605M36	1.5MnMo	480–850												
608M38	*1.5MnMo*	480–850	817M40	1.5NiCrMo	635–1240	817M40	1.5NiCrMo	635–1240				817M40	1.5NiCrMo	635–1240
709M40	1CrMo	480–950										897M39	3.25CrMoV	640–940
816M40	*1.5NiCrMo*	540–850												
945M38	1.5MnNiCrMo	480–850												
526M69	0.05Cr	620–740	830M31	3NiCrMo	635–940	823M30	2NiCrMo	635–1235				823M30	2NiCrMo	635–1235
817M40	1.5NiCrMo	635–1240				826M31	2.5NiCrMo	635–1235				826M31	2.5NiCrMo	635–1235
												897M39 (85min.)	3.75CrMoV	1110–1235
830M31	3NiCrMo	635–940	823M30	2NiCrMo	635–1235							826M40	2.5NiCrMo	640–900
			826M31	2.5NiCrMo	635–1235									
823M30	2NiCrMo	635–1235	826M40	2.5NiCrMo	640–900	826M40	2.5NiCrMo	640–900	826M40	2.5NiCrMo	640–900	835M30	4NiCrMo	1125–1235
826M31	2.5NiCrMo	635–1235												
826M40	2.5NiCrMo	640–900												
826M40	2.5NiCrMo	640–900	826M40	2.5NiCrMo	640–900									
897M39	3.25CrMoV	640–940	976M33	3.25NiCrMoV	700–980									
976M33	3.25NiCrMoV	700–980	897M39	3.25CrMoV	640–940									

3 Steels with first three digits '905' have high aluminium contents and are specifically intended for surface-hardening by nitriding. However, steels 722M24 and 897M39 are also suitable for nitriding, as well as for general purposes.

4 As a general rule, steels quoted in any one block can be tempered down to the next lower tensile range. Equally, where a tensile range is quoted up to a certain maximum diameter, the properties can be attained on smaller diameters. but in practice this may be wasteful of the alloy content, and a cheaper steel may be satisfactory.

5 All steels in the second block in the 10 in. diameter blocks refer to heavy forgings as specified in BS 4670, but all steels may be used as smaller section forgings as well as rolled bar or billet.

6 This table is based on the 1970 edition of BS 970 and the 1971 edition of BS 4670 except that steels in italics have been eliminated from the 1983 edition of BS 970.

Table 1.7 Influence of added (and adventitious) elements in steel

Element	Dominant characteristic	Influence in ferritic steel	Influence in austenitic steel
Carbon	Strong austenite former	Strongly increases strength and hardenability. Decreases ductility	Causes weld decay unless stabilised. Stabilises austenite
Nitrogen	Strong austenite former	Increases strength. Decreases toughness	Increases strength. Stabilises austenite
Manganese	Austenite former	Strongly increases strength. Increases hardenability. Increases tendency to quench cracking. Neutralises harmful effect of sulphur	Stabilises austenite
Nickel	Austenite former	Refines grain. Increases toughness. Increases hardenability. Slightly increases strength	Stabilises austenite. Stress corrosion cracking peaks at 17%N
Chromium	Ferrite former, carbide former	Improves corrosion and scaling resistance. Improves hardenability. Slightly increases strength. Retards softening in tempering	Improves corrosion and scaling resistance. Destabilises austenite. In high concentration forms brittle σ phase with iron
Molybdenum	Ferrite former, carbide former	Strongly increases hardenability. Moderately increases strength. Retards softening and tempering. Strongly increases strength at high temperature. Alleviates temper embrittlement	Improves corrosion resistance. Strongly increases strength at high temperatures
Vanadium	Ferrite former, carbide former	Strongly increases hardenability. Moderately increases strength. Strongly increases high-temperature strength. Increases toughness. Alleviates embrittlement by nitrogen	
Silicon	Deoxidiser	Improves scaling resistance. Increases hardenability. Reduces toughness. Increases resistivity. Promotes decarburisation	Improves scaling resistance
Niobium	Strong carbide former	Increases strength of carbon steel by age hardening	Stabilises against weld decay. Increases strength at high temperature
Titanium	Strong carbide former (with aluminium)	Strongly increases strength by age hardening	Stabilises against weld decay. Increases strength at high temperature. Very strongly increases strength by age hardening
Aluminium	Deoxidiser	Increases toughness by combining with nitrogen. Increases scaling resistance. Renders steels suitable for gas nitriding	
Boron		In small amounts greatly increases hardenability. Improves strength at high temperatures	Greatly improves creep and rupture strength
Sulphur	Impurity, except when added to improve machinability	Reduces cleanliness. Reduces ductility. Approx. 0.3% added to improve machinability	Reduces cleanliness. Reduces ductility. Approx. 0.3% improves machinability
Lead	Improves machinability	Added to improve machinability	
Selenium	Improves machinability		Added to improve surface finish on machining
Phosphorus	Impurity	Reduces ductility and cleanliness. Can improve strength of carbon steel	Reduces ductility and cleanliness
Copper	Normally an impurity	Improves corrosion resistance. May improve strength but reduces ductility by ageing	Improves corrosion resistance. Can increase strength at high temperature
Tin, antimony, arsenic, bismuth	Tramp element impurities	Strongly reduce ductility. Promote temper embrittlement	Fortunately seldom encountered
Hydrogen	Impurity. Decarburiser	Strongly promotes rupture and fracture	Fortunately seldom encountered

ing basically of coke or charcoal particles and barium carbonate. The coke and barium carbonate react to produce carbon monoxide from which carbon diffuses into the steel. The process is simple, of low capital cost and produces low distortion, but it is wasteful of heat. It is also labour intensive because the boxes have to be packed and later emptied before heat treatment.

In liquid carburising the component is suspended in a molten salt bath containing not less than 23% sodium cyanide with barium chloride, sodium chloride and accelerators. The case depth achieved (which is proportional to time) is 0.3 mm in 1 h at 815°C and 0.6 mm in 1 h at 925°C.

The process is efficient and the core can be refined in, and the component hardened from, the salt bath, but the process uses very poisonous salts, produces poisonous vapours and maintenance is required.

In gas carburising, hydrogen gas is circulated around the work piece at between 870 and 925°C. The relationship between case depth temperature and time is the same as for liquid carburising. The process is clean, easy to control, suited to mass production and can be combined with heat treatment, but the capital cost of the equipment is high.

1.3.11.3 Carbonitriding

Carbonitriding is achieved by heating the steel in a bath similar to a liquid carburising bath but containing 30/40% sodium cyanide which has been allowed to react with air at 870°C (liquid carbonitriding) or in a mixture of ammonia and hydrocarbon (gas carbonitriding) at a lower temperature than is used for gas carburising. The case produced is harder and more wear and temper resistant than a carburised case but is thinner. Case depths of 0.1–0.75 mm can be produced in 1 h at 760°C and 6 h at 840°C, respectively.

Steels which are carburised can also be carbonitrided, but because the case is thinner there is a tendency to use steels of slightly higher carbon and alloy content so that the harder core offers more support to the thinner case. A significant advantage of carbonitriding is that the nitrogen in the case significantly increases hardenability so that a hard case may be obtained by quenching in oil which can significantly reduce distortion in heat treatment. Case hardnesses of 65 Rockwell C may be produced with the same range of core strengths as by carburising.

1.3.11.4 Nitriding

Nitriding may be achieved by heating steel in a cyanide bath or an atmosphere of gaseous nitrogen at 510–565°C. The steel component is heat treated and finish machined before nitriding.

Liquid nitriding This uses a bath of sodium and potassium cyanides, or sodium cyanide and sodium carbonate. The bath is pre-aged for a week to convert about one-third of the cyanide into cyanate.

Two variants of the process are: liquid pressure nitriding, in which liquid anhydrous ammonia is piped into the bath under a pressure of 1 to 30 atm; and aerated bath nitriding, in which measured amounts of air are pumped through the molten bath.

All the processes provide excellent results, the depth and hardness of case being the same as that obtained from gas nitriding. Unlike gas nitriding, carbon steels can be liquid nitrided and the case produced on tool steels is tougher and lower in nitrogen than a gas nitrided case. However, liquid nitriding uses a highly poisonous liquid bath at a high tempera-

ture and the process may take as long as 72 h. It is really only suitable for small components.

Gas nitriding This is achieved by introducing nitrogen into the surface of a steel by holding the metal at between 510 and 565°C in contact with a nitrogenous gas, usually ammonia. A brittle nitrogen rich surface layer known as the 'white nitride layer', which may have to be removed by grinding or lapping, is produced. There are two processes, single- and double-stage nitriding.

In the single-stage process a temperature between 496 and 524°C is used and about 22% of the ammonia dissociates. This process produces a brittle white layer at the surface. The first stage of the double-stage process is the same as for the single-stage process, but following this the ammonia is catalytically dissociated to about 80% and the temperature increased above 524°C. Less ammonia is used in the double-stage compared with the single-stage process and the depth of the brittle white layer is reduced and is softer and more ductile. Process times are of the order of 72 h.

Gas nitriding can only be used if the steel contains an alloying element (e.g. aluminium, chromium, vanadium or molybdenum) that forms a stable nitride at nitriding temperatures. The film produced by nitriding carbon steels is extremely brittle and spalls readily. In general, stainless steels, hot work die steels containing 5% chromium and medium carbon chromium containing low alloy steels have been gas nitrided. High speed steels have been liquid nitrided.

There are also a number of steels listed in AISI/SAE or BS 970 (or having the name 'Nitralloy') to which 1% aluminium has been added to make the steel suited for gas nitriding. AISI 7140 (BS 970, 905M39) is typical.

Nitriding can produce case hardnesses up to 75 Rockwell C, depending on the steel. This hardness persists for about 0.125 mm but depths of case with hardness above 60 Rockwell C of 0.8 mm may be produced.

The relatively thin case compared with other methods of case hardening makes it customary to use fairly strong core material. For ferritic steels a UTS of 850–1400 MPa is usual. Typical components nitrided are gears, bushings, seals, camshaft journals and other bearings, and dies. In fact, all components which are subject to wear. In spite of their relatively low hardness, austenitic stainless steel components are nitrided to prevent seizure and wear, particularly at high temperatures. Two considerations apply.

In the first place stainless steels must be depassivated by mechanical or chemical removal of the chromic oxide film before nitriding. Secondly, nitriding decreases corrosion resistance by replacing the chromic oxide film by a chromium nitride film and should not be employed when corrosion resistance is of paramount importance.

Ion implantation This is achieved by bombarding the surface of a steel with charged ions, usually nitrogen when the aim is to harden the surface. The cost is high, the quantity of nitrogen implanted small, and the process can only be carried out in a laboratory which has an accelerator such as, for example, AERE. Ion implantation is used for special applications which will probably increase in number.

1.3.12 Stainless steels

The addition of strong oxide forming elements, aluminium silicon and chromium, replaces the oxide on the surface of iron by a tenacious film, which confers corrosion and oxidation resistance. Alloys of iron with substantial proportions of aluminium and silicon have undesirable properties so that chromium additions which in progressively increasing quanti-

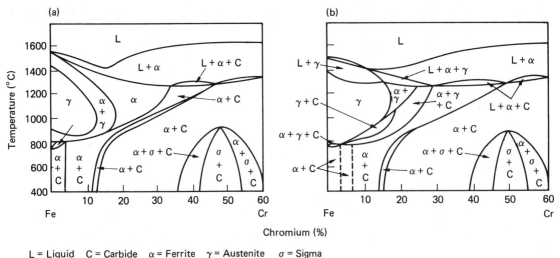

L = Liquid   C = Carbide   α = Ferrite   γ = Austenite   σ = Sigma

Figure 1.27 Iron–chromium–carbon phase diagrams: (a) at 0.10%C; (b) at 0.50%C

ties change the oxide film first to a spinel and then to chromium trioxide must be employed. Stainless steels are alloys with a minimum of 50% iron and a minimum of 12% chromium.

1.3.12.1   Metallurgy of stainless steels

The above comments are reflected in the phase diagram for the iron–chromium system (*Figure 1.27*). Of particular significance is the small austenite field known as the γ loop; alloys to the right of this loop are ferritic and undergo no allotropic changes in heating or cooling; consequently, grain refinement by such changes is not possible. The amount of chromium which closes this loop if no other element is present is 12.8%. Above this figure pure iron–chromium alloys are ferritic and subject to grain growth as temperatures are raised to the liquidus. Addition of austenite formers enlarge the γ loop so that, in the limit, the austenitic phase is stable over the entire range of temperature. Varying the proportions of chromium and nickel (and mangenese and nitrogen) produces the several types of stainless steel.

Ferritic stainless steels   These contain between 11 and 30% chromium, a minimum of austenite formers (see *Table 1.7*) such as carbon whose influence on the extent of the γ loop is shown in *Figure 1.27*, and often some other ferrite formers so that they always retain a ferritic structure.

The standard ferritic (and martensitic) stainless steels have '400' series AISI and BS 970 numbers. These numbers increase with the chromium content, low numbers, e.g. 403, denoting 12% chromium. Other things being equal, therefore, a higher numbered steel will have a better resistance to general corrosion than a lower numbered steel. The following numbers indicate a ferritic steel: 405, 409, 430, 434 and 436.

The non-standard steels include Carpenter 182 FM and four aluminium containing steels (Armco 18 SR and BSC Sicromal 9, 10 and 12).

Ferritic stainless steels are marketed only in the form of plate and strip and all have similar mechanical properties (UTS 415–460 MPa; yield strength 275–550 MPa; elongation 10–25%, depending on the thickness of the plate). They require no heat treatment beyond an anneal at about 800°C

followed by air or furnace cooling. They are easily drawn and pressed and their machinability is good, 430 FSe being naturally the best. They are prone to grain growth, particularly during welding, and this impairs toughness and ductility.

Ferritic stainless steels are virtually immune to chloride induced stress corrosion cracking at the relatively low temperatures at which they are used and have good resistance to scaling at elevated temperatures, the aluminium containing varities (e.g. the Sicromals) being some of the best available materials in this respect. They are significantly cheaper than austenitic steels and are used for chemical-plant components, domestic and catering equipment, automobile trim, domestic and industrial heater parts, exhaust systems, and fasteners. The higher numbers, which have greater resistance to general corrosion, are used for the more demanding applications.

'Low interstitial' grades   These are characterised by carbon and nitrogen contents below 0.03%, chromium contents between 17 and 30%, usually with molybdenum, and other additions in recently developed ferritic stainless steels. These include one standard steel, 444 (which in spite of its high number, contains only 18.5%Cr) and non-standard steels (Allegheny Ludlum 'E Brite 261', 'A129.4.4' and 'A294C', Nyby Uddeholm 'Monit', Crucible 'Seacure/SCI' and Thyssen 'Superferrit').

These steels, particularly the versions which contain 28% Cr and 4% or more molybdenum, are claimed to have exceptional resistance to general stress and pitting corrosions and to be suitable for the most aggressive environments obtaining in chemical plant and elsewhere.

Martensitic stainless steels   These contain 11–18% chromium and some austenite formers (see *Table 1.6*), such as carbon (see *Figure 1.27*) so that they can be hardened by cooling through the γ–α phase transformation.

The martensitic stainless steels also have 400 series numbers (403[B], 410[B], 414[B], 416[B], 420[B], 422, 431[B] and 440—the superscript B indicates a BS 970 version) with chromium contents increasing with specification number from 12 to 17% (the highest chromium content at which a steel can have a fully martensitic structure). They have, therefore, less general corrosion resistance than the ferritic stainless steels, but have fair resistance to stress corrosion. They can be hardened by quenching from above 950°C to form a hard and brittle structure which must be tempered. Tempering at 150–370°C improves ductility with little loss of strength, but above 500°C the strength falls off rapidly. Holding at temperatures between 370 and 600°C causes temper embrittlement which reduces impact resistance and must be avoided.

The martensitic high carbon grades are difficult to form and weld. They are particularly suited for operations requiring resistance to wear and manufacture of a cutting edge and their applications include valves, tools, cutlery, scissors, turbine blades, coal-mining equipment and surgical instruments. The most widely used (and, therefore, most commonly available) martensitic and ferritic stainless steels are listed in *Table 1.8*.

Austenitic stainless steels These contain 15–27% chromium and, in the case of the '300' series, 8–35% nickel. In the '200' series, for which there is no BS 970 eqivalent, some of the nickel is replaced by manganese and nitrogen which cost less

Table 1.8 Most readily available martensitic and ferritic stainless steels

AISI No.	Approximate composition (%)	UTS (MPa)	Additional information
403	C 0.08 max. Cr 12.0/14.0 Ni 0.50 max.	420	A low carbon stainless iron suitable for rivets, split pins, and lightly stressed engineering fittings. Nearest equivalent specifications: BS 970: 1970 403S17; BS 1449: 1970 403S17; BS 1501: 1973: Part 3 403S17
405	C 0.08 max. Cr 12.0/14.0 Ni 0.50 max. Al 0.10/0.30	420	Non-hardenable. Suitable for welded fabrications. Nearest equivalent specifications: BS 1449: 1970 405S17; BS 1501: 1973: Part 3 405S17
409	C 0.09 max. Cr 11.0/13.0 Ni 0.70 max. Ti 5 × C/0.60	420	Non-hardenable. Suitable for welded fabrications. Nearest equivalent specifications: BS 1449: 1970 409S17
410	C 0.09/0.15 Cr 11.5/13.5 Ni 1.00 max.	540/690	Martensitic stainless steel for general engineering applications. Nearest equivalent specifications: BS 970: 1970 410S21; BS 1449: 1970 410S21
420	C 0.14/0.20 Cr 11.5/13.5 Ni 1.00 max.	690/850	Surgical instruments, scissors, taper and hinge pins. General engineering purposes. Nearest equivalent specifications: BS 970: 1970 420S29
420	C 0.20/0.28 Cr 12.0/14.0 Ni 1.00 max.	690/850	Valve and pump parts (which are not in contact with non-ferrous metals or graphite packing), surgical instruments. Nearest equivalent specifications: BS 970: 1970 420S37
420	C 0.28/0.36 Cr 12.0/14.0 Ni 1.00 max.	690/930	Cutlery and edge tools. Nearest equivalent specifications: BS 1449: 1970 420S45; BS 970: 1970 420S45
430	C. 0.10 max. Cr 16.0/18.0 Ni 0.50 max.	430	Ferritic stainless. Domestic and catering equipment, motor car trim, domestic and industrial heater parts. Nearest equivalent specifications: BS 970: 1970 430S15; BS 1449: 1970 430S15
431	C 0.12/0.20 Cr 16.0/18.0 Ni 2.00/3.00	850/1000	General engineering. Pump and valve parts (in contact with non-ferrous metals or graphite packing). Nearest equivalent specifications: BS 970 431S29
434	C 0.10 max. Cr 16.0/18.0 Mo 0.90/1.30 Ni 0.50 max.	—	Ferritic stainless. Motor car trim. Nearest equivalent specifications: BS 1449 434S19
SF67*	C 0.70 Cr 13.0	—	Razor-blade strip

Free machining versions of 13%Cr steels are available to BS 970 416S21, 416S29, 416S37.
* BSC trademark.

than nickel. These steels can be cold worked to higher strengths than the '300' series steels.

Austenitic materials with much more than 30% nickel are known as 'nickel alloys'. If they contain age hardening aluminium and titanium additions, they are known as 'iron (or nickel) superalloys'.

The mechanical properties of austenitic steels are: UTS 490–680 MPa, yield strength 205–575 MPa, elongation 30–60%.

Some of the AISI specification numbers are followed by letters, these letters (and, where applicable, to the BS 970 numerical code) are:

H   (BS code 49)—these steels contain 0.006%B and 0.15%Nb (except 347 which already has a higher niobium content) and have creep resisting properties.

Se   this steel contains 0.15%Se and is free machining.

L   (BS code 11)—these contain a maximum of 0.03%C.

N   (BS code 6X)—these steels contain 0.2%N and, therefore, have proof stresses 50–130 MPa higher than non-nitrogen containing steels.

Ti or Cb (BS code 40)—these steels contain titanium or niobium to combine with the carbon and, thereby, prevent weld decay.

There are over 50 standard AISI and slightly less BS 970 austenitic stainless steels. *Table 1.9* lists those most commonly used and, therefore, the most readily available. (Steels suitable for use at elevated temperatures are listed in *Table 1.10*.)

There are, in addition, a very large number of non-standard austenitic steels of which the following list gives a small selection.

(1) Allegheny Ludlum 'A286'—this is really a superalloy but it is also used as a stainless steel because of its high yield strength.
(2) 'Nitronic'—a high nitrogen steel with high yield strength.
(3) Avesta 'SMO'—high molybdenum content steel with exceptional resistance to pitting corrosion.
(4) Carpenter '20 Cb3'—really a nickel alloy but generally known as a stainless steel, this has high resistance to sulphuric acid attack.
(5) BSC 'Esshete 1250'—a steel with exceptional creep resistance and high yield.

Austenitic stainless steels are chosen on account of their resistance to general corrosion which is superior to that of a ferritic steel of similar chromium content and also because of the high ductility of the face-centred γ structure which confers high hot and cold formability and high toughness down to cryogenic temperatures.

It is not possible to state exactly where the limits of stability of austenite steel lie at room temperature because transformation can be too sluggish to permit precise delineation of the phase fields and is influenced by further alloy addition such as molybdenum, silicon or nitrogen. The austenite should ideally be 'persistent', that is it should not transform under the temperature or working conditions encountered in fabrication and service.

The range of compositions with 'persistent' austenite at room temperature is shown in *Figure 1.28* (labelled 'A'). Austenite stability is increased by increasing the nickel, manganese, carbon and nitrogen content.

Partial transformation will cause the steel to lose its non-magnetic character, impair its deep drawing characteristics and reduce notch toughness at cryogenic temperatures. There may be other drawbacks but service performance is not usually impaired.

Substantial advantages—the prevention of fissuring on solidification and resistance to intergranular corrosion—are conferred by the presence of a proportion of ferrite.

Except in the case of welding (see Section 1.3.17) these advantages apply to cast rather than wrought austenic steels.

Many austenitic stainless steels, including 304, the typical 18.8 grade, are partially transformed by cold work and work harden appreciably. Steels such as these are air cooled in thin section, but thicker sections are water quenched. Besides promoting stability, this retains carbide in solution.

Duplex stainless steels These contain 18–27%Cr, 4–7%Ni, 2–4%Mo with copper and nitrogen in proportions which ensure that they have a mixed ferritic austenite structure which is not heat treatable (see *Figure 1.28*). Their mechanical properties are: UTS 600–900 MPa, yield strength 410–850 MPa, elongation 16–48%.

The one standard duplex stainless steel is AISI 329, but there are in addition, BSC 'SF22/5', Langley Alloys 'Ferralium 255', Sandvik '2RE60' and 'SAF2205/AF22' and Sumitomo 'DP3'.

The duplex stainless steels have outstanding properties. Their resistance to stress corrosion cracking is superior to that of comparable austenitic steels and they have good resistance to pitting corrosion. They have better toughness than ferritic steels and they are easily welded. Those containing nitrogen can be cold worked to higher strengths than can ferritic or austenitic steels, and are highly weldable provided that a welding consumable that will ensure the presence of ferrite in the weld metal is employed.

They have so far been used for tube plates, marine applications, sour-gas pipeline and acetic acid production. When they are better known and more widely available they should become used in preference to austenitic steels for the more demanding applications.

Precipitation hardening stainless steels These contain 12–28%Cr, 4–7%Ni, and aluminium and titanium to give a structure of austenite and martensite which can be precipitation hardened. The mechanical properties of precipitation hardening stainless steels are: UTS 895–1100 MPa, yield strength 276–1000 MPa, elongation 10–35%.

No precipitation hardening stainless steels are standardised by AISI or in BS 970, but Firth Vickers 'FV 520' is covered by BS 1501 460552 for plate and BS 'S' specifications S143, S144 and S145 for bars, billets and forgings. Non-standard steels include Armco '15–5PH', '17-4 PH' and '17-7 PM' and Carpenter 'Custom 450' and 'Custom 435'

The excellent mechanical properties and corrosion resistance has caused precipitation hardened stainless steels to be used for gears, fasteners, cutlery and aircraft and steam turbine parts. They can be machined to finished size in the soft condition and precipitation hardened later. Their most significant drawback is the complex heat treatment required which, if not properly carried out, may result in extreme brittleness.

1.3.13 Corrosion resistance of stainless steels

Corrosion resistance of stainless steels depends on surface passivity arising from the formation of a chromium containing oxide film which is insoluble, non-porous and, under suitable conditions, self-healing if damaged. Passivity of stainless steel is not a constant condition but it prevails under certain environmental conditions. The environment should be oxidising in character.

Other factors affecting corrosion resistance include composition, heat treatment, initial surface condition, variation in corrosion conditions, stress, welding and service temperature.

1.3.13.1 Composition

Those ferritic and martensitic steels with roughly 13%Cr are rust resisting only and may be used for conditions where corrosion is relatively light, e.g. atmospheric, steam and oxidation resistance up to 500°C. Applications include cutlery, oil cracking, turbine blades, surgical instruments, and automobile exhausts.

17% chromium (ferritic and martensitic) steels are corrosion and light acid resisting. They have improved general corrosion resistance compared with 13%Cr steels. Applications for the ferritic grade include domestic and catering equipment, automobile trim, and industrial heater parts. The martensitic grade is used in general engineering, for pump and valve parts in contact with non-ferrous metals or graphitic packings.

The addition of molybdenum significantly improves the integrity of the oxide film. The ferritic 434 and 436 grades can withstand more severe corrosion conditions and the martensitic 440 grades are used where wear and acid resistance is required such as in valve seats.

Additional amounts of nickel above 8% to form an austenitic structure in the '300' steels further improve resistance to corrosion and acid attack. Applications include domestic, shop and office fittings, food, dairy, brewery, chemical and fertiliser industries.

The stainless steels with the highest corrosion resistance are those with even higher chromium contents such as 310 with 25%Cr and the low interstitial steels with up to 30%Cr. The addition of molybdenum up to 6% is also highly beneficial. The resistance to sulphuric acid attack of Carpenter 20 Cb-3 which contains 3.5%Cu as well as 2.5%Mo has already been mentioned.

Note that in order to ensure an austenitic structure the nickel content of the molybdenum bearing steels increases above 8% as the content of molybdenum and other ferrite stabilising elements (titanium, niobium, etc.) increases. The 4.5%Mo alloy 317 LM is used in sodium chlorite bleaching baths and other very severe environments in the textile industry.

1.3.13.2 Heat treatment

Heat treatment has a significant influence on corrosion resistance. Maximum resistance is offered when the carbon is completely dissolved in a homogeneous single-phase structure. The 12–14%Cr steels are heat treated to desired combinations of strength, ductility and toughness and, because of their low carbon content, are generally satisfactory unless tempered in the temperature range 500–600°C.

Austenitic steels (18/8) are most resistant when quenched from 1050–1100°C, their normal condition of supply. A steel chosen for welded fabrications should be titanium or niobium stabilised (AISI, 'Ti' or 'Cb', BSI '40') or, better still, an extra-low carbon grade (AISI, 'L' or BSI 11). Quenching after welding is usually impracticable.

1.3.13.3 Surface condition

For maximum resistance to corrosion the passive film must be properly formed; this is ensured by removing all scale, embedded grit, and metal pick-up from tools and other surface contaminants. Polishing improves resistance. Passivating in oxidising acid (10–20% NHO_3 by weight) solution at 25°C for 10–30 min confers maximum resistance to austenitic steels. The ferritic–martensitic grades are passivated in nitric acid/potassium dichromate solution (0.5% nitric acid + 0.5% potassium dichromate at 60°C for 30 min).

1.3.13.4 Variation in corrosion conditions

In the absence of experience, samples of the proposed steels should be tested in the condition in which they are to be used (i.e. welded, if fabricated) in the intended environment, taking full note of any possible variation in service conditions. The effect of welding on corrosion resistance is considered in Section 1.3.17.

1.3.13.5 Service temperature

Because stainless steels other than those with very low carbon which are unstabilised or partly stabilised with titanium or niobium may show chromium carbide precipitation when subjected to service temperatures above 350°C (see Section 1.3.16), this should be the upper limit for service in corrosive environments. Fully stabilised steels are not restricted in this manner.

1.3.13.6 Localised corrosion of stainless steels

The considerations discussed in Sections 1.3.13.1 to 1.3.13.5 apply principally to general corrosion which progressively reduces the thickness of a component until it is completely dissolved or its strength is so reduced that it can no longer withstand imposed stress. More insidious attack mechanisms on stainless steels are the five varieties of localised corrosion: galvanic, crevice, pitting and stress corrosion and intergranular penetration. These are confined to isolated areas or lines on the surface, but penetrate through the thickness of a component to destroy its integrity without materially affecting its dimensions. Their incidence is less predictable and their onset more difficult to predict than is general corrosion, but their effect may be catastrophic.

Crevice and galvanic corrosion must be countered by designing to eliminate crevices and the juxtaposition of metals of different solution potential. Intergranular penetration is discussed in Section 1.3.17. Pitting and stress corrosion are composition dependent.

Pitting occurs in conducting aqueous liquid environments (usually halide solutions) when local penetration of the oxide film creates stagnant locations in which diffusion generates strongly acid environments which rapidly penetrate a component. *Figure 1.29* shows a pit in an early and in a very late stage. Resistance to pitting in low interstitial ferritic steels increases with increase of chromium content from 18 to 29% and molybdenum from 1 to 4%, while austenitic steels require at least 20%Cr and between 4.5 and 6%Mo. Typical austenitic steels with very high resistance are Allegheny Ludlum 'A129-92' Avesta '254 SMO' and Langley Alloys 'Ferralium' 255, all of which are claimed not to pit in stagnant seawater. (These steels are also claimed to resist crevice corrosion should this not have been eliminated by design.) The best standard austenitic stainless steel is 317 LM.

Stress corrosion cracking occurs when a material is stressed in tension in an aggressive aqueous environment, usually an alkali metal halide or hydroxide solution.

Cracks may be intergranular (see *Figure 1.30(a,b)*) or transgranular (see *Figure 1.30(c,d)*). The tendency to stress corrosion cracking of a material is measured by its K_{ISCC} value, which is the lowest value of stress intensity (in MN m$^{-3/2}$) at which a crack will propagate in a specific medium at a specific temperature.

The growth rate of stress corrosion cracks is highly temperature dependent, increasing about 500 times with an increase in temperature from 20 to 100°C. Most austenitic steels are resistant at ambient temperature, but if the temperature rises above about 40°C in a saline environment a change should be

Table 1.9 Most readily available austenitic stainless steels

AISI No.	Approximate composition (%)	UTS (MPa)	Additional information
202	C 0.07 max. Mn 7.00/10.0 Cr 16.5/18.5 Ni 4.00/6.50 N 0.15/0.25	630	Nearest equivalent specifications: BS 1449: 1970 284S16
301	C 0.15 max. Cr 16.0/18.0 Ni 6.00/8.00	540/1240	Readily hardens by cold working. Structural steels for applications where high strength is required. Nearest equivalent specifications: BS 1449: 1970 301S21
302	C 0.08 max. Cr 17.0/19.0 Ni 8.00/11.0	510/790	For spoons and forks, holloware, architectural and shop fittings, domestic catering, food manufacturing, dairy and brewery equipment. Nearest equivalent specifications: BS 970: 1983 302S25; Bs 1449: 1970 302S17, 302S25
303	C 0.12 max. S 0.15/0.30 Cr 17.0/19.0 Ni 8.00/11.0	510/790	A general purpose austenitic free-cutting steel. Nearest equivalent specifications: BS 970: 1983 303S21
304L	C 0.03 max. Cr 17.5/19.0 Ni 9.00/11.0	490	A low carbon version of 304, fully resistant to weld decay. For chemical plant, food manufacturing, dairy and brewery equipment. Nearest equivalent specifications: BS 970: 1983 304S12; BS 1449: 1970 304S12; BS 1501: 1973: Part 3 304S12
304LN	C 0.03 max. Cr 17.5/19.0 Ni 9.00/12.0 N 0.25 max	590	A high proof stress version of 304L. For cryogenic, storage, and pressure vessels. Nearest equivalent specifications: BS 1501: 1973: Part 3 304S62 (Hi-proof 304L*)
304	C 0.06 max. Cr 17.5/19.0 Ni 8.00/11.00	510/790	Holloware, domestic, catering, food manufacturing, dairy and brewery equipment. Recommended for stretch forming applications. Readily weldable. Nearest equivalent specifications: BS 970: 1983 304S15; BS 1449: 1970 304S15; BS 1501: 1973: Part 3 304S15, 304S49
304	C 0.06 max. Cr 17.5/19.0 Ni 9.00/11.0	510	As above. Preferable for deep drawing applications. Nearest equivalent specifications: BS 1449: 1970 304S16
304N	C 0.06 max. Cr 17.5/19.0 Ni 8.00/11.0 N 0.25 max.	590	A high proof stress version of 304. Cryogenic, storage and pressure vessels. Nearest equivalent specifications: BS 1501: 1973: Part 3 304S65 (Hi-proof 304*)
305	C 0.10 max. Cr 17.0/19.0 Ni 11.0/13.0	460	Dental fittings, thin walled deep drawn pressings. Low cold working factor and very low magnetic permeability. Nearest equivalent specifications: BS 1449: 1970 305S19
316L	C 0.03 max. Cr 16.5/18.0 Ni 11.0/14.0 Mo 2.25/3.00	520	A low carbon version of 316 fully resistant to weld decay. For chemical and textile plant, dairy and food equipment. Nearest equivalent specifications: BS 970: 1983 316S12; BS 1449: 1970 316S12; BS 1501: 1973: Part 3 316S12
316LN	C 0.03 max. Cr 16.5/18.5 Ni 11.0/14.0 Mo 2.25/3.00 N 0.25 max.	620	A high proof stress version of 316L. Cryogenic storage and pressure vessels. Nearest equivalent specifications: BS 1501: 1973: Part 3 316S62 (Hi-proof 316L*)

AISI No.	Approximate composition (%)	UTS (MPa)	Additional information
316	C 0.07 max. Cr 16.5/18.0 Ni 10.0/13.0 Mo 2.25/3.00	540	Chemical and textile plant. Dairy and food equipment. A lower ferrite content version is for use in special applications e.g. urea plant. Nearest equivalent specifications: BS 970: 1983 316S16; BS 1449: 1970 316S16; BS 1501: 1973: Part 3 316S16
316N	C 0.07 max. Cr 16.5/18.5 Ni 10.0/13.0 Mo 2.25/3.00 N 0.25 max.	620	A high proof stress version of 316. For cryogenic storage and pressure vessels. Nearest equivalent specifications: BS 1501: 1973: Part 3 316S66 (Hi-proof 316*)
317L	C 0.03 max. Cr 17.5/19.5 Ni 14.5/17.0 Mo 3.00/4.00	490	A low carbon version of 317 fully resistant to weld decay. For chemical plant. Nearest equivalent specifications: BS 970: 1983 317S12; BS 1449: 1970 317S12
317	C 0.06 max. Cr 17.5/19.5 Ni 12.0/15.0 Mo 3.00/4.00	540	For chemical plant. Nearest equivalent specifications: BS 970: 1983 317S16; BS 1449: 1970 317S16
320Ti	C 0.08 max. Cr 16.5/18.0 Ni 11.0/14.0 Mo 2.25/3.00 Ti 4 × C/0.60	520	Fully stabilised against weld decay. Nearest equivalent specifications: BS 970: 1983 320S17; BS 1449: 1970 320S17; BS 1501: 1973: Part 3 320S17
321	C 0.08 max. Cr 17.0/19.0 Ni 9.00/11.0 Ti 5 × C/0.70	540	Fully stabilised against weld decay. Chemical, dairy and brewing plant, food manufacturing and textile equipment. Domestic and catering equipment. Nearest equivalent specifications: BS 970: 1983 321S12, 321S20; BS 1449: 1970 321S12; BS 1501: 1973: Part 3 321S12, 321S49
Warm worked 321	C 0.08 max. Cr 17.0/19.0 Ni 9.00/11.0 Ti 5 × C/0.70	620	A high proof stress version of 321 obtained by controlled low temperature hot working. Nearest equivalent specifications: BS 1501: 1973: Part 3 321S87
325	C 0.12 max. Cr 17.0/19.0 Ni 8.00/11.0 Ti 5 × C/0.90 S 0.15/0.30	510/790	A free-cutting version of 321, fully stabilised against weld decay. Nearest equivalent specifications: BS 970: 1983 325S21
347	C 0.08 max. Cr 17.0/19.0 Ni 9.00/11.0 Nb 10 × C/1.00	510/540	Chemical, dairy and brewing plant. Food manufacturing and textile equipment. Domestic and catering equipment. Particularly suitable for use in welded plant in contact with nitric acid. Nearest equivalent specifications: BS 970: 1983 347S17; BS 1449: 1970 347S17; BS 1501: 1973: Part 3 347S17, 347S49
347N	C 0.08 max. Cr 17.0/19.0 Ni 9.00/12.0 Nb 10 × C/1.00 N 0.15/0.25	650	A high proof stress version of 347. Nearest equivalent specification: BS 1501: 1973: Part 3 347S67 (Hi-proof 347*)

* BSC trademark.

Table 1.10   Steels suitable for use at elevated temperatures showing 0.2% proof and creep rupture strengths near the top of their useful temperature ranges

BS 1501, 2 or 3 Designation	AISI equivalent	Type of steel*	Min. 0.2% proof stress (MPa) at temperature (°C)	10^5 h rupture strength (MPa) at temperature (°C)	Note
161 Grade 28	1025	Si killed carbon	147.5 at 450	133 at 450	
221 Grade 32	1527	Si killed carbon manganese	172 at 450	147 at 450	
223 Grade 32		Si killed carbon manganese Nb treated	173 at 450	142 at 450	
271		Mn, Cr, Mo, V	292 at 450	309 at 450	Used for boiler drums in heavy sections
620		1Cr, 0.5Mo	136 at 550	49.4 at 550	
622		2.25Cr, 1Mo	145 at 550	72.5 at 550	
625		5Cr, 0.5Mo	210 at 550	290 at 550	Used in refinery. Not in power plant
		9Cr, 1Mo	210 at 550	84 at 550	
660		0.5Cr, 0.5Mo, 0.25V	199 at 550	74 at 550	
Jessups H46		12Cr, 0.5Mo, V, Nb, N, B	181 at 600		Gas turbine disc or steam turbine blade material
BS 4882 B16A (Durehete 1055)		1Cr, 1Mo, 1.5V, 0.1Ti, 0.005B	Stress relaxation specification		Bolting materials for temperature range 500–565°C
304 S49	304H	18Cr, 12Ni, 2Mo, 0.15Nb, 0.005B	100 at 600	74 at 600	
316 S49	316H	18Cr, 10Ni, 0.5Ti, 0.15Nb, 0.005B	100 at 600	118 at 600	
321 S49	321H	18Cr, 10Ni, 0.5, 0.45, 0.005B	111 at 600	105 at 600	
347 S49	347H	18Cr, 12Ni, 1Nb, 0.005B	123 at 600	106 at 600	
	310	25Cr, 20Ni	120 at 550	120 at 550	
BSC Esshete 1250		15Cr, 6Mn, 10Ni, 1Mo, 1Nb, 0.5V, 0.006B	140† at 650	160 at 650	
Iron superalloy		15Cr, 25Ni, Mo, V, 3Ti, 0.3Al	150 at 700	79 at 700	Used in aircraft gas turbines

* Compositions are given as percentages.

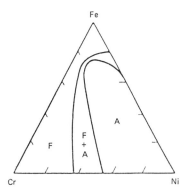

Figure 1.28   Iron–nickel–chromium phase diagram (room temperature) showing persistent austenite (A) and duplex ferrite–austenite (F+A) phase fields

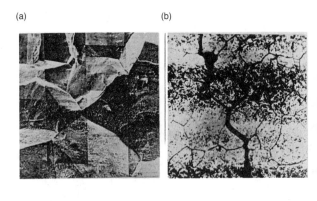

(a)   (b)

(c)   (d)

Figure 1.30   Stress-corrosion cracks. (a,b) Intergranular; (c,d) transgranular. (a) Continuous wall at interface with brittle fracture. (b) A network of cracks penetrating around grains. The cracks are thin and sharp and there is no distortion of grains. Cracks may be filled with corrosion product. (d) Multiply branching, thin, sharp cracks which may be filled with corrosion product. There is no distortion of grains. ((a,b) Courtesy of Berkeley Nuclear Laboratories; (c,d) courtesy of Dr D. Bagley, BNF Fulmer Contracts Research)

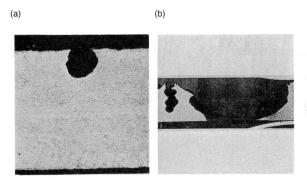

(a)   (b)

Figure 1.29   (a) An early stage of pitting corrosion showing a partially spherical pit. (b) Pitting corrosion has penetrated the wall causing a leak—pits can also initiate fatigue stress corrosion. (Courtesy of Dr D. Bagley, BNF Fulmer Contracts Research)

made to a high molybdenum steel, a duplex steel, a nickel free ferritic steel or a nickel alloy, depending on the severity of the conditions.

Materials are tested for resistance to stress corrosion by exposure to stress in a boiling aqueous solution of 42% $MgCl_2$. Very few materials will withstand this for long.

1.3.14 Heat resisting steels

The range of operating temperature of carbon steels is limited to about 400°C by decrease in resistance to deformation, and in oxidising atmospheres to about 500°C by diffusion of oxygen through the oxide film. Operation above these temperatures is achieved by the addition of alloying elements such as chromium, aluminium and silicon which render the oxide film more tenacious and limit diffusion (see *Figure 1.31*) and chromium, molybdenum, vanadium and niobium which in solid solution or as carbides impart stability and increase resistance to deformation.

The scope for providing high-temperature yield and creep strength in ferritic steels is limited so, for service above about 550°C, austenitic steels are used. Steels for use at high temperature must, in addition, be stable and capable of fabrication to the design shape. Material requirements vary with application.

Steam-power-plant materials, with minor exceptions, operate below 650°C and in an atmosphere which is frequently little more aggressive than air. However, they must last for at least 10^5 hours and preferably two or three times that figure. Chemical and refinery plant may be required to operate over a wider temperature range in very varied environments, but they have shorter lives of usually about $2-4 \times 10^4$ h. Aircraft propulsion turbines require materials to withstand high stresses at very high temperatures, but component operating lives are seldom above 10^3-10^4 h. The materials used to meet these requirements are basically as follows.

For components which must resist oxidation but are not stressed, ferritic chromium steels, preferably also containing silicon and aluminium, may be chosen. The choice will depend on cost, temperature, aggressiveness of environment and ability to fabricate from 405, 409 (the cheapest weldable stainless steel) 446 and BS Sichromal '9', '10' and '12'.

For stressed components operating at high temperatures a choice may be made from the steels listed in *Table 1.9*. This table lists only a few of the many alternatives but provides at least one steel that may be selected with confidence to operate over any part of the temperature range. At temperatures below 450°C silicon killed carbon or carbon manganese steels are used, except for heavy pressure vessels where bainitic steels such as BS 1501 271 or 281 which have a high proof stress in the normalised condition are used.

Time-dependent deformation becomes more important than yield at temperatures above 400°C and the design criterion changes from a factor of the proof stress to a factor of the creep rupture stress at the design life of the component. As the temperature increases above 400–450°C carbon steels start to give place to chromium–molybdenum bainitic or martensitic steels. In power plant, strength rather than corrosion resistance is the critical parameter and the lower chromium steels are preferred. In chemical or refining plant the environment may be hydrogenous and higher chromium contents are essential to prevent hydrogen, which diffuses into the steel, combining with carbon to cause internal ruptures (see *Figure 1.32*).

Increase in temperature beyond 550°C requires the higher creep resistance of an austenitic steel. One of the steels designated 'H' by AISI or coded 44 by BS 970 should be selected. These are stabilised by niobium and their creep rupture strengths and ductilities are improved by the addition of 0.006%B.

Standard steels are satisfactory up to about 600°C, but non-standard steels such as BS 'Esshete 1250' have an increased temperature range and allow the use of thinner sections. In critical locations the very high scaling resistance of 310 may be used as the corrosion resistant face of a laminated structure backed with 'Esshete 1250'.

For the higher temperatures and higher stresses in aircraft gas turbine engine blades or discs recourse must be had to superalloys such as Allegheny Ludlum 'A286' or 'Discoloy'. Where high-temperature strengths of these are inadequate, recourse must be made to nickel alloys (see Section 1.6).

Scaling resistance of aircraft gas turbine blade materials is provided by coating with aluminium.

1.3.14.1 Structural stability

Stainless steels heated to above 600°C in fabrication or operation are subject to embrittlement mechanisms which have in the past given rise to severe problems. These mechanisms are listed and the compositions over which they may occur are indicated in *Figure 1.33*. Better understanding of the problems

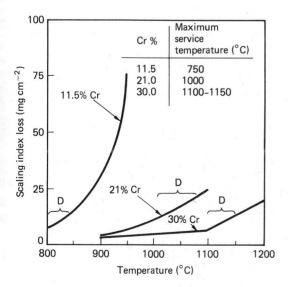

Figure 1.31 Relationship between scaling loss, temperature and chromium content of chromium steels

	Cr %	Maximum service temperature (°C)
	11.5	750
	21.0	1000
	30.0	1100–1150

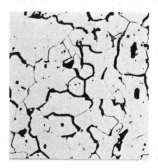

Figure 1.32 Microfissures caused by hydrogen in steels. (Courtesy of Dr D. Bagley, BNF Fulmer Contracts Research)

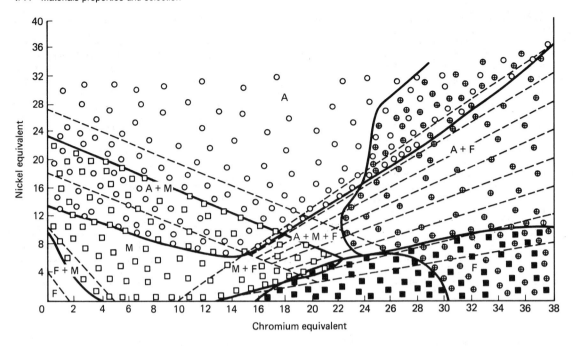

Figure 1.33 Embrittlement mechanisms in stainless steels related to composition. (□) Martensitic cracking between 0 and 290°C; (○) hot cracking above 1250°C; (⊕) σ phase embrittlement after heat treatment or service at 500–900°C; (■) cold brittleness after grain growth due to high temperatures (>1150°C)—ductile above 400°C. Where symbols overlap, the material shows the characteristics of both mechanisms

has resulted in a method of avoidance or the appreciation that they are not so serious as was originally considered.

Embrittlement due to carbide precipitation is avoided by using a low carbon or a stabilised steel (AISI 'L' or 'H' or BS 970 Code '11' or '44').

Straight chromium steels with chromium contents less than 27%Cr are not subject to σ phase embrittlement (precipitation of an intermetallic iron–chromium phase). Steels with more than 27%Cr should not be employed within the temperature range 520–700°C at which σ phase embrittlement occurs. The σ phase is dissolved by heating to 820°C. Straight chromium steels with more than 15%Cr suffer from '475°C' embrittlement if held in, or slowly cooled through, the range 525–425°C.

Austenitic steels, particularly 310 with silicon in excess of 1.5%, develop σ phase when heated in the range 590–925°C but, provided local stressing by differential expansion is prevented by design, the embrittlement has little effect on service performance, even though the steel has zero room-temperature ductility.

Austenite transformation to ferrite may be avoided by ensuring that the composition of the steel is such as to produce persistent austenite. There is little evidence of the transformation leading to problems in service even when this condition has not been met.

1.3.14.2   Valve steels

Internal combustion engine valves operate under severe conditions of fatigue, impact, high-temperature corrosion and wear. In the USA the SAE lists a special category which includes all types of steel which are used for valves (see the *SAE Handbook*[16]).

In the UK five steel types classified as stainless steels in BS 970 are described as 'valve steels'. These are

(1) Grade 401S45; 3%Si, 8%Cr (strictly not a stainless steel)—martensitic steel used for inlet valves in petrol engines and exhaust valves in medium-duty diesels. Limiting temperature 700°C.
(2) Grade 382S34; 21%Cr, 12%Ni—austenitic steel used for diesel exhaust valves, must be hard faced above 700°C.
(3) Grade 443S62; 2%Si, 20%Cr—martensitic steel used for exhaust valves in petrol engines. Limiting temperature 750°C.
(4) Grades 331S40 and 331S42 (KE965—common (or trade) name); 14%Cr, 14%Ni, 1%Si, 2.5%W—austenitic steel suitable (with hard faced seats) for temperatures up to 800°C.
(5) Grades 349S52, 349S54, 352S52 and 352S54 (the S54 types are free cutting with sulphur additions)—scaling resistance up to 900°C. Used for petrol engine exhaust valves.

1.3.15   Toughness in steels

Toughness is the property that prevents failure of a material when a load is either rapidly applied or generates a high stress intensity at the root of a discontinuity. Toughness is defined as the critical stress intensity resulting in fracture (K_{IC} in MN m$^{-3/2}$) or Charpy impact energy (in J). In the case of metals with a body-centred cubic structure (ferritic steels) or a hexagonal structure (magnesium), which decline sharply in toughness over a narrow temperature range, toughness is also defined by the impact transition temperature (FATT in °C). This section describes materials which meet the requirements for specific applications.

Other things being equal, fracture toughness bears an inverse relationship to the tensile strength, grain size, and carbon content of a steel.

Martensitic structures are tougher than bainitic which are themselves tougher than pearlite structures with the same hardness. Toughness is reduced by increase in the content of hydrogen, oxygen, nitrogen and sulphur and the so-called 'tramp elements' (phosphorus, antimony, arsenic and tin) which cause 'temper embrittlement'. Toughness is increased by increasing the content of nickel, manganese and appropriate amounts of aluminium, vanadium, niobium and molybdenum which specifically reduces temper embrittlement.

The face-centred cubic austenitic steels do not suffer from a ductile–brittle transition at low temperature.

Two examples of failures which were eliminated by a change to a tougher material are given below.

(1) Failures in the original welded ships which in some cases split in half. These failures all occurred at low temperature. In one specific case failure occurred at a weld start strake at 2°C in a steel with a ductile–brittle transition temperature of 30°C and a Charpy energy at failure temperature of 11 J. Failures were eliminated by deoxidising steel with 0.15–0.3%Si and 0.02–0.05%Al which refines grain size and combines with nitrogen. Ship plate is now specified to have a Charpy V notch value of 20 J at 4°C, a figure which is well within the capability of modern steels low in hydrogen, oxygen, sulphur and phosphorus.

(2) Failures in heavy section turbogenerator forgings, which operated at relatively low temperature, due to embrittlement by hydrogen combined with temper embrittlement. Hydrogen was eliminated by vacuum treatment of the molten steel. 'Lower nose' temper embrittlement is associated with the migration of 'tramp' elements such as phosphorus, arsenic, antimony and tin, which are taken into solution at the tempering temperature and reprecipitate at grain boundaries at temperatures around 500°C. The migration is promoted by carbon, silicon, nickel and manganese, but is retarded by molybdenum. The embrittlement could be avoided by quenching the steel from its tempering temperature, but the internal stresses so produced would be worse than the temper embrittlement. Temper embrittlement is minimised by reducing the content of 'tramp' elements and by using carbon vacuum deoxidation which obviates the need for silicon.

Figure 1.34 and *Table 1.11* show the properties available in modern large forgings in 3.5 nickel–chromium–vanadium steel. Control of embrittlement is also important to avoid the risk of failure in light water pressure vessels made from ASTM 533B MnMoNi and 508 NiCrMo steels and is achieved by a specification with limits of: Cu, 0.10; P, 0.012; S, 0.015 and V 0.05% which guarantees a K_{IC} value of 176 MN m$^{-3/2}$ at room temperature.

1.3.15.1 Cryogenic applications

Care must be taken when choosing steels for cryogenic applications for which, as a result of their ductile–brittle transition, normal ferric steels are unacceptably brittle.

All the common standard austenitic stainless steels have excellent toughness at temperatures down to −240°C, measured by Charpy impact values usually between 140 and 150. Ultimate tensile stress and 0.2% proof stress increase as the temperature is lowered to around 1500 and 456 MPa, respectively, and elongations decrease slightly but remain adequate at 40–50%. The 0.2% nitrogen grades, typically 316N, H (316S66), have higher proof stresses and are particularly suited to cryogenic applications, because the nitrogen ensures that the austenite is persistent. There is some evidence that the endurance limit of austenitic steels increases as temperature decreases.

If a non-stainless steel is preferred there are the French 'Afnor' specification steels (3.5%Ni, 5%Ni, 9%Ni) whose low-temperature properties improve with increasing nickel content. If high strength combined with high toughness at cryogenic temperatures is required, a maraging steel (see Section 1.3.16) should be specified.

1.3.16 Maraging steels

Maraging steels are supplied to ASTM A579. They are high nickel steels which are hardened by precipitation of an aluminium–titanium compound on ageing at 500°C. They have a number of advantages, including high strengths, normally ranging between 1100 and 1930 MPa (but a steel with a proof stress of 2400 and a UTS of 2450 MPa is available), excellent toughness even at −196°C, and good resistance to stress corrosion cracking.

Their greatest advantage, however, is ease of fabrication. They can be machined at their low solution treated hardness of 300 VPN (Vickers pyramid (hardness) number) and then aged to their optimum hardness at 500°C with minimal distortion and no risk of cracking. They have good weldability needing no preheating and their properties may be restored after welding by ageing.

Their main disadvantage is their high cost and the fact that, to obtain optimum toughness, they should be made by ESR or

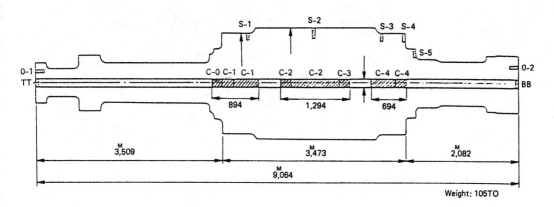

Figure 1.34 Dimensions of a typical large rotor forging, showing the location of test specimens

Table 1.11 Mechanical properties of the rotor shown in *Figure 1.34* after quality heat treatment*

Sample No.	Position and orientation of sample		Tensile test				Charpy impact test (notch: 2mm V)							
			$S_{0.2}$ (kg mm^{-2})	S_u (kg mm^{-2})	e (%)	A_R (%)	As received		De-embrittled†		Embrittled‡		$\Delta FATT_1$ (°C) AsRe-WQ	$\Delta FATT_2$ (°C) STC-WQ
							Energy (J)	FATT (°C)	Energy (J)	FATT (°C)	Energy (J)	FATT (°C)		
Desired properties														
—	Surface	R	82.4	84.5	15	45	4.1	≤+15.5						
—	Centre of core bar	L	81.0	84.5	15	45	—	—						
		T	—	—	—	—	4.1	≤+15.5						
Actual properties														
S—1	Surface radial	O	84.0	95.3	19.1	69.6	out. 11.0	<−75						
		I					in. 12.1	−65						
S—2		O	83.8	96.0	19.1	65.6	out. 12.7	−63						
		I					in. 12.1	−29						
S—3		O	83.7	94.8	20.5	68.3	out. 12.7	−75						
		I					in. 12.5	−37						
S—4		O	92.1	102.8	18.0	62.8	out. 9.3	<−75						
		I					in. 11.0	<−75						
S—5		O	90.9	101.6	19.1	68.8	out. 10.8	<−75						
		I					in. 11.5	<−75						
O-1	Ends	L	90.9	101.9	20.5	67.4								
O-2		L	92.8	103.6	20.5	67.9								
C-1	Centre of core bar	L	84.3	98.4	19.8	57.9	4.6	+15	10.5	+1	4.1	+38	+14	+37
		T	83.8	98.7	17.7	51.9	5.9	+12	6.9	+13	4.1	+43	−1	+30
C-2		L	85.0	98.5	18.3	60.5	6.7	0	12.5	−18	7.7	+6	+18	+24
		T	85.0	99.0	16.9	53.9	7.3	0	13.0	−20	9.2	+5	+20	+25
C-3		L	84.0	97.1	19.9	59.5	8.7	−8						
		T	83.8	96.4	18.2	57.8	8.7	−10						
C-4		L	83.4	96.0	20.8	62.6	8.1	+11	14.0	−4.5	7.1	+20	+15.5	+24.5
		T	83.4	95.7	18.7	59.9	10.6	+10	9.0	+8	5.8	+22	+2	+14

R. radical; T, transverse; L, longitudinal; I, inner; O, outer.
ΔFATT, increase in FATT; AsRe-WQ, s re-water quenched; STC-WQ, step cooled and water quenched.
* Reproduced by courtesy of Japan Steelworks.
† De-embrittled: 590°C × 1 h − WQ.
‡ Embrittled: step cooled.

vacuum arc melting. Also, because of the absence of hard carbides they are inferior in wear properties to hardened or tempered steels.

1.3.17 Weldability of steels

Steels may be welded by almost all varieties of electric arc welding methods, including gas shielded metal inert gas (MIG) and tungsten inert gas (TIG) with and without filler, flux shielded manual metal arc, suberged arc, electroslag, spot, projection and flash butt welding. Other fusion methods include the more recently developed electron beam and laser, and the relatively old-fashioned gas welding. Solid-phase methods, forge, diffusion, friction and explosive welding may also be used. Many of these procedures are concerned with relatively thin sections or special design and applications.

1.3.17.1 Weldability of non-stainless steels

The term 'weldability' as applied to steels usually implies the ability to make long runs in fairly large sections either by manual metal or submerged arc and is governed in ferritic steels by the 'carbon equivalent' (CE):

$$CE = C\% + N\% + \frac{Cr\% + Mo\%}{5} + \frac{V\% + Ni\% + Cu\%}{15}$$

(Boron is not taken into account in this equation but has a great influence on hardenability and, therefore, on weldability).

Steels with carbon equivalents below 0.14% are readily welded without special precautions in a wide range of thicknesses. Steels with carbon equivalents between 0.14% and 0.45% require the following precautions, depending on the value of CE and section size, to prevent the formation of austempered martensite cracking aggravated by hydrogen.

(1) *Specification of low hydrogen electrodes.* This is always desirable but requires operator skill to compensate for the more sluggish metal and slag flows compared with other electrodes.

(2) *Use of preheat before welding.* The preheat temperature required depends on the CE and the metal thickness; for a CE of 0.2, 40°C and 110°C are advisable for metal thicknesses of 25 and 225 mm, respectively, while for a CE of 0.45, 170°C and 260°C are advisable for the same thicknesses.

(3) *Control of heat input.* Other things being equal, a higher heat input gives less risk of formation of austempered martensite than does a lower heat imput, but care must be taken to limit distortion and the introduction of stress.

(4) *Use of post-heat after welding.* This is seldom required for CE values below about 0.35%, but high-duty components with restrained welds should be post-weld heat treated at between 600 and 650°C for 1 h per 25 mm thickness. Besides preventing immediate cracking (or making it obvious during inspection) post-weld heating improves dimensional stability. It is essential when welding thick and complex structures to post-heat-treat one weld before commencing to weld a cross-seam.

Steels with a CE above 0.45 present very severe problems in welding. Very high preheats ranging up to 340°C for CE = 0.6 and a 225 mm thickness, low hydrogen electrodes (preferably lower in CE than the parent material) and immediate post-heat-treatment at temperatures around 800°C are essential. Sample test welds are advisable.

Maraging steels with carbon contents around 0.3% have a soft martensite matrix and are highly weldable with no risk of decarburisation, distortion or cracking. These should be used where very high strength combined with weldability is required.

1.3.17.2 Weldability of stainless steels

Welding is the normal method of fabricating stainless-steel vessels, etc. The heat-affected zones are raised to incipient fusion temperature, but the time spent at this temperature varies with different welding processes. Argon arc and spot welding are most satisfactory in heating for minimum time; metal arc welding, inert-gas metal arc and submerged arc are less so in that order from this point of view.

The problems associated with welding stainless steels fall into two categories. The first, associated with carbide precipitation includes 'weld decay' and 'knife line attack', and affects mainly corrosion behaviour. The second includes those phenomena which may be assessed by means of the Schaffler diagram (see *Figure 1.33*).

Carbide solution and precipitation The solubility of chromium carbide in austenite decreases with decreasing temperature and increasing nickel content (*Figure 1.35*). At room temperature the solubility in 18%Cr, 8%Ni austenite (solid line) is approximately 0.03%. If an 18%Cr, 8%Ni alloy containing, say, 0.06%C is annealed at 1050–1100°C all chromium carbide is in solution and remains in unstable solution after quenching to room temperature. If the alloy is heated to an intermediate temperature excess carbide is precipitated.

The mode of precipitation depends on whether the austenite has been worked. If the quenched but unworked material is heated in the temperature range 450–750°C, chromium carbide is precipitated at the grain boundaries; and the lower the temperature the longer the time required. Thus at around 450°C the time taken for precipitation can be about 2 years, whereas at 700°C it is a matter of minutes.

Precipitation is effected by diffusion of carbon atoms to the grain boundaries where they each combine with approximately four times the number of chromium atoms. Diffusion of carbon is relatively fast at these temperatures, but that of chromium is extremely slow. Consequently, the chromium atoms are almost entirely supplied by the grain boundaries, so that the grain boundary chromium content is substantially lowered. This local depletion of chromium causes loss of passivity in acid corrodants with consequent attack along grain boundaries.

Weld decay The resultant 'intergranular penetration' in a casting or, if the heating has been caused by welding, 'weld decay' (see *Figure 1.36*) can completely disintegrate the material. Precipitation in cold worked material takes place along slip planes as well as grain boundaries. Consequently, the distance that the chromium atoms must diffuse is small. Hence, although the same amount of chromium is removed as carbide, the depletion is more uniformly distributed with a consequential lowering of general corrosion resistance, but a lower tendency to intergranular failure.

There are two alternative approaches to the problem of preventing intergranular corrosion. Either the carbon content of the steel is limited, by using an AISI 'L' or BS 970 code '11' steel, to 0.03% at which precipitation of carbide in sufficient quantity to cause trouble is impossible or an element such as titanium or niobium, which has a stronger affinity for carbon than chromium, is added to form the appropriate carbide by using an AISI 'Nb' or 'Ti' or a BS 970 code '40' steel. The theoretical amounts required to ensure that all carbon in excess of 0.02% is combined are: titanium = 4 × excess carbon; niobium = 8 × excess carbon. In practice, allowance must be made for nitrogen combining with the added element (particularly titanium) and for the efficiency of combination—carbon levels below 0.06% requiring a higher titanium or niobium to carbon ratio for complete combination than those above 0.08%C.

Knife-line attack When a stabilised (titanium or niobium treated) steel is heated to successively higher temperatures above 950°C up to 1250°C, the carbide enters solution and is broken down into its constituent elements to an increasing extent so that above 1100°C a relatively small amount of carbon remains combined. The free carbon is then available to form chromium carbide on subsequent reheating in the 450–750°C range. Combination of carbon with titanium occurs in the range 850–950°C, given adequate time. The whole

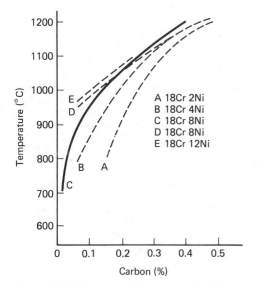

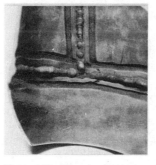

Figure 1.35 Variation in solubility of carbon in 18%Cr steel with temperature and nickel content

Figure 1.36 Example of 'weld decay' in an austenitic steel. (Courtesy of ICI)

question of 'stabilisation' is concerned with time, temperature and amount of free carbon.

Time at temperature affects the extent of re-solution of the titanium and niobium carbides present in stabilised steels; titanium carbide dissolves more rapidly than niobium carbide. Re-solution takes place at temperatures in excess of approximately 1200°C under welding conditions. The extent depends on carbide particle size as well as time at temperature. If reheated within the sensitisation temperature range (around 650°C) this narrow zone immediately adjacent to the weld metal precipitates intergranular chromium carbide, because combination of titanium or niobium with carbon cannot occur at this temperature.

Thus, although the stabilised steels will not precipitate chromium carbide in the region of the heat-affected zone raised to 650°C by welding, there is the possibility, in conditions where the edge of the weld metal is reheated to 650°C, that intergranular attack can occur. The existence of such conditions depends on the welding practice but, in most fabricated articles, as distinct from samples with single run welds, positions must arise at weld junctions where these conditions will obtain; welded samples should therefore always have crossed welds.

This particular type of intergranular attack at the weld metal edges is known as 'knife-line attack' (see *Figure 1.37*). It is most likely to be seen in boiling dilute nitric acid solutions. The composition of the steel affects its incidence; steels with lower nickel-to-chromium ratio, which produce a greater amount of δ ferrite in the knife-line zone are less susceptible. Fully austenitic titanium stabilised grades appear to be more susceptible than fully austenitic niobium stabilised ones.

Where corrosion conditions are known to offer a knife-line hazard, treatment of the fabrication at 870°C will promote precipitation of the carbon as titanium of niobium carbide with consequent resistance to attack.

The unstabilised 431 and 434 grades are susceptible to intergranular attack after welding. This can be prevented by heat treating for 2 h at 600–800°C which coalesces the carbide films.

Weld problems which may be assessed by means of the Schaefler diagram (Figure 1.33) Fully austenitic weld metal tends to crack on solidification because of inherent weaknesses at the boundaries of columnar grains. The composition of the filler metal is therefore adjusted to ensure that all the molten zone contains a small proportion of ferrite and also to ensure that its strength is not excessive compared with the parent metal (that is, it is adjusted to lie on the A + F side of the A/A + F line in *Figure 1.33*). So long as the weld-metal composition is maintained within the zone in which ferrite and austenite coexist austenitic steels have excellent weldability.

Figure 1.37 Example of 'knife-line' attack on stainless steel welds. (Courtesy of ICI)

Ferritic stainless steels are weldable but suffer from brittleness and grain growth problems (see Section 1.3.12).

Martensitic stainless steels suffer from the same brittleness problems as carbon and quenched and tempered steels unless the carbon content is below 0.12%.

Both precipitation hardening and duplex stainless steels (the compositions of which can be roughly estimated from *Figure 1.28*) are fully weldable without preheat, and the precipitation hardening steels may be hardened by precipitation after welding.

Brief summary notes on corrosion and welding aspects of stainless steels are given in *Table 1.12*.

1.3.18 Tool steels

The name 'tool steels' (BS 4659: 1971 and AISI/SAE 'Tool steels') covers a wide variety of steels used for forming and cutting materials which have as essential properties high hardness, resistance to wear and abrasion and adequate toughness. There are, or have been, some 82 AISI standard steels, 25 BS steels and many non-standard steels, but it should be possible to meet almost all requirements from the 10 steels listed here.

Carbon steels are used for hand tools and other applications where high levels of toughness are required and where some distortion in heat treatment can be tolerated. Recommended steels are:

(1) AISI 109; BS 4659 BWIA (0.9%C)—a steel with good combination of hardness and toughness, good general purpose steel; and
(2) AISI 210; BS 4659 BW2 (1%C, 0.25%V)—a steel which retains a sharp edge and withstands shock better than BWIA.

A carbon tool steel should be quenched in water or brine and tempered as soon as its temperature has been quenched to 'hand warm'. Carbon tool steels will soften and lose their edge if appreciable heat is generated by the cutting action.

High speed steels have a high content of carbide forming elements (tungsten, vanadium and chromium) and, therefore, retain their hardness at high temperatures (i.e. they have good 'red hardness'). Recommended steels are:

(1) AISI M2; BS 4659 BM2—for normal duty;
(2) AISI T4; BS 4659 BT4—for faster cutting and increased output; and
(3) AISI M42; BS 4659 BM42—for cutting hard materials.

'T' steels are tungsten steels and 'M' steels molybdenum steels which are cheaper but slightly more difficult to heat treat. Heating must be carried out in atmosphere-controlled furnaces to prevent decarburisation; heating done slowly to 825°C and then quickly to the manufacturer's recommended temperature (around 1300°C), followed by quenching in air blast, oil or salt bath at 525°C and air cooling. After an optional refrigeration treatment the steel must be tempered (secondary hardened) two or three times at about 545°C, again in controlled atmosphere.

For many purposes high speed steels are being replaced by sintered carbides (see Section 1.5) or ceramics such as sialons (see Section 1.6) which have exceptional wear and heat resistance even though they may not be as tough as high speed steels.

Hot work steels are used for forming (not cutting) hot materials. They must not soften at high temperature and they must have good wear resistance. They must also be able to resist thermal fatigue when heated and cooled (sometimes by

Table 1.12 Corrosion resistance and weldability of stainless steels

AISI No.	Nominal composition (%)					Corrosion notes
	C	Cr	Ni	Mo	Other	
410	0.08–0.4	12–14	—	—	—	Rust resisting. Higher carbon grades for engineering applications, turbine blades, cutlery, etc.
405	0.08 max.	13	—	—	Al	Weldable grade.
430	0.01 max.	16.5	—	—	—	Resists mild acids. Special feature is resistance to nitric acid. May require heat treatment after welding (600–800°C) to avoid intergranular attack. Forming of sheets up to 3 mm at room temperature; greater thicknesses at 200–350°C
430Ti	0.1 max.	17.5	—	—	Ti	Weldable grade not requiring heat treatment. Argon arc (gives minimum grain growth) preferred. Both grades, if welded, should not be applied under conditions of shock loading or vibration
304	0.08 max.	18	10	—	—	Rust and acid resistant. Suitable for welding in certain applications.
304L	0.03 max.	18	10	—	—	Extra low carbon. Very resistant to intergranular corrosion. Weldable for practically all applications.
309Cb 321	0.1 max.	18	10.5	—	Nb or Ti	Not susceptible to intergranular attack (but see reference to knife-line attack in text). Applicable above 300°C. Weldable
316	0.07 max.	17.5	11	2.2	—	Resistance to chemical attack better than 18/8 (e.g. severe acid attack). Resists intergranular attack up to 6 mm thickness. Applicable below 300°C. Weldable for most applications.
316L	0.03 max.	17.5	11	2.2	—	Superior resistance to intergranular corrosion, suitable for thicknesses greater than 6 mm
316Cb 316Ti	0.1 max.	17.5	11.5	2.2	Nb or Ti	Not susceptible to intergranular attack (but see reference to knife-line attack in text). Applicable above 300°C. Suitable for strong acids at elevated temperatures. Weldable
317	0.07 max.	17.5	12	2.8	—	Resists intergranular attack up to 6 mm thickness. Applicable below 300°C. Corrosion resistance superior to 2%Mo alloys. Weldable for most applications.
317Cb 317Ti	0.1 max.	17.5	12.5	2.8	Nb or Ti	For strong acids at high temperatures. Applicable above 300°C. Weldable
317LM	0.03 max.	17	13.5	4.5	—	Resistance to strong organic acids at elevated temperatures. Increased resistance to pitting. Applicable below 300°C. Resists intergranular attack. Weldable for most applications

water jets). Their metallurgy is similar to that of high speed steels. A recommended steel is:

AISI H13; BS 4659 BH13—this steel has the highest and deepest hardness of the hot work steels.

Cold work steels are used for forming cold materials and resistance to abrasive wear is of highest importance. In addition, they may have to be machined to very complex shapes and must therefore have very high dimensional stability during heat treatment.

Recommended steels are:

(1) AISI 01; BS 4659 B01 (0.95C, Mo, V)—a steel for light duties, simple to heat treat; and
(2) AISI D2; BS 4659 BD2 (1.5%C, 12%Cr, Mo, V)—a martensitic stainless steel with very high hardness and wear resistance for general application.

Shock resisting steels are used for tools which are subject to heavy vibration or hammering; they must be hard but also have reasonable toughness to avoid failure by brittle fracture. A recommended steel is:

AISI S1; BS 4659 BS1 (0.5%C, Si, Mn, Cr, W)—the metallurgy of this steel is relatively uncomplicated and heat treatment is straightforward.

1.3.19 Steels for springs

There are three different types of spring steel. 'Patented' and cold drawn carbon steel wire is used for small coil springs. 'Patenting' consists of heating the billet to roughly 1000°C to develop a coarse grain size so that after slow cooling the steel has a coarse pearlite–bainite structure which is readily drawn into wire.

The steels used and the properties of the wire are covered in BS 5216 and ASTM A227 and 228. The carbon content varies between 0.65% for 'hard drawn spring wire', which has the largest diameter (up to 9 mm), the poorest surface finish and the lowest tensile strength (less than 940 MPa), and 0.85% for 'piano or music wire', which has the smallest diameter (0.1 mm minimum), the best surface finish and the highest tensile strength (up to 3780 MPa).

Many ranges of tensile strength are available. Springs are cold coiled from the wire.

Carbon and alloy spring steels are made to the specifications given in BS 770: 1972: Part 5 and the corresponding AISI/SAE grades. Coil springs are usually made from hot rolled and ground bar of the diameter required for the final spring. The bar is heated to a temperature within the hardening temperature range, coiled on a mandrel, slipped off the mandrel, quenched and tempered to a tensile strength around 1650 MPa. Carbon steels are used for springs up to 13 mm diameter, more highly alloyed steels are used for larger diameters, the maximum diameter (around 80 mm) being made from BS 925 A60 manganese–molybdenum steel.

The purchase specification must strictly limit decarburisation of the surface (to which silicomanganese steel, which is popular for springs, is particularly prone), because fatigue cracking, which will propagate across the spring, may start in a soft decarburised surface layer.

The surfaces of all but the smallest springs are conditioned by shot peening which induces a compressive surface stress and increases fatigue strength by 25–30%. 'Scragging', which overloads the spring in the direction it will be used in service, produces residual stresses which oppose service stresses in the surface layers and, therefore, improves endurance.

Rust is harmful to spring performance and to prevent it a spring should be protected immediately after peening. Corrosion resistant steel springs are covered by an old British Standard (BS 2056: 1953) which uses the EN designations. In practice, stainless steel wire for springs is usually supplied to AISI number.

Martensitic steels are usually supplied softened and lightly cold drawn to a UTS of 820–850 MPa. They are hardened and tempered after forming. Austenitic steels are cold drawn to a UTS of 1800–2000 MPa for diameters below 2 mm and 1000 MPa for diameters up to 10 mm. One precipitation hardening stainless steel (DTD5086) can be supplied for forming in the softened condition and can then be precipitation hardened to 1800 MPa.

1.3.20 Cast steel

All the types of steel described earlier in this section can, in principle, be produced as castings. In practice, the steel grades listed in British Standards and by the several US standards authorities are confined to a limited number of types. *Table 1.13* gives:

(1) BS specifications;
(2) ASTM grades for carbon steels and for steels with alloy content up to 8% and a UTS of 482–827 MPa; and
(3) ACI (Alloy Castings Institute of the USA) grades for heat resistant and corrosion resistant steel castings.

While each grade in a steel-castings specification is the equivalent of a grade in BS 970 or SAE/AISI, they differ in important aspects. For example, a foundry is less likely to be equipped for carbon vacuum deoxidation than is a large steelworks. To allow for this the silicon content of steel castings is usually set at a higher level than for the corresponding wrought steel and the very low carbon grades are not included. This may require the content of other alloying additions to be adjusted. Also, an austenitic steel casting often contains more ferrite than the corresponding wrought steel to prevent fissuring during solidification (both in casting and welding) and to resist intergranular penetration. Therefore, when ordering a casting the designer should specify the BS 1504, BS 3100 or ACI grade rather than the BS 970 or AISI grade number for the corresponding wrought steel.

If a compelling reason exists for specifying a steel not listed in a standard casting specification, the casting will almost certainly be more expensive because the foundry may have to make experimental castings and will not be able to recycle scrap directly. In addition, it may be more difficult to obtain a guarantee of quality.

Design of castings is too complex a subject for detailed consideration here. The essential criterion is to make absolutely sure that nowhere within the casting is there a point where, through a local increase in section, metal is left to solidify surrounded by metal that has already solidified. Walls should be of as uniform thickness as possible, corners radiused, multiple junctions eliminated, changes of section tapered and, where large sections are inevitable, they should be so placed that they solidify progressively towards a feeding head. Where isolated large increases in section are unavoidable, chills may be used (see *The ASM Metals Handbook*[17]).

In principle, the properties of a casting should be identical to those of a forging of the similar composition and, in practice, castings are available with a UTS to match any forgings up to 827 MPa UTS, and corresponding yield strengths are available.

There are, however, significant differences in the cast and wrought structure, particularly in alloys with more than one phase present. Castings have a 'cast structure' which is effectively a skeleton of intermetallics which tend to limit and restrict slip. In a correctly worked wrought alloy this skeleton is broken up so that it becomes effectively a dispersion of fine particles rather than a network. Working refines the grain and renders the alloy more susceptible to heat treatment, and has two effects which are significant in design; i.e. ductility is increased and creep strength decreased.

The reduction in ductility of castings compared with wrought steel has a negligible effect on design with steels with a UTS of around 500 MPa. However, with strengths of 800 MPa and above, the fracture toughness of cast material is lower and more variable and the fatigue endurance limit is about 20% lower than that of wrought material.

In addition to this, the continuous casting process for manufacturing ingots from which wrought material is forged has much superior feeding and segregation characteristics than is possible in a large sand casting, so that the material is inherently superior. Furthermore, forging, if correctly programmed, can be made to align the grain (and any discontinuities) in the principal stress direction and thus make the component more resistant to both brittle fracture and fatigue. On the other hand, the transverse properties of a casting should be superior, and thorough inspection will reduce or eliminate dangerous defects.

The improved creep resistance of the cast structure is of considerable value in the case of large turbine castings, while the creep properties of the small lost wax castings listed in BS 3146 could not be obtained in forgings. Even higher creep properties could be obtained by directional solidification, but this is used for the more highly creep resistant nickel alloys rather than steels.

1.3.21 Cast irons—general

Cast iron is an alloy of iron with 1.7–4.5% carbon (1.7% is the eutectic composition). There are two basic types, one of which is a composite of steel and graphite, while the other (white cast iron) consists of cementite in a matrix of steel.

White irons (white cast iron, low alloy white cast iron, martensitic white cast iron and high chromium white cast iron) have special wear and environmental resistant properties.

The graphite containing cast irons, which include the flake graphite, nodular graphite and malleable grades, have been regarded as a cheap and brittle substitute for other engineering materials but, in addition to their relatively low cost, have very definite technological advantages. These advantages are particularly evident in the case of the newly developed austempered ductile irons.

The several grades of cast iron are listed according to BS specification in *Table 1.14*, which gives all (with the possible exception of damping capacity) relevant physical and mechanical properties.

1.3.22 Grey cast iron

Grey cast iron (flake graphite iron) can be 'non-alloyed', 'low alloy' or acicular. Design stresses, etc., given in *Table 1.14* (BS 1452 and ASTM A48 class 20–60) are for non-alloyed grey cast iron with carbon contents varying from 3.65 to 2.7%, silicon from 2.5 to 1.35%, phosphorus from 0.5 to 0.09% and manganese around 0.6%.

This is the cheapest engineering metal, not only because the raw materials—pig iron, cast iron and steel scrap, limestone,

Table 1.13 Standards for steel castings

Grade			UTS (MPa)	Steel type	Special requirements
BS 3100: 1976	BS 1504: 1976	ASTM*			
Non-stainless steels					
A1, 2 & 3		A27 & A148	430, 490, 540	Carbon steel for general purposes	
	430, 450, 540	A356	430, 480, 540	Carbon steel for pressure vessels	0.2% PS specified at temperature
AL1			430	Carbon steel for low temperature	Charpy 20 J at −40°C
A4, 5 & 6			540, 620, 690	Carbon–manganese steel for general purposes	
B1		A27	460	Carbon–molybdenum steel for elevated temperatures	
	26	A356	460	Carbon–molybdenum steel for pressure vessels	0.2% PS specified at temperature
	27		460	3.5%Ni steel for pressure vessels	Charpy 20 J at −60°C
BL1			460	0.5%Mo steel at low temperatures	Charpy 20 J at −50°C
BL2			460	3%Ni, 0.5%Mo steel	Charpy 20 J at −60°C
B2			480	1.25%Cr–molybdenum steel	
	28	A389	480	1.25%Cr–molybdenum steel	0.2% PS specified at temperature
B3			540	2.25%Cr–molybdenum steel	
	29		540	2.25%Cr–molybdenum steel	0.2% PS specified at temperature
B4			620	3%Cr–molybdenum steel	
	30		620	3%Cr–molybdenum steel	0.2% PS specified at temperature
B5			620	5%Cr–molybdenum steel	
	31		620	5%Cr–molybdenum steel	0.2% PS specified at temperature
B6			620	9%Cr–molybdenum steel	
	32		620	9%Cr–molybdenum steel	0.2% PS specified at temperature
B7			510	0.5%Cr, 0.5%Mo, 0.25%V	
	33		510	0.5%Cr, 0.5%Mo, 0.25%V	0.2% PS specified at temperature
Stainless and heat resisting steels		*ACI No.*			
302 C25		CF-20	480		
302 C35					
304 C12		CF-3	430		
304 C15		CF-8	480	Low carbon 18/8 type	
309 C30		CH-20			
309 C32			560		
309 C35			510		
309 C40		CK-20	450		
311 C11					
315 C15		CF-16F	480	1.5%Mo 18/8	
315 C16			480		
316 C12	316 C12	CF-3M	430	2.5%Mo 18/10 low carbon	
316 C16	316 C16	CF-8M	480	2.5%Mo 18/8	
316 C71	316 C71		510	2.5%Mo 18/8	
	317 C12		430	3.5%Mo 18/10 low carbon	
317 C16	317 C16	CG-8M	480	3.5%Mo 18/10	
318 C17		CF-12M	480	2.5%Mo–niobium 18/10	
347 C17	347 C17	CF-8C	480	Niobium stabilised 18/12	
364 C11		CN-7M	430	Chromium–nickel–copper	
410 C21		CA-15	540	13%Cr martensitic steel	
420 C29	34	CA-40	690	13%Cr martensitic steel	
425 C11	35	CA-6NM		13%Cr, 4%Ni	
452 C11				28%Cr, 1.5%Mo ferritic steel	
452 C12				28%Cr, 0.5%Mo ferritic steel	
BS 3146: Part 2				Corrosion and heat resisting investment castings	

Most grades are covered by ASTM A743 and A744. PS, proof stress.
* The figures indicate carbon content.

coke and air—are all relatively cheap, but also because melting costs in a cupola are relatively low. Casting is very easy because cast iron is more fluid, has a narrower solidification range and a lower in-mould shrinkage than steel. Machinability is excellent because graphite acts both as a chip breaker and as a tool lubricant.

Grey cast iron has good dry bearing qualities and its freedom from scuffing makes it a good material for automobile cylinder walls. Its wear resistance is assisted by slight chilling and a hard network of phosphide eutectic.

It also has an excellent damping capacity, particularly in the lower (higher carbon) grades and is particularly suitable for machine tool bases and frames.

However, it is brittle because the graphite flakes reduce strength and the maximum recommended tensile design stress is only 25% and its fatigue loading limit is 11–16% of the

Table 1.14 Cast irons, classes, grades and properties

Safe design stresses and other properties	Flake graphite iron, BS 1452: 1977 Grade 150	180	220	260	300	350	400	Compacted graphite iron	Nodular graphite (SG) iron, BS 2789: 1973 Grade 370/17	420/12	500/7	600/3	700/2	800/2	Austempered ductile iron	BS 309: 1972 W340/3 Whiteheart	W410/4 malleable iron	BS 310: 1972 B290/6 Blackheart	B310/10 malleable	B340/12 iron	BS 3333: 1972 P440/7	P510/4 Pearlitic	P540/5 malleable	P570/3 iron	P690/2	BS 1591: 1975 10–14% Si High-silicon	16% Si. Fe	4% Si, 0.5% Mo	BS 4844: 1972–1974 Grades 1A to 1C Abrasion resisting white iron	Grades 2A to 2E	Grades 3A to 3E	Austenitic cast iron, BS 3468: 1974 Flake 15% Ni, 6% Cu, 2% Cu	20% Ni, 2% Cr	13% Ni, 7% Cu, Mn	2% Ni, 20% Cr	23% Ni, 4% Mn	35% Ni, 3% Cr; Nodular	BCIRA Broadsheet 63 3% Cr, 15% Mo High-chromium	23–28% Cr iron	30–35% Cr iron	Hematite, high-C low-P iron
Tensile stress (Nmm⁻²)	38	45	55	65	75	88	100	92, 103	129	138	145	156	173, 198, 270,	270,	270,	81, 88,	102	110	121	120, 130, 134, 162,	257		140	15, 10				54 49, 57	90	86	77	90	150	35		
Elongation (%)	<1	<1	<1	<1	<1	<1	<1	4, 2	25, 17	20, 12	15, 7	15, 3	5, 2	5, 1	10, 1	>3 >4	>6	>10	>12	>4 >5	>3	>2						15 7	25							
Compressive stress (Nmm⁻²)	156	187	229	270	312	364	416	150	152	173	204	216	238, 271, 350,	850		93, 116,	120, 140,	130	138	149	172, 184, 188, 238,	353														
Unnotched-fatigue stress (Nmm⁻²)	23	27	33	39	45	50	51	60	63	67	75	83	93 101 130,	130,		41, 51,	53, 62,	58	62	68	71, 85, 88, 91	93						35	50							
Hardness (HB)	130, 150, 160, 180,	200,	225,	250,	305			140, 150, 165 155,	115, 140,	170,	170, 215, 250,	265, 305, 500			180, 190, 240, 270,			130, 140, 170	180, 190	240	170, 180, 190, 240, 270		140, 180, 180, 200,					120, 130, 140,	170	200	112, 120,	150 140 123	140		550, 550, 250, 90,	750 340 120
Young's modulus (GNm⁻²)	100	109	120	128	135	140	145	155, 165	169	169	169	174	172, 172, 155, 170		176 176 176	176		169	169	169	172, 172, 172, 176 176		124	200			200 200	140, 150, 130	105		85, 105			217	224	80, 100
Notched-impact values (J) at 20 °C				24, 20				7, 3	15, 13	10, 5	5, 2	5, 2	5, 5, 8, 4		17, 13	13, 12	10 5	13, 5	17, 13	10, 5 5, 2	5, 5, 2		11, 2	30, 25		20 25		27, 15	27,	62, 86	24, 13	7				59, 36
Notched-impact transition temperature (°C)								–10, +30	–10, +30	–10, +30	–10, +30						–10, +30	–10, +30	–10, +30	40, 40, 40, 0,* 0,*																
Fracture toughness K_IC (MNm⁻³ᐟ²)	12 20	20 24							46	46	100, 120	120	100, 100, 100, 120																							
Service temperature Max. (°C)	500	500	500	500	500	500	500	500	500	500	500	500	500, 500, 800, 70, ‡			500	500	500	55, 50	55, 35	50, 500, 250, 250, 800,			30, 25	20	25	700, 700, 500, 700, 700, 800,					900,1050,900, ‡	900 1050 500 §
Service temperature Min. (°C)	–50	–50	–50	–50	–50	–50	–50	–40	–40	0	0	20	–40, +30			0	0	–40	35	35	40 100 100 100 100			0	0	0	–40 –80 –80 –80 –200 –80								
Design stress +450 °C (Nmm⁻²)	17, 18, 30, 30,						100	22, 23,	22	23	27, 32, 173// 198//				19, 19, 19, 23, 28,	28,		19, 23,	19	85 102 120// 130// 134// 162 257			23,	35,	39 57	62, 77 86	70, 70, 70 70 70								
Design stress –50 °C (Nmm⁻²)	45	55	65	75	88	100		129	138	145	156			72																							
Density (gcm⁻³)	7.0	7.1	7.2	7.2	7.2	7.3	7.3	7.2	7.1	7.1	7.1, 7.2	7.2	7.2 7.4		7.4	7.3	7.3	7.3	7.3	7.3	7.3 7.3 7.3 7.3 7.3		7.0	6.85	7.7	7.7 7.7	7.3 7.3 7.4 7.6 7.67		7.3	7.3		7.69 7.43 7.0				
Thermal expansion 20 °C (10⁻⁶ K⁻¹)	10, 12.5	10, 12.5	10, 12.5	10, 12.5	10, 12.5	10, 12.5	10, 12.5	12.5	12.5	12.5	12.5	12.5	12.5		11, 13	11, 13	11, 13	11, 13	11, 13	11, 13 11 13 11 13	12, 13	12, 15.9	11, 12, 12, 15.9 13.5					18, 18, 18, 18, 18.	18	18	18	18 18 18				
Thermal conductivity 100 °C / 400 °C (Wm K⁻¹)	53, 50	52, 49	50, 47	49, 46	47, 44	46, 43	44, 41	41, 38	37, 36	37, 36	37, 35	35, 33	33, 31 31, 31		36, 35	36, 35	36, 35	38, 36	38, 36	35, 36, 35, 34, 33, 32 34 33 34 33 32		27, 25														
Electrical resistivity Max. / Min. (μΩ-m)	0.85 0.50	0.78	0.76	0.73	0.70	0.67	0.64	0.50	0.50	0.50	0.51 0.53	0.54	0.54 0.24,0.24, 0.26 0.26		0.37 0.37	0.40 0.40	0.40 0.40	0.40 0.40	0.40 0.40	0.40 0.40 0.50 0.50												0.26				
Maximum magnetic permeability (μHm⁻¹)	310, 380	310, 380	310, 380	310, 380	310, 380	310, 380	310, 380	2140 2140	1900 1820	910, 1820	1600 870	500	500		1900 1900	1900 1900	1280, 900, 900 750	750, 650	650 400	400 380						Non-mag.									
Hysteresis loss (B = 1 T), J/m⁻³	2500,2500, 3000 3000	2500, 3000	2500, 3000	2500, 3000	2500, 3000	2500, 3000	2500, 3000	600 600	1300 850	2200	1300, 2200	2700	2700 1500,1500, 2700 850		450, 450	450, 450	1300,1900,2200,2400,3800, 1900 2200 3800 4500																			
Poisson's ratio	0.26							0.275										0.26														T				

tensile strength. (It should be remembered that tensile stress is measured by bend—see Section 1.1)

There are (or were) two variants with better fatigue properties. Compacted graphite iron or mechanite is made by inoculating an iron which would otherwise solidify white. Haematite high carbon low phosphorus iron was originally made from haematite pig iron. Its low phosphorus content improves its fatigue properties (while reducing fluidity).

Low alloy and acicular cast irons made by adding nickel, copper, chromium, molybdenum, vanadium or titanium (and in the case of acicular cast iron reducing the phosphorus content) enables grey cast iron to be used in higher duty applications without re-design or technological change.

1.3.23 Nodular graphite iron

Nodular or spheroidal graphite (SG) cast irons are available in grades corresponding to those of grey cast iron, but are produced by inoculation of the melt with nickel, magnesium and caesium compounds which change the form of the graphite to near spheroidal nodules (see *Figure 1.38*). This produces material which has strength, ductility and thermal shock resistance more typical of steel but castability, damping capacity and machinability more typical of cast iron.

The recommended design, tensile and fatigue stresses are a much higher proportion of the UTS than is the case with cast iron. Steel castings, fabrications and sometimes forgings may be replaced with considerable economic advantage.

Matrix structures can be varied (by changing the cooling rate or alloying) between ferrite, pearlitic carbide and acicular structures for higher duty applications.

The development of nodular and other higher duty irons detailed here has accelerated the trend to modern melting practice. Casting from a cupola is not amenable to the precise composition control which is possible with an electric or gas furnace. Even where a cupola is used for the actual melting, final control of composition requires a holding furnace.

1.3.24 Austempered ductile iron

Austempered ductile iron is SG iron with added alloying elements (usually molybdenum, nickel and/or copper) sufficient for a bainitic structure, usually with retained austenite, to be produced in the section size by austempering. Such material can have yield strength and UTS up to 1150 and 1400 MPa, respectively, with elongations of 6% and fatigue limit up to 33% of the UTS. Wear resistance because of the graphite and retained austenite is superior to steel of the same hardness and components such as gears are quieter in operation. The potential of austempered ductile iron, which is substantially cheaper than forged steel and can be cast closer to shape than steel is usually forged, exceeds that of any other recently developed material.

Obtaining the required properties requires dedication to process control in foundries and heat-treatment departments.

The most economically rewarding application for austempered ductile iron is as a material for gears which can be made quieter, lighter and cheaper than equivalent steel gears. One disadvantage, for the highest rated gears, is the lower fatigue strength of austempered ductile iron compared with that of steel, but this is being overcome by shot peening the teeth of the gears.

Austempered ductile iron has been used successfully for tracks for off-the-road vehicles, pump bodies, agricultural equipment, friction blocks and drive shafts.

(a)

(b)

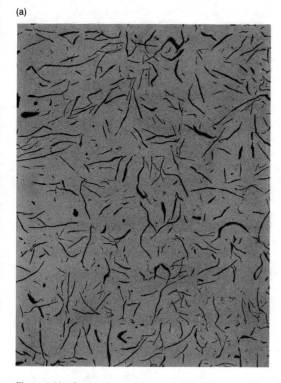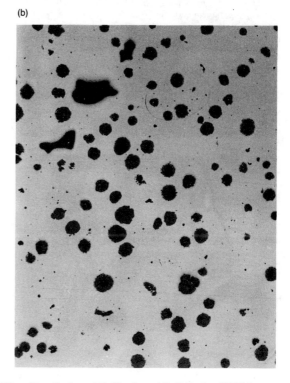

Figure 1.38 Contrast in graphite morphology: (a) grey cast iron; (b) 'SG' iron. Magnification ×100. (Courtesy of Mr R. Davies, BNF Fulmer Contracts Research)

1.3.25   Malleable iron

Malleable iron is cast with a white cementite structure which is converted to a steel–graphite composite by annealing. The requirement for the as-cast structure to be graphite free, limits the maximum section to about 38 mm and the general run of castings weigh under 5 kg and have a maximum section of 25 mm. There are three varieties.

Whiteheart malleable   This has a carbon content of about 3.5% which improves castability compared with the other varieties. Other alloying elements are silicon (0.6%), manganese (0.25%), sulphur (<0.3%) and phosphorus (<0.1%). It is heat treated for 5–6 days at 875°C packed in an oxidising medium to produce spidery graphite aggregates in a pearlite–ferrite matrix. This long heat treatment increases cost, limits rate of production and decarburises the surface layer.

Blackheat malleable   This has compositions varying between 2–2.65%C, 0.9–1.65%Si, 0.25–0.55%Mn, 0.05–0.18%S and <0.18%P. It is heat treated in a neutral atmosphere for 40–60 h at 860°C, cooled to 690°C, held for 4–5°C/hour and air cooled. It has graphite aggregates in a ferrite matrix (no decarburisation) and, although not so easy to cast as whiteheart, has rather better properties and the best combination of machinability and strength of any ferrous material.

There are two ASTM A47 grades (32510 and 35018) and one A197 'Cupola' grade of lower quality.

Pearlitic malleable   This usually has a higher manganese content (0.25–1.25%) and may be cooled rapidly after annealing. It has higher strength than the other malleables and, unlike them, has good wear resistance and is difficult to weld.

Because of their low cost and excellent shock resistance the malleable irons have been used extensively in power train, frame, suspension and wheels of motor vehicles, rail, agricultural and electrical equipment. The market for malleable iron has contracted with the sole exception of galvanised pipe fittings.

1.3.26   Austenitic cast irons

Austenitic cast irons have an austenitic matrix containing either flake or nodular graphite. They are non-magnetic, have thermal expansion coefficients similar to low-expansion aluminium alloys (with which they can be used as wear and thermal fatigue resistant inserts for pistons) and are available in a wide range of grades including:

(1)   Ni resist (14–32%Ni, 20%Cr)—for resistance to medium concentration acids; and
(2)   Nirosilal (Ni + Si)—for resistance to high temperature oxidation and growth up to 950°C.

High nickel nodular irons have excellent ductility and are suitable for cryogenic applications.

1.3.27   High silicon cast irons

High silicon cast irons, composition (10–17%Si, <3.5%Mo) have a silicoferritic solid solution matrix with dispersed graphite, exceptional corrosion resistance to mineral oxidising acids and, although extremely brittle, are used as pipes, stills and vats where strength is not needed.

The 4%Si, 0.5%Mo grade has good resistance to oxidation and acids and better strength than the high silicon grades.

1.3.28   White cast irons (abrasion resisting white irons)

There are four types of white cast iron: unalloyed, low alloy, martensitic and high chromium.

Unalloyed white cast irons have a reduced content of silicon so that on fairly rapid cooling after casting no graphite is formed and the carbon is in the form of cementite or pearlite.

Chill in white cast iron is increased by increasing the content of carbon and manganese, but is reduced by increasing sulphur and phosphorus.

White irons with carbon contents above 3.5% can have Brinell hardness of up to 600. However, increased carbon decreases transverse breaking strength and *low alloy white cast irons* have added elements (usually chromium and nickel) that increase chill and improve toughness and wear resistance (but are insufficient to produce a martensitic structure).

Martensitic white cast irons (e.g. Nihard) have sufficient alloying elements (usually chromium and nickel) to produce a cementite–martensite structure with higher hardness (up to 90 schleroscope) and toughness than other cast irons, and are also stable at temperatures up to 550°C. Martensitic white cast iron should preferably be stress relieved.

The white cast irons can be machined only with difficulty using carbide tools and should be cast as nearly to size as possible. They have higher solidification shrinkage than other cast irons and require careful running and feeding.

White cast irons are used for grinding and ore crushing equipment, mill liners, tables, rollers and balls and other applications requiring wear resistance.

The selection of the correct wear-resistant material for any application depends on relative life and relative cost. Martensitic white irons cost more than low alloy which cost more than unalloyed irons but, depending on the application, the life in wear of the most expensive material may be between 50 and 400% longer than the cheapest. Also, the more ductile more expensive material should be less prone to fracture, but this may depend more on events in the mill or on the technique of the supplying foundry. It is advisable to carry out comparative trials on different materials including, where appropriate, forged martensitic steel and deposited carbides and to standardise on that material which proves to be most economical for the specific application.

1.3.29   High chromium iron

Irons having a chromium content of 15–35% have a partially austenitic structure with higher toughness and strength than nickel based irons. They also have high corrosion and oxidation resistance.

They have a higher resistance to strong acids than silicon cast irons and can be used for heat-treatment equipment, melting pots for lead, zinc and aluminium, other parts exposed to corrosion at high temperature and for wet grinding operations.

1.4   Non-ferrous metals

1.4.1   Copper and its alloys

1.4.1.1   General

Copper is basically more expensive than iron but has important advantages for special applications, the most significant of which is conduction. The electrical conductivity of pure copper is superior on a volume basis to all metals other than silver, and on a weight (and specific cost) basis to all metals other than aluminium. The same relationships apply to its thermal conductivity. Both properties are reduced by alloying, but the conductivities of copper alloys are superior to those of steels.

Copper's second most important characteristic is its resistance to natural environments. Where iron rusts, copper remains bright or develops an attractive patina, and this characteristic is improved by appropriate alloying. In marine environments the toxicity of copper prevents fouling. There is, however, a temperature limitation on the use of copper alloys compared with steels.

Copper and a high proportion of its alloys are highly ductile so that they are eminently suited to forming operations. Some, particularly the leaded brasses, are also highly machinable so that the finished cost of a brass component may well be competitive with that of any other material, when allowance is made for the value and easy recovery of scrap.

The mechanical properties of copper (tensile and fatigue strength and creep resistance) can be improved by alloying, but without achieving the strengths of steels or approaching the specific strengths of the light metals. The good mechanical properties are retained at cryogenic temperatures but are inferior to steels at elevated temperatures. Other properties also benefit from alloying. The influence of specific additions is indicated in *Table 1.15*.

Copper alloys are by no means the easiest to cast or weld and their toxicity, although having useful biocidal applications, prohibits contact with foodstuffs.

Copper alloys are divided into classes, the main classes having traditional names. The main classifications together with their British Standard designations are listed in *Table 1.16*. The British Standards for product forms are listed in *Table 1.17* and the material condition codes are given in *Table 1.18*.

1.4.1.2 Copper

'Copper' is an alloy of copper and oxygen. The oxygen content of the conductivity grades is not such as to affect their electrical conductivity, but unless an 'oxygen free' grade is used would cause problems in welding and also when heated in a reducing atmosphere. Non-conductivity grades are deoxidised, usually with phosphorus which reduces electrical conductivity.

The suitability of copper for electrical conductors depends on its high conductivity combined with a high resistance to atmospheric corrosion and ease of drawing and fabrication. Material for conductors should be selected from grades C100 to C104 or C110, all of which have electrical conductivities of at least 101 IACS annealed, 97 IACS cold drawn (100 IACS = 0.019 $\mu\Omega$-m). If the conductor is to be heated in a reducing atmosphere the oxygen free grade C103 should be used. The high conductivity grades have UTS 385 and proof stress 325 MPa hard and 220 and 60 MPa annealed.

Additions of silver between 0.02 and 0.14% improve creep strength and resistance to annealing without impairing conductivity and should be used for rotating machinery or where a component must be heated during manufacture.

For general engineering and building operations where conductivity is not significant any of the grades C101 to C107 may be used. Arsenical grades have slightly better strength at high temperature, phosphorus deoxidised grades are better to braze or weld.

1.4.1.3 High conductivity copper alloys

There are a number of alloys containing a high percentage of copper which balance the minimum possible reduction in conductivity against some other desirable property, machinability (C109, C110) strength (CB101) wear resistance (C108) or strength at high temperature (CC101, CC102). Strengths range from 495 to 1346 MPa and conductivities from 90 to 20 IACS. Several of these alloys are available in cast form.

1.4.1.4 Brass

The range of composition of copper–zinc alloys is illustrated in *Figure 1.39*. There are two classes: α brasses containing 24–37% zinc; and duplex brasses with 40–47% zinc.

Table 1.15 Influence on copper of alloying additions

Improved property	Alloying addition
Strength	Aluminium, beryllium, chromium, zirconium, zinc, tin, phosphorus, silicon, nickel, manganese, iron
Corrosion resistance	Nickel, aluminium, silicon, tin, arsenic, manganese, iron
Wear resistance	Tin, cadmium, silicon, aluminium, silver
Bearing properties	Lead
Colour	Zinc, tin, nickel
Cost	Zinc
Machinability	Lead, tellurium, sulphur, zinc
Castability	Zinc
Resistance to annealing (conductivity not impaired)	Silver

Table 1.16 Classes of copper alloy

Common Name	Description	British Standard: wrought	British Standard and ASTM: cast	ASTM designation: wrought
Copper	Alloy of copper and oxygen. Sometimes accompanied by deoxidant. Sometimes contains silver	C100–107, 110	HCC1 (BS 4577 A1/1)	OF. ETP. OLP.
High conductivity copper alloy	Alloy of copper with additions which improve strength at a minimum loss in conductivity	C108, 110, 111–113, CB 101, CC 101/2	A2/1–4/2	
Brass	Alloy of copper and zinc to which other elements (usually with their name as a prefix) may have been added	CZ 101–137	SCB1–5 DCB1–3 PCB1 HTB1–3	2XX 3XX (leaded)
Bronze Phosphor bronze Leaded bronze Leaded phosphor bronze Copper–tin	Traditionally alloys of copper and tin Copper, tin and phosphorus Copper, tin and lead Copper, tin, phosphorus and lead Copper and tin	PB101–104	PB 1–4 LB 1–5 LPB 1 CT	5XX
Silicon bronze	Copper, silicon and manganese	CS 101		
Gunmetal Nickel gunmetal Leaded gunmetal	Copper, tin and zinc Copper, tin, zinc and nickel Copper, tin, zinc and lead		G1, G2 G3 LG 1–3	
Nickel–silver Leaded nickel–silver	Copper, nickel and zinc Copper, nickel, zinc and lead	NS 103–109 NS 101, 103, 111		7XX
Cupronickel	Copper, nickel, manganese (sometimes iron)	NC 102–108	NC 1, 2	7XX
Aluminium bronze	Copper aluminium (iron, nickel and/or manganese–silicon)	CA 101–107	AB 1–3, CMA 1	6XX
Copper–lead	Bearing alloy. Copper, 20–40% lead		(ISO 20%Cu, 40%Pb)	
Constant resistivity alloys		(ISO Cu, 13%Mn, 3%Ni; Cu, 13%Mn, 2%Al)		
Memory alloys	Copper, zinc and aluminium	No standard		

Table 1.17 British and some ASTM standards for copper alloys in product form

British Standard	ASTM*	
BS 6017	Depends on material	Copper refinery shapes
BS 1400	B176, 584, 148	Cast copper alloys
BS 2870	B248	Copper and copper alloys—sheet, strip and foil
BS 2871	B251	Copper and copper alloys—tubes (3 parts)
BS 2872	B124	Copper and copper alloys—forging stock and forgings
BS 2873	B250	Copper and copper alloys—wire
BS 2874	B249	Copper and copper alloys—rods and sections (other than forging stock)
BS 2875	B248	Copper and copper alloys—plate
BS 1432		Copper for electrical purposes—strip with drawn or rolled edges
BS 1433		Copper for electrical purposes—rod and bar
BS 1434		Copper for electrical purposes—commutator bars
BS 1977		High conductivity copper tubes for electrical purposes
BS 4109		Wire for general electrical purposes and for insulated cables and flexible cords
BS 4608		Copper for electrical purposes—rolled sheet, strip and foil

* These are a selection of equivalent ASTM standards many of which refer to particular material applications.

Table 1.18 Material condition coding for copper and copper alloys

Code	Description
0	Annealed
¼H, ½H, H, EH	Harder tempers, produced by cold working (or part annealing)
SH, ESH	'Spring hard' tempers produced by cold rolling thin material
M	'As manufactured'
W	Solution treated. Will precipitation harden
W(¼H), W(½H), W(H)	Material solution treated and then cold worked to progressively harder tempers
WP	Solution and precipitation treated
W(¼H)P, W(½H)P, W(H)P	Material solution treated, cold worked to progressively harder tempers and then precipitation treated

All brasses can be hot worked; α brasses are readily cold worked and cast, but duplex brasses are significantly more workable at elevated temperatures and can be extruded and forged into complex sections and shapes. This formability has the result that brasses are available in a very wide range of shapes. They are intrinsically easy to machine and machinability is improved even more by the addition of low percentages of lead. Brasses have a very useful strength range, (330–810 MPa), they are resistant to atmospheric and natural water corrosion, and the incorporation of zinc lowers their cost appreciably compared with copper.

The most suitable alloy for high speed machining is the leaded CZ1214PB and for hot stamping CZ122. These alloys have an ultimate tensile strength (UTS) of 450 MPa and they are, because of their high zinc content, the least expensive of the wrought brasses.

Where a higher tensile strength is required there are a number of high tensile brasses, with additions which include aluminium, iron, manganese and silicon. The wrought alloys CZ114 to CZ116 and CZ135 have UTS values of between 430 and 770 MPa, the cast alloys between 470 and 810 MPa.

Corrosion resistance may be improved by adding tin, aluminium arsenic and nickel.

Two important considerations must be observed when introducing brass components into service in aqueous environments.

(1) In some potable, marine or industrial waters containing sulphur compounds, many brasses (particularly duplex brasses) are prone to 'dezincification' (the preferential dissolution of zinc leaving a weak copper sponge). This can usually be prevented by choosing an α brass inhibited with arsenic. Recently, a duplex alloy (CZ132) has been

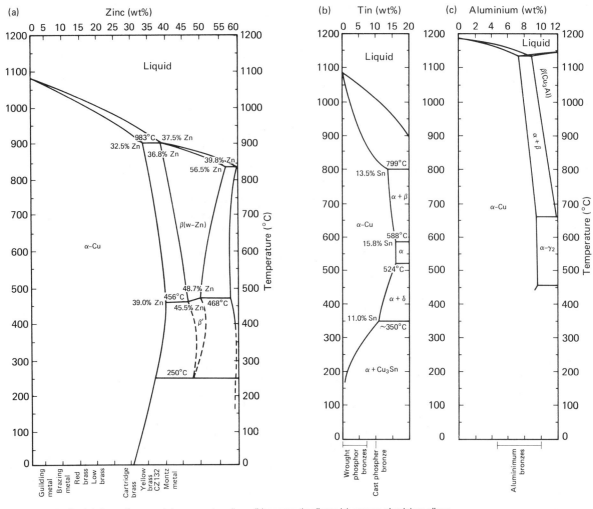

Figure 1.39 Partial phase diagrams: (a) copper–zinc alloys; (b) copper–tin alloys; (c) copper–aluminium alloys

developed which, when correctly heat treated, is claimed to resist potable waters (see *Figure 1.39(a)*).

(2) Stress corrosion may be caused by ammonia derived from organic refuse, by chlorine, or by mercury. Brass which may be exposed to these environments should be given a stress relief heat treatment.

Where there is a possibility that either of these phenomena will occur, trials should be made of the behaviour of the brass in the chosen environment. Alternatively, a bronze or aluminium bronze (but *not* a manganese bronze) should be used.

1.4.1.5 Bronze and gunmetal

Brasses all contain zinc but not all bronzes, which traditionally are copper–tin alloys, contain tin. 'Aluminium bronze', 'silicon bronze' and 'lead bronze' contain no tin and 'manganese bronze' is a brass.

The true wrought bronzes contain up to 8%Sn and are deoxidised with phosphorus. They have a single phase solid solution structure at the annealing temperature, but the solubility of tin falls with decreasing temperature (see *Figure 1.39(b)*). Cold working the resultant supersaturated solution gives them excellent elastic properties to which, coupled with

a high resistance to corrosion, they owe the majority of their applications.

The phosphor bronze most generally used is PB102 which contains 5%Sn but higher and lower tin contents with UTS values ranging between 590 and 680 MPa (hard) and 320 and 380 MPa (annealed) are available. They are used for springs, instrument components and bearings.

Cast phosphor bronzes, PB1, PB2 and PB4 contain 10%Sn, which is more than the wrought alloys, and are also used for bearings with hardened steel where load-carrying capacity at low speed is required. Gunmetals G1 (a single-phase alloy of copper, tin and zinc) and LG1, LG2 and LG4 (similar alloys which also contain lead), which combine modest strength (UTS ranging between 180 and 440 MPa) with good corrosion resistance, are more widely used because of their better castability. Applications include pumps, valves, bearings and statuary.

Also used for bearings are the leaded bronzes (LB1 to LB5) which have better plasticity and are particularly suitable for mating with soft steel journals which would be scored by harder materials. The addition of lead (which does not dissolve in solid or liquid copper but forms a composite) confers machinability and very considerably improves bearing characteristics.

1.4.1.6 Aluminium bronze

Aluminium bronzes CA 102 and CA 104 to CA107 and AB1-3 contain between 6 and 10% aluminium together with iron, nickel or silicon (see *Figure 1.39(c)*).

They all have excellent corrosion resistance properties, due basically to the combination of the electrochemical properties of copper with the tenacious oxide film of aluminium, and good mechanical properties (UTS 430–820 MPa for wrought and 460–700 MPa for cast types).

They can be used in a very wide range of environments, including dilute acids, with the exception of nitric acid. They are particularly suited to marine environments where, unlike most steels, they resist pitting and crevice corrosion and, because of the biocidal nature of copper, are not subject to biodeterioration.

The alloys with higher aluminium contents (see *Figure 1.39(c)*) have a duplex structure of α solid solution and β-Cu_3Al and develop high strengths when worked.

Although aluminium bronzes have quite good wear resistance they should not be used in applications which involve relative movement in contact with other metals, because the oxide film causes galling, fretting and seizure. This restriction apart, aluminium bronze has a very high potential as an engineering material which has yet to be realised fully.

1.4.1.7 Silicon bronze

There is one silicon bronze CS101 (Cu, Si, 3%Mn) used mainly as wire for marine fasteners.

1.4.1.8 Copper–nickel alloys

The copper–nickel alloys (CN101, CN102, CN104, CN105 and CN107 (wrought) and CN1 and CN2 (cast)) have even better corrosion resistance than the aluminium bronzes. They have moderate strengths, the UTS ranging from 300 to 390 MPa (annealed) and 360–650 MPa (hard). They may be used with confidence in the most severe conditions of marine pollution, and corrosion. However, no copper alloy is completely resistant to corrosion/erosion. Where, as is the case with those marine condenser tubing installations from which sand or silt cannot be excluded, corrosion/erosion is the principal cause of failure, titanium or a pitting resistant stainless steel performs better.

1.4.1.9 Nickel silvers

Nickel silvers (NS102 to NS106) are copper alloys with 10–25%Ni and 18–28%Zn. They have good corrosion resistance and attractive colour and are available as strip and wire

with proof stresses between 600 and 620 MPa, which makes them very suitable for relay springs at ambient and moderately elevated temperatures.

Leaded nickel silvers (NS101 and NS111) which have improved machinability are also available as bar.

1.4.1.10 Miscellaneous alloys

Copper alloys with between 20 and 45% lead are used for bearings whose loadings are too high for white metals. They were originally cast into a steel backing but are now usually made using powder metallurgy techniques to improve consistency and produce a more favourable structure.

Alloys of copper with manganese and aluminium or manganese and nickel can have very small temperature coefficients of electrical resistance. They are used in instruments and are usually supplied as annealed wire. Other copper–manganese alloys have high damping capacities.

Typical compositions and mechanical properties of these alloys (which are not covered by British Standards) are listed in *Tables 1.19* and *1.20*.

Alloys of copper with zinc (12–40%) and aluminium (2–8%) are capable of existing in two distinct configurations above and below a critical temperature which may be selected to lie between $-100°C$ and $+100°C$. This 'shape-memory' characteristic enables the manufacture of a component that will generate a force 200 times that which can be obtained from a bimetallic strip of similar size and is used for controlling temperature.

1.4.1.11 Selection of copper alloys

Choice of the alloy of copper that will most satisfactorily fulfil most appropriate engineering applications may be made by consulting *Tables 1.21* to *1.24*.

1.4.2 Aluminium and its alloys

1.4.2.1 General

Although aluminium has been used commercially for little more than 100 years, on a volume basis it has become the most widely used non-ferrous metal. It has the essential characteristic of high corrosion resistance in normal atmospheres and conditions and its cost per unit volume is usually lower than that of any other metal except steel.

One of the most important characteristics of pure aluminium is its high electrical conductivity, 61 IACS on a volume basis; because of its low specific gravity, aluminium conductors are usually cheaper than copper conductors.

Table 1.19 Composition and mechanical properties of copper–lead alloys for bearings (lead bronzes)

Type	Composition (wt%)					Tensile Strength (MNm^{-2})	Characteristics
	Cu	Pb	Sn	Ag	Other		
80/20 Cu–Pb	78	20	—	—	1.2–5Ni	140	Bonded to steel shells for bearings. Alloys up to 45% lead are in use
74/24 Cu–Pb	74	24	2	—	—	140	Good conductivity
70/30 Cu–Pb	69	30	—	0.6	—	140	Withstands 'pounding'
60/40 Cu–Pb	59	40	—	1	—	—	Lead distribution improved by powder metallurgical processing
55/45 Cu–Pb	55	45	1	—	—	120	

* Reproduced by permission of the heirs of Dr E. A. West.

Table 1.20 Composition and mechanical properties of copper–manganese alloys*

Type	Composition (%)				Condition	Tensile properties		Characteristics
	Cu	Mn	Al	Ni		Strength (MNm^{-2})	Elongation (%)	
Cu–Mn–Ni ('Manganin')	Balance	12	—	2–4	Drawn—soft	420	30	High specific resistance alloys with low temperature coefficient of electrical resistance
Cu–Mn–Al	Balance	13	2	—	Drawn—soft	430	30	
					Drawn—hard	680	10	
General purpose, high damping alloy ('Incramute')†	58	40	2	—	Hot and cold worked			High capacity for vibration damping and good corrosion resistance. Good casting and hot and cold forming properties, but difficult to machine. Costly
Marine high damping alloy ('Sonoston')‡	30	63.5	4	2.5	Cast			High capacity for vibration damping and good marine corrosion resistance. Casting requires skill. Poor machinability. Costly
	60	20	—	20	Heat treated			High strength suitable for springs

* Reproduced by permission of the heirs of Dr E. A. West.
† International Copper Research Association Inc.
‡ Stone Manganese Marine Ltd.

Table 1.21 Initial guide to selection of copper and copper alloys for electrical purposes*

Application	Wrought	Cast
Power cables, overhead lines, industrial and domestic wiring	C101, C102	
Telecommunication and coaxial cables	C101, C102	
Overhead lines and electric-traction catenaries	C101, C102, C108	
Bus-bars	C101, C102	
Flexible cables	C101, C102, C108	
Generator windings and transformers at low and normal temperatures	C101, C102	
Generator windings at raised temperatures	CC101, CC102 Cu + 0.1% Ag	
Commutators	C101, C102, A2/1 Cu + 1% Ag	
Slip rings	CZ108	PB4, HTB1, SCB4, SCB6
Electronic components in vacuum	C103, C110	
Switch blades, etc.	C101, C102, CB101 Cu + 0.1% Ag	HCC1, CC1
Machined components for fittings (terminals, etc.)	C111, C109	
Cast components for conductors		HCC1
Spring contacts, etc.	PB101, PB102, PB103, C108, CB101, A3/2, NS103, NS105, NS106, NS107	

* Reproduced by permission of the heirs of Dr E. A. West.

Table 1.22 Initial guide to selection of copper and copper alloys for applications in mechanical and chemical engineering*

Application	Wrought	Cast
Bellows and diaphragms	PB101, PB102	
Valve bodies for high duties	CA104, CA105	AB1, AB2
Pressure vessels	C106, C107, CS101	
Wear-resistant cans, guides, etc.	CB101, CA105	AB1, AB2
Pump components	CA106, PB103, PB104	AB1, AB2, CMA1, G1, LG2, LG4
Springs	CB101, PB102, PB103, NS104, NS106, NS107 CZ106, CZ107, CZ108, C101	
Bearings		
Heavy duty, rolling mills	CA106, CA105, Copper lead	AB2
Marine		G1
Non-critical, low loads		LG2, LG4, LB4
Light duties	CZ120, CZ121, CZ122	DCB1, DCB2, DCB3
Average duties, good lubrication	PB104, CZ124	LG2, LG3
Average duties, poor lubrication		LB2, LB3, LB4, LPB1
For hard shafts	PB104, CZ124	PB1, PB2, PB4, G1
For soft shafts, low loads	PB103, PB104, CZ121, CZ124	LB1, LB4, LB5
Plates for bridges	CA104, PB104	PB1, PB2, PB4
Ball and roller cages	A/3/2	
Automobile radiators		
Tubes	C101, C102, C106, CZ105, CZ106	
Strip	Cu + 0.1%Ag	
Tanks	CZ105, CZ106, CZ107	
Gears		
Light duty		G1, LG2, LG3, LG4, DCB1
Moderate duty	PB103, PB104, CA103, CA104, CZ105	DCB2, DCB3, AB1
Moderate duty	PB103, PB104, CA103, CA104, CA105	PB1, PB2, PB4
Very heavy duty, low speed		PB2, AB2
High loading	CA103, CA104	PB1, PB2, PB4, AB2
High abrasive loading	CA104	AB1
Pinions	PB103, PB104	PB1
Clocks and similar	CZ118, CZ120, CZ122	
Instruments (high precision)	CZ120, CZ122	
Bushings for sleeves	PB103, PB104, CZ120, CZ121, CZ122	
Deep drawn and pressed items	CZ105, CZ106, CN104, NS104, NS105, C104	
Repetition machined items	C109, C111, CZ118, CZ119, CZ120, CZ121, CZ122	
Brazed assemblies	C106, C107, CZ103	
Non-sparking tools, etc.	CB101, CA103, CA104, CA105	AB1, AB2
Chains	CB101, CA104	AB1
Hot forgings and stampings	CZ109, CZ120, CZ122, CZ123	

* Reproduced by permission of the heirs of Dr E. A. West.

Table 1.23   Initial guide to selection of copper and copper alloys for corrosion resistance*

Environment and Application	Wrought	Cast
Rural and industrial		
Roofing sheet and cladding, flashing and gutters	C101, C102, C104	
External decorative items, formed strip and sections	C102, C104, CZ107, CZ121, CZ125, NS101, NS102, NS111, NS112	
Lightning conductors	C101, C102, C104	
Wall ties and masonry fittings	PB102, CA103	PB1, AB1, AB2
Window frames and statuary	CZ121	
Statuary	C101, C102, C104, C106	Special tin bronzes
Solar heating panels	C106, C10	
Damp-proof course and weather strip	C104, C107, CZ101, CZ102	
Tubes for water and gas	C106	
Water cylinders, calorifiers	C106, CS101	
Tubes for soil and waste systems	C106	
Hinges and butts	CZ108, CZ121	
Water fittings, taps, etc.		SCB1, SCB3, SCB6, DCB1, DCB3, PCB1
Valves for water		SCB1, SCB3, SCB6, DCB3, LG2, LG4, G1
Nails and screws	C102, CS101, CZ106, CZ108, CZ121, CZ124	
Marine		
Condenser tubes	CN101, CN107, CZ109, CZ110, CZ112, CA105	
Tubes for seawater	CN101, CN107, CZ110, CS101, CA104, CA105	
Oil tank heaters	CZ110	
Fittings for seawater	CZ112, HTB1, CN107, CS101, CA104, CA105	PB1, PB4, G1, LG2, LG4
Valves and pump components	CA104, CA105	G1, LG2, LG4, PB1, PB4, AB2
Boiler feed water fittings	CT1, PB1, PB3	AB2, G1, LG2, LG4, PB1, PB4
Propellors		CMA1, (CMA2), AB2, HTB1
Portholes, deadlights, windows	CA104	G1, AB2
Non-magnetic fittings	CZ112, CA103, CA104	G1, LG2, LG4, AB1, AB2
Small fittings and cleats	CZ110, CZ111, CZ112, CS101, CA104	HTB1, LG2, LG4
Nails, clouts and screws	CS101, CZ112, CA103, CA105	

* Reproduced by permission of the heirs of Dr E. A. West.

Table 1.24   Initial guide to selection of copper and copper alloys for applications at raised temperatures*

Application	Wrought	Cast
Superheated steam valves and fittings	CA103, CA104, CA105	LG2, LG4, AB2, CMA1
Oxygen lance heads	CC101, CC102	
Arc furnace electrode holders	C108	CC1-TF
Disc brakes	CC101, CC102	
Aircraft brakes	CC101, CC102	
Spot, seam and flash butt welding electrodes and dies	C108, CB101, CC101	

* Reproduced by permission of the heirs of Dr. E. A. West.

A most important point to consider in the selection of aluminium is the cyclical price variation. The supply of aluminium is inelastic because furnace capacity cannot be expanded quickly. Therefore, while aluminium components cost less than copper or magnesium components when supply exceeds demand, magnesium components and occasionally even copper conductors may become superior economically when aluminium is in short supply. In extreme cases a correct choice made when a component is designed may prove to be incorrect when it comes to be manufactured.

Aluminium is ductile and easily fabricated and, although its mechanical properties are not outstanding, the specific strength and modulus of its alloys are excelled only by titanium and magnesium among common engineering metals because of its low specific gravity (2.7).

The tensile properties of aluminium improve with reduction in temperature and, because it has a face centred cubic structure, it does not embrittle at cryogenic temperatures. It is, therefore, suitable for components for operation at cryogenic temperatures.

The casting and fabrication of aluminium are easy due to its low melting point and high ductility. Because of this, and because of its wide usage, aluminium is readily available in all wrought forms, shapes and finishes, and all varieties of castings. British Standards specifications covering aluminium alloys are listed in *Table 1.25*.

Most wrought alloys can be hardened by cold working and are available in a variety of tempers.

Some alloys, both cast and wrought, are amenable to heat treatment by solution treatment and ageing. The several heat treatments and conditions are denoted by letter and number codes which are listed in *Table 1.26*. All except some heat-treated alloys are very easy to weld.

Aluminium is easily machined and, in spite of its ductility, is capable of a good finish. It is non-sparking and non-magnetic.

Drawbacks are poor fatigue and poor elevated-temperature properties. Aluminium has no endurance limit, the fatigue strength of the pure metal at 10^8 cycles is only about 60% of its tensile strength and this ratio is lower the higher the tensile strength of the alloy. Because of its low melting point, its high-temperature capability is very limited and the tensile strength of wrought alloys falls very sharply above 220°C and that of cast alloys above 270°C.

Table 1.25 British Standard specifications covering aluminium alloys*

BS 1490: 1970	Aluminium and aluminium alloy ingots and castings
Applications	
BS 1470: 1985	Wrought aluminium and aluminium alloys for general engineering purposes—plate, sheet and strip
BS 1471	Drawn tube
BS 1472	Forging stock and forgings
BS 1473	Rivet, bolt and screw stock
BS 1474	Bars, extruded round tubes and sections
BS 1475	Wire
BS 3087: 1974	Anodic oxide coatings on wrought aluminium for external architectural applications
BS 1615: 1972	Anodic oxidation coatings on aluminium
BS 5762: 1979	Methods for crack opening displacement (COD) testing
BS 8118: 1985	Code of practice for the design of aluminium structures

* See also BS L series and DTD specifications for defence materials.

Table 1.26 BS codes for aluminium product forms and conditions

Wrought material
Prefixes indicating product form:

B	Bolt and screw stock
C	Clad
J	Longitudinally welded tube
R	Rivet stock
E	Bars, extruded round tube and sections
F	Forging stock and forgings
G	Wire
S	Plate, sheet and strip
T	Drawn tube

Suffixes indicating condition:

F	As fabricated
O	Annealed
H	Strain hardened (i.e. strengthened by cold working—non-heat-treatable materials)
H2	Approximately equivalent to previous BS designation 1/4 H
H6	Approximately equivalent to 1/2H
H8	Approximately equivalent to 3/4H

Heat treatable:

T4	Solution treated and naturally aged (formerly TB or W)
T3	Solution treated, cold worked and naturally aged (formerly TD or WD)
T5	Precipitation treated (formerly P or TE)
T8	Solution treated, cold worked and precipitation treated (formerly TH or WDP)
T6	Solution treated and precipitation treated (formerly TF or WP)

Electrical grade:

E	A special suffix indicating an electrical grade of material

Cast material
Code indicating heat treatment:

M	As cast
TB	Solution heat treated and naturally aged (formerly W)
TB7	Solution heat treated and stabilised (W special)
TE	Precipitation treated (P)
TF	Solution heat treated and precipitation treated (WP)
TF7	Solution heat treated, precipitation treated and stabilised (WP special)
TS	Thermally stress relieved

Ingot material
Ingot materials have the same designations as castings, but no code letters for condition

Although corrosion resistance in normal atmospheres is good, aluminium is prone to galvanic corrosion when coupled to most engineering metals (except magnesium and zinc) and certain alloys in high heat treatment conditions are prone to localised corrosion.

1.4.2.2 Wrought aluminium alloys (see *Table 1.27*)

All aluminium alloys are collectively known as 'aluminium'. 'Pure aluminium' is an alloy of aluminium, iron and silicon

Table 1.27 US and UK wrought aluminium alloy designations

Composition (%)	USA AA	UK BS (where different)	Heat treatable	UTS (MPa)	Weldability		Notes
					Fusion	Resistance	
99.99Al	1099, 1199	—	No	50–100	Very good	Good	
99.8Al	1060		No	50–150	Very good	Good	
99.5Al	1050	IE	No			Very good	
99.5Al	EC	IE	No			Very good	
99.0Al	1200		No	100–150	Very good	Very good	
Al, 4Cu, Si, Mg	2014		Yes	400–500	No	Very good	
Al, 4Cu,Mg	2024	L97, L98	Yes	400–500	No	Good	
Al, 12Cu, 1.5Mg, 1Fe, 1Ni	2618		Yes	400–500	No	No	
Al, 12Cu, 1Ni, 1Mg, Fe, Si	—				No	No	RR50
Al, 15.5Cu, Cu, Pb, Si	—	FC1			No	No	free cutting
Al, 1Mn	3103		No	200	Good	Good	
Al, Mn, Mg	3105		No	200	Good		
Al, 5Si	4043/4543	—	No				
Al, 12Si	4047	—	No				
	4032	DTD324B					
Al, 1Mg	5005				Excellent	Excellent	
Al, 2Mg	5052		No	150	Excellent	Excellent	
Al, 3Mg	51.54A	5154	No	200	Excellent	Excellent	
Al, 5Mg	5056A	5056	No	200–300	Excellent	Excellent	
Al, 3Mg, Mn	5454		No	250–300	Excellent	Excellent	
Al, 4.5Mg, Mn	5083		No	300–400	Excellent	Excellent	
Al, Mg, Si	6063		Yes	150–200	Good	Good	
	6363		Yes	150–200	Good	Good	
Al, 1Mg, Si, Cu	6061		Yes	200–300	Difficult	Good	
Al, Mg, Si, Mn	6082		Yes	200–300	No	Good	
Al, 14.5Zn, 1Mg	7005		Yes	350	No	Good	
	7075	DTD5074A					
Al, 12.5Li, 1.3Cu, 0.8Mg, 0.1Zr	8090		Yes	400–500	Not stated	Not stated	Low density,
Al, 12.3Li, 2.7Cu, 0.12Zr	2090		Yes	400–500	Not stated	Not stated	high modulus

derived from the reduction process, in contents varying according to grade up to 1% with minor quantities of other metals. 'Bright trim' grades have slightly higher impurity contents and small additions of magnesium. In 'super pure' aluminium, which is double refined, the impurities (copper, silicon and iron) total less than 0.01%.

The tensile strength of the pure aluminium grades varies from about 55 MPa (temper designation 'O') annealed to about 160 MPa strain hardened full hard (temper designation 'H.8'). Increases in strength are obtained by alloy additions. Manganese and magnesium in amounts up to 1.25%Mn and 4.5%Mg increase the UTS up to about 300 MPa annealed and 375 MPa strain hardened half hard.

None of these alloys owes its tensile properties to heat treatment, and corrosion resistance is reduced only marginally compared with that of pure aluminium. All of the alloys are readily welded.

The lower strength alloys of this group are used for domestic and culinary equipment and automobile trim; the medium strength alloys are used for architectural, marine and commercial road vehicles; and the higher strength alloys are used for marine, welded structural applications and aircraft tubing.

Higher strengths are obtained by quenching suitable alloys from elevated temperatures and then ageing by reheating at lower temperatures.

There are several groups.

(1) Alloys containing amounts of magnesium and silicon in the ratio of the compound Mg_2Si with manganese or chromium. These have UTS up to 330 MPa without any real deterioration in corrosion resistance. They are used for applications similar to those of the non-heat-treated alloys, but which require slightly higher strength.

(2) Alloys containing copper as well as Mg_2Si with one or more of manganese, iron, nickel, chromium, titanium, zirconium or niobium may age harden at room temperature after quenching or may require precipitation heat treatment to give UTS up to 450 MPa. The corrosion resistance of these alloys is impaired so that they must be protected from weathering. These alloys may be used for aircraft structures, other miscellaneous structural applications and, in some cases, forged aircraft engine parts, including forged pistons.

(3) Alloys containing zinc and magnesium, and sometimes chromium, some of which give UTS around 450 MPa, while others give UTS up to 600 MPa. Alloys heat treated to the higher levels may have both their fracture toughness and their resistance to stress corrosion impaired. These alloys are mainly used for aircraft structural applications, but other structural applications are becoming important.

(4) The addition of lithium in amounts of around 2.5% to copper–magnesium–zirconium alloys gives an increase in modulus and a decrease in specific gravity of 10% in alloys with 0.2% proof stress and UTS of 500 and 420 MPa, elongations around 7%, and plane strain fracture toughness values around 35, which are to be compared with the properties of 2324 or 2014 for which the

Table 1.28 UK and US aluminium casting alloys designations

Alloy type	Composition (%)	UK BS	USA AA	UTS* (MPa)	Notes
Pure Al	99.5Al	LM0		80	
Al, Si	Al, 5Si	LM18	443	120–150	
	Al, 12Si	LM6	—	170–200	Linear expansion coefficient 20×10^{-6}
Al, Si, Mg	Al, 15Si, 1Cu, Mg	LM16	355	210–310†	
	Al, 7Si, Mg	LM25	356	140–310†	
	Al, 11Si, Mg, Cu	LM13	A332	150–290†	
	Al, 12Si, Mg	LM9	A360	180–310†	
	Al, 19Si, Cu, Mg, Ni	LM28		130–200†	
	Al, 23Si, Cu, Mg, Ni	LM29		130–210†	Linear expansion coefficient 16×10^{-6}. Low ductility
Al, Si, Cu	Al, 10Cu, 2Si, Mg	LM12	222	180	
	Al, 5Si, 1Cu	L78		200–310†	
	Al, 5Si, 3Cu	LM4	319	150–310†	
	Al, 5Si, 3Cu, Mn	LM22	—	260	
	Al, 6Si, 4Cu, Zn	LM21	319	180–200	
	Al, 7Si, 2Cu	LM27	319	150–180	
	Al, 8Si, 3Cu, Fe	LM24	A380	200	
	Al, 9Si, 3Cu, Mg	LM26	F 331	230	
	Al, 10Si, 2Cu, Fe	LM2	—	180	
	Al, 12Si, Cu, Fe	LM20	413	220	
	Al, 17Si, 4Cu, Mg	LM30	390	160	
Al, Mg	Al, 5Mg	LM5	314	170–230	
	Al, 10Mg	LM10	520	310–360	
Al, Cu, X	Al, 2Cu, 1.3Si, 1.3Ni, 1Mg, 1Fe	L51	—	215†	Maximum impact 250
	Al, 4Cu	L91/92		280–355†	Maximum impact 250
	Al, 4Cu, 2Ni, 2Mg	L35		220–310†	Maximum impact 270
	Al, 4.5Cu, 0.7Ag, 0.3Mg, 0.10Si, 0.14Fe, 0.4Mn, 0.4Cr	Not in specification		450†	High technology sand casting
Al, Zn, Mg	Al, 5Zn, Mg	DTD5008		215	
	Al, 7.7Mg., 1.2Zn	DTD5018		275–305	

* Weakest sand casting—strongest die casting.
† Solution heat treated and precipitation treated.

lithium alloys might be substituted. Fatigue crack growth rates also appear comparable or better than the standard alloys as does the range of general corrosion. These alloys are, however, prone to stress corrosion. Samples of these alloys have been used in aviation applications for 2 years and qualifying tests for major usage are proceeding.

1.4.2.3 Aluminium casting alloys (see Table 1.28)

The alloying additions for casting alloys differ significantly from those for wrought alloys because of the role of silicon in improving castability and reducing thermal expansion. The coefficient of thermal expansion is reduced by 33% for a silicon content of 23% and proportionately for other contents—an effect which can be important in components such as pistons. All aluminium casting alloys contain approximately 0.2% titanium as a grain refiner. In addition, aluminium–silicon alloys must be 'modified', i.e. the acicular silicon phase in the aluminium–silicon eutectic must be converted to a fine spheroidal particle by the addition of sodium, either as a metal or as a fluoride compound, if the alloy is to have a reasonable ductility. The eutectic composition of modified silicon–aluminium is approximately 12.6%. If, to reduce the coefficient of thermal expansion, the silicon content is raised above this value, the elongation will become very low because of the presence of hypereutectoid silicon.

Five major casting alloy groups can be identified:

(1) In castings, additions of manganese or up to 12% silicon increase the UTS up to 170 MPa without heat treatment. The straight aluminium–silicon alloys are used where castability is essential but only moderate strength is required.

(2) Moderate increases in tensile strength without heat treatment are obtained by additions of copper, sometimes accompanied by iron, magnesium or zinc, to the basic silicon alloys. These alloys are used where good casting characteristics, weldability and pressure tightness are required with moderate strength. Applications include manifolds, valve bodies, ornamental grills and general-purpose castings.

(3) Heat-treated aluminium–silicon alloys, either aluminium–silicon–magnesium or aluminium–silicon–copper may have UTS ranging up to 310 MPa, but ductility is low with elongation values ranging around 1% or less, particularly with the higher silicon content materials. These alloys are used for automobile and diesel pistons and pulleys.

COSWORTH PROCESS

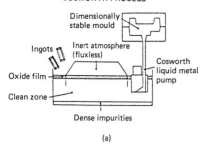

(a)

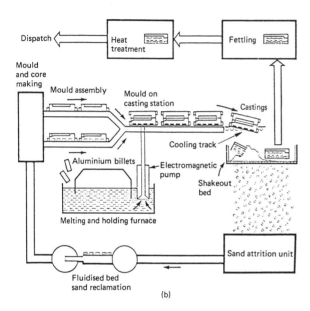

(b)

Figure 1.40 The Cosworth Foundry: (a) the casting unit; (b) flow diagram of the Cosworth Process. (Courtesy of *Metals and Materials*)

(4) UTS ranging up to 300 MPa with elongation between 4 and 10% may be obtained by heat treating aluminium–magnesium, aluminium–copper, aluminium–zinc–magnesium or more complex alloys of low silicon content. These alloys are used for more highly stressed components such as aircraft fittings, internal combustion engine pistons and cylinder heads where higher strength and hot strength are important but castability is not so critical.

(5) There is one, silver containing, casting alloy (Al, 4.5%Cu, 0.7%Ag, 0.34%Mg, 0.10%Si, 0.14%Fe, 0.4%Mn, 0.4%Cr) not in any standard which, when manufactured by a high-technology casting process, may be heat treated to UTS 450 MPa, elongation 10%; this is equivalent to many forging alloys. It is expensive and unlikely to have a high resistance to corrosion, but it offers great advantages for complex highly stressed components such as may be used in fighter aircraft.

1.4.2.4 Fabrication of aluminium and its alloys

Aluminium has a face centred cubic structure and its consequent ductility combined with its relatively low melting point make it easy to cast and work. It is available in the form of sand, gravity and pressure die castings, a very wide range of shapes and sizes of extruded sections, forgings, plate, sheet and strip, many sheet metal products, rod, wire and tube. Most of its alloys can be welded by inert gas shielded metal arc or electric resistance processes. Powder processes have been developed, principally for producing composites usually of an alumina or other ceramic in an aluminium matrix.

High quality aluminium components depend on the production of a fine-grained defect-free casting. Aluminium has no allotropic modification and cannot be refined by heat treatment. While aluminium alloys can be recrystallised after working, their fatigue characteristics depend on the production of an inherently fine-grained casting with a very fine dispersion of intermetallic particles. Furthermore, because of the high solidification contraction, aluminium alloys depend for their soundness on very efficient feeding during solidification.

These problems are compounded by the tendency of molten aluminium to absorb hydrogen which is evolved to form cavities in the metal on solidification and by the tenacious oxide film which forms on the surface of the molten, as of the solid, metal and may be carried under the surface of the casting by turbulence to form discontinuities and interrupt feeding.

Hydrogen must be minimised by melting practice and eliminated by fluxing shortly before pouring.

A fine cast structure and satisfactory feeding of ingots for mechanical working can be achieved by continuous casting and this process is amenable to smooth liquid metal transfers which minimise or eliminate oxide inclusions.

The production of high quality conventional sand and gravity die castings has, in the past, depended to a very large extent on the experience and skill of the foundry. Recent developments in the technique, both of sand and die casting, have gone a long way to ensure rapid and efficiently fed solidification of a melt free from oxide inclusions. The most modern techniques of sand and die casting consist of causing molten alloy to rise from the centre of the melt into a superimposed die or mould.

The liquid metal (which has no oxide skin) rises with an unbroken surface during all stages of filling with metal, transferred in non-turbulent conditions, from a large quiescent enclosed bath of molten alloy. The pressure head developed to raise liquid smoothly into the mould is maintained to feed the casting during solidification. The more even flow of metal itself increases the chilling effect of the mould on the solidifying metal, but this effect is augmented by using mould materials with greater heat capacity.

Cosworth foundry, whose casting unit is represented diagrammatically in *Figure 1.40* uses zircon sand which, besides producing a greater chill, has a lower thermal expansion than silica sand and significantly improves accuracy.

The newly developed casting processes improve tensile and fatigue strength by about 30% and double the elongation compared with conventional sand casting.

1.4.2.5 Identification and standards of aluminium alloys

Wrought products The British Standards Institution (BSI) the US Government and the standards organisations oof many countries use the four-digit system of the Aluminium Association of the USA to identify wrought aluminium alloys, as listed in *Table 1.27*.

Table 1.29 Guide to exposure of aluminium alloys to rural, marine and industrial environments*

Castings (BS 1490)	Wrought (BS 1470—1475)	Notes†
Group 1		
LM5-M	1060, 1050, 1200	Weathers to pleasant grey colour, deepened
LM10-TB		to black in industrial atmospheres.
LM6-M, LM9-TE, LM9-TF	3103, 3105	Superficial pitting occurs initially and
LM18-M, LM20-M	5005, 5251, 5154, 5454	gradually ceases. Seldom needs painting,
	5083	except for decoration. May be anodised for
	6060-T4, 6061-T4,	appearance, but some alloys (e.g. LM6) give
	6082-T4, 6463-T4	dark films
	2014C-T4, 2024C-T4	
Group 2		
LM4-M, LM2-M	6060-T6, 6061-T6,	Weathers as above. Is normally painted in
LM16-TF, LM13-M	6065-T6, 6082-T6,	severe industrial environments and for
LM21-M, LM22-TB, LM24-M,	6463-T6	marine service. May be anodised as above
LM27-M, LM31-M, LM20-M		
Group 3		
LM12-TF	2014-T4	Painting needed in marine and industrial
	2024-T4, 2024-T6	atmospheres, but coatings need only
		infrequent renewals. Sprayed aluminium
		coatings, or cladding, give excellent
		protection. Seldom anodised

* Reproduced by permission of the heirs of Dr E. A. West.
† Aluminium retains its initial appearance if it is washed periodically (3- to 12-month intervals), depending on the severity of the pollution in the air. For many applications the weathered surface is satisfactory, but anodising is often undertaken, particularly for architectural items, to preserve a smooth appearance, but periodic washing is desirable, especially when the anodic film is coloured. Anodising should be in accordance with BS 1615: 1972 and BS 3987: 1974.

Table 1.30 Initial guide to corrosion resistance of aluminium alloys at normal temperatures*

Exposure/application	Wrought products (BS 1470–1477, 4300)	Castings (BS 1490)
Inland atmosphere/ building components, roofs	1st choice: 1050, 1090, 1200, 3103, 3105, 5005, 5251	1st choice: LM0, 5, 6, 9, 10, 18, 25
	2nd choice: 5083, 5154, 5454, 6061, 6063, 6082, 7020	2nd choice: LM2, 41, 13, 16, 20, 21, 22, 24, 26, 27, 28, 29, 30
Marine/boats and ships, fittings	1st choice: 5005, 5083, 51540, 5251, 54540, 6061TB, 6063TB, 6082TB	1st choice: LM5, 10
	2nd choice: 5154H, 5454H, 6061TF, 6062TF, 6082TF, 7020TB, 7020TF	2nd choice: LM0, 9, 18, 25
Chemical and food plant	1st choice: 1080, 1050, 3103, 3105, 51040	1st choice: LM0, 5, 10
Structural items	1st choice: 6061TB, 6063TB, 6082TB	1st choice: LM5, 10
	2nd choice: 6061TF, 6063TF, 7020, 6082TF, 2014A (clad)	2nd choice: LM6, 9

* Reproduced by permission of the heirs of Dr E. A. West.

In most cases the British Standard numbers are identical to those of the US ones, but a few British alloys, for which there is not always a US equivalent, retain the old 'L' system of numbering and a few newer alloys are so far covered only by Department of Technical Development specifications.

Castings Casting alloy numbers have not been homologised. Most British Standard alloys have 'LM' numbers, but a few 'L' numbers survive and some alloys are covered by DTD specification. At least three distinct number classifications are used in the US Aluminium Association casting alloy numbers and

Table 1.31   Initial guide to selecting cast aluminium alloys according to static strength*†

UTS (MPa)	Elongation (%)	Alloy reference (BS 1490)	Form of casting	Condition
120–140	1–5	LM4, LM5, LM18, LM25, LM27	Sand	M
		LM28, LM29	Sand	TF
		LM28	Chill	TE
150–180	1–5	LM2, LM4, LM5, LM12, LM18, LM21, LM24, LM25, LM27, LM28, LM30	Chill	M
		LM16	Sand	TB
		LM6	Sand	M
		LM25	Sand	TE
190–210	1–5	LM6, LM9, LM20	Chill	M
		LM13, LM25, LM26, LM29	Chill	TE
		LM28, LM29	Chill	TF
210–245	1–8	LM4, LM9, LM6, LM25	Sand	TF
		LM9	Chill	TE
		LM16, LM22, LM25	Chill	TB
250–310	1–2	LM4, LM9, LM13, LM16, LM25	Chill	TF
	8–12	LM10	Chill	TB
450	10	Special silver alloy‡	High technology sand	TF

* Reproduced by permission of the heirs of Dr E. A. West.
† Figures are those obtained on standard test bars.
‡ Referred to in Section 1.4.2.3.

Table 1.32   Initial guide to selecting wrought aluminium base materials according to static strength*†

UTS (MPa)	Elongation (%)	Alloy reference (BS 1470–1475, 4300)	Condition range
55–95	35–20	1080A, 1050A, 1200, 1350	0
90–155	25–16	3103, 3105, 5005	0
	18–16	6082	0
	10–3	1080A, 1050A, 1200, 1350	H4–H8
145–185	8–1	3103	H2–H4
	15	3103, 5005	H4–H8
	12	6082, 6463	TB/TF
		6061, 6063	TF
160–200	20–18	5251	0
215–285	18–12	5154A, 5454	0
	8–3	5251	H3–H6
	12	7020	TB
	9	6061, 6063	TF
275–350	16–12	5083	0
	8–3	5154A, 5454	H2–H4
	8	6082	TF
	10	7020	TF
	12	2031	TB
375–400	14–8	2014A, 2014 (clad), 2618A	TF
	10–4	5083	H2–H4
400–480	8–5	2014A, 2014 (clad), 2618A, DTDXXXA	TF
480–580	4	7075, DTDXXXB	T6

* Reproduced by permission of the heirs of Dr E. A. West.
† As there are variations depending on form and thickness, details must be obtained from the relevant British Standard.

British Standard Product Specifications are listed in *Table 1.25* and the condition codes are given in *Table 1.26*.

1.4.2.6   Design using aluminium

The attention of the design engineer is called to three matters which are of great importance in the design and operation of aluminium components.

Creep buckling in compressed struts and sheets can occur under loads far smaller than those calculated from the normal Euler formula using the modulus of elasticity.[18] Creep buckling is not unique to aluminium alloys, but the loading conditions in aircraft struts are often such that it is critical.

When a typical aluminium alloy is compared with a steel, the *fatigue crack growth rate* per cycle da/dN for the aluminium alloy is 40 times that for the steel. When compared with a titanium alloy of twice the yield stress, da/dN for the aluminium alloy is 20 times that of the titanium alloy. K_{Ic} for the aluminium alloy is typically 20 MN m$^{-3/2}$ compared with 160 MN m$^{-3/2}$ for the steel and 60 MN m$^{-3/2}$ for the titanium alloy (see Chapter 8, Section 8.3). This is to some extent offset by the lower stresses used in design with aluminium. Even so, fatigue crack growth calculations play a much more significant part in the design process for aluminium alloys than for medium strength steels.

The fatigue crack growth versus stress intensity curve shows less tendency to turn down to the vertical with aluminium alloys than with steel. Therefore aluminium alloys do not have a well defined *endurance limit*. This is not so serious as it might seem because aluminium alloy components are seldom de-

Table 1.33 Initial guide to selecting aluminium alloys for use at elevated temperatures*†

UTS (MPa)	Temperature (°C)	BS references	
		Castings	Wrought forms
100–155	100	LM4, LM5, LM6, LM10 TB, LM12 TF, LM13 TF, LM25 TF, L35, 2L92 TF	1200 H8, 2014 T6, 2618 T6, 3103 H4–H8, 5083 0, 5251 0 H4–H8, 5454 0/H2–H4, 6063 T6, 6082 T6, 7075 T6
	150	LM4, LM5, LM6, LM10 TB, LM12 TF, LM25 TF, L35, 2L92	1200 H8, 2014 T6, 2618 T6, 3103 H4–H8, 5083 0, 5251 0/H4–H8, 5454 0/H2–H4, 6063 T6, 6082 T6, 7075 T6
	200	LM4, LM5, LM6, LM10 TB, LM12 TF, LM13 TF, L35, 2L92	2014 T6, 2618 T6, 5083 0, 5251 0/H4–H8, 5454 0/H2–H4, 7075 T6
	250	LM4, LM5, LM10 TB, LM12 TF, LM13 TF, L35, 2L92	5083 0, 5454 0/H2–H4, 6082 T6, 7075 T6
	300	LM10 TB, LM12 TF, L35, 2L92	—
150–200	100	LM4, LM5, LM6, LM10 TB, LM12 TF, LM13 TF, L25 TF, L35, 2L92	1200 H8, 2014 T6, 2618 T6, 3103 H8, 5083 0, 5251 0/H4–H8, 5454 0/H2–H8, 6063 T6, 6082 T6, 7075 T6
	150	LM5, LM10 TB, LM12 TF, LM13 TF, LM25 TF, L35, 2L92	2014 T6, 2618 T6, 3103 H8, 5083 0, 5251 0/H4–H8, 5454 0/H2–H4, 6082 T6, 7075 T6
	200	LM10 TB, LM12 TF, LM13 TF, L35, 2L92	2618 T6, 5083 0, 5251 H4–H8, 5454 0/H2–H4
200–250	100	LM10 TB, LM12 TF, LM13 TF, LM25 TF, L35, 2L92	2014 AFT, 2618 T6, 5083 0, 5251 H4–H8, 5454 0/H2–H4, 6063 T6, 6082 T6, 7075 T6
	150	LM10 TB, LM12 TF, LM13 TF, L35, 2L92	2014 T6, 2618 T6, 5083 0, 5251 H4–H8, 5454 H2–H8, 7075 T6
	200	LM10 TB, LM12 TF, LM13 TF, L35, 2L92	2618 T6
	250	LM12 TF, L35	—

* Reproduced by permission of the heirs of Dr E. A. West.
† Based, in general, on static tests undertaken at temperature after soaking at the same temperature.

signed to withstand cyclic loading in the region of 10^8 cycles. It should be noted that the endurance limits normally quoted for sand castings refer to those produced by conventional methods and not the high technology methods described in Section 1.4.2.4.

The choice of alloy to meet specific requirements will be assisted by reference to *Table 1.29* for resistance to environmental corrosion, *Table 1.30* for resistance to aqueous corrosion, *Table 1.31* for resistance of castings to stress at room temperature, *Table 1.32* for resistance of wrought products and *Table 1.33* for resistance to stress at elevated temperatures.

1.4.3 Titanium and its alloys

1.4.3.1 General

The importance of titanium as an engineering material depends on alloys which have specific strengths which (except for beryllium which is costly and toxic) are greater than those of any other metal alloys—excellent creep resistance up to 600°C and exceptional corrosion resistance to oxidising media.

The high specific strengths derive from alloys with tensile and yield strengths up to 1380 and 1230 MPa, respectively, and a specific gravity of 4.51 (for the pure metal) rising to a maximum of 4.85 when alloyed, i.e. roughly half that of steel.

The corrosion and oxidation resistance is due to a tenacious oxide film on a metal above silver and nickel and below only Hastelloy and Monel in the electrochemical series for structural metals in seawater. One research programme on coupled specimens in seawater has indicated titanium to be the most noble of structural metals. Titanium alloys are very resistant to cavitation erosion.

These advantages are reinforced by a high fatigue resistance (endurance limits are roughly half the UTS) and a thermal expansion (8×10^{-6} °C^{-1} between 0 and 500°C), roughly half that of austenitic stainless steels and aluminium and two-thirds that of ferritic steels. Most titanium alloys may be welded using gas shielded or diffusion methods, and some are capable of superplastic forming.

The main disadvantage of titanium is its cost. Ore separation and reduction to the metal is expensive and the metal reacts readily when hot with oxygen and nitrogen, so that melting and casting must be done in vacuo and welding must be done under a protective atmosphere.

Further features are a modulus of 105–125 GN m^{-2} compared with 180 GN m^{-2} for steel and 65 GN m^{-2} for aluminium alloys, and low electrical and thermal conductivities only about 4% of that of copper. (The higher modulus compared with aluminium can have advantages for use in springs, ultrasonic devices, surgical implants, etc.) In addition, in spite of its generally excellent performance, the corrosion reactions of titanium with very strongly oxidising media can be catastrophic and it is attacked by uninhibited reducing media.

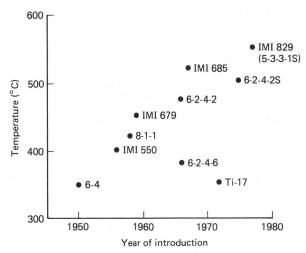

Figure 1.41 High-temperature alloys of titanium used at present in production aeroengines: years of development and temperature capability under optimum stress conditions. (Reproduced by permission of IMI Titanium Ltd)

There are essentially five major groups of engineering applications of titanium: process plant, marine condenser tubing, electrochemical fitments, aerospace frames and aeroengines. The most exciting applications are in the compressors (and other comoponents) of aircraft turbine engines. The development of these engines has depended on the parallel development of alloys with progressively improved creep resistance without sacrificing fatigue resistance. The development of titanium alloys has kept pace with these requirements, as illustrated diagrammatically in *Figure 1.41*.

1.4.3.2   Titanium metal and alloy structure

The structure of titanium metal is transformed from the low temperature close packed hexagonal α phase to the higher temperature body centred β phase on heating above 850°C. Alloying elements influence the properties of one or both phases, and their relative stability.

Titanium with α stabilising and strengthening elements Zirconium and tin are two additions that strengthen the α phase considerably, particularly at high temperature. Unfortunately they also raise the density.

Oxygen and nitrogen which may be introduced during reduction and processing stabilise the α phase and increase strength at the expense of toughness. Their content must

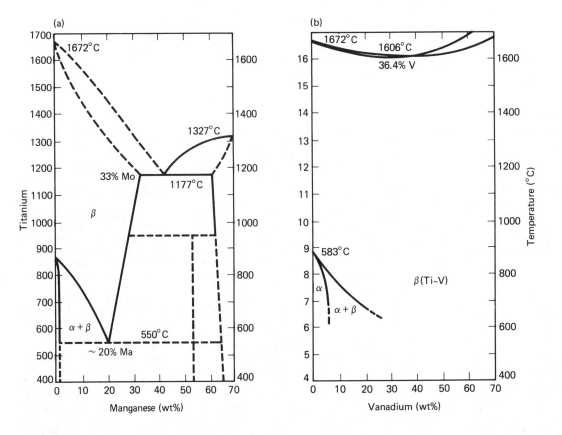

Figure 1.42 Part phase diagrams of β stabilising additions to titanium. (a) Complex structure of manganese alloys. (b) Favourable simpler structure of vanadium alloys

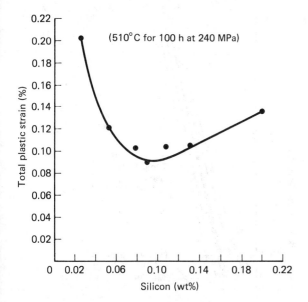

(510°C for 100 h at 240 MPa)

Figure 1.43 Effect of silicon on the creep strain in 6-2-4-2. (Reproduced by permission of IMI Titanium Ltd)

tensile properties up to 300°C and is weldable but difficult to cold form.

Unfortunately attempts to increase strength still further by increasing the alloy content lead to embrittlement due to ordering and formation of Ti_3Al according to:

$$Al + \frac{Sn}{3} + \frac{Zr}{6} + 10\,O_2 < 9\%$$

Titanium with β stabilising and strengthening elements Manganese, iron, copper, chromium, vanadium and molybdenum additions tend to stabilise the β phase. The most favourable elements are those that do not have a peritectic reaction (see *Figure 1.42*).

Very high strengths (up to 1500 MPa) can be achieved in metastable β alloys but, although alloys such as Ti, 8%Mn (which was processed in the α + β phase region) and Ti, 13%V, 11%Cr, 3%Al (which was strengthened by precipitation of the α phase) have been developed the alloys can embrittle on heating above about 300°C and none have found high temperature applications in aeroengines.

Titanium with both α and β stabilising and strengthening elements Titanium alloys with added elements that stabilise both the α and β phases simultaneously provide significantly increased room and high temperature strengths.

Ti, 6%Al, 4%V is rated up to 350°C and has found widespread application. The addition of silicon significantly increases creep strength and Ti, 4%Al, 2%Sn, 4%Mo, 0.5%Si (IMI 550) is not only 10% stronger than Ti, 6%Al, 4%V but, in addition, can be operated up to 400°C. (It was later found that the optimum benefit of silicon additions occurs at a lower level of addition to near α titanium alloys (see *Figure 1.43*).)

therefore be limited. The addition of small quantities of palladium (IMI grades 260 and 262) markedly improves resistance to reducing environments; the addition of molybdenum and nickel (ASTM grade 12) is a cheaper but less effective means of obtaining a similar result.

Aluminium stabilises the α phase and also reduces density. The Ti, 5%Al, 2.5%Sn alloy (ASTM grade 6) has useful

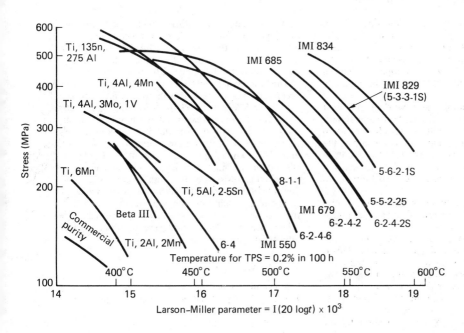

Figure 1.44 Creep resistant alloy development. Larson–Miller plot showing improvements achieved over the last 30 years. (Reproduced by permission of IMI Titanium Ltd)

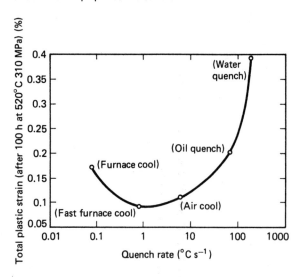

Figure 1.45 Effect of cooling rate on the creep strain in IMI 685. (Reproduced by permission of IMI Titanium Ltd)

Near α alloys have also been developed to have high strength for moderate temperature applications (up to about 400°C). These include Ti, 2.25%Al, 11%Sn, 4%Mo, 0.2%Si (IMI 680) and Ti, 6%Al, 5%Zr, 4%Mo, 1%Cu, 0.2%Si (IMI 700) and in the US Ti, 6%Al, 2%Sn, 4%Zr, 6%Mo and Ti, 5%Al, 2%Sn, 2%Zr, 4%Mo, 4%Cr.

Near α titanium alloys The near α titanium alloys have the maximum amount of α stabilisers with sufficient amounts of β stabilising elements to give medium strength levels but low enough amounts to avoid weldability or high temperature creep resistance.

Ti, 11%Sn, 2.25%Al, 5%Zr, 1%Mo, 0.2%Si (IMI 679) can be operated up to 450°C and the US alloy Ti, 8%Al, 1%V, 1%Mo up to 400°C although it suffers from ordering embrittlement and needs careful control in processing. Ti, 6%Al, 2%Sn, 4%Zr, 2%Mo ('6-2-4-2') was developed to have a temperature capability of about 470°C and has been widely used in the USA.

All these alloys had previously been worked and heat treated in the $\alpha + \beta$ phase region. β heat treatment significantly improved creep strength but reduced tensile ductility to an unacceptably low level. This problem was overcome by the development of near α Ti, 6%Al, 5%Zr, 0.5%Mo, 0.25%Si (IMI 685), which is β heat treated to produce an acicular microstructure and in this temper has creep capability up to 520°C (see *Figure 1.44*).

β heat treatment by itself may not be sufficient to maximise creep resistance. Cooling rate is also critical and an example of this is shown in *Figure 1.45*.

Fracture toughness and fatigue crack growth rate are improved by changing from $\alpha + \beta$ to β heat treatment. This is illustrated in *Figure 1.46*. Unfortunately, room temperature tensile ductility and fatigue crack initiation are worsened. The effective initiation unit is the prior β grain or the α colony which, in a β treated structure, can be from 0.5 to 2 mm in size compared with 10–15 μm in an $\alpha + \beta$ structure.

Reduction in the size of structural features while retaining an acicular structure can be achieved by using thermomecha-

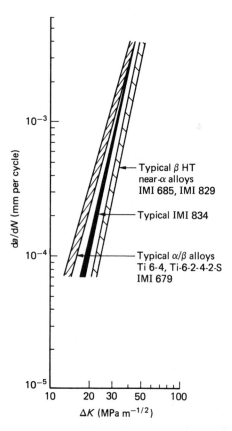

Figure 1.46 Crack propagation in high-temperature alloys. (Reproduced by permission of IMI Titanium Ltd)

nical processing with multirecrystallisation stages, and by alloy development. This procedure has resulted in β grain sizes of 0.5–0.75 mm in the alloy Ti, 5.6%Al, 3.5%Sn, 3%Zr, 1%Nb, 0.3%Mo, 0.3%Si (IMI 829). Improvements have also been carried out in the USA on Ti, 6%Al, 2%Sn, 4%Zr, 2%Mo.

The best creep properties (see *Figure 1.44*) together with an acceptable fatigue crack growth rate are given by Ti, 5.8%Al, 4.4%Sn, 3.5%Zr, 0.7%Nb, 0.5%Mo, 0.35%Si, 0.06%C (IMI 834). This alloy is characterised by a more gradual change in α phase content with heat-treatment temperature, as shown in the comparison of the β transus approach curves for this alloy and 829 shown in *Figure 1.47*. This allows the alloy to be heat treated in the high α field for optimum creep and crack propagation resistance, but with β grain size controlled to about 0.19 mm for improved fatigue performance.

The relationship between heat-treatment temperature and the balance of creep and fatigue properties is illustrated schematically in *Figure 1.48*. The optimum balance can be adjusted for a particular application.

1.4.3.3 Nomeclature and standards

The chemical composition, nomenclature, (where available) UK and US specification numbers, tensile properties and weldability are listed in *Table 1.34*.

In Europe, titanium alloys are often referred to by their IMI number. Some of the more recent titanium alloys had not at

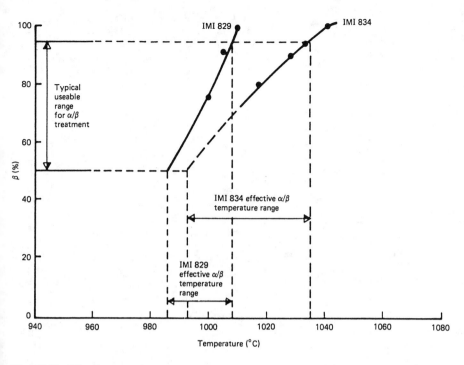

Figure 1.47 β Transus approach curves

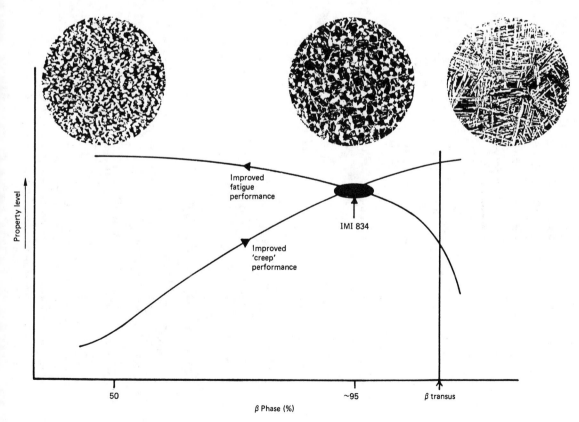

Figure 1.48   Effect of α/β phase proportions on creep and fatigue properties

Table 1.34 Alloys of titanium

Nominal composition (%)	IMI No.*	BS	ASTM	Typical mechanical properties		Weldability
				0.2% PS† (MPa)	UTS (MPa)	
α-phase alloys						
Ti, 0.05O₂/N₂	110	—	Grade 1 (B265, 337, 338, 348, 367, 381)‡	190	310	
Ti, 0.07O₂/N₂	115	2TA.1		250	375	
Ti, 0.13O₂/N₂	125	2TA.2, 3, 4, 5	Grade 2 (B265, 337, 338, 348, 367, 381)	330	465	
Ti, 0.02O₂/N₂	130		Grade 3 (B265, 337, 338, 348, 367, 381)	410	540	Excellent
Ti, 0.28O₂/N₂	155	2TA6	Grade 4 (B265)	540	650	
Ti, 0.30O₂/N₂	160	2TA7, 8, 9	Grade 4 (B348, 367, 381)	500	670	
Ti, 0.15Pd, 0.07O₂/N₂	260		Grade 11 (B265, 337, 338, 348, 381)	220	330	
Ti, 0.15Pd, 0.13O₂/N₂	262		Grade 7 (B265, 337, 338, 348, 381)	350	465	
Ti, 2.5Cu	230	2TA 21–24 TA 52–55, 58		530(a)§ 667(STA)§	620(a) 770(STA)	
Ti, 0.3Mo, 0.8Ni	(Grade 12)	—	Grade 12 (265, 337, 338, 348, 367, 381)	400 400	500 500	Very good/ excellent
Ti, 5Al, 2.5Sn	317	TA14–17		850 850	950 950	
α/β-phase alloys						
Ti, 6Al, 4V	318	2TA11–13, 28	Grade 5 (B265, 337, 338, 348, 367, 381)	930(a) 1070(STA) 1040(STA)	1030(a) 1130(STA) 1170(STA)	Good
Ti, 4Al, 4Mn, 2Sn, 0.5Si	550	TA 56, 59		1200(STA)	1310(STA)	Poor, not normally welded except by diffusion
Ti, 4Al, 4Sn, 0.5Si		TA45–51, 57		960(a)	1020(a)	
Ti, 6Al, 6V, 2Sn	551	TA38–42		1080(STA)	1150(STA)	
Ti, 6Al, 2Sn, 4Zr, 6Mo	(662) (6246)			1145(STA)	1260(STA)	Poor, not normally welded
Ti, 5Al, 2Sn, 2Zr, 4Mo, 4Cr	(T.17)			1090(STA)	1170(STA)	
Ti, 4.5Al, 5Mo, 1.5Cr, 0.13 O	(Corona 5)			890(a) 1172(STA)	956(a) 1255(STA)	
Near-α-phase alloys						
Ti, 2.25Al, 11Sn, 5.2Zr, 1Mo, 0.25Si	679	TA 18–20, 25, 26, 27	—	1025	1230	Poor
Ti, 6Al, 2Sn, 2Zn, 2Mo, 0.1Si	(62425)		—	950	1050	Poor
Ti, 8Al, 1Mo, 1V	(811)		—	975	1095	Fair
Ti, 6Al, 5Zr, 0.5Mo, 0.25Si	685	TA 43, 44	—	900	1020	Very good/excellent
Ti, 5.6Al, 3.5Sn, 3Zn, 1Nb, 0.25Mo, 0.3Si	829		—	850	1000	Very good/excellent
Ti, 5.8Al, 4.4Sn, 3.5Zn, 0.7Nb, 0.5Mo, 0.35Si, 0.06C	834		—	950	1090	Very good/excellent
β-phase alloys						
Ti, 15V, 3Cr, 3Sn, 3Al	—	—	—	1450	1500	Fair
Ti, 10V, 2Fe, 3Al, <0.16O	—	—	—	1115	1170	—

* Designations in brackets are not IMI numbers.
† Proof stress.
‡ In US Specifications: B265 Plate sheet and strip; B337 Seamless and welded pipe; B338 Seamless and welded tube; B348 Bar and billet; B367 Castings; B381 Forgings.
§ (a) annealed. (STA) Solution treated and aged.

Table 1.35 Properties of IMI 829 and 834

Alloy No.	Temperature (°C)	0.2% PS* min. (MPa)	UTS min. (MPa)	Elongation (min.)	Stress for 0.1% total plastic strain in 100 h (MPa)	Modulus of elasticity (GPa)	Endurance limit of UTS (%)	Fracture toughness, K_{Ic} (MN m$^{-3/2}$)	Specific gravity	Coefficient of thermal expansion at 20–1000°C (K^{-1})
829	20	820	950	10	200	~120	50	78	4.54	10.4×10^{-6}
	540			12		~ 93				
834	20	910	1030	6	150	~120	50	37	4.59	11.3×10^{-6}
	600	450	585	9		~ 93				

* Proof stress.

the time of writing appeared in national material specifications. The properties of the more attractive of these (IMI 829 and IMI 834) are, therefore, listed in *Table 1.35*.

Such is the pace of titanium alloy development that the designer is recommended to approach the manufacturer for information on materials available at the time before designs are finalised.

1.4.3.4 Fabrication

Titanium, and where appropriate alloying metals, are compacted and welded into electrodes for vacuum arc consumable electrode furnaces and double or triple melted to ensure homogeneity. The resulting ingot is forged and then rolled. Bar, rod, wire, sheet, plate, seam welded tube and bored or extruded and drawn seamless tube are available commercially.

Almost all titanium alloys can be forged. The technique which can produce near-to-size components is isothermal press forging in which the metal and dies are heated to the same temperature (900–950°C). Some fine grained duplex alloys including IMI 318, 550 and 6242 have high strain-rate sensitivity at temperatures between 900 and 950°C and will deform superplastically at low strain rates to strains of over 1000%.

Titanium will dissolve its own oxide and can therefore be diffusion bonded; this process can be combined with superplastic forming to make a variety of hollow components economically.

Most titanium alloys are weldable by electron beam, inert gas shielded arc, resistance, flash butt, pressure or friction welding and, even in those alloys which are considered to deteriorate when welded, the properties can usually be augmented by heat treatment.

The titanium alloys IMI 110 to 160, 260, 262, 318, 685 and 829 can be centrifugally cast in rammed graphite or investment moulds using a consumable electrode skull melting furnace.

Tensile and yield strengths are about 5% below those of forgings and dimensional tolerances vary from 0.6 to 2%, depending on process and size.

It is becoming increasingly common for castings to be hot isostatically pressed to give properties comparable to those of forgings.

1.4.3.5 Applications

So-called 'commercially pure' titanium (IMI 110 to IMI 160) is primarily a chemical-plant vessel and piping material resisting seawater, halogen compounds, oxidising and organic acids and many gaseous environments. It is also used for non-consumable anodes, sometimes coated with precious metal, and fitments for metal finishing and other electrochemical operations. IMI 260 or IMI 262 may be substituted if there is any possibility that the environment may become reducing in nature.

A coarse-grain faceted structure can be produced which can be selectively anodised and used for decorative and jewellery products.

The UTS of these alloys ranges up to 650 MPa, so that they can be used for less highly stressed aircraft components.

Aircraft engine casings and bypass ducts require stronger and more creep-resistant material, and for these applications the higher strength more creep resistant readily formed and welded α strengthened IMI 230 or, if there is no severe forming operating involved, ASTM grade 5 (IMI 318) may be used.

The α + β alloys (for example IMI 550) have much higher tensile and fatigue strengths (which are exceeded only by the less stable β strengthened alloys) and are used for discs, blades and highly stressed air-frame components.

The creep properties of the near α alloys have already been described. They are used for the higher temperature compressor discs, rings, blades and impellers at temperatures approaching 600°C. Above this temperature the limitation for titanium appears to be in the field of oxidation rather than strength. Oxidation resistant noble-metal-ion plated coatings and titanium–aluminium and titanium–niobium–aluminium alloys or compounds are showing promise.

1.4.4 Magnesium and its alloys

1.4.4.1 Introduction

Magnesium is the least dense of the engineering alloys and, because of this, the specific strengths of its alloys are superior to those of aluminium and to medium strength steels. More importantly, although the specific rigidities $E\rho^{-1}$ of the alloys differ little from those of aluminium and steel, the stiffness in bending of a section of equal weight (which is related to $E\rho^{-3}$) is far superior. Substantial savings in weight are possible for designs for which rigidity is a significant criterion.

A further advantage of magnesuum is its high damping capacity which helps to minimise vibration fatigue.

Further scope for weight reduction in comparison with aluminium arises from the excellent castability of some alloys for components where section thickness is determined by casting considerations. Die casting benefits from the fact that

Table 1.36 Casting alloys of magnesium*

Typical chemical composition—major alloying elements (%)	Elektron alloy	Tensile properties†			Compressive properties		Fatigue endurance values§		Hardness
		0.2% proof stress (MPa)	Tensile strength (MPa)	Elongation‡ (%)	0.2% proof stress (MPa)	Ultimate strength (MPa)	Unnotched (MPa)	Notched (MPa)	(Brinell)
Y5.25, Nd and other heavy rare earth metals 3.5, Zr0.5	WE54 Solution and precipitation treated:								
	Sand cast	185	255	2	—	—	95–100	—	80–90
	Chill cast	185	255	2	—	—	—	—	—
Rare earth metals 3.0, Zn2.5, Zr0.6	ZRE1 Precipitation treated:								
	Sand cast	95	140	3	85–120	275–340	65–75	50–55	50–60
	Chill cast	100	155	3					
Zn4.2, rare earth metals 1.3, Zr0.7	RZ5 Precipitation treated:								
	Sand cast	135	200	3	130–150	330–365	90–105	75–90	55–70
	Chill cast	135	215	4					
Zn5.8, rare earth metals 2.5, Zr0.7	ZE63 Solution and precipitation treated:								
	Sand cast	170	275	5	190–200	430–465	115–125	70–75	60–85
Th3.0, Zn2.2, Zr0.7	ZT1 Precipitation treated:								
	Sand cast	(85)	185	5	85–100	310–325	65–75	55–70	50–60
	Chill cast	(85)	185	5					
Zn5.5, Th1.8, Zr0.7	TZ6 Precipitation treated:								
	Sand cast	155	255	5	150–180	325–370	75–80	70–80	65–75
	Chill cast	155	255	5					
Ag1.5, Nd rich rare earth metals 2.0, Zr0.6, Cu0.07	EQ21A Solution and precipitation treated:								
	Sand cast	175	240	2	165–200	310–385	100–110	60–70	70–90
	Chill cast	175	240	2					
Ag2.5, Nd rich rare earth metals 2.5, Zr0.6	MSR-B Solution and precipitation treated:								
	Sand cast	185	240	2	165–200	310–385	100–110	60–70	70–90
	Chill cast	185	240	2					
Ag2.5, Nd rich rare earth metals 2.0, Zr0.6	QE22(MSR) Solution and precipitation treated:								
	Sand cast	175	240	2	165–200	310–385	100–110	60–70	70–90
	Chill cast	175	240	2					
Al8.0, Zn0.5, Mn0.3	A8 As cast:								
	Sand cast	(85)	140	2	75–90	280–340	75–85	58–65	50–60
	Chill cast	(85)	185	4					
	Solution treated:								
	Sand cast	80	200	7	75–90	325–415	75–90	60–70	50–60
	Chill cast	80	230	10					
Al9.0, Zn0.5, Mn0.3, Be0.0015	AZ91 plus Be Die cast	(150)	(200)	(1)					
Al9.5, Zn0.5, Mn0.3	AZ91 As cast:								
	Sand cast	(95)	125	—	85–110	280–340	77–85	58–65	55–65
	Chill cast	(100)	170	2					
	Solution treated:								
	Sand cast	80	200	4	75–110	185–432	77–92	65–77	55–65
	Chill cast	80	215	5					
	Solution and precipitation treated:								
	Sand cast	120	200	—	110–140	385–465	70–77	58–62	75–85
	Chill cast	120	215	2					
Al7.5–9.5, Zn0.3–1.5, Mn0.15(min.)	C As cast:								
	Sand cast	(85)	125	—	65–90	278–340	73–80	58–65	50–60
	Chill cast	(85)	170	2					
	Solution treated:								
	Sand cast	(80)	185	4	75–90	330–415	77–85	62–73	50–60
	Chill cast	(80)	215	5					
	Solution and precipitation treated:								
	Sand cast	(110)	185	—	90–115	340–432	62–73	58–62	70–80
	Chill cast	(110)	215	2					
Zn5.5–6.5, Cu2.4–3.0, Mn0.25–0.75	ZCM630-T6 Sand cast	125	210	2			94	57	55–65

* Reproduced by courtesy of Magnesium Elektron Ltd.

† The tensile properties quoted are the specification minima for the first specification listed for that alloy and condition. The ranges given are the specified minima; bracketed values are for information only. The values quoted are for separately cast test bars and may not be realised in certain portions of castings.

MoD, Ministry of Defence.

| Description | Specifications | | | ASTM | |
	MoD Procurement Executive (D.T.D. Series)	British Standards Aircraft	General Engineering	Alloy designation and temper	Specification
Excellent retention of strength after long exposure at 250°C. Good castability, weldable. Good corrosion resistance	— — —	— — —	— — —	WE54A-T6	—
Creep-resistant up to 250°C. Excellent castability. Pressure tight and weldable	— —	2 L.126 2 L.126	2970 MAG6 TE 2970 MAG6-TE	EZ33A-T5	B80-76
Easily cast, weldable, pressure tight, with useful strength at elevated temperatures	— —	2 L.128 2 L.128	2970 MAG5-TE 2970 MAG5-TE	ZE41A-T5	B80-76
Excellent castability, pressure tight and weldable with high developed properties in thin wall castings	5045	—	—	ZE63A-T6	
Creep-resistant up to 350°C. Pressure tight and weldable	5005A 5005A	— —	2970 MAG8-TE 2970 MAG8-TE	HZ32A-T5	B80-76
Stronger than, but as castable as RZ5, weldable, pressure tight	5015A 5015A	— —	2970 MAG9-TE 2970 MAG9-TE	ZA62A-T5	B80-76
Heat-treated alloys with high yield strength up to 200°C. Pressure tight and weldable	— —	— —	— —	—	—
	5035A 5035A	— —	— —	—	—
	5055 5055	— —	— —	QE22A-T6	B80-76
General purpose alloy. Good founding properties. Good ductility, strength and shock resistance. Also available as a high purity grade	— —	— —	2970 MAG1-M 2970 MAG1-M	AZ281A-F	B80-76
	— —	3 L.122 3 L.122	2970 MAG1-TB 2970 MAG1-TB	AZ81A-T4	
General purpose pressure diecasting alloy. Draft ISO specification	—	—	—	AZ91B-F	B94-76
General purpose alloy. Good founding properties. Suitable for pressure die castings	—	—	2970 MAG3-M 2970 MAG3-M	AZ91C-F	—
	— —	3 L.124 3 L.124	2970 MAG3-TB 2970 MAG3-TB	AZ91C-T4	—
	— —	3 L.125 3 L.125	2970 MAG3-TF 2970 MAG3-TF	AZ91C-T6	
General-purpose alloy with good average properties	— —	— —	2970 MAG7-M 2970 MAG7-M	—	
	— —	— —	2970 MAG7-TB 2970 MAG7-TB	—	
	— —	— —	2970 MAG7-TF 2970 MAG7-TF	—	—
Good founding properties. Good creep resistance	—	—	—	—	—

‡ Elongation values are based on a gauge length of $5.65\sqrt{A}$ except in the case of thin material where a gauge length of 50 mm may be used (see BS 2L.500, 3370 and 3373). With the latter gauge length, elongation requirements for sheet and plate depend on thickness and a range of minima is quoted.
§ Endurance values for 50×10^6 reversals in rotating bending-type tests; semi-circular notch, radius 1.2 mm; stress concentration factor approx. 2.0. Reversed bending for sheet.

dies are not attacked by molten magnesium as they are by molten aluminium.

Magnesium is readily extruded and forged at elevated temperatures. It is highly machinable but suffers from a risk of fire and even explosion unless precautions are taken to prevent accumulation of swarf.

These advantages are (or have been in the past) offset by three major drawbacks.

The major drawback compared with aluminium is the less protective nature of the oxide. The early magnesium alloys had very poor corrosive resistance and, although this disability has to a large extent been overcome in more recently developed alloys, the liability to galvanic corrosion has not been lessened and care must be taken to prevent it.

A second drawback arises from the hexagonal structure of the magnesium crystal which prevents fabrication by cold working.

In the past a third drawback was the higher price of magnesium compared with aluminium which, even allowing for the lower specific gravity, confined its use to applications in which weight reduction is at a premium. These include aerospace, components for high performance automobiles, materials handling, portable tools, high speed machinery and high value portable consumer goods. Recently, however, the relative costs of the two metals have reversed and magnesium components may cost less than aluminium components.

1.4.4.2 Alloy designation and standards

Magnesium alloys are usually referred to by their ASTM designations which, together with their temper designations, are given in a four-part code.

The alloy designation consists, first, of *letters* which identify the major alloying elements, given in order of decreasing concentration. The letters that refer to the alloying elements are:

A	aluminium	N	nickel
B	bismuth	P	lead
C	copper	Q	silver
D	cadmium	R	chromium
E	rare earth	S	silicon
F	iron	T	tin
H	thorium	W	yttrium
K	zirconium	Y	antimony
L	lithium	Z	zinc
M	manganese		

The letters are followed by *numbers* which give, to the nearest percentage, the weight percentage of the alloying elements.

In the ASTM system only the *two most concentrated* alloying elements are designated. Thus MEL alloy ZCM711, which contains 6.5 wt% Zn, 1.2 wt% Cu, 0.7%Mn has the ASTM designation ZC71.

The *third* part of the alloy designation consists of a letter which refers to a standard alloy within the broader composition range specified by the first two parts of the designation.

The *temper* designation, which is separated from the first two or three parts of the code by a hyphen, is the same as that adopted for aluminium alloys. This is given in full in ASTM specification B296-67.

Thus a complete designation might be QH21A-T6. The alloy contains approximately 2 wt% silver and 1 wt% thorium; its exact composition is covered by the specification for the 'A' version of the alloy; and it is in the solution treated and artificially aged condition. The alloy designation is given in full in ASTM specification B275.

ASTM specification numbers comprise: B92 magnesium ingot; B93 magnesium alloy ingot; B94 die castings; B50 sand castings; B199 permanent mould castings; B403 investment castings; B90 sheet and plate; B91 forgings; and B107 extruded bar shapes and tubes.

BS 2970 lists cast and wrought alloys and has a code for conditions similar to ASTM with the addition of 'M' for 'as cast'.

The designations of the major British supplier, Magnesium Elektron, are roughly similar to those of the ASTM.

Tables 1.36 to *1.38* list the cast and wrought alloys, their Magnesium Elektron designations, British and ASTM specifications, and principal mechanical properties and characteristics.

1.4.4.3 Development of magnesium alloys

The first magnesium alloys contained aluminium up to 10% and sometimes also zinc up to 6%. These alloying additions increase strength and give precipitation hardening properties. The addition of manganese reduced iron pick up and so improved corrosion resistance. Further development was aimed at improving creep and corrosion resistance, both of which were very poor in the earlier alloys.

Addition of zirconium refines the grain structure and improves the strength and ductility, both hot and cold. About 0.5% of this alloying addition, together with additions of thorium, rare-earth metals, silver and/or copper, provide a combination of strength, castability and high-temperature tensile and creep strengths comparable with the high-temperature aluminium alloys.

Improvements in corrosion resistance have resulted from an understanding of the effect of heavy-metal impurities. Aluminium and zinc containing casting alloys are now produced with the following impurity limits: Fe <0.005%, Ni <0.001%, Cu <0.015%, with 0.15%<Mn<0.25% to suppress iron pick up. As a result of this the high purity versions (ASTM designations AZ1D and AM60B) have corrosion rates in salt solution that are only about 2% of those of the original alloys. Alloy WE54 is unique among the creep resisting alloys in that its corrosion resistance is excellent and on a par with that of the high-temperature aluminium casting alloys.

1.4.4.4 Casting of magnesium alloys

A major problem in the manufacture of magnesium castings has been the prevention of oxidation of the molten metal which has, in the past, been achieved by melting under a layer of flux. It has always been difficult to prevent entrainment of flux, and the possibility of flux inclusions in castings has worsened the problems arising from corrosion.

New techniques of fluxless melting and stirring have been developed and the introduction of low-pressure casting as used for aluminium alloys should bring further improvements.

1.4.4.5 High temperature strength of magnesium alloys

The variation in ultimate tensile and proof stress of the creep resisting cast alloys of magnesium is shown in *Figure 1.49* and the stress to produce 0.2% total strain at 1000 h is given in *Table 1.39*. These properties are adequate for many aerospace and automobile engine components. A creep resistant forging

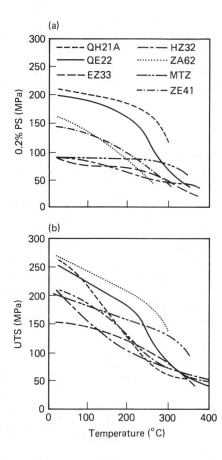

(a)

(b)

Figure 1.49 Effect of temperature on (a) 0.2% proof stress, and (b) ultimate tensile strength of various magnesium cast alloys. (Reproduced by permission of Mr. W. Unsworth, Magnesium Elektron Ltd)

alloy (ZT) is available but its application is restricted compared with that of castings.

1.4.4.6 Corrosion of magnesium alloys

The improved corrosion resistant alloys of magnesium have been referred to in Section 1.4.4.3.

The high purity casting alloys have performed at least as well as high temperature aluminium alloys in salt fog corrosion tests. Painted die cast high purity AX91D has shown negligible scribe corrosion creepages and excellent stone chip resistance.

Unfortunately, the galvanic corrosion of magnesium is not affected and great care must be taken in design not to place magnesium in electrical contact with a metal with a significantly more positive electrode potential. Where juxtaposition is unavoidable for service in a corrosive atmosphere, reliable insulation should be inserted between the two components, or the magnesium component should be protected in some other way. This applies particularly to rivetted joints. Because of the limitation on ductility imposed by the hexagonal crystal structure magnesium cannot be used for rivets. Rivets of aluminium, which has a relatively small positive electrode potential compared with magnesium, are used. Steel or copper rivets must not be used.

1.4.4.7 Applications of magnesium alloys

Improvements in high temperature and corrosion resistance have increased the possible applications of cast magnesium alloys in automobile, aerospace and other applications, where reduction in weight is significant. These applications include helicopter gearboxes, aircraft-engine casings, high performance car and motorcycle components (including wheels), computer parts, military equipment and video and conventional cameras.

The applications of wrought alloys are more restricted because of the problems of cold forming and rivetting, but include railings, ladders, brackets and cans for 'magnox' reactors.

Table 1.37 Physical properties of magnesium casting alloys

Alloy	Specific gravity (20°C)	Coefficient of thermal expansion (20–200°C) (10^{-6} K^{-1})	Thermal conductivity (20°C) (W m^{-1} K^{-1})	Electrical resistivity (20°C) (nΩ-m)	Specific heat (20–100°C) (J kg^{-1} K^{-1})
WE54	1.9	27	107	70	1000
ZRE1	1.80	26.8	100	73	1050
RZ5	1.84	27.1	109	68	960
ZE63	1.87	27.1	113	56	960
ZT1	1.83	26.7	105	72	960
TZ6	1.87	27.1	113	66	960
EQ21A	1.81	26.7	113	68.5	1000
MSR-B	1.82	26.7	113	68.5	1000
QE22	1.82	26.7	113	68.5	1000
A8	1.81	27.2	84	134	1000
AZ91	1.83	27.0	84	141	1000
C	1.81	27.2	84	134	1000

Table 1.38 Wrought magnesium alloys*

Typical chemical composition—major alloying elements (%)	Elektron alloy	Tensile properties‡			Compressive properties		Fatigue properties*		Impact value		Hardness (VPN)	Description	Specifications				
		0.2% Proof stress (MPa)	Tensile strength (MPa)	Elongation§ (%)	0.2% Proof stress (MPa)	Compressive strength (MPa)	Unnotched (MPa)	Notched (MPa)	Unnotched (J)	Notched (J)			MoD Procurement Executive (DTD series)	British Standards Aircraft	British Standards General engineering	ASTM Alloy designation	ASTM Standard No.
Zn6.5, Cu1.2, Mn0.7	**ZCM711**											The highest strength magnesium wrought alloy when fully heat treated.				ZC71A-T6	
	Extruded bars and sections: 0–13 mm diameter	160	240	7	—	—	—	—	—	—	—		—	—	—		
	As extruded: Precipitation treated	200	250	5	—	—	—	—	—	—	—	Weldable	—	—	—		
	Fully heated	300	325	3	—	—	—	—	—	—	—		—	—	—		
	Forgings**	—	—	—	—	—	—	—	—	—	—		—	—	—		
Zn3.0, Zr0.6	**ZW3**															ZK30A	—
	Extruded bars and sections: 0–10 mm	200	280	8	200–250	385–465	110–135	85–95	23–31	—	65–75	High strength extrusion, and forging alloy. Weldable under good conditions	—	2 L.505	3373 MAG-E-151M		
	10–100 mm	225	305	8	—	—	—	—	—	9.5–12	65–75		—	2 L.505	3373 MAG-E-151M		
	Extruded forging stock: 0–10 mm	195	280	8	—	—	—	—	—	—	65–75		—	L.514	3372 MAG-E-151M		
	10–100 mm	205	290	8	—	—	—	—	—	—	65–75		—	L.514	3372 MAG-E-151M		
	Forgings**	205	290	7	165–215	370–340	—	—	6–27	4.7–9	60–80		—	L.514	3372 MAG-F-151M		
Al6.0, Zn1.0, Mn0.3	**AZM**															AZ61A-F	B107-76
	Extruded bars and sections and extruded forging stock: 0–75 mm	180	270	8	130–180	370–420	125–135	90–95	34–43	6.7–9.5	60–70	General purpose alloy. Gas and arc weldable	—	L.512&3	3373 MAG-E-121M		
	75 mm	160	250	7	115–165	340–400	—	—	—	—	55–65		—	L.512&3	3373 MAG-E-121M		
	Extruded tube: 75–150 mm	150	260	7	130–180	—	—	—	—	—	60–70		—	2 L.503	3373 MAG-F-121M		
	Forgings**	160	275	7	130–165	340–400	115–125	80–90	16–23	3.4–4	60–70		—	L.513	3372 MAG-F-121M		
Al8.5, Zn0.5, Mn0.12(min.)	**AZ80**															AZ80A	B91-72
	Forgings—precipitation treated	200	290	6	—	—	—	—	—	—	60	High strength alloy for forgings of simple design	—	—	—		
Al3.0, Zn1.0, Mn0.3	**AZ31**															AZ31B-O	B70-90
	Sheet—soft: 0.5–6.0 mm	(120)	220–265	10–12	—	—	—	—	—	—	50–65	Medium strength sheet and extrusion alloy. Good formability. Weldable	—	—	3370 MAG-S-1110		
	Extruded bars and sections: 0.10 mm	150	230	8	—	—	—	—	—	—	50–65		—	—	3373 MAG-E-111M	AZ31B-F	B107-76
	10–75 mm	160	245	10	—	—	—	—	—	—	50–60		—	—	3373 MAG-E-111M		
Th0.8, Zn0.6, Zr0.6	**ZTY**															HZ11A	—
	Extruded forging stock: 0–25 mm	130	230	6	—	—	75	45	—	—	50–60	Creep resistant up to 350°C. Fully weldable	5111	—	—		
	25–50 mm	110	200	8	—	—	—	—	—	—	50–60		5111	—	—		
	>50 mm	95	200	8	—	—	—	—	—	—	50–60		5111	—	—		
	Forgings**	130	230	6	—	—	—	—	—	—	50–65		5111	—	—		

* Reproduced by courtesy of Magnesium Elektron Ltd.
‡ Larger sizes than those given in the table are available: when required property levels will be by agreement.
† The tensile properties quoted are the specification minima for that alloy and condition. Where a range is quoted the specification requirements depend on section thickness. Bracketed values are for information only.
§ Elongation values are based on a gauge length of 5.65√A except in the case of thin material where a gauge length of 50 mm may be used (see BS 2L.500, 3370 and 3373). With the latter gauge length, elongation requirements for sheet and plate depend on thickness and a range of minima is quoted.
* Endurance values for 5×10^8 reversals in rotating bending-type tests: semi-circular notch, radius 1.2 mm: stress concentration factor approx. 2.0. Reversed bending for sheet.
** Forging properties quoted are those in the most favourable direction of flow: the manufacturer should be consulted on directionality.
VPN, Vicker's pyramid number. MoD, Ministry of Defence.

Table 1.39 Stress (MPa) to produce 0.2% total strain in 1000 h in creep resisting cast magnesium alloys*

Alloy	Temperature (°C)			
	150	204	260	315
WE54	88	40	23	
HK31		55	24	6.7
EZ33		38	18	5.5
HZ32		44	28	12
QE22		31	12	
ZCM630	61	41		

* Reproduced by courtesy of Mr W. Unsworth, Magnesium Elektron Ltd.

1.4.5 Nickel and its alloys

1.4.5.1 General

Although the major proportion of nickel mined is used as an alloying agent for ferrous metals, nickel as a major constituent forms many alloys which have a very wide range of outstanding properties.

The face centred cubic lattice of nickel persists without allotropic change from very low temperatures to its melting point. This confers ductility down to cryogenic temperatures and strength up to 70% of the melting point, which makes nickel an outstanding base for creep-resistant materials.

Pure nickel has excellent corrosion resistance to non-oxidising media and this property is enhanced by the addition of copper, chromium and molybdenum. Oxidation and scaling resistance is conferred by the addition of chromium and adherence of the chromic oxide surface layer is improved by rare-earth additions.

The range of properties available in nickel alloys includes very low thermal expansion coefficients and almost constant elastic moduli over limited ranges of temperature. Other alloys have a wide range of electrical resistance and in some this is almost constant over a range of temperature. Excellent magnetic properties are available, but only at low magnetic induction.

The major drawback of nickel alloys is their cost. Nickel is not widely distributed and some of the more abundant deposits are expensive to process. The cost of the metal is augmented by that of some of the alloying additions. The embrittling action of some impurities requires their content to be kept very low. This complicates the refining and scrap segregation procedures and further increases cost.

The manufacturing cost of components from nickel alloys is high, partially because of their poor castability, partially because of their high resistance to deformation at elevated temperatures and partially because of their high rate of work hardening. Very powerful equipment operating at high temperatures is required to work nickel alloys and they are hard to machine.

The resistance of nickel to oxidising gases is poor and the action of sulphidising gases is catastrophic.

There are six main groups of nickel alloys: corrosion resistant; high temperature; electric; magnetic; controlled physical property; and hard facing. There are also a few miscellaneous types.

1.4.5.2 Nomenclature and standards

Some of the more important alloys are covered by national specifications, but among engineers they are normally referred to by the designation given by their supplier, usually a subsidiary of the International Nickel Corporation. Sometimes an identical material is available from another supplier and in this case it is usually referred to by the same number but without the tradename, e.g. 'Alloy 800' instead of 'Incoloy 800'.

Most wrought alloys are available in all the semifinished forms and may also be available as castings. Cast alloys are usually available only in that form.

Such British and ASTM Standards as exist are listed in Table 1.40.

1.4.5.3 Corrosion-resistant alloys

The corrosion-resistant alloys include:

(1) 'Nickel'—a name which covers a range of general-purpose corrosion-resistant materials;
(2) 'Monels'—essentially nickel alloys with 30% copper which have exceptional resistance to aqueous environments;
(3) 'Hastelloys' (a Cabot corporation tradename)—nickel–molybdenum alloys which have excellent resistance to acids at high temperature;
(4) 'Iliums'—nickel–chromium–molybdenum–copper alloys which resist sulphuric acid;
(5) 'Inconels' (nickel–chromium–iron alloys) and 'Incoloys' (nickel–chromium–molybdenum alloys)—these are really high temperature alloys but Inconel 625 and Incoloy 800 have exceptional resistance to stress corrosion and

Table 1.40 British and ASTM Standards for nickel and nickel alloys

Refined nickel (principally cathodes, briquettes and pellets)
BS 375—5 grades (99.5–99.95) of refined nickel
ASTM B39—99.8% refined nickel

Depending on grade, nickel may contain 0.005–1.5%Co, 0.05–0.1%C and 0.002–0.15%Cu. Minor impurities are restricted to 0.00002–0.002%

Nickel alloy castings
BS 3071 Nickel copper (N30%) castings
ASTM 494 Chemical composition and tensile requirements for 11 casting alloys

Wrought nickel alloys
BS 3072 Nickel and nickel alloy sheet and plate
BS 3073 Nickel and nickel alloy strip
BS 3074 Nickel and nickel alloy seamless tube
BS 3075 Nickel and nickel alloy wire
BS 3076 Nickel and nickel alloy bar

ASTM B564 Nickel alloy forgings
ASTM B161 Nickel seamless pipe and tube
ASTM B162 Nickel plate sheet and strip
ASTM B168 Nickel chromium iron alloys—plate sheet and strip
ASTM B169 Welded nickel alloy pipe
ASTM B670 Precipitation hardening nickel alloy—plate, sheet and strip
ASTM B335 Nickel–molybdenum alloy rod

Incoloy 825, with additional copper resists strong mineral acid.

The corrosion-resistant alloys are listed, with their standard designations, tensile strengths and dominant characteristics in *Table 1.41*.

1.4.5.4 High temperature alloys

The high temperature alloys of nickel (with the exception of the dispersion strengthened composites) can be regarded either as based on Nimonic 75, (80/20 nickel chrome) or as an extension of the austenitic steels with progressively reducing contents of iron and progressively increasing strengthening additions. There are two classes.

High temperature corrosion resistant alloys These have relatively low hot strength (for a nickel alloy) but good scaling resistance (see *Table 1.41*). They include:

(1) Nimonic 75 (originally developed as Brightray wire) but available in other forms;
(2) Inconel 600 and 601 with higher scaling resistance are similar to Brightray but contain some iron; and
(3) 50/50 nickel chrome available as castings or as a cladding

material (Incoclad) and IN657 (with niobium) available as castings are the materials with the highest resistance to fuel ash corrosion (except for precious metals).

Creep and corrosion resistant alloys (nickel superalloys) These are alloys of nickel and chromium, or nickel, chromium and iron with strengthening additions. They are employed anywhere that strength and oxidation resistance are required at high temperature, but their development as materials to withstand progressively more severe conditions in gas turbine engine components has stimulated research both in alloy composition and in manufacturing technique. The progress and results of this development are summarised in the following section. The relative rupture strengths of the several alloys produced in different ways are shown in *Figure 1.50*.

1.4.5.5 Nickel alloys for gas turbines[19]

Materials for forged blades[20-23] The original nickel alloy turbine blades were forged and their creep and creep rupture strengths were improved by increasing the proportion of ordered precipitating phase γ' ($NiAl_3$) and solid solution strengthening additions, chromium, cobalt, molybdenum, tungsten and tantalum. These elements also dissolve in the γ' phase and may have a two-fold hardening effect.

Alloy compositions must be optimised to obtain the most favourable balance of creep and thermal fatigue strength and oxidation resistance. Unfortunately, increasing content of γ' must be balanced by a reduction in chromium (which reduces the solubility of aluminium and titanium), but the effect of this on scaling resistance is offset to some extent by the increase in aluminium. At the highest temperatures, oxidation resistance is obtained by coating with aluminium. The increased temperature capacity of the improving forged nimonic alloys is illustrated in *Figure 1.51* and shown in *Table 1.42*.

Increase in γ' content reduces the range of temperature available for hot working between softening and incipient melting. This can be offset to some extent by the addition of cobalt, which increases the solvus temperature. Even so, the small margin between the maximum preheating temperature (1100°C) and the minimum working temperature (1050°C) of Nimonic 115 indicates that this is likely to remain the forged alloy with the best high temperature properties.

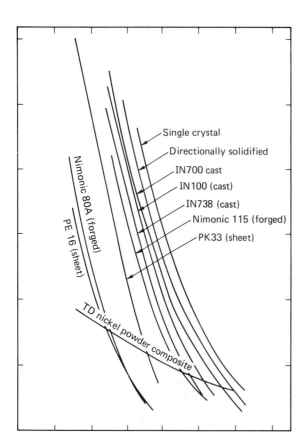

Figure 1.50 Comparative 1000 h rupture strengths of nickel alloy gas turbine blading materials produced by a variety of techniques

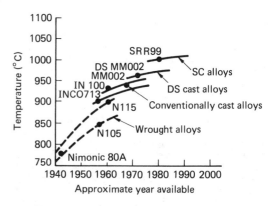

Figure 1.51 Increase in temperature capability for turbine-blade alloys based on creep rupture in 1000 h at 150 MPa. (Reproduced by permission of *Metals and Materials*)

Table 1.41 Corrosion resistant nickel alloys

Alloy	Composition* (%)													BS	Form	Tensile strength (MPa)	Corrosion resistance†
	Ni	C	Mn	Fe	S	Si	Cu	Cr	Co	Mo	Al	Ti	Other				
Low/medium temperature																	
Nickel 200	99.5	0.08	0.18	0.2	0.005	0.18	0.13	—	—	—	—	—	—	3072/6 NA11	International Nickel / Annealed → cold worked	380–550	Aqueous and general
Nickel 201	99.5	0.01	0.18	0.2	0.005	0.18	0.13	—	—	—	—	—	—	NA12	Annealed → cold worked	340–410	Stress corrosion and general
Nickel cast	98	0.1	1.0	0.2	—	1.0	—	—	—	—	—	—	Mg0.1		Cast bar	360–420	Aqueous and general
MONEL alloy 400	63.0	0.15	1.0	2.5 max.	0.024 max.	0.5 max.	31.0	—	—	—	—	—	—	3072/6 NA13	Annealed bar	480–620	Seawater, polluted water and acids
MONEL alloy 410	66.0 min.	0.2	0.8	1.0 max.	0.008	1.6	30.5	—	—	—	—	—	—	3071 NA1	Cast bar	450–580	Seawater, polluted water and acids
MONEL alloy K-500	63.0 min.	0.15	1.5 max.	2.0 max.	0.010 max.	0.5	30.0	—	—	—	2.9	0.6	—	3072/6 NA18	Heat-treated bar	620–760	Higher strength alloy resists seawater
HASTELLOY alloy B	Bal.	0.09	1.0	5.0	0.03	1.0	—	1.0	2.5	28.0	—	—	VO.3		Cabot Corporation / Solution-treated bar	880	Hydrochloric acid
HASTELLOY alloy C276	Bal.	0.10 max.	1.0	5.5	0.03	1.0	—	16.0	2.5	16.5	—	—	VO.3		Solution-treated bar	780	Chlorides and hypochlorites
HASTELLOY alloy D	Bal.	0.12	1.0	2.0	—	10.0	3.0	1.0	1.5	—	—	—	—		Solution-treated cast plate	790	Sulphuric acid
HASTELLOY alloy G	Bal.	0.05 max.	1.5	19.5	—	1.0 max.	2.0	22.0	2.5 max.	6.5	—	—	W1 max Nb+Ta2		Annealed sheet	710	Hot sulphuric and phosphoric acids
HASTELLOY alloy N	Bal.	0.06	0.8	5.0 max.	—	1.0 max.	0.35 max.	7.0	0.2 max.	16.5	0.5 max.	—	B 0.01 max		Solution-treated sheet	700	Molten fluorides
ILLIUM alloy B	Bal.	0.05 max.	1.0 max.	1.5	—	3.5	5.5	28.0	—	8.0	—	—	—		International Nickel	415–485	All concentrations sulphuric acid to b.p.
ILLIUM alloy G	56.0	—	—	—	—	—	6.5	22.5	—	6.5	—	—	—		—	470	
ILLIUM alloy R	68.0	—	—	—	—	—	3.0	21.0	—	5.0	—	—	—		—	780	60% sulphuric acid to 50°C
ILLIUM alloy 98	Bal.	0.07	1.5 max.	1.5	—	1.25	5.0	28.0	max.	8.5	—	—	—		—	370–540	60% sulphuric acid to b.p.
INCONEL alloy 625	60.5	0.10 max.	0.25 max.	5.0 max.	0.015 max.	0.5 max.	—	21.5	—	9.0	0.25	0.25	Nb+Ta3.65	3072/6 NA16	Annealed cold	830–1040 worked	Seawater
INCOLOY alloy 825	42.0	0.05 max.	1.0 max.	Bal.	0.03 max.	0.5 max.	2.25	21.5	—	3.0	0.20	0.9	—		Annealed bar	590–730	Strong mineral acids
High temperature																	
80Ni. 20Cr	Bal.	0.13	1.0 max.	5.0 max.	0.02	1.0 max.	0.5 max.	19.5	—	—	—	0.40	—		Cast and extruded clad	300	Resists fuel ash corrosion
50Ni. 50Cr	50	0.1 max.	0.3 max.	1.0 max.	—	0.5 max.	—	50.0	—	—	—	—	—		Cast	600	
IN 657	Bal.							48.52					Nb1.5				
INCONEL alloy 600	Bal.	0.15 max.	1.0 max.	8.0	0.015 max.	0.5 max.	0.5 max.	15.5	—	—	—	—	—	3072/6 NA14	Annealed bar	550–690	
INCONEL alloy 601	60.5	0.05 max.	0.5 max.	14.1	0.007 max.	0.25 max.	0.25 max.	23.0	—	—	1.35	—	—		Annealed bar	740	
INCOLOY alloy 800	32.5	0.10 max.	1.5 max.	Bal.	0.015 max.	1.00 max.	0.75 max.	21.0	—	—	0.38	0.38	—	3072/6 NA15	Annealed bar	590	Originally developed for electric-kettle elements
INCOLOY alloy DS	37.0	0.1 max.	1.2 max.	Bal.	—	2.3 max.	0.5 max.	18.0	—	—	—	—	—		Annealed bar	730	

* Bal., balance.
† b.p., boiling point.

Table 1.42 Nimonic alloys

Basic grade	Improved corrosion resistant grade	Operating temperature (°C)
80A	81	815
90 (Inconel 750)	91	920
105		940
115		1010

Cast blades Vacuum casting has provided an alternative manufacturing route. Problems met with in the earlier alloys included cracking due to the separation of massive carbides on solidification and the formation of σ-like intermetallic compounds. However, in one specific alloy, IN100, up to 70% γ' fraction was achieved. This had the immediate effect of eliminating the need for solid solution strengthening so that a low density high creep resistance material became available. Early casts of IN100 embrittled after prolonged exposure at 850°C due to the formation of intermetallic compounds σ, μ, χ, π or Laves phases which had to be avoided by computerised phase control (Phacomp) techniques.[23,24]

The achievement of yet higher temperature capability required the addition of a high-melting-point solid solution strengthener such as tungsten. The further addition of 2% hafnium improved ductility and minimised cracking by increasing the amount of low-melting-point eutectic in (Martin Marietta) MarM002. This and the alloy IN738, which although it lacks the best creep rupture properties is extremely well suited to coating with aluminium and is therefore used where the maximum creep resistance is not essential, were standardised.

The structure of the grain boundaries is important. Too little strengthening impairs creep resistance, too much impairs creep ductility and, therefore, thermal fatigue. The grain boundary structure is controlled through precipitation of carbides, usually chromium carbides (but substituted with other carbides) and by boron and zirconium additions. Compositions of wrought and cast superalloys are listed in *Table 1.43*. Two Martin Marietta alloys which are similar in many respects are referred to in this section. MarM002 is suitable for conventional casting, whilst MarM200 is better for directional solidification.

Directional solidification[25–28] The next improvement in temperature capability came with directionally solidified blades made in a plant of the type shown in *Figure 1.52*. The mould is enclosed within a hot zone and the heat extracted through a chill plate. The blade consists effectively of a bundle of crystals each with the (100) direction longitudinal and having random rotational orientations transverse to this direction.

This structure has little effect on creep but, because there are no transverse grain boundaries, very significantly increases rupture ductility, particularly in the highly creep resistant alloys such as, for example, MarM200 (see *Figure 1.53*). This produces greatly increased thermal fatigue life, which is further enhanced by a low elasticity modulus in the longitudinal (100) direction. This allows a greater elastic extension in a thermal stress cycle and reduces $\Delta\varepsilon_p$.

Single crystals Even greater improvements accrue from casting blades as single crystals either by restricting the cross-section of the solidifying metal in advance of the moulded blade (see *Figure 1.54*) or by using a seeding crystal. There are no grain boundaries in a single crystal and, consequently,

elements introduced to strengthen grain boundaries (e.g. carbon, boron, zirconium and hafnium) may be omitted. Higher solution treatment temperatures can therefore be used and stronger more uniform γ' precipitates produced. Thus, in the UK, SRR99 will replace MarM002, low density RR2000 will replace IN100 and RR2060 will be used for nozzle guide vanes.

The kind of performance that may be expected from directionally solidified and single-crystal materials is shown in *Figure 1.50*. The compositions of superalloys that are suitable for directional solidification are listed in *Table 1.44*. Any significant further advance in gas turbine blade materials must come from the use of composites.

Composite blades Composites can be produced by casting eutectic alloys in directionally solidifying furnaces or by powder techniques.

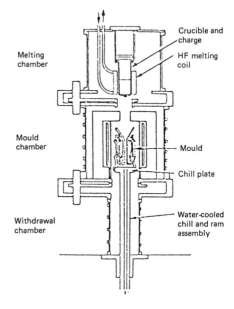

Figure 1.52 Schematic of Rolls Royce directional solidification plant. (Reproduced by permission of *Metals and Materials*)

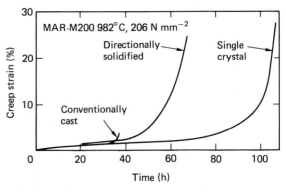

Figure 1.53 Comparison of creep curves for MarM200 tested at 1255 K and 206 MPa in conventional cast directionally solidified and single-crystal forms. (Reproduced by permission of the Metals Society)

Table 1.43 Typical composition of some wrought, cast and powdered nickel based superalloys

Alloy	Co	Cr	Al	Ti	C	Ta	Mo	W	Nb	Fe	Zr	B	Other	Developer/user
Turbine-blade alloys: wrought														
Inconel 600	—	15.5	—	—	0.08	—	—	—	—	8.0	—	—	—	International Nickel
Inconel 722	—	15.0	0.6	2.5	0.04	—	—	—	—	7.0	—	—	—	
Inconel X-750	0.5 max.	15.0	0.9	2.5	0.04	—	—	—	1.0	7.0	0.03	—	—	
Nimonic 75	—	19.5	—	0.4	0.10	—	—	—	—	—	0.05 max.	0.001 max.	—	
Nimonic 80A	2.0 max.	19.5	1.4	2.5	0.10 max.	—	—	—	—	5.0 max.	0.10 max.	0.005 max.	—	
Nimonic 90	18.0	19.5	1.5	2.4	0.13 max.	—	—	—	—	1.0 max.	0.10 max.	0.005 max.	—	
Nimonic 105	20.0	14.8	4.7	1.2	0.16 max.	—	5.0	—	—	1.0 max.	0.14 max.	0.008 max.	—	
Nimonic 115	14.8	15.0	5.0	4.0	0.15	—	4.0	—	—	1.0 max.	0.20 max.	0.06 max.	—	
Nimonic 120	10.0	12.5	4.5	3.5	0.08 max.	—	5.7	—	—	1.0 max.	0.05 max.	0.025	—	
Nimonic 81	2.0 max.	30.0	0.9	1.8	0.05	—	—	—	—	1.0 max.	0.06	0.002	—	
EPK 55	20.0	28.5	1.2	2.3	0.05	—	—	—	0.75	0.5 max.	0.07	0.006	—	
EPK 57	19.7	24.3	1.4	3.0	0.05	—	1.5	—	0.95	0.5 max.	0.05	0.012	—	
Udimet 500	16.5	17.5	2.9	2.9	0.15 max.	—	4.0	—	—	4.0 max.	—	0.010 max.	—	
Udimet 700	17.0	15.0	4.3	3.4	0.15 max.	—	5.3	—	—	4.0 max.	—	0.05 max.	—	
Turbine-blade alloys: cast														
IN 713 LC	15.0	12.5	6.1	0.8	0.12 max.	—	4.2	—	2.2	0.5 max.	0.10	0.012	—	International Nickel
IN 100	10.0	10.0	5.5	4.7	0.18	—	3.0	—	—	0.5 max.	0.06	0.014	1.0 V	Pratt and Whitney
B 1900	10.0	8.0	6.0	1.0	0.025	4.0	6.0	—	—	—	0.04	0.015	—	
MarM200	10.0	9.0	5.0	2.0	0.15	—	—	12.5	1.0	—	0.05	0.15	—	Martin Marietta
MarM002	10.0	9.0	5.5	1.5	0.15	2.5	—	10.0	—	—	0.05	0.15	1.5 Hf	
MarM247	10.0	8.3	5.5	1.0	0.15	3.0	0.67	10.0	—	—	0.05	0.015	Hf	
IN 738	8.5	16.0	3.4	3.4	0.17 max.	1.8	1.8	2.6	0.9	0.5 max.	0.10	0.010	—	International Nickel
René 77	15.0	14.5	4.3	3.3	0.07	—	4.2	—	—	—	0.04	0.010	—	General Electric
René 80	9.5	14.0	3.0	5.0	0.17	—	4.0	4.0	—	—	0.03	0.015	—	
Mechanically alloyed powder														
MA 6000	—	12.0	4.5	2.5	—	2.0	2.0	4.0	—	—	—	—	$1.1Y_2O_3$	
Sheet alloys (rolled from cast billet)														
PE 16	18.0	16.5	1.2	1.2	0.06	—	3.3	—	—	2.2	0.06 max.	0.05 max.	—	
PK 33	14.0	18.5	2.9	2.2	0.07	—	7.0	—	—	1.0	—	—	—	
Turbine-disc alloys (forged from cast billet)														
Nimonic	90.1	12.5	0.3	2.9	0.04	—	6.7	—	—	3.0	—	—	—	
Waspaloy	13.5	19.5	1.3	3.0	0.08	—	—	—	—	—	0.06	0.006	—	
Astroloy	17.0	15.0	4.0	3.5	0.06	—	5.25	—	—	—	—	0.03	—	
Turbine-disc alloys (forged from powder)														
AP1	17.0	15.0	4.0	3.5	0.025	—	5.0	—	—	—	0.04	0.025	—	

Figure 1.54 Single-crystal blade with spiral construction. (Reproduced by permission of *Metals and Materials*)

Directionally cast eutectics normally have a laminar structure but can, theoretically, be seeded to produce fibre reinforcement in a matrix. Typical compositions are listed in *Table 1.45*. Unfortunately, some directionally solidified eutectics have limited oxidation resistance, some delaminate on thermal cycling, and some have uneconomically long production times. None have so far been exploited commercially.

Dispersion strengthened alloys resist creep at very high temperatures (see TD Nickel in *Figure 1.50*), but it is extremely difficult to produce a material with high strength at lower temperatures. One such alloy, MA 6000 (see *Table 1.43*), is said to be strong enough for stubby tapered unshrouded blades which can be run at higher temperatures than other nickel based alloy blades.

Turbine disc materials　Turbine disc materials require good tensile strength (to prevent bursting in the event of an overspeed) and good low cycle fatigue life (to withstand the changes of stress undergone in the flight cycle) rather than creep resistance at high temperatures. The alloys adopted, Astroloy, Waspaloy and Nimonic 901 (see *Table 1.43*) are similar in composition to turbine blade alloys, but are given low temperature thermomechanical treatments which generate a homogeneous dislocation structure, improve tensile strength by up to 40% and produce a fine-grained microstructure which has good fatigue properties.

Ingots cast from more highly alloyed disc materials are prone to segregation and difficult to forge. It should be possible to overcome this by powder processing, but so far the presence of defects too small to be detected by non-destructive testing (NDT) has prevented the application of powder techniques to turbine discs. The mechanically alloyed powder API (*Table 1.43*) is an example of this type of alloy.

Table 1.44　Composition of high temperature alloys that have been directionally solidified*

Alloy	*Composition†* (%)												
	Ni	*Co*	*Cr*	*Al*	*Ti*	*C*	*Ta*	*Mo*	*W*	*Nb*	*Zr*	*B*	*Others*
IN713LC	Bal.	—	12.0	5.9	0.6	0.05	—	4.5	—	2.0	0.1	0.01	—
IN713C	Bal.	—	12.5	6.1	0.8	0.12	—	4.2	—	2.0	0.1	0.012	—
IN100	Bal.	15.0	10.0	5.5	4.7	0.18	—	3.0	—	—	0.06	0.014	1.0 V
MarM200	Bal.	10.0	9.0	5.0	2.0	0.15	—	—	12.5	1.0	0.05	0.015	—
MarM002	Bal.	10.0	9.0	5.5	1.5	0.15	2.5	—	10.0	—	0.05	0.015	1.5 Hf
MarM246	Bal.	10.0	9.0	5.5	1.5	0.15	1.5	2.5	10.0	—	0.05	0.015	—
B1900	Bal.	10.0	8.0	6.0	1.0	0.10	4.0	6.0	—	—	0.10	0.015	—
IN738LC	Bal.	8.5	16.0	3.5	3.5	0.11	1.6	1.75	2.5	0.7	0.08	0.008	—
IN 939	Bal.	19.0	22.5	1.9	3.7	0.15	1.4	—	2.0	1.0	0.09	0.009	—
MarM247	Bal.	10	8.4	5.5	1.05	0.15	3.3	0.65	10.0	—	0.05	0.015	1.4 Hf
UTRCMMT 143	73.9	—	—	5.8	—	—	6.0	14.3	—	—	—	—	—
X 40	10	Bal.	25.0	—	—	0.5	—	—	7.5	—	—	—	—
MarM509	10	Bal.	23.5	—	0.2	0.6	—	—	7.0	—	—	—	—
Single-crystal alloys													
P & W 444	Bal.	—	8.6	5.1	1.98	—	—	—	11.1	—	—	—	—
P & W 454	Bal.	5.0	10.0	5.0	1.5	—	12.0	—	4.0	—	—	—	—
NASAIR 100	Bal.	—	9.0	5.75	1.2	<0.01	3.3	1.5	10.5	—	—	—	—
RR SR99	Bal.	5.0	8.5	5.5	2.2	<0.015	2.8	—	9.5	—	—	—	—
RR 2000	Bal.	15.0	10.0	5.5	4.0	0.02	—	3.0	—	—	—	—	1.0 V
RR 2060	Bal.	5.0	15.0	5.0	2.0	0.02	5.0	2.0	2.0	—	—	—	—

* Reproduced (with addition) by permission of the Metals Society.
† Bal., balance.

Table 1.45 Composition of directionally solidified eutectic composites*

Alloy	Structure		Composition† (wt%)												
	Matrix	Reinforcement	Ni	Co	Cr	Al	Nb	Ta	C	W	Re	V	Mo	Y	Fe
NITAC3116A	γNi + γ'Ni₃Al	TaC	Bal.	3.7	1.9	6.5	—	8.2	0.24	—	6.3	4.2	—	—	—
COTAC 744	γNi + γ'Ni₃Al	NbC	Bal.	2.0	10.0	4.0	4.9	—	0.6	10	—	—	—	—	—
γ'-γ'-Cr₃C₂	γNi + γ'Ni₃Al	Cr₃C₂	Bal.	—	12.3	6.9	—	—	1.8	—	—	—	—	0.5	—
(Co,Cr)-Cr₇C₃	Co,Cr solid solution	Cr₇C₃	—	Bal.	41.0	—	—	—	2.4	—	—	—	—	—	—
γ-γ'-δ	γNi + γ'Ni₃Al	δ Ni₃Nb	Bal.	—	6.0	2.5	20.1	—	0.06	—	—	—	—	—	—
γ-γ'-α	γNi + γ'Ni₃Al	α Mo	Bal.	—	—	6.0	—	—	—	—	—	—	32	—	—

* Reproduced by permission of the Metals Society.
† Bal., balance.

Casing materials Sheet materials have compositions similar to blade materials, but their creep resistance is usually lower, partly because it is not needed, partly because the high technology casting procedures are not applicable to sheet, and partly because they usually have to be welded. The creep rupture properties of two sheet materials, PE16 and PK33, which span the available range are indicated in *Figure 1.50*. The compositions of a selection of available (and potential) nickel gas turbine alloys are listed in *Tables 1.43 to 1.45*.

The above account is based on experience at Rolls Royce, but similar developments, using alloys of marginally different compositions tttttailored to meet specific design requirements, have been achieved at General Electric and United Technologies.

1.4.5.6 Electrical alloys

Nickel and nickel alloys are used in electrical applications, usually in the form of wire or strip, because of their individual characteristics of resistivity, electron emission, thermoelectric properties or corrosion resistance. *Table 1.46* lists these materials giving composition, supplier, tensile strength, and property of major interest. They are usually supplied in the form of annealed bar for drawing into wire or strip.

1.4.5.7 Magnetic alloys

The magnetic alloys of nickel are principally those with high magnetic permeability in low or medium strength magnetic fields or some special form of magnetic hysteresis loop. They are used mainly in the form of tapes or sheet, or as powder for cores for electronic equipment. Very careful control is needed in production. Pure nickel or nickel rich cobalt alloys are used as magnetostriction transducers. Compositions, tensile strengths and an indication of magnetic characteristics are listed in *Table 1.47*. Magnetic alloys can be obtained from Telcon Metals, ITT (Hounslow) and International Nickel.

1.4.5.8 Alloys with special dimensional and elastic properties

Certain alloys of nickel have minimal or controlled coefficients of expansion over certain ranges of temperature. Alloys with minimal expansion coefficients are used in instruments and those with controlled coefficients for sealing to glass or ceramics are used in vacuum devices. Alloys with small positive temperature coefficients are used for temperature-insensitive vibrating instrument devices. There is also a low-temperature shape-memory alloy (Nitinol). The compositions, ultimate tensile strengths and an indication of the characteristic properties of nickel alloys with dimensional property applications are listed in *Table 1.48*.

1.4.5.9 Hard facing materials[29,30]

A large number of commercial hard facing alloys depend on a nickel matrix, usually containing a dispersion of tungsten and chromium carbides, boron and, sometimes, molybdenum and silicon. The hardness of the deposited material varies according to composition (300–720 VPN) and wear resistances can be obtained superior to other materials at room temperature, but inferior to cobalt alloys at high temperature or in resistance to aqueous environments.

1.4.6 Zinc and its alloys

1.4.6.1 Introduction and standards

Zinc owes its commercial applications to three characteristics:

(1) Zinc components are essentially low priced. The metal requires less energy to produce and to cast than any of its competitors and is cheap to machine.
(2) It has quite exceptional corrosion properties. Only aluminium, magnesium and the alkali metals are cathodic to it so that it will protect all other metals sacrificially and can be used as the cathode in an electrolytic cell, but in normal atmospheres it forms a protective film which is penetrated only about 0.01 mm per year in the worst industrial atmospheres.
(3) Its low melting point makes it exceptionally suited as a material for die casting.

The drawbacks of zinc are that it is relatively soft and weak, has poor creep resistance and its hexagonal structure makes it brittle at room temperature.
Its major uses are:

(1) as a coating to protect steel;
(2) as a base for casting alloys; and
(3) as a sheet material for battery cases and roofing.

Standards covering zinc metal include:

(1) BS 1004 and ASTM B669—Zinc alloys for casting
(2) BS 3436 Ingot zinc; and ASTM B6 Zinc
(3) BS 5338 Code of practice for zinc alloy pressure die casting

Table 1.46 Electrical alloys of nickel

Alloy	Composition† (%)													BS	Supplier	Form	Tensile strength (MPa)	Properties and uses
	Ni	C	Mn	Fe	S	Si	Cu	Cr	Co	Mo	Al	Ti	Other					
Nickel 205	99.5	0.08	0.18	0.10	0.004	0.08	0.08	—	—	—	—	0.03	Mg0.05	—	International Nickel	Annealed rod	460	High damping. Electronic valve electrodes
Nickel 212	97.7	0.10	2.0	0.05	0.005	0.05	0.03	—	—	—	—	0.03	—	3504	International Nickel	—	476	Resists sulphur embrittlement. Support wires
Nickel 222	99.5	0.01	0.02	0.04	0.0025	0.01	0.01	0.01	0.06	—	0.01	0.01	Mg0.08 Zr0.15	3504	International Nickel	—	340	Emits electrons. Valve cathode sleeves
Nickel 240	95.0	—	0.01	0.04	—	2.0	0.45	1.7	—	—	0.01	0.3	—		International Nickel	—	—	Resists lead and sulphur. Spark plug electrodes
Nickel 270	99.98	0.01	0.003	<0.001	<0.001	<0.001	<0.001	<0.001	<0.001	—	—	<0.001	Mg0.001		International Nickel	Annealed strip	340	Powder product, highly deformable. Valves. etc.
Constantan (Ferry)	45	—	—	—	—	—	55	—	—	—	—	—	—		ITT (Harlow)	Annealed bar	415	Low temperature coefficient of resistivity. Resistors and thermocouples
Brightray B	59.0	0.1	1.0 max.	Bal.	—	0.35	—	16.0	—	—	—	0.17	—			Annealed bar	686	Heating elements up to 750°C
Brightray C	Bal.	0.1 max.	0.25	1.0 max.	—	1.5	—	19.2	—	—	0.26	—	Rare earth metals 0.5		Wiggin alloys	Annealed bar	735	Heating elements up to 1150°C
Brightray S	Bal.	0.1 max.	0.4	1.0 max.	—	1.0 max.	—	20.0	—	—	—	—	—		Wiggin alloys	Annealed bar	734	Strip heating elements up to 1150°C
Brightway 35	37.7	0.1 max.	1.2	Bal.	—	2.2	—	18.0	—	—	—	—	—			Annealed bar	740	Resists carburisation. Heating elements up to 1050°C
Chromel alloy P	Bal.	—	—	—	—	0.4	—	10.0	—	—	—	—	—		British Driver Harris	—	—	Thermocouples up to 1110°C
Alumel	Bal.	—	1.75	0.2	—	1.2	—	—	—	—	1.6	—	—		British Driver Harris	—	—	Thermocouples up to 1110°C
Nicrosil	Bal.	—	—	0.1	—	1.5	—	14.3	—	—	—	—	—		British Driver Harris	—	—	Thermocouples up to 1110°C
Nisil	Bal.	—	—	—	—	4.5	—	—	—	—	—	—	Mg0.1		British Driver Harris	—	—	Thermocouples up to 1110°C

† Bal., balance

Table 1.47 Magnetic alloys of nickel*

Alloy*	Composition† (%)													BS	Form	Tensile strength (MN m⁻²)	Properties and uses
	Ni	C	Mn	Fe	S	Si	Cu	Cr	Co	Mo	Al	Ti	Other				
Nickel 205	99.5	0.08	0.18	0.10	0.004	0.08	—	—	—	—	—	0.03	Mg0.05	—	—	—	High magnetostriction. Ultrasonic transducers
Nickel, 4Co	Bal.	—	—	—	—	—	—	—	4.0	—	—	—	—	—	—	—	High magnetostriction. Ultrasonic transducers
Nickel, 18Co	Bal.	—	—	—	—	—	—	—	18.0	—	—	—	—	—	—	—	—
75Ni, 25Fe	70–80	—	—	Bal.	—	x	x	x	—	x	—	—	—	2875A	Annealed strip	540	—
50Ni, 50Fe	50	—	—	Bal.	—	x	x	x	—	x	—	—	—	2875A	Annealed strip	430	—
36Ni, 74Fe	36	—	—	Bal.	—	x	x	x	—	x	—	—	—	2875A	Annealed strip	530	—
JAE metal	70	—	—	30	—	—	—	—	—	—	—	—	—	—	Annealed bar	430	High temperature coefficient of permeability. Magnetic components
30Ni, 70Fe	30	—	—	70	—	—	—	—	—	—	—	—	—	—	Annealed strip	430	

* Tradenames and suppliers of magnetic alloys are: Mumetal and Radiometal, Telcon Metals Ltd; Permalloy, ITT Harlow; Nilomag and JAE metal. Wiggin Alloys Ltd.
† Bal., balance.
x, Small additions of one or more of these elements.

Table 1.48 Controlled-expansion and constant-modulus alloys of nickel

Alloy*	Composition† (%)													Form	Tensile strength (MN m⁻²)	Properties and uses
	Ni	C	Mn	Fe	S	Si	Cu	Cr	Co	Mo	Al	Ti	Other			
NILO alloy 36[1]	36.0	0.15 max.	0.5	Bal.	—	0.5 max.	0.5 max.	—	—	—	—	—	—	Annealed bar	460	Very low thermal expansion 20–100°C. Metrology. Chronometers
NILO alloy 42[2]	42.0	0.15 max.	0.5	Bal.	—	0.5 max.	0.5 max.	—	—	—	—	—	—	Annealed bar	525	Thermal expansion $c.5.5$ K⁻¹ (20–300°C). Thermostats for sealing into glass
NILO alloy 48[2]	48.0	0.15 max.	0.5	Bal.	—	0.5 max.	0.5 max.	—	—	—	—	—	—	Annealed bar	494	Thermal expansion $c.9$ K⁻¹ (20–400°C). Thermostats for sealing into alumina
NILO alloy K[2]	29.5	0.05 max.	0.3	Bal.	—	0.5 max.	0.5 max.	—	17.0	—	—	—	—	Annealed bar	525	Thermal expansion $c.6$ K⁻¹. Sealing into borosilicate glass
INCONEL alloy 903	38.0	—	—	Bal.	—	—	—	—	15.0	—	0.7	1.4	Nb3.0	Warm-worked bar	—	Thermal expansion $c.5$ K⁻¹. Components stressed at variable temperature
NI-SPAN alloy C-902[3]	42.25	0.1 max.	0.5	Bal.	—	0.6	—	5.3	—	—	0.55	2.5	—	Heat-treated bar	1240	Low positive coefficient of elastic modulus. For accurate vibration devices
55 Nitinol‡	55	—	—	—	—	—	—	—	—	—	—	45	—	Annealed bar	860	
50 Nitinol	60	—	—	—	—	—	—	—	—	—	—	40	—	Annealed bar, heat-treated bar	940 1070	

* These are Wiggin designations. Other designations are: [1]Telcon Metals 'Invar', British Driver Harris 'Thermo'; [2]Telcon Metals 'Telcoseal', British Driver Harris 'Therlo'; [2]Telcon Metals 'Telcoseal'; [2]Telcon Metals 'Elinvar'.
† Bal., balance.
‡ Property of the Naval Ordinance Laboratory, Washington, DC, USA.

(4) ASTM B69 Rolled zinc
(5) ASTM B418 Cast and wrought galvanic zinc anodes for use in saline electrolytics

1.4.6.2 Zinc based casting alloys

Zinc based alloys may be used for all types of sand, investment, gravity and pressure die casting. Melting is clean and easy, casting is less sensitive to problems of oxide inclusions, misruns and voids, finish is superior, wall thicknesses can be thinner, detail is sharper, and shapes are more complex than is possible with such competitive materials as aluminium and copper alloys and cast irons. Zinc alloys have excellent machinability and less machining is required than with castings in other metals. Complex shapes can therefore be produced more cheaply in them than in any other metallic material.

A further advantage of the zinc alloys is that for those low speed bearing applications for which copper alloys have traditionally been used there are indications that zinc casting alloys perform better than leaded gunmetal.

Casting alloys, which usually have aluminium as their main alloying element are commonly known by their 'ZA' number.

The composition and properties of zinc based casting alloys are listed together with those of competitive materials in *Tables 1.49* to *1.51*.

1.4.6.3 Alloy development

In early days, zinc based alloys developed a poor reputation because some castings disintegrated in service and electroplate sometimes showed poor adherence. The cause of disintegration was traced to the pressure of heavy metals (iron, lead, cadmium and tin) which are held strictly to acceptable limits in modern high-purity zinc and research on plating eliminated the problems of adherence.

The alloys ZA3 and ZA5 are very suitable for hot chamber high pressure die casting because the molten metal has a very low affinity for iron and there is very little wear on the dies. Castings in these alloys are therefore inexpensive to produce and very suitable for complex lowly stressed components.

Increasing the aluminium content very significantly increases strength but also increases aggressiveness to iron, so that above 8% (the aluminium content of ZA8) the alloys cannot be hot chamber die cast. Although more expensive in the form of components, the higher aluminium casting alloys

are more than competitive with cast iron, copper and aluminium alloys in situations where their low hot strength is not detrimental.

ZA12 should not be considered for use in stressed applications at or above 120°C, but the ASME boiler code design stress for ZA27 at 150°C (which is the stress required to produce a secondary creep of 1% in 100 000 h) is 69 MPa.

As long as these limitations are observed, ZA12 and ZA27 have a potential market for components for mass-produced cars equivalent to that of magnesium alloys for high performance sports cars.

1.4.6.4 Wrought zinc alloys

Zinc is available in the form of sheet, extrusions (and forgings) and wire.

Sheet zinc is used mainly for roofing and as anode cans in the common dry cell. Zinc for roofing can be the pure metal, or the more creep resistant zinc–copper or zinc–titanium alloys which are an economical replacement for copper or lead. Zinc strip for battery cases contains approximately 0.5%Pb and zinc plate for photoengraving (an important application) contains 0.2%Pb and 0.2%Cd.

Sheet containing 78%Zn and 22%Al can be made superplastic and drawn and stretched into shapes that require a very high degree of deformation.

Extruded zinc is available as the copper–titanium alloy. These alloys are used for architectural purposes in conjunction with roofing or walling sheet, but their applications have been largely superseded by plastics. A major remaining application for pure extruded (or sheet) zinc is as sacrificial anodes to protect steel ships or buried pipes.

Zinc wire is largely used for spray metallising, but is available for nails, hooks, gauze and similar wire products, and for solder.

1.4.6.5 Zinc coating for corrosion protection

The major use for zinc is as a coating on steel for corrosion protection. Zinc operates in two ways. Its room-temperature oxidation product is adherent (as opposed to that of steel in moist conditions) but, after it has eventually been penetrated, it protects the underlying metal electrochemically. It makes an excellent base for paint.

There are three types of zinc coating. 'Hot dip galvanising' is achieved by immersing the steel in a bath of molten zinc.

Table 1.49 Composition (%) of zinc casting alloys

| Component | Alloy | | | | |
	ZA3	ZA5	ZA8	ZA12	ZA27
Specification	BS 1004	BS 1004	—	ASTMB-669-82	ASTMB-669-82
Aluminium	3.9–4.3	3.9–4.3	8–8.8	10.5–11.5	25.0–28.0
Copper	0.10	0.75/1.25	0.8–1.3	0.5–1.25	2.0–2.5
Magnesium	0.025/0.05	0.05/0.06	0.015–0.030	0.015–0.030	0.010–0.020
Iron	<0.075	<0.075	<0.010	<0.075	<0.010
Lead	—		—	<0.004	—
Cadmium	—		—	<0.003	—
Tin	—		—	<0.002	—
Zinc	—		—	Bal.	—

Bal., balance.

Table 1.50 Mechanical and physical properties (20°C) of zinc casting alloys and competitive materials

Property	ZA3	ZA5	ZA8			ZA12			ZA27				Brass BS 1004 SCB3	Aluminium LM6M	Cast iron	
	Pressure die	Pressure die	Sand cast	Gravity die	Pressure die	Sand cast	Gravity die	Pressure die	Sand cast	Sand H/T	Gravity die†	Pressure die			Blackheart malleable	Grey
UTS (MPa)	283	324	248–276	221–255	365–386	275–317	310–345	392–414	400–440	310–324	424	407–441	185–240	160–185	290–345	160–345
Elongation in 2 in. (%)	15	9	1–2	1–2	6–10	1–2	1–2	4–7	3–6	8–11	1	1	15–30	5–7	6–12	<0.5
Young's modulus (GPa)	—	—	88	85	—	83	—	—	78	79	—	—	83	71	169	75–145
Hardness*	83	92	82–89	85–90	99–107	90–110	85–95	95–105	110–120	90–100	110–120	116–122	45–65	55–60	110–149	200–250

* Brinell hardness number.
† Preliminary data.

Table 1.51 Physical properties (20°C) of zinc casting alloys and competitive materials

Property	ZA3	ZA5	ZA8			ZA12			ZA27				Brass BS 1004 SCB3	Aluminium LM6M	Cast iron	
	Pressure die	Pressure die	Sand cast	Gravity die*	Pressure die	Sand cast	Gravity die	Pressure die	Sand cast	Sand HT	Gravity die*	Pressure die			Blackheart malleable	Grey
Density (g cm^{-3})	6.7	6.7	6.3	6.3	6.3	6.0	6.0	6.0	5.0	5.0	5.0	5.0	8.5	2.6	7.3	7.3
Electrical conductivity (%IACS)	26	26	27.7	27.7	27.7	28.3	28.3	28.3	29.7	29.7	29.7	29.7	20	37	—	—
Thermal conductivity (W m^{-1} °C^{-1})	113	110	115	115	115	116	116	116	125.5	125.5	125.5	125.5	90	142	49	42–50
Melting range (°C)	382–387	379–388	375–404	375–404	375–404	380–430	380–430	380–430	380–490	380–490	380–490	380–490	920–1000	580–640	1450–1550	1090–1260

* Preliminary data.

The zinc and steel form an alloy at the interface. 'Electrogalvanising' is achieved by electroplating zinc through an aqueous electrolyte. The thickness of coating is more uniform but usually thinner than that produced by other methods. 'Sheradising' (heating the component to 370°C in zinc powder) produces a layer of uniform thickness.

'Zalutite' (a trademark of British Steel Corporation), a 55%Al, 43.5%Zn, 1.5%Si silicon coated steel combines the protection of zinc and aluminium and in many environments is superior to galvanising.

1.4.7 Low-melting-point metals, lead, tin and their alloys

1.4.7.1 Introduction

Lead and tin are soft metals with low melting points. Lead is used pure or strengthened by alloying with antimony for roofing, cable sheathing, radiation shielding, battery electrode and chemical plant materials. Both lead and tin alloys are used for bearings.

Alloys of lead and tin have melting points lower than those of either metal and may be further alloyed with antimony and bismuth to reduce the melting point still more or to confer other properties. Tin is used for coating steel and may be alloyed with lead for this purpose.

1.4.7.2 Nomenclature and standards

The standards and nomenclature for chemical lead are:

BS 334 Type A Pure lead
       Type B1 Copper lead
       Type B2 Copper tellurium lead
       Type C Antimonial lead
ASTM B 29 Pig lead

Standards for alloys and products are:

BS 3332 and ASTM B23 White metal bearing alloys (Babbitt metal)
ASTM B 32 Solder metal
BS 2920 Tin plate

1.4.7.3 Lead in corrosion service

Lead forms adherent corrosion products which may be sulphate, oxide, carbonate, chromate, or more complex compounds. These coatings protect the base alloy and have led to its use for roofing, underground pipes and products in contact with sulphuric, sulphurous, chromic and phosphoric acids.

The two main drawbacks of lead are its toxicity which has caused its withdrawal from all applications associated with potable liquids and its low creep strength.

Antimony additions increase both the ultimate tensile strength (UTS) and the creep strength of lead. The tensile strength of pure lead is about 15 MPa and the stress to cause 1% creep per year at 30°C is about 2.1 MPa. Lead containing 8% antimony has a tensile strength of 60 MPa and a stress to cause 1% creep per year at 30°C of 2.7 MPa.

1.4.7.4 Bearing metals (babbitts)

'Babbitt' bearing metals contain up to 90% tin with antimony, copper and in some cases lead, or up to 90% lead with antimony and tin. Their important characteristics are 'antiseizure' properties and 'fatigue resistance'.

Anti-seizure properties render the material readily wetted by oil but, should the oil film break down, the material will not adhere to a steel journal but flow out of the way locally and embed and cover any hard particle which may have gained access. Both lead and tin have these characteristics.

Fatigue resistance in a bearing implies that the material will, when the bearing is subjected to an alternating load, resist the formation of cracks that initiate at right angles to the bearing surface, propagate almost to the backing and then turn at right angles so that 'loose tiles' will erode away. Traditional tin based babbitts were considered to have greater fatigue resistance than lead based ones, but it is now appreciated that lead is superior if it is made thin enough, preferably down to a thickness of 0.125 mm.

Both tin and lead based babbitts are easy to cast, easy to bond to the backing and easy to machine. They are useful for low duty bearings made in short production runs. They have, however, been supplemented by other types of bearing for high duty service and mass production. Often a very thin layer of pure lead bonded to the surface of a stronger support material is used.

1.4.7.5 Low-melting-point alloys

Solders are alloys of tin and lead in varying proportions, for example 38%Pb, 62%Sn for tinman's solder and 66%Pb, 34%Sn for plumbers' solder.

There are a number of lead–tin alloys with bismuth or antimony the applications of which depend on their low melting points. The most important are the type metals which contain 12–30% antimony. Antimony expands on solidification and an alloy containing 20–30% of this metal has a negligible contraction and produces a clear type face.

1.4.7.6 Tin and terne plate

Electrolytic or immersion tin plate has been used to protect steel sheet; this is used mainly in the food industry. For products other than those associated with food, terne plate (which contains 10–25% tin, with the remainder being lead) is supplanting tin plate, largely due to cost but also because the lubricity of lead assists drawing and forming operations.

1.4.8 Cobalt and its alloys

1.4.8.1 Introduction

Cobalt is a soft silvery metal which is readily corroded by aggressive environments, but properties developed by alloying include a high coercive force, exceptional resistance to corroding and oxidising environments at ambient and elevated temperatures, exceptional hardness and wear resistance and high hot strength.

Cobalt can also be transmuted by irradiation with neutrons to ^{60}Co which emits 7×10^6 eV γ-rays; this is used for high-penetration radiography.

The major drawbacks of cobalt are its high cost and the distribution of its ores which are found in politically unstable countries. In addition, its alloys compare unfavourably in high temperature strength to those of nickel (see below).

1.4.8.2 Applications of cobalt alloys

The major application of cobalt is as a permanent-magnet material. Alnico 5 (24%Co, 14%Ni, 8%Ai, 3%La, 51%Fe) has, for example, a retentivity B_r of 12 500 G and a coercive force H_c of 550 Oe. Cobalt is also a constituent of high permeability magnets. Permendar has a saturation induction B_s of 24 500 G and a coercive force H_c of 2 Oe.

The next most important group of applications comprises those which depend on strength, hardness and corrosion resistance at room and elevated temperatures. These include gas turbine materials, furnace hardware and wear resistant and spring alloys. Almost all of these materials are based on an alloy containing 20–30% chromium to which are added other constituents conferring specific mechanical properties.

Because cobalt is a high-melting-point metal with no allotropic modifications, its prospects as a high creep strength material for aircraft turbines would appear highly favourable. Unfortunately, no strengthening mechanism comparable with the γ' precipitation process in nickel has been found. This deficiency might be overcome by the development of a creep resisting composite, but there is little incentive to develop such a material on a base whose supply might fail in an emergency. Nickel has therefore supplanted cobalt as a gas turbine material.

The deformation, corrosion and oxidation resistant characteristics of the cobalt superalloys, LG05, HS188, UMCo50 and stellite 250 have led to their extensive application in furnace hardware.

There are a number of dental and prosthetic alloys specified in ANSI/ASTM F75-76 for cast alloys and ANSI/ASTM F90-76, ANSI/ASTM F562-78 and ASTM F563-78 for wrought alloys. BS 3561: 1980: Part 2 covers both cast and wrought alloys. Cobalt alloys are superior for dental purpose to gold because of their higher strength and lower specific gravity. The decision as to which cobalt alloy to use for prosthetic purposes, or whether to use surgical stainless steel, titanium or tantalum instead depends on the preference of the surgeon.

Cobalt based wear resistant products are, in effect, alloys of Co, 20%Cr with tungsten carbide. They are available to a number of designations, depending on the wear and other characteristics required, and in forms which include castings, forgings, powder for compaction and hard facing consumables (see Bibliography).

The most comprehensive classification of cobalt alloys is that of the Australian Welding Research Association (AWRA).

Cobalt is also the basis for a number of alloys for springs ('Elgiloy' or 'Cobenium'), low expansion corrosion resistant alloy ('Stainless Invar') and alloys with low temperature coefficients of modulus of elasticity.

(See also Section 1.5.12.)

1.4.9 Other non-ferrous metals

There are some 28 non-ferrous metals not considered individually here. Many of them are used as alloying additions to the metals described in Sections 1.3.1 to 1.3.7. Only those which form the major constituents of alloys are considered here.

1.4.9.1 Metals used for alloying steels

Tungsten This has the highest melting point (3410°C) of any metal and the highest strength at elevated temperatures. The ultimate tensile strenth (UTS) (of 1 mm diameter wire) is approximately 2600 MPa at room temperature and 30 MPa at 2800°C. It has good corrosion resistance, but oxidises in air at temperatures above 500°C. It is normally fabricated by powder metallurgical techniques, but can be cast by vacuum arc or electron beam.

Its applications depend on its high temperature properties in protective atmospheres. They include the following.

(1) Electric light filaments—These are made by a powder metallurgy technique incorporating an oxide which restricts grain growth so that the filament consists of a bundle of single crystals each of which is continuous along the length of the filament.

(2) Electrodes in electron tubes.
(3) Electrodes for inert gas welding—for this and the previous application, thorium oxide may be incorporated to promote ionisation and the smooth striking of an arc.
(4) Electrical contact materials for highly repetitive and continuous arcing applications—Tungsten has outstanding resistance to arcing welding or sticking. It has, however, a tendency for the positive terminal to oxidise which can be prevented by substituting palladium or platinum for this terminal only.

Molybdenum This has many of the high temperature characteristics of tungsten, but its melting point (2610°C) is lower and its high temperature strength, although good, does not compare with that of tungsten. Like tungsten it oxidises in air above 500°C unless protected. It is, however, more ductile and more easily fabricated than tungsten. It is highly resistant to liquid media, including glasses and molten metals, but is attacked by oxidising agents. It is fabricated by consumable arc melting.

Its applications include the following:

(1) electrical and electronic parts,
(2) high temperature furnace parts (particularly for vacuum furnaces),
(3) glass-melting furnaces,
(4) hot working tools, and
(5) dies and cores for die casting.

Chromium This is a light silvery metal which is highly resistant to oxidation and to many corroding media, but not to hydrochloric acid. It is, however, extremely brittle (the production of ductile chromium has been reported but not followed up) and this restricts its use.

Applications include the following:

(1) mirrors,
(2) X-ray targets, and
(3) decorative and wear resistant electroplate.

The hardness of electroplated chromium depends on its hydrogen content and may be up to 1200 VPN as plated reducing to 70 VPN annealed. Chromium has a low coefficient of friction (0.12 sliding, 0.14 static) and can be plated in a porous form to retain oil.

Tantalum and niobium These have excellent fabricability, high hot strength and a low ductile brittle transition, but oxidise in air above 300°C. Tantalum is used for chemical-plant liners, surgical implants and, because of the electrical properties of its oxide film, for rectifiers and capacitors.

1.4.9.2 Precious metals

The precious metals have very high resistance to oxidation and corrosion. They include:

(1) the platinum group metals (platinum, palladium, iridium, rhodium, osmium and ruthenium) which have high melting points (1760–3050°C), high strengths and excellent dimensional stability at elevated temperatures; and
(2) gold and silver which have low melting points (1063 and 961°C, respectively) and are soft and ductile.

Precious metals are specified for applications which demand extreme reliability or absolute freedom from corrosion and those in which high recovery value and long trouble-free service offset a high original cost. Applications of specific precious metals include the following.

Silver:
(1) electrical conductors which, unlike those of copper, do

not oxidise at elevated temperatures and contacts;

(2) corrosion resistant containers for food processing (largely superseded by stainless steels);
(3) bearings;
(4) ornaments and jewellery (sometimes as plates); and
(5) high-melting-point solder.

Gold:
(1) instruments requiring corrosion resistance;
(2) conductors for transistor circuitry;
(3) dentistry;
(4) high-melting-point solder; and
(5) coinage, jewellery and ornaments.

Platinum (and to a lesser extent palladium):
(1) catalysts, particularly for automobile exhaust systems and chemical synthesis;
(2) furnace windings;
(3) laboratory crucibles and containers;
(4) resistance thermometers and thermocouples;
(5) linings for optical glass and fluoride process plant; and
(6) electrical contacts.

Iridium and rhodium are used mainly as alloying agents for platinum, but pure iridium has been used for the manufacture of high quality glass and rhodium as an electroplate of even higher resistance than platinum.

Osmium and its alloys have very high hardness (approximately 800 VPH), resistance to wear and corrosion and moduli. They are used for fountain pen nibs, record-player needles, instrument pivots and electrical contacts.

1.4.9.3 Nuclear metals

Nuclear metals are divided into 'fissile', 'fertile', 'canning' and 'control' materials.

Fissile metals undergo fission when irradiated by neutrons, disintegrating into two major fission products, a number of neutrons which serve to carry on the chain reaction, other particles and energy (including γ radiation). Fissile materials include ^{235}U a constituent of natural uranium, ^{233}U a product of neutron capture by thorium, and *plutonium* a product of neutron capture by ^{238}U the major constituent of natural uranium. They constitute the fuel in nuclear reactors.

'Fertile' metals include ^{238}U and *thorium*. They are incorporated into nuclear-reactor fuel or used separately in 'blankets' to absorb neutrons and produce additional fissile material.

Canning metals are used to contain nuclear fuel in a reactor, maintain its integrity and dimensions, protect it from attack by the coolant, retain fission products so that they do not contaminate the coolant and, through it, the environment, transfer the heat produced efficiently and absorb a minimum proportion of neutrons. Canning and core structural materials now in use include *stainless steel* for sodium cooled and high temperature gas-cooled reactors, *magnesium alloy* for the original 'Magnox' reactors, *zirconium* for pressurised water and boiling water cooled power reactors and *aluminium* for water-cooled research reactors.

Zirconium occurs naturally together with *hafnium*, which has high neutron-absorbing properties. These must be separated by a complex chemical process before they can be used in water reactors, zirconium as a core structural material and hafnium as a control rod material. Both have excellent resistance to pressurised water attack if they are suitably alloyed and satisfactorily pure.

Beryllium combines a very low nuclear capture cross-section with good strength and hardness at moderately high temperatures. It appeared to have great promise as a canning and core structural material, but the promise has not been fulfilled,

mainly because of its lack of ductility and resistance to environmental attack and partly because of doubts concerning the effect of helium which is produced when beryllium is irradiated by neutrons.

The applications of alloys based on beryllium are confined to those such as spacecraft where its high specific strength outweighs its high cost and hazard to health. (Its oxide causes 'beryllicosis', which is similar to silicosis but when inhaled beryllium is more virulent.)

1.4.9.4 Metals used in integrated circuits

Silicon and *germanium*, which when pure are very poor electronic conductors of electricity, can be transformed by 'doping'. Introducing into the lattice pentavalent elements (phosphorus, arsenic or antimony) introduces free electrons and gives rise to negative or n-type conductivity. Introducing trivalent elements (boron or aluminium) reduces the number of electrons and forms 'holes', giving rise to positive, or p-type, conductivity.

Junctions between regions of these two conductivity types are called p–n junctions. These junctions are at the heart of most semiconductor devices (diodes, transistors, solar cells, thyristors, light-emitting diodes, semiconductors, lasers, etc.).

By taking a slice of highly pure single-crystal silicon, diffusing into it p- and n-type atoms in a geometrical pattern controlled photographically, and then insulating or interconnecting regions by metallisation, circuits with millions of components can be formed on one silicon chip.

Highly pure, zone-refined single-crystal silicon has completely superseded germanium for the manufacture of transistors and silicon integrated circuits. The quantity used is small, amounting to only tens of tons per annum, but its technological importance is enormous.

1.5 Composites

1.5.1 Introduction

A 'composite' is a combination of two or more constituents to form a material with one or more significant properties superior to those of its components. Combination is on a *macroscopic* scale in distinction to alloys or compounds which are *microscopic* combinations of metals, polymers or ceramics.

Those properties that may be improved include:

(1) specific gravity;
(2) elasticity and/or rigidity modulus;
(3) yield and ultimate strength and, in the case of ceramics and concrete, toughness;
(4) fatigue strength;
(5) creep strength;
(6) environmental resistance;
(7) hardness and wear resistance;
(8) thermal conductivity or thermal insulation;
(9) damping capacity and acoustical insulation;
(10) electrical conductivity;
(11) aesthetics (attractiveness to sight, touch or hearing); and
(12) cost.

Not all these properties can be, or should be, improved at the same time, but the consideration which governs the choice of a composite is that a critical property has been adequately improved, while deterioration in other properties has not been significant.

Usually, but not always, a composite consists of a matrix which is relatively soft and ductile containing a filler which is harder but may have low tensile ductility.

The use of composites has persisted ever since tools of wood or bone (which are naturally occurring composites) were used by primitive humans. The earliest human-made composite was probably straw-reinforced mud for building. The Egyptians invented plywood, an early example of the improvement (which continues to the present day) of the natural composite, wood.

There are two ways of classifying composites: either according to the material of the matrix or according to the geometrical distribution of the components. Composites classified according to geometry include:

(1) particulate composites which are distributions of powder in a matrix;
(2) laminar composites which comprise layers of two or more materials; and
(3) fibrous composites which comprise a matrix which is usually relatively soft and ductile surrounding a network of fibres which are usually stronger but may be brittle relative to the matrix. The fibres may be short and their orientation effectively random or they may be long and carefully aligned. Composites with long fibre reinforcement are known as 'filamentary composites' and the fibres may be aligned on one, two or three directions.

Composites classified according to matrix include:

(1) fibre-reinforced or powder-filled polymers;
(2) concrete, reinforced concrete and prestressed concrete;
(3) wood and resin-impregnated wood;
(4) metal-matrix composites;
(5) fibre-reinforced ceramics and glasses;
(6) carbon-fibre-reinforced carbon.

Of these, reinforced concrete probably has the greatest industrial importance, but fibre-reinforced polymers have the greatest technological and engineering interest and the major part of this section is devoted to them. Some other classes of composite are described briefly and their properties and applications outlined..

1.5.2 Reinforcing fibres

Reinforcing fibres may be long (even continuous) and carefully aligned, or short with only the limited degree of alignment that is produced by flow during fabrication. The ability to control directional properties by the use of long fibres has opened up new fields in design, particularly for aerospace. Fibre material may be glass, ceramic, high-modulus polymer, carbon, boron or metal. Most types of fibre are manufactured and used in long lengths, but the cheaper materials may be used in short lengths. In the case of glass, the fibres are manufactured long but may be chopped. Some ceramic fibres, notably alumina, can be produced short or long, depending on the technique of manufacture.

The most widely used glass fibres are 'E glass', a non-alkaline aluminoborosilicate of average composition 54% SiO_2, 14% Al_2O_3, 8.5% B_2O_5, 18.5% CaO, 4% MgO and 0.8% rare-earth oxide. This material was originally developed for electrical insulation and represents a compromise betwen mechanical-property considerations and ability to form fibres from the melt. Fibres are made by feeding the liquid glass through a multitude of orifices to form up to 4000 filaments of approximately 20 μm diameter in one strand of roving. E-glass fibres have a specific gravity of 2.5 and a breaking strength of 2.6 GPa, which is higher than that of massive glass, notably because discontinuities are less severe. However, their modulus (86 GPa) is low and this makes it difficult to utilise

the high tensile strength in design because of the danger of buckling and the high elastic deformation.

Glasses with higher strength and slightly higher modulus are available, but it is more difficult to manufacture fibre from them. Quartz fibres have excellent high-temperature strength, but must be drawn individually from heated rod, and are, therefore, expensive.

E glass cannot be used to reinforce concrete because it is attacked by alkali. Glass in which the alumina is replaced by zirconia is more difficult to convert to fibre, but retains its strength much better in concrete and is marketed by Pilkington under the name of 'Cemfil', as a substitute for asbestos, which may no longer be used. The specific gravity of Cemfil is 2.5, its UTS is 3.6 GPa and its elastic modulus is 75 GPa.

The most extensively used oxide fibre is aluminia prepared by spinning or extruding an aqueous solution of a precursor, aluminium hydroxide/chloride, with a slurry of aluminia in suspension. Spinning produces relatively short staple fibre 2–4 cm long in blanket form; extrusion produces continuous filament. Small amounts of silica may be present. The spun or extruded fibre is converted into aluminia by heating and may have UTS up to 1.8 GPA and elastic modulus up to 380 GPa. Aluminia fibres have no application for resin-matrix composites, but their high-temperature strength makes them an attractive constituent of metal-matrix composites and they have been used both in the staple form and as whiskers.

Carbon and graphite fibres are made by spinning a precursor filament, usually polyacrylonitride (PAN), which is then heat treated while being subjected to tensile stress. The heat treatments include stabilisation, oxidation, pyrolysis and graphitisation at temperatures increasing from around 200 to 3000°C. Fibres with tensile properties ranging between 3.4 GPa UTS, 235 GPa elastic modulus and a breaking strain of 1.5% (for Courtaulds Graphil AXS) and 2.5 GPa UTS, 340 GPa elastic modulus with a breaking strain of 0.6% (for Graphil HMS) may be obtained in the more expensive grades by varying the temperatures and stress applied. The production of carbon fibres is illustrated schematically in *Figure 1.55*. Cheaper grades are available with inferior properties. Cheaper precursors for carbon fibre include cellulose and pitch of high aromatic low impurity content.

Boron filaments (boron/tungsten (B/W) fibres, 95–200 μm diameter) are manufactured by the chemical deposition of boron from a gaseous mixture of boron trichloride and hydrogen at between 1000 and 1300°C on a 12.5 μm diameter tungsten wire. Boron/carbon (B/C) fibres have been made

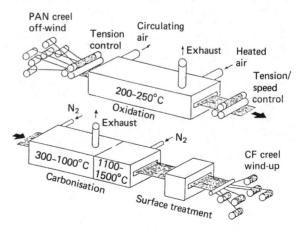

Figure 1.55 Schematic to show the method of production of carbon fibre. (Reproduced by permission of North Holland Publishing Company)

experimentally by deposition on carbon filaments. The tensile strength of B/W fibres ranges from 3.4 to 3.9 GPa, their elastic modulus is 390 GPa, and their specific gravity is 2.5. Although the ductile properties of boron fibres are no better than those of a ceramic they do not suffer long-term loss of strength and their creep properties are superior to those of tungsten. They are the strongest and stiffest fibres at present used for composite reinforcement.

Silicon carbide fibres are manufactured by spinning high-molecular-weight polycarbon silane fibres at between 200 and 300°C, curing in air at 200°C and heat treating at 1250°C. Depending on the grade of precursor, fibres of 10 μm diameter may have a UTS between 2 and 3 GPa and elastic moduli between 150 and 200 GPa. They have excellent high-temperature strength and useful electrical properties.

Paraorientated aromatic polyamide (Aramid) fibres are manufactured by spinning a liquid crystalline solution (or 'dope') of their sulphuric acid ester to form highly ordered domains of extended polymer chains. This structure confers a much higher modulus than is common in polymers and is the basis of the commercial exploitation of poly-p-phenylene teraphthalamide (Kevlar) fibres.

After spinning, the 'dope' is spun into fibres, coagulated in an aqueous bath at low temperature, stretched, washed, dried and heat treated at temperatures between 250 and 550°C. The resultant fibres have specific gravities around 1.45, a UTS of 2.64 GPa, elastic moduli of 59–127 GPa and breaking strains of 2.8–4.0%. Fibres with moduli up to 60 GPa can be made from polyolifins and polyoxymethylene.

Metal reinforcing fibres include carbon, alloy and stainless steels, tungsten, niobium and tantalum. The characteristics and properties of these materials are described in Section 1.3.

The properties of fibres and whiskers appropriate for use in composites are listed in *Table 1.52*.

1.5.3 Polymer matrices

The essential qualities of a matrix material are the ability to infiltrate among and bond strongly to reinforcing fibres. After these criteria have been met the matrix should set as quickly as possible to a strong heat- and environment-resisting solid. Polymers share with concretes the advantage over other possible matrix materials that they fulfil these requirements at a relatively low processing temperature. There are two classes of polymer: thermosetting and thermoplastic.

Thermosetting resins compounded with a hardener may be infiltrated between fibres while liquid and allowed to harden at room or elevated temperature. They include unsaturated polyesters, which are relatively cheap and easy to work, but do not bond well to fibres and have a relatively high shrinkage. They are used for large and comparatively low-duty composites, usually with glass reinforcement.

Epoxide resins are the most extensively used matrix materials for high duty carbon, boron and aramid fibres. They perform excellently at temperatures up to the range 160–200°C.

Thermosetting resins which have been used as matrices operating at higher temperatures include phenolics, phenol arakyls and the recently developed polyphenylene quinoxialine. Resins which are beginning to replace epoxides for high-temperature service with carbon reinforcement are bismaleides (BMI) and polyimides (PI) which have continuous-service capabilities of 200 and 300°C, respectively. (Some polyimides have survived short exposures to 760°C.) These polymers are however difficult to handle and polyimides in particular are expensive and require high cure temperatures.

Thermoplastic matrix materials are tougher than thermosets, have an indefinite shelf-life, the semifinished composite can be hot formed and, in some cases, thermoplastics have better high temperature and solvent resistance. However, the molten polymer has a higher viscosity than an uncured thermoset, fabrication temperatures are high and some are expensive. Many thermoplastics have been used ranging from the cheapest (nylon) to the highly expensive polyamide imide (PAI) and polyether–ether ketone (PEEK). PEEK composites have a maximum service temperature of 250°C, a work of fracture up to 13 times that of epoxide composites and significantly better fatigue resistance, but are expensive.

1.5.4 Manufacturing procedures for filamentary polymer composites

Filamentary composites are manufactured by 'layup' a term used for the positioning of the fibres and matrix to form the shape of the final component. Layup may be accomplished by 'pultrusion', 'winding' or 'laying', 'tow', 'tape', 'cloth' or 'mat'. In none of these forms are the fibres twisted to form a yarn. All forms of subassemblies can be obtained as 'prepregs' saturated with the resin which is later to form the matrix.

In *pultrusion* (see *Figure 1.56*) the reinforcing fibres are used to pull the material through a die.

In *winding*, impregnated single filaments, rovings or tapes are wound onto a former or mandrel. *Figure 1.57* shows a winding machine which may be computer controlled to produce any convex shape from which the mandrel can be removed. Filaments may be orientated according to the pattern of stresses that are to be withstood.

Cloth winding or laying utilises preimpregnated cloth which is deposited in the desired form and orientation. The bidirectionality and convolutions of the fibres in cloth make for lower precision in strength and stiffness. Cloth laying is therefore often used for filling where strength and stiffness are not critical.

Moulding can start with a deposition of precut layers of prepreg fibres which are compressed at elevated temperature to form the final laminate.

Continuous lamination is the application of pressure by rolling to bond layers of prepreg cloth or mats.

1.5.5 Properties of filamentary polymer composites

Filamentary composites consist in principle of 'laminae' which are assembled into 'laminates'. A 'lamina' is a flat or curved assembly of unidirectional fibres in a matrix. It is highly anisotropic having high stiffness and strength in the fibre direction and very low stiffness and strength transverse (see *Table 1.53*).

Laminae of varying orientations are therefore superimposed in a stack to form a 'laminate' with directional properties tailored to match the stress. Laminates are therefore essentially two-dimensional structures (the 'dimensions' may be curved when the component is a cylinder or sphere) and the mechanical properties in any of the principal directions of a laminate are inferior to those in the principal direction of one of the constituent laminae. In addition, the thermal stresses which arise on cooling from the curing temperature may impair strength.

Three-dimensional reinforcement, such as is employed in carbon–carbon composites (see Section 1.5.8) and reinforced concrete (see Section 1.5.11), is not normally applied to laminated plastics, and shear and transverse tensile stresses can result in delamination.

The matrix supports, protects, and distributes load among and transmits load between the fibres. If a fibre should break, the matrix, stressed in shear, transmits load from one broken end to the other and to adjacent fibres. Because boron or graphite fibres in a polymer matrix provide by far the greater proportion of strength and stiffness, composites with these

Table 1.52 Properties of fibres and whiskers appropriate for use in composites

Fibre	Type of filament	Whisker	Density (kg m⁻³)	Elasticity modulus, E (GPa)	Tensile strength, σ_u (MPa)	Specific strength, σ_u/ρ (km)	Specific stiffness, E/ρ (Mm)	Elongation, ϵ (%)	Composites in which used
		Alumina	38.80	380	18000	4.5	9.8		Metal matrix
Asbestos	Chrysotile		25.50	164	1000	0.4	6.5	2.5	Previously cement
	Crocidolite		33.70	196	3500	1.1	6	2.5	
Boron			25.20	390	3400	1.35	15		Polymer matrix
		Boron carbide	24.70	450	6700	2.70	18		Metal matrix
Carbon	High modulus		19.00	340	3500	1.8	18	0.5	Polymer
	Low modulus		19.00	235	2350	1.37	12	1.0	Polymer, metal and ceramic matrix
		Carbon	16.30	980	21000	13.00	60		
Cellulose			12.00	10	400	0.33	0.8	15	Concrete
		Copper	87.40	124	3000	0.34	1.4		Polymer matrix
Glass	E		25.00	86	3200	1.37	3.4	4.8	Polymer matrix
	Alkali resistant		27.00	75	2500	0.95	3	3.6	Polymer and concrete matrix
		Iron	76.80	200	13000	1.70	2.6		Polymer matrix
Kevlar	High modulus		14.40	133	2900	2.01	9	2.1	Polymer matrix
	Low modulus		11.40	69	2900	2.63	6	4	Polymer matrix
		Nickel	87.90	215	3900	0.44	2.4		Polymer matrix
Nylon			11.40	>4	850	0.75	0.35	13.5	Concrete matrix
Polypropylene			9.00	>8	400	0.44	0.9	18	Concrete matrix
		Silicon carbide	31.20	840	11000	3.50	27		Polymer and metal matrix
Steel	High tensile		78.60	200	2000	0.25	3.1	3	Concrete matrix
	Stainless		78.60	160	1700	0.22	2.5	3.5	Polymer matrix

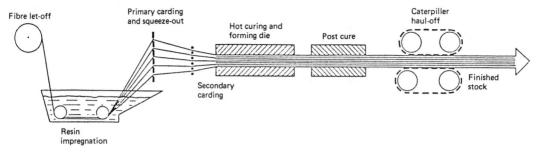

Figure 1.56 Pultrusion process in which the reinforcing fibres are used to pull the material through the die. (Reproduced by permission of *Metals and Materials*)

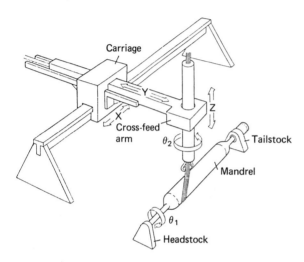

Figure 1.57 Gantry-type five-axis filament winding machine. (Reproduced by permission of *Metals and Materials*)

fibres can, in most cases, be considered to be linear elastic materials. In composites with glass or aramid fibres the lower modulus results in the matrix bearing a higher proportion of the load and the stress–strain relationship may depart from linearity.

Elastic and physical properties may, in the case of high-strength composites, be calculated using classical theory.[31]

Strengths are more difficult to calculate because the secondary stresses induced in a composite may exceed the transverse shear strengths and may themselves cause failure.

The parameters which must be taken into account in design include the following.

(1) Elastic properties:
    longitudinal stiffness $E_{1,1}$,
    Transverse stiffness $E_{2,2}$,
    In-plane shear modulus $G_{1,2}$,
    Poisson's ratio $\nu_{1,2}$.
(2) Strength properties:
    Longitudinal tensile strength $\sigma_{1,T}$,
    Transverse tensile strength $\sigma_{2,T}$,
    Longitudinal compressive strength $\sigma_{1,C}$,
    Transverse compressive strength $\sigma_{2,C}$,
    Yield strength σ_y,
    In-plane shear strength $T_{1,2}$.
(3) Physical properties:
    Specific gravity SG,
    Longitudinal thermal expansion coefficient α_1,
    Transverse thermal expansion coefficient α_2,
    Longitudinal thermal conductivity k_1,
    Transverse thermal conductivity k_2.

The specific strengths and moduli of fibrous composites and other engineering materials are illustrated diagrammatically in *Figure 1.58* (in this figure specific properties are derived by dividing the modulus or strength by the density and a gravitational term of 9.81).

The fatigue processes which occur in composites differ fundamentally from those in metals and, providing that they are well understood, offer very significant advantages to the designer. High modulus fibres such as carbon and boron confer excellent tension/tension fatigue properties, the fatigue stress at 10^7 cycles of longitudinal boron epoxy being only 15% less than the tensile stress. This is because the high modulus fibres limit the stress in the lower modulus matrix and so protect it from fatigue damage. However, in those plies in which fibres are orientated transverse to the principal cycle stress, the matrix is subjected to transverse tensile and shear stresses which cause cracking parallel to the plies and delamination. The effect of fibre orientation on carbon-fibre-reinforced polymer (CFRP) laminates is shown in *Figure 1.59*.

Only glass composites have a steep *S–N* slope, presumably caused by the diffusion of moisture which causes cracks to initiate in the glass fibres. Even so, the *specific* fatigue resistance of longitudinal fibreglass is far superior to that of any metal.

A further advantage of composites subjected to fatigue is that, whereas in metal fatigue there is, during the greater part of the life of a component, no superficial evidence of deterioration, there is in filamentary reinforced plastics a slow and progressive deterioration revealed at an early stage by a decrease in modulus or an increase in cracking in specific plies which is easily detectable by non-destructive examination.

This reduction in modulus could, if allowed to continue, lead to failure by buckling but, because of both the higher specific fatigue strength and the more obvious incidence of failure, catastrophic fatigue failures in filamentary composites are much less likely than in metals.

The assessment of the influences of impact on filamentary composites is more complex than metals because of their anisotropy and large number of failure mechanisms. Where, for example, in a jet engine a titanium blade will shear undamaged through the body of an intruding bird, a composite blade of equivalent strength will shatter. It can be stated that, in terms of impact strength, for composites the common fibres may be ranked in order of superiority: Kevlar 29, glass, Kevlar 49, boron, high tensile carbon, high modulus carbon.

The resistance to attack of polymers depends on the specific polymer and its environment. Traditional matrices based on polyesters, vinyl esters and epoxides perform very successfully in atmosphere, soil and many items of chemical plant. Protection may, however, be needed against degradation by ultraviolet radiation from sunlight. Some polymers, including the

Table 1.53 Properties of 60% fibre plies in epoxide laminates*

Property	E glass	S glass	Kevlar 49	HT-CFRP	HM-CFRP	Boron
Elastic moduli (GPa)						
E_{11}	37–50	55	77–82	140–207	220–324	210
E_{22}	12–16	16	5.1–5.5	9.8–10.0	6.2–6.9	19
G_{12}	4.5–6	7.6	1.8–2.1	5–5.4	4.8	4.8
ν_{12}	0.20	0.26	0.31	0.25–0.34	0.20–0.25	0.25
Strengths (MPa)						
σ_{1T}	1100–1200	1600–2000	1300–2000	1240–2300	783–1435	1240
σ_{2T}	40	40	20–40	41–59	21	70
σ_{1C}	620–1000	690–1000	235–280	1200–1580	620	3300
σ_{2C}	140–220	140–220	140	170	170	280
τ_{12}	50–70	80	40	80	60–70	90
ILSS	60	80	60	90–100	60–90	90
Strains to failure						
ϵ_{11}	2–3	2.9	1.8	1.1–1.3	0.5–0.6	0.6
ϵ_{22}	0.4	0.3	0.5	0.5–0.6	<0.7	0.4
ϵ_{11C}	1.4	1.3	2.0	0.9–1.3	—	1.6
ϵ_{22C}	1.1	1.9	2.5	1.6	2.8	1.5
Thermophysical						
SG	1.9–2.1	2.0	1.35–1.38	1.5–1.6	1.63	2.2
α_1 $(\times 10^{-6}\ K^{-1})$	6.3	3.5	−4 to −4.7	+0.4	−0.4 to −0.8	4.5
α_2 $(\times 10^{-6}\ K^{-1})$	30	29	60–87	25	27–32	23
k_1 $(W\ m^{-1}\ K^{-1})$	1.26	1.58	1.7–3.2	10–17	48–130	—
k_2 $(W\ m^{-1}\ K^{-1})$	0.59	0.57	0.15–0.35	0.7	0.8–1.0	—
Specific heat $(J\ kg^{-1}\ k^{-1})$	840	840	1260	840	840	1260

* Reproduced by courtesy of *Metals and Materials*.

fluoroplastics PTFE and PDF and polyether–ether ketone (PEEK), have exceptional resistance to radiation damage and may be used as matrices and as coatings.

1.5.6 Applications of filamentary polymer composites

The cost of glass-fibre-reinforced polymer (GFRP) is of a similar order to that of steel, aluminium or timber, and it is used where its lightness and corrosion resistance are advantageous, and where its fabrication methods are suitable for the specific component. Applications include small boats (and not so small minesweepers), roofing and cladding for buildings and many components for road and rail transport.

Other uses of GFRP are promoted by one or more specific property parameters. For example, GFRP is displacing steel for vehicle leaf springs on account of its lightness and fatigue resistance, and it is replacing porcelain and glass for electrically insulating components on account of its strength and insulating properties. In addition, GFRP is replacing steel for aqueous liquid vats, tanks and pipes because of its lightness, strength and corrosion resistance.

High performance composites are used in aerospace or sport applications, where the requirement for the specific stiffness and/or specific strength justifies the increased cost. The aerospace applications of carbon-fibre-reinforced polymer (CFRP) include the basic structures of spacecraft and commenced with ancillary fittings, floors and furniture of aircraft, but is now extending to major structural items such as stabilisers, tail-planes and fins. Future fighter aircraft will probably contain a high proportion of CFRP and will benefit from a reduced sensitivity to radar.

High performance sports goods are also increasingly made of CFRP because the reward of coming first in a race (or a fishing contest) far outweighs the additional cost of a CFRP racing car skin or a CFRP fishing rod compared with any conceivable alternative material, save possibly boron-fibre-reinforced composites.

The combination of a specific tensile strength of around 0.8 and a modulus of around 105 GPa m³ kg⁻¹ can only be obtained from boron-fibre-reinforced plastics. Boron fibres may be used by themselves or as a hybrid composite, part boron fibre, part carbon fibre for horizontal and vertical stabilisers, control surfaces, wing skins, flaps, slats, tail surfaces, spars, stringers, fuselage reinforcement tubes, spoilers, airhole flaps, doors, hatches, landing gear struts, helicopter rotor shafts and blades for military and civil airplanes and space shuttles.

The use of such materials (including aluminium matrix composites) can reduce weight by 12–45%, almost double service life, and reduce fuel usage and maintenance by about 10%.

Boron-fibre-reinforced composites are also used for the pick-up arms for hi-fi record-playing decks where specific stiffness is paramount.

The relative cost of glass, carbon, hybrid and boron reinforced plastics is 1, 10, 20 and 30, but the cost of the high strength, high modulus fibres is reducing with time.

The use of 'aramid' para orientated aromatic polyamids fibres has been restricted because their relatively low moduli (58.9–127.5 GPa) makes it difficult to take advantage of their high UTS (up to 2.64 GPa) in designs which may be buckling critical. They have been used for golf shafts, tennis racquets

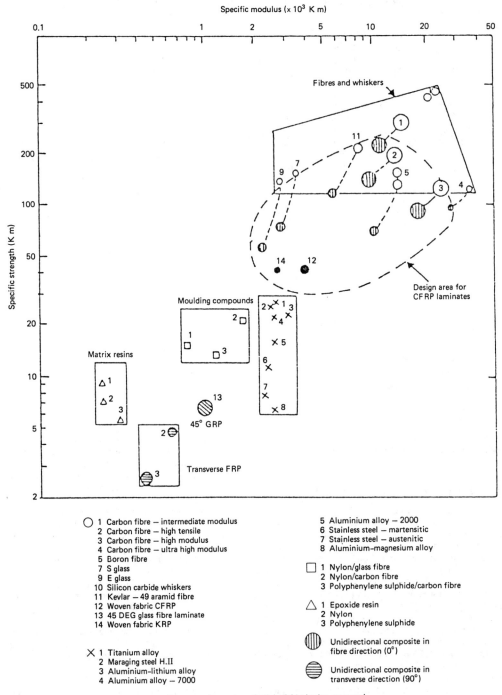

Figure 1.58 Specific strengths and moduli of composites and competing materials. (Reproduced by permission of *Metals and Materials*)

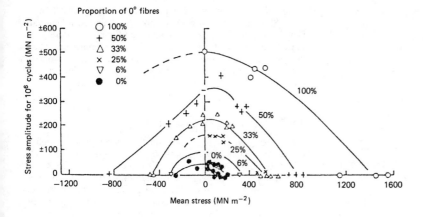

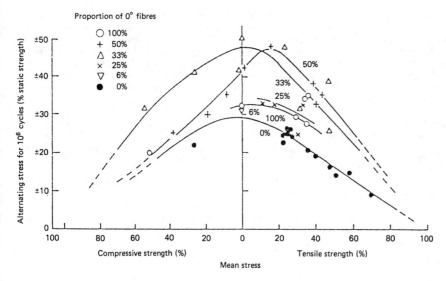

Figure 1.59 Maximum and minimum stress in fatigue cycling causing failure at 10^6 cycles in various CFRP laminates. (Reproduced by permission of *Metals and Materials*)

and boat hulls; Kevlar T950 has been used for tyres and Kevlar T956 for other rubber components.

1.5.7 Discontinuous fibre reinforced polymer components

1.5.7.1 General

Discontinuous fibres of an average length in the region of 380 μm may be incorporated in proportions up to about 25% by volume in mouldable polymers to enhance their stiffness, strength, dimensional stability and elevated temperature performance.

Reinforced thermoplastic materials (RTPs) may be shaped by melt fabrication techniques, injection moulding, extrusion, blow moulding and thermoforming. The material is melted or plasticised by heating, shaped in the plasticised condition and cooled to resolidify. Reinforced thermosets may be made to flow in the precured state and cured or cross-linked to an infusible mass in the hot mould.

'Commodity' thermoplastics, polyolefins, polystyrene, polyvinyl chloride, etc., are used mainly in the non-reinforced form but are marketed in the fibre-reinforced form. A much higher proportion of engineering thermoplastics, polyamides, polyacetyls and thermoplastic polyesters are reinforced, usually with short glass fibre and the specialised high performance thermoplastics such as polysulphones are also reinforced often with short carbon fibre. Short-glass-fibre reinforcement is used for thermosets such as phenolic, amino and melamine formaldehyde resins which may be injection moulded, although the curing time lengthens the manufacturing cycle.

An important class of composite is the long-fibre-reinforced sheet-moulding compounds (SMC) and the dough-moulding compounds (DMC) based on unsaturated polyester, vinyl ester and epoxide resins. These materials are normally compression moulded (see *Figure 1.60*) and have to compete with steel pressings. Similar composites are based on thermoplastics which are produced as sheets which are heated and then pressed between cold dies.

Two materials are used for discontinuous fibre reinforcement: short and long staple glass fibre, and short staple carbon

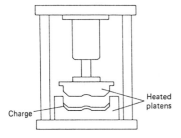

Figure 1.60 Press moulding arrangement for discontinuous fibre reinforced plastics. (Reproduced by permission of *Metals and Materials*)

fibre. Aramid fibres have the required properties but polymers compounded with them are not yet obtainable commercially.

Discontinuous fibre reinforced plastics cost less to fabricate than the corresponding filamentary reinforced materials, but their mechanical properties are significantly inferior. This is because the rule of mixture that is obeyed precisely insofar as modulus is concerned, and approximately insofar as yield strength and UTS is concerned, for high modulus continuous fibres is not obeyed for discontinuous fibres. The strength of short fibre reinforced polymers is controlled by a complex series of interactions between the fibres and the matrix.

The fibre–matrix interface is usually the weakest link. In aligned fibres the end becomes debonded at quite low loads and the debonding spreads along the fibre as the load increases. Debonding reduces the stiffening efficiency of the fibre and constitutes a microcrack which may extend into the matrix.

The mechanical strengths of typical short and woven fibre reinforced thermosets are listed in *Table 1.54*. *Table 1.55* lists the mechanical properties of short-fibre-reinforced thermoplastics, which includes some carbon-fibre-reinforced ma-

terials. The superiority of filamentary reinforcement is evident. Short-fibre reinforcement shows to even less advantage in fatigue, creep and impact loading and is not to be recommended for highly stressed parts. Short-fibre reinforcement is, however, much cheaper than filamentary reinforcement and is used extensively for a wide variety of domestic, architectural, engineering, electrical and automotive components.

1.5.8 Carbon–carbon composites

Carbon–carbon composites retain their strength to a higher temperature than any competitive material (see *Figure 1.61*). They are unique in that the matrix is identical in composition to that of the reinforcing fibres. They differ from the polymer composites already described in that the matrix, which can exist in any number of quasi-crystalline forms from 'glassy' or amorphous carbon to graphite, has low strength and negligible ductility.

While, therefore, single and bidirectionally reinforced carbon–carbon composites are manufactured, the need to avoid delamination has promoted three-dimensional reinforcement.

Complex weaving equipment has been developed to achieve multilayer locking by means of structures such as those shown in *Figure 1.62*. Even more complex patterns are employed. As an alternative to three-dimensional or eleven-dimensional weaving, the directional reinforcement may be produced by fabric piercing. Arrays of layers of two-directional fabric are pierced with metal rods or needles. The metal needles are withdrawn and replaced by yarns or by precured resin yarn rods. Fabric piercing is versatile and can produce a higher overall fibre volume and a higher preform density than weaving.

Other techniques for producing multidirectional structures involve the assembly of rod consisting of yarns prerigidised with phenolic resins by pultrusion. These can be used to form 'four-directional' tetrahedral structures or, by incorporating a filament winding operation, a cylindrical structure.

Table 1.54 Properties of short-fibre- and woven-fibre-reinforced thermosets*

Property	DMC†	SMC‡	Glass fibre polyester	Glass fibre epoxide	Woven CF epoxide	Woven Kevlar epoxide
Stiffnesses (GPa)						
E_{11}, E_{22}	11	12–13	17–21	23–26	70	31
ν_{12}	0.11	0.11	0.11–0.12	0.12–0.16	0.08	—
Strengths (MPa)						
σ_{1T}, σ_{2T}	60–69	75–120	303	379–517	586–620	517
σ_{1C}, σ_{2C}	138	179–193	276	345–413	690	83
σ_{f}	103	138–172	214	517–624	841–1034	345
τ_{1LSS}	13.8	17–28	24	28	55–67	55
Izod impact (J m⁻¹)	430–640	640–850	750–960	1600	—	—
SG	1.65–1.80	1.7	1.7–1.8	1.8–1.9	1.59	1.33
$\alpha_1, \alpha_2 (\times 10^{-6}\ \mathrm{K}^{-1})$	18–31	22–36	10–16	10.6	3	0
$k_1, k_2 (\mathrm{Wm}^{-1}\ \mathrm{K}^{-1})$	0.1–0.23	0.6–0.22	0.16–0.20	0.16–0.33	—	—
Specific heat (J kg⁻¹ °C⁻¹)	850	850	850	850	—	1260

* Reproduced by courtesy of *Metals and Materials*.
† 15–25% glass.
‡ 30–40% glass.

Table 1.55 Mechanical properties of short-fibre-reinforced thermoplastics*

1B Polymer	2B Glass fibre content (w%)	(v%)	9 Water absorption (max.) (%)	10 Flexural modulus (GPa)	11 UTS (MPa)	12 Tensile elongation (%)	13 Notched Izod impact (J m⁻¹)
Polyethylene (HD)	20	9	0.1	4.0	55	2.5	50
Polyethylene (HD)	40	20	0.3	7.5	80	2.5	70
Polypropylene	20	8	0.02	4.0	63	2.5	75
Polypropylene (chemically coupled)	20	8	0.02	4.0	79	4	90
Polypropylene (chemically coupled)	40	19	0.09	7.0	103	4	100
Nylon 6	40	23	4.6	10.5	180	2.5	150
Nylon 6.6	20	10	5.6	9.0	130	3.5	100
Nylon 6.6	40	23	3.0	15	210	2.5	136
Nylon 6.10	40	22	1.8	9.0	210	2.5	170
Nylon 11	30	15	0.4	3.2	95	5	—
Acetal homopolymer	20	12	1.0	4.3	60	7	40
Acetal copolymer	30	19	1.8	9.0	90	2	40
Acetal (chemically coupled)	30	19	0.9	9.7	135	4	95
Polystyrene	40	22	0.1	11.3	103	2.5	60
SAN	40	22	0.28	13.4	128	2.5	60
ABS	40	22	0.5	7.6	110	3.5	70
Modified PPO	40	22	0.09	8.6	135	3.5	80
PETP	30	18	0.24	8.3	130	4	85
PBTP	40	26	0.4	9.6	150	4	155
Polysulphone	40	26	0.6	11.0	138	2	100
Polyethersulphone	40	26	—	11.0	205	—	80
PPS	40	26	0.06	12.5	160	3	80
Polycarbonate	20	10	0.19	5.8	110	6	180
Polycarbonate	30	17	0.18	8.2	127	5	190
Polycarbonate	40	24	0.16	10.3	145	4	200
Carbon fibre filled materials							
Nylon 6.6	30	21	2.4	20.0	240	3.5	75
PETP	30	24	0.3	13.8	138	2.5	60
Polysulphone	30	24	0.4	14.0	158	2.5	60
PPS	30	24	0.1	16.9	186	2.5	55

* Reproduced by permission of North Holland Publishing Co.

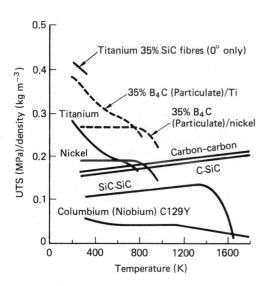

Figure 1.61 Strength-to-density ratio for several classes of high temperature materials with respect to temperature. (Reproduced by permission of *Metals and Materials*)

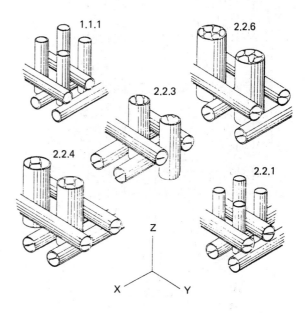

Figure 1.62 Three-dimensional orthogonal weaves for carbon–carbon composites. (Reproduced by permission of North Holland Publishing Company)

Densification of the structure with carbon is achieved either by impregnation with pitch, a thermosetting phenol or furfural type resin or by depositing carbon from a hydrocarbon (CVD process). The preform may be impregnated with liquid by a vacuum process, carbonised at 650–1100°C at low pressure and then graphitised within the range 2000–2750°C. The cycle is repeated until the desired density is achieved.

Alternatively, the preform may be impregnated with pitch, carbonised and then graphitised at high pressure and temperature in a furnace; and the cycle repeated as required. In this process the work piece must be isolated from the pressure vessel in a furnace of the type shown in *Figure 1.63*.

Impregnation by carbon by the CVD process is carried out by feeding hydrocarbon gas through and into the pores of the preform, isothermally, under a thermal gradient or under differential pressure. Carbon is deposited at 1100°C and, in this case, as in impregnation with a thermosetting resin, a carbon rather than a graphite matrix is formed.

The tensile properties of carbon–carbon composites with various matrices are listed in *Table 1.56*.

The application of carbon–carbon composites has so far been restricted by high cost and their susceptibility to oxidation at temperatures above 400°C. Coatings to protect against oxidation are under development. Their most important application so far has been as rocket nozzles, thrust chambers, ramjet combustion lines and heat shields for space vehicles.

They are used commercially for aircraft brake systems for Concorde and military aircraft. They are also used for hot-pressing moulds. They can be used for very high temperature heat shields and elements for vacuum furnaces.

The high temperature strength of carbon–carbon composites will favour a large number of uses if their cost is reduced.

1.5.9 Fibre-reinforced metals

The potential of fibre-reinforced metals is so great that they have been declared a strategic material in the USA. The fibres may be:

(1) whiskers, usually silicon carbide, made by pyrolysis of rice hulls;
(2) discontinuous fibres, of alumina and alumina silica, often packed together as insulation blankets;
(3) continuous fibres, of boron, silicon carbide, alumina, graphite, tungsten, niobium zinc and niobium titanium.

Most engineering metals could be used as a matrix for a composite. Matrices of titanium, magnesium, copper and superalloys are the subject of investigation, but almost all the applications so far recorded have used an aluminium matrix.

Typical properties of metal-matrix composites so far investigated are shown in *Table 1.57*. The fabrication techniques used include conventional and squeeze casting, powder methods including hot moulding and isostatic pressing, diffusion bonding and vapour deposition. A major problem is to prevent damage or dissolution of the fibres during the manufacturing process, while producing a good metallurgical bond with the matrix. Surface treatments are sometimes employed to promote one or both of these objectives.

Silicon carbide particles and whiskers are given a special surface treatment which promotes wetting by aluminium. After casting, the composite may be worked by any of the conventional methods.

Powder metallurgy techniques are best suited to the manufacture of particle-filled composites but, provided the pressing operation is designed to avoid damage to the fibres, either by carefully controlled direction of the pressure, or by hot hydrostatic pressing it can be used to make fibrous metal matrix composites.

In 'squeeze casting', pressure is used to force molten metal into the interstices of fibre preforms which have preferably been evacuated. The production of a piston by this process is illustrated in *Figure 1.64*.

Diffusion bonding has been used to fabricate boron fibre reinforced aluminium by the process illustrated in *Figure 1.65*, but it can also be used for magnesium or titanium. Fibres of silicon carbide or aluminium may be coated with aluminium by vapour diffusion or by passing through molten aluminium, and bundles of coated files may then be compacted by rolling, swaging or hot hydrostatic pressing.

Almost all the manufacturing processes for metal matrix filament composites are expensive and most of the applications have so far been limited to space technology. The aluminium–boron composite described earlier has been used for tubular spars for the space shuttle. An antenna boom is being built for the NASA space telescope from graphite fibre reinforced aluminium which has the advantage of a very small thermal expansion coefficient and a good thermal conductivity.

1.5.10 Fibre reinforced glasses and ceramics

Fibre reinforcement of glasses and ceramics is an attractive concept because, in theory, it should eliminate the problems

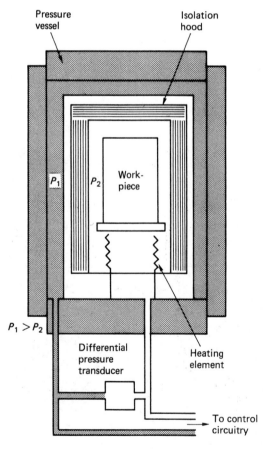

Figure 1.63 Isolation hood and differential pressure equipment for HIPIC processing. (Reproduced by permission of *Metals and Materials*)

Table 1.56   Typical mechanical properties of carbon–carbon composites*

| Property | CVD pyrocarbon | Matrix | | |
		Pitch	Pitch	Phenolic resin
Heat-treatment temperature (°C)	2000–2500	1700	2500	1000
Bulk density (g cm^{-3})	1.6	1.6–1.65	1.6–1.9	1.73
Flexural strength (MN m^{-2})	250	200–240	150–200	220
Tensile strength (MN m^{-2})	180		240	300
Young's modulus (GN m^{-2})	90	70–90	75–95	65
Interlaminar shear strength (MN m^{-2})	8	12–15	12–15	12

* Reproduced by courtesy of *Metals and Materials*.

Table 1.57   Representative properties of metal-matrix composites*

| Matrix | Reinforcement | Reinforcement (vol. %) | Modulus (GPa) | | UTS (MPa) | |
			Longitudinal	Transverse	Longitudinal	Transverse
Aluminium	None	0	70	70	280–490	280–490
Epoxy	High-strength graphite fibres	60	147	10.5	1260	35
Aluminium	Alumina fibres	50	203	154	1050	175
Aluminium	Boron fibres	50	203	126	1530	105
Aluminium	Ultrahigh modulus graphite fibres	45	350	35	630	35
Aluminium	Silicon carbide particles	40	147	147	560	560
Titanium	Silicon carbide monofilament fibres	35	217	188	1750	420

* Reproduced by courtesy of *Metals Engineering*.

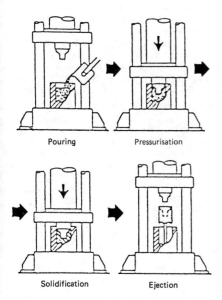

Figure 1.64   Production of a piston with fibrous inserts by squeeze casting. (Reproduced by courtesy of *Metals Engineering*)

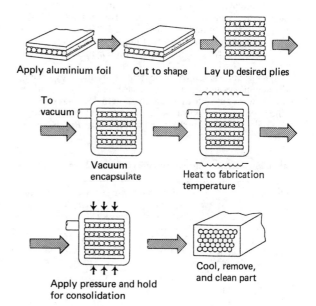

Figure 1.65   Aluminium–boron fibre composite fabrication by diffusion bonding. (Reproduced by courtesy of *Metals Engineering*)

associated with 'static fatigue' and allow significant tensile stresses to be imposed.

In practice, however, the behaviour of a not very ductile fibre such as carbon or boron in a brittle matrix is not such as to confer significant advantage, and the high fabrication temperature required restricts the choice of fibre and increases cost.

For these reasons, the only composite of this class to have achieved significant commercial application has been glass reinforced with steel-wire mesh which continues to present a barrier after the glass has been shattered.

Further information on the principles and technique of fibre reinforcement of ceramics is to be found in references 32 and 33.

1.5.11   Reinforced concrete

Concrete and mortar constitute, in terms of volume, the most important matrix materials for human-made composites. They are particulate composites.

'Concrete' comprises a matrix of hydrated Portland cement (other cements have been used) surrounding mineral particles, usually silica in the form of sand and aggregates. 'Fine' aggregates are limited in size to 5 mm and 'coarse' aggregates to between 5 and 20 mm. 'Mortar' is hydrated Portland cement and sand.

Portland cement should conform to BS 12. The sampling and testing of aggregates is described in BS 812 in conjunction with BS 882.

The strength of concrete and mortar depends on the proportion of cement to aggregate and sand and the ratio of the volume of water added to the volume of cement, because this controls the proportion of voids.

Depending on arriving at the best optimisation, the compressive strength of concrete may vary up to 120 MPa; 50 MPa is readily obtainable. The tensile strength of concrete is affected by slow crack growth and must therefore be assessed by the Weibull techniques referred to in Section 1.6. A reasonable working estimate is 5 MPa.

A material, 'macro defect free (MDF) cement' has been made by removing macroscopic flaws during preparation of cement paste having flexural strengths between 60 and 70 MPa and compressive strengths greater than 200 MPa.[34]

The addition of polymers further improves concrete. There are two types: polymer-impregnated and polymer-added concrete.

Polymer-impregnated concretes (PICs) are made by drying and vacuum/pressure impregnating hardened concrete with a liquid monomer such as, for example, methylmethacrylate to fill the voids and then polymerising the monomer by radiation, thermal or promoter catalysis. The strength of PIC is about four times that of normal concrete, 200 MPa compressive and 20 MPa tensile strength being obtained. However, the material is more prone to brittle failure. Water permeability, water absorption and chemical attack are also reduced, but PIC is expensive and its commercial application is therefore limited.

Polymer-added concrete is prepared by the addition of a polymer or monomer during the mixing stage. The increases in strength are not as great as those of PIC, compressive strengths being limited to about 100 MPa, while tensile strengths up to 18 MPa are reported.

Polymer-added concretes have increased resistance to abrasion and chemical attack and bond well to existing concrete. Since the increase in cost is only the cost of the materials, polymer-added concretes have extensive applications for items such as floors which are subject to heavy wear.

Evidence is accumulating as a result of studies on biomechanics[35] that comparatively high values of modulus and toughness can be achieved in such materials as nacre (mother of pearl) and antler by geometrical arrangements of calcium carbonate and small quantities of organic material.

It is conceivable that high modulus, high strength and high toughness might be achieved at relatively low cost by combining high strength high modulus fibres, an MDF cement matrix and a thin, tough, flexible polymer or elastomer interlayer which would bond both to the fibres and the matrix.

Concrete, like carbon, is a brittle material and benefits from three-dimensional reinforcement by both short and long fibres, as shown (for both concrete and mortar) in *Table 1.58*.

Chopped steel, glass and polypropylene have been used for precast concrete sections and steel for *in situ* concrete. Flexural reinforcement can be achieved by adding 2% of random steel fibre to concrete. The main problem is to obtain an adequate dispersion of this concentration. The mix must contain 50% of fines and if vibration compacting is used care must be taken to avoid unacceptable fibre alignment.

Polypropylene (0.44%) has been used successfully for pile sections. Glass fibre and steel bar and wire are used for filamentary reinforcement. Concrete pipes are manufactured with 1% wound glass fibre concentrated at the inner and outer sections.

Reinforced concrete contains a three-dimensional network of steel bar and/or wire aligned to resist tensile stress. Compression stresses are resisted by the concrete, but those regions of the concrete near to a steel bar which is stressed in tension are subjected to tensile stress during operation.

The versatility of reinforced concrete is illustrated by its use in the Thorpe railway suspension bridge. This bridge actually contains more steel than would have been used in a steel-girder bridge, but it was cheaper steel, and does not require painting.

Table 1.58   Fracture strengths and work of fracture of cement–mortar composites*

Material	Flexural cracking strength ($MN\,m^{-2}$)	Ultimate flexural strength ($MN\,m^{-2}$)	Work of Fracture, γ_F ($kJ\,m^{-2}$)
Carbon fibre/cement	30–50	130–185	2–8
Steel fibre/cement	6–12	6–17	2–4
Asbestos/cement	—	17–38	—
Glass/mortar	5–8	7–17	0.6–1.0
Polypropylene/mortar	~5	~5	—
Plain concrete	~5	~5	~0.03

* Reproduced by courtesy of Applied Science Publishers Ltd.

In prestressed concrete steel tendons may be stressed, while the concrete is poured over them, and released to compress the concrete after it has cured. More frequently, the tendons pass through channels or holes in the concrete and are stretched, and the ends secured after the concrete has cured. In correctly designed structures of prestressed concrete the whole of the concrete should be in compression. Prestressed concrete structures can be designed to have much lighter sections than reinforced concrete.

Corrosion of the steel reinforcement can be a serious problem with both reinforced and prestressed concrete. A sound layer of concrete (25 mm thick) will protect steel from corrosion, but cracks which may form under either tensile or compressive loading may allow water, which is very likely to contain salt, to gain access to the steel reinforcement. If the steel rusts it will increase in volume and eventually cause disintegration of the concrete.

The techniques for improving the fracture strength of concrete (discussed earlier) should prove extremely beneficial, not only in preventing failure due to corrosion, but also by allowing reinforced and prestressed concrete structures to be designed to higher stresses, in opening up new applications in the field of mechanical as well as civil engineering.

1.5.12 Particulate composites

The most important metal-matrix particulate composites are cemented carbides or hard metals. These are cermets, consisting of finely divided hard particles of carbide of tungsten, usually accompanied by carbides of titanium or tantalum, in a matrix, usually cobalt, but occasionally nickel and iron.

Hard metals have the high elastic moduli, low thermal expansions and low specific heats of ceramics combined with the high electrical and thermal conductivities of metals. They are hard, have high abrasive wear and corrosion resistance, good resistance to galling, and good friction properties. In addition, compared with ceramics, hard metals are ductile, having fracture strengths about 1000 MPa and work of fracture γ_f of about 250 J m^{-2}.

Hard metal dies and tools are manufactured by a powder route. A 'green' or partially sintered compact is machined into shape and then sintered in hydrogen atmosphere at a temperature approaching (or even reaching) the melting point of the matrix metal. After sintering (or solidification) cobalt occupies the interstices between the grains as an almost pure metal with its original ductility. If nickel or iron is used as binder, more tungsten carbide is dissolved than with cobalt as binder and the ductility of the resulting composite is impaired. Increasing the percentage of cobalt increases ductility but decreases hardness, modulus, resistance to wear, galling and crater formation.

Cemented carbides with 3% cobalt have a hardness (HV 500 g) above 1900, a flexural strength of 2200 MPa and an elasticity modulus of 675 GPa. Increasing the cobalt percentage to 25% decreases hardness to 950 and the modulus to 462 GPa, but increases flexural strength to 3200 MPa.

In general, high carbide versions, particularly those with added titanium carbide, are used for finishing cuts. Medium carbide content materials are used for roughing cuts and low carbide content materials for high impact die applications. Tantalum carbides are used for applications involving heat.

Tungsten carbides may be used in oxidising conditions up to about 550°C and in non-oxidising conditions up to about 850°C. Titanium carbides can be used at temperatures up to about 1100°C.

The application of hard metals for wear resistant cutting tools is now being challenged (in the absence of shock) by ceramics such as alumina and sialon.

British Standards for hard metals include:

BS 3821: 1974 Hard metal dies and associated hard metal tools.
BS 4193: 1980 Hard metal insert tooling.
BS 4276: 1968 Hard metal for wire, bar and tube drawing dies.

A new and important development in metal-matrix particulate composites is the reinforcement of aluminium by silicon carbide particles. A level of 40 vol.% silicon carbide in aluminium doubles the modulus, halves the coefficient of thermal expansion, increases tensile strength and greatly reduces frictional wear. This composite is available as sheet, extrusions, forgings and castings.

The field of metal-in-metal particulate composites is, by comparison, restricted. Examples are the additions of lead to steel to promote machinability and to copper to produce a bearing material. These applications are both giving place to other materials. In particular, bearings are made of porous sintered bronze impregnated with PTFE.

1.5.13 Laminar composites

The number of laminar composites is vast and defies classification.

Metal–metal laminates usually comprise a substrate that provides strength but reduces cost, with a surface material that resists environment or improves marketability. Examples include rolled gold, Sheffield plate and electroplate, tin plate, galvanised iron or titanium sheathed steel.

Alternatively, the core may be ductile and the surface hard enough to provide a cutting edge, e.g. a Damascus sword blade.

Wrought iron consists of layers of iron and slag which confers corrosion resistance (and solid phase weldability).

Glass is laminated with transparent plastic for automobile wind shields.

A most important class of laminar composites is the sandwich (lightweight) structure which comprises two high-strength skins which may be metal, wood, plastic or cardboard, separated by a core which may be basically lightweight such as balsa wood or may be foam or of honeycomb construction.

1.5.14 Wood and resin-impregnated wood

Wood is a natural composite and is one of the oldest, if not the oldest, composite used by humans. It is reinforced by a system of parallel tubes constructed of cellulose fibres which confer longitudinal properties such as those shown in *Table 1.59*. This structure has developed by natural selection in such a way as to ensure that failure of one element does not interact with an adjacent element in such a way as to lead to progressive failure. The low specific gravity of wood gives it specific strengths comparable with steel. Its transverse properties are very poor and its very easily splits longitudinally. Provided, however, that this is allowed for in design, timber structures, such as, for example, the Lantern Tower at Ely, may be designed to support heavy loads for many centuries.

Plywood was developed by the ancient Egyptians to provide strength in two directions and to prevent warping.

The next advance was to impregnate laminated wood with resin. This can be achieved in two ways.

(1) Veneers of softwoods such as Douglas fir are impregnated with epoxy resin, laid up in a female mould and cured under a vacuum bag. *Figure 1.66* shows a section near the root end of an aerogenerator blade made by joining two half sections made in this way. Design

Table 1.59 Mechanical properties of raw materials used for wood–resin composites

	Douglas fir	Dry yellow birch	West system epoxy
Specific gravity	0.52	0.62	1.14
Compressive strength (MPa)	48.2*	56.8*	96.5
UTS (MPa)	103.4*	138*	62
Elastic modulus (GPa)	14	14.3	2.06

* Parallel to grain.

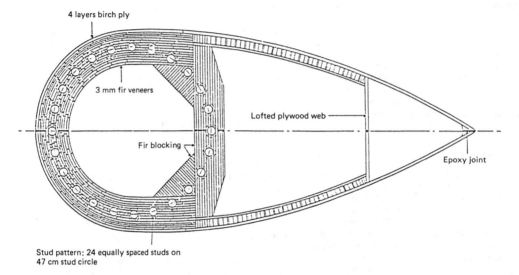

Figure 1.66 Section near the root end of an impregnated wood aerogenerator blade

allowables (based on wooden propeller blade experience) are given in *Table 1.60*. Tests on prototype aerogenerator blades have indicated that the mechanical properties are adequate and, in series production, the blades would be cheaper than any other material (with the possible exception of prestressed concrete).

(2) Compressed impregnated wood (as manufactured by Permali, Gloucester) involves laying up birch or beech veneers interleaved with phenolic resin and bonding at high temperature and pressure. This product is a highly weather resistant electrical insulating material with mechanical properties as listed in *Table 1.61* and a substantially flat *S–N* curve with fatigue strength better than 90 MPa at 10^9 cycles.

Table 1.60 Design allowable for wood laminates*

Strain type	Static allowables		4×10^8 cycles Fatigue allowables	
	'One time'	Working	$R = -1$	$R = 1$
Parallel to grain				
Flexural (MPa)	39.25	32.87	16.69	39.25
Tensile (MPa)	34.43	28.84	16.69	34.43
Compressive (MPa)	30.21	25.30	16.69	30.21
Shear (MPa)	4.71	3.94	2.11	4.71
Perpendicular to grain				
Tensile (MPa)	1.15	0.97	0.47	1.15
Compressive (MPa)	9.92	8.25	3.81	9.92
Rolling shear (MPa)	0.70	0.59	0.32	0.70

* Douglas fir laminae; values corrected to 8% moisture, 49°C except $R = -1$ (room temperature).

Table 1.61 Mechanical properties of permali impregnated compressed wood

Property	Longitudinal grain	75% longitudinal grain	Equal longitudinal cross grain
Cross-breaking strength (MPa)	190	150	85
Compressive strength			
Parallel to laminae (MPa)	170	170	140
Perpendicular to laminae (MPa)	105	190	205
Shear strength			
Endwise (MPa)	20	34	48
Flatwise (MPa)	62	55	48
Elastic modulus (GPa)	17.2	15.2	13.8

1.6 Engineering ceramics and glasses

1.6.1 Introduction

Ceramics These are inorganic crystalline materials, low in ductility and high in melting point which are usually fabricated not by melting but by processes involving powder compacting and sintering at very high temperatures. They are usually compounds of metals with non-metals and they owe their high temperature resistance to deformation to directed covalent or electronic bonds between the constituent atoms which do not permit plastic flow at operating temperatures too low for diffusion to occur.

Glasses These have analogous compositions to ceramics but have lower melting points and retain their amorphous liquid structure on cooling. They have, instead of a melting point, a glass transformation temperature above which their viscosity is low so that they can be melt formed, and below which the viscosity is high so that they have effective shape retention equivalent to a solid.

Glass ceramics Also known as 'melt-formed ceramics', these can be melt formed as glasses but crystallise on a micrometer or submicrometer scale at a lower temperature.

Single-crystal materials These are special products with a uniform structure.

Machinable ceramics These fall into two types. One type has plate-like structures which can be readily cleaved and the other consists of particles that are only weakly bonded together so that they can be readily chipped apart. These, unlike other ceramics that have to be ground with an abrasive, can be machined with conventional tools.

1.6.2 Standards

Most national standards are highly oriented to testing or application. Only in the field of electrical insulation has any systematic classification been attempted in IEC 672: 1980, which is to be the basis for BS 6045. So far as is possible, the IEC classification is followed in this section.

1.6.3 Clay based ceramics

1.6.3.1 Technical porcelains (IEC classes 1XX)

These are ceramics based on clay and other silicates with added alumina. Fine-grain versions (IEC 1XX, 2XX and 4XX) are used primarily for electrical purposes, but coarser grain size chemical stoneware and kiln furniture are also included. These are well vitrified and, therefore, non-porous.

IEC class C110—porcelains with <30% alumina These consist of quartz grains and massive mullite surrounded by a matrix of glassy feldspar, and the shrinkage of the quartz on transformations at 530°C leads to microcracking. They have low strengths (50–100 MPa) and the lowest thermal conductivities (1.7–2.1 W m^{-1} K^{-1}). They are good insulators at ambient temperature, but resistivity falls off as temperature increases. They are cheap and used for internal and external insulators.

IEC class C111—pressed porcelains with <30% alumina These are similar in composition to class C110 but, because the pressing process, which is carried out with a clay containing less water, allows trapping of air between the granules, pressed porcelains are more porous and have lower strengths. They must be glazed for outdoor applications and cannot be used at high voltages.

IEC class C112—crystobalite porcelains These contain crystobalite instead of quartz, but are otherwise similar to class C110.

IEC class C120—aluminous porcelains with 30–50% alumina In these porcelains most or all of the quartz in class C110 is replaced with alumina giving improved strength (c. 100 MPa).

IEC class C130—aluminous porcelains with 75% alumina These are feldspar porcelains in which all the quartz has been replaced by alumina improving modulus and strength (up to 200 MPa).

1.6.3.2 IEC class C2XX

These are steatite and forsterite ceramics where the quartz or alumina in class C1XX is replaced with enstatite or forsterite talc. These materials comprise the following.

IEC class C210—low voltage steatites These are porcelains with 80–90% talc, 5–10% plastic clay and 5–10% feldspar. They have low strength (50–100 MPa) and, because they have about 0.5% open porosity, are limited to low-voltage insulation.

IEC class C220—normal steatites In these the electric losses of the glassy phase are reduced by the substitution of barium or calcium carbonate for the feldspar and the addition of magnesia and zirconia. Class C220 has higher strength, resistivity and lower dielectric dissipation than class C210.

IEC class C221—low loss steatites Alkali levels are lower than those of class C220 to reduce dielectric loss by minimising the glassy phase.

IEC class C230—porous steatites These are low-loss steatites with c. 30% open porosity which improves machinability and thermal shock resistance but limits the material to low-voltage applications.

IEC classes CO240 (porous forsterites) and CO250 (dense forsterites) In forsterites the magnesia level is raised so that forsterite is formed instead of steatite to give a coefficient of expansion of c. $10 \times 10^{-6}\,K^{-1}$, which is close to that of titanium and some nickel–iron alloys for ceramic to metal assemblies in high-frequency power devices and microwave tubes.

IEC class C410—dense cordierite Cordierite is a magnesium–aluminium silicate which is made synthetically from clay, talc and sillimanite and has a low coefficient of thermal expansion. Ceramics consisting of this material with 5% vitrifying feldspar are readily fired and can be made dense. Although their strength is only 50–100 MPa, their low thermal expansions (c. $3 \times 10^{-6}\,K^{-1}$) confer shock resistance.

IEC class C420—dense celsian and zircon porcelains (unclassified) These are little used.

1.6.3.3 Porous ceramics (IEC class G5XX)

These may be based on clay, cordierite or alumina up to about 80%. (Porous ceramics with higher alumina contents are included under high aluminas, class C7XX.) They are used primarily as electrical heating and thermocouple insulations where their porosity confers resistance to thermal shock and machinability.

These ceramics are compounded from fine aggregates for electrical applications, but coarse aggregates may be used for crucibles and tubing for high-temperature processing.

IEC class C510—porous aluminosilicates These are manufactured primarily from refractory clays with additions of quartz, silimanite or alumina to increase refractoriness. They contain very little fluxing material and their 30% porosity gives them the thermal shock resistance required for such applications as gas fire radiants.

IEC classes C511, C512 and C520—porous magnesium aluminosilicates These are developed from class C510 by the addition of magnesia containing minerals to produce cordierite which greatly enhances shock resistance and the distinctions between classes are based on performance. The highest performance class (C520) will withstand severe thermal cycling and is used for experimental heat exchangers and automobile exhaust catalyst supports.

IEC class C530—porous aluminous materials These ceramics have alumina contents up to 80% and are more refractory but less shock resistant than the other classes of porous ceramic.

1.6.3.4 Impermeable aluminosilicate and mullite ceramics (IEC class C6XX)

These are impermeable fine-grain materials with high strength (100–200 MPa) and high dielectric strength. They are used for

gas-tight tubing and high temperature electrical insulation. They are classified according to alumina content.

IEC class C610—50/60% alumina ceramics These are made from clay and alumina (or aluminosilicates) and consists of mullite in a siliceous glassy phase.

IEC class C620—60–80% alumina ceramics This specification covers materials with a wide range of properties and crystalline compositions because mullite which is $71.8\%\,Al_2O_3$ and $28.2\%\,SiO$ is highly refractory and compositions containing more Al_2O_3 are also highly refractory (and require a high firing temperature). Compositions with more silica than mullite are less refractory and the addition of other oxides such as CaO reduces refractoriness considerably. If a highly refractory material is required this must be specified closely. However, all materials in this class are fine grained with no open porosity and have fair shock resistance.

1.6.4 Oxide based ceramics

Oxide based ceramics (IEC classes C3XX C7XX and C8XX and non-classified materials) consist of synthetic oxides together with bonding materials which sometimes form a glassy phase. The amount and composition of the glass has a significant influence on the engineering properties. In general, a low glass content gives a high refractoriness, a large grain size, high electrical resistance and low dielectric loss. On the other hand, strength and wear resistance are favoured by a low grain size which requires more glass to reduce the firing temperature.

1.6.4.1 IEC class C7XX—high alumina ceramics

In these ceramics the second two digits denote the minimum percentage of alumina. This degree of classification is presumably adequate for electrical purposes, but not for all applications.

IEC class C799—high purity aluminas ($>99.9\%$ Al_2O_3) These are single-phase ceramics which may be sintered to give a large grain size or hot pressed to give low porosity and a small grain size. If sintered in a reducing atmosphere they can be made translucent or transparent and may be used, for example, for sodium vapour lamp envelopes.

Aluminas with over 99.7% Al_2O_3 may be made from relatively lower purity aluminas or may have added magnesium oxide which restricts grain growth. The magnesia free products are more refractory but weaker (100–200 MPa tensile strength), while those containing magnesia have tensile strengths of 200–400 MPa and are more suitable for thin-walled tubing or small-diameter rods.

Aluminas with 99–99.7% Al_2O_3 usually have a small amount of glassy phase which facilitates sintering and controls electrical properties.

High purity aluminas are the most widely used of the alumina ceramics for engineering and electrical purposes, those with fine grains having strengths up to 400 MPa, while those with coarse grains are used for electrical insulation.

IEC class C795 These aluminas all have a deliberately formulated glassy phase which facilitates metallising by manganese/molybdenum.

IEC class C786 This is a somewhat heterogeneous class, some having a glassy phase while others have additions of manganese and titanium dioxides which impair electrical properties and refractoriness. Those with lower alumina contents are non-refractory and are used for low temperature electrical insulators and mechanical components.

IEC class C780 There are two types of material in this class, one is refractory consisting of mullite and silica, and the other is non-refractory and limited to low temperatures and non-critical uses.

Other There are also aluminas which do not fall within the IEC classification. Porous alumina ceramics are available in the same composition ranges with high resistance to corrosion or slag attack. Unlike the non-porous materials these can be machined (or rather chipped) by hard metal tools.

The addition of 10–20% of zirconia to alumina ceramics enhances strength to over 400 MPa making them suitable for cutting-tool tips. The addition of 20–40% TiC also increases strength and stiffness for similar applications.

The addition of TiO (a high permittivity material) significantly increases the dielectric constant of alumina ceramics where this is desirable.

Synthetic alumina crystals such as sapphire and ruby are used for high temperature and scratch resistant windows and for bearings and laser elements.

1.6.4.2 IEC class C8XX

IEC class C830—zirconia ceramics Zirconia is more refractory than alumina, it can be machined readily to give a low coefficient of friction against metals and it is a useful oxygen ion conductor above about 700°C. It is, however, more expensive than alumina which is therefore used preferentially where either will suffice, and its use is complicated by phase instability. Pure zirconia is monoclinic at room temperature, transforms to tetragonal form at about 1200°C and to cubic form at about 2370°C. Its use as a ceramic depends on stabilisation either of the cubic phase, of a fine tetragonal phase within cubic crystals or of an all tetragonal material.

Truly stable zirconia would be very strong (above 400 MPa) and highly refractory. Development is proceeding and if manufacturers' claims are fully justified zirconia ceramics have wide applications. In the meantime they are used for oxygen probes, small crucibles, dishes, etc., and heating elements for operation in air at high temperature.

IEC class C3XX—titania ceramics Ceramics based on TiO_2 have been used for applications requiring high permittivity and clay bonded substochiometric materials are used for thread guides in the textile industry because they are semiconducting and discharge static electricity.

IEC class C820—magnesia ceramics Magnesia possesses few advantages over alumina as a ceramic but, considered as a refractory, it has a higher thermal conductivity, greater refractoriness and greater resistance to basic slags. Magnesia ceramics are therefore used in applications requiring resistance to corrosion at high temperatures and, specifically, for electrical purposes as an insulator in mineral insulated cabling.

IEC class C810—beryllia ceramics Beryllia is much more expensive than alumina and is highly toxic. However, it has a thermal conductivity at room temperature of 500 W m^{-1} K^{-1}, i.e. one order of magnitude greater than other electrical insulating material, and it is therefore used for some electronic devices. Its widely heralded future as a nuclear material has not yet been realised.

Other Other oxide ceramics not in the IEC classification include: lime, thoria, uranium dioxide (used as a nuclear fuel),

ferrites (which have very extensive magnetic applications) and titanates.

Ferrites, compounds of Fe_2O_3 with other oxides such as ZnO, are available with very high magnetic permeability and can be made in powder form to have very low losses so that they may be used for cores for high-frequency inductors. Certain ferrites will form strong permanent magnets.

Titanates are used for their electrical properties. Some, usually barium titanates, have permittivities up to 5500 and are used as dielectrics for electronic circuitry. Magnesium titanates have permittivity-compensating properties. Lead zirconate–titanates have strongly piezoelectric properties.

1.6.5 Non-oxide ceramics

Non-oxide ceramics differ essentially from oxide ceramics in that they oxidise at high temperatures in air and cannot therefore be fabricated by means of a conventional oxide sintering process. They must therefore be fabricated by the alternative routes listed in Section 1.6.11 and their compositions and microstructures are adjusted accordingly.

The most important are carbides, silicides, borides and carbons.

1.6.5.1 Silicon carbide (carborundum)

This owes it application to its high hardness, very high refractoriness, high thermal conductivity and its semiconducting properties. Its electrical conductivity is low at low temperature and low currents, but increases by orders of magnitude at high temperatures and high currents.

Although silicon carbide is oxidised in air at high temperature, many of the commercial materials form a protective layer of silica.

The name 'silicon carbide' covers a number of products manufactured in different ways.

The conventional method of manufacture is the reduction of silica sand by carbon in the arc furnace which produces the high temperature stable α phase. The phase is produced by low temperature gas reactions and is converted to the β phase by heating above 2000°C.

Commercial products include the following.

(1) Clay bonded silicon carbide containing 10–15% clay which bonds chemically to the surface film of silica on sintering.

(2) Other additives which enhance sintering, but permit a higher proportion of silicon carbide than is possible with clay bonding, have been used with reported success.

(3) Very high temperature sintering of prime silicon carbide produces a coarse grained, porous and highly refractory product.

(4) Hot pressed silicon carbide requires the addition of a secondary material, usually about 2% alumina. It is very strong (above 400 MPa) fine grained and non-porous, but very expensive.

(5) Reaction bonded silicon carbide is made by two distinct processes. In one process a compact of silicon carbide, carbon and an organic binder is infiltrated by liquid silicon which bonds the particles by the formation of additional silicon carbide. The free silicon remaining may, if required, be leached out with acid to leave an open porous refractory. The other process achieves the same object by firing a compact of silicon carbide and silicon in nitrogen to produce a silicon nitride bond.

(6) Coatings of thin-walled components of pyrolytic silicon carbide can be produced by decomposing a gas mixture containing both silicon and a hydrocarbon on a heated substrate. This process has found an important applica-

tion in the manufacture of fuel elements for high-temperature gas-cooled reactors. In these fuel elements a silicon carbide shell contains fission products produced in small spheres of fissile material.

(7) Graphite can be 'case hardened' with silicon carbide produced from solid (or gaseous) silicon.

Applications of silicon carbide The original application of silicon carbide was as an abrasive for grinding steel, but it has been replaced for this purpose by fused alumina and its use as an abrasive is now restricted to shaping materials of lower tensile strength.

It is used extensively for refractories, particularly where the shock resistance conferred by its high thermal conductivity is an asset both in vacuum or reducing conditions and in air.

A major use is in electrical resistance heating elements and it is also used for susceptors in high frequency induction fields.

Its high hardness makes it suitable for abrasion-resistant components, in particular for non-lubricated bearings.

1.6.5.2 Silicon nitride based materials

Silicon nitride and its derivatives have properties which augur a great engineering potential because of its high hardness, high strength, refractoriness, low coefficient of thermal expansion (which confers thermal shock resistance) and its resistance to oxidation which, like that of silicon carbide is due to the formation on its surface of a silica film.

The application of silicon nitride ceramics has, however, been hindred by fabrication difficulties. Like silicon carbide it cannot be readily sintered, tending to dissociate at above 1850°C and alternative methods of manufacture are difficult to control or expensive. Two products are available.

Reaction bonded silicon nitride Reaction bonded silicon nitride can be produced by converting a silicon powder compact to nitride by sintering in a nitrogen atmosphere. The resulting compact is porous and the product variable.

Hot pressed silicon nitride Hot pressed silicon nitride requires the addition of an oxide, typically 1–5% MgO which provides a liquid phase at the sintering temperature. This produces a fine, strong product but limits the temperature at which the high strength is retained. Hot pressing is expensive and the shapes which can be produced are limited.

Applications of silicon nitride The applications of silicon nitride have so far been limited to high temperature industrial applications. It is extremely valuable for such components and operations as furnace supports, heating tubes, jigs and printer's saggars, silver soldering and vacuum and copper brazing.

1.6.5.3 Sialons

The difficulties experienced with silicon nitride have, to a large extent, been overcome by the partial substitution in the molecule of aluminium for silicon and oxygen for nitrogen. The resulting 'SiAlONs' can be hot pressed from mixtures of silicon nitride, alumina, silica and aluminium nitride or, if additional densifying constituents are added, they can be sintered. These materials merit, and have received considerable study and, as a result, porosities as low as those in hot pressings and strengths above 400 MPa may be achieved with sintered materials.

An alternative route to sintering is the nitriding (and subsequent high temperature sintering) of a mixture of oxides and silicon.

Sialons and related products are finding wide and increasing applications in manufacturing processes, where resistance to heat and abrasion are required.

1.6.5.4 Boron carbide

Boron carbide is one of the hardest ceramic materials available and can be hot pressed to achieve tensile strengths above 400 MPa. It is used for very abrasive conditions, e.g. for shot blast nozzles and ballistic armour. Its main drawback is its cost, which is high. The sintered product is more porous and has a coarser grain size.

1.6.5.5 Boron nitrides

Boron nitride in its most commonly available form differs from the ceramics already considered in that it has a laminar structure and is therefore soft and machinable.

The properties of the boron nitride crystal are highly anisotropic and shapes deposited pyrolytically on a substrate have thermal and electrical conductivities and coefficients of thermal expansion which may be up to 100 times greater parallel to the substrate than perpendicular to it. This property and the fact that many molten metals do not wet boron nitride has led to its use for crucibles for metal melting.

There is also a cubic form made under high pressure, which is used as an abrasive.

1.6.6 Carbons and graphites

Three allotropic modifications give rise to a very wide range of carbons and graphites with a correspondingly large range of applications. The three allotropes are:

(1) *Diamond*, a cubic crystalline material with a very stable structure. It is the hardest substance known, highly refractive, highly transparent when pure and very expensive. It can, with difficulty, be manufactured, but the manufactured product, which can be used as an abrasive, is barely competitive with the mined product.

(2) *Graphite* has a laminar hexagonal ring structure. Each ring is strongly bonded to six other rings in approximately the same plane but the bonding between planes is weak. Graphite has, therefore, excellent lubricating properties, is highly refractory and, if the layers of hexagonal rings are continuous and aligned, very high specific strength and specific elasticity modulus in a direction in the plane of the rings.

(3) *Amorphous carbon* is, as its name implies, amorphous to X-rays, but this structure is probably a large scale statistical phenomenon rather than a true liquid-type structure. The available amorphous carbons are probably random assemblies of graphite platelets, the size and distribution of which govern the properties.

Solid carbon and graphite products are usually made from a mixture of graphite (which occurs naturally) or carbon particles (usually derived from coal or oil cokes) pressed with a carbonaceous binder to form a solid block. The mass is converted to carbon on firing. Very high temperature firing graphitises both the amorphous carbon and the carbon derived from the binder.

Both industrial carbons and graphites are highly refractory materials which must be protected from oxidation at high temperatures. They have strengths around 45 MPa and electrical resistivities around 30 Ω-m. The thermal conductivity of graphite is higher than that of carbon.

Carbons may be impregnated with resin to improve strength and soundness or with metals.

Carbons and graphites manufactured by polymer carbonisation are assuming great technical importance. They include the following.

(1) Carbon fibres, made by carbonising polymer fibres which are subjected to tension during carbonisation and graphitesation. These are highly oriented graphite fibres with high specific tensile strengths and specific moduli exceeded only by boron fibre. They are used for the manufacture of the composites described in Section 1.6. Carbon–carbon fibre combines the refractory and electrical conducting properties of graphite with high specific strength.
(2) Vitreous carbon, produced by controlled carbonisation of a cross-linked polymer, is glassy in appearance and has no open porosity. It is available in plate, dish and crucible forms and, according to grade, will withstand temperatures of 1000–2500°C in inert atmospheres.
(3) Graphite foam and graphite or carbon felts are produced by carbonisation or graphitisation of the appropriate polymer foam or mat.

The refractoriness, electrical conductivity, lubricity and (in the form of fibre) high mechanical properties of carbon and graphites together with their reasonable cost render them suitable for many and varied engineering and consumer applications.

1.6.7   Miscellaneous ceramics

A number of other ceramics have specific applications. They include refractory metal (usually tungsten or titanium) carbides which are used extensively for cutting or forming tools. These are usually bonded with cobalt, nickel or nickel molybdenum and are described in Section 1.5.

1.6.8   Glasses

All commercial glasses are based on silica (SiO_2) to which may be added other oxides which progressively reduce the softening point, increase the thermal expansion and affect other characteristics.

99.8% silica glasses include:

(1) fused silica, which contains small bubbles, and is used for applications that require good thermal shock and corrosion resistance;
(2) fused quartz, which is transparent and used for high-quality tubing; and
(3) vitreous silica, which is used for high-quality optical components.

Aluminosilicate glasses have service temperatures close to those of silica glasses and are resistant to alkalis. They are used for high performance electronic applications and other industrial applications.

Borosilicate glasses have lower softening temperatures than aluminosilicate glasses and, because of the lower cost of fabrication, have more widespread applications.

Sodalime silicate glasses have still lower softening temperatures and high coefficients of expansion. They are widely used for windows and other domestic applications because of their relatively low cost.

Alkali–lead silicate glasses have, besides the lowest softening points commonly available, refractive indices and dispersive power, which makes them useful for optical and similar applications.

Other glasses include photochromic glasses, semiconducting glasses, solder glasses for glass–metal seals and ophthalmic glasses.

1.6.9   Glass ceramics

The properties of glass ceramics can be tailored to have specific properties to fit the desired application. The properties concerned include:

(1) thermal expansion, which can be made to match that of the material to which the glass ceramic is to be fused, or possibly to approach zero;
(2) refractoriness;
(3) transmission of light;
(4) colour; and
(5) machinability.

Where an application is foreseen, manufacturers or specialists should be consulted for advice on manufacturing procedure and practicability.

1.6.10   Mechanical properties

The applications of ceramics are strongly influenced by their differences in mechanical behaviour compared with metals.

1.6.10.1   Elastic properties

The elastic properties of ceramics are influenced by the level of porosity and also by the proportion of different phases present. The effect of small, closed, randomly distributed spherical pores on the modulus of a material of Poisson's ratio $\nu = 0.3$ has been shown to be approximately:

$$E = E_o(1 - 1.9v + 0.9v^2)$$

where E_o is the modulus of the non-porous material, and v is the proportion of pores by volume.

Each phase in a multiphase material contributes, to a first approximation, according to its volume fraction.

Where a high-modulus major phase is bonded by a low modulus minor phase, as is common in engineering ceramics, the modulus reduces rapidly with an increase in the content of the minor phase.

The modulus usually decreases with increasing temperature, but glasses and complex ceramics are available with moduli which increase with increasing temperature (over a limited range). The range of moduli for ceramics is indicated in *Table 1.62*.

Table 1.62   Typical room-temperature elastic moduli of glasses and ceramics

Material	Young's modulus ($GN m^{-2}$)
Glass	63–73
Glass ceramics	78–120
Carbons and graphites	9–13
Porous ceramics	54
Porcelains	69–138
Oxides	132–155
Zirconias	230–580
Aluminas ≮99%	280–350
Silicon carbide	385–470
Boron nitride*	44–104
Silicon nitride	160–310
Sialon	280
Boron carbide	440

* Anisotropic.

1.6.10.2   Strength and ductility

The tensile strength values for ceramics quoted here and elsewhere are not comparable with those quoted for metals. In a metal the strength of the material obtained from a simple mechanical test adjusted by a safety factor chosen on the basis of previous experience may be used to determine a safe loading stress.

In ceramics and other brittle materials, measurements on nominally identical materials will show scatter in strength results with standard deviations around 20% of the mean strength. Catastrophic failure occurs by rapid propagation of a crack as soon as the stress at the most severe flaw reaches a critical value. (For more detailed discussion of this see references 36 and 37.)

the consequences of this are as follows.

(1)   As specimens contain critical flaws of varying severity, nominally identical specimens exhibit scatter in strength.
(2)   Large components and components with larger surface areas are, on average, weaker than smaller ones because there is a higher chance of finding a flaw of given severity. (It is also likely that a manufacturing method will produce a larger flaw in a larger component, but this is secondary to the argument.)
(3)   Because a severe flaw at a stress less than the maximum stress in a component may reach a critical value for fracture before a less severe flaw subjected to the maximum stress, a component will not necessarily fail at the point of maximum stress. A thorough stress analysis of a component must therefore be carried out.

There is, therefore, no single measure of strength applicable to both a test specimen and a component, and no absolute guarantee can be given that a component will not fail under its design load.

Design must be based on a probabilistic approach which will keep the frequency of failure below an acceptable figure depending on the seriousness of the consequences.

1.6.10.3   Methods of testing

The standard uniaxial tensile and compression tests used for metals are not applicable to ceramics because of the difficulties of gripping the specimen and of aligning it axially. The tests used apply specific parts of the specimen to tensile stress by loading other parts in compression. A number of tests have been devised.[38] The most straightforward method is the three-point bend test in which the tensile stress rises to a maximum on a line on one surface of the specimen.

There is no recognised international standard for the testing technique, specimen size, surface finish or shaping procedure; therefore small differences in strength quoted by a number of manufacturers should not be regarded as significant where it is desired to select the strongest.

Procedures have been developed for determining the load which may be applied to a component to give a specific probability of failure in the short term. It is necessary to undertake a programme of tests on at least 20 specimens from which a statistical distribution known as the 'Weibull modulus' m is determined, which gives an estimate of the degree of scatter of the data (see *Figure 1.67*). From the value of m, and

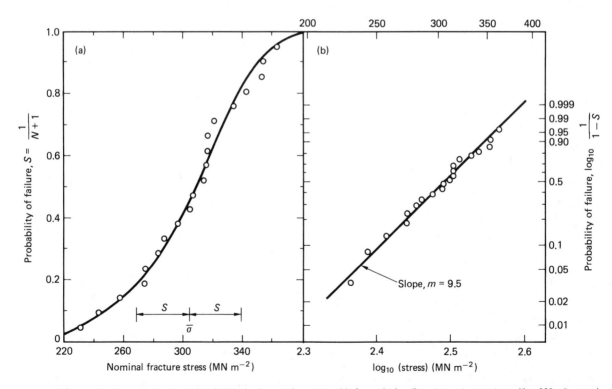

Figure 1.67   The fracture strengths of a batch of 95% alumina specimens tested in four-point bending at room temperature with a 320 grit-ground surface finish are plotted (a) as a cumulative distribution and (b) as a Weibull plot to get a straight-line fit to the points. Using the Weibull parameters calculated the solid line in (a) represents the Weibull distribution calculated. The slope m of the straight line is the 'Weibull modulus'. (Reproduced by permission of Dr R. Morrel and the National Physical Laboratory)

the effective volume or surface of the component the probability of failure at a given load may be calculated. Working stresses for a 10^{-4} failure probability may be of the order of 25–50% of the mean bend stress.

The procedures are described in detail in references 12 and 36.

1.6.10.4 Long-term strength

The long-term strength of many ceramics, in particular silicate bonded materials, is lower than the short-term strength. This phenomenon has been described, usually referring to the behaviour of glasses, as 'static fatigue' but is more correctly termed 'slow crack growth'.

The crack-like defects present in all ceramics (except whiskers) extend by a thermally activated atomic process, driven by the applied stress, influenced by the environment at the crack tip and strongly dependent on temperature.

In oxide ceramics hydroxyl (OH) ions can break metal–oxide bonds; therefore clay-bonded materials may have lower long-term strengths in high-humidity atmospheres and, particularly, in alkaline environments.

With most non-oxide ceramics, water vapour has little influence so that little effect may be observed at stresses well below the fracture stress.

Most ceramic components are subject to low values of tensile stress in service, but where a guarantee of survival for a specific period at stresses in excess of 10% of the nominal breaking stress is required the only solution is a short-term overload proof stress which rejects components in the low strength tail of the distribution.[36]

1.6.10.5 Fatigue under cyclic loading

Little systematic work has been carried out on the effect of cyclic loading on ceramics. If cyclic loading cannot be avoided the effect may best be estimated by integrating the varying stress intensity factor raised to the power n (obtained from reference 36) with time.

1.6.10.6 Strength at elevated temperatures

As described in Section 1.6.10.4, slow crack growth increases with increasing temperature; consequently, increasing temperature may initially reduce long-term strengths. However, further increases in temperature may result in flaw rounding which increases strength to a maximum. At still higher temperatures strength is reduced by creep and creep rupture.

Few data are available beyond statements on 'upper use temperature' usually under no-load conditions, which implies suitability for short-term use without disastrous changes in microstructure.

Temperatures at which ceramics are suitable for long-term use under load are indicated in *Table 1.63*.

Table 1.63 Approximate temperatures at which ceramics and glasses are suitable for long-term operation under load

Type of ceramic	Temperature (°C)	Notes
Sodalime glass	300	
Silica glass	800	
High purity alumina	1400	
Silicon nitride	1600	Oxidises in air above 900°C
Graphite	>1500	Oxidises in air above 400°C

1.6.11 Manufacturing processes

The manufacture of ceramics includes shaping, with or without the application of pressure, and consolidation by firing. The characteristics of the ceramic and its cost depend on the process which must be used. The simplest (and cheapest) method is compaction and sintering.

Compaction This may be carried out by a variety of processes which may involve the addition of a binder or lubricant or may include the application of pressure.

Sintering In the pure sense, 'sintering' implies the densification of an assembly of particles at high temperatures by solid-state-diffusion processes. This process is relatively slow, needs very high temperatures and a dense impervious body is achieved only with difficulty. Ceramics other than oxides may require a protective atmosphere.

Vitrification Also known as liquid-phase sintering, vitrification is achieved by incorporating a component which melts to a viscous liquid during sintering and solidifies (in the case of a clay-based ceramic) to a glassy phase. The term 'well vitrified' indicates that a dense impervious body has been achieved.

Reaction bonding This is achieved by arranging for two or more components of a compacted body to react to form both a desired phase and a bond between particles. The reacting components may all be solid or one of them may be gaseous or liquid and introduced during the sintering process.

Self bonding This involves bonding the major constituent of a ceramic with its constituents formed by reaction during sintering.

Hot pressing This is achieved by the application of pressure during sintering. There are two variants. Hot pressing, as the term commonly understood, consists in applying pressure unidirectionally by a plunger acting on material in a rigid container. Hot hydrostatic pressing consists in submitting the material contained in a membrane to hydrostatic pressure during heating. Both processes produce material of a higher density and superior quality more quickly than can be achieved by sintering, but they are substantially more expensive and the range of possible component shapes is restricted.

Pyrolysis This is the process whereby a ceramic is deposited from a gaseous environment onto a (usually) heated substrate. This process is usually employed to deposit coatings or produce thin-walled components of graphite or silicon carbide from carbon or silicon containing gas.

1.6.12 The future prospects of engineering ceramics

Ceramics and glasses have a wide range of application in manufacturing technology and even wider application in domestic use. If they are to achieve the really widespread engineering application that their properties appear to merit, they must find this in some field such as the automotive industry.

Some components, including oxygen sensors, catalyst supports, diesel glow plugs, rocker arm pads, precombustion chambers and turbocharger rotors have already been standardised, usually for high efficiency and military engines. *Tables 1.64* and *1.65* and *Figure 1.68* list the properties and applications of the ceramics at present available for the required breakthrough into the mass market.

Table 1.64 Properties of selected metals and technical ceramics (properties are at room temperature unless otherwise stated*†)

	Density ($\times 10^3$ kg m^{-3})	Bend strength (MN m^{-2})	Young's modulus (GN m^{-2})	Fracture toughness (MN m$^{-3/2}$)	Thermal expansion‡ ($\times 10^{-6}$ K^{-1})	Thermal conductivity‡ (W m^{-1} K^{-1})
Alumina 99%	3.9	400	400	~3.0	9.0	24
Zirconia toughened alumina	4.1	450	340	~8	8.1	23
Aluminium titanate	3.0	~40	20	—	0	1.5
Cordierite (MAS)	2.5	120	110	~2.5	2.0	1.5
Silicon carbide						
Reaction bonded	3.1	~500	410	~4.5	3.8	100
Sintered	3.1	460	400	~4.5	4.0	90
Silicon nitride						
Hot pressed	3.2	800	310	6	3.2	~20
Reaction bonded	~2.5	200	170	~3	3.0	12
Sintered	3.2	~400–700	250	5	3.4	~16
Sialon	~3.2	~950	~290	~8	~3.1	~21
Zirconia						
Plasma sprayed	5.2	6–80	48	~2	8.0	1.0
Partially stabilised	5.6	500	205	8	9.5	1.7
PZT (Y)	6.05	1000	210	15	9.0	2.0
Fully stabilised	5.8	180–250	160	~4	10.0	2.0
Cast iron						
Nimonic (80A)§	7.2	—	117	—	12.0	54
	8.2	—	200	—	13.0	12
Aluminium alloy IM27	2.75	—	71	—	21.0	155

* Reproduced by permission of *Metals and Materials*.
† The values should be used as a guide only, since various ceramic grades exist under one generic name.
‡ Values for 300–600 K.
§ Inconel 751 has a K_{IC} of 80–100 Mn m$^{-3/2}$.

Table 1.65 Automotive ceramic components*

Component	Material	Car maker	Ceramic manufacturer	Year of introduction
Engine				
Diesel glow plug	Silicon nitride	Isuzu	Kyocera	1981
		Mitsubishi	Kyocera	1983
		Toyota	Toyota	1984
		Caterpillar	Kyocera	1986
IDI precombustion chamber	Silicon nitride	Isuzu	Kyocera	1981
		Mazda		1986
		Toyota		1986
Exhaust port liner	Aluminium titanate	Porsche	Hoechst	1986
Rocker arm pad	Silicon nitride	Mitsubishi		1984
Turbocharger rotor	Silicon nitride	Nissan	NTK	1985
		Buick		1987
Knock sensor	PZT Pb(ZrTi)O$_3$	Toyota		1980
Other				
Oxygen sensor	Zirconia/titania	Vehicles with three-way catalytic converters		
Catalyst support	Cordierite	Vehicles with three-way converters		
Warning alarm	PZT	Toyota plus other		
Sensors	NTC/PTC† thermistors			

* Reproduced by permission of *Metals and Materials*.
† Negative/positive temperature coefficient ceramics (Al$_2$O$_3$–Cr$_2$O$_3$/BaTiO$_3$).

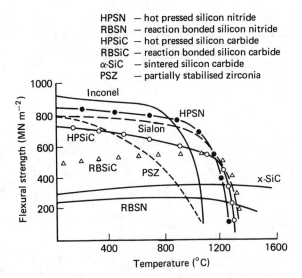

HPSN — hot pressed silicon nitride
RBSN — reaction bonded silicon nitride
HPSiC — hot pressed silicon carbide
RBSiC — reaction bonded silicon carbide
α-SiC — sintered silicon carbide
PSZ — partially stabilised zirconia

Figure 1.68 Flexural strength variations with temperature of several materials. (Reproduced by permission of *Metals and Materials*)

The major problem is that, depending on the fabrication procedure, the cost of a ceramic component is 5–15 times that of the presently used steel component, so that a substantial improvement in performance or a marked reduction in cost is required to justify the use of the ceramic.

The potential advantage of materials, such as sialons, which combine excellent properties and a relatively inexpensive manufacturing route (sintering), is evident.

References

1 GILLAM, E., *Metallurgist and Material Technologist*, **9**, 521–525 (1979)
CRANE, F. A. A. and CHARLES, J. A., *The Selection and Use of Engineering Materials*, Butterworth, London (1984)
2 EDELEANU, C., 'Information requirements and the chemical industry', *Metals and Materials*, **3**(3), pp. 43–44 (March 1987)
3 TURNER, M. E. D., 'Materials for the process industry', *Metals and Materials*, **3**(3), pp. 136–139 (March 1987)
4 *Proc. Workshop, Fairfield Glade, Tennessee, November 1982*, Steering Committee of the Computerised Materials Data Workshop (1983)
Proc. 9th International CODATA Conference, Jerusalem, June 1984, North Holland Physics Publishing, Amsterdam (to be published)
Proc. CEC Workshop of Factual Materials Data Banks, Petten, The Netherlands (to be published)
5 HAMPEL, V. E., BOLLINGER, W. A., GAYNER, C. A. and OLDANI, J. J., *UCRL Report No. 90942* (June 1984)
6 BREEN, D. H., WALKER, G. H. and SPONZILL, J. T., *Metal Progress*, **103**, pp. 83–88 (1973)
7 FARAG, M. M., *Materials and Process Selection in Engineering*, Applied Science, London (1979)
8 Matsel Systems Ltd., 14 Mere Farm Road, Birkenhead, Merseyside L43 9TT, UK
9 SWINDELLS, N. and SWINDELLS, R. S., 'System for engineering materials selection', *Metals and Materials*, **1**(5), pp. 301–304 (May 1985)
10 DIMAID, A. and ZUCKER, J. J., 'A conceptual model for materials selection', *Metals and Materials*, **4**(5), pp. 291–296 (May 1988)
11 AMERICAN SOCIETY FOR TESTING MATERIALS, *The ASTM Metals Handbook*, current edition, ASTM
12 THE FULMER INSTITUTE, *The Fulmer 'Optimizer'*, current edition, Elsevier, London
13 BRITISH STEEL CORPORATION, *Iron and Steel Specifications*, 5th edition, BSC
14 THE INSTITUTE OF METALS, *Secondary Steelmaking for Product Improvement*, London (1985)
15 THE INSTITUTE OF METALS, *Continuous Casting*, London (1982)
16 SAE, *The SAE Handbook*, current edition
17 AMERICAN SOCIETY FOR TESTING MATERIALS, *The ASTM Metals Handbook*, pp. 122–146, ASTM (1961)
18 COTTRELL, A. H., *Creep Buckling. The Mechanical Properties of Matter*, p. 337, Wiley, New York (1964)
19 DRIVER, D., 'Developments in aero engine materials, *Metals and Materials*, **1**(6), 345–353 (June 1985)
20 WHITE, C. H., *The Development of Gas Turbine Materials*, Chap. 4, Applied Science, London (1981)
21 BETTRIDGE, W. and HESLOP, J., *The Nimonic Alloys*, 2nd edition, Edward Arnold, London (1974)
22 BECLEY, P. R. and DRIVER, D., *Metals Forum*, **7**, 146 (1984)
23 WALLACE, W., *Metal Science*, **9**, 547 (1975)
24 ASHDOWN, C. P. and GREY, D. A., *Metal Science*, **13**, 627 (1979)
25 MEETHAM, G. W., *Metallurgist and Materials Technologist*, **9**, 387 (1982)
26 BECLEY, P. R. and DRIVER, D., *Metals Forum*, **7**, 146 (1984)
27 SIMS, C. T. and HAGEL, W. E. (Eds), *The Superalloys*, Wiley, New York (1972)
28 MCCLEAN, M., *Directionally Solidified Materials for High Temperature Service*, The Institute of Metals, London
29 The Wear Technology Division of the Cabot Corporation (Deloro Stellite) Stratton St Margaret, Swindon, Wilts, issue publications on their 'Deloro' series of wear-resistant nickel based alloys and 'Triballoy' intermetallic compounds for wear-resistant coatings.
30 Wall Colmonoy Ltd, Pontardawe, West Glamorgan, issue brochures listing their nickel alloys for wear-resistant coatings.
31 JONES, *Mechanics of Composite Materials*, Scripto Book Co. (1975)
32 KELLY, A. and MILEIKO, S. T., *Fabrication of Composites*, North Holland, Amsterdam (1983). This book is part of a series intended to cover all aspects of composites
33 JAYATILAKA, A. de S., *Fracture of Engineering Brittle Materials*, Applied Science, London (1979). This book deals with the mechanisms whereby fibre reinforcement improves strength and, sometimes, toughness
34 BIRCHALL, J. D., HOWARD, A. J. and KENDALL, K., 'Flexural strength and porosity of cements', *Nature*, **289**, 288, 289 (1981)
35 JACKSON, A. P., VINCENT, J. R. F. and TURNER, R. M., 'The mechanical design of nacre', *Proceedings of the Royal Society, London, Series B*, **234**, 415–440 (1988)
36 MORREL, R., *Handbook of Properties of Technical and Engineering Ceramics, Parts 1 and 2*, National Physical Laboratory (1985)
37 DAVIDGE, R. W., *Mechanical Behaviour of Ceramics*, Cambridge University Press, Cambridge (1979)
38 SHOOK, W. B., 'Critical survey of mechanical property test methods for brittle materials', *Technical Report No. ASD-TDR-63-491. AD 417620* (1963)

Bibliography

In addition to the publications specifically referred to in the text, the reader is referred to the following associations and publications. The addresses of relevant associations, etc., are given where appropriate.

General

AMERICAN SOCIETY FOR TESTING MATERIALS, *The ASTM Metals Handbook*, current edition, ASTM
SAE, *The SAE Handbook*, current edition, SAE
THE FULMER INSTITUTE, *The Fulmer 'Optimiser'*, current edition, Elsevier, Amsterdam
The Metals Reference Book, Butterworths, Oxford

Ferrous metals

BRITISH STEEL CORPORATION, *Plates—Steel Specification Comparisons, Part 1 Structural Steels, Part 2 Pressure Vessel Steels*, BSC, British Steel House, 31 Oswald Street, Glasgow

PLOCKINGER, M. E. and ETTERICH, C., *Electric Furnace Steel Production* (trans. Babler, E. B. and Babler, P. E. O.), Wiley, Chichester (1985)

THE BRITISH CAST IRON RESEARCH ASSOCIATION* publish information on all aspects of cast iron

THE DROP FORGING RESEARCH ASSOCIATION* publish information on all aspects of drop forgings

THE INSTITUTE OF METALS, *Continuous Casting*, The Institute of Metals, 1 Carlton House Terrace, London (1985)

THE INSTITUTE OF METALS, *Stainless Steels '84*, London (1985)

THE MOND NICKEL CO., *Transformation Characteristics of Direct Hardening Nickel Alloy Steels*, 3rd edition

THE STEEL CASTINGS RESEARCH AND TRADE ASSOCIATION*, 5 East Bank Road, Sheffield, publish information on all aspects of steel castings

THE WELDING INSTITUTE*, Abington, Cambridgeshire, publish information on all aspects of welding

Non-ferrous metals

Copper and its alloys

COPPER DEVELOPMENT ASSOCIATION, Orchard House, Mutton Lane, Potters Bar, Herts EN6 7AP, more than 20 brochures giving technical data, properties, manufacturing procedure, applications and suppliers of copper alloys are available without charge from this association

DAWSON, R. J. C., *Fusion Welding and Brazing of Copper and Copper Alloys*, Butterworths, Oxford (1973)

SMITH, C. S., 'Mechanical properties of copper and its alloys at low temperatures—a review', *Proceedings of the ASTM*, **39**, 642–648 (1939)

UPHEGROVE, C. and BURGHOFF, H. L., 'Elevated temperature properties of coppers and copper base alloys', *ASTM Special Publication No. 181*, ASTM (1956)

WEST, E. G., *Copper and its Alloys* (Industrial Metals Series), Ellis Horwood, Chichester (1982)

WEST, E. G., 'Copper manganese alloys'

WEST, E. G., 'Copper lead alloys for bearings (lead bronzes)'

Aluminium and its alloys

ALUMINIUM FEDERATION, 'The properties of aluminium and its alloys', *Publication No. A4*, Aluminium Federation (1983)

EVANS, B., MCDARMAID, D. S. and PEEL, C. J., 'The evaluation of the properties of improved aluminium–lithium alloys for aerospace applications', *SAMPE Conference, Montreux* (June 1984)

GRIMES, R., CORNISH, A. J., MILLER, W. S. and REYNOLDS, M. A., 'Aluminium–lithium based alloys for aerospace applications', *Metals and Materials*, **1**(6), 357–363 (1985)

HUFNAGEL, W., *Key to Aluminium Alloys—Designations, Compositions and Trade Names of Aluminium Materials*, Aluminium-Zentrale (1982)

LAVINGTON, M. H., 'The Cosworth process—a new concept in aluminium alloy casting production', *Metals and Materials*, **2**(11), 713–719 (1986). The Cosworth Process may be licensed through the International Mechanite Co.

PEEL, C. J., EVANS, B. and MCDARMAID, D. S., 'Development of aluminium–lithium alloys in the U.K.', *Metals and Materials*, **3**(8), 449–455 (1987)

WOODWARD, A. R., 'The use of aluminium for stressed components. Selection of materials in machine design', *Institute of Mechanical Engineers Conference Publication 22*, IME (4–5 Dec. 1973)

* Some, but not all, of the information published by these bodies is restricted to members of the particular institute or association.

Titanium and its alloys

IMI TITANIUM LTD, P.O. Box 704, Witton, Birmingham B6 7UR, publish a series of brochures listing the properties, fabrication procedures and applications of their standard range of alloys

LUTJERING, G., ZWICKER, U. and BURRK, W., *Titanium Science and Technology*, 4 Vols, Deutsch Gesellschaft fur Metallkunde EV (1984)

Magnesium and its alloys

ALICO, J., *Introduction to Magnesium and its Alloys* (1945)

BECK, A., *The Technology of Magnesium and its Alloys*, Magnesium Elektron, Twickenham (1940)

GRENFIELD, P., *Engineering Applications of Magnesium*, Mills and Boon Limited, London (1972)

MAGNESIUM ELEKTRON LTD, Royal House, London Road, Twickenham TW1 3QA, is the main British supplier who publishes brochures detailing the properties of the principal magnesium alloys, all of which they supply

RAY, M. S., *The Technology and Applications of Engineering Materials*, Prentice-Hall (1987)

UNSWORTH, W., 'Meeting the high temperature aerospace challenge', *Light Metal Age* (Aug. 1986)

UNSWORTH, W., 'Developments in magnesium alloys for casting applications', *Metals and Materials*, **4**(2), (1988)

Nickel and its alloys

MESSRS INCO ALLOYS INTERNATIONAL, Holmer Road, Hereford, publish a series of brochures detailing the properties and applications of the nickel alloys, which they supply free of charge

Zinc and its alloys

BARBER, N. I. and JONES, P. E., 'A new family of foundry alloys', *Foundry Trade Journal* (17 Jan. 1980)

GERVAIS, E., LEVENT, H. and BESS, M., 'Development of a family of zinc base foundry alloys', *84th Casting Congress of the American Foundrymen's Society*, St Louis, Missouri (Apr. 1950)

INTERNATIONAL LEAD ZINC RESEARCH ORGANISATION, INC., *Engineering Properties of Zinc Alloys*, ILZRO (Apr. 1981)

LYON, R., 'High strength zinc alloys for engineering applications in the motor car', *Metals and Materials*, 55–57 (Jan. 1985)

MARTIN, S. R., *The Technology and Application of Engineering Materials*, Prentice-Hall (1987)

MORGAN, S. W. K., *Zinc and its Alloys*, Macdonald and Evans, Estover, Plymouth (1972)

Low-melting-point metals

CARVIS, J. H. and GILBERT, P. T., *The Technology of Heavy Non-Ferrous Metals and Alloys Copper, Nickel, Zinc, Tin and Lead*, George Newnes Ltd, London (1967)

THE INTERNATIONAL TIN RESEARCH INSTITUTE, Fraser Road, Perivale, Greenford UB6 7AQ, will provide information on the properties of tin and its alloys

THE LEAD DEVELOPMENT ASSOCIATION (LDA), 3 Berkeley Square, London W1X 6AS, publish a number of brochures on lead and its alloys

RAY, M. S., *The Technology and Application of Engineering Materials*, Prentice-Hall (1987)

Cobalt and its alloys

CENTRE D'INFORMATION DU COBALT, *Cobalt Monograph*, Brussels (1960) gives a general account of cobalt alloys

DELORO STELLITE, Stratton St Margaret, Swindon, and WALL COLMONOY, Pontardawe, Swansea, market wear-resistant cobalt alloys as 'Stellites' and 'Triballoys' and as 'Wallex', respectively. Both these organisations publish brochures giving compositions and properties of their products

SIMS, C. T. and HAZELL, W. C. (Eds), *The Superalloys*, Wiley, New York (1972) gives a full account of the cobalt superalloys

Other non-ferrous metals

HARWOOD, J. S., *The Metal Molybdenum*, Symposium
  Proceedings of the ASM, Cleveland, OH (1958)
MCGACHIE, R. O. and BRADLEY, A. G., *Precious Metals*,
  Pergamon Press, Oxford (1981)
MILLER, G. L., *Zirconium*, Butterworth, Oxford (1957)
SIMON, E. W., *Guide to Uncommon Metals*, Frederick Muller,
  London (1967)
SULLY, A. H. and BRANDES, E. A., *Chromium*, Butterworth,
  Oxford (1967)
WICKERS, R. F., *Newer Engineering Materials*, Macmillan,
  London (1969)

Composites

LUBIN, G., *Handbook of Composites*, Van Nostrand Rheinhold,
  New York (1982)
Metals Engineering (May 1986) provides information on
  metal-matrix composites

Metals and Materials Vol. **2**, Nos 4, 6, 7, 9, 10 and 12 contain a
series of articles contributed by members of the Materials
Development Division of AERE dealing with all aspects of
filamentary and some short-fibre composites (with the exception
of cement based composites). Vol. **4**, No. 5 contains additional
information on filament-winding techniques. Vol. **3**, No. 11 deals
with some aspects of automotive applications of composites and
Vol. **4**, No. 7 with aircraft applications. Vol. **4**, No. 9 contains
two articles dealing with carbon–carbon composites. Vol. **2**, No.
3 contains an article dealing with metal-matrix composites

Engineering ceramics and glasses

KIRK, J. N., 'Ceramic components in automotive applications',
  Metals and Materials, **3**(11), 647–652 (1987)
NATIONAL ENGINEERING LABORATORY, Engineering
  Ceramics as Applied to Reciprocating Engines (1987)

2 Polymers, Plastics and Rubbers

John Brydson

Contents

2.1 Introduction

2.1.1 Polymers, rubbers and plastics

Rubbers and plastics are important examples of materials based on chemicals known as *high polymers*. Other materials which belong to the same family include adhesives, surface coatings and fibres.

Such polymers are characterised by being made up of a large number of repeating units. One such example is polyethylene which is usually made by joining together many molecules of ethylene, a gas consisting of two atoms of carbon and four of hydrogen and which has the structure shown on the left-hand side of the following diagram. (In this diagram 'C' represents a carbon atom and 'H' one of hydrogen, a single link represents a single bond and the double link between the two carbon atoms a double bond.)

$$
\begin{array}{cc}
H & H \\
| & | \\
C\!\!=\!\!C & \rightarrow \quad -CH_2\!-\!CH_2\!-\!CH_2\!-\!CH_2\!-\!CH_2\!-\!CH_2\!- \\
| & | \\
H & H
\end{array}
$$

Ethylene is an example of a *monomer* and the process of joining up the molecules is known as *polymerisation*. In this case chemical reactions are initiated that lead to the double bonds opening up so that the particular process is often known as *double-bond polymerisation*. Many other important plastics such as polyvinyl chloride (PVC), polypropylene and polystyrene are made in a very similar way.

In the above examples only one monomer is used to make the polymer and the products may be referred to as *homopolymers*. Where more than one monomer is used the process is known as *copolymerisation* and the products as *copolymers*. In the case of two monomers the products are strictly known as 'binary copolymers' (although the adjective is commonly dropped), whilst with three monomers the products are known as 'ternary copolymers' (or more commonly simply as *terpolymers*).

Other ways of making polymers also exist. Without going into detail, the polyamides (nylons) are made either by opening up small molecules which have a ring structure (*ring-opening polymerisation*) or by reacting small molecules which become joined together with the elimination of some small molecule. This is an example of *condensation polymerisation*, in this case the small molecule being split out is that of water.

To say that a chemical is a 'high polymer' means that it has a high relative molecular mass (which is also known as 'molecular weight'). This is approximately the number of times the molecule is heavier than an atom of hydrogen. Common laboratory chemicals usually have molecular weights of less than 300. For example, that of water is 18, benzene 78, ethyl alcohol 46 and sulphuric acid 98. Typical commercial polymers may have molecular weights ranging from a few thousand to over a million. A typical grade of polystyrene, for example, has a molecular weight of about 200 000.

Although it is the high molecular weight which confers on polymers important properties it is as well to bear in mind that, since the weight of an atom of hydrogen is only about 1.7×10^{-24} g, then the absolute weight of a polymer molecule is still very small.

A further point that should be made at this stage is that rubbers and plastics do not solely consist of polymers. It is usual to incorporate a number of *additives* into the polymer in order to adjust the properties. For example, PVC plastics may contain plasticisers (to increase flexibility), stabilisers (to improve heat and light resistance), lubricants (to aid flow or to stop sticking to processing equipment), fillers (such as china clay, usually primarily to reduce cost) and colorants.

2.1.2 Thermoplastics and thermosetting plastics

All of the polymers mentioned in the previous section are long chain-like polymers (generally referred to by chemists as *linear polymers*). Let us consider what happens when we heat up and subsequently cool one of these, namely polystyrene. At usual ambient temperature polystyrene is hard and rigid. The long-chain molecules are intertwined and are generally fixed in space, rather like a frozen mass of spaghetti. If the polystyrene is then heated up to above 100°C the molecules acquire energy which enables them in effect to 'wriggle about'. If subjected to an external stress such as a shear stress or simply the effect of gravity, the mass starts to move or, in common terminology, we say that the material has melted. At this stage the material may be shaped by such processes as moulding or extrusion. On cooling down the mass it will harden again. Materials that behave in this way are known as *thermoplastics*. In theory, heating and cooling processes can be repeated over and over again. However, polymers can be damaged to a greater or lesser extent each time they are heated and sheared, so that there is a limit to the number of times the process may be repeated.

Until about 1960, thermoplastics were less important than *thermosetting plastics*. As supplied to the processor, these materials contain fairly small polymer molecules. However, during and after shaping operations, reactions occur which cause the molecules to join up into a three-dimensional network and the polymer becomes hard or *set* at the processing temperature. This process is generally irreversible and is frequently referred to as *cross-linking*. (It may be noted at this point that, although the term 'thermosetting' was originally used to denote materials that were set at elevated temperatures a number of similar materials may now be hardened at normal ambient temperatures. Strictly, it is probably better to use the term *cross-linked plastics* to cover these materials, irrespective of the reaction temperature used.) Important thermosetting plastics include the phenolic resins, urea formaldehyde and melamine formaldehyde resins, and epoxide resins. Some polyesters are thermoplastic and some are thermosetting (and, for that matter, some are rubbers and some are fibres).

2.1.3 Amorphous thermoplastics and orientation

It is important to distinguish between two types of thermoplastic: the amorphous and the crystalline types. In the former type, exemplified by polstyrene and acrylic plastics such as polymethyl methacrylate, the polymer molecules will, given time, coil up into a random structure in the melt state. When the melt is cooled in the absence of external stresses this random structure will be retained. This leads to the following characteristics:

(1) the solid material will be isotropic in its properties (i.e. they will be the same in every direction);

(2) the shrinkage that occurs on solidification will be small (typically about 0.005 cm per centimetre in an injection-moulding process); and

(3) in the absence of powdery additives or bubbles of air or other gas the product is likely to be transparent.

In practice, the molecules are often subject to extensive shear in the melt during processes such as injection moulding and extrusion, and the melt freezes in an oriented or partly oriented state so that the product is anisotropic. The following types of orientation may be distinguished.

(1) Monoaxial orientation—this occurs when a molten rod is elongated into a filament. The molecules tend to align with the filament axis leading to increased strength and modulus in the axial direction.
(2) Biaxial orientation—this occurs when a sheet or film is stretched in two directions in the plane of the sheet. This can give a toughening effect to the sheet.
(3) Triaxial orientation—because solid thermoplastics are very resistant to volume change triaxial orientation leading to general dilatation is virtually impossible but locally may cause fracture.
(4) Complex orientation—this commonly occurs during injection moulding where orientation tends to follow the lines of melt flow and thus varies from place to place.

Orientation may or may not be a desirable feature as will be discussed in Section 2.8.

2.1.4 Crystalline thermoplastics

Several thermoplastics in the solid state show many of the characteristics of crystalline materials. In such cases individual molecules pass through several regions where segments of many molecules are arranged or packed in a highly ordered way. Such ordered regions tend to be spatially mingled with less ordered and virtually amorphous regions. Examples of crystalline polymers are polyethylene, polypropylene, the nylons and the polyacetals. As a result of this ability to pack, crystalline polymers show the following general characteristics:

(1) there is a greater shrinkage on cooling from the melt (typically 0.015–0.060 cm per centimetre), which depends on polymer type and the processing conditions;
(2) properties can be affected by seeding to encourage initiation of crystallinity;
(3) because the different zones will have different densities and hence different refractive indices crystalline polymers tend to be opaque although there are some exceptions; and
(4) since levels of crystallinity vary with temperature and with processing conditions, the physical properties of products made from crystalline thermoplastics are more dependent on the manner in which they have been processed and on the ambient temperature than are those of amorphous thermoplastics.

As with the amorphous thermoplastics, when cooled from the molten state without subjection to external stress crystalline thermoplastics are isotropic in that the crystalline structure tends to be randomly dispersed. Furthermore, orientation of the various types indicated with amorphous thermoplastics can also occur. In the case of crystalline thermoplastics, however, orientation of the crystal structures is more important than orientation of molecules. Most synthetic fibres are made by drawing crystalline thermoplastic materials. Because of the layering effect that takes place it is interesting to note that biaxial stretching tends to enhance the transparency of crystalline materials.

2.1.5 The glass transition and the crystalline melting point

Successful processing and use of plastics and rubbers are facilitated by an understanding of two properties: the *glass transition temperature* and the *crystalline melting point*.

At room temperature an amorphous thermoplastic material such as polystyrene is hard and rigid because the constituent molecules are also effectively frozen in space. If, however, polystyrene is heated up to above 90°C the material becomes more flexible and further heating takes it into a rubbery state. If the molecular weight is not too high, raising the temperature above about 150°C causes the material to melt but, in the case of very high molecular weight materials (e.g. over 500 000), the material may decompose before melting as the temperature is raised. The change at about 90°C arises because the polymer molecules start to 'wriggle' and on application of a stress they tend to be elongated. However, entanglements and limited motion stop the chains from sliding past each other so that, on release of a stress, the molecules tend to coil up randomly and the sample recovers, at least in part, its original shape. In other words its exhibits *rubberiness* (also known as 'reversible high elasticity'). This transition from being glass like to rubbery is known as the *glass transition temperature*, universally given the abbreviation T_g. Many properties other than mechanical ones change at the glass transition temperature whilst processing behaviour is very dependent on the relationship between processing temperatures and T_g. It may be noted that T_g does depend slightly on molecular weight and also on the method by which it has been measured, so that any quoted figure can only be considered as approximate. However, for practical purposes, this does not matter too much.

In the case of crystalline polymers the changes at T_g are less, the extent depending on the degree of crystallinity. Indeed, with highly crystalline materials T_g is often difficult to identify. This is because many of the polymer molecules are incorporated in whole or in part in crystalline structures which still exist above T_g, thus still effectively freezing the mass. In the case of lightly crystalline structures the crystal structures present act more like cross-links between which the molecules are able to coil and uncoil on application of a stress so that a measure of rubberiness may be observed.

If, however, the crystalline polymer is progressively heated to higher temperatures, it may be observed that the degree of crystallinity will fall. Eventually a temperature is reached where the last traces of crystallinity disappear and this is known as the *crystalline melting point* (T_m). (In practice, it is becoming more common to obtain this value from peaks observed in differential scanning calorimetry which may be slightly lower.)

It is frequently found that the following rule-of-thumb relationship exists between T_g and T_m

$$T_g = 2T_m/3 \text{ (when expressed in degrees Kelvin)}$$

In many common instances this means that the crystalline melting point is about 100 K above the T_g. Providing the molecular weight is not so high that entanglements prevent flow above this temperature, the polymer rapidly becomes molten rather than rubbery. The various relationships are illustrated schematically in *Figure 2.1*.

At this point it is also instructive to show graphically how the stiffness or modulus of the various types of material considered so far are affected by temperature. This is shown in *Figure 2.2*.

2.1.6 The elastic state and vulcanised rubber

It has already been indicated that high elasticity in polymers arises from the fact that, on application of a stress, molecules above T_g are able to deform from a randomly coiled up state. Where there is some entanglement, flow or slippage of molecules past one another is difficult so that when the stress

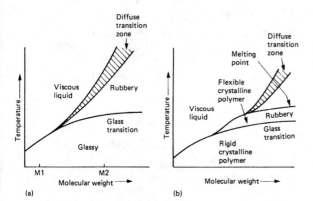

Figure 2.1 Temperature vs. molecular weight diagrams for (a) amorphous and (b) moderately crystalline polymers (with highly crystalline polymers the glass transition is less apparent). (Reproduced by permission from Brydson, *Plastics Materials*, Butterworth)

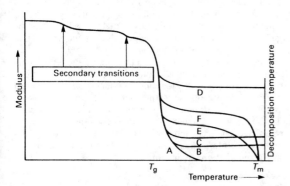

Figure 2.2 Scheme to show the dependence of the modulus of a polymer on a variety of factors. A is an amorphous polymer of moderate molecular weight, whereas B is of such a high molecular weight that entanglements inhibit flow. Similar effects are shown in C and D, where the polymer is lightly and highly linked, respectively. In E and F the polymer is capable of crystallisation, F either being more highly crystalline than E or containing fibre reinforcement. (Reproduced by permission from Brydson, *Plastics Materials*, Butterworth)

is removed the molecules recover a coiled up state. Simply relying on entanglements or small crystalline structures acting as knots is somewhat inefficient and substantial slippage or creep may occur. This may be very much reduced by introducing a small amount of cross-linking, as with thermosetting plastics, but to a much lower extent than prevents chain slippage. In rubber technology this process is known as *vulcanisation*. It is interesting to note that the process was first discovered by Charles Goodyear in the USA as long ago as 1839 when natural rubber was heated with sulphur, a process exploited first by Thomas Hancock in London. Today the use of 1–2 parts of sulphur per 100 parts of rubber is still by far the most common way of vulcanising both natural and most of the large tonnage synthetic rubbers.

In effect the cross-links tend to lock the polymer chains relative to each other in space so that on release of any deforming stress the mass recovers its original shape. Unfilled

vulcanised rubbers may be stretched more than 1000%, hence the term 'high elasticity' as distinct from the 'elasticity' of metals. Commercial rubbers, however, usually have much lower breaking elongations (150–550%) in order to obtain improvements in other properties such as stiffness, tear resistance and abrasion resistance.

2.1.7 Thermoplastic rubber

One of the more interesting developments in the past 30 years has been that of thermoplastic rubbers. One of the problems with conventional vulcanised rubber is that, once cross-linking or vulcanisation has taken place, the material is, like a thermosetting plastic, 'set' and cannot be melted and reprocessed. Thus material used in a process but not incorporated into the final product or defective products cannot be re-used as can a thermoplastic material. Thus many attempts have been made to produce a rubbery material in which effective cross-links exist at normal use temperature but which disappear (become *heat fugitive*) at elevated temperatures. To some extent entanglements and crystalline zones fulfil this role and rubbers have been used that are simply very high molecular weight polymers (but thus difficult to process) or slightly crystalline. Polymers have also been made in which ionic cross-links exist at low temperatures, but which lose their force at elevated ones. A number of true covalent-bonded systems have been devised in which these bonds break down at elevated temperature, but which reform at low temperatures. Such systems have been of limited use. Far more successful has been the use of *block copolymers*.

Such block copolymers differ from the more common *random copolymers* in that the monomers of each type are grouped together in one chain. One such material consists of a block of butadiene molecules (forming a segment of a rubbery polybutadiene block) set between two blocks of styrene molecules (forming glassy polystyrene blocks). Such as system is known as a *styrene–butadiene–styrene (SBS) triblock copolymer*. At room temperature the polystyrene ends congregate into domains effectively forming cross-links between many triblock molecules at the chain ends. However, above T_g these domains tend to break up and, because the overall molecular weight is quite low, the whole system melts and is capable of flow. When the melt is cooled the domain structures and thus the cross-links reform. Slightly different are the polyether–polyester block copolymers in which amorphous polyether zones are separated by crystallisable polyester blocks. At room temperature these latter blocks do crystallise together to produce small crystal structures which act as cross-links. These latter materials are available in a variety of polyether/polyester ratios and thus vary in stiffness and rubberiness. Because of the high melting point of the polyester blocks these materials have good heat resistance, and because of their chemical nature they have good oil resistance. They have become important engineering rubbers.

Blends of polypropylene (which is a crystalline polymer) with rubbery ethylene–propylene rubbers, which are very slightly crystalline have also been prepared. Appropriately modified by a heat treatment these blends have also been successful as thermoplastic rubbers.

2.1.8 Additives

The polymers used for making rubbers and plastics are very rarely used on their own without the incorporation of additives. In the case of a rubber it is usually necessary to incorporate a *vulcanising agent* such as sulphur. Since the reaction between most rubbers with sulphur is slow and inefficient, a chemical known as an *accelerator* is often used, an additive which often works better in the presence of an *accelerator activator* such as zinc oxide and a fatty acid such as stearic acid. To retard ageing, *antioxidants* and *antiozonants* may be used. To increase stiffness, tear resistance and abrasion resistance *reinforcing fillers* such as fine particle carbon blacks may be incorporated, whilst coarser, so-called *inert fillers* may be used primarily to reduce cost. Adjustment of stiffness may be made by softeners (plasticisers), whilst blowing agents may be used to give a cellular structure. Non-black compounds may incorporate colorants. Thus, for example, a bridge bearing based on natural rubber could consist of the following formulation:

Rubber	Natural rubber	100
Vulcanising system	Sulphur	2.5
	Zinc oxide	5
	Stearic acid	2
	Accelerator (e.g. CBS)	0.6
Reinforcing filler	General purpose furnace type (GPF) carbon black	50
Protective system	Antioxidant (TMQ)	3
	Antiozonant (IPPD)	2
	Paraffin wax	2

where CBS is *N*-cyclohexylbenzothiazole-2-sulphenamide, IPPD is *N*-isopropyl-*N'*-phenyl-*p*-phenylene diamine, and TMQ is polymerised 1,2-dihydro-2,2,4-trimethylquinoline.

As one example with thermoplastics, a compound for PVC wire covering insulation flex would contain, in addition to the polymer, *plasticiser(s)* to make the material flexible, china clay as a *filler* which also improves electrical insulation properties, a *stabiliser* to reduce degradation on heating and exposure to light, a *colorant* and a *lubricant* (to prevent sticking to processing machinery). For both rubbers and PVC other additives may also be used for other purposes and similar comments also apply for other rubbers and plastics.

2.1.9 Polymer blends

In recent years increasing use has been made of blends of polymers. In many cases the reason for this is to produce tough polymers and frequently this involves blending a glassy amorphous thermoplastic with a rubbery material as, for example, with toughened polystyrene, a blend of glassy polystyrene with rubbery polybutadiene, and the acrylonitrile–butadiene–styrene (ABS) plastics, originally blends of glassy acrylonitrile–styrene with butadiene–acrylonitrile, but today rather more complex structures. Other important blends include blends of polystyrene and the crystalline polymer 2,6-dimethyl-*p*-phenylene oxide (the produce marketed as Noryl) and blends of polypropylene and ethylene–propylene rubbers to give a type of thermoplastic rubber.

In principle, these materials, often known as *polyblends*, are a physical mixture of different polymers and are thus distinct from copolymers where the different monomers are on the same polymer chain. Therefore a blend might be expected to have quite different properties from a copolymer. For example, a random copolymer usually has too irregular a structure to enable it to crystallise, whereas it is often possible to make a polyblend containing one or more crystalline polymers so that the polyblend retains a measure of crystallinity. In practice this situation may be more complicated with some blends involving copolymers, whilst in some instances there may be chemical reaction between the components of a blend.

2.1.10 Composites

By definition a *composite* consists of two or more physically distinct materials which are combined in a controlled way. Three classes of composite may be distinguished:

(1) where one material (component) forms a continuous phase (*continuous matrix*) and the other components are embedded within this matrix;
(2) *laminar* structures where the components are in separate layers (alternating where there are just two components); and
(3) *interpenetrating* systems where both components form a continuous matrix.

Strictly speaking any polymer compound containing an additive or a polyblend could be considered a composite, although common usage of terminology excludes these materials.

Whilst all three types are used the most typical in the case of rubbers and plastics is the first group and, specifically, the *structural composites* which are characterised by the matrix material being enhanced in stiffness and strength by the presence of a *reinforcing material* embedded within it. Where the continuous phase is a polymer, such materials are known as *polymer-based structural composites*.

Such systems have long been used in the rubber industry where, for example, the strength of a tyre is greatly enhanced by incorporating fibrous filament or cloth which may be organic (e.g. cotton) or a synthetic fibre, or inorganic (e.g. glass fibre or metal filament).

In the plastics industry, laminates using paper have long been used for electrical insulation, whilst cloth-based phenolic laminates have been used for constructional materials for most of the twentieth century. Rather better known is the use of glass-fibre reinforcement in polyester resins for sports cars, boats and railway engine cabs and in epoxide resins in the aircraft and chemical industries. For extreme operating demands the reinforcement may be carbon fibre or an *aramid* fibre (aromatic polyamide fibre).

2.2 General properties of rubbery materials

In this section the general properties of rubbery materials relevant to their use in engineering applications are considered, followed in the next section by a short review of the types available. Unless otherwise stipulated, the use of the word 'rubber' implies vulcanised rubber.

Today, the bulk of manufactured rubber articles is used in engineering applications such as tyres, springs, seals, bearings, mine belting, transmission belting, chemical plant liners and electrical cable and wire insulation in severe environments. Successful use clearly requires an understanding of relevant properties.

2.2.1 Stress–strain properties in tension

Probably the best known property of rubbers is that they are capable of large elastic deformations. It is also well known that this deformation is highly recoverable on release of stress. Two other properties may be noted at this stage:

(1) rubbers have a low modulus or stiffness;
(2) rubbers are virtually incompressible in bulk (Poisson's ratio ν is about 0.5).

In tension an unfilled natural rubber vulcanisate (the product of vulcanisation) may be extended in tension by more than 1000%. Commercial products which more commonly contain large amounts of filler will have lower *elongations* at break, but for many engineering applications values do exceed 400%. The tension stress–strain curve is not linear but is somewhat S shaped (*Figure 2.3*).

A number of attempts have been made to derive theoretical expressions to fit this curve. The so-called 'Gaussian theory' yields the expression

$$f = \tfrac{1}{2}G(\lambda_1 - 1/\lambda_1^2)$$

where f is the tensile stress, G the shear modulus and λ the strain (strained length/original length).

This gives fair fits up to about 500% elongation, but is poor at higher elongations. A rather better fit is given by the Mooney–Rivlin equation

$$f = 2(\lambda - 1/\lambda^2)(C_1 + C_2/\lambda)$$

where C_1 and C_2 are constants.

In practice, most uses of rubber in engineering applications involve elongations of less than 50%. In this range the rubber is almost Hookean in behaviour and it is possible to use a Young's modulus derived from the slope within this range.

Compared with metals the modulus of rubbery materials is very low. Some typical figures are given in *Table 2.1*.

2.2.2 Modulus and hardness

Rubber technologists make wide use of *hardness measurements*. With rubbers this involves the measurement of a reversible deformation produced by a specified indentor,

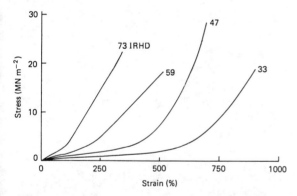

Figure 2.3 Tensile stress – strain curves for four natural rubber compounds of different hardness: 73 IRHD contains 50 parts of a reinforcing black; 59 IRHD contains 50 parts of a non-reinforcing black; and different vulcanising systems account for the different curves of the two gum compounds (47 and 33 IRHD). (Reproduced by permission from Lindley, *Engineering Design with Natural Rubber*, Malaysian Rubber Producers' Research Association)

Table 2.1 Comparison of the modulus and Poisson's ratio of rubbers with those of mild steel

	Natural rubber		
	Gum*	Filled†	Mild steel
Young's modulus (MPa)	1.8	5.9	210 000
Shear modulus (MPa)	0.54	1.37	81 000
Poisson's ratio	0.4997	0.4997	0.29

* A vulcanisate of an unfilled natural rubber compound.
† A vulcanisate.

unlike metal hardness measurements which are concerned with an irreversible plastic indentation. There is a good correlation between hardness and Young's modulus. The most widely used hardness test is that which yields 'international rubber hardness degrees' (IRHD). Also often quoted in the literature are British Standard (BS) hardness degrees and Shore A units. All three give similar numerical values. *Table 2.2* shows the relationship between various types of modulus and IRHD.

2.2.3 Bulk modulus, Poisson's ratio and incompressibility

The bulk modulus of rubber is in the range 1000–2000 MPa and Poisson's ratio is almost exactly 0.5. This means that solid rubber vulcanisates are virtually incompressible (in bulk). *It is therefore essential that if rubber is being used in an application which involves deformation, there should be some room for the rubber to deform.* For example, if a rubber disc was placed at the base of the inside of a cylinder and a close-fitting piston was dropped into the cylinder, the piece of rubber would be unable to deform and would behave as a hard solid. Failure to realise this fact of incompressibility even with very soft rubbery articles has led to embarrassing design failures. (This restriction does not, of course, apply to cellular (sponge) rubber articles.)

2.2.4 Stress–strain properties of rubber in shear

At constant rates of shear deformation the stress–strain relationship in shear is linear up to about 100% shear strain, i.e. Hookean behaviour may be assumed in shear. In this case it is possible to postulate the relationship

$$\tau = G \tan \gamma$$

where τ is the shear stress and $\tan \gamma = x/t$ where x is the displacement in the direction of shear and t is the separation between shearing surfaces (*Figure 2.4*).

It is also possible to define the shear stiffness K of a mounting being sheared by

$$K = F/x = GA/t$$

where F is the shearing force and A the area over which the force is applied.

It may be noted at this stage that inserting plates into a rubber mounting parallel to the shear direction does not affect the stiffness of the plate, providing that the total thickness of rubber in the mounting is unchanged. The importance of this will become apparent later.

2.2.5 Stress–strain properties of rubber in compression

The compression stress–strain properties of vulcanised rubber are usually measured by compressing a block of rubber between two metal plates. Unless special measures are taken to lubricate the interfaces between rubber and metal, the rubber at the interface will be constrained. Thus, on compression the interfacial area will not change but, since rubber is incompressible in bulk, there must be a lateral expansion between the plates which is at a maximum at the midplane of the sample. In turn this will mean that the free surface of the rubber block will increase, setting up tensile strains on the surface (*Figure 2.5*).

This tensile strain will be in addition to the forces required simply to compress the sample and will, therefore, increase the stiffness. It may be appreciated that this effect is greater

Table 2.2 Hardness and elastic moduli*†

Hardness‡ (IRHD ± 2)	Young's modulus§, E_0 (MN m^{-2})	Shear modulus//, G (MN m^{-2})	k**	Bulk modulus, E_∞ (MN m^{-2})
30	0.92	0.30	0.93	1000
35	1.18	0.37	0.89	1000
40	1.50	0.45	0.85	1000
45	1.80	0.54	0.80	1000
50	2.20	0.64	0.73	1030
55	3.25	0.81	0.64	1090
60	4.45	1.06	0.57	1150
65	5.85	1.37	0.54	1210
70	7.35	1.73	0.53	1270
75	9.40	2.22	0.52	1330
Shore A (approx.)	(lbf in.$^{-2}$)	(lbf in.$^{-2}$)		(lbf in.$^{-2}$)
35	168	53	0.89	142 000
45	256	76	0.80	142 000
55	460	115	0.64	154 000
65	830	195	0.54	171 000
75	1340	317	0.52	189 000

* Reproduced by permission from *Engineering Design with Natural Rubber*, 4th edition, Malaysian Rubber Producers' Research Association.
† Based on experiments on natural rubber spring vulcanisates containing (above 48 IRHD) SRF black as filler. Note that hardness is subject to an uncertainty of about ±2 IRHD.
‡ The majority of springs are in the hardness range 40–60 IRHD.
§ Theoretically, with a Poisson's ratio of ½, E_0 should equal $3G$. This is so for soft gum rubbers, but for harder rubbers containing a fair proportion of non-rubber constitutents, thixotropic and other effects increase E_0 to about $4G$.
// Average design limits: 15% compression, 50% shear.
** k is used in the calculation of compression characteristics.

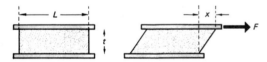

Figure 2.4 Rubber mounting subjected to shear deformation

Figure 2.5 Rubber block in compression.

with thin blocks and it is necessary to introduce a *shape factor* S in order to determine the compression modulus E_c effective in any particular sample, test piece or mounting. This shape factor is given by

$$S = LB/2t(L + B)$$

for a rectangular block of length L, breadth B and thickness t.
For a block of square section (i.e. $L = B$) or circular section (diameter L) then

$$S = L/4t$$

The compression modulus depends on the shape factor by the relationship

$$E_c = E_Y(1 + 2kS^2)$$

where E_Y is Young's modulus and k is a numerical factor the values of which are given in *Table 2.2*.
The compression stiffness K_c may be defined as

$$K_c = F/x = E_c A/t$$

where x is the deflection.
The load (F) vs. deflection (x) curve for vulcanised rubber is non-linear. Providing there is no slip at the interface it approximates to

$$F = E_c Ae(1 + e)$$

where e is the compressive strain ($= x/t$). The non-linearity is usually ignored for strains up to 10% (*Figure 2.6*).
If plates parallel to the metal–rubber interfaces are inserted into the blocks, one thick sample will in effect be replaced by a number of thinner samples with a higher shape factor (*Figure 2.7*). In effect we have a stacked array of rubber springs. If these springs have deflections x_1, x_2, x_3, etc., the total deflection is

$$x_1 + x_2 + x_3 + \ldots$$

Since the deflecting force will be the same for each element, then the total stiffness for the unit K_{tot} is given by

$$1/K_{tot} = 1/K_1 + 1/K_2 + 1/K_3 + \ldots$$

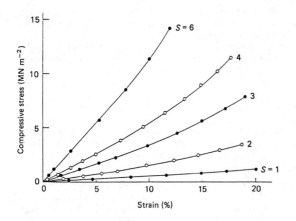

Figure 2.6 Effect of shape factor: experimental stress–strain curves for 6.3 mm thick discs of rubber (47 IRHD) in compression. The shape factor is shown alongside each curve; the diameter (in millimeters) is 25.4 times the shape factor. (Reproduced with permission from Lindley, *Engineering Design with Natural Rubber*, Malaysian Rubber Producers' Research Association)

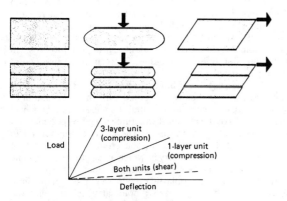

Figure 2.7 The vertical stiffness of a rubber block can be increased by inserting into the block horizontal metal spacer plates which reduce the freedom of the rubber to bulge. The shear stiffness is not altered by the presence of these horizontal plates. (Reproduced by permission from Lindley, *Engineering Design with Natural Rubber*, Malaysian Rubber Producers' Research Association)

If all the rubber 'springs' between the plates are equal and there are n separating metal plates, i.e. $(n + 1)$ rubber springs, then the expression simplifies to

$$K_{tot} = K/(n + 1)$$

where K is the compression stiffness of each element.

Simple calculations show that insertion of such metal plates markedly increases the compression stiffness. For example, *Table 2.3* illustrates how the insertion of a number of metal plates increases the stiffness of a rectangular mounting of loaded face area L^2 and thickness $t = L/4$. This has the effect of *increasing the compression stiffness but not affecting shear stiffness*. This is of considerable importance as a means of producing rubber components with differing levels of stiffness in different directions. This is discussed in Section 2.7.1.3.

Table 2.3 Effect of the insertion of metal plates on the compression stiffness of a rubber mounting of loaded face area L^2 and thickness $L/4$

No. of plates	S	Increase in stiffness,* K/K_o
1	1	3
2	2	9
3	3	19
5	5	51
7	7	99
10	10	201

* K is the stiffness of the composite and K_o is the stiffness of the block with no inserted metal plates.

2.2.6 Resilience

The resilience of a rubber may be considered as the energy recoverable after deformation. A simple example is where a metal ball is dropped onto a slab of rubber from a height H. If the ball rebounds to a height $0.7H$ it would be said that the resilience of the rubber is 70%. Energy not recovered is converted to heat. Since no rubber is 100% resilient it follows that if a piece of rubber is constantly flexed or otherwise deformed there is a build up of heat which, unless dissipated, causes a temperature rise and possibly failure of the part. For this reason such components as sidewalls of lorry tyres need to be made from high resilience compounds.

Above T_g, resilience increases with temperature up to about $(T_g + 100)°C$. Since the rate of change of resilience with temperature is similar for many, but not all, rubbers, this also tends to mean that the lower the T_g of the rubber the higher the resilience at normal ambient temperature (*Figure 2.8*).

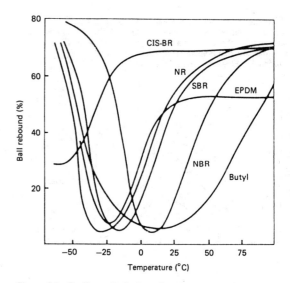

Figure 2.8 Resilience (ball rebound) vs. temperature relationships for butyl rubber compared with other major rubbers. (Reproduced by permission from Gunter, *Developments in Rubber Technology*, 2nd edition, Applied Science)

2.2.7 Creep, set and dynamic properties

When a polymer is subjected to a tensile stress the total deformation may be considered as the sum of three components:

(1) an instant deformation due to bond bending and stretching (D_{OE}) similar to that of ordinary elastic bending in metals;
(2) chain uncoiling which is not instantaneous and leads to high elastic deformation (D_{HE}); and
(3) chain slippage leading to viscous deformation (D_{visc}).

Hence,

$$D_{total} = D_{OE} + D_{HE} + D_{visc}$$

These various responses and the total response are shown in *Figures 2.9* and *2.10*.

The relative proportions of each component vary with the type of polymer, but with a vulcanised rubber the D_{HE} component predominates. Even so there will be some irreversible *creep* which will progress with time. Similarly, if a piece of rubber is held under constant strain, the stress required to maintain this strain decreases with time. This phenomenon is known as *stress decay*.

Total deformations in other strain modes have similar relationships and, in practice, since rubber is more often used in compression and shear, these modes can be more important. In particular it is important that, when rubber is being used as a compression seal, the stress decay is not so great that seal fails. Also of concern is the *compression set*, i.e. the irrecoverable component of a compressive deformation.

Because D_{HE} is not instantaneous, strain will lag behind stress in a cyclic deformation. The mathematics are similar to those in a.c. theory and at this point just two points will be made.

(1) If the stressing frequency is progressively increased there will be less time for the strain to reach equilibrium so that the response in each cycle will decrease. This will effectively increase the modulus.
(2) By paying regard to the natural vibrating frequency of a rubber component it is possible to isolate vibrations. This technique is useful, for example, for protecting delicate instruments from machinery vibrations and buildings from earthquakes.

2.2.8 Non-rubbery properties of technical importance

Clearly there are many properties other than stress–strain properties that are often important in the use of a rubbery material. These may include heat resistance, oil resistance, electrical insulation behaviour, general chemical resistance, permeability to gases, colour and, always importantly, cost. In this subsection brief consideration is given to the first two of these properties.

In the case of a cross-linked (vulcanised) rubber, the heat resistance will be mainly concerned with the degradation in structure that occurs at elevated temperatures. This will depend on both the chemical structure of the raw polymer *and* on the vulcanising system used. Therefore it is important to understand that any statement concerning the relative heat resistance of various rubbery polymers needs to be given some latitude depending on the vulcanising system employed.

In the case of thermoplastic rubbers the upper service limit is determined by the softening point at which the material loses its rubberiness and begins to flow.

In many applications rubbery components are often in contact with hydrocarbon oils. As a general rule the hydrocarbon rubbers (which form the bulk of the market) have less oil resistance than do non-hydrocarbon rubbers.

2.2.8.1 Classification system for vulcanised rubbers

The American Society for Testing and Materials (ASTM) and the Society of Automotive Engineers (SAE) have issued a report entitled *Classification System for Rubber Materials for Automotive Applications* (SAE J200/ASTM D2000); BS 5176 issued by the British Standards Institution is almost identical. The classifications do not at present apply to thermoplastic rubbers.

The classification provides basic classification codes based on test values obtained in standard laboratory tests. Each code is made up of two parts: the first part based on *basic requirements*; the second on *supplementary* or *suffix requirements*. These provide, for a given specification, a *line call-out*.

The first part of the line call-out results from the basic requirements and consists of:

(1) a letter 'M' to indicate that SI units are being used (this may be followed or preceded, according to which specifi-

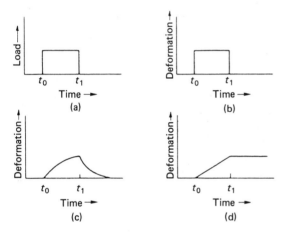

Figure 2.9 Types of deformational response as a result of a fixed load being imposed between times t_0 and t_1 (a). (b) Ordinary elastic material. (c) Highly elastic material. (d) Viscous material. (Reproduced by permission from Brydson, *Rubbery Materials and their Compounds*, Elsevier/Applied Science)

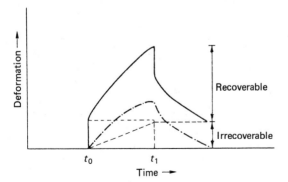

Figure 2.10 Deformation vs. time curve for material showing substantial ordinary elastic, high elastic and viscous components of deformation. (Reproduced by permission from Brydson, *Rubbery Materials and their Compounds*, Elsevier/Applied Science)

cation is being used, by a number indicating supplementary requirements);

(2) a letter indicating the *type*—assigned on the level of heat resistance required;

(3) a letter indicating the *class*—assigned on the level of mineral oil (hydrocarbon oil) resistance required;

(4) a single figure indicating the hardness range; and

(5) a double figure indicating the tensile strength.

The *type* is based on changes in tensile strength of not more than 30%, an elongation change of −50% maximum and a hardness change of not more than 15 points. The temperatures at which the materials are tested for determining type are shown in *Table 2.4*.

The *class* is based on the resistance of material to swelling in ASTM Oil No. 3 (or in the case of BS 5176, Oil No. 3 of BS 903). In test the immersion time shall be 70 h and the oil temperature shall be the type temperature or 150°C, whichever is the lower. Limits of swelling are listed in *Table 2.5*.

The basic mechanical properties are indicated by a single figure for hardness followed by a double figure for tensile strength. Thus the set '810' indicates a Shore hardness (a commonly used hardness scale) of 80 ± 5 and a tensile strength of 10 MPa minimum. (The BS specification uses IRHD and with slightly different tolerance limits of +5 to −4.)

An example of an ASTM D2000 line call-out is: M6 HK 810 A1-11 B38 EF31. This indicates the use of SI units, an elastomer of type H (resistant to 250°C) and of class K (10% swell), a hardness of 80 ± 5 and a minimum tensile strength of 10 MPa. The number '6' after 'M' provides a key to supplementary suffix requirements which are detailed in the second part of the call-out (A1-11 B38 EF31). Interpretation of suffix requirements requires reference to details in the appropriate standards, but it may be noted that in the above example call-out these refer to particular heat resistance, compression set and fuel resistance requirements.

Figure 2.11 shows in graph form the relative positions of selected rubbers in respect of type and class. *Table 2.6* provides a summary of properties of most commercial rubbery materials.

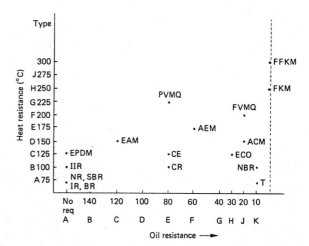

Figure 2.11 SAE J200 ASTM D2000 classification for cross-linked elastomers. Whilst specific points are shown for clarity, there will be a spread for each rubber, the exact position depending on the grade of polymer and compounding formula used. (Reproduced by permission from Brydson, *Rubbery Materials and their Compounds*, Elsevier/Applied Science)

2.3 Survey of commercial rubbery materials

There are about 30 types of rubbery material on the market, although just two (natural rubber and styrene–butadiene rubber) account for about 70% of total tonnage and three for about 80%. These materials are listed together with their recommended ISO/ASTM abbreviations in *Table 2.7*.

In addition to the nomenclature based on ISO and ASTM recommendations, several other abbreviations are widely used. Those most likely to be encountered are listed in *Table 2.8*.

During World War II the US Government introduced a system of nomenclature which continued in use, at least partially, until the 1950s and is used in many publications of the period (see *Table 2.8*).

Rubber usage continues to be dominated by the tyre industry, which consumes about two-thirds of total new rubber each year. Hence it is not surprising that those rubbers widely used in the industry tend to dominate the market, not simply because they are used in tyres but also because the large-scale production tends to give these materials a competitive price structure. The term *tyre rubbers* is given to these materials.

In the non-tyre industry, tyre rubbers are used alongside other rubbers and four such materials are used in this area in quantities similar to those of the tyre rubbers. These materials are commonly known as *special purpose rubbers*, although in reality they may be used in a wider range of applications than some of the tyre rubbers.

There are a number of applications, particularly in engineering, where more severe demands are put on the rubber than can be met with the tyre rubbers and special purpose rubbers, particularly in terms of heat and oil resistance. The term *speciality rubber* is commonly used for these materials. Some extremely specialised materials, available only in small quantities, and at high cost are sometimes separately classed as *exotics*.

Table 2.4 Classification of rubber by type—assigned according to the level of heat resistance

Type	Test temperature (°C)	Type	Test temperature (°C)
A	70	F	200
B	100	G	225
C	125	H	250
D	150	J	275
E	175		

Table 2.5 Classification of rubber by class—assigned according to the level of resistance to mineral oil

Class	Maximum volume swell (%)	Class	Maximum volume swell (%)
A	No requirement	F	60
B	140	G	40
C	100	H	30
D	100	J	20
E	80	K	10

Table 2.6 Summary of properties of principal rubbery materials*

Property	ASTM D1418 class							
	NR	SBR	BR	IR	EPDM	IIR	CR	NBR
ASTM D2000 rating	AA	AA/BA	AA/BA	AA	CA	BA, AA	BC, BE	BF, BG, BK, CM
Specific gravity	0.92	0.94	0.93	0.93	0.86	0.92	1.23	1.00
Tensile strength (gum) (MPa (psi))	>20.7 (3000)	<6.9 (1000)	<6.9 (1000)	>17.2 (2500)	<6.9 (1000)	>10.3 (1500)	>20.7 (3000)	<6.9 (1000)
Tensile strength (black loaded) (MPa (psi))	>20.7 (3000)	>13.8 (2000)	>20.7 (3000)	>20.7 (3000)	>20.7 (3000)	>13.8 (2000)	>20.7 (3000)	>13.8 (2000)
Tear resistance	Good–v.good	Fair	Fair	Good–v. good	Fair–good	Good	Good	Fair
Abrasion resistance	Excellent	Good–excellent	V. good	V. good	Good–excellent	Good	Excellent	Good
Compression set	Good	Good	Good	V. good	Good	Fair	Fair–good	Good
Rebound (cold)	Excellent	Good	Excellent	Excellent	V. good	Poor	V. good	Good
Rebound (hot)	Excellent	Good	Excellent	Excellent	Excellent	Good	V. good	Good
Dielectric strength	Excellent	Excellent	V. good	V. good	Excellent	Excellent	Good	Poor
Electrical insulation	Good–excellent	Good–excellent	V. good	V. good	Excellent	Good–excellent	Fair–good	Poor
Environmental resistance								
Oxidation	Good	Fair	Fair	Fair	Excellent	Excellent	V. good–excellent	Good
Ozone	Poor–fair	Poor	Poor	Poor	Outstanding	Excellent	V.good	Fair
Sunlight	Poor	Poor	Poor	Poor	Outstanding	V. good	V.good	Poor
Heat ageing (upper limit at constant service)(°C)	85	90	90	85	145	120	95	120
Low-temperature flexibility	Excellent	V. good	Excellent	Excellent	Excellent	Good	Good	Fair–good
Fluid resistance								
Gas permeability	Fair	Fair	Fair	Fair	Fair	V. low	Low	V. low
Acid resistance (dilute)	Good	Good	Good	Good	Excellent	Excellent	Excellent	Good
Acid resistance (conc.)	Good	Good	Good	Good	Excellent	Excellent	V. good	Good
Swelling								
In aliphatic hydrocarbons	Poor	Poor	Poor	Poor	Poor	Poor	Fair–good	Excellent
In aromatic hydrocarbons	Poor	Poor	Poor	Poor	Poor	Poor	Fair	Good
In oxygenated solvents	Fair–good	Good	Good	Fair–good	Good	Good	Poor	Poor

Property	ASTM D1418 class							
	ACM	ECO/CO	AEM	CSM	FKM	PVMQ	FVMQ	YBPO
ASTM D2000 rating	DJ	CM	EF	CE	HK	GE	FJ	Not assigned
Specific gravity	1.10	1.27–1.36	1.08–1.12	1.12–1.28	1.85	1.06–1.42 (compound)	1.41 (compound)	1.17–1.25
Tensile strength (gum) (MPa (psi))	—	<6.9 (1000)	—	>17.2 (2500)	>12.4 (1800)	—	—	25–45 (3500–6500)
Tensile strength (black loaded) (MPa (psi))	>13.8 (2000)	17.2 (2000)	>17.2 (2500)	>20.7 (3000)	>13.8 (2000)	5–10 (725–1450)	>6.9 (1000)	—
Tear resistance	Fair	Fair–good	Good	Fair	Fair	Poor	Poor	Outstanding
Abrasion resistance	Fair	Fair–good	Good	Excellent	Good	Poor	Poor	Outstanding
Compression set	V. good	Poor	Poor	Fair	Fair–good	Poor–fair	Poor–fair	Fair–good
Rebound (cold)	Low	Good	Poor	Fair–good	Fair–good	Poor	Poor	Fair–v. good
Rebound (hot)	—	Good	Fair	Good	Good	—	—	Good–excllent
Dielectric strength	Good	Good	Good	V. good–excellent	Good	Good	Good	Fair–good
Electrical insulation	Fair	Good	Fair–good	Good	Fair–good	Excellent	V. good	Fair–good
Environmental resistance								
Oxidation	Good	Good	Excellent	Excellent	Outstanding	Excellent	Excellent	Excellent
Ozone	Excellent	Excellent	Outstanding	Outstanding	Outstanding	Excellent	Excellent	Excellent
Sunlight	Good	Good	Outstanding	Outstanding	Outstanding	Excellent	Excellent	V. good
Heat ageing (upper limit at constant service)(°C)	177	135	170	135	205	235	230	100–110
Low-temperature flexibility	Fair–v. good	Good–v.good	Good	Good	Fair–good	Excellent	Excellent	Excellent
Fluid resistance								
Gas permeability	Fair	Low	V.low	V. low	V. low	High	High	Low
Acid resistance (dilute)	Fair	Fair–good	Good	Excellent	Excellent	Excellent	Excellent	Excellent
Acid resistance (conc.)	Poor	Fair	Poor	V. good	Excellent	Fair	Good	V. good
Swelling								
In aliphatic hydrocarbons	Excellent	Excellent	Good	Good	Excellent	Poor	Excellent	Excellent
In aromatic hydrocarbons	Fair–poor	Good	Fair	Fair	Excellent	Poor	Good–excellent	Good
In oxygenated solvents	Poor	Poor	Poor	Poor	Poor	Fair	Good–excellent	Fair

* Reproduced by permission from Brydson, *Rubbery Materials and Their Compounds*, Elsevier/Applied Science.

Table 2.7 Standard abbreviations for rubbery materials (based on ISO Recommendation and ASTM D 1418)

ABR	Acrylate–butadiene rubber
ACM	Copolymers of ethyl or other acrylates and a small amount of a monomer which facilitates vulcanisation
AECO	Terpolymers of allyl glycidyl ether, ethylene oxide and epichlorohydrin
AEM	Copolymers of ethyl or other acrylate and ethylene
AFMU	Terpolymer of tetrafluoroethylene, trifluoronitrosomethane and nitrosoperfluorobutyric acid
ANM	Copolymers of ethyl or other acrylate and acrylonitrile
AU	Polyester urethanes
BIIR	Bromo-isobutene-isoprene rubber (brominated butyl rubber)
BR	Butadiene rubber
CFM	Polychlorotrifluoroethylene
CIIR	Chloro-isobutene-isoprene rubber (chlorinated butyl rubber)
CM	Chlorinated polyethylene
CO	Epichlorhydrin rubber
CR	Chloroprene rubber
CSM	Chlorosulphonated polyethylene
ECO	Ethylene oxide and epichlorhydrin copolymer
EAM	Ethylene–vinyl acetate copolymer
EPDM	Terpolymer of ethylene, propylene and a diene with the residual unsaturated portion of the diene in the side-chain
EPM	Ethylene–propylene copolymer
EU	Polyether urethanes
FFKM	Perfluoro rubbers of the polymethylene type having all substituent groups on the polymer chain either fluoro, perfluoroalkyl or perfluoroalkoxy groups
FKM	Fluoro rubber of the polymethylene type having substituient fluoro and perfluoroalkoxy groups on the main chain
FVMQ	Silicone rubber having fluorine, vinyl and methyl substituient groups on the polymer chain
GPO	Polypropylene oxide rubbers
IIR	Isobutene–isoprene rubber (butyl rubber)
IM	Polyisobutene
IR	Isoprene rubber (synthetic)
MQ	Silicone rubbers having only methyl substituent groups on the polymer chain
NBR	Nitrile–butadiene rubber (nitrile rubber)
NIR	Nitrile–isoprene rubber
NR	Natural rubber
PBR	Pyridine–butadiene rubber
PMQ	Silicone rubbers having both methyl and phenyl groups on the polymer chain
PNR	Polynorborene rubber
PSBR	Pyridine–styrene–butadiene rubber
PVMQ	Silicone rubbers having methyl, phenyl and vinyl substituent groups on the polymer chain
Q	Rubbers having silicon in the polymer chain
SBR	Styrene–butadiene rubber
T	Rubbers having sulphur in the polymer chain (excluding copolymers based on CR)
VMQ	Silicone rubber having both methyl and vinyl substituent groups in the polymer chain
XNBR	Carboxylic–nitrile butadiene rubber (carboxynitrile rubber)
XSBR	Carboxylic–styrene butadiene rubber
Y	Prefix indicating thermoplastic rubber
YBPO	Thermoplastic block polyether–polyester rubbers

Table 2.8 Miscellaneous abbreviations used for rubbery materials

ENR	Epoxidised natural rubber
EPR	Ethylene–propylene rubbers (either EPM or EPDM)
EVA	Ethylene–vinyl acetate copolymers (instead of EAM)
SBS	Styrene–butadiene–styrene triblock copolymer
SEBS	Hydrogenated SBS
SIR	Standard Indonesian rubber
SIS	Styrene–isoprene–styrene triblock copolymer
SMR	Standard Malaysian rubber

US Government nomenclature

GR-A	Government Rubber	Acrylonitrile (modern equivalent NBR)
GR-I	Government Rubber	Isobutylene (IIR)
GR-M	Government Rubber	Monovinyl acetylene (CR)
GR-P	Government Rubber	Polysulphide (T)
GR-S	Government Rubber	Styrene (SBR)

Two further groups of rubbery materials which have a quite different processing technology to the above are the *thermoplastic rubbers* which are processed on equipment used in thermoplastics processing and the *liquid processing rubbers* such as the cast polyurethanes and the liquid silicones which have their own specialist technology.

2.3.1 Tyre rubbers

This term is used for three diene rubbers widely used in the tyre industry. Although capable of making products with good performance in respect of toughness, strength, abrasion and tear resistance these materials have a lower resistance to hydrocarbon oils, oxygen, ozone, heat and light than do most of the rubbers described in the following sections.

All of the tyre rubbers are usually vulcanised using conventional sulphur based systems.

2.3.1.1 Natural rubber

Natural rubber (NR) has about 40% of the total new rubber market in tonnage terms. Although rubbery materials have been obtained from the latices (plural of latex) of some 500 plants, nearly all commercial rubber comes from *Hevea brasiliensis*, originally from Brazil but now grown widely in South-East Asia, West Africa and the Indian subcontinent. The material is obtained by coagulating the latex which has been tapped from the bark of the tree. The raw polymer has a high molecular weight (>1 million) and, because of entanglements, is quite elastic and incapable of shaping by melt processing. It is therefore necessary to subject the rubber to an intensive shearing process (mastication) to break down the molecular weight to a state in which the material may be blended with additives, shaped and then vulcanised. Newer grades may be treated at the latex stage to produce raw material which needs little or no mastication (known as 'constant viscosity (CV) rubbers').

Unfilled vulcanisates (gum stocks) are highly elastic and resilient and, although this may be reduced by incorporation of fillers, NR vulcanisates tend to be better than comparable filled vulcanisates from most synthetic rubbers. On ageing, the rubber tends to soften.

Because of its high resilience NR is widely used in tyre sidewalls and, more generally, in large tyres such as truck tyres. It is also an engineering rubber, where it is used, for example, in engine mountings and bridge bearings. Unlike other tyre rubbers, gum stocks show high vulcanisate strengths and this leads to many uses in stationery and medical applications.

The two major disadvantages of NR are its poor resistance to hydrocarbon oils (vulcanisates swelling extensively) and moderate heat resistance. However, penetration by oil and by oxygen (the main agent for limiting heat resistance) is slow so that thick articles or articles in which a large portion of the surface is in contact with metal and thus not accessible to oil and/or oxygen may be satisfactorily used at high temperatures or in oily environments for long periods of time.

2.3.1.2 Styrene–butadiene rubber

For many years styrene–butadiene rubber (SBR) was the leading rubber in tonnage terms, NR recovering its preeminence about 1980. Currently SBR has about 30% of the market. It differs from NR in the following ways.

(1) It does not break down on mastication and so has to be supplied at the correct molecular weight. Unvulcanised compounds also have inferior strength (green strength) and lack natural tack (a property useful when assembling or plying up pieces of rubber to make an article).
(2) Carbon black reinforcement is necessary for reasonable strength.
(3) Vulcanisation is slower than with NR with similar vulcanising (curing) systems.
(4) Abrasion resistance of reinforced compounds is generally better than with natural rubber. Wet grip and rolling resistance of tyres is also better.
(5) It hardens, rather than softens, on oxidation.

Mainly used for tyres, SBR is also used in such areas as mechanical goods, shoe soling and carpet underlay.

2.3.1.3 Polybutadiene rubber

Polybutadiene rubber (BR) is superior to both NR and SBR in terms of:

(1) abrasion resistance and groove cracking of tyres;
(2) low-temperature flexibility;
(3) high resilience at low deformations;
(4) heat ageing resistance;
(5) ozone resistance (under both static and dynamic conditions); and
(6) ability to accept higher levels of fillers and oil with less deterioration in properties.

It does, however, exhibit the following limitations:

(1) poor tack;
(2) poor road grip of tyre treads; and
(3) poor tear and tensile strength (compared with NR and SBR).

With regard to (3), in some ways BR vulcanisates may be considered as brittle with the possibility of products being subject to chunking and gouging, that is pieces of rubber breaking off when operating under severe conditions.

For these reasons BR is seldom used alone but mainly as a blending material (e.g. with SBR in passenger-car tyres and NR in truck tyres). It is also widely used as a toughening rubber in the manufacture of high-impact polystyrene. A small application is for very highly resilient solid playballs. BR has about 10% of the new rubber market.

2.3.2 Special purpose rubbers

This term is something of a misnomer but is commonly employed for those rubbers that find fairly widespread use but which are not normally used in any quantity as tyre rubbers.

2.3.2.1 Ethylene–propylene–diene terpolymers

These materials are now widely used in the general rubber-goods field. They have the best heat resistance of the tyre and special purpose rubbers as well as good resistance to oxygen, ozone and to light. As with the diene tyre rubbers (BR, NR and SBR) they do not have good resistance to hydrocarbon oils. Unlike NR, but like SBR, they require reinforcement with fine fillers in order to obtain good strength and abrasion resistance.

Ethylene–propylene–diene (EPDM) polymers contain just a small amount of some diene component which allows the rubber to be vulcanised using conventional sulphur based curing systems.

2.3.2.2 Ethylene–propylene copolymers

With properties similar to the above, the absence of a diene component in ethylene–propylene (EPM) copolymers prevents the use of conventional sulphur cures and require such materials as peroxides for vulcanisation. These rubbers therefore have limited use compared with EPDM.

2.3.2.3 Isobutylene–isoprene rubbers (butyl rubbers)

Isobutylene–isoprene rubbers (butyl rubbers; IIR) have the following particular characteristics:

(1) very low air permeability for a rubber (but not as low as many plastics materials);
(2) low resilience over a wide temperature range, although resilience does increase with temperature; and
(3) low unsaturation (i.e. low level of double bonds in the polymer) giving good resistance to oxygen, ozone and weathering, although not so good as with EPM and EPDM.

As with EPDM, a small amount of a diene component (from isoprene) allows conventional sulphur vulcanisation.

Comparatively recent developments are the brominated and chlorinated butyl rubbers (BIIR and CIIR). They are faster curing and capable of vulcanisation bonding. The main uses for IIR and derivatives are tyre inner tubes and inner linings, tyre curing bags, sealants, pipe-wrapping systems, pharmaceutical bottle closures, acid-resistant tank linings, reservoir roofing membranes, pond liners, and even chewing gum.

2.3.2.4 Polychloroprene rubbers

Available since the early 1930s polychloroprene (CR) rubbers are well known under the tradenames 'Neoprene', 'Baypren' and 'Butachlor'. They are well established as engineering rubbers of moderate heat, oxygen and ozone resistance with useful fire retardancy characteristics and better hydrocarbon oil resistance than any of the hydrocarbon based rubbers described earlier in this section.

They cannot be vulcanised using conventional sulphur based systems and require their own specialist curing technology.

2.3.2.5 Butadiene–acrylonitrile rubbers (nitrile rubbers)

The principal use for butadiene–acrylonitrile rubbers (NBR) is where oil, grease and, particularly, petrol resistance is required. Recently, more severe demands for under-the-bonnet applications have required NBR to be replaced by some of the speciality rubbers.

Nitrile rubbers may be vulcanised using conventional sulphur based systems.

2.3.2.6 Polyisoprene rubber

Polyisoprene rubber (IR) is a synthetically produced version of NR. Although generally purer than the natural product the polymer structure is less perfect. This means that the mechanical properties are generally not so good but, because of the lower level of impurities, IR finds use in pharmaceutical goods and as the raw material for producing chlorinated and isomerised rubber for the surface-coating industry. Other specialised uses include golf-ball thread (because of the low modulus) and some engineering applications where the low compression set (compared with NR) is an important prerequisite.

At one time IR was extensively used in tyre blends, but in recent years this application has proved uneconomical (rather than technically unsatisfactory).

2.3.3   Speciality rubbers

2.3.3.1   Acrylic rubbers

Acrylic rubbers (ACM) were developed for use where it is necessary to operate at higher temperatures than is possible for nitrile rubbers or where oils contain sulphur bearing additives which tend to react with nitrile rubber. They are also superior in heat resistance to all the rubbers considered above.

2.3.3.2   Fluoro rubbers

Many types of fluoro rubber are now available but, in general, they are used where a high level of oil resistance coupled with a high level of heat resistance is required.

2.3.3.3   Silicone rubbers

These materials are used because of their excellent heat resistance and excellent electrical insulation properties, their physiological inertness, their high permeability and low compression set. Besides conventional materials, special grades include liquid silicones rubbers and room temperature vulcanising systems that may be used for flexible mould-making, for seals and for caulking applications.

2.3.3.4   Polysulphides

Polysulphides (T) have been in production for longer than any synthetic rubber, having been introduced in the 1920s. They have poor heat resistance but outstanding resistance to hydrocarbon oils.

2.3.3.5   Epichlorhydrin rubbers

Epichlorhydrin (CO and ECO) rubbers have excellent resistance to ozone and weathering, an even lower gas permeability than butyl rubber, and good resistance to both heat and swelling in hydrocarbon oils. They have replaced nitrile rubber in some applications where specifications have become too stringent for the latter.

2.3.3.6   Polynorbornenes

Polynorbornenes (PNR) are hydrocarbon rubbers, of interest because they can be blended with large amounts of plasticiser and thus are available in very soft forms. They can also be formulated with interesting damping characteristics.

2.3.3.7   Chlorinated polyethylene and chlorosulphonated polyethylene

Chlorinated polyethylene (CM) and chlorosulphonated polyethylene (CSM) have good resistance to heat, ozone, weathering and a wide range of chemicals. Their main uses are in coated fabric and sheet form and for wire covering.

2.3.3.8   Ethylene–acrylic rubbers

Ethylene–acrylic (AEM) rubbers have intermediate heat resistance to more expensive rubbers and also may be formulated to have good flame retardance coupled with low emission of smoke and toxic gases under fire conditions. Uses are split between cable and automotive applications.

2.3.4   Thermoplastic rubbers

All the materials discussed in this subsection can be processed on conventional thermoplastic processing machinery such as injection-moulding machines.

2.3.4.1   Styrene–butadiene–styrene triblock rubbers

The high level of creep with styrene–butadiene–styrene (SBS) triblock rubbers limits their use in conventional rubber applications. They are important as bitumen additives, in pressure-sensitive and contact adhesives and for plastics modification. They are also used in shoe soling. Related SIS rubbers using isoprene instead of butadiene are also useful in adhesives.

2.3.4.2   Thermoplastic polyester rubbers and similar materials

These materials have better oil resistance and heat resistance (both in terms of thermal stability and heat softening resistance) than SBS materials. They are also available in a range of hardness from a stiff rubber to that of a more leather-like plastics material. They are important specialised engineering rubbers. Somewhat similar in properties are the thermoplastic polyurethanes and the thermoplastic polyamide rubbers.

2.3.4.3   Thermoplastic polyolefin rubbers

These materials are blends of EPDM with polypropylene, and sometimes also polyethylene. Being hydrocarbons they have limited oil resistance. They are of extensive interest in automotive applications, window profiles, cable insulation and footwear. Although they have somewhat limited levels of high elastic recovery after deformation, there have been considerable improvements with these properties in recent years.

2.3.5   Liquid processing rubbers

Both polyurethane and silicone rubbers are frequently processed in the liquid (low-molecular-weight state). Their processing technology is, therefore, highly specialist. Polyurethanes made using these techniques may have, if suitably formulated, outstanding oil resistance and abrasion resistance coupled in many cases with high values of tensile strength. Silicones tend to retain the properties mentioned in Section 2.3.3.3.

2.3.6   Exotics

The term 'exotics' is applied to highly specialised materials often only produced to specific order. They include the fluoropolyphosphazenes which have a very wide operating range (-65 to $+175°C$) coupled with good oil resistance and the totally non-inflammable carboxy–nitroso rubbers (AFMU).

Rubbery materials are discussed in greater detail in *Rubbery Materials and their Compounds*[1] with more recent developments discussed in more detail in *Handbook of Elastomers*.[2]

2.4   General properties of plastics

Since plastics may be regarded as those polymers that are not rubbers, fibres, adhesives or surface coatings, the range of properties possible is very wide. Some materials such as low-density polyethylene and plasticised PVC are flexible; the polyurethane foams are flexible and, in some cases, rubbery;

glass-reinforced polyester laminates are hard and rigid; whilst the nylons are less rigid but tough. Some plastics dissolve in water, others have excellent resistance to a wide range of chemicals. Other properties such as electrical-insulation characteristics and heat resistance also vary widely.

Since this handbook is concerned with engineering, this section is restricted to the consideration of certain selected properties of particular importance when plastics are used in engineering applications. The next section then provides a brief review of the main types of plastics material.

The properties considered here are:

(1) softening point (T_g and T_m);
(2) heat resistance;
(3) fire retardancy;
(4) toughness;
(5) creep resistance; and
(6) solvent resistance.

2.4.1 Softening point

The maximum temperature that may be used with a given plastics material depends on two independent factors:

(1) the softening behaviour of the polymer; and
(2) the thermal stability of the polymer, particularly in air.

Whilst the softening behaviour will put a short-term limit on the maximum temperature, the thermal stability is more time dependent.

The softening point of a plastics material is an arbitrary property, being the point at which the modulus of the material drops to some arbitrary, and usually unmeasured, value. Two tests are now widely used.

The Vicat softening point test A sample of the material is heated at a specified rate of temperature increase and the temperature is noted at which a needle of specified dimensions indents into the material a specified distance under a specified load. Most commonly a load of 10 N is applied, the needle indentor cross-section is 1 mm², the specified penetration distance is 1 mm and the rate of temperature rise is 50°C per hour. (The relevant standards are ISO 306; BS 2782: Method 120; ASTM D1525 and DIN 53460.)

The deflection temperature under load test Also widely known by its earlier name of the *heat distortion temperature test*, the temperature is noted at which a bar of material, subjected to a three-point bending stress, is deformed a specified amount. The load F applied to the sample will vary with the thickness t and width w of the sample and is determined by the maximum stress specified at the midpoint of the beam P which may be either 0.45 MPa or 1.82 MPa. (The formula used for the calculation is $F = 2Pwt^2/3L$, where L is the distance between the outer supports.) (The relevant standards are ISO 75, BS 2782: Method 121, ASTM D648 and DIN 53461.)

It can be seen from *Figure 2.2* that, in the case of amorphous polymers such as polystyrene and polymethyl methacrylate, the modulus will eventually reach a temperature where the modulus drops catastrophically. In these circumstances the softening points are very similar whatever test method is used and are all quite close to T_g (*Table 2.9*).

In the case of crystalline polymers, modulus does not change so sharply with temperature and a wide spread of softening point values are obtained. In this case Vicat temperatures are often quite close to T_m (*Table 2.10*).

Table 2.9 Glass transition temperature (T_g), Vicat softening point (VSP) and deflection temperature under load (DTL) of selected amorphous plastics

Plastic	T_g (°C)	VSP (°C)	DTL (at 1.82 MPa) (°C)
Polycarbonate*	149	155	127–135
Polymethyl methacrylate	104	114	85–100
Polystyrene	90–100	90	80
Polysulphone	190	—	174
Polyethersulphone	230	226	203
UPVC	80	73–77	76–80

* Polycarbonate is very slightly crystalline.

Table 2.10 Relationship between T_g, T_m and softening points for selected crystalline plastics

Plastic	T_g (°C)	T_m (°C)	VSP (°C)	DTL (at 1.82 MPa) (°C)
Nylon 6	50	215	—	80
Nylon 66	60	264	240	80
Polyacetal (copolymer)	−13	163	162	110
PBT	22–43	225	169–180	65

Many of the so-called *engineering thermoplastics* (see below) are supplied in fibre-reinforced form. It is interesting to note here that fibre reinforcement has a greater effect on the softening point of crystalline than on amorphous polymers (*Figure 2.12*).

It is of further interest to note that with such polymers the deflection temperature of reinforced polymers is often close to T_m, whilst that of the unreinforced materials is closer to T_g (*Table 2.11*).

2.4.2 Thermal stability

As with assessment of softening point, the assessment of thermal stability is also somewhat arbitrary. One method that

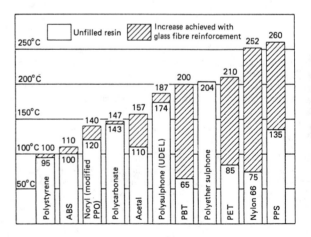

Figure 2.12 Heat deflection temperatures under a load of 1.82 MPa for selected polymers. Note that incorporation of glass fibre has a much greater effect with crystalline polymers than with amorphous ones. (Reproduced by permission from Whelan and Craft, *British Plastics and Rubber*, 29 (Nov. 1982))

Table 2.11 Comparison of T_g, T_m and heat deflection temperatures of polymers with and without glass reinforcement*

	T_g (°C)	T_m (°C)	Heat deflection temperature (°C)	
			Unfilled	Filled
Polyether ether ketone	143	334	150	315
Polyether ketone	165	365	165	340
Polyphenylene sulphide	85	285	135	260
Polyethylene terephthalate	70	255	85	210
Polybutylene terephthalate	22–43	225	54	210

* Reproduced by permission from Brydson, *Plastics Materials*, 5th edition, Butterworth-Heinemann.

has become increasingly used in recent years is the test leading to the Continuous Use Temperature Index of the Underwriters Laboratories.

In order to obtain this rating a large number of samples are subjected to oven ageing at a variety of temperatures for periods up to 1 year. In this time samples are withdrawn and tested and values of the property tested are plotted against time. The time taken for the property to lose half its value is noted for each temperature tested. This time is, quite arbitrarily, referred to as the *failure time*. A plot of the logarithm of the failure time against $1/K$ (where K is the temperature in kelvin) is made and the temperature corresponding to a failure time of 10 000 h is referred to as the *temperature index* (*Figure 2.13*).

Clearly, the temperature index will depend on the property being measured. Amongst those properties widely used are tensile strength (where the results are often quoted as 'mechanical, without impact'), Izod impact strength and electrical properties. Results also depend on sample thickness. Some representative data are given in *Table 2.12*.

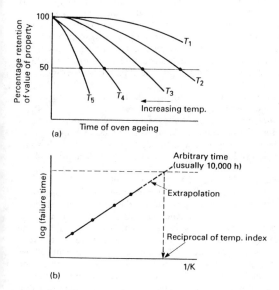

Figure 2.13 Method for determining the UL temperature index. (Reproduced by permission from Brydson, *Handbook for Plastics Processors*, Butterworth-Heinemann)

Table 2.12 Continuous Use Temperature Index (mechanical, without impact unless otherwise stated)*†

Polymer	Temperature (°C)
ABS	60–80
Nylons	75
Polyacetal (homopolymer)	90
Polyacetal (copolymer)	90
Polybutylene terephthalate	105–140
Polyethylene terephthalate	140 (glass reinforced)
Polyarylate (Ardel D-100)	130
Polysulphones	160
Polyetherimide	170
Polyethersulphone	180
Polyphenylene sulphide‡	200
Aromatic polyester (Ekkcel I-2000)	220
Liquid crystal polyester	220
Polyether ether ketone	240

* Reproduced by permission from Brydson, *Handbook for Plastics Processors*, Butterworth-Heinemann.
† The data given in this table have been obtained using procedures related to the UL Temperature Index method. The data given here do not, however, necessarily imply that the Underwriters Laboratories have awarded the temperature ratings quoted here either for the class of materials or for specific grades. Unless specified the figures are for standard unreinforced polymers.
‡ Glass reinforced, mechanical with impact.

2.4.3 Flame retardancy

For many applications resistance to burning is an important requirement. Whilst in many cases it is the fire reistance of a total structure that is of prime concern, the inherent flame-retardancy properties of the plastics materials used are also important. Two tests, described in the next two subsections, are now widely used for this purpose. It is also to be noted that in a fire more deaths occur through inhalation of smoke and toxic gases than through direct burning. For this reason assessments should also be made of smoke evolution and of toxicity of fire products.

2.4.3.1 Limiting oxygen index

The 'limiting oxygen index' (LOI) is the minimum concentration of oxygen, expressed as a percentage, in a slowly rising stream of a mixture of oxygen and nitrogen which will support combustion. The percentage of oxygen in air is 20.9%, so that polymers with a higher figure than this will not burn easily, if at all. Typical LOI figures are given in *Table 2.13*.

2.4.3.2 Underwriters Laboratory UL 94 Vertical Burning Test

Where plastics materials exhibit a measure of flame retardancy it is more common to classify them according to the Underwriters Laboratory UL 94 Vertical Burning Test.

In this test a rectangular bar is held vertically and clamped at the top. The burning behaviour of the bar and its tendency to form burning drips when exposed to a methane or natural gas flame applied to the bottom of the specimen is noted. Methods of specimen preparation and details of the test method are laid down in Underwriters Laboratory Inc. Standard UL 94. Materials are rated 94 V-0, 94 V-1 or 94 V-2 according to the following characteristics:

Table 2.13 Typical limiting oxygen index (LOI) values for polymers*†

Polymer	LOI (%)
Polyacetal	15
Polymethyl methacrylate	17
Polypropylene	17
Polyethylene	17
Polybutylene terephthalate	18
Polystyrene	18
Polyethylene terephthalate (unfilled)	21
Nylon 6	21–34
Nylon 66	21–30
Nylon 11	25–32
PPO	29–35
ABS	29–35
Polycarbonate of bisphenol A	26
Polysulphone	30
Polyethylene terephthalate (30% G.F.)	31–33
Polyimide (Ciba-Geigy P13N)	32
Polyarylate (Solvay Arylef)	34
Polyether sulphone	34–38
Polyether ether ketone	35
Phenol–formaldehyde resin	35
Polyvinyl chloride	23–43
Polyvinylidene fluoride	44
Polyamide imides (Torlon)	42–50
Polyphenylene sulphide	44–53
Polyvinylidene chloride	60
Polytetrafluoroethylene	90

* Reproduced by permission from Brydson, *Plastics Materials*, 5th edition, Butterworth-Heinemann.
† Where only a single figure is given it applies to raw polymer, unless otherwise indicated; where there is a range of figures the result is dependent on the grade of material, which may be dependent on the polymer structure or the presence of additives such as glass fibre and flame retardants. The relevant standards are ASTM D-2863 and BS 2782: Method 141B.

94 V-0 Specimen burns for less than 10 s after application of test flame. The wad of surgical cotton beneath the specimen is not ignited. When the specimen stops burning the flame is replaced for a further 10 s; after removal of this flame glowing combustion should die within 30 s. The total flaming combustion time for 10 flame applications on five specimens should be less than 50 s.

94 V-1 Specimen burns for less than 30 s after either test flame application: cotton not ignited. Glowing combustion dies within 60 s of second flame removal. Total flaming combustion time for 10 flame applications on five specimens should be less than 250 s.

94 V-2 As for 94 V-1, except that there may be some flaming particles which burn briefly but which ignite the cotton.

The UL 94 ratings of some selected engineering thermoplastics are given in *Table 2.14*.

The UL codings apply to a *specific grade* of material. The ratings may be dependent on the sample thickness, conditions being more stringent with thinner samples. Many of the so-called engineering thermoplastics now meet the V-0 rating and there has been an increasing tendency to quote minimum sample thickness for which the material will pass the V-0 test.

Table 2.14 The UL 94 ratings of some selected engineering thermoplastics*

Polymer	Tradename	UL 94 rating
Polycarbonate	Lexan 101	V-2
Nylone 66	Maranyl A100	V-2
Nylon 66 G.F.	Maranyl AD 447	V-0
Polysulphone	Udel P-1720	V-0
Polyethersulphone	Victrex 200P	V-0
Polyethylene terephthalate G.F.	Rynite 530 FR	V-0
PPO	Noryl HS 1000	V-0
	Noryl GFN-2	HB
Polyacetal	Delrin 500	HB
Polyphenylene sulphide G.F.	Ryton	V-0
Polyarylate	Ardel D-100	V-0
Polyether–imide	Ultem	V-0
Polyether ether ketone	Victrex PEEK	V-0
ABS Standard Grades		HB
ABS/polycarbonate alloy	Cycovin KMP	V-0

* Reproduced by permission from Brydson, *Plastics Materials*, 5th edition, Butterworth-Heinemann.

Mention should also be made of two other UL ratings; the 94HB rating for which the sample passes a less severe burn test than the vertical test and the 94 5-V rating which is more severe than the V-0 rating and involves a test in which the sample is subjected to a 5 in. flame five times at 5-s intervals with each exposure lasting 5 s.

2.4.4 Toughness

An important prerequisite of plastics materials in most applications is that they should have an acceptable level of resistance to sudden impact. It is quite common to classify plastics into two types: brittle plastics and tough plastics.

A *brittle* plastic material is one that can only withstand a very small strain before fracture (e.g. 1–2%) so that the area under a stress–strain curve is small. Such materials break cleanly with little permanent distortion of the broken surfaces and such parts may sometimes be repaired by welding or by adhesives. Just because a material is classified as brittle does not mean that it is unsatisfactory. This will depend on the expected operational conditions.

Tough plastics extend by several per cent before fracture. In this case the area under the stress–strain curve is quite large, this being a measure of the energy required to break. In some cases a product may yield irreversibly before or without fracture, rendering it unsatisfactory for its purpose so that a simple consideration of toughness is not enough.

For general comparative purposes toughness is measured by standard impact tests, in particular the Izod impact strength. In this test half of a notched sample is supported and then struck at a fixed point above the notch, on the same side as the notch, by a swinging pendulum. The energy required to break is obtained by considering the amount the pendulum follows through after striking the sample relative to the height of the pendulum before its release. Results may be expressed in a number of ways, but it is to be noted that the impact strength will depend on the cross-sectional area of the sample and also on the angle of the notch. Some materials are very sensitive to the notch-tip radius (*Figure 2.14*).

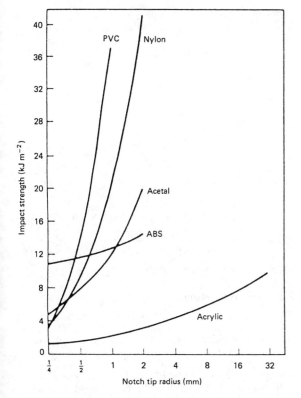

Figure 2.14 Impact strength as a function of notch-tip radius for samples of five different polymers. (Reproduced by permission from Vincent, *Impact Tests and Service Performance of Plastics*, Plastics and Rubber Institute)

Considerable interest has been given to the application of fracture mechanics to plastics since this can help to predict the service life of articles containing small cracks or crack-like defects. For the comparatively simple case of a crack subject to opening forces in a linear brittle elastic material the stress at the crack tip in the situation illustrated in *Figure 2.15* is given by

$$\sigma_Y = K_I/(2\pi r)^{1/2}$$

where r is the distance from the crack tip along the direction of crack growth and

$$K_I = Y\sigma(\pi a)^{1/2}$$

i.e. the local stress is proportional to the applied stress and varies with the square root of the crack size. Y is a modifying factor depending on the location of the crack relative to the plate geometry and is obtainable from standard texts. K_I is independent of the material, provided that it is linear elastic. A material will fail by very rapid crack growth when K_I reaches a critical value for the material; typical values of K_{IC} are given in *Table 2.15*.

The crack growth rate is given by

$$da/dt = CK_I^n$$

for materials under constant stress where C and n are constants. These 'constants' may vary with the stage of crack growth. The time for a crack to grow may be found by integration of the crack growth equation between the original size of the crack and the new size of the crack.

2.4.5 Creep behaviour

Under long-term load plastics materials will continue to deform or *creep*. The extent of creep will depend considerably on the type of polymer and also on whether or not it has been reinforced, e.g. with glass fibre. Creep will also depend on the temperature (*Figure 2.16*) and on the load (*Figure 2.17(a)*). A consequence of this creep is that, after an interval of time, the amount of deformation may reach unacceptable levels and this has to be built into design calculations (considered in Section 2.8.1).

It is possible to derive from a family of creep curves such as those shown in *Figure 2.17(a)* two other types of curve. In the first case the strain corresponding to various levels of stress at some specified time, e.g. 1000 h, is noted. The resultant plot is known as an isochronous stress–strain curve (*Figure 2.17(b)*). Any number of such curves may be so derived, simply by changing the specified time.

Alternatively, the amount of stress that will give a specified strain at various times will yield an isometric stress vs. log(time) curve (*Figure 2.17(c)*). Finally, a creep modulus vs. time curve may also be derived (*Figure 2.17(d)*). Although no additional information is actually provided by the derived curves they are sometimes more convenient to use.

It is interesting to note that acceptable upper strain limits are frequently used for plastics materials. Whilst there is some level of arbitrariness, some typical figures are suggested in *Table 2.16*.

On release of a load recovery of plastics material is time dependent and also incomplete, since some of the creep will be irreversible. Data are sometimes provided in *reduced* or *dimensionless form* using the following:

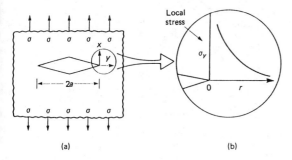

(a)   (b)

Figure 2.15 Stresses near crack tip in infinite thin plate

Table 2.15 Typical values for fracture toughness in air at 20°C

Material	K_{IC} (MN m$^{-3/2}$)
Epoxy resin	0.6
Polystyrene	1.0
Cast acrylic sheet	1.6
Polycarbonate	2.2
PVC pipe compound	2.3
High density polyethylene	3.0
Nylon 66	3.6

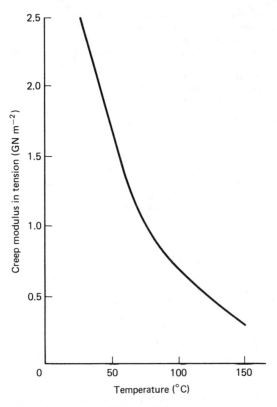

Figure 2.16 Example of the dependence of the creep modulus on the temperature for a crystalline thermoplastic: data typical for a polyacetal copolymer after 100 s loading and 0.2% strain

Table 2.16 Upper strain limits for plastics

Plastic	Strain (%)
Glass-reinforced nylon	1
Acetal copolymer	2
Polypropylene	3
ABS	1.5
Polyethersulphone	2.5
Polycarbonate	0.6

Fractional recovered strain

$$= \frac{\text{Strain recovered}}{\text{Total creep strain at moment load is removed}}$$

$$\text{Reduced time} = \frac{\text{Recovery time}}{\text{Duration of creep period}}$$

Typical data are given in *Figure 2.18*.

2.4.6 Solvent resistance

Amorphous polymers tend to dissolve in solvents which have a solubility parameter within 2 $MPa^{1/2}$ of that of the polymer. In the case of crystalline polymers the situation is more complicated and two main effects may be discerned.

If the polymer is not capable of specific interaction with a solvent (as with PE, PP and PTFE) then there are no solvents until temperatures approaching the melting point are reached.

If the polymer is capable of a specific interaction with a solvent then it may be possible to find a solvent if it shows complementary specific interaction. Such interaction is possible with polycarbonates, PVC and the nylons.

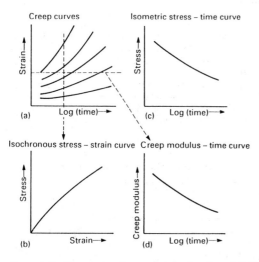

Figure 2.17 Presentation of creep data: sections through the creep cruves at constant time and constant strain give curves of isochronous stress–strain (b), isometric stress–log(time) (c) and creep modulus—log(time) (d). (Reproduced by permission of ICI)

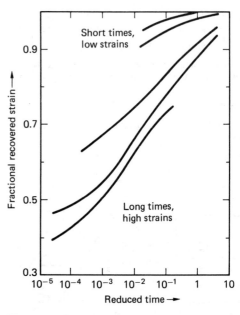

Figure 2.18 Recovery from tensile creep of an acetal copolymer at 20°C and 65% relative humidity. (Reproduced by permission from *ICI Technical Service Note G117*, ICI)

Some solubility parameter data for polymers and solvents are given in *Tables 2.17* and *2.18*. Crystalline polymers marked with a ‡ are not capable of specific interaction and do not dissolve at room temperature, whilst those marked with a § are capable of specific interaction with some solvents and, where these solvents have similar solubility parameters to the polymers, dissolution may occur. In most of these cases it is necessary that one of the components (i.e. polymer and solvent) be what is known as a 'proton donor' and the other a 'proton acceptor'. Examples of proton donors are PVC, dichloromethane, formic acid, cresol and dimethyl formamide, whilst examples of proton acceptors are the nylons, polyethylene terephthalate and other polyesters, the polycarbonates and cyclohexanone.

Cross-linked polymers including vulcanised rubbers will not dissolve but will swell in solvents of similar solubility parameter. The extent of swelling also depends on the level of cross-linking.

Table 2.17 Solubility parameters of polymers*†

Polymer	Solubility parameter, δ	
	(cal cm^{-3})$^{1/2}$	(MPa$^{1/2}$)
Polytetrafluoroethylene‡	6.2	12.6
Polychlorotrifluoroethylene	7.2	14.7
Polydimethyl siloxane	7.3	14.9
Ethylene–propylene rubber	7.9	16.1
Polyisobutylene	7.9	16.1
Polyethylene‡	8.0	16.3
Polypropylene‡	8.0	16.3
Polyisoprene (natural rubber)	8.1	16.5
Polybutadiene	8.4	17.1
Styrene–butadiene rubber	8.4	17.1
Poly-t-butyl methacrylate	8.3	16.9
Poly-n-hexyl methacrylate	8.6	17.6
Poly-n-butyl methacrylate	8.7	17.8
Polyethyl methacrylate	9.0	18.3
Polymethylphenyl siloxane	9.0	18.3
Polysulphide rubber	9.0–9.4	18.3–19.2
Polystyrene	9.2	18.7
Polychloroprene rubber	9.2–9.4	18.7–19.2
Polymethyl methacrylate	9.2	18.7
Polyvinyl chloride§	9.5	19.4
bis-Phenol A polycarbonate§	9.5	19.4
Polyvinylidene chloride‡	9.8–12.2	20.0–25.0
Ethyl cellulose	8.5–10.3	17.3–21.0
Cellulose dinitrate	10.55	21.6
Polyethylene terephthalate§	10.7	21.8
Acetal resins§	11.1	22.6
Cellulose diacetate	11.35	23.2
Nylon 66§	13.6	27.8
Polymethyl-α-cyanoacrylate	14.1	28.7
Polyacrylonitrile	14.1	28.7

* Reproduced by permission from Brydson, *Plastics Materials*, 5th edition, Butterworth-Heinemann.
† Because of the difficulties in their measurement the published figures for polymer solubility parameters range by ±3% from the average figure quoted.
‡ See text.
§ See text.

Table 2.18 Solubility parameters of common solvents

Solvent	Solubility parameter, δ	
	(cal cm^{-3})$^{1/2}$	(MPa$^{1/2}$)
n-Pentane	6.3	12.8
Isobutylene	6.7	13.7
n-Hexane	7.3	14.9
Diethyl ether	7.4	15.1
n-Octane	7.6	15.5
Methyl cyclohexane	7.8	15.9
Ethyl isobutyrate	7.9	16.1
Di-isopropyl ketone	8.0	16.3
Methyl amyl acetate	8.0	16.3
Turpentine	8.1	16.5
Cyclohexane	8.2	16.7
2,2-Dichloropropane	8.2	16.7
sec-Amyl acetate	8.3	16.9
Dipentene	8.5	17.3
Amyl acetate	8.5	17.3
Methyl-n-butyl ketone	8.6	17.6
Pine oil	8.6	17.6
Carbon tetrachloride	8.6	17.6
Methyl-n-propyl ketone	8.7	17.8
Piperidine	8.7	17.8
Xylene	8.8	18.0
Dimethyl ether	8.8	18.0
Toluene	8.9	18.2
Butyl cellosolve	8.9	18.2
1,2-Dichloropropane	9.0	18.3
Mesityl oxide	9.0	18.3
Isophorone	9.1	18.6
Ethyl acetate	9.1	18.6
Benzene	9.2	18.7
Diacetone alcohol	9.2	18.7
Chloroform	9.3	19.0
Trichloroethylene	9.3	19.0
Tetrachloroethylene	9.4	19.2
Tetralin	9.5	19.4
Carbitol	9.6	19.6
Methyl chloride	9.7	19.8
Dichloromethane	9.7	19.8
Dichloroethane	9.8	20.0
Cyclohexanone	9.9	20.2
Cellosolve	9.9	20.2
Dioxane	9.9	20.2
Carbon disulphide	10.0	20.4
Acetone	10.0	20.4
n-Octanol	10.3	21.0
Butyronitrile	10.5	21.4
n-Hexanol	10.7	21.8
sec-Butanol	10.8	22.0
Pyridine	10.9	22.3
Nitroethane	11.1	22.6
n-Butanol	11.4	23.2
Cyclohexanol	11.4	23.2
Isopropanol	11.5	23.4
n-Propanol	11.9	24.2
Dimethyl formamide	12.1	24.7
Hydrogen cyanide	12.1	24.7
Acetic acid	12.6	25.7
Ethanol	12.7	26.0
Cresol	13.3	27.1
Formic acid	13.5	27.6
Methanol	14.5	29.6
Phenol	14.5	29.6
Glycerol	16.5	33.6
Water	23.4	47.7

* Reproduced by permission from Brydson, *Plastics Materials*, 5th edition, Butterworth-Heinemann.

2.5 Survey of commercial plastics materials

As with the rubbery materials the market for plastics materials is dominated, in tonnage terms, by a small number of polymers. Today, four polymers, namely polyethylene, polypropylene, PVC and polystyrene, account for about 80% of the total market tonnage. There are additionally several hundred other materials which could be classified as plastics that are on the market. Most of these have been considered at greater depth by the writer elsewhere[3] and this section can only give an outline indication of the more important ones, particularly in respect of engineering applications.

Abbreviations for the most well-known materials are given in *Table 2.19*.

2.5.1 Major tonnage thermoplastics

2.5.1.1 Polyethylene

In tonnage terms polyethylene (PE; also known as polyethene or polythene) is the world's leading plastics material. Besides its relatively low cost, other outstanding features include excellent electrical-insulation characteristics at both low and high frequencies, excellent resistance to chemicals, absence of solvents at normal ambient temperature (although it is swollen by solvents of similar solubility parameter), negligible water absorption, toughness and flexibility, ease of processing, and clarity in thin-film form.

It does, however, have a number of limitations such as a low softening point, susceptibility to oxidation, susceptibility of low-molecular-weight grades to cracking under stress in many environments (particularly oils, detergents, silicones and ethers), wax-like appearance and poor scratch resistance, lack of rigidity and low tensile strength and stiffness.

Three main types of polymer are recognised.

HDPE (high density polyethylene) This is a virtually unbranched polymer. Having the most regular structure it is the most crystalline and has the highest strength, stiffness and softening point of the polyethylenes.

LLDPE (linear low-density polyethylene) A fairly recent development in which occasional short hydrocarbon side-chains are attached to the main chain. These reduce regularity and ability to crystallise. The main interest in these materials is for film applications.

LDPE (low-density polyethylene) The original commercial material which, because of the method of manufacture, has both long and short branches attached occasionally to the chain. Because of the limited ability to crystallise this material is quite flexible.

Also to be mentioned here is the copolymer *EVA (ethylene–vinyl acetate)* in which the presence of vinyl acetate may reduce further, and even eliminate, the ability to crystallise according to the amount of vinyl acetate used. Grades range from materials similar to but softer than LDPE to polymers that are rubbery.

About 75% of LDPE use is for film with the rest used for a variety of applications.

2.5.1.2 Polypropylene

Polypropylene (PP) can exist in a number of forms, depending on the way in which the monomer molecules are joined together. The most important is the so-called 'isotactic polypropylene' and the following comments apply to this form.

Compared with HDPE, polypropylene is stiffer, has a higher softening point (it may be sterilised in boiling water), is less wax-like, and mouldings can have a better finish. It has a lower density (0.90 g cm^{-3}) and is substantially free from stress cracking problems, but it is more susceptible to oxidation. Copolymers containing small amounts of ethylene can show improved impact strength and reduction in brittle point, but with a small loss in hardness and softening point.

Whilst the main application is in fibres and filaments the polymer is also widely used in injection moulding for such diverse products as stacking chairs, beer bottle crates, washing-machine tubs and car parts. It is also an important film material. It does not generally have the finish to make it suitable for housings of domestic electrical equipment.

2.5.1.3 Polyvinyl chloride

Polyvinyl chloride (PVC) is also known as polychloroethene or poly(1-chloroethylene); it is also often identified by the term *vinyl* but this should be discouraged because of ambiguity.

PVC differs from polyethylene in possessing a chlorine atom attached to alternate carbon atoms on the main polymer chain. This makes the polymer harder, raises T_g, increases resistance to hydrocarbons and gives a measure of flame resistance. On the other hand, electrical-insulation properties are not so good (particularly at high frequencies), the polymer is more susceptible to degradation during processing, there are some solvents at room temperature and, because the commercial materials are amorphous and therefore do not possess a crystalline melting point, the softening point is lower than for HDPE.

There are two main forms of PVC.

UPVC (unplasticised PVC) Commercial grades are blends of polymer with fillers, stabilisers, processing aids (lubricants) and colorants. It is widely used in building and pipe applications and for chemical plant.

PPVC (plasticised PVC) In this case the compound also contains substantial amounts of a compatible liquid known as a 'plasticiser' (20–100 parts per 100 parts of polymer). Increasing the level of plasticiser increases the flexibility and, at the higher levels, the material may be quite rubbery. Well-known applications include domestic flex, leathercloth, sheeting and hose piping.

Mention should also be made here of the vinyl chloride–vinyl acetate copolymers which are used for gramophone records and for flooring materials.

2.5.1.4 Polystyrene and related materials

Polystyrene (PS) is an amorphous hydrocarbon plastic material. It is rigid; has low water absorption; it is transparent (with a high refractive index) when unfilled; and is an excellent electrical insulator over a wide frequency range. As with the material described above it is also a low-cost material.

Disadvantages include a low impact strength, low softening point, solubility in a wide range of solvents and the need for care in processing in order to optimise mechanical properties. Polystyrene in a cellular form (expanded polystyrene, XPS) is an excellent thermal insulator and shock absorber widely used in packaging and insulation applications.

Blending PS with small amounts of a synthetic rubber yields a material with enhanced toughness. This is known either as

Table 2.19 Some common names and abbreviations for plastics*

Abbreviation	Material	Common name
ABS	Acrylonitrile–butadiene–styrene polymer	ABS
ACS	Acrylonitrile–styrene and chlorinated polyethylene	
AES	Acrylonitrile–styrene and ethylene propylene rubber	
ASA	Acrylonitrile–styrene and acrylic rubber	
CA	Cellulose acetate	**Acetate**
CAB	Cellulose acetate–butyrate	CAB, butyrate
CAP	Cellulose acetate–propionate	CAP
CN	Cellulose nitrate	Celluloid
CP	Cellulose propionate	CP, propionate
CTA	Cellulose triacetate	Triacetate
CS	Casein	Casein
DMC	Dough-moulding compound (usually polyester)	
EP	Epoxide resin	Epoxy
ETFE	Tetrafluoroethylene–ethylene copolymer	
EVAC	Ethylene–vinyl acetate	EVA
EVOH, EVAL	Ethylene–vinyl alcohol	
FEP	Tetrafluoroethylene–hexafluoropropylene copolymer	
FRP, FRTP	Thermoplastic material; reinforced, commonly with fibre	
GRP	Glass-fibre reinforced; plastic based on a thermosetting resin	
HDPE	High-density polyethylene	
HIPS	High-impact polystyrene	
LDPE	Low-density polyethylene	
LLDPE	Linear low-density polyethylene	
MBS	Methacrylate–butadiene styrene	
MDPE	Medium density polyethylene	
MF	Melamine–formaldehyde	**Melamine**
PA	Polyamide	Nylon (some types)
PBTP, PBT, PTMT	Polybutylene terephthalate	Polyester
PC	Polycarbonate	Polycarbonate
PETP, PET	Polyethylene terephthalate	Polyester
PCTFE	Polychlorotrifluoroethylene	
PE	Polyethylene	Polythene
PEBA	Polyether block amide	
PEEK	Polyether ether ketone	
PEG	Polyethylene glycol	
PEK	Polyether ketone	
PES	Polyether sulphone	
PF	Phenol–formaldehyde	Phenolic
PMMA, PMM	Polymethyl methacrylate	Acrylic
POM	Polyacetal, polyoxymethylene, polyformaldehyde	**Acetal**
PP	Polypropylene	Propylene, polyprop
PPG	Polypropylene glycol	
PPO	Polyphenylene oxide	
PPO	Polypropylene oxide	
PPS	Polyphenylene sulphide	
PS	Polystyrene	**Styrene**
PS, PSU	Polysulphone	**Polysulphone**
PTFE	Polytetrafluoroethylene	**PTFE**
PUR	Polyurethane	Polyurethane, urethane
PVA	Polyvinyl acetate	
PVA	Polyvinyl alcohol	
PVA	Polyvinyl acetal	
PVAC	Polyvinyl acetate	PVA
PVB	Polyvinyl butyral	
PVC	Polyvinyl chloride	PVC, vinyl
PVDC	Polyvinylidene chloride	
PVDF	Polyvinylidene fluoride	
PVF	Polyvinyl fluoride	
PVF	Polyvinyl formal	
PVP	Polyvinyl pyrrolidone	
RF	Resorcinol–formaldehyde	
SAN	Styrene–acrylonitrile	SAN
SI	Polysiloxane	Silicone
SMC	Sheet moulding compound (usually polyester)	
TPS	Toughened polystyrene	
UF	Urea–formaldehyde	**Urea**
UP	Unsaturated polyester	Polyester
UPVC	Unplasticised PVC	
VLDPE	Very low-density polyethylene	
XPS	Expanded polystyrene	

* Those marked in bold indicate that they are in the main schedule of BS 3502: 1978 *Common names and abbreviations for plastics, Part 1.*
Reproduced with permission from Brydson, *Handbook for Plastics Processors,* Butterworth-Heinemann, 1990.

high-impact polystyrene (HIPS) or toughened polystyrene (TPS). Such materials are widely used for a variety of industrial, household and automotive applications. Such materials are normally opaque.

2.5.1.5 Acrylonitrile–butadiene–styrene polymers

These materials (ABS polymers) may be considered as related to polystyrene but have been considered under a separate heading because of their importance. They are complex blends of polymers produced from acrylonitrile, butadiene and styrene monomers. Suitably formulated, they have better heat resistance than other polystyrene polymers, have a high level of toughness and scuff resistance and a finish of moulded articles acceptable for the demanding household-appliance and car-component markets. These materials are normally opaque.

2.5.2 Engineering thermoplastics

This rather loose and inexact term is frequently used to describe thermoplastics used in load bearing engineering applications. PTFE, whilst not load bearing in the usual sense, is also included here because of its widespread use as an engineering material.

2.5.2.1 The nylons

The nylons are a form of *polyamide* which are fibre-forming or derived from fibre-forming polymers. They are the most well established of the engineering thermoplastics, well known for their toughness, reasonable rigidity, very good abrasion resistance, good hydrocarbon resistance and reasonable heat resistance. Glass-fibre-filled grades are widely used because of their higher rigidity, resistance to creep, low coefficient of friction and high heat deflection temperatures. Glass bead, carbon fibre and rubber modified grades are also used.

The principal disadvantages of the nylons are their: high water absorption (as high as 10% with some grades), tendency to oxidise at elevated temperatures and thus discolour; opacity; poor electrical insulation characteristics at high frequencies; high coefficient of friction; and special care needed in processing because of their low melt viscosity and water absorbing tendencies.

The nylons differ from polyethylene in that an amide (–CONH–) group occurs regularly along the chain backbone. The more frequently it occurs the higher the softening point, the hydrocarbon resistance, the hardness, the yield strength and the water absorption. A number of types are available in which the concentration of such amide groups varies. Without going into details of the chemistry, the following sequence gives the descending order of amide concentration: nylon 46 > nylon 66 > nylon 6 > nylon 610 > nylon 11 > nylon 12. Nylons 11 and 12 have only about half the concentration of amide groups as the most commonly used nylons 6 and 66 and thus are intermediate in properties between these two nylons and polyethylene.

Most applications of nylons in engineering involve quite small parts such as cams, gear wheels, bearings bushes and valve seats. The ability to run silently and without lubrication in many applications are attractive features. The polymers are also used in medical applications, in low-frequency electrical insulation applications; extruded nylon applications include packaging film, filaments, petrol tubing and chemical plant.

Mention may be made here of the so-called *glass-clear polyamides*. These materials are based on aromatic materials and have an irregular structure that prevents them from crystallising. These are clearly useful where transparency is required in conjunction with properties more generally characteristic of the polyamides.

2.5.2.2 Polyacetals

Also known as 'acetal resins', 'polyoxymethylenes' and 'polyformaldehyde resins', these polymers have $-CH_2O-$ as a repeating unit in the chain, i.e. they resemble polyethylene in which an oxygen atom is inserted between each $-CH_2-$ group.

In many respects the acetal resins are similar to the nylons but may be considered superior in fatigue endurance, creep resistance, stiffness and water resistance. The nylons (except under dry conditions) are superior in impact toughness and abrasion resistance. As with the nylons these materials are opaque. Two particular points should also be made. Firstly, the density of the polyacetals is about 30% higher than that of the nylons so that the weight of material used in an article of a given volume will be that much higher, a factor to be considered when making comparative costings. Secondly, the polymer is somewhat unstable at elevated temperature, particularly in the presence of PVC. It is therefore essential that cross-contamination of PVC and acetal resin should not be allowed to occur since this may lead to explosive decomposition.

Uses of polyacetals include gears, bearings, cams, fan blades, conveyor hooks, electric kettles, plumbing valves, pump housings, lawn sprinklers and ballcocks.

2.5.2.3 Polycarbonates

The outstanding features of the polycarbonates (PC) are their rigidity up to about 140°C, toughness, transparency, very good electrical-insulation characteristics (although limited resistance to tracking), measure of flame resistance and physiological inertness. Disadvantages include a strong notch sensitivity and a tendency to crack if subject to long-term strain exceeding about 0.75%. They also have a somewhat limited resistance to chemicals and some solvents may cause crazing and cracking. They are therefore not commonly used in the same areas as the nylons and polyacetals but rather where a combination of several of the desirable features listed above is required.

2.5.2.4 Polysulphones and polyether sulphones

Polysulphones (PSU) and polyether sulphones (PES) are expensive, highly specialist engineering materials usually only considered when polycarbonates do not meet specifications. They have a high temperature resistance coupled with a high softening point and resistance to chemical change on heating. In addition, creep resistance for an unfilled material is exceptional and this is combined with high rigidity, transparency (when unfilled) and a measure of fire resistance. Smoke evolution is particularly low. As with the polycarbonates the polymers are tough but notch sensitive, and electrical properties are broadly similar. There are two main types; the so-called 'polyether sulphones' being considerably more expensive but having the highest level of performance. Terminology is unsatisfactory as all types may be considered both as polysulphones or polyether sulphones. Glass fibre and carbon fibre filled grades are also available.

2.5.2.5 Polyphenylene sulphide

Polyphenylene sulphide (PPS) is another special-purpose material with excellent heat resistance for a thermoplastic with UL (Underwriters' Laboratories) temperature indices of the order of 240°C. Filled grades have high deflection temperatures whilst polymers also show excellent flame resistance without the use of additives. PPS is also attractive because of precision mouldability (low moulding shrinkage) and high dimensional stability (low water absorption).

2.5.2.6 Polyether ether ketone and polyether ketone

Polyether ether ketone (PEEK) and polyether ketone (PEK) are high-cost polymers that are usually considered when polyether sulphones do not meet specifications. They have a high softening point, are suitable for long-term use above 200°C, have very low flammability and exceptionally low smoke emission. Other properties include excellent hydrolytic stability, very good stress cracking resistance and good radiation resistance. PEK has the higher softening point and heat resistance.

2.5.2.7 Modified PPO (e.g. Noryl)

This is a blend of poly-p-2,6-dimethylphenylene oxide with polystyrene or modified polystyrene. Self-extinguishing, glass-reinforced and structural foam grades are important variants.

The polymers are rigid and may be used for precision work because of low moulding shrinkage, coefficient of expansion and water absorption. They have moderately high deflection temperatures. The polymers are particularly useful for components used in contact with water such as washing machines, dishwashers and water pumps as well as marine components.

2.5.2.8 Polytetrafluoroethylene and other fluoroplastics

Polytetrafluoroethylene (PTFE) is a white opaque crystalline material with exceptionally good electrical insulation properties over a wide frequency range, exceptional chemical inertness, low coefficient of friction, good non-stick properties and excellent heat resistance with a crystalline melting point of 327°C. It does not burn but may decompose at elevated temperatures giving off toxic gases. Disadvantages include low creep resistance (for an engineering thermoplastics material), difficulty of processing by conventional techniques and high cost. It is widely used in a variety of industries.

Several other fluorine containing polymers are also available. These do not have quite the same exceptional heat resistance of PTFE but are generally more processable.

2.5.2.9 Polyimides and related polymers

The original polyimides were introduced as materials with unexcelled heat resistance for a plastics material. They are, however, expensive and require special processing techniques. This has led to the availability of a range of modified polyimides, such as the polyamide imides and the polyether imides which may be processed on conventional equipment but at some sacrifice to the exceptional physical properties of the original polymers.

2.5.3 Other thermoplastics

In this section brief mention is made of some other thermoplastics which do not quite fit into the above categories.

2.5.3.1 Polymethyl methacrylate

This is best known as the transparent sheet material, first introduced in the UK by ICI as 'Perspex'. It is also available in moulding and extrusion grades. The material is outstanding for its high light transmission and excellent weathering properties. (This material is often referred to simply as *acrylic*. This

should be discouraged since quite different acrylic polymers are also important in the fields of rubbers, surface coatings, fibres and adhesives.)

2.5.3.2 Cellulose plastics

Materials such as cellulose acetate and celluloid still find some specialist applications, but are no longer of great importance and are of negligible interest for use in engineering applications.

2.5.3.3 Polyesters

The term 'polyester' simply implies that there are many ester groups in the polymer molecule without saying anything about what other chemical groups may be present. There are thus many very different polyesters used not only in the field of thermoplastics, but also as thermosetting plastics, fibres, rubbers and surface coatings.

Two particular thermoplastic polyesters are of importance.

Polyethylene terephthalate (PET) This was originally introduced as a fibre-forming material, but today is also a useful injection moulding material and of great importance for disposable bottles for carbonated drinks.

Polybutylene terephthalate (PBT) Also known as 'polytetramethylene terephthalate' (PTMT) this is usually used in glass-fibre-filled form, to give a rigid material with excellent dimensional stability, particularly in water, and resistance to hydrocarbon oils without showing stress cracking.

2.5.4 Thermosetting plastics

Until about 1960 these were in, tonnage terms, more important than thermoplastics. They now have a lesser but still important role.

2.5.4.1 Phenolic plastics

Originally introduced as 'Bakelite', these materials have been widely used because of their heat resistance, good electrical-insulation properties at low frequencies and low thermal conductivity. Limitations include a tendency to surface electrical tracking, odour, limited colour and limited durability of a good finish.

2.5.4.2 Aminoplastics

There are two main types of aminoplastic.

Urea–formaldehyde (UF) plastics These are available in a wide colour range but do not have the heat resistance of PFs. They also absorb water and are stained by many liquids. Uses include plugs and switches and toilet seats.

Melamine–formaldehyde (MF) plastics These have better heat and stain resistance than UFs and are used mainly in tableware applications.

2.5.4.3 Thermosetting polyester resins

Used mainly with glass fibres for structural use in sports-car bodies, lorry and railway engines, boats and aircraft as well as many other applications. More mouldable materials (dough-moulding compounds (DMC) and sheet-moulding compounds (SMC)) are widely used in applications requiring a high level of strength and rigidity (e.g. crash helmets, car headlamp fittings, and microwave cooking equipment).

2.5.4.4 Epoxide resins

Epoxide (epoxy) resins are similar to polyesters but with improved heat and chemical resistance. They are used with carbon fibres where exceptional performance is required. Epoxide resins are also widely used for encapsulation of electronic components, for surface coatings and as adhesives.

2.6 Processing of rubbers and plastics

2.6.1 Shaping and setting

Whilst rubbers and plastics may be converted into a variety of products using a wide range of different processes, these processes may be considered as requiring two stages:

(1) getting the shape; and
(2) setting the shape.

The shaping operation may be carried out with the polymer in one of the following states:

(1) as a melt (as in compression, injection and blow moulding and in extrusion);
(2) in the rubbery state (as in vacuum forming);
(3) in solution (as in casting film or fibre spinning);
(4) as a suspension (as in rubber latex technology or in PVC processes);
(5) as a liquid monomer or low-molecular-weight polymer (as in casting and laminate production); or
(6) as a rigid solid (in machining operations).

In all but the last example a special operation is needed to set the shape. The most important approaches are:

(1) cooling of thermoplastics and thermoplastic rubbers from the melt;
(2) cross-linking of thermosetting plastics and of rubbers (vulcanisation);
(3) polymerisation of monomer as in casting of polymethyl methacrylate and nylon 6;
(4) evaporation of solvents as in film casting;
(5) gelation or coagulation of rubber latex which separates the rubber from the continuous (usually aqueous) phase; and
(6) gelation of PVC paste in which continuous-phase plasticiser is absorbed on heating into the PVC particles to form a single-phase blend of polymer and plasticiser.

The bulk, by far, of rubbers and plastics processing is carried out in the melt state.

2.6.2 Processing in the molten state

Amongst techniques available for shaping melts are squeezing in a mould (compression moulding), injecting into a mould under pressure (injection moulding), forcing through a hole (extrusion) and forcing through an annular hole and inflating the resultant tube in a shaping mould (extrusion blow moulding).

When processing in the melt state it is important to consider the following:

(1) the water absorption of the raw materials;
(2) the physical form of the raw material;
(3) the thermal stability of the polymer;
(4) the flow properties of the melt;
(5) the adhesion of melt to metal;
(6) the thermal properties affecting heating and cooling of the melt;
(7) the compressibility and shrinkage; and
(8) the cross-linking (curing, vulcanising) characteristics (in the case of products the shapes of which are set by cross-linking).

This topic has been dealt with in the case of thermoplastics in greater detail elsewhere, but the following points may be made here.

Traces of water in a compound (perhaps more associated with an additive than the polymer) may turn to steam at elevated processing temperatures causing porosity and possibly affecting cross-linking processes. The higher the processing temperature the worse the effect so that the level of water tolerable varies between materials. The water may simply be deposited on the surface of the polymer or may be absorbed into its mass. Different levels of action may thus be necessary to remove water, although prevention of contamination in the first place will reduce many problems.

Polymer compounds may be in a variety of forms before melt processing, the most common being as pellets, as powder or as slab. The pellet form is most common with thermoplastics and thermoplastic rubbers, facilitating handling, storage and feed. There has been some interest in powder feed with rubbers and thermoplastics, but this may cause problems in feed, in explosive hazards and in porosity of the products. Powder feed is common with thermosetting plastics used for compression moulding. With both pellet and powder, regularity of size is important to facilitate feed but, for some purposes (e.g. where a pelleting process is involved), a less regular particle size distribution may be required. For most rubber processing operations it is common to handle the polymer compound in slab or sheet form; this approach also being used when feeding PVC into a calender. Glass-fibre-reinforced polyester and epoxide moulding compounds may also be supplied in the form of both doughs (dough-moulding compounds) and sheet (sheet-moulding compounds).

All polymers will be adversely affected to a greater or lesser extent on heating. It is important to ensure that the compound has sufficient stability for the processing operation envisaged. Within the processing equipment, points where melt may be held up or allowed to 'stagnate' must be avoided. It should be realised that use of reworked material may adversely affect the level of stability. In the case of rubbers, overheating at early stages of processing may cause premature curing (*scorching*) before shaping is complete.

Flow properties of polymer melts vary enormously between type of polymer, polymer molecular weight and melt temperature. Polymer melts also differ from simple liquids in that the *apparent viscosity* decreases as the rate of shear increases, i.e. small increases in shaping forces can cause large increases in flow rates (*Figure 2.19*).

For most operations a level of polymer-to-metal adhesion is desirable; for example, in order to provide a positive pumping action by the screw in a single-screw extruder. However, with some polymers adhesion may be so high that if a polymer is allowed to cool from the melt inside an injection-moulding machine barrel it can pull pieces of metal away from the barrel during contraction of the melt.

The amount of heat needed to bring a polymer compound up to processing temperature will vary according to the melt temperature required, the specific heat of the polymer, the latent heat of melting of crystallites (in the case of crystalline polymers) and, of course, the mass of the polymer compound. (Note: since most articles are of a definite *volume* then, when

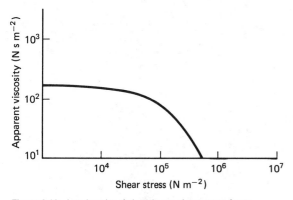

Figure 2.19 Log–log plot of viscosity vs. shear stress for an injection-moulding grade of nylon 66. All polymer melts show a decrease in viscosity with increase in shear—this is known as 'pseudoplasticity'

comparing materials, it is necessary to also take into account the density of the compound.) Such differences may not just be of the order of a few per cent, but sometimes two- or three-fold.

The amount of heat that must be removed from a compound before it can, for example, be removed from a mould will also vary from one polymer compound to another. In this case it will be necessary to know the temperature at which a moulding may be removed from a mould.

The melt and mould temperatures quoted in *Table 2.20*, which is designed primarily for injection moulding, are taken from a paper by Whelan and Goff.[4] Using data obtained by differential scanning calorimetry methods, the authors measured the amount of heat required to cool unit mass of polymer from the melt temperature to the mould temperature and used this as a measure of the heat required to be removed from unit mass of a moulding before the moulding was extracted. In practice, the mould is opened before the polymer reaches the mould temperature but the data do provide useful relative values. Since a moulding is made by volume, the present author feels that it is more useful to compare the heat required to be removed per unit volume and this has been obtained by using specific gravity data.

In their paper Whelan and Goff[4] estimated the specific heat averaged between the melt and mould temperatures by the

Table 2.20 Heat required (enthalpy required) to raise polymers to their processing temperatures from an ambient temperature of 20°C and the heat required to be removed in cooling a polymer from the melt to the mould temperature

Polymer	Melt temp. (°C)	Mould temp. (°C)	SG	specific heat (J kg^{-1} K^{-1})	Heat required to melt		Heat required to cool	
					(J g^{-1})	(J cm^{-3})	(J g^{-1})	(J cm^{-3})
FEP	350	220	2.2	1600	528	240	240	109
Polyether sulphone	360	150	1.37	1150	391	285	242	177
Polyether ether ketone	370	165	1.3	1340	469	361	275	212
Polyethylene terephthalate (crystalline)	275	135	1.38	2180	556	403	305	221
Polystyrene	200	20	1.05	1720	310	295	310	295
Polyacetal	205	90	1.41	3000	555	394	345	245
Polycarbonate	300	90	1.2	1750	490	408	368	307
ABS	240	60	1.04	2050	451	434	369	355
Polymethyl methacrylate	260	60	1.18	1900	456	386	380	322
Polyphenylene sulphide	320	135	1.4	2080	624	446	385	275
PPO (Noryl type)	280	80	1.06	2120	551	520	434	409
Polysulphone	360	100	1.24	1675	570	459	436	351
Polyethylene terephthalate (amorphous)	265	20	1.34	1970	483	360	483	360
Nylon 11/12	260	60	1.03	2440	586	568	488	474
LDPE	200	20	0.92	2780	500	543	500	543
Nylon 6	250	80	1.13	3060	703	623	520	460
Nylon 66	280	80	1.14	3075	800	701	615	539
Polypropylene	260	20	0.91	2790	670	736	670	736
HDPE	260	20	0.96	3375	810	843	810	843

* Data taken from Whelan and Goff.[4] The table is modified from Brydson, *Plastics Materials*, 5th edition, Butterworth-Heinemann.

method described above. It has been assumed by the present author that this value is valid between the melt temperature and 20°C and has then been used also to estimate the amount of heat required (both per unit volume and unit mass) to raise the polymer from 20°C to the melt temperature. Whilst this will be in some error, it once again provides some useful relative information.

The considerable difference between polymers should be noted, particularly noting that the polymers with the highest processing temperatures do not necessarily require the greatest amount of heat to raise to processing temperature and, in addition, often require much less heat to be removed before a moulding can be extracted.

When a polymer solidifies from the melt it shrinks, the shrinkage being greater with crystalline polymers. However, where a melt has been subjected to high pressures during shaping some of the cooling may cause decompression rather than shrinkage. As a general guide, the linear shrinkage of amorphous thermoplastics is about 0.005 cm per centimetre, but crystalline polymers may shrink by anything from 0.007 to 0.060 cm per centimetre.

It is generally desirable with thermosetting plastics and vulcanisable rubbers that the cross-linking system be such that it is not activated during shaping but only after the material is shaped. It is then normally desired for the reaction to be as fast as possible, i.e. the activity should increase rapidly above a certain temperature and/or after a delay period. This is not always satisfactory when shaping thick articles because of the time taken to heat up the centre of the articles to the required temperature.

2.6.3 Compression moulding

In principle, compression moulding consists of heating a polymer compound in a mould and then closing the mould under pressure forcing the melt to the shape of the mould cavity, excess material being squeezed out as *flash*. The main features of a simple compression mould are shown in *Figure 2.20*.

The process is of only limited value with thermoplastics and thermoplastic rubbers because it is necessary to cool the moulding each cycle before the moulding can be removed. On the other hand, the process is very suitable for thermosetting plastics and vulcanising rubbers which are cross-linked in the mould and thus harden *in situ* and may be removed whilst still hot. A typical moulding cycle, applicable to both rubbers and thermosets, is shown in *Figure 2.21*.

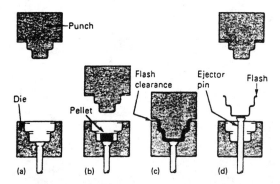

Figure 2.20 The moulding sequence in compression moulding. (a) Mould open, cavity cleaned; (b) pellet loaded, mould closing; (c) mould closed, curing stage; (d) mould open, moulding ejected. In this figure a pellet (possibly preheated) is used but, frequently, and particularly with automatic processes, powder is used. (Reproduced by permission of the Open University, Milton Keynes)

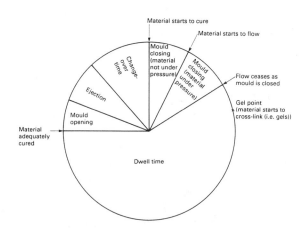

Figure 2.21 The compression-moulding cycle. In many instances the dwell time will comprise a larger fraction of the moulding cycle than indicated here. (It is instructive to compare this diagram with that for the injection-moulding cycle—see *Figure 2.25*). (Reproduced by permission from Brydson, *Handbook for Plastics Processors*, Butterworth-Heinemann)

Rubbery compounds are usually fed in the form of a slab, whilst thermosetting plastics are normally put into the mould either in powder or pellet form. Whilst pelleted material is less suitable for automated operations, it is useful where preheating is required and where rapid weighing and minimal contamination is desired. Mention should, however, be made of *dough-moulding compounds* based on polyester resins and which are widely used for such diverse applications as microwave cookware, headlight reflectors and electricity meter boxes, which are supplied in the form of a dough which is fed to the mould in a lump form. A variation of this is *sheet-moulding compound* which is supplied in sheet form for moulding.

An important variation of the compression-moulding process is used for pneumatic tyre manufacture. In this case the various parts of a tyre (tread, sidewalls, bead, etc.) are assembled on a drum into the form of a cylinder. This cylinder is then turned into the general shape of a tyre by using an inflating bladder which is actuated as the mould closes. The pressure exerted by the bladder forces the rubber to adopt the contours of the mould surface and this is immediately followed by cross-linking (vulcanisation).

2.6.4 Transfer moulding

In this process the raw material is placed in a transfer pot from which it is forced by means of a plunger through a runner and gate into a mould cavity. A simple transfer-moulding system using an *integral pot mould* is shown in *Figure 2.22*. More complex systems are also used which may be considered as intermediate between the integral pot system and injection moulding (see below).

The process has been useful in helping to ensure clean surfaces of mouldings (rather difficult when compression moulding from slab feedstocks) and because of the heating effect when passing down the runners makes it easier to mould thick articles. The process has been widely used for rubber moulding, for example O-rings, complex thermoset compression mouldings and, at one time, unplasticised PVC. The development of injection-moulding techniques for vulcanising rubbers and thermosetting plastics has in recent years reduced the use of transfer moulding, although it has found special

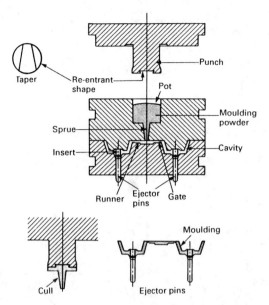

Figure 2.22 Transfer mould of the integral-pot type. (From Bown and Robinson, *Injection and Transfer Moulding and Plastics Mould Design*, Business Books)

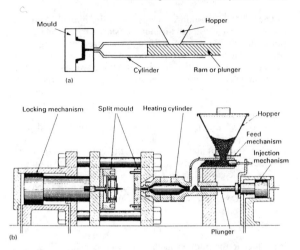

Figure 2.23 (a) Scheme to show basic ram injection moulding. (b) Ram machine incorporating a torpedo or spreader inside the heating cylinder

applications for encapsulating electrical components with epoxide resins.

2.6.5 Injection moulding

This process may be considered as the workhorse of the
plastics industry (although in tonnage terms more material is
probably extruded than injection moulded). In principle, the
process consists of heating the material in a cylinder (barrel)
until it softens and then forcing the material into a mould in
which the material sets. In the case of thermoplastics and
thermoplastic rubbers the mould is cooler than the temperature at which the material hardens (e.g. $<T_g$ in the case of an
amorphous thermoplastic), but with thermosets and vulcanising rubbers the mould temperature is kept high to enable rapid
cross-linking to occur.

At one time most injection machines were of the ram type,
as illustrated in *Figure 2.23*. The diagram shows a spreader
inserted into the barrel to ensure that polymer comes close to
the heated barrel wall in order to give more uniform and more
rapid heating. However, the resulting constriction leads to
undesirable pressure loss between the front of the ram and the
melt in the mould cavity. There will also be undesirable
pressure losses in the granular material which is present at the
hopper end of the cylinder. These opposing requirements of
efficient heat transfer and efficient pressure transmission led
to the development of the *in-line single-screw preplasticising
injection moulding machine* by BASF and first introduced
commercially by Ankerwerke. The process is illustrated in
Figure 2.24.

In the first stage of the cycle the screw rotates, pumping
polymer from the hopper end towards the front of the barrel,
melting it during the process. In order to accommodate the
material the screw moves backwards at the same time that it
rotates until such time that there is a sufficient amount of
polymer ahead of the screw to fill the mould cavity (or
cavities). The screw then ceases to rotate and moves forward
as a ram, injecting preplasticised material into the mould
which then sets either by cooling or cross-linking.

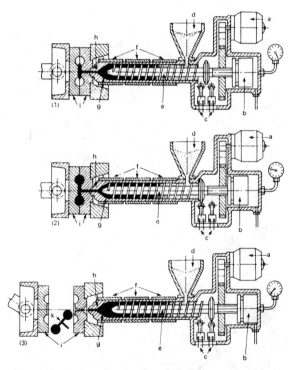

Figure 2.24 Screw injection-moulding cycle. (1) Molten polymer
being injected into mould by screw acting as a ram (usually
non-rotating) under injection pressure. (2) Screw in forward position
under holding pressure until the material in the gate (point of entry into
the mould cavity) has frozen. (3) Screw has rotated and moved axially
away from mould to charge sufficient material to front of cylinder for
next shot. Mould opens when moulding is sufficiently hard. a, Motor
for rotation of the screw; b, drive for reciprocating movement of the
screw; c, left limit switch for setting holding pressure, right limit switch
for limiting the screw stroke; d, feed hopper; e, conveying, plastifying
and injection screw; f, heater bands; g, hot plastified material; h,
injection nozzle; i, cooled or heated mould with sprue brush (attached
on the locking unit); k, moulded part with sprue. (Reproduced by
permission from Saechtling, *Plastics Handbook*, 2nd edition, Hansen)

The injection-moulding cycle is shown in *Figure 2.25*. In this diagram it is easiest to consider the start of the cycle at the point the mould starts to close. The preplasticising stage whereby melt is fed to the front of the barrel takes place during the stage labelled 'further cooling of moulding'.

A simple two-cavity injection mould is shown in *Figure 2.26* illustrating some of the principal features of such moulds.

A typical pressure profile occurring during the injection moulding cycle is shown in *Figure 2.27(a)*.

It is desirable that the pressure in the cavity should drop to zero at the point that the material sets (not a very exact concept, since not all the melt sets at the same time). If there is residual pressure after the material sets then the moulding may be difficult to eject from the mould, whilst if the pressure drops to zero whilst the material is still molten further cooling may lead to sink marks and voids (*Figure 2.27(b)*).

2.6.5.1 Injection unit specifications

The injection-moulding machine can be considered as comprising two active components: the injection unit based on the barrel and screw, and the locking unit which operates the mould.

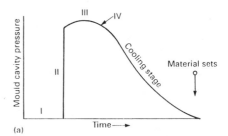

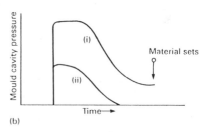

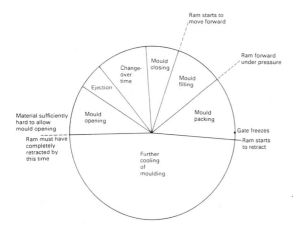

Figure 2.25 The injection-moulding cycle. (Reproduced by permission from Brydson, *Handbook for Plastics Processors*, Butterworth-Heinemann)

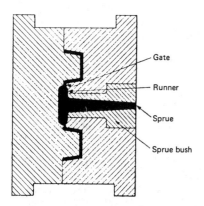

Figure 2.26 Simple two-cavity injection mould. (Reproduced by permission from Pye, *Injection Mould Design*, 2nd edition, Plastics and Rubber Institute)

Figure 2.27 (a) Preferred trace of mould cavity pressure with time during an injection-moulding cycle. The cavity pressure should drop to zero at the moment when the moulding has reached the settling point. I, Mould filling; II, pressure build up; III, consolidation; IV, gate freezing point. (b) Undesirable cavity pressure vs. time traces. (i) Residual pressure when the moulding sets causes sticking problems. (ii) Cavity pressure is at zero while the polymer is still molten. Further cooling leads to sink marks, voids and shrinkage defects. (Reproduced by permission from Brydson, *Handbook for Plastics Processors*, Butterworth-Heinemann)

Injection capacity The maximum amount of material that may be injected per cycle of operations (per shot). This is now usually quoted as the 'swept volume' (usually in cubic centimetres) that the injection unit can deliver in one stroke at a nominated injection pressure (e.g. 100 MPa). The old tendency to quote the weight of polystyrene that could be injected per shot is now obsolescent.

Length/diameter ratio (L/D ratio) This term is only relevant to in-line screw machines. The figure quoted is based on the *effective* length rather than the actual length of the screw, i.e. the distance between the rear of the hopper when the screw is in the fully forward position to the end of the screw flights. A typical value of L/D is 20:1, but may be smaller for large diameter screws (e.g. 14:1 for a 90 mm diameter screw) and larger where a vented barrel is used.

Screw stroke The maximum linear movement of the screw during the plasticisation stage. As the screw moves back, the effective L/D ratio is reduced and, if this is overdone, the polymer will become insufficiently plasticised. To avoid this problem the maximum screw stroke is usually $<3.5D$.

Plasticising capacity The maximum amount of material that may be plasticised to an acceptable melt quality per unit time. Clearly this depends on the material. It is most commonly expressed in terms of weight of polystyrene plasticised per hour (this material presumably being chosen by the machine manufacturers as it gives the highest value for any common polymer). Plasticising capacity is, however, also affected by shot size, injection rate, screw design, barrel specification and melt temperature. Comparison between machines should therefore be made using standard procedures such as those of the Society of the Plastics Industry Inc. (SPI) or the European Committee of Machinery Manufacturers for the Plastics and Rubber Industries (EUROMAP).

Injection rate The maximum rate at which melt may be ejected from the barrel during a single shot. It is usually measured when no mould is present (i.e. an *air shot*). The value depends on the material used and may be expressed in either volume or weight terms.

Injection pressure This is the maximum pressure that may be exerted by the ram (screw) on the molten polymer during the mould-packing stage of the injection cycle. It is not the maximum line pressure of the hydraulic oil between the pump and the back of the ram (screw) nor is it the maximum pressure exerted by the ram (screw) on the melt before the mould cavity is filled.

2.6.6 Extrusion

In tonnage terms more plastics materials are extruded than processed by any other method. This is because the process is not only used for making important products such as film, sheet, piping, ducting, cable, hose and so on, but also it is an important intermediate operational stage for such purposes as the preparation of sheet for vacuum forming, for mixing operations and reworking of waste material. The injection unit of an in-line screw injection-moulding machine and the corresponding part of a blow-moulding machine may also be considered as captive extrusion operations. Although perhaps not quite so dominant in the rubber industry it is widely used not only to produce end-products but also for intermediate operations such as producing tyre tread material prior to tyre building and moulding. A modification of the process has been used for thermosetting plastics using a ram extruder, but this is a highly specialised process.

The process consists, in principle, of forcing fluid material through an orifice (generally known as a *die*) to give an extrudate of constant cross-section. It is not confined to polymers, but is also used for such diverse materials as pasta and metals. The pumping may be brought about by gas pressure (as in some rheometers), by the use of gear pumps or rollers (at one time used with rubber and gutta percha) or by means of a ram. However, the use of screw pumps is today virtually universal, mainly single-screw but occasionally of the twin-screw type.

2.6.6.1 Single-screw extrusion

The extrusion may be considered in four stages:

(1) material feed,
(2) extruder barrel and screw,
(3) extrusion die, and
(4) haul off.

Thermoplastics and thermoplastic rubbers are usually fed in granule form, although powder is occasionally used. Rubbers are usually fed in strip form, whilst some extruders are designed to take large lumps of material that have been dumped directly from mixing equipment.

The essential features of the barrel and screw are shown in *Figure 2.28*, whilst some common terms associated with screw design are given in *Figure 2.29*.

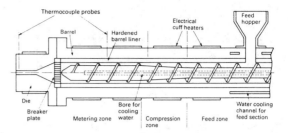

Figure 2.28 The essential features of a single-screw extruder. (Reproduced by permission of the Plastics and Rubber Institute)

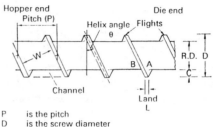

P	is the pitch
D	is the screw diameter
W	is the channel width
C	is the channel depth
L	is the land width
R.D.	is the root diameter
θ	is the helix angle
A	is the screw leading edge
B	is the screw trailing edge

Figure 2.29 Common terms associated with screw design. (Reproduced by permission of the Plastics and Rubber Institute)

It is convenient to consider the barrel in three zones:

(1) the *feed zone* accepts granules from the hopper, pumps them up the barrel and commences the heating and plasticisation of the polymer;
(2) the *compression zone* compresses the polymer, squeezing air which had been between the granules back out of the system via the hopper (the actual bulk compression of solid polymer is very small); and
(3) the *metering or delivery zone* ensures that the material is in the correct state for feeding to the die.

The relative lengths of each zone may vary from polymer to polymer. With some materials the zones may be of similar length with a gentle transition from zone to zone, whereas with other materials the compression zone may be very short. This will affect the type of screw used; some typical screw designs are shown in *Figure 2.30*.

In addition to those shown in *Figure 2.30*, important screw characteristics are:

(1) the *length/diameter ratio* (defined as in Section 2.6.5.1);
(2) the *compression ratio*, which may be defined by the expression

Compression ratio

$$= \frac{\text{Swept volume of one turn of the channel at feed}}{\text{Swept volume of one turn of the channel at exit}}$$

It should also be noted that commonly the screw pitch is numerically the same as the screw diameter.

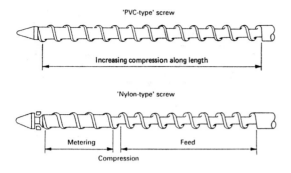

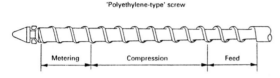

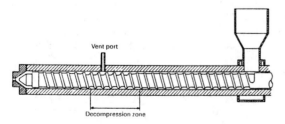

Figure 2.30 Typical screw profiles for three types of thermoplastic

For materials that can volatilise or which may give off volatiles on heating, a screw with a decompression zone may be used in conjunction with a vented barrel. Such a design is shown in *Figure 2.31*.

2.6.6.2 Extrusion dies

On leaving the barrel the melt is fed into a die that determines the shape of the cross-section of the extruded product. Many types of die exist. Simple designs for making tube, lay-flat film, wire covering and sheet are shown in *Figures 2.32* to *2.35*.

2.6.6.3 The overall extrusion process

The extruder and die form only a part of the total extrusion process. *Figures 2.36* to *2.39* show typical layouts for tube, lay-flat film, wire covering and sheet using dies similar to those described in the previous section.

2.6.6.4 Multi-screw extrusion

In a single-screw extruder the forward pumping action caused by screw rotation is partially offset by a reverse pressure flow caused by restrictions at the die. In the case of a twin-screw

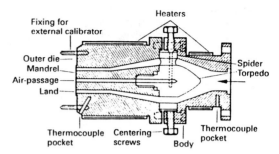

Figure 2.32 Tubing die. (Reproduced by permission of the Plastics and Rubber Institute)

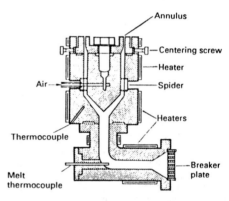

Figure 2.33 Tubing (film) die involving a 90° turn in the flow path via an elbow adaptor. (Reproduced by permission of the Plastics and Rubber Institute)

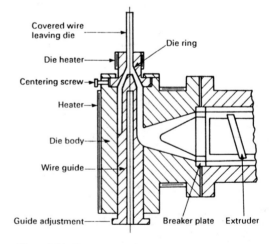

Figure 2.34 Wire-covering die. (Reproduced by permission of the Plastics and Rubber Institute)

extruder, in which the screws intermesh, the polymer becomes trapped between C-shaped spaces between the screw flights and is positively pumped up the barrel. In this case the output is independent of the die head pressure. Twin-screw extruders are expensive and are limited to low extrusion speeds. They

Figure 2.31 Decompression-type screw used in conjunction with a vented barrel

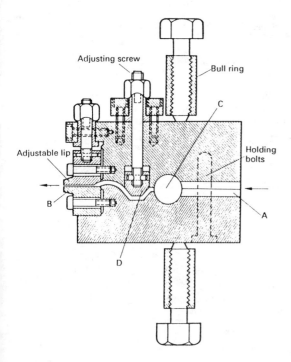

Figure 2.35 Manifold die for sheet manufacture. In this design the melt passes to the die down an inlet tube (A) into the tubular manifold (C) which is transverse to the cylinder barrel axis. (In many designs the manifold may be slightly bent so that the distance between the manifold and the die lip is less on the edge of the die than at the centre, in order to compensate for the pressure drop down the manifold.) Between the manifold and the die lip there is also an adjustable restrictor bar (D) which can 'fine tune' the flow rate across the width of the die. The melt then flows between the fixed die lip (B) and an adjustable lip. (Reproduced by permission of the Plastics and Rubber Institute)

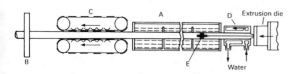

Figure 2.36 Extrusion-line layout for hollow tubing. A, water tank; B, cutting screw for cutting tubing into lengths; C, caterpillar haul off; D; sizing disc; E, floating plug to prevent pressurising air from escaping out of end of pipe. (Reproduced by permission of the Plastics and Rubber Institute)

are, however, useful with difficult materials such as unplasticised PVC.

2.6.7   Blow moulding

The term *blow moulding* is used to describe a process for making hollow articles by inflation of a tube (known in this context as a *parison*). There are three main variants of the process:

(1)   extrusion blow moulding,
(2)   injection blow moulding, and

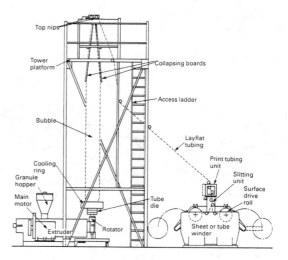

Figure 2.37   A blow film line for low-density polyethylene

(3)   stretch blow moulding (which may be based on either extrusion or injection methods).

2.6.7.1   Extrusion blow moulding

The principle of extrusion blow moulding is illustrated in *Figure 2.40*. In the first stage a tube (parison) is extruded between the halves of an open mould. When a sufficient length has been extruded the mould closes round the parison which is then inflated by compressed air to the shape of the mould cavity. After cooling the mould is opened and the moulding is ejected. Variants of the process depend on whether extrusion of the parison is continuous or intermittent, how many blow stations exist, how many mouldings are made at a time, the position of the blowing pins, and methods of controlling parison dimensions. *Figure 2.41* shows the outline of a *two-station* system using two four-cavity moulds, making eight mouldings per cycle.

2.6.7.2   Injection blow moulding

The principle of injection blow moulding is illustrated in *Figure 2.42*. In this process the parison is formed by injecting directly onto a blow stick. This is then transferred, together with the molten parison, to the blowing cavity. In turn the parison is blown to the shape of the cavity by means of compressed air which passes through the blowing stick.

The process is useful where parison weight control is critical and where the dimensions of the threaded neck may be more accurately controlled. The process has been used particularly for making polystyrene containers for talcum powder and other toiletries.

2.6.7.3   Stretch blow moulding

In this process, arrangements are made for controlled stretching of the parison. This will increase the level of crystallinity (in crystallising thermoplastics) and will give products that are lighter for a given rigidity, enhanced transparency and gloss, reduced permeability, greater precision and improved burst strength. The process has been of particular use in the manufacture of polyethylene terephthalate (PET) bottles for beer and other carbonated drinks. The process may be based

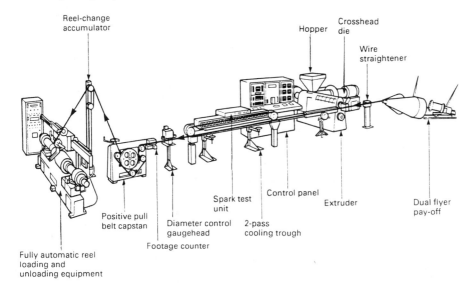

Figure 2.38 A typical wire-covering extrusion line

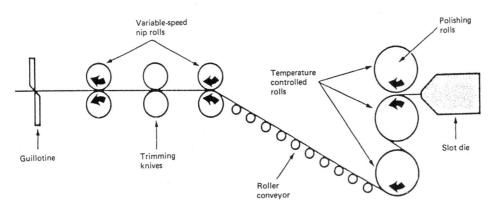

Figure 2.39 An extruded-sheet line

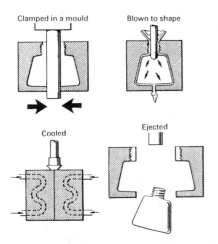

Figure 2.40 The principles of extrusion blow moulding, showing the four main stages of the process. (Reproduced by permission from the Plastics Processing Industry Training Board)

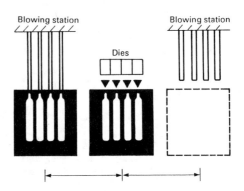

Figure 2.41 Extrusion blowing system with two blowing stations and four cavity moulds. (Reproduced by permission from the Plastics Processing Industry Training Board)

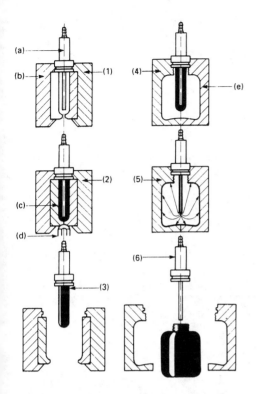

Figure 2.42 Principles of injection blow moulding. (1) Injection mould (b) containing transferable blowstick (a). (2) Material injection via hot nozzle or runner (d) to form molten parison. (3) Mould opens—blowstick removed with hot parison (c). (4) Blowstick placed in blowing cavity (e). (5) Air blown through holes in blowstick inflates parison. (6) Blow cavity opens—moulding removed

on either an injection moulding or an extrusion blow moulding technique. *Figure 2.43* shows a schematic outline of a process based on injection-moulded parisons.

2.6.8 Calendering and frictioning

The calendering process was originally used for applying a sheen onto fabrics and paper but has since been adopted by both the rubber and plastics industries. In the plastics industry it is used mainly for making PVC sheeting employing highly expensive equipment. In the rubber industry it is also used

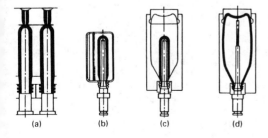

Figure 2.43 Injection stretch–blow moulding process sequence. (a) Multiple injection moulding of amorphous preform. (b) Reheating stage. (c) Axial extension with telescopic mandrel. (d) Blow moulding

both for making sheet (e.g. butyl rubber pond liners) and for impregnating fabrics (frictioning) with rubber prior to their use in the manufacture of such products as tyres, belting and hose. The rubber industry also uses a similar process for coating rubber onto fabric (skin coating).

The calender consists of an arrangement (stack) of rolls (bowls) mounted in bearing blocks supported by side frames (gables) and equipped with roll drives, nip adjusting gear and feed, heating and haul off arrangements. Some typical configurations are shown in *Figure 2.44*.

2.6.9 Thermoforming

All the processes described above involve processing of the melt. The term *thermoforming* is used to describe processes in which the material is shaped in an elastic state and then frozen. It is particularly important with sheet shaping, but may also be used for hot stamping (warm forging) operations. In the case of sheet shaping three main variations are known:

(1) mechanical shaping,
(2) vacuum forming, and
(3) compressed-air techniques.

Mechanical shaping involves forcing the material to shape either by stamping between the halves of a matched metal mould or by the use of skeleton tooling which only contacts the sheet at key points but leaves the sheet unmarked elsewhere. Such processes find some use with thicker grades of sheet (e.g. acrylic sheet) where the material may be preheated in an oven and then transferred to the press and shaped before it has had time to lose too much heat.

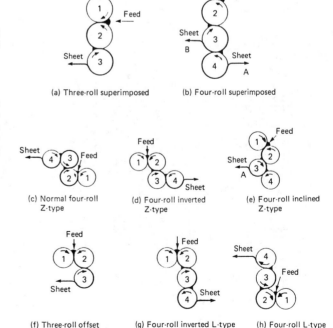

Figure 2.44 Some common calender configurations. (Reproduced by permission of the Plastics and Rubber Institute)

Far more important is vacuum forming. In this process sheet is clamped over the shaping mould, the sheet is heated by infra-red heaters until it is rubbery and the air is then evacuated from the space between the sheet and the mould. The atmospheric pressure then forces the sheet down onto the mould and after cooling the shaping is then removed. A simple form of vacuum forming known as 'female moulding' is shown in *Figure 2.45*.

A somewhat more complicated process known as 'male drape forming' is shown in *Figure 2.46*. In this case the sheet is first clamped in a frame and heated. The male mould then moves upward into the sheet and this is followed by application of the vacuum. Whereas the female mouldings are thinner at the base of the moulding, the corresponding part of a male moulding will be the thickest part of the moulding.

There are a number of variations of the vacuum forming process. The method chosen depends on such factors as thickness control and material saving. One disadvantage of sheet-shaping methods is that scrap material may not be almost directly reprocessed as is often possible with melt-processing methods such as injection moulding.

Vacuum-forming techniques are also limiting in that the only pressure available for shaping is that of the atmosphere (c. 0.1 MPa). This has led to a number of techniques by which positive air pressure is used to force sheet to shape. In its simplest form softened sheet may be clamped over a blowing table by a ring. Air is introduced below the sheet which is then free blown to the shape of a bubble. Numerous variations on this process, some of which simultaneously involve vacuum and mechanical pressure, are used.

2.6.10 Other processing methods

The processes described above are widely used with many different polymers. There also exists a large number of other processes which, although of considerable importance, tend to be restricted to a few polymers. Some of these are mentioned briefly below.

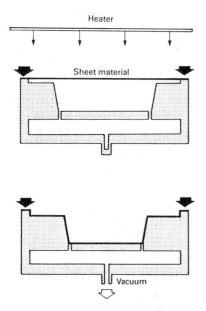

Figure 2.45 Vacuum forming with female moulding. (Reproduced by permission from *ICI Technical Note G109*, ICI)

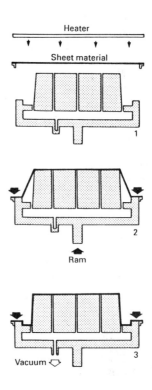

Figure 2.46 Vacuum forming with drape forming over a male mould. (Reproduced by permission from *ICI Technical Note G109*, ICI)

2.6.10.1 Polymerisation casting

In this process monomer or low-molecular-weight material or a mixture of monomer and polymer is mixed with a polymerising agent and is then poured into a mould and allowed to polymerise. This method has been used to make such widely differing products as very large gear wheels from nylon, artificial dentures from polymethyl methacrylate and hydrophilic contact lenses, also from acrylic materials.

2.6.10.2 Fibre-reinforced thermosetting plastics laminates

In some ways this may be considered a development of polymerisation casting. Liquid material (usually low-molecular-weight polymer and generally known as *resin*) is blended with a polymerising catalyst and then applied to glass-fibre material. The resin then hardens with the fibre *in situ*. Very often the process, as used to make boat hulls, sports-car bodies and lorry cabs, is basically very simple. The mould may be of wood, plaster, metal or even plastic. It is draped with glass fibre which is then impregnated with resin, often simply by means of a brush, and the air is removed using ribbed rollers. When used with polyester laminating resins it is not necessary to apply external pressure and formulations are commonly employed that allow for curing (cross-linking) at room temperature. More elaborate methods are, however, available; for example, using fibre preforms or preimpregnated sheet (sheet moulding compound) with matched metal moulds, processes widely used for microwave cookware, car light fittings and crash helmets.

Whilst the market is dominated in tonnage terms by polyesters, epoxide resins are widely used in the aerospace industry (sometimes using carbon fibre or aramid fibre rather than glass fibre) and with more sophisticated techniques such as filament winding. Furan resins are sometimes used for chemical plant.

2.6.10.3 Polyurethane foams

This may also be considered as a modification of poly-merisation casting. In this process a liquid polyether or polyester (generally known as a 'polyol') is blended with an isocyanate, some water, a blowing agent, a cross-linking agent and other more specialised components. The highly agitated mix is then run into containers (either open trays or enclosed moulds) where various reactions occur causing the material to cross-link and give off gases causing the material to form a cellular structure more or less simultaneously. By varying the components the products may be hard (as used in insulation), semi-rigid (for shock absorption) or soft (as in upholstery). Much effort has been expended in recent years to make the products less liable to burn or to give off gases that may harm the ozone layer. The variation of the process in which the reacting mixture is pumped into an enclosed mould, known as *reaction injection moulding* has been subject to considerable development over the years.

2.6.10.4 Rotational moulding and centrifugal casting

In rotational moulding polymer is put into a mould which is then closed and rotated in an oven. In most operations the polymer melts and spreads evenly over the inner wall of the mould. The mould is then cooled and the moulding extracted. The process may be used for such diverse products as large container bins and traffic cones.

In one variation, PVC polymer, plasticiser and other ingre-dients are mixed together at room temperature to give a paste with the solid components supended in the liquid ones. On heating the liquid components are absorbed by the PVC particles and the whole mass fuses into a homogeneous material during the rotational operation. The process is used to make balls and toys.

Pipes may also be made by centrifugal casting of polymer in a long cylinder which rotates about its axes whilst being heated.

2.6.10.5 Spreading and dipping

PVC pastes may also be spread onto fabrics and then gelled by heating in an oven to give leathercloth. In the rubber industry unvulcanised rubber compounds are dissolved in solvents and spread onto cloth. The solvent is then removed and the rubber vulcanised in an oven. The process is important for making waterproof goods and gas-impermeable fabrics.

Products may be made by dipping formers into polymers in fluid form. These could be PVC pastes which are then gelled on heating, rubber solutions (where the solvents are removed before vulcanisation), rubber latex (where water is first removed) or even fine particles of polymer suspended in a gentle air stream (fluidised-bed techniques).

2.7 Design of rubber components

Since the greater part of manufactured rubber goods are used for engineering purposes and because of the nature of this handbook, this section is confined to a short discussion of the design of selected rubber components used in engineering applications.

The severe technical demands of such products are often underestimated. For example, a car tyre has an area of contact with the road only of the order of the size of the palm of the hand and yet this is expected to give sufficient grip for the car to move forward without skidding. Similarly, the tread and sidewalls are expected to have good abrasion resistance so that the tyre will be long lasting whilst the danger of blow-outs due to overheating, often in most severe conditions, has to be negligible. In many other applications less well known to the general public such as in springs, seals, bearings and mine belting, the specification requirements are no less demanding but, using properly formulated compounds and well-designed products, rubbery materials have been successfully used in a wide range of applications.

In Section 2.2 some of the more well-known features of vulcanised rubbers were discussed. In particular mention was made of:

(1) capability of large elastic deformations,
(2) low modulus or stiffness,
(3) non-linearity of the tensile stress–strain curve,
(4) bulk incompressibility, and
(5) damping capacity.

It is pertinent to make some comment here on the relevance of the above to design considerations.

Although an elongation at break as high as >1000% is possible and 400% is common, this is seldom of immediate importance. This is because, often for fail-safe considerations, rubber components are usually stressed in compression, shear or torsion with deformations usually less than 50%. In such circumstances the stress–strain relationship is almost linear and the modulus may be estimated by measuring the slope of the curve at the intercept. This is a useful simplification in calculations.

In many applications it is desirable that the stiffness of a component be anisotropic. This is achieved by incorporation of fabric or of metal plates; the effect of inserting metal plates on compression stiffness without affecting shear stiffness was discussed in Section 2.2.5.

Mention was also made in Section 2.2.3 of the importance of taking into account the incompressibility of rubber in bulk. It bears repeating that if a rubber component is required to be deformed in operation then there must be some room for it to deform. *Figure 2.47* provides an illustration of this. (This problem of course applies primarily to solid rubber compo-nents and not those of cellular (foam, sponge, or expanded rubber) consistency.)

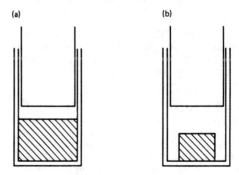

(a)   (b)

Figure 2.47 Because of its bulk incompressibility the rubber pad in (a) will have no cushioning effect if struck by the piston, whereas in (b) deformation of the pad is possible, thus allowing some cushioning

Damping by a rubber results from the fact that response to a stress is not instantaneous and this enables such features as shock absorption and vibration isolation to be designed into a product.

Before discussing some selected applications it is also pertinent to mention the causes of failure in rubber–metal bonded units which are of considerable engineering importance. It is found that, providing a good rubber compound has been used and the method of bonding is appropriate, failure normally occurs where there is splitting or tearing at a point of stress concentration. Rubber is also easily cut when under stress so that it is important to avoid:

(1) conditions which start cuts such as sharp metal edges; and
(2) conditions that result in high local stress concentrations.

Some examples of good and bad design are given in *Figure 2.48*.

2.7.1 Rubber springs

A steel coil spring has a number of undesirable features, notably:

(1) it has a limited capacity—it may only be compressed to the point where the coil is 'closed';
(2) it is easily buckled on misalignment; and
(3) unless made of stainless steel it is liable to corrosion.

On the other hand, a rubber spring has an infinite capacity, if squat in shape it will not buckle, is not subject to corrosion and is silent in operation. Rubber springs take many forms and may be used in a variety of applications, some examples of which are give below.

2.7.1.1 Bushes

Bushes consists of two concentric metal sleeves separated by a hollow rubber cylinder which may be either bonded to the sleeves or a push fit (*Figure 2.49*). Such a bush may be subjected to a variety of deformation modes such as torsion, shear (axial deformation), tension–compression (radial deformation) or tilt (conical deformation) (*Figure 2.50*).

Rubber bushes are widely used for purposes such as shaft couplings in order to accommodate misalignments, and for a host of vibration isolation purposes.

2.7.1.2 Inclined shear mountings

In the simple design shown in *Figure 2.51* the two inclined metal plates are separated by two rubber pads which will be subjected to compression and/or shear stresses according to the direction in which the load is imposed and, therefore, *the stiffness will vary in each of the three perpendicular directions*.

Calculation shows that the stiffness in the three mutually perpendicular directions x, y and z (the direction normal to the plane of the paper) is given by

$$K_x = 2(K_c \sin^2\beta + K_s \cos^2\beta)$$

$$K_y = 2(K_c \cos^2\beta + K_s \sin^2\beta)$$

$$K_z = 2K_s$$

In the above example it is assumed that in the calculation of horizontal stiffness in the plane of the paper K_x there is a couple preventing rotation under the action of the horizontal load. Where there is no restraining couple the equation is:

$$K_x = 2K_c K_z / K_y$$

The relative values of the three mutually perpendicular stiffnesses may be varied further by inserting metal plates into the rubber pads (as discussed in Section 2.2.5) since these will affect K_c greatly but have only a small effect on K_s. Many industrial mountings may be considered as developments of this type of mounting.

2.7.1.3 Bridge bearings

A typical modern road bridge comprises decks supported on piers (*Figure 2.52*). Such bridges suffer from the well-known phenomenon of deck expansion and contraction due to seasonal and diurnal temperature variations. If the deck was to sit directly on the piers the expansion of the deck, if transmitted directly to the piers, could cause dangerously high bending moments. To accommodate this the deck could be mounted on metal rollers, thus reducing the transmission of bending forces to the pier. Such metal rollers are, however, subject to corrosion and, because of high localised stresses, also cause wear on the top of the piers. The introduction of rubber pads (in effect a form of spring and generally known as 'bridge bearings') has overcome this problem and in the second half of this century became widely accepted by civil engineers.

The main requirements for the rubber pad are that

(1) it must support the weight of the deck giving a controlled and predictable deflection; and
(2) it must be able to accommodate thermal expansions and contractions of the deck without transmitting an excess horizontal force to the top of the pier.

For (2) to be possible in practice *it is essential that vertical and horizontal stiffness are independently controllable*. This clearly requires a consideration of the effect of shape factor and of the use of metal plates to control compression modulus independently of shear modulus.

In order to design a bearing it is necessary to establish from the civil engineer the following data:

(1) the mass of the deck W;
(2) the maximum horizontal force allowable on the pier H;
(3) the maximum vertical stress allowable to avoid damage to the concrete σ;
(4) the maximum vertical deflection allowed by the bridge authority d;
(5) the maximum allowable shear strain on the rubber γ;
(6) the maximum horizontal movement due to thermal expansion m; and
(7) the number of bearings proposed for each deck X.

From these data it is possible to calculate, in turn:

(1) the area of the pad or bearing A;
(2) the thickness of the pad t;
(3) the shear modulus of the rubber E_γ, and hence its formulation; and
(4) the vertical stiffness required, and hence the number of plate inserts.

The area of the pad If we take 300 000 kg as the mass of the deck W, the maximum vertical stress considered as being acceptable without adversely affecting the concrete piers (σ) as 5 MPa and the number of bearings per deck (X) as 4, then by the definition of stress

$$\sigma = W/4A$$

i.e.

$$A = W/4\sigma = 3000\,000/(4 \times 5 \times 10.2) = 1470 \text{ cm}^2$$

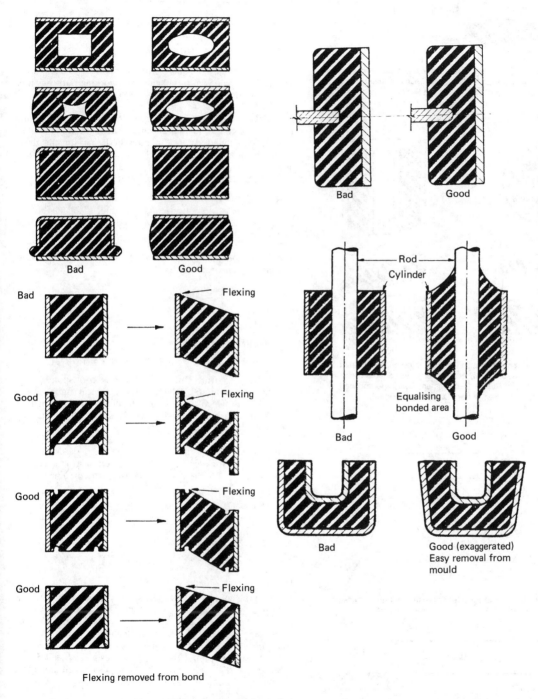

Figure 2.48 Examples of good and bad design in rubber–metal bonded units. (From Naunton, *What Every Engineer Should Know About Rubber*, Natural Rubber Development Board. Reproduced by permission of the Malaysian Rubber Development Board)

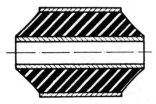

Figure 2.49 Rubber bush unit. (From Naunton, *What Every Engineer Should Know About Rubber*, Natural Rubber Development Board. Reproduced by permission of the Malaysian Rubber Development Board)

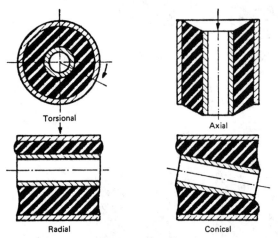

Torsional

Axial

Radial

Conical

Figure 2.50 Principal deformation modes of the bush illustrated in *Figure 2.49*. (From Naunton, *What Every Engineer Should Know About Rubber*, Natural Rubber Development Board. Reproduced by permission of the Malaysian Rubber Development Board)

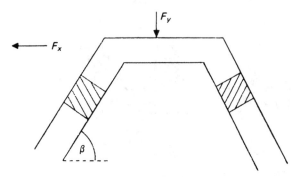

Figure 2.51 Inclined shear mounts. The stiffness in three mutually perpendicular directions may be controlled independently

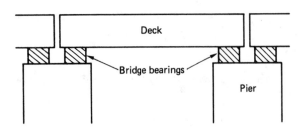

Figure 2.52 Terms used in discussing bridge bearings

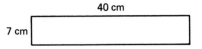

40 cm

7 cm

Figure 2.53 Overall pad dimensions (rubber component only)

If the pad is to have a square cross-section then the length of one side L is 38.35 cm ($\simeq$ 40 cm). For working purposes it is assumed in the following that $L = 40$ (*Figure 2.53*).

The thickness of the pad (t)   In order to prevent bond failure at rubber–plate interfaces a maximum shear strain (γ) of 0.35 may be postulated. It is also estimated that for the deck in question the maximum horizontal thermal expansion–contraction (m) is likely to be ± 2.5 cm.

From the definition of shear strain it follows that

$$\gamma_{max} = m/t$$

i.e.

$$t = m/\gamma_{max} = 2.5/0.35 \simeq 7 \text{ cm}$$

The shear modulus of the rubber   This is obtained from the definition of shear modulus

$$E_\gamma = \text{Shear stress}/\gamma$$

i.e.

$$E_\gamma = (H/A)/\gamma$$
$$= 60/1600 \times 0.35 \text{ kN cm}^{-2}$$
$$= (60 \times 10\,000)/(0.35 \times 1600 \times 1000) \simeq 1.07 \text{ MPa}$$

It can be seen from *Table 2.2* that this corresponds to a hardness of about 60 IRHD, a fairly typical figure for a vulcanisate in which the rubber is loaded with about half its weight of a carbon black and which is, therefore, easy to formulate.

Adjustment of vertical stiffness   This requires a determination of the number of plates required to be inserted into the pad. In Section 2.2.5 it was pointed out that the compression stiffness K_c can be defined as

$$K_c = F/x = E_c A/t$$

It was also noted that

$$E_c = E_\gamma (1 + 2kS^2)$$

Thus, combining these two equations,

$$F/x = (A/t)E_\gamma (1 + 2kS^2)$$

Since, in the context of this exercise, F is equal to the force on each pad ($= 300\,000g/4 \simeq 0.75$ MN), A is the area of each pad (1600 cm^2), $x \le d \le 0.3$ cm and the value for E_γ corresponding to $E_\gamma = 1.01$ is approximately 4.5 with $k = 0.57$ (see *Table 2.2*), the only unknown is the value of the shape factor S. Substitution of the known values gives $S \ge 4.5$.

Since $S = L/4t_i$ (where t_i is the thickness of each individual rubber spring) and as $L = 40$ cm, then

$$t_i = L/4S \le 40/(4 \times 4.5) \le 2.2 \text{ cm}$$

Furthermore, as the overall thickness of the pad (ignoring the metal plates) is 7 cm, then it follows that the correct stiffness is strictly achieved by the use of four separate rubber springs sandwiched between a total of five metal plates. However, the use of just three rubber springs would be

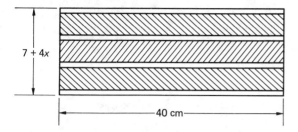

Figure 2.54 Pad with reinforced metal plates (x is the thickness of each metal plate). In practice the whole assembly may be wrapped in a rubber sheet

satisfactory if the calculation had yielded a value of 2.33 instead of 2.2 for d and, taking into account the various assumptions made in the calculation, this should be an acceptable simplification (*Figure 2.54*)

2.7.2 Rubber seals

Rubbery materials are useful for seals because their low rigidity enables them to fill the gap, including surface roughnesses, between the faces to be sealed whilst the elasticity helps them to resist extrusion by the pressure of any enclosed fluid.

The most commonly used seal is the O-ring which fits between two faces of circular cross-section. The ring itself is of solid rubber with a circular cross-section and it fits into a circumferential groove cut into one of the faces to be sealed. The dimensions of groove and ring are chosen to minimise any cutting of the rubber during fitting or use.

One feature of the O-ring is that the tightness of the seal increases with an increase in the pressure exerted by any fluid

being enclosed by the seal. Care must however be taken to ensure that this pressure is not so high, the gap between the surfaces so great, or the modulus of the rubber so low that the ring becomes extruded between the faces (*Figure 2.55*).

Experience has shown that extrusion will not occur if the combinations of clearance, pressure and modulus do not exceed the values indicated in *Figure 2.56*.

Where high sealing pressures are required, the rubber rings may be backed up by thermoplastic anti-extrusion rings which may be further reinforced by a metal casing. O-rings generally are only suitable for static applications or those involving small reciprocating movements.

For applications where a low friction between shaft and seal is required a U-ring may be preferred. In the absence of a fluid sealing pressure the sealing force will be very small. However, in the presence of the fluid sealing pressure the effectiveness of the seal is instantly enhanced (*Figure 2.57*).

Modification of a U-ring enables a design to be used as a rotary seal (where there is relative movement of shaft and housing). The knife edge contact with the shaft reduces friction, but a garter spring (a tight coil of wire formed into a spring) enhances the sealing stress (*Figure 2.58(a)*). Such a design is only suitable for low-pressure work, but somewhat higher pressures may be accommodated by the use of a more rigid pressure support ring (*Figure 2.58(b)*).

The choice of rubber clearly depends on the fluid that is to be retained by the seal and on the operating temperatures. Thus a wide range of synthetic rubbers including speciality and exotic types (see Section 2.3) may be employed.

2.7.3 Vibration isolation

Rubber pads are widely used for vibration isolation. Examples are the isolation of buildings from earthquake vibrations, isolation of delicate equipment from general vibrations in a building or the isolation of machinery vibrations from the adjacent parts of a building.

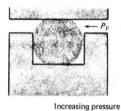
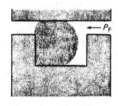
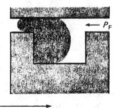

Increasing pressure

Figure 2.55 Extrusion of a seal under pressure. (Reproduced by permission of the Open University, Milton Keynes)

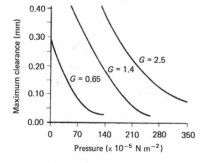
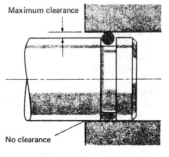

Figure 2.56 Relationship between pressure, extrusion gap and shear modulus G (in MN m^{-2}). (Reproduced by permission of the Open University, Milton Keynes)

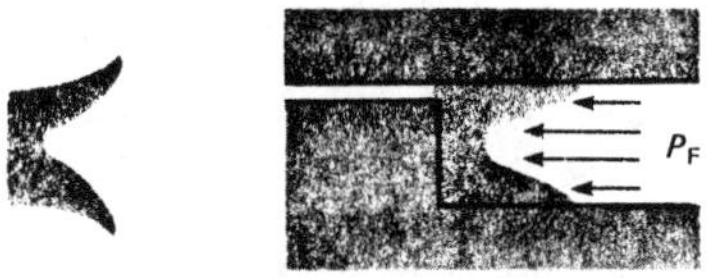

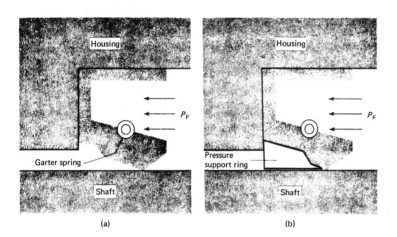

Figure 2.57 A typical U-ring: (a) unstressed; (b) installed. (Reproduced by permission of the Open University, Milton Keynes)

Figure 2.58 (a) Rotary lip seal. (b) Rotary lip seal with pressure support ring. (Reproduced by permission of the Open University, Milton Keynes)

In principle, insertion of a spring between a source of vibration and the mass to be protected *can* reduce the vibration of the latter. This may easily be demonstrated by connecting an object to a length of elastic thread. If the other end of the thread is held in the hand and the hand then moved up and down at increasing frequency it will be noticed that a frequency will be reached at which the vertical movement of the object becomes negligible. (The reader is recommended to try this for him/herself.)

The relevant equation for an ideal spring is

$$T = x_2/x_1 = 1/[1 - (\omega/\omega_o)^2]$$

where T is the transmissibility (the ratio of the amplitudes of motion of the mass x_2 and the source x_1), ω is the vibration frequency and ω_o is the natural frequency of the spring given by

$$\omega_o = (K/m)^{1/2}$$

where K is the spring stiffness and m is the mass being isolated.

Calculations show that as ω increases from zero the vibration increases up to $\omega = \omega_o$ at which point it tends to infinity. Between this frequency and $(2\omega_o)^{1/2}$ the amplitude decreases but is still greater than the disturbing amplitude. Above this value the displacement of the mass falls rapidly with increasing frequency.

Although it is intended that anti-vibration springs should operate at these higher frequencies, the very high transmissibility around the natural frequency ω_o may be a severe problem, particularly as most vibration sources have a spectrum of vibrating frequencies. In such circumstances the viscoelastic properties of rubbers come into their own. This enables them to act simultaneously as both spring and damper

and this ensures that the transmissibility at the natural frequency becomes finite. The dependence of transmissibility of a simple system on the type of rubber used is indicated in *Figure 2.59*.

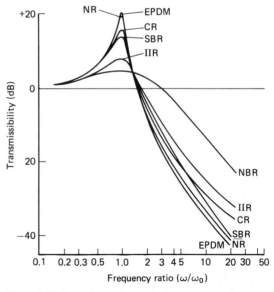

Figure 2.59 Dependence of transmissibility of a simple linear system on type of rubber. (Reproduced by permission from Freakley and Payne, *Theory and Practice of Engineering with Rubber*, Applied Science)

2.8 Design of plastic components

Successful use of a plastic material requires that both the product designer and the fabricator have an understanding of the properties of plastic materials. In particular there are a number of characteristics of plastic materials that need specifically to be appreciated including the following items which are particularly relevant to thermoplastics.

Anisotropy The long-chain polymer molecules of thermoplastics become oriented during shaping operations and recoiling is only partially complete before the melt cools after shaping. This will result in the mass of material being anisotropic, i.e. its properties will differ in different directions, the level of anisotropy depending on the amount of relaxation (recoiling) that has occurred, this in turn depending on the processing conditions. There may also be a differential shrinkage on cooling and this may lead to warping of the product. The anisotropy effect may be aggravated by the use of anisotropic fillers such as glass and carbon fibres. Furthermore, because an oriented molecule is not in a stable state (it will always try to coil up in a random fashion), internal stresses may be set up in the product. This may mean that the additional level of external stress that may be imposed before the product breaks will be less than if there is no 'frozen-in' internal stress. (However, it should be noted that in some instances orientation *may* increase the strength, particularly where this is measured along the direction of orientation.)

Shrinkage Shrinkage on cooling can be quite substantial in the case of crystalline polymers, being typically in the range 0.01–0.05 cm per centimetre, i.e. 1–5% (linearly) for unfilled polymers. This figure will be much less for glass-filled polymers and for amorphous thermoplastics (0.005 cm per centimetre). Clearly important for precision products, this also has more general implications. If a product, e.g. an injection moulding, has a thick section the surface will tend to cool much earlier than the core and, therefore, will be the first to shrink. This may then cause the moulding to shrink 'outward' leaving a void in the core and/or the surface to sink.
    Problems of such shrinkage may be avoided by using thin-section mouldings, suitably stiffened by the use of ribs and, where possible, domed surfaces. Where a thick section is absolutely necessary then shrinkage problems such as sink marks and voids may be overcome by incorporating a gas into the melt, thus generating an internal pressure to counteract any tendency to shrink.

Bulk compression Whilst it is correct to think of polymers as being incompressible in bulk in terms of normal conditions of service, bulk compression is not negligible under the very high pressures that exist during injection moulding and may be up to 8%. This can lead to overpacking of polymer into a mould and this may cause problems when ejecting mouldings and may also cause severe stresses around the gate of the mould.

Stress concentrations High stress concentrations can occur at sharp edges and at sudden changes in cross-section. This will provide a weak point in the product which may, if allowed to be strained in directions that increase the stress, lead to premature failure, particularly in some aggressive environments. Aggressive environments include liquids that have solubility parameters just outside the range enabling solution of the polymer but sufficient to soften the surface.

Flow-path ratio Since in injection moulding the melt is cooling as the mould cavity fills, a moulding may be incomplete if the flow-path ratio (flow length/flow cross-section)

is too high. Whilst the maximum flow-path ratio may be increased by raising injection pressures this may lead to too high a cavity pressure near the gate end causing both release problems and problems due to excess stressing.

Temperature It should be appreciated that the properties of most plastics materials vary with temperature even though this may be below T_g (for an amorphous material) or T_m (for a crystalline polymer).

Creep Many plastic materials will tend to creep under load and this should be taken into account when designing parts for long-term service.

Ageing Many polymers change their properties with time due to ageing which may be accelerated by heat, oxygen or ultraviolet light.

2.8.1 Designing under creep loading

For many engineering applications plastic materials may be used under conditions of long-term loading which will give rise to creep. For design purposes this may most conveniently be taken into account by the use of an *ad hoc* system known as the *pseudo-elastic design approach*. This method uses classical elastic analysis but employs time- and temperature-dependent data obtained from creep curves and their derivatives. In outline the procedure comprises the following steps.

(1) Ascertain the function and the operating conditions for the component part in question including both the expected lifetime and the maximum service temperature.
(2) Assume a 'worst-case' scenario before commencing calculations. For example assume the component will be operating continuously at the maximum temperature and under the maximum load encountered during its service life.
(3) Select appropriate formulae from classical elastic analysis.
(4) Obtain the appropriate figures for stress, modulus, etc., from a creep curve or a derivative curve. Insert this into the appropriate formula.

Example 1 A blow-moulded container, cylindrical in shape but with one spherical end, is prepared from the polysulphone material for which (at 20°C) are shown in *Figure 2.60*. The cylindrical part of the container has an outside diameter of 200 mm and is required to withstand a constant internal pressure of 7 MPa at 20°C. It is estimated that the required lifetime for the part will be 1 year and that the maximum allowable strain is 2%. What will be the minimum wall thickness for satisfactory operation?

    The appropriate formula from classical elastic analysis is

$$t = pd/2\sigma$$

where t is the thickness, p is the internal pressure, d is the outside diameter and σ is the hoop stress.
    Figure 2.60 shows that the stress that will lead to a strain of 2% after 1 year is about 39 MPa. Substituting this into the equation as the hoop stress gives

$$t = (7 \times 200)/(2 \times 38) = 18.4 \text{ mm}$$

Example 2 A straight rectangular beam of the same polysulphone material as in Example 1 is clamped at one end but is loaded at the other end by a load of 50 N. If the distance between the clamp and the point of loading at the other end is 150 mm and the width of the beam is 12 mm what should be

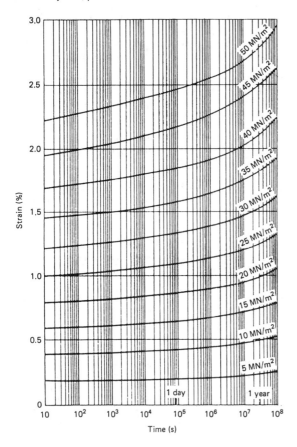

Figure 2.60 Curves for creep in tension of a commercial polysulphone (Polyethersulphone 300P-ICI) at 20°C. (Reproduced by permission from *ICI Technical Note PES 101*, ICI Plastics Division)

the depth of the beam if the deflection of the free end is not to exceed 3 mm after 1 year at 20°C?

The appropriate formula from classical elastic analysis is

$$I = bh^3/12 = PL^3/3E_c\delta$$

where I is the second moment of area, P is the load (50 N), L is the span (150 mm), δ is the free-end deflection (3 mm), b is the width of the beam (12 mm), h is the thickness of the beam, and E_c is the creep modulus.

Rearranging this in terms of the unknown value h

$$h^3 = 4PL^3/b\delta E_c$$

There is a possible problem in determining the actual modulus at the appropriate strain, but it will be noted by inspecting the curves that up to a strain of 1.5% the creep modulus at 1 year (obtained by dividing the stress by strain at 1 year) is virtually constant at about 2000 MPa (= 2000 N mm^{-2}) and it would seem reasonable to use this figure.

Hence

$$h^3 = (4 \times 50 \times 150)/(12 \times 3 \times 2000) = 4.167$$

or

$$h = 161 \text{ mm}$$

2.8.2 Stiffness of structural foam mouldings

The stiffness of a piece of plastic material is related to the cube of the thickness. On the other hand, if a cellular polymer is produced then the stiffness only decreases linearly with the proportion of solid polymer in a foam.

If the *blow-up ratio* (e) is defined by

$$e = \rho/\rho_f$$

where ρ and ρ_f are the densities of solid polymer and foamed polymer, respectively, then it may be shown that for a bar of polymer clamped at one end and loaded at the other (i.e. as in the previous example) the thickness required of the foamed product in order to keep the deflection constant is given by

$$h_f = e^{1/3}h$$

where h and h_f are the thicknesses of solid polymer and foam, respectively.

Similarly, it may be shown that the mass of the foamed part as a fraction of the mass of a solid part of equal stiffness is given by $e^{-2/3}$.

The data in *Table 2.21*, derived from these expressions, give the relative thickness and mass for bars of solid and cellular polymer of varying blow-up ratio which will give the same deflection under load. In effect this means that, if solid polymer is replaced with a cellular polymer with a blow-up ratio of 2, then a product will be made of equal stiffness with a saving of almost 40% in material. It will, however, require an increase in thickness of about 25%.

The use of cellular polymers as replacements for solid materials has further advantages that the internal gas pressures help to reduce sink marks and large voids in the product, but usually at the cost of an inferior surface finish. A number of techniques is available such as co-injection, sandwich moulding, etc., in which cellular material is sandwiched between solid skins (of good finish).

2.8.3 Design of plastics parts

Because plastics are used in such a wide variety of applications, any comments concerning design procedures must be very general. Although the remainder of this section is basically a statement of the obvious, it is probably worth listing the following sequence of procedures involved in introducing a new plastics component or product.

(1) Establish the working conditions (including stress loadings, temperature and environmental conditions) to be demanded of the component.
(2) Establish the number of components to be manufactured and the financial constraints (e.g. maximum acceptable cost per part).
(3) Consider possible polymers and provisionally select the material that will meet the specification requirements

Table 2.21 Relative thickness and mass for various blow-up ratios that will give the same deflection under load (solid and cellular polymer bars)

	Blow-up ratio					
	1	1.2	1.4	1.6	1.8	2.0
Relative thickness	1	1.06	1.12	1.17	1.22	1.26
Relative mass	1	0.886	0.800	0.738	0.676	0.63

and which will be the cheapest to produce the component in question (taking into account both raw-material and production costs).

(4) Select the most appropriate processing method. (Let us now assume that the method selected is that of injection moulding. With other processes the procedures would be similar but nevertheless peculiar to the process chosen.)

(5) Arrange for design of suitable mould and refine the component shape taking into account such features as avoiding undue stresses, providing stiffness without making the part excessively heavy and bearing in mind the maximum flow-path ratio for the plastic material selected.

(6) Prepare mouldings and subject them to a testing programme to establish their fitness for use.

(7) Modify the design of the mould or part or change the material if indicated as a result of testing. Repeat the testing procedures.

(8) Prepare a quality-control programme for use during manufacture.

References

1 BRYDSON, J. A., *Rubbery Materials and their Compounds*, Elsevier/Applied Science (1988)

2 BHOWMICK, A. K. and STEPHENS, H. L., (Eds), *Handbook of Elastomers*, M. Dekker, New York (1988)

3 BRYDSON, J. A.. *Plastics Materials*, 5th edition, Butterworth Heinemann, Oxford (1989)

4 WHELAN, A. and GOFF, J. P., Paper presented at *PRI Mouldmaking '86*, Solihull (Jan. 1986).

Further reading

Much of the material used in this chapter is drawn from more detailed publications by the author. For example, the material covered in Sections 2.1, 2.4 and 2.5 is covered more deeply in *Plastics Materials*.[3] Similarly, Sections 2.2 and 2.3 are condensed from *Rubbery Materials and Their Compounds*.[1] The material for Sections 2.6 and 2.8 is covered in greater depth in *Handbook for Plastics Processors* (Butterworth-Heinemann, Oxford (1990)).

For design with vulcanised rubber see:

FREAKLEY, P. K. and PAYNE, A. R., *Theory and Practice of Engineering Design with Rubber*, Applied Science, London (1978)

GÖBEL, E. F., *Rubber Springs Design*, Newnes-Butterworth, London (1974)

LINDLEY, P. B., *Engineering Design with Natural Rubber*, Malaysian Rubber Producers' Research Association, Brickendonbury, Hertford SG13 8NL, UK

For design with plastics see:

POWELL, P. C., *Engineering with Polymers*, Chapman and Hall, London (1983)

BECK, R. D., *Plastic Product Design*, Van Nostrand Reinhold, New York (1970)

3

Metal Casting and Moulding Processes

J Glownia

Contents

3.1 Economics of casting and moulding

Casting may be defined as 'metal shaping obtained by the solidification of molten alloy in the mould cavity of a required shape'. Of course, there are also other methods of shaping metal, i.e. forging, stamping, rolling, machining or welding. In practice, wrought iron is cast as ingots, and then plastically worked into the required shape. This means that all primary alloy metals are cast first.

The process of casting has certain advantages over the other processes of shaping:

(1) it can be adapted to the requirements of mass production;
(2) castings have more uniform properties from the directional point of view than does wrought iron;
(3) heavy pieces (100–200 ton) can be cast, while it would be difficult to make these in any other way;
(4) the design of parts can be simplified so that machining or forging is reduced to a minimum;
(5) some metals or alloys cannot be hot worked from ingots into other shapes.

The predominant factor in every process is its economic aspect. The design of the product may dictate whether a single process of shaping or a combination of several methods is suitable.

Developments in the production of castings are related to the implementation of new foundry technologies and with achievements in the many branches of industry. A new development requires the use of a material of strictly determined properties that corresponds to the operating conditions that it will experience. Examples of well-defined materials are steels for service at low temperatures, e.g. transcontinental pipelines in Siberia and Alaska, and the new grades of corrosion-resistant cast steel with a carbon content of less than 0.03% which eliminates intercrystalline corrosion. In the latter case interest is focused on materials which are used in aviation and space applications, e.g. composites and monocrystals. Each of the above-mentioned operating conditions requires the use of special-purpose alloys.

Remaining competitive in what is now a world market for cast products requires continuous study and evaluation of the production of the main metals and alloys.

For more than 100 years the production of steel has been systematically growing all over the world (*Figure 3.1*). In 1960, the total production of steel in the world amounted to 341 million ton and in 1987 it increased to 734.7 million ton.

It is interesting to note that in highly industrialised countries such as the USA and the EEC countries, the rate of steel production decreased, while it increased rapidly in countries with new industrial traditions such as Brazil, Australia and South Korea (*Table 3.1*).

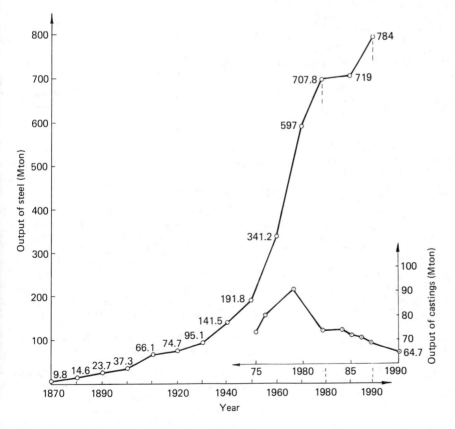

Figure 3.1 World output of steel and castings, 1870–2000

Table 3.1 Production of steel castings (SC) and raw steel (RS) in the years 1970–1987

Country	1970 SC	1970 RS	1975 SC	1975 RS	1980 SC	1980 RS	1982 SC	1982 RS	1983 SC	1983 RS	1985 SC	1985 RS	1987 SC	1987 RS
Australia	—	6900	60	8100	74	7900	75	6400	75	5700	75	6600	—	6100
Austria	26.5	4100	34.5	4500	28.3	5100	—	4300	18.1	4400	25.81	4700	—	4300
Belgium	97.5	42600	88.7	11600	80.7	12300	74.8	10000	76.6	10200	70.5	10700	—	9800
Brazil	—	—	116	—	163.5	15300	109.4	13000	103	14700	133.7	20500	—	22200
Canada	174	11200	217	13000	198.2	15900	117	11900	36.4	12800	80.4	14600	—	14700
Czechoslovakia	300	11500	349	14300	363.2	15200	363.6	15300	342.6	15000	342.6	15000	—	15400
France	288	23800	263	21600	218.2	23700	180.4	18400	142.4	17600	139.6	18800	—	17700
FRG	390	45000	369	40400	293.7	43000	251.8	35900	206.6	35700	228.2	40500	—	36300
GDR	—	5100	204	6500	233.2	7300	237.4	7200	—	7200	228	7900	—	8200
Great Britain	263	28300	284.5	20200	174.1	11500	147.8	13700	114.9	13000	118.7	15700	—	17200
India	—	6300	68	7800	74	9500	74	11000	—	10200	—	11900	—	12600
Italy	154	17200	168	21800	132.9	26500	115.3	24000	84.5	21800	93.8	23900	—	22800
Japan	897	93300	644	102000	732.6	111000	612.9	99500	518.8	97200	544	105300	—	98500
Poland	286	11795	342	15000	354.4	19500	355	14800	266	16200	258	16100	—	17000
Spain	127	7400	142	11100	140	13000	105	13200	99.5	13000	105	14200	—	11900
Sweden	38	5500	37.4	5200	41	4200	12	3900	13.5	4200	14.7	4800	—	4600
USSR	—	116000	5582	141000	323	140000	—	147200	—	152500	—	154700	—	161400
Republic of Korea	—	—	—	—	—	—	92	11800	—	11900	105	13500	—	16800
China	—	—	—	—	—	—	683	37200	—	40000	1000	46700	—	55300
South Africa	—	—	—	—	—	—	141.7	8300	—	720	1077	8500	—	8800
Taiwan	—	—	—	—	—	—	23	4200	—	5000	34.7	5100	—	5600

In recent decades some important changes have occurred in the structural system of steel manufacture. The Bessemer process, discovered in Germany in 1856, and the Thomas process dominated the manufacture of steel in the last century, but are now very little used. The Siemens–Martin process discovered in France in 1865, and the extension to the basic lining in 1875 have also become less important. The steel-manufacturing processes used today are aimed at achieving a higher melting rate with a simultaneous reduction in cost and improved quality of the steel obtained. Currently, attention is focused on the oxygen converter processes. A small but constant growth in popularity has been gained by the electric arc process. The share of these processes in the general production of steel is shown in *Figure 3.2*. Different production processes are used in different countries, depending on the availability of raw materials and the level of technical knowledge.

Future forecasts are based on the use of three processes: Siemens–Martin, electric arc, and oxygen converter. It is anticipated that in the year 2000 the share of the oxygen-converter processes will account for about 70% of the steel manufactured, the electric-arc process for 20%, and the Siemens–Martin process for the remaining 10%. It is expected that Siemens–Martin furnaces will be used only for the manufacture of large castings (50–100 ton).

Figure 3.3 shows the trends in the amount of various alloys used world-wide. In the last decade there has been a decrease in the production of grey cast iron and cast steel, but an increase in the share of ductile iron and aluminium alloy casting.

3.2 Sand casting

Making castings in sand moulds is a multistage process consisting of the following operations:

(1) pattern making—including core boxes, preparation of the equipment such as drags, flasks and patterns of the gating system;
(2) core and mould making;
(3) melting and pouring of the alloy;
(4) solidification; and
(5) cleaning and inspection.

3.2.1 Pattern making

The main objective of using a pattern is to reproduce the external shapes of a casting. The core reproduces the internal shape of a casting. The pattern equipment also includes a set of gate or riser patterns and patterns of the core boxes in which cores are made.

Three main groups of patterns are used:

(1) *Loose patterns* These are unsplit or split patterns with or without loose pieces. When making patterns allowance must be made for the contraction of the alloy during solidification. Patterns are mainly made of either wood, plastics or alloys such as aluminium and cast iron. They are used in the manufacture of a small number of castings.
(2) *Loose patterns with core prints* The loose part remains in the mould when the pattern is lifted from the mould, and is taken out at a later stage. This enables the pattern to be lifted from the mould without damage.
(3) *Gated patterns* Compared with the above two types, gated patterns are designed in such a way that, in

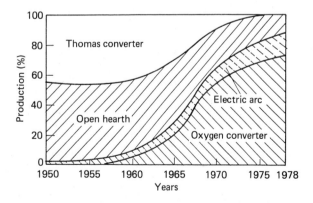

Figure 3.2 Production of steel by various processes. (Reproduced by permission from Graf[15])

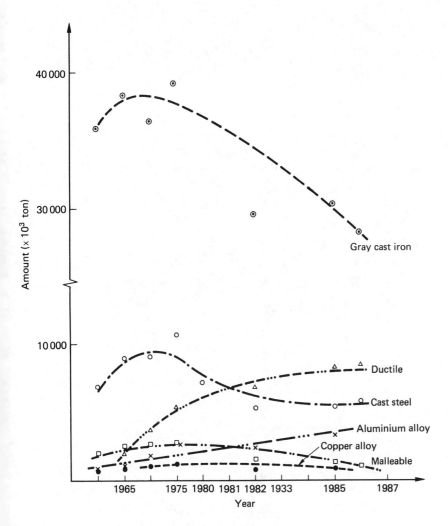

Figure 3.3 Trends in the amount of various alloys used world-wide

addition to reproducing the proper shape of a casting, they also reproduce the gating system and core prints. They are usually made of aluminium alloy or cast iron. They are used to reproduce small castings made in large batches. They are usually fixed to the cope and drag pattern plates. The cope and drag half moulds are prepared separately. This method is more expensive than those described above. However, it is necessary to use patterns of this type in a mechanised moulding shop.

Although a pattern reproduces the shape of a casting, its dimensions vary from those of the casting itself. The differences arise due to contraction during the solidification process and the allowance made for machining the cast surface.

In view of this contraction in the volume of an alloy in the liquid, solid–liquid or solid state, the pattern has larger dimensions than the casting. The amount of contraction depends on the type of alloy cast, the degree of intricacy of the casting shape, the range of cooling temperatures used, and the rate of cooling. Mathematically, the relative contraction (ε) of an alloy is given by

$$\varepsilon = \alpha(T_1 - T_2) \tag{3.1}$$

where α is the coefficient of thermal expansion and T_1 and T_2 are the temperatures (in kelvin) from and to which the casting is cooled, respectively.

The percentage linear casting contraction of an alloy is defined as

$$\varepsilon_f = \frac{l_p - l_c}{l_c} \times 100\% \tag{3.2}$$

where l_p is the length of the pattern and l_c is the length of the casting.

Values of the casting contraction for various cast alloys are given in *Table 3.2*.

3.2.2 Moulding and casting processes

The process of making a casting consists of the following steps:

(1) preparation of charge and moulding materials;
(2) preparation of pattern equipment such as patterns and core boxes;
(3) melting;

Table 3.2 Casting contraction in the main cast alloys

Alloy	Contraction		$T_L{}^*$ (K)	$T_p{}^\dagger$ (K)
	Restrained	Free		
Grey cast iron			1403	
Small castings	0.9	1.0		1573–1623
Medium-size castings	0.8	0.9		1523–1573
Heavy castings	0.7	0.8		1493–1523
Cast steel			1783	
Carbon cast steel	1.3–1.7	1.6–2.0		1693–1743
High-alloy chromium	1.0–1.4	1.3–1.7		1713–1753
Austenitic	1.7–2.0	2.0–2.5		1613–1653
Ferritic–austenitic	1.5–1.9	1.8–2.2		1713–1743
Malleable cast iron			1773–1848	
Whiteheart	0.8–1.5	1.5–2.5		1653–1723
Blackheart		<1.2		
Brass			1188	
Zinc	1.5–1.7	1.8–2.0		
Silicon	1.6–1.7	1.7–1.8		1323–1423
Manganese	1.8–2.0	2–2.3		
Bronze				
Tin	1.2	1.2–1.8	1253	1273–1423
Silicon				1323–1423
Aluminium				
Silumin	0.8–1.0	1–1.2	843	973–1003
Aluminium–copper	1.4	1.6	903	

* Temperature of melting.
† Temperature of pouring.

(4) proper preparation of the foundry mould and casting process, i.e. pouring of mould and solidification;
(5) cooling of the casting; and
(6) cleaning and final treatment which includes heat treatment and control of the surface and of mechanical properties, cutting of risers and machining.

The term 'mould making' refers to a method of making moulds and to the materials used in the process.

The following guidelines are recommended in the design of parts to be produced by casting:

(1) large surfaces of a casting which are to be machined should be located in a horizontal position in the drag;
(2) the position of a casting in the mould should be such as to ensure easy placing of the cores, and hanging cores should be avoided if possible;
(3) the size of moulding boxes and the position of a casting in the mould should ensure the lowest possible consumption of the sand, hence reducing labour cost;
(4) castings made from alloys which have a high shrinkage value should be moulded in a position that ensures their directional solidification;
(5) the pattern should have one parting plane and no loose parts;
(6) the machining datum should be a rough surface or a surface with the smallest possible machining allowance;
(7) the core box should have sufficent strength to counteract

any changes in its position on pouring and allow for the vents and the stability of its shape during mould assembly; and
(8) the placing of several cores in one core print seat should be avoided.

3.2.3 Making a mould

In a traditional process the mould is made by pouring the sand and ramming it around the pattern. Packing of the sand can also be done by hand, a pneumatic rammer, a slinger, jolt squeezing, or squeezing and shooting.

There are three types of mould, the type used depending on the life required:

(1) expendable, single-use type moulds;
(2) semi-permanent moulds; and
(3) permanent moulds, or dies.

The life of a mould is expressed as the number of pourings for which the mould can be used before its destruction.

Expendable moulds are made of natural or synethetic sands. The basic material is silica sand, mixed with a binder which is bentonite, kaolinite clay with an addition of oils, dextrine, waste sulphite liquor, or a resin. *Green sands* are sands with a moisture content of up to 6% and a compression strength of $2–8 \times 10^4$ Pa, and *skin-dried and dried sands* have compression strengths above 1×10^6 Pa.

Semi-permanent moulds are made of highly resistant materials such as *chamotte*, graphite, or plaster. Their life corresponds to several dozen pourings, and for every pouring only the surface layer of the mould cavity is renewed. These moulds are used only for large and medium-sized castings of simple shape that are manufactured in batches.

Permanent moulds are made of cast iron, alloyed cast steel or copper. Their life ranges from several tens of pourings for steel castings up to several thousand pourings for aluminium alloys. They are used in the mass batch production of small and medium-sized castings.

In foundry practice single-use moulds are most popular. Some different types of these moulds are shown in *Figure 3.4*. These types of mould can be made in a box or on the foundry floor. The selection of the type of mould depends on the number of castings, the dimensional accuracy and the surface finish required.

When designing single-use moulds the following considerations should be taken into account.

(1) A mould should have sufficient resistance to the impact of the stream of metal and to the metal static pressure; a resistance of this type is ensured through using a sand of appropriate compression and bending strength.

(2) A mould should preserve its refractory abilities from the moment of pouring until knocking out of the casting. The time of solidification for large castings may be 20–40 h and during this time reaction between the molten metal and the moulding sand is avoided by using protective coatings applied to the surface of the mould cavity, or by using more refractory sands for direct contact with molten alloy.

(3) A mould should possess appropriate permeability to enable escape of the gas evolved from the sand during the solidification of an alloy; the permeability is controlled during sand preparation.

(4) A mould should possess sufficient flexibility to control the stresses and shrinkage on solidification.

(5) There should be good collapsibility during the separation of a casting from the mould and crushing of the moulding sand.

The chemical compositions of moulding sands are given in *Table 3.3*.

In modern technologies synthetic sands based on cement, water glass and furan resins are widely used. The main advantage of using such sands is that it is possible to control their properties. Disadvantages include poor collapsibility, difficult reclamation and pollution of the environment.

Self-setting liquid and loose sands, which do not necessitate drying of moulds and cores are currently gaining in popularity. Liquid self-setting sands are poured into a moulding or core box, where they harden without any ramming. Loose self-setting sands, on the other hand, require some packing.

The choice of self-setting sands is based on four criteria:

(1) reducing the time required (and thus the cost) to prepare the mould so that liquid sand can simply be poured into the mould only;

(2) the rate of setting;

(3) ease of knocking out of casting; and

(4) reducing the cost of the sand components used.

A comparison of the relative costs of the manufacture of moulds and cores in various types of sand is given in *Figure 3.5*.

The main operations of the mould-making process consist of the following stages:

(1) placing the pattern and box on the pattern plate;

(2) coating the pattern with a material such as powder, kerosene or paraffin which prevents the sand from sticking to the pattern;

(3) filling the moulding box with a facing and backing sand;

(4) packing the sand;

(5) removing the excess sand and placing the vents;

(6) turning over and lifting the pattern;

(7) operations (1) to (6) for the second half-mould; and

(8) blowing the mould cavity and gating system with compressed air, assembly of the two half-moulds, and clamping.

In practice, all of the above operations can be mechanised. The mechanised method of sand packing consists of shooting and squeezing the sand, as shown in *Figure 3.6*. The advantage

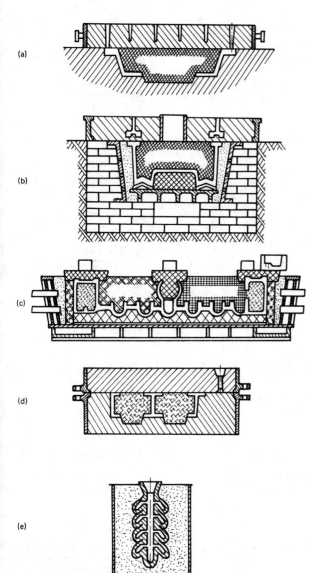

(a)

(b)

(c)

(d)

(e)

Figure 3.4 Types of single-use mould: (a) foundry floor; (b) caisson; (c) core assembly; (d) boxes; (e) shells. (Reproduced by permission from Skarbiński[8])

Table 3.3   Composition of moulding and core sands

Type of sand	Application	Composition (wt%)	Binder (wt%)	Additives (wt%)
System sand	Machine moulding of castings in aluminium alloys	Sand: 80–95 Silica sand: ≤20	Bentonite: 6–8	—
	Machine moulding of castings in copper alloys	Sand: 70–90 Silica sand: ≤25	Bentonite: 6–8	—
	Machine moulding of castings in grey iron (weight 5–100 kg)	Sand: 85 Silica sand: 10–15	Bentonite: 1.5	Coal dust: 0.6
	Machine moulding of steel castings	Sand: 55–65 Silica sand: 25–30	Bentonite: 7–12	Dextrine: 0.3
Facing sand	Iron castings	Sand: 50–90 Silica sand: ≤38	Bentonite: 7–10	Coal dust: ≤5
	Iron castings (weight ≤2 ton, wall section 30 mm)	Sand: 50–60 Half-strong sand: 30–50	—	Sawdust: 10
	Responsible steel castings of medium weight	Chamotte: 85 vol%	Kaolinite clay: 15 vol%	—
	Hadfield steel castings	Chromite sand: 98–100	Kaolinite clay: 2 vol%	—
	Large castings in aluminium alloy	Sand: 50–60 Half-strong sand: 40–50	—	—
Core sand	Heavy iron castings	Sand: 35–45 Half-strong sand: 35–45	—	Sawdust: 10–15 Coal dust: 5–6
	Steel castings susceptible to cracking	Sand: 20–30 Half-strong sand: 50–60	—	Sawdust: 20 Waste sulphite liquor: 1
	Shell moulds and cores	Fractionated sand: 90–92	Resin: 7–10	Kerosene: 1
	Steel castings, CO_2 hardening	Silica sand: 90–95	Water glass ($M = 2.7$): 5–10	—
Loose sand	For heavy castings up to 2 ton and wall section above 30 mm	Sand: 20–30 Silica sand: 55–65	Clay: 3–5	Cement: 8–10
	Furan sand	Silica sand: >97	Resin: <2	Orthophosphoric acid
Liquid sand	Self-setting good collapsibility; Synflo sand	Silica sand: <95	Hydrophilic agent: <1 Resin: <3.5	Orthophosphoric acid: 0.5 Sapogen: 0.05

of this method is a high moulding output: up to 150 moulds per hour for horizontal mould joints and 300 moulds per hour for vertical mould joints.

Another way of making the mould stable is the 'V' process, which utilises the physical effect of vacuum (*Figure 3.7*). In essence, this method consists in 'sealing' the pattern with foil of thickness 0.05–0.1 mm and making it fit the pattern and plate through the application of negative pressure using a vacuum pump. A special moulding box with a collector and sucking pumps is put into place and filled with binder-free sand. After a preliminary packing by means of vibration, the second foil is put into place and the air is sucked off from the sand mould. Atmospheric pressure then squeezes the sand to the required hardness. The negative pressure is maintained

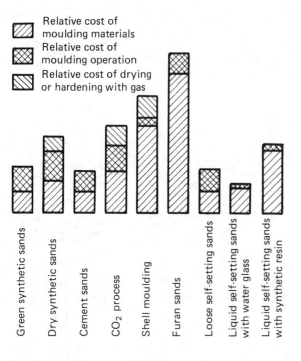

Figure 3.5 Relative cost of manufacturing moulds from various types of sand

until the casting solidifies. When the vacuum is removed the mould spontaneously disintegrates. The main advantage of this process is the total elimination of the use of a moulding-sand mixture. Compared with traditional methods, the vacuum method reduces both the cost of manufacturing a casting and environmental pollution.

3.2.4 Gating system

After the foundry mould has been made, the next stage is to pour the liquid alloy from tilting ladles.

The alloy, previously melted in a furnace, is subjected to a treatment of deoxidation, desulphurising and inoculation, and is then poured into the mould cavity through a gating system. The main objectives of using a gating system are to:

(1) ensure uniform feeding of the molten metal to all parts of the mould cavity;
(2) obtain a proper distribution of temperature over the entire casting during its solidification (simultaneous or directional) and cooling; and
(3) arrest the slag and protect the metal stream from oxidation.

A typical gating system (*Figure 3.8*) is composed of a pouring cup, down-sprue, runner, ingates, risers and flow-off. The place where the metal is fed to the casting is very important for the casting quality. The general rules that should be observed are as follows.

(1) For castings which do not require feeding during solidification (e.g. grey cast iron, bronzes and light alloys), the

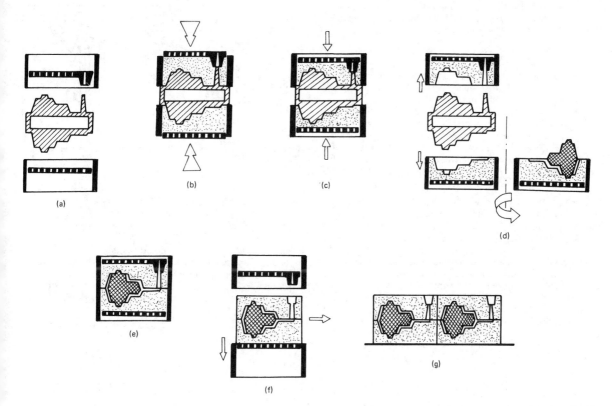

Figure 3.6 Mechanical method of compacting sand in a moulding box. (a, b) Placing of pattern and box; (c) packing of the sand; (d) removal of the pattern; (e) assembly of the two half-moulds; (f, g) before pouring on the moulding line. (Reproduced by permission from Jeancolas *et al*.[6])

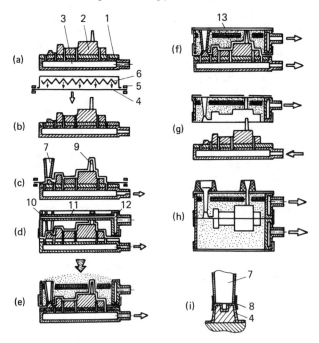

Figure 3.7 Making sand moulds using the vacuum process. (1, 2) Pattern elements; (3) vacuum chamber; (4, 13) foils; (5) foil holder; (6) heating; (7) sprue; (8) sand; (9) riser; (10) box; (11) tubes; (12) collector. (a) Preparation of pattern plate and boxes; (b) heating of the foil; (c) fitting of the foil; (d, e) filling with sand; (f) vacuum turned on; (g, h) assembly of the moulds; (i) moulding of the sprue. (Reproduced by permission from Lewandowski[7])

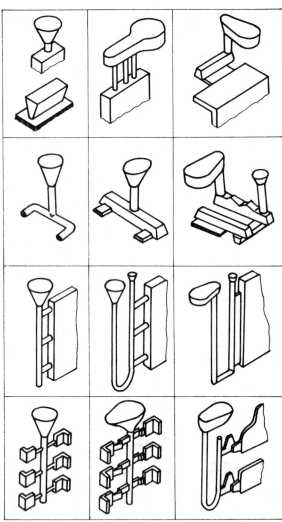

Figure 3.8 A typical gating system. (Reproduced by permission from Skarbiński[5])

metal is fed to the thin sections; a great number of the ingates of small cross-section and quick pouring ensure simultaneous solidification.

(2) For heavy castings which require intensive feeding during their solidification (e.g. cast steel, malleable and ductile cast iron, brasses, and copper–aluminium and copper–nickel alloys) the metal must be fed to the thick walls of a casting; thus it is advantageous to feed the metal through feeders and to ensure slow pouring of the mould. This will give directional solidification.

Calculations associated with the gating system consist in determining the pouring time, the pouring temperature, and the sprue–runner–gate ratios. The time of pouring for ferrous alloys is given in *Table 3.4*. The results reported therein were obtained assuming that optimum conditions for filling the mould cavity exist only for a shorter time than the time for alloy cooling in the narrowest cross-section of the mould cavity, in the range of the pouring and liquidus temperatures.[6]

The starting point for the calculations associated with gating systems is to determine the optimum time of pouring (t) or the optimum speed of the metal rising in the mould (V_f). In the case when moulds are poured from tilting ladles there are numerous formulae, both theoretical and practical; for example:

$$t = s \cdot \delta^2 \text{ for cast iron}^1$$

where δ is the wall thickness, s is a coefficient ($s = 12.6$ for green-sand moulds; $s = 23.0$ for dried-sand moulds).[2]

The calculated time of pouring is used to calculate the speed of the metal rising in the mould cavity:

$$V_f = \frac{C}{t}$$

where C is the height of the mould cavity.

The calculated value of V_f is checked with the values of V_f given in the literature (*Table 3.4* and *Figure 3.9*).

Once the pouring time has been determined, the smallest cross-section of all the elements of a gating system is calculated using

$$A_i = \frac{Q}{tm} \text{ for cast iron}$$

and

$$A_s = \frac{Q}{tm_1} \text{ for non-ferrous alloys}$$

where A_i is the cross-section of the ingates (cm), A_s is the cross-section of the sprue (cm), Q is the mass of alloy in the mould (kg), and m and m_1 are mould-pouring-rate coefficients (kg cm^{-2} s^{-1}).

The value of A_i or A_s is used as the basis for calculating the remaining dimensions of the gating system. For this purpose the practically tested values of the sprue–runner–gate ratios $A_s : A_r : A_i$ are used. Values of this ratio are given in *Table 3.5*.

Steel castings are characterised by different conditions of pouring. The use of bottom-stoppered ladles instead of tilting ones is characterised by a low flexibility in the outflow of the liquid steel from the ladle. The discharge of liquid steel is limited by the diameter of the tapping hole in the ladle and by the height of the column of liquid steel in the ladle. Therefore, the size of the ladle and the diameter should be adjusted to the size of the castings.

The formulae used to determine the pouring time are:

$$t = a \sqrt{G} \text{ for castings of weight } G \geqslant 15 \text{ ton}$$

$$t = b^3 \sqrt{\delta G} \text{ for castings of weight } G \leqslant 15 \text{ ton}$$

Table 3.4 Optimum values of the linear mould pouring rate (cm s^{-1}) for a height of $C = 10$–150 cm[6]

Alloy	δ (mm)				
	4	4–7	7–10	10–40	>40
Cast iron	3–10	2–3	2–3	1–2	0.8–1
Cast steel	—	—	1	1	0.8
Light alloys	3–10	2–7	1.5–4	0.8–2.8	0.5–1.4

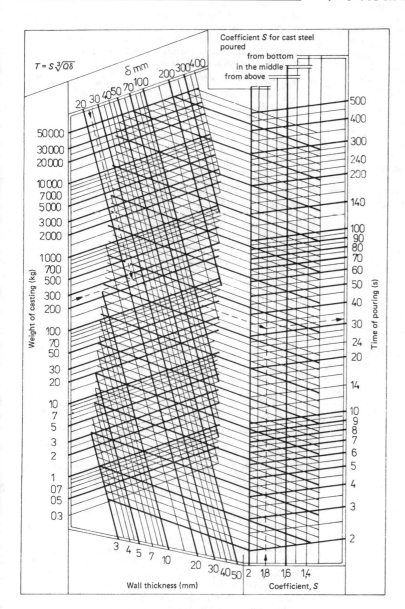

Figure 3.9 Nomogram for calculating the dimensions of a gating system. (Reproduced by permission from Baranowski[14])

Table 3.5 The ratio of the cross-sectional areas of the sprue to that of the runners and ingates

Alloy	A_i	A_r	A_s
Grey cast iron	1	1.1–1.4	1.2–1.6
Malleable cast iron	1	1–1.2	1–1.3
Ductile cast iron	1	1.4	1.5–1.8
Cast steel	1	1–1.2	1.1–1.4
Copper alloy	1	2	1
Copper–aluminium alloy	2–4	2–4	1
Aluminium alloy	2–6	1.2–2	1

where a, b, and p are coefficients and G is the casting weight (kg).

Due to the low flexibility of the system, the down-gate must receive the whole volume of the alloy flowing out from the ladle. Hence the discharge of the liquid steel from the ladle (Q_l) and the volume received by the mould (Q_f) are equal:

$$Q_l = Q_f = \frac{G}{t} = \mu_l\sqrt{2gH_a}A_l\rho = \mu_s\sqrt{2gH_g}A_s\rho \qquad (3.3)$$

where A_l and A_s are the cross-sectional areas of the tapping holes in the ladle and the down-gate, respectively, μ_l and μ_s are coefficients of the flow resistance in the ladle (0.8–0.95) and the mould (0.25–0.5), H_a is the mean metalostatic pressure of the alloy in the ladle, and H_g is the static pressure of the metal in the mould.

After calculating the diameter of the tapping hole in the ladle (25–55 mm) and rearranging equation (3.3), the cross-sectional area of the down-gate is given by

$$A_s = A_l\frac{\mu_l}{\mu_f}\sqrt{\frac{H_a}{H_f}}$$

In the further course of the calculations the sprue–runner–gate ratios (*Table 3.5*) are used.

3.2.5 Risers

During solidification of an alloy, shrinkage cavities are formed in those parts of a casting which solidify at the lowest rate (hot spots). These parts must, therefore, be fed with molten metal. This is possible through proper control of the solidification process in such a way as to locate the shrinkage cavities outside the hot spots in the feeding part. This is done by means of risers connected directly to the hot spots. Risers fulfil their task only when:

Table 3.6 Maximum feeding distance of risers*

Alloy	Wall thickness (mm)	A	K	A + K
Carbon cast steel	<25	2.5d	3.5d	6d
< 0.3%C	50–200	2.5d	2d	4.5d
0.3–0.4%C	15–40	2.5d	3d	5.5d
Malleable cast iron	6–10	—	—	11d
Bronze	10–40	2d	2d	4d
Aluminium alloy	25	—	—	5d

* For definition of *A* and *K* see *Figure 3.11*.

(1) there is directional feeding (solidification) from that part of the casting which is most remote from the foot of riser;
(2) the time of action of the riser (time of solidification) is longer than the time of solidification of a hot spot; and
(3) the liquid alloy in the riser absorbs gas and non-metallic inclusions.

In addition, risers must:

(1) be easy to remove and not hinder the transfer of heat from the casting (overheating of the moulding sand); and
(2) not hinder the casting contraction.

The parameter which characterises in a consistent way the process of casting solidification is called the *modulus of solidification* (M). This is determined[3] as the casting volume/cooling surface ratio M_o or as the ratio between the casting cross-sectional area (hot spot) and the circumference of this cross-section, M_j. The solidification modulus is sometimes replaced by the concept of 'reduced wall thickness'.

For intricate castings the calculation of the modulus M appears at first to be somewhat complicated. However, the problem is simplified because:

(1) the value of M for the whole casting is only calculated for large, simple castings; and
(2) in the majority of cases M is calculated only for that part of a casting which acts as a hot spot.

The casting solidification modulus M_o is given by

$$M_o = \frac{V_o^2}{F_c^2} \text{ or } M_j = \frac{F_j}{O_j}$$

where V_o is the casting volume, F_c is the cooling surface of the casting, F_j is the cross-sectional area of the hot spot, and O_j is the circumference of the hot spot.

To preserve directional solidification, the solidification modulus of a riser M_r should have values higher than $M_o \times M_r = 1.2M_o$ to $1.3M_o$ for cast steel, non-ferrous alloys and nodular graphite cast iron, and $M_r = 0.2M_o$ to $0.8M_o$ for grey cast iron, where M_r is the ratio of the riser volume to the cooling surface of the riser.

The calculation of most riser volumes can be done using the reduced formula

$$V_r = M_pF_c + V_{rg} \qquad (3.4)$$

where V_r is the riser volume, M_p is the modulus of the protected part (the riser is divided into two parts—the protected part and the fed part), and V_{rg} is the volume of the fed part of the riser.

Equation (3.4) can be used to calculate the riser volume required for an arbitrary alloy and for an arbitrary shape of riser, providing that the riser cooling rate is low. The calculation of riser dimensions has now been simplified by the availability of nomograms or computer programs,[4] which can be used to calculate and check in practical application the volume of risers. Results are compiled in the form of tables giving the shape and size of risers. An example for risers having a height/diameter (h/d) ratio of 1.5 is shown in *Figure 3.10*.

The next operation is to calculate the number of risers. Although the riser should feed each hot spot which occurs in a casting with metal, for flat and long walls use of a riser does not ensure the elimination of shrinkage cavities since the range of its effect is limited. It is assumed that the range of effect of a single riser is equal to $4M_o$. Section A in *Figure 3.11* is free of any shrinkage defects, due to the boundary effect (heat transfer through a mould) that acts there. The range of effect of the riser depends on the cast wall thickness and on the type

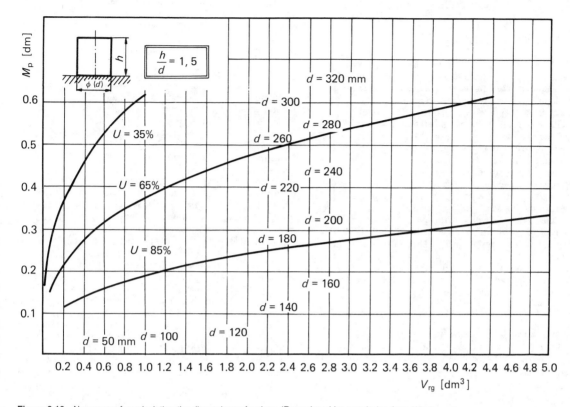

Figure 3.10 Nomogram for calculating the dimensions of a riser. (Reproduced by permission from Wlodawer[4])

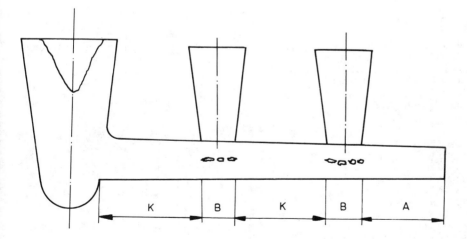

Figure 3.11 The range of effect (*K*) of a riser. (Reproduced by permission from Skarbiński[8])

of alloy; the height of the riser does not play any part in its range of effect (*Table 3.6*).

3.3 Pressure die casting

Pressure die casting is one of the few technologies which in recent years have shown definite growth in usage. This is due to advantageous market conditions in the automotive, electronics and household goods manufacturing industries.

Pressure die casting involves exerting pressure on the liquid metal. Filling of the metallic mould, referred to as a die, is achieved due to a positive gauge pressure (10–70 MPa) of air or nitrogen in the case of low pressure die casting, or by means of a mechanical plunger, which can exert up to 200 MPa, in the case of high pressure die casting.

Pressure-casting dies are most often used to manufacture aluminium alloy, zinc alloy or copper alloy castings; they are also used for casting steel. In the majority of cases dies are made of steel. The die cavity is made of steel characterised by a strength of 700–1100 MPa for low-melting-point zinc alloys or 1200–1600 MPa for copper alloys. These are usually steels containing 1–5%Cr and 0.3–1.1%Mo or 11–12.5%Cr and 0.3–1.2%Mo, in addition to nickel and vanadium.

Compared with other foundry technologies, the pressure die casting process has a number of advantages:

(1) a very high production rate (up to 500 pourings per hour);
(2) good dimensional tolerances (0.03–0.07 mm);
(3) it is possible to cast thin-walled (0.3 mm) details; and
(4) high surface finish and high yield of metal (up to 95%) are obtained.

These characteristics of the process enable elimination of machining and fettling of castings and as much as 40% reduction in the weight of castings. However, the process also has some disadvantages, e.g. high capital costs of the pressure die casting machine and of the new dies, and its application is limited to low-melting-point alloys and castings of low weight.

3.3.1 Low pressure die casting

This process is used for the manufacture of castings which would require the use of large risers in gravity die casting and for thin-walled intricate castings. It is most widely used in the automotive industry (engine cylinders, heads and cylinder blocks) and in the electronics industry (parts of motor rotors).

The advantages of low pressure die casting are:

(1) higher productivity than gravity die casting;
(2) easy mechanisation;
(3) lower investment outlays for machining and fettling; and
(4) a reduced gating system (risers).

There are two types of low pressure die casting machine. In the first type gas pressure is exerted only on the liquid alloy and the walls of the crucible. The machine includes the following elements: a furnace, a timing device, and a gravity die casting machine. Usually, resistance or main frequency induction furnaces are used as the crucible. The timing device is an auxiliary unit which serves to feed compressed air from the cylinder or central supply system, to reduce and stabilise its pressure and, after solidification of the casting, to remove the applied gas. Modern designs are equipped with automatic pressure-controlling devices and appropriate time relays. They control the whole casting cycle and facilitate the operation of machines.

The crucible in a furnace is sealed with a plate on which the casting machine is placed. Under the effect of positive gauge pressure, the liquid alloy is squeezed out of the crucible and fills the die cavity. Since the pressure does not drop, contraction during solidification is compensated for by permanently supplied liquid alloy—this either eliminates or reduces the need for risers.

The metallic mould (die) is made of cast iron and the cores may be made from sand mixture or steel. Since it is possible that cores made of moulding sands might suffer some deformation, the velocity of flow of the alloy in the gate (*Figure 3.12*) should not exceed 1.5 m s^{-1}. The die cavity is coated with a layer of TiO_2 or Al_2O_3 with an addition of $CaCO_3$, while the die is preheated to a temperature of 573–723 K. The temperature used depends on the wall thickness of the casting and on the type of alloy cast.

In the second type of machine, gas pressure is not only exerted immediately above the liquid alloy, but also inside the whole furnace. An example of such a design is shown in *Figure 3.13*. The furnace is equipped with two crucibles. The external cast iron crucible encloses a graphite crucible in which the liquid alloy is held. The temperature is measured at two places: with an immersion thermocouple in the graphite crucible, and in the space between the cast iron crucible and the furnace lining (4 in *Figure 3.13*). The molten alloy flows through a cast-iron pipe (5) heated from the outside by a heating coil. The furnace chamber is sealed by means of a lid (6) and the asbestos lining which adheres to a flange of the cast-iron crucible (2). The cycle of pouring is mechanised and is based on automatic adjustment of alloy temperature and on a control system which includes the time relays.

Automation of the manufacturing cycle usually covers the assembly and removal of a die and filling of the die cavity. The process is controlled through measuring the pressure above the level of molten alloy, the temperature of the molten alloy, and the level of the alloy in the crucible.

Modern machines for low pressure die casting are equipped with a die pouring rate control system. The principle of operation of these machines is by casting with counterpressure. A characteristic of this method is the possibility of pouring the alloy into a die at a pressure higher than atmospheric pressure. A variation of this process is shown in *Figure 3.14*. At the starting position in this process, the compressed air is fed to an area above the surface of the molten metal (P_1) as well as to the die (3). By turning the device to position II, the die cavity is filled with molten metal due to the effect of gravity.

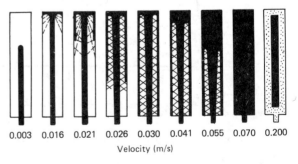

| 0.003 | 0.016 | 0.021 | 0.026 | 0.030 | 0.041 | 0.055 | 0.070 | 0.200 |

Velocity (m/s)

Figure 3.12 The velocity of flow of an alloy in a gate. (Reproduced by permission from Frommer and Lieby[11])

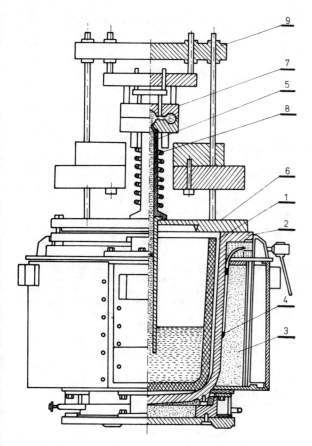

Figure 3.13 Low pressure die casting machine. (1) Crucible; (2) external crucible; (3) housing of the furnace; (4) thermocouple; (5) heat resistant tube; (6) cover; (7) chill; (8) heating; (9) moveable plate. (Reproduced by permission from Green[9])

3.3.2 High pressure die casting

Unlike in the gravity die casting process, in high pressure die casting machines the liquid metal flows into the die cavity at a pressure of 7–70 MPa. There are two types of machine: hot chamber machines for low-melting-point lead, tin and zinc alloys, and cold chamber machines used for aluminium, magnesium and copper alloys. In the former type of machine it is possible to exert a pressure of 7–15 MPa on the liquid metal, while in cold-chamber machines the pressure can exceed 200 MPa. The design of a die is determined by the temperature of pouring, the alloy castability, the shape and dimensions of a casting, and the type of pressure die casting machine used. Nowadays, the most popular machines are of the hot chamber plunger type in which a high pressure (10–60 MPa) is used.

The plunger is usually hydraulically driven, which makes it possible to exert a pressure of up to 300 MPa on the liquid alloy with the die locking force amounting to 100–5000 ton.

In cold-chamber machines the liquid alloy is kept in a separate furnace outside the machine. An exact quantity of liquid metal is carried in a bucket and poured into the pressure chamber.

In hot-chamber machines the liquid alloy is kept in a crucible directly connected to the pressure chamber. Pressure is exerted on the liquid metal in the crucible, and the metal is then transported under a pressure to the die chamber. In these machines the liquid alloy remains in contact with the plunger and the chamber, which affects the machine operation. This restricts the application of hot-chamber machines, mainly to the production of zinc castings.

The principle of operation of cold-chamber machines is shown in *Figure 3.15*. The liquid alloy is poured into the pressure chamber located in the horizontal position. Under the effect of pressure the plunger pushes the excess alloy out of the chamber in the form of a disk.

Horizontal-chamber machines have the following advantages compared with vertical-chamber machines:

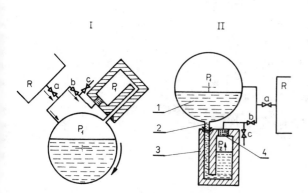

Figure 3.14 Casting of molten metal by counter-pressure. (1) Crucible with molten alloy; (2) bottom outlet; (3) mould. (a–c) Valve system for the compressed air. (P_1, P_2) Pressure ($P_1 > P_2$), (R) pressure vessel. (Reproduced by permission from Balewski and Dimov[10])

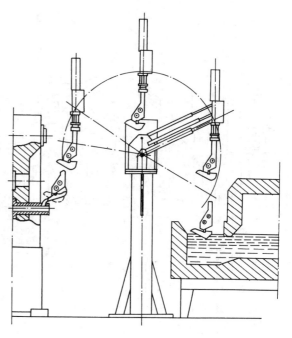

Figure 3.15 Principle of operation of a cold-chamber machine. (Reproduced by permission from Dańko[12])

(1) the molten alloy has a shorter path to pass through from the pressure chamber to the die cavity;
(2) the plunger has better guidance because it does not leave the chamber;
(3) the lubrication of the plunger is easy to mechanise; and
(4) the yield of metal is higher because of the smaller gating systems.

The limitations of horizontal-chamber machines are their greater overall dimensions, and the necessity to provide a better venting system for the die cavity.

A typical pressure die casting machine is shown in *Figure 3.16*. The technological process of making a pressure die casting consists of the following stages:

(1) preparation of the die cavity, including venting and lubrication;
(2) filling the die with liquid alloy; and
(3) consolidation and ejection of the casting.

With regard to casting quality, the consolidation stage is most important. The main task of the squeezing system is to control the increase in pressure during the filling of the die cavity. This factor is very important in view of the way in which the die is filled and the application of pressure.

The following is a typical cycle in the pressure die casting process:

(1) the molten alloy strikes the opposite die wall and its stream is divided into streams flowing in the opposite direction;
(2) the stream of molten alloy strikes the opposite wall and loses its kinetic energy, resulting in a whirling motion of the metal;
(3) the filling of the die proceeds from the opposite wall with intensive whirling of the molten alloy.

Such a die-filling mechanism requires certain operating conditions of the machine and the squeezing system. The most important factors which determine the operation of a pressure die casting machine are:

(1) the locking force and unit squeeze pressure, as they determine the choice of machine power according to the size of the casting;
(2) the ratio between the casting volume and the volume of the liquid alloy introduced into the chamber, as this determines the dimensions of the die and the pressure chamber;
(3) a constant pressure during pouring and solidification.

The three successive stages in an operating cycle of the squeezing mechanism are shown in *Figure 3.17*. The vertical axis in this figure refers to the pressure (P) and the plunger travel (S) and the horizontal axis to time (t).

Stage I: taking the air off from the chamber of consolidation when the pressure exerted on the liquid alloy is still low (1 in *Figure 3.17*).
Stage II: increase in pressure and increase in plunger travel speed (1–3 in *Figure 3.17*). Point 3 denotes the end of filling and the termination of plunger travel.
Stage III: this stage begins at point 3 with the turning on of the hydraulic pressure multiplicator and ends at the moment when the required pressure P is reached (point 4 represents static pressure). This is the consolidation stage (point 5).

An example of a system that controls the operation of the machine is shown in *Figure 3.18*. The system controls the plunger travel speed, the shot time, the consolidation time and the pressure level.

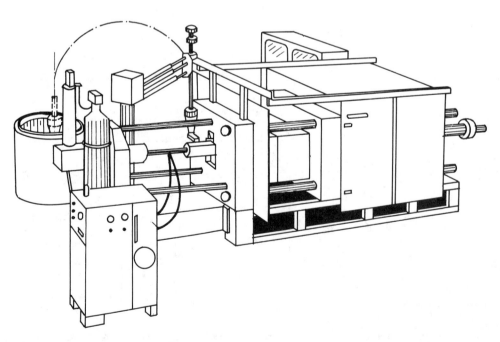

Figure 3.16 A typical pressure die casting machine. (Reproduced by permission from the Bühler Catalog[16])

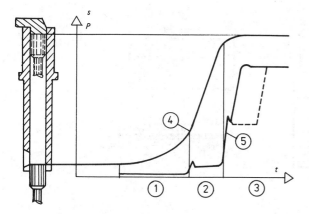

Figure 3.17 A typical operating cycle. For further details see text

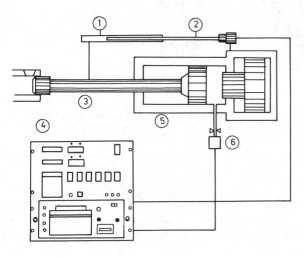

Figure 3.18 Control system for a pressure die casting machine. (1) Cylinder; (2) speed indicator; (3) plunger rod; (4) control unit; (5) shot cylinder; (6) valve

3.4 Investment casting

The investment, or 'lost-wax', process is widely used in the manufacture of intricate castings of low weight. This technology is particularly suitable for batch production. Surface reproducibility is so good that castings need no further machining. With this method it is possible to obtain castings of wall thicknesses as thin as 0.5 mm.

The investment casting process has the following features:

(1) the use of expendable patterns;
(2) moulding in a slurry;
(3) moulds are bonded with an inorganic ceramic binder which ensures exact reproduction of the pattern;
(4) the wax is melted out of the hardened mould;
(5) the mould is heated to remove gases; and
(6) pouring is done into a preheated mould.

The patterns are removed from the wax together with the gating system and risers. The wax patterns with gating systems

are made in metallic moulds by injection or pouring. The gating system allows for shrinkage of both the wax and the alloy, and for the expansion of the mould on preheating.

Patterns are made of a mixture of beeswax, paraffin, ceresin or other resins, usually in the form of blends, e.g. 50% paraffin and 50% stearin.

The wax mixture is injected into the chill at 333–343 K. After 3–10 min the pattern is lifted from the chill. When the whole pattern assembly is to be intricate in shape, the patterns and the gating system are made separately. After lifting out the pattern, the flashes are removed. The single patterns of the castings are joined with the gating system to form a cluster (*Figure 3.19*). The patterns and the gating system are joined together by heating and surface melting of the parts which are to be stuck to each other.

The manufacture of a mould consists of the following steps:

(1) preparation of the slurry;
(2) coating of the pattern cluster with a few layers of the ceramic coating;
(3) melting out of the pattern;
(4) moulding of a ceramic shell; and
(5) preheating of the mould.

The slurry, of a refractory coating material, is prepared from ethyl silicate solution, distilled water, acetone (or ethyl alcohol) and hydrochloric acid. Examples of such compositions used for coated-wax assemblies are given in *Table 3.7*.

The pattern assembly is immersed in the slurry several times in order to obtain a uniform coating. After the pattern has been removed from the slurry it is coated by sprinkling with coarse sand and allowed to dry for 4 h in air. During drying the coating hardens due to coagulation of silicic acid and evaporation of acetone and water.

To confer upon a mould certain durability, it is necessary to apply further coatings. The number of coatings applied depends on the weight, shape and degree of complexity of the casting and on the material cast. Usually 3–8 layers are applied. The second and subsequent coatings are of the materials listed in *Table 3.8*.

Successive layers of coating can also be applied by immersing the coated-wax assembly in the slurry on a vibrating table. As the table vibrates, the moulding material surrounds the patterns. Each new layer is sprinkled with sand and then allowed to set in air.

The next operation consists of melting the wax out of the assembly. This is done at a temperature of 353–363 K for 15–30 min. If melting out is done in water, the moulds must subsequently be dried. For drying and preheating, empty moulds are put into containers, surrounded with clean sand and heated to 1150–1320 K (ferrous alloys) or 920 K (aluminium alloys).

The temperature of preheating of moulds is determined by the pouring temperature required for a given alloy and the casting design. The heating and melting out operation must completely remove wax and gas-forming particles from the mould.

Casting is carried in by means of either a gravity feed process or a centrifugal method. After cooling, the casting is cleaned.

3.4.1 The Shaw process

The Shaw process is a variation of the investment casting process. Its features are:

(1) excellent reproduction of the casting shape, and
(2) very good accuracy and surface finish.

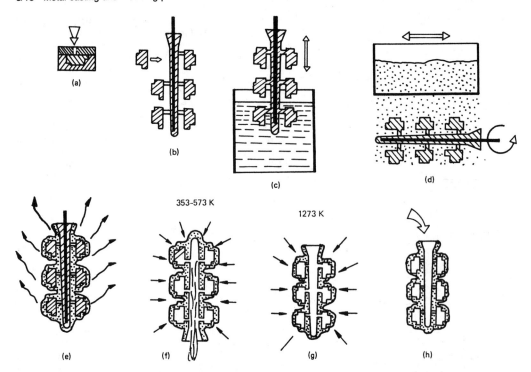

Figure 3.19 Investment casting. (a) Casting of pattern in the chill; (b) collecting of patterns with pouring system; (c) making of the shell in the slurry (up to 8–10 shells); (d) covering each shell with powder; (e) drying or hardening of shells; (f) removal of patterns by melting; (g) annealing; (h) pouring

Table 3.7 Coating mixtures used in the 'lost-wax' process

Materials	Application
94 parts of 325-mesh silica, 56 parts of 325-mesh alumina, 37 parts of 40-mesh silica, 80 parts of liquid (4 parts of sodium silicate, 1 part of 2% polyvinyl alcohol)	High-melting-point alloys
90% silica, 6% magnesia, 3% ammonium phosphate, 1% sodium phosphate, water or 1% hydrochloric acid	Cast steels
52% SiO_2, 45% Al_2O_3 + TiO_2, 0.87% Fe_2O_3 Molichite*	

* From English China Clays Sales Ltd.

The moulds are expendable, and are made using permanent patterns. This process is particularly suitable for the manufacture of castings made of high-melting-point alloys such as cast steel. The dimensional accuracy of steel castings of dimensions up to 25 mm is ±0.08 mm, whilst for dimensions above 370 mm the accuracy is ±1.14 mm.

Table 3.8 Second and subsequent coatings used in the 'lost wax' process

95% sand, 5% alumina cement; then 30% water additionally

90.6% sand, 7.1% calcium phosphate, 2.3% MgO; then 51% water additionally

1460 g of clay/silica flour (3 : 17), 800 ml of ethyl silicate (No. 40)/denatured ethyl alcohol/HCl in water (37.5 : 60 : 2.5)

The moulding mixture is composed of a fine-grained material (e.g. MgO, $MgO \cdot Al_2O_3$ or $ZrO_2 \cdot SiO_2$), a binder (e.g. hydrolysed ethyl silicate) and a hardener, which is a gelling agent. The hardener causes a change in the consistency of the moulding sand: it changes from a dense liquid to a plastic composition and then to a ceramic shell of high strength at the final stage.

Moulds are made by the following techniques.

(1) *Block (monolith) moulds*—the whole mould is made of the Shaw composition; these moulds are used in unit production; they are expensive.
(2) *Two-layer moulds*—the mould is made of two types of the Shaw composition; these moulds are used for art castings.

(3) *Shell moulds*—these are manufactured by means of pressing; they are used for medium-sized castings in an automated production cycle.

(4) *Metallic moulds*—these are cast iron moulds lined with the Shaw composition.

(5) *Composite moulds*—one layer is made of the Shaw composition and another layer is made of a filler (i.e. a mixture of chamotte and water glass); this method enables savings of up to 60% total costs.

The ceramic material used to prepare the Shaw composition is usually mullite ($3Al_2O_3 \cdot 2SiO_2$) or zirconium silicate ($ZrO_2 \cdot SiO_2$). The Shaw composition is characterised by high refractoriness, low expansion, absence of molten alloy, mould reaction, and low price.

The binder is hydrolysed ethyl silicate of pH 1.5–3.0. Alcohol, water and acid are used for hydrolysis. The acid is usually concentrated HCl; sulphuric or orthophosphoric acids are occasionally used.

The gelling material is an ammonium salt or carbonate. The binding rate of the Shaw composition depends on the amount and type of hardener introduced into the mixture of binder and ceramic material.

The process of making two-layer moulds is shown in *Figure 3.20*. Two patterns are used here. One is an oversize pattern made of wood or plastic and characterised by a low grade of accuracy. It is 8–10 mm larger than the mould, i.e. it is larger by the thickness of the facing sand layer made of the Shaw composition. After placing this pattern on a plate, the box is filled with filler (e.g. a chamotte mixture) and hardened with CO_2.

After hardening of the mould, the oversize pattern is removed, and the mould is placed on another pattern plate with a pattern exactly reproducing the shape of a casting. This pattern determines the accuracy of the casting and, therefore, must be made with great precision. The mould cavity is then filled with the Shaw slurry. After hardening, the mould is turned over, the pattern lifted off and the alcohol is burnt out from the Shaw composition. During this burning out process water is removed from $Si(OH)_4$ and the sand loses its elasticity, becomes rigid and acquires proper strength. If burning out begins immediately after lifting off the pattern, the surface of the mould is covered with small grazes. These are small enough to keep the molten metal from penetrating them, but they increase mould permeability and its resistance to temperature variations. Further increase in the mould strength and stability takes place during baking at temperatures of 1173–1273 K.

If the mould is not burnt, the surface of the mould cavity is hardened using a special mixture. If burning is not applied, volatile matter is not removed from the mould and microcracks are not formed. Hence, with no burning, mould permeability is lower and the mould is suitable for the manufacture of cores.

3.5 Shell moulding

This is a variation of investment moulding which can be fully mechanised. The process, invented by J. Croning in 1944, is used very widely because of its high dimensional accuracy (± 0.003 mm). This accuracy is due to the use of pattern plates with stabilised dimensional deviations (≤ 0.025 mm). Other advantages of this process include:

(1) reduced consumption of moulding materials;
(2) higher productivity per square metre of foundry shop;
(3) long bench life of moulds;
(4) reduced costs of knocking out, fettling and labour; and
(5) elimination of machining.

The disadvantages of the shell moulding process are the high cost of pattern plates and resin, and limited weight of possible castings.

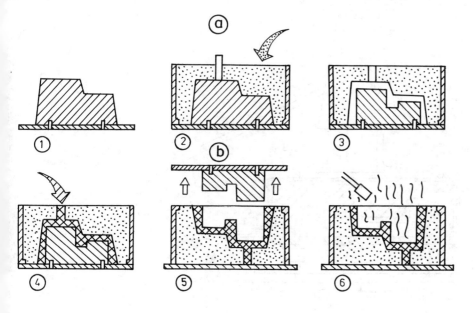

Figure 3.20 The making of two-layer moulds. (1) Enlarged pattern; (2) making of the mould with backing sand; (3) proper pattern; (4) facing sand (with ethyl silicate); (5) drawing of pattern; (6) drying. (a) First stage: making the mould; (b) second stage: making the proper shell

The process is, therefore, applied in large production runs in the automotive and precision parts industries.

In the Croning process, the main materials used are washed silica sand (97% silica), resin (phenolic and aldehyde) and a hardener (usually hexamine).

After mixing the sand and resin the process of hardening occurs only after heating the sand to a certain temperature. Thus it is possible to store the sand in the place where it was prepared and then to carry it to the moulding stand.

The sand used in this process should have a grain diameter of 0.1—0.15 mm and a sintering point of 1620 K. The sand may contain up to 0.5% iron oxides and up to 0.5% carbonates.

The binder for the sand is a resin which, during the formation of a shell, will first form a thin layer and then undergo hardening. Resins used for this purpose are furfuryl or polyester (saturated or unsaturated) resins.

Shell moulds are made in the following way. The pattern plates are coated with a material which facilitates the separation of the shell. These are usually silicon oils, wax or kerosene.

The steps involved in the shell moulding process consist of:

(1) coating the pattern plate or core box with a parting agent;
(2) heating the plate to 500–530 K; and
(3) pouring the sand onto the plate and waiting until a 5–8 mm thick shell is formed.

The excess sand is then removed and the shell heated to 620 K for further hardening. Both mould halves prepared in this way are joined together, either mechanically or with a glue, placed in a container and surrounded with coarse sand. At this stage the mould is ready for pouring. Similarly, cores can also be made as a solid or in halves.

A typical machine for making shell moulds is shown in *Figure 3.21*. The machine comprises a blower fitted with a membrane squeezing system. The resin-coated sand (1) is introduced under pressure through ducts (2) into a space between the pattern plate (3) and the contour plate (5). Attached to the contour plate is an elastic membrane (4). The sand is blown into cavity B, and then the compressed air is blown into cavity A. The air presses the sand to the pattern plate through the membrane. The space between the heated pattern plate and the box is sealed with a rubber which is resistant to the prevailing temperature.

Sand with 3.5% resin is fed from the tank (8). The blowing time is 1–1.5 s at a pressure of 6×10^5 Pa. After filling the box, squeezing is done for 5 s at 1.5×10^5 Pa pressure.

Another technique of making shell moulds is the hot-box process. This process consists of shooting or blowing the sand mixture into a hot core box (480–500 K) and holding it there until hardening takes place.

The core boxes used in the hot-box process are made of cast iron or cast steel; they are heated either electrically or with gas. The heating elements are a part of the shooting machines and their operation is automatically controlled. After heating to 450–550 K the core box is coated with a parting agent (water emulsion of silicon oils, or wax), and the sand is then shot inside. After hardening of the external layer of sand the cores are removed automatically.

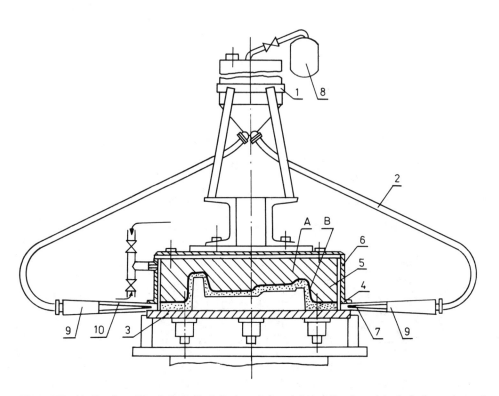

Figure 3.21 Machine for making shell moulds. 1, Resin coated sand; 2, duct; 3, pattern plate; 4, elastic membrane; 5, contour plate; 6, bottom flask; 7, nozzle outlet; 8, reservoir with resin coated sand; 9, nozzle; 10, air valve. A and B, cavities. (Reproduced by permission from Dańko[12])

3.6 Sintering

As one of the processes of powder metallurgy (powder production, mixing and blending, compaction, sintering and testing), sintering has found wide application in the manufacture of new types of material.

The use of powder metallurgy has enabled the fabrication of alloys which cannot be produced by melting and casting and the development of high-strength materials, composites, refractory compounds, and dispersion strengthened metals.

The final properties of a sintered material depend on the properties of the materials used for the manufacture of the powders and on the properties of the mass powder. The latter are determined by structure, the density, the melting point of the particles, the average size and shape of the particles, the particle size distribution, the specific surface of the particles, and the friction conditions.

The powders are manufactured by milling, grinding and crushing of solids. If the metals are in the liquid state, powder is produced by atomisation, i.e. by breaking up the molten stream into gases or liquids. The particles are mixed to obtain the required powder characteristic. The essence of the process is shown in *Figure 3.22*. The desired particle size and size distribution are obtained by crushing, milling (ball or rod) and grinding.

The next stage in powder metallurgy is the mixing of powders. The aim of this process is to introduce the alloying elements, binders (glycerine, camphor, or spirit) or sliding agents (e.g. graphite, oils, wax, or paraffin in amounts of 0.3–1.0%). Alloying elements are added to obtain certain predetermined technological and mechanical properties of the sinters. Sliding agents are used to obtain a more advantageous distribution of pressure during the subsequent process of consolidation. Binders are added when the alloying elements differ in terms of density or particle size from the base powder.

During mixing of the powder changes occur in the particle size (disintegration) and surface structure (oxidation). In addition, during mixing there is also abrasion of the particles due to the presence of the introduced additives and due to the walls of the mixer.

The next stage in powder metallurgy is the consolidation of the powders. Moulding of powders into objects of a given shape is carried out with or without the application of pressure. Without the application of pressure compaction involves unidirectional and isostatic pressing, rolling, forging and explosive compacting. When pressure is applied, compaction also involves slip casting, sintering of powder in the mould and vibrational compacting.

One of the most popular methods of compaction is isostatic pressing of powders. The method is applied in the case of sinters of complicated or oblong shape. In sinters of this type there are some local differences in density.When uniaxial pressing is replaced by isostatic pressing these differences in density are removed.

Isostatic pressing essentially consists of placing an elastic mould (made of neoprene rubber, urethane or polyvinyl chloride) and a powder in a chamber with compressed air or liquid. This causes the same pressure to act simultaneously on all parts of the mould (*Figure 3.23*) and enables the application

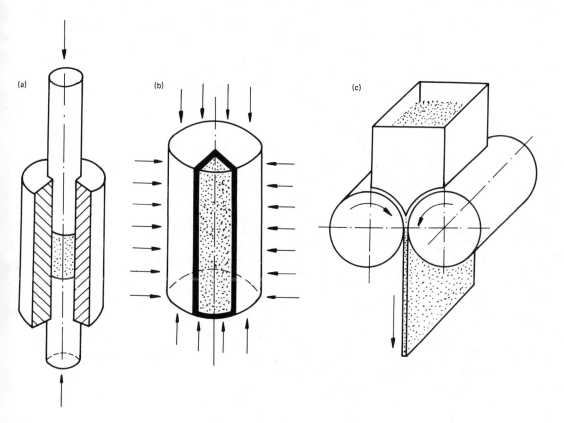

Figure 3.22 Compacting of powder in the sintering process: (a) pressing; (b) in an elastic mould; (c) by rolling

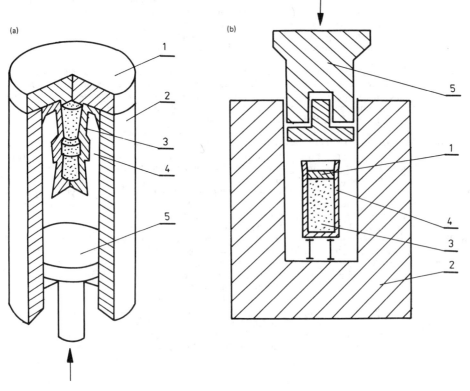

Figure 3.23 Compacting a powder: (a) with stable mould; (b) with stable mould removed. 1, Stopper; 2, press chamber; 3, powder; 4, mould; 5, plunger

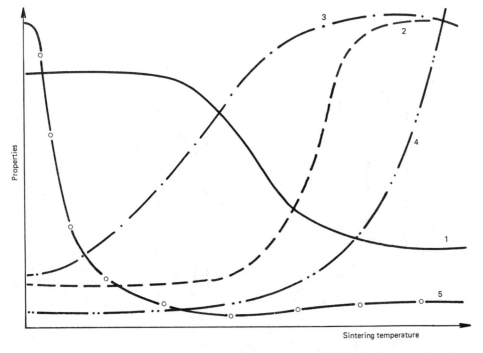

Figure 3.24 The effect of temperature on the properties of powder compacts. (1) Porosity; (2) density; (3) tensile strength; (4) grain size; (5) electrical resistivity

of lower pressing pressures. Owing to these effects, isostatic pressing can be used to consolidate hard and brittle powders as well as to mould large and heavy sintered compacts.

The moulded powder compacts are subjected to temperature in order to strengthen the powder mass. During sintering the powder compacts attain the required properties (*Figure 3.24*).

Sintering is a complex process and is accompanied by phenomena such as:

(1) a change in the surface and crystal lattice at the grain boundaries;
(2) a change in porosity;
(3) recrystallisation;
(4) stress formation; and
(5) changes in foreign constituents (gases and oxide films).

References

1 TRENCKLE, C., *25th International Foundry Congress*, Brussels (1958)
2 CLEGG, A. J., *Foundry Trade Journal*, **3197**, 429, 436 (1980)
3 CHVORINOV, N., *Giesserei*, **74**(14), 461 (1987)
4 WLODAWER, R., *Gelenkte Erstarrung von Stahlguss*, Giesserei-Verlag, Dusseldorf (1961)
5 SKARBINSKI, M., *Poradnik inżyniera odlewnictwo*, WNT, Warsaw (1986)
6 JEANCOLAS, M., *Le Remplissage des Empreintes de Moules en Sable*, Editions Techniques des Industries de la Fonderie, Paris (1966)
7 LEWANDOWSKI, L., 'Sand mixtures' (in Polish), *Publication No. 820*, Academy of Mining and Metallurgy, Cracow (1986)
8 SKARBINSKI, M., *Uruchomienie produkcjiw odlewni*, WNT, Warsaw (1972)
9 GREEN, R. E., *Machinery*, **2528**, 953 (1961)
10 BALEWSKI, A. and DIMOV, T., *The British Foundryman*, **LVIII**(7), 280 (1965)
11 FROMMER, G. and LIEBY, S., *Druckgusstechnik*, Springer-Verlag, Berlin (1965)
12 DANKO, J., 'Particular machines for foundries' (in Polish), *Publication No. 521*, Academy of Mining and Metallurgy, Cracow (1976)
13 LONGA, W., *Risers for Sand Castings* (in Polish), Slask, Katowice (1976)
14 BARANOWSKI, A., HAJKOWSKI, M., IGNASZAK, Z. STODULNY, C. and WASOWICZ, H., *Poradnik inzyniera odlewnictwo*, WNT, Warsaw (1986)
15 GRAF, H., *Stahl u. Eisen*, **96**, 117 (1976)
16 Bühler Catalog, *Technical Information No. 4*, Switzerland (1985)

4

Metal Forming

T Z Blazynski

Contents

4.1 The origin, nature and utilisation of plastic flow

4.1.1 Introductory observations

The initiation of plastic flow owns its origin to a variety of micro-effects associated with the application of an external loading system, the level of which cannot be contained within the elastic recovery capabilities of the material. The dimensional change (or changes) becomes permanent and leads to either a desirable accommodation of the metal within the confines of the forming pass or, in extreme conditions, to failure resulting in fracture.

There are two basic mechanisms of flow. Usually, but not always, the initiation of flow takes place when a sufficiently high shearing stress forces the atoms of a crystal to move to a new position of equilibrium. Since an individual atom cannot change its position without affecting the neighbouring structure, the movement will be possible only if a whole layer of atoms moves along. Slipping of the crystalline structure is thus effected and occurs along the planes of highest atomic density. If the load associated with the shearing stress does not exceed a critical value, return to the original position of equilibrium takes place on its removal and the elastic recovery is complete. Generally, stretching is produced by interfacial slip which, in turn, is independent of the value of the normal stress. Slip of the material is made easier by the presence of dislocations or metal-lattice defects. These are instrumental in producing consecutive atom movement which allows plastic deformation to occur.

Smaller plastic deformations can be produced by twinning or by a shift, of constant atomic distance, in a specified direction. A twinned crystal becomes distorted (in a mirror-like fashion) along the twinning plane.

Although, on the microscopic scale, the mechanism of yielding and the subsequent plastic flow are fairly well understood, it is nevertheless very difficult to extrapolate from the micro-condition to the macro-state with which engineers are constantly concerned. It is for this reason that the concept of continuum of matter is generally adopted and the macro-effects are used as criteria of design and general operational purposes.

Even with the acceptance of the continuum approach, it is still clear that a differential in flow is likely to occur. Although most materials are regarded as isotropic, or as continua in which mechanical properties are independent of direction, in a polycrystalline conglomerate lattice faults are usually present and even on this level perfect isotropy is unlikely. Forming of a metal—which takes place prior to the final manufacture of the component—will impose a directionality of flow and a consequent orientation of the grains. Clearly, therefore, directionality in properties must be expected and will be reflected in the level and type of the material response when the specimen is tested in two mutually perpendicular directions. Anisotropy becomes a recognisable property which has to be accommodated in the computation of the relevant numerical values.

A particular case of a micro- rather than macro-anisotropy is that of the *Bauschinger effect* which manifests itself in the reduction of the level of the yield strength of the material on reversal of the direction of plastic deformation imposed.

The Bauschinger effect is associated with the anistropy of the individual grains which, on application of a specific load, find themselves either in the elastic or plastic state. The subsequent removal of the load produces a conglomerate that varies in its properties and remains residually stressed. If the type of stress imposed thereafter corresponds to the prevailing residual, yielding will occur at a lower stress level than the original condition of the material would suggest.

Although the realisation of the presence of these phenomena and the understanding of their effect on plastic deformation are of relatively recent origin, the lack of this knowledge has never, in recorded history, prevented the human race from utilising the empirically established facts.

Metal forming, i.e. changing the shape of the material without actually removing any part of it, was practised at least 3000 years ago in Egypt, where hammer forging to produce gold sheet, cut subsequently to make wire, is recorded in the Bible to have taken place. Rolling in wooden mills was employed to manufacture papyrus. Manual swaging and wire drawing were well established in the Middle Ages but, naturally, were limited in scope by the power available. It was only with the advent of the Industrial Revolution that progress was made and processes like extrusion and cross and longitudinal rolling became available. But even here, the restrictions imposed by the low quality of tool materials, lubrication problems and the lack of understanding of the basic precepts of plasticity impeded progress until, in some cases, well into the twentieth century.

The ever-increasing demand for high quality products—often of sophisticated shape in difficult to process materials—economically produced, fabricated or semi-fabricated, combined with the rising cost of metallic engineering alloys has focused attention on metal-forming processes and techniques.

The emphasis here lies on the 'chipless' approach to shaping. This provides an economical, direct means of converting a cast ingot to slab, plate, billet or bloom and then—in another chipless operation—of changing these basic shapes into profiled finished or semi-finished products. The avoidance of the removal of the material during a forming operation enhances the economics of the process by reducing wastage associated with the swarf-producing machining. Whereas the latter has, of course, a very considerable and necessary role to play in the range of manufacturing activities, its indiscriminate use (a feature of the early years of plentiful supply of cheap labour and materials) is no longer acceptable when high tonnage of accurately manufactured product can be obtained at a much lower cost.

In the most simplistic terms, the desired change in shape is effected either in the cold, warm or hot state (the latter below the melting point of the material) by the application of external forces, pressures or torques of sufficient magnitude to induce plastic flow, and thus a permanent set, of the material through the forming pass. Depending on the operation, the material is forced to flow between driven rolls, through (or into) open or closed dies, or between sets of dies and rolls. Solid or hollow sections are thus produced from the initially solid blocks of metal.

The *standard* basic operations are:

(1) rolling (flat, oblique or longitudinal),
(2) extrusion (axisymmetrical or asymmetrical),
(3) drawing (solid or hollow components),
(4) sheet forming (deep drawing, bending, pressing or bulging),
(5) forging (solid and hollow sections), and
(6) cropping (shearing and piercing).

Within the compass of any of these operations, a number of variants exists which reflects not only a variety of manufacturing routes and subroutes, but also the nature, properties and characteristic responses of the processed materials. Modern metal-forming technology makes use of solid and semi-solid ('mashy' state), and superplastic, as well as explosively prewedled metallic composites and dynamically compacted particulate matter. Mixtures of metallic and/or ceramic and polymeric materials are formed to manufacture composites of very specific properties. The problem of forming these into desirable shapes presents the engineer with new and often difficult situations to solve. Selection of the appropriate forming process, the tool design, the effects of the pass geometry on the final physical and mechanical properties of the product, the dimensional accuracy, and the achievement of the as near as possible final shape in the minimum of operational stages have to be faced.

The apparently simple sequence of ingot–slab–semifabricate–finished product becomes complex unless there is good understanding of the basic characteristics of the individual processes and an appreciation of the principles of the theory of plasticity, as well as that of the concepts of tool and process design. The bases for and fundamentals of the major processes and technological developments are discussed in the following sections of this chapter, but detailed treatment of the individual topics is only indicated by reference to the appropriate literature.

4.1.2 Classification of processes

For a given application, the selection of the correct process necessitates the introduction of a criterion of process classification. Since hot working homogenises and refines the crystallographic structure of the material and thus, ultimately, improves its strength and toughness, whereas cold working increases strength, hardness, dimensional tolerances and improves surface finish, these temperature-induced effects are often used to differentiate between the various manufacturing methods.

Important as the processing temperature is, in some circumstances other criteria of classifying metal-forming processes may well be more appropriate. From a purely manufacturing point of view, quantity and shape may have to be considered, while the likely response of the processed material to the level and/or rate of stressing, as well as the manner of application of the forming load system, may offer a better clue

to the desirability or otherwise of using a particular technique or operation.

The parameters that characterise forming operations give rise to the following possible classification systems:

(1) operational temperature (hot, warm or cold forming),
(2) shape effect (bulk or sheet forming),
(3) operational stress system,
(4) operational strain rate,
(5) starting material (ingot, slab, billet, bloom, slurry, or powder).

4.1.2.1 Operational-temperature criterion

The idea behind the subdivision into hot, warm and cold processing of materials is not only to indicate the nature of the operation, but also to draw attention to the plant and ancillary equipment needed, to the level of force parameters required, and to the likely metallurgical response of the processed material.

An outline of this classification scheme, including only the basic operations, is given in *Figure 4.1*.

Starting with a cast ingot, the primary hot operations of flat, billet and slab rolling, and slab forging will produce the starting stage for the secondary, further processing of the slab into plate, billet or a large forging. These, in turn, will form the first step in the manufacturing route of a more sophisticated, profiled product. Hot operations are carried out at elevated temperatures exceeding annealing and normalising ranges and, consequently, yield a hot-finished product showing a relatively low level of flow stress. However, the force parameters required match the mechanical properties of the material and are also relatively low. It follows that the rate of wear of the tooling can be kept at an economical level especially if the lubrication problems are well under control.

To improve the mechanical properties of the product, while at the same time keeping the loading at a moderate level, warm processing is used. Here, the temperatures are well above ambient but, equally, well below the hot-processing range, and usually slightly less than for recrystallisation. The

Table 4.1 Classification of dynamic regimes

Characteristic time (s)	10^6 10^{-8}	10^6 10^{-6}	10^2 10^{-4}	10^0 10^{-2}	10^{-2} 10^0	10^{-4} 10^2	10^{-6} 10^4	10^{-8} 10^6
Strain rate (s^{-1})								
	Creep rates		Quasi-static rates		Intermediate strain rates		High strain rates	Very high strain rates
Primary load environment	High or moderate temperatures		Slow deformation rates		Rapid loading or low velocity impact		High velocity impact or loading	Very high velocity or hypervelocity impact
Usual method of loading	Constant load or constant stress machines		Conventional hydraulic or mechanical machines		Fast-acting hydraulic or pneumatic machines, cam plastometers, low impact devices		High velocity impact devices, expanding-ring technique, high-speed metal cutting	Light gas gun or explosively accelerated plate or projectile impact
Dynamic considerations in testing	Strain versus time		Constant-strain-rate test		Machine stiffness wave effects in specimen and testing machine		Elastic–plastic wave propagation	Shock wave propagation, fluid-like behaviour

————————Isothermal———————— ←————————Adiabatic————————→

←————Inertia forces neglected————→ ←————Inertia forces important————→

increased material ductility is sufficient to reduce the power requirement of the plant. Cold-working conditions are confined to ambient temperature and are characterised by a high energy requirement—necessitated by large operational forces and/or torques—but result in very high quality final product displaying both good dimensional tolerances and mechanical properties.

A rough guide to the temperature ranges can be obtained by considering the operational temperature/melting point ratio. On this scale, hot working takes place when the ratio is >0.6, warm working when the ratio is 0.3–0.5 (the latter corresponds to recrystallisation conditions), and cold when the ratio is <0.3.

4.1.2.2 Shape-effect criterion

The effect of shape reflects the geometry of both the initial and final component and, consequently, the nature of the change imposed on it by the forming operation.

A process in which a component of a relatively small initial surface area/thickness ratio is deformed in such a way that the ratio is increased, is often classed as a 'bulk deformation operation'. On the other hand, the component of an initially high surface area/thickness ratio, shaped in a process which does not impose any change in the thickness but effects shape changes only, is said to be 'sheet formed'. Any change in the thickness of such a component can easily lead to tensile plastic instability and incipient, localised yielding.

Bulk processes are those of rolling, extrusion, forging and solid- and/or hollow-section drawing. Bending, pressing, deep drawing, spinning and shearing are the main sheet-forming operations.

4.1.2.3 Operational-stress system

Because of the inherent severity of many forming processes, particularly the rotary ones, a consideration of the type and property of the induced stress field is of primary importance. The success of the operation may well depend on its compatibility with the properties of the processed material.

The presence of tensile and compressive stress fields results in the appearance of shearing stresses which, in turn, lead to the sliding of molecular planes and, eventually, to the yielding and plastic flow of the metal. Stress systems containing these components are most likely to give rise to plastic flow which, if it is controlled, will produce the desired amount of deformation.

Purely compressive or tensile systems create conditions of hydrostatic pressure in a triaxial field (absence of shear), or produce shearing stresses in uni- and bi-axial conditions. Clearly, since it is the configuration of the individual stress system that is indicative of the type of deformation which can be expected, its assessment prior to choosing a forming system is imperative. These various possibilities are illustrated, diagrammatically in *Figure 4.2*.

As an indication of the incidence of any of the stress systems, the following, non-exhaustive, list can be considered:

Tensile–compressive systems
Biaxial tension/uniaxial compression:
(1)   under a roll of a two-roll piercer,
(2)   under a roll of a two- and three-roll piercer, and
(3)   under a roll in the helical rolling process.

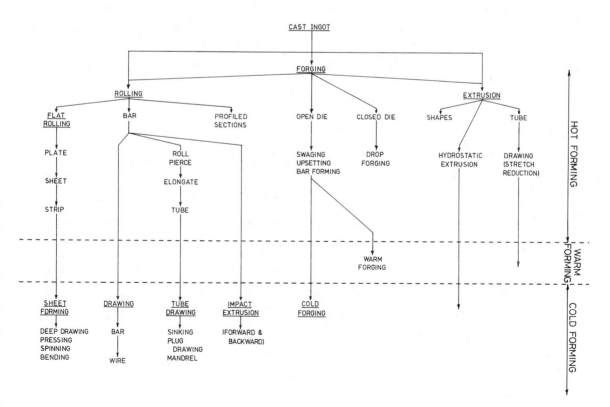

Figure 4.1   Process classification system based on operational temperature

Uniaxial tension/uniaxial compression:
(1) between the rolls in roll forming, and
(2) in the flange in deep drawing.

Uniaxial tension/biaxial compression:
(1) in the drawing die.

Compressive stress systems
Triaxial stresses:
(1) In the oblique zone of a three-roll rotary plug piercing mill,
(2) in the closed forging die,
(3) near the die throat in extrusion of bar, and
(4) under the punch in tube extrusion.

Biaxial stress:
(1) between the rolls of a longitudinal rolling mill with no front and/or back tension, and
(2) in the upsetting, open dies.

Tensile stress system
Biaxial stress:
(1) stretch forming, and
(2) bulging.

4.1.2.4   Operational strain rate

A number of engineering alloys and even some practically 'pure' materials, e.g. commercially pure aluminium, are susceptible to the changes in the rate of straining. Modern technological techniques have either 'speeded up' conventional processes—for instance, wire can be drawn at some 120 m min^{-1} —or have introduced new ones that operate in truly dynamic conditions. Impact extrusion, explosive forming, welding and compaction, and mechanically and electrically induced discharges of energy producing high strain rates, have all combined to introduce an entirely new field of high-energy rate fabrication, known commonly as HERF.

The range of possibilities arising in this context are listed in *Tables 4.1 and 4.2* which provide a detailed insight into the effect of different strain rates and the means of producing them in an industrial environment.

In terms of the mechanical properties of the material, an increase in the operational rate generally leads to a rise in the yield and flow stress and to an accompanying reduction in the value of the strain to fracture. The engineering alloys usually employed fall, in this respect, into two main groups (*Figure 4.3*). One (*Figure 4.3(a)*) will initially display little change in the value of the strain to fracture, but this will be reduced rapidly as the flow stress increases. The other may even show some initial improvement in the strain value, but the deterioration sets in as soon as the optimum strain rate is exceeded (*Figure 4.3(b)*).

Table 4.2 Mass–velocity relationship contributing to total kinetic energy

Forming system	Mass contribution (kg)	Velocity contribution (m s^{-1})	Velocity/mass
Drop hammer	144	122	0.84
Trapper rubber	144	122	0.84
Floating piston	4.54	3050	672
Explosive in gas	0.004	2.75×10^6	687×10^6
Explosive in liquid	0.008	1.52×10^6	190×10^6
Electric discharge	0.008	1.52×10^6	190×10^6

The toughness of these materials will be correspondingly affected (*Figure 4.4*). Consequently, if dynamic working conditions are likely to prevail in the selected process, a classification based on the strain rate may offer a more realistic means of assessment, particularly in the case of cold-forming operations.

4.1.2.5   Starting material

Since some modern processes do not require bulk solids as starting materials, but utilise particular matter and semi-solid substances, a classification based on the initial physical state of the material offers an interesting alternative to the more conventional approach.

Typical examples of unconventional starting materials are: 'mashy' state processing, leading to conventional rolling of composite sandwich components; the Conform-type extrusion, starting with a powder or granulated material, or an explosive compaction of powders.

4.1.3   Characteristics of the basic groups of processes

Of the major processes listed in Section 4.1.1, forging is the most diverse and cannot therefore be described in more general terms. For this reason, the basic characteristics of only four groups of processes are indicated here and those of forging, sheet forming, cropping, etc., are discussed later.

All rolling processes rely on the forces transmitted through the rolls to the material to effect deformation and on the rigidity of the roll system for the dimensional accuracy of the product.

Sheet and plate are initially obtained from a slab by rolling the slab in a relatively simple system (*Figure 4.5*). Driven rolls introduce the material into the roll gap, or working zone of the pass, and reduce the thickness. The success of any further processing to obtain strip rather than a sheet or large area of plate, depends on the ability of the system to maintain a constant width of the processed metal and on reduction of the thickness (this being equivalent to the reduction in the cross-sectional area). These requirements call for a plane strain operation which is possible only if the lubrication of the pass is very efficient. Processing in this mode can proceed in either cold or hot conditions.

A much more complex rolling system is that of longitudinal rolling, which is employed in the production of axisymmetrical billets, bars and hollows (*Figure 4.6*). A train of suitably shaped rolls, mounted on stands (either in pairs or in three-roll configurations) inclined at right angles (between the successive stands) is used, as shown diagrammatically in *Figure 4.6(a)*. A gradual reduction in the cross-sectional area of the material takes place (*Figure 4.6(b)*) as the specimen moves axially forward through the sets of driven rolls. While fully engaged in the train, the processed material experiences, additionally, axial tensions resulting from a differential distribution of successive stand velocities. The ovality of the early passes is slowly reduced along the train until the last stand is reached. Here, the final, circular cross-section is expected to be achieved.

An alternative to longitudinal rolling is offered by the oblique-rolling system in which a single set of two or three driven rolls produces tractive, frictional forces which propel the specimen axially while, at the same time, causing it to rotate. The motion of an element of the worked material is thus forward, but helical.

Figure 4.7 illustrates, using an example of tube rolling, the basic principle involved. In this case, three profiled, driven rolls, disposed at 120° to each other, and inclined at an angle α

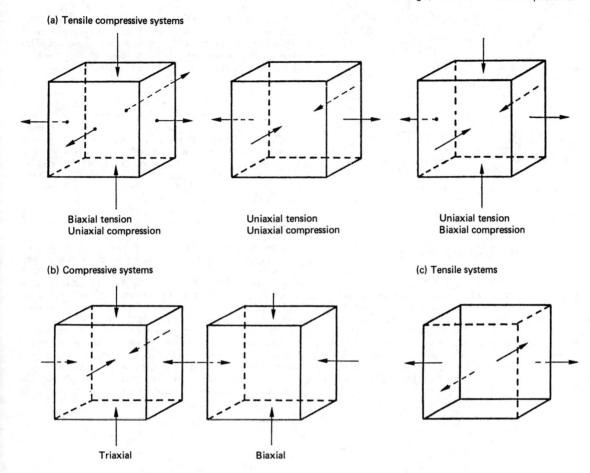

(a) Tensile compressive systems

Biaxial tension
Uniaxial compression

Uniaxial tension
Uniaxial compression

Uniaxial tension
Biaxial compression

(b) Compressive systems

(c) Tensile systems

Triaxial

Biaxial

Figure 4.2 Process stress classification system

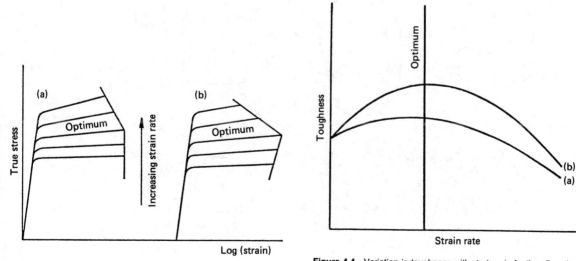

Figure 4.3 Stress–strain (to fracture) relationships

Figure 4.4 Variation in toughness with strain rate for the alloys in *Figure 4.3*

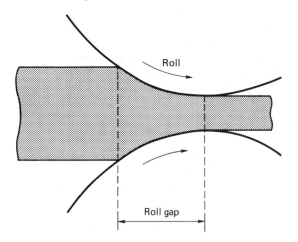

Figure 4.5   The principle of sheet and plate rolling

These basic characteristics of oblique rolling operations (the variants of which are discussed later) are common to all operations, as indicated, for instance, in *Figure 4.8*. This shows, diagrammatically, the operation of the so-called 'secondary piercing', or 'oblique plug rolling' of a tube — a process in which a long cylindrical mandrel is replaced by a short profiled plug.

On the other hand, processes of profiling by rolling can take various forms, two of which are indicated in *Figures 4.9* and *4.10*. A stepped shaft, required to acquire a series of specific profiles, can be manufactured by oblique rolling in a single three-roll stand (*Figure 4.9*). An operation in which the billet is rotated and fed through a system of driven rolls produces this effect. In another variant of oblique rolling, a two-roll system of helically ribbed rolls (*Figure 4.10*) will produce metal balls out of a solid cylindrical billet.

These few examples illustrate the versatility of rolling operations, a more full discussion of which is given in Section 4.5.

When the initial shape of the work piece has been imposed on it by one of the processes described above, there often arises the problem of how to achieve a degree of further deformation leading, possibly, to the final product. Drawing processes answer this need by providing a means of producing either solid (bar, rod or wire) or hollow tubular sections, either circular or non-circular in shape. The drawing operation is carried out in a die—or a set of consecutive dies forming a tandem drawing system—into which the work piece, with a swaged leading end, is introduced (*Figure 4.11*). An axial force is applied through a gripping device (as indicated by the arrow in *Figure 4.11*) and the work piece is pulled through the die. In the case of a solid specimen, the outer dimension only

(the feed angle) to the horizontal mill axis, and an angle β (the cone angle) in the vertical plane, introduce the bloom (supported internally in the bore by a mandrel) into the forming pass. The bloom is 'sunk' onto the mandrel in the zone AB and has its wall thickness reduced on the roll 'hump' BC. Slight elastic recovery takes place along DE. The bloom is thus elongated and its wall is thinned. The amount of deformation imposed depends on the size of the inter-roll opening or the 'gorge'.

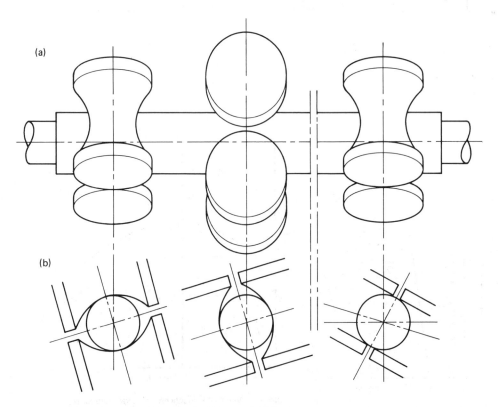

Figure 4.6   Longitudinal rolling: (a) a roll train; (b) successive roll passes

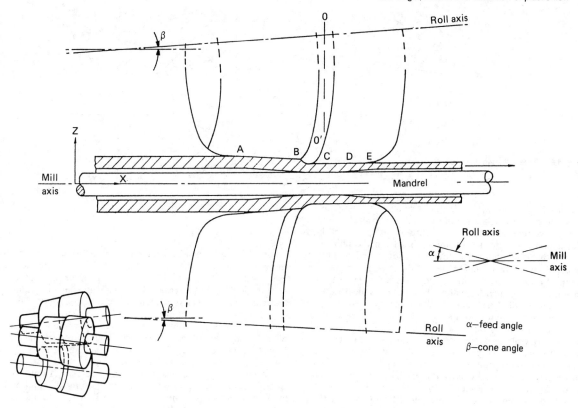

Figure 4.7 Oblique, three-roll tube rolling on a mandrel

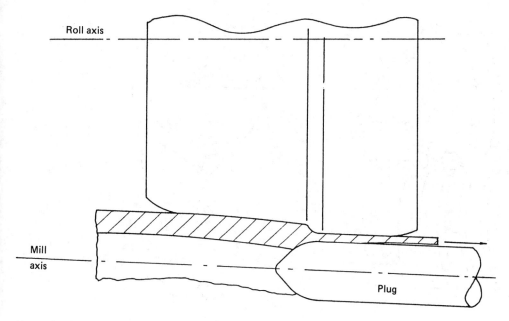

Figure 4.8 Oblique, tube rolling on a plug (secondary piercing)

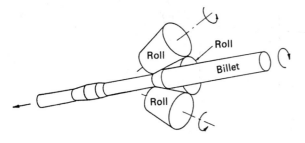

Figure 4.9   Three-roll shaft shaping

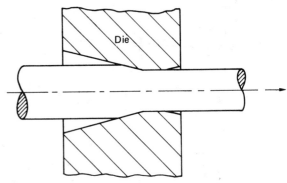

Figure 4.11   Open die drawing

is reduced, whereas with a hollow section there is also a change in the wall thickness. Lubrication of the working zone of the pass (the part of the die surface along which the deformation is effected) is of importance from the point of view of the magnitude of both the drawing load and the induced drawing stress, and in view of the surface finish.

Similar results can be obtained in extrusion, a process in which the starting billet (sometimes referred to as the 'slug') is inserted into a cylindrical container and is then pushed mechanically through a suitably profiled die (*Figure 4.12*). there is a number of variants of this process (see Section 4.7), but the two basic operations are those of forward (or direct) and inverse (or backward) extrusion. In the forward extrusion a solid moving ram is brought into direct contact with the billet and activates the latter by moving it axially forward through the die. In inverse extrusion a hollow ram is in contact with a movable die which bears onto the billet, firmly held in the container. When the pressure exerted by the tooling is sufficiently high to exceed the yield stress of the material, plastic flow is initiated and backward extrusion into and through the hollow ram takes place.

Considerable control over the dimensional accuracy can be exercised in such systems but, again, solution of the lubrication problem is of importance. In this latter context, hydrostatic extrusion (to be described later) provides an important alternative to the conventional arrangements indicated here.

A large group of 'unorthodox', dynamic processes introduces a number of new elements and opens new operational possibilities of using materials which are sometimes difficult to process and of reducing manufacturing costs by dispensing with heavy plant and equipment.

The high-energy-rate processes stem essentially from the usually overlooked fact that the working of metal requires energy and not merely the application of force, and that, in addition, the rate of dissipation of energy is of importance. A simple consideration of the basic equation for kinetic energy shows that a comparatively small change in the velocity of a body will have a more pronounced effect than will a change in

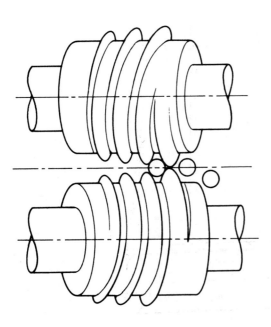

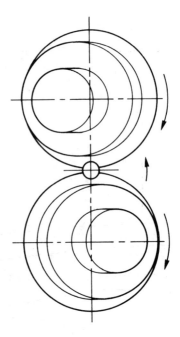

Figure 4.10   Two-roll ball manufacture

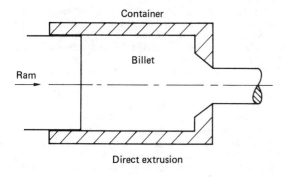

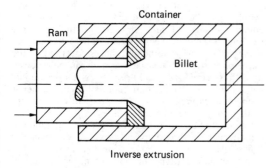

Figure 4.12 Direct and inverse extrusion

its mass. A typical conventional system approaching the conditions of high-energy forming, i.e. drop-hammer forming, is limited in its usefulness by the necessity of using large masses and, therefore, unwieldy and costly equipment.

In a high-velocity system, the mass contribution to the available energy can be negligible, as indicated in *Table 4.2*. The forces generated in a system of this type are easily accommodated by the inertia of the forming dies.

The sources of energy used in the high-velocity systems are chemical explosives, electrostatic and magnetic fields, and pneumatic–mechanical devices.

The basic processes are those of forming (shaping), welding and powder/particulate-matter compaction. A variety of forming systems exists, each displaying specific characteristics associated with either sheet or tube forming, for which it is intended. Consequently, a more detailed description of such processes is given later, but to appreciate the principles involved, an example (*Figure 4.13*) which refers to the explosive bulging of a tube is used here.

A stainless-steel (in this case) tube to be shaped is inserted into a profiled, vented die which is then closed. An explosive, linear charge, capable of delivering the required level of energy, is then placed axially in the tube and either air or water is used as the energy-transmitting medium. On detonation of the charge, high-pressure, high-temperature gases are generated and these propel the tube radially outwards, until it fills the die and acquires the required shape. The two-part die is then opened and the product extracted. The finish of the die surface is of importance because the outer surface of the tube comes into intimate contact with it and will bear the imprint of any machining, etc., marks.

The action of electrical discharge, whether capacitance in type or magnetic, is similar, but the forming system employed will, of necessity, be more complex and sophisticated. Because of the inherent cost of electrical equipment of this type, only relatively small sized components can be formed economically.

4.2 Process assessment

4.2.1 Force/torque estimate

The efficiency of a metal-forming operation can be assessed on two different levels. From the point of view of plant design and its subsequent rational utilisation, it is necessary to have a means of predicting the values of the externally applied forces and torques. This calls for the development of a suitable mathematical model of the operation and for a purely theoretical approach to the determination of numerical values. A theory of plasticity and methods of modelling based on it must be established. These are reviewed in Section 4.3.

However, the characteristics of metal flow, i.e. its incidence and pattern, cannot be neglected in an assessment because they reflect both the chosen geometry of the pass and the

Table 4.3 Summary of the characteristics of laboratory stress–strain testing methods

	Tension	Simple compression	Plane compression	Torsion (solid test piece)	Torsion (hollow test piece)
Definability of stress conditions	Not definable. Dependent on metal at high strain	Fairly definable, but dependent on friction at high strain	Definable (plane strain)	Not definable when axial stresses present	Definable (simple shear)
Amenability of features of test to correction	Corrections doubtful, complex measurements (contours) needed	Sound correction methods, but multiple tests needed	No correction needed	Strain rate corrections needed for very strain-rate sensitive materials	—
Information at high strains	Stress–strain curve doubtful	Information below 2.3 strain only	Information below 2.3 plane strain only	No limit*	Information obtained at strains >6
Ductility indication	Obtained	None	None	Obtained	Obtained
Suitability for tests at high temperatures	Suitable	Suitable	Doubtful	Suitable	Suitable
Suitability for tests at high strain rates	Suitable, but some severe experimental difficulties	Suitable	Unsuitable owing to need to change tools	Suitable	Suitable

* Results obtained by the present author up to strains of 10 000.

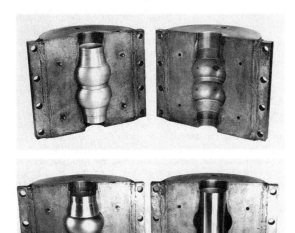

Figure 4.13   Explosive forming of a profiled 18/8 stainless-steel tube in a die

prevailing lubrication conditions, and the response of the processed material to the imposed mechanics of the operation.

The response of a material depends, primarily, on its physical and mechanical properties, but it also depends on the presence, or otherwise, and the type of constraints imposed on the flow. These parameters govern the degree of severity of the operation and, consequently, its efficiency.

Constraints imposed by the geometry of the pass and the initial state of the processed material not only affect the quality of the final product, but also influence the magnitude of the operational force parameters. The economics of the operation will thus be affected from the point of view of both the power consumption and tool wear, and the volume of imperfect or even defective product.

Once the expected characteristics of the flow have been anticipated, the forces, stresses and power parameters can be estimated theoretically.

4.2.2   Flow characteristics

The desired change in shape of the work piece, i.e. the basic objective of the metal-forming operation, normally involves elements of both useful and redundant or unnecessary straining of the material. In this context, useful or homogeneous deformation is associated with the absolutely necessary change in shape, from the initial to the final shape, but not with any other change in the material that does not contribute to this situation. In general, homogeneous deformation is defined by three mutually perpendicular strains and its magnitude and pattern are related to the external forces applied and to the geometry of the system. It does not, however, depend on the mechanical properties of the processed material.

To varying degrees, every forming operation will produce shearing strains associated with the angular and/or longitudinal distortion of the material. Since this type of deformation does not contribute anything to the required change in shape, but merely produces undesirable additional straining, it is clearly redundant to the basic objective of forming.

The pattern of this deformation depends only on the geometry of the system, but its magnitude is governed by the mechanical properties of the material, as reflected by the stress–strain curve, and also frictional effects produced in the working zone of the pass.

In some forming operations, e.g. rotary processes, friction constitutes the only tractive force and its elimination or reduction below a certain critical level results in break down of the operation. Frictional effects can, therefore, be viewed from two different angles. As shears, distorting the structure of the material, they are detrimental to the operation by increasing the magnitude of unnecessary straining, whilst not contributing to the change in shape. As effective tractive forces, they must be tolerated in certain cases in spite of the introduction of additional strain.

Thus the redundant deformation consists of two components: the frictional component, and the deformation produced by the geometry of the pass. The latter manifests itself as shearing at the entry to and exit from drawing dies, longitudinal warping and circumferential twisting in rotary processes, etc. It is mostly the magnitude of the latter component of inhomogeneous deformation that is governed by the mechanical properties of the material and can be assessed from the stress–strain curve. For a given tool–material–lubricant system, the specific contribution of friction depends mostly on temperature and is more pronounced for non-metallic than for the metallic materials.

Detailed study of inhomogeneity calls for the separation of frictional effects from those imposed purely by the geometry of the pass. However, for comparison between, say, the performance of different types of tool profile, this separation is not strictly necessary. If the work piece and tool materials, lubrication and temperature conditions are kept constant, the contribution of friction, although dependent on tool profile, is minimised. The level of inhomogeneous deformation, as measured or calculated on this basis, reflects the performance of the tools and serves as a measure of their efficiency in changing the shape of the work piece.

Typical examples of the incidence and likely pattern of inhomogeneous deformation are given in *Figures 4.14* and *4.15*.

The former refers to the rotary piercing of a solid billet in an oblique rolling operation (similar to those in *Figures 4.7* and *4.8*, but not starting with a hollow). If, using a model material that is capable of being marked internally, the operation is stopped before the whole of the billet has been pierced, and the product is sectioned, the angular distortion along the pass becomes obvious. In the absence of the redundant shear, the originally vertical and horizontal banks or markers retain their angular positions and only change dimensions (homogeneous strains). The distortion seen here is due to a differential shearing of notionally, infinitesimally thin cylindrical elements with respect to each other in tangential directions.

In non-rotary drawing or extrusion processes (*Figure 4.15*), an internally marked billet (*Figure 4.15(a)*) shows that distortion takes place in the axial direction; the distortion can, however, be quite severe. If the distortion is neglected when estimating the strain for the purpose of calculating extrusion pressures or drawing loads, considerable error can be introduced. For instance, if the evaluation of redundancy is based on the distortion of the marker AB (*Figure 4.15(b)*) to its steady-state position A′B′, then it could be concluded that the longitudinal shear strain is practically uniform across the section and may be represented by, say, $A_1A_2 = B_1B_2$. If, however, the variation in strain along a number of flow lines, such as aa′, bb′, etc., is considered, the distribution is non-uniform and corresponds to, say, A_3B_3.

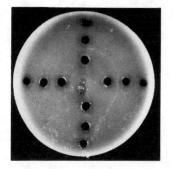

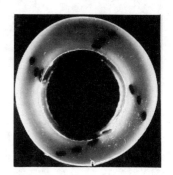
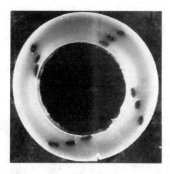
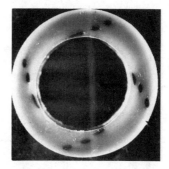

Figure 4.14 Development of redundant, shearing circumferential distortion and distortion due to twist strain in the rotary piercing of a billet

Clearly, the material experiences different levels of straining in different internal sections, and its potential suceptibility to overstraining and, possibly, fracture must be taken into account when designing the pass.

The effect of the additional, redundant shearing strain on the response of the product depends on the type of operation. In cold processes, the presence of shear increases the level of flow stress and, therefore, tends to introduce an element of brittleness into the material. This effect is indicated in *Figure 4.16*. The curve OA'B'C represents the stress–strain relationship in the original, unworked metal. When this material is cold formed and the homogeneous strain imposed amounts to, say, OA, the stress–strain curve obtained will not coincide with the 'basic' curve, but will show a stress level difference of A'A''. This reflects the amount of redundant shearing. A shift of the curve to the right, until the two coincide (the Hill–Tupper principle), gives the true value of the strain imposed, by including the redundant component AB.

In hot processes, the level of stressing is of less importance, but the presence of shear affects recrystallisation, gives rise to internal faults, and further weakens the structure which may already be affected by the presence of inclusions, etc. These observations are summarised schematically in *Figure 4.17*.

To assess *a priori* the suitability of the given material for processing in a specific forming system, the material properties must be either known or will have to be determined. Standard, tensile, compressive and torsional laboratory tests may have to be conducted, but these, in turn, must be chosen carefully since not every test will necessarily be suitable. For example, rotary operations proceed mainly in torsion and,

consequently, material properties established in the torsional mode will give a better idea of material response than, say, tensile or compressive testing. The selection of the correct type of test may be based on the information provided in *Table 4.3*.

4.3 An outline of the theory of plasticity

4.3.1 Basic concepts

Once the initial yield point of the material has been exceeded and the load is allowed to continue to increase, plastic flow is initiated. In strain-hardening materials this produces an ever-increasing level of the current flow stress and, also, an internal consolidation of the material. The latter physical fact provides the basic law of plasticity, i.e. constancy of volume. If the three, mutually perpendicular strains are ε_x, ε_y, ε_z, the law is given in the form:

$$\varepsilon_x + \varepsilon_y + \varepsilon_z = 0 \tag{4.1a}$$

The characteristic behaviour of the material in the plastic regime can differ from its behaviour in the elastic range in that a reversal of strain (in direction) is possible and the strain path need not be straight since a rotation of the successive planes of the loaded specimen can take place. To account fully for this possible pattern, it is *insufficient* to consider the initial and final stages of the deformation only, but it is necessary to follow the deformation of an element through the zone of

(a)

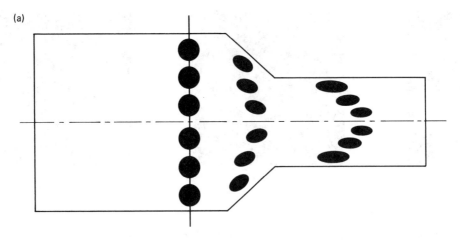

(b)

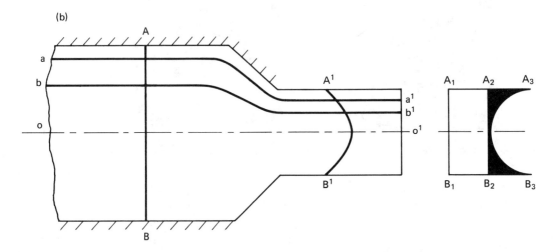

Figure 4.15   The pattern of shearing strains expected in bar drawing or extrusion

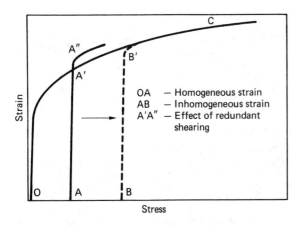

Figure 4.16   The effect of redundancy on the flow stress of a strain-hardening material

deformation in an incremental manner. In other words, analysis of the pattern of flow has to be based on incremental and *not* finite changes.

Equation (4.1a) is therefore often quoted in the form:

$$d\varepsilon_x + d\varepsilon_y + d\varepsilon_z = 0 \qquad (4.1b)$$

where the strain increments have to be defined, in each individual case, in terms of the variable dimensions of the worked specimen. For instance, in the rod or wire drawing, shown in *Figure 4.11*, the radial strain increment ($d\varepsilon_y$) would reflect the change in the radius (R) varying through the zone of deformation, the axial increment ($d\varepsilon_x$), the variation in the cross-sectional area (A), the circumferential increment ($d\varepsilon_z$) and the change in the circumference (S). Hence

$$d\varepsilon_x = \frac{dA}{A}; \ d\varepsilon_y = \frac{dR}{R}; \ \varepsilon_z = \frac{dS}{S} \qquad (4.2a)$$

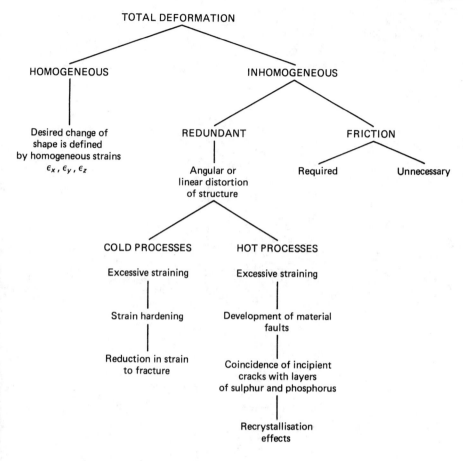

Figure 4.17   Components of total deformation

But, $A = \pi R^2$, and $dA = 2\pi R dR$, and $S = \pi R$ and so $dS = \pi dR$. On substituting these values into equation (4.2a), and accounting for the constancy of volume (equation (4.1a)), the strain increments become:

$$d\varepsilon_x = \frac{2dR}{R}; \ d\varepsilon_y = -\frac{dR}{R}; \ d\varepsilon_z = -\frac{dR}{R} \qquad (4.2b)$$

This approach is adopted in all metal-forming processes, but naturally the expressions obtained are generally more complex. *Table 4.4* gives the values of strain increments for a selected number of more common processes.

On integrating individual expressions of the type given by equation (4.2a) between the initial and final values of the considered sections, finite values of strains are obtained in natural-logarithmic form. Strains so evaluated are known as 'natural', 'logarithmic' or 'true' since, unlike engineering percentage strains, they define the actual deformation in any given section.

The true strains correspond to specific loading conditions which result in the presence of stresses. In the plastic regime, changes in the cross-sectional area, material thickness, etc., are so substantial that the concept of 'nominal' stress, based on the original dimensions becomes untenable and its use introduces considerable numerical error. All the stresses must be calculated on the basis of the current, actual dimensions.

Although the onset of plastic flow is usually the result of the interaction of different stress fields, it is quite possible to visualise this situation as arising from the application of a single force of sufficient magnitude to cause the material to flow. To have the desired physical effect, the stress associated with this hypothetical force must assume the same value as the yield or flow stress Y. This particular stress, known as the 'generalised stress' or 'equivalent stress', is often denoted by the symbol $\bar{\sigma}$. It is used in calculations concerned with the plastic regime of flow and, in terms of the principal stresses, is given by:

$$\bar{\sigma} = \frac{\sqrt{2}}{2}[(\sigma_1 - \sigma_2)^2 + (\sigma_2 - \sigma_3)^2 + (\sigma_3 - \sigma_1)^2]^{1/2} \qquad (4.3)$$

Similarly, in order to combine the effect of the individual strains—to be able to consider the total deformation of the specimen—the concept of the generalised or equivalent strain is introduced. In terms of homogeneous strains only, the homogeneous strain increment is given by:

$$d\varepsilon_H = \frac{\sqrt{2}}{3}[(d\varepsilon_x - d\varepsilon_y)^2 + (d\varepsilon_y - d\varepsilon_z)^2 + (d\varepsilon_z - d\varepsilon_x)^2]^{1/2}$$

$$(4.4a)$$

Table 4.4 Strain and strain-rate relationships

Process	Homogeneous strain increments	Generalised homogeneous strain	$\dot{\varepsilon}_m$
Rotary piercing	$d\varepsilon_x = -\dfrac{D_0 - 2h}{h(D_0 - h)}\, dh$ $d\varepsilon_y = -\dfrac{dh}{h}$ $d\varepsilon_z = -\dfrac{dh}{d_0 - h}$ $d\varepsilon_H = \dfrac{2\sqrt{3}}{3}\left[\dfrac{3h^2 - D_0 h + D_0^2}{h^2(D_0 - h)^2}\right]^{1/2} dh$	$\varepsilon_H = \ln\left[\left(1 - \dfrac{8BC}{(C - 3B)(C - B)}\right)^{1/\sqrt{3}}\left(\dfrac{D_0}{C - D_0 - \sqrt{3}B}\right)\right]^2$ $B = (2h_1 - D_0);\ C = (3h_1^2 - 3D_0 h_1 + D_0^2)^{1/2} + D_0$	$\dot{\varepsilon}_m = \dfrac{aV_0}{2\sqrt{3}[-b + (G + 4ah)^{1/2}]}\dfrac{2B}{(1 - B^2)}$ $+ \ln\dfrac{(1 + B)}{(1 - B)}$ $h = ax^2 + bx + c;\ B = \dfrac{2h - D_0}{(3h^2 - 3D_0 h + D_0^2)^{1/2}};$ $G = b^2 - 4ac$
Secondary rotary piercing	$d + h = \text{constant}$ $d\varepsilon_x = -\dfrac{dh}{h}$ $d\varepsilon_y = -\dfrac{dh}{h}$ $d\varepsilon_z = 0$	$\varepsilon_H = \dfrac{2\sqrt{3}}{3}\ln\left(\dfrac{h_0}{h_1}\right)$	—
Assel elongation	$d\varepsilon_y = -\dfrac{dh}{h}$	$\varepsilon_H = \ln\left\{\dfrac{(1 + B_0)(1 - B_n)}{(1 - B_0)(1 + B_n)}\right.$ $\left.\left[\dfrac{\left(1 - \dfrac{2}{\sqrt{3}}B_0\right)\left(1 + \dfrac{2}{\sqrt{3}}B_n\right)}{\left(1 + \dfrac{2}{\sqrt{3}}B_0\right)\left(1 - \dfrac{2}{\sqrt{3}}B_n\right)}\right]^{2/\sqrt{3}}\right\}$ $B = \dfrac{\sqrt{3}(2h + d)}{2(3h^2 + 3hd + d^2)^{1/2}}$	$\dot{\varepsilon}_m = (\beta \text{ or } \gamma \text{ or } \delta)\left\{\dfrac{2(B_n - B_0)(B_0 B_n + 1)}{(1 - B_0^2)(1 - B_n^2)}\right.$ $\left. + \ln\dfrac{(1 + B_n)(1 - B_0)}{(1 + B_0)(1 - B_n)}\right\}V_0$ $\beta = \dfrac{D_0^2 - d^2}{4\sqrt{3}\,d^2(x_n - x_0)};\ \gamma = \dfrac{2\sqrt{3}\,d^2[-b + (G + 4ah_n)^{1/2}]}{a(D_0^2 - d^2)};$ $\delta = \dfrac{(D_0 - h_0)h_0 \tan\alpha}{\sqrt{3}\,d^2(h_0 - h)};\ B = \dfrac{2h + d}{(3h^2 + 3hd + d^2)^{1/2}}$ a, b and G as in rotary piercing
Tube extrusion	$d\varepsilon_x = \dfrac{d + 2h}{(d + h)h}\, dh$		
Mandrel and plug tube drawing	$d\varepsilon_z = -\dfrac{dh}{d + h}$ $d\varepsilon_H = \dfrac{2}{\sqrt{3}}\left[\dfrac{3h^2 + 3hd + d^2}{(d + h)^2 h^2}\right]^{1/2} dh$	$B = \dfrac{\sqrt{3}(2h + d)}{2(3h^2 + 3hd + d^2)^{1/2}}$	
Tube sinking	$d\varepsilon_x = \dfrac{dD}{D - h_0}$ $d\varepsilon_y = -\dfrac{dh}{h} = 0$ $d\varepsilon_z = -\dfrac{dD}{D - h_0}$ $d\varepsilon_H = \dfrac{2}{\sqrt{3}}\left(\dfrac{dD}{D - h_0}\right)$	$\varepsilon_H = \dfrac{2\sqrt{3}}{3}\ln\left(\dfrac{D - h_0}{D_0 - h_0}\right)$	$\dot{\varepsilon}_m = \dfrac{4V_0 \tan\alpha}{3(D - h_0)}$
Rod extrusion	—	—	
Bar and wire drawing	—	—	$\dot{\varepsilon}_m = \dfrac{2(D_0 + D)V_0 \tan\alpha}{D^2}$

In many actual operations in which rotation of the axes of strains is either absent or negligibly small (such as in wire and tube drawing and extrusion, or strip rolling) and in which the strain path is clearly defined, equation (4.4a) can be integrated directly to give:

$$\varepsilon_H = \sqrt{\frac{2}{3}} \, (\varepsilon_x^2 + \varepsilon_y^2 + \varepsilon_z^2)^{1/2} \qquad (4.4b)$$

The final, generalised strain, including the redundant strain component, is given by:

$$\varepsilon_T = \sqrt{\frac{2}{3}} \, [\varepsilon_x^2 + \varepsilon_y^2 + \varepsilon_z^2 + \tfrac{1}{2}(\varepsilon_l^2 + \varepsilon_c^2 + \varepsilon_r^2)]^{1/2} \qquad (4.4c)$$

where ε_1, ε_r and ε_c are the inhomogeneous longitudinal, radial and circumferential strains, respectively. Since the shears produce only internal distortion of the structure and do not influence dimensional changes, the principle of constancy of volume does not apply to them and they can thus appear singly, in pairs, or all three may be present with the mathematical linkage missing. These strains must be assessed experimentally in each individual case. The values of the generalised, homogeneous strains for a number of the forming processes, are given in *Table 4.4*.

The relationship between the generalised, or flow stress and true, generalised strain is known as the 'constitutive equation' of the material and is characteristic of the material behaviour in specified working conditions. Equations of this type are determined experimentally, using standard laboratory test techniques (*Table 4.3*). Generally, the engineering alloys used in metal forming fall into five distinct groups according to their properties. In terms of their constitutive relations, they can be classified as:

(1) perfectly elastic (no plastic ductility);
(2) elastic, but linearly strain hardening in the plastic range (approximation to, say, ferrous alloys);
(3) elastic–perfectly plastic (a critical and constant value of stress produces flow, until fracture occurs);
(4) rigid–perfectly plastic (negligible elasticity); and
(5) rigid–linear strain hardening (negligible elasticity).

These properties are illustrated in *Figure 4.18*.

The usual, empirical equations representing the properties of an actual material are of the power-law type, and the functional relationship between the flow stress Y and the total, generalised strain is either in the form

$$Y = A\varepsilon^n \qquad (4.5)$$

or

$$Y = B(C + \varepsilon)^n \qquad (4.6)$$

where A, B and C are constants related to the material properties, and n ($0 \leqslant n \leqslant 1$) is the exponent indicating the strain-hardening characteristics.

In the condition of increasing strain rate, the response of the material is conditioned by the properties of the forming system (both in terms of geometry and kinematics) and the influence of frictional and internal shearing parameters. The effect of the rate of strain can be represented either by the instantaneous or the mean strain rate. Using the concept of generalised homogeneous strain and considering the time in which it takes place, the instantaneous rate ($\dot{\varepsilon}_H$) is given by

$$\dot{\varepsilon}_H = \frac{d\varepsilon_H}{dt} \qquad (4.7)$$

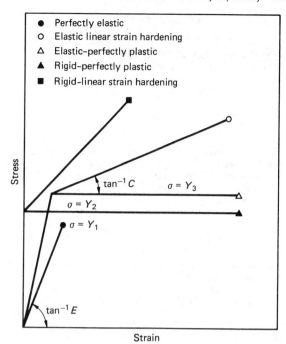

● Perfectly elastic
○ Elastic linear strain hardening
△ Elastic–perfectly plastic
▲ Rigid–perfectly plastic
■ Rigid–linear strain hardening

Figure 4.18 Idealised representation of basic stress–strain relationships

and the mean strain rate $\dot{\varepsilon}_m$ is given by

$$\dot{\varepsilon}_m = \frac{1}{t} \int_0^t \dot{\varepsilon}_H dt \qquad (4.8)$$

Since the strain rate is defined in the manner of equation (4.7), it does not reflect the variation in the velocity of the strained elements as they proceed along the pass. Furthermore, the mean strain rate (equation (4.8)) refers solely to the material element under consideration, but not to the volume of the material undergoing instantaneous deformation.

In metal forming, it may be more important to consider the strain rate prevailing in the whole zone of deformation, rather than just locally, since this will account for the variation in velocity. This objective is attained if the mean strain rate is referred to the physical boundaries of the pass, rather than to the time elapsed (t). In this case,

$$\dot{\varepsilon}_m = \frac{1}{x} \int_0^x \dot{\varepsilon}_H dt \qquad (4.9)$$

The values of the latter strain rate are quoted in *Table 4.4*.

4.3.2 Yielding and criteria of yielding

The elastic 'failure' or the onset of plasticity are determined for a given material by using an appropriate criterion of yielding which defines the value of the critical flow stress as a function of the total strain.

The simpler of the two criteria, and that most commonly used, is the maximum shear stress (or the Tresca criterion). This ignores the effect of the third principal stress and is concerned solely with the influence that the maximum possible stress difference has on the stability of the molecular planes.

The Tresca criterion is defined as:

$$| \sigma_3 - \sigma_1 | = Y(\varepsilon) = k(\varepsilon) \tag{4.10}$$

where k is the yield stress in shear.

The more general Huber–Mises criterion is based on the critical value of the shear strain energy that can be absorbed by the material within the elastic range of response, and is given by:

$$(\sigma_1 - \sigma_2)^2 + (\sigma_2 - \sigma_3)^2 + (\sigma_3 - \sigma_1)^2 = Y'(\varepsilon) = k'(\varepsilon) \tag{4.11}$$

The use of a mathematical function will account for the changes in the level of the yield stress, but the function itself may be difficult to handle in actual calculations. In recognition of this difficulty, a concept of the *mean* yield stress, operating over the working zone of the pass, can be used.

A mean yield stress Y_m is defined as:

$$Y_m = \frac{1}{\varepsilon} \int_0^\varepsilon Y(\varepsilon) d\varepsilon \tag{4.12}$$

where ε is the total, generalised strain of equations (4.5) and (4.6). Alternatively, the value can be obtained by employing the Hill and Tupper approach and utilising the information available in a plot of the type represented in *Figure 4.16*. When the translation of the actual stress–strain curve is effected, the values of the mean flow stresses, in addition to those of the component strains, can be determined by reference to the areas enclosed under the respective curves. Thus, in the absence of any redundant effect, the mean flow stress would be Y_H which is obtained by measuring the area OA′A and dividing it by ε_H (OA). In the presence of redundancy, the mean flow stress (Y_T) is defined as the area OB′B/ε_T, and the proportion of the stress induced by redundancy Y_R = (area AA′B′B)/ε_R. Where ε_R = AB.

4.3.3 Redundancy and geometry factors

The level of redundancy in an operation has to be assessed in order to provide a means of determining the efficiency of the process. This, as already indicated, depends on the geometry of the pass and the properties of the material.

For most metal-forming processes, when the total strain is plotted against the homogeneous deformation, the former increases at a higher rate than would be expected for an efficient forming process. The relationship is practically linear, but the ratio of the two quantities, for any given value of the homogeneous deformation, can be high. This type of relationship leads to the definition of the 'strain redundancy factor' ϕ:

$$\varepsilon_T = \phi \varepsilon_H \tag{4.13}$$

The factor is clearly a function of both the strain and the geometry of the pass and so can be defined further as:

$$\phi = \phi(\varepsilon, \Delta) \tag{4.14}$$

where Δ is the geometry factor.

Therefore the factor can be represented by both a linear function (a reasonable approximation for most bulk metal-forming processes) and, by a modified expression that involves the geometry factor. Hence,

$$\varepsilon_T = K_1 \varepsilon + K_2 \sin \alpha$$

or

$$\phi = K_1 + K_2 \frac{\sin \alpha}{\varepsilon} \tag{4.15}$$

and

$$\phi = C_1 + C_2 \Delta \tag{4.16}$$

where K_1, K_2, C_1 and C_2 are constants depending on the material and the frictional conditions.

Equation (4.15) is appropriate for drawing and extrusion. Similar expressions can be obtained for rotary and longitudinal rolling processes but, of necessity, can be, and often are, more complex.

The equivalence of the expressions given by equations (4.15) and (4.16) must be proved in each individual case and will depend on the somewhat arbitrary definition of the geometry factor.

The case of drawing represented in *Figure 4.11*, can again be taken as an example. On substituting equation (4.2b) into (4.4a), and integrating the latter within the limits of the initial and final diameters D_0 and d, the generalised homogeneous strain is obtained as:

$$\varepsilon_H = 2 \ln(D_0/d) \tag{4.17}$$

The geometry factor for a conical die is often defined as:

$$\Delta = \frac{D_0 + d}{D_0 - d} \sin \alpha \tag{4.18}$$

where α is the die semi-angle. Consequently, equation (4.16) in conjunction with equation (4.17) gives:

$$\phi = C_1 + C_2 \left(\frac{e^{\varepsilon/2} + 1}{e^{\varepsilon/2} - 1} \right) \sin \alpha$$

$$= C_1 + C_2 \left(\frac{1}{\tan h (\varepsilon/4)} \right) \sin \alpha$$

For small die angles and normal bar-drawing practice, expansion of this function gives

$$\tan h(\varepsilon/4) \simeq \varepsilon/4$$

and so

$$\phi = C_1 + 4C_2 \frac{\sin \alpha}{\varepsilon_H} \tag{4.19}$$

It is clear that equations (4.15) and (4.19) are equivalent, with $K_1 = C_1$, $K_2 = 4C_2$, and $\Delta = (\sin \alpha)/\varepsilon_H$.

Similar arguments are applied when considering other metal-forming processes.

The strain redundancy factor is the more often used in practice since it is based on measurable quantities. However, occasionally a *stress* redundancy factor Φ is introduced. This is defined as:

$$\sigma = \Phi \sigma_H \tag{4.20}$$

where σ and σ_H represent the operational stress levels referred to the total and homogeneous force requirements, respectively. Since stresses can only be calculated, their numerical assessment depends on the type of mathematical model or theory selected and will, therefore, be more arbitrary than strain values. The two factors are not numerically equivalent unless, of course, there is no measurable redundancy in the system in which case their value amounts to unity.

Definitions of geometry factors are somewhat arbitrary but, even so, they take into account the influence that the individual process parameters are likely to have on the conduct of the operation and the quality of the processed product. A selection of the expressions defining the individual geometry factors is given in *Table 4.5*.

Table 4.5 Geometry factors in basic processes

Process	Δ
Rotary piercing*	$\dfrac{1}{2}\left(\dfrac{D_0/2 + h}{D_0/2 - h}\right)\cot\alpha$
Wire and bar drawing and extrusion	$\dfrac{D_0 + d}{D_0 - d}\sin\alpha;\ \text{or}\ \dfrac{1}{2}\left(\dfrac{D_0 + d}{D_0 - d}\right)\left(\dfrac{1 - \cos\alpha}{\sin\alpha}\right)$
Tube drawing	
Sinking	$\dfrac{4h_0[(D_0 + d) - h_0]}{(D_0 + d)(D_0 - d)}\sin\alpha$
Plug or mandrel	$\dfrac{1}{2}\left(\dfrac{h_0 + h}{h_0 - h}\right)\sin\alpha$
Tube extrusion†	$\dfrac{1}{2}\left(\dfrac{h_0 + h}{h_0 - h}\right)\left(1 - \dfrac{1}{b}\right)\sin\alpha_e$
	$\sin\alpha_e = \left[1 - \dfrac{1}{24a^2\sin^2\alpha}\right]\sin\alpha$

* α is the feed angle as in *Figure 4.7*.
† $a = \rho(h_0 - h)$; $b = (D_0 + d)/(h_0 + h)$; ρ is the radius of die curvature.

4.3.4 Stress–strain relationships

Unlike the elastic regime, the relationship between stress and strain is not necessarily unique once the yield point of the material has been exceeded. The two parameters are firmly linked at any instant, but the nature of this can change and does not therefore constitute a characteristic material property equivalent to, say, the Young's, rigidity or bulk moduli which define the conditions in the elastic range.

When the effect of elastic strain is negligible, as in the forming of metals, the general relationship between the two parameters is governed by the Levy–Mises rule of low, defined as:

$$d\varepsilon_1^{\text{p}} = \lambda[\sigma_1 - \nu(\sigma_2 + \sigma_3)]$$
$$d\varepsilon_2^{\text{p}} = \lambda[\sigma_2 - \nu(\sigma_3 + \sigma_1)] \quad\quad (4.21a)$$
$$d\varepsilon_3^{\text{p}} = \lambda[\sigma_3 - \nu(\sigma_1 + \sigma_2)]$$

where the superscript 'p' indicates purely plastic conditions, subscripts 1, 2 and 3 refer to the mutually perpendicular, principal directions, and $\lambda = d\bar{\varepsilon}/\sigma$ is the *instantaneous* constant. This brings in both the differential nature of strain and the generalised stress and strain—as defined by the appropriate constitutive equation. Note that, again unlike the elastic range, an increment of strain is proportional to finite stress.

Since the material is incompressible, Poisson's ratio ν has a single value of 0.5, irrespective of the metal considered.

By eliminating λ from equation (4.21a), the following standard expression is obtained:

$$\frac{d\varepsilon_1^{\text{p}}}{\sigma_1 - \frac{1}{2}(\sigma_2 + \sigma_3)}\ \frac{d\varepsilon_2^{\text{p}}}{\sigma_2 - \frac{1}{2}(\sigma_3 + \sigma_1)}\ \frac{d\varepsilon_3^{\text{p}}}{\sigma_3 - \frac{1}{2}(\sigma_1 + \sigma_2)} = \lambda$$

$$= \frac{d\bar{\varepsilon}}{\sigma} \quad\quad (4.21b)$$

The relationship between and interdependence of the stresses are defined by the selected criterion of yielding (equations (4.10) and (4.11)) or by a suitable mathematical model of the operation.

4.3.5 Mathematical modelling of metal-forming processes

A number of analytical methods have been developed for assessing the various parameters of metal-forming processes, showing a varying degree of accuracy and complexity. All these methods aim ultimately at relating the required deformation of the specimen to the necessary external deforming agent that must be supplied by the plant.

At present, five basic methods of analysis are usually employed. In chronological order of development these are:

(1) the equilibrium technique,
(2) the slip-line approach,
(3) the upper-bound method,
(4) the visioplasticity technique, and
(5) the finite-element method.

The equilibrium method is based on the stress field that develops in the processed material, whereas the other four are associated with phenomena related to metal flow. Irrespective of the method of analysis adopted, the basic characteristics of the plastic state of the material (i.e. the condition of constancy of volume, the yield criterion and the functional yield, flow or constitutive stress equation) will have to be assumed.

4.3.5.1 The equilibrium technique

This approach, sometimes referred to as the 'stress' or 'slab' technique, consists basically of isolating a representative volume element in the body of the material undergoing plastic deformation, and observing its behaviour as it moves along the working zone of the pass. Since in a normal physical situation the element does not cease to form an integral part of the whole of the material body, it must clearly remain in the state of force equilibrium throughout its period of deformation. The behaviour of the element, reflecting that of the whole of the worked specimen, can therefore be analysed by considering the equilibrium of forces acting on it at any instant of deformation.

A volume element of a material undergoing axisymmetric-type deformation is shown in *Figure 4.19*. The equation of equilibrium of forces in one direction is

$$\frac{\sigma_r - \sigma_\theta}{r} + \frac{\partial\sigma_r}{\partial r} - \frac{2\tau}{t} = 0 \quad\quad (4.22)$$

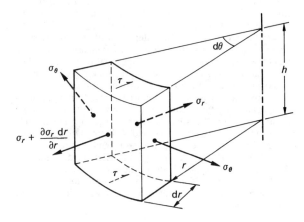

Figure 4.19 An element of material in axisymmetric equilibrium analysis

Basically, the equilibrium method requires that conditions of force equilibrium in three directions be established, i.e. $\Sigma F_x = \Sigma F_y = \Sigma F_z = 0$, or in, say, cylindrical co-ordinates that $\Sigma F_r = \Sigma F_\theta = \Sigma F_z = 0$. The analysis usually leads to the formulation of one or more differential equations involving the stress system produced in the body of the material. The traction or friction effect, originating at the tool–work piece interface, is normally allowed for by introducing a friction term incorporating a value of the coefficient of friction. The relationship between the stresses produced is established by means of a suitable criterion of yielding, ultimately related to the tensile/compressive or shear yield stress of the material. Finite values of the operative stresses are obtained by integrating differential equations between the conditions obtaining at the entry to and exit from the working zone of the pass or, if this is complex in nature, by splitting the pass into a number of interconnected zones and summing the effects in the individual zones.

The method is only approximate and underestimates the magnitudes of the forces involved by ignoring the effects of redundancy and of the pattern of flow and, therefore, by taking an oversimplified view of the mechanics of the process in question. Since, however, the analysis involved is usually reasonably straightforward, the technique is widely used whenever approximate solutions are acceptable. It is, for instance particularly suitable for wire and tube drawing, hot and cold rolling of strip and sheet, and also gives an acceptable degree of approximation for rotary piercing and elongation of seamless tubing.

If the effects of redundancy in any given situation are either known or assessable, the accuracy of force predictions given by the equilibrium approach can be improved. In addition, a separation of the homogenous effect from the inhomogeneous one can be carried out, and the respective components of work done in changing the shape and in distorting the structure can be determined. A deforming 'efficiency' of the process can thus be established.

To improve the accuracy of calculations a modification of the value of the operative stress resulting from the integration of the equation of equilibrium can be made, either by introducing the stress redundancy factor Φ (equation (4.20)) or by modifying the value of the yield stress. In the latter case, equation (4.13) (by giving the value of the total strain) will either provide the limits of integration of the yield-stress function or will give the mean operative value of the stress Y_m. Either of these can be introduced into the analysis when a criterion of yielding has been selected.

An estimate of the deforming 'efficiency' of the operation can be made in the following manner. The total operative stress, and therefore by implication the magnitude of the total deforming agent, can be calculated from the final expression given by the analysis. If, then, the friction term is omitted from the equation of equilibrium, the equation will define the stress of total deformation only. To separate the effect of the homogeneous deformation from that of the redundant one, the equation of equilibrium for frictionless forming is used once again. This time, however, the 'lower' value of the yield is introduced, i.e. the value that corresponds to the magnitude of the homogeneous strain only and is obtained from equation (4.12). The two expressions give the magnitudes of the loads necessary to produce the total homogeneous deformation. The numerical difference between the two will, of course, give the load of the redundant deformation. Work done or energy dissipated in effecting these physical changes can, therefore, be computed.

4.3.5.2 The slip-line analysis

The technique used here is based on the physical observation that plastic flow occurs predominantly as a result of microscopic slip (on an atomic scale) along crystallographic planes. A polycrystalline metal will normally contain a high number of planes of close-packed atoms, along which slip is likely to occur, a sufficient number of these being orientated in such a way that, on a macroscopic scale, the general direction of slip will coincide with that of the maximum shear stress. The directions of the maximum shear stresses thus become characteristic properties of the system and, when established, lead to a numerical assessment of stress distribution throughout the deforming region including, of course, the boundaries on which loads exerted by tools are acting. Calculations of such loads becomes possible.

Detailed descriptions of the theoretical bases of the technique, its scope and limitations, the practical methods of application to a variety of problems and the derivation of the appropriate equations can be found in the standard textbooks on the subject (see Bibliography). It is sufficient to note here that this particular method of analysis applies rigorously only to plane strain deformations of a rigid, perfectly plastic isotropic solid, and that, unless artificially modified, it cannot account for strain hardening, strain rate, or frictional and redundancy effects.

The method forms a useful tool in the general field of metal forming, mainly by giving some insight into the pattern of flow of the material. The basic difference between a slip-line approach and that of, say, equilibrium-type solution lies, however, in the fact that a specific slip-line field has to be postulated a priori, and that its feasibility must then be checked against the existing experimental data. It is clear that in the general course of events the method cannot give a unique solution, and that a number of fields could be proposed in a specific case that would fit the specified system of boundary conditions equally well. In fact, a criterion of minimum work must be invoked to help select the solution that is most likely to be the probable one.

To obtain a solution, families of curvilinear or straight lines are constructed that intersect orthogonally and correspond to the directions of maximum and minimum constant shear lines. Static equilibrium, the yield condition and the pattern of flow in the plastic zone must be satisfied by the assumed network of these lines. The mutually perpendicular lines in any x–y plane are designated as α and β.

Deformation in a plane strain situation is that of pure shear and, consequently, it is normally assumed at this point that the deformation is the result of pure shear stress. Yielding will occur when the maximum shear stress reaches the appropriate value k, which is limiting for the given material. Since no linear strain can be experienced along the slip lines, an element formed between two α and β lines, say that shown in *Figure 4.20*, can only distort and will neither extend nor contract. If σ_1, σ_2 and σ_3 are principal stresses, then in plane strain $\sigma_2 = \frac{1}{2}(\sigma_1 + \sigma_3)$. This can be expressed in terms of hydrostatic compressive pressure $(-p)$ which is appropriate in most metal-forming operations as $\sigma_2 = -p$. The principal stresses are then given by:

$$\sigma_1 = -p - k; \; \sigma_2 = -p; \; \sigma_3 = -p + k \tag{4.23}$$

The equations for the equilibrium of forces acting on the element shown in *Figure 4.20* are:

$$\frac{\partial p}{\partial \alpha} + \frac{2k}{r_\alpha} = 0 \text{ (along an } \alpha \text{ line)}$$

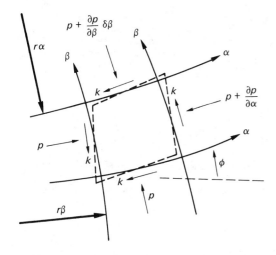

Figure 4.20 A material element between two slip lines

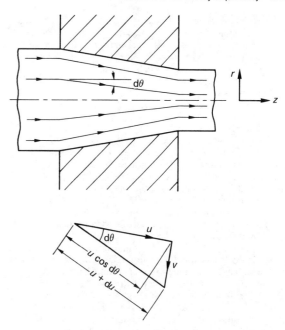

Figure 4.21 Diagrammatic representation of the velocity field of a particle in tube extrusion

and

$$\frac{\partial p}{\partial \beta} - \frac{2k}{r_\beta} = 0 \text{ (along a } \beta \text{ line)}$$

Expressing the radii of curvature in terms of θ, it is found that $1/r_\alpha = \partial\theta/\partial\alpha$ and $1/r_\beta = \partial\theta/\partial\beta$, and therefore on integration

$$p + 2k\theta = C_1 \text{ (along an } \alpha \text{ line)} \tag{4.24}$$

$$p - 2k\theta = C_2 \text{ (along a } \beta \text{ line)} \tag{4.25}$$

For a hardening material, the yield stress in shear k remains constant along a slip line, but the pressure p may vary. The stress configuration can be determined at any point on the deforming material if the magnitude of p and the direction of k are known. The latter is defined by the direction of the considered slip line, whereas the former can be estimated by considering the changes in p produced by an angular rotation of the slip line between two points in the field. The actual finite value of p can be established from knowledge of the appropriate boundary conditions.

As already indicated, the validity of a slip-line field which satisfies the stress boundary conditions must be verified by means of continuity equations. In other words, a check must be made on whether a velocity field that satisfies equilibrium conditions actually exists.

In the absence of any linear changes in the dimensions of an element along a slip line, there can be no change in the direction of shear between two points on a slip line; the velocity component perpendicular to the line will vary from point to point. Denoting the velocity component parallel to the α line as u, and that parallel to the β line as v (*Figure 4.21*) it is found that the following equations hold:

$$du - vd\theta = 0 \text{ (along an } \alpha \text{ line)} \tag{4.26}$$

$$dv + ud\theta = 0 \text{ (along a } \beta \text{ line)} \tag{4.27}$$

If a material point within the deforming body suffers a change in the magnitude of its absolute velocity on crossing a slip line, the conditions of continuity demand that the components of velocity normal to the slip line be the same. However, no such stipulation is made with regard to the tangential velocity components, i.e. velocity discontinuities are possible in directions tangential to the slip lines. These conditions, used in

conjunction with equations (4.26) and (4.27), provide the required means of testing the feasibility of the proposed slip-line field. It should, however, be remembered that a good approximation to this set of conditions can be expected only within the fully developed plastic region, and not when crossing from rigid to plastic regions of deformation where the change takes place over a finite transition zone.

To establish a slip-line field, the stress and slip boundary conditions have to be clearly specified. The following four possibilities exist:

(1) free surface,
(2) frictionless interface,
(3) Coulomb friction condition (μ is constant), and
(4) perfectly rough surface.

Free surface In the absence of a normal stress at the surface, or with $\sigma_3 = 0$, $p = k$ (equation (4.23)) if σ_1 is compressive, and $\sigma_1 = -2k$. If σ_1 is tensile, $p = -k$ and $\sigma_1 = k$. Slip lines meet at the free surface at 45°.

Frictionless interface With the resultant shear on the surface being zero, slip lines meet again at 45°, but in the presence of a normal stress $\sigma_3 \neq 0$ and, consequently, $p \neq k$.

Correction for Coulomb friction Taking the vertical pressure at the point of contact as q, it is clear that the frictional effect μq on the surface must be balanced by stress $k \cos 2\theta$, related to the plane of maximum shear or the slip line inclined at angle θ to the interface, and consequently $k \cos 2\theta = \mu q$. The problem that immediately arises in this connection is that of knowing the value of q. Since q cannot be determined before the field has been drawn, it is necessary to adopt an iterative technique of calculation by taking the first value of q as that obtained from the consideration of a frictionless surface.

Perfectly rough surface This normally represents the condition of sticking friction or absence of interfacial movement. The frictional stress τ is equal to k, and is independent of the normal stress. The necessary slip-line configuration gives one line normal to and the other tangential to the surface.

The use of the slip-line method of analysis in the examination of problems involving large redundancies presents certain difficulties. Although the effect of friction can be incorporated and a fair degree of accuracy can be attained, the problem of strain hardening is more difficult. Hencky's equations ((4.24)–(4.27)) can be modified by the addition of another term to account for hardening, but the method is cumbersome and has so far shown promise only for machining operations. The concept of the mean yield stress is, therefore, again introduced, and an equation of the type given by (4.12) is used with substitution of $2k$ for $Y(\varepsilon)$.

Effects of redundancy are usually incorporated by correcting the value of the operative stress, determined from the field, by means of factor Φ.

The most serious limitation of the analysis, of course, is its dependence for accuracy on the conditions of plane strain. Attempts have been made to extend the technique to axisymmetric problems, and have met with some limited success in application to the extrusion of bars. A similarity between the results obtained from slip-line-theory considerations, applied to plane strain extrusion through square dies, and experimentation with axisymmetric extrusion (see Johnson in the Bibliography) led Johnson to propose a semi-empirical method of solution based on the slip-line-field approach. This method is discussed later (see Section 4.4) in the context of extrusion operations. It should be noted here that a correction for hardening was made on the assumption that the total equivalent strain is independent of hardening characteristics.

4.3.5.3 The upper-bound method

Within the scope of its assumptions, the slip-line-field technique produces an exact solution and, whilst this seems desirable, difficulties are often encountered in finding suitable slip-line fields for new problems. In addition, axisymmetric problems cannot be accommodated. The real power of the technique lies in the derivation of general solutions to plane strain problems.

Limit analysis has been proposed as a means of obtaining solutions to many otherwise intractable metal-forming problems. In its complete form, this approach seeks to establish two expressions for forming loads, one (the so-called 'lower bound') which is definitely an underestimate. Since it is generally sufficient to ensure that the calculated load for a forming operation is indeed sufficient to perform the process, then interest naturally centres on the upper-bound technique. If used correctly, realistic overestimates are obtained.

Upper-bound solutions involve consideration of the conditions which must be fulfilled by the strain increments in a plastically deforming medium and are not concerned with stress equilibrium. An important concept involved is that of a kinematically admissible velocity field. This is a distribution of generalised particle velocities which is kinematically compatible with itself and with the externally imposed velocities at boundaries. The total amount of work done by the externally applied forces in a metal-forming operation is used to overcome the resistance to plastic deformation and the frictional resistance to relative motion at the tool–work piece interface. These two components can be calculated from a velocity field and the upper-bound theorem that states that the kinematically admissible velocity field which minimises the work done is the actual velocity field. From such a field the actual work done can thus be derived, although in practice the actual field may never be completely attained. When formulating a solution the general procedure is to divide the deformation zone into one or more assumed zones throughout each of which the velocity is continuous. In adajcent zones a different velocity distribution may exist, whilst across these interfaces and at the tool–work piece interfaces a tangential velocity discontinuity may occur. The existence of velocity distributions in each zone implies that plastic strain and, therefore, energy must be dissipated:

$$P_1 = \int \bar{\sigma} \dot{\bar{\varepsilon}} \, dV \tag{4.28}$$

where V is the volume of the deforming metal. P_1 is often referred to as the 'power of deformation'. Similarly, energy must be dissipated when material is sheared across velocity discontinuities. This term, known as the 'shear power', is given by:

$$P_2 = \int k \, | \dot{s} | \, dA \tag{4.29}$$

where A is the area over which the velocity discontinuity of magnitude $\dot{s}$ occurs.

Energy dissipated due to frictional resistance along tool–work piece interfaces can be accounted for in a similar manner. A tangential velocity discontinuity $\dot{s}$ compatible with a shear stress of maximum value k for perfectly rough interfaces will occur. For Coulomb friction, a proportion mk of the shear stress can be used. Thus the friction power p_3 is given by:

$$p_3 = \int \text{c}k \, | \dot{s} | \, dA \tag{4.30}$$

The upper-bound theorem can be stated as:

$$P = P_1 + P_2 + P_3 \tag{4.31}$$

where P is the total power required to perform the operation.

Frequently, a characteristic parameter of the velocity field is retained as a variable when calculating the right-hand side of equation (4.31). This parameter can then be optimised to minimise the total power.

4.3.5.4 The visioplasticity technique

The visioplasticity technique makes use of experimental observations of flow patterns and provides a means of deriving complete solutions. However, solutions are obtained after the experimental data have been obtained. The technique cannot therefore be used to predict the likely behaviour of the processed medium, but it is very useful in a detailed analysis of the distribution of stress, strain and strain rate in any section within the deforming zone of a plastic body. Furthermore, because the analysis is based on observed experimental fact and does not, therefore, require any assumption other than the usual one regarding the stress–strain relationship, it gives an exact solution and thus can be used as a 'standard' to assess the accuracy of results obtained by other means.

The technique is based on the examination of a velocity field developed incrementally within the deforming body. In a general axisymmetric case a marking grid pattern is imprinted on the meridian plane of a cylindrical work piece undergoing a metal-forming operation and the latter is conducted by imposing on the material a unit strain at a time. A record of the change in the grid, after an increment in strain has been imparted, will give a sufficiently clear picture of the instananeous pattern of flow.

As an example of the determination of stress and strain-rate fields, consider *Figure 4.21*. If V is a known or measurable instantaneous velocity of a particle at A, whose position is defined by co-ordinates r, θ and z, and u and v are the components of V in directions z and r, respectively, then plots of u and v against, in turn, r and z can be constructed. Strain rates at point A, and any other point in the deforming material, can then be determined from the slopes of these curves:

$$\dot{\varepsilon}_z = \frac{\partial u}{\partial z}; \ \dot{\varepsilon}_r = \frac{\partial v}{\partial r}; \ \dot{\gamma}_{zr} = \frac{\partial u}{\partial r} + \frac{\partial v}{\partial z} \tag{4.32}$$

If it is assumed that the material is homogenous and isotropic, the directions of the principal stresses are given by:

$$\tan 2\alpha_z = \frac{\dot{\gamma}_{zr}}{\dot{\varepsilon}_z - \dot{\varepsilon}_r}$$

where α_z is the angle between the z axis and the direction of a principal stress.

The strain-rate component $\dot{\varepsilon}_\theta$ can be found from the condition of incompressibility (constancy of volume):

$$\dot{\varepsilon}_z + \dot{\varepsilon}_r + \dot{\varepsilon}_\theta = 0$$

The effective (generalised) strain rate is given by

$$\dot{\varepsilon} = \frac{\sqrt{2}}{3} \left[(\dot{\varepsilon}_z - \dot{\varepsilon}_r)^2 + (\dot{\varepsilon}_r - \dot{\varepsilon}_\theta)^2 + (\dot{\varepsilon}_\theta - \dot{\varepsilon}_z)^2 \right.$$
$$\left. + \frac{3}{2} \dot{\gamma}_{zr}^2 \right]^{1/2} \tag{4.33}$$

and the generalised stress by

$$\bar{\sigma} = \frac{1}{\sqrt{2}} \left[(\sigma_z - \sigma_r)^2 + (\sigma_r - \sigma_\theta)^2 + (\sigma_\theta - \sigma_z)^2 \right.$$
$$\left. + 3\tau_{zr}^2 \right]^{1/2} \tag{4.34}$$

In its final form, the expression for σ_z, when the velocity field upstream is known, is given by:

$$\sigma_z = \frac{2}{3} \int_{r1}^{r} \left[\frac{\partial}{\partial r} \left(\frac{\dot{\varepsilon}_z - \dot{\varepsilon}_r}{\dot{\varepsilon}/\bar{\sigma}} \right) \left(\frac{\dot{\varepsilon}_r - \dot{\varepsilon}_\theta}{r\dot{\varepsilon}/\bar{\sigma}} \right) - \frac{1}{2} \frac{\partial}{\partial z} \left(\frac{\dot{\gamma}_{zr}}{\dot{\varepsilon}/\bar{\sigma}} \right) \right] dr$$
$$- \frac{1}{3} \int_{z0}^{z} \left[\frac{\partial}{\partial r} \left(\frac{\dot{\gamma}_{rz}}{\dot{\varepsilon}/\bar{\sigma}} \right) + \frac{\dot{\gamma}_{rz}}{r\dot{\varepsilon}/\bar{\sigma}} \right] dz + \sigma_{z0} \tag{4.35}$$

where σ_{z0} is a known stress related to the conditions obtaining at the entry to the working zone of the pass.

It is clear that, because of its complexity, equation (4.35) must be integrated either numerically or graphically. In fact, the use of a computer ensures a high degree of accuracy. Equation (4.35) is, of course, easily reducible to conditions of plane strain if these become characteristic of the operation considered.

Although the visioplasticity method gives very satisfactory results, it has been applied mainly to various extrusion problems. This is due partly to the labour involved in the measurements necessary to establish various relationships, and partly to the actual integration of the final expression.

4.3.5.5 Finite-element method

With the wide availability of powerful computers, a new tool in the analysis of metal-forming processes is offered by the finite-element method (FEM). This requires that the deformation zone be divided into a number of elements that form a mesh and are connected at a specified number of nodal points.

When this has been done, and the assumed mesh pattern has influenced the outcome of the exercise, the velocity

distribution for each element is assumed. The unknown velocity vectors form part of a set of simultaneous equations which are solved to give the velocity distributions and the associated stresses actually present.

Both frictional and material property elements of the working system can be accommodated, but boundary conditions—pertaining to real physical situations—must be carefully assessed *a priori*.

Considerable computing time is required and each individual process must be treated in isolation. Consequently, no general solution can be offered and details of the analysis should be obtained from specialised literature (see Bibliography).

4.3.6 Simulation of metal flow

The development of various techniques of analysis of metal flow and, in particular, of controlled tool design, has led to considerable improvements in the performance of metal-forming plant. These improvements, as already indicated, are associated with the reduction in the incidence and magnitude of inhomogenous strains in the given operation. The resulting reduction in the severity of the process leads to an increase in the output and to an improvement in the quality of the product. The success of an attempt at optimising tool design can, therefore, be assessed on the basis of a comparison between the levels of the inhomogeneous deformation.

In many instances, however, the unavailability of industrial plant for long term experimental work on the effects of varying tool profiles makes it necessary to employ model materials in laboratory conditions. Internally marked specimens (see *Figure 4.22*) can be easily made using a model material and can then be processed. The development of the flow pattern can be studied by stopping the operation before it is completed, withdrawing the partially processed product,

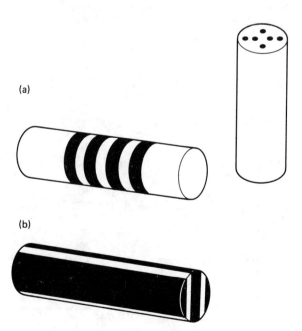

(a)

(b)

Figure 4.22 Internally marked wax billets: (a) billets for the investigation of longitudinal and circumferential shearing strains; (b) a different, more simple way of marking and sectioning the product within the range of the working pass.

The basic requirement of a model material is that its mechanical behaviour should resemble as closely as possible that of the material which it is used to simulate. In general, no model material will comply completely with this requirement, and consequently the observed behaviour must be interpreted bearing in mind the differences existing between the model and the prototype.

If only qualitative analysis of the problem is undertaken, the discrepancies between the two materials are of less importance since they do not affect the observed pattern of flow but only the relative magnitudes of the measured quantities. In the quantitative analysis, however, it is essential to be able to 'translate' the meaning of the measurements made on the model material in terms of the properties of the prototype of the simulated material.

The magnitude of the divergence between the behaviour of a model and a prototype medium depends on the difference between the geometrical shapes of the basic true stress–strain curves, on the susceptibility of the considered materials to the effects of strain rate and/or temperature, on frictional effects, and on the work-hardening function.

For a long time, Plasticine and lead have been used as the most common model materials. It is only fairly recently that waxes and clays have been added to the list, and the possibility of using aluminium and copper has been investigated.

One of the difficulties arising from the use of non-metallic materials is their usually high degree of thixotropy. In this respect, Plasticine is less reliable since the amount of inbuilt thixotropy in the original material is unknown and the cumulative effect of further processing is impossible to determine. Waxes, however, can be standardised easily by melting and recasting.

The susceptibility of non-metallic materials to strain rates is often underestimated and considerable error may be introduced in an analysis which is based on an incorrectly selected stress–strain curve. Examples of the dependability of non-metallic materials on the conditions of testing are given in *Figures 4.23* to *4.27*.

In the case of, say, Plasticine (which is a mixture of oil and a filler), a slight change in the contents will be reflected in the different stress–strain curves (*Figure 4.23*). Equally, chemical changes produced by the dye, and the temperature level used during the test will influence the characteristic properties of the material (*Figure 4.24*).

A mixture of waxes (paraffin, ceresin, carnauba and beeswax) that has satisfactory frictional and strength properties and can simulate rotary rolling processes (which proceed in the condition of shear) can be tested in torsion. This will establish the basic properties of the mixture. In turn, these properties can be converted to the effective tensile or compressive stress, and the stress–strain curves can thus be obtained. Matching of such curves with those of the prototype or 'real' material will provide a means of selecting the correct conditions of the model.

Figure 4.25 shows the different amounts of deformation (in shear) imposed on a number of wax specimens. Shear stress–strain curves for these specimens (obtained at different shear strain rates) are shown in *Figure 4.26*. Finally, these relationships converted to the effective, tensile stress–strain quantities are given in *Figure 4.27*.

Although the incidence and pattern of the redundant deformation is independent of the material, its magnitude does depend on the constitutive relationships. Therefore the difference in response of the various materials to strain rate, temperature and frictional effects will be responsible for either a magnification of or a reduction in the effect of the shearing strain present in a given operation. A comparison between the stress–strain curves of a number of materials (*Figure 4.28*) indicates that, in general, this effect, as shown by either hot or cold wax, is likely to be exaggerated when compared with the actual effects present in the prototype. To achieve a realistic interpretation of results when using a model material which is as near as possible to the prototype in its characteristics, its stress–strain curve is matched with that of the prototype in terms of strain rate.

A convenient method of converting measurements made on the model material to the prototype has been developed and can be explained with reference to *Figure 4.29*. In this example, wax is regarded as the model (M) and lead as the prototype (P), since their respective stress–strain curves show a sufficient degree of geometrical difference to illustrate the approach. OE is the value of homogeneous strain ε_H and Y_{HP} and Y_{HM} are the mean yield stresses in the prototype and model, respectively. $Y_{HP} = OACE/OE$ and $Y_{HM} = OBDE/OE$. The inclusion of inhomogeneous strains will alter the values of the respective generalised strains. ε_T will be different for each material. Let OF and OG be the strains in the prototype and the model, respectively. The mean yield stress will then be given by $Y_{TP} = OAC'F/OF$ and $Y_{TM} = OBD'G/OG$. Provided that the stress–strain curves of the two materials are geometrically similar, the following approximate relationships will be true:

$$(Y_{TP}/Y_{HP}) \approx (Y_{TM}/Y_{HM}) \tag{4.36}$$

Once the mean yield stress, corresponding to the total strain, has been determined for the model material, the value for the prototype can be estimated from an equation similar to equation (4.36), since Y_{HP} can be found directly from the appropriate stress–strain curve.

For the sake of both convenience and clarity, it can be assumed that the mean yield stress of a material is represented by two components: one that corresponds to the homogeneous deformation and the other that is induced by the presence of the inhomogeneous deformation. Using this assumption, it becomes possible quantitatively to convert inhomogeneous

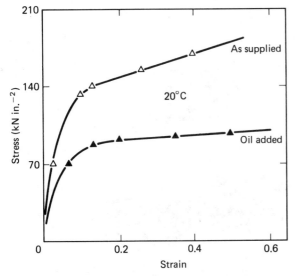

Figure 4.23 The effect of the addition of oil on the properties of Plasticine

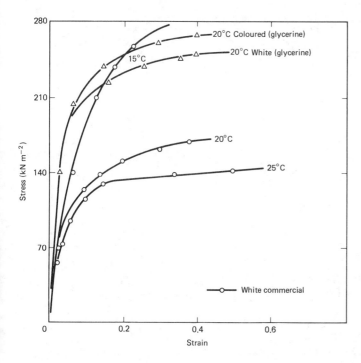

Figure 4.24 The combined effects of dying and temperature on the properties of Plasticine

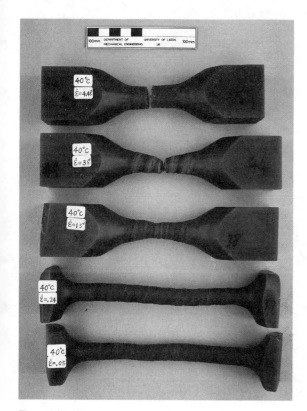

Figure 4.25 Effective strain effects in wax subjected to torsion testing at 40°C

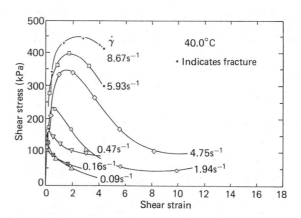

Figure 4.26 Shear stress–strain curves for the wax specimens shown in *Figure 4.25* at different shear strain rates

strains in the model to those in the prototype. Such a concept is, from the point of view of physical reality, purely artificial, but it provides a means of accounting directly for the effects of these two types of deformation.

The inhomogeneous component of the yield stress is given by $Y_{IP} = ECC'F/EF$ and also $Y_{IM} = EDD'G/EG$. In general, for a material having a stress–strain curve that does not vary considerably with strain, the following relationship holds:

$$(Y_I/Y_T) = m(Y_T/Y_H) \tag{4.37}$$

where m is a correction factor determined experimentally and used if greater numerical accuracy is required. In the ordinary range of materials the value of m varies from 5% to 11%.

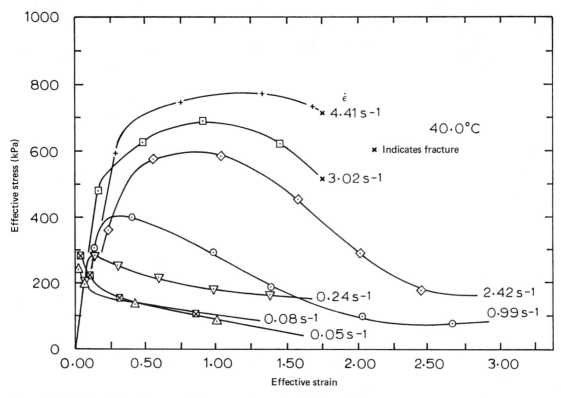

Figure 4.27 Effective stress–strain curves for the wax specimens shown in *Figure 4.25*

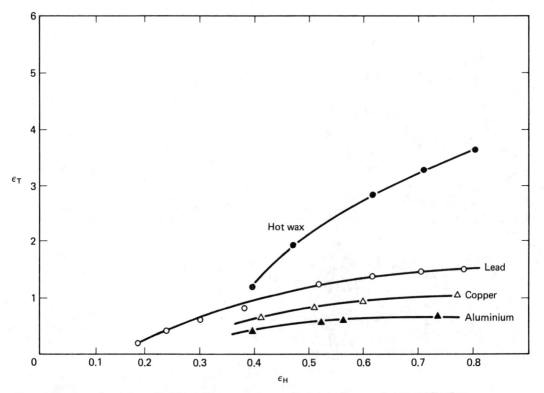

Figure 4.28 Comparison between the total and homogeneous generalised strains in wax and some metallic alloys

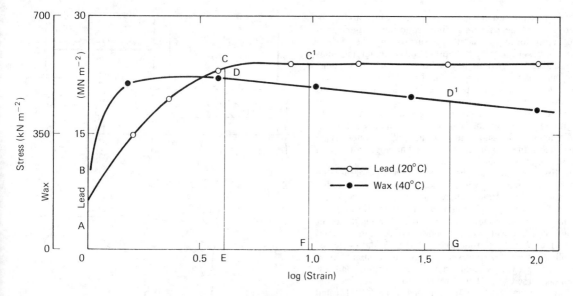

Figure 4.29 Basic and particular stress–strain curves for a strain-hardening material

Since Y_T for a material can be determined from equation (4.36), Y_I is fully defined by equation (4.37).

The convertibility of measurements made on a model material, under certain working conditions, to those likely to occur in a prototype processed under the same set of conditions can be made using equations (4.4b) and (4.4c), then

$$\varepsilon_T = \sqrt{2}[3\varepsilon_H^2/2 + \tfrac{1}{2}(\varepsilon_{xy}^2 + \varepsilon_{zx}^2 + \varepsilon_{yz}^2)]^{1/2}\sqrt{3} \qquad (4.38)$$

To determine the value of the expression in paranthesis (which represents inhomogeneity) for the prototype, the value of ε_T must be evaluated. *Figure 4.16* shows that $\varepsilon_T = \varepsilon_H + \varepsilon_I$. This is numerically equivalent, in this case, to OE for both materials. The value of ε_I is, for instance, EF for the prototype and EG for the model. In general, therefore, if ε_I is measured or calculated, then if ε_H is known the total strain can be ascertained and the true value of the total inhomogeneous strain can be calculated using equation (4.38).

For any material $Y_T/Y_H = W_T/\varepsilon Y_H$ and the total work done per unit volume $W_T = W_H + W_I$. Also, in the general case, $W = Y\varepsilon$ and so

$$\varepsilon_1 = \varepsilon_H(Y_H - Y_T)/(Y_T - Y_1) \qquad (4.39)$$

Again Y_I is known from equation (4.37) and, consequently, equation (4.39) can be solved. Accordingly, the quantity $(\varepsilon_{xy}^2 + \varepsilon_{yz}^2 + \varepsilon_{zx}^2)$ can be determined or, in a simple case, the actual inhomogeneous strain can be calculated.

The values of the ratios of the components of the yield stress for a few model materials are given in *Table 4.6*.

4.4 Tool design

4.4.1 Concepts of design

The problem of rational design of tools in metal-forming processes is comparatively new. Tools have been, and in most cases still are, designed on the basis of experimental selection of profiles, followed by the equally experimental 'trial and error' methods of testing their suitability. Changes in the

Table 4.6 Redundant strains in lead, aluminium and copper

ε_H	Lead		Aluminium		Copper		Hot wax
	$\dfrac{Y_I}{Y_T}$	$\dfrac{Y_T}{Y_H}$	$\dfrac{Y_I}{Y_T}$	$\dfrac{Y_T}{Y_H}$	$\dfrac{Y_I}{Y_T}$	$\dfrac{Y_T}{Y_H}$	$\dfrac{Y_T}{Y_H}$
0.4	1.07	1.11	1.27	1.04	1.19	1.07	1.015
0.5	1.29	1.24	1.09	1.02	0.97	0.99	1.010
0.6	1.25	1.16	1.13	1.02	0.89	0.96	0.960
0.7	1.16	1.09	1.24	1.03	0.87	0.96	0.960
0.8	1.11	1.05	1.11	1.01	0.85	0.96	0.935

lubrication, the range of processes materials, etc., will often call for further modifications to be made and, consequently, for the extension of the laborious and unreliable testing technique.

In addition, the lack of appreciation of the fundamentals of the mechanics of metal flow and deformation in a given process very often results in the adoption of tool profiles which, although apparently working, impose unduly severe conditions. Tool profiles, so established, can become complicated and basically unsound.

The last decade or so has seen a number of serious attempts to remedy the situation, but these developments have been mostly concerned with either the drawing/extrusion or cross-rolling groups of processes. For historical as well as practical reasons, systematisation of work in these fields has been regarded as both more important and more productive because it offers a variety of possibilities. Eventually, these could be made applicable to more than one operation. This contrasts, of course, with the situation arising in, for instance, strip and plate rolling where the basic shape of the roll cannot be altered and improvements have to be sought through the medium of plant design. Forging constitutes a problem of its own since the almost infinite variety of shapes required precludes the possibility of formulating generalised solutions which are possible for axisymmetric processes.

Rotary piercing and elongating processes, which are representative of the most severe forming conditions, offer considerable scope for improvement. Not surprisingly, therefore, a number of possible design concepts of rolls and piercing plugs has emerged. These are usually based on the idea of organising or manipulating the flow of metal in such a way that the change along the pass in a specific parameter of the process is controlled. The most usual parameter considered is either that of a constant reduction in cross-sectional area or the longitudinal homogeneous strain, or that of a constant reduction in the diameter (in a solid billet) or the wall thickness (in a hollow billet). In the latter case the radial homogeneous strain plays the dominant role.

The first approach is synonymous with considering the effect of axial strain only while neglecting the effects of radial and circumferential strain, whereas the second concept reverses the problem by neglecting axial and hoop effects. Since even in the virtual absence of one of these strains the other two must be numerically equal (condition of constant volume) it is obvious that the consideration of a single parameter is, generally, inadequate. However, because of the simplicity of such an approach, it is often used for the 'simpler' and less demanding processes. There are two concepts of tool design—the constant rate of homogeneous strain (CRHS) and the constant mean strain rate (CMSR)—which satisfy the three-dimensional nature of a general situation and are, therefore, more representative of real forming conditions.

The CRHS method is concerned only with the rate of deformation which is not related to the time necessary to strain the material. With the instantaneous strain rate, which operates in the zone of deformation, in some processes as high as three orders of magnitude in comparison with the mean, the time element is taken into consideration in the CMSR concept. Incorporation of the CRHS idea into the CMSR is possible and offers the advantage of fulfilling automatically both conditions.

Therefore, the five main principles of tool design are:

(1) the constancy of the reduction in area,
(2) the constancy of the reduction in diameter or thickness,
(3) the constancy of the rate of homogeneous strain (CRHS),
(4) the constancy of the mean strain rate (CMSR), and
(5) the conformity with the flow or stream-line pattern.

To illustrate the use of these concepts (discussed more fully in Section 4.5), consider *Figures 4.30* and *4.31*. *Figure 4.30* refers to a secondary piercing operation, and *Figure 4.31* to the extrusion or drawing of a solid cylindrical product.

4.4.1.1 Reduction of area

In this case, the development of the tool profile is related directly to the imposed mode of flow defined, in turn, by the change in the cross-sectional area of the work piece as it proceeds through the working zone of the pass. The latter is subdivided into an arbitrary number of equispaced sections, as for instance in *Figures 4.30* and *4.31*. The reduction in area R_n in any section n is given by

$$R_n = (A_0 - A_n)/A_0 \qquad (4.40)$$

where A is the cross-sectional area of the work piece. The generally accepted condition of constant ratios of the increments in area reductions can be defined as

$$(R_2 - R_1)/(R_1 - R_0) = (R_3 - R_2)/(R_2 - R_1)$$
$$= (R_n - R_{n-1})/(R_{n-1} - R_{n-2})$$
$$= c \qquad (4.41)$$

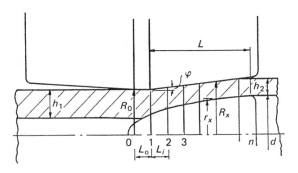

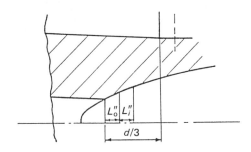

Figure 4.30 Rational design of a repiercing plug for the secondary tube piercing shown in *Figure 4.8*

With no deformation in section 0 (at the entry to the pass), $R_0 = 0$ and equation (4.41) becomes

$$A_n = A_{n-1} - (A_0 - A_1)c^{(n-1)} \qquad (4.42a)$$

Once the actual mode of deformation, given by the value of c, has been decided upon, and the number of sections has been stipulated, the value of A_n can be calculated since A_0 is simply the cross-sectional area of the undeformed specimen. In a simpler version of this approach the concept is related to the condition of constant volume V, so that

$$V_0/V_1 = V_1/V_2 = \ldots V_{n-1}/V_n = 1 \qquad (4.43a)$$

or, in order to reduce the deformation uniformly over the whole length of the tool, the following relation must hold:

$$A_0/A_1 = A_1/A_2 = \ldots = A_{n-1}/A_n = a \qquad (4.42b)$$

This means that the length of the individual material element is given by

$$L_0/L_1 = \ldots = L_{n-1}/L_n = b \qquad (4.43b)$$

Naturally, $b = a^{-1}$, and the total length L of the tool is

$$L = \sum_{x-0}^{n-1} L_x \qquad (4.43c)$$

where $L_x = L_0 a^x$, and x takes the values $0, 1, 2, \ldots, (n-1)$. The relationship between the cross-sectional area at the entry to the pass and any section of it is given by

$$A_x = A_0 a^{-x} \qquad (4.43d)$$

where $x = 0, 1, 2, \ldots, n$.

In a single-tool process, the value of A in successive sections will lead to the development of the required tool profile by giving the variation in, say, the diameter or the wall thickness. In a two-tool forming system, one of the tool profiles must be

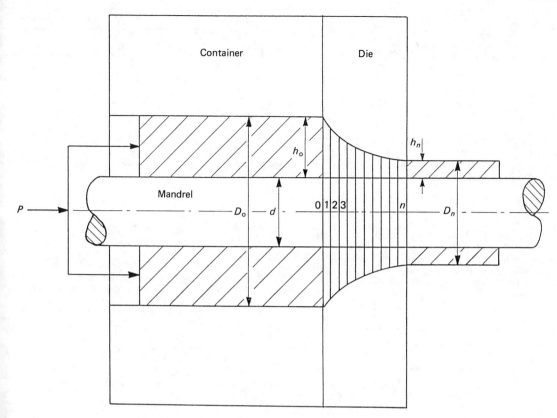

Figure 4.31 Rational die design for bar extrusion or drawing

assumed *a priori* and then used as a datum for setting off the calculated dimensional specimen variations. The selection of the 'basic' profile is usually dictated by a consideration of the economical aspects of tooling. Unnecessary and perhaps complex profiling of large and expensive-to-make tools, which are also likely to suffer a high degree of wear, should be avoided. Such tools should be as simple as possible geometrically and shaping should be confined to smaller tool(s).

4.4.1.2 Reduction in thickness

Although originally proposed in connection with hollow tubular semifabricates, this particular concept can be extended to solid sections if the overall radial or width dimensions are considered. Applying the precept of constant simple wall or overall thickness ratios in consecutive sections of the pass, the required condition is:

$$h_o/h = C \qquad (4.44a)$$

where C is the coefficient of the overall thickness reduction. Equation (4.44a) can be rewritten in the form of, say,

$$h_o/h_n = C_{av}^n \qquad (4.44b)$$

so that $h_o/h_1 = h_1/h_2 = \ldots = h_{n-1}/h_n = C_{av}$, where C_{av} is the ratio averaged over two consecutive sections. The application of this concept to a specific design problem is made on the lines discussed above for the reduction of area.

4.4.1.3 CRHS concept

The concepts described so far, fall short of providing reliable methods of designing tools because they do not regard a metal-forming process as a complex, but well-integrated entity. In fact the above-described approaches are normally based on the consideration of only one or, at the most, two facets of the given process and, therefore, they may give a very inaccurate picture. To introduce the concept of generality and, therefore, of applicability to any metal-forming operation, the CRHS condition is imposed on the flow of the metal. This makes use of the generalised homogeneous strain and, consequently, accounts fully for the straining of the metal within the range of useful shape-forming deformations.

On substituting the expressions for strain increments appropriate to the process, and integrating within the boundary conditions, the following equation for the finite homogeneous strain between sections 0 and n is obtained:

$$\varepsilon_n = \ln(Z_n) \qquad (4.45)$$

The natural or logarithmic strain ε_n is then defined by a function Z_n that is related to the dimensions of the work piece in section n. The concept itself is defined simply as

$$(\varepsilon_2 - \varepsilon_1)/(\varepsilon_1 - \varepsilon_0) = (\varepsilon_n - \varepsilon_{n-1})/(\varepsilon_{n-1} - \varepsilon_{n-2}) = s \qquad (4.46)$$

where s is an arbitrary numerical constant. On substituting into equation (4.46) from (4.45). The general expression is obtained in the form:

$$Z_n/Z_{n-1} = Z_1^{s^{(n-1)}} \qquad (4.47)$$

Solution of this equation gives the dimensions of the work piece in any section, and thus defines the required geometry of the pass.

The value of s (equation (4.47)) decides the rate at which deformation takes place. If $s = 1$, the rate is uniform and with $s \leqslant 1$ or $s \geqslant 1$ the rate is decelerated or accelerated, respectively. The choice depends entirely on the severity of the process, but it is possible to combine two different modes of flow in a single pass. Although there is clearly an infinite number of values that s can take, experience indicates that for most practical purposes the range 0.8–1.2 is quite sufficient.

Since equation (4.47) cannot automatically provide the value of the function Z_1, this must be assumed by taking an acceptable (from a practical point of view) amount of reduction in that section.

Using as an example the profile of a rotary piercing plug shown in *Figure 4.32*, it can be seen that the choice of the design concept can be all important. Plug profiles based on equations (4.42a) and (4.47) (using the same values for c and s) differ substantially in that the use of the CRHS concept results in smooth surfaces, whereas that of the reduction in area introduces discontinuities. The problem of tool wear is apparent in the latter case.

4.4.1.4 CMSR concept

This concept is particularly applicable to the processes performed at higher speeds which can affect the material properties of the product.

The concept of constant mean strain rate is defined analytically as:

$$\dot{\varepsilon}_1 = \dot{\varepsilon}_2 = \ldots \dot{\varepsilon}_n = \text{constant} \qquad (4.48)$$

where $\dot{\varepsilon}_n$ is the mean strain rate between section 0 and any section n, and $\dot{\varepsilon}$ is obtained from the appropriate equations ((4.7) and (4.9)).

However, it is obvious that various values of either h_n or D_n can be obtained by an arbitrary choice of the distance x. Such a procedure, although satisfying the equation, will inevitably produce discontinuities in the tool profile and a lack of uniformity in the mode of reduction. A simple solution to this problem lies in the incorporation of the CRHS condition. This ensures that the variation in the chosen dimensional parameter is governed by a specific mode of deformation and, therefore, the distance x is evaluated with respect to each value of h or D in a predictable manner. Equation (4.48) is then satisfied automatically.

4.4.1.5 Flow-line concept

Optimised die profiles, based on visioplasticity studies of the patterns of flow, must be developed in each individual case. This approach is particularly useful for extrusion and drawing processes. In the former case the concept shows that, should a dead-metal zone be formed, the material will flow past it in such a way that a 'natural' die is formed. The zone of deformation is then radically limited to a more homogeneous flow and the redundancy factor is substantially reduced. The practical difficulty in using this very useful concept lies in the necessity of making rather complex calculations, which follow from the use of the theory of visioplasticity.

Some simplification of this problem is provided by the slip-line-field theory rather than a full scale visioplasticity approach. *Figure 4.33* shows, for example, a method of designing an extrusion die.

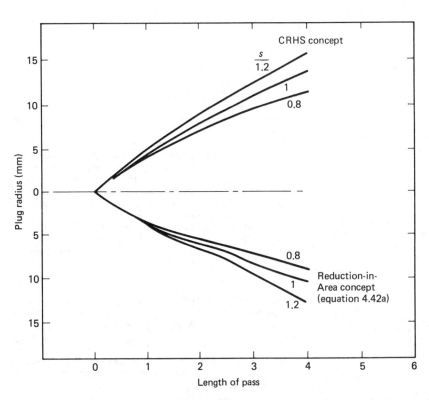

Figure 4.32 Comparison between the effects of different design concepts on the radius of a rotary piercing plug

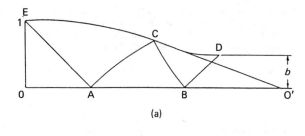

(a)

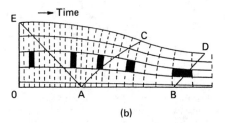

(b)

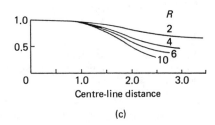

Centre-line distance

(c)

Figure 4.33 Flow-line- and slip-line-field design concepts

A bound on redundancy is imposed here by defining the conditions of homogeneous flow. An analysis of this particular problem leads to the slip-line field shown in *Figure 4.33(a)*. The average extrusion pressures, calculated by means of the proposed velocity–stress field, will constitute the bounds to the redundant work. In practice, therefore, a die profile corresponding to curve ECD, in the figure, is likely to produce conditions of flow approximating closely to those of homogeneous deformation.

A die profile of this type (for a typical extrusion ratio $R = 4$) together with the calculated flow lines and orthogonal time lines, for a constant volume flow, is shown in *Figure 4.33(b)*. This also shows how an element of the material is likely to change its shape when passing through the die. It is noticed that the high degree of shearing and 'unshearing' normally expected, appears to be considerably minimised in this case.

The effect of the extrusion ratio R on the shape of the tool is indicated in *Figure 4.33(c)*.

4.4.2 Lubrication

The effects of friction in a typical metal-forming process, bearing in mind the exceptions mentioned earlier, are three-fold. Friction is responsible for, or adds to, the wear of tools and affects the surface finish of the product. By locally increasing the level of the internal stress system, it also affects the loading on the plant and may cause internal material defects in the work piece. Consequently, deformation pat-

terns, temperature rise, surface pressure on tools, and total force requirements may be influenced.

The specific effects depend entirely on the process considered. In drawing and extrusion, friction increases forces or pressures, raises the level of inhomogeneity by causing a high degree of differentiality of flow, and affects the pattern of residual stresses. In extrusion, friction is likely to be instrumental in the formation of a dead-metal zone and will thus again affect the pattern of flow by forming the natural internal die profile already mentioned. The metal is forced to flow along this surface in preference to the actual surface of the tool.

Friction hills occur in both sheet and plate rolling and in forging. In these processes the effect on tooling can be considerable because of a large increase in pressure on the roll or dies. When the pressure increase exceeds the yield stresses of the tool material employed, plastic deformation of the latter will take place and dimensional accuracy of the product will be lost. Costly tool replacement or, at least, remachining will be necessary.

The importance of lubrication (where appropriate) becomes obvious. The basic function of an efficient lubricant is, naturally, to reduce friction, but it also has a number of other roles to play. It should act as a thermal insulator, it must prevent pick up of material on the tool–work piece interface, and should not adversely chemically affect either surface. This implies that contamination, such as carbon pick up or corrosion, cannot be tolerated, and that efficient removal of the lubricant must be possible.

Clearly, a wide range of lubricants is required to provide optimal conditions in a metal-forming operation, because the characteristics of these operations vary widely. The lubricants are either solid or fluid at the time of application, but can change their original phase in the course of processing. The general classification of lubricants is given in *Figure 4.34*.

Metallic films that are employed normally consist of lead, copper and, exceptionally, indium. All of these are used in complex and severe working conditions, in which it is particularly important to ensure that additional shearing is confined to the metal film and the processed specimen. The coefficient of friction is then given as the ratio of the shearing stress in the film to that in the specimen.

Non-metallic films are used mainly at elevated temperature and include graphite, glass, polymers, oils and molybdenum

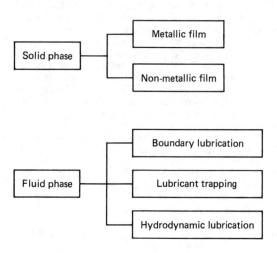

Figure 4.34 General classification of lubricants

disulphide. The latter, although expensive, is of particular value since it is operative over a wide temperature range and gives very good surface finish combined with very low frictional forces in cold-forming operations.

The inherent difficulty in processing certain materials such as stainless steels requires even more efficient and, to a certain extent, more sophisticated lubricating methods than those needed for a wide range of ordinary engineering alloys. Although graphite type lubricants are still widely used, a significant step forward was made when the Ugine–Sejournet process employing glass as lubricant was introduced. This involves not only using glass, but also introducing the novel idea, insofar as hot working of metals is concerned, of lubricating only the work piece and not the tools. When a tool is lubricated, but not the work piece, the danger of break down of the lubricant film is considerable. The conversion of solid or liquid lubricant to a liquid or gas, respectively, usually results in a substantial loss through mechanical ejection. To counteract this tendency it is necessary either to trap the lubricant in the grooves or recesses of the tools, thus complicating their design, or to maintain the lubricant by artificially increasing the pressure. Both methods introduce additional complexities and increase the cost of production. The Sejournet method of lubricating the billet obviates these difficulties but poses two main engineering problems which can be solved only by careful manipulation of the system. The two problems are as follows.

(1) The considerable heat gradient between the hot billet and the relatively cold tool leads to the rapid cooling of the former and, consequently, to an increased load requirement; and
(2) there may be a reduction in the resistance of the tool caused by an increase in its temperature.

To remove these difficulties, the lubricant chosen must have very good thermal insulating properties. It should have a low coefficient of friction in order to minimise load requirements and oxidation during handling and it should afford some protection in the form of a semipermanent coating to the billet. The application of the lubricant to the billet must be uniform throughout the operation, both in the sense of the constancy of the film thickness and in terms of chemical stability. The viscosity and chemical composition of the lubricant are, therefore, of importance.

All these conditions are satisfied by a variety of glasses but only some of them are met by graphite type lubricants. In addition, glass is even more attractive due to its low cost compared with, say, graphite greases.

Comparison between the properties of glass and graphite clearly shows the basic difference between the two lubricants. Glass is almost solid at tool temperature, but viscous at the hot billet temperature. Rapid heat exchange between the two glass and the billet is prevented. During processing most of the shear strain due to friction takes place, as required, within the layer of glass and not on the surface of the billet. As a result, tool wear is reduced and die life is increased. The effectiveness of glass in reducing shear forces on the billet surface also reduces the possibility of intergranular failure at elevated temperatures.

In the extrusion of stainless steel, for example, it has been shown that a reduction of up to 35% is possible when a graphite type lubricant is replaced by glass (the coefficient of friction of glass is <0.01 and can be as low as 0.001) and, also, that the danger of carbon pick up is non-existent.

When fluid-phase lubricants are used, the situation is changed. With boundary lubrication, a chemical reaction between the lubricant and metal takes place and results in the appearance of, say, a metal soap which then forms a film of low-friction properties. As already indicated, trapping of the lubricant (caused either by the condition of the surfaces or deliberate entrapment due to tool design) may occur and, although low friction can be expected, a matt finish and indented surface may result if the pressures generated in the pools of lubricant are sufficiently high.

A different situation arises with hydrodynamic types of lubrication. The function of such lubricants is to separate permanently the two surfaces, but always at high pressure (of the order of that of the yield stress of the processed material). Break down of the film should not occur under these conditions, but the efficiency of the operation will still depend on the sealing of the working zone of the pass with a view to preventing the escape, or even slight leakage of lubricant. Loss of lubricant lowers, or even destroys, the build up of pressure. The plant system required is more complex and expensive and the process is, therefore, suitable only for steady-state operations such as wire drawing.

Specific lubricants, selected for a range of normally processed metals such as steels, copper alloys and aluminium, are indicated under the headings of the actual processes in the following sections.

4.4.3 Tool materials

The importance of the correct selection of tool material cannot be overemphasised but, at the same time, it must be realised that, unlike machining operations in which standardisation is relatively easy, metal forming presents very considerable difficulties.

Because of the range of conditions encountered, the problem of selecting appropriate tool materials calls for specialised knowledge and the solution cannot, therefore, be presented here in a simple form. An outline of the requirements and possibilities is given below, but detailed information should be sought in the literature listed in the Bibliography.

The basic requirements of the tool material and the tool are that it should fulfil a number of often mutually exclusive conditions. To resist deformation (which if imposed on the tool will result in the loss of accuracy) the material should possess high tensile or compressive strength but, because of the economics of the production line, it should also have a high degree of resistance to wear, mechanical and thermal shock and respond well to cycling loads. The latter, in particular, implies resistance to fatigue and fracture.

A conflict between the 'theoretical' requirements is immediately obvious since, for instance, resistance to abrasive wear depends on the ability of the material to resist plastic deformation, whereas the latter will assist in sealing an incipient crack that is usually responsible for the onset of fatigue. A compromise is, therefore, necessary.

Additional problems may arise as a result of purely local difficulties such as misalignment of tool settings and inefficient lubrication. These aspects of the conduct of the process will subject the tool to disadvantageous loading conditions which may lead to failure. Since the failure of a tool calls not only for its possibly costly replacement, but also means an interruption of production, the main causes of fracture and wear should be identified. These are:

(1) fracture due to either mechanical shock or material fatigue;
(2) fracture due to either loading or thermal stress cracking;
(3) fracture caused by faulty machining;
(4) wear due to friction, oxidation and corrosion; and
(5) wear caused by impact loading.

The standard materials used in metal-forming tooling are steels, cemented and titanium carbides, ceramics and special surface coatings, sometimes combined with an appropriate surface treatment.

4.4.3.1 Steels

The generally accepted classification of steels is that of the American Iron and Steel Institute which subdivides tool steels into seven groups:

(1) *high speed steel (HSS)*—mostly molbydenum based M type steel possessing higher toughness than the tungsten T type;
(2) *hot-work steel*—chromium, tungsten and molybdenum based (used mainly for extrusion and forging tooling);
(3) *cold work steel*—high carbon/high chromium medium alloy, air or oil hardening;
(4) *shock-resisting steel*—combinations of the above with suitable heat treatment, and maraging steels;
(5) *mould steels*;
(6) *special purpose steels*; and
(7) *water hardening steels*.

Of these, (5) to (7) are of least importance, but the mould steels (5) are used as numerically controlled machined inserts in otherwise cast-iron stretch-forming dies.

4.4.3.2 Cemented carbide composites

These are primarily cermets consisting of crystalline tungsten carbide in a cobalt matrix. Although very much more expensive than any tool steel, carbides are characterised by a long tool life and a high degree of accuracy. More efficient production lines can thus be planned using tooling made of this material.

The percentage of cobalt binder (4.5–25%) in the composite is instrumental in defining the response of the tool. Composites containing 4.5–6%Co are used for drawing and ironing dies that can withstand light shocks and have satisfactory anti-wear properties. The medium 'cobalt range' ($\leqslant$13%Co) composites are used for the manufacture of wire, tube and deep drawing dies, static and dynamic extrusion tooling, punches, and a variety of blanking tools. Medium shocks can be sustained by these tools which also remain wear resistant. Higher percentages of cobalt ($\leqslant$25%) product materials suitable for heading punches and dies which are subjected to heavy shocks under normal working conditions.

4.4.3.3 Ceramic materials

These are new materials that offer interesting possibilities but, as yet, they have not been fully developed. This is partly because of the high cost of production and partly because they are confined to steady-state processes operating in compression. The response of these materials to tensile stress is generally unsatisfactory. A typical application of their use is in wire drawing dies (or as shaped ceramic inserts in metal die holders). The materials normally used are alumina (Al_2O_3), silicon nitride (Si_3N_4), partially stabilised zirconia, and polycrystalline diamond.

4.4.3.4 Titanium carbide composites

Titanium carbide/HSS cermets are of particular interest because when subjected to suitable surface preparation they display very good anti-wear properties. They are used as replacements for steels in extrusion and forging dies, in blanking and rolling tools and for hot-working applications.

The cermets can be machined to specific profiles by wire spark erosion.

From the point of view of economy, these materials are very much cheaper than tungsten based composites.

4.4.3.5 Surface treatment

The properties of the base tool material can often be enhanced by an additional surface treatment which increases surface hardness, wear and scuffing resistance, and lowers the effect of friction. All these effects are obtained by using either conventional or unconventional surface-treatment processes. The former include carbon case hardening, salt bath or gas carbonitriding or siliconising, oxidation and, sometimes, sulphonising. The more recent, unconventional methods consist of ion implantation, plasma nitriding and physical and chemical vapour deposition.

Surface coating is used for drawing and deep drawing dies where the dimensional accuracy of the product is particularly important and where good properties and low friction are required. These techniques are relatively expensive and, consequently, the economics of the whole operation should be considered carefully before any particular technique is chosen.

4.4.4 Residual stresses

Removal of the operative stress fields, the presence of temperature gradients, and the phase changes taking place in the processed material can result in the appearance of residual stresses. Their incidence, pattern and level depend primarily on the geometry of the pass and, therefore, on the design of tool(s), but also on the properties of the metal in question.

Although the build up of these stresses due to phase changes is of interest in hot- and warm-working processes, the main problem which they present lies in the area of cold working where warping of a section can occur together with a permanent internal stress field that may well be incompatible with the operational conditions under which the processed material is expected to work.

When a material body is subjected to a full cycle of loading in which is passes from the elastic, through the elastoplastic, into the fully plastic state, the effective external force system is then removed, the body will behave as though it was subjected to a loading system that is equal numerically but opposite in direction. In many cases the relaxation will be purely elastic in nature but, even so, it will not be sufficient in value to equal the level of the original plastic stress field. Consequently, tensile and/or compressive stresses are locked (on the macroscopic scale) in the material. These 'remnant' or residual stresses have very little effect, if any, on the state of the material when it is altered, even slightly, by mechanical means or heat treatment that may follow a forming operation. Very simply, the relationship between the loading (L), unloading (U) and residual stress systems (R) is given by

$$\sigma_R = \sigma_L - \sigma_U \tag{4.49}$$

A typical development of these stresses together with the complex pattern that they may form can be illustrated using the example of the bending of a strip of metal by an applied external moment. With reference to *Figure 4.35*, the effect of the loading (forming) phase is shown in the sequence of the parts of the figure (a) to (d):

(a) in the elastic state, elastic stresses are distributed linearly through the cross-section of the strip (tensile on the upper and compressive on the lower surface);
(b) onset of plasticity is confined to the outer fibres;

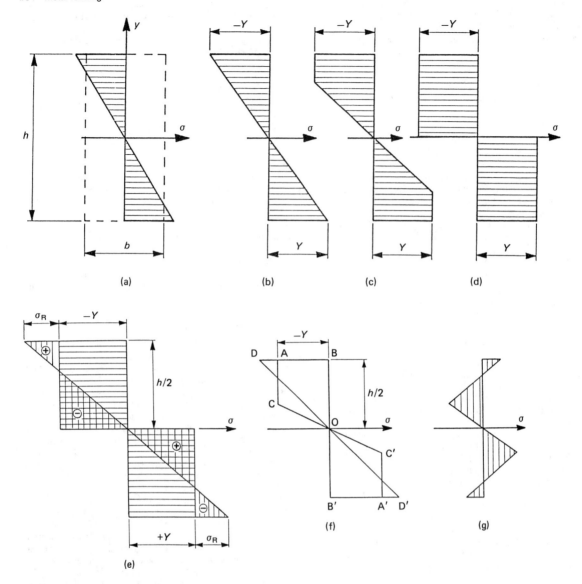

Figure 4.35 Sequence of events in the build up of a residual stress field during cold bending of a metal strip. For a description of (a) to (g), see text

(c) increasing the operative moment causes penetration of the plastic zone, compatible with the current value of the moment; and

(d) the advancing bands of the plastic material approach the neutral axis, and the section becomes (theoretically) fully plastic.

On completion of the forming operation and removal of the bending moment, the strip recovers elastically (with the material normally strain-hardened and possessing a higher flow stress). The stress system created by the recovery is superimposed on the existing, initially plastic system (horizontal hatching in part (e)) and thus generates a residual stress field (vertical hatching in (e)). Both tensile and compressive stresses are generated.

If the partial elastoplastic situation is the final state reached (equivalent to part (c)), then the residual stress field will be as in (f). In general, the final pattern in such a case like this is as shown in (g).

The presence of residual stresses is not always detrimental. Compressive stresses on the surface are desirable and are often imposed deliberately on the material by operations such as rolling or shot peening. Tensile stresses should be avoided as they affect the strength of the material and its fatigue life, and can easily lead to stress corrosion and associated cracking, as well as to the purely mechanical stress cracking.

If necessary, relief of these stresses is obtained either by post-processing heat treatment or by further plastic deformation.

4.5 Rolling processes and products

4.5.1 Classification of processes

With the exception of special steels and/or profiled sections rolling processes follow one of two main routes:

(1) production of plate and strip; or
(2) production of billets, bars, rod, sheet, tube or sections.

Traditionally, both routes start with a cast ingot (*Figure 4.36*) which is then rolled down to slabs (route 1) by cogging. In route 2, cogging again leads to the production of a bloom (a product of over 10 cm × 10 cm, or equivalent, in cross-section), and then to either a variety of small flats or large rounds or, through a billet mill, to a billet (a product of cross-sectional area less than 10 cm × 10 cm). However, very satisfactory developments in the area of continuous casting have led to the introduction of casting machines into these cycles. In the new, fully automated and computer-controlled, high-productivity works, continuous casting of slabs has to a great extent eliminated the cast ingot.

In route 2, in a modern mill the stress is on the use of continuous billet casters (in preference to bloom casters), thus eliminating one stage of the production line. Where blooms are still required, normal practice is to employ two or three strands of material which are then rolled in two or three passes to produce blooms. With smaller sizes of billet, up to six strands can be cast.

It can be seen from *Figure 4.36* that the manufacture of a wide range of either semifabricates or finished products calls for a variety of mills and plant settings. A very brief review of these is provided here but, again, detailed information can only be obtained from the references cited in the Bibliography.

Basically, the process, whether hot or cold, begins with the preparation of stock such as an ingot (in older plant) or continuously cast bloom or billet. In hot operations this is followed by heating in a strictly controlled atmosphere and temperature, and then rolling proper. Finishing of the work peice includes a number of operations such as cutting, cooling and, very often, straightening. In cold operations, which are used to enhance the mechanical properties of the material and improve dimensional accuracy, the ancillary equipment consists of furnaces for heat treatment and plant for surface finishing. Whereas modern plant comprise not only the rolling mill(s) proper, but also a number of pieces of ancillary equipment concerned with the preparation of the material prior to and post rolling, interest centres mainly on the actual mill since the dimensional quality of the product will depend mainly on its performance.

According to their actual functions, rolling mills are subdivided into the following classes:

(1) cogging mills (production of blooms, billets and slabs from ingots, where these are still used);
(2) plate, strip and sheet mills;
(3) tube mills (longitudinal rolling and cross-rolling);
(4) section mills (production of profiled sections, rounds, flats and strip); and
(5) special mills (production of machine parts).

An individual mill is characterised by the function for which it is intended and also by the type, position and number of rolls. These are located in stands, which in some cases are incorporated in a stand train. Depending on the position of the stands, mills can be further classified as linear (where the stands are in a line and there may be one or two lines driven by a single motor), or continuous (where a number of stands is

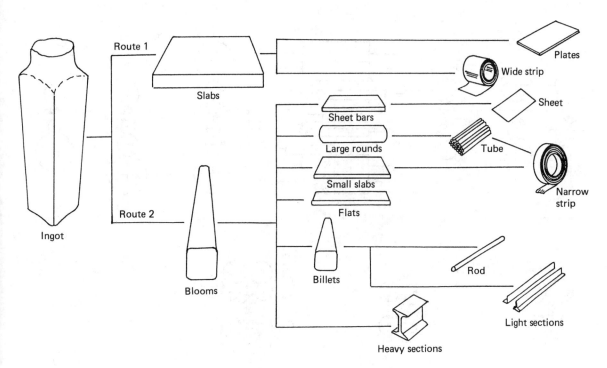

Figure 4.36 Schematic, simplified representation of the sequences of the basic rolling processes. Route 1 leads to the production of plate and wide strip. Route 2 results in sheet, narrow strip, tube and sections

placed in tandem and, at any given time, rolling takes place in more than one stand).

In each stand the number of rolls involved defines the stiffness of the structure and, therefore, reflects on the dimensional accuracy of the product, whereas the disposition of the rolls specifies the function of the system. Irrespective of the complexity or otherwise of the roll arrangement, the actual work of deformation is normally performed by a pair (or a set) of innermost rolls which are backed up (for the mentioned structural stiffness) by a number of other usually larger diameter sets of rolls. The number of these gives rise to the name of the mill which is described as *n*-high.

A selection of basic arrangements for flat, horizontal rolling is given in *Figure 4.37*. With reference to the figure, the following types of mill can be distinguished.

(1) *Two-high*—either (i) reversing or (ii) non-reversing.
    (i) Used for slabbing, large rail, profiled sections and thick plate rolling.

    (ii) Used for continuous rolling of billets, rods, rounds and plate, linear rolling of sheet, and for dimensional calibration of the product.

(2) *Double duo*—used mainly in older mills for rolling small and medium-sized rods, strip and profiled sections.

(3) *Three high*—used for rolling medium and large billets, rails, rods and profiled sections.

(4) *Three-high with the middle roll oscillating between the two outer rolls*—all the rolls are driven; used for blooming or slabbing of ingots of around 3.5 ton in weight.

(5) *Three-high Lauth, with an idle middle roll*—used for rolling thick plate and as a first stage in linear hot rolling of thin sheet.

(6) *Four-high*—either (i) reversing or (ii) non-reversing.
    (i) Used for cold rolling thin strip and hot rolling thin and thick plate.

    (ii) Used for continuous hot rolling strip and plate and cold rolling strip.

(7) *HC rolling mill (Hitachi)*—used mainly for cold rolling plate and strip.

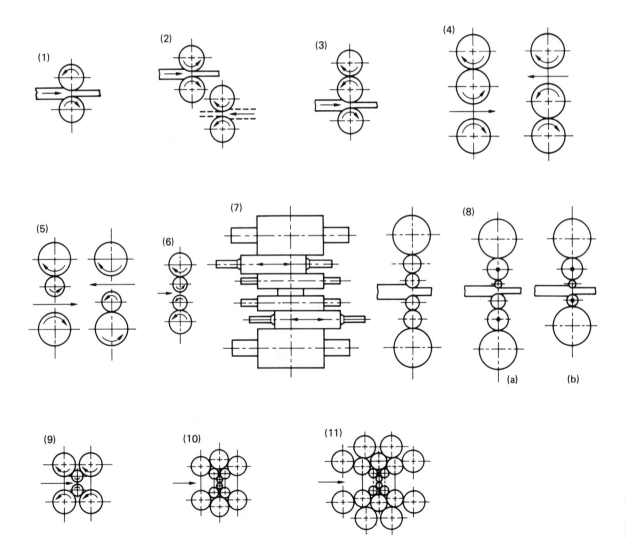

Figure 4.37 Flat rolling: scheme to show the position of horizontal roll systems. For a description of (1) to (11), see text

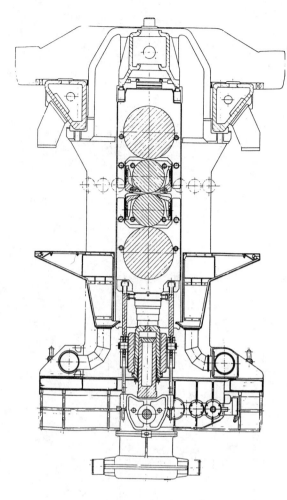

Figure 4.38 Arrangement of a four-high horizontal stand for plate rolling

(8) *Taylor mill*—used for precision rolling of thin sheet and strip.
(9) *Six-high*—used for cold rolling thin sheet and foil.
(10) *Sendzimir twelve-high mill*—used for cold rolling thin sheet and foil.
(11) *Twenty-high, planetary Sendzimir mill*—used for cold rolling thin sheet and foil.

For general information, the arrangement of a four-high stand mill, which is representative of a standard mill structure, is shown in *Figure 4.38*.

For more complex shapes and/or sophisticated components, the disposition of the rolls has to depart from the horizontal scheme. Vertical roll systems on their own, or vertical roll systems in combination with horizontal roll systems are used, thus enabling the rolling of solid rounds and hollow tubular components. A few basic systems are shown diagrammatically in *Figure 4.39*. With reference to the figure, the following systems can be identified.

(1) *Vertical system*—used for the continuous rolling of plate and strip, and linear rolling of very narrow strip. Dimensional control of the strip width is maintained.
(2) *Intermediate vertical–horizontal system*—used for the continuous rolling of billets and small- to medium-sized profiled sections.
(3) *Universal slabbing system*—used for rolling slabs.
(4) *Universal structural mill*—used for rolling I-sections; the usual size range is 30–120 cm in height. Only the horizontal roll sets are driven, the vertical ones remain idle.
(5) *Intermediate horizontal–vertical system*—used for longitudinal tube rolling.

The arrangement of a selection of these mills in a production line, supplemented by ancillary equipment, is not standard and depends entirely on the objective of the operation.

For reasons of expediency and well-established rolling practice, individual mills are often referred to by the size(s) of rolls which they employ. In plate and sheet rolling mills, rolls are specified by their length since this governs the width of the product that can be rolled. The range of lengths is substantial, lying between 5000 mm for plate mills and 100 mm or so for single-strand applications. For other rolling mills it is the diameter of the roll that is used as a specification. Again

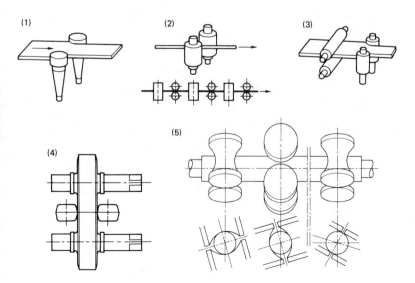

Figure 4.39 Flat and cross-rolling: scheme to show the position of vertical and inclined roll systems. For a description of (1) to (5), see text

depending on the system, large rolls (about 1450 mm in diameter) are used in universal structural mills ((4) in *Figure 4.39*) and very small rolls (down to 10 mm diameter) may be used in cluster or planetary mills ((11) in *Figure 4.37*).

The roll speed of every driven stand can be adjusted within specified range. This provides a means of introducing iner-stand tensions (affecting the pattern of flow of the metal and the mechanics of the operation) and of creating conditions suitable to the rolling of a specific product. A rough guide to the effect of velocity is given in *Figure 4.40*.

The quality of product in terms of material properties depends considerably on whether processing takes place at an elevated or at ambient temperature. Rolling hot is not only intended to minimise the forces and torques necessary to effect deformation, but also to improve the structure through recrystallisation. Unless recrystallisation takes place, strain hardening (produced by deformation) will adversely alter the structural characteristics of the metal, even at temperatures much higher than ambient. The processing temperature should, therefore, lie above that of recrystallisation, in order to produce new, small grains (*Figure 4.41*), but below the value at which oxidation takes place and affects the surface of the product.

The main function of cold rolling, as already mentioned, is to improve the mechanical properties of the alloy, to impart a surface finish suitable for the future use of the product, and to enable the manufacture of dimensionally accurate very thin sheet, strip and foil (0.001–2 mm thick) and 'medium' thick sheet (1.0–1.5 mm thick). Depending on the quality of the surface, the rolled sheet can be used for stamping when the material displays fine-grain finish, for plating when the surface is smooth, or for further artificial surface coating when the surface is dull.

4.5.2 Roll design

Apart from the straightforward rolling of flats through ordin-ary sets of smooth rolls (*Figure 4.42(a)*) conversion of blooms, billets and slabs into rounds, rods, tubes and profiled sections calls for the use of grooved rolls of the type shown in *Figure 4.42(b)*. The grooved roll profiles fall into five main groups (shown diagrammatically in *Figure 4.43*). Combinations of these profiles are often used in a particular system, but the main production lines consist of the following sequences.

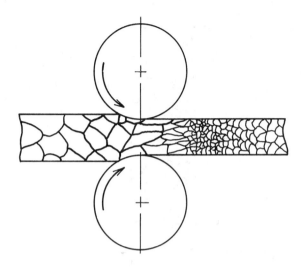

Figure 4.41 Recrystallisation in hot rolling

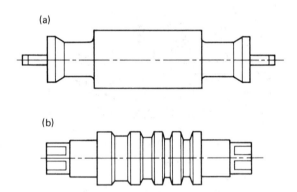

Figure 4.42 Basic types of roll: (a) smooth surface; (b) grooved surface

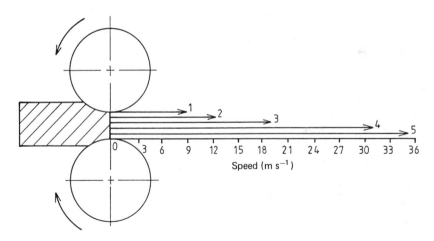

Figure 4.40 Speed of rolling and type of product. (1) Billet, bloom; (2) sheet; (3) profiled sections; (4) rod; (5) plate

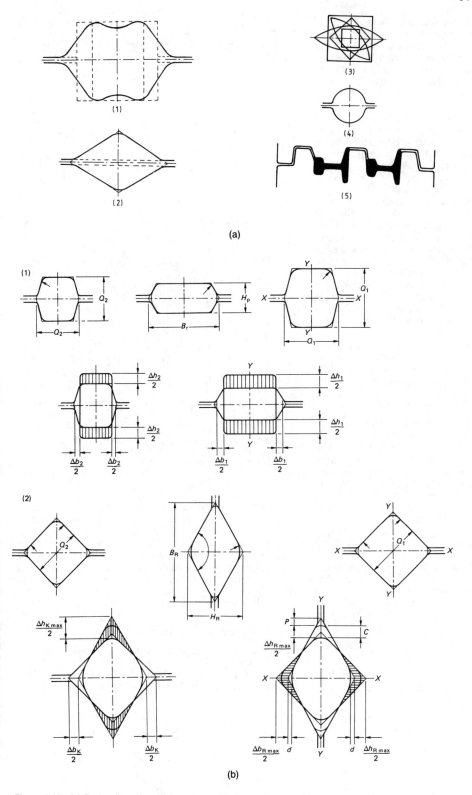

Figure 4.43   (a) Basic roll profiles: (1) box groove; (2) diamond groove; (3) oval groove; (4) circular groove; (5) an example of a special pass for rolling rails. (b) Typical profile sequences: (1) box/rectangle/box (Δh is the reduction in the roll, Δb is the axial expansion, and a_1 and a_2 are the initial and final box dimensions); (2) box/rhombus/box

(1) *Box pass*—used in blooming mills producing billets, and in the production of smaller items. Several passes through stands (utilising box-type rolls) disposed at 90° to each other are necessary before the desired amount of deformation is achieved. By their very nature and small size, box grooves occupy relatively little room and, therefore, a number of them can be accommodated on a roll.

(2) *Diamond passes*—used either in pure diamond sequences or in conjunction with square grooves, for the production of square sections. These profiles are used as drawing passes for large components. Unlike box-type rolls in which the sides of the specimen are not worked on by the rolls, the diamond and diamond/square groove sequence tools operate on all sides of the material being processed. A large variation in the size of the finished bars can be obtained in these systems by the simple expedient of regarding each, alternate square groove pass as the finishing one. The advantage of the purely diamond groove system over the mixed sequence lies in the greater stability of the product, combined with a greater degree of uniformity in the rate of deformation.

(3) *Oval grooves*—square and hexagonal cross-sections can be obtained by employing oval/square and oval/hexagon sequences. These are capable of producing large deformation, thus increasing considerably the length of the finished bars.

(4) *Circular grooves*—these are used either on their own (*Figure 4.44*) or in conjunction with oval-grooved rolls to produce circular cross-sections for rounds and billets required in the production of either solid bars and rods or seamless tubing.

(5) *Special profiles*—used for rolling rails and sections.

The proximity of the rolls in any *n*-high system gives rise to either an open or a closed type of pass. In the first case, typical profiled rolls are positioned as shown in *Figure 4.45(a)*, with the result that deformation takes place not only within the confines of the roll grooves but also in the inter-roll gaps. Fins are thus formed and any misalignment of the neighbouring rolls will result in lining and scoring of the surface of the product. An actual, open pass system in a three-high, box/square/diamond blooming sequence (shown in *Figure 4.45(c)*) illustrates this point—the metal can flow easily through the intervening horizontal gaps and produce unnecessary fins.

In the second case, a closed pass is formed by 'fitting' a forming part of one roll into the groove of the other, thus bringing the two rolls close together and eliminating, to a degree, the sideways flow of metal. *Figure 4.45(b)* shows how the sections shown in *Figure 4.45(a)* can be obtained by bringing the rolls together.

4.5.3 Basic mill layouts

Although no standard plant arrangement is recommended for any specific production line, basic layouts have been deve-

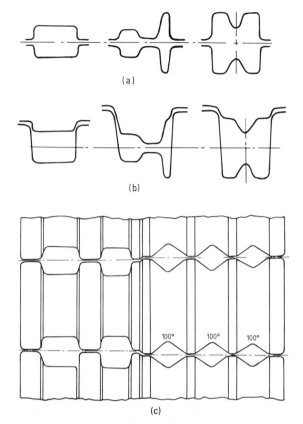

Figure 4.45 Open roll passes: (a) typical examples; (b) closed passes; (c) passes in a three-high blooming stand consisting of a box/square/diamond sequence

loped and serve the purpose of building up the most economical rolling set-ups that may be required in a given situation. Each set-up includes specially selected rolling units and items of ancillary equipment such as pickling, drying, lubricating, heat treatment, and cutting plant. A selection of basic rolling-mill systems is shown in *Figure 4.46*. This figure shows rolling-plant arrangements only, without any auxiliary items. Insofar as the terminology is concerned, 'continuous mills' take their name from the fact that, at any given time, rolling takes place simultaneously in more than one stand.

4.5.4 Rolling of plate, strip and sheet

The definition of each of these products is somewhat vague, but it is usually taken that, in hot processes, a rolled item is referred to as 'plate' when its thickness is in the range 4–160 mm and its width is 600–3800 mm. Sheet that is

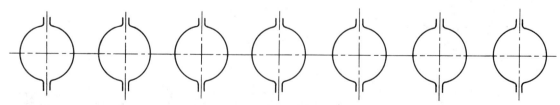

Figure 4.44 Sequence of circular grooved passes

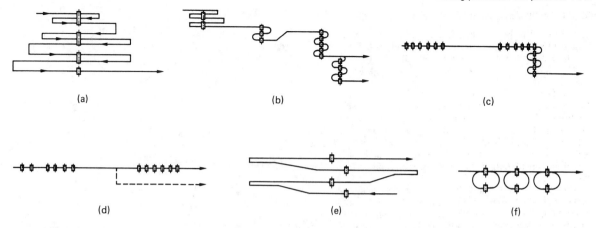

Figure 4.46 Basic production rolling-mill systems: (a) single line; (b) multiline; (c) semicontinuous; (d) continuous; (e) staggered; (f) loop

4–50 mm thick and 200–1500 mm wide is also sometimes called 'universal', whereas 'thin sheet' or 'thin strip' is material of thickness 0.2 to <4 mm with a width of 600–2400 mm. 'Plate', intended as a constructional material for bridge structures, general building purposes, boilers, ship hulls, car and railway rolling stock bodies, etc., is rolled up to a length of some 36 m, depending on the final use of the product. Rolling is done either from slabs or flat billets. Thicker plates are usually produced in linear, two-stand, two-high reversing systems ((1) in *Figure 4.37*) followed by a train of four-high finishing stands. Typically, thinner plates may be rolled from slabs in a single stand, Lauth linear system ((5) in *Figure 4.37*) or, for greater accuracy, in a planetary mill.

More recently, thick and thin products have been rolled in semi-continuous systems ((6) in *Figure 4.46*) employing two universal stands ((4) in *Figure 4.39*) in tandem, followed by a continuous mill arrangement.

Strip and medium heavy sheet are produced in lengths of 5–20 m on either linear or planetary mills; with the latter, in particular, high thickness reductions of up to 95% are obtained.

Thin sheet, intended as structural material, is often rolled, as separate items, to a length of approximately 1800 mm, or alternatively in long lengths in continuous mills ((d) in *Figure 4.46*) and is then coiled. The car and electronic industries, in particular, make considerable use of steel and aluminium alloy sheets and strips coated with thin layers of plastic material. The metallic product used in such applications varies in thickness from some 0.3 to 2 mm.

Cold rolling operations have a two-fold objective: to improve the mechanical properties and dimensional accuracy of the product and to enhance the economic feasability of the operation (hot rolling of materials to less than about 2 mm thickness is uneconomical). Due to economic considerations hot-rolled strip or sheet is further processed in a variety of mills (very similar to those shown in *Figure 4.37*), but particularly in planetary or cluster mills if very thin material is required. A typical arrangement employed in the latter case is shown in *Figure 4.47*. The usual working sequence consists of pickling the hot-rolled material, rolling to the required size, followed, where necessary, by heat treatment and, finally, by general finishing operations.

The range of thicknesses that can be obtained depends, to a degree, on the properties of the starting material, since the roll forces and torques required can be very high. It is often because of this that cluster mills, characterised by their high rigidity, are used. Low carbon steel can be rolled down to a thickness of about 0.2–0.4 mm. Similar strip thicknesses are obtained in Inconel, Monel, Nimonic and bimetallic combinations, rolled in Sendzimir mills which use rolls of 450–1300 mm in length. Consecutive rolling operations are employed to produce very thin foils.

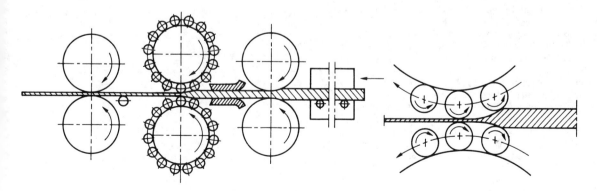

Figure 4.47 System incorporating a planetary mill for cold strip rolling

4.5.5 Rolling of structural sections

Semifinished or hot-finished structural sections are rolled in semicontinuous ((6) in *Figure 4.46*) continuous ((d) in *Figure 4.46*), Universal ((4) in *Figure 4.39*), linear ((b) in *Figure 4.46*), or H mills ((7) in *Figure 4.37*). Although a variety of shapes and components can be made on such mills, of particular engineering interest is the rolling of medium and heavy sections. Depending on the actual sizes and weight per length run, these comprise channels, I- and T-beams, rails and angular sections, as well as semiproducts intended for further processing, such as rounds, squares and flats. Without being specific about the type of mill, it is generally understood that 'bar mills' are used for rolling heavy sections and 'rod mills' for light sections.

Channels (50–450 mm high), designated as heavy sections when the height exceeds 200 mm) are usually processed in continuous mills, as are T-sections (30–220 mm high). The method of manufacture of I-beams is less uniform. Smaller sizes (below, say, 200 mm in height) are produced in continuous mills, whereas heavy sections (up to 600 mm and, occasionally, up to 1 m in height, with flanges of 200–400 mm in width) are rolled in universal-type mills. The method of rolling steel rails (shown by (5) in *Figure 4.43(a)*) depends on the weight per metre run. In general, sizes range from 8 to 75 kg m^{-1}, the 40–75 kg m^{-1} sizes representing heavy sections.

In most cases, two-line, two- or three-high, two- to seven-stand linear systems are used but, more recently, special universal and H-type mills have been introduced.

The semifabricates (rounds 6–200 mm in diameter; square sections 8–200 mm side) are produced on semicontinuous and continuous mills, as are rods (5–8 mm in diameter) intended for wire manufacture. These products are rolled in up to 39 stands.

4.5.6 Rolling of seamless tubing

Seamless tubing is produced by either longitudinal or cross (oblique) rolling of hot finished rounds. Both manufacturing routes start with pickling, dressing, heating and piercing the billet to provide a roughly made tube which can then be further hot and, if necessary, cold processed. The latter stages are intended to give dimensional tolerances to the product and, in the case of cold processing, to impart both better mechanical properties and, often, better surface finish.

The two routes comprise the following intermediate processes:

(1) *Longitudinal rolling* consits of rotary or punch piercing, plug or mandrel rolling. It occasionally may incorporate the Erhardt or push-bench process, followed by the finishing sequences of sinking and stretch reducing.
(2) *Cross-rolling* involves rotary piercing, Assel elongating (possibly a combined piercing–elongating process) or Diescher elongating.

Irrespective of the route followed, the piercing stage precedes further operations and consists either of rotary piercing of a solid round or hydraulic punch piercing of a square billet.

Of these two methods, cross-rolling (*Figure 4.48*) is carried out in a two- or three-roll system in which the rolls are inclined at an angle (feed angle α) to the central axis of the mill, and a plug (situated on a long bar) is inserted between the rolls. The plug bar can either be free to rotate or driven.

With the rolls set obliquely to each other, the billet is drawn into the pass by frictional tractive forces, the axial component of the forces being responsible for the forward movement and

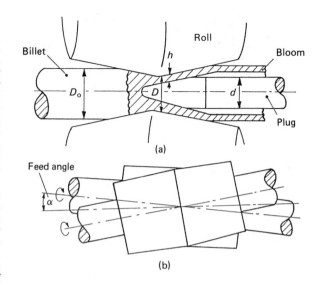

Figure 4.48 Rotary piercing of rounds

the tangential component for the rotation. The physical conditions that develop in the pass are equivalent first to oblique rolling and then to cross-rolling. The cross-rolling zone extends from the plug nose to the exit from the pass, and the billet is worked between the roll and plug surfaces.

The process is not continuous, in the sense that the successive arcs of the billet circumference come, in turn, into contact with the rolls. The set-up clearly tends to develop ovality in the billet, unless horseshoe-type adjustable guides are introduced between the rolls. Up to a point, these guides rectify this undesirable condition and help to preserve circularity of the tube.

Three basic roll profiles are used (*Figure 4.49*). The original Mannesmann barrel-type roll has been supplemented by the cone and Stiefel disc rolls.

In a barrel-roll system the conditions of pure rolling do not obtain and, therefore, there is a 'mismatch' of the tangential

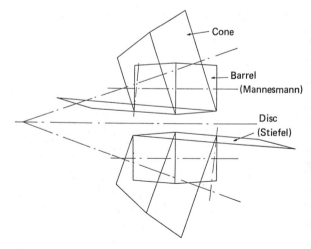

Figure 4.49 Basic profiles of piercing rolls

velocities of the roll and the billet at various sections along the pass. This is compensated for by the slipping and twisting of the billet and finally results in the introduction of a considerable degree of inhomogeneity of flow.

In the Stiefel system, the matching of the roll and billet surface velocities over a finite length of the tool–billet engagements is not easy and, owing to the shape and required setting of the tools, the rigidity of the roll assembly is low. The range of tube sizes that can be produced is therefore limited, and the system is not often used on its own although it is incorporated in the Diescher elongator.

The cone-type roll does not solve the problem of velocity matching either, but rather shifts the emphasis. Whereas a match between the two velocities is possible in the cross-rolling zone of the Mannesmann pass, it is in the entry, i.e. in the oblique zone, that the matching can be effected when using cone rolls.

The pierced billet, which in tube-making terminology is called the 'bloom', can now be elongated to reduce both its outer diameter (this remains almost unchanged in the piercing operation) and the wall thickness.

4.5.6.1 Longitudinal rolling

The main elongation of the bloom, to produce a hot finished hollow, is carried out in either a plug or a mandrel (continuous) mill and is then completed in one or two finishing stages.

In the plug-mill system, the plug, a short cylindrical, detachable tool, is maintained in position by a plug bar, an arrangement similar to that in the rotary piercing plant, and controls the dimensions of the bore of the hollow. The work piece passes through grooved, oval rolls that are usually positioned in two stands and maintained at 90° to each other. The mill is

generally used as a braking-down stage and operates on tubing of 3–40 mm wall thickness and 20–400 mm diameter.

In the mandrel-mill system the hollow bloom is threaded on a long mandrel and is fed into a continuous train of 7–10 pairs of rolls arranged in tandem. The tube is rolled out on the mandrel without interruption to approximately the finished thickness. The axes of the rolls are alternately at right angles to one another. The earlier breaking-down passes have considerable ovality. The finishing passes are designed to leave the tube loose on the mandrel and more or less circular so that no reeler is required to enable the mandrel to be withdrawn. A typical sequence of reductions is shown in *Figure 4.50).*

The method is no more continuous than any other seamless process which starts with a billet of finite length and rolls it into tube. The continuous mill is outstanding for producing long lengths of all thickness at a very high rate of output.

The final rolling stage uses a sizing or reducing mill to give the tube its final outside diameter with, preferably, little change in wall thickness. Reducing mills fall into three broad categories:

(1) sizing,
(2) stretch reducing, and
(3) sinking.

Sizing mills These are used to produce an accurately sized finished tube without appreciably reducing the outer diameter. They are usually equipped with three to seven stands, each containing two rolls, and are normally designed for an overall diameter reduction of 3–8%. In general, the diameter reduction is accompanied by a thickening of the tube wall.

Sizing mills are usually installed in plants designed to produce the larger tube-size range, but can be used with advantage in the manufacture of smaller tubes when following main seamless processes that are reasonably flexible in the sizes that they can produce.

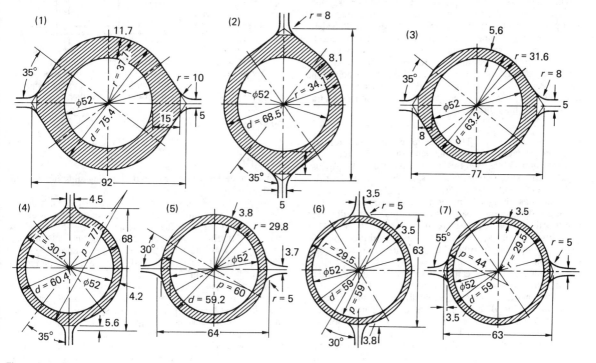

Figure 4.50 Typical reduction sequence in seven stands

Reducing sinking mills These comprise a continuous train of passes designed to reduce a larger tube to a smaller one without a mandrel or plug to support the bore. The amount of reduction per pass is usually 3–5% according to the type of mill.

A reducing mill is usually employed in conjunction with a main unit to improve the performance of the latter. This may be done by:

(1) reducing the number of size changes to be made in the main mill;
(2) increasing the weight of the billet worked by the main mill and, therefore, the rate of output; and
(3) increasing the maximum length available, and decreasing the available minimum diameter.

In obtaining these advantages the following disadvantages must be taken into account:

(1) A reheating furnace is necessary;
(2) an additional process is thus introduced increasing costs of fuel, power, labour and maintenance;
(3) tubes become thicker during reduction and, therefore, the main mill must roll thinner gauge; and
(4) the quality of the bore of the tube deteriorates for various reasons and this limits the amount of reduction possible.

These mills are usually designed with up to about 24 two-roll stands, and are capable of overall diameter reductions of up to 55%. The stands are driven through bevel gear boxes from common line shafts, and the behaviour of the tube-wall thickness depends on the relationship between the gearbox reductions. It is normal for sinking mills to operate with no axial tension between the stands so that the wall thickens freely. If, however, some tension is applied, wall thickening can be restricted to about half of that which would occur without tension.

Whilst heavy wall ends result from the sinking process, they are usually short and do not greatly influence the yield.

Of particular interest, from the point of view of roll design, is the effect that slipping in the rolls can have on the circularity of the tube. The work of Blair (see the Bibliography) concerned with two-, three- and four-roll passes, introduces the concept of the 'effective radius' and emphasises the point that this is the most important dimension to be considered when designing rolls. The salient points of this statement are summarised, diagrammatically, in *Figure 4.51*.

Stretch reducing mills These are primarily used for the economic production of small diameter thin-walled tubes. As the name implies, the wall thickness is reduced in the process and, to achieve this condition, considerable tension must be applied between the stands (*Figure 4.52*).

The mills are designed with up to 22 stands, each with two or three rolls, and are capable of overall diameter reductions of more than 75%, and wall reductions of up to 40%. To obtain the control required over the interstand tensions, the mills are designed with either d.c. motor drives to each stand or hydraulically regulated drives.

As the ends of the tube cannot be subjected to full tension in the mill, they are inevitably heavy walled and of a length that is proportional to the elongation, stand spacing and percentage wall reduction. The end crop can have a significant effect on the yield and, therefore, it is necessary to put the longest possible length into the mill to minimise the percentage loss.

In addition to the essential change in the outer diameter of the tube in the reducing operation, there is always some change in wall thickness. The thickness change plays a vital

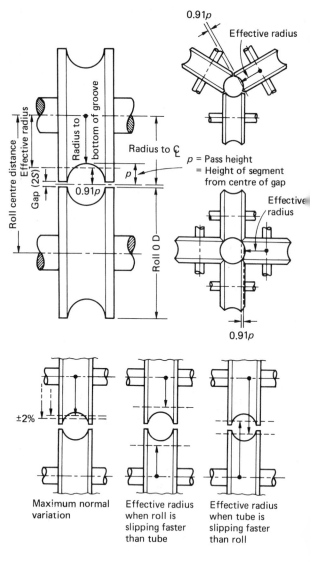

Figure 4.51 Effective radius in roll design

role in the design and operation of reducing mills and depends on the rolling condition.

No-stretch conditions prevail when a tube passes through a single stand, so that its diameter is reduced without any external axial forces being applied to it; the thickness also increases. In this case, where the change in wall thickness occurs purely as a result of the compressive action of rolls on the tube, the reduction is said to take place under the 'no-stretch condition'.

In a mill consisting of a number of stands that are all reducing the tube simultaneously, this condition can arise providing that the roll speeds in successive stands are set such that there is no tension or compression in the length of tube between each pass. Since there is always an increase in the linear speed of a tube as it is reduced, the roll speeds in successive stands must be increased in relation to the elongation that occurs. It follows that if a tube is being continuously

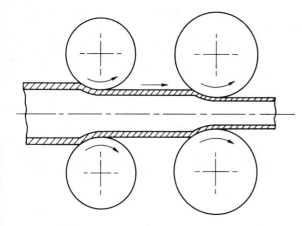

Figure 4.52 Rolling with interstand tension in a stretch-reducing mill

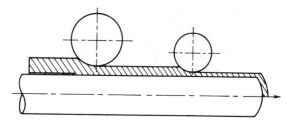

Figure 4.53 Push-bench process

processed through a number of reducing mill stands, the volume of flow must be constant. The cross-sectional areas multiplied by the linear speed must, therefore, be the same at all points in the mill, and as the tube area is reduced its speed must correspondingly increase.

Under no-stretch conditions, the whole tube thickens uniformly throughout its length, the amount of thickening being dependent on the diameter to thickness ratio of the in-going tube and the total diameter reduction given. It is unaffected by the reduction given in each stand or by the number of stands working on the tube at any time.

Conversely, in no-stretch conditions, if tension occurs between the stands, wall thickening can be partly or fully compensated for, or the wall thickness can be reduced. Tension is applied by arranging for each pass to run at a higher speed, relative to the preceding pass, than is required for no-stretch conditions.

With *stretch conditions*, not only is the thickness reduced between the stands but the diameter is also reduced so that the actual reduction in diameter brought about by the compressive action of the rolls is slightly less than that which takes place with no stretch, resulting in slightly less thickening of the tube wall between the rolls. Stretching can occur only on a section of tubing which is between two passes. Consequently, the extreme ends of the tube can never be subjected to stretch nor can an appreciable length at the ends suffer the full stretching action. Therefore, a mill that is designed to stretch or in which stretching occurs, produces a tube that is thinner in the centre than at the ends, the thickness at the extreme ends being that which would arise under no-stretch conditions and the centre thickness being that arising from the stretching action. The thickness change is not abrupt but gradual and there is normally an appreciable length of uniform thickness between the thickened ends.

Push-bench process This process (*Figure 4.53*) forms an alternative to the above routes. The standard push-bench system incorporates punch-piercing of the hot billet, while retaining a solid base on the leading end, elongating on the bench proper and hot-reducing on an appropriate mill. To retain a reasonable degree of concentricity, the depth of penetration of the piercing punch must not exceed seven times the punch diameter. The 'bottle' which is fed into the bench is relatively short and the process may not be fully economical.

Modern developments have led to the introduction of roller-type dies (the Manfred–Weiss process) and, even more recently, to the incorporation of an elongator in the cycle. The elongator increases the depth of penetration to nine times the punch diameter while retaining good concentricity through the medium of equalisation of the wall thickness. Since the elongator will also reduce the diameter of the bottle, the amount of deformation to be obtained on the bench is lower and the quality of the manufactured tube is higher.

Some of the advantages of the elongator are, however, offset by the fact that the gauge correction cannot be carried out at the ends of the bottle and that some reduction in output must therefore be expected.

When the elongator forms an integral part of the cycle, reheating of the bottle, after punch piercing, will be necessary and will, of course, add to the cost of the operation.

The modern push-bench is normally used for the manufacture of tubing of outer diameter of 6–15 cm. Thin-walled tubing can be obtained in lengths of up to 10 m when an elongator is used, and in lengths of up to 6 m without the elongator.

The advantage of the modern roller-die beds of the bench proper over the old ring-die system lies in the reduction of thrust required by about 30%, and a reduction in the size of the die bed by about 25%. Although the provision of three or four rolls in a cluster is more expensive than the installation and maintenance of ring dies, the considerable increase in speed of the working cycle, resulting from the possibility of using more passes on a given bed, and from the actual increase in the axial speed of the operation, offsets the apparent economic disadvantage. High speeds of operation are, in fact, required in the case of alloy steels which harden rapidly with a drop in temperature.

4.5.6.2 Cross rolling

The two major rotary production processes employed are those of Mannesmann and Assel. In the Mannesmann cycle a cold centred round billet is heated in a roller-type furnace, pierced in an oblique rotary mill, possibly repierced on another rotary piercer, hot forged, reheated in a continuous furnace and stretch reduced. The piercing and repiercing operations are of primary importance from the point of view of the incidence of redundancy.

In the Assel cycle (*Figure 4.54*), the billet is heated in a roller-type or rotary-hearth furnace, pierced in an oblique rotary mill, elongated in the Assel mill, reheated in a continuous furnace, and then either sized or sunk on a longitudinal rolling mill. The first two stages, i.e. piercing and elongating, account for most of the redundant shears induced in the worked material.

To increase the range of possible deformations, it is occasionally necessary to follow the piercing stage by a repiercing or secondary piercing operation in which the bore of the pierced bloom is increased while its wall thickness is reduced. A purely elongating operation in which the bore remains

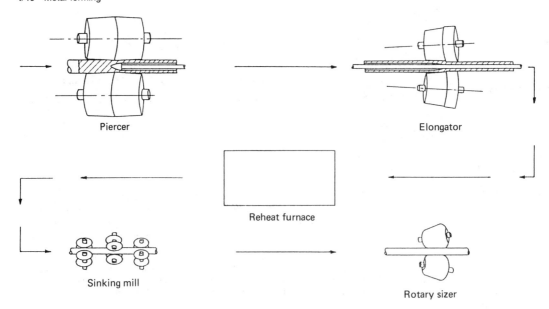

Piercer

Elongator

Reheat furnace

Sinking mill

Rotary sizer

Figure 4.54 The Assel cycle

unchanged and the wall is further deformed will follow. Secondary piercing operations have been developed from the basic Assel elongating technique and are considered in that context.

An Assel mill is shown diagrammatically in *Figure 4.7*. In this process, the previously pierced tube, with a cylindrical floating mandrel inserted in the bore, is drawn by frictional forces into the pass defined by the setting of the rolls. In the entry zone of the pass, the diameter of the tube is slightly reduced. The hump of the roll then reduces the wall thickness, and a smoothing operation takes place in the reeling zone of the pass to remove the triangulation of the vertical section of the tube. Final rounding of the tube proceeds in the exit zone of the roll. Elongation occurs primarily as a result of wall thinning, but it is also slightly influenced by the reduction in tube diameter.

To obtain the required feed angle, the rolls are rotated about the mill axis by applying the 'housing angle' or, in other words, by rotating the plate containing the roll entry end chocks about the mill axis. The variation in the housing angle gives a range of feed angles.

The roll axis is inclined to the mill axis in such a way that the axes of all three rolls always intersect at a point on the entry side, forming what is known as the 'cone angle'. The cone angle for the zero housing angle is normally about 3° and remains so on all modern mills.

With this arrangement, the roll diameter increases along the pass and the roll surface velocity and, in particular, its axial component, increase accordingly. This permits the rolling of thinner walled tubes than was possible with the original Assel mills.

The main advantage of the Assel mill over the push-bench process is its flexibility. This shows itself in the ability of the mill to cope with a wide range of tube sizes, necessitating only minor changes in the tool setting.

The process cannot operate when the tube diameter/thickness (D/h) ratio exceeds 11.5 and, therefore, its field of application is for medium and heavy wall tubing. For reasons which, as yet, are not clear, but which appear to pertain to

strain-rate effects, the conventional Assel process fails in the case of stainless steels.

Triangulation is a problem usually associated with the Assel elongating. It is particularly pronounced in the thinner walled tubes where $D/h > 9$. At the leading end of the work piece, triangulation is constrained by the bloom behind the deforming section and shows itself only as a 'belling' out of the tube end. In the middle portion of the tube, triangulation is constrained by the lengths of the tube in front and behind the regions in which it originates. At the trailing end, however, if the amount of distortion is too great, the rolls may fail to round up the tube in the reeling region and a 'fishtail' end is produced (*Figure 4.55*).

The bulging of the tube between the rolls also results in a redundant, alternate bending of the wall in opposite directions as the latter passes from the rolls to the gap between them. This may give rise to wall laminations or even serious cracking.

Irrespective of the zone of the pass considered, a material element undergoing deformation in an Assel elongator will be subjected to a biaxial tension and uniaxial compression with the attending emergence of shears.

Economics of tube making demand the final ratio to be of the order of 8–12. However, to reduce both the excessive loads and the degree of triangulation, the reduction in wall-thickness obtainable in the Assel elongating stage is not normally allowed to exceed the height of the roll hump. The highest elongations are therefore obtained with small-diameter and thin-walled tubes, but even in this case, elongation factors are not in excess of 2. An increase in the size of the hump does not solve the problems associated with the processing of larger, thick-walled tubes. The hump constitutes a discontinuity in the Assel pass and, therefore, affects redundancy. An increase in the effects of such a discontinuity would result in an increase in the degree of the severity of the operation.

It is for these and other practical reasons that an intermediate stage of secondary piercing is used in some tube-making cycles. If a piercing plug of the type shown in *Figure 4.48* is

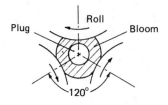

Three-roll system

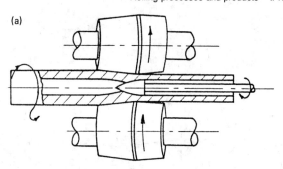

(a)

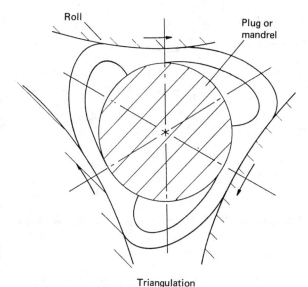

Triangulation

Figure 4.55 Triangulation in Assel elongation

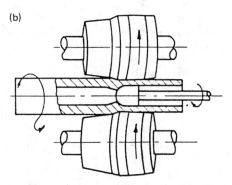

(b)

Figure 4.56 (a) Two-roll barrel mill. (b) Calmes elongator

used between Assel-type rolls, the maximum possible reduction in the wall thickness will no longer be limited by the height of the hump since the total reduction will be accomplished by both the roll and the shoulder of the plug. Typical of these arrangements are the two-roll barrel-type machine and the Calmes elongator (*Figure 4.56*). In the latter (used also in the push-bench cycle) the hump is used to correct eccentricity and does not reduce the wall thickness. The elongation of the tube is achieved between the plug and the parallel portion of the roll.

Surface faults are common, especially so in the Calmes elongator. It seems, however, that a similar degree of redundancy exists in both these secondary piercing operations. Stress systems present in either case are similar to those encountered in the Assel elongated bloom.

A more recent development in the field of secondary piercing consists in the introduction of three Assel-type rolls in conjunction with a profiled plug (*Figure 4.8*). The plug is positioned in such a way that both its shoulder and the roll hump lie in the same vertical plane. An increase in the bore and a reduction in the outer diameter of the bloom take place in the same section of the pass. The 3° Assel cone angle is retained in the system. Again, stress conditions are basically those of biaxial tension and uniaxial compression.

A possibility also exists of combining rotary piercing and elongating operations into a single process to be carried out on a rotary mill equipped with suitably designed tools.

For the combined process to be industrially successful, large deformations coupled with soundness of the product must be achieved. The selection of the roll profile depends, on the one hand, on the available length of the working contact and, on the other, on the possibility of accommodating physically both the piercing and elongating zones. The first condition precludes the disc-type roll and the choice is, therefore, limited to either the Mannesmann or cone roll. The use of either profile will introduce a mismatch of surface velocities and this will result in tangential slip and twisting of the billet. The presence of the plug and the variation in the frictional tractive force will produce axial slip. Since both the tangential and axial slips, especially the latter, alter the stress system in the working zone of the pass by increasing the magnitude of stresses, and particularly that of shearing stresses, it is essential to try to minimise their magnitudes. This is done effectively by matching the surface velocities of the roll and of the billet and, therefore, reducing, in the first instance, the magnitude of the tangential slip. The matching of velocities is more important in the cross-rolling part of the pass than it is in the entry zone, because it is the cross-rolling zone that determines the initial rolling conditions. The barrel-type roll, is therefore, preferred.

Essentially, the combined process consists of a three-roll system incorporating a barrel-type piercing and Assel type elongating roll (*Figure 4.57*). The process is basically that of the Mannesmann piercing type, i.e. the consideration of the piercing operation is of greater importance than that of the elongating operation. The mill is run as a Mannesmann piercer

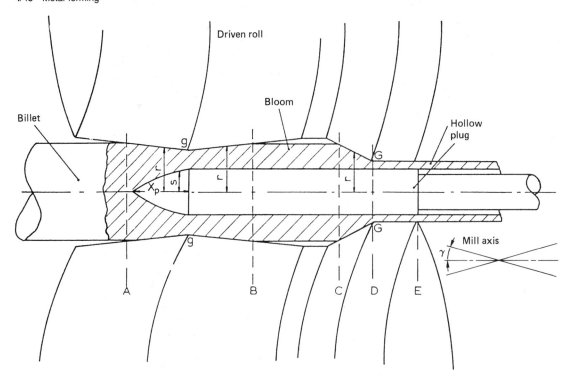

Figure 4.57 A combined rotary piercing–elongating process: AB, piercing; CD, elongation; DE, reeling; gg, piercing gorge; GG, elongating gorge; X_p, plug advance

and no use is made of the cone angle that would be used in an Assel elongator. The amount of deformation in the elongating zone is conditioned by the setting of the piercing zone of the pass.

Triangulation is a feature of the tube made in a combined process, but a general improvement in the quality of the product, as compared with a two-mill production, is observed. An element of the processed material is subjected to a system of triaxial compressive stresses in the oblique zone of rolling in front of the plug, and to biaxial tensile–uniaxial compressive system in the cross-rolling zones of the piercing and elongating portions of the pass. The effect of this stress system will manifest itself in the presence of axial, circumferential shearing strains, as well as that due to twisting of the billet. The three components of redundant deformation are shown in *Figure 4.58*.

Of special interest in tube making is the production of stainless-steel tubing. This, as will be seen later, is generally manufactured in extrusion, but rotary processes are also used, since the rate of yield is likely to be higher.

The production routes used in this case are somewhat complex and reflect both the cost of the material and basic processing difficulties. The main process lines are:

(1) punch piercing, Calmes elongator, rotary hot forging;
(2) punch piercer, push-bench, polishing mill;
(3) disc rotary piercer, plug mill; and
(4) Mannesmann rotary piercer, hydraulic push elongator, rotary hot forging.

A possible alternative to asseling is offered by the Diescher elongator. This mill incorporates two oblique barrel rolls and two discs (set at right-angles to the rolls) which replace the

horseshoe guides necessary (see piercing) to correct the ovality of the tube. Elongation is carried out between the rolls and a bore-supporting mandrel. The mill improves concentricity of the tube and, by effecting elongation in both forward and backward directions relative to the gorge, imparts a burnishing finish to the tube.

4.5.7 Tooling for tube rolling

The design of tooling, consisting of plugs, rolls and mandrels, for tube rolling processes must take into account the inherent severity of these processes, particularly in the oblique, cross-rolling systems. The problems arising here depend very much on whether the mill in question uses two or three rolls.

Three major faults associated with a two-roll system are central cavitation (or rupture), wall lamination and ovality. Cavitation occurs axially in front of the piercing plug and causes longitudinal uneven rupture and opening in the body of the billet (*Figure 4.49(a)*). Wall lamination is equivalent to material failure and introduces an unacceptable lowering in the quality of the product. Ovality, inherent in the geometry of the system, introduces an element of redundancy and has to be controlled by means of rigid horseshoe guides positioned in a vertical plane between the rolls, or by discs of the type employed in the Diescher process.

More recent developments in the design of rotary piercers have led to the adoption of a three-roll system in which driven rolls are placed at 120° with respect to each other. The introduction of three rolls changes drastically the stress system in the oblique rolling zone, converting it from a biaxial tensile–uniaxial compressive to a triaxial compressive one. The stress system existing in the cross-rolling zone remains,

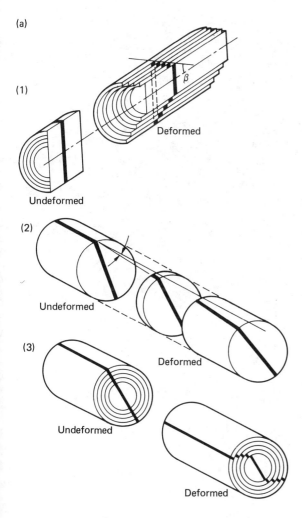

Figure 4.58 Development of longitudinal (1), due to twist (2) and circumferential (3) shearing strains in rotary piercing of billets

however, unchanged. Cavitation cannot occur in a three-roll piercer or elongator. Some wall lamination is, nevertheless, possible and ovality is replaced by triangulation.

The incidence of these faults in rotary piercing and the attending inhomogeneity of flow (of varying degrees) are responsible for an increase in the redundancy. An example of this is shown by *Figure 4.59(b)* which clearly indicates the much higher levels of redundancy present in a two-roll system.

In the three-roll system, the problem of cavitation is replaced by the difficulties created by the presence of triangulation. In addition to the appearance of a triangular 'fishtail' at the trailing end of the tube, which can cause jamming, surface and annular cracking of the wall can also occur. The latter is, to some degree, of lesser importance, since with a reasonable quality material cracking will tend to occur after the size of the fishtail has become excessive. It is unlikely, therefore, that in normal production runs this situation will be allowed to develop. It should be realised that the appearance of a fishtail is indicative of the incidence and severity of triangulation, in the sense that the same effect would have been noticed along the whole length of the processed tubing if the smoothing and

ironing action of the rolls was stopped. In effect, therefore, a fishtail represents the only visible sign of triangulation.

The appearance of triangulation is of much greater importance in the elongating processes than in three-roll piercing because it effectively imposes an upper limit on the diameter/wall-thickness ratio of the tube that can be attained.

The real effect that triangulation has on the shape of the tube is shown in *Figure 4.60*. The figure clearly shows two characteristic aspects of the phenomenon: i.e. an appreciable thickening of the wall between the roll and mandrel or plug and before entry to the pass, and the fact that the maximum curvature of the bulge between the rolls is displaced away from the central position towards the inlet to the roll.

The thickening of the wall at the entry to the roll becomes more prominent as the triangulation and elongation increase. The effect appears to be caused by the piling up of material at the entry to the interface between the tube and the inner tool. At the inner surface of the tube the metal is rolled back out of the gap formed by the roll and the plug or mandrel. This backward flow of metal, relative to the remainder of the bloom, will tend to produce the triangular shape in the specimen. The velocity pattern developed in the wall of the tube causes bending stresses which, in turn, produce the bulging of metal between the rolls.

A confirmation of this hypothesis seems to lie in the fact that the apex of the bulge is, as already indicated, displaced towards the inlet to the roll.

Quantitative determination of the magnitude of triangulation presents some practical difficulties, in the sense that the shape factor can be ambiguous. To simplify the matter, the triangulation factor S, which determines the departure from a cricular cross-section to that of an almost equilateral triangle, has been defined as:

$$S = \frac{x}{D} \qquad (4.50)$$

where x is the mean of the three sides of the triangular cross-section of the trailing end of the tube, and D is the diameter of the finished circular tube obtained in the considered pass. Effectively, therefore, in the absence of triangulation $S = 1$.

While cavitation is removed by the use of a three-roll system, axial and tangential internal shearing persists in both systems unless the geometry of the pass is rationally designed.

The use of the constant rate of homogeneous strain (CRHS) and constant mean strain rate (CMSR) concepts will produce suitably profiled passes and will minimise the severity of the operation. In view of the cost of large rolls, it is more customary to concentrate on profiling the plugs in piercing and repiercing operations, since the wear on the relatively small tool is low and, in any case, the cost of replacement is insignificant. The two design concepts produce considerably different plug profiles, as shown in *Figures 4.32* and *4.61*.

In the elongators, and especially in an Assel mill, with a cylindrical mandrel of a somewhat basic shape supporting the bore, the only possibility of rationalising pass design lies in profiling the hump of the roll. Such a set of profiles is shown in *Figure 4.62*.

Adoption of this approach results in a substantial reduction in the redundancy factor in the accelerated (ACRHS), uniform (UCRHS) and decelerated (DCRHS) passes as compared with a standard industrial (INDST) system. For a specific 8° feed angle Assel pass, an average difference can amount to as much as 180%, as shown in *Figure 4.63*. A further advantage of using a correctly designed pass is the improvement in the degree of triangulation (see *Table 4.7*).

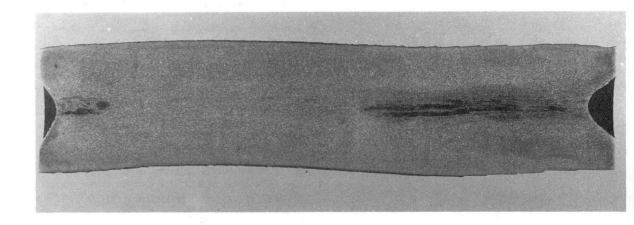

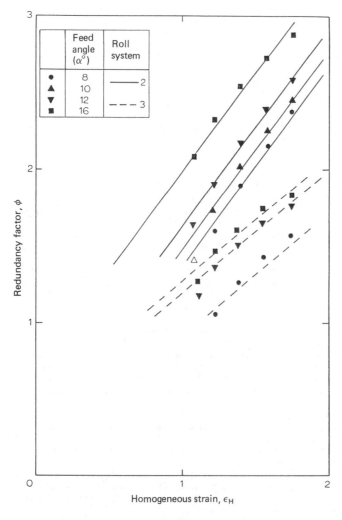

Figure 4.59 (a) Axial cavitation in front of plug nose in two-roll, oblique piercing system. (b) Redundancy factor and homogeneous strain in two- and three-roll oblique piercing systems

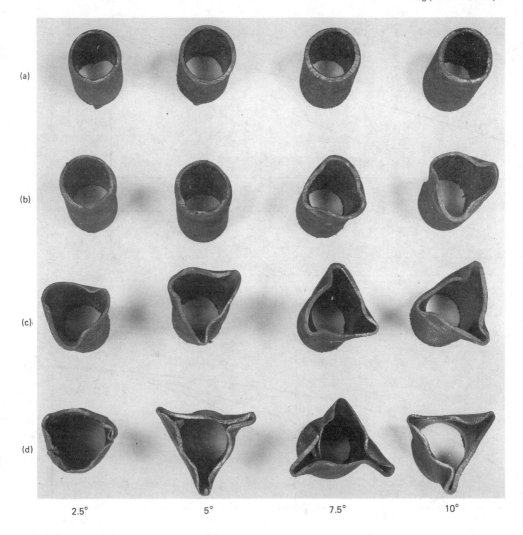

Figure 4.60   The incidence and effect of triangulation in a three-roll system: (a) $D/h = 10.3$, (b) $D/h = 11.6$, (c) $D/h = 13.8$, (d) $D/h = 18.6$

4.5.8   Flow forming

Practical difficulties of producing large diameter, thin-walled long tubing in a variety of materials, including stainless steel, has introduced a new approach because the ordinary manufacturing techniques could not always cope with the sizes and materials involved and, even where this was possible, the cost of production was very high.

A solution to this problem was first proposed, on a modest scale, in the early 1950s when the horizontal flow-turning lathe was introduced. This was followed by a vertical cold 'floturning' machine in which three stands, carrying five rollers each, were introduced. In this, short thick-walled bloom is supported in the bore by a solid cylinder of hardened tool steel machined to the exact tolerances and to the degree of surface finish which is required of the bore. Each station, capable of angular adjustment to obtain optimum forming conditions, is powered hydraulically. Rollers are advanced into the forming positions and retracted at the end of the cycle. The wall thickness of the bloom is easily reduced by one-sixth of the

total reduction by each roller. As an example of the possibilities, in one pass a steel cylinder of 1.25 m outer diameter, 1.25 mm thickness and 3.5 m in length can be formed from a hollow of 8 mm thickness and about 0.75 m in length. In effect, therefore, the machine reduces the wall thickness of the hollow, elongates the work piece between the rollers, but retains the original diameter of the bore. The control of the mechanical properties of the material can be maintained through annealing, and that of the surface finish can be maintained through the finish of the roller surfaces.

Some limitations on the length of the tubing obtained on a vertical machine can be expected and this appears to be the reason for a parallel development of horizontal 'floturning' equipment. The floturn process does, in fact, utilise, in general, the horizontal-type machine.

The development of horizontal machines was initially stimulated by the demand for large diameter, long tubing required for the boosters on space vehicles. The equipment designed for this purpose is capable of turning out precise tubing of up

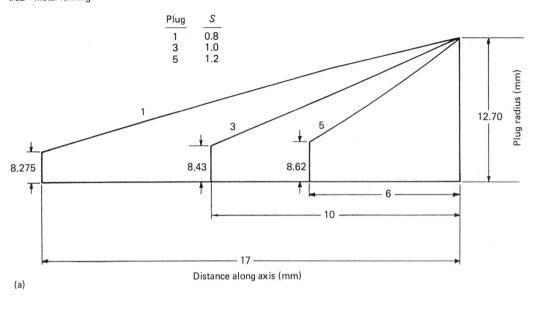

(a)

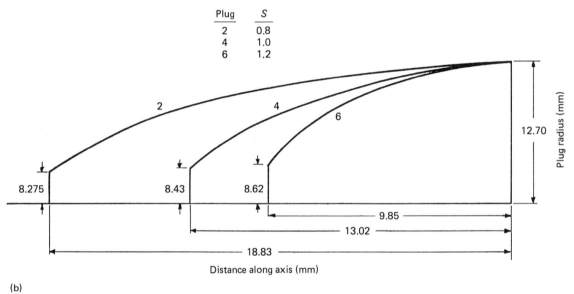

(b)

Figure 4.61 (a) CRHS and (b) CMSR piercing plug profiles for decelarated (0.8), uniform (1.0) and accelerated (1.2) deformation rates S

to 635 mm in outer diameter and of 30 m length in alloy steels, titanium, Monel, aluminium and stainless steel.

Essentially, the process consists of reducing the diameter of an initially thick-walled, short hollow by threading and clamping the latter on a rotating mandrel, and driving a carriage with the forming rolls along the work piece. The carriage contains three rollers located at 120° to each other which, when driven slowly over the surface of the hollow, will reduce its thickness. This particular process, known as 'par-forming', can retain wall-thickness tolerances to within ±0.51 mm on, say, a 655 mm outer diameter, 2.95 mm thick and 13 m long aluminium tube, and on a 508 mm outer diameter and 2.15 mm thick stainless-steel component.

To achieve these high tolerances, a certain amount of preparatory dressing of the hollow is necessary. Initially, the hollow is turned, faced and machined in the bore. The bore is usually honed to remove tool marks and to impose the required tolerances. The tube should fit the mandrel to within 200 μm and, for the final tube sizes stipulated above, it should be about 1.5 m long. Mandrels are made of high quality chromium–molybdenum–tungsten steels and are ground to give tolerances of the order of 51 μm.

The actual forming operation takes place in two stages. Again, for the sizes discussed, something of the order of five passes of the forming rolls over the hollow will be needed, these will be followed by annealing and, finally, by two further finishing passes.

Machines of this type are fully capable of forming tubular components with shoulders, tapers and partial or full closures. The range of tube sizes varies from about 75 to 700 mm in

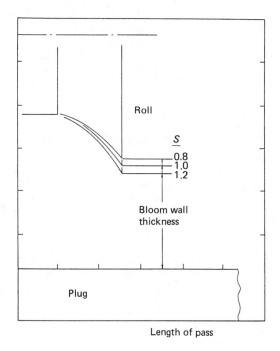

Figure 4.62 The CRHS-type Assel roll hump design

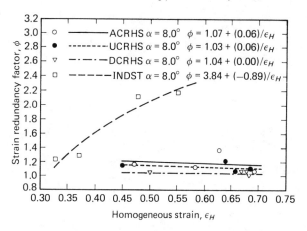

Figure 4.63 Redundancy factor and homogeneous strain in an Assel CRHS elongating pass (feed angle 8°). Rate: A, accelerated; U, uniform; D, decelerated

outer diameter, the range of wall thickness being 0.38–3.8 mm. Surface finish can be controlled to within 380 μm on the bore and 760 μm on the outer surface.

Floturning machines are often incorporated in mechanised plant and this facilitates cleaning of tubing prior to adhesive bonding, where required, of two dissimilar metallic components. Cleaning sequences depend on the material processed but, in general, will include emulsion, water and alkaline cleaning for non-ferrous materials (e.g. aluminium) and pickling in nitric hydrofluoric acid for, say, stainless steels.

The capital cost of the equipment involved in these operations is obviously high, as is the cost of hollow dressing, and the cost increases with the increase in the outer diameter of the tube. Clearly, the existing industrial units serve very specific purposes and are not necessarily intended for ordinary commercial purposes. In their own field of application, i.e. large diameters, thin walls, and substantial lengths, they are well advanced compared with other methods.

The possibility, however, of using the same principle for processing small diameter tubing has not been neglected. One of the main problems arising here is that of buckling. In the nose forming of steel tubes, where the components are compressed on a mandrel by conical rollers, the effect of wall thickness on the stability and formability of the tube is noticeable. The reduction in the tube diameter depends only on the applied axial stress and is essentially independent of the cone angle and wall thickness. Up to a reduction in diameter of about 30%, no change in wall thickness is reported to occur, and for specimens of 2–4 mm thickness no buckling occurs over a range of reductions. For specimens of 1 mm wall thickness, buckling of the tube occurs at a cone angle of 30° and a compressive stress of about 770 kPa, and at an angle of 45° and a stress of 310 kPa.

4.5.9 Rolling of machine parts

Although developed primarily for tube manufacture, cross-rolling processes are now often employed, for reasons of economy, in the production of a variety of machine parts.

These range from a three-roll, oblique rotary grooved shaping of a shaft (*Figure 4.9*) to the helically two-roll manufactured special hubs (*Figure 4.64*). Three-roll (inclined at 120° to each other) finned-roll systems are used in the production of finned tubing of high dimensional accuracy, whereas grooved two-roll mills are used to produce ball bearings (*Figure 4.10*). Bevel and spur gears, as well as bearing races and similar components are manufactured by

Table 4.7 Critical values of feed angles producing triangulation

| Process* | Feed angle, α (°) | | | | | | | | | |
| | 6 | | 8 | | 10 | | 11 | | 12 | |
	D/T = 5.4	D/T = 10	D/T = 5.4	D/T = 10	D/T = 5.4	D/T = 10	D/T = 5.4	D/T = 10	D/T = 5.4	D/T = 10
ACRHS	0	0	0	▽	0	▽	▽	—	▽	—
UCRHS	0	0	0	▽	0	▽	▽	—	▽	—
DCRCHS	0	0	0	▽	0	▽	▽	—	▽	—
INDST	0	▽	0	▽	▽	▽	▽	—	▽	—

* See text for details.

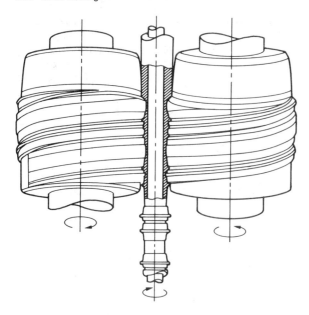

Figure 4.64 Cross-rolling shaped-hub manufacture

cross-rolling with a very substantial saving in material as compared with many machining production routes in a variety of rolling systems.

4.5.10 Sheet roll bending

An interesting development in the area of cold-rolling is that of using especially profiled rolls to bend metal sheet (usually aluminium, but also hot or cold rolled steel) sheet into simple or even complex shapes that serve as structural components in the car and aircraft industries.

Bending is carried out in machines incorporating roll trains of driven and idle tools, which can produce long length (up to 30 m) suitably profiled components, without changing the initial thickness of the sheet.

4.5.11 Basic rolling theory

The problems involved in any mathematical modelling of rolling processes are complex and cannot possibly be discussed in a brief review. Specialised monographs and texts are available on this subject (see Bibliography).

An indication of the methods used is provided here with reference to the three most basic modes of flat, longitudinal tube, and cross-tube rolling.

4.5.11.1 Cold rolling of strip

In the process of cold rolling of strip between smooth cylindrical rolls, the frictional force between the stock and tools is low, but at the standard speeds of these mills, the flow stress characterisitcs of the material are unaffected by the rate of deformation. The direction of frictional forces in the pass changes from providing tractional effects at the entry (the roll velocity is higher than that of the strip) to opposing the delivery at the exit (the strip moves faster than the rolls). A neutral plane does exist in the pass and may change its position if there is a variation in the relative velocities. In the normal

course of events, the strip will be subjected to back p_1 and front p_2 tensions and, in addition, the original roll radius R will in many cases become elastically distorted to, say, R' and thus alter the assumed circular arc of contact between the strip and the rolls.

Although a certain degree of inhomogeneity will be present, as indicated schematically in *Figure 4.65*, this is normally neglected in a basic analysis of the rolling operation. The analysis leads to the following final expressions:

Normal roll pressure, s   On the exit side the pressure is

$$s^+ = \frac{2kh}{h_2}\left(1 - \frac{p_2}{2k_2}\right)e^{\mu H} \tag{4.51}$$

and on the entry side the pressure is

$$s^- = \frac{2kh}{h_1}\left(1 - \frac{p_1}{2k_1}\right)e^{\mu(H_1 - H)} \tag{4.52}$$

where subscripts 1 and 2 refer to the entry and exit, respectively, $k = Y/2$ is the yield stress in pure shear, h is the strip thickness, μ is the coefficient of friction, and

$$H = 2\sqrt{\frac{R'}{h_2}}\tan^{-1}\left(\sqrt{\frac{R'}{h_2}}\,\phi\right) \tag{4.53}$$

where ϕ is the angle subtending the arc of engagement considered.

Both of the equations above are obtained by using the equilibrium of forces analytical method, and by considering the geometry of the system.

The roll force, P   The roll force per unit length is given by

$$P = \int_0^{\phi_1} sR'\mathrm{d}\phi \tag{4.54}$$

Roll torque, T   This is given by

$$T = RR'\left[\left(\int_{\phi_2=0}^{\phi_1} s\phi\mathrm{d}\phi\right) + \frac{s_1h_1 - s_2h_2}{2R'}\right] \tag{4.55}$$

The coefficient of friction, μ   This is given by

$$\mu = T/PR \tag{4.56}$$

4.5.11.2 Longitudinal rolling of tube

For analytical considerations, the processes of sinking and stretch reducing can be treated in the same way. The effect of ovality is often included, and in one possible approach this gives the following set of conditions.

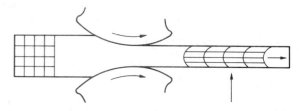

Figure 4.65 Inhomogeneous deformation in strip rolling

In the entry zone, the specific roll pressure p' is defined by

$$p' = \frac{yh}{r_1 + y \tan \varphi \cos \theta - 0.5h)(\cos \varphi + \mu \sin \varphi \cos \theta)}$$

$$\left[\frac{I}{A + I} + \left(\frac{r_1 + L \tan \varphi \cos \theta - 0.5h}{r_1 + y \tan \varphi \cos \theta - 0.5h}\right)^{(A+1)}\right.$$

$$\left.\left(\frac{A}{A + I} - \frac{\varphi_0}{F_0 y}\right)\right] \tag{4.57}$$

where Y is the flow stress, h is the wall thickness of the tube, r_1 is the radius of roll groove, y is the axial coordinate of the considered point, $\tan \varphi = \Delta h/2L$, Δh is the absolute reduction, L is the projected length of the arc of contact, θ is the position of the considered point with respect to a zero datum at the root of the groove $[(r_1^2 - x^2)^{1/2}/r_1]$, and μ is the coefficient of friction.

$$A = \frac{\cos \theta (\mu - \cos \theta \tan \varphi)}{\sin(I + \mu \tan \varphi = \cos \theta)}$$

$$L = \sqrt{R_x^2 - (R_x - h)^2} = \frac{\sqrt{(R_I + E_1 - \sqrt{r^2 - x^2})^2}}{-[R_I - \sqrt{r_0^2 - (E_0 + x)^2}}$$

$$\Delta h = \sqrt{r_0^2 - (x + E_0)^2} - \sqrt{r_1^2 - x^2} + E_1$$

where R_x is the radius of roll at the considered section with the abscissa x, R_1 is the ideal radius of roll, E_0 and E_1 are the eccentricities of stock and groove, respectively, φ_0 is the back tension, and F_0 is the cross-section of stock.

In the exit zone, the pressure p'' is

$$p'' = \frac{Yh}{(r_1 + y \tan \varphi \cos \theta - 0.5h)(\cos \varphi - \mu h - \varphi \cos \theta)}$$

$$\left[-\frac{I}{B - I}\left(\frac{r_1 + y \tan \varphi \cos \theta - 0.53}{r_1 - 0.53}\right)^{B-1}\right.$$

$$\left.\left(\frac{B}{B - I} - \frac{\varphi_1}{F_1 Y}\right)\right] \tag{4.58}$$

where

$$B = \frac{\cos \theta(\mu + \tan \varphi \cos \theta)}{\sin \varphi(I - \mu \tan \varphi \cos \theta)}$$

φ_1 is the front tension and F_1 is the cross-section of tube.

4.5.11.3 Cross-rolling of tube

A combined rotary piercing–elongating pass is shown in *Figure 4.66*, in which zones I to III refer to a Mannesmann-type piercer, and zones IV and V to an Assel elongator. For an ordinary system in which piercing and elongating operations constitute separate stages, the determination of roll-separating force, roll torque and axial plug load, in the case of piercing, is carried out separately. The expressions derived for the combined operation can obviously be used.

The five zones refer to the following operations:

I —oblique rolling between the entry to the pass and plug point;
II —piercing and cross-rolling between the plug and the narrowest part of the piercing pass (gorge);
III—cross-rolling with a small amount of diametral expansion;
IV—elongating and cross-rolling; and
V —reeling.

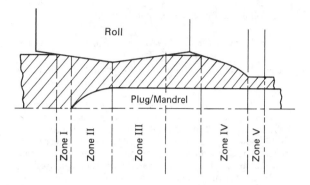

Figure 4.66 Combined rotary piercing–elongating pass

With reference to *Figure 4.57*, it can be shown that in zones I to V the relationship between the radial pressure p and the circumferential stress σ_3 is given by

Zone I:

$$\sigma_3 r \lambda = -p[r\lambda - R\Omega(1 + \mu \tan \gamma \cos \delta) + 2\mu r] \tag{4.59}$$

Zone II:

$$\sigma_3(r - s) = -p[s\lambda(1 - \mu \tan \theta \cos \delta) - RB\Omega + 2\mu(r - s)] \tag{4.60}$$

where $B = 1 + \mu \tan \gamma \cos \delta$.

Zone III:

$$\sigma_3(r - s)\lambda = -p[s\lambda - RB\Omega + 2\mu(r - s)] \tag{4.61}$$

Zone IV:

$$\sigma_3(r - s)\lambda = p\left[R\Omega\left(1 - \mu \cos \delta \frac{dr}{dx}\right) - s\lambda - 2\mu(r - s)\right] \tag{4.62}$$

Zone V does not contribute substantially and can therefore be neglected in calculations.

The relationship between p and σ_3 is defined by the Tresca criterion:

$$p - \sigma_3 = 2k \tag{4.63}$$

To determine the roll-separating force it is necessary to sum the vertical components of the forces in the relevant zones. To a first approximation

$$P_R = \sum_I^n P_z \tag{4.64}$$

where

$$P_z = \eta \int_{r_0}^r pR\Omega dx$$

The roll torque is supplied in zone I in the case of simple piercing, and in zone IV in the case of simple elongating, and can be obtained from the following equation

$$T_R = \mu \eta \sin \delta \sec \gamma \int_{r_0}^r pR^2 \Omega dx \tag{4.65}$$

The plug load for simple piercing is given by

$$L_P = \nu \int_{r_0}^{r} ps\lambda(\tan\theta + \mu\cos\delta)\mathrm{d}x \qquad (4.66)$$

The values of angles λ and Ω and the variation in r along the pass are determined from the geometry of the pass. In these expressions the following notation is used: σ_3 is the circumferential stress, k is the yield stress in shear, L_p is the axial force on the plug, p is the radial pressure, P is the radial force, r is the radius of billet or bloom, R is the radius of roll at a section, s is the radius of plug at a section, T is the torque on a roll, α is the feed angle, γ is the slope of the piercing roll, δ is the angle of inclination of the resultant friction force to the axis rolling, η is the coefficient of trangential slip, θ is the variable slope of the elongating part of the roll, λ is the angle of engagement of billet, μ is the coefficient of friction, ν is the coefficient of axial slip, and Ω is the angle of engagement of roll.

Equations (4.59) to (4.66) are obtained from consideration of the equilibrium of forces acting on a deforming element.

Considering the geometry of the pass, it is seen that in the piercing part, at any transverse section distant x_0 from the gorge of a Mannesmann-type roll, the coordinate y_0 is given by:

$$ay_0^4 + by_0^3 + cy_0^2 + dy_0 + f = 0 \qquad (4.67)$$

where

$a = (1 + E)^2 E$

$b = -2K(1 + E)(2E + 1)$

$c = (1 + E^2)H + (4 + 5E)K + r_b^2 E$

$d = -2K[K^2 + (1 + E)H - r_b^2 E]$

$f = K^2(H - r_b^2)$

$E = \sin^2\alpha\sec^2\gamma - 1$

$H = R_g(R_g - 2x_0\cos\alpha\tan\gamma)$
$\quad + (\cos^2\alpha\sec^2\gamma - 1)x_0^2$

$r_b = r_0 + R_g$

$K = (R_g\tan\gamma - \cos\alpha\sec^2\gamma\, x_0)\sin\alpha$

The angles Ω and λ are given by

$$\Omega = 2\tan^{-1}\left[\frac{y_0\cos\alpha - x_0\sin\alpha}{z_0 - r_b}\right] \qquad (4.68)$$

and

$$\lambda = 2\tan^{-1}\left(\frac{y_0}{z_0}\right) \qquad (4.69)$$

In the elongating part of the pass, for the conical part of the roll profile, equation (4.67) is valid if the roll radius in the Assel gorge (R_G) is substituted for R_g in the equations giving the two angles.

In the cylindrical, reeling part of the roll, the constants in equation (4.67) will be:

$$E = -\cos^2\alpha; \; H = R_G^2 - x_0^2\sin\alpha; \\ K = -x_0\sin\alpha\cos\alpha \qquad (4.70)$$

4.6 Forging operations

4.6.1 Basic concepts

Forging, one of the oldest known metal-forming operations, relies entirely on the application of compressive forces to effect the change of shape. In its simplest form, forging is used to preform a billet by changing its dimensions (preparatory to further shaping), and in its more sophisticated form it produces complex shapes to a very high degree of accuracy. Depending on whether the operation is carried out hot, warm or cold, forging affects the structure and properties of the forged component to varying degrees.

Whilst being essentially simple in concept, forging processes, in their many varied forms, are in fact extremely complex. By far the most common group of processes is concerned with the forging of ferrous alloys and a break down of the costs involved in producing the average ferrous forging illustrates the reason for the introduction of new methods, techniques and ideas. The cost can be apportioned as follows:

Material in final forging	35%
Material wasted in forging	15%
Labour	10%
Overheads	30%
Tools	10%

It is clear from these figures that material usage is an area in which savings should be introduced and, consequently, innovative preforming processes of powder forging, transverse rolling and cast preform forging, aimed specifically at reducing material usage, must be considered in addition to the standard techniques normally employed. In this context, the idea of forging is of particular interest.

In the hot forging process, the work piece preheat temperature is usually chosen to be as high as possible, consistent with the production of a sound forging, thus exploiting the benefit of minimum flow stress. Any accompanying consideration of economic and technical feasibility must make reference to the work piece preheating costs, corresponding handling costs and the effects of surface oxidation and decarburisation. Cold forging of steel, on the other hand, is conducted at room temperature with a consequent saving in heating costs and material wastage, but against this must be set the higher tooling costs. The fact that some work hardening usually takes place is only infrequently exploited.

It is not surprising, therefore, that the possibility of forging at intermediate warm temperatures is advocated in order to obtain the benefits of both hot and cold forging. In this connection, two approaches are adopted.

The first seeks to improve the accuracy of the hot forging process by selecting preheat temperatures high in the spectrum between hot and cold forging. If, for a typical mild (low carbon) steel a forging temperature of, say, 800°C is chosen, then oxidation will be reduced, thus permitting a lower machining allowance and benefit will be gained from the low flow stress. Tolerances of ±0.25 mm are claimed for warm forging in this temperature region. In addition, the small amount of strain-hardening which occurs may be beneficial in some cases. It is important, however, to realise that the heating and attendant costs are still present. As the temperature is reduced further, the flow stress begins to increase rapidly and the process becomes less attractive.

The second approach is from the cold-working end of the spectrum. Due to strain-hardening and strain-ageing effects, lower grade steels may be taken to higher final strengths. This is also in competition with heat treatment over which it has the advantage of being more controllable. A further refinement is to start the process with a quenched and tempered structure on top of which warm forging will produce even more strength and ductility.

In its classical form, forging is of three main types:

(1) open die,
(2) impression die, and
(3) closed die.

In addition, the already mentioned modern developments have given rise to a whole series of processes of hot and cold rotary forging (including tube making), orbital forging, high-energy-rate (dynamic) forging of both shapes and particulate matter, and cast-preform forging.

The objective of the 'standard' operations is to produce, in stages, machine parts and components such as gears, wheels, compressor and turbine discs, crankshafts and connecting rods, small tooling, screws and bolt heads, and coins and medals, and to assist in the conduct of other operations by providing simple preforms. Punch piercing of billets is a typical example of the latter application of forging.

4.6.2 An outline of open-, impression- and closed-die operations

Open-die forging, also known as 'upsetting', is concerned mainly with reducing the height of a cylindrical billet. This is generally done between two flat dies, although the dies can be profiled in a simple manner to impart a specific shape to the ends of the upset specimen.

The outcome of an upsetting operation, in terms of the shape of the preform, depends on the frictional effects that develop between the dies and the faces of the billet. With efficient lubrication, the reduction in height produced by the application of compressive forces is accompanied by an increase in the diameter of the billet, but with the billet retaining its original sharp edges. In an unlubricated or poorly lubricated operation, the deformation becomes inhomogenous and barrelling occurs. The amount of barrelling depends on the value of the width/height (d/h) ratio and the reduction in height r (*Figure 4.67*). Since, for a cylinder of given diameter, the degree of barrelling depends on its height and is lowest for the longest cylinder, a cylinder of infinite height ($d/h = 0$) would be expected to be free of barrelling. This observation is

helpful is assessing conditions leading to barrelling. If a graph of applied stress is obtained experimentally for a series of different aspect ratios and is then extrapolated to zero, the intercept on the stress axis should correspond to the ideal, non-barrelling condition.

Whereas upsetting of a basic cylinder leads to a build up of compressive stresses only in the state of either pure, homogeneous deformation (*Figure 4.68(a)*) or inhomogeneous flow (*Figure 4.68(b,c)*), the situation is more complex when a circular section is forged by using flat dies. Here, the differential in flow is much higher and depends on the degree of height reduction. When this is small, only the layers adjoining die surfaces are deformed plastically, but as d/h changes, the degree of deformation of the inner material begins to improve, but less so than in corresponding square-sectioned specimens. This is because with circular contours only part of the section is in contact with the dies and, consequently, additional resistance to flow is offered by the remaining material of the section. As a result, tensile stresses are created in the central plastic zone. The tensile stress generated is a function of the height reduction and diminishes with an increase in the latter. In practice, the preforms obtained may well display a degree of brittleness in their central core. These particular patterns of flow are shown in *Figure 4.69*.

To effect an improvement, upsetting of circular sections can be carried out to advantage in shaped, rather than flat, dies which create conditions approximating to those given by hydrostatic pressure (*Figure 4.70(a)*). Optimal conditions are reached with circular dies and a large angle of engagement of 150° (*Figure 4.70(b)*). In this case σ_2 is compressive across the whole of the section.

Although open-die forging offers a simple and relatively inexpensive means of producing small components, it calls for a high degree of manipulative skill on the part of the operator, since acquisition of the basic shape can only be achieved by

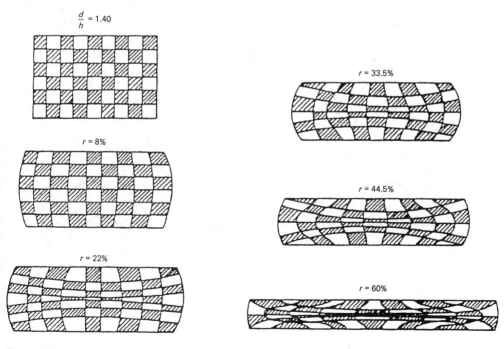

Figure 4.67 Barrelling in unlubricated, open-die forging

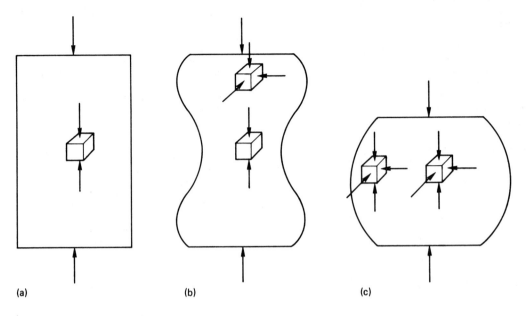

Figure 4.68 Stress systems induced during upsetting of a cylindrical billet: (a) pure homogeneous deformation; (b, c) inhomogeneous flow

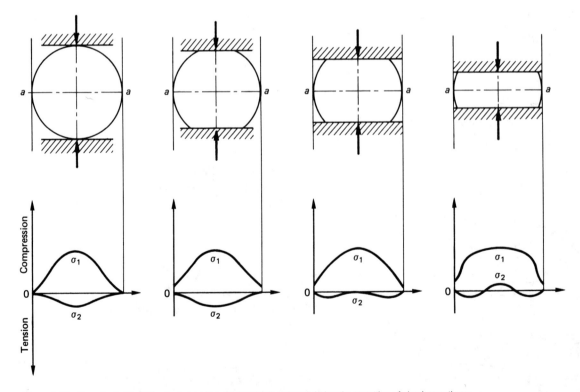

Figure 4.69 The effect of height reduction on the stress pattern induced during the upsetting of circular sections

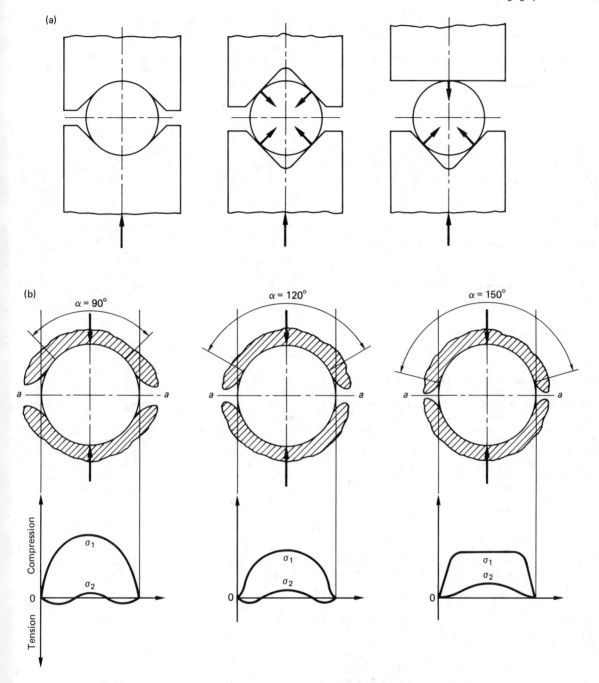

Figure 4.70   (a) Direction of forces and pressures in shaped upsetting dies. (b) Stress patterns and angles of engagement in shaped dies

turning the work piece to different positions between success-ive blows. The deforming forces are applied either mecha-nically by means of powered hammers or manually. In either case, the rate of yield of the component is low and the process is unsuitable for mass production. The additional difficulty, which has an additional cost associated with it, is the require-ment to machine the specimen in order to obtain both the desired final shape and the required dimensional tolerances, neither of which is likely to result from the forging operation on its own. The machining stage naturally introduces an element of wastage of material and an additional labour requirement.

To introduce a high degree of dimensional accuracy and better material usage, impression and closed-die forging ope-rations are used. These operations are shown in their most basic form in *Figure 4.71*. The fundamental requirement is

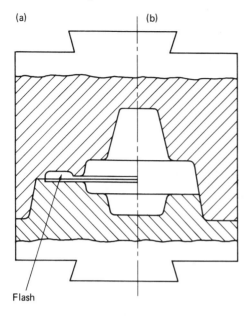

(a) (b)

Flash

Figure 4.71 (a) Impression-die and (b) closed-die forging

that the die cavity, whether that of an impression die (*Figure 4.71(a)*) or of a closed die (*Figure 4.71(b)*), is completely filled with the forged material and, most importantly, that the material remains structurally sound. The more complex the shape to be forged the more involved the production line becomes since, clearly, it is impossible to convert a starting cylindrical or square billet to the desired shape in a single blow. Not only are preforming operations required, but several actual forging stages may also be needed.

The basic characteristics of impression dies is the fact that, when activated by a mechanical hammer or hydraulic press, they do not close completely, and thus some metal is allowed to escape and forms a flash (*Figure 4.71(a)*) between the flat surfaces. The presence and magnitude of the flash are of considerable importance in impression-die forging because they influence the mechanics of the operation. By the nature of its geometry, the flash is subjected to high pressures and, consequently, it experiences a high degree of frictional resistance as it propagates radially outwards. This constrains the tendency of the bulk of the material to flow between the dies and, therefore, creates better conditions for the filling of the die cavity. This action is further assisted by rapid cooling of the flash (as compared with the cooling of the material in the die), which results in a further increase in the resistance to flow. These characteristics form the basis of die design.

In the normal sequence of operations, preforming will be necessary and may consists of the following operations:

(1) simple upsetting,
(2) blocking, and
(3) rolling.

Upsetting and blocking in blocker dies are the more usual routines, although rolling (see previous sections) of profiled shapes is also used.

A blocker die belongs to the impression-die group, but provides only a general outline of the work piece. The final stage is not achieved using a blocker die, but the material is distributed in a way that ensures a more acceptable uniform-

ity. On trimming the flash, the work piece can undergo further processing in a finishing die that will impart to it its final shape.

A set of impression dies will produce a parting line in the specimen and, unless the die design takes into account the most likely response of the material, in terms of its pattern of flow and formation of fibres in its structure, problems may arise because, on trimming, the fibres are severed.

To facilitate the removal of the forging on completion of the operation, dies are provided with draft angles (a situation similar to that in casting).

Reduction in height, piercing and bulging are the basic operations associated with impression-die forging. However, their sequence and intensity depend on the required shape of the forging. *Figure 4.72* illustrates a selection of the possibilities arising in these operations. If the required forging is of simple shape, the arrangement shown in *Figure 4.72(a)* may be sufficient. If one- or two-sided indentations are needed (*Figure 4.72(b,c)*), the final shape is obtained by height reduction and piercing in the die. Formation of more complex shapes with side bulges, etc., depends on the dimensions of the starting material. If the diameter of the billet is less than that of the elongated part of the axisymmetric forging, then the cavity is filled simply by reducing the billet height (*Figure 4.72(d)*). On the other hand, when the starting diameter is larger than that of the forging, height reduction and flow into the die cavities are required (*Figure 4.72(e)*). A more complex operation is required (*Figure 4.72(f)*) when the forging has both indentations and side bulges, and the starting billet diameter is larger than that of the side of the forging. Height reduction, piercing and bulging must occur to give the final impression.

Four basic stages can be distinguished in any of the above-mentioned cases. These are represented in *Figure 4.73*, in which the right-hand half of each figure refers to the initial stage, and the left-hand side to the current stage shape. The first stage (*Figure 4.73(a)*) begins when the upper die contacts the metal specimen positioned in the cavity of the lower die. On application of a compressive force to the system, free flow of the metal takes place and culminates when contact with the die side is established.

In the second stage (*Figure 4.73(b)*) further flow is generated, with the die cavity filling more fully, but leaving the corners free of metal. Flash also starts forming at this stage.

The third stage (*Figure 4.73(c)*) involves both the full formation of the flash (with the attending mechanics of flow coming into operation) and complete filling of the die cavity. However, the height of the forging is still greater than the required final height. It is at the fourth stage that the final reduction Δh_4 takes place to give the required dimensions. The excess material flows into the flash which undergoes plastic deformation, as does the central core of the forging itself.

More recent investigations of the impression-die-forging process have shown that the speed of the operation has considerable influence on the quality of the product. Forging at speeds between those of slow presses (say, 0.75 m s^{-1}) and those of fast hammers benefits the material in a number of ways. The reduced effect of cooling and improved lubrication which accompanies an increase in speed create a higher degree of homogeneity of flow and, therefore, a better chance of successful filling of the die cavity. In most cases, although a slight increase in pressure will occur in the die, the pressure on the flash will be reduced. To obtain optimal conditions, the flash is kept within a width/thickness ratio of 2–5. When this value is exceeded, deformation of the flash will take place and part of the available forging load will be used up unnecessarily. Dimensionally, a much better final product is obtained

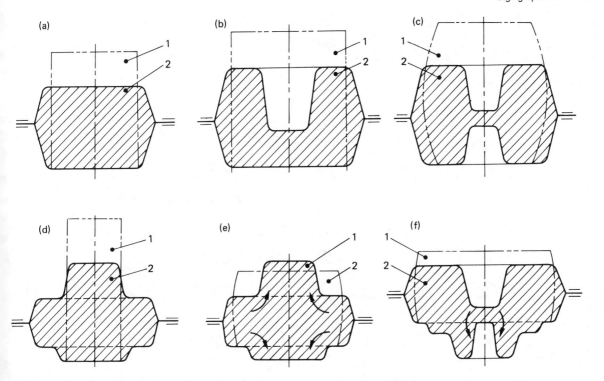

Figure 4.72 Different routes of filling the cavity of a finishing die: 1, starting shape; 2, final shape of forging

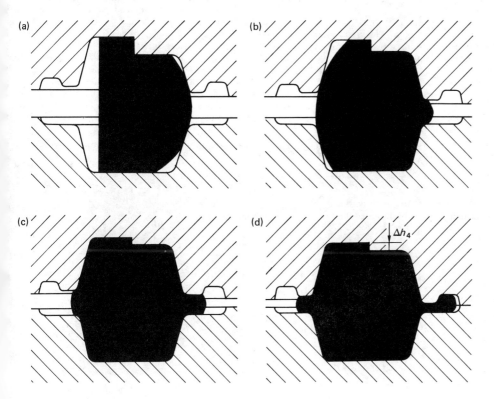

Figure 4.73 Typical sequence of forging in impression dies

at higher speeds, but the total forging load increases by up to 25%.

The basic advantage of high-speed forging lies in its ability to reduce the amount of preforming needed when working on more complex shapes.

However, in the more conventional approach, blocking precedes impression forging, with blocker dies having the advantage of low cost and high production rate but, as with open-die processing, still requiring additional machining operations.

In true closed-die forging, the operation produces a completely flashless forging (*Figure 4.71(b)*). To achieve this highly desirable situation, the die/billet relationship must be carefully worked out, since incorrect dimensions of the billet lead to either incomplete filling of the die cavity or damage to the dies and punches.

If the forging system is correctly designed and the die (or dies) and the punch are machined to the required degree of accuracy, a precision, or near-net-shaped forging will be obtained. Zero draft tooling is used and, because of the high loads required, softer materials like aluminium or magnesium are used in preference to steels.

In the course of normal operations, the blank or billet will be forged by the compressive force of the punch transmitted to the material. Any shape can thus be obtained, with or without piercing in the die.

A standard, non-piercing operation can be divided into three stages (see *Figure 4.74*). If the material is initially in the condition indicated in *Figure 4.74(a)*, the first two stages (*Figure 4.74(b,c)*) will be identical with those in impression forming. Unless precision forming is considered, stage three (*Figure 4.74(d)*) will be completed when the die cavity is completely filled and a vertical flash (resulting from the difference between the punch diameter D_s and the die diameter d_m) begins to form.

To reduce the flash, or even to prevent its formation, the dies designed for more complex profiles, which are also associated with piercing in the die, are often provided with a compensator which allows excess material to accumulate without flowing axially onto the outer circumference. The compensator may remain partially unfilled (the usual arrangement ((1) in *Figure 4.75(a)*), or be designed accurately to form part of the profile of the forging ((2) in *Figure 4.75(a)*). The pattern of flow of the metal can be seen clearly in the sequence shown in *Figure 4.75(b)*, wich starts with the initial blank and ends with the finished product. The effect of the shape of the forging on the crystallographic structure of the material is clearly visible and demonstrates the difference in the proper-

ties between the various parts of the section, as reflected in the grain size and grain orientation.

One of the characteristics of closed-die forging is the provision of indentation which either forms part of the desired profile as, for instance, in the case of the component shown in *Figure 4.75*, or becomes the first step in, say, rotary forging or extrusion of a seamless tubular component. In either case, this part of the forging process depends on the shape and dimensions of the blank or billet. In its simplest form, punch piercing of this type is carried out either on circular billets or square blanks (*Figure 4.76(a)*).

If, initially, the material has a circular cross-section and its diameter is smaller than that of the die ((1) in *Figure 4.76(a)*), the process can be divided into two stages. In the first, the billet is reduced in height and is bulged sideways until it fills the die. In the second stage, the material is forced to flow between the punch and side of the die and so piercing proper commences.

However, if the initial blank is square (usually with $a = 0.7d_m$), as in (2) in *Figure 4.76(a)*, the first stage of piercing is different. The blank already touches the side of the die at four corners and, therefore, initially it can only flow in the direction perpendicular to its square sides. When the cavity is eventually filled, the piercing operation proceeds as with a circular billet.

A third possibility is that of piercing a cylindrical billet of a diameter close to that of the die ((3) in *Figure 4.76(a)*). In this case, the piercing operation begins almost immediately and is identical with stage two.

The first stage of piercing is of particular importance from the point of view of the development of stress fields and their intensities. With reference to *Figure 4.76(b)*, it can be seen that immediately under the punch (case (i)) there develops a state of triaxial compression; but, in the annular ring, biaxial compression is accompanied by uniaxial tension. The presence of tensile stress is undesirable because forging of low ductility steel may easily result in the presence of axial cracks (2). Furthermore, folding of the upper face (1) of the billet into a meniscus takes place radially inward and is reflected in a similar semi-spherical shape of the lower face (3). It is only in stage two, when the effect of the side surfaces of both the die and punch comes into operation, that the triaxial compressive system reestablishes itself everywhere in the bulk of the pierced material (case (ii) in *Figure 4.76(b)*).

4.6.3 Special applications of closed-die forging

A number of special operations associated with closed-die

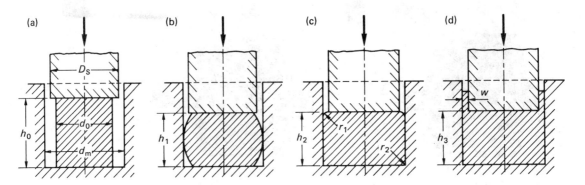

Figure 4.74 Sequence of forging operations in a closed-die system

(a)

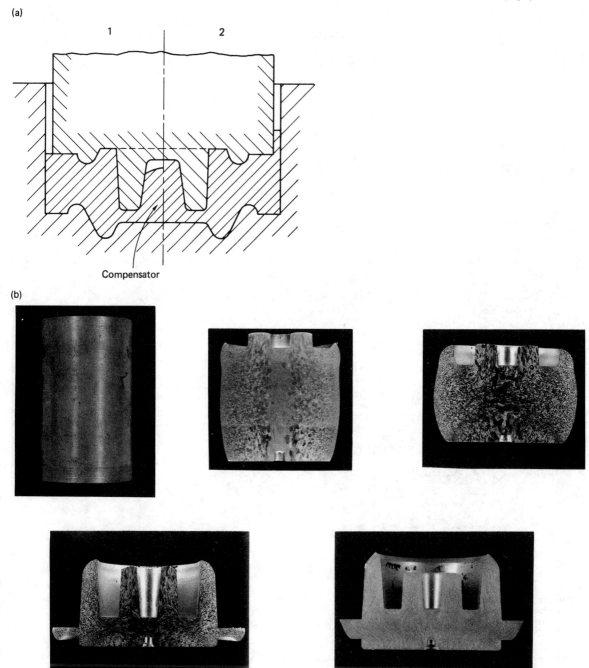

(b)

Figure 4.75 Piercing and forging in a closed-die system. (a) Basic operation. (b) Sequence of forging operations showing the pattern of flow of the metal

forging are carried out cold, i.e. at a temperature below the recrystallisation temperature of the metal in question. However, some difficulty is experienced in the case of steels, since these alloys recrystallise at temperatures above 600–700°C, temperatures that are too high to be called 'cold'. Consequently, forging at room temperature is often referred to as 'cold', and forging at elevated, but below recrystallisation, temperatures is called 'warm'. Generally, the purpose of cold forging is to produce a finished part with high dimensional accuracy.

Highly ductile materials, such as aluminium, lead and tin, have been cold forged for a long time, mainly in the form of

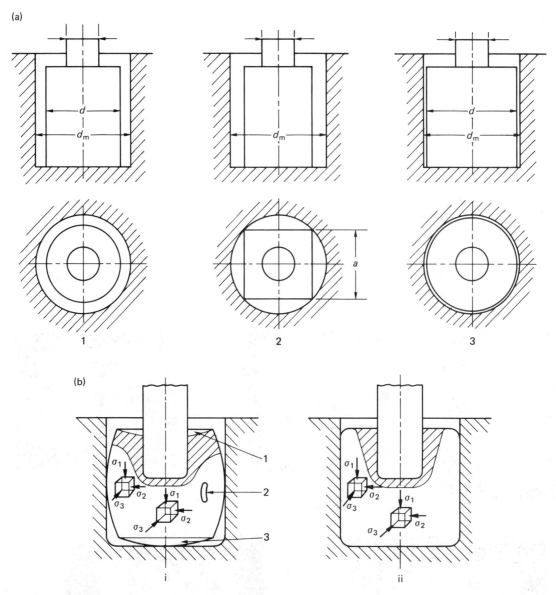

Figure 4.76 Piercing of cylindrical and square sections in the die. (a) Schematic representation of the operation. (b) Stress fields at different stages of piercing

extrudates, but 'proper' cold forging is normally limited to either small parts in low and medium carbon steels (such as bolts and nuts) or larger parts of up to 10 kg in mass which require good dimensional tolerances.

Of the better known applications in this area are the coining and embossing operations (*Figure 4.77*) used for the production of coins and medals, and for the improvement in dimensional accuracy of other preforms. The coining operation is one that is actually carried out in a closed-die system, sometimes with embossing forming, a stage that includes an open-die system. In these processes, three-dimensional details are reproduced in the material giving not only a faithful impression of the punch and/or die surface, but also good surface finish. It is not surprising that the forging loads used are high,

the punch pressures necessary to emboss being of the order of $3Y$, and those needed to produce fine-detailed coins reaching five to six times the value of the flow stress of the material. The design of the dies is critical since lubrication cannot be used in coining operations because of the danger of entrapment and the consequent damage of the forging.

A completely different type of application of closed-die hot forging is that of powder forming, also known as 'sinter forging'. This process, an offshoot of conventional powder metallurgy processes, involves the following sequence of operations. Cold powder is compacted in a press to produce a powder preform. The preform is subsequently sintered in a controlled-atmosphere furnace but, instead of being allowed to cool in that atmosphere (as in conventional powder metal-

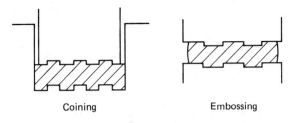

Coining   Embossing

Figure 4.77 Coining and embossing

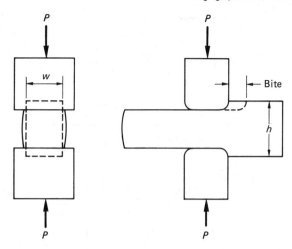

Figure 4.78 Cogging of a rectangular specimen

lurgy processes), the preform is removed from the furnace whilst still hot and forged in closed dies to produce the final shape. Since only the exact amount of powder required to make the final shape is actually used, there is an obvious elimination of waste.

The same final shape could also be made in the first compaction operation, but the introduction of the forging stage produces a superior product. Improvements in strength, ductility and density are expected, with accuracy increasing to some 0.025 mm. Against these advantages must be balanced the drawbacks of additional processing and the possibility of deleterious oxidation occurring immediately before and after the forging operation.

Commercial applications of powder forging range from forged-powder connecting rods, through material properties approaching those of commercial forged steels, to valve spring cage components. In the latter, for instance, the density (as compared with normally sintered material) increases by some 30%, the elongation quadruples, and the ultimate tensile strength almost trebles.

More recently, the process has been extended to the cold forging of polymeric powders which, although they show some post-forging eleastic recovery, can be quite successfully compacted in this way.

4.6.4 Subsidiary forging operations

These operations are often associated with preforming and comprise:

(1) cogging,
(2) fullering,
(3) heading, and
(4) hubbing.

Cogging or drawing (*Figure 4.78*) of a rectangular section specimen involves reduction in height by successive blows or bites, leading to a gradual elongation of the original blank. The successive bites reduce the force requirement, but they must be grouped closely together to produce an even surface. The operation will not cause plastic deformation of the bulk material to occur, unless bites of sufficient length are initiated. To avoid possible buckling of the specimen, the height/width (h/w) ratio should be kept below 2.5.

Fullering (*Figure 4.79*) is another preform operation in which the original bar is shaped by profiled open dies into an outline suitable for more detailed, further forging. Thus the basic function of fullering is to redistribute the bar material along its length, prior to cropping of the individual preforms.

Heading is an operation that combines forging and extrusion. In the forging part an upsetting operation carried out between flat dies will flatten (by bulging) the end of a cylindrical billet, thus forming a bolt head. The remaining portion of the billet can then be extruded to form the bolt.

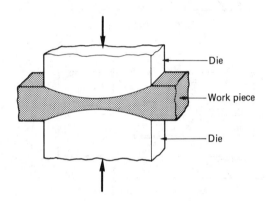

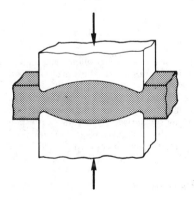

Figure 4.79 Fullering

Hubbing is used as a means of forming die cavities by indenting the material of the blank with suitably profiled punch heads. The operation is thus very similar to in-die punch piercing (discussed above). It generally calls for punch pressures of about three times the flow stress of the material to be forged.

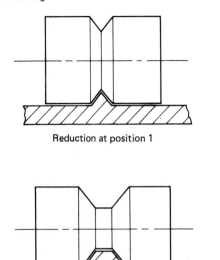

Reduction at position 1

Reduction at position 2

Reduction at position 3

Figure 4.80   An example of transverse roll forging

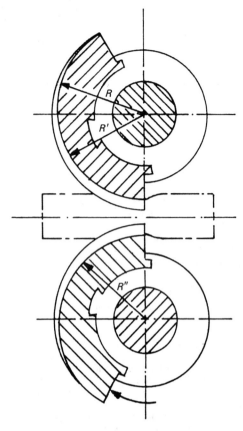

Figure 4.81   Profiled roll system in rotary forging

4.6.5   Rotary forging

Rotary forging, also known as 'transverse rolling', is used as a preform operation in which profiled tools are positioned on driven rolls (see *Figure 4.80*).

The bar stock is located transversely between three rolls. The rolls make one revolution and the bar makes several. The wedge impressions on the roll surface (*Figure 4.81*) are then progressively imparted to the work piece to give the required preform shape. Subsequent flash discards of 15–20% are usually expected.

4.6.6   Rotary tube forging

Of very considerable industrial importance is the use of the rotary-forging concept in hot and cold seamless tube manufacture. The operation is carried out on a pilger mill.

In hot pilgering, the already rotary-pierced bloom is threaded onto a cylindrical floating mandrel and is steadily advanced forward and retracted in a reciprocating motion in which the forward stroke always exceeds the length of the backward one (hence the name of the operation, which is derived from the movement of the chorus of pilgrims in an opera) and is executed by two profiled, driven rolls. Each successive movement is accompanied by a 90° turn of the tube–mandrel assembly.

Figure 4.82 shows a sequence of successive 'bites' culminating in a finished tube. Irrespective of whether the working roll surfaces are parabolic, hyperbolic or form logarithmic spirals (*Figure 8.83*), each roll (being basically a cam) is characterised by three distinct zones. The working zone AB consists of the reducing portion that produces the highest degree of wall reduction. It is this deformation which, in turn, converts a bloom to a tube and is accomplished by a change in roll radius. The finishing zone BC imparts the required dimensions to the tube; the roll radius remaining constant along this part of the profile. The third zone is the idle intercept CA during which reciprocating and rotational motions of the tube–mandrel assembly are performed. To obtain the final geometry of the tube up to six passes may be required, with each successive bite reducing the wall thickness of the bloom further.

The working zone of the roll constitutes some 45–50% of the revolution, while the finishing zone accounts for 30–35%. The maximum elongation attainable in a pilger mill is of the order of 15:1, and the process is used to rotary forge heavy walled blooms to thinner walled tubing.

To obtain even thinner walled seamless tubing displaying high dimensional accuracy and good mechanical properties, cold pilgering (also known as 'cold reducing') is employed. The cold-reducing process is a step-by-step operation which uses tapered grooved rolls (*Figure 4.84*). Several dissimilarities exist between hot and cold pilgering. The most obvious of

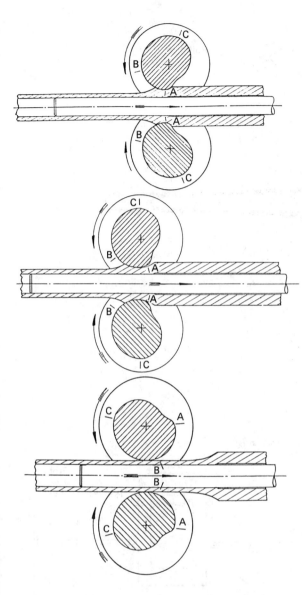

Figure 4.82 Sequence of events in hot pilgering of a seamless tube. AB, working zone; BC, finishing zone; CA, idle zone

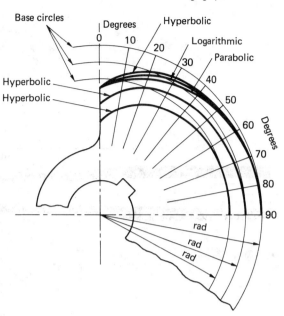

Figure 4.83 Basic profiles of pilger mill rolls

which is that, whilst in the hot process the rolls revolve completely and continuously, in cold reducing they rotate backwards and forwards through approximately 180°. In the hot process the rolls are mounted on a fixed stand, and the work piece reciprocates, whilst in cold reducing the roll stand reciprocates and the work piece remains stationary.

The machine operates as follows. Each time the saddle is at the end of the stroke, so that the large ends of the die grooves are presented to the in-going hollow, the cross-head pushes the hollow forward by a fixed amount (or the feed). The dies then roll the material thus inserted down the cone, simultaneously thinning the wall and reducing the diameter, with the consequent elongation of the tube. At the far end of the saddle stroke, the tube and mandrel are rotated through approximately 60°, and the saddle returns. This is repeated at

frequencies of 60–120 cycles/min, depending on the size and type of machine.

At each end of the saddle stroke, because of the clearance in the dies and also because the rolls turn through slightly more than 180°, there is a short period of time when the dies are not in contact with the tube being rolled. It is during this time that feed has to occur at one end of the stroke and turning of the tube at the other. The arrangement of a mill is shown diagrammatically in *Figure 4.85*.

Tool design and preparation of the hollows are of particular importance in this process. Use is made of rolls (sometimes called 'dies') containing a tapered groove which is divided into two sections. The first section, comprising four-fifths of the groove length, does the work of deforming the material of the work piece. The second section, known as the 'sizing section', irons out any irregularities produced in the first section. Modern mills contain rolls having the grooves profiled in the form of a continuous curve. The curve is designed to give an equal percentage reduction in area per stroke of the machine. In other words, if, for instance, each increment of feed requires 15 machine strokes to pass from hollow, down the cone until it emerges as a finished tube, then the percentage reduction in area it undergoes during each stroke is the same, thus helping to equalise the torque required along the working stroke.

The preparation of hollows, to optimise forging conditions, includes spheroidising and pickling prior to dressing. The leading end of each hollow is radiused with a view to eliminating any stress raisers remaining in the tube end from the hot-rolling process. Failure to carry out this procedure results in the incidence of splitting as the front ends of tubes pass through the cold reducer, with a consequent risk of serious damage to the rolls. After rounding, the hollows are dipped in a lubricant bosh, drained and stored in racks until required.

In general, this cold forging process is associated with deformation of tubing in the 10–230 mm wall thickness range, and is employed whenever the specified wall-thickness tolerances of the finished product are required to be below ±10%,

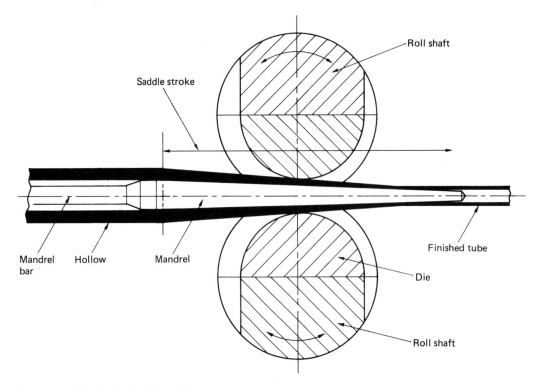

Figure 4.84 Cold pilgering (reducing) tube forging process

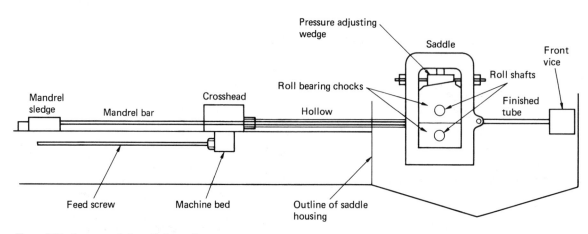

Figure 4.85 Arrangement of a cold pilger mill

and the quality of the tubing, with regard to surface or structure, calls for cold deformation.

Tube obtained in this manner can either be used directly, in industrial applications, or can form an intermediate stage of production requiring a cold-drawing operation. An interesting point emerges here regarding welded tubing. After cold pilgering, such tubing almost acquires the properties and appearance of the genuine seamless product, having undergone a considerable amount of material deformation and consolidation. The most important advantages of the process, however, are the following. A negligible waste of material is involved and, since cold deformation is done simply by means

of the forging pressure, a large reduction in wall thickness can be achieved which, in the case of high alloy and carbon steels and condenser tubing, can amount to some 75%, whereas for copper tubing the reduction can be as high as 93%. Furthermore, it is possible to employ particularly thick walled hollows, which means that heavier billets can be used in the preceding hot-working operations and, also, that only a few sizes of hot-produced hollows are necessary in any instance to satisfy the production programme.

The use of straight-tapered or concave-tapered mandrels makes it possible to achieve diameter reductions of considerably more than 50% in the outer diameter of the hollow and, because of these large wall and diameter reductions, the wall tolerances and any eccentricity of the hollow are improved by approximately 50%.

As a result of the overall reduction that occurs in the cold pilgering process, ball-baring tubing can be elongated 2.8-fold, high alloy, medium and low carbon steel tubing five- to six-fold, and copper tubing 12.5-fold. Consequently, intermediate annealing, degreasing, pickling and bonderising processes can be dispensed with.

4.6.7 Orbital forging

The orbital forging process is intended to allow smaller machines to be used for suitable forging operations. This process is based on the simple concept that the axial force required to effect a desired deforming zone between the platens is confined to a small region. The plastic zone is then moved through the work piece, thus resulting in progressive deformation.

The plastically deforming region is formed by initially indenting with a conical platen into the work piece which is supported by a lower platen capable of axial movement only. The plastic region is then moved through the material by the upper conical platen which is capable of rotation about the central axis of the machine and, also, about its own geometric axis. In order to produce a flat top surface on the work piece, the geometric axis of the upper conical platen is inclined to the central axis of the machine.

The process, known also as 'Rotaform', can be applied successfully to the hot, warm and cold forging of a wide range of materials and axisymmetric shapes. The Rotaform is a completely automatic machine provided with automatic feeder and ejection equipment. *Figure 4.86* illustrates the principle involved.

The die uses a pair of cooperating dies, one of which is adapted to perform a wobbling motion, relative to the other, about a centre at or near the axial centre-line of the dies. The wobbling die, which has a circular rocking motion without actual rotation about its axis, is actuated at high frequency. The complementary half-die is secured to a hydraulic ram which conveys the work piece to, and presses against, the wobbling die. The cycle of operations commences with the introduction of a billet into the receiving station of the automatic feeder which places the billet accurately into the non-wobbling die. The hydraulic ram raises the non-wobbling die at a speed programmed for the most suitable deformation rate throughout the forging stroke. When the forging has assumed its final shape, the ram retracts rapidly and the forging is automatically ejected and blown clear of the dies.

The cycle time is very short and outputs of 15 forgings per minute have been achieved.

4.6.8 Cast-preform forging

The material of all metal components usually originates from a cast and subsequently undergoes a considerable degree of processing before being finally transformed by, say, forging into a component. Since all the intermediate mechanical working improves the properties of the structure, this improvement, to a level necessary for most components, can be achieved by cast-preform forging.

From a technological point of view, the structure of a reasonably homogeneous casting can be transformed to a forged structure by quite small amounts of hot working,

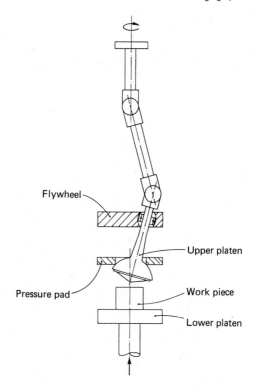

Figure 4.86 Orbital forging arrangement

amounts which could quite easily be achieved in the forging process itself. Internal cavities can be closed up and welded, providing they are located in regions of high deformation. Due to oxidation, however, cavities extending to the surface are likely to remain as faults. The economics of such a process require some consideration, but it becomes obvious that the operation is viable when the casting can be produced without a lot of waste material in the form of runners and risers. In addition, shapes normally forged from a cropped billet in single impression dies, where material usage is high, are not amenable to this process. Essentially, success rests on the ability to design an appropriate preform shape, to produce low-quality or low-cost casting of it, and to transform this into a sound forging. For certain materials, e.g. cast iron, the potential benefits of improvements in the tensile, impact and dimensional properties together with the existing corrosion and wear characteristics are considerable.

Major economic advantages can be gained if the casting and forging operations are combined. Several techniques are available for this; for example, a conventional steel-casting plant is situated alongside the forging press and reheating is done between the two operations. This technique has been developed as the 'Auto-forge' process which consists of a rotary table with a number of stations at which the various operations are performed. Casting is carried out at the first station and the moulds are water cooled while the table rotates to the next station where the forging operation is performed. The work piece then rotates to the clipping station. Another type of process, the 'Auto Forcast', consists of a melting plant coupled with a continuous casting plant. The various sections of cast steel produced are straightened and hot cropped into billets before being fed into the forging press.

4.6.9 Isothermal forging

A more expensive and complicated process is that of iso-thermal forging in which the dies are heated to the tempera-ture of the blank prior to the operation itself.

The advantages of this technique include low stress fields and, consequently, improved flow of the material which, combined with the absence of a temperature gradient at the tool–work piece interface, reduces the inherent difference in the bulk forging material.

4.6.10 Tooling in standard forging operations

Since the main objective of an impression or closed-die forging operation is to ensure that the die cavities are completely filled with the processed metal, the complexity of the die profile plays a considerable role in tool design. Whereas simple shapes are clearly associated with a reasonably uniform flow of metal, parts with thin, long sections in the form of fins, webs, ribs or projecting flanges belong to the category which, in terms of the quality of the final forging, is related directly to the geometry of the system. The latter provides means of quantifying the difficulty of forging, because a clear relation-ship exists between the volume of the processed material and its surface area of contact with the tools.

The shape classification of forgings, indicating somewhat subjectively the degree of difficulty in processing, give the following, main groups of forgings:

(1) compact, bulky shapes;
(2) thin components with elongated parts; and
(3) flat, disc shapes.

To quantify the effect of these geometries on the character-sitics of the process in which rounds are forged, shape factors have been introduced. These refer either to longitudinal (α) or lateral effects (β).

The longitudinal factor is defined as

$$\alpha = X_f/X_c \tag{4.71}$$

where

$$X_f = P^2/A; \; X_c = P_c^2/A_c \tag{4.72}$$

P is the circumference of the axial cross-section, A is the surface area of that cross-section, P_c is the circumference of the axial cross-section of the cylinder circumscribing the forging, and A_c is its cross-sectional area. The longitudinal factor thus gives a comparison between the simple, cylindrical shape and that of the forging.

When thin and elongated parts exist, the degree of difficulty in filling them completely increases with their distance from the core of the bulk material. This is accounted for by the lateral shape factor, defined as

$$\beta = 2R_G/R_c \tag{4.73}$$

where R_G is the radial distance from the axis of symmetry to the centre of gravity of one-half of the cross-section of the forging, and R_c is the maximum radius of the forging (equiva-lent to the radius of the circumscribing cylinder).

In more complex geometries, both factors must be taken into account and are involved in the form of a general shape difficulty factor S, as

$$S = \alpha\beta \tag{4.74}$$

The factor S reflects the complexity of the geometry of a half-section of the forging, with respect to that of a simple circumscribing cylinder.

Whereas the factor S is of importance in any die-forging operation, the effect of flash (as indicated earlier) is of

particular significance in an impression-die process, in which its formation is preceded by other operations. In this context, the forging operation is often examined as a sequence of events, each of which contributes part of the total applied load and is responsible for specific features of the process.

Therefore, a general approach to tool design and process planning requires the consideration of the following para-meters:

(1) the shape complexity and, hence, the factor S;
(2) the geometry and number of blocking, preforming opera-tions;
(3) an estimate of the flash characteristics;
(4) an assessment of the total power requirement; and
(5) a decision on the point(s) of load application in each sequential operation.

Again, in the case of complex geometries, the finishing impression-die operation will have to be preceded by preform-ing in blocker dies. The basic requirements concerned with the design of these are that the blocker die should be slightly smaller than the finishing one (so that it can fit into the latter) but, unlike the finishing tool, the blocker die should tend to enhance a more uniform distribution of the material by having larger radii and fillets.

The reduction in the wear rate of a blocker die must be considered, particularly when ribs and webs are to be formed. Thus the web thickness in the blocker will have to be larger than in the finishing die and, furthermore, in order to improve the flow of metal into the ribs, an opening taper in the web may have to be provided, with the heights of the ribs being higher than those in the intended finishing system.

In a combined process involving the three separate stages of upsetting, blocking and finishing, the choice of any particular route to obtain a given shape depends on the relative reduc-tion involved. An example of this is given in *Figure 4.87*, in which the forging of a webbed and ribbed H-shaped compo-nent by three different routes is shown.

A very much simpler example of using blocker dies is (*Figure 4.73*) where the left-hand sides of individual stages represent blocker preforms.

When the impression-die forging is carried out in a single die, the above-mentioned apportioning of load contributions becomes a necessity. A notional load–stroke curve (*Figure 4.88(d)*) indicates the importance of the individual loading phases. The actual load levels are very difficult to estimate since, among other parameters, flash dimensions, tempera-tures and frictional conditions must be known, but the use of computer modelling is becoming more common in such cases. A computer-aided-design (CAD) approach is very expensive, if applied to the whole cycle of an operation, and, conse-quently, for simpler shapes (for which the expense might be unjustified) only a few stages are considered and the process parameters are deduced from these. This type of simplified treatment is illustrated in *Figure 4.88(a–c)*.

For steel forgings the flash dimensions of thickness t and width w can be estimated from the following empirical formu-lae:

$$t = -0.09 + 2\sqrt[3]{Q} - 0.01Q \tag{4.75}$$

and

$$w/t = -0.02 + 0.0038\frac{D_oS}{t} + \frac{4.93}{Q^{0.2}} \tag{4.76}$$

where D_o is the diameter of the initial round stock, Q is the weight of the forging (in newtons), and S is the factor given by equation (4.74).

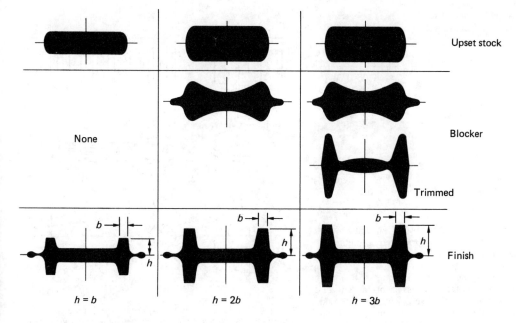

Figure 4.87   Three different routes of upsetting, blocking and finishing a specific shape of forging

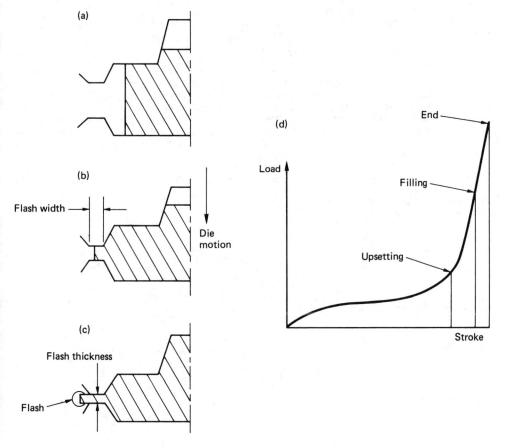

Figure 4.88   A simplified approach to impression-die design based on the consideration of a notional load–stroke curve: (a) upsetting, (b) filling, (c) finishing

Table 4.8 Speed range of forging equipment

Forging machine	Speed range (m s^{-1})
Hydraulic press	0.06–0.30
Mechanical press	0.06–1.5
Screw press	0.6–1.2
Gravity-drop hammer	3.6–4.8
Power-drop hammer	3.0–9.0
Counterblow hammer (total speed)	4.5–9.0
HERF machines	6.0–24.0

HERF, high-energy-rate forming.

The choice of the appropriate forging machine will, naturally, depend on both the expected load requirement and the type of forging. *Table 4.8* provides an indication of the available plant.

Computer simulation of metal flow, leading to a better assessment of the parameters of the process of forging axisymmetric components, is based on the use of the upper bound elemental technique (UBET). The process is approached in an incremental manner (*Figure 4.88*) and the flow inside the die cavity is predicted by establishing a velocity field that minimises the rate of energy consumption. A more recent analytical development in this area can be used to design preforms using UBET in reverse, i.e. by starting with the finished forging and working backwards.

4.6.11 Estimation of forging loads

Approximate values of required forging loads can be obtained by using a simple equilibrium approach. When considering a specific component, say that shown in *Figure 4.89*, it is necessary to establish, first of all, the planes in which flow will take place (*Figure 4.89(a)*) and then the direction of flow (*Figure 4.89(b)*). On completing this exercise, the axisymmetric section of the forging is further subdivided into simple geometrical shapes and the stress system in each is considered. The method is illustrated below using the example of the forging shown in *Figure 4.90*.

Taking the representative flow stress as Y (but bearing in mind that this will vary, to certain extent, across the section) and the frictional shear factor for the material (a notional equivalent of the coefficient of friction) as m, the stress at the parting line of the die where the flash begins to form is

$$\sigma_{ea} = \left(\frac{2mw}{t\sqrt{3}} + 1\right)Y \tag{4.77}$$

On integrating the stress over the region of the flash, the flash forging load P_{fa} is given by

$$P_{fa} = 2\pi Y\left\{-\frac{2m(R^3 - r^3)}{3t\sqrt{3}} + \left(1 + \frac{2mR}{t\sqrt{3}}\right)\left(\frac{R^2 - r^2}{2}\right)\right\} \tag{4.78}$$

where $R = r + w$, and r is the radius of half-width of the cavity.

The load acting on the die cavity P_{ca} is then

$$P_{ca} = 2\pi r^2\left(\frac{mrY_c}{3H\sqrt{3}} + \frac{\sigma_{ea}}{2}\right) \tag{4.79}$$

where Y_c is the flow stress in the cavity.

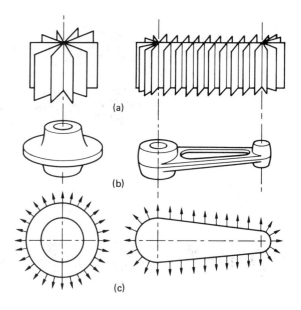

Figure 4.89 Examples of metal flow in the forging of two different components. (a) Planes of flow. (b) Finished forging. (c) Directions of flow

The total load on the section P is the sum of the component loads:

$$P = P_{ca} + P_{fa} \tag{4.80}$$

A similar approach is adopted for other shapes which can always be geometrically simplified, or idealised, for the purpose of these computations. It must be emphasised, however, that the load values thus obtained can only be approximate.

When the profiles of the forgings are relatively simple and, in particular, when the coefficient of friction is low in the operation, the slip-line-field technique can give fairly satisfactory results.

The slip-line method has been applied in the analysis of the bulging (in upsetting) and elongation of both flat and profiled

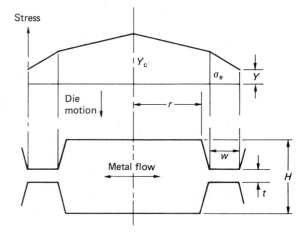

Figure 4.90 Force equilibrium approach to the calculation of stresses in a simple forging

dies, and also in the analysis of impression and closed-die forging. However, in all these cases, the problem of accounting for frictional effects remains, and this imposes severe limitations of the use of the technique.

4.7 Extrusion

4.7.1 General concepts

One of the very widely used metal-forming processes is that of extrusion. In its basic form, the operation involves preparation of a, generally, cylindrical billet (sometimes referred to as the 'slug'), insertion of it into a container (which holds a suitably profiled die), and applying pressure (by means of a punch or a ram) to the trailing end of the billet. The billet is thus pushed, or extruded, through the die.

Since extrusion normally constitutes a more or less direct step from the billet to the finished product, tool design (including its material) and lubrication are of primary importance.

The term 'extrusion' is sometimes applied to forging operations that involve elongation of the product, but it is used here only for the case of 'pure' extrusion described above.

The extrusion process is used in the manufacture of round or profiled bars and tubes, profiled sections, simple and shaped containers, finned or ribbed components and, more recently, spur gears.

Actual extrusion operations are carried out using either mechanical or hydraulic presses; the former being of the vertical type and the latter, as a rule, of the horizontal type. These machines provide extrusion ratios (initial to final cross-section ratios) varying from 10 to 100, with punch speeds of up to 500 mm s^{-1}. Slower speeds are used for most of the lighter non-ferrous alloys of aluminium, magnesium and copper, and higher velocities of deformation are needed when processing ferrous alloys, refractory metals and titanium.

Mechanical presses show a slight advantage over hydraulic machines in terms of speed. The speed of operation precludes almost completely the possibility of using glass as a lubricant which plays an important role in extrusion of stainless steels. Consequently, the process becomes more severe and the billet to hollow ouput can be slightly lower than that for hydraulic presses. A further disadvantage of mechanical presses lies in the fact that the punch velocity varies during the working stroke, and so very considerable difficulties can be created when processing alloys suceptible to strain-rate variations.

In addition, the maximum length of the product is limited by the length of the crankshaft. Hydraulic presses remove these difficulties because the speed of extrusion is controllable and can be maintained almost constant while the length of the extrudate depends on the load capacity of the press and not on its mechanical characteristics.

A distinct, additional and practical advantage of this type of press is the ease of removal of long tubing, or other extrudate, as opposed to the vertical, mechanical presses where a provision for, say, a pit must be made.

It is, however, advisable to remember that, in general, alignment is better on vertical presses than on horizontal ones, and that loading conditions are less severe on the former.

With the great advantages of glass as a lubricant, the number of horizontal presses has inevitably increased, being at present about three times that of vertical presses.

Because of its very wide range of application, the extrusion process is looked upon in a variety of ways that reflect the particular use to which it is put in any specific circumstance.

As a consequence of this, the classification of extrusion operations follows diverse courses.

The most basic classification recognises four extruding techniques:

(1) forward or direct extrusion (*Figure 4.12*),
(2) backward or inverse extrusion (*Figure 4.12*),
(3) side extrusion (*Figure 4.91*), and
(4) continuous extrusion (conform).

However, depending on the shape of the product, a further subdivision is often used (*Figure 4.91*):

(1) solid (circular or non-circular rod) extrusion,
(2) hollow (tubular) extrusion, and
(3) can extrusion.

In view of the importance of die design and, therefore, the die profile, the process is sometimes classified with reference to the tooling:

(1) square (flat faced) die extrusion,
(2) conical (linearly converging) die extrusion, and
(3) profiled die extrusion.

Naturally, extrusion processes can be carried out at elevated temperatures (hot extrusion), or below the recrystallisation temperature of the alloy (warm or cold extrusion).

Square-die processing is used normally in the hot extrusion or profiled light metal (aluminium or copper alloys) components. Continuous profile dies are mainly used for lubricated cold or warm extrusion to enhance material properties and improve surface finish. Streamlined dies are employed because of the higher degree of homogeneity of flow that can be obtained.

Depending on the properties of the starting material and the complexity of the shape of the extrudate, cold extrusion processes (which clearly require high extrusion pressures) involve three distinct techniques which, under specific conditions, ameliorate to some degree their severity:

(1) conventional extrusion,
(2) impact extrusion, and
(3) hydrostatic extrusion.

As in other cold-forming operations, the mechanical properties of the metal are considerably improved if, of course, the increase in temperature associated with the operation does not exceed the temperature of recrystallisation. Dimensional tolerances are of a very high order and, as already mentioned, efficient lubrication gives good surface finish, which is also made possible by the absence of oxidising effects that are present in hot working conditions.

4.7.2 Basic extrusion operations

The three most basic extrusion operations of forward, backward and side flow are shown in *Figure 4.91*. Although, as shown in the figure, these operations refer to axisymmetric solid or hollow components of circular cross-section, profiled or asymmetric shapes can also be obtained in this way. Again, the techniques are equally applicable to hot, warm or cold conventional processing.

In a forward extruding operation involving the manufacture of a solid rod or hollow tube (*Figure 4.91(a,b)*) the previously prepared billet is placed in the container and is either directly extruded to form the rod or, if prepierced, is threaded onto a cylindrical mandrel which supports its bore and is then extruded. The deformation of the tube results in the thinning of its wall and the consequent elongation of the extrudate. If there is no prepiercing, a piercing punch is positioned in the

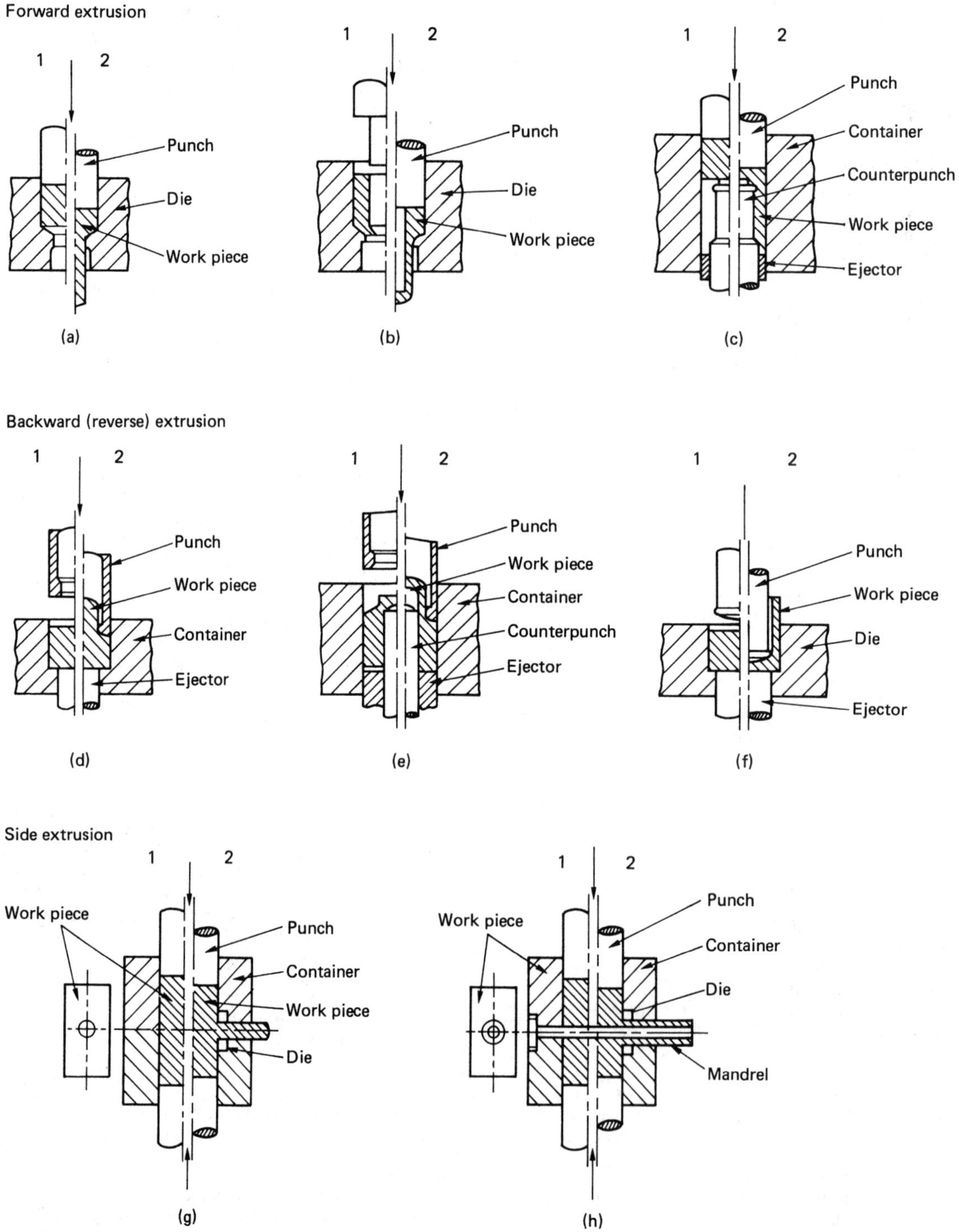

Forward extrusion

(a) (b) (c)

Backward (reverse) extrusion

(d) (e) (f)

Side extrusion

(g) (h)

Figure 4.91 Forward extrusion of (a) solid, (b) tube, and (c) can. Backward (reverse) extrusion of (d) solid, (e) tube, and (f) can. Side extrusion of (g) solid, and (h) tube. 1, Pre-extrusion condition; 2, post-extrusion condition

container to provide, on the one hand, the initial hole and, on the other, to act as a mandrel.

In the extrusion of a can (*Figure 4.91(c)*), a counter punch in the container acts as a die by causing the material to flow between the container walls and its own outer surface. At the conclusion of the operation and retraction of the punch an ejector removes the formed can from the container.

In the backward extrusion of rod (*Figure 4.91(d)*), the billet is placed in the bottom of the container and a hollow punch, of bore diameter corresponding to the outer diameter of the rod to be extruded, is forced into the material, causing it to flow upwards.

A similar arrangement is used when extruding tubular components (*Figure 4.91(e)*), but here either a prepierced hollow is used or piercing in the container forms the first stage which is followed by extrusion. Again, an ejector is necessary to remove the product.

When extruding a can (*Figure 4.91(f)*), a reverse of the forward process is effected. The billet is placed in the die and a solid punch is moved axially into the metal. An ejector is necessary to remove the can.

Side extrusion, either one or two sided, is limited to rod (*Figure 4.91(g)*) or tube (*Figure 4.91(h)*) manufacture. In the first case, the die is situated in the side of the container with the billet positioned at right angles to it, but supported at its lower end. The punch moves axially downwards forcing the metal to flow through the die. When making a tube, the billet must be pierced first, inserted into the container and the

mandrel threaded through it. It is only then that the punch can be actuated and cause plastic flow to commence.

Although, as indicated above, the arrangement of tooling in the actual extrusion process varies with the component to be made and the technique used, the basic features of the press are the same. Thus the arrangement shown in *Figure 4.92*, for backward can extrusion, is representative of the scheme often adopted.

As in other forming processes, extrusion, in any form, produces a number of material defects which are associated either with the characteristics of the process itself, or with the selected geometry of the forming pass. In more general terms, these can be summarised in the form given in *Figure 4.93*. Defective items in extrusion may reach a proportion as high as 10–15% of the total volume of the product and, although inspection of extrudates can prevent the use of defective components, rejection of parts increases production costs since, in addition to the expense of full processing, the inspection itself is expensive.

Defects that may occur are usually due to any, or a number, of the following:

(1) defective billets,
(2) defective or unsuitable tooling, and/or
(3) processing technique.

Irrespective of their origin, all these defects can be reduced or even eliminated by correct design of the extrusion tooling.

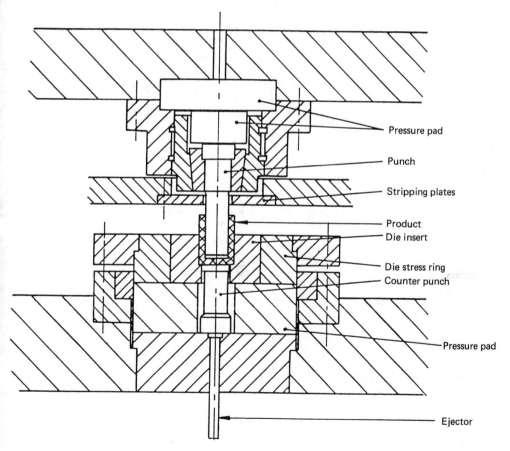

Figure 4.92 Arrangement of tooling in backward can extrusion

Pressure pad
Punch
Stripping plates
Product
Die insert
Die stress ring
Counter punch
Pressure pad
Ejector

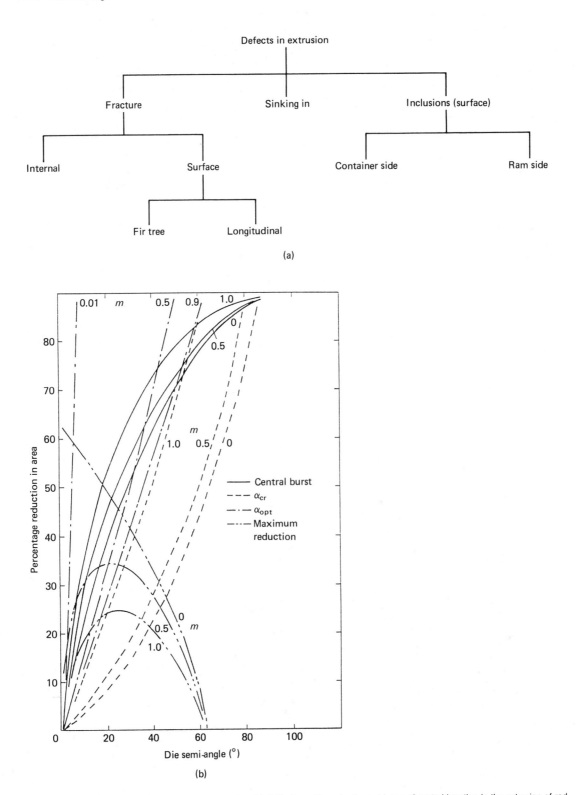

Figure 4.93 (a) Possible defects caused by extrusion. (b) Critical conditions for the incidence of central bursting in the extrusion of rod

Defects arising from unreasonably high stress fields manifest themselves in fracture. Fracture-type defects form two different classes, depending on their location with respect to the tooling. Consequently, two specific types of fracture may occur:

(1) die contact fracture (fir-tree cracking), and
(2) internal fracture (central bursting).

Fracture may be considered as a limiting quantity in the formability of a material. Bulk metal forming processes produce a state of stress in which major stresses are generally compressive. However, in extrusion, hydrostatic pressure is a decisive factor in the incidence and growth of a ductile fracture. The magnitude of hydrostatic pressure decreases along the die axis towards the exit from the die and the pressure effects may become tensile in nature, thus leading to cavitation. The defect produced by this phenomenon is known as 'central bursting'. Central bursting is rare and analysis based on the slip-line approach suggests that it may occur when the extrusion ratio is less than 4, since at ratios greater than this all the stresses in conventional extrusion tend to be compressive.

Different criteria are used to predict central bursting defects and use forming variables and/or die geometry. For instance, the so-called 'Indian feather' criterion involves the die semi-angle and extrusion ratio. It assumes different degrees of strain hardening and considers the effect of friction. The criterion leads to the establishment of a safe zone. Parts produced within the zone are likely to be defect free. According to this approach, an increase in the strain-hardening capacity of a metal increases the safe zone. The incidence of central bursting is favoured by large die angles, small reductions per pass, and a low fracture strain of the worked material. An increase in the reduction ratio for a given die, or a decrease in the die semi-angle at a constant reduction ratio, may shift the process from a dangerous area into the safe zone.

However, forming-limit criteria predict that fracture will occur when the levels of stresses and strains reach critical values. It follows, therefore, that the existence of redundant strain may cause the effective strain to approach its critical value.

Inhomogeneous deformation in extrusion is synonymous with non-uniformity of flow of the material. For instance, the material near the centre of the specimen moves faster than that near the walls. The result of this is that, on leaving the die, the extrudate is subject to the compressive residual stresses at the centre and tensile stresses near the surface.

While the temperature is an important factor in surface cracking, such cracks do not normally appear in the absence of longitudinal or hoop stresses on the surface of the product. Surface cracking can be eliminated by using a properly designed die which reduces residual stresses. The stresses can be related to the geometry factor Δ in rolling, extrusion and drawing because, as the geometry factor increases to above unity, the residual stresses begin to build up and increase in magnitude with the factor.

Formability in extrusion is not, therefore, seen as a unique property of the material, but depends on localised conditions of stress, strain rate and temperature, in combination with material characteristics. Processing parameters are associated with die design and lubrication control the local stress state and strain rate in the processed material and are, therefore, instrumental in determining the quality of the finished product.

A very characteristic feature of extrusion, especially when using flat dies, is the phenomenon of the 'dead-metal zone'.

Figure 4.94 Formation of a dead-metal zone. (a) Mechanism of formation and the associated ram-force displacement curve. (b) Separation in the dead-metal zone of 99.5% pure aluminium bar (lubricant methylolstearamide)

It can be seen from *Figure 5.94(a)* that in flat dies (90° die semi-angle) part of the billet material becomes trapped in the corner of the die/container space and does not participate in extrusion. The bulk of the material moving through the die shears past the trapped annular ring of stationary metal which thus effectively forms a new, curved die surface that merges with the proper die.

Depending on the point of view adopted, the formation of a dead-metal zone can be regarded either as a defect or as a desirable phenomenon which may enable the material to adopt the optimal flow path.

In support of the latter viewpoint is the ram (punch) load–displacement curve (*Figure 4.94(a)*) which suggests that, although when the zone is first formed (A) the pressure peaks, the extrusion load actually drops within the dead-metal zone and increases sharply again when the metal begins to extrude through the 'real' die.

The dead-metal zone can be sizeable in volume and can, therefore, constitute serious loss when more expensive materials are processed. An example of the size of this discard, as well as of the pattern of flow which it creates, is provided by the extrusion of an aluminium bar through a 90° die (*Figure 4.94(b)*).

Flat dies are employed for the extrusion of a variety of materials and shapes although, as will be seen later, contoured tools are often essential if the efficiency of the process (in terms of the quality of the product) is to be enhanced.

In simple, conventional hot extrusion, working conditions are well defined; these are summarised briefly in *Table 4.9*.

4.7.3 Conventional tube extrusion

One of the most important applications of the extrusion process is the manufacture of seamless tubing and, in particular, the manufacture of the strainless-steel variety. This is because rotary, longitudinal rolling processes often give unsatisfactory results, and operations such as Assel elongating fail altogether. The importance of extrusion becomes obvious in these circumstances, and its applicability to the processing of ferrous alloys cannot be stressed too strongly.

Although extrusion was used for non-ferrous metals since about the middle of the last century, serious interest in its application to steels was not shown until the mid-1920s when the first experiments carried out in France, the USA and Germany showed the distinct feasibility of the operation. The problem encountered was the high rate of tool wear, in particular of the dies. The slow development of tool materials and manufacturing techniques prohibited any extensive use of extrusion until almost the outbreak of World War II, but rapid progress was then made in Germany and the USA, whereas in France, Sejournet continued the development of his glass lubricating technique, which became widely accepted in the 1950s.

Since the extrusion of stainless steel offers wide scope for discussion, these materials are used in the following as the basis for a review of techniques and practices.

The current practice of extruding stainless-steel and tubing and sections, either by direct or reverse methods, is based on the following sequence of operations:

(1) preparation of billet,
(2) heating,
(3) lubricating,
(4) providing a pilot hole,
(5) reheating and relubricating,
(6) extruding,
(7) removing the lubricant, and
(8) straightening whenever required.

Reheating and relubricating (5) are not necessary in some processes.

In general, four main variants of the process are in operation. These involve the following techniques:

(1) prepiercing, extruding;
(2) drilling the billet, expanding, extruding;
(3) drilling the billet, extruding; and
(4) staving (dumping) the billet, piercing, extruding.

The choice and application of a given technique are examined in the following detailed discussion of the general sequence of operations.

4.7.3.1 Preparation of billet

Irrespective of the process involved, all billets are faced at least on one end. Furthermore, they are normally machined, or sometimes ground, since the surface finish of the billet determines to a great extent the surface quality of the hollow. A surface finish of about 7.5 μm is often required, thus increasing considerably the cost of production.

In the case of direct extrusion (using drilling as the means of providing the pilot hole), billets are drilled centrally at this stage. Holes of up to 50 mm in diameter are machined out on twist drilling and vertical boring machines. Larger holes are made by trepanning. Trepanning is also used on smaller size holes in an effort to increase the initial length of billets.

Invariably, billets are either radiused or chamfered externally at the leading ends to offset the tendency to cracking in this region during extrusion. In some cases, provision for internal radius at the trailing end of the billet is made. The radius is usually smaller than the leading-end radius and serves to prevent flash and breaks when extruding to thin discards. The magnitudes of the radii depend on the initial sizes of billets but, on average, they are of the order of 10 mm for the leading ends and about 5 mm for the trailing ends.

Table 4.9 Materials and die geometries in simple conventional hot extrusion

Metal extruded	Billet temperature (°C)	Extrusion ratio	Comments
Aluminium (soft architectural alloys)	400–500	10–400 (630 for indirect extrusion)	Flat dies, hollows using bridge, spider, porthole types. Highly complex shapes
Brass	650–750	10–400 (600 for indirect rod extrusion)	Flat dies, hollows using mandrels. Complex shapes
Copper	750–950	10–250	Flat dies, hollows using mandrels. Simpler shapes
Steels	1000–1300	10–25	Extrusion ratios apply to sections, not tubes. Profiled or bell-mouthed dies
Nickel alloys	1050–1200	7–50	are essential for the Ugine–Sejournet process. Hollows using mandrels.
Titanium alloys	850–1150	9–100	Simple shapes

4.7.3.2 Heating

Heating of billets is of great importance, not only from the point of view of the time involved, but also in terms of the economics of the possible descaling operation.

The general trend appears to be towards the low frequency inert gas atmosphere heating of billets. Theoretically, the economic use of a low frequency induction furnace puts severe limitations on the use of smaller diameter billets below, say, 100 mm diameter; the range above 150 mm being considered economical. The advantage of obtaining a scale-free billet may sometimes outweigh the disadvantage of higher heating costs and, consequently, the intermediate range between 100 and 150 mm is occasionally used.

When using induction furnaces, the general practice is towards preheating billets to about 1020°C followed by further induction heating to about 1200°C for staving and prepiercing, and expanding processes.

Other techniques are also in use, e.g. preheating to about 820°C in a gas-fired furnace to avoid heavy scaling, followed by heating in a salt bath to extrusion temperature. A practice widely adopted in the USA consists of preheating in a Selas radiant heat slot furnace to about 1200°C, followed by heating in a barium or sodium chloride bath (to dissolve scale) to extrusion temperature.

A new heating system that makes use of dual fuel furnaces is also in use. In this system, steel is preheated to 900°C in a gas-fired furnace and is brought rapidly to 1250°C in an induction furnace. Scaling appears to be negligible.

In the case of billet drilled for direct extrusion, heating is usually carried out in a salt bath, with the exception of the revolving Balestra type furnace. The latter combines the heating and lubricating operations, being a drum type furnace in which the refractory lining is coated with a thick layer of molten glass. Prior to heating, glass wool is inserted into each end of the hole drilled in the billet in order to prevent oxidation. Equiverse type furnaces, which have proven very successful in the case of low carbon steels, are not often used for stainless steels.

Essentially, the final heating operation is carried out either in an induction- or salt-type furnace. Each of these processes has distinct advantages and disadvantages. A brief survey of the available information leads to the following conclusions.

Advantages of induction heating

(1) The temperature can be easily regulated, ensuring a correct rate of heating for the given material.
(2) The rapid increase in temperature means that the required heat level is reached in a short time. In the case of a serious interruption of production, heating of surplus billets can be terminated quickly.
(3) The furnace can be easily adjusted to heat single billets and, if required, to heat the leading billet end to a higher temperature in order to facilitate the start of the working stroke of the press.
(4) Induction heating is scale-free, even if the operation is not carried out in an inert atmosphere. In the latter case, the rate of heating to the final temperature is higher than the rate of scale formation.
(5) The loss of material from the billet is practically zero.
(6) Induction furnaces are compact and can be fitted easily into the existing plant. Furthermore, they are well suited to fully automatic handling of billets.

Disadvantages of induction heating

(1) There is a danger of overheating some parts of a hollow billet.

(2) The length and diameter of successive billets should be kept fairly constant in order to reduce possible changes in the pattern of the heat level.
(3) The measurement of temperature is not easy.
(4) It is necessary to change the coil when changing either the type of steel or the billet size.

Advantages of salt bath heating

(1) The rate of heating is fairly high and the level is uniform.
(2) The salt bath affords protection against scaling of billets and any scale previously acquired is dissolved.
(3) The film of salt remaining on the billet after its removal from the bath affords protection against scaling when the billet is transferred to the next stage of the cycle.

Disadvantages of salt bath heating

(1) The salt bath is, in general, inefficient and the running costs are high.
(2) The bath must be serviced regularly in order to prevent accumulation of sludge.
(3) The surfaces of billets become corroded if, for any reason, they are left too long in the bath.

4.7.3.3 Lubrication

The inherent difficulty in extruding steels and, particularly stainless steels, requires more efficient and, to a certain extent, more sophisticated lubricating methods than those used for the processing of ordinary engineering alloys. Originally, graphite type lubricants were widely used mainly because of the possibility of lubricating the mandrel when extruding small-bore hollows. The Ugine–Sejournet process, using glass as lubricant, has changed this situation significantly.

The innovation consists not only in using glass, but also in introducing the novel idea (in hot working) of lubricating only the work piece and not the tools. (Details of glass lubricants are given in Section 4.4.2.)

However, the possibility of increasing the range of extrudable materials and the amount of deformation in a given operation when using glass, can be predicted from a theoretical analysis of the process. The relevant equations, often used in industry, are as follows.

Solid sections:

$$P = \pi R^2 \rho \exp(2\mu l/R) \ln(a) \tag{4.81}$$

Tubular sections:

$$P = \pi(R^2 - r^2)\rho \exp(2\mu l/(R - r) \ln(a)) \tag{4.82}$$

where P is the required load, R is the radius of the container, r is the radius of the tube bore, l is the length of the extrudate, a is the extrusion ratio, ρ is the resistance to deformation, and μ is the coefficient of friction.

Glass is applied by means of pads of fibre, cloth or pressed powder. The only disadvantage of glass is the slight difficulty in removing it from the extrudate. In a properly controlled process, however, the layer of glass is very thin and on cooling there is a tendency for iron oxides to form. These dissolve glass slightly and facilitate its removal.

4.7.3.4 Providing a pilot hole

An initial hole is made either by drilling or machining, or by hot piercing. Drilling and machining, followed directly by extrusion, are used where the finished bore is small (usually up to about 30 mm). The cost of the waste material and labour involved is considerable, but the lack of eccentricity in the finished tube outweighs to a certain extent this distinct disadvantage. This technique is also used whenever the formed metal is difficult to hot pierce.

To obtain bore sizes of 32–115 mm, small pilot holes are drilled and then expanded. Pilot holes for this operation are 20–25 mm in diameter.

The technique is particularly useful when lack of eccentricity is of primary importance, and also when, subject to length limitations imposed by the extrusion container, longer billets are preferred to those that can be used in direct hot piercing.

For bore sizes larger than 115 mm in diameter, hot piercing followed by extrusion is employed. Two different techniques are used: the billet is either pierced directly using a mandrel and extruded in the same press; or it is staved (dumped) in a slightly tapered container, hot pierced in it, and then extruded on a separate press. In the first case, eccentricity presents a problem, and to minimise its incidence and magnitude the billet length/diameter ratio is kept low and should not exceed 5:1. This technique, however, has the distinct advantage of being contained in a single unit, thereby introducing savings in both the capital cost and reheating.

The staving operation is used mostly to reduce the possible occurrence of eccentricity. This is achieved by both ensuring that the initial indentation of the punch is concentric with the billet being seized, and by reducing the length to diameter ratio by shortening the billet prior to piercing. The main disadvantage of staving lies in the introduction of an additional press.

4.7.3.5 Extrusion

The techiques of extrusion are discussed in Section 4.7.2 but, in the case of stainless steel and the Ugine–Sejournet process, tube sizes and the corresponding required press capacities need careful assessment. The standard practice adopted is summarised in *Table 4.10*.

The inherent weakness in the extrusion process lies in the production of discard associated with the formation of the dead-metal zone which, naturally, represents the total loss of material. With the high cost of stainless steels, the problem of discard is more serious than for other materials. The weight of discard is approximately proportional to the cube of the mandrel diameter. The weight is not affected by the introduced variations in the diameter of the die. Although a reduction in the die angle gives some improvement, this is small in comparison with the improvement achieved by completing the piercing–extruding operation in two stages. In this operation, the billet is first pierced against a solid plate which temporarily replaces the die. On completion of piercing, the plate is removed and the extruding die is introduced into its place.

The advantage of reducing the weight of the discard in this way must be set against the cost of the labour involved and the provision of back-plates which, obviously, can be damaged easily.

4.7.3.6 Removal of lubricant

Graphite based lubricants are easily removable by conventional methods. When glass is used as lubricant, techniques vary slightly but, essentially, they all depend on the application of some type of pickling bath. Thus, glass is removed by pickling in a mixture of 4% hydrofluoric acid and 14% nitric acid, or in a mixture of hydrofluoric acid and sodium sulphate.

Small amounts of the mixture of partly decomposed glass and scale are sometimes removed by sand blasting.

The increase in the use of the extrusion process in the last 25 years has been due partly to economical and partly to technical considerations. Extrusion becomes economical and competitive with rolling processes for a range of sizes of up to 150 mm outside diameter, and for comparatively short lengths and runs. The cost of plant required for the production of bigger hollows, both in terms of presses and ancillary equipment, increases rapidly when this limit is exceeded without there being, at the same time, any possibility of increasing the length. At present, the intermediate range (say, 150–250 mm outer diameter) is manufactured using extrusion presses but the process becomes rather expensive owing to considerable tool wear and increased rate of scrap. The additional, inherent disadvantage of the process is its comparative slowness with, on average 50–60 extrusions per hour.

A more important reason for the use of extrusion is the fact that, from the point of view of formability, the process is capable of dealing with very difficult materials. This is possible due to the lower incidence of redundant strains than in rotary processes, and the consequent reduction in the severity of the operation.

4.7.4 Cold extrusion processes

Conventional cold extrusion operations are based on the same three basic systems of forward, backward and side operations (see *Figure 4.91*). The range of materials usefully employable in engineering applications is limited to steels and aluminium and copper alloys. Although easily extrudable, materials like tin, lead and magnesium show no benefit from strain hardening and, in any case, are of no great industrial importance.

Although hollow sections are produced by extrusion in non-ferrous alloys, the bulk of the products consists of profiled sections (particularly in aluminium) used extensively in the car and aircraft industries, as well as in domestic situations.

Very high extrusion pressures, combined with extensive frictional effects that occur in the container and in the die, make conventional extrusion unsuitable for processing less ductile materials or composites. Although the individual demand for either is not yet very high, developments in modern technology make it imperative that processes capable of coping with such materials be developed. Two processes fulfilling the demand are hydrostatic and impact extrusion. Both of these rely on high pressure, but reduced friction, forming.

The process of hydrostatic extrusion differs from conventional extrusion in that it employs a pressurised liquid instead of an extrusion ram. A diagram of the type of hydrostatic extrusion systems available is shown in *Figure 4.95*. The

Table 4.10 Range of stainless steel tube sizes

Wall thickness (mm)	Capacity of press (ton)					
	3000		4000		5000	
	OD (mm)	Length (m)	OD (mm)	Length (m)	OD (mm)	Length (m)
2.3	81.3–241	16–6.1	81.3–249	19.8–10	91.4–254	19.8–10
5.1	91.4–279	16–5.02	71.1–320	19.8–5.03	119–355	15–5.03
10.2	51–250	14.9–5.03	61–279	19.8–5.03	101–457	17–5.03
15.2	61–223	14.9–5.03	61–254	18.6–5.03	119–317	14.9–5.03
20.3	61–198	12–4	71.1–241	15–5.03	119–279	10–5.03
25.4	—	—	—	—	109–211	8.84–5.03
30.5	—	—	—	—	101–200	6–5.03

OD, outer diameter.

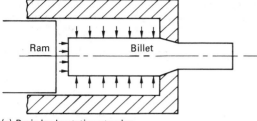

(a) Basic hydrostatic extrusion

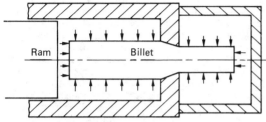

(b) Differential extrusion

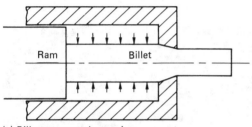

(c) Billet augmented extrusion

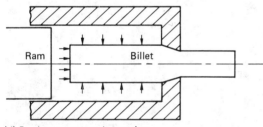

(d) Product augmented extrusion

Figure 4.95 Different hydrostatic extrusion systems

extrusion die has, in this case, a relatively small cone angle, but the high bursting stresses associated with this are offset by the fact that the fluid pressure acts as a containing element around the die circumference.

In practice, this operation has a number of advantages over conventional extrusion, most of which are related to the fact that there is no contact between the billet and the extrusion container wall. This results in:

(1) lower extrusion pressure,
(2) improved lubrication from the pressurising liquid,
(3) a reduction in redundant work since smaller die angles can be used, and
(4) the extrusion of longer billets—the limit being the length of the container.

A further advantage of hydrostatic extrusion is associated with the improvement in ductility which most materials undergo when deformed under hydrostatic pressure. In practice, this means that the extrudability of materials is improved when the hydrostatic pressure component is present and under such conditions many nominally brittle materials have been satisfactorily cold extruded.

As in other processes, it is found that for a given amount of deformation and specific frictional conditions, there is an optimum die angle which gives the best balance between friction and inhomogeneous strain and, hence, there is a minimum extrusion pressure.

In spite of considerable research and development work on hydrostatic extrusion, the process has not been as widely accepted as an important industrial process as was originally expected. There are various reasons for this. On occasion, the process can be unstable because the lubrication between the die and the work piece varies as extrusion proceeds and this can lead to rapid changes in the velocity of the extrudate. At the end of the operation, the billet can be completely ejected which, although avoiding the existence of a discard, can actually be a dangerous event. There is still a lack of knowledge about the fatigue life of thick-walled containers of the dimensions required in hydrostatic extrusion work and, consequently, there is some reluctance to use this technique unless absolutely necessary.

Although, in some cases, it has proved to be advantageous to perform warm or even hot hydrostatic extrusion (in fact the hot extrusion of steels (using glass as lubricant) is often quoted as the most successful production-type hydrostatic extrusion process), one of the major advantages of the process is that it can be used successfully as a cold-forming operation.

Since the fluid surrounding the work piece isolates it from the walls of the container, thus eliminating friction, conditions exist for hydrodynamic lubrication to occur. The consequent reduction in frictional effects makes it possible to use narrow die angles which reduce the level of redundancy. In general, therefore, longer components can be extruded at a given pressure than would be possible in a conventional operation. Under certain circumstances, hydrostatic extrusion offers the possibility of avoiding interstage annealing which can be of particular interest in the extrusion of wire.

The four basic systems available are shown in *Figure 4.95*. Simple hydrostatic extrusion (*Figure 4.95(a)*) relies on the hydrostatic pressure exerted by the fluid (usually oil) on the billet. The billet must be preshaped to fit the conical die exactly and thus provide a seal which enables the liquid medium to be pressurised when the ram is actuated. The ram does not come into contact with the billet.

To inhibit cracking, the billet can be extruded into another, pressurised chamber (*Figure 4.95(b)*). In this case the extrusion pressure is provided by the differential liquid pressure across the die. It is possible to extrude the billet completely, i.e. without discard.

Two further variants of the process involve both extrusion and, to provide higher deformation combined with good dimensional control, drawing. These systems are usually referred to as either 'billet-augmented' or 'product augmented' extrusion.

In the billet augmented operation (*Figure 4.95(c)*), part of the hydrostatic pressure is replaced by the use of a ram, with the latter being in direct contact with the billet. In the product-augmented system (*Figure 4.95(d)*), forward pull is provided to supplement the hydrostatic pressure existing around the billet and in the die.

In all these cases the effect of the hydrostatic pressure is to inhibit and, indeed, to reduce the magnitude of the tensile stress, and thus to approximate the conditions of triaxial compression.

Special industrial needs, particularly those of the atomic energy and chemical industries, occasionally require the production of reasonably long and geometrically complicated components capable of combining strength with good protective and/or heat-transfer characteristics. In the former case, refractory metals are likely to be involved, whereas in the latter situation extended surface areas of specified shapes and anticorrosive properties may be required. These, at times opposing, requirements can be fulfilled if a satisfactory technique of manufacturing assemblies of mono- or bi-metallic multilayered, coaxial rods is developed.

Conventional metal-forming processes which would normally provide a means of elongating the initially short refractory metal combinations are often difficult to operate. This can be demonstrated using the example of the production of composite, remote control rods.

Drawing or extrusion of loose arrays, held in position by locating outer sheaths (*Figure 4.96(a)*) may lead to unsatisfactory results associated with the flow characteristics of perhaps widely different constituent materials. Differential elongation will be aggravated further by the relative slipping of the individual rods, or of whole layers with respect to each other. A possible solution to this type of problem is provided by initial explosive welding of the constituent elements of the

assembly to each other, followed by a conventional forming operation. In this approach the welding part ensures the integrity and rigidity of the composite and, therefore, facilitates, its processing.

As a specific example, a variety of low carbon steel–copper rod arrays is considered, with a view to comparing their behaviour under conventional and hydrostatic extrusion. The two materials differ sufficiently in their properties to provide a realistic basis for assessing the problems involved.

The process of manufacturing integral assemblies of bimetallic rods consists of two distinct steps, i.e. implosive welding and subsequent extrusion. The primary objectives of the former are to obtain a good quality weld between the two elements of the array, to ensure that the basic shape of the assembly is affected as little as possible, and at the same time to provide a billet suitable for extrusion. The welding can be accomplished in two slightly different basic ways. The rods can be held either in a plastic outer shield, or in a metallic one. The material of which the shield is made is normally the same as one of the constituent alloys of the assembly. In the first instance, the cavities existing between the individual rods or layers of rods remain unfilled by the metal, whereas in the latter such initial cavities and voids will contain the materials of the shield deposited in them as a result of jetting. In

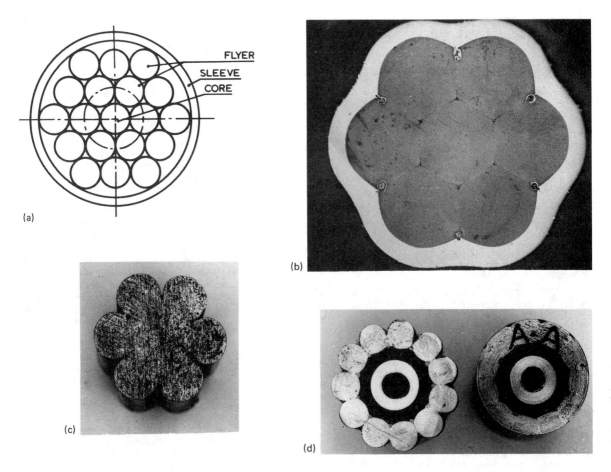

Figure 4.96 Explosive-welded array of rods for remote-control systems operating in toxic conditions. (a) Basic arrangement and definition of terms. (b) An array welded explosively in an outer metal sheath. (c) An array welded explosively in a plastic sheath. (d) A heat-exchanger unit after welding and after hydrostatic extrusion

general, the plastic shield system is preferred but, from the viewpoint of further processing, the irregular outline of the outer billet surface makes it directly suitable for conventional extrusion only. The metallic shield system, although giving a slightly irregular cross-sectional outline, is by virtue of possessing an additional layer of metal, more suitable for initial machining and, therefore, for hydrostatic extrusion. The problem of jetting of the shield material remains, but may not be critical in a number of applications.

Figure 4.97 Failure of an explosively welded array of rods in convetional extrusion

The basic requirement of the cold-extrusion operation is that it produces as homogeneous deformation of the constituent elements of an array as possible and, furthermore, that the integrity, and therefore the weld quality, of the assembly remains unimpaired after the operation. This, in turn, implies the use of as low an extrusion pressure as is compatible with the required deformation and, consequently, the effect of the die profile becomes important.

In an axisymmetric bimetallic combination, consisting of materials which differ in their mechanical properties, the sequence in which they are used (i.e. which material constitutes the core and which the outer shield (the flyer)) plays an important role and must be carefully assessed before the system is designed.

The presence of tensile stresses, generated in a conventional extrusion operation, leads to the failure of the specimen (*Figure 4.97*) when larger deformations are attempted. In hydrostatic extrusion of specimens such as, for instance, the heat exchanger assembly shown in *Figure 4.96(d)*, a degree of inhomogeneity of deformation is discernible in that the flyer always overflows the core, but its level varies. The percentage reduction in the core depends on the material from which it is made, being slightly higher for a more ductile material.

In general, the inner-rod cavities and voids, often containing resolidified material of the sleeve will substantially close during extrusion. The resulting structure presents a more homogeneous appearance than that of the unprocessed billet.

The basic soundness of extruded specimens is illustrated in *Figure 4.98* in which the polished leading ends of the specimens correspond to the initially shaped ends of the billets.

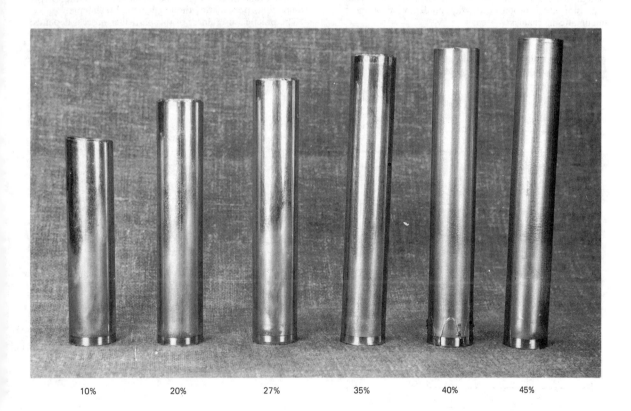

| 10% | 20% | 27% | 35% | 40% | 45% |

Figure 4.98 Hydrostatically extruded prewelded arrays of rods

Another group constituting successful processing of other-wise difficult to extrude composites is that of simple shaped, bimetallic solid core/tubular sleeve systems.

Irrespective of the method by which the two materials are initially arranged with respect to each other, i.e. whether they are metallurgically bonded or mechanically interlocked, the success of further processing depends on the preservation of the condition of homogeneity of deformation or of identical elongation of both components. If this requirement is not fulfilled, fracture may easily result.

Essentially, but not entirely, the fracture of either the core or the sleeve is caused by the failure of the harder material to elongate to the same extent as the softer one. It is occasionally found that a soft sleeve will, for instance, fracture over a hard core or that a soft core will fail in a harder matrix. In other words, the incidence of fracture cannot be attributed entirely to the mechanical properties of the materials involved, but is also conditioned by the geometry of the pass (and therefore redundancy), by frictional conditions, and by the relative dimensions of the two components.

Although common faults, such as, for example, 'fish skin' can also be present, it is basically the condition for incidence of fracture that is of immediate interest. As indicated earlier, the existing analyses are concerned with the determination of practical limits within which sound flow can be expected.

Away from the conventional operations, including hydrostatic extrusion, it is the high-energy-rate operations that are steadily finding application in specialised industrial areas, and extrusion is no exception to this development.

Impact extrusion can be employed in the case of either bar or tube forming. The forming energy is supplied through the medium of a ram which, in turn, is actuated by a sudden expansion of gas produced by, say, an explosive charge or a pneumatic–mechanical system.

In the forward extrusion of bar (*Figure 4.99*), immediately after impact the ram and billet travel together at high speed with the extrusion of the billet taking place in the die. Relatively high strain rates can be attained and in some alloys

this will lead to a substantial increase in the value of the yield stress Y, combined with a reduction in the strain to fracture. This is particularly noticeable towards the end of the operation when a high degree of deceleration is reached. It is also at this stage that high tensile stresses are reached at the base of the extruded bar and result in either the necking or breaking off of the product.

The maximum possible deformation depends mainly on the ability of the tool materials to withstand shock conditions.

Impact extrusion of tubular components is more conveniently carried out in a reverse system (*Figure 4.100*). This employs a ram of either flat or contoured face and a means of applying additional carefully controlled external pressure p to the walls of the extruded product. Pressure p is required to minimise the effect of two major faults which characterise the operation, i.e. cavitation and fish skin. Fish skin occurs throughout the operation and is equivalent to the appearance of circumferential tearing of the surface. Cavitation is associated with the flow of the material upwards, leaving the corners of the container unfilled. Rounded instead of sharp edges are thus produced. The determination of the levels of energy required to allow this type of flow to occur, in preference to one which would give a full deformation, is of practical importance and lends itself to a satisfactory investigation by means of the upper-bound theory. This is made possible by the fact that the incidence of defects is associated with a lower energy level than that required for a sound metal flow.

In general, an investigation of the impact extrusion process is best approached from the viewpoint of the energy balance. This can be done either in the form of an overall equilibrium-type assessment, or by detailed examination of the energy components provided by the upper-bound technique.

The practical usefulness of this particular process depends, to a considerable degree, on whether the incidence of faults such as break-offs in a bar or cavitation and fish-skin in, say, tubular bottle manufacture, can be avoided. Although to some extent all these faults are characteristic of dynamic conditions,

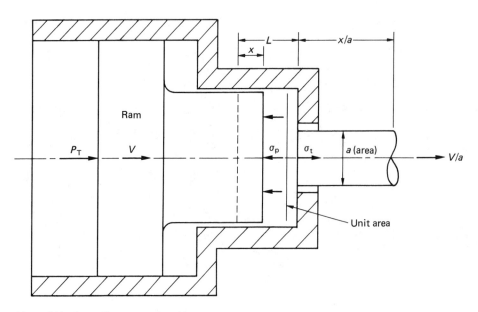

Figure 4.99 Forward impact extrusion of bar

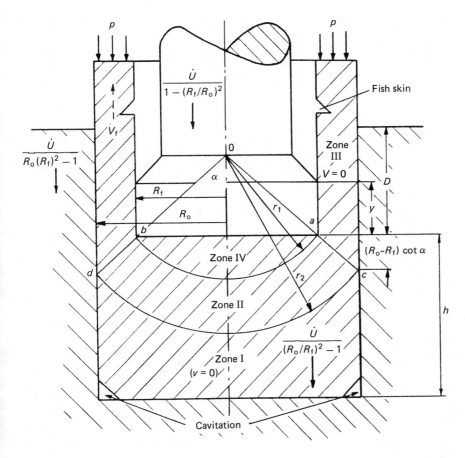

Figure 4.100 Reverse impact tube extrusion

it should be remembered that cavitation, for example, can also occur in a conventional operation.

In bar extrusion, the incidence of these faults is attributed to the magnitude of the tensile stress (σ_t) appearing at the instantaneous base of the bar, and supporting the instantaneous acceleration of the material. The effect of the stress, in conjunction with the mechanical properties of the metal, dynamics of the system, and its redundancy, will determine the conditions for breaks.

A measure of redundancy is introduced by a work redundancy factor Φ' defined as

$$w = \Phi' Y \ln(1/a) \tag{4.83}$$

where w is the work done per unit volume, and a is the cross-sectional area. In this expression the work done per unit volume in deforming and shearing in the pass is related to that expended in the homogeneous deformation associated with the extrusion ratio only. The work redundancy factor is, therefore, based on both stress and strain, but does not necessarily present a complete picture of what happens in the pass. To account fully for the strain-hardening effect on deformation, it would be necessary either to adjust directly the value of Y, while retaining the simple definition of homogeneity given by the logarithmic term, or to use the concept of the generalised homogeneous strain in its place. In its present form the factor is not equivalent to either the stress or the strain redundancy factors (defined previously in equations (4.20) and (4.13), respectively).

The relative importance of the various process parameters for an ideal situation and the simplified case when the ram is in free flight, is illustrated in *Figure 4.101*. Considering the variation in the limiting stress ratio σ_t/Y with the ram position $\bar{x}$, it is seen that for a range of values of a and v ($v = M/\rho L$, where M is the mass of moving parts per unit area of billet, and ρ is the density), the possibility of a break is increased when a low value of a is used in conjunction with low v. The influence of redundancy on the incidence of breaks is shown in *Figure 4.102* for a variety of combinations of v, θ and a (where $\theta = MV^2/2gLw$). Safe operating conditions lie above the plots in the figure, whereas the onset of faults is expected when operating below the curves.

The conditions of flow and the incidence of faults in the reverse extrusion of a tubular bottle depend mainly on the shape of the ram face.

A flat-faced tool will produce a dead-metal zone, the spread of which is determined by the angle α. The variation in the required ram pressure with α gives the optimum value of the angle of the dead-metal zone for which the pressure p_r is a minimum. If, by virtue of using a suitably contoured ram, the dead-metal zone is not allowed to develop, the optimum angle of the ram nose, corresponding to the minimum ram pressure, will influence the relative thickness of the wall.

A comparison between the extrusion with a flat and contoured ram, based on the upper-bound analysis, is given in

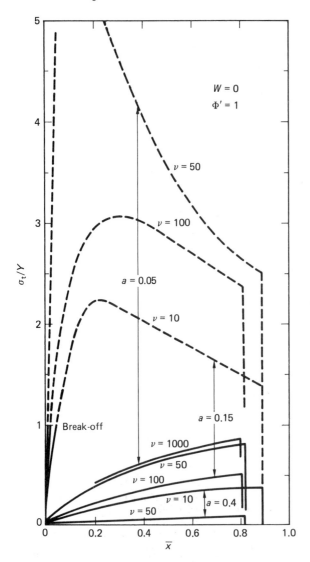

Figure 4.101 The incidence of breaks in impact extrusion as a function of the tensile stress

Figure 4.103, which shows the relationship between the power components and the wall thickness. Irrespective of the system used, the ram pressure increases with increasing penetration (ratio D/R_f), since the resulting increase in the area affects the magnitude of friction. Consideration of the inertia term, omitted in the figure, would of course result in a further increase in the numerical value of p_r. Essentially, both shear and frictional losses are relatively independent of the wall thickness if this is substantial, but the increase in p_r for low values of R_o/R_f is rapid and numerically significant because of the increase in both shear and frictional losses caused, in the case of shear, by greater values of Δv at the surfaces of the velocity discontinuity. An almost uniform increase in the power of internal deformation, with the wall deformation, is thus accompanied by a decrease in shear and friction. Since, in the absence of a dead-metal zone, friction losses at the boundary *ab* (*Figure 4.100*) are much lower than the cor-

responding shear loss would be, it follows that the total ram pressure required when using a contoured ram is less than that when using a flat face tool.

In a similar manner, the incidence of faults can be predicted. Application of an external pressure p is required to inhibit the onset of cavitation after the billet has been reduced to a thickness critical for a given geometry. The analysis of the problem indicates that cavitation lowers the power of internal deformation, shear losses and friction loss along the bottom of the container, but that it increases the frictional loss along the container walls and ram surface. The interplay between these two power contributions determines the incidence of the fault.

It is recognised that the critical value of the R_o/h ratio corresponds to the situation in which a balance between the gains and losses in the supplied power is achieved. As the billet becomes thinner, i.e. as h decreases, the friction loss along the bottom wall of the container will increase and predispose the system towards cavitation. To prevent the onset of cavitation an increase in pressure is required. The relationship between the minimum pressure p and the R_o/h ratio, neglecting the effect of inertia, that would cause cavitation is shown in *Figure 4.104*.

Flow with fish skin is characterised by an increase in frictional losses along the ram surface and container walls, and by an increase in the power of internal deformation. However, shear losses are reduced. The minimum values of the external pressure p required to prevent the incidence of fish skin are shown in *Figure 4.105*.

In general, in a given physical situation the pressure required to prevent cavitation is higher than that needed to stop the formation of fish skin. Fish skin is, therefore, unlikely to occur before cavitation and the prevention of the latter will automatically remedy the situation with regard to the former.

Impact extrusion is performed in high-speed presses, either specially built presses for specific hot or cold operations, or more general presses, such as the Petro-forge machine. The power capacity of these installations varies between 3 and 30 kJ and the ram velocities between 9 and 15 m s^{-1}.

The process, in its various forms, is particularly suitable for the production of tubular sections (simple or finned) of very thin walls in comparision with their diameters. The h/D ratio can be as low as 0.005.

Steels are extruded using phosphate coating combined with commercial soaps as lubricants, whereas for aluminium alloys and copper and brass proprietory lubricants are used.

4.7.5 Continuous extrusion

Continuous extrusion processes have been explored in some depth, but it appears that, at present, the Conform process is the only one that is used extensively.

The process relies on frictional forces generated between the billet and container. These are sufficiently high to effect extrusion through the die. The operational system is illustrated in *Figure 4.106*. The product can be delivered either axially or radially (as in side extrusion) as shown in *Figure 4.107(a)*.

The tooling consists of a rotating wheel with a circumferential groove, a shoe which overlaps a portion of the wheel surface and includes a grip segment and an abutment containing the die.

Solid and tubular sections can be extruded in aluminium and copper through single- or twin-port dies (*Figure 4.107(b)*). Both solid and particulate matter can be used as starting material, but very high extrusion ratios of up to 200 produce temperatures of up to 500°C and die pressures of up to 1000 MPa. Consequently, good tool materials must be used and the system must be cooled efficiently. With the groove of

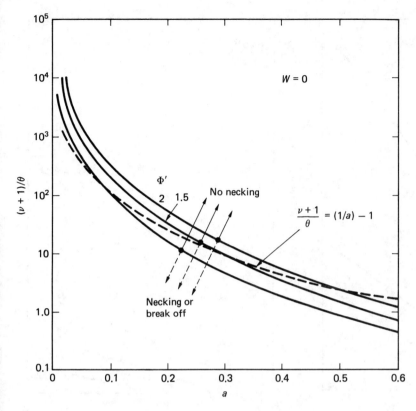

Figure 4.102 Criteria of necking or break off in impact extrusion of bar

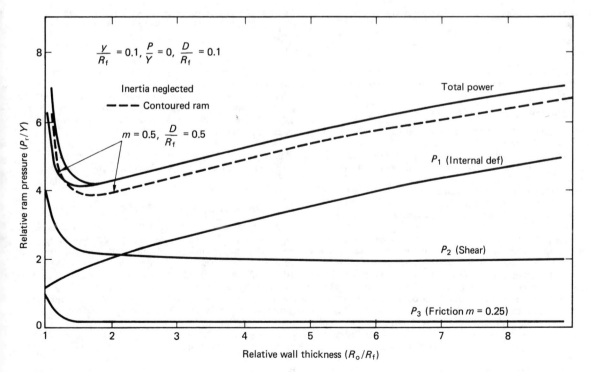

Figure 4.103 Comparison between the variation in ram pressure with wall thickness for impact extrusion with flat and contoured rams

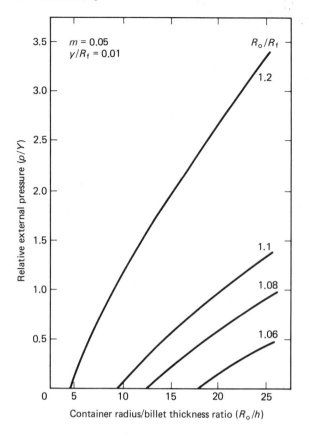

Figure 4.104 Minimum extrusion pressure required to inhibit cavitation in reverse impact extrusion

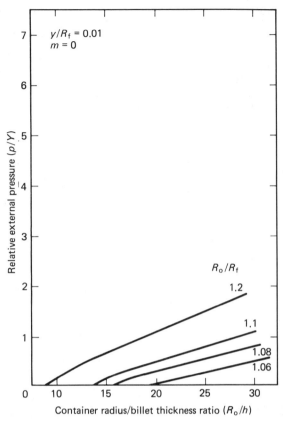

Figure 4.105 Minimum extrusion pressure required to prevent the incidence of fish skin in reverse impact extrusion

the wheel undergoing cyclic thermal stressing, the possibility of fatigue must be considered when designing this part.

4.7.6 Section and sheet extrusion

Although interest often centres on the extrusion of rod and wire, a large proportion of extrudates in light-metal alloys is manufactured as profiled sections ranging from simple curtain rails to very complex structural shapes a selection of which are shown in *Figure 4.108*.

In such cases the flow of metal during the operation is far from uniform and the incidence and level of redundancy can be high unless die profiles are designed correctly (see the next section).

The requirements of modern industry demand more and more composite bi- or tri-metallic components which posses sufficiently good mechanical and physical properties to be suitable for forming to specific shapes. Examples of such applications involve the use of bimetallic sheets in the chemical industry where, say, a thin layer of stainless steel on a low carbon steel base will provide both structural strength and anticorrosive properties, and can be used in the fabrication of containers and pressure vessels, or in the electrical industry for the manufacture of bimetallic strip conductors.

Bimetallic strip is often produced by rolling but, more recently, an extrusion method has been developed for small width and thickness strip in which billets of different metals

are extruded simultaneously from two or more containers to form a composite strip.

This manufacturing method provides a means of obtaining good dimensional tolerances and results locally in sufficiently high stresses to produce pressure welding. High deformations exceeding 50% are needed for this to occur.

Extrusion pressures, experienced by the metals, are only slightly higher than the corresponding pressures occurring with similar extrusion ratios in circular sections.

4.7.7 Die design and metal flow

Studies of the actual patterns of flow and of the incidence of faults as opposed simply to the determination of stress fields and the loads associated with them, are of primary importance because of the high level of redundancy in extrusion operations. The comparative ease of internally marking a processed specimen by imprinting grids on the median surfaces of split specimens, or by using laminate and composite model and prototype material billets, has made it possible to obtain a far better insight into the pattern of flow than is possible in many other processes. The use of the visioplasticity technique is of particular value here since it describes and defines fully the mechanics of the operation at any instant in time and at any point in the pass. It is thus possible to separate the different effects, to identify their causes and, therefore, if necessary, to

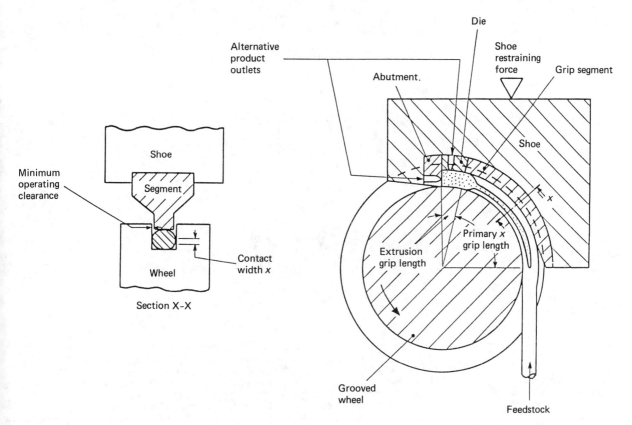

Figure 4.106 Arrangement of a Conform extrusion machine

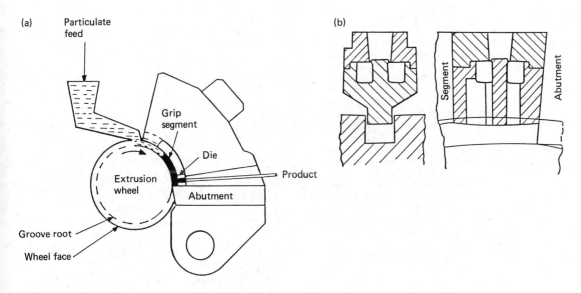

Figure 4.107 (a) Detail of the Conform machine. (b) Twin-port dies

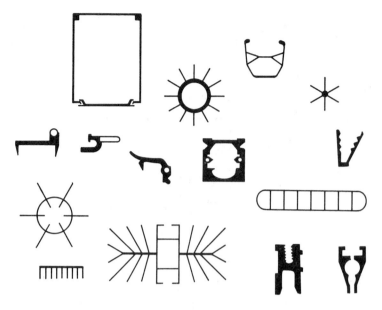

Figure 4.108 A selection of profiled sections extruded in light-metal alloys

be in a position to take remedial action by suitably altering the geometry of the pass.

The patterns of flow which develop in the extrusion passes strongly depend on the geometry of the system and, consequently, the levels of redundant deformation are affected. In normal industrial processes of rod and tube extrusion, the tendency is to use flat 90° die semi-angled tools or straight conical dies. These do not necessarily provide the best results because of the formation of the dead-metal zone and, sometimes, excessive shearing. An examination of these effects in rod extrusion indicates the problems arising.

The effects of friction and shearing distortion vary in intensity depending on the die angle. It is possible to obtain optimum conditions which will minimise the required extrusion pressure, but it must also be remembered that the distortion effects are large. Shearing distortion increases with the increasing die angle and may also be slightly influenced by the extrusion ratio R. An example of these trends is given in *Figure 4.109*. For the radial reduction of $R = 2$, the increase in shear with the die angle is clearly visible. For an angle of 60° (*Figure 4.109*) the originally vertical lines become almost parallel to the horizontal axis, particularly in the vicinity of the rod surface where frictional effects are superimposed on the shear effect. For large die angles, where shear is already high, the increase in the homogeneous deformation may marginally affect the situation. It can also be seen from *Figure 4.109* that an increase in the ratio R from 3.5 to 5 causes the originally vertical lines to become practically parallel to the longitudinal axis. It must not be forgotten that individual variations in the pattern of flow will be introduced by variations in the friction conditions and in characteristics of the materials processed.

Figure 4.109 indicates clearly that shearing distortion varies across the section of the rod and that the redundant effects will, therefore, vary accordingly. This is reflected in the variation in the generalised strain across a section. For instance, for commercially pure aluminium extruded through a die of a semi-angle of 45° with an extrusion ratio of 3 and lubricated with fluorocarbon, the generalised strain distribution at various sections varies considerably in the radial

direction. The generalised strain increases rapidly near the die surface, while remaining reasonably constant nearer to the centre of the billet. This is indicative of a state of almost homogeneous deformation within a certain radial zone that can be identified with a certain flow line.

In view of the potential danger of producing faulty extrudates, optimisation of die design must clearly be considered.

The use of the constant rate of homogeneous strain (CRHS) and constant mean strain rate (CMSR) concepts (equations (4.47) and (4.48)) in the design of rod extruding tools offers the possiblity of minimising the effects of internal macroshearing by facilitating the flow of the metal. An example of this is given in *Figure 4.110*. *Figure 4.110(a)* shows the CRHS die profiles for the accelerated (A), decelerated (D) and uniform

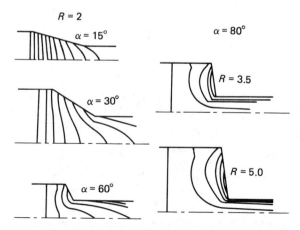

Figure 4.109 Shearing distortion in forward extrusion of rod. R, extrusion ratio

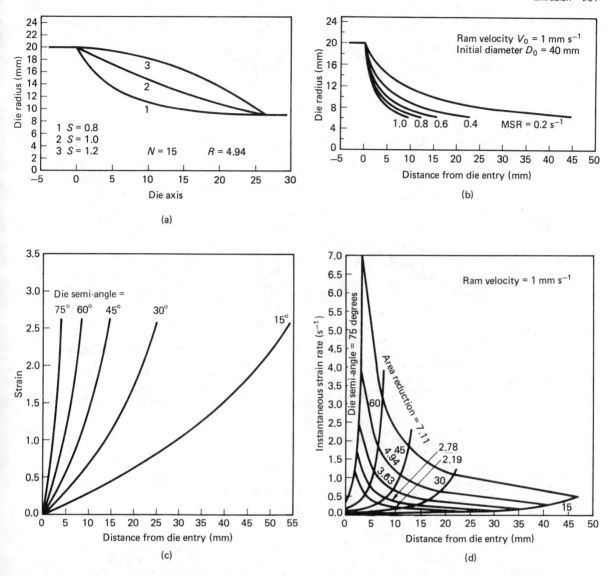

Figure 4.110 CRHS and CMSR die profiles and their effect on metal flow in rod extrusion. (a) CRHS die profiles: *N*, number of sections considered; *R*, extrusion ratio; *S*, rate of deformation. (b) CMSR die profiles. (c) Homogeneous strain distribution in conical U-type dies. (d) The effect of die semi-angle on the strain rate in U-type dies

(U) rates of deformation, all three of which start with the same billet size and produce the same final diameter. The total homogeneous strain developed is, therefore, the same in each case. However, it is clear that, whereas the A-type die (3) will produce relatively little work of deformation at the entry to the pass, but most of it nearer to the exit, the D-type die (1) will have an almost opposite effect, with the effect of the U-type die (2) being in between. The stress fields generated in the material will therefore be different and, in cold processing, the residual built-in stress systems will reflect the patterns of flow and will be associated with the effects of its inhomogeneity.

A similar situation arises with the CMSR tools (*Figure 4.110(b)*), which affect the distribution of work done in the pass, depending on the strain rate generated.

The effects of the CRHS passes on the distribution of homogeneous strain and strain rate is shown in *Figure 4.110(c,d)*. In *Figure 4.110(c)* for a given die land of a U-type die, but differing die semi-angle, the rate of deformation is highest for large die angles and slowest for small angles; the severity of the operation is thus controllable and can be adjusted depending on the material and lubricant used. However, if this criterion of design is adopted, the strain rate along the pass will vary (*Figure 4.110(d)*) and, again, in cold processing will affect the mechanics of the operation.

In tube extrusion, for the same rates of deformation as those considered above, the homogeneous strain distribution along the pass shows very similar effects (*Figure 4.111*). The patterns of strain, in the accelerated and decelerated rate extrusion dies, show almost mirror-like inverted reflections, thus indi-

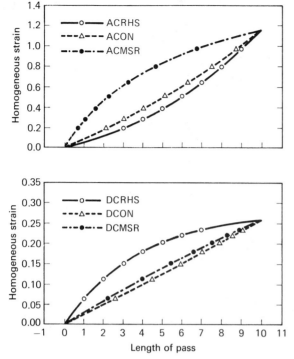

Figure 4.111 Distribution of homogeneous strain in accelerated (A) and decelerated (D) CRHS, industrial conical (CON) and CMSR tube extrusion dies

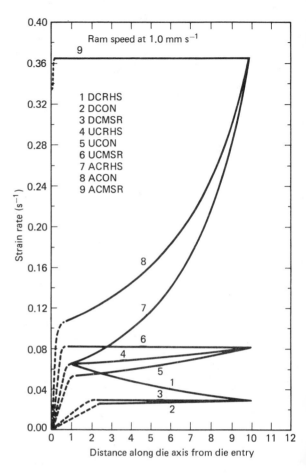

Figure 4.112 Variation in the mean strain rate in accelerated (A) and decelerated (D) CRHS, industrial conical (CON) and CMSR tube extrusion dies

cating once gain the inherent differences in the manner in which deformation occurs.

Figure 4.112 shows that, unless tube extrusion dies are designed with a view to maintaining constant strain rates, the strain rates will vary along the pass just as in rod extrusion.

The effect of the various design concepts is summarised briefly in *Table 4.11* by reference to the values of the redundancy factors.

Dies can be designed using the information provided in Section 4.4 and in *Table 4.4*.

Unlike the extrusion of axisymmetric parts (the products of cylindrical rods obtained from cylindrical stock), the extrusion of sections includes a wide variety of configurations. Flat-faced dies are usually employed in shape extrusion, but inhomogeneity, incidence of defects and the limitations on the speed of processing are major problems arising with these dies. Since the achievement of a minimum extrusion pressure is possible when smooth curved dies are used, interest often centres on this approach, although a general solution for the optimal curved tools has yet to be obtained.

Becuse of the three-dimensional nature of flow occurring in the extrusion of sections, it is very difficult to obtain analytically complete solutions. The existing numerical solutions are confined to the simple shapes only, but the upper-bound approach has produced information on the extrusion of polygonal sections derived from similar polygonal shapes and for drawing of shapes from round-rod stock. An upper-bound solution for the extrusion of square sections from round bars is in existence and can be used to obtain die profiles for the extrusion of more complicated regular shapes and with higher efficiency. This analysis considers curved streamlines, the envelope of which forms the surface of the deformation zone and, therefore, of the die.

From the viewpoint of the uniformity of flow and the level of the forming load, the superiority of profiled dies over flat ones has now been established and the general conclusion drawn is that in the hexagonal, square, rectangular and elliptical sections, the deformation patterns and tendencies that develop during extrusion are similar to those observed in a circular solid rod of the equivalent (circumscribing) diameter. The only exception is that of a triangular section in

Table 4.11 Performance of industrial and theoretical tools in extrusion

Process	Material	Die type	α (°)	Redundancy factor, ϕ	Range of ϕ
Bar extrusion	Low carbon steel	CRHS	60	U 0.88 ± 0.12	1.4–1.6
		CON	60	U	1.6–2.0
Tube extrusion	Non-metallic	CRHS		A	1.02–1.12
		CRHS		U	1.01–1.15
		CRHS		D 1.0 ± 0.05	1.03–1.05
		CON		D	1.4–1.7
		CMSR		A	1.025
		CMSR		U	1.02–1.06
		CMSR		D 1.01 ± 0.009	1.01–1.04
		CON		D	1.4–1.7

which the flow pattern is asymmetric with respect to the axis of the rod.

Apart from the use of flat-faced dies, the first and fundamental step in converting the initial, cylindrical billet to a polygonal section is often carried out in profiled tools. Again, the optimum profile is expected to correspond to the minimum expenditure of deformation energy (maximum uniformity of flow) and should be produced as economically as possible. Very complex shapes are avoided because of the obvious problems of machining, on the one hand, and high rates of wear in operational conditions, on the other. Instead, a step-by-step gradual approach is adopted which, although requiring a number of tools and passes, reduces the severity of a round-to-polygon transformation.

Conical and exponential dies for the extrusion of rectangular bars from round stock are used (*Figure 4.113*) because the pressure curves obtained for exponential and conical dies are very different from those for the flat-faced tools, thus reflecting the differences between the deformation rates.

Three possible profiles involving plane surfaces for the conversion of round bars to square products are shown in *Figure 4.114(a–c)*. Despite the non-uniform change in cross-sectional planes, continuous transformation takes place in the axial direction. In some cases, the profiled dies are made in such a way that the die lands are inclined at a certain angle towards the axis of the die. This reduces the parallel land length and also results in a further progressive reduction in area. Such dies are used in the extrusion of I-sections.

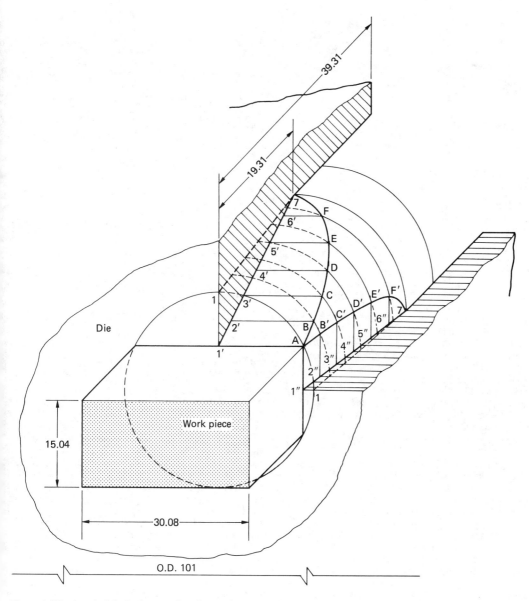

Figure 4.113 A conical die for the extrusion of a round to rectangular shape

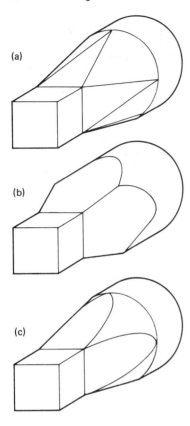

(a)

(b)

(c)

Figure 4.114 Possible die profiles for the transformation of a round to a square section

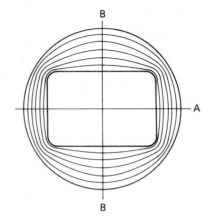

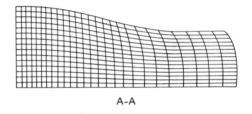

A-A

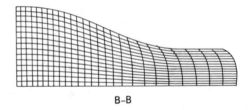

B-B

Figure 4.115 Difference in the patterns of flow in the mutually perpendicular planes when processing through a rectangular extrusion die

The difference in flow patterns in the planes inclined at right angles to each other (*Figure 4.115*), is the origin of the possible general non-uniformity of deformation and, therefore, internal shearing. The observed severity in the level of deformation (section B–B), as compared with that of section A–A, is indicative of the redundancy effects that set in.

A gradual reduction of a circular cross-section to a polygonal one is therefore recommended, in spite of the existence of frictional effects, at least in the early stages, when the die–billet surface of engagement is large.

The problems should be alleviated to some extent if the intended section is square, in which case symmetry with respect to the two mutually perpendicular axes should eliminate planar differences in flow.

The conversion of a circular billet to an elliptical profile is, of course, much simpler, because a gradual reduction and change in shape are easier to impose on the material. *Figure 4.116* shows that, unlike the polygonal-shape extrusion, this can be achieved in a single pass, thus decreasing the cost of the operation.

Current developments in streamlined dies will offer a complete solution to the forming-die design of the future. The analysis of the streamlined dies is a CAD operation in which the streamlines, within the deformation zone, are imposed in such a way that some prerequisite conditions that agree with the real flow are satisfied. The particle velocities are calculated, and the constraints preventing free flow of the material are identified. The possible methods of modification are then studied. However, an important problem that still remains is that of the determination of conditions under which particles will follow the desired flow lines. One advantage of this approach is that the flow of material can be illustrated on the screen of a monitor without any need to operate the real process. The uniformity of the flow and the deformation forces can, therefore, be evaluated and optimised. The method can provide accurate information, provided that the parameters are selected properly.

An illustration of this approach is given in *Figure 4.117*. External mapping of the section is performed first, as in (*a*), followed by a development of successive stages of the profile (*b*). Conversion from the round to the final shape (*c*) is envisaged, and the mathematical functions bounding the shape for the geometrical control (*d*) and the degree of reduction in area (*e*) are established and are reproduced pictorially to enable final assessment to be made.

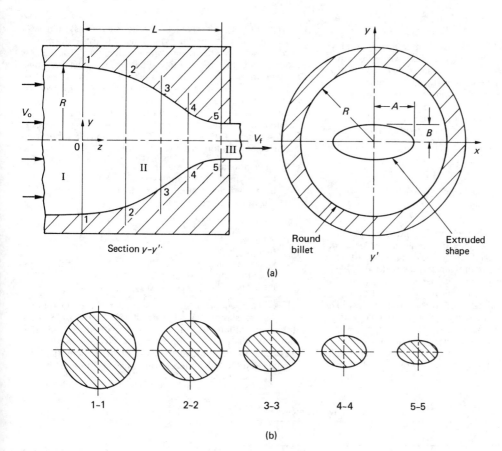

Section y-y'

Round billet

Extruded shape

(a)

1-1   2-2   3-3   4-4   5-5

(b)

Figure 4.116 Gradual conversion of a round shape to an elliptical shape

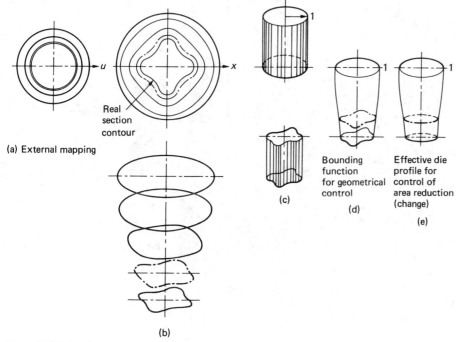

(a) External mapping

Real section contour

Bounding function for geometrical control

Effective die profile for control of area reduction (change)

(c)

(d)

(e)

(b)

Figure 4.117 Development of streamline dies

4.7.8 Die materials

For the hot extrusion of aluminium and similar light alloys, H13 steel dies are normally used. α/β brass extrusions require H10A and H21 steel dies or stellites and cobalt-based alloys. Copper components are extruded through Nimonic 90, or other nickel based alloy dies, whereas steels are usually processed in H13 and H21 steel dies, or in TZM molybdenum alloy. Nickel alloys are invariably extruded through Nimonic 90 dies, and titanium alloys through hot-work refractory oxide coated steel tools.

4.8 Cold drawing of wire and tube

4.8.1 Introduction

As pointed out earlier, hot processing normally produces low-strength materials and uneven dimensional properties of the product. If high-quality fabricates are required, cold processing must follow the preliminary hot method of shape acquisition. In axisymmetric components, such as wire (rod) and seamless tubing, this is achieved by cold drawing the hot-finished product through a die (or a series of dies). This treatment imparts good mechanical properties and effectively regulates dimensional tolerances.

Basically, a nozzled (in the case of a tube) specimen is inserted in the die and is then gripped by a suitable device which can pull it forward on a mechanical or hydraulic bench.

A reduction in the diameter (solid specimens) and wall thickness (tubular specimens) results. Multi-die drawing is necessary in wire production, whereas single-die processing is more usual in the case of a tube although, for special reasons, tandem drawing is sometimes employed.

Coiled rod, obtained in a hot-rolling or extruding process, forms the starting stage of wire drawing, whereas a hot-finished tube is used for a cold-drawing operation.

There are a number of variants of wire-drawing processes. Processes are usually based on the method of lubrication adopted, but the four major routes are:

(1) conventional dry or wet drawing,
(2) hydrodynamic lubrication system,
(3) hydrostatic lubrication system, and
(4) ultrasonic vibration system.

The main operational tube drawing processes are:

(1) sinking,
(2) floating plug,
(3) stationary plug,
(4) mandrel, and
(5) ultrasonic vibration.

The main characteristics of some of these operational systems are shown diagrammatically in *Figure 4.118*. In a sinking operation, the tube is drawn without any internal support of the bore and, therefore, the change in wall thickness cannot be controlled. Because of the bending and unbending that the wall undergoes in its passage through the die,

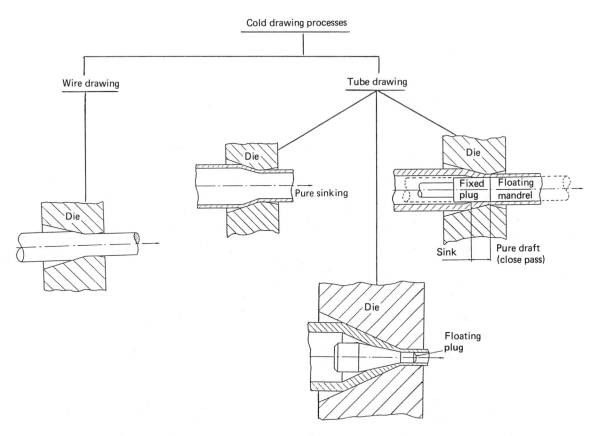

Figure 4.118 Cold drawing of wire and tube

Figure 4.119 Plug drawing of seamless tubing

thinning or thickening of up to 4–11% can take place. The process is generally used only as a preliminary operation to further manipulation.

The normally adopted route is that of drawing over a cylindrical plug. Either a stationary or a floating tool can be used. In the former case the tool is positioned in the die throat and is held there by a rigid plug bar. The tube, of a bore slightly larger than the plug diameter, is pulled over the plug and the plug bar (*Figure 4.119*). Initially the tube is 'sunk' onto the plug and is then drawn in what is known as 'close pass' or 'pure draft'. Both the diameter and the wall thickness of the tube are reduced in a controlled manner.

For the drawing of long, small bore, small diameter, thin-walled tubes, floating plug or semi-floating 'captive' processes are used. Here, an initially hot-processed or cold-annealed tube is inserted into a die which contains, in one case, an unsupported conical free plug. The tube is drawn over the plug and is then coiled on a drum for ease of storage and transportation. In the other variant of the operation, a conical plug that is free to float is prevented from moving in one direction by a bar.

The older method of drawing over a long mandrel, which supports the bore, is now seldom used, mainly because the tube drawn in this way clamps tightly onto the mandrel. Thus, to remove the tube after drawing, the tube and mandrel may have to be reeled in a cross-rolling reeler which expands the tube and thus frees it from the mandrel. The reeling operation can impart helical markings to the outer surface of the tube, thus necessitating another drawing operation, and is also likely to affect the uniformity of dimensions along the whole length of the tube. In some cases, where highly polished and dimensionally accurate, but relatively short tubing is needed, mandrel drawing can be used if withdrawal of the tool is possible without reeling. This is the case for tubing with sufficient wall thickness that buckling does not occur when the tube is freed by pulling the mandrel/tube assembly through a 'gate' of a diameter only fractionally larger than that of the mandrel. This results in a stripping operation.

In ultrasonic vibratory systems, used either in wire or tube drawing, the die is vibrated at an appropriate frequency to increase the efficiency of the process by affecting the rate of feed of lubricant and the mechanics of the drawing.

The techniques described above are equally applicable to the manufacture of ferrous and non-ferrous tubing, but conventional cold and warm drawing of stainless steels and cold drawing of square sections are of particular interest.

In the case of stainless steel, an increase in temperature to about 300°C reduces drawing forces by up to 35% and increases the attainable deformation by about 55%. Warm drawing thus constitutes an important development in this type of processing.

The development of the floating-plug technique, with its saving in tool material, better carry-through of lubricant, and saving of space when coil drawing, has been rapid and, in some case, has overtaken the use of more conventional fixed-plug operations. Again of particular interest is the ever-increasing use of this process in the coil drawing of stainless and carbon steels. Irrespective of the material being drawn, the rate of production is increased; some 40% saving in time is possible when, for example, coil drawing on 0.9 m diameter drums.

Although the spring back or the final diameter of the drawn coil depends on the material processed, it is easily assessable for aluminium, brass, copper and alloy steels.

Wire drawing, often thought of only in terms of steel and copper, is in fact used to produce satisfactory lengths of rods and wires in a number of more exotic metals and alloys. On the single-metal side, brittle materials such as molybdenum and beryllium can be drawn, whilst tungsten wires must be manufactured at high temperature in order to counteract the unfavourable mechanical properties of this material. Bimetallic wires are manufactured by coaxial drawing of solid cores surrounded by hollow tubular sheaths, including ultrafine composites of niobium and copper, and copper and aluminium. Transformation-induced-plasticity steel has been processed, as has ausformed silicon–chromium steel.

Clearly, the range of application of cold drawing is large and it continues to increase bringing in non-circular section rods and wires.

4.8.2 Basic concepts of wire drawing

Although it is customary when considering theoretical aspects of wire drawing to refer to a single pass, it must be made clear that drawing from a coiled rod to the finished product must be carried out in a multistage, normally automated, machine or draw bench. Successive passes call for the correct design of dies and the provision of suitable lubricating conditions. Die design requires minimisation of the degree of redundancy and size of the drawing load, leading to the choice of the optimum (for the given conditions) effective die angle, while at the same time allowing a protective lubricating film to develop and be maintained constantly throughout the operation. Failure in lubrication results in impaired quality of the surface finish and an increase in the rate of wear of the die. The cost of remachining or replacement of the tool is then added to the cost due to the loss of dimensional accuracy of the product.

Since successive drawing passes are not able to rectify original surface faults or the presence of scale created by annealing the rod, the preparation of the rod is all important. Here, either 'dry' or 'wet' processing is adopted.

In a typical 'dry' in-line system for rod drawing, the specimen undergoes the following sequence of operations:

(1) grit blasting by three or four sets of guns at, say, 90° to each other;
(2) grit extraction through filters in a chamber;
(3) air blasting to remove the dust;
(4) lubricating in an enclosed chamber; and
(5) drawing.

A dry blasting operation results in pitting of the surface or, at least, a matt finish, either of which assists in trapping the lubricant and creating local conditions of hydrostatic lubrication. This, in turn, promotes the possibility of applying single heavy passes. Various alloys, including stainless steels, can be treated in this way without any additional surface preparation.

The 'wet' descaling processes include the well-established acid pickling, improved, and sometimes accelerated by, the passage of an electric current of some 7 A cm^{-2} and ultrasonic vibration, both of which help to dislodge and precipitate the scale.

However, environmental considerations weigh against the atmospheric and effluent pollution associated with the use of acids and, therefore, attention has been focused on reducing the original amount of scale and on using molten salt bath heating in place of acid pickling.

A controlled carbon oxide atmosphere is suitable for high-speed, low-scale-forming annealing, and is sometimes preferable to vacuum annealing which may cause strand welding. Depending on the alloy processed, resistance heating can be used for, say, brass annealing, or fluidised beds can be used to reduce the time of operation and the amount of scale.

Electrochemical lubrication, resulting from the use of suitable molten salts of, for instance, potassium and lithium clorides is the direct result of descaling in a non-acidic environment. Descaled and cleaned rods are then coiled and fed into drawing machines. Drawing of the wire itself is carried out at speeds ranging from 30 to 2500 m min^{-1}, depending on the material, with resulting reductions in cross-sectional area of 15–25% in the case of narrow diameter wires and 20–45% for coarser wires.

The two basic techniques employed are, again, the 'wet' and 'dry' processes. In the former, the entire production line is usually immersed in the lubricating liquid, whereas in the latter the wire 'picks up' the lubricant on passing through a container.

Copper and copper alloy, some aluminium, and very fine diameter wires are normally processed in the wet condition, while ferrous alloys and all other matrials tend to be drawn dry.

In consequence, a wide range of lubricants is employed. The lubricant used must account for not only the specific material requirements, but also for the effects of cooling that are now recognised as being of major importance.

For most ferrous materials, the preferred method of lubrication is that of precoating the wire. Ordinary carbon steels are phosphate coated, but stainless steels require either oxalate compounds or borax as a lubricant. Although oxides protect the surface from die damage during drawing, they do not necessarily act as lubricants and so two-stage lubrication is still, reluctantly, used. In addition to oxide films, crystalline lubricants are required, and chlorine and sulphur type additives are used.

Electrochemical deposition of lubricants is still in the development stage but, if proven sufficiently economical, it will provide a relatively easy answer to standard processing lubrication problems in that the deposition and removal (combined with temperature control) of substances such as molybdenum sulphide is relatively simple by this means.

The removal, and possibly reuse (for reasons of economy), of lubricant is of importance. Cleanliness of the lubricant as achieved by, say, filtration of debris, can be high and will thus affect the 'brightness' of the surface when the wire undergoes the final series of passes. When drawing very fine wires, ultrasonic cleaning is necessary.

4.8.3 Non-standard wire-drawing techniques

These techniques include principally hydrodynamic lubrication systems and ultrasonic drawing. The basic idea behind hydrodynamic lubrication is to provide a continuous, but sufficiently thick, layer of pressurised lubricant that will separate the tool from the wire. In this way tool wear is considerably reduced.

An effective technique of achieving these conditions is that of drawing through a sealed tube containing oil as lubricant (although occasionally soap can be employed) and terminating in a constriction or nozzle and, eventually, in a die. As the speed of drawing increases (and high speeds are necessary) so does the oil pressure until, theoretically, it reaches the value of the yield stress of the processed material. The difficulties experienced in operating the system are related to the problems of sealing at high pressures and velocities, and to those of precision nozzle design which is 'adjustable' to materials and conditions.

Copper and aluminium wires can be drawn, but the pressures required for 'harder materials' such as, for example, steels, are too high to maintain successfully over a period of time. Nevertheless, since very good lubrication is generated with, consequently, low friction and tool wear, the technique should be considered where economy of operation is important.

In an ultrasonic multidie system, one or more dies when vibrated in the direction of drawing create conditions of back-pull which, in turn, alter the force requirement and directly effect its magnitude. As a result of die vibration, an oscillatory force is induced in the wires, between the consecutive dies and, eventually, the coiler drum. In consequence, greater reductions in cross-sectional area are achievable without either an increase in the degree of plasticity of the material or any decrease in interface friction. The level of this reduction depends on the value of the back-pull exerted and increases with decreasing back-pull factor. The surface finish and mechanical properties of the drawn wire remain unaffected by the oscillatory nature of the force system.

4.8.4 Mechanisms of wire drawing

Drawing through a single die is accomplished by generating a tensile stress, due to the applied forward pull, and a compressive stress produced by the die. However, in a real multidie system, conditions of back-pull are easily produced by suitable matching of the amounts of deformation in the subsequent dies. The generation of back-pull increases the tensile stress induced and lowers the compressive stress, thus reducing frictional effects and improving the efficiency of the process. The total drawing load is, naturally, increased since it must overcome the effect of the back-pull, but in a continuous process some of this energy can be recovered and the net power expenditure may turn out to be slightly lower than in a conventional operation.

The efficiency of back-pull drawing is assessed by means of the back-pull factor which is related to the rate of decrease in the die load produced in this method of drawing. The maximum possible back-pull is limited by the yield stress of the drawn material. This will vary with the rate of strain hardening and pass geometry, but the limiting back-pull/drawing load ratio is:

(1) highest for very ductile alloys undergoing small deformations,

(2) lower for strain-hardening alloys, and

(3) reduces as the pass deformation increases.

for a given back-pull condition, a high back-pull factor results in a smaller increase in the drawing load and a larger saving in power.

Irrespective of the manufacturing technique adopted, the efficiency and soundness of a wire-drawing process depends, as in the other operations discussed, on the geometry of the pass and the associated levels of redundancy.

The die design ranges from simple conical (*Figure 4.118*) in which a die land is provided as the calibrating device, through streamline, sigmoidal to double-reduction and control relief-angle tools. The tool more often used industrially is the conical die which approximates closely to the U-rate CHRS-type die. The relationships between redundancy and geometry factors (defined in *Table 4.5*) and the levels of stress and strain redundancy factors for a variety of conical dies and materials are listed in *Table 4.12*. The information given in the table is approximate in nature and should be treated accordingly. Changes in die geometry and lubrication will affect the numerical values quoted in the table.

Die materials include tungsten carbide inserts, drilled diamond for small diameter wires, boronised carbides, ceramic coatings of chromic oxide and solid ceramics. Drilling is effected by electrospark or discharge machines, electrolytic capillary piercing and ultrasonic means.

A number of material defects may originate as a result of either incorrectly designed passes or misalignment, and incorrect operating practices. The three major faults are:

(1) the dead-metal zone;
(2) chip formation, with or without a built-up edge; and
(3) central burst.

These faults are illustrated in *Figure 4.120*.

The dead-metal zone, similar to that encountered in extrusion, is associated with large die angles. If the dead-metal zone fails to adhere to the die and begins to roll backwards producing a chip, the core of the wire will not deform but will merely travel through the die, retaining a constant and practically unchanged diameter.

Periodic fracture of the core of the work piece, referred to as 'central bursting', occurs occasionally with large die angles and low deformations. The fault is serious and is comparable with central cavitation in rotary piercing.

The criteria for the incidence of these faults, proposed by Avitzur (see the Bibliography) are summarised in *Figure 4.121*. To avoid the occurence of any of these faults, and to minimise the drawing load, the optimum effective die angle should be used. This will provide a balance between frictional

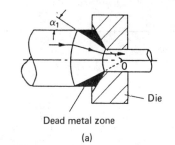

Dead metal zone

(a)

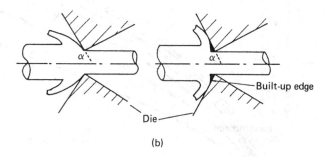

Built-up edge

Die

(b)

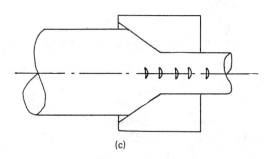

(c)

Figure 4.120 Patterns of flow and fault formation in wire drawing. (a) Dead-metal zone. (b) Chip with or without a built-in edge. (c) Central burst

and redundancy effects, and can be estimated from the following empirical expressions (see the Bibliography):

Hill and Tupper:
$$\alpha_{opt} = \sqrt{\left(0.87\,\frac{a}{1-a}\right)} \qquad (4.84a)$$

Siebel:
$$\alpha_{opt} = \sqrt{\left(\frac{3\mu a}{2-a}\right)} \qquad (4.84b)$$

Avitzur:
$$\alpha_{opt} = \sqrt{\frac{3m}{2}\ln\!\left(\frac{R_i}{R_f}\right)} \qquad (4.84c)$$

where a is area reduction, μ is the coefficient of friction, m is the friction factor, and R_i and R_f are the initial and final radii of the wire, respectively.

4.8.5 The drawing stress

The magnitude of the drawing stress depends not only on the pass geometry, general drawing conditions, and the material used, but also on the mathematical model of the process adopted. In its simplest form, using the equilibrium-of-forces approach, the drawing stress σ is given by

Table 4.12 Wire drawing—industrial tools

Material	Equation	Φ or ϕ
Strain-hardened copper	$\Phi = 0.89 + 0.76\Delta$	1.08–1.46
Annealed 60/40 brass	$\Phi = 0.94 + 0.61\Delta$	1.09–1.40
Strain-hardened brass	$\Phi = 0.87 + 0.76\Delta$	1.06–1.44
Annealed aluminium	$\Phi = 0.87 + 0.69\Delta$	1.04–1.38
Strain-hardened aluminium	$\Phi = 0.87 + 0.84\Delta$	1.10–1.52
Annealed 0.12%C steel	$\Phi = 0.90 + 0.73\Delta$	1.08–1.45
Strain-hardened steel	$\Phi = 0.81 + 0.89\Delta$	1.03–1.48
Annealed copper		2.62
70/30 brass		2.62
Annealed low carbon steel	$\Phi = 0.88 + (0.19 + 0.22)\Delta$	1.34
Annealed aluminium		1.34
Annealed copper		1.95
Annealed 70/30 brass		1.94
Annealed 0.11 steel	$\phi = 0.88 + 0.12\Delta$	1.30
Annealed aluminium		1.31

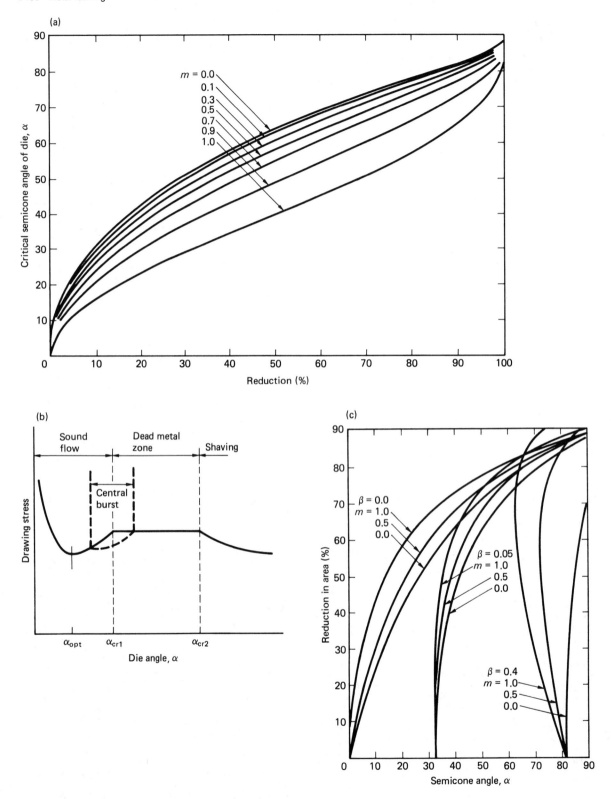

Figure 4.121 Critical conditions for the incidence of (a) dead-metal zone, (b) chip (shaving) formation, and (c) central burst. β, slope of the stress–strain curve of the material; m friction factor

$$\sigma = Y_m(1 + B) \ln\left(\frac{1}{1 - a}\right) \qquad (4.85)$$

where Y_m is the mean value of the operative yield stress (incorporating the effect of redundancy), $B = \mu \cot \alpha$, α is the effective die semiangle, and a is the reduction in area. The drawing force is obtained by multiplying the stress by the cross-sectional area of the drawn wire.

The upper-bound approach gives rise to much more complicated expressions.

4.8.6 Tool design in tube drawing

With the very considerable deformations (up to 50% reduction in area) produced in tube drawing, the problem of tool design and tube preparation assumes even greater importance than in the manufacture of wire.

Hot-finished tubing is usually the starting material in seamless tube drawing. Pickling or hot-sand blasting, followed by lubrication of the type described in the preceding sections, are the preliminary steps taken. However, as the percentage of internal defects caused by metallurgical phenomena is lower when cast rather than rolled material is used, there is increasing interest in the use of cast billets and tubes. The possibility of continuously casting steel tubes, to be rotary repierced, has been investigated and black hollow sections of 175 mm diameter and 3.75 m length are obtainable. Centrifugally cast tubes are equally suitable for further processing.

However, to initiate the process, the tube must be tagged, or nozzled, on its leading end to enable it to be passed through the die and be gripped by the drawing carriage (*Figure 4.119*). This operation is carried out either in a rotary swager, in which rotating hammers forge the tube end to the required diameter, or by push-pointing dies. Both operations necessarily reduce the yield by rendering part of the tube dimensionally unacceptable and, also by strain hardening the tag before actual drawing, they introduce an element of non-uniformity in mechanical properties. Post-nozzling annealing is, of course, possible, but can be expensive.

The majority of uncorrected ('standard') industrial dies used in sinking, plug and mandrel drawing operations are either straight–concial or slightly curved. This gives rise to high redundancy factors, which for low carbon steel can be as high as 20, depending on the die semiangle. A representative set of data is given in *Table 4.13*. A more general relationship between the geometry factor, for the three basic processes is given in *Figure 4.122*.

The redundancy factors are particularly high in sinking, where the shearing stresses due to bending and unbending of the unsupported tube wall reach very high levels. Intermediate values are obtained in plug drawing, where the control of macroshearing is much better, but reach their lowest values in mandrel drawing where the mechanics of the operation is changed to a favourable one by virtue of the relative movement of the mandrel with respect to the bore of the tube which it supports.

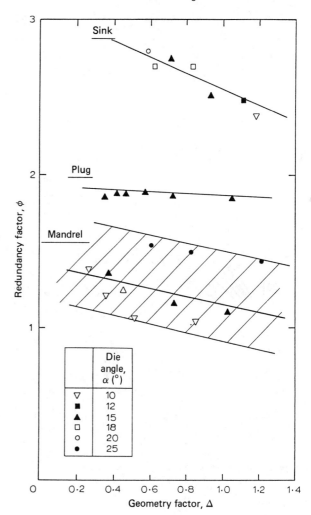

Figure 4.122 Correlation between the redundancy and geometry factors in cold drawing of steel

Optimisation of drawing conditions can be achieved by designing the dies according to the CRHS or CMSR concepts, and by avoiding as far as possible the presence of sink. As indicated earlier, die profiles, whether used in conjunction with short, fixed cylindrical plugs, long cylindrical mandrels, or even purely sinking tools, depend not only on the rate of deformation imposed, but also on the total general homogeneous strain suffered. Sets of die profiles developed for plug drawing of copper tubing are shown in *Figure 4.123*. The

Table 4.13 Tube drawing—industrial tools

Process	Material	Equation	α (°)	ϕ
Mandrel	Low carbon steel	$\phi = 1.35 - 0.167\Delta$	10–25	1.2–1.3
Plug	Low carbon steel	$\phi = 2.0 - 0.132\Delta$	15	1.7–1.8
Sinking	Low carbon steel	$\phi = 2.4 + 0.06(D_0/h_0)\Delta$	12–20	1.95–2.7

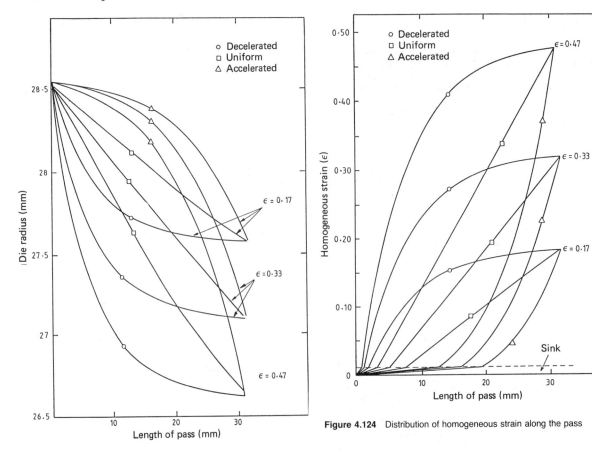

Figure 4.123 Die profiles in plug drawing of copper tube

Figure 4.124 Distribution of homogeneous strain along the pass

distribution of homogeneous strain along the pass, produced by such tool configurations (*Figure 4.124*) is of vital importance from the point of view of the mechanical properties of drawn tube.

Since most of the cold-drawn tubing is used either as structural machine and equipment parts, or as pressurised fluid containers, the mode and level of residual stresses induced in the drawn material, together with the effects of stain hardening imposed, may well determine the suitability, indeed the safety, of using the tube manufactured in any given manner.

An indication of this is provided by *Figure 4.124* which shows how the work of deformation is distributed along the die and, therefore, the effect this may have on residual stressing. A comparison between D- and A-type dies shows that two basic differences in the pattern of flow exist. In the former case, most of the work of deformation is done at the entry and forward parts of the die and very little towards the exit. The exit part of the die resembles closely a standard straight-tapered tool of a relatively small included angle. The pressure distribution in the die reflects the geometry and is higher in this zone of the pass. The increase in pressure, combined with the consequent increase in frictional effects, is likely to result in the presence of compressive stresses on the outer surface and tensile stresses of similar magnitude in the bore.

The accelerated tool displays the opposite charactersitics. The work of deformation being concentrated in the exit part of the die, thus creating mechanical constraints to the flow opposite in effect to those described above. The uniform rate falls, in its effect, somewhere in between these two situations.

These tendencies are illustrated in *Figure 4.125* (based on measurements made on plug-drawn copper tubes) which refers to longitudinal residual stress distributions. The levels and patterns of the maxima of residual stresses developed in these conditions, are shown in *Figure 4.126*.

If optimisation of the drawing load only is sought, the range of optimal die semiangles must be established. As in the case of wire, a balance between frictional and redundancy effects must be achieved. This will vary with the mode of drawing and the, sometimes unavoidable, presence of sink. An example of this situation is given in *Figure 4.127*, which indicates not only the required die angles, but also the effect of increasing the proportion of sink in a pass.

The two drawing operations normally employed for the processing of tubing that varies in outer diameter from 5 to 50 mm, are sinking and plug drawing. Drawing on a cylindrical plug results in a reduction in wall thickness and a good bore finish. Since this method of operation is only applicable to relatively short lengths of tubing (the length being limited by the length of the draw bench), the cost of production considering, among other things, swaging operations, increases, while the overall yield remains low.

The sinking of short lengths of tubing on ordinary benches, and the sinking of long but small diameter tubes on rotary

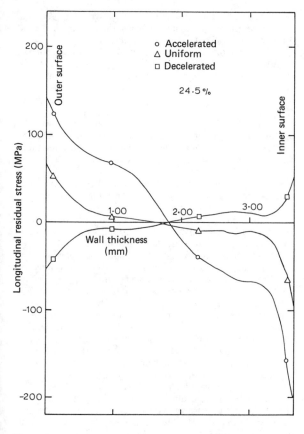

Figure 4.125 Distribution of longitudinal residual stress in plug-drawn copper tubing (24.5% reduction in area)

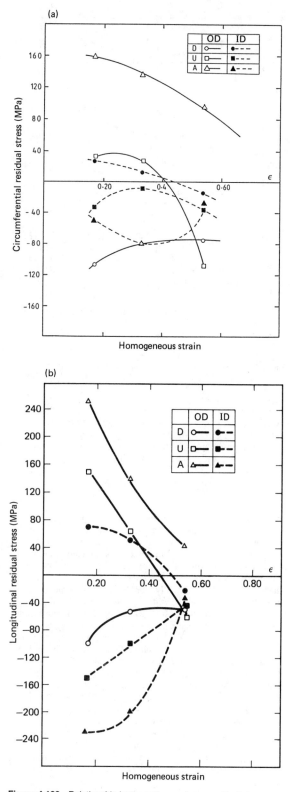

Figure 4.126 Relationship between the maximum residual stresses and maximum homogeneous strain in plug drawing of copper tubing: (a) circumferential stress, (b) longitudinal stress

benches, enables a considerable reduction in diameter to be achieved, but usually results in very poor bore finish. For these reasons the use of floating plugs has increased in recent years. It is found that satisfactory tube drawing on rotary benches with reduction in wall thickness is easily achieved. The life-span of tools is longer and regrinding of plugs is easy. Saving in tool material is, therefore, assured. Better carry-through of lubricant on the tube bore reduces the possibility of scoring, particularly in the case of alloy and stainless steels. In the case of low carbon steels this reduction in the incidence and magnitude of faults is often equivalent to an increase in output, per interpass anneal, of up to three good passes.

Elimination of plug bars, which are the feature of fixed-plug drawing, reduces both the cost of the material and the labour involved in maintenance and tool changes, and the chatter caused by the oscillation of plug bars is no longer a problem.

The new plug design has eliminated bore fissures and radial cracking in thick-walled tubing, which is subject to these faults in conventional operations. However, the possibility of using a small diameter plug means that tubes of very small outer diameter can be drawn to high dimensional tolerances, instead of having to be sunk in the last stages of production.

When the technique is used for multiple tube drawing on straight benches, the only modifications required are those made to the pulling carriage and the die plate.

The floating-plug technique is particularly attractive when coil drawing is adopted. Coil drawing of very long tubing at high speeds gives a saving in loading time. For instance, a 40%

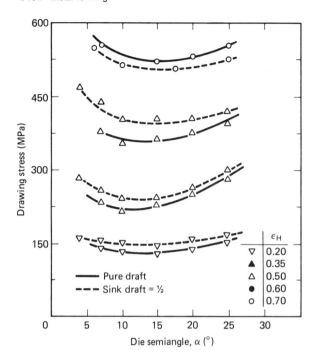

Figure 4.127 Drawing stress and homogeneous strain in mandrel-drawn 0.12%C steel tubing

saving in the cost of converting a thick bloom to 6 mm diameter tubing is normally observed. In addition, there is always the possibility of saving on storage. Coiled tubing can be easily stored and equally easily redrawn on straight benches for final use.

Mechanical properties of most varieties of steel tubing seem to be better, for a given amount of deformation, than the properties of materials processed in a fixed-plug system. An important point observed in the coil production technique is the more gentle release of the back end of the tube from the die. This removes the possibility of the usual forward jerk, which on straight benches and fixed plugs can cause bending in the long length of small-diameter tubing. A coiling operation also has the unsuspected advantage of showing quite clearly whether the tube has actually been drawn or has merely been sunk. In conventional operations this problem arises fairly frequently when drawing thick-walled small-diameter tubing. it is of some importance because this type of tubing is usually intended for heavy-duty service and, consequently, its quality should be high. Coil drawing with a floating plug determines, to a high degree of repeatability, the spring-off diameter of the coil according to the conditions of the draw. For instance, the diameter of the coil of a sunk tube coming off a 90-cm drum is of the order of 150 cm, while that of a truly drawn tube is no more than 100 cm.

An advantage of correctly designed floating plugs is that they can be used on machines which drawn from coiled to straight material. In the case of large quantities of a given size of tubing, the operation becomes continuous since the machine will part off and straighten the cut-off length at a high rate.

Automation of plant is easier when the floating-plug technique is used. It is possible to design a space-saving layout for delivery, drawing and discharging by using a fixed carriage and a moving die. This set-up is much more easily automated than is the fixed- plug bench system in which the carriage, the most complicated drawing tool, moves on the bench.

For most ferrous materials the improvement obtained in the eccentricity of tube is better with floating plugs and, as a result of the improved mechanical properties of the material, large deformations are possible. For small-diameter tube, well designed floating plugs give a 45% reduction in area as compared with 20% obtainable with fixed plugs.

For the same pass size, drawing loads are lower with the floating plug, and it is possible to stop and restart the operation. This is of importance when coiling and when tubes drawn on straight chain benches exceed the lengths of the bench.

The obvious disadvantage of the process lies both in correct tool design and in the critical lubrication. The latter in particular is quite restricting.

The dynamic equilibrium of a plug floating in the zone of deformation with a mixture of linear and axial movements, and also angular rotations, depends almost entirely on the magnitude of the coefficients of friction. The magnitude of friction is dependent on and defined by the surface finish of the tube and the tools, the quality and quantity of the lubricant used, the mechanical properties of the tube material, etc. The effect of the speed of drawing must be taken into account when considering the general magnitude of frictional forces, and consequently, the assumption that the coefficient of friction remains constant is not realistic. It is, however, possible to analyse the variation in this parameter that is likely to occur in a clearly defined drawing situation. The basic advantages of the floating-plug process are the possibility of operating at relatively high speeds (up to 120 m min^{-1}) and the fact that large elongations are obtainable. To realise these conditions, it is imperative to devise a rational plug profile and to ascertain that the plug dimensions are correctly geared to the given situation.

The selection of the dimensions of the individual ements of the plug profile must be based on the requirements of the dynamic equilibrium and self-stabilisation of the plug in the zone of deformation. Since the effect of the plug profile should be such as not to hinder the reduction in either the tube diameter or its wall thickness, the plug cone angle must be less than the die cone angle.

The theory behind correct plug design is rather complicated and cannot be summarised briefly. The reader is, therefore, referred to the appropriate literature listed in the Bibliography.

4.8.7 Special tube-drawing operations

Two completely diverse drawing systems now in operation are: ultrasonic drawing of ferrous and non-ferrous alloys; and fixed-plug drawing of explosively prewelded bi- or tri-metallic tubing.

Ultrasonic drawing is based on either the volume or surface effects produced by the oscillatory vibration of tools. The volume effect is subdivided into:

(1) the superposition mechanism,
(2) metallurgical effects, and
(3) the swaging effect

The surface effects comprise:

(1) the change in friction between the tool and the work piece, and
(2) the friction vector.

If only the plug is vibrated at about 20 kHz, friction is reduced by surface effects, i.e. by a thickening of the film of lubricant between the tools and the work piece. This reduces the drawing load and/or eliminates pick-up and chatter. However, practical difficulties are experienced in designing the plug and plug bar that will give the required resonance.

If, however, the die is vibrated ultrasonically in the radial direction, rather than remains stationary as in a conventional drawing operation, a swaging effect is produced in the tube which, although conforming to the subdivision listed above, indicates that, industrially, the division between volume and surface effects is not all that important.

Again, the details of the design of suitable equipment cannot be reproduced here, but information is available in the publications listed in the Bibliography.

A general view of a commercially available, standard 3.2 kW tube drawing unit is given in *Figure 4.128*.

The process has the following basic characteristics:

(1) the reduction in the drawing stress increases with increasing ultrasonic energy density,
(2) a reduction in area of more than 54% is achievable,
(3) the ultrasonic energy density can be increased by reducing the speed of drawing rather than by increasing the energy input,
(4) surface finish is improved when radial die oscillations are used, and
(5) pick-up and chatter are low or absent.

The high degree of sophistication required in the rapidly developing technology of the manufacture of petrochemicals, electronic devices (particularly cybernetics, cryogenic systems, and of atomic pile and toxic metals remote control systems) calls for the development of a new range of engineering, tubular composites. Some of these cannot be produced by conventional techniques alone or can only be manufactured at high cost. A typical example of the first group is a multilayer, multimetallic cylindrical pressure vessel, whereas an example of the second is a semiconductor system enclosed permanently in a protective tubular metal sheath. In between these extremes there is a wide range of components such as special bi- or multi-metallic heat exchangers, that combine structural strength with anticorrosive properties and ensure a rate of heat flow as good as that shown by the individual metals of the composite.

Conventional codrawing or coextruding of such assemblies gives less satisfactory results, because even with a high degree of process control it is practically impossible to ensure that no lubricant or debris is trapped on the surfaces. Furthermore,

lack of cohesion between the original components of the assembly is likely to lead to very high differential deformation and the associated in-built shearing stress.

Many of these limitations can be eradicated if the integrity of the component is assured *a priori* by, say, explosive welding. In this respect, the manufacture of duplex or triplex, bimetallic or multimetallic cylindrical pressure containers is of practical importance and this has been accomplished successfully. Because of the technical problems involved when long composite cylinders are explosively welded, the usual technique is to weld short, large diameter combinations, and then to obtain the required dimensions by means of cold-plug drawing.

A set of drawn triplex low carbon steel/brass/copper tubing is shown in *Figure 4.129*. It is noticeable that even with prior welding some differentiality in drawing is present and increases with increasing deformation.

The characteristic features of drawing are the changes effected in the distribution of hardness across the tube section, and in the quality of the weld. While the former simply reflects the effect of strain hardening of the composite associated with the imparted deformation, the latter represents the effect of shearing at interfaces. Unlike hydrostatic extrusion, discussed earlier, where the adhesion of implosively welded elements is improved, in drawing weakening and even failure of the weld can occur. Of course, the failure, often only local, of the weld does not reduce the strength or tightness of the cylinder, since the function of the weld is only to promote more 'homogeneous' drawing conditions.

4.8.8   Drawing loads

Although the three fundamental modes of cold drawing, i.e. sinking, plug and mandrel drawing, have not undergone any major development, the simplified theoretical equilibrium-of-forces approach to the problem of drawing has become more established.

In its basic form, the treatment offers the possibility of estimating the magnitude of drawing loads and the influence of the lubricant on the operation.

The efficiency of a drawing operation can be determined by considering the variation in the components of the total drawing load with the conditions of the draw. These parameters are of practical interest to the plant and tool designer, and to the production planner.

4.8.8.1   Tube sinking

The total drawing stress at the exit from the die is given by

$$\left(\frac{r_1}{r_2}\right)^b = \frac{\sigma_1 b - c}{\sigma_2 b - c} \tag{4.86}$$

where $b = \mu \cot \alpha$ and $c = Y_0(I + b)$.

The stress associated with the load which is required to produce the desired change in shape and to overcome redundant deformations, excluding friction, is given by

$$\sigma_2' = \sigma_1 + Y_0 \ln\left(\frac{r_1}{r_2}\right) \tag{4.87}$$

and the frictional component of the load can be obtained using

$$F = \mu \pi D \int_0^x p\,dx \tag{4.88}$$

The solution of this equation requires knowledge of the variation in the length of contact between the tools and the tube.

Figure 4.128   General view of a standard 3.2 kW ultrasonic tube-drawing unit. (Courtesy of Technoform-Sonics, Brierley Hill)

Figure 4.129   Cold-plug drawn, explosively prewelded low carbon steel/brass/copper tubing. Area reductions: (a) 17%, (b) 21%, (c) 23%

4.8.8.2   Mandrel drawing

In mandrel drawing, the condition of back-pull, found useful in the manufacture of wire, can be imposed by using a tandem die system, with a sinking die (*Figure 4.130(b)*) providing back-pull, and a pure draft (No. 2) die imposing the required dimensional conditions (*Figure 4.130(d)*).

Again, the stress related to the total drawing load is given by

$$\sigma_2 = Y_0(1 + C)\ln\left(\frac{t_1}{t_2}\right) + \sigma_1\left[1 - c\left(\frac{t_1}{t_2} - 1\right)\right] \quad (4.89)$$

where $C = \mu \cot \alpha$.

The stress of deformation is given by

$$\sigma_2' = \sigma_1 + Y_0 \ln\left(\frac{t_1}{t_2}\right) \quad (4.90)$$

The frictional component is given, as before, by equation (4.88).

4.8.8.3   Plug drawing

The total stress can be calculated using the expression:

$$\left(\frac{t_1}{t_2}\right)^c = \frac{\sigma_1 c - a}{\sigma_2 c - a} \quad (4.91)$$

where $a = Y_0(1 + c)$ and $c = 2\mu \cot \alpha$.

The stress of deformation is defined by equation (4.90), and the frictional component of the total load is given by

$$F = 2\mu\pi D \int_0^x p\,\mathrm{d}x \quad (4.92)$$

In the above expressions, D is the outer tube diamter, σ is the true flow stress, σ_1 is the back-pull stress at entry to the die, σ_2 is the drawing stress, F is the frictional load, p is the radial die pressure, r_n is the outer tube radius, and t_n is the wall-thickness at any section.

4.8.8.4   Drawing load

The total drawing load P_T required is composed of two major components: the load of deformation, and the frictional load F. In turn, the load of deformation P_D serves the dual purpose of shaping the tube to the required size and of overcoming the redundant shearing deformation. The load can, therefore, be split into two components: the load of useful deformation P_U, and the load of redundant distortion P_R.

In any given case,

$$P_T = \sigma_2 A_2 \quad (4.93)$$

where A_2 is the cross-sectional area of the tube at the exit from the die, and the load of useful deformation can be obtained from equation (4.87) or (4.90) by using the lower flow stress value which corresponds to the apparent (not the generalised) strain.

$$P_U = \sigma_2' A_2 \quad (4.94)$$

Then, since

$$P_D = P_T - F \quad (4.95)$$

the redundant load can be obtained from

$$P_R = P_D - P_U \quad (4.96)$$

4.9   Sheet-metal forming

4.9.1   Forming processes

The considerable success and continuous expansion of the sheet-forming industry are due to a number of unrelated but important developments.

The introduction of the continuous tandem rolling mill has made it possible to produce wide strip which can either be coiled or cut to suitable operational sizes. At the same time,

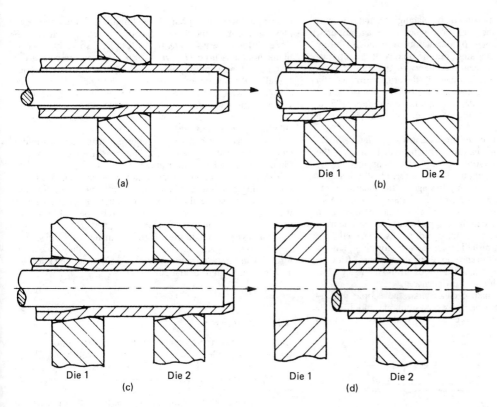

Figure 4.130 Tandem mandrel tube drawing. (a) Combined sinking–pure draft standard pass. (b) Pure sink system: die 1, sinking. (c) Tandem system: die 1, sinking; die 2, pure draft. (d) Die 2, pure draft

cold forming of sheet ensures greater resistance of the formed section to corrosion, since the preformed material has good surface finish due to treatments such as pickling. Surfaces cleaned in this way form suitable bases for the application of protective layers of, for instance, paint or plastic coating and, furthermore, they can be galvanised and then rolled or drawn.

The good dimensional qualities of preformed sheet make it a good starting material for operations such as bending, piercing, dimpling and drawing, by ensuring general uniformity of thickness. The exception here is the change in sheet thickness on transition sections where thinning is likely to occur. However, in a cold operation, strain hardening accompanied by changes in thickness will tend to counteract the effect of the actual variation in thickness.

The five basic processes effecting a change in shape are:

(1) shearing,
(2) bending,
(3) spinning,
(4) stretching, and
(5) drawing.

A wide range of sections and shapes, and a varied range of products for industrial, domestic or general use can be produced.

Industrial applications are found in the following areas:

(1) computer, nuclear and space industries;
(2) the chemical industry;
(3) the aircraft industry;

(4) the car industry; and
(5) the food industry.

Cold-formed sections are used industrially as components, plant, equipment, containers, panels, radiators, body parts, chassis, frames, racks, stiffeners, etc.

Domestic applications include, among others:

(1) components,
(2) cookers,
(3) panels,
(4) refrigerator bodies,
(5) washing machines, and
(6) shower cabinets.

General applications range from building constructions and components through structural elements, to rail and road transport, civil and highway engineering, agricultural machinery and equipment, and architectural and shop fittings. A major part of the output is intended for the electrical heating and ventilating applications represented by conduits, casings, cable supports, electrical appliances, heating panels, and dust-extraction equipment.

4.9.2 Shearing operations

The shearing operation constitutes the first stage of any forming process by producing either the starting material (cutting out of a sheet) or preparing an existing work piece by punching a hole or a series of holes before forming.

Basically, the operation involves placing the material between the edges of a shearing tool, which serve as supports, and separating a part of it by the action of a punch. Purely shearing stresses are generated, but the quality of the separating cut depends on the clearance between the tool and the specimen. If the clearance is too small, the cracks produced by the tool do not coalesce and tearing occurs. At the other extreme, too large a clearance results in considerable plastic flow which inhibits formation of the cracks and produces a burr at the upper edge.

The usual sequence of events is shown in *Figure 4.131*. Three stages of shearing can be distinguished. In stage I, the deformation of the material increases the shearing force by producing strain hardening; the load is carried by the material away from the plane of shearing. In stage II, the shearing force reaches its maximum and begins to decrease in magnitude. The displacement U_s of the free surface reaches its maximum value and a slight change in the displacement h of the shearing blade (due to the increase in speed v_h) is observed. An increment of the burnished zone Δ_{BB} is formed. This is determined by the difference in displacements U_f (due to fracture) and U_1 (due to localisation of strains). In stage III, ductile fractures develop and propagate.

To produce good quality, clean cuts, the simple punch–shearing tool assembly and stress system are modified by introducing an additional compressive stress. The effect of this is to suppress, up to a point, crack initiation and to encourage plastic shearing of the sheet. This objective can be achieved in three different ways. (*Figure 4.132*). If a specially shaped blank holder is used (*Figure 4.132(a)*), a compressive-stress system will be superimposed on the pure shear-stress one. Equally, the introduction of a negative clearance (*Figure 4.132(b)*), which produces part extrusion and part shearing, will have the same effect. The third method consists in using a tight clearance to finish an already blanked specimen by shearing (*Figure 4.132(c)*).

When very fine blanking is required, triple-action tooling can be employed (*Figure 4.133*). This consists of an additional blank holder and a counter-punch.

The sequence of operations is shown in *Figure 4.133*. From the starting position (*a*) the blank holder is moved up to press the sheet against the die (*b*). Shearing then takes place against the resistance of the ejector (*c*), followed by the withdrawal of the tools (*d*, *e*). Finally, the blank is ejected (*f*).

To assess the magnitude of the shearing force, a simple approximate expression can be used, giving the force P in terms of ultimate tensile strength (UTS), sheet thickness h, and the total length of the cut L:

$$P = 0.7hL(\text{UTS}) \qquad (4.97)$$

4.9.3 Bending

Bending is executed either by pushing the strip into a die of the desired profile or, more often, by cold-roll bending (see Section 4.5.10). Either method can be used to produce a variety of shapes, a selection of which are shown in *Figure 4.134*.

The bending process applied here is basically the same in character as that shown in *Figure 4.35*, in that tensile and compressive stresses are formed, with a single fibre of the neutral axis remaining unaffected. With the thinning of the edges and stretching of the material, one of the criteria limiting the amount of deformation is the localised necking, or plastic, tensile instability. This is likely to occur when the strain in the material exceeds a critical value or, approximately, when

$$\frac{h}{2R + h} \le (\exp n) - 1 \qquad (4.98)$$

where h is the sheet thickness, R is the radius of the die, and n is the power exponent of the constitutive stress–strain equation.

The minimum permissible bending radius which should be used to avoid fracture can be assessed, in terms of the reduction in area a, as

$$R = h\left(\frac{1 - 2a}{2a}\right), \text{ for } a < 0.2$$

and $\qquad\qquad\qquad\qquad\qquad\qquad\qquad (4.99)$

$$R = h\frac{(1 - a)^2}{a(2 - a)}, \text{ for } a > 0.2$$

Since the bent strip is rarely fully plasticised during the operation, the residual elastic material will produce spring-back when the tooling is removed. This, of course, impairs the dimensional accuracy of the finished part. The tendency to spring-back can be reduced either by overbending or by imposing a compressive stress or hydrostatic pressure during the operation. An approximate expression for the required bending force P is

$$P = Lh^2(\text{UTS})/w \qquad (4.100)$$

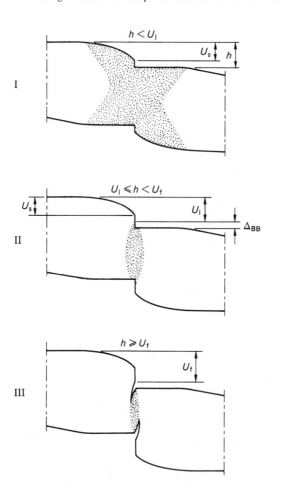

Figure 4.131 Development of strains in pure sheet shearing

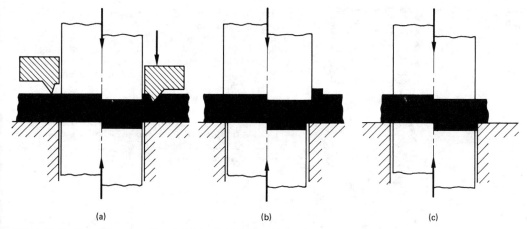

(a)   (b)   (c)

Figure 4.132   Three techniques of sheet shearing: (a) blanking, (b) negative clearance, (c) shaving

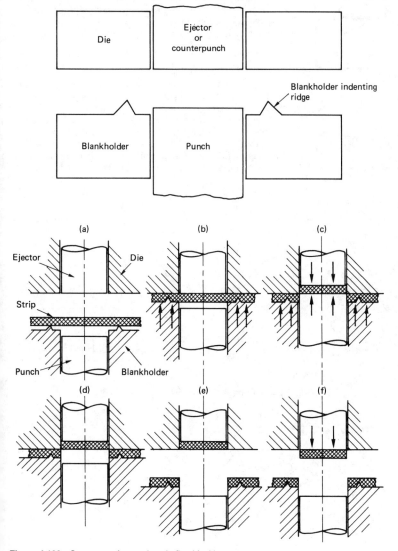

Figure 4.133   Sequence of operations in fine blanking

Figure 4.134 A selection of sheet sections obtained in bending

where L is the length of the bent section, and w is the width of the opening of the die.

Since the final quality depends on the material used, it may well be that the phenomenon limiting the degree of bending allowable in any given case will not be simply that of fracture or instability, but also one that produces an unsatisfactory finish. A selection of possible limiting factors, including the basic fracture and necking, is given in *Figure 4.135*. The maximum deformation may be limited by indentation caused by the die edges, galling, wrinkling and crease forming.

In addition to actual sheet bending, the operation is often extended to tube or blank flanging. Severe compressive stresses are imposed in these processes and may lead to buckling in a tube.

4.9.4 Spinning

Sheet forming by spinning is applicable to axisymmetric shapes only. The operation involves a driven, rotating system to which the male forming die is attached. The sheet to be formed is held against the die. Either manual or power, cold or hot spinning operations are used. The limit of formability is imposed by the ductility of the material and can be assessed from a simple tensile test.

4.9.5 Stretch forming

Stretch forming is often used to produce either simple components, comparable to cylindrical cups, or more complicated pressing-type shapes. Basically, stretch-forming operations depend on the use of conventional punch and die systems or a dieless-punch system. In either case, the rim of the blank is clamped (*Figure 4.136*) and the stretching, with consequent thinning of the material, can proceed on loading the punch. The thinning is particularly pronounced on the radiused rim of

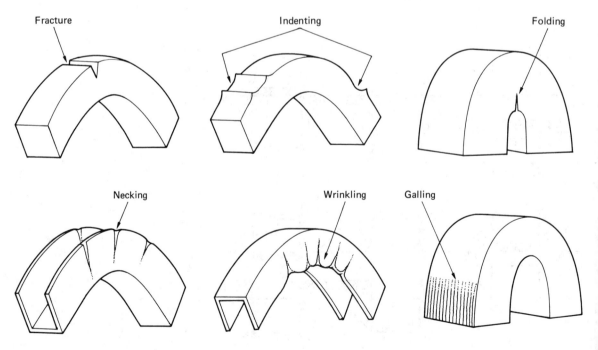

Figure 4.135 Sheet bending limiting product-quality phenomena

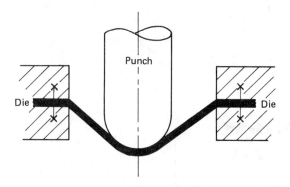

Figure 4.136 Conventional stretch forming

the supporting die or anvil. Instability can set in and lead to failure through localised necking.

Dynamic or explosive stretch forming is also used (see Section 4.10.3).

4.9.6 Deep drawing

Deep drawing is normally associated with the manufacture of cups, cans and similar containers. The operation is usually divided into two main groups: first-stage drawing, in which a flat circular metal blank is made into a cup; and a redrawing stage (or stages) in which the cup reaches its final size. The latter operation is necessary because first-stage drawing cannot normally produce a higher degree of deformation than that defined by the ratio of the diameter of the blank and the die throat (drawing ratio) of about 2.2, or a cup height/diameter ratio of about 1.

Figure 4.137 shows first-stage drawing and indicates the instantaneous stress distribution and changes in the thickness of the material likely to occur during drawing.

The sequence of operation is as follows. Initially, the specimen held in position by a blank holder, is partly in contact with the die, partly with either the die or punch, and partly with the punch only. The downward movement of the punch initiates drawing. The outer rim of the blank is then subjected to pure radial drawing (i.e. drawing towards the vertical axis of the system) between the die and blank holder. A part of the material bends and slides over the die and is further stretched between the punch and the die, whereas the material initially in the vicinity of the punch head and actually in contact with it bends and slides over the radiused part of the punch and stretches over the punch head.

The redrawing systems often used are shown in *Figure 4.138*. Parts (*a*) and (*b*) in the figure show direct redrawing systems with and without blank holders, respectively, while a reverse system is shown in (*c*). In (*a*), the wall of the cup undergoes double bending and unbending, the severity of which is expected to be high because the respective directions of deformation are at right angles to each other. System (*b*) shows less severity because of the tapered wall support, although double bending is involved. This system can be used only for relatively low cup diameter/wall thickness ratios which do not require the use of a blank holder. In comparison with the direct methods, system (*c*), having a generously radiused die profile, tends to reduce the degree of (or with a semicircular profile to eliminate completely) one bending and unbending effect. Whether there is significant advantage to using any system depends on the balance between the reduction in redundancy and practical production considerations.

The definition of 'redundancy' in deep drawing is not easy since redundancy is not necessarily associated with the effects of macroshear. The nature of the processes is such that portions of the blank material undergo some phases of deformation which in themselves induce redundant effects and yet are physically unavoidable if the process is to be completed. It is therefore the degree of severity imposed rather than the avoidance of a certain phase of the operation that matters. In this respect, the process differs significantly from the bulk forming operations discussed previously.

The three main sources of unnecessary strain in and/or distortion of the blank or cup material are flange wrinkling, the already discussed bending and unbending, and, partly, ironing. The latter is used to eliminate the increase in cup wall thickness which can be as much as 30% in the first stage of drawing. If this is followed by a further substantial rise in successive processing stages and is accompanied by wrinkling, an additional drawing operation becomes necessary. As far as redundancy is concerned, ironing is the only operation that brings back the 'standard' features of shearing.

The formability of a material depends on the blank-holder pressure and, consequently, the deep drawing ratio $R = D/d$ may be limited either by wrinkling of the flange, tearing of the cup bottom, or by galling. *Figure 4.139* shows diagrammatically the boundaries of these conditions and indicates the presence of a 'safe window' within which deep drawing is likely to be successful.

In determining the drawability, the criterion to be adopted is that relating to the first incidence of any fault.

4.9.7 Hydroforming and hydromechanical forming

To increase the depth of the draw, while reducing local stress concentrations, the techniques of fluid-backed or fluid-mechanically augmented processing have been introduced.

In a pure hydroforming operation (*Figure 4.140(a)*), a rubber diaphragm pressurised by the fluid acts as both a blank holder and a flexible tool. The only rigid tool is the punch. Since hydraulic pressure is exerted on the blank, the stressing is uniform and is accurately controlled. With reduced sheet thinning and considerable suppresion of cracks, high draw ratios (up to $R = 3$) are possible. The method is suitable for the forming of both jet-engine components (some 1 m × 1 m in size) and a variety of car components, including side and roof panels. A selection of parts so formed is shown in *Figure 4.140(b–e)*.

In the Hydromec, or mechanically augmented system (*Figure 4.141*), the sealing diaphragm is omitted and the blank comes into direct contact with the fluid. The sealing of the fluid is effected through the use of sealing rings. The blank is pressed firmly against the punch and thus possible bulging between the die ring and punch is avoided; a feature of the process that is of particular importance in the deep drawing of parabolic components. Complex parts can be drawn in a single operation, thus reducing operational costs. Sizing tools are claimed to be unnecessary.

4.10 High-energy-rate operations

4.10.1 Introduction

The last three decades have seen the development of industrial forming techniques based on the utilisation of high-energy-rate dissipation. The dynamic aspects and mechanisms of these operations make it possible to manufacture, usually fairly economically, either large components, such as radar

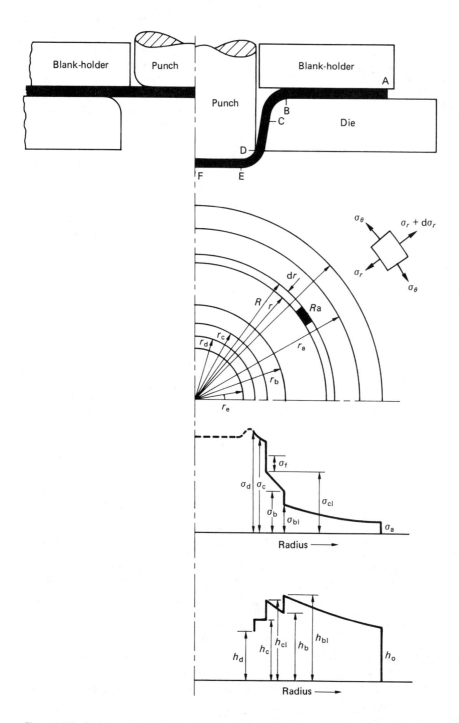

Figure 4.137 First-stage of cylindrical cup drawing showing instantaneous stress and deformation distributions

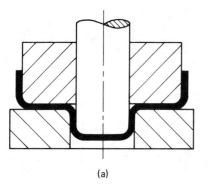
(a)

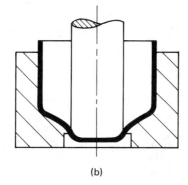
(b)

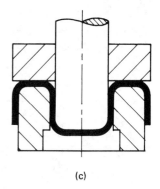
(c)

Figure 4.138 Direct cup redrawing with (a) and without (b) a blank holder. (c) Reverse drawing

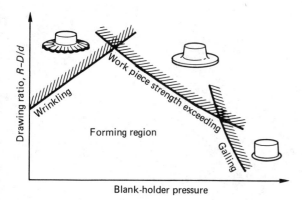

Figure 4.139 The effect of blank-holder pressure on the cup drawability

dishes, rocket nose cones and pressure vessels, or to produce smaller semifabricates (in sheet or tubular form) to be processed further in conventional operations. Some examples of this approach were given in Sections 4.7 and 4.8.

The usual sources of energy available are:

(1) chemical explosives,
(2) electrostatic fields (hydroelectric forming), and
(3) magnetic fields (magnetic forming).

These sources of energy are used in:

(1) forming (sheet and tube),
(2) welding (sheet sandwich semifabricates, and multimetallic tubular components),
(3) powder compaction (to form prefabricates),
(4) forging, and
(5) hardening.

The potential use of any of these operations is conditional on the formability of the given metal in the dynamic condition. An indication of this, based on 1100-0 Al as standard, is given in *Figure 4.142*.

4.10.2 Sources of energy

The proper utilisation of the energy evolved when an explosive charge is detonated depends on the degree of understand-ing of both the properties of explosives and of the detonation phenomena.

From a practical point of view, commercial chemical explosives will generally fulfil the requirements of the welding, forming and compacting processes. On detonation, the detonating front travels through the explosive converting its mass to a high temperature, high pressure gas which is then used in the generation of a stress wave and, depending on the conditions, in the appearance of shock and release waves.

Chemical explosives are subdivided into 'high' and 'low' (deflagrating) materials. High explosives are characterised by very high detonation pressures and high rates of reaction. Because of this, high explosives are subdivided further into two groups.

(1) Primary or detonating explosives which are sensitive and may be detonated by slight impact, flame, static electric charge, or simple ignition. They are normally used in detonators, but seldom as a source of energy in metal working.
(2) Secondary high explosives are used mainly in metal working and other industrial applications. They require a detonator to intiate the reaction and sometimes a booster charge to reinforce the detonation wave. They have a higher energy content than primary explosives.

Deflagrating or low explosives burn rather than detonate when the reaction is initiated and produce much lower pressures. They usually contain their own oxygen supply and, therefore, burn easily, but in some materials the reaction is difficult to initiate. Their low rate of burning makes them excellent propellants, but fire risk is considerable when handling such chemicals.

The charges are initiated by detonators. Electric detonators are widely used and are safe to handle if reasonable precautions are taken. Commercial detonators normally consist of a thin metallic container protecting the contents of an initiating primary high explosive, and a small amount (about 1 g) of a sensitive secondary explosive, e.g. PETN or tetryl. Initiation is achieved electrically, using an exploding bridge wire. Some types contain a slow-burning material to provide a time delay when many charges are fired at different time intervals.

The following explosives are commercially available and are used in industrial applications.

Primary high explosives
(1) mercury fulminate,
(2) lead azide,

(a)

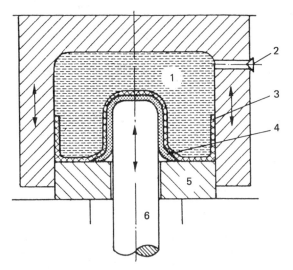

(b)

(c)

(d)

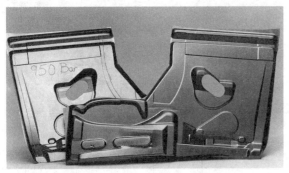

(e)

Figure 4.140 (a) Pure hydroforming operation: 1, hydraulic cushion; 2, pressure regulator; 3, cup diaphragm; 4, blank material; 5, pressure pad; 6, punch. (b, c) A range of jet engine and aircraft components. (d) Side panel of a car body. (e) Car parts. (Parts (b) to (e) courtesy of Saab-Scania AB Car Division, Sweden)

(3)   diazodinitrophenol,
(4)   lead styphnate, and
(5)   nitromannite.

Secondary high explosives
(1)   TNT (trinitrotoluene),
(2)   tetryl (trinitrophenylmethylnitramine),
(3)   RDX (cyclotrimethylenetrinitramine),
(4)   PETN (pentaerythritol tetranitrate),
(5)   ammonium picrate,
(6)   picric acid
(7)   ammonium nitrate,
(8)   DNT (dinitrotoluene),
(9)   EDNA (ethylenediamine dinitrate),
(10) NG (nitroglycerine), and
(11) nitro starch.

Low explosives
(1)   smokeless powder;
(2)   nitrocotton;
(3)   black powder (potassium nitrate, sulphur, charcoal); and
(4)   DNT (dinitrotoluene ingredient).

Explosive materials are availabe in different forms and some of them in more than one. Many of them can be melted allowing other explosives to be added in the form of slurries. For instance, cyclotol is made by slurrying granular RDX in molten TNT or Composition B. Powdered, granular, solid, liquid and plastic explosives can be used. One of the most useful types is so-called 'Datasheet' or 'Metabel', which is essentially a PETN explosive combined with other ingredients to form a tough, flexible waterproof sheet that can be cut and

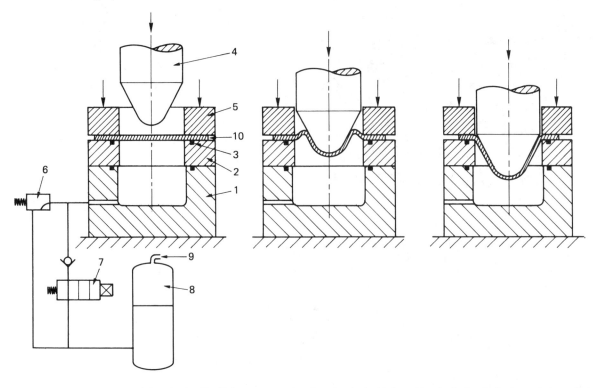

Figure 4.141 Hydromechanical deep drawing (the Hydromec process). 1, Container for hydraulic cushion; 2, die ring; 3, die ring seal; 4, punch; 5, blank holder; 6, pressure regulator; 7, electromagnetic control valve; 8, fluid container; 9, compressed air supply; 10, circular blank

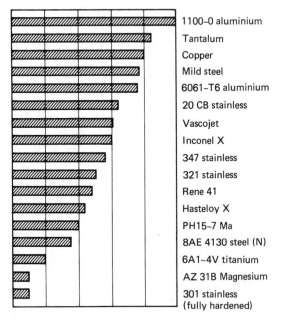

1100-0 aluminium

Tantalum

Copper

Mild steel

6061–T6 aluminium

20 CB stainless

Vascojet

Inconel X

347 stainless

321 stainless

Rene 41

Hasteloy X

PH15–7 Ma

8AE 4130 steel (N)

6A1–4V titanium

AZ 31B Magnesium

301 stainless
(fully hardened)

Figure 4.142 An assessment of the formability of metals

shaped to the required size for contact and stand-off operations. It is available in different thicknesses and can be glued together or used as shaping back-up material if a shape charge is required. Another very useful type is powder explosive, particularly various mixtures of TNT with aluminium powder, which can fill a container of any form and then be compacted to attain higher densities. Sealed containers are not required, thus offering an advantage over liquid explosives. Cord explosives, e.g. Cordtex, are also available. These consist of a flexible cord containing a core of explosive. Cords are very useful when continuous long charges are needed and give reasonable accuracy in a number of forming operations.

The effectiveness of a charge depends on the characteristics of the explosive, as reflected by the pressure–time function, velocity of detonation, explosive/specimen mass ratio, stand-off distance and the transmitting medium.

'Detonation' is a term used to describe the process in which an explosive charge undergoes a chemical reaction accompanied by a characteristic type of shock wave (or detonation wave). Depending on the properties and type of the explosive material, the velocity and intensity of the characteristic shock wave varies, but remains constant for a given type of explosive and for a charge of uniform geometry and density. This simplifies the mathematical solution of the hydrodynamic theory which applies to the process. The general behaviour of a primary explosive during reaction is characterised by a slow combustion process at the beginning, and then by deflagration, up to a point of sudden transition to detonation. The whole process is completed in a few microseconds. However, the rate of build up of the reaction and the transfer to detonation when a secondary explosive is detonated without a detonator is much slower and burning before detonation may occur.

Low explosives are characterised by the absence of the transition period. They react at rates which are proportional to the build-up pressure which, in turn, increases as the chemical reaction speeds up. This cycle leads to explosion within a fraction of a second, but the rate of reaction is usually much slower than 1% of that in detonation, and peak pressures are also lower. However, energy comparable with that obtained from high explosives can be generated by low explosives when they are adequately confined or used in sufficient quantity, and pressure distribution is easily controllable.

As far as metal-working processes are concerned, the most important parameters of the detonation process are:

(1) the energy released by the detonation,
(2) the detonation velocity (i.e the velocity of propogation of the detonation front), and
(3) the pressure exerted by the gaseous products of detonation on the specimen.

Other aspects of the process, such as the thermal stability and sensitivity of the explosive, the temperature, the heat generation and the ionisation phenomena in the gaseous products, as well as their composition, are of no importance.

To carry out a successful metal-working operation the optimum strain rate for the metal must be ascertained, and the amount of energy dissipated must be adjusted to give the desired rate.

The adjustment of the amount of energy available for a given operation depends on the source used and the method of application. Regardless of the type of source used, the energy can be delivered in one of two ways: through a transmitting medium, or directly to the metal. Apart from some special applications, delivery through a transmitting medium is more usual.

4.10.2.1 Transmitting through a medium

Although most of the comments given below are directly applicable to explosives, they also apply to hydroelectric and magnetic systems.

Considering a theoretical case of an explosive charge detonated in a transmitting medium, it is found that as the high-pressure and high-temperature gaseous products of explosion begin to expand, the surrounding medium suddenly becomes compressed and the major part of the impulse produced is propagated as a shock wave. The velocity of propagation is considerably higher than the velocity of the products of explosion. The physical state of the medium is thus altered from its original density ρ_0, pressure p_0, and temperature t_0 to one of high density ρ, pressure p and temperature t. In addition, particles of the medium acquire a translational velocity u, i.e. the velocity with which a particle in the compressed medium moves behind the shock front. The following relationships then apply:

$$\rho(U - u) = \rho_0 U \tag{4.101}$$

$$p - p_0 = \rho_0 u U \tag{4.102}$$

$$E - E_0 = \tfrac{1}{2}(P + p_0)\left(\frac{1}{\rho_0} - \frac{1}{\rho}\right) \tag{4.103}$$

where U is the velocity of the shock front, and E_0 and E are the internal energies before and after compression, respectively. The intensity and configuration of the shock wave depend on the weight and shape of the charge. For instance, for a spherical charge the peak pressure will be proportional to the cube root of the charge weight

$$\frac{p_1}{p_2} = \left(\frac{w_1}{w_2}\right)^{1/3} \tag{4.104}$$

where w_1 and w_2 are the weights of charges 1 and 2.

In addition to the weight and shape of the charge, peak pressures are also influenced by the type of transmitting medium. When water is used, a reasonably good approximation for the pressure p at any point in time is given by the expression $p = p_m(\exp - t/\theta)$, where p_m is the peak pressure, θ is a constant depending on the characteristics of the charge, and t is the time. The total impulse delivered can be expressed as an integral of pressure and time ($I = \int p \, dt$).

4.10.2.2 Transmitting directly to the metal

A contact operation of this type applies only to chemical explosives. When a charge is detonated against the surface of a metal, the detonation front is propagated through the, as yet, unexploded part of the charge, eventually reaching the surface of the specimen. A transient high pressure shock wave is then partly reflected onto the medium of the products of explosion and partly transmitted through the metal. The situation within the metal is somewhat analogous to that in a transmitting medium. The original metal boundary becomes depressed in the direction of the travelling wave and the region behind the shock wave undergoes severe compression. A further reflection of the wave takes place at the outer boundary of the metal specimen. The high-pressure wave is transmitted through the metal in the form of a transient stress wave which is capable of deforming and, if it exceeds certain magnitude (depending on the properties of the metal), of fracturing the work piece.

The effect of the passage of a shock wave through the metal is complex and produces a response which depends entirely on the material structure. A comparison of the effects produced by high velocity, high temperature or large mass systems is given in *Figure 4.143*.

4.10.2.3 Hydroelectric forming

The principle of operation is based on the rapid dissipation and transmission of energy evolved when an electrostatic field is suddenly discharged. A basic circuit is shown in *Figure 4.144*.

Two different techniques of using the energy stored in a bank of condensers are employed: underwater discharge, and exploding wire.

In the former case, the discharge across two submerged electrodes produces a shock wave in the transmitting medium, accompanied by heating and vaporising of the adjacent layers of the medium. The plasma created by the spark expands as a gas bubble, transmitting the force of explosion to the work piece. The efficiency of the operation depends on the conductor material, losses of energy in the circuit, and the geometry and surface conditions of the electrodes.

The second method consists of connecting submerged electrodes with an initiating wire. The transmitted energy vaporises the wire and converts it into plasma, creating a pressure wave. The increase in volume of the vaporised wire is of the order of 25 000 times its original volume. The exploding-wire method possesses certain distict advantages over the spark-discharge method in that the process can be more rigidly controlled. The shape of the shock wave can be determined by the shape of the wire, and a long arc discharge can be obtained as opposed to a point source one. The amount of energy can be controlled by the dimensions and material of the wire. For instance, tungsten produces more energy than tantalum, niobium, molybdenum, titanium, nickel or aluminium (in that order).

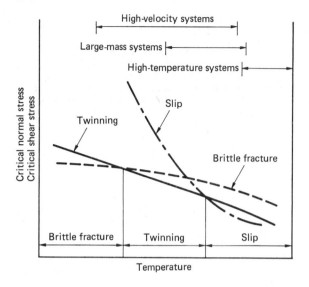

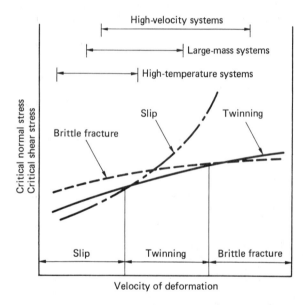

Figure 4.143 Effect of various parameters on the mechanism of metal working

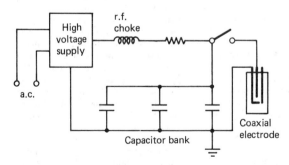

Figure 4.144 Hydroelectric forming

4.10.2.4 *Electromagnetic forming*

The principle of electromagnetic forming is, basically, the same as that of an electrohydraulic operation. The energy stored in a bank of condensers is rapidly discharged through a magnetic coil, which surrounds, is placed inside, or is in the proximity of the work piece (*Figure 4.145*). A high intensity magnetic field is thus created and, providing that the material of the work piece is conductive, electric current is induced in the specimen. The current interacts with the coil field and produces high transient forces. The specimen thus acts as a secondary short-circuited coil. The energy level produced by the magnetic field depends on the conductive properties of the formed metal, its shape and mass and the duration of the initial current pulse. These factors can quite easily be controlled by suitable choice of the materials and geometry of the coils. Materials of low conductivity are sometimes lightly coated with copper. The shape of the impulse wave can be modified by using 'field shapers' which consists of shaped beryllium–copper pieces inserted in the coils. Electrically, shapers help to depress or concentrate the intensity of the magnetic flux in those sections of the work piece which may require a lower or higher degree of deformation.

The life of a coil depends on the magnitude of the force to which it is subjected. This being equal to the force generated on the surface of the work piece, the pressures can be very high. The usual practice consists, therefore, of using one-shot disposable coils for more complicated operations involving only a few parts, and of limiting the pressures to the value of the compressive strength of the coil material for mass-produced parts.

4.10.3 Forming systems

4.10.3.1 *Explosive forming*

Explosives are used primarily for shaping sheet, plate and tubes, and for sizing and flanging, all these operations being of the stand-off type. Close-contact operations are used to a

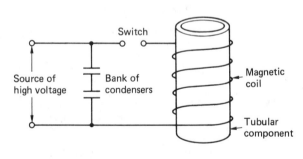

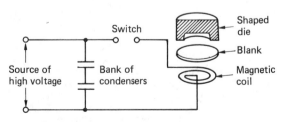

Figure 4.145 Some circuits used for magnetic forming

lesser extent and for more sophisticated processes such as extruding, cladding (welding), powder compaction and controlled surface hardening of Hadfield steels. The particular applicability of explosives to forming arises from the fact that virtually any shape in any size can be obtained without recourse to plan or machinery. The forming tool consists of a die which is comparatively light and does not require any foundations since the inertia of the tool mass is sufficient to counteract the applied force. It is essential, however, to realise that explosive forming is not economically viable when a very large number of components is required. Its advantage over the conventional methods lies in the possibility of producing complex parts very accurately and with very little or no machining, but in small quantities. The saving is due to the fact that there is no capital cost of presses, tooling, etc. The dies for stand-off operations are very often made from cheap cast materials such as epoxy or concrete.

The range of metals successfully worked includes: aluminium and its alloys, stainless steels, magnesium and some of its alloys, titanium and its alloys with aluminium, vanadium and manganese, refractory metals, copper and its alloys, and special alloys such as stellite, iron–nickel, nickel–copper, chromium–nickel and cobalt–iron. Carbon and low-alloy steels are less often used because of their low formability in dynamic conditions.

Three essential types of techniques characterise the stand-off operations: (i) free forming (cups, flanging and deep drawing), (ii) cylinder forming, and (iii) bulkhead forming (sheet and plate). These techniques are shown diagrammatically in *Figure 4.146*. Cylinder shaping is done in an open system (*Figure 4.13*) using high explosives. A transmitting medium other than air is used to sustain the pressure for a longer period, thus increasing the impulse delivered to the work piece. In general, the forming of metals can be carried out either in a tank sunk in the ground and filled with water as the transmitting medium, or in an empty tank with the water being contained in a polythene bag (in the case of a cylindrical component) or in any suitable, disposable container. The first method is used for large components that require large charges, where the confinement of explosion within a large volume of water serves both as a safety measure and as a means of reducing noise. In such a case, the space between the underformed metal and the die must be sealed-off and evacuated to enable full deformation to take place. In the second case, it may not be necessary to evacuate the air, providing that the die is fitted with a suitable system of ventilating holes.

A very important aspect of this method of forming is correct die design. The near absence of spring-back depends on the proper balance of tool profile, avoidance of sharp edges and deep narrow grooves and the provision of reasonably smooth transition sections. The difficulties encountered in this respect with both spring-back and wall thinning are illustrated for an 18/8 titanium stainless steel component in *Figure 4.147*.

Bulk forming of sheet is carried out by using either a single blank subjected, possibly, to a series of shots until the die is filled completely, or by using a mechanically shaped preform which is then given its final accurate dimensions in an explosive operation (*Figure 4.148*).

Forming without a die is also possible when relatively simple shapes are required, e.g. the impeller shown in *Figure 4.149*. An initially circular specimen can be formed to the shapes shown in *Figure 4.150*.

Although explosive forming cannot be regarded as a substitute for existing processes, it is an extension of the techniques now in use in the sense that it enables a number of difficult materials to be formed to a high degree of accuracy and at low cost. In the general field of metal shaping, the process is very advantageous when large parts are formed,

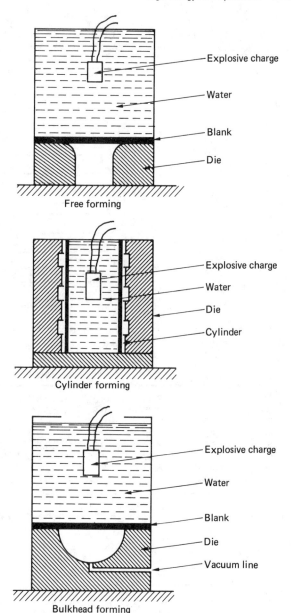

Figure 4.146 Basic explosive forming systems

often in a single operation, as costly heat treatment, tooling and machining are not required. The process is not competitive when parts can be produced in bulk using conventional equipment in a small number of operations. The use of explosives requires special precautions when storing and handling, and for safety reasons forming cannot be done in a congested enclosed area.

4.10.3.2 Hydroelectric forming

In the application of this controlled and repeatable electrical force, the most obvious difference from explosive forming is

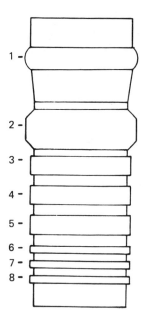

Section	Die diameter (cm)	Section diameter (cm)			Final thickness (mm)		
		Max.	Min.	Mean	Max.	Min.	Mean
1	9.84	9.84	9.82	9.830	1.42	1.29	1.36
2	9.85	9.85	9.84	9.847	1.29	1.19	1.24
3	8.95	8.95	8.94	8.945	1.37	1.27	1.32
4	8.95	8.95	8.94	8.945	1.37	1.32	1.34
5	8.95	8.95	8.94	8.945	1.39	1.32	1.35
6	8.95	8.70	8.66	8.682	1.47	1.39	1.43
7	8.95	8.74	8.63	8.688	1.55	1.42	1.48
8	8.95	8.61	8.49	8.549	1.52	1.47	1.49

Figure 4.147 The effect of spring-back and wall thinning in an explosively formed 18/8Ti steel tube (initial wall thickness 1.62 mm)

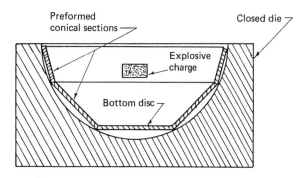

Figure 4.148 Preform bulk forming

the design of the forming apparatus. The dies are completely enclosed, since the volume of plasma generated is very small compared with the gas bubble created by a chemical explosive. By closing the die, full advantage can be taken of the pressure wave and, in the case of tubular parts, the shock wave is reflected from the ends, thus giving rise to additional energy.

Each die has its own set of electrodes. In either technique the discharge takes place in a matter of microseconds, and is repeatable within 30 s. Operations can be automated and, unlike explosive forming, are eminently suited for mass production. Industrial forming machines are commercially available that have a basic output of, for example, 15 kJ, the amount of energy sufficient to form ordinary commercial materials using blanks of 12–100 mm diameter and wall thicknesses of 0.2–2.5 mm. The output can be increased to 60 kJ.

For most purposes, dies used in hydroelectric forming are made of aluminium or cheap castable materials. The serious disadvantage of this technique is the comparatively high cost of the electrical equipment needed.

4.10.3.3 Electromagnetic forming

The dies used in this process are made from cheap materials and, in some cases, are not required at all since the magnetic fields produce uniform pressures sufficient to expand cylindrical parts.

The technique is very suitable for shaping sheet and tubular parts, swaging and the production of finned components. The

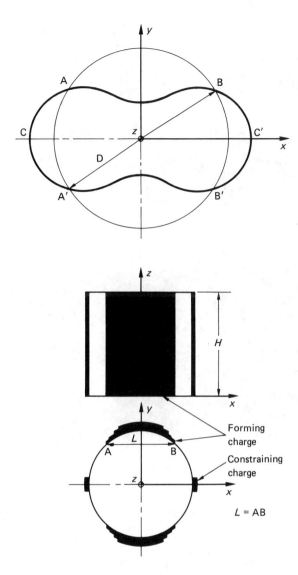

Figure 4.149 Forming an impeller without using a die

(tubular) surfaces of metallic alloys which are either incompatible (from the point of view of conventional welding) or impossible to join because of the geometry involved.

Explosive welding is a pressure operation in which two initially separate elements are joined together as a result of the metal jetting action which originates at the point of contact between them. The contact is achieved by detonating a charge covering the surface of one component (the flyer) which then gradually collapses onto the other (the base) component. The latter may be rigidly supported on an anvil. The shearing stresses generated at the point or line of contact are too high for either material to sustain and liquefaction takes place. A jet containing the elements of the flyer and the base is created and is propagated along the surface of the base. Since the detonation of the charge continues in a finite, albeit a short, time of microseconds, the flyer collapses in stages, creating new points of contact with the base and thus giving rise to further jetting. An indenting wave is produced (*Figure 4.151*) which, in addition to molecular welding, also creates a mechanical bond between the two surfaces.

Elements welded in this way are usually made of the materials listed in *Table 4.14*, but more 'exotic' combinations can be produced easily. The success of the operation depends on the correct matching of the dynamic angle of collision between the welded elements and the collision velocity. For any material there is a 'welding window' (*Figure 4.152*) within which welding will take place.

The quality of the weld is tested using a number of standard techniques (see *Figure 4.153*).

Explosive welding is used to produce large sandwich (multimetallic) plates which can then be processed further by rolling, shearing, bending and/or deep drawing, and to manufacture duplex or triplex cylinders, assemblies of rods and tubes which can be further cold drawn or hydrostatically extruded.

Special welding–forming operations and processes are employed and range from the manufacture of honeycomb panels to multilayered foil cylinders. The latter can be used as light but strong pressure vessels and are made either from layered foils or, if required, can be reinforced by metallic mesh interleaved with foil.

Typically, a flat sandwich of either variety is built up, containing up to 10 layers of materials, and is then wrapped on a bore-supporting mandrel to form a cylinder. This is surrounded by a layer of suitable explosive which, on detonation, will effect the welding of the individual material layers to each other. *Figure 4.154* shows diagrammatically an assembly of this type intended for the welding of a triplex cylinder.

A cross-section of a copper/brass non-reinforced multilayer foil cylinder is shown in *Figure 4.155*.

If mesh reinforcement of the foil matrix is required, four basic combinations can be used (*Figure 4.156*). A copper foil matrix, stainless steel mesh reinforced, explosively welded section of a cylinder is shown in *Figure 4.157*. This figure clearly shows the reinforcing wires welded to the surrounding foil, as well as the welding of adjacent layers of the foil matrix to each other.

4.10.5 Powder compaction

Various dynamic powder-compaction techniques involving the use of chemical explosives, high pressure gas guns, or electromagnetic or mechanical pneumatic power sources are used to compact metallic powders into semifabricates and finished machine parts. Either cold or preheated powders are used and the products, produced at pressures of 100–1000 kbar, amount to many thousands of tonnes annually. Like explosive-welding operations, explosive powder compacting is an exact and very complex process which requires detailed in depth study. The

process can be fully automated and forming machines are available. These are capable of producing magnetic flux densities of the order of 3×10^5 G, corresponding to pressures of over 340 MPa. The duration of the pulse is short, because of the permeability of metals, and is usually limited to 10 ms. Magnetic fields are quite sufficient for forming copper, aluminium and similar materials, but for metals like tungsten and for large parts, fields of the order of 5×10^5 to 5×10^6 G are needed.

The major disadvantage of the process is, again, the relatively high cost of the electrical equipment, but this is offset to some extent by the very high reliability, repeatability, speed of production and accuracy obtainable.

4.10.4 Explosive welding

The use of explosives for welding is now a well-established practice and provides a means of joining either flat or curved

Figure 4.150 Steel impellers formed without using a die

Figure 4.151 Explosively welded stainless steel (lower part) and carbon–steel interface

Table 4.14 Standard cladding combinations

Cladding metals	Base metals
Stainless steels Copper and copper alloys Nickel and nickel alloys Hastelloy Titanium and zirconium Tantalum Aluminium	Carbon and low alloy plates and forgings. Stainless steel
Copper	Aluminium

reader is, therefore, referred to the Bibliography for detailed information, but an indication of the use of dynamic compaction in forming is given by the following examples.

A collection of tools, made from repressed sintered tool material is shown in *Figure 4.158*. A planar flow, cast ribbon is comminuted into powder (both shown in the figure) and the powder is then compacted to form a selection of tools.

In another example (*Figure 4.159*), an aircraft nose wheel is manufactured from a metal powder of composition of 18.5%Al, 1.3%Fe, 1.7%V, Si (residue).

A different type of product is made of a metallic/non-metallic composite, e.g. a metallic/polymeric or ceramic/polymeric compound obtained by explosive compaction of particulate matter. A steel/polymer composite of interest in the electronics industry and capable of further conventional processing is shown in *Figure 4.160*.

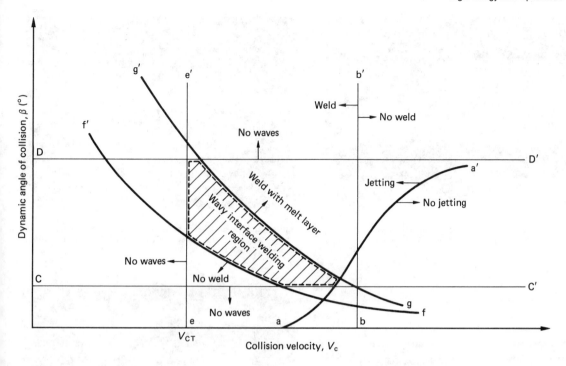

Figure 4.152 A typical 'welding window'

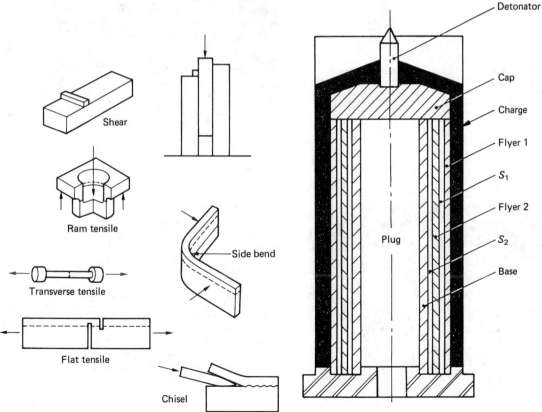

Figure 4.153 Methods for testing explosively produced welds

Figure 4.154 Explosive welding of a triplex cylinder. S_1 and S_2, original stand-off distances

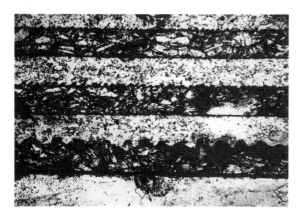

Figure 4.155 Cross-section of an explosively welded and formed multilayer copper/brass foil cylinder. Magnification ×110

Figure 4.158 A selection of tools produced by powder compaction of a repressed sintered tool material (Courtesy of Dr D. Raybould, Allied Signal Inc., USA)

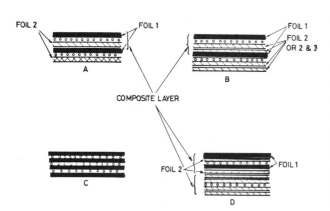

Figure 4.156 Four basic mesh-reinforcing systems

Figure 4.159 An aircraft wheel nose obtained by compaction of 18.5%Al, 1.3%Fe, 1.7%V, Si powder. (Courtesy of Dr D. Raybould, Allied Signal Inc., USA)

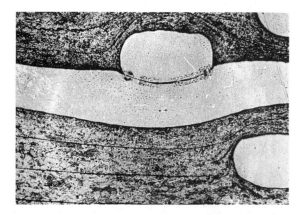

Figure 4.157 A section of a copper matrix/stainless steel mesh reinforced explosively welded cylinder. Magnification ×100

4.10.6 Explosive forging

The important characteristic of high-energy-rate forming (HERF) as far as hot forging is concerned is the higher than normal speed of deformation. The interdependence of the principal parameters of the hot-forging process with respect to the effect of the change in the speed of deformation or strain rate is shown in *Figure 4.161*.

The energy and load requirements are the two principal process parameters, their magnitude being determined by the yield strength of the work piece and the frictional resistance at the tool–work piece interface. Increasing strain rate and reducing temperature act to increase the yield strength. Friction at the tool–work piece interface is also found to reduce with increasing strain rate. Increasing the strain rate reduces the duration of the deformation period and, since the tools are usually at a much lower temperature than the work piece, chilling is reduced. Conversely, the effect of internal heating due to plastic working is to cause a faster rate of heating and, therefore, a larger temperature rise.

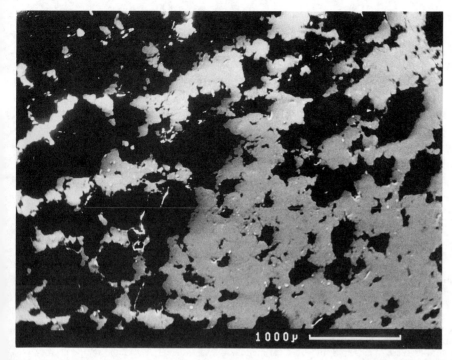

Figure 4.160 Explosively compacted steel (dark areas) and polymer powder composite for use in the electronics industry

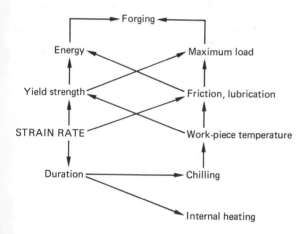

Figure 4.161 The interdependence of the principal parameters of the hot-forging process

Simple forging tests suggest that the net result of this interdependence of parameters, with respect to an increase in strain rate, is to reduce substantially the maximum loads required to increase the amount of energy and to improve forgeability. The latter feature is a result of the reduced heat loss which enables the work-piece temperature to remain high with a corresponding low flow stress.

In terms of production forging, these characteristics of HERF manifest themselves in the ability to form components prossessing thin sections and fine detail. These advantages can, however, be invoked when the part is forged in a single blow. The manipulation of the more exotic alloys is not always successful and, although the details are by no means clear as yet, it seems that if in conventional processing these materials present difficulties due to overheating, then these will not be improved in dynamic conditions. Converesely, if excessive cooling is the problem, then HERF can probably be used to advantage.

Ultimately, the choice of the process for a particular component is invariably dictated by economic considerations. Of prime importance in this context is the question of tool life; reduced heat transfer, lower friction and lower maximum loads are expected to give increased tool life in HERF conditions. Comparative die trials do not necessarily provide conclusive evidence, but are nevertheless considerably enlightening. Studies suggest that the different dwell times of the different types of HERF machine can have an important effect but, in general terms, tool performance is comparable. An example illustrating this argument is provided by, for instance, the manufacture of pipe reducers.

The conventional methods of hot-forging pipe reducers are inherently expensive, usually requiring multiple forging operations. In addition, subsequent heat treatment is frequently required to restore physical properties. Surface scaling and grain growth are often problematic. The concept of explosively forging pipe reducers is both practical and economical. Considerable advantage is offered over conventional techniques because virtually any size reducers can be explosively forged provided that the ratio of the diameters required is not too large.

A variety of engineering alloys have been successfully formed in this way, including carbon steel, 304 stainless steel, 5038 and 6061-T6 aluminium and Monel 400. The only requirement for successful explosive forging is that the alloy possesses sufficient ductility (under impulsive loading) to conform to the die. The process is particularly attractive for forging the corrosion-resistant alloys used in modern ship-board piping, including Monel, Inconel and copper–nickel. Conventional hot-forging techniques require close control of forging temperatures in order to prevent excessive grain growth with these alloys.

In general, forgings in aluminium, chromium and stainless steel have been made with zero draft angles, tolerances of ±0.02 mm on diameters of up to 75 mm, and surface finishes of less than 0.45 μm. In some forgings, rapid stiffners (25 mm high, 1.3 mm thick) have been formed in a 25 mm × 25 mm waffle pattern on 1.6 mm thick plate.

4.10.7 Explosive hardening

High pressures of short duration, produced by detonating a high-explosive charge against the surface of a specimen, are often used to effect controlled hardening of steels. A contact operation in which a sheet of chemical explosive is detonated on the surface of the metal considerably increases the hardness of the material and promotes an advantageous distribution of hardness with depth, while giving rise to only a small plastic deformation of the material.

When subjected to the detonation of the charge, austentitic manganese and low carbon steels, experience instantaneous shock pressures of the order of 270 GPa. The pressure wave, when travelling through the metal, causes severe lattice deformation. The wave decays exponentially in a matter of microseconds, but is of sufficient intensity to effect the formation of additional dislocations which, in turn, increase the hardness of the metal.

Providing that the magnitude and configuration of the high-pressure wave are carefully controlled and that, therefore, no spalling, cracking or gross damage to the metal takes place, two direct practical applications are possible. Firstly, large metal components can be hardened *in situ*. For example, railway points and blades of stone-crushing plant can be hardened in a manner that combines simplicity of operation with a high degree of qualitative uniformity. Austenitic manganese (Hadfield) type steels treated in this manner attain their maximum possible hardness after a single shot, thus producing a saving in the cost of both specialised equipment and labour.

Secondly, it is possible to autofrettage or prestress metal components, because of the imposition of controlled plastic deformation. This application is even more attractive if it is remembered that it can be done without the use of any machine tools by simply shaping the explosive charge to the pattern required and by controlling the level of energy which it contains.

Both applications, but particularly explosive hardening, have been used on an industrial scale for the last two decades or so.

4.11 Superplastic and mashy state forming

4.11.1 Superplasticity

The definition of the term 'superplasticity' is related to the fact that 'normal' alloys rarely show a greater degree of elongation than, at most, 60% when cold worked and only very occasionally exceed 100% when hot worked. In a few cases (*Table 4.15*) metallic alloys elongate at elevated temperatures to as much as 10 times their original length—these are called 'superplastic' materials.

A number of alloys behave superplastically at temperatures ranging from ambient to 500°C but, because they are prone to corrosion and have poor creep properties, their industrial application is extremely limited. Tin/lead alloys are typical examples of this group.

Table 4.15 Superplastic alloys used in practical applications

Alloy	Superplastic temperature (°C)	Elongation (%)	Strain rate (s⁻¹)	m*	Flow stress (MPa)
Titanium					
Ti, 6%Al, 4%V	927	1000–2000	2×10^{-4}	0.8	10
Aluminium					
Supral 100	450	600–1000	10^{-3}	0.38	9
08050	565	500	10^{-3}	0.3	2.8
7475 (fine grain)	516	1200	2×10^{-4}	0.75	2
Zinc					
Zn, 22%Al	200	2000	10^{-2}	0.5	10
Iron					
Fe, 1.6%C(+1.5%Cr)	650	1200	10^{-4}	0.46	45
Fe, 26%Cr, 6.5%Ni (IN 744)	900	1000	5×10^{-4}	—	28
Nickel					
IN 199 (PM)	1010	1000	—	0.5	35

* The strain-rate-sensitivity index.

The introduction of modern manufacturing techniques, e.g. diffusion bonding, enables the production of materials characterised by high strength (e.g. the titanium alloy Ti, 6% Al, 4%V) which are of considerable practical interest, although they require temperatures of up to 900°C.

Although welcome in general terms, the ability of a material to extend by several hundred per cent, may not be acceptable from the point of view of the quality of the final product, since it could imply considerable thinning. Consequently, practical considerations limit the extent to which the superplastic tendency of a material can be utilised.

A practical guide defines a desirable material as one that will deform at very low stress, with a high rate sensitivity and freedom from damage, up to an elongation of some 100%.

The usefulness of these materials depends on the grain-growth response to strain rate, because only fine-grain structure is acceptable. Practically acceptable superplastic alloys should, therefore, display the following characteristics:

(1) they must be stable at high temperatures (0.3–0.5 T_m);
(2) the grain growth must be limited at high strain rates;
(3) the flow stresses must be lower than the corresponding stresses in the same, but coarse grain, alloy; and
(4) the acceptable properties must be obtainable at normal operational strain rates of up to 10^{-2} s^{-1}.

Since superplastic materials can be formed at low stresses, they are ideal for first-step operations in which blanks are preformed to non-uniform thickness, but are subsequently formed to the required shape exhibiting a very high degree of thickness uniformity.

4.11.2 Mashy state operations

When a metal is heated to a temperature higher than its solidus temperature, its structure changes and consists of both solid and liquid elements, the latter generated by partial melting. The 'mashy state' is thus created the properties of which enable easy processing of the material in both the conventional forming processes and composite manufacturing techniques.

In the mashy state, the liquid component is normally present at grain boundaries and separates the solid component consisting of the incompletely melted grains (*Figure 4.162*). The properties of mashy state metals are as follows:

(1) low flow stresses;
(2) behaviour like that of a slurry when the solid fraction φ is less than 65%;
(3) when φ is low a mashy state metal can be mixed with other metals; and
(4) two different mashy state materials can be bonded together.

The following operations can be carried out on mashy state metals:

(1) extrusion,
(2) rolling of sheet metal,
(3) production of particle-reinforced metals,
(4) production of particle-reinforced cladding metals, and
(5) the manufacture of composite sheet.

4.11.2.1 Extrusion

Extrusion of wires, bars and tubes is carried out in a conventional manner by using an electric furnace and an extrusion press. Billets are heated to the required temperature to obtain solid fractions of at least 70–80% and are extruded using

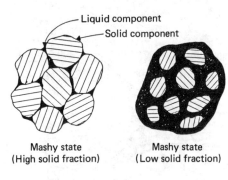

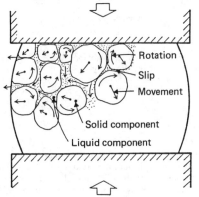

Figure 4.162 Diagrammatic representation of the mashy state

preheated tools. A typical example of directly extruded aluminium alloy tubing is given in *Figure 4.162*.

The characteristics of mashy state extrusion are:

(1) a low extrusion pressure (20–25% of that used in hot extrusion),
(2) a high extrusion ratio obtainable in a single pass,
(3) the liquid component provides lubrication, and
(4) complex sections are easy to extrude.

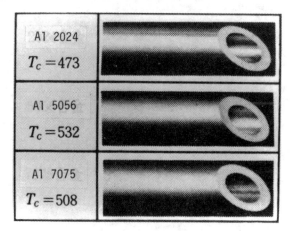

Figure 4.163 Aluminium alloy tubes extruded directly in the mashy state. (Courtesy of Prof. M. Kiuchi, Tokyo University)

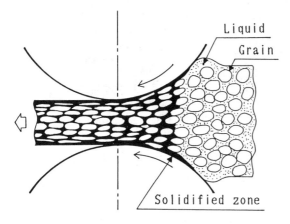

Figure 4.164 Diagrammatic representation of mashy state rolling. (Courtesy of Prof. M. Kiuchi, Tokyo University)

4.11.2.2   Rolling of sheet metal

The success of this operation depends on steady flow and homogeneous deformation, which can only be achieved if the state of hydrostatic pressure is maintained along the pass. The process has the advantage of enabling metals with complex internal structures to be processed, where a conventional cold- or hot-rolling operation is inappropriate.

A diagrammatic representation of the forming system is given in *Figure 4.164*. The liquid component is cooled by the rolls, solidifies and is forced into the roll gap, whereas the grains of the solid component are compressed, deformed and then drawn into the gap. In the gap each grain is deformed and elongated.

4.11.2.3   The manufacture of particle-reinforced metals

Since additives, in the form of particulate matter, can be mixed effectively with a mashy metal, composites of different metals, or metals and ceramics, can be made and these are particle-reinforced structures. These can then be processed conventionally by extrusion and rolling to produce the desired shapes and dimensions.

Two possible production routes and applications are shown in *Figure 4.165*. A metallic matrix and, say, reinforcing ceramic particles can be stirred and mixed together and then be extruded directly to form bars, wires or tubes, or be rolled to form sheet. On the other hand, on cooling the metal matrix gradually during stirring, metal powder is obtained. This can then be mixed with reinforcing particulate matter, formed into billets and then be extruded. Both routes are indicated in *Figure 4.165*.

Particle reinforced and extruded aluminium bars and wires are shown in *Figure 4.166*, and a forged gear wheel is shown in *Figure 4.167*.

4.11.2.4   Particle-reinforced cladding metals

Although particle-reinforced metals possess good hardness and antiwear properties, they are basically brittle and so have low formability. To compensate for this disadvantage, particle reinforced cladding metals obtained in the mashy state can be used. The particle reinforced cladding metal consists of a layer of particle-reinforced material and a metal base which provides the required ductility. The bonding between the two layers is obtained in mashy state forging.

The various manufacturing routes used are shown in *Figure 4.168*, which shows the possibilities of preforming the reinforcement to sheet followed by forging, cold forming following by forging, or starting with a mixture of powders taken directly to mashy forging.

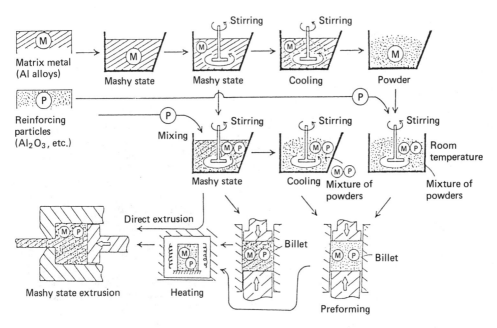

Figure 4.165 Production of particle-reinforced metals in the mashy state. (Courtesy of Prof. M. Kiuchi, Tokyo University)

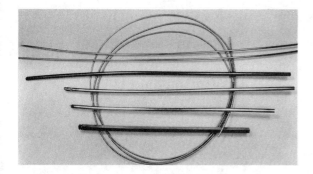

Figure 4.166 Bars and wires of a particle reinforced aluminium alloy extruded in the mashy state. (Courtesy of Prof. M. Kiuchi, Tokyo University)

Figure 4.167 A gear wheel of particle reinforced aluminium alloy forged in the mashy state. (Courtesy of Prof. M. Kiuchi, Tokyo University)

4.11.2.5   The manufacture of composite sheet

Either metal/ceramic-fibre reinforced or metal/metal combinations can be produced. An example of the adopted practice is given in *Figure 4.169* which refers to the manufacture of a particle- or fibre-reinforced cladding sheet. The new manufacturing processes, briefly outlined here, are beginning to be used in normal industrial routines but, naturally, a great deal of further development work is still necessary.

Bibliography

General

ALTING, L., *Manufacturing Engineering Processes*, Marcel Dekker, New York (1982)

AVITZUR, B., *Metal Forming: Processes and Analysis*, McGraw-Hill, New York (1968)

BLAZYNSKI, T. Z., *Metal Forming: Tool Profiles and Flow*, Macmillan, London (1976)

BLAZYNSKI, T. Z., *Applied Elasto-Plasticity of Solids*, Macmillan, London (1983)

BLAZYNSKI, T. Z., *Design of Tools for Deformation Processes*, Elsevier/Applied Science, London (1986)

BLAZYNSKI, T. Z., *Plasticity and Modern Metal-Forming Technology*, Elsevier/Applied Science, London (1989)

BURKE, J. J. and WEISS, V., *Advances in Deformation Processes*. Plenum Press, New York (1978)

JOHNSON, W. and MELLOR, P. B., *Engineering Plasticity*, Von Nostrand Reinhold, London (1973)

JOHNSON, W., SOWERBY, R. and HADDOW, J. B., *Plane Strain Slip-Line Fields,* Edward Arnold, London (1977)

KALPAKJIAN, S., *Manufacturing Processes for Engineering Materials*, Addison-Wesley, London (1984)

PITTMAN, J. F. T., WOOD, R. D., ALEXANDER, J. M. and ZIENKEIWICZ, O. C., *Numerical Methods in Industrial Forming Processes*, Pineridge Press, Swansea (1982)

ROWE, G. W., *Principles of Industrial Metalworking Processes*, Edward Arnold, London (1977)

SCHEY, J. A., *Introduction to Manufacturing Processes*, McGraw-Hill, New York (1977)

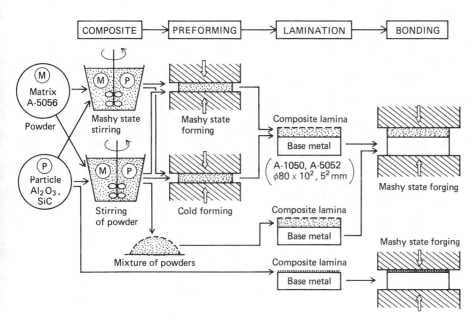

Figure 4.168   The manufacturing techniques for particle reinforced cladding metals. (Courtesy of Prof. M. Kiuchi, Tokyo University)

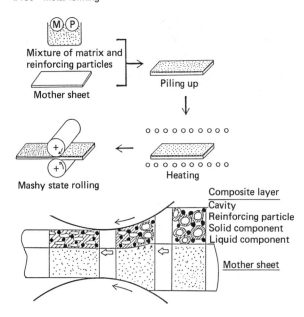

Figure 4.169 The mashy state manufacture of composite sheet. (Courtesy of Prof. M. Kiuchi, Tokyo University)

SCHEY, J. A., *Metalworking Tribology: Friction, Lubrication and Wear*, American Society for Metals, Metals Park, OH (1983)

Rolling

JAVORONKOV, V. A. and CHATURVEDI, R. C., *Rolling of Metals*, Yantrik, Bombay (1981)

LARKE, E. C., *The Rolling of Strip, Sheet and Plate*, Chapman & Hall, London (1963)

ROBERTS, W. L., *Cold Rolling of Steel*, Marcel Dekker, New York (1978)

ROBERTS, W. L., *Hot Rolling of Steel*, Marcel Dekker, New York (1983)

STARLING, C. W., *The Theory and Practice of Flat Rolling*, University of London Press, London (1962)

WUSATOWSKI, Z., *Fundamentals of Rolling*, Pergamon Press, New York (1969)

Forging

ALTAN, T., OH, S. and GEGEL, A., *Metal Forming: Fundamentals and Applications*, American Society of Metals, Metals Park, OH (1983)

FELDMAN, H. D., *Cold Forging of Steel*, Hutchinson, London (1961)

SABROFF, A. M. *et al.*, *Forging Materials and Practices*, Reinhold, New York (1968)

WATKINS, M. T., *Metal Forming I, Forging and Related Processes*, Oxford University Press, New York (1975)

Extrusion

ALEXANDER, J. M. and LENGYEL, B., *Hydrostatic Extrusion*, London (1971)

JOHNSON, W. and KUDO, H., *The Mechanics of Metal Extrusion*, Manchester University Press, Manchester (1962)

LANE, K. and STENGER, H., *Extrusion—Processes, Machining, Tooling*, American Society of Metals, Metals Park, OH (1981)

PEARSON, C. E. and PARKINS, R. N., *The Extrusion of Metals*, Chapman & Hall, London (1960)

Drawing

BERNHOEFT, C. P., *The Fundamentals of Wire Drawing*, The Wire Industry Ltd., London (1962)

CAMERON, A., *Principles of Hydrodynamic Lubrication*, Longman, Harlow (1966)

IRBRAHIM, I. N. and SANSOME, D. H., *An Experimental Study of the Mechanics of Ultrasonic Tube Bending*, Conf. Ultrasonics International, Canada (1983)

SANSOME, D. H., 'Ultrasonic tube drawing', *Journal of Tube International*, 219 (Dec. 1985)

TASSI, O. J., *Non-Ferrous Wire Handbook*, Wire Association Int., Branford, CN (1977, 1981)

Sheet forming

GRAINGER, J. A., *Flow Turning of Metals*, The Machinery Publishing Co., Brighton (1969)

KOSTINEN, D. P. and WANG, N.-M., *Mechanics of Sheet Metal Forming*, Plenum Press, New York (1978)

WATKINS, M. T., *Metal Forming II: Dressing and Related Processes*, Oxford University Press, New York (1975)

WATKINS, M. T., *Source Book on Forming of Steel Sheet*, American Society for Metals, Metals Park, OH (1976)

WATKINS, M. T., *Developments in the Drawing of Metals*, Metals Society, London (1983)

High-energy-rate forming

BLAZYNSKI, T. Z., *High Energy-Rate Fabrication*, Int. Conf. Proceedings, University of Leeds, Leeds (1981)

BLAZYNSKI, T. Z., *Explosive Welding, Forming and Compaction*, Applied Science, London (1983)

BLAZYNSKI, T. Z., *Materials at High Strain-Rates*, Elsevier/Applied Science, London (1987)

CROSSLAND, B., *Explosive Welding of Metals and its Application*, Clarendon Press, Oxford (1982)

SCHROEDER, J. W., and BERMAN, I., *High Energy Rate-Fabrication '84*, American Society of Mechanical Engineers, New York (1984)

5

Large-chip Metal Removal

D B Richardson

Contents

5.1 Large-chip processes

All the large-chip processes use cutting tools of defined geometry which are applied in a controlled manner to remove metal at a predetermined rate. The processes could be classified in many ways, but it is convenient to consider them in terms of the kinematics of the machine tools. With this in mind they have been separated into four main machine groups:

(1) turning (rotating work),
(2) shaping (reciprocating tool or work),
(3) milling (rotating tool), and
(4) drilling and boring (rotating tool).

Turning machines embrace the wide variety of lathes and vertical boring machines which can be controlled manually or automatically. Automatic control can be achieved using cams, sequential controllers, hydraulic copying devices or numerical programming. All machines in this group are capable of performing six basic operations, as shown in *Figure 5.1*. In addition, copying lathes and numerically controlled lathes can generate non-parallel forms by traversing the tool simultaneously in two planes.

Most turning processes use tools with a single cutting edge, where the cutting action is characterised by a relatively uniform section of material being presented to the cutting zone, resulting in a continuous chip when cutting ductile materials, or a repetitive form of short discontinuous chips when cutting brittle materials. Although the production of continuous chips indicates an efficient cutting action, the chip streamer itself presents disposal problems, frequently wrapping itself round the work piece, the cutter, or parts of the machine tool, creating a hazard to both the process and the operator. Chip breakers are extensively used to induce continuous chips to break into short lengths which are relatively safe and can be easily disposed of. These push against the underside of the chip and cause it to curl into a tight spiral, the free end of which strikes against the tool, and the resulting

bending stress causes fracture. The earliest form of chip breaker, still extensively used with flat-top tools, consists of a hard wedge-shaped block of sintered carbide clamped to the rake face of the tool about 2 or 3 mm from the cutting edge (*Figure 5.2(a)*). The introduction of disposable sintered carbide inserts has allowed more complicated rake-face geometries to be used which act as built-in chip breakers (*Figure 5.2(b)*). Effective chip breaking is largely a matter of trial and error, being influenced by the feed, tool bluntness and cutting speed, as well as by the material being machined.

The development of new and improved cutting materials has enabled a 100-fold reduction in cutting time to be achieved since the beginning of the twentieth century. Unfortunately, the reduction in idle time—caused by the need for tool adjustment—and in the tool approach and retraction times before and after cutting has not been of a comparable order. In achieving lower production costs the emphasis has now rightly moved away from further reducing cutting time to attacking the disproportionately large amount of non-cutting time.

Disposable sintered cutting inserts are made to a high level of precision which allows them to be indexed or replaced in tool holders in a few seconds, usually without the need for sizing cuts. When all the cutting edges are worn the inserts are discarded, obviating the need for time-consuming regrinding which is common practice when using high speed steel tools. When resetting lathes between work batches, a substantial time saving can now be achieved by using preset tools mounted in holders which can be replaced as cartridges in the tool post. No doubt further improvements will be achieved in the development of cutting-tool materials, but any further reduction in manufacturing time achieved in the future will be determined mainly by reducing idle time, both by better tool-changing mechanisms and by improved machine tool design to facilitate chip disposal and to reduce the tool approach and retraction times.

The shaping group of machine tools produce chips by a relative linear motion between the cutting tool and the work. This group includes shaping machines, planing machines and slotting machines, all of which are used mainly for tool manufacture or maintenance work and have little application in modern production. They operate on a reciprocating principle, cutting on the forward stroke and idling on the return stroke. Although they have quick return mechanisms the cutting time is only of the order of half the reciprocating cycle time. Swarf disposal is usually no great problem due to the intermittent nature of the cut. *Figure 5.3* shows typical configurations of these three machine types, which have changed little in recent years.

Other machines in the shaping group are gear shapers and gear planers, outlines of which are shown in *Figure 5.4*. In gear

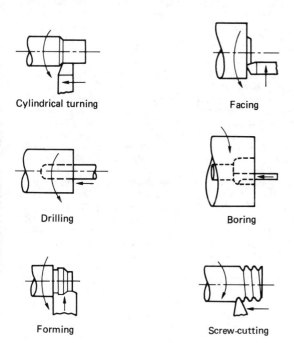

Figure 5.1 Basic lathe operations

Cylindrical turning

Facing

Drilling

Boring

Forming

Screw-cutting

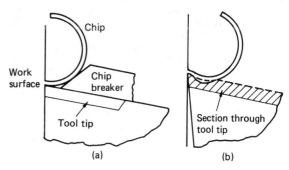

Figure 5.2 Chip breakers: (a) clamped; (b) built-in

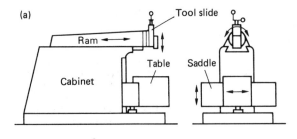

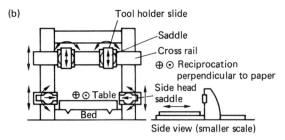

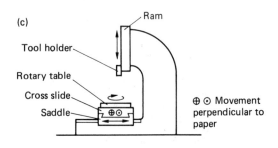

Figure 5.3 Outlines of reciprocating machines: (a) shaping machine; (b) planing machine; (c) slotting machine

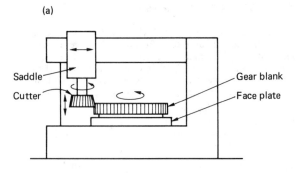

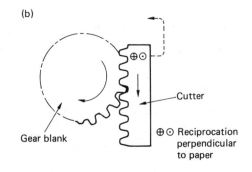

Figure 5.4 Kinematics of gear generating machines: (a) gear shaper; (b) gear planer

shaping the cutter resembles a side-relieved spur gear, the involute profile being generated by rapid reciprocation of the cutter whilst slowly revolving the cutter and gear blank in synchronism. Helical gears can be generated using a cutter with helical teeth and applying an appropriate helical motion to the spindles. Gear shaping is used for producing gears when hobbing would be impossible due, for instance, to a turned shoulder close to the involute profile. Gear planers have little modern use, particularly in a production environment. The cutter is in the form of a straight tooth rack, suitably relieved, and the gear is generated by reciprocating the cutter and moving the gear blank and the cutter at a constant speed. To enable a short rack to be used the cutter can be removed from the cut and indexed back at intervals.

Broaching machines also belong to the shaping group, but these produce the required form in a single pass. Internal broaching is for opening circular holes to produce non-circular forms. The cutter is a broach which has a number of cutting edges along its length and which is usually drawn, but sometimes pushed, through the hole by means of hydraulic pressure. Each cutting edge is larger than its predecessor by about 0.05–0.08 mm, so that the number of cutting teeth is determined by the form to be produced. Push broaching is limited to broaches with a small number of teeth which have a low length/cross-section ratio and which would not buckle under compression. Surface broaching is a more recently introduced technique, and is used as an alternative to milling for the production of external surfaces. Surface broaches are rigidly clamped to a machine slide and traversed against the component being machined, producing a surface in a shorter time than is required for milling, and usually giving a superior finish. Whereas internal broaching is usually the only feasible method for producing the desired shapes, surface broaching is an alternative to milling and is usually justified only if the quantities required are sufficient to offset the high equipment and tooling costs.

The milling group comprises a large range of manually operated or numerically controlled machines, many of which can perform operations such as drilling, reaming and boring as well as the accepted milling operations. Milling cutters generate surfaces either by means of cutting edges on the periphery or the face of the cutter. Peripheral milling is now seldom used for generating large plane surfaces, its main use being for machining slots or profiles. Although peripheral cutters can be fitted with carbide cutting edges the majority are of high speed steel and, except when used for machining the more exotic hard materials, will probably continue to be so in the foreseeable future. Frequently, such cutters also have shallow teeth on the cutter face, although these teeth usually contribute little to the total metal removed. A range of typical peripheral cutters is shown in *Figure 5.5*. Face milling cutters are essentially for generating plane surfaces. They are fastened to the end of stub arbors in the machine spindle and their configuration makes them suited to the use of specially designed carbide inserts (*Figure 5.6*). The cutting edges of both peripheral and face milling cutters are in contact with the uncut part of the component for, at most, half a revolution. Since the chip length is of the order of one-third of the length

Figure 5.5 Peripheral cutters: A, high radial rake cutter; B, helical cutter; C, side and face cutter; D, end milling cutter; E, slot drill

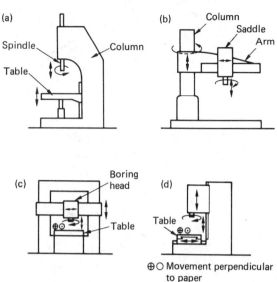

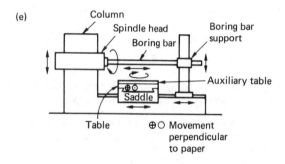

Figure 5.7 Machines in the drilling and boring group: (a) single spindle drill press; (b) radial drilling machine; (c) gantry type jig boring machine; (d) open front jig boring machine; (e) horizontal boring machine

Figure 5.6 Face milling cutters: A, zero corner angle; B, 15° corner angle

of the uncut surface, the chips have a maximum length approximating to the cutter radius, so chip breaking poses few problems.

Gear hobbing machines also belong to the milling group. Hobs are in the form of a screw with a straight side rack-form thread, gashed to give cutting edges and relieved to provide cutting clearance. Gears are generated by synchronously rotating the blank and the hob, and the hob is fed parallel to the axis of the arbor on which the blank is mounted. When a number of identical gears is required, several blanks can be fitted to the same arbor and machined at a single pass.

The machines in the drilling and boring group can be subdivided into drill presses, radial drilling machines, jig boring machines and horizontal boring machines. They have rotating spindles which hold drills, stepped cutters, taps, reamers or single-point boring tools, and the cut is applied by feeding either the spindle or the work table. Typical configurations of cutters in this group are shown in *Figure 5.7*.

Increased metal-removal rates, made possible by the development of new cutting materials, have forced machine-tool manufacturers to design new machines capable of large ranges of spindle speeds and feeds. This has necessitated more powerful motors and structures having high rigidity to resist the increased cutting forces and to reduce the likelihood of self-induced vibrations, giving rise to chatter. With increased rates of metal removal, the problem of swarf disposal has become more acute. The magnitude of this problem can be realised by considering that swarf occupies about 100 times the volume of the metal removed, so a 10 kW motor running at full power can generate about 1 m^3 of mild steel swarf or about 3 m^3 of aluminium swarf per hour.

Modern high production metal cutting machines commonly cost in excess of £100 000. If they are amortised over a period of 5 years the depreciation cost of such a machine when used continuously on a double-shift basis is more than £5 per hour. It follows that such plant requires high utilisation, efficient programming to produce at optimal metal-removal rates, an effective system of tool management to reduce the non-cutting time and intelligent application of terotechnology to minimise lost time due to maintenance.

Where large-scale production justifies continuous or large-batch manufacture, the achievement of these objectives becomes a feasible possibility. Unfortunately, few products are marketed in such large quantities, and a large-batch approach usually results in uneconomically high stocks. In the 1970s, the British manufacturing industry became notorious for its in-

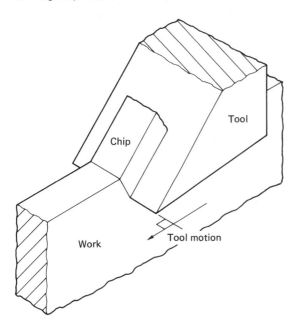

Figure 5.8 Orthogonal machining

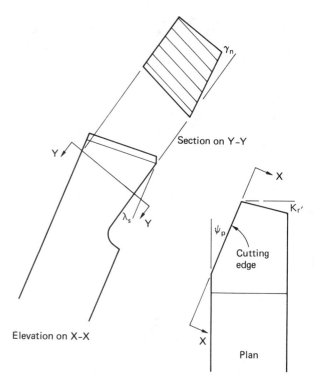

Figure 5.9 Angles in the normal rake system

flexibility and for its disproportionately high stocks, which resulted from the pursuit of large-batch policies. Present policies are directed towards small batches and this manufacturing philosophy has highlighted the need for rapid change-overs and for manufacturing systems accommodating large numbers of tools which can be called into use in response to the demands of small batches.

5.2 Cutting-tool geometry

Most of the research into chip formation has been based on orthogonal cutting, a simplified situation which is seldom met with in practice. The tool approaches the work with its cutting edge parallel to the uncut work surface and at right angles to the direction of cutting. To prevent end effects, the tool is wider than the work, as shown in *Figure 5.8*. Orthogonal cutting can be achieved only in a planing or shaping operation, although a close approximation can be obtained when turning on the end of a thin-walled tube.

In practice, the cutting tools usually approach the work obliquely and have rake angles in both directions on the rake face, together with a nose radius at the end of the cutting edge. The direction in which the chip flows across the tool surface is determined by this complicated geometry. BS 1296: 1972 defines the angles on single-point cutting tools in terms of the normal rake system (*Figure 5.9*), based on two coordinate rake angles.[1] The back rake or cutting edge inclination λ_s is measured parallel to the cutting edge in the vertical plane and the normal rake γ_n is measured in a plane at right angles to the cutting edge and perpendicular to the rake face. ψ_r is the tool approach angle and $\kappa_{r'}$ is the horizontal clearance angle, or the tool minor cutting edge angle. In addition, the tool is relieved to give vertical clearance angles of about 5°.

Other systems of tool nomenclature relate the rake angles to the coordinate axes of the tool shank, or to the cutting edge, measuring the angles in each case in the vertical plane. Although these systems are conceptually simpler, they are of little use in deducing the direction of chip flow. The British Standard relates to single-point tools but it can be applied also to multi-point tools and is generally preferable to the other systems.

Experimental work has shown that the direction in which the chip passes across the rake face of the tool can be expressed using Stabler's law.[2] This is best understood by reference to *Figure 5.10*, which shows an oblique shaping operation. The law simply states that $\gamma = \lambda_s$, where λ_s is the cutting-edge inclination and γ is the angle measured on the rake face between the normal to the cutting edge and the direction of chip flow. In this example the tool approach angle is zero, but the result is equally applicable for turning or face milling cutters where the tool approach angle is generally not zero.

The chip geometry and the principal cutting force are determined by a rake angle measured in the direction of chip flow. This is known as the 'effective rake' and is shown as β_e in *Figure 5.11*, which relates to a lathe tool, where

$$\sin \beta_e = CE/AE = (BF + CD)/AE$$

$$= BG/AE \cos \lambda_s + BD \sin \lambda_s/AE$$

$$= BG \cos \gamma/AG \cos \lambda_s + (EG - FG) \sin \lambda_s/AE$$

$$= \sin \gamma_n \cos \gamma/\cos \lambda_s + \sin \gamma \sin \lambda_s$$

$$\quad - BG \tan \lambda_s \sin \lambda_s \cos \gamma/AG$$

$$= \sin \gamma_n \cos \gamma/\cos \lambda_s + \sin \gamma \sin \lambda_s$$

$$\quad - \sin \gamma_n \tan \lambda_s \sin \lambda_s \cos \gamma$$

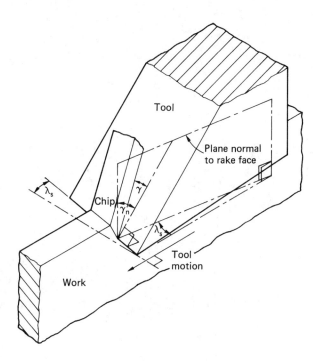

Figure 5.10 Oblique machining

Substituting $\gamma = \lambda_s$ (Stabler's law),

$$\sin \beta_e = \sin \gamma_n + \sin^2 \lambda_s - \sin \gamma_n \sin^2 \lambda_s$$
$$= \cos^2 \lambda_s \sin \gamma_n + \sin^2 \lambda_s \qquad (5.1)$$

Figure 5.12 shows how β_e varies with changes in cutting-edge inclination and normal rake.

If the normal rake is not specified it can be calculated from the vertical rake angles coordinate to the cutting edge, as shown in *Figure 5.13*, from which it is seen that

$$\tan \gamma_n = BG/AB = BG/BF \cdot BF/AB$$
$$= \tan \beta' \cos \lambda_s \qquad (5.2)$$

Figure 5.14 shows graphically how the normal rake is affected by variation in cutting-edge inclination and side rake.

The angles specified for a roller milling cutter are the radial rake and the helix angle (*Figure 5.15*), where the helix angle is the cutting edge inclination λ_s and the radial rake is β'.

Face milling cutters usually have a small nose radius at the end of the cutting edge, which helps to reduce the roughness of the machined surface. The cutting edge is usually inclined to the direction of travel by an angle θ, known as the corner angle, which is analogous to the tool-approach angle in *Figure 5.9*. This angle effectively increases the length of the cutting edge and reduces the uncut chip thickness t (*Figure 5.16*). By distributing the cut along a larger length, tool life can be increased. However, large corner angles can give rise to self-induced vibration, and they are normally restricted to values of $\leqslant 45°$.

The basic rake angles on face milling cutters are defined in the plane of rotation as the radial rake r_r and the axial rake r_a. To obtain the cutting edge inclination and the normal rake the following transformations (based on *Figure 5.17*) are necessary:

$$\tan \lambda_s = (CC' - BB')/B'C'$$
$$= (AC' \tan r_a - AB' \tan r_r)/B'C'$$
$$= \cos \theta \tan r_a - \sin \theta \tan r_r \qquad (5.3)$$

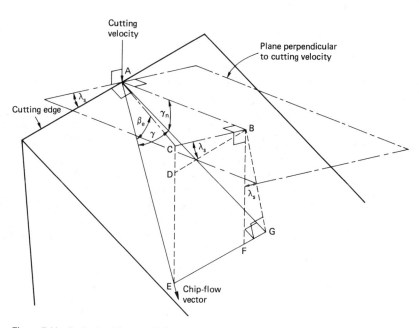

Figure 5.11 Angles in oblique machining

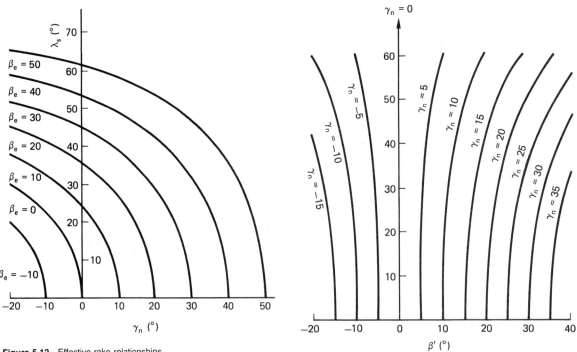

Figure 5.12   Effective rake relationships

Figure 5.14   Rake-angle relationships for oblique cutting

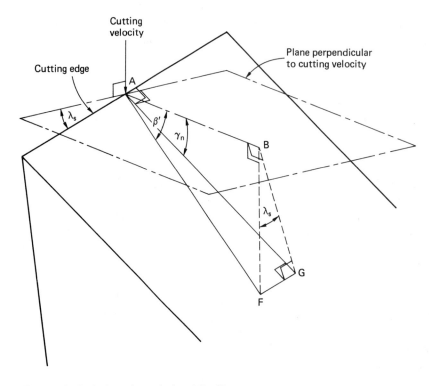

Figure 5.13   Back rake and normal rake relationship

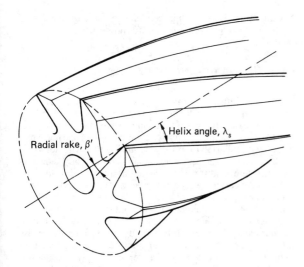

Figure 5.15   Rake angles on a helical milling cutter

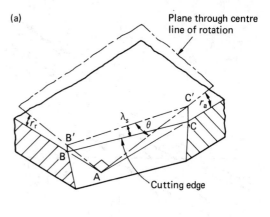

(a)

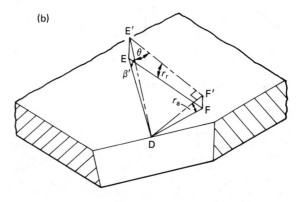

(b)

The side rake β' is evaluated by reference to *Figure 5.17(b)* as follows:

$$\tan \beta' = EE'/DE'$$

$$= (DF' \tan r_a + EF' \tan r_r)/DE'$$

$$= \sin \theta \tan r_a + \cos \theta \tan r_r \qquad (5.4)$$

The normal rake γ_n is evaluated from equation (5.2):

$$\tan \gamma_n = \tan \beta' \cos \lambda_s$$

and the effective rake angle is evaluated from equation (5.1):

$$\sin \beta_e = \cos^2 \lambda_s \sin \gamma_n + \sin^2 \lambda_s$$

Figure 5.17   Geometry of a face milling cutter

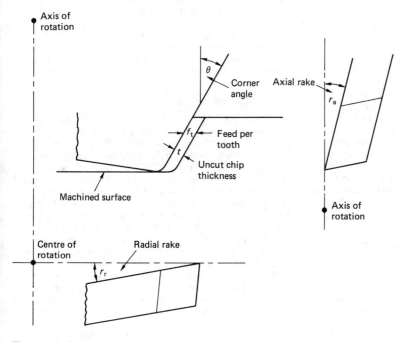

Figure 5.16   Basic angles on a face milling cutter

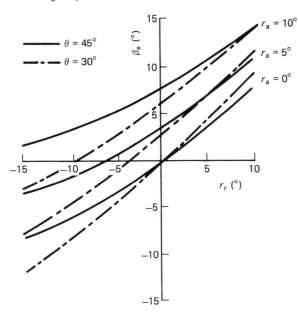

Figure 5.18 Cutting-angle relationships on a face milling cutter

Figure 5.18 shows families of curves for cutters having corner angles of 30° and 45°, where β_e is plotted against r_r for three values of r_a. It can be seen that the benefits of negative rake angles can be obtained by selecting suitable combinations of angles whilst retaining the lower axial thrust associated with positive rakes. In addition, as demonstrated in Section 5.7, a large corner angle gives a longer tool life.

5.3 Cutting-tool materials

The range of available cutting-tool materials has increased rapidly in recent years due to the development of more difficult to machine materials and to the demand for higher productivity. Any attempt to summarise these materials is unlikely to be completely successful due to the rate at which improvements and innovations are occurring. However, this is no excuse for ignoring the published state of the art at the time of writing.

When selecting a tool material for a particular application, it is necessary to measure its rating against the following list of properties, some of which are mutually opposed. For instance, in most cases hardness and impact resistance of competing tools tend to be inversely related. The essential properties are:

(1) high hardness at elevated temperatures,
(2) high compressive strength,
(3) adequate impact resistance,
(4) insusceptibility to large variation in local temperature,
(5) chemical inertness at working temperatures,
(6) low interface friction, and
(7) good abrasion resistance.

Temperatures at the chip–tool interface can be about 1000°C when machining steel, and can be considerably higher when machining some of the more exotic materials, particularly if heat is applied to help soften the work material. As the cutting temperature is largely dependent on cutting speed, it follows

that the pursuit of higher productivity creates increasing demands on the high-temperature properties of the tool materials.

5.3.1 High speed steel

High speed steels are likely to continue to be used in the foreseeable future for many applications such as drilling, reaming, tapping and dieing, forming, broaching and milling due to the ease with which they can be shaped in both the soft and hardened state. Typically, they consist of carbon steel alloyed with tungsten or molybdenum, together with additions of chromium, vanadium and cobalt. The alloying elements raise the temperature at which tempering occurs, allowing high speed steels to be used at temperature up to about 650°C. Their hardness is limited to 750 H_V (Vicker's hardness number), which is adequate for machining most of the common metals, including alloy steels in their unhardened forms. Cutting speeds are necessarily limited to prevent excessive rise in temperature. When machining mild steel, cutting speeds of about 1.5 m s^{-1} are possible if a plentiful supply of coolant is provided. A recent development is the coating of high speed steel drills with a deposit of about 3–5 μm of titanium nitride which allows rotational speeds to be increased, resulting in a 50% increase in penetration rate and longer tool life. With a few exceptions of this sort, it is unlikely that high speed steels will ever again pose a threat to the supremacy of sintered carbides for heavy-metal removal.

Despite their relatively low hardness and susceptibility to softening at high temperatures, high speed steel tools are tougher than most of the competing materials, enabling them to be used for interrupted cuts without fear of fracture. They can also be reground, giving a number of cutting lives before they must be finally discarded.

5.3.2 Cast non-ferrous alloys

These alloys consist of cobalt, chromium, tungsten and carbon. Although less versatile than high speed steels, they enable cutting to be performed at higher temperatures. Their main use is for drilling, where their superior hardness at elevated temperatures is an advantage when the application of fluids is frequently a problem. Their cutting performance is generally superior to high speed steel, but inferior to that of sintered carbides, so their use is unlikely to grow in popularity.

5.3.3 Sintered carbides

The introduction of sintered carbides for cutting has been the most important single contribution to increased productivity during the past 50 years. Sintered carbides are essentially cermets, which consist of hard carbide ceramic particles embedded in a metal matrix. Early carbide tools were usually of tungsten carbide and cobalt. Their brittleness encouraged the use of negative rakes to promote compressive stresses and restricted their use to continuous cutting. Subsequent improvements in the sintering process, and the introduction of alternative ceramic and metallic components, has enabled a range of carbide tools to be produced which withstand the rigours of thermal and mechanical shock, making them suitable for interrupted cutting in turning and milling.

The first generation of carbide tools consisted of sintered tips brazed to steel shanks. This arrangement provided a fairly rigid cutting system and the tools could be reground when blunt. More recently, the brazed tip has been almost completely superseded by disposable tips, which are mechanically clamped into steel tool holders. These tips are polygonal, with three or more cutting edges which can be indexed when worn to expose new edges. The negative-rake types can be inverted to double the number of edges. They are a throw-away concept, regrinding being uneconomical. Their introduction has forced a reappraisal of metal cutting economics, making tool lives of the order of 15 min a desirable objective. Practical cutting speeds are about three times as great as could be achieved with high speed steel, 5 m s^{-1} being a typical maximum when machining mild steel.

Depending on their composition, the hardness of sintered carbides varies between 1200 and 1600 H_V, impact resistance being obtained at the cost of hardness. Thermal softening becomes a problem only if the cutting temperature is allowed to exceed 1000°C, enabling most cutting operations to be performed without the use of a coolant. Improvements in cutting performance are now obtained by applying a variety of vacuum deposited coatings to the surface of the carbide tools. Although the coatings are usually only 1.0–5.0 μm thick, when used singly or in combination they can improve tool life by a factor of about 2. Thus, a coating of titanium carbide reduces wear due to cratering on the rake face, aluminium oxide increases the surface hardness, and titanium nitride reduces rake face friction.

5.3.4 Ceramics

Sintered ceramic tools based on aluminium oxide (Al_2O_3) have been available for more than 30 years. Their brittleness and poor thermal shock resistance can be improved by additions of zirconium oxide and titanium carbide, but they are generally unsuited to interrupted cutting. Having hardness values in the range 1500–1800 H_V, which does not significantly decrease over the normal metal cutting temperature range, they are suitable for machining hardened steels and chilled cast irons. They are similarly suitable for heat-assisted machining of hard materials such as the nimonic alloys and Stellite. However, their brittleness has proved a severe limitation for general-purpose machining of steels, an area in which their high metal removal potential would have been an advantage.

Mixed ceramics, based on carboxides with dispersed titanium carbide, have achieved better impact resistance without significant loss of hardness. They can be used at high cutting speeds of the order of 15–20 m s^{-1} when operated at low feeds, making them suitable for finish turning and finish milling hard materials.

A recent addition to the range of commercially available ceramic cutting tools uses silicon nitride (Si_3N_4) with various levels of aluminium and oxygen substitution. The silicon nitride ceramics have good resistance to thermal and mechanical shock, enabling them to be used for discontinuous cutting. They can be operated at higher cutting speeds than carbides together with higher feeds than can other types of ceramics. Their main applications to date are for rough turning and rough milling grey cast iron and for turning nickel alloy steels. Due to their high thermal shock resistance they can be used to cut dry or with coolant, the latter method being inadvisable with other ceramics. Chemical reaction occurs at cutting temperatures, causing rapid wear when machining most steels. Thus, at present, there is little likelihood of silicon nitride supplanting carbides in this important area of manufacture.

5.3.5 Cubic boron nitride

Polycrystalline cubic boron nitride is another comparatively recently introduced addition to the range of metal-cutting materials. It has hardness considerably in excess of that of ceramic tools and retains its hardness at temperatures well in excess of 1000°C. The main application is for machining hard ferrous materials at very high cutting speeds, giving a surface finish comparable to that obtained by grinding. It has relatively good impact resistance, allowing it to be used for interrupted cutting. It is also of use for hot machining of refractory metals such as Stellite, where a 90% reduction in machining time compared with carbide has been claimed. Due to the high cost of cubic boron nitride tools, manufacturers do not claim great cost savings, but the time saving is very significant.

5.3.6 Diamonds

Diamond is the hardest material known to man and it has found a limited cutting application where this is an important attribute. Natural diamonds, brazed to steel holders, have been used for many years for producing fine finishes on copper and aluminium. Being monocrystalline, natural diamonds have planes of weakness which render them unsuitable for anything but fine finishing cuts.

The production of polycrystalline synthetic diamonds has extended the usefulness of diamonds by improving their impact resistance. Polycrystalline synthetic diamond cutting tools are now used extensively in machining abrasive aluminium–silicon alloys, fused silica, and reinforced plastics. They are chemically reactive at high temperatures, so they are of little use for machining ferrous materials.

5.3.7 Limitations imposed by machine tools

Industry has, until recently, been very reluctant to replace machine tools whilst they continue to perform the function for which they were purchased. Machine-tool manufacturers market machines which adequately utilise the cutting tools available at the time of purchase. Inevitably, with the rapid development of new cutting materials, the existing machine tools cease to provide a service which uses the cutting tools in an economical manner.

Self-induced vibration, giving rise to chatter, is undesirable in any cutting operation. It is particularly undesirable when using brittle cutting tools where catastrophic failure can become a very real possibility. The problem of self-induced vibration becomes more acute as metal removal rates increase at high cutting speeds. Ideally, resonant frequencies should be as high as possible, but this requires high structural stiffness and low mass. Unfortunately, dynamic stiffness tends to be directly related to mass, so a simple scale-related solution does little to reduce chatter. The solution, if it exists, lies in structural redesign to enhance stiffness without a proportional increase in mass.

Optimal cutting speeds using modern cutting tools require a large increase in the rotational speed of spindles and a corresponding increase in input power. Manufacturers of machine tools recognise this need, which is reflected in their latest designs, together with improved provision for handling the greater volumes of swarf which are produced.

5.4 Chip formation and cutting parameters

It is convenient to assume that a chip shears across a plane and that this is the sole mechanism of deformation in the cutting process (see *Figure 5.19*). Much useful research in metal cutting has been developed using this simple analogy. Unfortunately, it is a gross oversimplification due to two factors: the effects of temperature and strain rate on the yield behaviour of work materials. The shear plane analogy, literally interpreted, implies an infinite rate of shear strain at the shear plane, giving

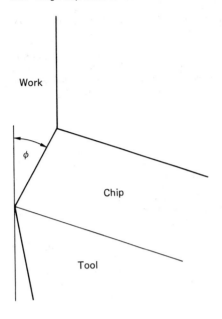

Figure 5.19 Ideal chip formation

rise to an instantaneous change in velocity and, presumably, an infinite force to effect it.

In practice, primary shearing occurs in a roughly triangular zone originating from the cutting edge. Friction effects along the initial length of the rake face cause a secondary deformation zone due to retardation of the underside of the chip (*Figure 5.20*). When the clearance flank of the tool wears, a further deformation zone may occur on the cut surface of the work. This zone is generally small and can be ignored, unless the surface integrity of the component is of particular impor-

tance. Measurement of the deformation of fine square grids scribed or photographed on the sides of the work piece has shown that, although the deformation zone departs dramatically from the shear plane concept, the very high strain rates are concentrated in a narrow band which follows closely the hypothetical shear plane.[3] In the following analysis of chip formation, the shear plane analogy is used extensively to demonstrate how the parameters of cutting speed and feed affect the chip root configuration.

A further complication affecting the understanding of chip formation is the existence of a built-up nose, usually referred to as a 'built-up edge', on the underside of the chip adjacent to the cutting edge. This results from tool bluntness and rake face friction, and its formation can be visualised as follows. Assume that the tool has a slight wear radius along the cutting edge. A particle of work material at the bottom extremity of the shear plane reaches this radius which acts as a stagnation point, the particle being constrained equally to join the cut work surface and the underside of the chip. Succeeding particles, impinging on the first, become stationary and a built-up nose is initiated. Adjacent particles slightly higher on the shear plane do not stagnate, but proceed along the rake face where the high compressive stress causes a friction force of sufficient intensity to initiate sticking friction. In this way a velocity gradient is created, varying from zero at the rake face to the full chip velocity at a line located part way into the chip. Eventually, an equilibrium is achieved where the effect of the temperature gradient produces a shear flow stress large enough to prevent further secondary deformation. This determines the boundary of a persistent, if unstable, built-up edge (*Figure 5.21*). The front of the built-up nose is constantly breaking away and being replaced, the debris being transported by the cut surface, producing a characteristically poor finish. Elevated temperatures at the interface between the tool and the built-up edge can cause diffusion welding to occur, leading to tensile failure of the surface on carbide tools.

Obviously-built-up edge is an undesirable phenomenon which should be avoided if possible. It occurs commonly when using high speed steel or carbide tools, but causes little problem with ceramics. Although the reason for this is not fully known, it is probable that the very low thermal conductivity of ceramics causes a high interface temperature and a

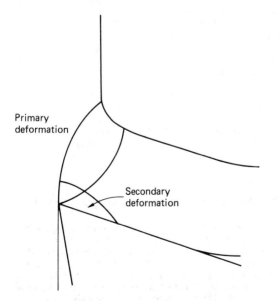

Figure 5.20 Actual chip formation

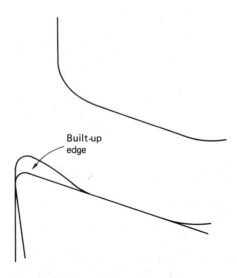

Figure 5.21 Typical built-up edge formation

consequent reduction in shear flow stress on the underside of the chip which, in turn, localises the secondary deformation. When cutting steel at impractical cutting speeds below about 0.050 m s^{-1}, there is no significant built-up edge. Such an edge assumes troublesome proportions at cutting speeds between about 0.2 and 1.2 m s^{-1} and then ceases to be significant at cutting speeds above about 2.0 m s^{-1}. The effect can be observed by performing cutting tests at increasing speed, where surface roughness improves progressively prior to a dramatic improvement at speeds in excess of about 2.5 m s^{-1}. Cutting speeds of this order are economically possible when cutting mild steel with carbides, but may not be economical when cutting some high carbon and alloy steels.

Some aluminium alloys promote a built-up edge which persists at very high cutting speeds. This accounts for the fact that, although these alloys are easily machined, it is frequently extremely difficult to produce a good surface on them.

5.4.1 Influence of cutting parameters

Heat generated in the primary shear zone is mainly carried away by the chip. However, some can be back-conducted into the work piece. This raises the temperature ahead of the shear plane and reduces the shear flow stress of the incoming material. In turn, the shearing action occurs at an earlier point, causing the shear plane to rotate and reducing the shear angle ϕ in *Figure 5.19*.

Since heat conduction is time dependent, the amount of heat conducted to the body of the work material at low cutting speeds is likely to be greater than at high speeds, even allowing for the greater total amount of heat generated at higher speeds. This belief is consistent with the observed fact that a surface machined at high speeds is usually cooler to the touch than one machined at low speeds. Furthermore, chips produced at high speeds are noticeably thinner than those produced at low speeds, a result of shear plane rotation. The principal cutting force decreases as cutting speed increases, due to a shorter shear plane, although this effect is frequently obscured by the effect due to a built-up edge.

When cutting with low feeds the shear angle is small, but it increases as the feed increases. This, no doubt, is also attributable to the effect of heat conduction, although the argument is more tenuous than for speed variation.

Intuitively, the maximum shear angle could be half the angle between the work surface and the rake face. In practice, it is always appreciably smaller. Sintered tools are operated over a small range of rake angles, varying between about +10° and −10°. When using negative rakes the shear angle is correspondingly reduced, giving very low angles when cutting at low speeds. However, negative rake cutting is essentially a high-speed operation, and at high cutting speeds the shear angle is not much less than that using positive rakes.

If the shear-angle concept is modified to assume that primary deformation occurs across a narrow band (*Figure 5.22*), the primary shear strain γ depends on the rake angle β_e and the shear angle ϕ. A particle entering the shear zone at O changes direction under the influence of shear and leaves the shear zone at B:

$$\gamma = AB/OC = (AC + BC)/OC$$

$$= \cot \phi + \tan (\phi - \beta_e) \qquad (5.5)$$

It can be seen that primary shear strain, and consequently the energy of primary deformation, are reduced if the effective rake angle β_e and the shear angle ϕ are large.

The ratio of the uncut chip thickness t divided by the actual chip thickness t' is generally denoted by r_t (*Figure 5.23*),

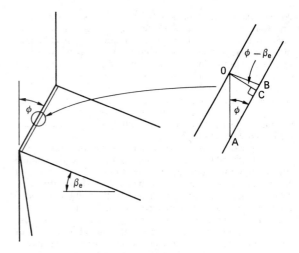

Figure 5.22 Rectangular shear zone model

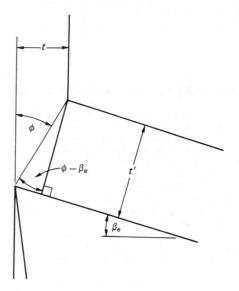

Figure 5.23 Chip thickness relationship

where it is easily deduced that:

$$\tan \phi = (1 - r_t \sin \beta_e)/(r_t \cos \beta_e) \qquad (5.6)$$

High values of r_t, i.e. thin chips, are sometimes taken as an indication of efficient cutting, giving rise to reduced primary shear strain and, consequently, to lower cutting forces.

5.4.2 Cutting fluids

Cutting fluids are used for three main purposes: as a lubricant at low cutting speeds, to cool the tool and work, and to assist in clearing the swarf.

At cutting speeds in excess of about 0.7 m s^{-1}, there is little noticeable lubricating effect. Below this speed extreme pressure (EP) mineral oils containing sulphur or chlorine additives can be used to reduce friction in the latter stages of chip–tool contact due to the formation of low-shear-strength sulphides or chlorides. The balance of the cutting forces is affected, giving rise to a larger shear angle and reduced contact length, and encouraging the thinner chip to curl, making for more effective chip breaking. High compressive stresses near the tool point prevent lubricant penetrating in this area, so the lubricating effect is limited to the latter part of the chip–tool interface where sliding friction occurs. The chemical reaction giving rise to low-strength compounds is both temperature and time dependent. Hence, at higher cutting speeds the lubricant rapidly loses its effectiveness.

High speed steel cutting tools start to soften at temperatures above about 650°C. When using these materials, the cooling effect of cutting fluids enables the tools to be operated at higher cutting speeds than would be possible when cutting dry. Water soluble oils, having high specific heat and good metal wetting properties, are better coolants than the mineral oils when used as lubricants. With carbides and ceramics the poor resistance to thermal shock makes the use of cutting fluids inadvisable, except in special circumstances. Fortunately, these materials can be used satisfactorily at high temperatures, and coolants are therefore not usually required.

The purely mechanical function of using cutting fluid to assist swarf disposal is sometimes of prime importance. An example of this application is in deep-hole drilling where cutting fluid is pumped to the cutting edges at high pressures of about 6 N mm^{-2}.

5.5 Forces and power in metal cutting

Most lathes and milling machines lack the power to exploit the cutting tools in an economical manner. This shortcoming is usually aggravated by a natural trepidation on the part of operators to run machine tools near their power limits, for fear of stalling the drive motor. Surprisingly, few machine tools are fitted with wattmeters, so the operator usually has no idea how near he/she is to causing an overload.

Among the more sophisticated numerically controlled machines, very few are fitted with adaptive control devices which cause feed or cutting speed to respond to excessive power demands. The vast majority operate from a predetermined programme which has been based on safety considerations, where the power requirements are well within the rated output of the motors.

There is an ill-founded belief that the cutting forces, and hence the power required, increase significantly as the tools wear, increases of 40% sometimes being quoted. In fact, cutting power seldom increases by more than 10% over the life of the tools.

At rated power, transmission losses usually account for about 30% of the input power, with a correspondingly greater percentage loss when operating at lower energy levels. Transmission losses are higher when the machine tool is cold, and drop significantly over the first half hour of operation. It is desirable, therefore, to record the transmission power over the full range of cutting speeds and feeds on a machine tool in both the cold and warmed-up conditions. Only then is it possible to know the available power which can be used for cutting.

5.5.1 Forces in turning processes

Although a knowledge of cutting forces is desirable to prevent excessive structural loads, the main reason for requiring such knowledge is to use it as a basis for estimating power. The power in watts is simply the product of the peripheral speed in metres per second and the tangential cutting force measured in newtons.

Due to the formation of built-up edge when cutting steel at low speeds, forces on the tool vary in an unpredictable manner, but above about 2.5 m s^{-1} become relatively constant when built-up edge ceases to have a significant effect. At cutting speeds below about 0.7 m s^{-1}, the lubricating effect of cutting fluids can do much to inhibit build up, and on the rare occasions when such low speeds are used the cutting force can often be reduced by this means. When using sintered tools it is usually possible to operate at cutting speeds high enough for the forces to be considered constant.

As would be expected, the effective rake has a significant effect on cutting force but, as mentioned in the previous section, the shear angle disparity between positive and negative rake tools becomes less at high cutting speeds. *Figure 5.24* shows approximately how the principal cutting force varies with effective rake at both high and low cutting speeds. High speed steel tools typically operate at the lower speed, where the cutting force increases by about 90% as the rake decreases from +30° to zero. Carbide tools, which operate typically at the higher speed, experience a rise in cutting force of only about 15% across the range of rakes from +10° to −10°.

This would seem to indicate a decided advantage in using high rakes when cutting at low speeds. However, tool-life considerations militate against high rakes when machining hard materials, because the restricted heat conduction path from the cutting edge leads to high interface temperatures. Typical rake angles for high speed steel tools are shown in *Table 5.1*.

As discussed previously, the shear angle increases as the uncut chip thickness increases, so that the shear plane length does not increase linearly with feed. This causes the cutting force to be related non-linearly to feed f, or to uncut chip

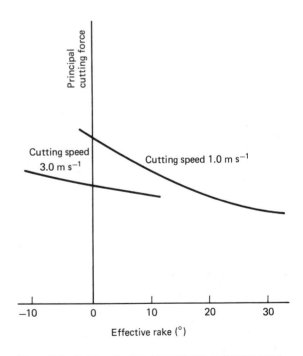

Figure 5.24 Variation of cutting force with rake and cutting speed

Table 5.1 Typical rake angles for high speed steel tools

Work material	Rake angle (°)
Aluminium and soft alloys	30–40
Free cutting steels	20–25
Medium carbon steels	10–15
Hard alloy steels, hard cast iron	0–5

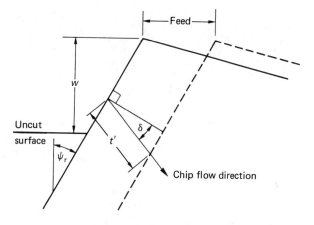

Figure 5.25 Plan geometry of a pointed tool

thickness t. When cutting orthogonally, where $f = t$,

$$F = Ct^x w \qquad (5.7)$$

where F is the principal cutting force (in newtons), C is a material constant at a given cutting speed and rake angle, t is the uncut chip thickness (in mm), w is the width of the cut (in mm) and x is a constant for the material being machined.

The exponent x varies slightly with rake angle, but for practical purposes can be assumed to be 0.85 when machining the whole range of steels and about 0.7 when machining cast irons. Average values of C when using positive or negative rake carbide tools at typical cutting speeds are given in *Table 5.2*.

In the more practical case of oblique cutting, equation (5.7) must be modified to allow for the tool approach angle and cutting edge inclination. The plan geometry of chip flow for a pointed tool, with an approach angle ψ_r and cutting edge inclination λ_s, is shown in *Figure 5.25*, where the angle δ is angle BAC in *Figure 5.11*.

The equation for cutting force becomes modified as

$$F = Ct'^x w' \qquad (5.8)$$

where t' is the uncut chip thickness in the direction of chip flow, and w' is the length of cutting edge in engagement.

It can be shown that,

$$\cos \delta = \cos \lambda_s \cos \gamma_n / \cos \beta_e$$

and

$$t' = f \cos \psi_r / \cos \delta$$
$$= f \cos \psi_r \cos \beta_e / (\cos \lambda_s \cos \gamma_n) \qquad (5.9)$$

where f is the feed per revolution, and ψ_r is the tool approach angle.

Allowing for the effect of tool approach angle and cutting edge inclination:

$$w' = w / (\cos \psi_r \cos \lambda_s) \qquad (5.10)$$

Expressed in terms of feed and width of cut, equation (5.8) becomes

$$F = Cf^x w (\sec^{1-x} \psi_r \cos^x \beta_e \sec^{1+x} \lambda_s \sec^x \gamma_n) \qquad (5.11)$$

Considering a practical range of values for the angles used in industrial practice, the force calculated from equation (5.11) seldom differs from that calculated from equation (5.7) by more than 6%. Allowing for the effect of tool wear, it would seem reasonable to multiply the result obtained from equation (5.7) by a factor of 1.2 to give a simple and slightly overvalued estimate of cutting force.

However, when applied to the peripheral milling operation, where helix angles exceeding 45° are not uncommon, the effect of obliquity significantly affects the cutting force and the cutter geometry cannot be ignored.

5.5.2 Forces and power in milling operations

Peripheral milling removes metal by means of teeth on the circumference of the cutter. It is seldom used to produce large flat surfaces, which are more effectively generated by face milling. Mostly, peripheral milling cutters are used for end milling slots or for producing slots or stepped surfaces by using one or more horizontally mounted side and face cutters or helical slab milling cutters.

End milling seldom requires the rated power of the drive motor. The limiting factor is usually the maximum recommended feed per cutting tooth which will prevent damage to the cutter. Horizontal peripheral milling, however, can be limited by the power of the drive motor, and it is useful to consider the way in which the cutting parameters affect the power required.

Table 5.2 Average values of C at typical cutting speeds for positive and negative rake carbide tools

Work material		Specific angle (°)		Specific power
		+10	−10	
Free cutting carbon steel	120HB	980	1080	2.6
	180HB	1190	1310	3.1
Carbon steels	125HB	1620	1780	4.2
	225HB	2240	2470	5.8
Nickel–chromium steels	125HB	1460	1610	3.8
	270HB	2100	2310	5.4
Nickel–molybdenum and	150HB	1600	1760	4.1
chromium–molybdenum steels	280HB	1960	2160	5.1
Chromium–vanadium steels	170HB	1820	2000	4.7
	190HB	2380	2620	6.2
Flake graphite cast iron	100HB	635	700	1.6
	263HB	1330	1470	3.5
Nodular cast iron	Annealed	1110	1220	2.9
	As cast	1240	1370	3.2

Although the changing chip section precludes the possibility of orthogonal milling, it is possible to consider the instantaneous situation when using a zero helix cutter as quasi-orthogonal. Due to the motion of the work the profile of the surface generated by a cutter tooth is not strictly circular, but there is not significant loss of accuracy in assuming it to be so. It then follows, from *Figure 5.26*, that,

$$\cos \theta_m = 1 - d/R \tag{5.12}$$

where d is the depth of cut (in mm), and R is the radius of the cutter (in mm).

Figure 5.27 shows the uncut chip geometry and considers the instantaneous uncut chip thickness when the cutting edge has rotated through an angle θ from the vertical,

$$t \simeq f_t \sin \theta$$

where f_t is the feed per cutting tooth (in mm) and t is the instantaneous uncut chip thickness (in mm).

From equation (5.7),

$$F = Ct^x w$$

and the work done in removing one chip is

$$W = RwC \int_0^{\theta_m} t^x d\theta/1000 \text{ watts}$$

$$\simeq RwCf_t^x \int_0^{\theta_m} (\sin \theta)^x d\theta/1000 \text{ watts}$$

Since θ is usually less than 1 rad,

$$W \simeq RwCf_t^x \theta_m^{1+x}/[1000(1 + x)] \text{ watts}$$

Substituting from equation (5.12),

$$\cos \theta_m \simeq 1 - \theta_m^2/2 = 1 - d/R$$

$$\theta \simeq (2D/r)^{1/2} \tag{5.13}$$

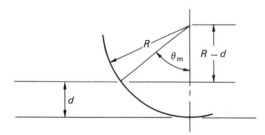

Figure 5.26 Approximate path swept by a cutter

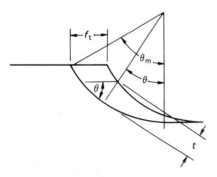

Figure 5.27 Variation of uncut chip thickness with angle of rotation

If the peripheral cutting speed is v m s^{-1} and s is the number of teeth on the cutter, the number of tooth impacts per second will be $1000vs/(2\pi R)$ and, substituting for θ_m, the cutting power with zero helix angle is,

$$P_0 = 2^{(1+x)/2} vsw \ Cf_t^x d^{(1+x)/2}/(2\pi(1 + x)R^{(1+x)/2}) \tag{5.14}$$

The volume of metal removed per second is,

$$V = 1000vf_t swd/(2\pi R) \text{ mm}^3$$

Hence, the specific cutting power is

$$P_s = 2^{(1+x)/2} \ CR^{(1-x)/2}/[1000(1 + x)d^{(1-x)/2} f_t^{1-x}] \tag{5.15}$$

Since d is predetermined, it follows that doubling f_t decreases the specific power by 5%, and halving R can reduce it by 5% when cutting steel, assuming $x = 0.85$.

When the cutter has a non-zero helix angle, the quasi-orthogonal model can no longer be applied. The angles by which a helical cutter is specified are the helix angle, which corresponds to the cutting edge inclination λ_s, and the radial rake which corresponds to β' in equation (5.2), from which the normal rake can be calculated.

Cutting power is now modified by including the trigonometric terms in equation (5.11), i.e.

$$P = P_0(\sec^{1-x} \psi_r \cos^x \beta_e \sec^{1+x} \lambda_s \sec^x \gamma_n)$$

For a cylindrical cutter, ψ_r is zero. Thus the formula simplifies to:

$$P = P_0(\cos^x \beta_e \sec^{1+x} \lambda_s \sec^x \gamma_n) = P_0A \tag{5.16}$$

Consider the three milling cutters of identical size and number of teeth, but with very dissimilar tooth geometry, listed in *Table 5.3*. These cutters are used to machine steel.

The product of the trigonometric terms A for the first two cutters is almost identical, whereas the helical cutter has a product about 60% greater. If the cutting speed is about 1 m s^{-1}, it can be seen from *Figure 5.24* that the cutting force constant C for the second and third cutters is approximately the same, but for the first cutter is about 50% greater. It follows that the cutting efficiencies of the helical and the side and face cutters are roughly similar, but the cutter with the high radial rake is almost 40% more efficient.

Peripheral cutting can be performed in either the upcut or climb mode (*Figure 5.28*). In upcut milling the cutting edge must penetrate the previously cut surface before chip generation commences. This causes a high radial force at the commencement of the cut which does not happen with climb milling. Cutters with large radial rakes have a weak tooth form which results in rapid wear when subjected to the high radial forces associated with upcut milling. Radial forces experienced in climb milling are much lower and wear is usually not a severe problem.

In summary, the most economical performance in terms of specific power can be achieved when climb milling with a high radial rake cutter operated at large feeds. It is preferable to operate at high feeds rather than high cutting speeds since the index of feed in the cutting power equation is less than unity, whereas power is directly proportional to cutting speed.

Table 5.3 Three milling cutters (see text)

	λ_s	β'	γ_n	β_e	A
Side and face cutter	0	10	10.0	10.0	1.0
High radial rake cutter	10	30	29.6	30.6	1.03
Helical cutter	45	10	7.1	34.2	1.63

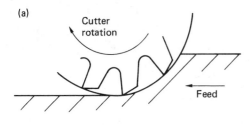

(a)

Cutter rotation

Feed

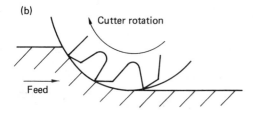

(b)

Cutter rotation

Feed

Figure 5.28 Climb and upcut milling: (a) climb; (b) upcut

The foregoing analysis provides useful indicators to the selection of cutting parameters, but it is unlikely that industry would adopt such a complicated method to calculate power requirements. It would be even more difficult to assess the power required for face milling where the work–cutter configuration can vary from total immersion of the cutter face to small arcs of contact in either the upcut or climb modes. A more simple method of power assessment is, therefore, required. This is provided by using the concept of specific energy which is the energy considered necessary to remove unit volume of material. Unfortunately, the influence of cutting speed and rake angle on cutting force, together with the non-linear effect of feed, makes the use of specific energy a very rough guide to cutting power assessment. It is obviously essential to over-predict if a value is to cover all contingencies, so the prediction must be based on very low feeds and negative rakes if it is to be universally applicable to a given work material. The values shown in *Table 5.2* give specific power in watts per cubic millimetre per second, based on the conservative assumption that feeds are of the order of 0.1 mm and that the milling process is about 60% as efficient as turning.

5.5.3 Hot machining

Some work materials pose machining problems which cannot readily be solved by conventional methods. These include the nimonic alloys and Stellite cast alloys. For example, some of the nimonic alloys when machined with carbide tools on a 100 mm diameter bar, necessitate cutting speeds as low as 0.15 m s^{-1} and tool failure commonly occurs after machining a 100 mm length of the bar. When the surface of the bar is preheated with a gas tungsten arc or a transferred plasma arc struck between the electrode and the work surface and using ceramic or cubic boron nitride tools, cutting speeds of about 2.5 m s^{-1} are possible, and the tools remain serviceable after machining a considerable length of bar.

This technique is not one which would be advised if alternatives are possible but, with some of the more refractory metals now in use, the hot machining process is frequently the only practical solution. The surface preheat temperature is about 600°C, giving such high interface temperatures that carbide tools cannot be used.

5.6 Surface-finish considerations

Built-up edge is one of the main factors contributing to poor surface finish. When machining most materials this can be reduced, if not eliminated, by operating at high cutting speeds. Where finish cuts are required, the uncut chip area is relatively small, so the cutting power is never likely to be an important consideration, even at very high cutting speeds. If ceramic tools are used the speed limitation is usually that imposed by the available spindle speeds, but when using carbides or high speed steels the speed constraint is usually that imposed by tool wear.

The theoretical surface roughness in turning is determined by tool plan geometry, a pointed tool operated at a given feed producing a rougher surface than one having a nose radius (*Figure 5.29*). In Britain, the most popular measure of surface roughness is the centre line average, which is the average deviation from the mean line drawn through the surface. For a pointed tool the theoretical surface roughness is given by

$$R_a = f/(4 \tan \psi_r + \cot \kappa_{r'}) \tag{5.17}$$

and for a tool of nose radius r

$$R_a = f^2/(31.2r) \tag{5.18}$$

In equation (5.18) the surface roughness is proportional to the square of the feed and inversely proportional to the nose radius. A large nose radius is obviously desirable but, since this is likely to induce chatter, most disposable tips are limited to a radius of about 1 mm.

Due to imperfect chip formation and surface roughness of the tool, the theoretical value is never achieved in practice. Rake angle and width of cut have very little effect on the value obtained. *Figure 5.30* shows how surface roughness varies with cutting speed and compares it with the theoretical value for a tool having a feed of 0.075 mm and a nose radius of 1 mm.

The surface generated by a peripheral milling cutter is directional in property. In the direction of feed the theoretical surface is geometrically similar to that for a turned surface, the cusps having a radius equal to that of the cutter, and the pitch between cusps being equal to the feed per cutting tooth. Due to the almost inevitable lack of straightness of the arbor on a horizontal milling machine, the contour generated by the cutter teeth varies as some teeth take a greater depth of cut than others. In severe cases, one tooth may take such a disproportionately deep cut that the surface generated has a periodicity corresponding to the feed per revolution rather than the feed per tooth, and the cusps are correspondingly deeper.

Face milling usually produces a finish superior to that generated by peripheral cutters. The geometry of the cutter is specially designed so that the combination of corner angle, end cutting edge angle and nose radius produce very flat cusps. In addition, due to the use of carbide cutters, cutting speeds are much higher than those achievable using high speed steel peripheral cutters.

5.7 Tool-life assessment

It is fortunate that, with few exceptions, tool wear occurs in a predictable manner. Although it takes different forms, each form of tool wear is associated with a known cause which happens within particular ranges of feed and cutting speed. The production engineer's initial task in optimising the metal cutting process is to decide which form of wear is to be accepted, and then to work within the constraints which such wear imposes. Before pursuing this line of thought further, it

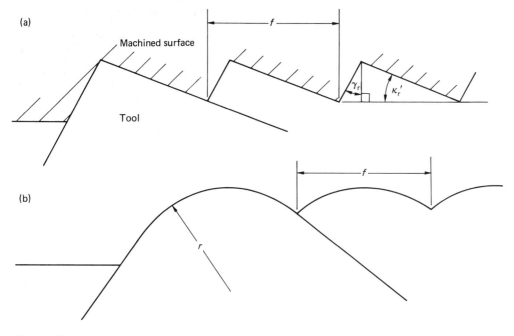

Figure 5.29 Surfaces generated by single-point tools: (a) pointed tool; (b) round-nosed tool

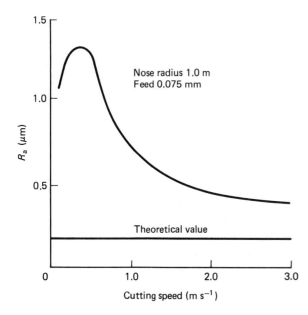

Figure 5.30 Comparison of experimental and theoretical values of surface roughness

is desirable to consider the various wear patterns which can occur.

5.7.1 Crumbling of the cutting edge

As mentioned earlier, when cutting steel at low speeds, built-up edge is a persistent problem. The build up is initiated by purely mechanical causes but, if it is allowed to remain *in situ* for sufficient time, diffusion welding occurs between the tool and the stationary built-up edge. If a carbide tool is used, the coefficient of expansion is less than that for a steel tool, so a subsequent drop in temperature, as for instance when the tool is taken out of cut, causes tensile stresses in the surface of the tool which can result in flakes of carbide being pulled from the surface. After this has been repeated a few times, the cutting edge will have deteriorated, giving the appearance of crumbling.

The solution when machining steel is fairly straightforward. An increase in cutting speed, if necessary at the expense of feed, will reduce the problem of build up and prevent crumbling.

5.7.2 Cratering

When tungsten carbide was used as the main component of sintered carbide tools, it was common for a crater to form near the hottest portions of the rake face due to diffusion between the tool and the chip, tool material being carried away on the underside of the chip. Since the high temperatures were largely the result of high cutting speeds, and to a lesser extent large feeds, the obvious solution was to restrict the cutting speed.

This problem has been reduced considerably by applying a surface coating of titanium carbide to the tool which diffuses less readily. Crater wear still occurs, but it is seldom the main wear source and judicious selection of cutting speed and feed can usually be made to reduce its effect.

5.7.3 Plastic deformation of the tool

Most materials subjected to high stresses and high temperatures start to creep. On a cutting tool this can result in the cutting edge deforming, causing the tool geometry to change and producing high cutting forces. These forces, in turn, raise the temperature and cause greater deformation to occur.

If the interface temperature and stress levels are reduced, creep can be prevented. A surface coating which reduces friction, such as titanium nitride, reduces the rise in interface temperature and reduces plastic deformation. Usually, unless the cutting speed and feed are very high, creep is not a severe problem.

5.7.4 Thermal and mechanical shock

Although these are different phenomena, they usually occur in combination. Mechanical shock occurs with a sudden change in chip formation, such as a turning tool crossing a keyway or a milling tool coming into or leaving engagement. Inevitably, discontinuities of this sort also cause thermal shock.

Mechanical shock, inducing tensile stresses, can cause small particles to break away from the cutting edge. Thermal shock causes a temperature gradient to occur along the cutting edge. Tensile stresses resulting from this can cause hair-line cracks which act as stress concentrations and result in breakdown of the cutting edge.

A main contributor to thermal shock is the application of inadequate coolant. For this reason it is usually advisable when using brittle tool materials such as carbides or ceramics to cut dry. In general, the limitation of mechanical shock may be relaxed by reducing feed and, to a lesser extent, cutting speed.

5.7.5 Flank wear

When all other forms of wear are eliminated or substantially reduced, tools will exhibit a progressive form of attritive wear, causing a scar on the front clearance face or flank. This is the mechanism by which most tools fail, and a study of metal-cutting economics must start from a consideration of what constitutes a reasonable amount of wear before replacing the cutting edge.

5.7.6 Catastrophic failure

Although cutting tools have been known to fail catastrophically for no obvious reason, such failures in modern carbide tools are so rare that they can be ignored. Generally, when catastrophic failure occurs it is a direct consequence of excessive wear due to one of the other modes of failure. If the tools are regularly inspected for wear they can almost invariably be withdrawn from service before catastrophic failure becomes a significant probability.

5.7.7 Tool-life relationships

The ranges of feed and speed within which the various wear mechanisms predominate will vary considerably with different tool/work material combinations.[4] Figure 5.31 shows a typical set of boundaries within which cratering, plastic deformation, crumbling of the cutting edge or flank wear predominate. These boundaries are usually so well separated that the area due to flank wear provides a wide range of both feeds and cutting speeds from which the cutting parameters can be chosen, these ranges being shown dotted in the figure.

If cutting tests are conducted under constant conditions, the wear scar will be seen to increase in three fairly well-defined stages. In the first stage wear is fairly rapid as the tool loses its initial sharpness. This is followed by a secondary stage of slow, almost constant wear rate. The third stage, which commences when the wear scar approaches 1 mm in length, is due to rapidly rising temperature at the cutting edge caused by

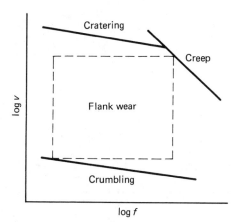

Figure 5.31 Tool-wear envelope

secondary heating on the rake face together with a third heat source due to rubbing on the flank face.

When tests are conducted at different cutting speeds, a family of curves similar to the one shown in Figure 5.32 will be produced. The third stage develops rapidly and, if allowed to continue, will result in catastrophic failure. It is therefore desirable to specify a wear scar near the boundary between the second and third stages of wear. This usually occurs with a flank wear scar of about 1.0 mm when using high speed steel tools and 0.75 mm when using carbides.

If the time to produce scars of this magnitude is plotted against cutting speed on logarithmic scales, a straight line relationship results (Figure 5.33), giving rise to a tool-life equation of the form[5]

$$vT^n = \text{constant} \tag{5.19}$$

It is fortunate that the exponent n varies very little across the whole range of steels, taking values of about 0.13 for high speed steel tools, 0.20 for negative rake carbides and 0.25 for positive rake carbides. Since most high production cutting is performed with carbides, the second and third values are of greatest importance. Values of about 0.38 have been quoted for ceramic tools, but these appear to be based on scanty information and are probably unreliable.

Equation (5.19) is important in that it shows the influence of cutting speed on tool life. It is very restrictive in that it does

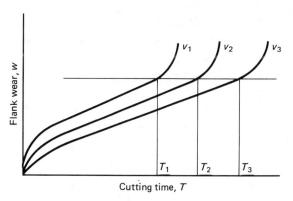

Figure 5.32 Flank wear related to cutting time

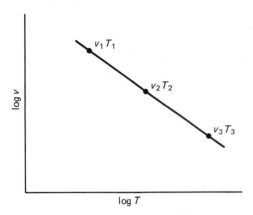

Figure 5.33   Cutting speed vs. tool life relationship

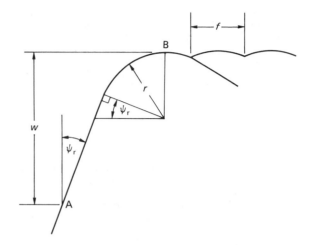

Figure 5.34   Tool plan geometry

not allow for variation in feed, width of cut or tool geometry. Well documented evidence exists which enables this equation to be extended to embrace feed variation, as follows,

$$vT^n f^m = \text{constant} \tag{5.20}$$

Again, it is fortunate that the exponent m does not appear to vary appreciably when cutting any of the steels and is unaffected by rake angle. A value of 0.37 can be assumed.

The next stage in generalising the equation is far more tentative, being based on intuition rather than experimental data. Several researchers have suggested a somewhat similar approach, but the main significance of the generalised equation can be attributed to Brewer.[6] He reasoned that, since tool wear is a function of interface temperature, a temperature equilibrium must exist which is determined by the ratio of heat generation and heat conduction from the cutting edge. The rate of heat generation is assumed to be proportional to the uncut chip area (fw) and the rate of heat dissipation is assumed to be proportional to the length of cutting edge in engagement.

If a tool taking a width of cut w and a feed f has an approach angle ψ_r and a nose radius r the effective length of the cutting edge can be shown to be approximately equal to the line AB in *Figure 5.34*, i.e.

$$l_e = [w - r(1 - \sin \psi_r)]/\cos \psi_r + (90 - \psi_r)\pi r/180$$

and the interface temperature θ is a function of fw/l_e. This ratio has the dimensions of length and is referred to as the 'equivalent chip thickness', b_e. It includes the cutting variables f, w, ψ_r and r, of which f is readily separable,

$$b_e = fw/l_e = fG$$

G has a maximum value of unity for a pointed tool with zero tool approach angle, and in practical situations is usually about 0.9.

A generalised equation including these variables could, therefore, be of the form

$$vT^n b_e^m = vT^n (fG)^m = \lambda \tag{5.21}$$

where λ is constant for any given work and tool material and

$$G = w/[(w - r(1 - \sin \psi_r))/\cos \psi_r + (90 - \psi_r)\pi r/180] \tag{5.22}$$

Purists may argue that the assumptions underlying this equation are so sweeping as to make it unrealistic. Its intuitive appeal lies in the fact that a longer cutting edge due to a non-zero tool approach angle and a nose radius would be

expected to reduce the effective value of uncut chip thickness, which is what the value of G achieves.

5.8   Economics of metal cutting

The main financial objective of a manufacturing company is to maximise the return on capital expenditure. This implies a knowledge of profitability, which in turn requires a knowledge of cost and selling price. Production engineers are concerned with the processes needed to make components which are eventually assembled into finished products. Selling price, therefore, is not usually a very useful statistic for the process planner. He/she must settle for subobjectives such as minimum cost or, sometimes, maximum output which, although not synonymous with profitability, at least contribute to its achievement. When discussing manufacturing economics, these are the objectives which must be addressed.

Metal cutting is an intrinsically wasteful operation, involving the removal of large quantities of material. Although there are no reliable figures to support this contention, it is probable that only about 70% of the material purchased is contained in finished parts, the balance being expensively converted to swarf which has a very low resale value. Intelligent design can do much to increase material utilisation, but material wastage will always be a significant proportion of the total component cost. In spite of the attractiveness of contending production options such as metal forming, it is inevitable that cutting processes will continue to be used extensively. The subsequent analysis assumes that due cognisance has been taken at the design stage of the importance of material utilisation, and the cost factors include only the direct cost of manufacture and its associated overhead. The operating cost, taking into account direct labour, machine depreciation and factory overheads, may well be of the order of £20 per hour.

5.8.1   Minimum-cost production

The cost of manufacture can be divided into five parts:

(1)   set-up and idle time cost per component, K_1;
(2)   machining cost per component, K_2;
(3)   tool changing cost per component, K_3;

(4) tool depreciation cost per component, K_4; and
(5) tool regrinding cost per component, K_5.

Assuming that disposable inserts are used, there is no regrinding cost, so this can be ignored.

The setting cost can be substantially reduced by using preset tooling. However, with the current trend towards small batch sizes, the setting cost ascribed to each component will increase proportionally. The idle time per cutting cycle is composed of loading and unloading time in addition to the tool approach and tool retraction times before and after machining has taken place. Set-up and idle time can, therefore, contribute significantly to production cost, and its reduction is frequently the largest single factor in cost minimisation.

Machining cost is directly related to cutting time which, in turn, is dependent on cutting speed and the size of feed. The use of large feeds and high cutting speeds reduces machining cost but reduces tool life and, consequently, increases both the unit tool depreciation and unit tool changing costs. It is therefore the minimisation of the total of these three costs which determines the minimum production cost for any given set-up.

The simplest analysis, and in fact the only analysis which has been exhaustively attempted, is that for a plain bar turning operation.[7] Fortunately, the conclusions to be drawn from the bar turning analysis provide good indicators when seeking to optimise other turning and milling operations.

If a metal bar of diameter D mm is machined along a length L mm with a cutting speed v m s^{-1} and a feed f mm, the machining time,

$$t_m = L\pi D/(60\ 000fv) \text{ min}$$

from equation (5.21), the tool life is

$$T = \lambda^{1/n}/[v^{1/n}(fG)^{m/n}]$$

and the number of tool changes per component is

$$t_m/T = L\pi Dv^{(1/n)-1}f^{(m/n)-1}G^{m/n}/(60\ 000\lambda^{1/n})$$

The set-up and idle time cost per component is

$$K_1 = k_1 t_s$$

where k_1 is the cost per minute of direct labour, overhead and machine tool depreciation, and t_s is the set-up and idle time per component.

The machining cost per component is

$$K_2 = k_1 L\pi D/(60\ 000fv)$$

The tool changing cost per component is

$$K_3 = k_1 t_c L\pi Dv^{(1/n)-1}f^{(m/n)-1}G^{m/n}/(60\ 000\lambda^{1/n})$$

where t_c is the time needed to change a tool.

The tool depreciation cost per component is

$$K_4 = EL\pi Dv^{(1/n)-1}f^{(m/n)-1}G^{m/n}/(60\ 000\lambda^{1/n})$$

where E is the tool cost per cutting edge.

The total manufacturing cost per component is

$$
\begin{aligned}
K &= K_1 + K_2 + K_3 + K_4 \\
&= k_1 t_s + k_1 L\pi D/(60\ 000fv) + L\pi Dv^{(1/n)-1}f^{(m/n)-1}G^{m/n} \\
&\quad (k_1 t_c + E)/(60\ 000\lambda^{1/n})
\end{aligned}
\tag{5.23}
$$

The three variables in this equation over which the production engineer has some control are f, v and G. By inspection it is seen that a low value of G is desirable. This can be achieved by having a large tool approach angle and a large nose radius. In practice, these may be limited by geometrical requirements of the work piece or by the likelihood of chatter, and they are

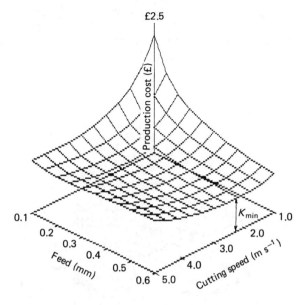

Figure 5.35 Production cost related to cutting speed and feed

usually limited to 15° and 1.0 mm, respectively, when using disposable tips.

Figure 5.35 illustrates by way of a carpet plot how K varies with both f and v. The main point worth noting is that, within the broad range of feeds and cutting speeds selected, the minimum cost occurs inside the speed range and at the maximum value of feed.

It would be useful to possess a set of values for λ covering all work materials and up to date cutting materials, but the few values which exist relate to carbide tools which were commercially available several years ago. Thus, *Figure 5.35* relates to a test using positive rake carbide tools to cut EN24 steel bar having a Brinell hardness of 280, for which the value of λ is 3.5. Using currently available coated tools this value would be expected to be somewhat larger. The other variables were given the following values:

Tool cost per cutting edge $E = £1.0$
Diameter of bar $D = 100$ mm
Length of cut $L = 150$ mm
Machine rate per minute $k_1 = £0.25$
Set-up and idle time per component $t_s = 2.0$ min
Tool changing time $t_c = 1.0$ min

This graphical model was used to predict the effects on manufacturing cost of varying the parameters singly and in combination. The most significant observations obtained from these simulations, together with the graphs obtained are detailed below.

(1) The value of G has a very small effect on the optimal cutting speed or on minimum cost across the practical range of values for G of 1.0–0.8. This would seem to suggest that the effects of tool plan geometry and of width of cut are so slight that they can be ignored for practical purposes. *Figure 5.36* shows cost graphs obtained using the two extreme values of G.

(2) Minimum cost is almost directly proportional to the machine rate k_1 (see *Figure 5.37*). Since this parameter comprises the sum of depreciation rate, direct labour cost

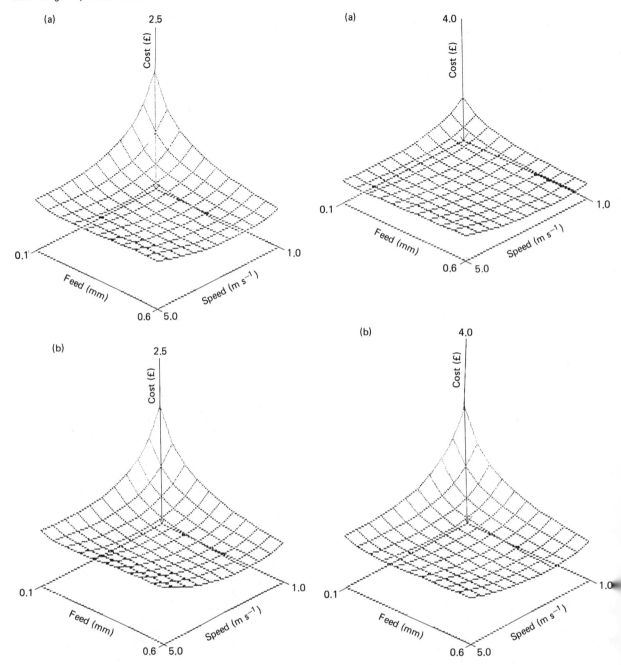

Figure 5.36 Effect on cost of varying G: (a) G = 0.8; (b) G = 1.0

Figure 5.37 Influence of machine rate per minute on cost: (a) k_1 = £0.15; (b) k_1 = £0.40

and factory overhead it is obvious that, when using expensive equipment, round-the-clock operation and high utilisation are very important.

(3) The value of λ has a dramatic effect on the shape of the cost curve and on the minimum cost values, as shown in *Figure 5.38*. Low values of λ give a very cost-sensitive curve, with low optimal cutting speeds. Large values of λ, associated with the newer high performance cutting materials, produce curves which are very flat and insen-

sitive to cost near the minima, which occur at high cutting speeds. High values therefore have the double advantage of low costs and of considerable flexibility in selecting cutting speeds, which makes for high productivity with little penalty in cost.

(4) Tool-changing time has relatively little effect on minimum cost but, as shown in *Figure 5.39*, long changing times cause a greater cost sensitivity to variations in cutting speed.

(a)

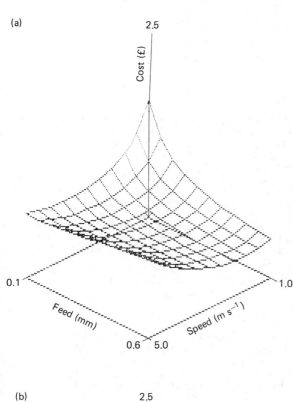

(b)

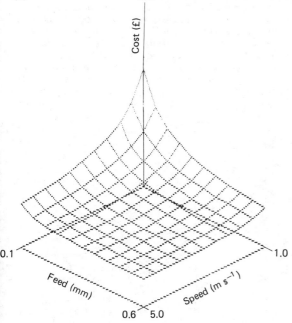

Figure 5.38 Variation of cost due to varying λ: (a) $\lambda = 3.0$; (b) $\lambda = 7.0$

(a)

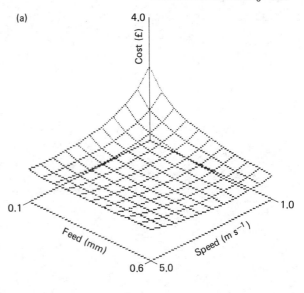

(b)

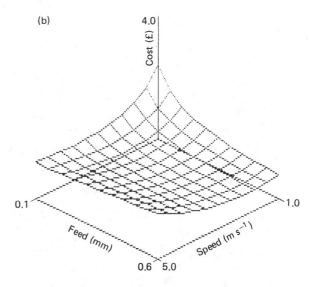

Figure 5.39 Effect of tool changing time on cost: (a) $t_c = 0.4$ min; (b) $t_c = 4.0$ min

(5) Total idle time has marked effect on minimum cost (*Figure 5.40*). This comprises setting time attributed to each unit of output, and to the non-cutting time in each manufacturing cycle (due to loading and unloading, and also to tool approach and retraction time). A strong case exists for streamlining batch changeovers, for improving materials handling, and for more rapid approach and withdrawal speeds.

Although reliable values of λ are not available from which to calculate optimal cutting speed, it is possible, as shown below, to establish the optimal tool life. Knowing the maximum feed which can be selected the appropriate cutting speed can be arrived at by experiment, i.e. by varying the speed to

(a)

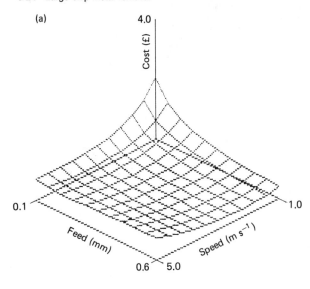

(b)

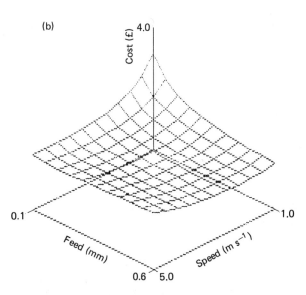

Figure 5.40   Influence of idle time on cost: (a) $t_s = 0.4$ min; (b) $t_s = 5.0$ min

give the desired tool life. The maximum practical feed may be determined by the largest available on the lathe, by the maximum permissible cutting force, or by surface roughness considerations.

Differentiating equation (5.23) with respect to cutting speed,

$$dK/dv = -k_1 L\pi D/(60\,000 f v^2) + (1/n - 1)$$
$$L\pi D v^{(1/n)-2} f^{(m/n)-1} G^{m/n}(k_1 t_c + E)/(60\,000\lambda^{1/n})$$

Equating to zero for minimum-cost conditions:

$$k_1 = (1/n - 1)v^{1/n}(fG)^{m/n}(k_1 t_c + E)/\lambda^{1/n} \quad (5.24)$$

From equation (5.21),

$$T = \lambda^{1/n}/(v^{1/n}(fG)^{m/n})$$
$$T_{opt} = (1/n - 1)(k_1 t_c + E)/k_1 \quad (5.25)$$

It is clear that T_{opt} could vary considerably depending on the values of k_1, t_c and E. Taking the values used in *Figure 5.35*, i.e. $k_1 = £0.25$, $t_c = 1.0$ min and $E = £1.0$ per cutting edge,

$$(k_1 t_c + E)/k_1 = 5.0$$

When cutting steel with positive rake tools, $n = 0.25$, giving an optimum tool life of 15 min. Even if the cost term in equation (5.25) increased by a factor of 2 the optimal tool life would still be only 30 min, which is considerably less than that normally used in industry.

This presupposes that sufficient power is available to permit the use of feeds and cutting speeds high enough to cause the tools to wear out in the optimal time. Consider a fairly high powered lathe with an available spindle power of 10 kW, capable of operating at large feeds and high spindle speeds, taking a 5.0 mm width of cut on EN24 steel, for which the cutting force constant C is 2000 when using positive rake carbide tools:

Power P = Principal cutting force × Cutting speed

$$P_{max} = 10\,000 = 2000 \times 5f^{0.85}v_p$$
$$v_p = 1/f^{0.85}$$

where v_p is the maximum speed constrained by spindle power. Assuming the values taken to produce *Figure 5.35*,

$$v_m = 1.849/f^{0.37}$$

where v_m is the cutting speed giving minimum cost.

Figure 5.41 shows that for any feeds in excess of 0.3 mm the available power constrains the maximum cutting speed to a value less than that which would produce a 15 min tool life. In most practical situations of heavy metal removal the power constraint is even more severe, so the case for operating at maximum power assumes added importance.

What conclusions can be drawn from this analysis when considering the other main heavy metal removal process, face milling? Obviously, with a multipoint cutter the optimal tool life would be greater than for a single-point cutter. Although

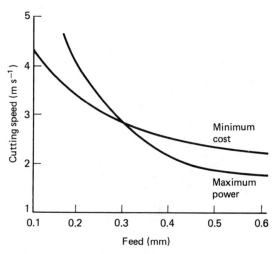

Figure 5.41   Typical minimum cost and maximum power constraints

no comparable analysis exists, it would seem reasonable that, if the optimum life for a single-point cutter is 15 min, that for an eight-tooth cutter would be of the order of 2–3 h.

With multipoint cutters the problems associated with catastrophic failure are less acute, since the failure of one cutting edge will not pose a dangerous threat. Usually, cutting is allowed to proceed until the generated surface is deemed to be unsatisfactory. This happens after incurring more severe wear than would be permissible with a lathe tool. To approach an optimal solution to the face milling process it would appear, therefore, that the maximum recommended feed per cutter tooth should be selected and the spindle speed be set to provide a cutter life of about 3 h if this is possible. More likely, the power constraint will override the economic constraint, in which case the cutter should be run at a spindle speed approaching that requiring maximum available power.

5.8.2 Machining for maximum output

Whereas minimum-cost machining is based on optimising the equation for unit cost, maximum productivity is achieved by minimising the unit manufacturing time.

The time taken to produce one component is given by

$$M = t_s + t_m + t_c \times \text{Number of tool changes per component}$$

$$= t_s + L\pi D/(fv) + t_c L\pi D v^{1/n}(fG)^{m/n}/(fv\lambda^{1/n})$$

Differentiating with respect to cutting speed, and equating to zero:

$$v_0 = \lambda/[(1/n - 1)^n(fG)^m t_c{}^n] \tag{5.26}$$

For minimum cost, from equation (5.24),

$$v_m = k_1^n \lambda/[(1/n - 1)^n(fG)^m(k_1 t_c + E)^n]$$

Since $1/t_c > k_1/(k_1 t_c + E)$, it follows that, for a given maximum feed, the cutting speed giving maximum output v_0 will always be greater than that giving minimum cost v_m. This would demand very short tool lives, having disproportionately large standard deviations, and a high probability of catastrophic failure. Even if the drive motor possessed sufficient power to operate under these conditions it is not normally to be advised. The sole purpose in extending the analysis to encompass maximum output is to show that the short tool lives associated with minimum-cost machining assist also in improving productivity.

5.8.3 The influence of chatter

A great deal of research has been published on the stability boundary between width of cut and spindle speed.[8] The boundary envelopes vary with the machine tool and the process parameters, but a typical boundary is shown in *Figure 5.42*, where the maximum width of cut under stable machining conditions decreases as the spindle speed increases. A large tool approach angle and nose radius increase the effective length of the cutting edge which is tantamount to increasing the width of cut and decreasing the uncut chip thickness. If chatter is likely to occur it is obvious the tool approach angle and nose radius should be kept as small as possible, a recommendation which conflicts with recommendations as regards tool life.

The influence of feed on the stability boundary is more significant, but this does not appear to have received the attention it deserves. *Figure 5.43* shows how the critical width of cut increases as feed increases, making the case for using large feeds to oppose the onset of chatter.[9]

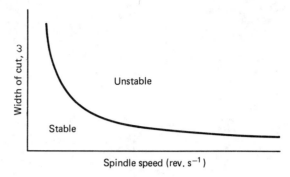

Figure 5.42 Variation of stability threshold with spindle speed

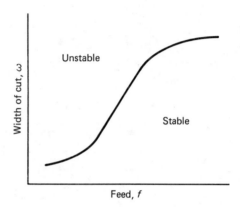

Figure 5.43 Variation of stability threshold with feed

In conclusion, it is pertinent to stress again the importance of large feeds in achieving not only chatter-free performance, but also low specific cutting power and minimum-cost machining.

References

1 BRITISH STANDARDS INSTITUTION, *BS 1296 Specification of single point cutting tools*, Milton Keynes (1972)
2 STABLER, G. V., 'The fundamental geometry of cutting tools', *Proc. Inst. Mech. Eng.*, **165**, 14 (1951)
3 CROOKALL, J. R. and RICHARDSON, D. B., 'Use of photographed orthogonal grids and mechanical quick stopping techniques in machining research', *Conference on Photography in Engineering*, Institute of Mechanical Engineers, London (1969)
4 TRENT, E. M., *Metal Cutting*, Butterworth, London (1977)
5 TAYLOR, F. W., 'On the art of cutting metals', *Trans. ASME*, **28**, 31 (1907)
6 BREWER, R. C., 'On the economics of the basic turning operation', *Trans. ASME*, **80**, 1479 (1958)
7 BREWER, R. C. and RUEDA, R. A., 'Simplified approach to the optimum selection of machining parameters', *Eng. Digest*, **24**, 9 (Sep. 1963)
8 TOBIAS, S. A., *Machine Tool Vibrations*, Blackie, London (1965)
9 PEARCE, D. F. and RICHARDSON, D. B., 'Improved stability in metal cutting by control of feed and tool/chip contact length', *Joint Polytechnic Symposium on Manufacturing Engineering*, Leicester (1977)

6

Non-chip Metal Removal

D Koshal

Contents

6.1 Introduction

There are applications where metal removal by chip formation is not satisfactory or economical for the following reasons:

(1) the tensile strength and hardness of the material is very high;
(2) the shape of the part is too complex;
(3) better surface finish than that possible with chip removal processes is required;
(4) temperature gradient or residual stresses in the work piece are undesirable; and
(5) miniaturisation in the electronic industry.

These requirements led to the development of a number of non-traditional machining processes. The most commonly used processes are described in this chapter.

6.2 Electrical processes

6.2.1 Electrical discharge machining

This process is sometimes referred to as 'electro-erosion' or 'spark machining'. It may be described briefly as the removal of material by means of repetitive short-lived electric sparks which occur between the tool (i.e. electrode) and the work piece.

The principle of the process is shown in *Figure 6.1*, in which the tool is mounted on a chuck or other clamping device attached to the spindle of the spark machine. The down feed of this spindle is controlled by a servodrive through a reduction gearbox. The work piece (i.e. cathode) is immersed in a tank filled with a dielectric fluid, usually paraffin; in order to reduce the risk of fire a depth of several inches is maintained over the top of the work piece. A dielectric fluid is circulated under pressure by a pump and filtered to remove vaporised work-piece particles.

The circuit shown in *Figure 6.1* is a d.c. relaxation circuit fed usually from a mercury arc or selenium type rectifier. In order to achieve erosion, a gap (approximately 0.025–0.05 mm) between the tool and the work piece has to be maintained by the servodrive. The frequency of spark production is of the order of 10 000 per second. As the current is discharged across the gap, temperatures in the region of 10 000–50 000°C are developed, this evaporates parts of the surrounding dielectric fluid and vaporises the metal, thus forming a small crater on the work surface.

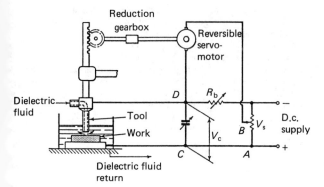

Figure 6.1 Electrical discharge machining apparatus

6.2.1.1 Tool wear and tool materials

The fact that wear occurs upon the tool (i.e. electrode) is one of the major difficulties in electrical discharge machining (EDM). The wear ratio, i.e. the metal lost from the tool divided by the volume of metal removed from the work, will depend on the tool and work materials used. Typical examples are that of a brass tool used on a brass work piece, where the ratio is of the order of 0.5, whereas for a brass tool used on a tungsten carbide work piece the ratio may be as high as 3.

Electrode wear will produce inaccurate machining, the wear taking place predominantly at the leading edges of the electrode, thus producing tapered cavities. Advantage may be taken of this feature when producing press tool blanking dies, the electrode being put through the die from the reverse side, thus providing natural die clearance. In most cases, however, electrode wear is not an advantage as it may be necessary to use several electrodes in order to achieve acceptable accuracy.

In order to minimise the cost of electrode manufacture, much work has been done to obtain suitable materials for the electrodes. For many applications brass or copper is used, although copper–graphite and tungsten carbide have been used to advantage. The higher the melting point of the electrode material the less wear occurs. Thus, by using materials such as copper–graphite and, more recently, with the pulse generator circuit, graphite, lower wear rates are obtained.

6.2.1.2 Metal removal

If the highest quality surface finish is required, then lower metal removal rates must be used. Thus it will be necessary to achieve a compromise, depending on whether rough or finish machining is required.

Using modern pulse generator machines, metal removal rates of up to 8 mm^3/min can be achieved when rough machining. The metal removal rate is affected by the spark duration for a given energy input. For a given power input an optimum discharge time exists, thus giving maximum metal removal per discharge. If the discharge time is too short, tool wear becomes excessive with a resulting loss of accuracy.

6.2.1.3 Accuracy

In addition to errors occurring due to incorrect discharge times, the accuracy with which the electrode is produced is clearly an important factor. Allowance must be made for the spark gap, and parallel-sided cavities can be produced by using several electrodes or by relieving the tool behind the tip. As the depth of the cavity increases it becomes more difficult for the dielectric fluid to wash the disintegrated work particles clear of the tool–work gap. This has the effect of reducing the breakdown voltage of the dielectric fluid with the result that sparking occurs between the sides of the electrode and the work cavity.

6.2.1.4 Surface roughness

The surface finish in EDM depends on the rate of metal removal. Good finish requires low power discharges that leave small craters, but the rate of metal removal is slow. It is generally advisable to use roughing electrodes first and then finish off. When high production rates are required, the surface roughness R_a may be as high as 1.75–3.75 μm, but by reducing the spark rate R_a values in the region of 0.25 μm may be achieved.

6.2.1.5 Applications

The application of EDM is limited to the machining of electrically conducting work piece materials, but the process has the capability of cutting these materials, regardless of their hardness or toughness.

Spark machining has its greatest application in the manufacture of press tools, mould forgings and extrusion dies. The material can be spark machined after it has been hardened, thus eliminating the effects of distortion due to heat treatment. Tools manufactured from cemented carbide can be machined when sintering has been completed, thereby eliminating the need for an intermediate partial sintering stage and thus eliminating the errors resulting from final sintering after the cavities, etc., have been machined.

The successful application of this method to the drilling of holes with a large length/diameter ratio, as in the case of the nozzles of diesel engine fuel injectors, has been carried out.

In addition to the conventional applications of EDM, the process has been used for turning, grinding, and precision band sawing. In the former case the work piece is rotated as in turning, but the cutting tool is replaced by the electrode. The dielectric fluid is flooded into the spark gap.

When precision band sawing hard materials, a fine wire wound onto two reels is used, with the dielectric fluid flooding into the work. To prevent the wire from breaking and wearing it is continually wound from one reel to the other.

6.2.2 Electrochemical machining

Electrochemical machining (ECM) is a process of metal removal in which electrolytic action is utilised to dissolve the work piece material. It is, in effect, the reverse of electroplating. The term 'ECM' is sometimes used for the electrolytic grinding process, which is a modification of electrochemical machining.

The basic elements of an electrochemical machine are shown in *Figure 6.2*. The work piece (which must be a conductor of electricity) is placed in a tank on the machine table and connected to the *positive* terminal of a d.c. supply. The tool electrode, which is shaped to form the required cavity in the work piece, is mounted in the tool holder and connected to the negative terminal of the supply. An electrolyte flows through the gap between the tool and work piece and is then pumped back to the working zone, either through the tool or externally, depending on the application.

The action of current flowing through the electrolyte is to dissolve the metal at the anode, i.e. the work piece. The

electrical resistance is lowest (and hence the current highest) in the region where the tool and work are closest together. Since the metal is dissolved from the work most rapidly in this region, the form of the tool will be reproduced on the work.

There is no mechanical contact between work and tool and any tendency of the work metal to be plated on the tool (the cathode) is counteracted by the flow of electrolyte which removes the dissolved metal from the working zone. Hence there is neither tool wear nor plating of the work material on the tool, so that one tool can produce a very large number of components in its life.

6.2.2.1 Surface roughness

The fact that metal removal in ECM is not achieved by mechanical shearing or by melting and vaporisation of the metal means that no thermal damage occurs and no residual stresses are produced on the worked surfaces. The only heat generated is that due to electrical resistance, and the temperature cannot be allowed to rise above the boiling point of the electrolyte.

Surface finishes of 0.75–1 μm are readily obtained, while surface finishes better than 0.25 μm have also been achieved. A great advantage of ECM, which is contrary to conventional machining, is that, as the metal removal rate is increased, the surface finish and accuracy are improved.

6.2.2.2 Features

These can be listed as follows.

(1) Stress-free surface on work piece.
(2) No work-hardening of material.
(3) No thermal cracking of work surface.
(4) High surface finish.
(5) No softening of work surface.
(6) Freedom from burrs.
(7) High metal removal rates on hard materials.
(8) No electrode wear.

6.2.3 Electrolytic grinding

This is a modification of the ECM process. The tool electrode consists of a rotating abrasive wheel which can conduct electricity; this is usually a metal bonded diamond wheel. The electrolyte is fed between the wheel and work surface in the direction of movement of the wheel periphery, so that it is carried past the work surface by the wheel rotation. The abrasive particles help to maintain a constant gap between wheel and work. The current flows between work and wheel as shown in *Figure 6.3*.

In this process the predominance of the electrolytic action reduces wear to a negligible amount and makes it possible to grind hard materials rapidly. Furthermore, the wheel can be used for long periods without dressing.

In applying electrolytic grinding it is desirable to design the operation in such a way that the area of contact between the wheel and work is as large as possible. This gives the highest rate of removal for a given current density and allows full use to be made of the available current capacity. The most notable application of this method is in the grinding of carbide tool tips where surface finishes of the order of 0.1 μm have been obtained.

6.2.3.1 Applications

The main application of electrolytic grinding is in machining hard metals such as those used in high temperature service.

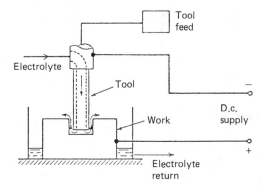

Figure 6.2 Electrochemical machining apparatus

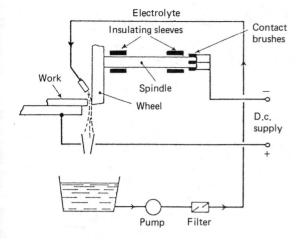

Figure 6.3 Electrolytic grinding apparatus

The removal rate is much higher than in conventional machining.

ECM has been applied successfully to the following machine operations:

(1) machining of through holes of any cross-section;
(2) machining blind holes with parallel sides;
(3) machining shaped cavities, such as forging dies;
(4) wire cutting of large slugs of metal; and
(5) machining complex external shapes, e.g. turbine blades.

6.2.4 Other electrochemical processes

A number of ECM processes find application in industry. These include:

(1) electrochemical turning (ECT);
(2) electrochemical honing (ECH);
(3) Shaped tube electrolytic machining (processes developed by General Electric Co., Aircraft Engine Group)—this process is used to drill small shaped or round holes (0.64–6.35 mm diameter), in electrically conducting materials it is usually difficult to machine alloys;
(4) Electro-stream (process developed by General Electric Co., Aircraft Engine Group)—this process is used to 'drill' extremely small holes (0.21–1.02 mm diameter);
(5) electrolytic polishing;
(6) electrolytic end milling; and
(7) electrolytic belt grinding.

6.3 Mechanical processes

6.3.1 Water-jet machining

Water-jet machining removes work-piece material and produces a narrow slit by the cutting action of a fine, high pressure, high velocity stream of water or water based fluid with additives. The water-jet nozzle can be designed to mix abrasive particles with the water and hence provide abrasive jet machining.

High pressure is needed to cut large thicknesses, while low pressure is used for deburring or finishing processes. The machined surface is generally unaffected by stress and heat.

6.3.1.1 Applications

Complex profiles may be produced using sophisticated nesting techniques when the nozzle is carried on a computerised numerical control cutting system. Computer control of the abrasive feed, water pressure and cutting speed ensures precise cutting.

Water jets are used for slitting and profiling of many non-metallic materials such as wood, paper, glass, marble, asbestos, plastics, leather, rubber, nylon and fibreglass. To cut large thicknesses of steel sheets, high pressure water jets are required.

6.3.2 Ultrasonic machining

Machining by the electrochemical or electrical discharge processes is confined strictly to metals, whereas ultrasonic machining may be applied to a much wider range of materials. Typical examples of the range of work covered by this process are:

(1) machining of ceramics and porcelain;
(2) cutting glass, including thread cutting;
(3) cutting precious and semiprecious stones;
(4) machining silicon, germanium and other semiconducting materials;
(5) machining ferrite;
(6) machining small punches and dies for presswork;
(7) machining accurately cavities of limited depth in cemented carbide; and
(8) drilling holes of varied internal configuration.

The development of this process has opened up several possibilities for the extended use of materials such as ceramics and semiconducting materials which would hitherto have been impractical. It must be recognised, however, that at this moment in time there are limitations to the process which have been not been overcome, namely:

(1) relatively small machining area;
(2) low cutting rates for hard metals relative to other work materials;
(3) relatively high power consumption;
(4) high rate of tool wear when cutting hard metals relative to other work materials; and
(5) relatively shallow depth of cut.

The cutting of brittle materials by the application of repeated blows of a concentrated nature has long been applied. This principle is applied in ultrasonic machining where the machining of a wide range of brittle materials can be carried out.

In ultrasonic machining the material is removed by the action of an abrasive which is driven into the work surface by a tool oscillating normal to the surface at high frequency. The tool which is machined to the shape of the desired hole or external profile is made from a tough metal, such as tool bits of alloy steel. A slurry of abrasive powder in liquid is fed to the tool which is oscillated at frequencies between 2000 and 3000 Hz with amplitudes in the region of 0.025–0.075 mm. The tool is pressed onto the work with a force of a few newtons, feeding down as the profile is cut on the work.

Excitation of the tool at these frequencies may be produced by an electromechanical transducer of the piezo-electric or magnetostrictive design.

6.3.2.1 Basic principle

The basic principle of ultrasonic machining is shown in *Figure 6.4* and, in order to increase the amplitude of vibration of the

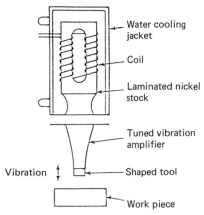

Figure 6.4 The principle of ultrasonic machining

transducer, a velocity transformer is fixed to the radiating face of the transducer. The resonant movement of the radiating face of the transducer being in the region of 0.012 mm, by attaching a velocity transformer the longitudinal amplitude can be increased to 0.025–0.075 mm. The transformer is made from materials having high fatigue strength and low energy loss and a material such as brass meets this specification. The profile of the transformer is tapered exponentially (*Figure 6.5*), the increase in the ratio of the amplitude being inversely porportional to the ratio of the areas of the two ends.

The toolholder (or tool cone) is fixed to the end of the transformer and to give minimum damping the vibrating components are clamped at the nodes and are acoustically matched. A transducer–transformer–tool system is thus produced which is a resonant mechanical transmission line that is two or more half-wavelengths long. As tool wear produces shortening of the combined length of tool cone and tool, it is

usual practice to make these slightly longer than a half-wavelength and, thereby, increase their effective life.

When machining components by this method the tool is held against the work by a light static force of about 18 N and an abrasive slurry is applied to the work area either by hand or by a circulating pump. As the tool tip is oscillated by the transducer, abrasive particles are trapped between the tool and the work and a chipping action ensues which removes small pieces of material from the work piece. The vibration of the tool tip accelerates the abrasive at very high rates, thus imparting the necessary force to produce the cutting action. Cavitation which occurs in the gap assists removal of the chips and circulation of the abrasive.

The abrasive is suspended in a liquid which serves several functions; it provides an acoustic bond between vibrating tool and work and, therefore, produces an efficient means of transferring energy between tool and work. It also acts as a coolant at the tool face and assists in carrying the abrasive to the cutting area and to wash the spent abrasive and swarf away.

A number of different types of abrasive have been used for ultrasonic machining, but those most commonly used are aluminium oxide, silicon carbide, boron carbide and, to a lesser extent, diamond dust. The grit size varies between 200 and 2000 with the coarse grade being used for roughing and the finer grades (say 600–700) used for finishing. The very fine grades (1000–2000) only being used for the final pass on work of the highest accuracy.

The liquid used to carry the abrasive must have the following properties:

(1) its density must be approximately equal to that of the abrasive;
(2) it must have good wetting properties;
(3) in order to be able to carry the abrasive down between tool and work, it must have low viscosity; and
(4) for the efficient removal of heat from the cutting zone it must have high thermal conductivity and specific heat.

Water will satisfy most of these requirements with such additives as rust inhibitors and wetting agents.

6.3.2.2 Surface finish and accuracy

The abrasive grit size is a major factor influencing surface finish. When using, for example, boron carbide on glass, the roughness may vary from 5 μm for 100-mesh grit to 1–25 μm when a 600 mesh is used. If tungsten carbide is machined, however, the roughness values are approximately one-quarter of those listed above. It has been found that the roughness obtained on the face of the tool is generally less than that on the sides.

Tolerances as close as ±0.013 mm may be readily attained.

6.3.2.3 Advantages

The advantages of ultrasonic machining are as follows.

(1) Accuracy and good surface finish are easily obtained.
(2) No heat is generated and, therefore, no changes occur in the physical structure of the material.
(3) Equipment is safe to handle and little skill is required in its operation.
(4) Cheap abrasives are used.
(5) The process is applicable to all brittle materials and is not limited by any other physical property.

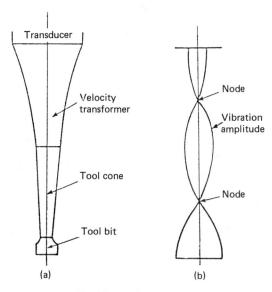

Figure 6.5 Profile of the transformer

6.4 Thermal processes

Thermal machining processes use thermal energy as the primary source of energy for metal removal. The energy is of such a magnitude as to evaporate the material being machined.

6.4.1 Electron-beam machining

Electron-beam machining (EBM) (*Figure 6.6*) is based on the principle that it is possible to accelerate electrons to very high velocities, i.e. upwards of 160 000 km/s. The electrons are then focused into a narrow beam by an electric field which can be bent by electrostatic and electromagnetic fields in a similar way to light rays being bent by glass lenses. The high-velocity stream of electrons are so focused as to impinge upon a very small spot on the work material, the kinetic energy of the electrons then being transformed into thermal energy which vaporises the material in the local area of the focused spot. In order to prevent the electrons from colliding with gas molecules in the atmosphere, which would scatter the electron beam, the operation is carried out under vacuum.

6.4.1.1 Advantages and limitations

This process provides what is perhaps the most precise cutting tool available. It can be used to cut any known material, metallic or non-metallic, which can exist in a high vacuum. As all cutting must be done under vacuum, the degree of vacuum restricts the component size, and the time must be allowed to evacuate the chamber. This method is very suitable for micromachining, as holes down to 0.05 mm diameter can be machined with no cutting-tool pressure or wear. Hole diameter/depth ratios of the order of 1:200 can be achieved and, due to the small beam diameter, close tolerances can be held (±0.012 to ±0.05 mm).

Distortion-free machining of thin or hollow parts is possible with no physical or metallurgical damage to the work material. Only relatively small cuts are economically possible, as the metal removal rate is only approximately 0.1 mg/s. The holes produced by this method have a slight crater where the beam enters the work and the hole is tapered by about 2° per side. The profile of the hole varies with its depth, due to the fact that the beam fans out above and below its focal point, thereby tending to produce an hourglass shape.

High capital cost of the equipment plus the need for considerable operator skill and the fact that the process is applicable to thin material rather limits its range of application.

6.4.1.2 Applications

The process has been applied to such operations as drilling either round or profile shaped holes for metering flow on sleeve valves, and rocket fuel injectors of injection nozzles for diesel engines. Further applications have been concerned with producing wire-drawing dies, spinnerets for manufacturing synthetic fibres and drilling gas orifices for pressure-differential devices.

6.4.2 Laser-beam machining

Laser-beam machining comprises the conversion of electrical energy into light energy and then into thermal energy. The highly focused, high density energy melts and evaporates a portion of the work piece in a controlled manner (*Figure 6.7*).

The term 'laser' is an acronym for 'light amplification by stimulated emission of radiation'. Lasers which are capable of producing highly directive beams of visible, infra-red and ultraviolet radiation, can be classified as solid state, gas, or liquid lasers. The process is usually automated by using computerised numerical control systems to control the movement of the machine table under the laser beam.

6.4.2.1 Applications

The laser machining process does not require a vacuum, and most engineering materials can be successfully laser processed. These include mild, stainless and special steels, aluminium, brass, nimonics, titanium, acrylics, ABS and other plastics, rubber, leather, textiles, Kevlar, advanced composites, quartz glass and fibre glass.

The process is suitable for small-scale cutting operations such as slitting and drilling of holes which may be as small as 0.005 mm with depth/diameter ratios of 50:1.

The following is a guide to the thicknesses of materials which can be cut.

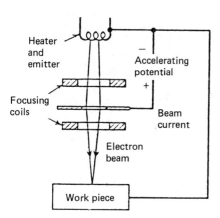

Figure 6.6 Electron-beam machining apparatus

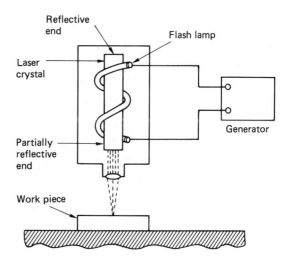

Figure 6.7 Laser-beam machining apparatus

Material	Maximum cutting thickness (mm)
Mild steel	6
Stainless steel	5
Aluminium	2
High temperature alloy	4
Brass	2
Acrylics	15

The main disadvantage in laser applications is the cost of the equipment.

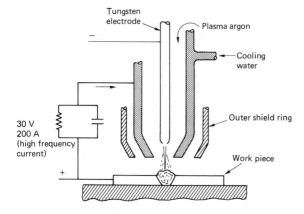

Figure 6.8 Plasma-arc machining apparatus

6.4.3 Plasma-arc machining

A 'plasma' is defined as a gas that has been heated to a sufficiently high temperature to become partially ionised and, therefore, electrically conducting. The temperature of the plasma may reach as much as 30 000°C. A torch is generally used to provide an electric arc between an electrode and the work piece. A typical plasma torch is shown in *Figure 6.8*.

6.4.3.1 Applications

The stream of ionised particles from the nozzle can be used to perform a variety of industrial jobs, including welding, cutting and machining.

Welding   Plasma arc welding can be employed to join almost all metals. The process is generally applied manually and requires some degree of welding skill.

Cutting   With appropriate equipment and techniques, the plasma arc can be employed to make cuts in electrically conducting metals. It can be used for cutting very hard materials and for machining of holes and profiles.

Machining   Plasma arc methods are employed in special applications to replace conventional machining operations such as the removal of metal from the surface of a rotating cylinder to simulate the turning operation.

7

Electronic Manufacture

D Koshal

Contents

7.1 Introduction

Electronic manufacture can be classified as the manufacture of printed circuit boards (PCBs) and electronic components, the placement of components onto PCBs, and the assembly of boards into a plastic/metal box or case. For several decades, through-hole components technology has been used to achieve great advances in the electronics industry, but now there is a growing trend whereby these components are being replaced by a method called 'surface mounting' in which the component leads or terminals are soldered to the top surface of the board.

This trend has accelerated recently due to the advantages offered by surface-mount technology (SMT) in terms of reduced cost, size and improved reliability. The new volume efficient packages for integrated circuits (ICs) and the use of passive chip components allow more dense circuit packing, permitting either an increase in functions for a given equipment size or a reduction in equipment size and lower system costs.

It has been found that approximately 40% of the cost of an electronic product is in the assembly of components onto a PCB. Therefore, an engineer needs to select the appropriate assembly method to minimise the cost of manufacture of a product. This chapter describes the various component assembly techniques that are currently used in industry.

7.2 Manual assembly of through-hole components

Figure 7.1 shows the stages involved in the manual assembly of PCBs with through-hole components. The types of assembly method currently used in industry depend on the quantity and the variety of boards required. There are three types of manual production which are commonly used. In the case of high throughput, automatic machines are available.

7.2.1 Low throughput

In this type of assembly an operator completes the entire board, including soldering. The board may consist of through-hole and surface-mount components. The reason for this type of production line is usually that either the board to be produced is very densely populated with components, or the board contains delicate or sensitive components.

7.2.2 Medium throughput

A cell is set up consisting of several operators and each person is trained to insert about 10 components manually into the PCB. When the board reaches the end of the line, the board is complete and, after visual inspection, is ready for wave soldering. To aid insertion of components a master board is usually made available to each operator.

7.2.3 Manual assembly with the aid of an assembly director

In the above methods the operators are required to memorise the component locations and it is inevitable that mistakes may be made. To reduce misinsertion of components, a machine called an 'assembly director', is employed (see *Figure 7.2*).

In the normal mode of operation of the assembly-director machine, the beam deflection system of the helium–neon laser displays on the surface of the board to be assembled one or more component locations. At the same time various component-dispensing machines can be driven to dispense the appropriate component. To advance to the next preprogrammed location a foot switch is incorporated, thus leaving the hands free for the insertion of components.

The method probably does not substantially increase the production rate, but it does reduce rework.

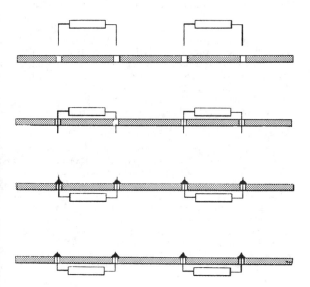

Figure 7.1 Stages in the assembly of through-hole components onto printed circuit boards

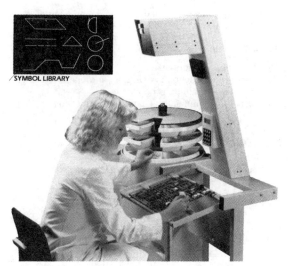

Figure 7.2 Assembly of components using an assembly director

7.2.4   Automated assembly of through-hole components

Automatic component insertion machines are available that are capable of inserting axial, dual in line (DIL), and radial components at a rate of about 1200 components per hour. Machines which are dedicated to inserting one or two types of the above components can operate at a much faster rate.

7.3   Surface-mount technology

Surface-mount technology (SMT) is defined as the placement of components on a printed circuit planar surface with component leads or terminals being soldered to that surface.

Surface-mount components (SMCs) are employed in various electronic products, including disc drives, personal computers, digital watches, communication equipment, test instrumentation, television, videos, and cameras.

7.3.1   Benefits of surface-mount technology

Circuit design

(1)   An increase in board packaging density is achieved. The use of SMT components can result in a 70% improvement in density. Savings of 35–50% are routine.
(2)   An improvement in frequency response is obtained. Conventional component leads act as antennas which generate electrical noise, thus the reduction in lead length afforded by surface mounting gives SMT a major edge in terms of external noise immunity.
(3)   A reduction in size and weight for the same function.
(4)   An improvement in shock and vibration characteristics due to lower mass.

Board manufacture and cost

(1)   A reduction in the number of boards required.
(2)   A reduction in material costs.
(3)   A reduced number of holes is needed, i.e. the drilling cost is reduced.
(4)   A lower number of conductor layers is required in the board structure.

Assembly of the PCB

(1)   This is easy to automate.
(2)   There is a reduction in inventory and manufacturing.
(3)   Improved reliability is obtained.
(4)   Increased manufacturing control is achieved.

Electronic components

(1)   SMCs are potentially cheaper than the leaded equivalent.
(2)   Surface-mounted devices are available in smaller overall dimensions.
(3)   Surface-mounted devices are more reliable.
(4)   The packaging of components makes storage easy.

7.3.2   Limitations of surface-mount technology

(1)   There are some components which cost more in surface-mount packaging than in through-hole technology.
(2)   Although the range of components available in SMT is wide, there are some families of components which are either difficult to find or not offered at all in SMT.
(3)   The cost for test and repair is higher.
(4)   There is a higher requirement for the assembly process.
(5)   A larger number of process steps is required in assembly.

7.4   Surface-mount components

Surface-mount components are classified as follows:

Passive components
Resistors
Capacitors
Inductance
Connectors

Active components
Diodes
Transistors
Thyristors
Integrated circuits

7.4.1   Active components

Small outline integrated circuit (SOIC) packages (Figure 7.3(a))   These are currently available in 6, 8, 10, 14 and 16 pin versions with a body width of 4 mm and in 16, 20, 24 and 28 pin versions with a wider body of 7.6 mm. The leads of these devices are of the gull-wing style.

Small outline transistors (SOT) (Figure 7.3(b))   These are used for discrete semiconductors such as transistors and diodes. Three standard outlines are available. The SOT-23 and SOT-143 have leads on opposite sides, while the SOT-89 has all leads on the same side.

Leadless ceramic chip carriers (LCCC) (Figure 7.3(c))   These are the most commonly used type of package in surface-mounted applications. The package consists of a small piece of plastic or ceramic with etched circuit traces on both the top and the bottom with over-the-edge interconnections on all four sides. The wire bonded IC chip is on the top and the terminating pads are on the bottom where final attachment to the medium takes place. The LCCCs need no holes to be drilled through the PCB. They are commonly available in 18, 20, 28, 32, 44, 52, 68, 84, 100, 124 and 156 terminal versions.

Plastic leaded chip carriers (PLCC) (Figure 7.3(d))   These low-cost plastic leaded chip carriers have leads (gull-wing profile or J shaped) on all four sides of the package. They are square in outline and commonly have between 18 and 84 pins.

Quad flat packs (Figure 7.3(e))   These offer surface-mount packages for devices requiring 64 to 196 pins. The high pin count is achieved by reducing the pitch between the adjacent leads. The quad flat pack devices have one of three different lead pitches (0.65, 0.8 and 1.0 mm), depending on the number of pins required. The leads are formed to a gull-wing mode to bring the ends of the leads level with the bottom of the package.

Cylindrical diode packages (Figure 7.3(f))   These components are available in metal electrode face bonded packaging.

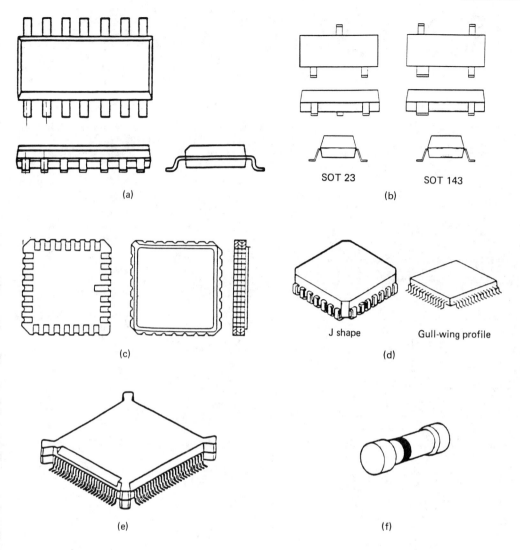

SOT 23

SOT 143

(a)

(b)

(c)

(d)

J shape

Gull-wing profile

(e)

(f)

Figure 7.3 Active surface-mount components. (a) Small outline integrated circuit package. (b) Small outline transistors. (c) Leadless ceramic chip carriers. (d) Plastic leaded chip carriers. (e) Quad flat pack. (f) Metal electrode for bonded packages

Discrete diode components are frequently packaged in SOT-23 encapsulations, but one of the three contacts is then redundant. Several two-terminal packages have been developed especially for diodes, the two most popular (the SOD-80 and the MELF diode) both being cylindrical.

The SOD-80, sometimes referred to as the 'MiniMelf', is less costly than the SOD-23, requires less board space and weighs less. However, the power dissipation of the SOD-80 is limited to 250 mW. When more power dissipation is required, a larger cylindrical encapsulation is used.

MELFs lead to placement problems because of the curve of the surface which will make contact with the board. The small area of contact means that the components can become detached relatively easily. The main advantages of MELFs is that they are less expensive than rectangular chips.

7.4.2 Passive components

Chip resistors The most commonly used sizes are the 1206 with height of 0.6 mm and the 0805 with a height of 0.5 mm. The typical power dissipation is 0.125 W at 70°C. The mechanical data for these packages are shown in *Figure 7.4(b)*.

Chip capacitors Chip capacitors are described by the same size notation as used for chip resistors. The following types are available:

(1) ceramic chip capacitors,
(2) cylindrical ceramic capacitors,
(3) plastic capacitors,
(4) tantalum chip capacitors, and
(5) aluminium electrolytic capacitors.

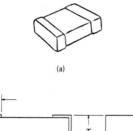

(a)

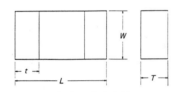

Size code	Dimensions (mm (in.))			
	L	W	T	t
0805	2.00 ± 0.20	1.25 ± 0.20	0.05 ± 0.20	0.40 ± 0.25
	(0.079 ± 0.008)	(0.049 ± 0.008)	(0.020 ± 0.008)	(0.016 ± 0.010)
1206	3.20 ± 0.20	1.58 ± 0.18	0.56 ± 0.15	0.50 ± 0.25
	(0.126 ± 0.008)	(0.062 ± 0.007)	(0.022 ± 0.006)	(0.020 ± 0.010)
1210	3.20 ± 0.20	2.50 ± 0.20	0.56 ± 0.15	0.50 ± 0.25
	(0.126 ± 0.008)	(0.098 ± 0.008)	(0.022 ± 0.006)	(0.020 ± 0.010)

(b)

Size code	Dimensions (mm (in.))			
	L	W	T_{max}	t_{max}
0805	2.0 ± 0.2	1.25 ± 0.2	1.3	0.7
	(0.079 ± 0.008)	(0.049 ± 0.008)	(0.051)	(0.028)
1206	3.2 ± 0.2	1.6 ± 0.2	1.5	0.7
	(0.126 ± 0.008)	(0.063 ± 0.008)	(0.059)	(0.028)
1210	3.2 ± 0.2	2.5 ± 0.2	1.7	0.7
	(0.126 ± 0.008)	(0.098 ± 0.008)	(0.067)	(0.028)
1812	4.5 ± 0.3	3.2 ± 0.2	1.7	0.75
	(0.177 ± 0.012)	(0.126 ± 0.008)	(0.067)	(0.030)
1825	4.5 ± 0.3	6.4 ± 0.4	1.7	0.75
	(0.177 ± 0.012)	(0.252 ± 0.016)	(0.067)	(0.030)

(c)

Figure 7.4 Passive surface-mount components. (a) Chip component. (b) Standard resistor sizes. (c) Standard capacitor sizes

The mechanical data for these packages are shown in *Figure 7.4(c)*.

7.4.3 Component packaging

In surface-mount assembly, the components must be packaged such that they can be handled by automatic equipment. The components must be presented to the assembly machine in a way that the machine tooling can locate and pick up the part. The commonly used types of packaging suitable for SMC placement machines are described below.

Sprocket driven tape (Figure 7.5(a)) 8, 12, 16, 24, 32, 44 and 56 mm sizes are available in tape which is either punched and formed cardboard or moulded plastic.

Bulk Bulk feeding is only possible with linear or vibratory feeders, specially tooled to handle a given part size. By means of a stop device fitted at a known position, the feeders can provide a source of components for the pick-up tool of a placement machine. Bulk purchasing is best suited to non-polarised components which are to be hand placed.

Sticks (Figure 7.5(b)) The components in flexible and rigid 'sticks' remain oriented and protected once loaded onto the machine. The machines are designed to select and remove devices from the sticks.

Matrix trays and waffle packs (Figure 7.5(c)) The internal waffle type nests (50 mm × 50 mm and larger) are made to conform to the part size. Each recess in the tray accommodates a component.

Figure 7.6 shows the body shapes and contact or lead arrangements, the circuit functions available, typical packages for each, and the general handling systems available.

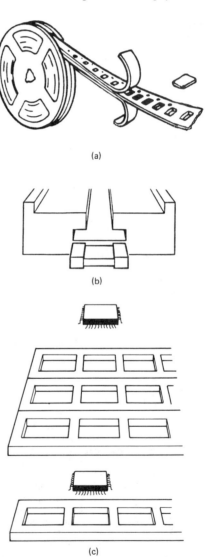

(a)

(b)

(c)

Figure 7.5 Component packaging. (a) Sprocket driven tape. (b) Sticks. (c) Waffle packs

Device function	Capacitor		Diode		Integrated circuit			
		Resistor		Transistor				
Available package forms	Polarised chip	Chip	Cylindrical	SOT	Bare IC or flip-chip	SOIC	Plastic leaded chip carrier	Flat pack	Leadless chip carrier
Available handling systems									
Tape—8 mm and larger	●	●	●	●		●	●	●	●
Sticks — rigid and flexible			●			●			●
Waffle trays					●			●	
Bulk — vibratory feeder	●	●	●						

Figure 7.6   Typical component packages. SOT, small outline transistor; IC, integrated circuit; SOIC, small outline integrated circuit

7.5   Assembly of printed circuit boards

A typical assembly sequence of a printed circuit board (PCB) is as follows:

(1)   insert pins/connectors,
(2)   insert electronic components,
(3)   solder PCB,
(4)   clean PCB with solvent,
(5)   inspection, and
(6)   test.

Depending upon whether SMCs are used exclusively or in combination with through-hole components gives various configurations and assembly procedures (*Figure 7.7*).

The possible configurations are described below.

7.5.1   Single-sided PCB with SMCs

Figure 7.8 shows the steps involved in the assembly of a PCB by wave soldering or by a reflow technique. Depending on the production volume of the boards, the adhesive or solder paste may be either screen printed or dispensed by the machine.

In the case of a sequential placement machine, the board is located on the table and the adhesive dispensing head places small drops of adhesive in the centre of the pads on which components are to be located. The pick-up head then places the SMCs on top of the adhesive. The adhesive is tacky enough to handle the populated board, but it should not be subjected to excessive vibration.

After visual inspection of the board it is placed in an oven to cure the adhesive. The PCB can now be soldered using a wave soldering machine. If solder paste is used it may be screen printed onto the board. After placing the SMCs the board is heated in an infra-red heater to melt and flow the solder to form a joint.

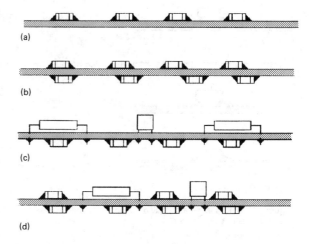

(a)

(b)

(c)

(d)

Figure 7.7   Various types of surface-mount assembly. (a) Single-sided printed circuit board (PCB) with surface-mount components (SMCs). (b) Double-sided PCB with SMCs on both sides. (c) Mixed assembly double-sided PCB with SMCs on one side only. (d) Mixed assembly double-sided PCB with SMCs on both sides

7.5.2   Double-sided PCB with SMCs on both sides

Considerable savings can be gained by mounting SMCs on both sides of the board using through-plated vias to make the interconnections between the opposite sides.

The steps involved are to screen print solder paste onto one side of the board, mount SMCs, and reflow the solder. Then, on the other side of the board apply the adhesive, place the SMCs, cure the adhesive, invert the board, and flux and wave solder (*Figure 7.9*).

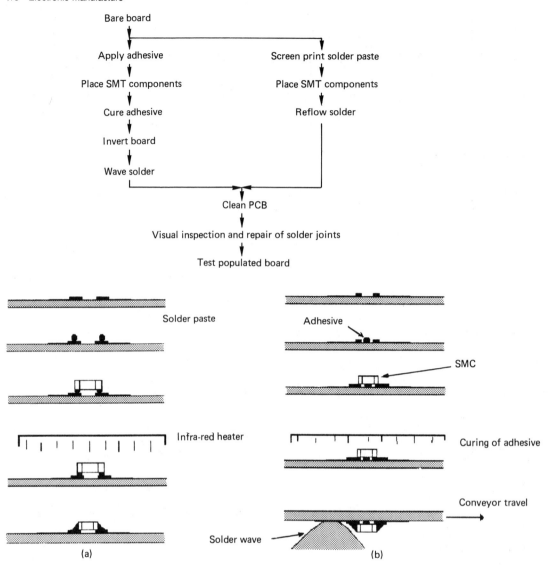

Figure 7.8 Stages in the assembly of a single-sided printed circuit board (PCB) with surface-mount components (SMCs) using (a) solder paste and (b) adhesive attachment

7.5.3 Mixed assembly, double-sided PCB with SMCs on one side only

In this case it is a common practice to place the SMCs before the through-hole components. The adhesive is dispensed, SMCs are placed, and the adhesive is cured. The board is turned over, through-hole components are inserted manually or by machine, and the board is fluxed and soldered (*Figure 7.10*). In this case the SMCs will be fully submerged in the solder paste. Therefore temperature-sensitive components will need to be assembled manually after soldering.

However, it is possible to assemble the through-hole components prior to the SMCs, but the above-described technique has the advantage that the adhesive can be screen printed, which is a much faster process than dispensing adhesive by means of a sequential placement machine.

7.5.4 Mixed assembly, double-sided PCB with SMCs on both sides

Figure 7.11 shows the steps involved in the assembly of SMCs and through-hole components on both sides of the board. Solder paste is screen printed onto the top surface of the PCB and the SMCs are placed using a sequential placement machine. Solder paste is then reflowed in an infra-red heater. The through-hole components can then be inserted manually or by means of a machine. The board is turned over, the adhesive applied, the SMCs placed, and the adhesive cured. The board is then turned over again and wave soldered.

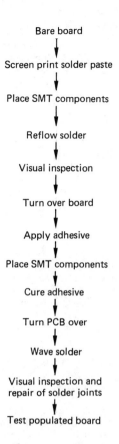

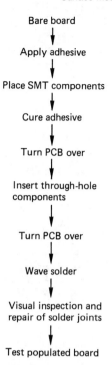

Figure 7.10 The steps involved in the mixed assembly of a double-sided printed circuit board (PCB) with surface-mount components (SMCs) on one side only

Figure 7.9 The steps involved in the assembly of a double-sided printed circuit board (PCB) with surface-mount components (SMCs) on both sides

7.6 Surface-mount-component attachment methods prior to soldering

For flow soldering, SMCs must be held in place by an adhesive until they are soldered. The adhesive is applied to the component sites prior to SMC placement, and is cured immediately after placement. Manufacturing speed is dependent on the adhesive being quickly cured. SMT flow soldering subjects the SMCs to one or more passes through molten solder while hanging from the bottom of the board (*Figure 7.12*). Therefore, it is important that the cured adhesive has sufficient holding power for the task in hand, and that it retains the holding power at the elevated temperatures present during the soldering process.

Adhesives are generally left in place after the soldering process. Thus, long-term reliability depends on the adhesive being fully cured.

7.6.1 Adhesive dispensing

There are three methods by which a small amount of adhesive can be applied at each site where a surface-mounting device is to be placed.

Screen printing This method employs a polyester or stainless-steel screen consisting of a rigid frame on which is a stretched mesh. The screen is held at a distance from the substrate to

which the adhesive or cream is to be applied and forced across the screen. Screen printing has the advantage of speed, but it cannot be used after conventional components have been inserted through the board.

Syringe A preset quantity of adhesive can be applied via a dispenser. The quantity of adhesive can be controlled by the adjustment of air-supply pressure, pressure-pulse duration, viscosity of adhesive, and the internal diameter of the nozzle. Dispensing the adhesive from a syringe is feasible for small boards with an assembly rate of less than abut 5000 components per hour. The syringe can either form part of the programmed pick-and-place machine or can be hand held for prototype manual assembly or repair work.

Pin transfer A pin is raised from below the surface of the adhesive and the small amount of adhesive at the tip of the pin is transferred to the substrate. With this method, either single pins can be used sequentially around the board or, for larger boards, an array of pins can be used for simultaneous application.

7.6.2 Curing the adhesive

Spots of adhesive are applied to the board followed by the placement of components upon it. At this stage the adhesion must be adequate to hold the component in place while the board is being handled. The adhesive is then hardened by curing.

To cure the adhesive the following methods are commonly used.

(1) Heating in an oven—a typical curing time for epoxy formulations might be about 6 min at 150°C.

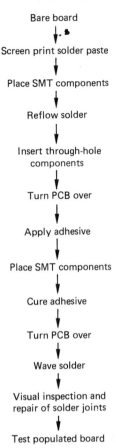

Bare board

↓

Screen print solder paste

↓

Place SMT components

↓

Reflow solder

↓

Insert through-hole
components

↓

Turn PCB over

↓

Apply adhesive

↓

Place SMT components

↓

Cure adhesive

↓

Turn PCB over

↓

Wave solder

↓

Visual inspection and
repair of solder joints

↓

Test populated board

Figure 7.11 The steps involved in the mixed assembly of a double-sided printed circuit board (PCB) with surface-mount components (SMCs) on both sides

(2) Ultraviolet radiation—in an ultraviolet conveyor system the typical curing time may be about 3 min at 150°C for acrylic formulations.

7.6.3 Solder pastes

Solder paste or solder cream consists of a powdered solder alloy bound in a flux medium. The paste holds the surface-mount components in place until the solder is reflowed, flux cleans the surface and promotes wetting, and the solder forms the final joint.

It can be applied in one of several ways. The most common method for volume production is screen printing, but other methods such as stencilling, dispensing from a syringe, roller coating and pin transfer are used.

7.6.4 Flux

Flux is used to prevent the formation of oxides or other compounds of the metal with components of air. Fluxes for soldering electronic assemblies have three main functions:

(1) to make the surfaces to be joined wettable and active;
(2) to protect those surfaces from contact with air, in order to prevent oxidation of the molten solder; and
(3) to promote wetting of the surfaces to be joined by

modifying the surface forces which control the wetting process.

Flux is applied to the surfaces to be soldered either by conventional wave soldering methods such as foaming and waving or as a flux paste.

Fluxes currently used for soldering fall into the following three main classes:

(1) rosin-based fluxes,
(2) synthetic fluxes, and
(3) organic fluxes.

7.7 Soldering

7.7.1 Wave soldering

Wave soldering can be done by the single-wave, dual-wave or the conventional drag-soldering approach. In dual-wave systems, the first wave is a turbulent wave which more evenly distributes the solder around the component. The second wave is a smoother wave which finishes the formulation of fillet and removes the excessive solder.

Wave soldering machines comprise a conveyor which transports the populated board either continuously or in a stepwise fashion from a loading position to, in turn, a fluxing station, a preheat stage, the solder wave and a cooling station, before removal from the conveyor at the unloading point. A gradual preheating profile is required to minimise the damaging effect of thermal shock (*Figure 7.12*).

Soldering of surface-mounting assemblies is carried out at a temperature in the range 235–260°C with a contact time of 1–4 s. The SMCs which move through the molten solder must be able to withstand this treatment, and remain unaffected either by the high temperature or the temperature gradients involved.

Advantages

(1) Suitable for soldering both through-hole and surface-mount components.
(2) Low technical risk when used on passive components.
(3) Especially suitable with a high percentage of conventional components.

Disadvantages

(1) SMCs are immersed in solder and hence exposed to large thermal gradients. They are not therefore suitable for temperature-sensitive surface-mount ICs.
(2) Difficulty in wetting all joints when components are closely spaced.
(3) Adhesive dispensing is required prior to placement.
(4) Dross is produced.

Figure 7.12 Wave soldering system

7.7.2 Reflow methods

Advantages

(1) Suitable for soldering a wide variety of surface components.
(3) Thermal gradients are easily controlled.
(3) The process is insensitive to component geometries.

Disadvantages

(1) Requires heavy investment in new equipment and process.
(2) There is a greater difficulty in adding conventional components.

7.7.2.1 Infra-red soldering

The process consists of a tunnel constructed from emitter lamps and refractory material. Both the mesh- and meshless-type conveyor holds the board and travels through the tunnel. The first segment of the system incorporates a short preheating zone which draws off flux volatiles in the solder paste. The next segment consists of a vent to extract these volatiles. The main area comprises the heaters with various controlled zones of top and bottom emitters. After a proper soldering profile has been established, the system is very repetitive (*Figure 7.13*).

Advantages

(1) Progressive and programmable heating rates.
(2) Adjustable solder temperature.
(3) Limited tombstoning.
(4) Low-cost operation.

Disadvantages

(1) For SMCs only.
(2) Solder paste dispensing required.
(3) Uneven temperature across the board.
(4) Temperature-sensitive components may require manual soldering.
(5) Some components such as J-bend and leadless chip carriers are not transparent to infra-red radiation.

7.7.2.2 Vapour phase reflow method

The vapour phase reflow method (*Figure 7.14*) involves passing the assembly to be soldered through a heated vapour. When a vapour comes into contact with the PCB below its condensation point, the vapour releases its latent heat of

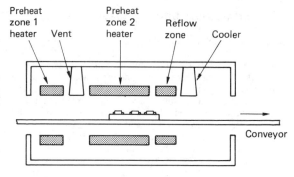

Figure 7.13 Infra-red reflow soldering system

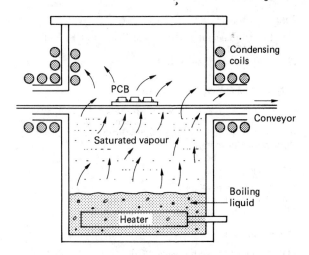

Figure 7.14 Vapour phase reflow soldering system. PCB, printed circuit board

vaporisation and condenses to form discrete droplets on the surface. As the vapour condenses on the cooler assembly it raises the temperature of the assembly to the point where the solder paste melts into solder. This process typically takes 15–30 s.

The fluid used in this process is usually one of a series of vaporised fluorocarbons with boiling points ranging from 174°C to 253°C.

Advantages

(1) Can be used for high density boards.
(2) The entire assembly is brought up to a well-controlled, uniform temperature where all solder connections are reflowed almost simultaneously.
(3) There is no large thermal gradient during heating.
(4) Components on both sides are soldered at the same time.
(5) Less flux is required because the soldering process takes place in an inert atmosphere.
(6) Components on both sides of the board can be soldered in one operation.

Disadvantages

(1) For SMCs only.
(2) Solder-paste dispensing required.
(3) The use of expensive chemical fluids is required.
(4) The high heating rates may damage some components.
(5) Mild acid formation occurs due to the decomposition of the working fluid.
(6) Tombstoning occurs.

7.7.2.3 Laser reflow soldering

With laser reflow soldering, each joint is made sequentially instead of simultaneously. Carbon dioxide lasers are used to heat the solder joints. Because the energy beam of the laser is highly focused, it is well suited to the soldering of temperature-sensitive components as well as narrow-spaced leads.

Advantages

(1) Only the lead solder paste is heated.
(2) Can be used for temperature-sensitive components.
(3) Produces high quality solder joints.

Disadvantages

(1) Low throughput because each joint must be reflowed individually.
(2) Very high equipment cost.

7.7.2.4 Conductive belt reflow soldering

In this process a flat heat-conducting belt is drawn over a controlled heated surface. The PCBs must have a very flat bottom surface and they are placed on the belt and moved through the heated area.

Advantages

(1) Low equipment cost.
(2) Simple operation.
(3) Moderate control over temperature profile.

Disadvantages

(1) Incompatible with organic PCBs.
(2) Small maximum board size.

7.8 Cleaning

The main function of cleaning an electronic assembly after soldering is to remove flux residues. Pure non-active resin flux should, by preference, be left unwashed on the board, but more usually an activated flux is required for good soldering and consequently washing is normally required. If water soluble flux is used, washing is mandatory. If washed with water, the subsequent drying process is expensive. Resin fluxes are washed away with solvents that are capable of removing both ionic and non-ionic contaminants from the assembly.

The type of flux used in wave soldering or the type of flux contained in the solder paste is very important in deciding whether or not to clean the assembly.

7.9 Automatic testing equipment

With the use of SMT there are fewer joints or solder joints to probe. The approach adopted has been to divide the board into 'modules'. These modules are tested on their testers in a quasi-functional test. The module is made small enough so that the fault can be narrowed down to a specific component in the module.

It is important to test the bare board as a large number of faults are due to faults in the board itself. It is less expensive to detect and correct such a fault in an unpopulated board than in an assembled one.

7.10 Surface-mount component placement machines

7.10.1 Selection criteria for placement machines

In selecting a placement machine, the following criteria must be considered.

(1) the PCB size to be assembled;
(2) the number of different PCB designs and batch sizes;
(3) the number of component types;

(4) the ease of programming the machine;
(5) the supply packing of components;
(6) the layout restrictions on the PCB;
(7) the testing of components for identity, polarity and faults;
(8) assembly performance speed;
(9) placement accuracy and reliability;
(10) assembly costs;
(11) adhesive dispensing; and
(12) flexibility.

7.10.2 Types of machine

There are four different placement methods used for SMCs.

In-line placement The PCB moves on a conveyor under several placement stations. Each module is dedicated to the placement of one type of package. Boards are loaded sequentially as they progress down the line. As a separate machine is required for each component, this approach is economical only for boards with a small number of components. Reconfiguration of the machine is difficult. In-line placement is best suited to the manufacture of a single product in very high volumes.

Sequential placement One or two software controlled pick-and-place heads pick up a SMC from a feeder or a feeding station and place it on the PCB. On some machines, the head moves in both x and y directions, while others employ a movable table to place the PCB under the fixed head. The latter technique is currently the most commonly used placement method. The accuracy of placement tends to be ± 0.1 to ± 0.2 mm for the x and y directions and $\pm 1°$ to $\pm 2°$ for rotation.

Simultaneous placement Multiple heads transfer an array of devices onto the board at one time. A board, or part of a board, is populated in a single operation. In this approach strict rules must be observed in the design of the board. The components must be laid out on a grid format which corresponds to the available feeder locations on the machine. The maximum number of components cannot exceed the number of feeder locations. Reconfiguration of the machine is usually very time consuming.

Sequential/simultaneous placement These machines incorporate an x–y table and multiple heads in succession. The heads are placed on a horizontally rotating turret. Each head picks up components at the feeding station and places them on the PCB.

7.11 Printed-circuit-board layout

The following factors affect the PCB layout.

Orientation The types of component used determine the choice of soldering methods available and thus influence the PCB layout. All components should be laid out parallel to the grid lines. This can result in an increased placement rate because of the reduction in the need for rotation of the pick up tool. The physical orientation of and the spacings between the components are primarily dictated by the soldering method to be used. Reflow soldering has no directionality and so components can be placed at any orientation to each other with a proximity restricted only by the footprint pad size requirements. Wave soldering, however, is directional and access to the pads by the solder wave is required. Thus both the relative orientation of components and their spacing are of prime importance to the reliability of the soldering process.

Component density The component placement equipment that is to be used can impose restrictions on the component density, since some machines require vacant space for placement tools and space for rotating components. Other factors include:

(1) the component placement accuracy of the machine; and
(2) the clearance required for cooling the components, visual inspection and the use of test and rework equipment.

Soldering method The soldering method employed is also a major consideration in the design of the PCB artwork. When adhesive is used to attach a component, a dummy land is placed under the component between the pads on which the component is to sit. This land reduces the effective component standoff height and controls the spread of adhesive.

Test points Testing requirements may require special considerations such as test pads in the circuit and spacing between components that is sufficient to allow the use of test probes. It is general practice to add a solder-coated test pad to the conductor adjacent to the component.

7.12 Possible defects during manufacture

Skip This arises when there is no continuous solder between the substrate and the component and is due to a shadowing effect of the solder pads by a component body as the board passes through the solder wave.

Blow holes This results from the volatilisation and evolution of moisture absorbed in the epoxy of the board laminate. These faults are manifested as visible blow holes in the surface of the solder fillets, internal hidden voids within the fillets and considerable, or even total, blow-out of the solder from the holes.

Solder bridges These are unwanted solder paths between two or more conductors that should not be short circuited. Bridging is caused by excessive solder deposit, poor component placement, inadequate reflow time, or poor screening alignment. The occurrence of bridging can be reduced by changing the mechanism associated with the backwash region of the ware or by a change in the conveyor speed.

Poor solder Poor solder joints arise due to the incorrect setting of the wave height for the land. The solder does not come through the hole in a through-hole board.

Tombstone The 'tombstone', 'Manhattan', 'Stonehenge' or 'drawbridge' effect arises when a chip component is reflow soldered using solder paste, and one end lifts off the substrate during the reflow process. It is caused by using an excessive heating rate, uneven solder deposit, or poor component–pad solderability.

Misalignment of components In this defect the component is displaced sideways or rotationally with respect to its footprint on the PCB. Misalignment can be caused by a placement machine, or a subsequent movement prior to, or during, soldering.

7.13 Introducing surface-mount technology

The introduction of SMT must be part of a valid business strategy aimed at achieving a competitive edge. The impact of SMT is felt in every department. Therefore, teamwork and cooperation is essential from all concerned. The design and development requires computer-aided design systems with new capabilities. New design guidelines for SMCs need to be developed. Material management must develop new procurement methods and meet new goods inspection, storage and stock-control requirements. The manufacturing department must select, purchase and install new production lines for prototype and volume production, gaining new expertise in placement, reflow soldering, and cleaning. The manufacturing and test department must learn quickly and aim for zero defects. There is need to work together with people who will be required to control the successful introduction of the new technology.

The following three approaches have been used by many companies in introducing SMT products.

(1) *Minimum-risk approach*—this involves redesigning an existing board. The board often results in using both through-hole and surface-mount components.
(2) *High-risk approach*—a new product is introduced that is entirely surface mounted. This method is likely to give rise to a successful economic marketing edge.
(3) *Contracting out*—contracting out the surface-mount design to specialised companies is often adopted by smaller companies to evaluate the product in the market before investing in the SMT equipment. This option results in long-term benefits rather than short-term ones.

8 Metal Finishing Processes

T Buttery

Contents

8.1 Introduction

In many applications, the functional behaviour of a component will depend critically on its surface and the regions immediately below. Factors contributing to surface performance are geometric shape, surface roughness R_a, materials properties, surface treatments and residual stresses. The total quality of a surface is, therefore, a function of the machining process by which it was produced and any surface treatment which may have been applied before or after machining to improve specific properties such as fatigue, wear or corrosion resistance.

Invariably, the approach to obtaining a 'perfect surface' involves a high cost penalty due mainly to the low rates of stock removal and high machining costs associated with abrasive machining processes (*Figure 8.1*). Continual efforts are therefore being made to improve metal removal rates and reduce floor-to-floor times for processes such as grinding, linishing and honing. Improved machining techniques, enhanced process control (by microprocessor or computer), in-process gauging and the application of automation or robotics increase the effectiveness of abrasive machining methods. Even though grinding gives good geometric shape, excellent dimensional tolerances and low values of surface roughness, it may not be ideal for some applications. For example, a ground surface has an approximately Gaussian height distribution (*Figure 8.2*), which in bearing applications will wear rapidly during the running-in period. Processes such as plateau honing are therefore applied after grinding to form a 'composite surface' having a topography similar to that of a component which has already been run in. In addition to the machining processes used to manufacture a component surface, treatments can also have major effects on surface quality and performance. A wide variety of processes is available to the engineer, ranging from the purely decorative to methods

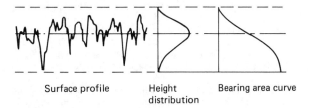

Surface profile   Height   Bearing area curve
                         distribution

Figure 8.2 Surface profile of a ground surface together with its height distribution and bearing area curve

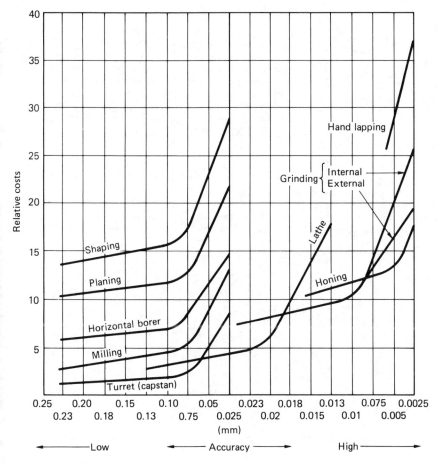

Figure 8.1 Relationship between the relative cost and accuracy for a number of machining processes. (Courtesy of British Standards Institution)

of increasing hardness and improving wear, fatigue, corrosion and scaling resistance. These methods vary in both complexity and cost.

8.2 Abrasives

In order for a material to act as an abrasive it must be significantly harder than the work piece. Richardson[1] has suggested a similarity with the Moh hardness scale, which places materials in 10 groups (0, soft, talc; to 10, hard, diamond) a material in a higher numbered group being able to scratch all materials in lower groups. Richardson concluded that an abrasive must be at least 10% harder than the work piece if it is to be effective. Furthermore, for an abrasive to act as a cutting tool it must have a suitable geometric shape, e.g. a sharp grit will cut more effectively than a blunt one, and as a hardened steel can act as a cutting tool or as a part of a bearing pair.

Many materials can be used as abrasives, but the most well known are silicon carbide (corundum), aluminium oxide (alumina), chromium oxide, iron oxide, magnesium oxide (magnesia), tungsten carbide, diamond (natural and synthetic) and cubic boron nitride. Abrasives are marketed in a variety of forms: solid abrasives such as wheels and sticks, coated abrasives such as belts and discs and loose abrasive in a range of particle sizes. For a particular application, the appropriate form must be chosen.

Although abrasives were originally naturally occurring, most of the abrasives in current use are man made. Silicon carbide and aluminium oxide were developed towards the end of the last century; more recently synthetic diamond (1950s) and cubic boron nitride (1960s) have become available. *Table 8.1* gives the hardness of the most widely used abrasive materials together with their chemical compositions.

There are two types of abrasive operation: two-body abrasion and three-body abrasion.

Two-body abrasion This type of abrasion involves interaction between the abrasive and the work piece. The abrasive is in solid form (grinding wheel, disc, stick or belt) and the work piece is forced against the abrasive making the second constituent of the abrasive pair (*Figure 8.3(a)*). Material is removed from the work piece as fine particles. Shot and vapour blasting are also examples of two-body abrasion; the abrasive pair consisting of the work piece and the abrasive particles propelled at its surface (*Figure 8.3(b)*).

Table 8.1 The hardness, chemical composition and applications of the most widely used abrasive materials

Material	Chemical composition	Knoop hardness (kg mm^{-2})	Applications
Diamond	C	7000	Grinding tungsten carbide, glass and some die steels
Cubic boron nitride	BN (cubic)	4500	Grinding tool steel
Silicon carbide	SiC	2400	Used on cast iron, copper alloys, aluminium alloys and some steels
Aluminium oxide	Al_2O_3	2100	Softer but tougher than silicon carbide; used to grind steels
Sand/quartz	SiO_2	850	Abrasive belts and papers, sand blasting

Three-body abrasion Loose abrasive in a carrier, such as oil or paraffin, is sandwiched between two surfaces (*Figure 8.3(c)*) and, as the surfaces move, the abrasive rolls between them occasionally removing material. Cutting rates are therefore low and the method is used mainly to lap in mating surfaces or to produce very high quality finishes. When geometric shape is not important, a cloth may be used to carry the abrasive slurry; this method is widely used in the preparation of microspecimens.

8.3 Grinding wheels and grinding wheel selection

A conventional grinding wheel is effectively a composite material consisting of abrasive, bond and void; each of these components has a major influence on grinding-wheel perfor-

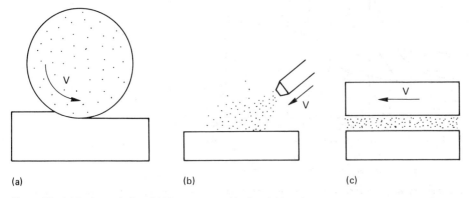

(a)   (b)   (c)

Figure 8.3 (a) Surface grinding. (b) Shot or vapour blasting. (c) Lapping

mance. The density of a grinding wheel (ρ_{wheel}) can be related to that of its constituent parts using the law of mixtures:

$$\rho_{wheel} = V_a\rho_a + V_b\rho_b + V_v\rho_v \tag{8.1}$$

Since $\rho_v = 0$, equation (8.1) can be rewritten as

$$\rho_{wheel} = V_a\rho_a + V_b\rho_b \tag{8.2}$$

where V_a, V_b and V_v are the volume fractions of the abrasive, bond and void, respectively, and ρ_a, ρ_b and ρ_v are the densities of the abrasive, bond and void, respectively.

All the major properties and the ultimate performance of a grinding wheel are related to the proportions of the constituent parts and their individual characteristics.

The convention for the specification and identification of grinding wheels is set out in BS 4481: 1969: Part 1. Grinding wheels are specified by the type of abrasive, grit size, grade, structure and bond (*Figure 8.4*).

Aluminium oxide and silicon carbide wheels are the abrasives most widely used in general engineering. Silicon carbide is the harder material, but suffers chemical wear when cutting steels. It is therefore used for grinding cast iron and non-ferrous materials. The softer but tougher aluminium oxide abrasive is used for grinding steels. For more specialist applications, superabrasives have been developed. These are natural and synthetic diamond and cubic boron nitride (CBN). Due to their high cost, superabrasives are used mainly on materials which are difficult to grind; typically tungsten carbide, glass and ceramics.

8.3.1 Grit size

The abrasive grains making up the wheel are accurately graded and a wheel will usually contain only one size of grit. Wheels having a coarse grit size are used for roughing operations where rapid metal removal is required and surface finish is not critical. Finer grit sizes are used for finishing operations, giving superior surface finish but with a reduction in metal removal rate. The relationship betwen the type of work and the grit size is shown in *Table 8.2*.

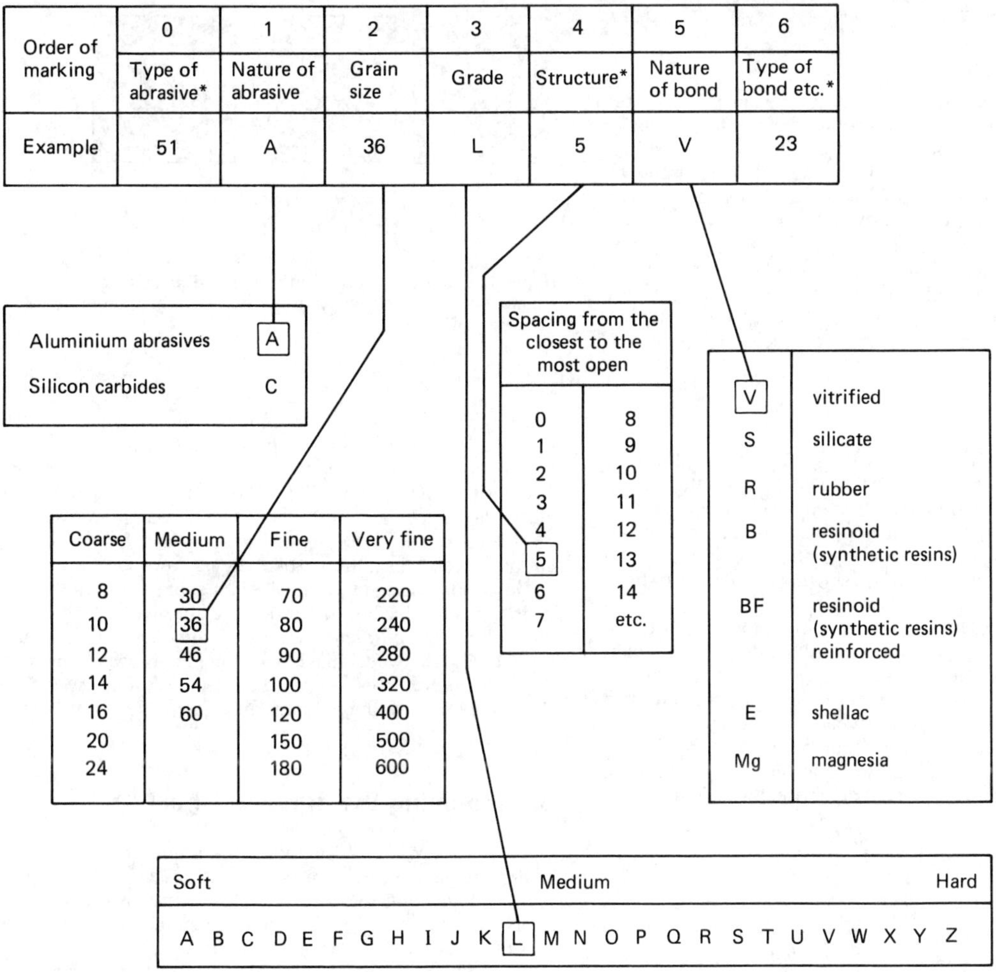

Order of marking	0	1	2	3	4	5	6
	Type of abrasive*	Nature of abrasive	Grain size	Grade	Structure*	Nature of bond	Type of bond etc.*
Example	51	A	36	L	5	V	23

Aluminium abrasives A
Silicon carbides C

Spacing from the closest to the most open

0	8
1	9
2	10
3	11
4	12
5	13
6	14
7	etc.

V	vitrified
S	silicate
R	rubber
B	resinoid (synthetic resins)
BF	resinoid (synthetic resins) reinforced
E	shellac
Mg	magnesia

Coarse	Medium	Fine	Very fine
8	30	70	220
10	36	80	240
12	46	90	280
14	54	100	320
16	60	120	400
20		150	500
24		180	600

Soft   Medium   Hard

A B C D E F G H I J K L M N O P Q R S T U V W X Y Z

*Optional symbols
The symbols 0 and 6 are the manufacturer's own

Figure 8.4 British Standard system for marking abrasive wheels (BS 4481: 1969: Part 1)

Table 8.2 Grit sizes used for different types of work

Type of work	Grit size
Roughing	8–24
General purpose	30–60
Fine finish	70–180
Extra-fine finish	220–600

8.3.2 Wheel grade

The wheel grade indicates the strength of bond holding the wheel together. If the bond is weak the wheel is said to be 'soft'. When grits become blunt in a soft wheel they are readily torn out to expose new sharp grits; a mechanism known as 'self-sharpening'. Since soft wheels are less likely to dull, they tend to give a cooler cut. Hard wheels are strongly bonded together and are less likely to exhibit self-sharpening; their use is favoured in finishing operations. Wheel grade is specified by a code letter (A, soft; to Z, very hard); the complete range of letters is shown in *Table 8.3*.

8.3.3 Wheel structure

The wheel structure refers to the overall arrangement of the abrasive grit, bond and void within the wheel. An open structure with a large amount of void would be preferred for roughing operations involving high rates of metal removal, since an open structure is less likely to load and has good self-sharpening characteristics. Wheels having a dense structure are favoured for finishing operations where a good surface finish is required but, as dense wheels are more likely to load, only low rates of stock removal can be used. Dense wheels are less likely to self-sharpen. Structure is indicated by a numerical scale ranging from 0 (dense) to 12 (open) as indicated in *Table 8.4*.

8.3.4 Bond

The type and quantity of bonding material used in the manufacture of a grinding wheel has a marked effect on its properties and performance. Strength, elastic modulus and flexibility are controlled by the bond, as is the ease with which grits can be dislodged by bond post-rupture (self-sharpening). *Figure 8.5* shows the effects of varying the proportion of bond on wheel structure.

For general applications, the most commonly used bonds are vitrified and resin. Less widely used bonds include shellac, rubber, oxychloride and metal.

Vitrified bonded wheels are used very extensively; this glass-type bond is strong, porous and unaffected by water, acids, oils or normal operating temperatures.

The resinoid bond is basically organic and wheels with this type of bond are more flexible than vitrified wheels. Resinoid

Table 8.3 Code letters used to specify wheel grades

Wheel grade	Symbol
Soft	A to I
Medium	J to Q
Hard	R to Z

Table 8.4 Numerical scale used to indicate wheel structure

Structure		Symbol	Applications
Dense		0 to 4	Finishing operations
Medium		5 to 7	General operations
Open		8 to 14	High rates of metal removal

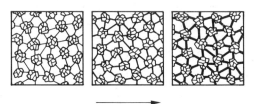

Increasing quantity of bond

Figure 8.5 Effects of varying the amount of bond on the structure of a grinding wheel. (Courtesy of Carborundum Abrasives GB Ltd)

wheels give high rates of stock removal and are widely used in foundries and fabrication shops for fettling and cutting-off operations. Shellac bonded wheels are capable of producing fine finishes on camshafts and mill rolls, but are unsuitable for heavy duty applications. Rubber bonded wheels are used for most feed wheels in centreless grinding and for specialised cut-off wheels which are particularly narrow (≤1 mm).

Metal bonding is used mainly with superabrasives such as diamond and cubic boron nitride. The abrasive is bonded to a metal disc using powder metallurgy techniques or electroplating. Finally, the wheel-maker can include his own identifying mark. Superabrasives are specified using a similar system to that used for conventional materials (*Figure 8.6*).

8.4 Mounting the grinding wheel

The Abrasive Wheel Regulations 1970 require that 'no person shall mount an abrasive wheel unless he has been trained to do so or is competent to do so'. Employers are also required to keep a register of persons appointed to mount grinding wheels and the class of wheels which they can mount.

Having selected the correct type and size of grinding wheel for the particular grinding machine or operation, the wheel should be examined carefully to ensure that it is free from defects. Visual inspection and a 'ring test' can find most defects. If the wheel is satisfactory it can be fitted to the

Superabrasive Marking System

Diamond Product Grading

MD	—	126	—	R	—	50	—	B	—	25	—	3 mm
ABRASIVE TYPE		*FEPA GRIT SIZE*		*GRADE*		*CONCENTRATION*		*BOND*		*SUFFIX*		*COAT DEPTH*
D		502 (Coarse)		I (Soft)		25 (Low)		B – Resinoid		Various		in millimetres
MD		to		to		to		M – Metal				
CMD		46 (Fine)		U (Hard)		100 (High)						

CBN Product Grading

CB	—	181	—	150	—	V	—	ST-H10
ABRASIVE TYPE		*FEPA GRIT SIZE*		*CONCENTRATION*		*BOND*		*SUFFIX*
XCB		252 (Coarse)		25 (Low)		B – Resinoid		Various
CB		to		to		M – Metal		
		46 (Fine)		200 (High)		V – Vitrified		

CBN – Cubic Boron Nitride

Figure 8.6 Superabrasive marking system. (Courtesy Carborundum Abrasives GB Ltd)

machine after a final check of wheel size, bore fitting, flanges and maximum operating speed. The wheel should fit easily onto the spindle and is clamped into position using the correct flanges and washers, as appropriate. Where a nut is used to clamp the wheel in position this should tighten up in the opposite direction of rotation of the wheel. Although the wheel should be held securely, over-tightening must be avoided. Where a torque figure is recommended, a torque spanner can be used.

8.5 Balancing and dressing

To ensure safe operation and to achieve the best quality of surface on the work piece, the wheel must be balanced and dressed. Several methods of balancing are available: static, dynamic and automatic balancing/self-balancing. In many applications, static balancing (which is simple and cheap) is quite adequate. With static balancing the balance weights are usually incorporated in the wheel flanges. Initially, the balance weights are set opposite each other, the wheel assembly is fitted to the grinding machine and then trued by passing a dressing diamond across its face. The wheel is now true but will be out of balance to some degree. To remove this 'out of balance', the wheel assembly is taken from the machine and mounted on a balancing arbor (*Figure 8.7*). The balance weights on the wheel flange are checked to ensure that they are positioned opposite each other (*Figure 8.8(a)*) and the wheel is placed on the balancing fixture. This fixture provides two parallel edges which can be levelled accurately. The wheel is allowed to rotate freely until it stops; the heaviest point is now at the bottom. Adjustments are now made to the balance weights which are moved until they are positioned as shown in *Figure 8.8(c)*). The balance weights are now moved a little at a time in the direction of the arrows (*Figure 8.8(d)*). After each small movement the wheel is allowed to rotate to see if it still stops with the marked position at the top. When the wheel is correctly balanced the mark will stop in a random position when the wheel is gently rotated. The wheel is now statically balanced and is replaced on the machine, dressed ready for grinding.

Dynamic balancing is performed with the wheel in place on the machine in a similar manner to wheel balancing on a car. Since this method is superior to static balancing it is used when machining critical components. Automatic balancing or self-balancing methods have been developed and have the advantage of correcting any imbalance which develops as the wheel wears.

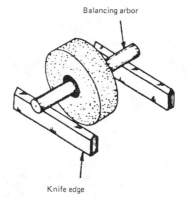

Balancing arbor

Knife edge

Figure 8.7 Grinding wheel mounted on an arbor ready for balancing

To prepare the wheel for use it must be dressed, an operation which ensures that the wheel is true and that the abrasive grits are in the correct condition for cutting. A profile can be dressed onto the wheel for manufacturing products such as threads and gears.

Dressing can be carried out in several ways, the most widely used methods being crush, diamond and roller dressing. Crush dressing uses a star wheel to ensure that the grinding wheel is clean, sharp and concentric on simple bench grinders and is unsuitable for precision grinding. For general dressing operations, diamond dressers are probably the most popular. Dressing is carried out by tracking the sharp dressing diamond across the face of the wheel which is rotated at normal speed. Shallow cuts are taken, removing a small layer from the surface of the wheel; loading and dulled grits are removed and concentricity maintained. Usually, several passes of the diamond are made progressively, reducing the depth of cut until the final pass may be at 'spark out'. Severity of dressing is one of the factors controlling wheel performance. Coarse dressing with a deep cut and high cross traverse speed gives a wheel with an open structure and well spaced sharp grits. A coarse dressed wheel is less likely to load, gives low cutting forces (the wheel cuts cool), high wheel wear and poor surface finish. Fine dressing where the diamond takes a small cut and is traversed slowly across the face of the wheel gives a wheel with more cutting edges, but these are blunt. Used in this condition the wheel will generate high cutting forces. Therefore, only

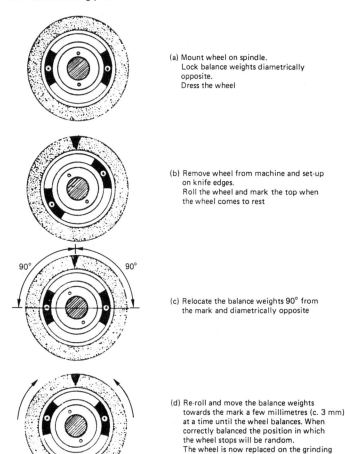

(a) Mount wheel on spindle.
Lock balance weights diametrically opposite.
Dress the wheel

(b) Remove wheel from machine and set-up on knife edges.
Roll the wheel and mark the top when the wheel comes to rest

90°   90°

(c) Relocate the balance weights 90° from the mark and diametrically opposite

(d) Re-roll and move the balance weights towards the mark a few millimetres (c. 3 mm) at a time until the wheel balances. When correctly balanced the position in which the wheel stops will be random.
The wheel is now replaced on the grinding machine and dressed ready for use

Figure 8.8 Procedure for balancing a grinding wheel using weights fitted to the wheel flange

shallow cuts can be made without the risk of thermal damage to the work piece. However, the larger number of grits and their relatively blunt shape produces a good surface finish. A summary of the effects of dressing on grinding performance is given in *Table 8.5*. If a profiled wheel is required, for example for the manufacture of gears or screw threads, this can also be produced by diamond dressing. The diamond is tracked across the face of the wheel using a proprietary system such as the diaform, special dressing attachment, or under computer numerical control (CNC). Crush dressing involves forcing the grinding wheel against a profiled dresser and crushing the grains, so that eventually the wheel takes up the shape of the dresser. The method is particularly suited to deep profiles and gives a sharp free cutting face to the wheel. The technique is often used for one-step manufacture of components for the automobile and aerospace industries by plunge or creep feed grinding.

The dressing/truing of superabrasives is minimised due to their hardness and cost. Since for most applications the abrasive is only a thin surface layer, it is best to take extra time in fitting the wheel to ensure the best concentricity. The wheel is then finally trued using an abrasive wheel dresser, removing the minimum of superabrasive material. Although truing and dressing are often seen to be overlapping in conventional

Table 8.5 Summary of the effects of dressing on grinding performance

Property	*Coarse dressing* (deep cut high cross-feed/rev.)	*Fine dressing* (shallow cut or spark-out small cross-feed/rev.)
No. of active grits	Small number	Larger number
Structure	Open	Close
Shape of grits	Sharp	Blunt
Cutting forces	Low	High
Risk of thermal damage	Low	High
Rate of stock removal	High	Low
Surface finish	Poor	Good

grinding, this is not true with superabrasives. Dressing conditions the wheel by removing bond and exposing the grits at the surface; a dressing stick of aluminium oxide is used for this purpose, the wheel being rapidly rotated. When properly carried out, dressing should only expose the superabrasive grains and make no alteration in wheel geometry.

8.6 Grinding mechanics

The mechanism of metal removal in abrasive machining operations is highly complex. Several factors distinguish grinding from other metal cutting operations; the abrasive grits in the wheel are very small and randomly orientated; the ratio of tangential to normal force is 0.5 compared with a ratio of 2 in single-point cutting; and the specific energy u is high, typically 50 J mm^{-3} compared with 3 J mm^{-3} for single-point cutting. These observations suggest that grinding processes are significantly different from other metal cutting operations. Examination of the grinding debris (*Figure 8.9*) shows some chips of conventional appearance plus rather more randomly shaped chips, metal globules and abrasive particles. It is only the outermost grits which interact with the work piece by cutting, ploughing or rubbing.

Grinding theory is discussed here using the following symbols.

F_n normal force on the work (N)
F_t tangential force on the work (N)
F_g average force per grit (N)
D wheel diameter (mm)
D_w work piece diameter (mm)
d wheel depth of cut (mm)
b width of cut or cross-feed (mm)
V wheel speed (m min^{-1})
v work speed (m min^{-1})
μ grinding coefficient F_t/F_n
u specific energy of grinding (J mm^{-3})
t grit depth of cut (mm)
G grinding ratio (volume of metal removed/volume of wheel removed)
C number of active grits per unit area (grits/mm^2)
r width/depth ratio of an average groove (equal to half the angle of an average groove)
Δ_a average slope of the ground surface

Any attempt to model the grinding process must of necessity involve a number of assumptions. From the earliest studies of Alden[2] and Guest,[3] grinding has been regarded as a form of micromilling, the abrasive grits acting as randomly orientated cutters. One of the most widely used parameters derived from the modelling procedures is the concept of grit depth of cut t. This is the depth of cut taken by an average grit (*Figure 8.10*). Backer *et al.*[4] have derived the following widely used expressions for grit depth of cut in surface and cylindrical grinding:

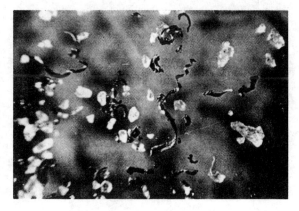

Figure 8.9 Grinding debris showing chips, metal globules and abrasive particles

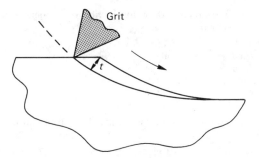

Figure 8.10 Diagram showing the grit depth of cut concept

Surface grinding $t = \left[\dfrac{4v}{VCr} \sqrt{\dfrac{d}{D}} \right]^{1/2}$ (8.3)

Cylindrical grinding $t = \left[\dfrac{4v}{VCr} \sqrt{\left(\dfrac{d(1 + D/D_w)}{D} \right)} \right]^{1/2}$ (8.4)

Equation (8.3) for surface grinding is a unique form of equation (8.4) with $D_w = \infty$.

All the functions in the above equations are readily available with the exception of the number of active grits per unit area C and the width/depth ratio r. One method of determining the value of C is to roll the wheel across a piece of smoked glass and then count the number of points where the glass is clear. This method has several disadvantages: the wheel must be removed from the machine to take a reading and the method only counts the outermost grits, giving no indication of any changes which may occur as the wheel penetrates the work piece. An alternative method uses a surface measuring instrument fitted with a chisel-edged stylus capable of giving a high-spot count p_c. The number of active grits per unit area is given by the high-spot count divided by the area traversed (width of stylus × traverse length). Sophisticated surface measuring equipment can indicate the high-spot count at varying levels, thus allowing the value of C to be determined at varying levels below the outer surface of the wheel. The value of C depends on the properties of the wheel, in particular the grit number, the dressing procedure and the grit depth of cut. Using the stylus method, the value of C can be monitored whilst the grinding wheel is mounted on the machine. In their original work, Backer *et al.*[4] determined the width/depth ratio r of an average scratch by taking a taper section of the work piece surface and measuring a number of typical scratches. More recently,[5] it has been suggested that the average slope Δ_a of the work piece surface could be used to derive the value of the width/depth ratio r. The value of r can be obtained using the expression:

$r = 2 \tan (90 - \Delta_a)$ (8.5)

The average slope of the work piece is measured using a stylus surface-measuring instrument of the appropriate type.

Taper sectioning is a destructive method, whereas stylus techniques are non-destructive and allow the process to be monitored continuously.

Although stylus methods can be carried out without removing either the wheel or the work piece from the machine and the values for C and r are more reproducible, the values obtained are only averages. Consequently, the value calculated for the grit depth of cut t will also be an average.

Once values have been determined for C and r, the grit depth of cut t can be calculated by substituting the grinding conditions in equation (8.3) for surface grinding or in equation (8.4) for cylindrical grinding.

For a surface grinding operation carried out under the following conditions

work speed v = 0.5 m s^{-1}
wheel speed V = 30 m s^{-1}
depth of cut d = 0.04 mm
wheel diameter D = 200 mm
No. of grits C = 2 grits/mm^2
width/depth ratio r = 18

the value of the grit depth of cut is calculated by substituting in equation (8.3):

$$t = \left[\frac{4 \times 0.5}{30 \times 2 \times 18} \sqrt{\frac{0.04}{200}} \right]^{1/2}$$

$$= 0.005 \text{ mm}$$

This calculation shows that the grit depth of cut is very much less than the wheel depth of cut, typically only about one-tenth. Despite the fact that it is difficult to determine a reliable value for the grit depth of cut t, the concept is very useful for predicting the effects of changes in grinding conditions on grinding performance.

The ability of a wheel to self-sharpen will depend to some extent on the average load per grit F_g. Assuming the average force per grit is proportional to the cross-sectional area of the undeformed chip, equation (8.3) can be rearranged to show that:

$$F_g \propto \frac{v}{VC} \sqrt{\frac{d}{D}} \tag{8.6}$$

As the value of F_g increases, the wheel will be more likely to self-sharpen and the wheel will thus act softer. Conversely, reducing the average load per grit will make the wheel act harder as it will be less likely to self-sharpen. The average load per grit F_g will increase as the work speed v and wheel depth of cut d are increased, and F_g is reduced as the wheel speed V, the number of active grits C and the wheel diameter D are increased. Dressing can affect the average load per grit in two ways: an open wheel will only have a few active grits and, therefore, a high average load per grit; and sharp grits will carry less load than blunt grits. The surface finish R_a produced in a grinding operation can be estimated from the grit depth of cut t using the equation

$$R_a \approx 0.25t \tag{8.7}$$

This equation assumes that the depth of an average groove on the work piece is equal to the grit depth of cut. By reference to equation (8.3) the factors favouring a good surface finish can be identified as those which give a small value for t. Thus in order to produce a good surface finish a low rate of metal removal (small values for v and d), a finely dressed wheel with a small grit size (large values of C and r) and a large diameter wheel (D) are required. In practice, grit depth of cut is not the only factor influencing surface finish; machine condition, work piece material, system stiffness and coolant can also have a significant effect on surface finish.

The mechanism of metal removal in the grinding process is complex and the grinding forces arise as a result of cutting, ploughing and rubbing. Indeed, as a freshly dressed wheel is used, wear flats form rapidly on the grits, increasing forces without contributing to the cutting processes. If, as in precision grinding, there is little self-sharpening, the blunting of the grits due to wear flats will control the frequency of dressing.

Grisbrook's[6] classic work on surface grinding shows the progressive increase in grinding forces with the number of passes of the work piece under the wheel (*Figure 8.11*). Hahn[7] has suggested that in controlled force grinding processes, the

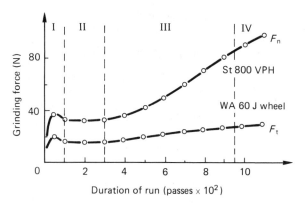

Figure 8.11 Relationship between grinding forces (normal F_n and tangential F_t) and duration of run

wheel develops optimum sharpness; when too sharp the wheel will blunt and when the wheel is dull it will self-sharpen (*Figure 8.12*).

The magnitude of the forces developed in grinding will depend on the machine settings, the work piece material and the condition of the wheel (dull or sharp). Grinding forces affect many aspects of the processes, including power consumption, flash temperatures, grinding ratio, specific energy and the risk of work piece damage.

The energy input to the processes is used in three ways: cutting, ploughing and rubbing. The specific energy u is, therefore, made up of three terms:

$$u = u_{\text{cutting}} + u_{\text{ploughing}} + u_{\text{rubbing}} \tag{8.8}$$

The overall value for the specific energy is given by

$$u = \frac{VF_t}{vbd} \tag{8.9}$$

Much of the energy expended in the grinding process is dissipated as heat, which is divided between the chip, abrasive grit and the work piece. Thus very high flash temperatures are

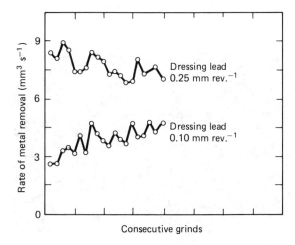

Figure 8.12 Effect of dressing lead on metal removal rate as grinding progresses

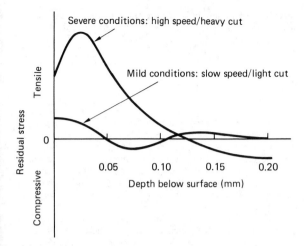

Figure 8.13 Effect of grinding conditions on residual stresses

generated at the chip–grit interface which will influence the wear of the grit. A proportion of the heat is carried away by the chip and the remainder goes into the work piece. Even when coolant is present the immediate surface of the work piece may be heated to temperatures which will produce changes in both structure and properties. Hardened steels are particularly prone to grinding damage. This can take the form of tempering where soft spots are produced on the surface of the component or, if conditions are very severe, the surface may be raised to the hardening temperature of the steel which on rapid cooling will produce martensite. Residual stresses are also produced as a result mainly of the temperature gradients associated with grinding. Typical stress distributions in the surface of ground components are shown in *Figure 8.13*. If conditions are exceptionally severe, the stresses may produce grinding cracks; these are generally at right angles to the direction of grinding. To avoid grinding damage a coolant should be used and the grinding conditions should not be too aggressive. Since grinding damage is often difficult to detect, critical components should be inspected.

Grinding cracks in steel components can be readily found using magnetic crack detection. Tempering and martensite formation are best detected by lightly etching the steel in the appropriate etching reagent. Tempered areas show up darker than the surroundings and martensite, burnt or glazed areas are lighter than the surroundings.

8.7 Wheel wear

Wheel wear is both an important and a complex parameter. Both the grinding ratio and the frequency of dressing are related to the rate at which the abrasive wears. Three major mechanisms of abrasive wear are: attritious wear, fracture and bond post-rupture.

8.7.1 Attritious wear

With this type of wear the abrasive grain is dulled due to a wear flat forming by a process called 'attrition'. The grit and the work piece react at the high flash temperatures associated with grinding, with wear taking place by a combination of physical and chemical processes. Attritious wear is lower

between materials which are chemically inert; aluminium does not react strongly with steel and so gives a lower rate of attritious wear than silicon carbide which will dissolve in molten iron.

8.7.2 Fracture

The grains of abrasive are hard and brittle so that if the forces on the grit build up as a result of wear-flat formation the grit may fracture and expose fresh cutting facets. The ease with which a grit fractures is called its 'friability'; good friability in an abrasive means that it is easily fractured. Aluminium oxide is less friable than silicon carbide and the grits in an aluminium oxide wheel tend to self-sharpen at a slower rate than those of silicon carbide.

8.7.3 Bond post-rupture

When a grit has become very dull and worn, the cutting forces may be sufficient to fracture the bond removing the grit and thus expose fresh abrasive. The ease with which a grit can be removed depends on the grade of wheel; soft wheels will self-sharpen easily maintaining a continuous supply of sharp grits and hence give low cutting forces, reducing the risk of thermal damage and residual stresses.

It must also be remembered that in many grinding operations it is the dressing operation which accounts for the bulk of wheel wear.

8.8 Grinding ratio

Grinding is no different to other metal cutting operations in that, as material is removed from the work piece, there is corresponding tool wear; in this case wheel wear. The grinding ratio G is a method of expressing this wear and is calculated using:

$$G = \frac{\text{Volume of metal removed}}{\text{Volume of wheel removed}} \qquad (8.10)$$

Many factors affect the grinding ratio, including work piece material, type of wheel, and machine conditions. Although there are some advantages in achieving a large value for G (typically 100–500 in fine grinding), such as a longer wheel life, it may lead to duller grits, higher forces and an increased possibility of damage to the work piece. The rate of wheel wear G is not constant throughout the grinding cycle. *Figure 8.14* shows volume of metal removed plotted against the volume of wheel removed. It shows that, initially, the freshly dressed wheel wears rapidly, but the rate of wear declines until a steady-state region is reached. It is in this region that the bulk of grinding takes place. Normally the wheel is redressed before the third stage where the wear rate begins to accelerate.

8.9 Grinding forces

Some estimate of grinding forces can be made by relating them to the specific energy. Equation (8.11) relates the power used in surface grinding to the tangential grinding force F_t, machine settings and specific energy u:

$$\text{Power} = F_t V = bdvu \qquad (8.11)$$

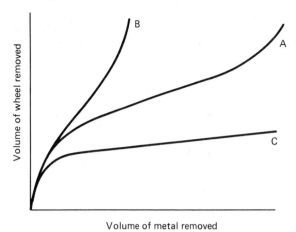

Figure 8.14   Volume of metal removed vs. volume of wheel removed; A, wheel satisfactory; B, wheel too soft; C, wheel too hard

hence

$$F_t = \frac{bdvu}{V} \qquad (8.12)$$

Unfortunately, the specific energy is not a constant even for one material as it depends on the condition of the wheel, the grinding conditions and the grit depth of cut. However, if an appropriate value is assumed for u, both F_t and the power can be estimated.

Similarly, the normal force F_n can be estimated using the expression for the grinding coefficient and rearranging:

$$\mu = \frac{F_t}{F_n} \qquad (8.13)$$

hence

$$F_n = \frac{F_t}{\mu} \qquad (8.14)$$

Estimated values of the forces in grinding are of considerable value since their practical determination requires very specialist equipment.

8.10 Coolant

Although grinding can be carried out dry in most applications, there are considerable advantages in using a coolant or lubricant. The most widely used coolants are water based soluble oils. Since the coolant is continuously recycled it must undergo a reconditioning treatment before re-use. The contaminants in the coolant (worn abrasive, and metallic and oxide particles) can be removed in several ways. The simplest and most economical method consists of a series of holding tanks and weirs; the grinding debris settles out by gravity. An alternative system is to filter the coolant through a continuously fed filter paper, and for very critical work this method can be enhanced by means of a centrifuge.

Having ensured an adequate supply of clean coolant, it is essential that it is applied correctly. Grinding damage can occur if coolant supply is insufficient and this damage is not always obvious, hence the need to inspect all critical parts after grinding.

8.11 Grinding processes

Grinding can be carried out on a wide variety of materials using a range of configurations. Conventional processes consist of surface grinding, cylindrical grinding (internal and external) and centreless grinding. Machines are designed to traverse the work past the wheel using cross-feed; alternatively, the work is fed directly into the wheel—a technique known as 'plunge grinding'. By dressing a profile onto the wheel, complex shapes such as screw threads and gear teeth can be produced.

Grinding is a relatively expensive method of machining and is generally used as a finishing process where good surface finish and accurate dimensions are required. Grinding is also applied to materials which are difficult to machine by conventional methods, e.g. hardened steels, aerospace materials, glass and ceramics.

8.11.1 Surface grinding

Surface grinding machines use a range of configurations, the most common being a reciprocating table and a horizontal spindle. The work reciprocates under the wheel, and most machines provide automatic cross-feed and automatic down-feed. There is also increasing demand for variable wheel speed. Components can also be plunge ground. Complex shapes can be manufactured by dressing a profile on the wheel using the appropriate attachment. Creep feed grinding is usually carried out in the surface-grinding mode. This process differs from conventional grinding in that the wheel depth of cut d is increased considerably (approximately 1000) and the work speed v reduced proportionally; hence the term 'creep-feed grinding'. Machines used for this type of grinding operation are normally special purpose if the process parameters are to be adequately controlled. Control of coolant application and wheel wear are particularly critical; often the wheel is continuously dressed throughout the process, appropriate adjustments being made for changes in wheel diameter. Creep-feed grinding is used to machine deep complex profiles into the work piece; typical products being the fir-tree serrations of gas turbine blades and connecting rod assemblies (*Figure 8.15*).

Surface grinding machines are available with computer numerical control (CNC), allowing the complete process to be carried out under computer control, including the dressing cycle, machining and gauging. In surface grinding CNC gives many benefits, including reduced cycle times, complete automation of the grinding cycle, more consistant quality and improved productivity. Further improvements in machine utilisation can be achieved by preparing programs off-line.

The complexity of product which can be produced by CNC grinding is shown in *Figure 8.16(a)*. In this example, the work piece is a semiconductor die used for the manufacture of integrated circuits; eight components, ganged in line, were machined at the same time. In each component there are 20 slots, each 1.5 mm wide by 2.5 mm deep. The slots are at 2.54 mm pitch, and the total cumulative pitch error between the first and last slot is limited to ± 30 μm.

The total process is fully automated, and starts with CNC wheel dressing to produce the profile shown in *Figure 8.16(b)*. Once the wheel has been formed the grinding cycle then deals with each slot in turn, progressively taking each slot to full

Figure 8.15 A connecting rod set up for creep-feed grinding using a formed wheel on an Elb grinding machine. (From Marsh[8])

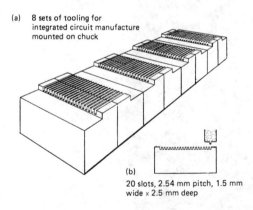

(a) 8 sets of tooling for integrated circuit manufacture mounted on chuck

(b)

20 slots, 2.54 mm pitch, 1.5 mm wide × 2.5 mm deep

Figure 8.16 (a) Eight sets of tooling for integrated circuit manufacture, ganged in-line. (b) Wheel profiled ready for grinding. (Courtesy of Jones and Shipman plc)

depth from the solid, and with 10 in-cycle dressings applied as an integral part of the programme.

There is automatic size compensation for reduction in wheel diameter, and automatic increase in wheel speed to maintain a constant peripheral speed of 30 m s^{-1}.

An increment in down-feed is automatically applied at each reversal of the table, and the total down-feed required to bring each slot to full depth is divided (in the programme) into coarse, medium and fine sections. Finally, there is a 15-stroke spark-out before the wheel head is automatically raised and the table indexed transversely to position it in readiness for grinding the next slot.

The total time to CNC grind the eight components is substantially less than that needed by skilled workers using manually operated surface grinders.

Gains resulting from CNC grinding are improved productivity, total automation of the process, improved quality and a reduction in batch-to-batch variation.

Vertical spindle reciprocating table (*Figure 8.17(a)*) and vertical spindle rotary table (*Figure 8.17(b)*) grinding machines are used to produce flat surfaces in quantity production applications. Rotary table horizontal spindle machines (*Figure 8.17(c)*) are capable of producing very flat surfaces, but are generally of limited capacity.

8.11.2 Work clamping

For surface grinding the work is usually held in position using magnetic, electrostatic or vacuum chucks. Obviously, magnetic chucks can only be used with ferromagnetic materials; electrostatic chucks will work on any electrically conducting material; and vacuum chucks will work on metallic and non-metallic materials. It is normally advisable to demagnetise components which have been held in a magnetic chuck.

Conventional jigs/fixtures are used for holding complex shapes requiring grinding.

8.11.3 Cylindrical grinding

Cylindrical grinding can be carried out on both external and internal cylindrical surfaces. External grinding provides the best grinding conditions with a small arc of contact (*Figure 8.18*), whereas internal grinding gives a larger arc of contact and, as a consequence, more severe conditions. External grinding is carried out between centres or in a chuck. The grinding wheel runs at normal speed (*c.* 30 m s^{-1}) and the work piece at 20–40 m min^{-1}. Wheel and work run in opposite directions at their point of contact. With cylindrical grinding there is often the facility to realign both wheel and table (universal grinding machine). Cylindrical grinding can be used to produce a wide range of products by traverse and plunge grinding with flat or profiled wheels. These products include straight cylinders, screw threads, tapers and complex shafts.

CNC has been applied to cylindrical grinding, giving similar benefits to those obtained when it is used with surface grinding. However, when grinding multidiameter shafts, additional advantages are obtained; in particular, all the operations are carried out without removing the component from the machine, thus ensuring concentricity of all diameters and giving very large reductions in floor-to-floor times.

A Jones and Shipman Ten[9] CNC cylindrical grinding machine is shown in *Figure 8.19* together with its control system. As an aid to programming the Series Ten[9] machine provides the operator with a series of coded 'canned cycles', examples of which are shown in *Figure 8.20*. Circular and linear interpolation are available for contour generation, both when dressing and when grinding (*Figure 8.21*). The grinding sequence for manufacturing a typical shaft is shown in *Figure 8.22*. Once the component has been placed in the machine, the grinding operations follow the sequence listed. In general, the sharpest curves/tapers are dressed onto the wheel, whilst other less sharp features are cut under CNC control by interpolation. In-process gauging allows adjustment of end stops and wheel speed to compensate for changes in wheel diameter. CNC grinding ensures that the machine spends a greater

(a)   (b)   (c)

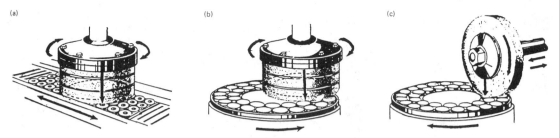

Figure 8.17 (a) Vertical spindle reciprocating table grinding machine. (b) Vertical spindle rotary table grinding machine. (c) Horizontal spindle rotary table grinding machine. (Courtesy of Carborundum Abrasives GB Ltd)

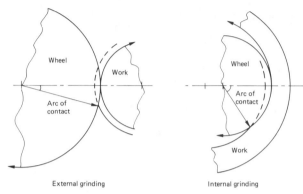

External grinding   Internal grinding

Figure 8.18 Effect of wheel–work configuration on arc of contact

CNC Grinding Machines

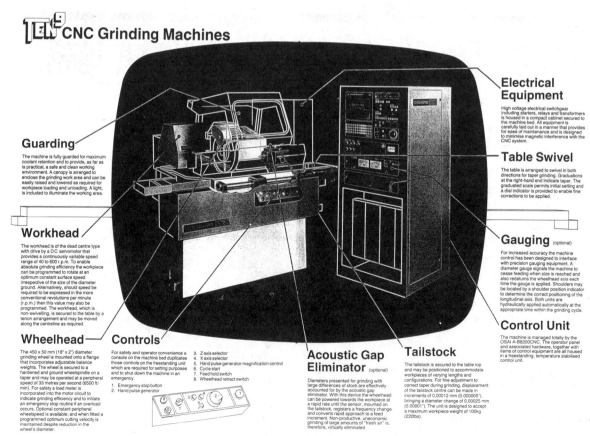

Guarding

The machine is fully guarded for maximum coolant retention and to provide, as far as is practical, a safe and clean working environment. A canopy is arranged to enclose the grinding work area and can be easily raised and lowered as required for workpiece loading and unloading. A light, is included to illuminate the working area.

Workhead

The workhead is of the dead centre type with drive by a DC servomotor that provides a continuously variable speed range of 40 to 600 r.p.m. To enable absolute grinding efficiency the workpiece can be programmed to rotate at an optimum constant surface speed irrespective of the size of the diameter ground. Alternatively, should speed be required to be expressed in the more conventional revolutions per minute (r.p.m.) then this value may also be programmed. The workhead, which is non-swivelling, is secured to the table by a tenon arrangement and may be moved along the centreline as required.

Wheelhead

The 450 x 50 mm (18" x 2") diameter grinding wheel is mounted onto a flange that incorporates adjustable balance weights. The wheel is secured to a hardened and ground wheelspindle on a taper and may be operated at a peripheral speed of 33 metres per second (6500 ft/min). For safety a load meter is incorporated into the motor circuit to indicate grinding efficiency and to initiate an emergency stop routine if an overload occurs. Optional constant peripheral wheelspeed is available; and when fitted a programmed optimum cutting velocity is maintained despite reduction in the wheel's diameter.

Controls

For safety and operator convenience a console on the machine bed duplicates those controls on the freestanding unit which are required for setting purposes and to shut down the machine in an emergency.

1. Emergency stop button
2. Hand pulse generator

3. Z axis selector
4. X axis selector
5. Hand pulse generator magnification control
6. Cycle start
7. Feed hold switch
8. Wheelhead retract switch

Acoustic Gap Eliminator (optional)

Diameters presented for grinding with large differences of stock are effectively accounted for by the acoustic gap eliminator. With this device the wheelhead can be powered towards the workpiece at a rapid rate until the sensor, mounted on the tailstock, registers a frequency change and converts rapid approach to a feed increment. Non-productive, uneconomic grinding of large amounts of "fresh air" is, therefore, virtually eliminated.

Tailstock

The tailstock is secured to the table top and may be positioned to accommodate workpieces of varying lengths and configurations. For fine adjustment to correct taper during grinding, displacement of the tailstock centre can be made in increments of 0.00012 mm (0.000005"), bringing a diameter change of 0.00025 mm (0.00001"). The unit is designed to accept a maximum workpiece weight of 100kg (220lbs).

Electrical Equipment

High voltage electrical switchgear including starters, relays and transformers is housed in a compact cabinet secured to the machine bed. All equipment is carefully laid out in a manner that provides for ease of maintenance and is designed to minimise magnetic interference with the CNC system.

Table Swivel

The table is arranged to swivel in both directions for taper grinding. Graduations at the right-hand end indicate taper. The graduated scale permits initial setting and a dial indicator is provided to enable fine corrections to be applied.

Gauging (optional)

For increased accuracy the machine control has been designed to interface with precision gauging equipment. A diameter gauge signals the machine to cease feeding when size is reached and also redatums the wheelhead axis each time the gauge is applied. Shoulders may be located by a shoulder position indicator to determine the correct positioning of the longitudinal axis. Both units are hydraulically applied automatically at the appropriate time within the grinding cycle.

Control Unit

The machine is managed totally by the OSAI A-B8200CNC. The operator panel and associated hardware, together with items of control equipment are all housed in a freestanding, temperature stabilised control unit.

Figure 8.19 Jones and Shipman Ten[9] CNC grinding machine. (Courtesy of Jones and Shipman plc)

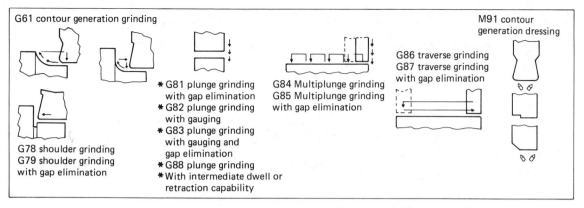

Figure 8.20 Examples of 'canned cycles' which enable more efficient programming. (Courtesy of Jones and Shipman plc)

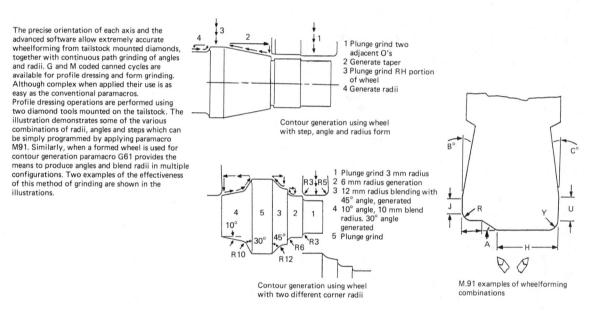

The precise orientation of each axis and the advanced software allow extremely accurate wheelforming from tailstock mounted diamonds, together with continuous path grinding of angles and radii. G and M coded canned cycles are available for profile dressing and form grinding. Although complex when applied their use is as easy as the conventional paramacros.
Profile dressing operations are performed using two diamond tools mounted on the tailstock. The illustration demonstrates some of the various combinations of radii, angles and steps which can be simply programmed by applying paramacro M91. Similarly, when a formed wheel is used for contour generation paramacro G61 provides the means to produce angles and blend radii in multiple configurations. Two examples of the effectiveness of this method of grinding are shown in the illustrations.

1 Plunge grind two adjacent O's
2 Generate taper
3 Plunge grind RH portion of wheel
4 Generate radii

Contour generation using wheel with step, angle and radius form

1 Plunge grind 3 mm radius
2 6 mm radius generation
3 12 mm radius blending with 45° angle, generated
4 10° angle, 10 mm blend radius. 30° angle generated
5 Plunge grind

Contour generation using wheel with two different corner radii

M.91 examples of wheelforming combinations

Figure 8.21 Circular and linear interpolation for contour generation. (Courtesy of Jones and Shipman plc)

proportion of its time cutting metal and gives very substantial reductions in cycle times. Components need no longer be moved in batches from machine to machine and a 'just-in-time' (JIT) processing approach becomes possible.

Internal cylindrical grinding is used in the manufacture of precision products such as the races of ball and roller bearings. The wheel work configuration gives a long arc of contact and, as the wheel is often a close fit in the bore, the application of coolant and the removal of grinding debris can be difficult. An added problem is the need to run the wheel at very high speed (*c.* 30 000 rev. min^{-1}), due to its small diameter, in order to maintain a normal cutting speed of 30 m s^{-1}. Frequently, the grinding spindle (quill) is also of small diameter and thus the grinding forces can readily deflect the wheel, giving problems with system stiffness and parallelism when grinding. The problem of parallelism has been overcome by an adaptation of the process called 'controlled force grinding' or 'constant force grinding'. By maintaining a constant force the deflection of the quill will be constant and this deflection can be allowed for in the machine settings. Machines using this technique are widely used in the manufacture of ball and roller bearing tracks.

8.11.4 Centreless grinding

Centreless grinding can be readily automated and is, therefore, a widely used production process; both external and internal surfaces can be ground by this process. Since the work piece does not require mounting between centres or holding in a chuck, fast cycle times can be achieved.

The work–wheel arrangement in centreless grinding is shown in *Figure 8.23*. A work rest blade carries the work, which is ground between two wheels; the larger wheel being the grinding wheel and the smaller one being a regulating wheel which is rubber bonded. Both wheels are wider than those used in conventional grinding in order to reduce the effects of wheel wear. The regulating wheel rotates at a much slower speed (20–60 m min^{-1}) than the grinding wheel, and controls both the rotational and longitudinal motion of the work piece.

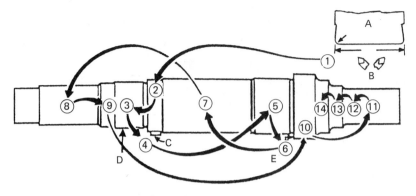

A Wheel B Diamonds C Steady D Diameter gauge
E Shoulder location gauge

Sequence to achieve EP7 class bearing shaft tolerances on diameters 3 and 5

1 Profile dressing of wheel
2 Plunge grind steady ϕ with gap elimination.
  (Program stop, apply steady)
3 Plunge grind with ϕ gauge
4 Shoulder grind with gap elimination
5 Plunge grind
6 Shoulder grind with gap elimination
7 Multiplunge grind

7 Traverse grind
8 Plunge grind
9 Plunge grind
10 Plunge grind
11 Plunge grind
12 Shoulder grind
13 Generate radius
14 Generate taper and blend radius

N	G	M	S/T	X	Z	Assigned/paramacro parameters
N10						(Spindle shaft)
N20						G71 G94 G97 G80 (CM.CAN)
N30		M36				
N40		M04	S100			(AP, P100 = 75)
N50		M60				
N60		M91		X20	Z145	B2 C2 D2 E.151.015 J5 P2 Q0
N70	G96		S15			
N80	G81			X56.900	Z145	A.015 P57.3 S10 Q0
N90		M00	S25			(Apply steady)
N100	G82			X50.000	Z166	A.015 P50.3 150.1 Q0
N110	G79				Z166	E.005 P50.02 S5 K-164.9 W-165.1
N120	G88			X60.000	Z1	A.015 P60.3 S5 Q0
N130	G79				Z1	E.005 P60.02 S5 K.1 W-.1
N140	G84			X57.000	Z45	A2 P57.3 S3 F1 K-150 Q1
N150	G86			X57.000	Z150	E.10 P57.004 S2 K45 Q100 R2 L2
N160	G88			X42.000	Z220	A.015 P42.3 S5 Q0
N170	G88			X49.800	Z197	A5 P50 S5 F1 Q0
N180	G88			X70.000	Z5	A.020 P70.3 S5 Q0
N190	G88			X38.000	Z55.3	A.015 P38.3 S5 Q1
N200	G78				Z56	E.003 P38.005 S5 K55.2 W55.32
N210	G61			X44.000	Z50	C.015 D.01 E1 P44.3 K40 R6 U56
N220	G61			X56.000	Z35	B60 C.015 D.01 E1 P56.3 K25 R4 U70
N230		M60				
N240		M02				

Figure 8.22 Sequence to machine a typical shaft using a CNC cylindrical grinding machine. (Courtesy of Jones and Shipman plc)

The work is forced against the work rest, the forward feed rate f being controlled by slightly inclining the regulating wheel. An approximate value of the feed rate is given by:

$$f = D_r N \sin \gamma \qquad (8.15)$$

where f is the feed rate (mm), D_r is the diameter of the regulating wheel (mm), N is the number of revolutions per minute of the regulating wheel, and γ is the angle of inclination of the regulating wheel.

External centreless grinding is particularly suited to the manufacture of cylindrical products of constant diameter by through feeding; almost any length can be produced and the

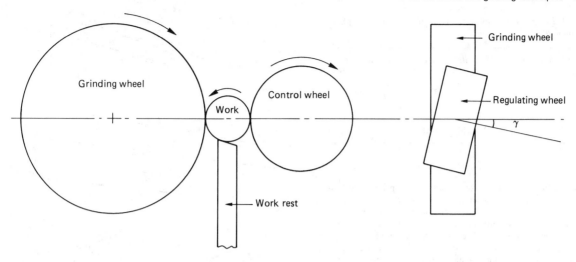

Figure 8.23 Wheel–work arrangement for centreless grinding

process is readily automated. However, the process can be configured to produce tapered or multidiameter cylindrical components. By profiling the wheel, spherical products can be made.

Internal centreless grinding is carried out by supporting the work piece between rollers and the grinding wheel moves into the work piece. To ensure that the internal and external diameters are concentric, the external aspect must be accurately machined before internal centreless grinding.

Centreless grinding processes are particularly suited to quantity production and are readily automated. However, there are restrictions on the shape of product which can be manufactured.

8.11.5 Tool and cutter grinding

Grinding is widely used in the manufacture and reconditioning of cutting tools. This work is usually carried out on universal tool and cutter grinders which are configured to grind complex tooling such as milling-cutters, single-point cutting tools, hobs and reamers. Specially shaped grinding wheels are normally used for this type of operation, and for the more exotic tool materials the use of superabrasives may be necessary.

8.11.6 Cutting off

Grinding is frequently used in fettling/snagging operations in the foundry to remove down runners and feeders from castings. Machines designed for this purpose achieve high metal removal rates using resinoid/rubber bonded wheels.

8.12 Newer abrasives and grinding techniques

Considerable research is being carried out on both abrasives and machining methods with the object of further improving grinding performance.

A new abrasive, known as SG, has been developed by Norton Abrasives. Chemically, the new material is aluminium oxide, but its unique structure gives it levels of performance similar to superabrasives at a price similar to conventional abrasives.

Slightly purer and about 10% harder than aluminium oxide used in the manufacture of ordinary grinding wheels, the new material is a ceramic produced by sintering the basic material produced by a seeded gel technique; hence the name SG. It is its structure which gives SG its enhanced properties. *Figure 8.24* shows an aluminium oxide wheel alongside a SG wheel of the same grain size. A traditional fused oxide grit is made up of a few, or possibly a single, grit. By contrast, the same single SG grain will contain millions of submicrometre particles. Each time a traditional aluminium oxide grit fractures in use, it can lose as much as one-fifth of its grinding surface, periodic dressing being needed to restore grinding performance. SG abrasive fractures on a much finer scale due to the submicrometre particle size of its constituents; cutting efficiency is maintained at significantly lower wear rates with SG wheels.

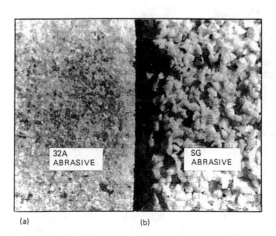

(a) (b)

Figure 8.24 (a) Aluminium oxide wheel and (b) SG abrasive of the same grain size[9]

Table 8.6 Improvement in performance obtained by using SG grinding wheels[9]

Wheel diameter (mm)	Wheel life increase (%)			Increase in grinding rate (%)		
	Oil	Water	Dry	Oil	Water	Dry
< 350	310	363	167	220	97	650
> 350	440	340	nil	140	100	nil

Since the grits remain sharp, even with minimal dressing, the risk of metallurgical damage to the work piece is much reduced. It is claimed that vitrified SG wheels give longer wheel life, increased productivity, require less frequent dressing and reduce the risk of metallurgical damage.

A summary of grinding test results comparing the performance of SG grinding wheels with conventional aluminium oxide wheels is given in *Table 8.6*. SG abrasive wheels are suitable for both ferrous and non-ferrous materials up to a hardness of $67R_c$. Cost savings due to reduced cycle times are proportionally greater when using semiautomatic and CNC grinding machines.

It is claimed[9] that prime sectors for the application of SG abrasives are the aerospace industry and for critical components particularly susceptible to metallurgical damage.

One obvious method of making the grinding process more competitive is to increase metal removal rates, and this is the main objective of high-speed grinding. Such high-speed-grinding machines require very careful design, paying particular attention to guarding the grinding wheel, frequently by the use of armour plate materials. Special grinding wheels have been developed for use at high speeds ($c.$ 60 m s^{-1}), and this has involved the building of special speed testing facilities.

Even with grinding there are some materials which can present machining difficulties. Electrochemical grinding was developed as a method of eliminating these problems. The process is carried out on a special-purpose machine, the bulk of the material being removed from the work piece by electrochemical action, the grinding wheel imparting surface finish and dimensional accuracy.

8.13 Honing

Honing should no longer be regarded as just a finishing process for cylindrical surfaces (internal and external). The process has been developed to the stage where it can frequently provide a cost-effective alternative to grinding and fine boring. Modern honing machines are capable of removing hard or soft materials at economical machining rates, whilst giving a product accurately to size, of good geometric shape (roundness and taper) and a controlled finish. Machines currently available range from manually operated machines to fully automated multispindle machines with microprocessor control.

8.13.1 Basic principles

Honing uses solid abrasives in the form of grinding blocks or sticks (*Figure 8.25*) and produces a highly characteristic surface which exhibits a cross-hatched pattern (*Figure 8.26(a)*). Honing is favoured for products requiring high accuracy and good surface finish and in bearing applications where the

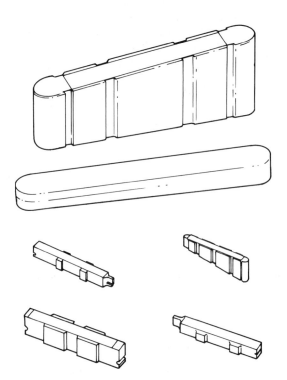

Figure 8.25 Examples of honing abrasives. (Courtesy of Jones and Shipman plc)

cross-hatched texture produced by the process improves oil-film retention.

Typically, the honing process would involve machining a bore using specialist tooling of individual design which forces the grinding blocks or strips against the work piece. The tooling is subject to a complex motion consisting of both rotation and reciprocation. The cutting velocity V_c and the cutting angle θ can be calculated if the reciprocating velocity V_r and the rotational velocity V_t are known.

With reference to *Figure 8.26(b)*, the cutting angle θ is calculated as:

$$\tan\left(\tfrac{1}{2}\theta\right) = \frac{V_r}{V_t}$$

or

$$\theta = 2\tan^{-1}\left(\frac{V_r}{V_t}\right) \tag{8.16}$$

Also with reference to *Figure 8.26(b)* the cutting velocity is given by:

$$V_c = (V_r^2 + V_t^2)^{1/2} \tag{8.17}$$

or

$$V_c = V_t \cos\left(\tfrac{1}{2}\theta\right) \tag{8.18}$$

or

$$V_c = V_r \sin\left(\tfrac{1}{2}\theta\right) \tag{8.19}$$

Unlike most grinding processes, where the grits only cut for a brief period per revolution, in honing the cutting grits are in continuous contact with the work piece. However, the reciprocating motion of the process causes the individual grits to

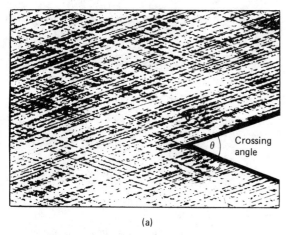

(a)

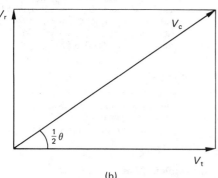

(b)

Figure 8.26 (a) Characteristic cross-hatched pattern produced by honing. (b) Diagram relating rotational velocity V_t and reciprocating velocity V_r to cutting velocity V_c

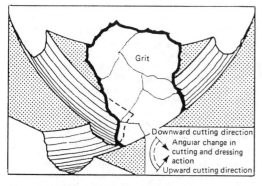

Figure 8.27 Effects of changes in direction on cutting facets. (Courtesy of Jones and Shipman plc)

describe a zig-zag path across the surface of the work piece, giving the characteristic cross-hatch pattern. These continual changes in direction alter the facets of the grit cutting at any one time, helping both self-sharpening and the cutting process (*Figure 8.27*). Honing is essentially a wear process in that a force is applied between the cutting stones and the work piece (similar to controlled force grinding) from which a cutting rate

evolves. For any given set of cutting conditions (product), the life of the stone will depend on the relative area of stone to work piece which is defined as the stone face value α:

$$\alpha = \frac{A_s}{\pi D^2} \tag{8.20}$$

When optimum honing conditions have been established, it is useful to calculate the stone face value α for future reference.

8.14 Honing practice

8.14.1 Abrasives

Honing uses the complete range of abrasive materials including silicon carbide, aluminium oxide, diamond and cubic boron nitride (CBN). In the honing process, the abrasive is in continuous contact over a large area of the component. In order to achieve both self-sharpening and free cutting requires very careful selection of the abrasive type, grit size and bond hardness. The combination of grit and bond should satisfy the above requirements and also give an economic abrasive life.

The honing action constantly changes the point of shear load on each cutting grit, encouraging grit fracture as the cutting edges dull, thus new cutting facets or fresh grits are presented to the work surface. This self-sharpening action depends critically on selecting the appropriate abrasive grit and bond. If the bond is too weak the abrasive will wear rapidly giving a short abrasive life. Conversely, too strong a bond prevents self-sharpening leading to glazing, overheating and, as a result, a reduced rate of stock removal and poor geometry.

8.14.2 Tooling

Honing tools should be designed to ensure two essential features: that feed is applied equally to all the abrasive stones, and that abrading is uniform on all areas of the work after all high spots have been removed. Expansion is applied to each of the abrasives by a central cone, whilst the tool is rotated and reciprocated in the bore (*Figure 8.28*). As the abrasives contact the work piece, both cone and abrasives form a rigid cutting unit (*Figure 8.28*). To maintain stability, the abrasives are normally spaced evenly around the tool.

The tools which carry the abrading hones demand specialist attention at the design stage to ensure that consistent results are produced throughout their life. An individual approach to the design of quality tooling is essential as the characteristics of each product can vary considerably. The length of a bore affects: the length of abrasive; the diameter and the number of abrasives; and component section and the width and spacing of abrasives. All these factors need to be balanced correctly if optimum results and maximum efficiency are to be achieved.

Honamould tools are an example of the individual approach; this type of tooling extends from very small diameter to multibanked tools, which are used in conjunction with one of the following types of abrasive.

(1) Honamould: abrasive is moulded into a plastic material which supports and protects the abrasives from the harsh dressing action that is prevalent in extremely rough or interrupted bores.
(2) Diamond or CBN hones: the abrasive medium is sintered onto a metal strip machined to the required configuration.

Both these types of cutting medium are precisely manufactured to allow immediate use without the need for mounting

(a)   (b)   (c)

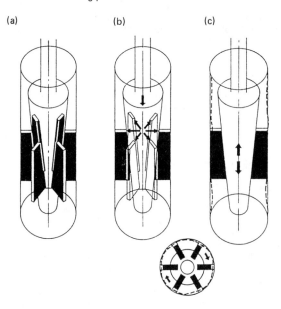

Figure 8.28 (a) Central cone with six abrasives. (b) Central cone applying pressure to the abrasives. (c) Abrasives acting to give round and parallel bore. (Courtesy of Jones and Shipman plc)

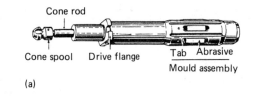

(a)

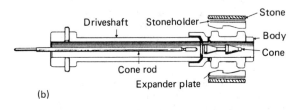

(b)

Figure 8.29 A typical Honamould tool (a) and a basic stone and shell assembly tool (b). (Courtesy of Jones and Shipman plc)

or predressing. A typical Honamould tool is shown in *Figure 8.29.*

Stone and shell assembly tools are more suited to larger bore sizes, and the design enables a greater versatility of its diameter range. Indeed, the range can be extended further by using oversize stone holders. The tool can be provided with either a single cone for abrasive expansion or, where necessary, a cone and tapered blades for dual expansion, allowing rough and finish or base and plateau honing. This feature allows for an even number of abrasives mounted around the tool body, 50% being used for the first honing operation and the remainder for the second operation; each abrasive has its own means of expansion (*Figure 8.30*). With this type of

Feed movement to outer tube and blades to expand finish abrasives

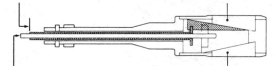

Feed movement to inner rod and cone to expand first abrasives

Figure 8.30 A double expansion honing tool. (Courtesy of Jones and Shipman plc)

tooling two operations can be carried out in the same cycle, thus reducing component handling.

8.14.3 Honing fixtures

Due to the wide variety of components that are honed and the high tolerances required, individual fixtures are generally necessary. In fixture design, basic principles are used to take advantage of the self-centring action that honing provides. Some fixture designs are shown in *Figure 8.31*.

8.14.4 Sizing systems

To ensure that sustained production with precise results is maintained, reliable automatic sizing methods are essential. Examples of such systems are shown in *Figure 8.32*. The actual method selected depends on the requirements of the component being machined.

8.14.5 Machine configurations

Typical machine configurations are shown in *Figure 8.33*. The single-spindle arrangement is ideal, provided production requirements can be satisfied and geometric and surface finish specifications can be achieved in one honing operation. The honing cycle is fully automatic and fixturing can be designed to provide either manual load or a completely automated method of component loading.

Where a honing machine is integrated in the production line, an in-line configuration is most desirable because the component leaves the machine at the opposite end to which it entered. Loading and unloading is usually automated and incorporated into the flowline, with the component passed through the machine by means of a transfer arm.

Rotary machines are most suited to manual loading because in this way the operator's movements are kept to a minimum. The rotary set-up is also of benefit when multioperations are necessary, as this arrangement allows the component to remain in the same fixture for all operations.

As with the other types of machine, the honing cycle will be completely automated.

Modern honing machines take full advantage of CNC control and are frequently incorporated into flexible manufacturing systems and unmanned manufacturing cells. Two machines are shown in *Figure 8.34* and examples of typical products, including production rates and tolerances, are shown in *Figure 8.35*.

8.14.6 Effects of honing

The honing process will improve both geometric shape and surface finish. When carried out correctly, honing will eliminate out-of-roundness by removing high spots first. Then the contact area will gradually increase until the whole surface is being machined, giving a fully machined surface and a round

Laterally floating component

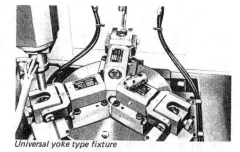
Universal yoke type fixture

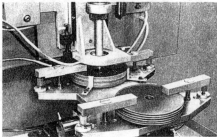
Rigid fixture with laterally floating tool

Universal gimbal type fixture

Rigid fixture with universally floating tool

Laterally floating fixture

Rigid fixture with rigid tool

Figure 8.31 A range of typical fixtures used for honing. (Courtesy of Jones and Shipman plc)

bore. Similarly, inaccuracies such as taper, waviness and bow can be corrected, provided there is sufficient machining allowance. The desired amount of stock removal can be calculated as:

$$\text{Total stock removal} = 2R + O + C + F \qquad (8.21)$$

where R is the roughness from the previous operation (R_a)

(μm), O is the out-of-roundness (μm), C is the taper (μm), and F is the maximum sum of other errors.

The preferred stock removal for honing a range of materials is given in *Table 8.7*.

If the process is not properly controlled, honing can produce errors such as bell, barrel and conical shaped bores. Selection of the correct over-run is particularly critical.

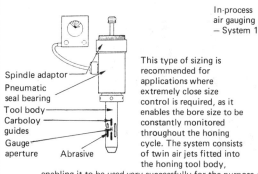

In-process
air gauging
— System 1

Spindle adaptor
Pneumatic
seal bearing
Tool body
Carboloy
guides
Gauge
aperture   Abrasive

This type of sizing is recommended for applications where extremely close size control is required, as it enables the bore size to be constantly monitored throughout the honing cycle. The system consists of twin air jets fitted into the honing tool body, enabling it to be used very successfully for the purpose of 'MATCH HONING' a sure way of guaranteeing consistent clearance between, for instance, a valve spool and the bore in which it operates.
Repeatability of 0.0025 mm (0.0001 in) is readily achieved.

A — Size signal from reference part
A1 — Size signal from workpiece
B — Air signal from reference part
B1 — Air signal from workpiece
C — Transducer, reference part signal
C1 — Transducer, workpiece signal
D — Electronics gauge — reference part
D1 — Electronic gauge — workpiece
E — Differential comparator (size signal to logic control)
F — Workpiece
G — Honing tool

In-process
air gauging
— System 2

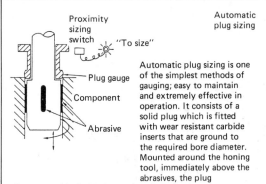

This type of sizing is recommended for applications where extremely close size control is required on bores which have a large number of interruptions, such as hydraulic or pneumatic valve bodies. The system consists of a separate air gauge mounted in line with, but opposite to, the honing tool. Size repeatability of 0.0025 mm (0.0001 in) can be readily achieved, which enables this system to also be used very successfully for the purpose of MATCH HONING.

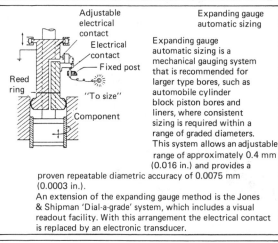

Adjustable
electrical
contact
                    Electrical
                    contact
                  — Fixed post
Reed
ring
      "To size"
                    Component

Expanding gauge
automatic sizing

Expanding gauge automatic sizing is a mechanical gauging system that is recommended for larger type bores, such as automobile cylinder block piston bores and liners, where consistent sizing is required within a range of graded diameters. This system allows an adjustable range of approximately 0.4 mm (0.016 in.) and provides a proven repeatable diametric accuracy of 0.0075 mm (0.0003 in.).
An extension of the expanding gauge method is the Jones & Shipman 'Dial-a-grade' system, which includes a visual readout facility. With this arrangement the electrical contact is replaced by an electronic transducer.

Proximity
sizing
switch   "To size"

Automatic
plug sizing

— Plug gauge
— Component

— Abrasive

Automatic plug sizing is one of the simplest methods of gauging; easy to maintain and extremely effective in operation. It consists of a solid plug which is fitted with wear resistant carbide inserts that are ground to the required bore diameter. Mounted around the honing tool, immediately above the abrasives, the plug reciprocates with the honing tool until the bore reaches the size required. At this point the plug will travel further downwards by entering the bore and operating a proximity switch, to terminate abrasive expansion.

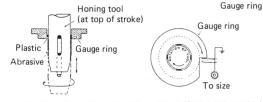

Honing tool
(at top of stroke)   Gauge ring

Plastic
Abrasive   Gauge ring

                    Gauge ring

                    To size

Gauge ring sizing is used in conjunction with vitrified abrasives which are either moulded to plastic (Honamould) or have a portion of plastic fixed at the top end of the abrasive as for a stone and shell assembly. This method, although particularly suitable for small diameters, can be successfully used on bores up to 100 mm (4 in.) diameter. A gauge ring, identical in diameter to the required bore size, is mounted in a position which allows only the plastic portion of the abrasive to enter the ring during reciprocation and rotation of the honing tool. When the bore reaches the required size, the plastic contacts the gauge ring on entry, causing it to rotate and actuate a small air valve or spring loaded contact, terminating the honing cycle.
Diameter repeatability of 0.0075 mm (0.0003 in.) is easily achieved.

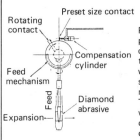

Rotating
contact   Preset size contact

                    Compensation
                    cylinder
Feed
mechanism

                    Diamond
                    abrasive
Expansion—

Positive stop
sizing

Positive Stop Sizing is particularly suitable where diamond hones are used as the cutting medium; and where diameters are too small to allow other sizing methods to be considered. The system uses two low voltage electrical contacts: one preset at size and the other that rotates as the diamond hones expand.
When the moving contact reaches the pre-set contact, expansion is terminated. Compensation for diamond wear can be manually or automatically initiated. Manual compensation is applied by depressing a button on the control panel. When this compensation feature is automatically initiated, it is usually part of an automatic transfer machining cycle.

Figure 8.32   A range of automatic sizing systems used in honing operations. (Courtesy of Jones and Shipman plc)

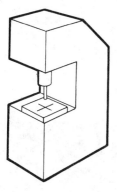 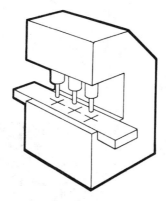 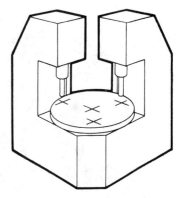

SINGLE SPINDLE
A single spindle arrangement is ideal, providing that production requirements can be satisfied and where geometric parameters and surface finish requirements can be achieved in the one honing operation. The honing cycle itself is fully automatic and fixturing can be designed to provide either manual load or a completely automated method of component placement.

IN-LINE
Where a honing machine is being integrated into a production flowline then the in-line formation is most desirable, because the component leaves the machine at the opposite end to which it entered. Direction of index can be arranged accordingly. Loading and unloading is normally automated and incorporated into the flowline, with the component passed through the machine by means of a transfer arm.

ROTARY
On a machine which, although automatic, does not form part of a transfer line then a rotary machine can be used to full advantage.
This is particularly apparent when the component is manually loaded, as the operator's movements are kept to a minimum.
The rotary set up is also of benefit when multi-operations are necessary, as this arrangement allows the component to remain in the same fixture for both operations until its processing is completed.

Figure 8.33   Typical machine configurations. (Courtesy of Jones and Shipman plc)

(a)

(b)

Figure 8.34   (a) Jones and Shipman JS784E vertical honing machine. (b) Jones and Shipman JS772-2 twin spindle honing machine with rotary indexing table, multiposition measuring equipment and automation gantry loader to and from carousel storage. (Courtesy of Jones and Shipman plc)

A major consideration for using honing, in addition to the production of good geometric shape, is the surface characteristics resulting from the process. Honing gives an excellent surface finish which has a cross-hatched pattern; a finish which supports an oil film particularly well and hence aids lubrication. A recent development in the process, called 'plateau honing', gives a surface texture similar to that of a run-in component. This is produced by base honing and then taking a further light cut to produce a surface with plateau; the talysurf traces given in *Figure 8.36* show run-in, base-honed and plateau-honed surfaces. Honing gives a typical production finish of 0.2–1.5 μm, although the process is capable of producing finishes down to 0.05 μm.

Internal honing is usually carried out using special-purpose machines; external honing is usually applied as a manual operation after finish turning, and is similar to superfinishing.

8.15 Superfinishing

This process is similar in principle to honing in that solid abrasive is used to generate a true geometric shape and an exceptionally good surface finish. As with honing, the abrasive sticks describe a zig-zag path over the surface of the work piece. However, the processes differ in that superfinishing automatically stops removing material at the end of the machining cycle.

The quality of surface produced by superfinishing could be achieved using other methods, and hence there is some confusion in name; 'finish honing' is probably a better description of the technique.

It is claimed that many of the problems associated with running-in are much reduced when superfinished components are used.

8.15.1 Machining set-up and machining mechanisms

The geometric arrangement for superfinishing a cylindrical component is shown in *Figure 8.37*. Low cutting speeds are used (10–45 m min^{-1}), completely eliminating the risk of

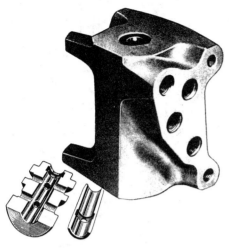

The functional design of the valves is such that it is necessary to produce accuracies of 0.0005 mm (0.00002 in.).

The sectioned insert sleeves are fitted into the aluminium casing of the respective valve assembly, before being mounted on to a floating fixture on the machine. A combination of the specially designed honing tool and the machine's inherent features, allow very close tolerances to be achieved.

(a)

Operational details

Bore diameter	6.3 mm (0.2485 in.)
Length of bore	22.2 mm (0.875 in.)
Material	Hardened steel
Stock removal	0.015/0.025 mm (0.0006 in.; 0.001 in.)
Geometric accuracy	0.0007 mm (0.00003 in.)
Surface finish	0.075–0.1 μm (3–4 μin.) Ra
Honing time	70 s

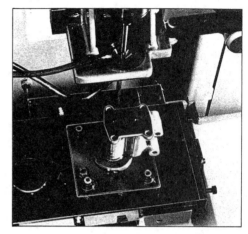

Operational details

	Piston Pin bore	Crank Pin bore
Bore diameter	20.63 mm (0.8125 in.)	51.35 mm (2.021 in.)
Material	Steel forging	Steel forging
Stock removal	0.064 mm (0.0025 in.)	0.064 mm (0.0025 in.)
Surface finish	0.75–1.0 μm (30–40 μin.) Ra	0.75–1.0 μm (30–40 μin.) Ra
Tolerances:		
Diameter	0.010 mm (0.0004 in.)	0.010 mm (0.0004 in.)
Geometry	0.005 mm (0.0002 in.)	0.005 mm (0.0002 in.)
Production rate	260 per h	260 per h
Sizing	Gauge plug	Gauge plug

The symmetrical cross-hatch pattern produced by the combined rotating and reciprocating motion of the honing operation, provides the ideal supporting surface in an automobile Connecting Rod assembly, for fitting of the soft metal shell bearing. Since the geometric and diameter tolerances are also vitally important it is essential for reliable honing equipment to be used.

(b)

Figure 8.35 Two examples of honed products: (a) hydraulic braking and suspension component; (b) automobile connecting rod. (Courtesy of Jones and Shipman plc)

Table 8.7 Preferred honing allowance (expressed as millimetres on the diameter) for a range of materials*

Material	One-off	Quantity production
Cast iron	0.06–0.15	0.02–0.06
Steel (soft)	0.06–0.15	0.02–0.06
Steel (hard)	0.03–0.08	0.005–0.03
Hard chromium	0.03–0.08	0.02–0.03
Cemented carbide	0.1–0.3	0.05–0.1
Light alloys	0.05–0.1	0.02–0.08
Copper alloys	0.04–0.08	0.02–0.05
Sintered metals	0.1–0.2	0.05–0.08
Glass	0.05–0.1	0.03–0.05
Bakelite	0.08	0.05

*From Rosenberger.[9]

(a)

(b)

(c)

Figure 8.36 (a) Worn cylinder bore showing plateau. (b) Base-honed cylinder bore. (c) Plateau-honed cylinder bore

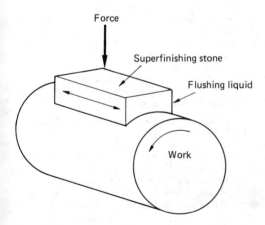

Figure 8.37 Basic arrangement for superfinishing a cylindrical component

thermal damage to the work piece. The superfinishing stone is usually silicon carbide or aluminium oxide, specially bonded and having a grit size in the range 400–1000. The superfinishing stone is held against the work piece under a constant force $(1 \times 10^5$ to 4×10^5 N m^{-2}) using springs or pneumatic pressure, and given a small oscillatory motion (4–6 mm) at frequencies ranging from 10 to 30 Hz. The process is carried out using a flushing liquid of appropriate viscosity.

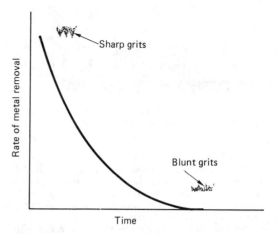

Figure 8.38 Superfinishing cycle showing the rapid fall in metal removal rate as the abrasive blunts/loads

The superfinishing cycle is a form of controlled running-in; this is shown diagramatically in *Figure 8.38*. At the start of the process, the grits cut efficiently and the flushing liquid carries away the debris produced. However, as the grits become blunted and worn down, little or no self-sharpening occurs because the process is carried out at constant force. The cutting rate progressively declines due to blunting and loading of the abrasive. Eventually, cutting stops completely, the flushing liquid giving fluid film lubrication. Since, unlike other machining processes, superfinishing will only remove a certain amount of material before the process stops, the surface finish prior to superfinishing must be carefully controlled. For this reason, components for superfinishing are generally ground.

A number of variables control the process, including the material being machined, the surface finish from the previous operation, machine condition, type of abrasive, machining speed, oscillating frequency, stone pressure and flushing liquid. The factors which control the cutting cycle (time to run-in) which must be optimised are cutting speed, oscillating frequency and stone pressure. To obtain the best results, these parameters may be altered during the machining cycle. Techniques have even been developed to change the viscosity of the flushing liquid, from high at the start of the cycle, to low at the end.

Most materials can be superfinished, but those which are difficult to grind generally give problems when superfinishing. External and internal cylindrical surfaces and flat annular surfaces can be superfinished. Since the process only removes a very thin layer of material, it only corrects microgeometrical errors within the original surface.

The process is favoured for bearing surfaces on certain automotive components which will be superfinished on special purpose machines. However, the process of superfinishing can be carried out on a normal lathe, using simple attachments.

8.16 Coated abrasives

Coated abrasives give two-body abrasion and are constructed by applying a thin layer of abrasive to a suitable backing material. A wide range of backing materials is used including paper, waterproof paper, cloth, waterproof cloth, polyester cloth and combination backings.

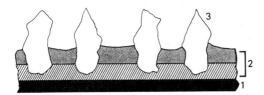

Figure 8.39 Structure of electrostatically coated abrasive cloth. 1, Backing; 2, primary and secondary bond; 3, grits

Coated abrasives are available in a range of materials, including emery, aluminium oxide, silicon carbide and zirconia alumina. Good bond strength is a vital factor in the performance of coated abrasives, current bond types include glue, resin over glue and full resin.

To obtain maximum cutting efficiency, the abrasive grains are aligned electrostatically during the manufacturing process to give the structure shown in *Figure 8.39*.

Since only a single layer of abrasive grains is available, and these are already aligned during manufacture to give the best performance, dressing is seldom used. Mackie[10] has suggested that a process called 'levelling' (very light dressing using a tungsten carbide stick) could improve the surface finish with little effect on metal removal rate.

Coated abrasives are available in the form of rolls, sheet, strips, belts, discs, cones and buffs.

8.16.1 Cutting mechanism

The cutting mechanism is very similar to that used in grinding, in that when the abrasive grit contacts the work piece it may rub, plough or cut the material. Similarly, depending on the operating conditions, the coated abrasive can self-sharpen, load or glaze. Theoretical studies of the behaviour of coated abrasives have regarded the mechanism of material removal as a wear rather than a machining process. It has been suggested that the cutting process is analogous to tracking a hardness indentor across a surface, producing a scratch. If it is also assumed that the load on the indentor is supported on its frontal facet, the geometry of the groove, and hence the wear rate, can be calculated using the equation:

$$\frac{V}{L} = \frac{K \times W \cot \theta}{p_m} \qquad (8.22)$$

where V is the volume loss, L is the sliding distance, K is a constant, W is the load, θ is the half angle of the scratch, and p_m is the material hardness (kg mm^{-2}).

The constant K gives an indication of the efficiency of the process; if the whole of the groove volume is removed $K = 1$. Values of K less than 1 indicate a mixture of cutting and ploughing, and $K = 0$ indicates 100% ploughing with no material removed. If the load on a blunt grit is very low it may rub the surface with no plastic deformation, the load being carried elastically. Obviously, the mechanism of material removal is complex and may involve microcutting, microploughing, microfatigue and microcracking. These physical interactions between abrasive particles and the surface of a material are shown in *Figure 8.40*.

The K factor is mainly a function of the grit geometry, the way in which the grit reacts with the work piece material, and the properties of the work piece material. Tough ductile materials have a lower value of K than do hard brittle materials.

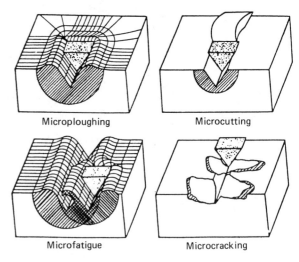

Microploughing   Microcutting

Microfatigue   Microcracking

Figure 8.40 Models of the interactions between abrasive particles and the surface (From Zum Gahr[11]).

Since coated abrasives consist of only one layer of abrasive material, there is a limit to the amount of material they can remove. When observing the rate of material removal versus time or sliding distance for coated abrasives, two types of behaviour have been observed.

(1) When the abrasive is cutting satisfactorily, the rate of metal removal is directly proportional to sliding distance and directly proportional to the load. The rate of material removal will gradually decline until all the abrasive has been used up; this decline is associated with some improvement in the surface finish produced.

(2) This type of behaviour is very similar to that observed in the superfinishing process. If conditions are such that the abrasive does not self-sharpen (light loads), the abrasive grits may blunt quite rapidly and, since in many applications the constant-force conditions are used, the rate of metal removal falls to zero. Associated with this decrease in metal removal is a progressive improvement in the surface finish of the abraded surface.

Behaviours (1) and (2) are shown graphically in *Figures 8.41(a)* and *8.41(b)*, respectively.

Surface finish is controlled by the grit size of the abrasive and the force applying the work piece to the coated abrasive. The greater the number of grits per unit area the less the penetration of each and the better the surface finish. Similarly, the greater the load per grit the further each will penetrate into the surface and the poorer the surface finish. Finally, the depth of penetration will depend on the shape of the grits; sharper grits will penetrate deeper into the surface for a given load and, therefore, produce a poorer surface finish.

Figure 8.42 shows the results of a controlled experiment where samples of SKF ball-bearing steel (1%C, 1%Cr; hardness 260 kg mm^{-2}) were abraded for 1 min under a force of 5 N against abrasive papers having a range of grit sizes (100–600 grit). *Figure 8.42(a)* shows that the weight of metal removed decreases as finer grades of paper are used. The reduced metal removal rate can be explained by the fact that the average slope Δ_a decreases as the grit number increases; therefore, θ in equation (8.22) increases, giving a reduced rate of metal removal. Surface roughness R_a also decreases as the grit number increases (*Figure 8.42(c)*).

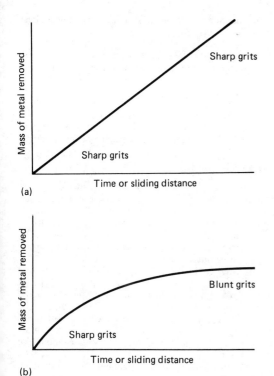

(a)

(b)

Figure 8.41 (a) Metal removal directly proportional to sliding distance as abrasive self-sharpens. (b) Abrasive blunts/loads in use and eventually stops cutting

8.17 Machining with coated abrasives

8.17.1 Abrasive belts

Abrasive belts are widely used to machine both metallic and non-metallic materials. A wide variety of machines has been developed to utilise the full potential of abrasive belts; these range from simple off-hand grinders to fully automated special-purpose machines.

8.17.2 Abrasive belt selection

An abrasive belt consists of four constituent parts: backing, bond, grit and joints. Each constituent has a significant effect on cutting performance.

Backing Both cloth and paper backings are available in varying stiffnesses.

Y weight — very stiff; used for heavy-duty grinding; coarse grits only.

X weight — stiff; heavy-duty grinding; coarse and medium grits.

J weight — semiflexible; heavy to fine grinding; all grits coarse to fine.

Flexiweight — flexible; medium to fine grinding; available in medium to fine grits; used where high conformability is required.

Bond The bond must be chosen to match the service conditions. If a too weak bond is used the grits will be stripped from the backing. If the bond is too strong and is used with a light cut, glazing and a consequent loss in cutting performance can result.

Grit type Aluminium oxide is the most widely used abrasive, being equally suitable for metals and wood. Silicon carbide is recommended for glass, ceramics and certain plastics. Zirconium based abrasives are frequently used when grinding stainless steels.

Grit size Abrasive grit sizes can be grouped as follows:

24, 36	— Very coarse, but give little better stock removal than 40 or 60 grit.
40, 50, 60	— Coarse; fast cutting and high stock removal.
80, 100, 120, 150	— Medium; general purpose.
180, 220, 240, 280	— Fine; suitable as a final finish or last operation before polishing.
320, 400, 500, 600	— Very fine; polishing non-ferrous metals and light deburring.

When grades are used sequentially (coarse to fine) steps of no more than 2 to 3 grit numbers should be made to ensure that all the coarser grit marks can be removed. Surfaces to be electroplated are usually polished down to a 320/400 grit finish

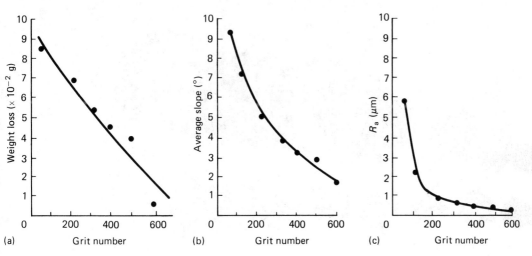

(a)

(b)

(c)

Figure 8.42 (a) Weight loss vs. grit number. (b) Average slope of the surface vs. grit number. (c) Surface roughness R_a vs. grit number

Type VL01

for normal belts and wide
belts except polyester.

Standard Lap splice for all kinds of application.

Type VL03

for normal belts and wide
belts (cloth only, no polyester)

Lap splice, grain removed completely or minus
tolerance for chatter mark free grinding of
plane or curved surfaces.

Type VS12

butt splice
for normal belts with
X-weight cloth backing and
wide belts with cloth backing

Type VS13

butt splice
for normal belt with F- and
J- weight cloth backing.

Type VS15

butt splice
for normal belts with
X-weight cloth backing

Type VS16

butt splice
for normal belts with F- and
J- weight cloth backing.

Figure 8.43 A range of standard splices used with abrasive belts.
(Courtesy of Carborundum Abrasives GB Ltd)

prior to plating. Surfaces to be polished after grinding are best
given a 240 to 320 grit finish.

Joints A range of standard splices is shown in *Figure 8.43*.
Most manufacturers will supply special splices if required.

8.17.2.1 Storage and handling of abrasive belts

Coated abrasives are sensitive to both humidity and temperature changes. Adverse storage conditions can result in curling
and brittleness, thus impairing the efficiency of the product.

8.17.3 Contact wheels

In many abrasive machining operations the work piece is
forced against the abrasive belt as it passes over a wheel called
the 'contact wheel'. Depending on the type of machine used,
the contact wheel can act as the drive wheel or it can be driven
by the belt and a separate drive wheel used. It is essential to
choose the correct type of contact wheel; the wrong wheel can
result in a poorer finish and a reduction in belt life.

The rim of the contact wheel is usually made from rubber
with a shore hardness of 30–90°. The depth and type of
serrations also affect performance; the wider and deeper the
serrations the more aggressively the abrasive belt cuts and the
higher the rate of stock removal. Plain wheels are most suited
to belts having a fine grit size. A complete range of contact
wheels is shown in *Figure 8.44* together with examples of
serration patterns.

The harder the contact wheel the greater the rate of stock
removal; therefore, for the best performance, the hardest
wheel to suit the shape of the work piece should be chosen.
Components having complex contours will require soft, flexible contact wheels.

Contact wheels are graded as follows:

30° shore — very soft;
40° shore — soft;
60° shore — medium;
80° shore — hard.

8.17.4 Factors controlling abrasive belt performance

(1) Coarser grits remove more stock.
(2) Higher belt speeds increase the rate of cut.

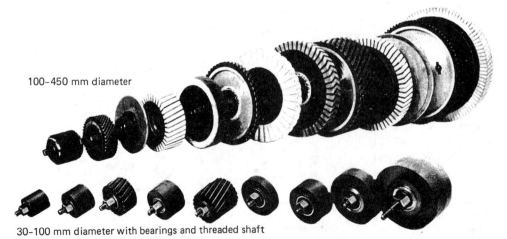

100–450 mm diameter

30–100 mm diameter with bearings and threaded shaft

Figure 8.44 A range of contact wheels for use on abrasive belt machines. (Courtesy of Surface Technology Products Ltd)

(3) Higher belt speeds give a finer finish.

(4) Small contact wheels remove more stock than do large contact wheels.

(5) Large contact wheels produce finer finishes than do small contact wheels.

(6) Small contact wheels produce shorter scratches than do large contact wheels.

(7) Hard contact wheels remove more stock than do soft contact wheels.

(8) Soft contact wheels produce finer finishes than do hard contact wheels.

(9) Wider grooving of the contact wheel results in higher stock removal.

(10) Lubrication prolongs belt life and improves finish, but reduces the rate of stock removal.

8.17.5 Abrasive belt machines

To obtain the best performance from the abrasive belt, the machine must be designed to operate at the correct cutting speed, have accurate belt tracking, positive tensioning and an economical belt length.

A number of well-established layouts are used in the design of abrasive belt machines.

Flat-bed linishers (*Figure 8.45*) are available with vertical or horizontal platens and are used for machining flat surfaces.

The abrasive belt runs over a metal platen and is tensioned between a drive wheel and idler wheel. The platen can be covered with a graphite cloth to reduce friction, increasing the power available and providing a cushioning effect. A stop or rest prevents the component being dragged along the belt.

Backstand machines (*Figure 8.46*) are another widely used type of abrasive belt machine. With this layout the contact wheel is also the drive wheel; therefore, any change in the diameter of the contact wheel results in a change in the surface speed of the belt. If the component requires a certain size of contact wheel, the cutting speed could be incorrect (either too fast or too slow).

Horizontal abrasive belt grinding machines have several advantages over backstand machines. A separate drive wheel is used and, therefore, the contact-wheel diameter can be chosen to suit the particular work piece, the diameter having

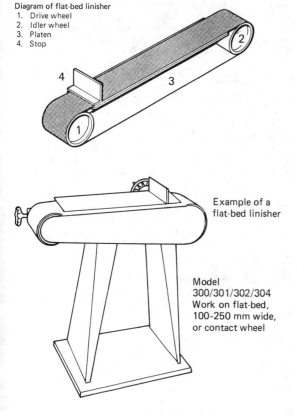

Available with vertical or horizontal platen. Recommended for components which must maintain dimensional flatness

Diagram of flat-bed linisher
1. Drive wheel
2. Idler wheel
3. Platen
4. Stop

Example of a flat-bed linisher

Model 300/301/302/304 Work on flat-bed, 100–250 mm wide, or contact wheel

Figure 8.45 Example of a flat-bed linisher together, with a schematic layout. (Courtesy of Surface Technology Products Ltd)

Backstands

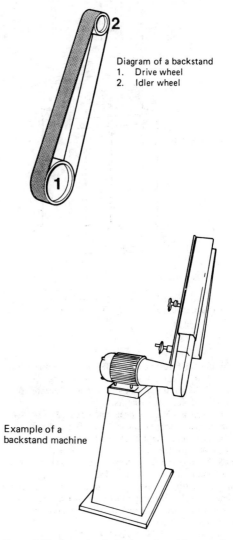

Diagram of a backstand
1. Drive wheel
2. Idler wheel

Example of a backstand machine

Figure 8.46 Example of a backstand machine, together with a schematic layout. (Courtesy of Surface Technology Products Ltd)

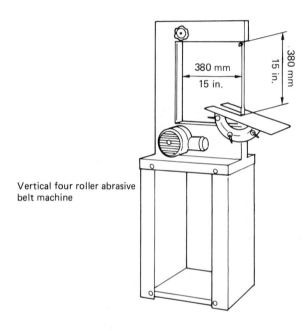

With platen   With contact wheel

1. Drive wheel
2. Idler wheel
3. Platen
4. Contact wheel

380 mm
15 in.

380 mm
15 in.

Vertical four roller abrasive
belt machine

Figure 8.47 Example of vertical four roller abrasive belt machine, with a schematic layout. (Courtesy of Surface Technology Products Ltd)

no effect on the surface speed of the belt. The machine configuration allows greater freedom to manipulate the work piece than with backstand machines. There is adequate space to handle large complex components.

Vertical four roller abrasive belt machines (*Figure 8.47*) have a similar arrangement to a bandsaw. The component is machined against a platen or a contact wheel and a variety of attachments can be used to hold or align the work.

Figure 8.48 shows a range of components being machined using off-hand grinding on basic machines.

The principles outlined have been used to develop more complex abrasive belt grinding machines, including plunge and centreless machines. A number of such machines are shown schematically in *Figure 8.49*.

Figure 8.50(a) shows a REMA FBM 4000 high performance abrasive belt grinding machine. It consists of a horizontal 250 mm wide by 4000 mm long abrasive belt head with

18.5 kW motor and an indexing rotary table with four jigs. The abrasive belt can be run wet or dry and can also oscillate. The height of the rotary table can be varied. A contact shoe applies a contact pressure which can be adjusted from 0 to 100 kg. The whole machining cycle can be carried out under automatic control; the operator is only required to load and unload components once the machine has been correctly set. *Figure 8.50(b)* shows the machining of an aluminium gear box cover.

8.18 Abrasive discs

A machine with an abrasive disc head can be used as an alternative to a flat-bed linisher. For the greatest stock removal, only the faster outer area of the disc can be used effectively. For most operations E-weight paper discs are adequate, cloth discs only being used for very heavy duty processes.

8.19 Lapping

Lapping is an example of three-body abrasion. The abrasive, carried in a suitable medium, is trapped betwen the lap and the work piece and, due to the relative movement, the abrasive removes material from both surfaces. As the rate of metal removal is low, the process is only used as a finishing operation, taking advantage of its ability to give good surface finish and high accuracy. Lapping can be applied to both flat and cylindrical surfaces. The technique is also used to bed in precision components such as plain bearings, gears and valves; effectively, a run-in surface is developed.

8.19.1 Abrasives

Lapping can be carried out using all the normal abrasives, the choice being a matter of experience. Having chosen the type of abrasive, the user must also select the appropriate grit size for the particular application. It may sometimes be necessary to use both a roughing and finishing grade of abrasive. In general, the harder the material being lapped, the harder the lapping medium chosen. The main categories of abrasive are described below.

Diamond This is a very hard, but expensive, material. It gives a high rate of stock removal for a longer period than do other abrasives and hence it is easier to maintain consistent results. It is used mainly for hard steels and cemented carbides.

Aluminium oxide (natural and synthetic) and silicon carbide These are widely used abrasives. They give adequate stock removal with steels, cast irons and non-ferrous metals. They are much cheaper than diamond.

Non-embedding compounds These compounds are relatively soft and do not become embedded in the work piece. They are particularly suited to the lapping of non-ferrous metals, such as white metal bearing materials, brass, bronze and aluminium alloys.

In addition to choosing the type of abrasive and its particle size, an appropriate carrying medium must be selected if the correct cutting mechanism is to be maintained. With the correct viscosity of cutting medium the film thickness allows

Weld grinding on metal window frame

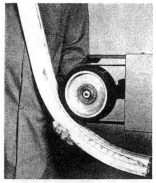
Removal of lines caused by bending of aluminium extrusion

Deburring of flame cut blank

Deburring of casting

Grinding of con rod

Grinding of forged spade

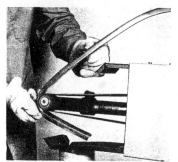
Finishing of car window trim

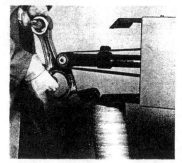
Grinding of tight radius

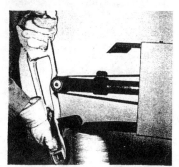
Abrasive belt polishing of small radii of aluminium casting

Figure 8.48 A range of products being off-hand ground on basic abrasive belt machines. (Courtesy of Surface Technology Products Ltd)

the abrasive grits to interact with the two surfaces. Too high a viscosity gives too thick a film and thus no cutting action; too thin a film and the grits are prevented from entering the gap between the lap and the work piece.

The arrangement used for flat lapping is shown in *Figure 8.51*. Cutting can occur only if the abrasive grits anchor themselves from time to time in the lapping disc. Cutting then occurs due to the motion of the work piece across the face of the lap. Since the grits may also attach themselves to the work piece, material is removed from both the work piece and the lap. Relative wear ranges from 1:0.3 to 1:1.5; the process is usually controlled to remove the greater amount of material from the work piece.

Lapping discs are frequently made from a pearlitic cast iron because this material gives good grain anchoring. An Archimedean spiral is often cut into the lapping disc; this helps the abrasive grains roll over thus presenting new cutting facets,

gives more even wear of the lapping disc, and a more uniform distribution of the lapping medium. By controlling the path of the work piece across the face of the lap, both uniform wear and a random cutting pattern will be achieved.

In lapping, the rate of stock removal depends on the pressure between the work piece and the lap, speed, type of grit, and shape of tool and work piece. Increasing speed and pressure will increase the stock removal rate. With each charge of lapping compound, the stock removal rate declines progressively with time, until a fresh charge of abrasive is required.

8.19.2 Lapping processes

Lapping can be carried out on flat or cylindrical surfaces by hand or by mechanical methods. When roughing and finishing cycles require two grades of lapping compound, separate

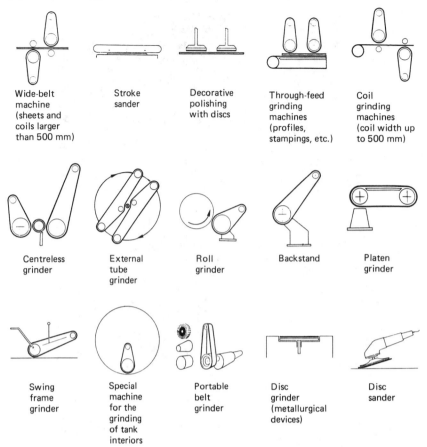

Figure 8.49 Schematic diagrams of the layouts of a range of machines using both abrasive belts and discs. (Courtesy of Carborundum Abrasives GB Ltd)

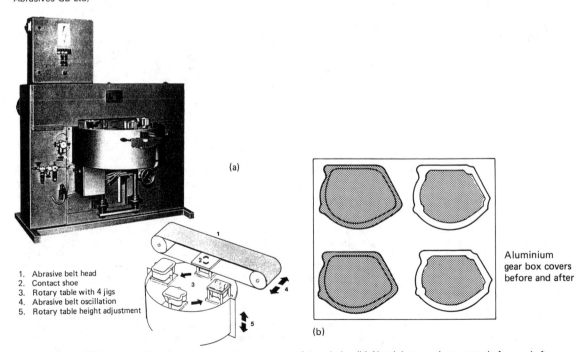

1. Abrasive belt head
2. Contact shoe
3. Rotary table with 4 jigs
4. Abrasive belt oscillation
5. Rotary table height adjustment

(a)

Aluminium gear box covers before and after

(b)

Figure 8.50 (a) REMA FBM 4000 high performance abrasive belt surface grinder. (b) Aluminium gearbox covers before and after machining. (Courtesy of Surface Technology Products Ltd)

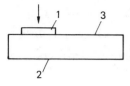
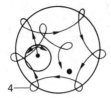

Figure 8.51 Lapping a plane surface. 1, Specimen; 2, lap; 3, abrasive paste; 4, path of specimen

Table 8.8 Recommended surface speeds for polishing and satinising

Process	Speed (m s^{-1})
Pre-polishing metals	36
Finishing polishing metals	30
Satinising with nylon buffs	15
Satinising with nylon wheels	15–25
Mush polishing	20
Polishing plastics	10–20

lapping discs should be used, the work piece being carefully cleaned between each stage.

Since the rate of stock removal is low compared with other abrasive processes, the amount of material to be removed should be kept to a minimum (typically 0.002–0.005 mm). The process can achieve an accuracy of 0.1 μm or better and a surface finish R_a of 0.0025 μm. The greater the accuracy and the better the surface finish required, the higher is the cost.

It is essential that all lapping compound is removed from the work piece at the end of the operation.

Errors which can occur in lapped surfaces include edge drop and roughness variations from the centre to the edges of the component.

8.20 Polishing

The term 'polishing' is used to describe processes the main objective of which is to produce a satin or mirror finish rather than to remove material. Polishing can be divided into three stages.

(1) When starting from a surface which is particularly rough, a relatively coarse abrasive (60–100 grit) in the form of belt, disc, cloth or flap wheel is used. This stage may be omitted if the initial surface has a good finish.
(2) Progressively finer grades of abrasive (down to 200 to 320 grit) are used in a preliminary polishing stage. If a satin finish is required, nylon buffs or nylon flap brushes (both products are made from abrasive impregnated nylon) are used. Where a highly polished surface is required, polishing is continued using a cutting compound (compo) on sisal, buffs or mops. Preliminary polishing or buffing is carried out using hard abrasives, such as aluminium oxide, in the form of bar, paste or emulsion. These are called 'cutting compositions' (compo) and produce a satin or semibright finish.
(3) Final polishing is done using cotton buffs and a polishing compo such as Tripoli or rouge. The compo is usually applied as bar when hand polishing; liquid compos are favoured for use in automatic polishing machines.

Recommended speeds for polishing and satinising are given in *Table 8.8.*

8.20.1 Micropolishing

Mechanical polishing of metal specimens for microscopic examination follows a similar procedure to that described previously. Initially, the specimen is polished on progressively finer grade (100–600 grit) papers in the form of rotating discs. The final mirror finish is produced by polishing on a rotating disc covered with a cloth impregnated with abrasive (aluminium oxide, magnesium oxide or diamond) and using an appropriate lubricant.

8.20.2 Barrel polishing/finishing

Barrelling is an inexpensive method of processing small to medium sized products in bulk, giving considerable savings on manual methods. Applications include cleaning small castings, deburring, and polishing prior to plating.

Barrels may be rotated vertically, horizontally or obliquely. Horizontal closed barrels can also be designed to reciprocate, which is advantageous when treating recessed components. Facilities must be provided for loading and unloading the barrels, separating the components from the barrelling medium, cleaning and drying. Barrelling is normally carried out wet using a barrelling medium and an aqueous solution of barrelling compound. The barrelling compound prevents the medium from glazing and keeps the articles being treated clean.

8.20.3 Barrelling medium

A wide variety of materials has been used as barrelling media, including rocks, stones, nut shells and steel balls. However, in order to get consistent performance, carefully graded materials are used in normal practice.

Barrelling media include bonded and fused aluminium oxide in random shapes, crushed and graded natural stone (limestone, granite and quartz), ceramic bonded shapes (generally triangular in form), and resin bonded shapes.

Soft or light materials require a light medium; heavy articles require a heavy medium. The particle size of the medium must be selected to avoid wedging of particles in any cavities or holes in the work piece. The medium is charged into the barrel to give a volumetric ratio of medium to product of between 2:1 to 4:1.

8.20.4 Barrelling compounds

Barrelling compounds are used to give greater control over the process and may include abrasives, wetting agents, detergents and rust inhibitors.

8.20.5 Equipment

The basic rotating barrel is cheaper than vibrating equipment, but requires longer cycle times. For maximum efficiency the load level for a rotating barrel is about 50%, and for vibrating equipment it is about 80–90%.

8.21 Blasting processes

In blasting processes a number of methods are used to project abrasive at the work piece. Methods include air, steam and mechanical systems.

Blasting processes are used for cleaning and fettling castings and stampings, preparation of surfaces prior to plating or painting and the work hardening of critical components to enhance their performance (shot peening).

8.21.1 Blasting abrasives

A range of abrasives is used in blasting operations, including sand, silicon carbide, aluminium oxide, and metal shot and grit.

References

1   RICHARDSON, R. C. D. *Proc. Inst. Mech. Eng.* **182** (3A), 410 (1968)
2   ALDEN, G. I. *ASME* **36**, 451 (1914)
3   GUEST, J. I. 'The Theory of Grinding', *Proc. Inst. Mech. Eng.*, 543 (1915)
4   BACKER, W. R., SHAW, M. C., and MARSHALL, E. R., *ASME* **74**, 61 (1952)
5   BUTTERY, T. C. 'Precision Engineering Surfaces—The Way Ahead', *Int. Conf. on Prod. Eng.*, Shoubra Faculty of Engineering, Cairo (1981)
6   GRISBROOK H. 'Precision Grinding Research', *Prod. Eng.*, **39** (5) (1960)
7   HAHN, R. S., *ASME*, **78**, 807 (1956)
8   MARSH, L. 'Abrasive Machining', *Prod. Eng.* **67**, 22 (1988)
9   ROSENBERGER, R. *Zusammenfassung verschiedener Forschungs und Versuchsarbeiten über das Honen*, Werkstsattstechnik, 1952, No.2, 1962
10   MACKIE, W. H., HAMED, M. S., and BUTTERY, T. C. 'Some Investigations into the Improvement of Abrasive Tool Behaviour', *2nd Joint Poly Symposium on Manufacturing*, Lanchester Polytechnic (1979)
11   ZUM GHAR, K. H. *Modelling of Two Body Abrasive Wear*, **124**, 87 (1988)

9

Fabrication

R D Cullum

E N Gregory

R Goss

Contents

9.1 Fasteners

Fasteners embrace a wide range of common and specialised components, but the emphasis for the designer of today is on fastening systems, largely because economy in production demands overall consideration of the assembly process. Fasteners inserted by hand are labour intensive and, therefore, make the assembly expensive.

Designers have to ask themselves which fastening system will serve the design purpose at the lowest overall cost. Does the design warrant fasteners and, if so, is it for assembly, maintenance, transport, etc? The reason for using a fastening device will lead to the choice of system, ranging from welding, through adhesives, bolts, rivets, etc., to highly specialised fastening devices. Despite the growing popularity of structural adhesives, mechanical fasteners are still widely used, but it must be emphasised again that the cost of inserting a fastener is likely to outweigh the cost of the individual fastener. Therefore, the designer must be convinced that a mechanical fastener is really the best solution for the job in hand. At this point it is convenient to enumerate questions such as:

(1) Is a fastener really necessary?
(2) Will the minimum number of fasteners be consistent with reliability and safety, as well as economic in production?
(3) Will the fastener specified perform its task efficiently?
(4) Will the fastener be simple to install and be capable of being placed by automatic methods?
(5) Will the fastener have to be removed during service for maintenance and repair, and will it be readily accessible with standard tools?
(6) If the fastener is to be removed during service, is it of a type that can be safely used again?
(7) Will the fastener material be compatible with that of the products?
(8) Bearing in mind the additional cost involved, will it be necessary to specify a special fastener instead of a standard one?

There are other questions that could be asked, and these depend on the type of assembly being considered, but the questions given above are basic to most designs.

The range of fastener systems currently available is bewilderingly wide, but it is the designer's responsibility to select from the right category, to assess performance from the supplier's data, and to consider in-place costs. Fasteners that fail in service are not only unreliable but uneconomical. The conditions under which the fastener operates must be known and the selected fastener must resist these conditions, which could well include extremes of temperatures, corrosive environments, vibration and impact loads.

9.1.1 Automatic insertion of fasteners

Depending on specific applications, most fasteners in use today can be installed automatically and, since some 80% of the total cost of fastening is on-assembly cost, a mere saving of 10% in assembly costs is more significant than a 40% saving in piece part costs.

Nevertheless, 80% of manufacturers assembling products with automated or robotic screwdriving equipment claim to have difficulties with their fasteners. This is the finding of a recent research survey carried out for European Industrial Services (EIS) who manufacture Nettlefolds screws. Faulty fasteners brought machinery to a halt and, in a significant number of instances, caused damage to machines or products.

For 23% of all users of automated equipment, the problem is a continuing one which production engineers appear pre-

pared to live with, either because they are unaware that a solution exists, or because they believe that any solution must be too expensive. It is against this background that, in 1987, European Industrial Services introduced a fault-free product known as 'Nettlefolds Gold Seal' to the robot user market. However, it appears that just as many problems are being experienced by people with automatic assembly equipment. Now, EIS are offering the Gold Seal products for use with automated screwdriving equipment.

Virtually all the screw products used in robotic and automated screwdriving are hardened, mainly fastening metal to metal, but with a growing proportion joining plastics, or plastics to metal. The Gold Seal error scanned products, therefore, cater for these requirements with a variety of self-tapping screws for metal applications and Polymate, a specially designed EIS patented screw for thermoplastics.

The Pemserter is a high volume automated fastener insertion press from PEM International for inserting any of the range of bushes in the PEM range. The press has an intelligent main cylinder containing an optical encoder which transmits absolute ram position to a microprocessor based logic control system. Operating and safety windows coupled with an end-of-ram sensor allow the machine to perform at its maximum rate while providing operator and tooling safety.

During a set-up procedure, these safety windows are automatically determined. Any variation from the initial parameters such as the operator removing the anvil, or the presence of a foreign object such as a finger, will cause the ram immediately to stop and return. The company claim that their press achieves new standards in performance, safety and ease of operation.

Another example of a modern assembly tool is the fully automated twin headed Heatserter, which is capable of installing up to 45 fasteners per minute in plastics mouldings (see *Figure 9.1*). The new CNC machine from PSM International plc, can be preprogrammed in three axes (x, y and z) to cater for plastics mouldings of a wide range of shapes and sizes. Preheating circuitry is used to facilitate high-speed post-

Figure 9.1 PSM Automated Systems new CNC Heatserter can install up to 45 fasteners per minute and can be preprogrammed in three axes (x, y and z)

moulding fastener installation, and the equipment features light guarding to provide easy tool changing. The Heatserter can be interfaced with moulding presses to form a link in an automated production line and is capable of installing the full range of PSM proprietary fasteners for thermoplastics.

The problems of automated assembly are not just handling and orientation, but are often also those of quality and consistency. Design for automated assembly will, therefore, embrace both component design and the manufacturing process.

9.1.2 Joining by part punching

A very effective way of fastening sheets together is by part punching the top sheet into the bottom sheet. Material is partly cut from the top sheet and pushed through the lower sheet. Simultaneously, the lower sheet is formed so that an overlap between top and bottom sheets occurs. As the partial cut is connected to the upper sheet by webs, a join of considerable strength is produced.

The Trumpf TF300 power fastener has been developed to cut and form the sheets at the rate of 2 strokes/s and consists of a three part tool set comprising punch, stripper and die. Developed for static or portable use, an optional coordinate table ensures the accurate positioning of joints. The power tool is particularly useful where box shape enclosures are being assembled or where flat strips are to be joined together.

9.1.3 Threaded fasteners

Where threaded parts are concerned, designers have been bedevilled by the numerous thread forms that have been available through history and many are still with us today in spite of the UK's declared policy to recommend the adoption of the International Organisation for Standardisation (ISO) metric screw thread system. Existing thread systems include Whitworth, British Standard Fine (BSF), British Association (BA) as well as a range of pipe threads, gas threads, oil industry threads and fire service threads. However, there has been a part change typified by the use of ISO inch series of threads (Unified Coarse and Unified Fine). Nevertheless, wherever possible, new designs and updates should call for ISO metric thread systems.

The choice between Unified Coarse and Unified Fine threads depends on the complete understanding of the assembly in which screws and bolts are used. In general terms, the coarse thread (UNC) can fulfill most design requirements. One small practical advantage with the fine thread (UNF) is when they are used in conjunction with a slotted nut and split pin. To align the slot in the nut with the hole in the bolt, less torque is required with UNF for a given bolt tension.

Where bolts are used in tapped holes, a failure under tension loads should be by breakage at the core of the bolt thread rather than by stripping of the thread. Where bolts are in shear, designers should ensure that the loads are carried on the bolt shank rather than on the threaded portion. Another important aspect is the clearance of the bolt in the hole, particularly when several bolts appear in a row. The tolerance for dimensions must be carefully considered, otherwise undue strain will be put upon a single bolt rather than being equally distributed among them all.

With nuts and bolts there will always be a problem of making certain that in field replacements, the bolt removed is replaced by one of similar strength. Members of the British Industrial Fasteners Federation (BIFF) all produce bolts of high quality and standards and these can be readily identified from their head markings. Additional markings also indicate tensile strength and thread forms. *Figures 9.2* to *9.4* show the most commonly used head forms for screws, bolts and nuts.

Cup head
Head dia. (2.2–2.5) × shank dia.
Head thick (0.54–0.67) × shank dia.

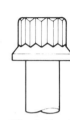

Square head

Raised hexagon washer head

12 point head

Raised countersunk head, also known as French, Oval, Instrument

Round heads, also known as Cup, Snap, Button

Cheese head

Connection head

Mushroom head, also known as Truss

Pan head (American type)

Raised cheese head also known as Filister

Pan head (English type), also known as ISO metric Cheese head

Countersunk head, also known as Flat

Hexagon head

Figure 9.2 Head forms for bolts and screws

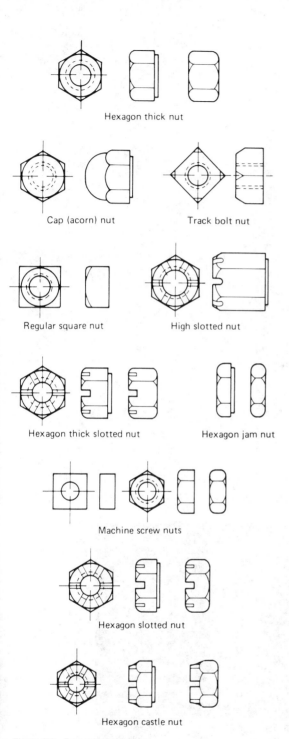

Machine screw nuts

Hexagon slotted nut

Hexagon castle nut

Figure 9.3 Free-spinning nuts

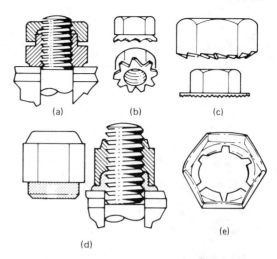

Figure 9.4 Free-spinning lock nuts. (a) The upper half of a two-piece nut presses the collar of the lower half against the bolt. (b) A captive tooth washer provides locking with a spring. (c) Ratchet teeth bite into a bearing surface. (d) A nylon insert flows around a bolt to lock and seal. (e) Arched prongs of a single thread lock the unit grip screw

the bolts. In cases like this, locking the nuts and bolts together may not always provide a satisfactory solution.

9.1.4 Load sensing in bolts

Much research has been carried out on problems relating to bolt tension, and in many critical applications, ranging from car engine cylinder heads to sophisticated chemical and nuclear plants, it is necessary to apply the right amount of torque and this is usually done through the use of a torque spanner.

A clever idea to indicate bolt tension is to be found in the new fastener which is the brainchild of John Hirst who heads Exotech, a design and development company based at the University of Warwick's Science Park. West Midlands based T. W. Lench, one of the UK's leading nut and bolt manufacturers, is currently setting up a special manufacturing unit at its Warley factory to produce the new Hi-Bolt fastener, as it is called (see *Figure 9.5*).

The system involves a structural bolt with an in-built load sensing device. It is identical to a standard bolt apart from the actual sensor—a steel pin running through the shank which is gripped as the nut is tightened. Before tightening the pin is free to move. As the nut is tightened, the bolt shank contracts and grips the pin. When this happens, the maximum preload in the bolt has been reached.

The RotaBolt uses the same principle in a different way. The internal pin is secured at one end within the shank of the bolt. The free end holds a Rota washer and cap clear of the bolt head. As the bolt stretches under the tension imposed by tightening the nut, it pulls the Rota washer into contact with the bolt head and locks it. The gap between the Rota washer and bolt head has been set previously at the manufacturing stage so when the Rota washer locks, the predetermined tension in the bolt has been achieved.

To determine the actual bolt tension required for a particular application is far more complicated and, as previously mentioned, particularly so when dealing with soft materials such as glass fibre reinforced plastics (GRP). A research team at Imperial College undertook an extensive test programme to

Threaded fasteners in general are used to provide a clamping force between two or more parts and where there is likely to be a need for dismantling at some time in the future, whether for access or for maintenance, etc. Clamping force can be a major problem, particularly when clamping soft materials which may shrink in service and cause loosening of

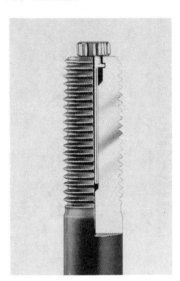

Figure 9.5 A close-up of the Hi-Bolt load sensing system from T. W. Lench

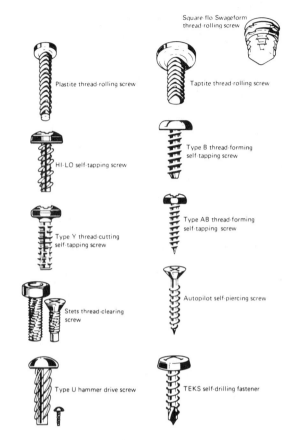

Figure 9.6 Self-tapping and thread-forming screws

determine the static strength of bolted joints in several forms of GRP. The experimental results were used to produce design charts for both single and multibolt joints subjected to tensile/shear loading. While design methods established for structural joints in metals are applicable in a general fashion to GRP joints, the physical nature of the material does introduce problems not encountered with metals. The anisotropic stiffness and strength means that unexpected failure modes may be introduced, the low interlaminar shear and through thickness tensile strengths being a particular difficulty in this respect. Joint strengths were found to be strongly dependent on choice of reinforcing fabric.

Many technical devices have been used to prevent the unscrewing of nuts and bolts. A selection of ideas is shown in *Figures 9.8* and *9.9*. Each has its own advantages and disadvantages. Some can only be effectively used once, and once removed should be discarded. Serrations which effectively lock the bolt head or nut to a surface could initiate cracks in sensitive parts. In other cases, although the nut is locked, the bolt can turn. The most effective way of preventing relative rotation is by the use of liquid adhesives or anaerobic cements.

9.1.5 Threadlocking

The selection of a chemical threadlocking compound will depend on the following criteria.

(1) The ultimate shear strength of the fastener must not be exceeded, although this is usually only of importance for screw sizes of 5/16 in. diameter or less.
(2) The severity of the loosening tendencies.
(3) The size of the threads and, therefore, the viscosity of the compound to ensure thread filling.
(4) The method of application and requirements for testing and putting into service.
(5) The environmental requirements of temperature and chemical resistance.

The viscosity should be such that the compound can be easily applied, will not run off and will fill threads which have the maximum clearance. Preapplied dry materials are avail-

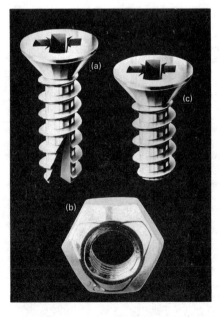

Figure 9.7 (a) Plasform K thread-cutting screw. (b) Reusable locking nut. (c) Plasform screw for plastics

able for application situations where liquids are not acceptable. They are easy to inspect and provide some reuse. The application is usually carried out by the bolt supplier and at least four formulations are available:

(1) low strength locking and sealing;
(2) medium strength plated fasteners;
(3) medium strength; and
(4) high strength.

PSM fasteners, for example, provide a comprehensive range of bolts and studs with the preapplied process for locking and sealing thread components. Their Scotch-Grip two-part epoxy system microencapsulates the resins while the hardener is freely carried in the coating substance. When the fastener is engaged with its mating thread, the capsules are crushed and the shearing action of the rotating fastener mixes the epoxy and hardener initiating the adhesive cure. An alternative but similar system uses microencapsulated anaerobic adhesives and sealants.

Fastening systems, whether they be adhesives, nuts and bolts, welding or any other method, are inherent in the design of the product or structure destined to be assembled as a one-off, by batch production or mass production. However, despite the popularity of adhesives, mechanical fasteners still have a prime role to play but, unless sufficient care is taken in their selection, the installation of such fasteners can lead to increased costs and possibly to a decrease in the required level of mechanical efficiency or in-service reliability.

To provide a comprehensive list of all of the various types of fastener available is impossible in this text, but it may be helpful to the reader to group the fastener types as follows:

(1) bolts, screws, studs, nuts and washers;
(2) self-tapping screws;
(3) locking and stiff nuts;
(4) blind bolts/screws;
(5) rivets including blind rivets;
(6) spring steel fasteners;
(7) plastics fasteners;
(8) quick release fasteners;
(9) self-sealing fasteners; and
(10) threaded inserts and studs.

Some of these are illustrated for general guidance, but each manufacturer produces its own wide range of products for similar applications. The reader would be wise, in the final analysis, to obtain specific data from these manufacturers, several of which are listed in Section 9.1.15.

9.1.6 Threaded inserts and studs for plastics

A method of securing a threaded bore in sheet is to use bushes designed to bite into the sheet and prevented from rotating either by being of hexagonal shape, or having a serrated shoulder or spigot. A large variety of these types of bush is available.

Whenever components must be fastened with screws to a plastics moulding, thread inserts which only require standard machine threads have distinct advantages:

(1) the assembly screw is completely reusable;
(2) unlike self-tapping or thread forming screws, which are applied directly into the plastics, there is no risk of thread stripping; and
(3) the assembly screw must often remain tight and maintain its clamping effect; threaded inserts allow this as they do not cause problems of stress relaxation in the material which is common when screws are driven directly into plastics.

The advantages of placing inserts after moulding instead of moulding-in are:

(1) reduced moulding time and, therefore, reduced unit costs;
(2) no expensive damage to tools due to misplaced inserts;
(3) no metal swarf from inserts to contaminate moulded surface; and
(4) availability of cost-effective assembly methods for post-moulded inserts.

Obviously, each job must be considered on its own merits, and post-moulded inserts tend to come into their own in long production runs. For small numbers, moulding-in can often be advantageous, in spite of the points made above.

9.1.7 Ultrasonic welding

Ultrasonics are widely used for joining plastics assemblies and for securing threaded inserts. The main limitation arises in joining parts made from different thermoplastics—a difference of only a few degrees in the softening temperature will cause one material to flow before the other. The pressure applied during ultrasonic welding is also a critical factor, and the most up-to-date ultrasonic welding equipment is microprocessor controlled to give the optimum pressure and heating conditions.

9.1.8 Self-tapping screws

These screws fall into two main categories: (a) true self-tapping screws which, like a tap, cut their own threads in a prepared hole; and (b) thread-forming screws which form their own threads by a rolling or swaging action (*Figures 9.6 and 9.7*). Typical of the latter are the proprietary types Taptite and Swageform. Taptite incorporates machine screw threads formed on a trilobular shank, and the Swageform has special lobes at 120° intervals along the tapered portion of the shank to impart a three-dimensional swaging action to form the threads.

Self-tapping screws are widely used in the assembly of sheet-metal components in which the thread-cutting action results in a good fit and is resistant to vibration and shock loads. Thread-forming screws are more applicable to thicker materials and blind holes. The self-drilling versions, for example Teks and Drill-kwick, are designed with a true drill point to produce the necessary pilot hole for the threads.

9.1.9 Stiff nuts

Stiff nuts, as opposed to free-spinning nuts, generally incorporate either a locking element to generate an elastic contact between the nut and bolt threads or, alternatively, a differential pitch or deformed threads to create the required friction between the mating threads. Examples are shown in *Figure 9.8*.

9.1.10 Washers

Washers are used to provide a seating for nuts and bolts in order to distribute the load over an area greater than that provided by the bolt head or nut. It also prevents damage to the surfaces being joined due to rotation of the nut or bolt. Washers can also seal, cover up oversize holes and act as a spring take-up between fastener and work piece. They also have the special task of preventing unwanted rotation of the nut or bolt. These include spring washers and serrated-tooth types as shown in *Figure 9.9*.

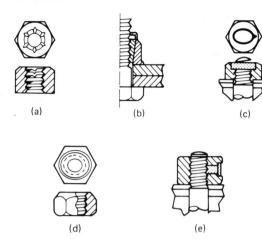

(a) (b) (c)

(d) (e)

Figure 9.8 Stiff nuts. (a) The deformed thread has depressions in the face of the nut to distort a few threads. (b) The slotted section forms beams that deflect inward and grip bolt. (c) The threaded elliptical spring steel insert produces locking. (d) The non-metallic plug insert grips bolt threads. (e) The out-of-round threads cause wedging action

9.1.11 Spring steel fasteners

Although these include certain standard parts, for example, caged nuts, most spring steel fasteners are designed for specific applications where speed of assembly is important and the cost of the fasteners is low. They are mass produced from hardened and tempered steel strip and are used in non-critical applications in cars, domestic electrical appliances and in the sheet metal industries. Some of the various shapes which can be produced are shown in *Figure 9.10*.

9.1.12 Plastics fasteners

These products fall into two categories: (a) standard thread forms made from plastics such as nylon, and (b) specially configured fasteners for specific applications. Plastics fasteners can be self-coloured and are non-corrosive. Some examples are shown in *Figure 9.11*.

9.1.13 Self-sealing fasteners

Bolts, screws, washers and nuts can be made into self-sealing elements by the addition of a sealing element in applications where it is necessary to seal the fastener hole against the leakage of liquids or gases (see *Figure 9.12*). The choice of the fastener and the sealing material is dependent on the specific application.

9.1.14 Rivets

Rivets are low cost, permanent fasteners suitable for manual and automatic setting. They can be classified as solid, tubular, semi-tubular, bifurcated and blind, all of which are available in several different materials and head styles. Correct joint design and preparation together with the right fastener will ensure a strong trouble-free and cost-effective assembly. A range of available types is shown in *Figures 9.13* and *9.14*.

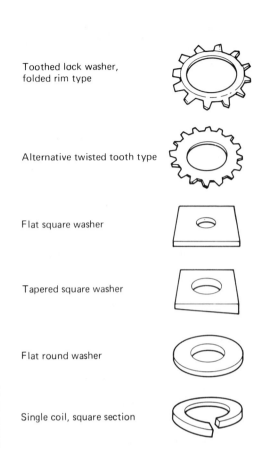

Toothed lock washer, folded rim type

Alternative twisted tooth type

Flat square washer

Tapered square washer

Flat round washer

Single coil, square section

Double coil spring washer

Tab washer

Figure 9.9 Washers

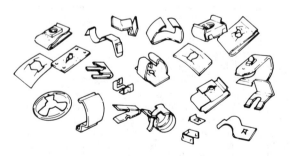

Figure 9.10 Selection of spring steel fasteners

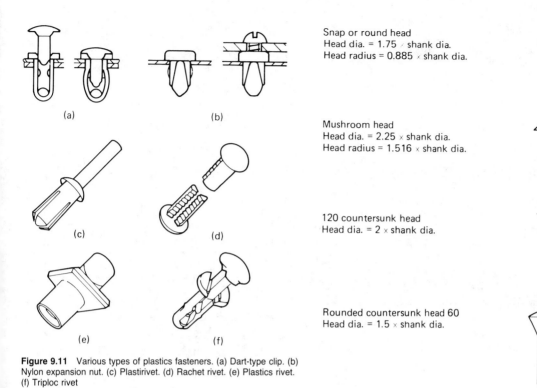

Snap or round head
Head dia. = 1.75 × shank dia.
Head radius = 0.885 × shank dia.

Mushroom head
Head dia. = 2.25 × shank dia.
Head radius = 1.516 × shank dia.

120 countersunk head
Head dia. = 2 × shank dia.

Rounded countersunk head 60
Head dia. = 1.5 × shank dia.

Figure 9.11 Various types of plastics fasteners. (a) Dart-type clip. (b) Nylon expansion nut. (c) Plastirivet. (d) Rachet rivet. (e) Plastics rivet. (f) Triploc rivet

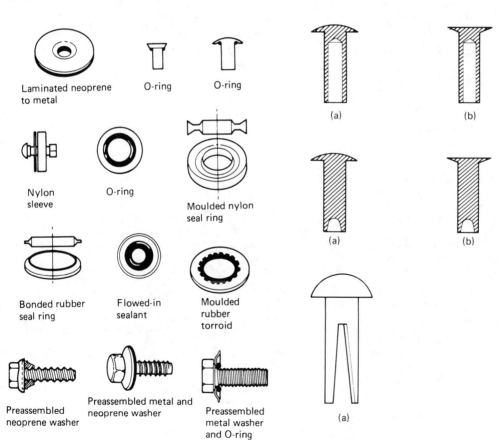

Laminated neoprene to metal

O-ring

O-ring

Nylon sleeve

O-ring

Moulded nylon seal ring

Bonded rubber seal ring

Flowed-in sealant

Moulded rubber torroid

Preassembled neoprene washer

Preassembled metal and neoprene washer

Preassembled metal washer and O-ring

Figure 9.12 Self-sealing fasteners and washers

(a)

(b)

(a)

(b)

(a)

Figure 9.13 Selection of rivets

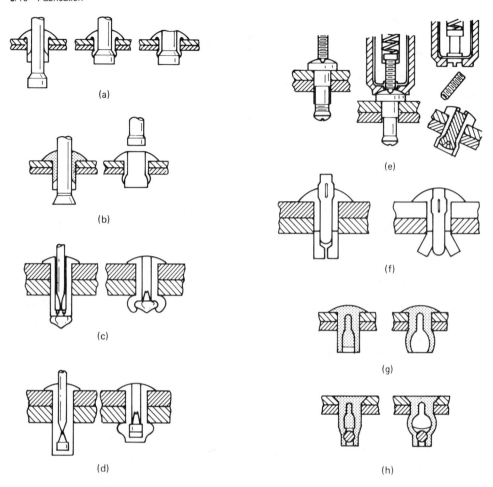

Figure 9.14 Blind rivets. (a) Break-stem mandrel. (b) Pull-through mandrel. (c) Break-stem mandrel (open end). (d) Break-stem mandrel (closed end). (e) Threaded mandrel. (f) Drive pin. (g) Chemical expanded (open end). (h) Chemical expanded (closed end)

9.1.15 List of suppliers of fasteners

Armstrong Screws and Fixings, 72 Great Barr Street, Birmingham B9 4BJ, UK

Bollhoff Fastenings Ltd, Midacre, The Willenhall Estate, Willenhall WV13 2JW, UK

British Industrial Fasteners Federation, Queens House, Queens Road, Coventry CV1 3EG, UK

Dzus Fasteners Europe Ltd, Farnham Trading Estate, Farnham, Surrey GU9 9PL, UK

European Industrial Services Ltd, Woden Road West, Kings Hill, Wednesbury, West Midlands WS10 7TT, UK

Evode Ltd, Como Road, Stafford ST16 3EH, UK

Fasteners Centre Ltd, The Hale House, Ghyll Industrial Estate, Heathfield, East Sussex TN21 8AW, UK

ISC Fasteners Europe Ltd, 180 Rooker Road, Waltham Abbey, Essex EN9 1JJ, UK

Jetpress Sales Ltd, Nunn Brook Rise, Huthwaite, Notts NG17 2PP, UK

T W Lench Ltd, P.O. Box 31, Excelsior Works, Rowley Regis, Warley, West Midlands B65 8BZ, UK

Micro Plastics International Ltd, Unit 2 Healey Road, Coventry CV2 1SR, UK

PEM International, Kirk Sandall Industrial Estate, Sandall Lane, Doncaster, South Yorkshire, UK

PSM Fastener Systems Ltd, Longacres, Willenhall, West Midlands WV13 2JS, UK

SEAC Ltd, 46 Chesterfield Road, Leicester LE5 5LP, UK

Spiral Industries Ltd, Princewood Road, Corby, Northants NN17 2ET, UK

Southco Fasteners Ltd, Unit E, Gregorys Bank Industrial Estate, Gregorys Bank, Worcester, UK

TR Fastenings, Trifast House, Framfield Road, Uckfield, East Sussex TN22 5AR, UK

Tappex Thread Inserts Ltd, Masons Road, Stratford on Avon, Warks CV37 9NT, UK

Trumpf Machine Tools Ltd, Lyon Way, Hatfield Road, St Albans, Herts AL4 0LB, UK

Unbrako, Gunns Lane, West Bromwich, West Midlands B70 9HF, UK

9.2 Welding, soldering and brazing

9.2.1 Welding

Welding is used for joining metals so that the physical and mechanical properties of the parent metal are reproduced in the joint.

The integrity of a welded component, which has metallurgical continuity across the joint, is also characterised by properties such as pressure tightness, or heat and corrosion resistance. These properties have contributed to the rapid development, both technical and economic, in all fields including nuclear power, chemical engineering, bridge building, offshore engineering, shipbuilding, and the manufacture of automobiles, railway locomotives and rolling stock, aircraft engines, domestic appliances, and military hardware from small arms to main battle tanks.

Metallurgical continuity across a joint can cause problems compared with bolted or rivetted connections. For example, under certain conditions, cracks originating from fatigue or brittle failure can propagate across welded seams, whereas they are arrested by mechanical joints. Nevertheless, extensive research has been carried out over many years on the factors controlling fatigue and brittle failure, and guidelines are available in the literature and in various codes of practice which are effective in avoiding cracking problems during the service life of welded structures and components.

A weld is, in effect, a miniature casting, and some of the defects that can occur in castings may also be present in welds. The heat-affected zone adjacent to a weld is heated to high temperature during welding and cools rapidly as heat is conducted into the body of the component.

This area may be heated again to a lower temperature when more weld metal is deposited on the top of the first weld to build up the joint. The sequence of heating, cooling rapidly, and reheating to a lower temperature is similar to the heat treatment of steel by quenching and tempering.

Some welding processes involve substantial plastic deformation of metal adjacent to the interface between the parts being welded, which is similar in many respects to the forging process which is used to form metal parts.

Thus, welding encompasses the whole field of metallurgy, albeit on a small scale, and the application of metallurgical science has successfully solved the major welding problems such as cracking of weld metal or parent plate, porosity and slag inclusions in welds, as well as problems that may occur in service such as cracking due to fatigue, brittle fracture or creep, and wear or corrosion.

It will be readily understood that welding technology has developed into a discipline in its own right, and a short section can only refer to the main points. However, it is important for the engineer to know what methods are available for joining the materials used because the most economical manufacturing route in many cases depends on selection of the most appropriate welding procedure that will produce welded joints having the minimum level of quality fit for their intended purpose. This will require decisions to be taken by an experienced welding engineer who will generally have a professional qualification, but to understand the logic of the welding engineer's decisions a brief knowledge of welding technology is advantageous for any engineer. The purpose of this section is to provide basic information on the principal methods of welding, brazing and soldering and their fields of application. For more detailed information the reader should consult the references cited. It should be noted that in the early 1990s most of the British Standards referred to will be revised and given a European Standard (EN) number, although the British Standard number will not be changed.

9.2.1.1 Types of joint

The types of joint used and their associated weld types are described in detail in BS 499: 1983: Part 1.[1] The commonest joint types are butt, T, corner and lap.

Butt joints These are joints between parts that are generally in line. If two plates are placed in contact (a close square-butt joint) they can be welded with full penetration by one run of weld metal deposited by a manual welding process from each side provided that the plate thickness does not exceed approximately 8 mm (*Figure 9.15*). However, plate above 6 mm thickness is generally bevelled, and the V-edge preparation formed is filled by depositing a number of runs of weld metal. If high current mechanised welding processes are used, penetration of the weld may be at least double the above dimensions and for electron beam welding may be many times as high.

Lap joints These are commonly used for sheet metal up to about 3 mm thick in which one sheet is overlapped by another. This type of joint is used for soldering, brazing, resistance spot or seam welding, and for arc spot welding, plug welding, as well as for adhesive bonding. For material of 3 mm or thicker, even up to 10 mm lap joints are occasionally used and fillet welds are deposited at the plate edges by arc welding.

T joints and corner joints The parts may be joined by fillet welds or butt welds made by an arc welding process (*Figure 9.16*).

9.2.1.2 Welding processes

The various welding processes can be used to join the majority of metallic materials whether in cast or wrought form, in thicknesses from 1 mm or less up to 1 m or more.

A simple classification of welding processes is given in *Figure 9.17*. A complete classification and definitions of the processes are given in BS 499: 1983: Part 1.[1]

A description of the various welding processes is given below.

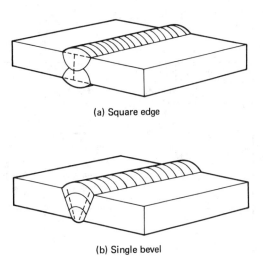

(a) Square edge

(b) Single bevel

Figure 9.15 Edge preparation for butt welds

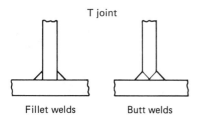

T joint

Fillet welds   Butt welds

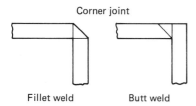

Corner joint

Fillet weld   Butt weld

Figure 9.16 Examples of welded joints

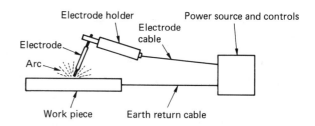

Figure 9.18 Welding circuit for manual metal arc welding

9.2.1.3 Manual metal arc welding

Manual metal arc welding, referred to as 'shielded metal arc welding' in the USA, is the most widely used welding process and accounts for approximately 50% of all the welding carried out in the world today. With this process, welding is carried out with flux coated electrodes which are connected via an electrode holder and length of cable to one terminal of a

welding power source, such as an a.c. transformer or a d.c. generator (*Figure 9.18*). The other terminal of the power source is connected to the work piece via the earth return or the ground cable, so that when the end of the electrode is placed in contact with the work piece, electric current flows through the circuit.

By withdrawing the tip of the electrode to about 3 mm from the work piece, an arc will be struck and current will continue to flow in the circuit and will pass through the arc which is electrically conductive.

If an arc is maintained between a rod type electrode and plates to be welded together, the tip of the rod becomes molten and so does a portion of the plates (the fusion zone). Gravity causes drops of molten metal to drip onto the plate and form a weld (*Figure 9.19*).

Apart from gravity, other forces caused by electromagnetic effects propel molten metal globules across the arc, and these forces always transfer metal from the rod to the plate whether

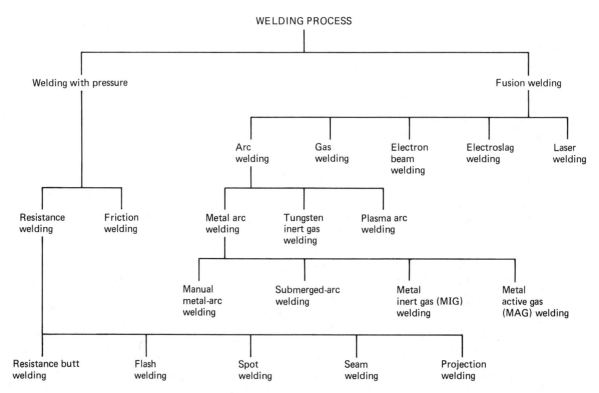

Figure 9.17 Classification of principal welding processes

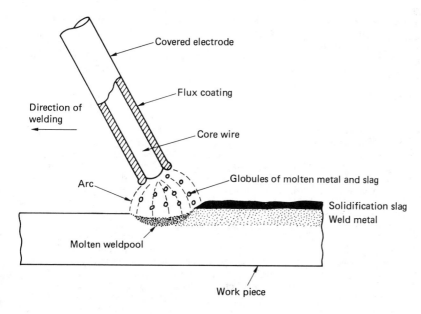

Figure 9.19 Manual metal arc welding with covered electrode

a.c. or d.c. is used and whether the polarity is electrode positive or negative. They will also transfer the metal against the force of gravity, so that vertical or overhead welding is possible.

Electrodes Electrodes have core wire diameters of 2.4–10 mm and are 300–450 mm in length. The deposition rate of weld metal, which governs the overall rate of welding, increases with increasing current and has a maximum value for each electrode length and diameter. Exceeding the maximum current causes overheating of the electrode core wire by resistance heating, which can damage the electrode coating. Welding currents vary from 60 A for the smallest electrodes, up to 450 A for the largest. The highest currents and deposition rates can only be used when welding downhand, i.e. in the flat position. Vertical and overhead welding can be used with electrodes having diameters up to 5 mm with maximum currents of approximately 170 A.

A bare wire can be used for welding, but the arc is mechanically unstable and the surface appearance of the weld is rough. The molten weld metal combines with nitrogen and oxygen from the air, resulting in poor mechanical properties.

The above problems can be overcome by coating wires with suitable fluxes and the principal purposes of this are:

(1) To facilitate the initiation or striking of an arc and to stabilise it so that it can be easily maintained.
(2) To provide a gas shield which protects the molten metal droplets from oxygen and nitrogen in the air as they are transferred through the arc.
(3) To provide a slag which protects the hot, solidifying metal from oxidation. The characteristics of the slag (e.g. melting point, surface tension, and viscosity) determine the shape of the weld bead and the suitability of the electrode for positional welding.
(4) To supply alloying elements to the weld metal; this means that an inexpensive rimming steel core wire can be used for many different weld metal compositions.

The constituents of the flux covering are mixed together in dry powder form and then binding agents are added. The weld paste is pressed into the form of slugs and loaded into machines which extrude the covering round electrode core wires as they pass at high speed through a die of appropriate size.

The electrodes are then dried as they pass through ovens and are stamped with identification marks before being packed.

The classification of flux coverings The development of flux coverings, consisting of mixtures of various minerals, has followed fairly well-defined lines with slight variations between different manufacturers and in different countries. Electrodes can be classified according to their coating types; for a full description BS 639: 1986[2] and the American standard AWS A5.1-81[3] should be consulted.

Steel electrode types are designated by letters in BS 639: 1986 and the chief characteristics of the different electrodes are as follows.

R (rutile)—Rutile coverings contain a high proportion of titanium dioxide in the form of the mineral rutile or ilmenite. The electrodes are easy to use but produce weld metal having high hydrogen contents which can cause cracking of the weld or parent metal heat-affected zone in heavily restrained joints.
RR (rutile, heavy coated)—The thick covering enables the electrodes to be used as contact electrodes which can be held in contact with the parent plate and dragged along the joint at high welding speed. Iron powder is often added to the coating to increase the deposition rate, and the RR type electrodes are not suitable for welding in the vertical and overhead position.
B (basic)—A basic covering usually has a high content of limestone (calcium carbonate) and fluorspar (calcium fluoride). Basic covered electrodes are often referred to as 'low hydrogen' because they were developed to produce weld metal having a low hydrogen content which reduces any tendency to hydrogen induced cracking. This covering decomposes to give a gas shield containing a large proportion of

carbon dioxide. These electrodes are used extensively because of their ability to weld medium and high tensile steels as well as high sulphur (free cutting) steels without solidification cracking of the weld metal and also because, by suitable drying treatment, the moisture content of the flux covering can be reduced so that the weld metal hydrogen content will be correspondingly low. This gives insurance against hydrogen-induced cracking of both the weld metal and the heat-affected zone (HAZ). Properly designed basic covered electrodes produce weld metal which has the highest fracture toughness properties, and they have the advantage over other types of electrodes in that high fracture toughness is maintained in all welding positions.

BB (basic, high efficiency)—These are similar to basic covered electrodes but have iron powder added to the coating so that the quantity of weld metal deposited is at least 130% of the weight of the core wire. The high deposition rate makes these electrodes unsuitable for welding in the vertical and overhead positions.

C (cellulosic)—This designation indicates a covering which has a high content of cellulosic material. These electrodes operate at a high arc voltage, which gives a deep penetrating arc and rapid burn-off. The covering forms a voluminous gas shield, consisting chiefly of carbon monoxide and hydrogen, and a small volume of slag which facilitates work involving changes in welding position such as pipe welding for which these electrodes are particularly suitable. They can also be used for the fast, vertical down welding of vertical seams in storage tanks up to about 12 mm thick. Because of the excellent penetration the root does not require gouging before making a sealing run on the reverse side. In pipe welding the close control of penetration is necessary because the deposition of a sealing run on the inside is generally impossible.

Power sources There are basically three types of power source:

(1) a.c. generators,
(2) d.c. rotary generators, and
(3) d.c. solid state.

The choice of power source is described by John and Ellis.[4]

A fourth type of power source of recent development is based on an invertor which is used to convert the mains frequency from 50 Hz to 5–25 kHz. Transformers for currents operating at these high frequencies are much lighter than those used in conventional a.c. generators. In the invertor-type power source, the a.c. mains input is rectified to give d.c. which is then fed to an invertor which converts it back to a high frequency a.c. The power is then reduced to the welding voltage by a lightweight transformer and it is rectified again to d.c. for welding.

Traditionally, in the UK transformers have been the most widely used type of power source for manual metal arc welding because of their relative cheapness, reliability and long life. The development of d.c. solid state power sources has eliminated the moving parts and maintenance costs associated with rotary generators, and has also reduced the capital costs.

Therefore, the use of d.c. welding is likely to increase in future, particularly in view of the fact that welding is easier because of the more stable arc and ability of d.c. to weld non-ferrous alloys and to produce the highest quality welds in stainless steel.

For use in repair work, for example in the garage trade, and for the 'do-it-yourself' market, a number of low-power welding sets are available which operate from the single phase 220/240 V supply. These power sources can be used with electrodes up to 3.25 mm diameter and both a.c. and d.c. units

are available. For the home market, video instruction is available.[5]

Applications The chief reasons for the popularity of manual metal arc welding are its versatility, simplicity of operation, and the relatively low cost of equipment.

The process can be used with equal facility in a workshop or on site. To weld on a remote part of a structure, it is possible, within reason, to lengthen the cables from the power source and, when the limits of extended cables are reached, the power sources are readily transported by crane or motor vehicle, or manually on level sites, and movement of the power sources is facilitated by their simplicity and robustness.

The range of thicknesses welded varies from less than 2 mm in the fabrication of sheet metal ventilation ducting to 75 mm and above in the production of nuclear containment pressure vessels. These two examples are indicative of the wide range of quality standards that may be required, from general sheet-metal work up to the highest possible standard of radiographic soundness and mechanical properties.

Metals that are most commonly fabricated by manual metal arc welding are carbon and carbon manganese steels, low alloy steels and stainless steels of both the corrosion- and heat-resisting types. By selection of suitable electrodes described in various standards,[2, 6–9] the mechanical properties of the weld metal in respect of strength, ductility and toughness match those of the parent plate at ambient temperature and also at elevated or subzero temperatures, as required. References to American Welding Society (AWS) specifications are included because of the world-wide use of these in the oil and petrochemical industries.

Non-ferrous metals such as nickel, copper and aluminium and their alloys are welded much more extensively with the gas shielded processes, although nickel and nickel alloys are readily welded by the manual metal arc process and a wide range of electrodes is available.[10–12]

Some tin–bronze (copper–tin) and aluminium–bronze (copper–aluminium) electrodes are manufactured, but their chief use is for repair work, particularly of castings, e.g. marine propellers. These alloys can also be used for welding pure copper, because the high conductivity of copper has prevented the successful production of a copper electrode. Pure copper is generally used for either its high thermal or electrical conductivity and, therefore, the application of copper alloy electrodes is strictly limited to those circumstances where weld metal, having low thermal or electrical conductivity, is satisfactory. Some non-ferrous electrodes based on nickel, nickel–iron or nickel–copper alloys are used for welding cast irons.[14, 15]

A wide range of electrodes is available for hard surfacing components to increase their wear resistance under conditions of abrasion, impact, heat or corrosion or various combinations of these factors. Electrodes for hard surfacing are manufactured from core wires of mild steel, carbon and alloy steels, stainless and heat-resisting steels, nickel–chromium and cobalt–tungsten–chromium alloys, and are also made from steel tubes containing granules of refractory metal carbides such as tungsten and chromium carbides.[14, 15]

Gravity welding This is a semimechanised welding process which is used principally in shipyards for fillet welding stiffeners to horizontal plates. Covered electrodes of the contact type, typically 600 mm long, are supported by an electrode holder which slides down one arm of a tripod. The other end of the electrode is positioned in the corner of a T joint to be welded and when the current is switched on an arc is initiated and the electrode moves along the joint line as the electrode holder slides down the arm of the tripod under the force of gravity. One man can operate three gravity welding units simultaneously, thus trebling the rate of manual welding.

Open arc automatic welding Although no longer used, it is appropriate to mention a mechanised welding process which was used extensively in the 1950s up to the 1970s in shipbuilding and bridge building. The engineer may find the process referred to in periodic inspection reports of these types of welded structures.

Mechanised welding with covered electrodes was carried out with a continuous coiled electrode having a core wire of 4–8 mm diameter. The core wire was wrapped helically with two thin wires (about 1 mm diameter) which anchored the extruded flux in place and also acted as a means of conducting electrical current from the jaws of the welding head to the core wire.

Automatic open arc welding with continuous covered electrodes has now been superseded by the submerged arc process.

9.2.1.4 Submerged arc welding

This is the most widely used mechanised welding process (*Figure 9.20*). A base wire (1.6–6.3 mm diameter, but usually 3.25 or 4 mm) is fed from a coil and an arc is maintained between the end of the wire and the parent metal. As the electrode wire is melted, it is fed into the arc by a servocontrolled motor which matches the wire feed rate to the burn-off rate so that a constant arc length is maintained.

The region of the joint is covered with a layer of granular flux approximately 25 mm thick, fed from a hopper mounted above the welding head. The arc operates beneath this layer of flux (hence the name 'submerged arc'). Some of the flux melts to provide a protective blanket over the weld pool and the unmelted flux is collected and reused. The electrode wire, welding head, wire drive assembly and flux hopper are mounted on a traverse system which moves along the work piece as the weld metal is deposited.

The traverse system may consist of a carriage mounted on a boom or it may be a motorised tractor either on rails or running freely with manual adjustment to follow the weld seam. Alternatively, the welding head can remain stationary while the work piece is moved. This method is used for welding the circumferential seams of a pressure vessel while it is rotated under the welding head.

Electrode wires The electrode for submerged arc welding is a bar wire in coil form usually copper coated. Two types are available: (a) solid wire, and (b) tubular wire. The solid wire is widely used for general fabrication of mild and low alloy steels, stainless steels and non-ferrous metals. For welding mild steel and low alloy steels, it is either a low carbon ultra-low silicon steel or a silicon-killed steel with manganese addition and sometimes low alloy additions, the selection of either type depending upon the type of flux to be used with it, i.e. a flux with manganese or manganese and alloy additions or a neutral flux, respectively. The tubular wire (made by forming narrow strip into a tube) carries alloy powders which permit the economical production of a wider range of weld compositions than is possible by using the solid wire type. Tubular wires are widely used for hardfacing.

Wire compositions for welding carbon steel and medium tensile steel are listed in BS 4165: 1984.[16]

With coated manual electrodes, wire and coating are one unit so that such electrodes can be classified according to the type of coating and its effect on weld mechanical properties. In submerged are welding, any wire may be used with a number of different fluxes with substantially different results in respect of weld quality and mechanical properties.

Consequently, BS 4165: 1984 grades wire flux combinations according to the tensile and impact strengths obtained in the weld metal.

A number of tubular wires are available, particularly for surfacing and hardfacing. These contain alloy powders which produce weld metals consisting of low alloy steels, martensitic and austenitic stainless steels, chromium and tungsten carbides, and various cobalt and nickel based heat- and corrosion-resistant alloys. Some corrosion-resistant alloys, including stainless steel are available in the form of coiled strips (100–150 mm wide, 0.5 mm thick) for high deposition

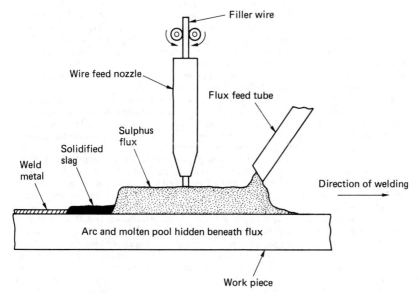

Figure 9.20 Submerged arc welding

rate surfacing by a submerged arc welding process known as 'strip cladding'.

Fluxes Two main types of flux are available, fused and agglomerated. Fused fluxes are manufactured by fusing together a mixture of finely ground minerals, followed by solidifying, crushing and sieving the particles to the required grain size. Fused fluxes do not deteriorate during transportation and storage and do not absorb moisture. Agglomerated fluxes are manufactured by mixing finely ground raw materials with bonding agents such as sodium or potassium silicates followed by baking to remove moisture. This type of flux is sensitive to moisture absorption and may require drying before use. Agglomerated fluxes are more prone to mechanical damage which can cause segregation of some of the constituents.

Fluxes are classified as acid, neutral, or basic, the latter being subdivided into semi-basic or highly basic. The main characteristics of the fluxes are as follows:

Acid fluxes—These have a high content of oxides such as silica or alumina. They are suitable for high welding currents and fast travel speeds. They are resistant to porosity when welding rusty plate, and have low notch toughness. They are not suitable for multipass welding of thick material.

Neutral fluxes—These have a high content of calcium silicate or alumina rutile. They are suitable for fairly high welding currents and travel speeds and also for multipass welding.

Basic fluxes—These have a high content of chemically basic compounds such as calcium oxide, magnesium oxide and calcium fluoride. They give the highest weld metal quality in terms of radiographic soundness and impact strength. Lower welding currents and travel speeds are suitable for multipass welding of thick sections. For further information on fluxes see references 16 and 17.

Power sources Either d.c. or a.c. may be used. D.c. may be supplied either by a motor generator or by a rectifier, either of which can have flat or sloping current voltage characteristics. These are referred to as 'constant-voltage' or 'constant-current' types. Generators will deliver up to about 650 A continuous output, and rectifiers deliver up to about 1200 A at 100% duty cycle. A.c. is supplied by welding transformers which are designed to give a drooping characteristic. Transformer output may range up to 2000 A.

Power sources are of rugged construction and are designed for 100% duty cycle. With the constant-current type source the arc voltage determines the wire feed rate which varies to maintain a constant arc length. The constant-voltage type power source produces a self-adjusting arc in which an increase in arc length or arc voltage causes a decrease in current and burn-off rate so that the original arc length is rapidly obtained. A decrease in arc length increases the current and burn-off rate and this self-adjusting effect occurs with a constant wire feed rate.

Application As the process operates with a continuous coil of electrode wire, butt welds in the flat position requiring multiple runs to fill the joint can be made with minimum stops and starts. Thus, circumferential joints in cylindrical bodies such as pressure vessels, pipes, etc., can be made with one stop and start per revolution of the work piece, this stop being necessary to reset the position of the welding head. Consequently, the possibility of stop and start defects is minimised; a most important consideration when reliability in costing is required. Although most widely applied to welding of joints in mild steel, low alloy high tensile steel, creep resisting steels, and to a lesser extent stainless steels, it is also widely used for building-up work, either for reclamation or replacement of defective parent metal or for hard surfacing.

Submerged arc welding is suitable for welding material of 5–300 mm thickness, and even thicker, but plates less than about 10 mm thick are generally welded by the gas shielded or flux cored arc welding process. A semi-automatic variant of the process is available in which the welder manually manipulates a welding gun on which is mounted a small hopper containing the flux. Electrode wire is fed to the gun from a coil by a wire feed unit. This process, which is only used to a limited extent, is sometimes referred to as 'squirt' welding.

Other variations of the submerged arc welding process are mainly concerned with increasing deposition rates and, therefore, welding speed and productivity. These include the following:

(1) Increasing the electrode extension or stick out by up to 150 mm by using an insulated guide tube. The resistance of the wire increases the burn-off rate by the I^2R heating effect.[18]

(2) Addition of iron powder to the joint which increases the weld volume.

(3) The use of multiple wire techniques in which two or more wires are used with two or three separate power sources.[18]

(4) Narrow gap welding of plates more than 100 mm thick in which a parallel gap of 14–20 mm between the square edges of the plates is used instead of a V or U groove. The saving in quantity of weld metal used is considerable, with a consequent increase in productivity.[19]

9.2.1.5 Electroslag welding

Electroslag welding is an automatic process for welding material 18 mm or thicker in the vertical position. Square edge plate preparation is used and joints are limited to butts and T butts. Heat for fusion is obtained by the current in the consumable electrode wire which passes through the molten slag formed by melting of a flux formulated to have high electrical resistance in the molten state.

The joint is set up with a wide gap (approximately 25–36 mm, depending on plate thickness), the very large weld pool being contained in the joint by water cooled copper shoes. The joint is filled in one pass. One, two or three wires are fed into the joint with or without a reciprocating motion to ensure uniform heat generation, the plate thickness determining the number of wires. One welding head with three wires can weld plate 450 mm thick. The copper shoes rise up the joint to prevent spilling of metal and slag. These shoes form part of the welding head.

A variation of the electroslag process is known as 'consumable guide welding'. Here the welding head remains stationary and feeds one or more wires down a tube which melts into the pool. The equipment is substantially cheaper than the conventional slag welding machines.

The consumable guide process, while theoretically suitable for very thick sections, is more usually applied to metal of thicknesses up to, say, 50 mm, because it is more manageable in these thicknesses. The equipment needed is much simpler than for the conventional process, consisting of a constant potential generator, rectifier or transformer with a flat characteristic and a wire feed unit. These are also the essential ingredients of submerged arc equipments and, by slight modification, they can be adapted for consumable guide welding.

The ultra-slow cooling rate of an electroslag or consumable guide weld minimises hydrogen cracking susceptibility in the parent steel and weld metal, but produces a large grain size weld. This tends to give a comparatively poor notch impact strength as determined by conventional tests. Consequently, for pressure vessel application, current codes require normalising after welding.

Fluxes Electroslag welding fluxes produce complex silicate slags containing SiO_2, MnO, CaO, MgO and Al_2O_3. Calcium fluoride is added to increase electrical conductivity and lower slag viscosity. Slags based on CaF_2–CaO have a strong desulphurising action which assists the welding of steels having carbon contents greater than 0.25%, without solidification cracking in the weld metal. Fluxes must be kept dry.

Economics of the process On heavy steel plate, for pressure vessels, boiler drums, etc., the actual welding speed (rate of filling the joint) is about twice as fast in 40 mm plate, four times as fast in 90 mm plate, and eight times as fast in 150 mm plate as compared with multirun submerged arc welding. The welding speed is 1–1.7 m h^{-1}. Plate-edge preparation by bevelling is also avoided. On this evidence the process would seem highly attractive economically speaking. However, 'setting up' the machine and selecting the correct welding parameters is a matter of experience or tests. Therefore the 'setting-up' time must also be considered and the cost of determining the correct procedure included in the cost estimate.

Thus, the process gives full economic benefit on repetition work where set parameters can be used based on experience. An example is the wide use of the process for the longitudinal seams of boiler drums in 125–150 mm thick steel. On one-off applications involving plate thicker than 150 mm, the process is economical on each joint and, providing sufficient work of suitable type is available, the high capital cost is recovered within a reasonable time.

Mechanical properties The weld metal mechanical properties are determined to a considerable degree by the composition and cleanliness of the parent steel. However, proper selection of wire and flux and suitable parent metal confer strength and ductility equal to or better than the parent metal.

In the as-welded condition, degradation of notched impact properties in the heat-affected zone results in values lower than the weld metal. Where notched impact requirements must be met, it is necessary to normalise the completed joints. In general, electroslag welding is an acceptable and economical method of welding thick steel. It finds application for ships' hulls, boiler drums, press frames, nuclear reactors, turbine shafts, rolling mill housings and similar heavy fabrications.

9.2.1.6 Gas shielded metal arc welding

This general term covers a group of welding processes in which no added flux is used. The molten weld pool is protected by a gas shield which is delivered to the welding gun through a flexible tube at a controlled rate, either from a gas bottle or from a bulk supply. The shielding gas may be inert (e.g. argon or helium or mixtures of these gases) or it may be active (e.g. carbon dioxide (CO_2) or mixtures of CO_2 with other gases such as argon). Sometimes small additions of oxygen or hydrogen are included in the shielding gas.

The wide variety of shielding gases, which may be completely inert (or non-reactive) or active (i.e. slightly oxidising or reducing), has led to the use of the terms 'metal inert gas' (MIG) and 'metal active gas' (MAG) to describe the principal gas shielded metal arc welding processes.

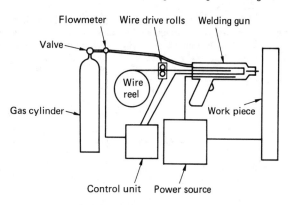

Figure 9.21 Metal inert gas (MIG) welding

9.2.1.7 MIG and MAG welding

In MIG or MAG welding, referred to as 'gas metal arc welding' (GMAW) in the USA (*Figure 9.21*), a small diameter (0.6–1.6 mm) wire is fed from a coil by a wire feed unit which contains an electric motor, gearbox and grooved drive rolls. The wire is fed to a welding gun which has a trigger which operates the wire feed drive, the current and the flow of shielding gas.

An arc is struck when the wire contacts the work and the arc length depends on the voltage which is preset by adjustment of a knob on the power source. Welding current is picked up by the wire from a copper contact tube through which the wire passes. The distance between the contact tube where current enters the wire and the end of the wire is usually a maximum of 25 mm, compared with 300–450 mm for a covered electrode. Therefore, overheating is not a problem, particularly as there is no flux coating. Higher currents can be used than those normally used for manual metal arc welding, e.g. 120–450 A for 1.6 mm diameter wire.

Therefore, deposition rates are generally higher than for manual metal arc welding. Another advantage of MIG/MAG welding is that the arc is automatically maintained at a length which depends on the arc voltage; thus the welder only has to move the welding gun along the joint line holding the nozzle of the gun at approximately the same distance from the joint. This is because the arc is self-adjusting because the voltage current characteristic of a MIG/MAG power source is flat or only slightly drooping.

If the welding gun is moved away from the joint the arc length and the arc voltage increase slightly. With a flat or slightly drooping characteristic, a small increase in voltage will cause a large decrease in welding current and the wire will burn off at a lower rate. The original arc length will be rapidly attained at which voltage the burn-off rate will once again match the wire feed speed.

Similar self-adjustment in the opposite sense will occur when the welding gun is moved towards the work piece.

Filler wires The commonest filler wires are 1.0, 1.2 and 1.6 mm in diameter, with 0.6 and 0.8 mm diameter wires less frequently used. Because of the wide range of current that can be used with each wire, it is only necessary to stock one or two diameters. This is in contrast with manual metal arc welding where a number of different diameters of electrodes and possibly two or more coating types may be required just for welding a single type of material such as carbon manganese steel.

However, a disadvantage of solid filler wires is the limited range of compositions available, because it would be too expensive for a steelmaker to produce small quantities of low alloy or stainless steel wires. Small batches of covered electrodes can readily be produced by introducing alloys in powder form through the coating. This situation is reflected by the number of electrodes and filler wires for welding low alloy and stainless steels listed in British Standards, which are as follows:

	Low alloy steels	*Stainless steels*
Covered electrodes	30	39
Solid wires	6	16

Any deoxidising elements such as silicon or aluminium required to refine or degas the weld pool are contained in the solid wire and compositions of wires available are listed in BS 2901: Parts 1–5[20] for ferritic steels, austenitic stainless steels, copper and copper alloys, aluminium and aluminium alloys and magnesium alloys, and nickel and nickel alloys.

Modes of metal transfer In MIG or MAG welding the operating conditions in terms of current and voltage determine the type of metal transfer which must be suitable for the application. There are four modes of metal transfer.

(1) *Short circuiting (dip transfer)*—This occurs when a low voltage and current are used which causes metal to be transferred from the end of the wire to the work piece by frequent short circuiting of the wire to the weld pool. This technique produces low heat input and a small controllable weld pool essential for welding steel sheet in all positions and thicker steel sections in the vertical and overhead positions. A disadvantage of the process is the production of spatter in the form of globules of metal expelled from the weld pool when each short circuit is broken. Spatter particles which adhere to the work piece can be reduced by fine tuning of the inductance of the power source.

(2) *Globular transfer (semi-shorting)*—This occurs when somewhat higher currents and voltages are used than for dip transfer welding of steel, but metal transfer still occurs by short circuiting of the filler wire to the weld pool. Because of the large droplet size and the larger weld pool, this mode of welding is not suitable for vertical or overhead welding. The production of spatter still occurs.

(3) *Spray transfer*—Free flight of metal droplets occurs with no short circuiting when the current and voltage are sufficiently high. This gives maximum deposition rates and deep penetration welding suitable for flat position welds in medium and heavy steel plate and for horizontal-vertical fillet welding (e.g. between a vertical and a horizontal plate). Spray transfer is used for welding aluminium and aluminium alloys in all positions because the spray transfer of droplets occurs at much lower welding currents than with other metals. Therefore, small weld beads can be deposited which solidify rapidly and enable welding to be carried out in the vertical and overhead positions.

(4) *Pulsed transfer*—This was developed to produce spray transfer at all current levels so that welding of all metal thicknesses in all welding positions could be carried out without the formation of spatter. In pulsed transfer the welding current is switched from a high pulse current to a low background current at a typical frequency of 50 Hz. The background current is sufficient to sustain the arc but it is insufficient for metal transfer. The pulse current is set above the critical level to produce sufficient electromagnetic force with each pulse to transfer one metal droplet from the tip of the wire. With the first pulsed arc power supplies the pulse frequency had to be a multiple of mains frequency and setting up welding conditions was difficult to the extent that it hindered the use of the process in industry. The average current, which depended on the background current, the pulse current and the frequency had to produce a usable burn-off rate at a constant arc length. The process was also sensitive to electrode stick out (the electrode extension beyond the contact tube) which could disrupt the balance between pulse energy and metal transfer, causing arc extension and spatter. The full advantages of the pulsed MIG process including stable low mean current operation, particularly when welding aluminium alloys or stainless steel and positional welding capabilities of all metal thicknesses were made readily available with the development of transistorised power sources referred to in the next section.

Power sources MIG and MAG welding are always carried out with d.c. and the principal types of power source are transformer rectifiers with constant potential or controlled slope characteristics and motor generators which are used for site work, e.g. welding pipelines.

Invertor-type power sources, described in Section 9.2.1.3, are also used for MIG and MAG welding. The main advantage of invertors is the considerable decrease in size and weight compared with conventional transformer rectifiers.

Electronic power control has had a considerable and beneficial influence on MIG/MAG welding enabling the process parameters to be preprogrammed which eliminates the complicated setting up operation and enables 'one-knob' control to be achieved.

Programmed control for both dip and spray transfer was originally developed in the late 1960s. The relationship between wire feed speed and voltage for any filler wire type and diameter could be programmed into the power source and a single control could be used to continuously vary mean current. The equipment contained preset resistors to store the fixed parameters but microprocessor-controlled units are now available.

A number of different electronic control systems have been developed for MIG/MAG welding in both the dip and spray transfer modes of operation, which overcome the setting-up difficulties mentioned above.

The setting-up difficulties were particularly acute with the pulsed arc mode of operation, but they have been overcome by the so-called 'synergic control technique' used in conjunction with a transistorised power source. With synergic control, precise independent regulation of the pulse shape, pulse current time, pulse frequency, and background current is obtained. The electronic power source can produce continuously variable 25–250 Hz pulse frequencies. With variable frequency pulsing, the correlation between pulse energy, burn-off rate, and arc characteristics can coincide with all electrode wire feed speeds to provide one metal drop transfer per pulse.[21]

With synergic control, all the pulse parameters are preprogrammed for a wide range of wire feed speeds. During welding the wire feed rate and pulse frequency automatically adjust together to produce one metal droplet transfer at a constant arc length. The welder only needs to adjust one control, i.e. average current.

Modern power supplies have control systems based on microprocessor technology. Memory chips in the control unit store process data and produce the optimum operating parameters if the user presses the appropriate switches to specify the types of filler wire and shielding gas. In some power sources the user can load his own operating programmes into the control unit. For further information the reader should consult references 22 and 23.

Similarly to manual metal arc welding, there are a number of 'hobby' sets on the market which can be used on the 13 A mains. These have a limited number of current settings for use with 0.6 or 0.8 mm diameter wire and can be used to weld carbon steel, stainless steel and aluminium in thicknesses up to 6 mm. Tuition by video is available.[24]

Applications MIG and MAG welding with the various modes of metal transfer can be used for similar applications to those that are fabricated by manual metal arc (MMA) welding. In addition, they are more suitable for welding some of the non-ferrous alloys such as aluminium and copper alloys and are probably equally suitable for welding stainless steel and nickel alloys. For sheet metal thicknesses which are welded by a single run of weld metal, MIG and MAG welding are generally up to 50% faster than MMA welding. In thicker materials, MIG/MAG welding and MMA welding used with the same duty cycle, i.e. the proportion of arcing time to total time, will have approximately the same overall welding rate, provided that full use is made of positioners to enable a large proportion of the welding to be carried out in the flat or horizontal–vertical positions. Claims made in the literature or in suppliers' brochures about the superiority of one process over the other in respect of productivity and economic advantages should be treated with caution, because they may only be valid for a specific application. One great advantage of MIG/MAG welding over MMA welding is the ease with which the process can be mechanised either by fitting the welding gun to a traverse unit, or by moving the work piece under a stationary gun either by linear motion or rotation.

Robotic and automated MIG/MAG welding has advanced with the developments in microcomputers and electronic power sources, the latter providing very stable arcing conditions in spite of mains voltage fluctuations. Automated MIG/MAG welding utilises seam-tracking devices which are necessary to compensate for inaccuracies of the component parts or distortion during welding.

Seam-tracking devices contain contact type sensors such as probes or guide wheels or the non-contact types such as electromagnetic, ultrasonic or video systems. For further information on robotic or automated welding the reader is referred to references 25–27.

9.2.1.8 Flux cored arc welding

Flux cored arc welding is similar in many respects to MIG/MAG welding, except that in one version of the process, no shielding gas is added. In this case the gas shield originates from the decomposition of minerals contained in the tubular cored electrode and this version of the process is sometimes referred to as 'self-shielded welding'.

Cored electrodes in coiled form are manufactured from steel strip which is first bent into a U section as it passes through forming rolls. The U-shaped strip is then filled with a metered quantity of flux and metal powders and the strip is passed through dies to form it into circular cross-section of 0.9–3.2 mm diameter.

Tubular cored electrodes Tubular cored electrodes may be gas shielded with CO_2 or Ar/CO_2 mixtures, or they may be self-shielded. Cored electrodes are classified according to the constituents contained in the core which influence the characteristics of the electrode.

For a full description of the different carbon and carbon–manganese steel types BS 7084: 1989,[28] AWS A5.20-79[29] and AWS A5.29-86[30] should be consulted. Many higher tensile and low alloy steels can be welded with flux-cored wires of matching strengths.

Stainless steel cored wires are available for use either with or without shielding gas, and many different types of cored wires are used for hard-facing applications in which a coating is applied to a steel base to confer resistance to wear, corrosion or heat.

Application Flux cored arc welding is used for similar applications to manual metal arc or MIG/MAG welding and, like MIG/MAG welding, the process can be mechanised.

Tubular cored electrodes are available in a wider range of compositions than are solid wires because of the ease of introducing alloying elements in powder form. Flux cored wires, particularly the gas-shielded types, meet the mechanical-property requirements of a range of applications and some grades give good low temperature impact properties. The mechanical properties attainable with self-shielded cored wires is more limited with maximum weld metal strengths of 700 N mm^{-2}.

Self-shielded wires are particularly useful for site work because no bottles of shielded gases, which are unwieldly, are required. Another advantage on site is that there is no externally added shielding gas which is susceptible to disruption by wind.

Flux cored wires can be used at higher maximum currents than can solid wires, resulting in high deposition rates.

9.2.1.9 Gas shielding tungsten arc (TIG) welding

In this process an arc is established between a tungsten electrode and the parent metal forming a weldpool into which filler rod is fed, generally by hand (*Figure 9.22*). Mechanised systems which feed the filler wire are available and movement of the welding head along the joint line can also be mechanised. The tungsten electrode is non-consumable and contamination of the weld pool by air is prevented by an inert shielding gas such as argon, helium, or mixtures of these gases.

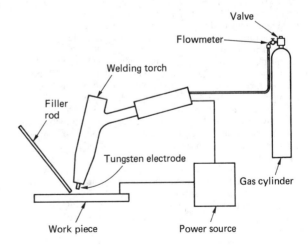

Figure 9.22 Tungsten inert gas (TIG) welding

A high level of skill is required by the welder to control penetration with great precision, which makes the process particularly suitable for the welding of thin sections and for the deposition of root runs in pipe.

Electrodes and filler rods Pure tungsten electrodes can be used, but improved arc initiation and stability are obtained by the use of electrodes containing additions of either thoria (thorium oxide) or zirconia (zirconium oxide). Thoriated electrodes are preferred for d.c. welding and zirconiated electrodes are used for a.c. welding. Electrode diameters vary from 1.2 to 4.8 mm, depending on the welding currents used, which can range from 75 to 450 A for thoriated electrodes and 50 to 200 A for the zirconiated types.

Filler rods which are specified in BS 2901: Parts 1–5: 1992 are of 1.2–5.0 mm diameter and are available in a wide range of compositions suitable for welding carbon and low alloy steels, stainless steels, copper and copper alloys, nickel and nickel alloys, aluminium and aluminium alloys, titanium and zirconium.

Power sources An a.c. or d.c. power source with standard generators, rectifiers or transformers is used. For stable operation the power source must have a 'drooping characteristic' so that when variations occur in voltage or arc length the current remains substantially constant. When changes occur in the arc length in which the welding torch is manually guided along the joint line, the power input remains within ±8% of the preset value.

If the arc is initiated by touching the tungsten electrode onto the parent metal, the electrode becomes contaminated and, to avoid this, a high frequency oscillator is incorporated in the power source. Alternatively, a spark starter using a high voltage coil similar to that in a car ignition circuit can be used. When the gas in the gap between the electrode and the parent plate is ionised by either the high frequency or the spark discharge the full welding current flows. With d.c., the high frequency is normally turned off automatically after arc initiation, but with a.c. it is operated continuously to maintain ionisation of the arc path when the arc voltage passes through zero.

Power sources are available for pulsed arc welding which enables a stable arc to be maintained at low currents down to 10 A. In pulsed TIG welding, the pulse frequency varies from 10 per second to 1 per second and each pulse forms a molten pool which solidifies before the next pulse. Pulsed TIG welding can be used to control penetration in thin sheet and in the root runs of pipes and positional welds in plate.

Applications TIG welding is particularly suited to welding light gauge carbon, alloy and stainless steels and all non-ferrous metals and alloys. A clear clean weld pool is formed with precise control of heat input and the ability to weld with or without filler metal in all positions makes the process attractive for critical applications where exceptionally high quality is essential. Examples are stainless steel piping for nuclear applications and the wide range of piping compositions used in chemical plant.

For such critical applications, fully mechanised orbital welding equipment has been developed in which the welding torch and wire feeding mechanism rotates round the pipe joint. Thin and thick section pipes can be welded with a narrow gap joint preparation and *in situ* fabrication of nuclear and chemical plant is now possible.

Other specialised TIG welding equipment is used for the mechanised welding of tubes to tube plates.

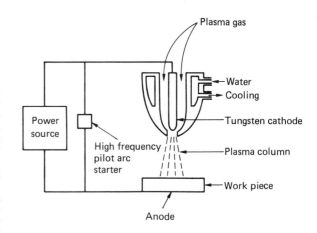

Figure 9.23 Plasma arc welding

9.2.1.10 Plasma arc welding and cutting

Plasma arc welding was developed from TIG welding by placing a narrow orifice round the arc and supplying a small flow of argon through the orifice (*Figure 9.23*). The constricted arc dissociates the argon gas into positive and negatively charged electrons to form a plasma. When the plasma gas flows away from the arc column, it forms neutral atoms again and gives up its energy in the form of heat.

A low current pilot arc is initiated between the tungsten electrode and the water cooled copper orifice. The argon gas flowing through the orifice is ionised and initiates the primary arc between the tungsten electrode and the parent metal when the current is increased.

The arc and the weld zone are shielded by a gas flowing through an outer nozzle. The shielding gas consists of argon, helium or gas mixtures of argon with either hydrogen or helium.

A normal tungsten arc has a temperature of approximately 11 000°C, but the constricted arc of a plasma torch can reach 20 000°C. The high temperature ionised gas jet gives up its energy when it contacts the parent metal and thus increases the energy of the tungsten arc.

This produces a deep penetration weld with a high depth-to-width ratio with minimum distortion of the parent metal. The term 'keyhole' is used to describe the shape of the hole formed in the parent metal when a close square edge butt joint is welded. As the torch is moved along the joint molten metal flows round the edges of the hole and solidifies at the rear of the hole. The molten metal at the sides of the hole is held in place by surface tension and the pressure of metal vapour in the hole.

The keyholing welding technique can be used on carbon, low alloy steels and stainless steels in thicknesses of 2.5–10 mm and in aluminium alloys up to 20 mm. Welding speeds are generally 50–150% higher than those possible with TIG welding.

A low current version of the process is microplasma arc welding which is used for precision welding of thin sheet (0.025–1.5 mm thick) at currents of 0.1–10 A. The plasma arc is much more stable than a TIG arc, which tends to wander from the joint line at low currents.

Plasma cutting If the current and gas flows are increased sufficiently the molten metal formed round the keyhole is ejected at the bottom of the hole and, as the plasma torch is traversed along the work piece, a cut is formed. Plasma cutting is especially suitable for cutting non-ferrous metals, such as aluminium, copper, and nickel and their alloys, which are not easily cut by oxy-fuel gas flames. Most non-ferrous metals are cut using nitrogen, nitrogen–hydrogen mixtures or argon–hydrogen mixtures as the plasma gas. A secondary shielding gas delivered through a nozzle that encircles the plasma gas nozzle is selected according to the material being cut. For mild steel and stainless steel it can be CO_2, and for aluminium it is an argon–hydrogen mixture. Sometimes water is used in place of the ancillary shielding gas and in another variety of the process water is injected round the end of the plasma gas nozzle which has the effect of concentrating the plasma flame and allowing higher cutting speeds.

Plasma cutting can be used for plate-edge preparation, i.e. bevelling and for shape cutting. The process can be used manually or the torch can be mounted on mechanised cutting equipment identical to that used for oxy-fuel gas cutting. For metal thicknesses up to 75 mm, carbon steels can be cut faster by plasma cutting than by oxy-fuel gas, and up to 25 mm thick the cutting speeds can be five times as fast.

An important variation of the process involves the use of compressed air for the plasma gas without the provision of any additional shielding gas. The use of compressed air in place of water for cooling enables the torch to be of simplified construction.

Small manual air plasma torches are available which are finding increasing application in sheet-metal cutting, e.g. motor repair shops.

For further information the reader is referred to reference 31.

9.2.1.11 Gas welding and cutting

Gas welding is carried out using a flame produced by burning approximately equal volumes of oxygen and acetylene which are delivered at equal pressures from gas bottles to a welding torch. The flame temperature is approximately 3100°C, which is high enough to melt steel and other metals. Filler metal, if required, is added by manually feeding a rod into the front edge of the weld pool while the torch is moved along the joint. The products of combustion provide sufficient protection from the atmosphere when welding steel. When welding other metals such as cast iron, stainless steel, aluminium alloys and copper alloys, fluxes are used to clean and protect the metal from oxidation.

Equipment The welding torch has two knurled control knobs which regulate the flow rates of oxygen and acetylene so that a neutral or slightly oxidising or reducing flame is obtained depending on the application. The torch has a screw-in nozzle from a set of nozzles having different diameter holes which produce the appropriate size of flame and, therefore, the required heat input for the particular metal and thickness to be welded.

The oxygen and fuel gas hoses are connected between the welding torch and the gas bottles, the gases passing through flash-back arresters and pressure regulators. Flash-back arresters are safety devices that prevent a flame travelling back into the cylinders in case a backfire occurs.

For workshop use, the gas bottles are generally mounted in pairs on a trolley which can be moved to the place where it is required.

Filler metal and fluxes Chemical compositions of filler metals are specified in BS 1453: 1972,[32] which includes ferritic steels,

cast iron, austenitic stainless steels, copper and copper alloys and aluminium alloys.

Ferritic steels do not require the use of a flux; proprietary fluxes are available for other materials.

Applications Gas welding is used mainly for repair and maintenance work, particularly in the repair of car bodies and agricultural implements, although it is slowly being replaced by small TIG and MIG welding equipment.

Gas welding is used to a certain extent for sheet-metal work, i.e. heating and ventilating ducting, and is still used for making the root runs in pipes where it is particularly useful for bridging gaps.

Two applications where gas welding has distinct advantages over other processes are in the welding and repair of grey iron castings and in hardfacing with expensive alloys.

Grey iron castings can be successfully welded by the use of high preheating temperatures of up to 600°C and gas welding with cast iron filler rods.

The deposition of high cost wear-resistant alloys such as the cobalt–chromium tungsten types or those based on chromium or tungsten carbides, can be carried out with minimum melting of the parent metal so that dilution of the deposited alloy and the consequent decrease in wear resistance is avoided.

Gas welding is also successfully applied in jewellery manufacturing with miniature torches and small gas bottles.

Acetylene is the only fuel gas suitable for gas welding because of its favourable flame characteristics of both high temperature and high propagation rates. Other fuel gases, such as propane, propylene, or natural gas produce insufficient heat input for welding but are used for cutting, torch brazing and soldering. They are also used for flame straightening of distorted components and for preheating before welding and post-heating after welding.

Gas cutting Gas cutting, sometimes referred to as 'flame cutting' or 'oxygen cutting', involves an active exothermic oxidation of the steel being cut when the material has been preheated by an oxy-fuel gas flame to the ignition temperature of around 900°C.

The equipment used for gas cutting is the same as that used for welding except that a special cutting nozzle is required. The nozzle has an outer ring of holes through which the preheating gas mixture is delivered and a central hole through which the oxygen jet flows. The exothermic reaction of oxidation of steel forms a fluid slag of iron oxide and, after a few seconds (depending on the metal thickness), the section is pierced. Iron oxide and molten metal are expelled from the cut by the oxygen stream. Movement of the cutting torch across the work piece produces a continuous cutting action and the torch can be operated manually or by a motorised carriage. Steel up to 300 mm thick can be cut by this process.

Oxidation resistant steels such as stainless steels may be cut by specialised methods including the introduction of iron powder or other proprietary powders into the oxygen stream. These powders react with the refractory chromium oxides and reduce their melting points and increase their fluidity, enabling cutting to take place. For further information the reader should consult reference 31.

Stainless steels and non-ferrous metals and alloys are usually cut by using the plasma cutting process which does not rely on an exothermic reaction.

Manual gas cutting is possible and the accuracy of cutting can be improved by the use of a small wheel mounted on the cutting torch. The wheel may be free running or motorised. For general cutting and profiling, including cutting bevels on plate edges, mechanised cutting is normally used. For mechanised cutting, electronic tracing devices are commonly used.

These consist of a photoelectric cell which follows the outline of a drawing and guides the cutting nozzle by means of driving motors which regulate the movement of a carriage and cross arm to which the torch is attached.

Numerically controlled cutting machines are available which use programs stored or punched on magnetic tape which send appropriate signals to the drive motors.

9.2.1.12 Welding and cutting with power beams

Electron beam welding and laser welding utilise high-energy beams which are focused on a spot on the work-piece surface about 0.2 mm in diameter. This intense heat source, which releases its kinetic energy when the beam hits the surface, is radically different from arc welding in which the arc melts an area of about 5–20 mm diameter, depending on the welding conditions.

When the power density of an electron or a laser beam at the focused spot is 10 kW mm^{-2} or greater, energy is delivered at a faster rate than can be conducted away in the form of heat in the work piece and the progressive vaporisation of metal through the section thickness forms a hole. If the beam is then traversed along the work piece, molten metal flows around the sides of the hole and solidifies at the rear of the hole. Molten metal at the sides of the hole is held in place by surface tension and the presence of metal vapour in the hole in the same manner as that described for the keyholing technique in Section 9.2.1.10 for making a square edge butt joint.

As with plasma welding, a deep penetration weld with a high depth-to-width ratio is formed with minimum distortion of the parent metal. Maximum penetration depths in steel are approximately 280 mm for electron-beam welding and 12 mm for laser welding, although developments in the latter process are likely to increase this to 25 mm or more.

Electron-beam welding In this process, a finely focused beam of electrons passes from a cathode and travels through a hole in an anode and is focused on a spot on the work piece 0.2–1 mm in diameter by means of a magnetic lens. Deflection coils are used to cause the beam to move in a circular pattern to increase the width of the weld, so that fusing two mating surfaces together in a close square butt joint is possible. The cathode in the electron gun is maintained at a negative potential of 60–150 kV and the gun is contained in a vacuum of 5×10^{-5} T. In the work chamber, a pressure of 5×18^{-3} T is suitable for welding most metals and for some applications a pressure of 10^{-1} T or less is used with the advantage of much shorter chamber excavation times resulting in increased production rates.

High vacuum welding High vacuum welding, in which the work chamber is maintained at a pressure of 10^{-3} to 10^{-5} T, depending on the application, has the following features.

(1) Maximum weld penetration and minimum weld width and shrinkage, enabling all thicknesses to be welded in a single pass.
(2) The highest purity weld metal is produced because of the absence of any contaminating gases such as oxygen or nitrogen.
(3) A high vacuum allows a long distance to be maintained between the gun and the work piece, which facilitates observation of the welding process.
(4) The pump-down time is lengthy (up to 1 h or more, depending on the size of the chamber) which lowers the production rate of small jobs with shallow welds but is

insignificant when welds in plate of 50 mm or more thickness are made with a single pass.

Welding speeds for steel are shown in *Table 9.1.*

Medium vacuum welding In medium vacuum welding the working chamber is maintained at a pressure within the range 10^{-3} to 10^{-1} T, although pressures of up to 25 T are reported in the literature[33] together with correspondingly short pump-down times of a few seconds.

Medium vacuum welding with small working chambers is used extensively for small repetition work such as welding of finish machined gear trains and similar high-volume mass-production applications for the motor industry. Many high precision semi-finished or fully machined components for aircraft engines are also welded by using the electron beam process.

Medium vacuum electron-beam welding is not suitable for welding reactive metals and alloys such as titanium and zirconium which require the high vacuum process to obtain sound welds.

Out of vacuum electron-beam welding Provided that the gun-to-work distance is less than about 35 mm, so that the greater dispersion of the electron beam compared with working in a vacuum is allowed for, it is possible to weld many materials out of vacuum. With 60 kW non-vacuum equipment, single-pass welds can be made in metal thicknesses up to 25 mm. Metals welded out of vacuum include carbon and low alloy steels, and copper and aluminium alloys. Because of the presence of air, which can cause contamination of the weld metal, it is usually necessary to provide an inert shielding gas to cover the weld zone.

Laser welding and cutting Two types of laser are used for welding and cutting: the solid state yttrium–aluminium–garnet (YAG) laser, and the carbon dioxide (CO_2) laser.

The term 'laser' is an acronym for 'light amplification by stimulated emission of radiation'. A laser is a device for producing monochromatic (single wavelength) light which is coherent, i.e. all the waves are in the same phase. A laser beam can be transmitted over a distance of many metres and can be focused to produce the high-energy density required for welding or cutting.

The solid state YAG laser The solid state laser is stimulated to emit coherent radiation by means of the light from one or more powerful flash tubes. The output is in the infra-red region (around 1.06 μm). Both input and output are generally pulsed, and the maximum power output is 500 W.

Pulsed laser welding is used extensively in the electronics industry where miniaturisation requires very precise position-

Table 9.1 Penetration depth and welding rates

Welding power (kW)	Plate thickness (mm)	Welding speed (mm min^{-1})
0.25	1	1500
2.0	10	750
4.5	10	750
15	50	150
75	280	50

ing of small welds. Typical examples are encapsulation of microelectronic packages, and the joining of fine wires by butt, lapped, or cross-wire joints. Spot welds can be made between overlapping sheets and seam welds can be formed by a series of overlapping spot welds. Most metals can be welded, including steel, copper, nickel, aluminium, titanium, niobium, tantalum, and their alloys.

Solid state lasers are also used for precision drilling of holes having very small diameters.

The carbon dioxide laser In the carbon dioxide laser the lasing medium is a gaseous mixture of carbon dioxide, nitrogen and helium at a reduced pressure of 2–50 T. The output of the carbon dioxide laser is in the mid-infra-red region (10.6 μm) and CO_2 lasers are available having powers up to 20 kW. The high power density produced (10^4 W mm^{-2} or more) forms a cavity or keyhole in the work piece which enables a deep penetration weld to be produced that is similar in appearance to an electron-beam weld.

Penetration is less than for an electron beam operating at the same power because of the plasma formed in the keyhole. The plasma gas escapes from the keyhole and interacts with the laser beam, restricting penetration to approximately 12 mm in mild steel. Methods have been developed for overcoming this problem,[33] either by the use of a pulsed output from the laser with pulses of shorter time than that required to generate the plasma, or by the use of a high velocity jet of helium to disrupt the plasma above the weld. Such developments are likely to increase the thicknesses that can be laser welded to 25 mm or more.

Carbon dioxide lasers are used in production in some automotive and aerospace applications, particularly for titanium and nickel alloys. The advantage of laser welding over electron-beam welding is that a vacuum is not required, which simplifies the welding operation. Other materials that can be welded using the carbon dioxide laser are steels, copper alloys, zirconium and refractory metals, but aluminium alloys are not readily weldable with this laser because of their reflectivity of the laser beam.

Laser cutting Cutting with the carbon dioxide laser is carried out by a combination of melting and vaporisation, with an auxiliary jet of gas to blow the molten metal from the cut. Various gases are used for this purpose, including oxygen, compressed air, inert gases, and carbon dioxide.

Oxygen produces an exothermic reaction with ferrous metals which increases the efficiency of cutting. The inert gases produce clean unoxidised surfaces, which are important when cutting readily oxidised metals such as aluminium and titanium.

9.2.1.13 Resistance welding

Spot, seam and projection welding are carried out by electric resistance heating of two overlapping metal parts which are pressed together by copper or copper alloy electrodes. Local melting occurs at the faying surfaces and an internal weld nugget is formed.

The welding cycle comprising current, pressure and time is readily controlled automatically, giving the following advantages:

(1) little skill is required for the operation;
(2) welding can be readily built into production lines;
(3) welding can be associated with automatic loading, unloading and transferring of components, so that resistance welding is the simplest process for automation or robotic welding;

(4) welding times are short so that output is high;
(5) no filler wires or fluxes are used; and
(6) distortion is a minimum, as heating is confined to a small area.

The high currents required for resistance welding are generally obtained from a single-phase a.c. transformer having a primary winding of several hundred turns. The secondary winding consists of one or two turns of thick copper which may be water cooled. The voltage is stepped down to a value between 4 and 20 V.

Three-phase welding machines are available; these are more expensive but have a better power factor. A three-phase transformer is sometimes used with a d.c. rectifier to carry out resistance welding with d.c.

For further information on power sources the reader should consult reference 33.

Spot welding Spot welding (*Figure 9.24*) is used for the fabrication of sheet metal of 0.6–3 mm thickness to produce lap joints that are intermittently welded and, therefore, not pressure tight. Typical applications are low carbon steel components for car bodies, cabinets, and general sheet-metal work. Spot welding is also applicable to stainless steel, aluminium and aluminium alloys, and copper alloys.

Because of its high electrical conductivity, pure aluminium is difficult to spot weld, and its softness results in heavy indentation by the electrodes. Similarly, the high electrical conductivity of copper makes it unsuitable for spot welding. Stainless and heat-resisting steels are spot welded for aircraft and gas turbine engines.

Seam welding Seam welding (*Figure 9.25*) is similar in principle to spot welding and uses rotating wheel electrodes which roll the overlapping components between them. The welding current passes intermittently through the electrodes forming a series of welds that overlap one another. The electrode wheels rotate continuously, and at least one is power driven to move the component along, in addition to carrying the welding current.

Pressure-tight seams can readily be made and the process is faster than spot welding. However, machines used for seam welding are heavier and more costly than spot welding machines. Seam welding is primarily suited to making long straight welds, although curved welds can be made; for example, welds which may occur at the corners when joining two half pressings together to form a fuel tank.

Typical applications are pressure-tight seams for oil drums and refrigerator parts.

Projection welding In this process the current is concentrated by the shape of the components themselves, small dimples

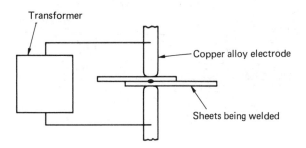

Figure 9.24 Resistance spot welding

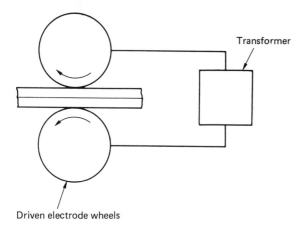

Driven electrode wheels

Figure 9.25 Resistance seam welding

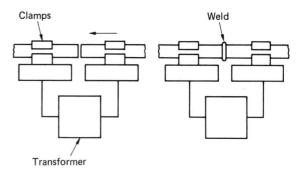

Figure 9.27 Resistance butt welding

being formed on one of the sheets to be joined, as shown diagrammatically in *Figure 9.26*.

The electrodes are of relatively large area compared with those used in spot welding and can, therefore, be made of a hard material having a comparatively high electrical resistance, such as a copper–tungsten alloy. As several welds can be made at the same time, the welding machine must be of high electrical capacity and must also be capable of applying the high total mechanical load to the components during welding. Three projections welded simultaneously give the best results, and up to five projections are common. Applications involving large numbers of projections are known, but as the number increases it becomes more and more difficult to ensure uniform and adequate welding of all the projections. The process can also be used for attaching studs, nuts and discs to flat plates.

The main features of the process are listed below.

(1)   Several welds can be made simultaneously in the time it would take to make a single spot weld.
(2)   Electrodes have longer life between dressings compared with spot welding electrodes, because of the relatively low current densities used.
(3)   Welds are generally of better external appearance than spot welds due to the absence of indentations.
(4)   T joints may be made as well as lap joints.
(5)   Machines are generally more costly than spot welding machines.
(6)   The cost of preparing components is greater because of the necessity of forming projections.

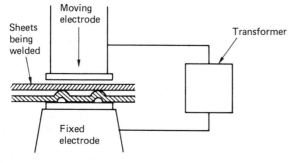

Figure 9.26 Projection welding

9.2.1.14   Resistance butt and flash welding

Resistance butt welding   The two ends of the parts to be joined are brought into contact and current is passed across the joint while a moderate mechanical force is applied to the components (*Figure 9.27*). As the joint is heated up by electrical resistance, the material softens and welding takes place. The finished joint shows a thickening or upsetting of the components in the joint area, and for some distance away from the joint. Resistance butt welding is essentially a solid state welding process in which no melting occurs.

Recrystallisation of the metal takes place across the faying surfaces, and the upsetting action has the effect of removing oxides from the joint.

The process has two main applications:

(1)   joining two components of the same cross-section end to end, e.g. wire, rod, bar or tubing, as used in rod and wire mills for joining the ends of coils for continuous processing; and
(2)   continuous welding of longitudinal seams in pipe or tubing formed from flat plate.

The process is applicable to the welding of carbon, alloy and stainless steel, aluminium alloys, copper alloys, and nickel alloys, and electrical resistance alloys.

Resistance butt welding machines are generally designed to weld a particular family of alloys because the current densities and pressures as well as the rate of application of the pressure differ widely between, for example, steels and aluminium alloys. The cross-sectional areas to be joined also determine the current capacity of the welding transformer.

The welding machine has two platens on which are mounted clamping dies which grip the components. One platen is stationary and the other moves to produce the pressure required.

Flash welding   Flash welding was developed from resistance butt welding; the principle is shown in *Figure 9.28*. The current is switched on before the ends of the components are in contact, with the result that small volumes of metal melt explosively at the points where contact is first made. This process continues as the two components are moved towards one another, causing a large amount of flashing and removal of metal, while at the same time heating the areas in contact and the material immediately behind. Once suitable temperature conditions have been established, the two ends of the components are forced together by a sudden increase in the

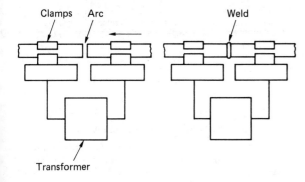

Clamps Arc   Weld

Transformer

Figure 9.28 Flash welding

load applied and the current is switched off. The effect of this sudden load is to squeeze out from the joint all the overheated and oxidised metal and to form a high quality pressure weld. The fin or flash of upset metal surrounding the joint is normally removed before the component is put into service.

Typical applications of flash welding are the production of continuous welded railway track, motor car wheel rims formed from flat steel stock and mitre joints in door and window frames. The materials that can be welded using this process are similar to those that can be resistance butt welded.

9.2.1.15   Friction welding

Friction welding is a solid state joining process, i.e. there is no fused metal involved. The heat required to make a forge weld is produced by moving one component relative to a mating component under pressure. The motion is usually rotational, although linear relative motion can be used and is under development.

A friction weld is similar in appearance to a resistance butt or a flash weld, with upset metal or flash which is generally removed after welding by a machining operation.

Butt welds can be made in rods or tubes, and rods or tubes can be welded to plates. One of the components is generally circular in cross-section, although square rods can be joined by friction welding. A wide range of sizes can be friction welded from small-diameter wires used in the electronics industry up to 150 mm diameter aluminium bus-bars.

Typical applications are found where high production rates are required, e.g. in the motor industry. Examples are axles, drive shafts, steering shafts and valves where a high alloy heat-resisting head is welded to a cheaper carbon steel shank. Bar stock can be welded to plates to produce parts that would normally be forged.

Most metals and alloys, with the notable exception of cast iron, can be friction welded. Dissimilar metals can also be welded without the formation of a brittle zone, which often occurs with arc welding, e.g. aluminium can be welded to steel.

Friction surfacing is an important development in which a corrosion or wear resistant alloy rod is rotated rapidly under pressure against a surface to be clad with the alloy. Under carefully controlled conditions a friction weld can be deposited over the surface and the weld metal is unchanged in composition because there is no dilution of the weld caused by melting of the parent metal as occurs in surfacing by arc welding. For further information on friction welding the reader is referred to reference 33.

9.2.2   Soldering and brazing

Brazing and soldering are carried out at temperatures below the melting points of the metals being joined. Brazing filler metals have melting points above 450°C and solders melt below this temperature. In both processes the molten filler metal flows by capillary action between closely fitted surfaces of the parts to be joined.

Unlike welding, soldering and brazing do not generally produce joints having mechanical properties matching those of the parent metal.

Typical applications, such as car radiators and electrical connections, require leak-tightness and good electrical conductivity, respectively. If high strength or ductility is required, correct alloy selection and joint design are essential. For further guidance on this subject references 34 and 35 should be consulted.

The low temperatures involved in soldering and brazing have two beneficial effects compared with welding: (1) parent metal properties are less affected, especially by soldering; and (2) residual stresses and distortion are lower.

For successful joining, the following procedures are required.

(1) For wetting of the metal by the solder or brazing alloy, the surfaces must be chemically clean. Degreasing and mechanical cleaning or pickling may be necessary.
(2) During the joining operation a flux is generally used to remove surface oxides and to prevent them from reforming. In some brazing processes a reducing or neutral atmosphere or a vacuum is used.
(3) Heating of the parent metal must be carefully controlled so that both mating surfaces reach the melting temperature of the filler metal before it is applied to the joint. When the filler metal is preplaced in the joint the same requirement applies because the molten filler metal will not wet the surface unless the surface is hot enough.
(4) All traces of flux residues which might lead to corrosion must be removed after soldering or brazing.

9.2.2.1   Soldering

Solders   The normal range of soft solders is listed in BS 219: 1977,[36] which also indicates typical uses. The commonest solders are the tin–lead alloys, some of which have antimony added. These solders are used in various applications, depending on the composition, including manufacture of food-handling equipment, tin-plated cans, domestic utensils, electronic assemblies, heat exchangers, and general engineering work. Tin–antimony solder is used for higher service temperatures, e.g. above 100°C, and in applications for joining stainless steels where lead contamination must be avoided. The alloy is also used for plumbing and refrigeration applications.

Tin–silver solder is used for fine instrument work and food applications while tin–lead–silver solder is used where service temperatures will be above 100°C or below −60°C.

Soldering methods   Many sources of heat can be used for soldering, including heated iron, flame or torch, hot-plate or oven, high-frequency induction, electric resistance, hot gas, as well as dip, wave or flow methods involving molten solder baths.

For further information on both manual and mechanised soldering techniques the reader should consult references 35, 37 and 38.

Manual soldering methods involve the use of a soldering iron or a torch, are fairly slow and require a reasonable degree of skill, although an experienced operator can achieve high production rates on repetitive work. Dip soldering is faster and the electronics industry uses specially designed equipment for wave soldering of printed circuit boards.

9.2.2.2 *Brazing*

Brazing alloys are specified in BS 1845: 1984[39] divided into eight groups based on aluminium, silver, copper–phosphorus, copper, copper–zinc, nickel, palladium and gold. The choice of brazing alloy for a particular parent metal can be fairly wide and a decision may be based on metallurgical considerations including corrosion resistance, the maximum temperature to which the parent metal can be heated to avoid deterioration in properties, or melting of previously applied brazing alloys. For further information the reader should consult references 34, 40 and 41.

A brief summary of the uses of the eight main groups of brazing alloys is given in *Table 9.2.*

Brazing alloys are available in the form of wires, rods, powders and inserts of various shapes for preplacing in the joint.

Heating methods A full description of the various heating methods and their advantages and disadvantages are described in BS 1723: 1986: Part 2.[40] The methods comprise hand torch brazing, mechanised flame brazing, induction brazing, furnace brazing including protective atmosphere, vacuum and open furnace brazing, and immersion brazing, including flux bath, dip bath and salt bath brazing. Special processes referred to are infra-red brazing and laser brazing. Reference 40 also contains details of design and location methods for brazing.

Applications Brazing is applicable to cast irons, steels, galvanised steel, aluminium, copper, magnesium, nickel and their alloys, stainless and heat-resisting steels, titanium, zirconium, ceramics and refractory metals and to dissimilar metal joints.

Brazing is selected in preference to other joining processes in the following cases:

(1) where heating must be restricted to avoid melting or distortion of one of the parts to be joined;
(2) where strength or dimensional accuracy of the assembly would be impaired by heating to high temperature;
(3) where a soft soldered joint would not be strong enough;
(4) where parts cannot be welded because of the properties of the parent materials involved, e.g. ceramics and refractory metals;
(5) where a number of joints must be made in a small complicated assembly; and
(6) where joints would be inaccessible to the welding processes.

9.2.3 Productivity and welding economics

Productivity, in the broadest sense, is what determines profit. Unfortunately, the cost of welding is one of the least documented topics in the whole field of welded fabrication. When a company decides to review its fabrication methods there is often a lack of understanding of the principles involved and only limited knowledge of the alternatives available. Reliance may be placed on equipment salesmen who may be fair and have an excellent product to sell but, quite naturally, may be biased in their approach.

There is a lack of information on welding compared with that readily available on machine tools, e.g. cutting speeds and depth of cut, which can enable a machine to be integrated into a manufacturing system to give a high utilisation or duty cycle and, therefore, maximise productivity. The same objective should be applied to welding as part of the manufacturing system.

The first task in reviewing fabrication methods is a simple one and concerns design. In the case of arc welding, fabrications should be designed to have the minimum amount of weld metal that will meet the service requirements of the components in terms of mechanical proeprties.[42] The total quantity of weld metal in a fabrication is, of course, only a very small proportion of the total weight (perhaps 2% or less), but nevertheless it is sensible to use the minimum amount necessary. This requires close attention to detail in design, shop floor supervision and inspection to ensure that welded joints are not overdesigned and weld metal is not being wasted.

Sometimes partial penetration butt joints will be adequate and intermittent fillet welds can replace continuous ones. Fillet weld profiles should not be excessively convex because

Table 9.2 Main uses of brazing alloys

Brazing alloy	Materials suitable for brazing
Aluminium	Pure aluminium and some aluminium alloys
Silver	Most ferrous and non-ferrous metals and alloys, except aluminium, magnesium and refractory metals and their alloys
Copper–phosphorus	Copper and copper alloys
Copper	Ferrous materials in a protective atmosphere or in a vacuum
Copper–zinc	Stainless steel and other heat- and corrosion-resistant alloys in a protective atmosphere or in a vacuum
Nickel	Stainless steel and nickel–chromium heat resisting alloys for high-temperature service
Palladium and gold	Metallised ceramics, copper, nickel and ferrous alloys in a protective atmosphere or in a vacuum

any weld metal in excess of a mitre fillet does not contribute to the strength of a joint.

For butt joints the plates are usually cut with bevel angles of 30° to give an included angle of 60° when the plates are placed together. With a gap at the root of 1–3 mm this is the optimum edge preparation to give full penetration and room to manipulate a covered electrode or MIG welding gun. If the included angle is increased to 70° the weld volume is increased by 20% which represents 20% waste. If a fillet weld having a size or leg length of 6 mm is adequate, then a leg length of 8 mm will result in 57% wasted weld metal because of the amount of overwelding. The increase in the cost of wasted consumables may be small, but the cost of depositing excess weld metal includes labour costs and overheads which can increase the overall cost considerably.

Having ensured that the optimum welded detail design is employed, there may be a desire to increase productivity still further. The next point to consider is the deposition rate, which, regardless of which arc welding process is used, depends on the welding current. As a simple example, in manual metal arc welding, if the electrode size is increased from 4 to 5 mm diameter and the current increased accordingly, the cost of welding can be reduced by approximately 25%. A similar increase in welding current has the same effect in other processes such as MIG or submerged arc welding. The above simple example relates to a particular job where the duty cycle was 30%, the duty cycle being the ratio of arcing time to total elapsed time for the welding operation. A welder is obviously not welding for 100% of the time, and typical activities for a manual welder in various industries are shown in *Table 9.3*.

Duty cycles may be considerably lower than those shown in *Table 9.3*. For example, in repair work where defects must be laboriously removed by mechanical means, the welding duty cycle may be less than 10%. In contrast to this the duty cycle of a robot MIG welding parts of car bodies may need to be 70% or higher to justify the capital investment.

Returning to the manual welder who increases his deposition rate by changing to a larger electrode and increasing the current, this gives a worthwhile cost saving provided that the duty cycle is at least approximately 20%. If the welder only deposits weld metal for 10% of his time, then even if his deposition rate is doubled, it will only have a slight effect on productivity. The duty cycle may be low because the welder has to fetch and carry components and set them up for welding. If a labourer is employed to deliver components and set them up at a second workstation or in a second jig, the welder can possibly double or treble his duty cycle and take full advantage of optimising the deposition rate of the welding process that he is using. It is necessary to carry out an exercise in method study to reveal the true cost of welding and to indicate how improvements in productivity may be accomplished with the equipment currently in use. Costing methods for arc welding must take account of the costs of consumables, weight or volume of weld metal, deposition rates and duty cycles, labour costs and overheads and amortisation of equipment. For further information on the calculation of welding costs, the reader should consult references 43 to 45.

A further important point to consider in economic surveys of welding operations is the welding position. Deposition rates can be maximised by welding in the flat position because a large molten pool of weld metal can be maintained, whereas in the vertical or overhead welding positions the size of the weld pool has to be restricted. Therefore, the provision of a welding manipulator or turning rolls can enable work to be rotated so that welding can be carried out in the flat position with maximum deposition rates. In some cases, depending on the shape and size of component, the rate of welding can be doubled or trebled by suitable manipulation of the work piece.

When a welding engineer has considered all the elementary principles of welding engineering and has determined the possible increase in productivity at minimum cost, the next question he/she must ask is whether this is satisfactory for the company. Management decisions are required on the rate of production required at present and in the foreseeable future and any plans for the manufacture of different products. New product lines may raise the question of what is technically required in terms of quality of welded joints, for example, static or fatigue strength, fracture toughness, corrosion or wear resistance. With this information, a choice can be made on the most appropriate welding processes, equipment and consumables capable of achieving the quality and production rates required. If a welding engineer is available he/she will obviously be expected to make recommendations on the choice of welding processes and equipment, otherwise a company may need to contact outside sources such as equipment suppliers, national advisory bodies, or a consultant.

In any of these circumstances it is advisable that the manager making investment decisions should have some basic knowledge of welding technology, if only to be able to ensure that all the possible alternatives have been considered.

For further information on the choice of manual, mechanised or robotic welding processes coupled with economic considerations the reader should consult references 27 and 46 to 48.

9.3 Adhesives

9.3.1 Introduction

Adhesives are used in almost every type of engineering industry, from automotives to lamp-shade manufacture, from heavy engineering to surface mount technology. Adhesives can simplify design and provide savings in manufacturing costs, thus making products more competitive in the marketplace. They are being increasingly proven as an engineering means of assembly, and in many circumstances offer significant benefits over traditional assembly techniques.

9.3.2 Modern engineering adhesives by category

Adhesives are essentially liquids that cure on demand to become solids. This is achieved by one of three means:

(1) chemical reaction,
(2) solidification on cooling, or
(3) solvent loss.

Adhesives are generally categorised as shown in *Table 9.4*.

Table 9.3 Percentage breakdown of the observed time spent on different activities by manual metal arc welders

Activity	Industry			
	Pressure vessel	*Site pipework*	*Light engineering*	*Heavy engineering*
Arcing (duty cycle)	42	36	62	35
Deslagging, setting up, etc.	22	30	25	32
Other activities (e.g. waiting/using crane, instructions, etc.)	14	23	5	14
Idle	22	21	8	19

Table 9.4 Categories of adhesive

Adhesive type	Cure mechanism	Typical use
Epoxy	Single part/dual part. Heat or room temperature cure	Structural assemblies where strong bonds are required
Anaerobic	Single part. Heat or room temperature cure	Machinery applications, threadlocking, gasketing, retaining, etc.
Cyanoacrylate	Single part. Room temperature cure	Bonding of plastics and rubber
Toughened acrylics	Dual part. Room temperature cure	Structural assembly where high peel loads are encountered
Ultraviolet curing acrylics	Ultraviolet light	Glass bonding and potting
Hot melts	Cure on cooling	Carboard-box manufacture, clothes
Phenolic	Heat cure to remove excess water	Aircraft industry, brake shoes
Polyurethane	Dual part	Window bonding, automotive industry
Silicones	Release of acetic acid or alcohol	Primarily sealing
Solvent based	Release of solvent	Laminates in furniture, paper

9.3.2.1 Chemical reaction

The adhesive changes from a liquid to a solid due to a chemical reaction. This may take the form of an activator applied to one surface and the adhesive to the other, or the mixing of a two-part adhesive, e.g. adhesive and hardener. Many adhesives also require heat to achieve full polymerisation. A more convenient means of curing is to use the substrate itself as the curing medium. Anaerobic adhesives, for example, use the activity of the metal substrate.

9.3.2.2 Solidification on cooling

These adhesives are initially solid and melt on heating, flowing to the required areas where they cool and resolidify to form a bond.

9.3.2.3 Solvent loss

A liquid adhesive contains a carrying solvent which evaporates after application to the joint faces. The adhesive becomes tacky and a bond is formed on closing the joint. This category also includes the emulsion-based adhesives (polyvinyl acetate) where the fluid carrier is water instead of solvent.

9.3.3 Designing with adhesives

To choose the correct adhesive requires, ideally, an understanding of cure and adhesion mechanisms. Factors which

affect cure speed and bond strength can then be identified and possible bond failure eliminated. In general, an adhesive must first flow and spread so that it thoroughly wets the surfaces to be bonded, penetrating into and filling all surface irregularities. It must then change from its liquid state into a load-bearing solid which links the surfaces together.

Adhesives, once cured, are essentially plastics (thermoset or thermoplastics) and, therefore, will have a maximum service temperature limitation of around 200°C, although certain metal filled epoxies and silicone sealants will withstand higher temperatures.

All adhesives are strongest in tensile or shear mode and weakest in peel or cleavage mode. Tension-loaded joints have the highest strength. The stress is distributed over the full bond area. Care should be taken that joint components have the rigidity to maintain this even loading because, if deflection under load occurs, the joint may be subjected to a cleavage stress resulting in early bond failure. (See *Figures 9.29* and *9.30*.)

Tensile loaded joints are very resistant to bond failure, and careful design can ensure an economical and successful bond.

In joints subjected to cleavage loading (*Figure 9.31*), the stress is concentrated at one end of the joint, and this can lead to failure.

Peel strength is usually the weakest property of a joint; a wide joint is necessary to withstand peel stress (*Figure 9.32*). Rubber toughened adhesives are recommended for this type of joint.

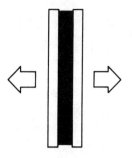

Figure 9.29 Tensile loading

Figure 9.30 Tensile shear loading

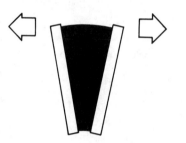

Figure 9.31 Cleavage loading

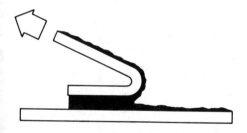

Figure 9.32 Peel loading

9.3.3.1 Bond line

The gap between the parts (adhesive bond line) has an important bearing on the characteristics of the joint. In general, the thinner the bond line the stronger the joint. The design of the joint can make a significant difference, but equally important is the choice of the adhesive applied to the substrate.

A thick bond line can, however, confer advantages. The adhesive is generally more flexible than the adherents or substrates, particularly in most engineering applications where metals or other rigid materials are bonded. A thick bond line can offer a capacity to absorb some impact energy, thus increasing the strength of the bond under this type of loading. Also, if dissimilar materials are bonded with differing coefficients of thermal expansion, a thick bond line may be able to accommodate this difference without being stressed to failure.

9.3.3.2 Surface preparation

Adhesion can be greatly affected by surface deposits. Dirt, grease, rust inhibitors, plating residues, plasticisers, mould release agents and similar contaminants will all separate the adhesive from intimate contact with the base material of the component.

For optimum conditions, substrates to be bonded should be cleaned with a chlorinated or Freon based solvent to remove grease and particulate contamination. A degree of surface roughness is often beneficial and for structural bonding it is recommended that plastic surfaces be lightly abraded and metal surfaces grit blasted. In some cases an activator or primer will be required to increase the surface activity of the substrate.

Correct cleaning and preparation of the surfaces is, therefore, vital for successful bonding. So, too, is the selection of the materials to be bonded, because not all surfaces have the same attraction for the molecules of adhesives. Non-polar surfaces, e.g. polyethylene, have little attraction for adhesives in general.

Some modern engineering adhesives will tolerate oil or other contaminants on the surface and, providing the constraints of the adhesive are realised, many successful designs can be completed using adhesive technology.

9.3.3.3 Environmental resistance

Adhesive bonds will always be susceptible to environmental attack and, therefore, thin bond lines are beneficial in applications where such attack is likely. However, the most important factor here is the choice of adhesive and the cleanliness of surface.

For specific environmental resistance characteristics of different adhesives the reader should refer to the manufacturers' data sheets.

9.3.3.4 Shelf life

The shelf life of an adhesive is defined as the time for which an adhesive can be stored in the unopened condition such that, when used, the adhesive will perform to specification. The shelf life is normally quoted in manufacturers' data sheets.

9.3.3.5 Approvals

Many adhesives are approved by the Ministry of Defence, and the major nationalised industries.

9.3.3.6 Testing

Thorough testing is always recommended for every adhesive application to ensure that all requirements are met.

9.3.4 Anaerobic adhesives

9.3.4.1 Cure mechanism and adhesive characteristics

Anaerobic adhesives cure in the absence of oxygen, but the activity of the metal part will accelerate cure. Anaerobic adhesives are generally more reactive on certain metals, e.g. copper, brass and mild steel. Other materials, e.g. aluminium and stainless steel, may require the use of an activator.

Fixturing strength is achieved in 10–30 min and full strength in 24 h.

Anaerobic adhesives can be thick or thin, strong or weak, and fast or slow acting, to suit a variety of engineering applications. Recent developments include oil-tolerant grades and ultraviolet (UV) curing grades for ultra-fast fixturing strength.

Anaerobic adhesives generally have an operating temperature of −50 to +150°C, although some can be used up to 220°C.

Typical applications of these adhesives are in threadlocking, bonding of cylindrical components, gasketting and thread sealing. These applications are described below.

9.3.4.2 Threadlocking

The locking of nuts and bolts to prevent loosening due to vibration is a typical and well-proven application of anaerobic adhesives.

The adhesive flows into the interstitial spaces between the threadforms where ideal conditions exist (metal-part activity and absence of air) for the adhesive to cure to a rigid thermoset plastic.

It is generally recognised that a pre-torqued nut and bolt will generally release at 75–95% of the seated torque. Any threadlocking material will add to or augment the put-down torque. This is known as *torque augmentation* (*Figure 9.33*).

A range of strengths of anaerobic adhesives is available for threadlocking materials. For example, a low-strength grade is normally selected for locking small precision screws or used where subsequent removal is required. A medium-strength grade is selected for typically M5 to M12 thread sizes to lock the thread and simultaneously seal against leakage and corrosion. For permanent locking of studs a high-strength product is required.

Preapplied anaerobic adhesives Threadlocking anaerobic products are also available in preapplied form. The adhesive is applied to the bolt in the form of a slurry containing microcapsules. The product is dried on and the bolt can be stored for at least 1 year prior to use. On assembly the microcaps burst, allowing the anaerobic adhesive to polymerise.

Application of the adhesive The adhesive can be applied to either the male or female thread. Note that thixotropic grades are available for non-run applications. Preapplied adhesives are applied to the male thread. Anaerobic adhesive dispensing equipment is available to ensure accurate adhesive application.

9.3.4.3 Bonding of cylindrical components

Engineering anaerobic adhesives are used to secure cylindrical parts such as bushes, bearings, liners, gears and commutators into, or onto other components. Traditional methods include the use of Woodruff keys, press or shrink fits, backing rings, splined shafts, collars, and stakes.

The use of an anaerobic adhesive can simplify design, improve fatigue resistance, increase dynamic and static joint strength and up-rate performance without the need for mechanical modification. Furthermore, an environmental barrier is formed which excludes corrosion.

The benefits of using anaerobic adhesives to join cylindrical parts are:

(1) *Faster assembly*—Anaerobic adhesives eliminate the requirement for press fitting and this can offer time savings for the production engineer.
(2) *Reduced machining*—Adhesives can simplify designs and negate the requirement for keyways, collars and stakes, resulting in cost savings.
(3) *Elimination of distortion*—Stresses in assemblies due to press or shrink fits can be reduced or eliminated.
(4) *Corrosion-free joint*—Anaerobic adhesives harden to become rigid plastics and this can be used to provide an environmental barrier which excludes corrosion.

Calculating the strength and torque of the joint The axial force required to break a joint can be expressed as

Axial force = Bonded area × Shear strength of adhesive

$$F = A \times [(\tau \times f_c) + (P \times \lambda)]$$

where F is the axial force (N), A is the bonded area (mm^2), τ is the shear strength of the adhesive (N mm^{-2}), f_c is a correction factor, P is the radial contact pressure between faces (N mm^{-2}), and λ is the coefficient of friction.

Note that the correction factor can be derived as a series of factors which takes into account the parameters that will affect the strength. The principal factors are:

(1) the materials to be bonded,
(2) the diametral clearance between parts,
(3) the geometry of components,
(4) the operating temperature, and
(5) the working environment.

Torque (N-m) = Axial force (kN) × radius (mm)

Note that consideration should be given to fatigue applications, i.e. the above formula will only give a maximum theoretical torque.

Loctite UK offer a complete computerised service to enable design engineers to select the correct grade of adhesive, the correct joint configuration and to calculate joint strength to suit the application.

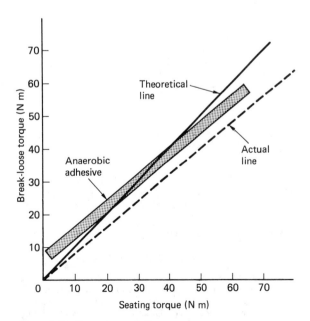

Figure 9.33 Torque augmentation

Press and shrink fits Anaerobic adhesives may be used to supplement an existing press or shrink fit, and also are often used to bond in dowel fins or keyways to prevent fretting between the components. Note that with press fits some scraping off of the adhesive is inevitable, and an allowance should be made for this.

A bonded shrink fit is actually extremely beneficial, as the adhesive (applied to the cold male part) is heat cured by the cooling female part and put under compression; both these conditions are ideal for retaining adhesives.

Differential thermal expansion In many cases, dissimilar materials with different coefficients of thermal expansion are used in co-axial assemblies. This can result in a tensile strain on the adhesive as the assembly reaches operating temperature. Whilst anaerobic adhesives can accommodate large compressive and shear strains, they generally cannot tolerate large tensile strains. However, three techniques can be employed to ensure that the adhesive bond can accommodate any thermal expansion which may occur.

(1) *Pre-heating of the female component*—The female component should be preheated to the operating temperature. The adhesive should be applied to the cold male component and the two components then assembled.
(2) *An adhesive augmented interference fit*—An adhesive should be used to supplement an interference fit. Provided that a small interference is maintained throughout the operating temperature range, the adhesive bond will operate successfully.

(3) *Specify tolerances*—The coefficient of thermal expansion for Loctite retaining products is $10 \times 10^{-5}{}^{\circ}\mathrm{C}^{-1}$ (for aluminium the value is $2.2 \times 10^{-5}{}^{\circ}\mathrm{C}^{-1}$).

Therefore tolerances can be specified to minimise tensile strain on the adhesive.

Application of the adhesive Retaining anaerobic adhesives are generally in the viscosity range 3–200 000 mPa (i.e. from the consistency of water to that of toothpaste). Adhesives in this range of viscosity can be readily dispensed. A common method of applying adhesive (up to 6000 mPa) to the female component is to use a spinning cup (see *Figure 9.34*).

Anaerobic adhesives may also be applied to the male component by brush or transfer roller.

The fluorescent characteristic of many anaerobic adhesives provides a useful means of inspection technique prior to assembly.

9.3.4.4 Gasketing

Sealing of flat faces with formed-in-place gaskets is another typical application area for anaerobic adhesives. Instead of sealing by compressing a malleable material, anaerobic gasketing products seal by filling the joint in liquid form, then curing rapidly to form a tough plastic which keys into and links the adjacent faces of the flanges.

This results in a number of benefits for using these products:

(1) *Increased structural integrity*—Anaerobic adhesives with high shear strength can be used to stop micromovements

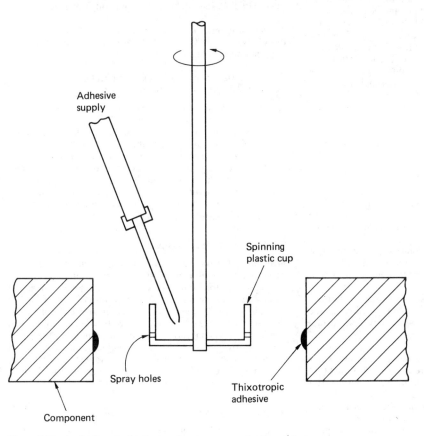

Adhesive supply

Spinning plastic cup

Spray holes

Thixotropic adhesive

Component

Figure 9.34 Application of adhesive to a female component using a spinning cup

due to side loads, and hence bolt loosening and fretting between flanges is eliminated. This feature is often taken advantage of when assemblies are subjected to external loads.

(2) *Non-shimming*—When anaerobic gaskets are used, flanges will mate together metal to metal. Because no allowance needs to be made for gasket thickness, tolerances can be more accurately maintained.

(3) *No retorquing*—Metal-to-metal contact ensures that bolt tension is maintained throughout the working life of the assembly. Therefore no retorquing of bolts is required.

(4) *Relaxed surface finish*—The use of anaerobic gaskets allows the surface finish and flatness tolerance of components to be relaxed. Stepped and scored surfaces can be sealed without the need for remachining.

(5) *Excess material remains liquid*—Unlike other types of liquid sealant, anaerobic gaskets only solidify between the flange faces. Excess adhesive can therefore be wiped away from the outside or flushed away from interiors, and air bleeds or oil channels will not be blocked.

(6) *Inventory cost reduction*—Preformed gaskets can lead to significant order, process and storage costs. A single container of anaerobic gasketing product will give a gasket of any size or shape.

(7) *Reduction of machining costs*—Static O rings replaced by anaerobic gaskets will result in the elimination of O-ring grooves.

9.3.4.5 Thread sealing

Medium viscosity anaerobic adhesives are suitable for sealing fine parallel threads of hydraulic and pneumatic connections. Higher viscosity anaerobics are suitable to seal coarse threads of pipes and fittings. When fully cured the sealant will withstand pressures in excess of 20 MPa. Many grades are approved by the British Water Council for use in contact with potable water. Anaerobic thread sealing grades have a number of world-wide approvals for use in gas appliances.

9.3.5 High-speed bonding

The main high-speed adhesives are UV acrylics and cyanoacrylates.

9.3.5.1 Ultraviolet acrylics

These products contain photoinitiators which will cure under ultraviolet light. UV acrylics generally have good adhesion to glass and most metals and are widely used in the cosmetic, jewellery and electronics industries.

The main advantage of these products is that the two components can be carefully aligned and no cure will commence until the product is held under UV light. Cure time is normally less than 1 min.

UV curing products are often used as potting materials. Under high intensity UV light the speed of cure can be seconds, whereas an epoxy based product may take several hours.

UV curing equipment UV acrylics require certain wavelengths of light to initiate the cure process. UV light of 365 nm wavelength will cure the bulk material, and shorter wavelengths (220 nm) are required to give a tack-free cure. The higher the intensity of UV radiation the faster the cure. UV acrylics are widely used in the electronics industry for bonding surface mount devices to printed circuit boards prior to the wave soldering process.

9.3.5.2 Cyanoacrylates

Modern cyanoacrylates will bond plastics, leather, wood, paper, fabric, rubber and metals, but are not satisfactory for bonding glass or glazed ceramics. Plastics with a low surface energy such as polyolefins and acetal plastics, cannot be bonded without pretreatment. Loctite UK has recently introduced 757 PRIMER for polyolefins.

The viscosity of cyanoacrylates ranges from 3 mPa for wicking applications to a gel form which will not migrate.

A wide range of cyanoacrylates is available to cater for almost every application and, whilst general guidelines are given here, the user should approach the adhesive manufacturer for specific applications.

Curing mechanism Cyanoacrylate adhesives solidify simply because of their reaction with moisture on the surfaces being bonded. They are kept liquid in the bottle by an acidic stabiliser which inhibits the adhesive molecules from linking. If this stabiliser is neutralised, the molecules link and the adhesive solidifies.

Cure speed The fixturing time for most substrates is 3–20 s, depending on the cyanoacrylate chosen and the gap between the components to be joined. The final strength of the joint is typically 10–16 N mm^{-2}.

Humidity and moisture resistance Cyanoacrylates are subject to attack from moisture. The degree to which they are attacked depends on joint geometry and the environmental conditions prevailing. Some rubber toughened grades such as Loctite PRISM 480 exhibit greater humidity resistance.

Temperature resistance Cyanoacrylates will generally perform between −55°C and +80°C, although certain grades are suitable for use at higher temperatures.

Cyanoacrylate activators In some applications activators are required to overcome possible problems.

(1) *Cure excessive adhesive*—Where a fillet of adhesive remains at the bond line, an activator can be used to prevent adhesive migration and reduce blooming. 'Blooming' is the name given to the white deposit due to excess cyanoacrylate molecules curing in the atmosphere and falling onto the surface adjacent to the bond line. Low-bloom products are available.

(2) *Incorrect humidity*—Low humidity conditions will slow the cure process and an activator can be used to regain fast cure speeds.

9.3.6 Toughened acrylics

Structural adhesives are used to bond different materials such as glass fibre reinforced plastics, prepared aluminium, and ceramics. The total coverage of the bonded joint achieved in these adhesives ensures in many applications that assemblies are simultaneously joined and sealed against electrolytic action and the ingress of corrosive fluids. Bonded assemblies can eliminate stress concentrations which are often inherent in a welded or riveted assembly. A bonded joint can also provide a cleaner, smoother joint with no requirement for remedial planishing. These adhesives have also been used to dampen noise or vibration.

9.3.6.1 Composition and cure mechanism

Toughened acrylic adhesives are two-part systems: the adhesive and an activator. The adhesive contains small rubber particles which are distributed evenly throughout the cured adhesive matrix. These particles act as a crack arrester and dissipate the energy, thus giving the adhesive excellent peel and cleavage properties.

Cure time is normally 5—15 min to handling strength and 12–24 h to full strength.

9.3.6.2 Temperature range

The service temperature range for toughened acrylic adhesives varies from one grade to the next, but a typical operating range is −50°C to +80°C.

9.3.6.3 Environmental resistance

Toughened acrylic adhesives generally have good environmental resistance, although they are not recommended for constant immersion in water. For specific data the manufacturer of the product should be consulted.

9.3.6.4 Typical applications

Toughened acrylic adhesives can be used to bond a variety of engineering materials, including 'as received' steel parts. The adhesive can tolerate a certain level of oil on the surface which is absorbed into the adhesive matrix. Many steel assemblies also pass through a paint bake process which can be used to cure the adhesive.

Toughened acrylic adhesives will bond some plastics, although certain grades are prone to crazing or stress cracking. Glass and ceramics may also be bonded but, if the assembly is to be subjected to high temperature, differential thermal expansion conditions may exist which could be detrimental to the strength of the bond.

Toughened acrylic adhesives are generally specified where high load conditions exist and, as with any adhesive application, trials should be conducted to ensure that all the specifications are met.

9.3.6.5 Typical application methods

Toughened acrylic adhesives may be applied by means of a simple hand gun or by using fully automated equipment.

The adhesive is generally applied to one surface and the activator to the other. It is normally advisable to leave the activator for about 1 min to allow the carrying solvent to evaporate. Handling strength is achieved in about 5 min and full strength in about 12 h (these times differ for individual grades). Whilst toughened acrylics will accept oily deposits, best results are achieved with clean surfaces.

With some grades of these adhesives the initiator or activator is formulated to be the same viscosity as the adhesive and static mixer nozzles mounted on the applicator gun can be used.

9.3.7 Epoxies

Epoxy adhesives will bond a wide variety of engineering materials and are widely used in industry. They can be single- or multi-part and can be cured at room temperature or heat cured.

The single-part epoxies generally need to be stored at low temperature to obtain optimum shelf-life and are heat cured at temperatures of 80°C or above. The room-temperature curing epoxies normally comprise a resin and a hardener and many require critical mix ratios.

Two-part epoxies are almost invariably exothermic and the cure time may depend on the mass of adhesive present. The larger the bulk of the adhesive the faster the cure and, in order to meet production requirements, the application of local heating may be required, especially where thin bond lines are present.

Epoxies are extremely tough and durable and certain grades will withstand quite high temperatures. Manufacturers' data sheets should be studied for selection of a suitable adhesive.

9.3.8 Dispensing equipment

Dispensing equipment is playing a more and more vital role for the production engineer as costs are analysed. The adhesive not only must bond the two components together but must also be easily applied. A number of manufacturers, including adhesive manufacturers, now offer dispensing equipment.

Many adhesives are packaged in bottles or tubes which lend themselves to easy application, but for more automated or precision processes standard or special-purpose dispensing equipment may be required. There are many factors that influence the choice of dispensing technique. These may include:

(1) adhesive type (e.g. anaerobic, epoxy, etc.; single or dual part);
(2) adhesive viscosity and temperature;
(3) packaging;
(4) health and safety considerations;
(5) volume of product required per assembly;
(6) accuracy of placement; and
(7) degree of automation.

There are many types of dispensing equipment available. Some of the more common types are described below:

(1) Pressure–time systems—the adhesive container is pressurised and the adhesive is forced down a pipe through to a dispense valve.
(2) Syringe dispensing—this is similar in principle to (1), but the product is dispensed directly from the syringe via a pulsed air line. Vacuum suck-back techniques are often used.
(3) Volume displacement systems—this relies on the filling of a chamber of known volume and ejecting out using a plunger.
(4) Drum pumps—this technique is normally used with the high viscosity products, e.g. silicones.

Many of the above dispensing systems can be incorporated into semi-automatic or purpose built robotic assembly systems with parts handling, clamping and adhesive curing facilities.

9.3.9 Health-and-safety considerations

With the Control of Substances Hazardous to Health (COSHH) regulations now in force it often falls to the Production Engineer to ensure that the adhesive on the production line is 'user friendly'.

By law adhesives must be labelled in accordance with the appropriate classification or category, e.g. flammable, toxic, corrosive, irritant, etc.

Full additional health-and-safety data must also be made available by the supplier to the end-user.

References

1 BRITISH STANDARDS INSTITUTION, *BS 499 Welding terms and symbols. Part 1 Glossary for welding, brazing and thermal cutting*, Milton Keynes (1983)

2 BRITISH STANDARDS INSTITUTION, *BS 639 Covered carbon and carbon manganese steel electrodes for manual metal arc welding*, Milton Keynes (1986)

3 AMERICAN WELDING SOCIETY, *AWS A.5.1-81 Covered carbon steel arc welding electrodes*, Miami, FL (1981)

4 JOHN, R. and ELLIS, D. J., 'A.c. or d.c. for manual metal arc', *Metal Construction*, **14**(7), 368 (1982)

5 THE WELDING INSTITUTE, *Manual Metal Arc Welding (Mild Steel with Low-current Sets)* (video), Abington, Cambridge

6 BRITISH STANDARDS INSTITUTION, *BS 2493 Low alloy steel electrodes for manual metal arc welding*, Milton Keynes (1985)

7 AMERICAN WELDING SOCIETY, *AWS A5.5-81 Low alloy steel covered arc welding electrodes*, Miami, FL (1981)

8 BRITISH STANDARDS INSTITUTION, *BS 2926 Chromium and chromium–nickel steel electrodes for manual metal arc welding*, Milton Keynes (1984)

9 AMERICAN WELDING SOCIETY, *AWS A5.4-81 Covered corrosion-resisting chromium and chromium–nickel steel welding electrodes*, Miami, FL (1981)

10 AMERICAN WELDING SOCIETY, *AWS A.5.11-83 Nickel and nickel alloy covered welding electrodes*, Miami, FL (1983)

11 AMERICAN WELDING SOCIETY, *AWS A.5.6-84 Covered copper and copper alloy arc welding electrodes*, Miami, FL (1984)

12 AMERICAN WELDING SOCIETY, *AWS A.5.3-80 Aluminium and aluminium alloy covered arc welding electrodes*, Miami, FL (1980)

13 AMERICAN WELDING SOCIETY, *AWS A.5.15-82 Welding rods and covered electrodes for cast iron*, Miami, FL (1982)

14 BRITISH STEEL CORPORATION, *Corporate Engineering Standards (CES) 23. Part 1. Supplement. Consumables, weld deposited surfaces* (1986)

15 AMERICAN WELDING SOCIETY, *AWS A5.21-8. Composite surfacing welding rods and electrodes*, Miami, FL (1980)

16 BRITISH STANDARDS INSTITUTION, *BS 4165 Electrode wires and fluxes for the submerged arc welding of carbon steel and medium-tensile steel*, Milton Keynes (1984)

17 DAVIS, M. L. E., *An Introduction to Welding Fluxes for Mild and Low Alloy Steels*, The Welding Institute, London (1971)

18 JONES, S. B., 'Variations on submerged arc welding', *WI Res. Bull.*, **15**(3), 67 (1974)

19 KENNEDY, N. A., 'Narrow gap submerged arc welding of steel, Part 1 Applications', *Metal Construction*, **18**(1), 687 (1986). 'Part 2. Equipment, consumables and metallurgy', *Metal Construction*, **18**(12), 765 (1986)

20 BRITISH STANDARDS INSTITUTION, *BS 2901 Filler rods and wires for gas-shielded arc welding. Part 1 Ferritic steels. Part 2 Austenitic stainless steels. Part 3 Copper and copper alloys. Part 4 Aluminium and aluminium alloys and magnesium alloys. Part 5 Nickel and nickel alloys*, Milton Keynes (1992)

21 LUCAS, W., 'Synergic pulsed MIG welding—process, equipment and applications', *FWP J.*, **25**(6), 7 (1985)

22 LUCAS, W., 'Microcomputer control in the control of arc welding equipment', *Metal Construction*, **17**(1), 30 (1985)

23 NORRISH, J., 'What is synergic MIG', *Welding Metal Fabr.*, **55**(5), 227 (1987)

24 THE WELDING INSTITUTE, *MIG Welding (Mild Steel With Low-Current Sets)* (video guide), Abington, Cambridge

25 WESTON, J., 'Arc welding robots—a welding engineering viewpoint', *Proc. Welding Institute Conference on Developments in Mechanical, Automated and Robotic Welding*, The Welding Institute, Abington, Cambridge (1980)

26 KENNEDY, N. A., 'Robotics for welding engineers. Part 1 Introduction', *WI Res. Bull.*, **26**(7), 221 (1985). 'Part 2a Programming methods', *WI Res. Bull.*, **26**(9), 302 (1985). 'Part 2b Programming languages', *WI Res. Bull.*, **26**(10), 334 (1985). 'Part 3a Hardware for robotic arc welding', *WI Res. Bull.*, **26**(12), 412 (1985)

27 WESTON, J. (Ed.), *Exploiting Robots in Arc Welding Fabrication*, The Welding Institute, Abington, Cambridge (1988)

28 BRITISH STANDARDS INSTITUTION, *BS 7084 Carbon and carbon–manganese steel tubular cored welding electrodes*, Milton Keynes (1989)

29 AMERICAN WELDING SOCIETY, *AWS A5.20-79 Specification for carbon steel electrodes for flux-cored welding*, Miami, FL (1979)

30 AMERICAN WELDING SOCIETY, *AWS A5.29-80 Specification for low alloy steel electrodes for flux-cored welding*, Miami, FL (1980)

31 AMERICAN WELDING SOCIETY, *Welding Handbook*, Vol. 2, 7th edn, 592, Miami, FL (1980)

32 BRITISH STANDARDS INSTITUTION, *BS 1453 Filler materials for gas welding*, Milton Keynes (1972)

33 AMERICAN WELDING SOCIETY, *Welding Handbook*, Vol. 3, 7th edn, 459, Miami, FL (1980)

34 BROOKER, H. R. and BEATSON, E. V., *Industrial Brazing*, 2nd edn, 263, Butterworth-Heinemann, Oxford (1975)

35 AMERICAN WELDING SOCIETY, *Soldering Manual*, 2nd edn, 149, Miami, FL (1977)

36 BRITISH STANDARDS INSTITUTION, *BS 219 Soft solders*, Milton Keynes (1977)

37 LOTTA, A. J., *Connections in Electronic Assemblies*, 277, Marcel Dekker, New York (1985)

38 WOODGATE, R. W., *The Handbook of Machine Soldering*, 224, Wiley, Chichester (1983)

39 BRITISH STANDARDS INSTITUTION, *BS 1845 Specification for filler metals for brazing*, Milton Keynes (1984)

40 BRITISH STANDARDS INSTITUTION, *BS 1723 Part 2 Guide to brazing*, Milton Keynes (1986)

41 AMERICAN WELDING SOCIETY, *Brazing Manual*, 309, Miami, FL (1976)

42 HINKEL, J. E., 'Joint designs can be both practical and economical', *Welding J.*, **49**(6), 449 (1970)

43 DOHERTY, J., 'Costing methods for arc welding', *WI Res. Bull.*, **1–3**, 7, 35, 63 (1968)

44 JACK, J. T., 'Controlling the cost of welding', *Austr. Welding J.*, **14**(4), 13 (1970)

45 MCMAHON, B. P., 'The price of welding', *Welding & Metal Fabr.*, **58**(1,2), 4, 58 (1970)

46 REYNOLDS, D. E. H., 'Decreasing welding costs in heavy fabrication', *Welding & Metal Fabr.*, **42**(3,5), 94, 185 (1974)

47 LINBLAD, L., 'The economics and methods of automatic welding', *Br. Welding J.*, **13**(5), 269 (1966)

48 AMERICAN WELDING SOCIETY, *Welding Handbook*, Vol. 5, 7th edn, 444 (1984)

Bibliography

CULLUM, R. D., *Handbook of Engineering Design*, Butterworth-Heinemann, Oxford (1988)

INSTITUTION OF ENGINEERING DESIGNERS, *Designers' Official Reference Book and Buyers' Guide*, Sterling Publications, London (1988)

10

Electrical and Electronics Principles

C J Fraser

Contents

10.1 Introduction

Within the confines of this chapter, it is impossible to include elaborate detail on the principles of electrical and electronics technology. A thorough depth of understanding can only come through numerous applications of the basic principles to the solution of practical problems. Since this is beyond the scope of this chapter, the content has been restrictively selected as appropriate to the practical requirements of manufacturing engineers. All the relevant basic principles are stated as matters of fact. For a more penetrating discourse readers must refer to the many standard texts on electrical and electronics technology which are readily available. An adequate number of references are cited for this purpose.

The graphical symbols used in the text are in accordance with BS 3939: 1986: Part 1.[1] Other symbols and abbreviations are consistent with the SI system of units as defined in BS PD 5686: 1972.[2]

10.2 Basic electrical technology

10.2.1 Flux and potential difference

The concept of flux and potential difference enables a unified approach to be adopted for virtually all 'field' type problems. In general, the flowing quantity is termed the 'flux' and the quantity which drives the flow is called the 'potential difference'. This consistency of method is equally applicable to problems in fluid flow, heat transfer, electrical conduction, electrostatics and electromagnetism, to name but a few. It is not particularly surprising, therefore, that thermodynamicists use an electrical analogy to describe many heat conduction phenomena, while electrical engineers often use heat and fluid flow analogies to enhance an understanding of electrical and electronics principles.

In general terms, the flux may be written as:

$$\text{Flux} = \frac{\text{Field characteristic} \times \text{Cross-sectional area} \times \text{Potential difference}}{\text{Length}}$$

(10.1)

In specific terms, for the flow of an electric current through a conducting medium, equation (10.1) takes the form:

$$I = \frac{\sigma a V}{l}$$

(10.2)

where I is the current (A), σ is the conductivity of the medium (S m^{-1}; i.e. the field characteristic), a is the cross-sectional area of the medium (m^2), l is the length of the medium (m), and V is the applied potential difference, or voltage (V). The group ($\sigma a/l$) is termed the 'conductance', denoted by G and measured in siemens, thus:

$$I = GV$$

(10.3)

The reciprocal of conductance is referred to as the 'resistance', denoted by R, and measured in ohms. Hence

$$I = V/R$$

(10.4)

Equation (10.4) is the familiar *Ohm's law*, which defines a linear relationship between voltage and current in a conducting medium. If the resistance R varies with the magnitude of the voltage, or the current, then the resistance is non-linear. Rectifiers constitute one particular class of non-linear resistors.

Comparing equations (10.4) and (10.2) gives:

$$R = l/(\sigma a)$$

(10.5)

It is more usual, however, to quote the 'resistivity' as opposed to the conductivity, and resistance is generally written as:

$$R = \rho l/a$$

(10.6)

where ρ is the resistivity of the conductor (in Ω-m).

The resistance of all pure metals is temperature dependent, increasing linearly for moderate increases in temperature. Other materials, including carbon and many insulators, exhibit a decreasing resistance for an increase in temperature.

10.2.2 Simple resistive circuits

The effective total resistance of the series arrangement shown in *Figure 10.1* is the algebraic sum of all the resistances in series, i.e.

$$R_t = R_1 + R_2 + R_3$$

(10.7)

where R_t is the total resistance of the circuit.

In *Figure 10.2*, the resistors are connected in parallel and each resistor is subject to the same applied voltage. The effective total resistance obeys an inverse summation law, i.e.

$$\frac{1}{R_t} = \frac{1}{R_1} + \frac{1}{R_2} + \frac{1}{R_3}$$

(10.8)

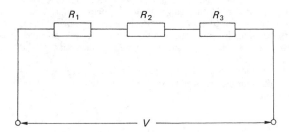

Figure 10.1 Resistors in series

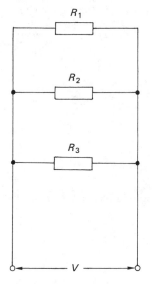

Figure 10.2 Resistors in parallel

Using equations (10.7) and (10.8), many series parallel resistive circuits can be reduced to a single effective resistance. Back calculation can then yield the current and potential differences across all the resistive elements which make up the circuit.

10.2.3 Electromotive force and potential difference

In a metallic conductor which has a potential difference applied across opposite ends, free electrons are attracted to the more positive end of the conductor. It is this drift of electrons which constitutes the electric current and the effect is simply Nature's attempt to redress an energy imbalance. Although the negatively charged electrons actually drift towards the positive end of the conductor, traditional convention gives the direction of the current flow from positive to negative. There is nothing really at issue here since it is only a simple sign convention which was adopted long before the true nature of the atom and its associated electrons was postulated.

A current of 1 A is associated with the passage of 6.24×10^{18} electrons across any cross-section of the conductor per second. The quantity of charge is the coulomb Q:

$$Q = It \qquad (10.9)$$

where 1 C of charge is passed when a current of 1 A flows for a period of 1 s.

The electromotive force (e.m.f.) is that which tends to produce an electric current in a circuit. Since e.m.f. is also measured in volts, the distinction between electromotive force and potential difference might at first appear rather subtle. However, e.m.f. is associated with the energy source. Potential difference is simply the product of current and resistance across any resistive element in a circuit, irrespective of the energy source.

For circuit elements other than purely resistive ones, the potential difference across the element becomes a time-dependent function.

10.2.4 Power and energy

Power is the rate at which energy is expended or supplied. The potential difference across any two points in a circuit is defined as the work done in moving unit charge from a lower to a higher potential. Thus the work done in moving Q coulomb of charge across a constant potential difference of V volt is:

$$W = QV \qquad (10.10)$$

Therefore

$$\text{Power} = \frac{dW}{dt} = \frac{dQ}{dt} V$$

From equation (10.9),

$$(dQ/dt) = I$$

Thus

$$\text{Power} = IV \qquad (10.11)$$

Using Ohm's law, the power dissipated across a simple resistive circuit element is:

$$\text{Power} = IV = I(IR) = I^2R \qquad (10.12)$$

10.2.5 Network theorems

A network consists of a number of electrical elements connected up in a circuit. If there is no source of e.m.f. in the

circuit, it is said to be 'passive'. When the network contains one, or more, sources of e.m.f., it is said to be 'active'.

Many practical resistive networks cannot be simplified to equivalent series and parallel arrangements. The unbalanced Wheatstone bridge provides one classic and frequently encountered example. For this reason a number of well-established theorems have been developed for the analysis of such complex resistive networks (see Hughes,[3] Bell and Whitehead,[4] or Bell[5]). The theorems are listed below.

(1) *Kirchhoff's first law*—The algebraic sum of the currents entering (positive) and leaving (negative) a junction is zero.
(2) *Kirchhoff's second law*—The algebraic sum of potential differences and e.m.f.s around any closed circuit is zero.
(3) *Superposition theorem*—In a linear resistive network containing more than one source of e.m.f., the resultant current in any branch is the algebraic sum of the currents that would be produced by each e.m.f. acting on its own while the other e.m.f.s are replaced with their respective internal resistances.
(4) *Thevenin's theorem*—The current through a resistor R connected across any two points in an active network is obtained by dividing the potential difference between the two points, with R disconnected, by $(R + r)$, where r is the resistance of the network between the two connection points with R disconnected and each e.m.f. replaced with its equivalent internal resistance.
(5) *Norton's theorem*—Any active network can be replaced at any pair of terminals by an equivalent current source in parallel with an equivalent resistance.

Although perhaps long-winded in definition, Thevenin's theorem has great practical application to complex resistive networks. An alternative statement of Thevenin's theorem is: 'Any active network can be replaced at any pair of terminals by an equivalent e.m.f. in series with an equivalent resistance'. The more concise version of Thevenin's theorem is perhaps a little more indicative of its power in application.

It might be apparent that Norton's theorem is complementary to Thevenin's theorem and both can be equally well used in the analysis of resistive networks.

Other useful network analysis techniques include 'mesh analysis', which incorporates Kirchhoff's first law, and 'nodal analysis' which is based on Kirchhoff's second law. Mesh and nodal analysis are also essentially complementary techniques.

A proficiency in the use of the network theorems can only be obtained through experience in their application. The books by Hughes[3] and Bell and Whitehead[4] provide a wealth of worked and tutorial examples with answers for practice.

10.2.6 Double-subscript notation

To avoid ambiguity in the direction of current, e.m.f. or potential difference, a double-subscript notation has been adopted. *Figure 10.3* shows a source of e.m.f. which is acting from D to A. The e.m.f. is therefore E_{da}. The current flows from A to B, by traditional convention, and is designated I_{ab}. From this simple circuit it is apparent that $I_{ab} = I_{bc} = I_{cd} = I_{da}$.

The potential difference across the load R is denoted by V_{bc} to indicate that the potential at B is more positive than that at C. If arrow heads are used to indicate the potential difference, they should point towards the more positive potential.

10.2.7 Electrostatic systems

Electrostatic systems are quantified by the physical behaviour of the somewhat intangible concept of 'charge'. Fortunately,

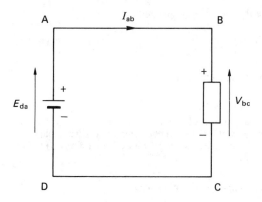

Figure 10.3 Double-subscript notation

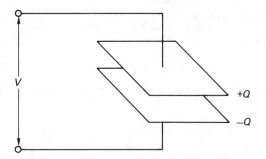

Figure 10.4 Electrostatic system

the unified field approach lends itself well to the quantification of electrostatic systems.

Figure 10.4 shows two parallel, conducting metal plates separated by an evacuated space. A potential difference is applied across the plates such that they become charged at equal magnitude, but opposite sign. For the electrostatic system, equation (10.1) is written:

$$Q = \frac{\varepsilon_0 a V}{l} \qquad (10.13)$$

where Q is the total charge (C), ε_0 is the permittivity of free space in (F m^{-1}; i.e. the field characteristic), a is the cross-sectional area of the plates, l is the distance separating the plates, and V is the applied potential difference.

The term $(\varepsilon_0 a/l)$ is called the 'capacitance' of the system. It is usually denoted by C, and is measured in units of farads. Thus

$$Q = CV \qquad (10.14)$$

Since the Farad is an unwieldy large number, it is more common to use the microfarad (μF) or the picofarad (pF) as the unit of measurement. (N.B. 1 μF = 10^{-6} F; 1 pF = 10^{-12} F.)

If the plates are separated by an insulating medium other than free space, then these so-called 'dielectric media' have a different value of permittivity. The actual permittivity is related to the permittivity of free space by the relative permittivity of the dielectric, i.e.

$$\varepsilon = \varepsilon_0 \cdot \varepsilon_r \qquad (10.15)$$

where ε_r is the relative permittivity of the dielectric.

Table 10.1 The permittivity of some typical dielectric materials

Material	Relative permittivity
Air	1
Paper	2–2.5
Porcelain	6–7
Mica	3–7

The permittivity of free space ε_0 is numerically equal to $(1/36\pi) \times 10^{-9}$. The relative permittivities of some of the more common dielectric materials are listed in *Table 10.1*.

10.2.8 Simple capacitive circuits

Figure 10.5 shows three capacitors connected in a simple parallel arrangement. The equivalent total capacitance is given as the algebraic sum of all the capacitances in the circuit, i.e.

$$C = C_1 + C_2 + C_3 \qquad (10.16)$$

where C is the total capacitance.

For the series capacitance arrangement shown in *Figure 10.6*, the total equivalent capacitance is related through the inverse summation:

$$\frac{1}{C} = \frac{1}{C_1} + \frac{1}{C_2} + \frac{1}{C_3} \qquad (10.17)$$

Equations (10.16) and (10.17) can be used to reduce series and parallel capacitor circuits to a single equivalent capacitor.

Composite capacitors, involving different dielectric media, may also be treated in the same manner as a series capacitor arrangement.

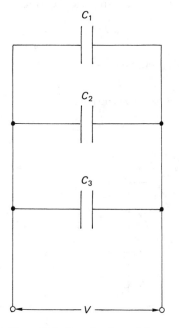

Figure 10.5 Capacitors in parallel

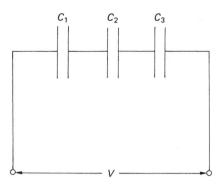

Figure 10.6 Capacitors in series

10.2.9 Charging a capacitor

Figure 10.7 shows a parallel plate capacitor which is connected in series with a resistor to a source of e.m.f., say a battery, through a switch. Initially, the capacitor is uncharged before the switch is closed. When the switch is closed a charging current will flow until such time that the potential difference across the capacitor is equal to the e.m.f. available from the source. The charging process consists of taking electrons from A and transferring them through the external wiring to plate B. The energy required to do this is derived from the battery. The build up of electrons from the negative terminal of the battery to plate B of the capacitor induces a dielectric flux between the plates and a balancing positive charge is developed on plate A. As long as the dielectric flux is changing, a current will flow externally. Eventually a state of equilibrium will be reached. Note that no electrons can pass through the dielectric because it is an insulator.

The instantaneous current during charging is:

$i = \mathrm{d}Q/\mathrm{d}t$

From equation (10.14), this may be rewritten for a capacitor as:

$$i = \mathrm{d}Q/\mathrm{d}t = C(\mathrm{d}v/\mathrm{d}t) \tag{10.18}$$

where v is the instantaneous voltage.

The instantaneous power is therefore:

$p = iv = Cv(\mathrm{d}v/\mathrm{d}t)$

The energy supplied over the time period dt is:

$Cv(\mathrm{d}v/\mathrm{d}t)\mathrm{d}t = Cv\mathrm{d}v$

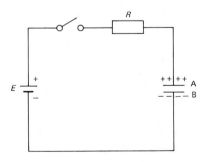

Figure 10.7 Charging a capacitor

Hence the total energy supplied is:

$$\int_0^V Cv\mathrm{d}v = \frac{1}{2}CV^2 \tag{10.19}$$

10.2.10 Types of capacitor

10.2.10.1 Air capacitors

These usually consist of one set of fixed plates and another set of moveable plates. The area between the capacitor plates is therefore variable and so also (from equation (10.13)) is the capacitance.

10.2.10.2 Paper dielectric capacitors

These consist of metal foil interleaved with wax, or oil impregnated paper and rolled into a compact cylinder form.

10.2.10.3 Mica dielectric capacitors

These consist of alternate layers of mica and metal foil, tightly clamped together. They are used mainly for high frequency applications.

10.2.10.4 Ceramic capacitors

The electrodes are formed by a metallic coating, usually silver, deposited on the opposite faces of a thin ceramic disc.

10.2.10.5 Polycarbonate capacitors

Polycarbonate is a plastic insulating material which is produced in a range of thicknesses, down to about 2 μm. The polycarbonate is bonded with aluminium foil and rolled to form the capacitor element.

10.2.10.6 Electrolytic capacitors

These generally consist of two aluminium foils, one with an oxide film and the other without. The foils are usually interleaved with paper which is saturated with a suitable electrolyte. Electrolytic capacitors have the advantage of having a large capacitance for a relatively small physical volume. They are used extensively for smoothing out the ripple voltage from rectified a.c. power supplies. They are only suitable, however, for use in circuits where the applied voltage across the capacitor can never reverse its direction.

10.2.11 Dielectric strength

If the potential difference across opposite faces of a dielectric material is increased above a particular value, then the material breaks down. The failure of the material takes the form of a small puncture, which renders the material useless as an insulator. The potential gradient necessary to cause breakdown is normally expressed in kilovolts/millimetre (kV mm^{-1}) and is termed the 'dielectric strength'. The dielectric strength of a given material decreases with increases in the thickness. *Table 10.2* gives approximate values for some of the more common dielectric materials.

10.2.12 Electromagnetic systems

The magnetic field can be defined as the space in which a magnetic effect can be detected, or observed. The standard school physics experiment involves sprinkling iron filings on a sheet of paper under which a bar magnet is placed. On tapping

Table 10.2 The dielectric strength of some common insulators

Material	Thickness (mm)	Dielectric strength (kV mm^{-1})
Air	0.2	5.75
	0.6	4.92
	1.0	4.36
	10.0	2.98
Mica	0.01	200
	0.10	115
	1.00	61
Waxed paper	0.10	40–60

the paper lightly, the filings arrange themselves along closed loops of equal potential and are clearly visible. An obvious magnetic field is also observable around a straight length of conductor carrying a current. In particular, the exact same magnetic field as that produced by the bar magnet is observed when the current carrying conductor is formed into a helical type coil. The equipotential loops describe the path of the magnetic flux ϕ and, although the flux lines have no physical meaning, they provide a convenient vehicle to quantify various magnetic effects.

The direction of the magnetic flux is governed by the so-called 'right-hand screw rule'. This states that the direction of the magnetic field produced by a current corresponds to the direction given by turning a right-hand screw thread. The direction of the current corresponds to the translational movement of the screw.

10.2.13 Magnetic field of a toroid

Figure 10.8 shows a toroidal coil, of N turns which is wound round an annular former. A resultant magnetic flux (shown as broken lines) is generated when the coil carries a current. For the magnetic field, equation (10.1) takes the general form:

$$\phi = \frac{\mu a F}{l} \tag{10.20}$$

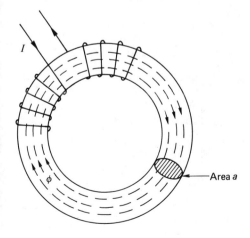

Figure 10.8 Toroid

where ϕ is the magnetic flux (W), μ is the permeability of the medium (H m^{-1}), a is the cross-sectional area of the flux path in the toroid, l is the length of the flux path, and F is the magnetic potential difference, or magnetomotive force (A).

The magnetomotive force (m.m.f.) is equal to the product of the number of turns on the coil and the current carried:

$$F = IN \tag{10.21}$$

Note that the m.m.f. is descriptively expressed in 'ampere-turns' (A-t). Since the number of turns is already a dimensionless quantity, the accepted unit of magnetomotive force is the ampere (A).

The term ($\mu a/l$) is called the 'permeance' and the inverse of permeance is the 'reluctance' S. Thus equation (10.20) may be rewritten as:

$$\phi = F/S \tag{10.22}$$

Equation (10.22) represents an electromagnetic version of Ohm's law.

Alternatively, equation (10.20) can be expressed as:

$$\frac{\phi}{a} = \mu \frac{F}{l}$$

or

$$B = \mu H \tag{10.23}$$

where $B = \phi/a$ is the magnetic flux density (W m^{-2}, or T) and $H = F/l$ is the magnetic intensity (A m^{-1}).

10.2.14 Permeability

The permeability of free space μ_0 is numerically equal to $4\pi \times 10^{-7}$. The absolute permeability of other materials is related to the permeability of free space by the relative permeability:

$$\mu = \mu_0 \mu_r \tag{10.24}$$

For air and other non-magnetic materials, the absolute permeability is the same constant. For magnetic materials, absolute permeability is not a fixed constant but varies non-linearly with the flux density. The non-linear variation of permeability is conveniently displayed as a functional plot of magnetic flux density B against magnetic intensity H. *Figure 10.9* illustrates a number of B vs. H curves for some common materials. From equation (10.23) it is apparent that the absolute permeability is given by the slope of a tangent to the B vs. H curve at any particular value.

Also shown in *Figure 10.9* is the B vs. H curve for air, the only straight-line relationship in the diagram. It is apparent that, for an applied magnetic intensity, the magnetic flux developed in a coil with a ferrous core is many times greater than that through a similar coil with an air core. In most practical systems, therefore, a ferrous core is normally used since it greatly facilitates the establishment of a magnetic flux.

10.2.15 Faraday's law

Perhaps the most fundamental law of electromagnetic systems, Faraday's law states that the e.m.f. induced in a magnetic circuit is equal to the rate of change of flux linkages in the circuit. In mathematical terms, Faraday's law is given by:

$$e = N(d\phi/dt) \tag{10.25}$$

where e is the instantaneous induced e.m.f.

Equation (10.25) forms the basis of all electrical power generating machines and is a statement of the fact that an

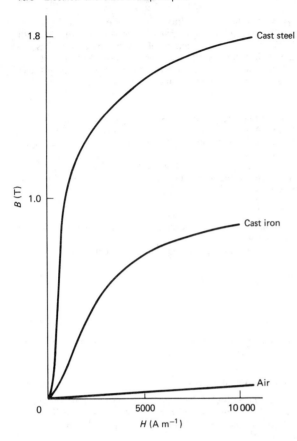

Figure 10.9 Plots of magnetic flux density B vs. magnetic intensity H for some common materials

electric current can be produced by the movement of magnetic flux relative to a coil. In all rotating electrical generators, it is actually the coil which is moved relative to the magnetic field. The net result, however, is exactly the same.

The direction of the induced e.m.f. is always such that it tends to set up a current to oppose the motion, or the change of magnetic flux which is responsible for inducing the e.m.f. This is essentially a statement of Lenz's law. In many texts, therefore, the right-hand side of equation (10.25) is often shown as a negative quantity.

The motion, or change of flux is associated with the application of a mechanical force which ultimately provides the torque required to drive the electric generator. *Figure 10.10* shows a single conductor of length l metres, carrying an induced current I and lying perpendicular to a magnetic field of flux density B tesla. The force applied causes the conductor to move through a distance dx metres. The mechanical work done is, therefore, $F dx$. The electrical energy produced is given as the product of the power developed and the time duration, i.e. $e I dt$. For no external losses, the mechanical work done is converted into electrical energy. Thus

$$e I dt = F dx \qquad (10.26)$$

Using Faraday's law (equation (10.25)) the induced e.m.f. is equal to the rate of change of flux linkage. For a single conductor, $N = 1$ and thus

$$e = (B l dx)/dt$$

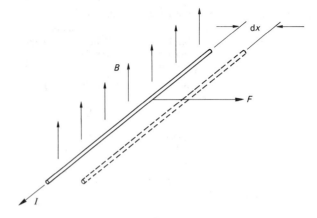

Figure 10.10 Generation of e.m.f.

Therefore

$$(B l dx/dt) I dt = F dx$$

i.e.

$$F = B l I \qquad (10.27)$$

Equation (10.27) relates the applied force to the corresponding current generated in a conductor moving through a magnetic field. The equation applies equally to an electric generator, or conversely to a motor in which case the electrical power supplied is converted into a mechanical torque via the electromagnetic effect.

10.2.16 Self-induced e.m.f.

If a current flows through a coil, a magnetic flux links with that coil. If, in addition, the current is a time varying quantity, then there will be a rate of change of flux linkages associated with the circuit. The e.m.f. generated will oppose the change in flux linkages.

When dealing with electric circuits it is convenient if the voltage across individual elements can be related to the current flowing through them. *Figure 10.11* shows a simple circuit comprising a coil having N turns and resistance R, connected in series with a time varying voltage.

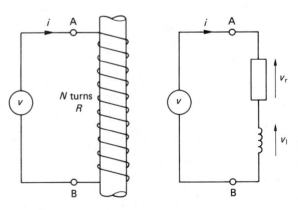

Figure 10.11 Self-induced e.m.f.

The voltage drop across the terminals A and B can be split into two components. First there is the voltage drop purely due to the resistance of the coiled element. Secondly there is a voltage drop which is a consequence of the self-induced e.m.f. generated through the electromagnetic effect of the coil. Thus

$$v = v_r + v_1$$

$$= iR + N\frac{d\phi}{dt} \tag{10.28}$$

From equations (10.20) and (10.21)

$$\phi = \frac{\mu a F}{l} = \frac{\mu a i N}{l}$$

therefore

$$v = iR + N\frac{d}{dt}\left(\frac{\mu a i N}{l}\right)$$

$$= iR + N^2\left(\frac{\mu a}{l}\right)\frac{di}{dt} \tag{10.29}$$

The group $N^2(\mu a/l)$ is called the 'self-inductance' of the coil and is denoted by L. The unit of self-inductance is the henry. Therefore,

$$v = iR + L\frac{di}{dt} \tag{10.30}$$

By comparing equations (10.28) and (10.30) it is apparent that

$$L\frac{di}{dt} = N\frac{d\phi}{dt}$$

Integration then gives

$$L = N\phi/i \tag{10.31}$$

The nature of the self-induced e.m.f., i.e. $(L di/dt)$, is such that it will oppose the flow of current when the current is increasing. When the current is decreasing the self-induced e.m.f. will reverse direction and attempt to prevent the current from decreasing.

10.2.17   Energy stored in an inductor

Instantaneous power $= vi$

Energy stored $= W = \displaystyle\int_0^t vi\,dt$

$$= \int_0^t L\frac{di}{dt}i\,dt$$

$$= L\int_0^I i\,di = \frac{1}{2}LI^2 \tag{10.32}$$

10.2.18   Mutual inductance

Two coils possess mutual inductance if a current in one of the coils produces a magnetic flux which links the other coil. *Figure 10.12* shows two such coils sharing a common magnetic flux path in the form of a toroid.

The mutual inductance (in henry) between the two coils is:

$$M = \frac{N_2\phi}{I_1} \tag{10.33}$$

where N_2 is the number of turns on coil 2, I_1 is the current through coil 1, and ϕ is the magnetic flux linking coils 1 and 2.

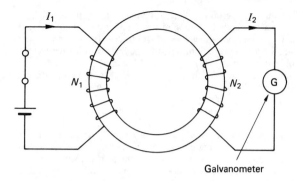

Figure 10.12   Mutual inductance

The mutual-inductance effect finds wide application both to electrical transformers and to rotating electrical machines.

10.2.19   Hysteresis in magnetic circuits

Hysteresis can be described with reference to a toroidal coil wound on an iron core (see *Figure 10.8*). The current applied to the coil can be imagined to be taken through a cyclic process where it is increased from 0 to $+I$ amperes, back through 0 to $-I$ amperes and again back through 0 to $+I$ amperes. Measurement of the flux density in the core, as the current varies, results in a B vs. H curve as depicted in *Figure 10.13*.

The behaviour of the B vs. H relationship is termed a 'hysteresis loop'. This behaviour is typical for ferrous cores and is an illustration of the fact that all of the electrical energy supplied to magnetise an iron core is not returned when the coil current is reduced to zero. The loss of energy is called 'hysteresis loss' and it is manifested as heat in the iron core. It can be shown that the hysteresis loss is directly proportional to the area enclosed by the hysteresis loop. It is to obvious

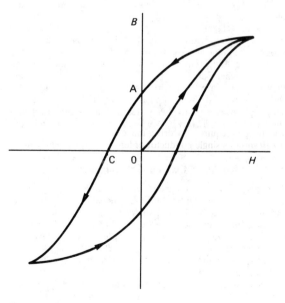

Figure 10.13   Hysteresis loop for an iron cored toroid

advantage, therefore, that any magnetic system which is subject to a cyclic variation of flux density should incorporate a magnetic core in which the hysteresis loop is minimally small. In practical applications a low-loss core of silicon iron, or a nickel–iron alloy such as Mumetal or Permalloy, is normally used. Some strain-relieving heat treatment process is also usually involved, after cold working, to restore the original magnetic condition of the material.

Hysteresis is characterised by two parameters: the 'remanent flux density' (or 'remanence') and the 'coercive force'. The remanent flux density is the flux density which remains in the core when the mangetic intensity, i.e. the coil current, has been reduced to zero. The remanent flux density is represented by line 0A in *Figure 10.13*. The coercive force is the magnetic intensity required to reduce the remanent flux density to zero and is represented by line 0C in *Figure 10.13*.

10.2.20 Eddy current loss

Faraday's law (equation (10.25)) shows that a time varying magnetic flux will induce an e.m.f. in a coil. If the ends of the coil are connected and form a closed circuit, then the induced voltage will circulate a current round the closed loop. Let us now consider an iron core, in which a time varying magnetic flux exists. Since iron is a conductor, then there will be a multitude of arbitrary closed paths within the iron matrix. These closed paths constitute effective conduction routes and the varying magnetic flux will generate a flow of current round them. The currents are called 'Eddy currents', and because of the ohmic resistance of the core, the end result is an energy loss as the eddy currents are dissipated as heat.

Eddy current losses can be greatly reduced by building the iron core in the form of laminations which are insulated from one another. The laminated assembly confines the path lengths for the eddy currents to each respective lamination. The cross-sectional area of the eddy current path is also reduced and the eddy current loss is approximately proportional to the square of the thickness of the laminations. A practical minimum thickness for any lamination is about 0.4 mm. Increasing manufacturing costs could not justify the use of much thinner laminations.

The eddy current phenomenon, while basically a detrimental effect, can be utilised to good purpose in the form of 'induction heating'. In this type of heater, a metallic work piece is enclosed within a coil which is powered with an a.c. supply. The induced eddy currents are dissipated as heat in the work piece. Induction heating is basically a surface heating technique where the 'depth' of heating is dependent on the depth of eddy current activity. This, in turn, is dependent on the frequency of the a.c. supplied to the coil. A high frequency supply produces a relatively low depth of heat penetration and this is used extensively in surface, or case hardening heat treatment processes.

10.2.21 Kirchhoff's laws and the magnetic circuit

Figure 10.14 shows a magnetic circuit in which a magnetising coil is wound on one of the limbs and another limb incorporates the usual features of an 'air gap'.

Using the analogy between the magnetic and the conduction circuits, the magnetic circuit can be represented in terms of an energy source, or m.m.f., and each limb of the magnetic circuit is written in terms of the appropriate reluctance S. This is illustrated in *Figure 10.15*.

Given all the relevant dimensions and material properties, the problem resolves to one of calculating the current required to establish a prescribed magnetic flux density in the air gap.

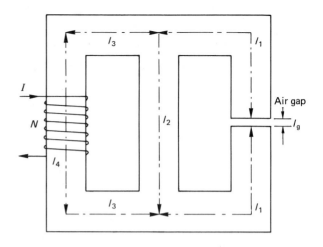

Figure 10.14 Magnetic circuit

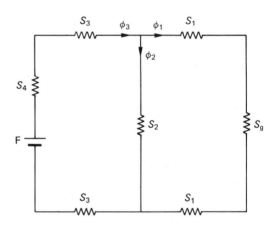

Figure 10.15 Representation of the magnetic circuit shown in *Figure 10.14*

The solution invokes the use of Kirchhoff's laws as they apply to magnetic circuits.

First law— At any instant in time, the sum of the fluxes flowing into a node is equal to the sum of the fluxes flowing out.

Second law— Around any closed magnetic circuit the total magnetomotive force is equal to the sum of all the m.m.f.s round the circuit.

Manipulation of equations (10.20) to (10.24) then yields the required solution. The self-inductance of the coil, if required, may be calculated from equation (10.31), or from the definition:

$$L = N^2(\mu a/l) = N^2/S \qquad (10.34)$$

It has already been shown that the lowest permeability is that of air and that the m.m.f. required to produce a flux density in air is many times greater than that required to produce the same flux density in a ferrous material. It may reasonably be questioned, therefore, why air gaps are used at all in iron cored magnetic circuits. The only function of the air gap is to provide a measure of linearity to the magnetic system such that the inductance remains reasonably constant over a range of operating currents.

10.2.22 Alternating quantities

If an electrical quantity varies with time, but does not change its polarity, it is said to be a 'direct current' (d.c.) quantity. If the quantity alternates between a positive and a negative polarity, then it is classified as an 'alternating current' (a.c.) quantity.

The period T is the time interval over which one complete cycle of the alternating quantity varies. The inverse of the period is the frequency, f (in cycles per second, or Hertz). The circular frequency ω in radians per second is also commonly used. (N.B. One cycle/second corresponds to 2π rad s^{-1}.)

Instantaneous values of the quantities encountered in electrical systems are usually denoted by lower-case letters. Since the instantaneous values are difficult to measure and quantify, a.c. quantities are usually expressed as 'root mean square' (r.m.s.) values. For a periodically varying a.c. quantity, the r.m.s. value is given by:

$$\text{r.m.s.} = \left[1/t \int_0^t (\text{Quantity})^2 dt \right]^{1/2} \quad (10.35)$$

If in a periodically varying a.c. quantity, the positive and negative half cycles are identical in all but sign, then the average value of the quantity is zero. The average value of the half cycle is sometimes useful, however, particularly when dealing with rectified signals.

Many electrical quantities vary in a sinusoidal manner and it can easily be shown that the r.m.s. value is simply related to the maximum value by:

$$\text{r.m.s.} = \text{max.}/(\sqrt{2}) = 0.707 \text{ max.} \quad (10.36)$$

10.2.23 Relationship between voltage and current in resistive, inductive and capacitive elements

For a simple resistive element, current is directly proportional to voltage. The current waveform will therefore be essentially the same shape as the voltage waveform.

For an inductive coil with negligible resistance, the relationship between voltage and current is given by equation (10.30), i.e.

$$v = L \frac{di}{dt}$$

Thus

$$i = \frac{1}{L} \int v \, dt \quad (10.37)$$

If, for example, the voltage is an a.c. square wave, then the current will take the shape of a 'sawtooth' waveform (see *Figure 10.16*).

The relationship between voltage and current for a capacitive element is given by equation (10.18), i.e.

$$i = C \frac{dv}{dt}$$

For the capacitive element it can be seen that a current will flow only when the voltage is changing. No current can flow if the voltage is constant since (dv/dt) will then be equal to zero. The capacitor then, will block any steady d.c. input and indeed is sometimes used for this express purpose.

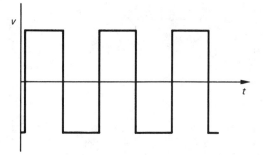

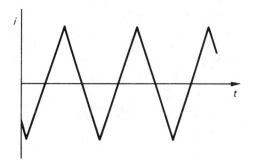

Figure 10.16 Voltage–current relationship for an inductive element

10.2.24 Resistive–inductance and resistive–capacitance circuits under transient switching conditions

Circuits involving a single resistor (R), capacitor (C) or inductance (L) are rare. It is more usual to find circuits involving some combination of these elements in both d.c. and a.c. applications. *Figure 10.17* illustrates two simple RL and RC circuits.

10.2.24.1 RL circuit

With the switch open there is no flow of current in the circuit. At the instant of switching, the current will rise and eventually reach a steady-state value of V_s/R. The transient period is governed by equation (10.30) which represents a first order, ordinary differential equation in i. The solution of equation (10.30) involves the technique of separating the variables to allow integration. The general solution is:

$$i = I[1 - \exp(-Rt/L)] \quad (10.38)$$

where I is the final steady-state current in the circuit (see *Figure 10.18*).

Equation (10.38) shows that the current growth in the circuit will rise exponentially to reach a steady-state value as time t increases.

It may also be shown that:

$$v = L \frac{di}{dt} = V_s \exp(-Rt/L) \quad (10.39)$$

If the initial rate of increase in current is maintained, then the steady-state current is achieved in a time T (seconds). From *Figure 10.18*:

$$\left[\frac{di}{dt} \right]_{t=0} = \frac{I}{T}$$

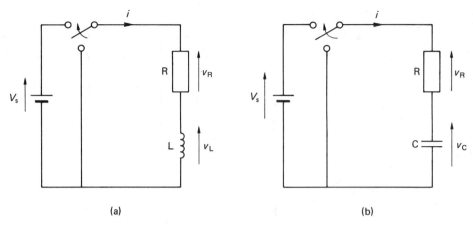

Figure 10.17 Simple RL (a) and RC (b) circuits under transient switching conditions

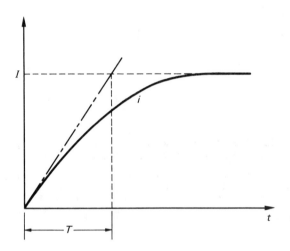

Figure 10.18 Transient current growth in an inductive element

At the instant of switching, $v_L = V_s$. Hence at $t = 0$

$$v_L = V_s = LI/T$$

Therefore

$$T = LI/V_s = L/R \qquad (10.40)$$

Equation (10.40) defines the 'time constant' for the RL circuit. The actual value of the current in the circuit when $t = T$ is $0.632 \times I$.

10.2.24.2 RC circuit

In *Figure 10.17(b)*, with the switch open there is zero potential difference across the capacitor. On closing the switch the voltage across the capacitor will rise in an asymptotic manner, reaching a steady-state value of V_s. From Kirchhoff's second law:

$$V_s = iR + v_c \qquad (10.41)$$

where v_c is the instantaneous voltage across the capacitor.

From equation (10.18) we can write:

$$V_s = RC\frac{dv_c}{dt} + v_c \qquad (10.42)$$

Equation (10.42) shows that the instantaneous voltage across the capacitor also conforms to a first-order system. The solution of (10.42) gives

$$v_c = V_s[1 - \exp(-t/RC)] \qquad (10.43)$$

Considering the initial rate of increase of voltage, it can be shown that the appropriate time constant for the simple RC circuit is:

$$T = RC \qquad (10.44)$$

Both the simple RL and RC circuits are first-order systems with a generalised form of transient behaviour.

In circuits containing both inductive and capacitive elements, the transient behaviour is governed by a second order ordinary differential equation. The transient behaviour of these circuits, however, is less important than their response to sinusoidally varying inputs.

An interesting additional feature of the RC circuit is its ability to function as an integrating, or a differentiating circuit. *Figure 10.19* depicts an RC integrator.

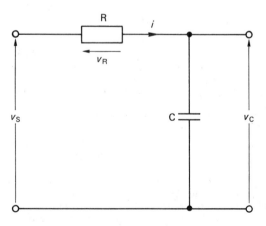

Figure 10.19 RC integrator

The components of the circuit are selected such that $v_R \gg v_c$. To a reasonable approximation $v_s = v_R$. Therefore

$$i = v_s/R$$

Since

$$v_c = 1/C \int i \, dt$$

Then

$$v_c = 1/RC \int v_s \, dt \qquad (10.45)$$

Equation (10.45) shows that the output voltage is proportional to the integral of the input voltage, with $(1/RC)$ the constant of proportionality.

For the RC differentiator (*Figure 10.20*) the components are selected such that the voltage drop across the capacitor v_C is very much greater than that over the resistor v_R, i.e. $v_s \gg v_c$. Therefore

$$v_s = 1/C \int i \, dt$$

Thus

$$i = C \frac{dv_s}{dt}$$

Taking the output across the resistor gives:

$$v_R = iR = RC \frac{dv_s}{dt} \qquad (10.46)$$

It is apparent that the output is proportional to the first derivative of the input.

10.2.25 Steady-state a.c.

In most practical applications in electrical engineering, the voltages and currents are sinusoidal. A simple series RLC circuit is depicted in *Figure 10.21*. Since the current is common to each of the circuit elements, it is used for reference purposes. The instantaneous current is defined as:

$$i = I_m \sin (\omega t) \qquad (10.47)$$

where I_m is the maximum value (or peak value) of the current, and ω is the angular, or circular frequency (in rad s^{-1}).

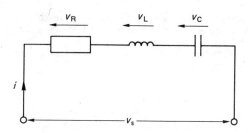

Figure 10.21 Series RLC circuit

The voltage drop across the resistor is:

$$v_R = iR = I_m R \sin (\omega t) \qquad (10.48)$$

Equation (10.48) indicates that the voltage drop across the resistor is in phase with the current. In other words, v_R reaches a positive maximum at the same instant as the current i.

The voltage drop across the inductor is:

$$v_L = L \frac{di}{dt} = L \frac{d}{dt} [I_m \sin(\omega t)]$$

$$= L I_m \omega \cos (\omega t)$$

$$= \omega L I_m \sin (\omega t + 90) \qquad (10.49)$$

The relationship between current and voltage drop across the inductor is shown in *Figure 10.22*. It can be seen that there is a phase difference between the voltage drop and the current through the inductor. In fact, v_L reaches a positive maximum 'before' i, and v_L is said to 'lead' the current by 90°.

For the capacitor, the voltage drop is given by:

$$v_c = 1/C \int i \, dt = 1/C \int I_m \sin (\omega t) dt$$

$$= -\frac{I_m}{\omega C} \cos (\omega t)$$

$$= +\frac{I_m}{\omega C} \sin (\omega t - 90) \qquad (10.50)$$

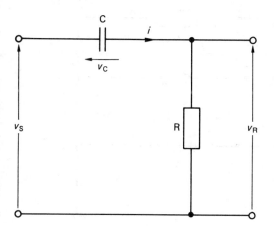

Figure 10.20 RC differentiator

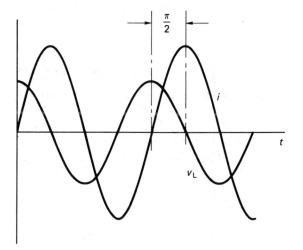

Figure 10.22 Current and voltage drop across an inductor

The voltage drop across the capacitor therefore reaches its positive maximum after that of i. In general terminology, v_c 'lags' i by 90°.

Equations (10.48) to (10.50) are all of a similar form in that they can be expressed as:

Voltage drop = Constant × Current

In equation (10.49), the constant ωL is termed the 'inductive reactance', and is denoted by X_L. In equation (10.50), the constant $(1/\omega C)$ is the 'capacitive reactance', which is denoted as X_c. Both these reactances are in units of ohms.

The total voltage drop across the three circuit elements is:

$$v = v_R + v_L + v_c$$
$$= iR + L\frac{di}{dt} + \frac{1}{C}\int i\,dt$$

Therefore,

$$v = I_mR \sin(\omega t) + \omega L I_m \sin(\omega t + 90°)$$
$$+ \frac{I_m}{\omega C} \sin(\omega t - 90) \tag{10.51}$$

While equation (10.51) defines the total instantaneous voltage drop in mathematical terms, it is a little cumbersome to deal with. To simplify the analysis, the addition of a.c. voltages is conveniently performed using a graphical technique involving 'phasors'.

10.2.26 Phasor diagrams

Any sinusoidally varying quantity can be represented as a phasor, which is a vector quantity. The length of the phasor is proportional to the magnitude of the product of the reactance and the maximum current. The direction of the phasor is determined by the phase angle and its relation to some common reference.

For the RLC circuit shown in *Figure 10.21*, the voltage drop across the inductance may be arbitrarily assumed to be greater than that across the capacitor. The total voltage drop in the circuit is then given as the phasor addition of the three individual potential difference components. This is illustrated in *Figure 10.23*.

The vector addition of the three phasors shows that the source voltage leads the current by an angle of ϕ degrees, i.e.

$$\bar{V} = V_m \sin(\omega t + \phi) \tag{10.52}$$

The circuit is therefore essentially inductive and, using the standard notation, the total phasor voltage is designated by a capital letter with an overscore.

10.2.27 Complex notation

Since inductive and capacitive elements in a.c. circuits involve a phase shift of +90° and −90°, respectively, the complex number notation is used extensively to manipulate phasor quantities.

The complex operator j, defined as $\sqrt{-1}$, is a unit operator which, when multiplying a phasor, shifts it by 90° in an anticlockwise direction. Thus for the series RLC circuit:

$$\bar{V}_R = \bar{I}R; \ \bar{V}_L = j\bar{I}X_L; \ \bar{V}_C = -j\bar{I}X_C$$

where $\bar{I}$ can be taken as the r.m.s. value of the current.

The voltage drop across the complete circuit can then be written as:

$$\bar{V} = \bar{I}R + j\bar{I}X_L - j\bar{I}X_C$$
$$= \bar{I}[R + j(X_L - X_C)] \tag{10.53}$$

The term within the square brackets is called the 'impedance' of the circuit and is denoted by $\bar{Z}$. Thus

$$\bar{V} = \bar{I}\bar{Z} \tag{10.54}$$

Equation (10.54) represents Ohm's law for a.c. circuits.

The phase angle between the source voltage and the current is:

$$\phi = \tan^{-1}[(X_L - X_C)/R] \tag{10.55}$$

10.2.28 The parallel resistive–inductive–capacitance circuit

A parallel RLC circuit is shown in *Figure 10.24*. The applied voltage is common to all of the circuit elements and it is therefore chosen as the reference.

Using Ohm's law, the currents through each of the circuit elements are:

$$\bar{I}_R = \bar{V}/R; \ \bar{I}_L = \bar{V}/X_L; \ \bar{I}_C = \bar{V}/X_C$$

Applying Kirchhoff's first law, the total current is the vector sum of the three currents $\bar{I}_R$, $\bar{I}_L$ and $\bar{I}_C$. The magnitude and phase of the total current may subsequently be determined

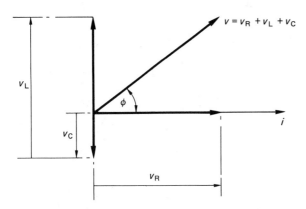

Figure 10.23 Phasor diagram for series RLC circuit

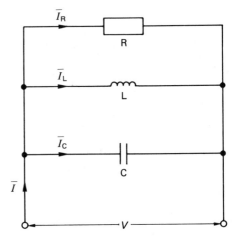

Figure 10.24 Parallel RLC circuit

from a phasor diagram, or calculated using the complex number notation. Using the latter, and noting that the current through an inductor lags the voltage while the current through a capacitor leads the voltage, it may be shown that:

$$\bar{I} = \bar{I}_R + \bar{I}_L + \bar{I}_C$$

$$= \bar{V}\left[\frac{1}{R} - j\left(\frac{1}{X_C} - \frac{1}{X_L}\right)\right] \tag{10.56}$$

and the phase angle is given by

$$\phi = \tan^{-1}\left[\frac{R(X_L - X_C)}{X_L X_C}\right] \tag{10.57}$$

10.2.29 Power and power factor in a.c. circuits

In general, for any a.c. circuit there will be a phase difference between the voltage and the current. If the circuit elements are purely resistive, the phase difference will be zero. In all other cases the phase difference will be a finite positive or negative value. Denoting the phase angle between the voltage and the current by ϕ, it may be shown (see Bell and Whitehead[4]) that the average power is:

$$P_{av} = \frac{V_m}{\sqrt{2}} \frac{I_m}{\sqrt{2}} \cos\phi$$

In terms of r.m.s. values:

$$P_{av} = VI\cos\phi \tag{10.58}$$

The term $\cos\phi$ is called the 'power factor'.

Power factor is an important parameter when dealing with electrical transformers and generators. All such machines are rated in terms of kilovolt-amperes (kV-A), which is a measure of the current-carrying capacity for a given applied voltage. The power than can be drawn depends both on the kV-A rating and the power factor of the load. *Figure 10.25* shows the relationship between kV-A, kilowatts (kW) and power factor, sometimes referred to as the 'power triangle'.

It can readily be seen that:

$$\text{kW} = \text{kV-A}\cos\phi \tag{10.59}$$

and

$$\text{kV A}_R = \text{kV-A}\sin\phi \tag{10.60}$$

where kV-A_R is the reactive power.

Thus, knowing the kV-A rating and the power factor of a number of various loads, the power requirements from a common supply may be determined.

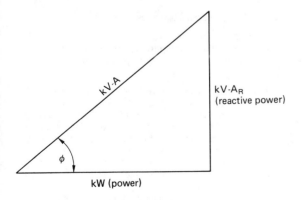

Figure 10.25 Power triangle

When quoting power factor in practical applications, it is usual to state the phase of the current with respect to the voltage. For an inductive load the current lags the voltage and the power factor is said to be 'lagging'. For a predominantly capacitive load the current leads the voltage and the power factor is leading.

If the power is supplied from, say, an alternator rated at 400 V and 1000 A, then these are the highest voltage and current that the machine can tolerate without overheating. The phase difference between the voltage and current is entirely dependent upon the load. Thus if the power factor of the load is 1 then the 400 kV-A alternator can supply 400 kW of power to the load. Neglecting losses, the prime mover which drives the alternator must also be capable of supplying 400 kW. If, on the other hand, the power factor of the load is 0.5, then the power supplied will only be 200 kW. This means that, although the generator will be operating at its rated kV-A, the prime mover which drives the generator will be operating at only half its capacity.

An alternative way of looking at this phenomenon is to consider a load of, say, 100 kW, with a lagging power factor of 0.75. If the supply voltage is 50 V, then the required current (from equation (10.58)) is 2.67 A. If, however, the power factor of the load were to be increased to unity, then the required current would be reduced to 2 A. This means that the conducting cables, in supplying a reduced current, may have a correspondingly reduced cross-sectional area. In general, the size of an electrical system including transmission lines, switchgear and transformers is dependent upon the size of the current. It is economically viable, therefore, to ensure that the current is minimised. As a further incentive to industrial consumers, the electricity supply authorities normally operate a two-part tariff system. The two-part system consists of a fixed rate depending on the kV-A rating of the maximum demand and a running charge per kilowatts consumed per hour.

For these reasons it is advantageous to try and increase the power factor such that it is close to, but not quite, unity. A unity power factor is in fact avoided because it gives rise to a condition of resonance (see Section 10.2.30). In practice, capacitors, connected in parallel, are often used to improve the power factor of predominantly inductive loads such as electric motors. For large scale power systems a separate phase advance plant is used.

10.2.30 Frequency response of circuits

Since both inductive and capacitive reactance are frequency dependent, it is clear that the output from circuits involving capacitive and inductive elements will also depend on the frequency of the input signal.

The 'frequency response' of a circuit is usually presented as a plot of output/input ratio against the frequency as base. The ratio plotted could be one of voltages, currents or powers. Since the range of frequencies involved may be quite large, a logarithmic scale is normally employed. A logarithmic scale is also usually adopted for the vertical axis and the output/input ratio quoted in decibels (dB), i.e.

$$\text{Voltage ratio (in dB)} = 20\log_{10}\left[\frac{V_{out}}{V_{in}}\right] \tag{10.61}$$

Considering the series RLC circuit shown in *Figure 10.21*, and taking the voltage across the resistor as an output:

$$V_{out} = IR$$

$$V_{in} = I[R + j(\omega L - 1/\omega C)]$$

Therefore

$$\frac{V_{\text{out}}}{V_{\text{in}}} = \frac{R}{R + j(\omega L - 1/\omega C)}$$

Using the complex conjugate and calculating the modulus of the voltage ratio gives:

$$\left| \frac{V_{\text{out}}}{V_{\text{in}}} \right| = \frac{R}{[R^2 + (\omega L - 1/\omega C)^2]^{1/2}} \tag{10.62}$$

The phase angle is given by

$$\phi = -\tan^{-1}\left[\frac{(\omega L - 1/\omega C)}{R}\right] \tag{10.63}$$

It can be seen from equation (10.62) that the voltage ratio will have a maximum value of 1 when the frequency is such that $(\omega L - 1/\omega C) = 0$. Equating this expression gives:

$$\omega = \frac{1}{\sqrt{LC}} \tag{10.64}$$

Equation (10.64) defines the so-called 'resonance' condition at which the inductive and capacitive reactances are equal and self-cancelling. The resonant frequency is usually denoted by ω_0 and it is the frequency at which the power transferred through the circuit is maximum. At any other frequency, above or below ω_0, the power transferred is reduced.

The impedance of the circuit is given by:

$$\bar{Z} = R + j(X_{\text{L}} - X_{\text{C}}) \tag{10.65}$$

At the resonant frequency the total reactance is zero and the circuit behaves as if only the resistive element were present.

The general variation in the voltage ratio, or amplitude ratio, and phase angle with frequency is illustrated in *Figure 10.26*. Also shown on *Figure 10.26* are the two frequencies ω_1 and ω_2 at which the amplitude ratio is −3 dB. The −3 dB amplitude ratio is chosen because it corresponds to a halving in the power transmitted.

The 'bandwidth' is the frequency range between ω_1 and ω_2. A quality parameter, used with respect to resonant circuits, is the so-called 'Q factor' which is defined as the resonant frequency/bandwidth ratio.

10.2.31 Semiconductors

The common materials used for semiconductors are germanium and silicon. In recent times silicon has all but replaced germanium as a semiconductor material. The semiconductor materials have a crystalline structure such that each atom is surrounded by equally spaced neighbours. The basic structure can be visualised as a two-dimensional grid where the node points represent the central nucleus and the inner shell electrons, while the connecting lines of the grid represent the four valence electrons associated with each nucleus. This grid concept is adequate to describe an intrinsic, or 'pure', semiconductor.

At absolute zero temperature the crystalline structure is perfect and the electrons are all held in valence bonds. Since there are no current carriers available, the crystal behaves as a perfect insulator. As the temperature rises above absolute zero, an increasing number of valence bonds are broken, releasing pairs of free electrons and their associated 'holes'. In the absence of an applied field the free electrons move randomly in all directions. When an electric field is applied the electrons drift in a preferential direction to oppose the field and a net flow of current is established.

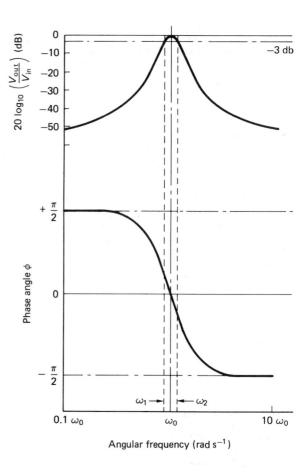

Figure 10.26 Voltage ratio and phase angle vs. frequency (series RLC)

The covalent bond, with a missing electron, has a large affinity for electrons such that an electron from a neighbouring bond may easily be captured. This will leave the neighbouring atom depleted of electrons and the flow of electrons is generally associated with a counterflow of so-called 'holes'. The mobile hole, to all intents and purpose, is essentially a simple positive charge.

10.2.32 Doped semiconductors

Doped semiconductors are those in which an impurity has been introduced into a very pure intrinsic silicon. The nature of the impurity depends on the type of semiconductor required.

(1) *n type*—Impurities with five valence electrons can be added to produce a negative type of semiconductor. These impurities are referred to as 'donors' since the additional electron is very easily freed within the matrix. In the n-type semiconductor the free electrons are the dominant current carriers.

(2) *p type*—The p-type semiconductor is one in which the added impurities have only three valence electrons. Such impurities are called 'acceptors' and they produce a positive type of semiconductor within which hole conduction is the dominant current carrier.

10.2.33   p–n Junction diode

A p–n junction is formed by doping a crystal in such a way that the semiconductor changes from p-type to n-type over a very short length, typically 10^{-6} m. The transition zone from p-type to n-type is called the 'carrier depletion layer' and, due to the high concentrations of holes on one side and electrons to the other, a potential difference exists across this layer. The diffusion of holes from p to n and electrons from n to p is the majority carrier movement, called the 'diffusion current'. The drift of electrons from p to n and holes from n to p is the minority carrier movement referred to as the 'drift current'. When there is no externally applied potential difference, the diffusion current and the drift current are balanced in equilibrium. If an electric field is applied across the device, then two situations can exist, as illustrated in *Figure 10.27*.

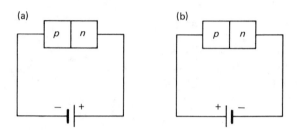

Figure 10.27   A p–n junction with applied potential difference. (a) Reverse bias; (b) forward bias

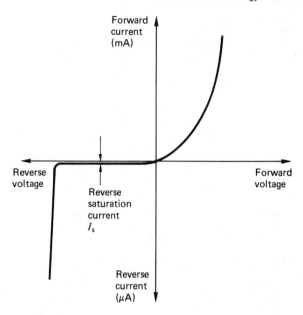

Figure 10.28   Current–voltage relationship for a p–n semiconductor diode

Figure 10.27(a) shows the reverse-bias mode in which the potential barrier is increased. The diffusion current is reduced while the drift current is barely altered. Overall, the current is negative and very small. When forward bias is applied (*Figure 10.27(b)*), the potential barrier is reduced and a large diffusion current flows. Overall, the current is positive and large. These general characteristics are the basis of a semiconductor diode which displays the typical current–voltage relationship depicted in *Figure 10.28*.

Figure 10.28 shows clearly that a very high impedance is presented by the diode to an applied voltage of reverse polarity. A low impedance is presented to a forward polarity voltage. In simple terms, the diode accommodates a forward flow of current, but greatly inhibits a reverse flow. The diode may be likened, therefore, to a switch which is activated 'on' for forward voltages and 'off' for reverse voltages. The reverse saturation current I_s is typically of the order of a few nano-amperes (nA) and can sensibly be regarded as zero.

The general characteristic also shows that the reverse voltage has a critical limiting value at which a 'breakdown' occurs.

Depending upon the diode construction, the breakdown, or 'Zener' voltage may range from as low as 1 V to as much as several thousand volts. Up to the breakdown voltage, the reverse saturation current is independent of the reverse voltage.

Since the current–voltage relationship for a diode is a non-linear exponential function, the analysis of circuits involving diodes can become complicated. A simple awareness of the diode's practical function as a rectifier is perhaps more important than is a proficiency in analysing circuits involving diode elements.

10.2.34   A.c. rectification

Figure 10.29 shows an a.c. circuit with a diode in series with a load resistor. When the diode is forward biased, a current will flow in the direction indicated by the arrowhead. No current can flow when the diode is reverse biased, provided that the applied voltage does not exceed the breakdown value. The resultant current waveform through the resistor, for a sinusoidal voltage input, will therefore consist of positive only,

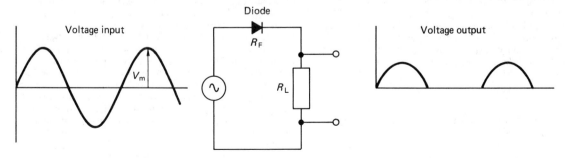

Figure 10.29   Half-wave rectification circuit

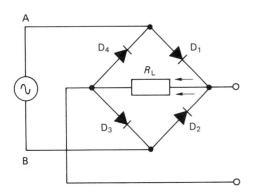
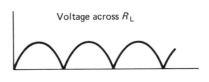

Figure 10.30   Full-wave rectification with a diode bridge

half sine waves. Since the output waveform is positive only, then it is by definition a d.c. voltage. It can be shown that the r.m.s. voltage across the resistor is:

$$V = \frac{V_m}{2} \frac{R_L}{R_L + R_F} \qquad (10.66)$$

where R_L is the load resistance, R_F is the diode forward resistance, and V_m is the peak input voltage.

Determination of R_F is problematic, however, and models of varying complexity are used to simulate the diode in the circuit.

The single-diode circuit results in half-wave rectification. To obtain full-wave rectification, a diode bridge circuit can be used. The diode bridge is shown in *Figure 10.30*. When A is positive with respect to B then diodes D_1 and D_3 are conducting. When B is positive with respect to A then diodes D_2 and D_4 are conducting. The circuit arrangement ensures that the current, which consists of a continuous series of positive half sine waves, is always in the same direction through the load R_L.

With full-wave rectification there are twice as many half sine pulses through the load as there are with half-wave rectification. In addition, there are always two diodes effectively in series with the load. The resultant r.m.s. voltage across the load resistor for the full-wave diode bridge rectification circuit is:

$$V = \frac{V_m}{\sqrt{2}} \frac{R_L}{(R_L + 2R_F)} \qquad (10.67)$$

The 'peak inverse voltage' (PIV) is defined as the maximum reverse biased voltage appearing across a diode. Thus when used as a rectifier, the diodes must have a sufficiently high reverse voltage rating in excess to the peak inverse voltage that the circuit can generate. For both the half- and the full-wave rectification circuits considered, the peak inverse voltage is equivalent to the maximum supply voltage V_m. Additional manufacturer's diode specifications would normally include the maximum power rating and the maximum allowable forward current.

10.2.35   The Zener diode

The diode breakdown effect is also used in a variety of circuits to provide a stabilised reference voltage. Special diodes which are designed to operate continuously in the reverse bias mode are called 'Zener diodes'. These diodes are manufactured with a range of breakdown voltages of 3–20 V. *Figure 10.31* shows

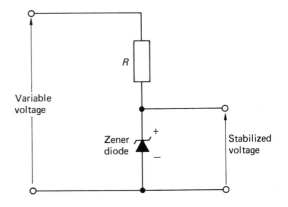

Figure 10.31   Zener diode as a reference voltage source

a Zener diode being used in a circuit to give a stable voltage which is essentially independent of the current flowing through the device. The series resistor in the circuit is included to limit the reverse current through the diode to a safe value.

10.3   Analogue and digital electronics theory

10.3.1   The bipolar, or junction, transistor

Originally developed in 1948, the transistor has revolutionised electronics and has influenced almost every aspect of life in the industrialised world. The term 'transistor', derived from 'transfer resistor', describes a device which can transfer a current from a low-resistance circuit to a high-resistance circuit with little change in current during the process.

The junction transistor consists of two p–n diodes formed together with one common section, making it a three-layer device (see *Figure 10.32*). Current flow in the transistor is both due to electron and hole conduction. The common central section is referred to as the 'base' and is typically of the order of 25 μm in length. Since the base can be made of either an n-type or a p-type semiconductor, then two basic configurations are possible. These are the n–p–n and the p–n–p types, as illustrated in *Figure 10.32*. The two other terminals are called

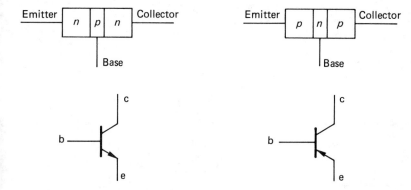

Figure 10.32 n–p–n and p–n–p junction transistors

the 'emitter' and the 'collector'. An arrowhead is traditionally shown between the emitter and the base to indicate the conventional direction of the current flow in that part of the circuit.

A brief description of the physical operation of the junction transistor can be made with respect to the n–p–n type. The mode of operation of the p–n–p type is the same as that of the n–p–n type, except that the polarities of all applied voltages, currents and charge carriers are reversed.

In normal use, as a linear amplifier, the transistor is operated with the emitter to base junction forward biased and the collector to base junction reverse biased. For the n–p–n transistor, the emitter is, therefore, negative with respect to the base while the collector is positive with respect to the base (see *Figure 10.33*).

The junction n_1p is forward biased such that the free electrons drift from n_1 to p. On the other hand, junction n_2p is reverse biased and it will collect most of the electrons from n_1. The electrons which fail to reach n_2 are responsible for the current at the base terminal I_B. By ensuring that the thickness of the base is very small and that the concentration of impurities in the base is much lower than either that of the emitter or the collector, the resultant base current will be limited to some 2% of the emitter current. The basic transistor characteristic is therefore:

$$I_C = h_{FB}I_E \qquad (10.68)$$

where I_C is the collector current, I_E is the emitter current, and h_{FB} is the current gain between the collector and the emitter.

Normally, h_{FB} would range between 0.95 and 0.995 for a good quality transistor.

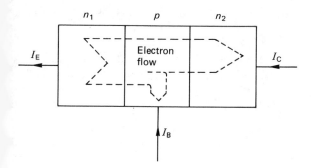

Figure 10.33 An n–p–n transistor in normal operation

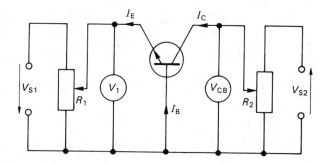

Figure 10.34 An n–p–n transistor in a common base circuit

10.3.2 Common base characteristics

Figure 10.34 shows an n–p–n transistor connected in a circuit to determine its static common base characteristics. The emitter current I_E is kept constant by varying R_1 and a range of values for I_C are imposed by varying R_2. The value of V_{CB}, the collector–base voltage, is noted. The test is repeated for another fixed value of I_E; the results obtained are as depicted in *Figure 10.35*.

It is found that, over a wide range of collector–base voltages, the collector current is essentially independent of the collector–base voltage. This is because most of the electrons entering the n–p–n junction are attracted to the collector. In effect, the collector circuit has a very high impedance and acts as a constant-current source. The actual value of the collector current is determined by the emitter current and the two are related through equation (10.68), which is the common base characteristic. The general characteristics also show that the collector–base voltage must be reversed, i.e. collector negative with respect to base, in order to reduce the collector current to zero. Finally, at a high collector–base voltage, the collector current increases rapidly due to the Zener effect. The same characteristics are observed with the p–n–p transistor, except that the signs are in the reverse direction to that shown in *Figure 10.34*.

10.3.3 Common emitter characteristics

Figure 10.36 shows the n–p–n transistor with its emitter terminal connected both to the base current circuit and to the collector current circuit. Using the same test procedure as

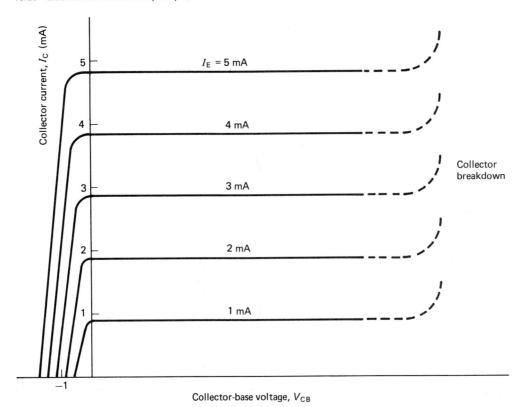

Figure 10.35 Common base characteristics

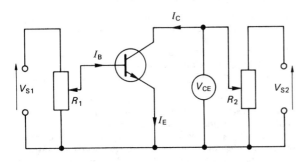

Figure 10.36 An n–p–n transistor in a common emitter circuit

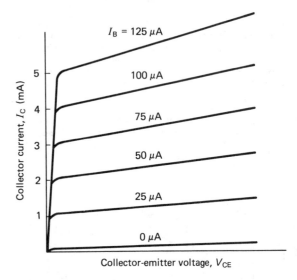

Figure 10.37 Common emitter characteristics

described above, the resulting characteristics are as shown in *Figure 10.37*.

The first significant observation is that the collector–emitter voltage V_{CE} must be positive to produce a positive collector current. At low values of V_{CE}, the collector current I_C is also low, but when V_{CE} exceeds the so-called 'knee' voltage, the characteristic assumes a linear relationship. The gradient of the linear region is generally much higher than that for the common base configuration and the collector impedance is, therefore, lower than that for the common base circuit. When the base current is zero, the collector current still has a positive finite value.

The common emitter characteristic is generally written as:

$$I_C = h_{FE}I_B \qquad (10.69)$$

where h_{FE} is the current gain between the collector and base.

Application of Kirchhoff's first law to the common emitter circuit gives:

$$I_E = I_C + I_B$$

Using equation (10.68) and eliminating I_E, it can be shown that:

$$\frac{I_C}{I_B} = \frac{h_{FB}}{1 - h_{FB}} = h_{FE} \tag{10.70}$$

For a transistor with a steady state current gain in common base of 0.95, the common emitter gain is:

$$h_{FE} = \frac{0.95}{1 - 0.95} = 19$$

If, due to some temperature effect, h_{FB} undergoes a minor change to, say, 0.96, then the new value of h_{FE} is 24. It is clear, therefore, that the common emitter gain h_{FE} is much more sensitive to small-order effects than the common base gain h_{FB}.

For a p–n–p transistor, the characteristics of the common emitter circuit are the same, except that the polarities of all voltages and currents are again in reverse order to that shown in *Figure 10.36*.

10.3.4 The transistor in a circuit

In most practical applications, transistors are operated in the common emitter mode where the emitter terminal forms the common connection between the input and output sections of the circuit (see *Figure 10.38*). The transistor collector characteristics are shown in *Figure 10.39*.

The load line for the resistor R_C is superimposed, and the operating point is given by the intersection of the load line with the collector characteristic. The operating point will, therefore, be dependent on the base current since this controls the collector characteristic. Also shown in *Figure 10.39* is the maximum power dissipation curve (broken line) which represents the locus of the product of collector current and collector–emitter voltage. The maximum power dissipation curve

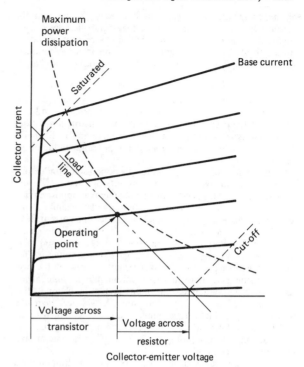

Figure 10.39 Common emitter characteristics with the load line superimposed

represents a physical limitation and the operating point must be constrained to lie below the curve at all times.

As the base current is reduced, the operating point moves down the load line. When I_B reaches zero, the collector current is minimised and the transistor is said to be 'cut-off'. Alternatively, as the base current is increased, the operating point moves up the load line and eventually reaches a maximum value at which the transistor is said to be 'bottomed', or 'saturated'. When saturated, the collector–emitter voltage is at a minimum of about 0.1–0.2 V and the collector current is a maximum. The two extremes between cut-off and saturation represent a very high and a very low impedance state of the transistor, respectively. These extremes have great practical application to rapid, low power switching, and transistors operating between cut-off and saturation are frequently used in digital electronics circuitry. The low impedance state represents a switch closed (or 'on'), and the high impedance state represents the switch open (or 'off'). When operating as a linear current amplifier, the operating point is ideally located in the centre of the active region of the characteristic.

The analysis of circuits involving transistors is conveniently dealt with by representing the transistor in terms of an equivalent circuit and using the conventional current-flow direction from positive to negative. Consideration of the charge carriers, i.e. holes or electrons, is only necessary to describe the internal physical operation of the transistor. Fully detailed worked examples are particularly informative and these are usually provided in all standard textbooks on electrical and electronics technology.

10.3.5 The field effect transistor

Field effect transistors (FETs) are a much more recent development than bipolar transistors and they operate on a

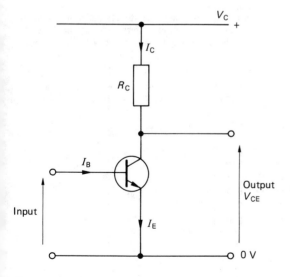

Figure 10.38 An n–p–n transistor in a practical, common emitter circuit

substantially different mechanism in achieving signal amplification. Operationally, FETs are voltage-controlled devices as opposed to the bipolar transistor which is a current-operated device. FETs are often described as 'unipolar' since conduction in the FET is the result of only one predominant charge carrier.

The junction field effect transistor (JFET) consists of a thin bar of semiconductor which forms a channel between its two end connections which are referred to as the 'source' and the 'drain'. If the semiconductor used in the construction of the FET is n-type, then the device is called an 'n-channel'. Conversely, a FET made from a p-type semiconductor is called a 'p-channel' device.

If the channel consists of a uniformly doped semiconductor, then the conductivity will be constant and the FET will function as a linear resistor. By introducing two opposite type semiconductor layers on either side of the channel, the effective thickness of the channel and hence the current flow can be controlled. The opposite type layers are denoted as 'gates', and in normal operation they are reverse biased by a d.c. potential V_{GS} referred to as the 'gate source voltage'. The reverse bias ensures that no current can flow between the two gates and the gate inputs have an extremely high impedance. By using a lightly doped semiconductor for the channel, the gate depletion layer, which is determined by V_{GS}, can be made to extend well into the channel width. This controls the resistance of the path between the source and the drain. The general characteristics of such a field effect transistor are shown in *Figure 10.40*.

For a given value of V_{GS}, an increase in drain–source voltage from zero initially gives a linear rise in drain current. Further increases in drain–source voltage result in a so-called 'pinch-off' in the drain current which then becomes independent of the drain–source voltage. Finally, at a particular limiting value of drain–source voltage, a breakdown is initiated. The similarity between *Figures 10.40* and *10.37* or *10.39* is clear, and it is evident, therefore, that the bipolar junction transistor and the unipolar FET can perform essentially a similar function in any given application. Many other types of transistor, for example the metal oxide semiconductor field effect transistor (MOSFET), use alternative means to control the resistance of the source to drain channel. However, the general characteristics of these devices are all very similar to those shown in *Figure 10.40*.

10.3.6 Integrated circuits

While transistor based amplifiers are still found as individual elements in many working circuits, the modern trend is towards the development of integrated circuits where all the circuit elements are housed within a single silicon wafer. MOSFET technology is predominant in this area since the number of components on a single silicon chip can be packed up to 20 times more densely than with bipolar technology.

The integrated circuit components include diodes and transistors which may be either bipolar junction type or FETs. Resistors can be deposited on top of the wafer in the form of tantalum, which is a poor conductor, or built into the wafer as 'pinch' resistors, which are partially turned-off FETs. Capacitors can also be produced within the silicon wafer. Capacitive elements may be formed when a p–n junction diode is reverse biased. The p-type and n-type layers form the plates of the capacitor, and the carrier depletion layer acts as a dielectric. However, the capacitance is limited to a few picofarads. There is no microelectronic equivalent for an inductor, but most circuit designs can generally avoid the requirement for coiled inductive elements.

When the integrated circuit is complete it is usually encapsulated as a 'dual-in-line' (DIL) package. This is the normal form in which the integrated circuit is sold. An eight-pin DIL package may contain a relatively simple circuit, but a 40-pin DIL could easily contain all of the electronics associated with a central processing unit (CPU) for a computer system. These latter devices contain an enormous number of transistors and diodes, approaching 10 000 on a chip of less than 10 mm². The technology to produce this density of integration is commonly called 'very large scale integration' (VLSI).

10.3.7 The thyristor

Both the bipolar transistor and the FET can be utilised for switching operations. These devices, however, are usually associated with low-power switching. For switching very large currents and voltages a special device called a 'thyristor', formerly known as a 'silicon controlled rectifier' (SCR), is normally used.

The thyristor is a four-layer, unidirectional semiconductor device with three connections referred to as the 'anode', 'cathode' and the 'control gate' (see *Figure 10.41*).

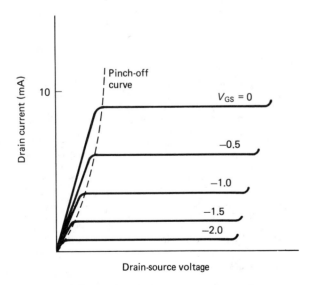

Figure 10.40 Characteristics of a field effect transistor (FET)

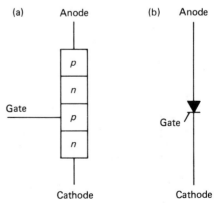

Figure 10.41 (a) Thyristor device. (b) Circuit symbol for a thyristor

The current flow is from the anode to the cathode only, and with the cathode positive with respect to the anode the device has a very high impedance. Under normal circumstances, the thyristor will fail to conduct current in any direction. If a voltage is applied such that if the thyristor were a diode it would conduct in the forward-biased direction, then application of a very small current between the gate and the cathode will cause the thyristor to change abruptly from non-conducting to conducting mode. The turn-on is rapid, within a few microseconds, and once turned on the thyristor will remain on, even if the gate current is removed.

Once triggered into conduction, the thyristor will turn off again only when the current flowing through it is reduced below a critical value. This minimum conducting current is called the 'holding current' and may range between a few microamperes to a few tens of milliamperes. Thyristors are additionally connected in series with a resistor which serves to limit the current to a safe value. The basic thyristor function is that of a power-control device and they are used extensively for switching mains electricity and as a speed controller for d.c. motors.

10.3.8 The triac

The triac, or bidirectional thyristor, is similar in operation to the thyristor, but differs in that it can be switched into conduction in either direction. In essence, the triac is equivalent to two thyristors mounted back to back. Triacs find application to switching in full-wave a.c. power supplies.

10.3.9 Amplifiers

In dealing with amplifiers, a 'systems' approach is normally adopted. The systems approach limits the considerations to the relationship between the input and output of the amplifier and the effect of feedback on the system. An amplifier may consist of a single transistor in a circuit, or be a complex arrangement of many transistors within an integrated circuit. In using the systems approach, the internal physical workings of the device are of no particular concern.

In general, electronic amplifiers are supplied with energy from a d.c. source. An input signal to the circuit controls the transfer of energy to the output and the output signal should be a higher power version of that supplied to the input. The amplifier does not, however, function as some magical source of free energy. The increased power across the amplifier is invariably drawn from the supply.

The term 'amplifier' is actually a shortened form for the complete specification 'voltage' amplifier. This has transpired

because most amplifiers are intended to magnify voltage levels. Any other type of amplifier is normally prefixed with the name of the quantity which is amplified, e.g. current amplifier, charge amplifier, and power amplifier.

Amplifiers may be broadly classified with reference to the frequency range over which they are designed to operate. In this respect there are two general categories, these being 'wide band' and 'narrow band' amplifiers. The names are self-explanatory in that the wide-band amplifier exhibits a constant power gain over a large range of input signal frequencies. The narrow-band, or 'tuned', amplifier, on the other hand, provides a power gain over a very small frequency range. The power gain is usually expressed in decibels and is defined by equation (10.61).

The bandwidth of an amplifier is used in the same context as described in Section 10.2.30, i.e. to define the operating frequency range. In this respect the -3 dB amplitude ratio is used consistently to define the upper and lower input signal frequencies at which the power transferred across the amplifier is halved.

Using the system model, the amplifier can be represented as shown in *Figure 10.42*. In the figure the amplifier is shown enclosed within the broken lines. There is a single input, a single output and one common connection. The amplifier also features an internal input impedance, shown as resistance R_i, and an internal output impedance, shown as resistance R_o. In actual fact the input and output impedances could have both inductive and capacitive components as well as the simple resistances as shown in the figure.

Connected to the input stage of the amplifier is a voltage source V_s and its associated internal resistance R_s. This could be taken to represent some form of transducer having a low voltage output in the millivolt range. At the output stage, the amplifier acts as a voltage source where A_v is the voltage gain. The output is shown connected to an external load R_L, which might be considered to be some sort of recording instrument such as a digital voltmeter.

Considering the input stage, it may be shown, from Ohm's law, that:

$$V_i = \frac{V_s}{(1 + R_s/R_i)} \tag{10.71}$$

Equation (10.71) indicates that the voltage applied to the amplifier input stage V_i will approach the source voltage V_s only when R_i tends to infinity. Therefore the amplifier should ideally have a very large input impedance to prevent serious voltage attenuation at the input stage. By a similar argument, the output impedance R_o should be very small in comparison with the load resistance R_L for maximum voltage gain.

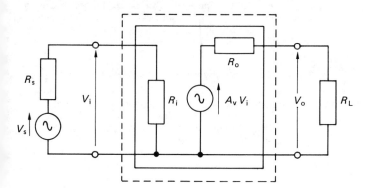

Figure 10.42 System representation of an amplifier

10.3.10 Effect of feedback on amplifiers

The amplifier illustrated in *Figure 10.42* is specified by its input and output impedances and its open-circuit gain A_v. The open-circuit gain being that gain obtained when the load resistance is infinite. These parameters are not fixed, but will vary with ambient temperature and power supply voltage variation and age. The adverse effects of these variabilities can be minimised through the application of 'negative feedback'.

One particular method of obtaining negative feedback is the so-called 'series voltage' method (see *Figure 10.43*). The feedback system shown in the figure is applied by connecting a potentiometer across the output terminals and tapping off a fraction β of the output signal. This fraction is connected in series with the input and with a polarity which will always oppose the input signal. Assuming both that the input impedance of the amplifier is very large in comparison with the internal resistance of the voltage source and the resistance of the potentiometer is very large in comparison with the output impedance of the amplifier, then:

$$V_i = V_s - \beta V_o \qquad (10.72)$$

Since

$$V_o = A_v V_i$$

Then

$$V_o = A_v V_s - \beta A_v V_o$$

The overall gain of the system with feedback A_f is:

$$\frac{V_o}{V_s} = \frac{A_v}{1 + \beta A_v} \qquad (10.73)$$

Equation (10.73) shows that the feedback loop has reduced the original gain by the factor $(1 + \beta A_v)$. If, in addition, the original gain A_v was in itself very large such that $\beta A_v \gg 1$, then:

$$A_f = A_v/(\beta A_v) = 1/\beta \qquad (10.74)$$

Under the above circumstances, the overall gain of the system with feedback is essentially dependent only on the feedback fraction β. Therefore any changes which alter the original gain A_v of the amplifier will not affect the gain of the overall system with feedback.

Consideration of the system with and without the feedback loop shows that the effect of series voltage negative feedback is to increase the input resistance by the factor $(1 + \beta A_v)$ and to reduce the output resistance by the same factor. Both these effects are of benefit to the operation of the system. These comments refer only to a negative-feedback system using the series voltage method. Other methods of obtaining negative feedback can be used, including series-current feedback, shunt-current and shunt-voltage feedback. These alternative methods have different effects on the overall gain and on the input and output impedances of the amplifier.

10.3.11 Noise and distortion in amplifiers

Noise is inherently present in all electronic amplifier systems. Sources of noise include the random charge movements within the solid-state devices, thermoelectric potentials, electrostatic and electromagnetic pick-up and interference from the standard 50 Hz (or 60 Hz) mains power supply. The noise is fairly evenly distributed across the whole frequency spectrum and appears superimposed on the amplified input signal. If the noise is generated at the input stage of the amplifier then the signal-to-noise ratio is not improved by feedback. The signal-to-noise ratio can be improved, however, if an intermediate amplifying stage, free from noise effects, can be included in the system.

Distortion is another undesirable feature which arises when the amplifier input–output characteristic (or transfer characteristic) deviates from an ideal linear relationship. If the transfer characteristic is linear then the output signal will be a faithful amplified replica of the input. A non-linear characteristic will give a distorted output and a non-sinusoidal output will be generated from a sinusoidal input. Distortion is usually associated with a high level of input signal which overextends the linear operating range of the amplifier.

10.3.12 Amplifier frequency response

The frequency response of an amplifier is usually illustrated as a plot of the gain in decibels against the input signal frequency. The graph is called a 'Bode plot' and the phase relationship between the output and input is also shown for completeness. *Figure 10.44* shows the frequency characteristics for a typical wide-band amplifier.

The bandwidth between the -3 dB cut-off frequencies is determined either by the characteristics of the active devices used to make the amplifier, or by other frequency-dependent elements in the amplifier circuit. The upper limiting frequency is fixed by the charge transit time through the active device. In practice, any stray capacitance, which is manifested as a parallel capacitance in the system, will reduce the upper limiting frequency considerably. In theory, the active device will respond to frequencies down to 0 Hz but, because of the variabilities due to ageing effects, a lower cut-off frequency is often imposed by including series capacitors on one, or both, of the input connections.

10.3.13 Positive feedback and stability

In *Figure 10.43* a negative-feedback signal is produced by using a series voltage. If the phase of the series voltage was changed such that the feedback signal augmented the input, then the nature of the feedback loop would become positive. With this positive feedback system, the overall gain would then become:

$$A_f = A_v/(1 - \beta A_v) \qquad (10.75)$$

Positive feedback therefore increases the overall system gain. If indeed the product (βA_v) is made equal to 1, then the

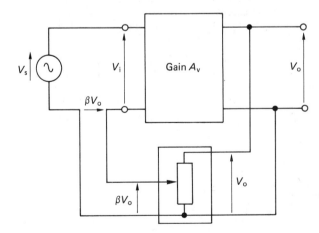

Figure 10.43 Series voltage method of negative feedback

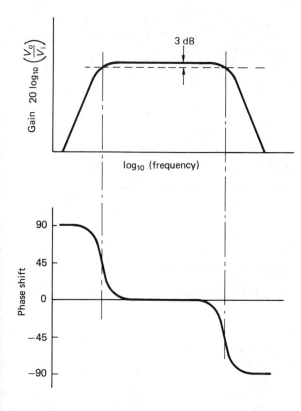

Figure 10.44   Frequency response for a wide-band amplifier

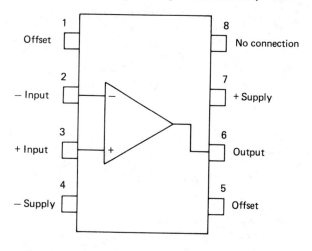

Figure 10.45   SN72741 operational amplifier

overall gain becomes infinite. Positive feedback, however, is inherently unstable since the output signal tends to increase indefinitely in an uncontrolled manner. Nonetheless, systems with positive feedback are found in oscillator circuits where the amplifier produces its own input signal via a positive-feedback loop.

10.3.14   The operational amplifier

Modern amplifier systems rely less on discrete active devices such as transistors and much more on the vast range of integrated circuits which are readily available. One of the most prevalent operational amplifiers based on integrated circuit technology is the generic type SN72741, or, as it is often abbreviated, the '741'. The 741 is available as an eight-pin DIL package and internally consists of 20 bipolar transistors, 11 resistors and one capacitor. The DIL package takes up less area than a small postage stamp and costs less than a cup of coffee. *Figure 10.45* shows the usual representation of the 741 operational amplifier, or 'op-amp', in its DIL form.

The internal circuitry is quite complex, but is conveniently reduced to the basic schematic form shown in the figure. The operational amplifier consists of an output, an inverting input and a non-inverting input. In addition, the integrated circuit requires a bipolar power supply which may range anywhere between ±3 and ±18 V. There is also provision for an offset null on connection pins 1 and 5. For the most part, the offset pins can be ignored.

The operational amplifier has a high input impedance, a low output impedance and a very high open circuit gain A. Ideally, the gain should be infinite. The bandwidth should also ideally

be infinite, but the 741, for example, has an effective bandwidth limited between 0 Hz and about 1 MHz.

For operational amplifiers such as the 741, there are a number of standard circuits which are used routinely to perform specific functions. These are described below.

10.3.14.1   Inverting amplifier

Figure 10.46 shows an op-amp wired up for an inverted output. The input current i_1 is given as V_1/R_1 and, because the amplifier input impedance is very high, the current flowing into the input terminal is approximately zero. This is equivalent to having the potential available at point E equal to zero. For this reason E is referred to as a 'virtual earth'. From Kirchhoff's first law then, it is apparent that $i_1 = -i_2$. Thus $V_1/R_1 = -V_0/R_2$, and the gain can be written as:

$$\frac{V_0}{V_1} = -\frac{R_2}{R_1} \tag{10.76}$$

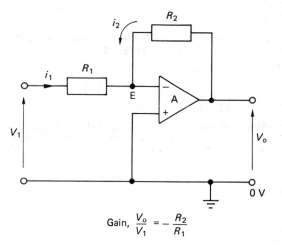

Gain, $\dfrac{V_0}{V_1} = -\dfrac{R_2}{R_1}$

Figure 10.46   Inverting amplifier

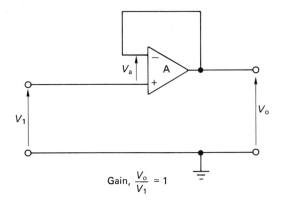

Gain, $\dfrac{V_o}{V_1} = 1$

Figure 10.47 Unity gain amplifier

Provided the open-circuit gain of the amplifier is very high, the overall gain with this negative-feedback system is given by the ratio of the two external resistors and is independent of the open-circuit gain.

10.3.14.2 Unity gain amplifier

Figure 10.47 depicts a unity gain amplifier in which no external resistors are wired into the circuit. The unity gain amplifier is also known as a 'voltage follower', or as a 'buffer amplifier'. This type of amplifier circuit is often used in instrumentation systems where the internal resistance of a voltage-generating transducer and that of the voltage-recording instrument are so poorly matched that the transducer voltage is seriously attenuated. This situation arises when the transducer internal resistance is large in comparison with that of the recording instrument. Since the buffer amplifier has a large input impedance and a low output impedance it can be interfaced between the transducer and the recording instrument to provide optimum impedance matching. This gives a low source impedance and high destination impedance between both the transducer and amplifier and also between the amplifier and the instrument.

Summing the voltages round the amplifier in *Figure 10.47* gives:

$$V_1 + V_a = V_0$$

Since the internal impedance of the amplifier is very large, then V_a is effectively zero and the gain is:

$$V_o/V_1 = 1 \tag{10.77}$$

10.3.14.3 Non-inverting amplifier

Figure 10.48 shows the operational amplifier connected up for a non-inverting output. Assuming that the currents through resistors R_1 and R_2 are equal and that point E is a virtual earth then:

$$\frac{V_i}{R_1} = \frac{V_o - V_i}{R_2}$$

Hence

$$\frac{V_o}{V_i} = \frac{R_2 + R_1}{R_1}$$

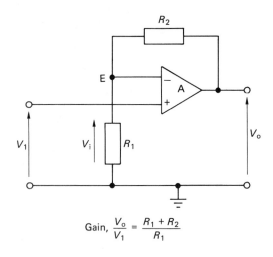

Gain, $\dfrac{V_o}{V_1} = \dfrac{R_1 + R_2}{R_1}$

Figure 10.48 Non-inverting amplifier

Since E is a virtual earth, then $V_i = V_1$ and

$$\frac{V_o}{V_1} = \frac{R_2 + R_1}{R_1} \tag{10.78}$$

If, in addition $R_2 \gg R_1$, then

$$\frac{V_o}{V_1} \gg \frac{R_2}{R_1} \tag{10.79}$$

10.3.14.4 Summing amplifier

The summing amplifier is shown in *Figure 10.49*. As point E is a virtual earth, then

$$-i_4 = i_1 + i_2 + i_3$$

Therefore,

$$\frac{V_o}{R_4} = -\left[\frac{V_1}{R_1} + \frac{V_2}{R_2} + \frac{V_3}{R_3}\right]$$

or

$$V_o = -R_4\left[\frac{V_1}{R_1} + \frac{V_2}{R_2} + \frac{V_3}{R_3}\right] \tag{10.80}$$

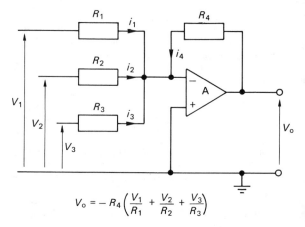

$$V_o = -R_4\left(\frac{V_1}{R_1} + \frac{V_2}{R_2} + \frac{V_3}{R_3}\right)$$

Figure 10.49 Summing amplifier

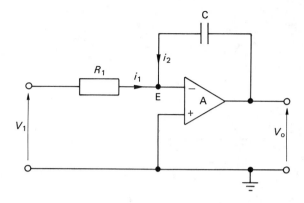

Figure 10.50 Integrating amplifier

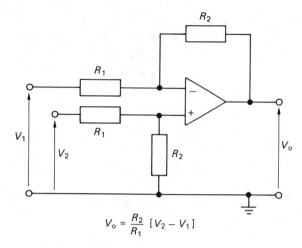

$$V_o = \frac{R_2}{R_1} [V_2 - V_1]$$

Figure 10.51 Differential amplifier

If the resistances used in the circuit are all of equal value, then the output voltage will be equivalent to the summation of all the input voltages and with a reversed sign. Subtraction of any of the voltages can be performed by reversing its polarity, i.e. by first of all passing the voltage through a unity gain inverting amplifier, before it is passed on to the summing amplifier.

10.3.14.5 Integrating amplifier

The integrating amplifier uses a capacitor, as opposed to a resistor, in the feedback loop (see *Figure 10.50*). The voltage across the capacitor is:

$$1/C \int_0^t i_2 dt$$

Since E is a virtual earth, then $i_1 = -i_2$. Therefore,

$$i_2 = -(V_1/R_1)$$

The voltage across the capacitor, which is in effect V_o, is:

$$V_o = -(1/C) \int_0^t (V_1/R_1)dt = -(1/CR_1) \int_0^t V_1 dt \qquad (10.81)$$

Thus the output voltage is related to the integral of the input voltage.

Apart from various mathematical processes, operational amplifiers are also used in active filtering circuits, waveform generation and shaping, as a voltage comparator and in analogue-to-digital (A/D) and digital-to-analogue (D/A) conversion integrated circuits.

10.3.15 The differential amplifier

The differential amplifier, or subtractor, has two inputs and one output (see *Figure 10.51*). The differential amplifier yields an output voltage which is proportional to the difference between the inverting and the non-inverting input signals. By applying the superposition principle, the individual effects of each input on the output can be determined. The cumulative effect on the output voltage is then the sum of the two separate inputs. It can be shown, therefore, that

$$V_o = (R_2/R_1)(V_2 - V_1) \qquad (10.82)$$

The input signals to a differential amplifier, in general, contain two components, these being the 'common-mode' and 'difference-mode' signals. The common-mode signal is the average of the two input signals and the difference-mode signal is the difference between the two input signals. Ideally, the differential amplifier should affect the difference-mode signal only. However, the common-mode signal is also amplified to some extent. The common-mode rejection ratio (CMRR) is defined as the ratio of the difference signal voltage gain to the common-mode signal voltage gain. For a good quality differential amplifier, the CMRR should be very large.

Although particularly important to the differential amplifier, the common-mode rejection ratio is a fairly general quality parameter used in most amplifier specifications. The 741 op-amp has a CMRR of 90 dB and the same signal applied to both inputs will give an output approximately 32 000 times smaller than that produced when the signal is applied to only one input line.

10.3.16 Instrumentation amplifier

Instrumentation amplifiers are precision devices having a high input impedance, a low output impedance, a high common-mode rejection ratio, a low level of self-generated noise and a low offset drift. The offset drift is attributable to temperature-dependent voltage outputs. *Figure 10.52* shows a schematic representation of a precision instrumentation amplifier.

The relationship between output and input is:

$$V_o = (R_4/R_3)[1 + 2(R_2/R_1)]V_1 \qquad (10.83)$$

The first two amplifiers appearing in the input stage operate essentially as buffers, either with unity gain or with some finite value of gain.

A number of instrumentation amplifiers are packaged in integrated circuit form and these are suitable for the amplification of signals from strain gauges, thermocouples and other such low-level differential signals from various bridge circuits.

The book by Kaufman and Seidman[6] gives a good practical coverage of the general use of amplifiers.

10.3.17 Power supplies

In Section 10.2.34, the use of p–n junction diodes was illustrated as a means of a.c. voltage rectification. Both the half-wave and full-wave rectification circuits give outputs which, although varying with respect to time, are essentially d.c. insofar as there is no change in polarity. These rectifica-

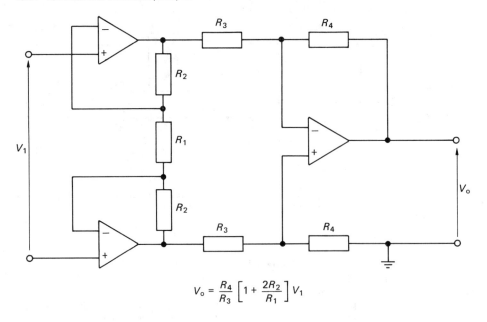

$$V_o = \frac{R_4}{R_3} \left[1 + \frac{2R_2}{R_1} \right] V_1$$

Figure 10.52   Precision instrumentation amplifier

tion circuits provide a first stage in the production of a steady d.c. voltage from an a.c. power supply. However, some further refinements are added to the circuits to reduce the variation, or 'ripple', in the d.c. output voltage. The ripple factor can be greatly reduced by adding a 'reservoir capacitor' (as shown in *Figure 10.53*) which is connected in parallel with the load.

While the supply voltage is positive and rising, the diode is forward biased and will conduct. The capacitor has a short charging time constant and the capacitor voltage initially follows the supply voltage. When the supply voltage reaches its peak value and starts to reduce, the capacitor voltage does not fall so rapidly. The diode therefore becomes reverse biased and no longer conducts. All of the load current is then supplied by the capacitor until such time that the supply voltage again becomes more positive than the capacitor voltage and the diode again conducts. The d.c. load voltage is

thereby 'smoothed', as shown in *Figure 10.53*, and the ripple is effectively reduced. A further reduction in ripple can be achieved by using a full-wave rectification circuit; as in this case there are then twice as many voltage pulses, the capacitor discharge time is halved. The reservoir capacitor is, of necessity, quite large and electrolytic capacitors are often used in this application. A leakage resistor is also frequently connected in parallel with the reservoir capacitor as a safety feature. In the event that the load is disconnected leaving the reservoir capacitor fully charged, the leakage resistor will dissipate the charge safely.

For applications where the reservoir capacitor still cannot reduce the ripple to an acceptable level, an additional ripple filtering circuit may be added. This consists of an additional series ripple resistor and a ripple capacitor (see *Figure 10.54*). The ripple filter is normally interfaced in the circuit between the reservoir capacitor and the load.

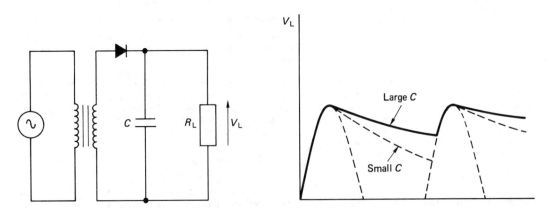

Figure 10.53   Half-wave rectification circuit with reservoir capacitor

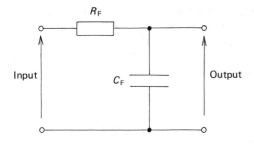

Figure 10.54 Ripple filter

The circuit shown in *Figure 10.54* acts as a low-pass filter where the low-frequency content, or d.c. component, of the input is largely unaffected, but the higher frequencies (i.e. the ripple components) are attenuated. This then provides further smoothing to the d.c. output voltage.

Further enhancements might include a variable resistor either in series, or in parallel with the load. The function of the variable resistor is to allow regulation of the voltage supplied to the load. The Zener diode discussed in Section 10.2.35 is often used in this capacity to provide a stabilised voltage supply.

For high-power systems, thyristors are used in place of diodes as the rectification element. The controlled conduction properties of thyristors allow close control to be exercised on the power supplied to the load.

10.3.18 Analogue and digital systems

Thus far this chapter has been concerned with purely analogue systems in which the circuit currents and voltages are infinitely variable. Digital systems, however, operate between one of two possible states. The two states are 'off' and 'on', conducting and not conducting and, as such, digital systems are essentially discrete in their operation. *Figure 10.55* illustrates the difference between an analogue and a digital system.

Figure 10.55 shows two possible arrangements for a liquid-level indicator. In both cases mechanical refinements have been omitted for clarity. *Figure 10.55(a)* shows an analogue

indicator in which a float is connected to a variable resistor and the output read on a voltmeter. As the liquid level rises, the output voltage v will increase linearly and continuously in direct proportion to the liquid level.

In *Figure 10.55(b)* a series of float switches, at equispaced heights, are connected to a resistor R and wired in parallel as shown. The current drawn from the battery is measured with an ammeter and this forms the basis of the output signal. As the liquid level rises the current drawn will exhibit a step increase every time a float switch is activated and connects another resistor into the parallel circuit. If the liquid level is rising uniformly, then a graph of current against time will exhibit a distinctive staircase form.

In this simple example, the analogue alternative is potentially the more accurate method of monitoring the liquid level. The accuracy of the discrete digital indicator could be improved with the addition of more level switches, but this would become prohibitively expensive. The digital system, however, is intrinsically more immune to 'noise' which might be manifested as surface waves and other disturbances in the liquid level.

10.3.19 Boolean algebra

The basic rules of Boolean algebra are conveniently described with reference to simple manually switched circuits. In binary notation, '0' denotes the switch as 'off', and '1' denotes the switch as 'on'. The '0' and '1' can also be taken to represent the absence, or presence, respectively, of a voltage or a current.

(1) *Logical AND—Figure 10.56* shows a simple AND circuit. Obviously the lamp will light only when both switch A AND switch B are closed. Writing this as a Boolean expression:

$$F = A \text{ AND } B \tag{10.84}$$

where A, B and F are Boolean variables denoting the switches A, B and the lamp, respectively.

The logical operator AND is denoted by a dot thus:

$$F = A \cdot B \text{ or } F = AB \tag{10.85}$$

(2) *Logical OR—Figure 10.57* shows the simple OR circuit. It is clear that the lamp will light in the OR circuit when either switch A OR switch B is closed. As a Boolean expression, the OR function is written:

$$F = A \text{ OR } B$$

i.e. $F = A + B \tag{10.86}$

The + sign is used to denote the logical OR and must not be confused with the arithmetical meaning.

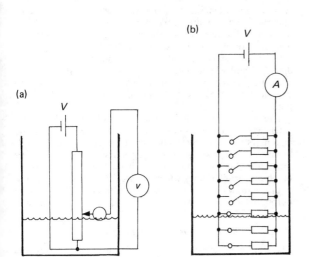

Figure 10.55 Analogue (a) and digital (b) liquid-level indicators

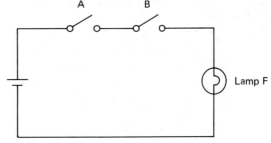

Figure 10.56 Simple AND circuit

The AND and the OR functions are the basic logical functions and quite complex switching circuits can be represented in Boolean form using them.

(3) *Logical NOT*—The NOT function is the inverse complement, or negation of a variable. The negation of the variable A is $\bar{A}$. Thus if $A = 1$, then $\bar{A} = 0$, and vice versa.

(4) *Logical NAND*—The NAND function is the inverse of AND.

(5) *Logical NOR*—In a similar fashion, the NOR function is the inverse of OR.

(6) *Exclusive OR*—In *Figure 10.57* it can be seen that the lamp will also light when both switches A and B are closed. The exclusive OR is a special function which does not enable an output when both switches are closed. Otherwise the exclusive OR functions as the normal OR operator.

The logical functions may also be represented in a tabular form known as a 'truth table'. The truth table indicates the output generated for all possible combinations of inputs. This is illustrated in *Table 10.3* for the AND and NAND operators with three inputs A, B and C.

Using the basic logical functions, the Boolean identities are specified in *Table 10.4*. In *Table 10.4* a '0' can be taken to represent an open circuit, while a '1' represents a short circuit.

Using a truth table, it is easy to prove the validity of various logical expressions by evaluating both sides, e.g.

$$A(B + C) = AB + AC$$

$$(A + B) \cdot (A + C) = A + BC$$

$$A + \bar{A}B = A + B$$

etc.

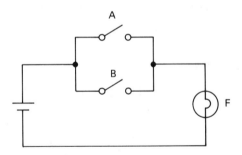

Figure 10.57 Simple OR circuit

Table 10.3 Truth table for AND and NAND operators with three inputs A, B and C

A	B	C	$A \cdot B \cdot C$	$\overline{A \cdot B \cdot C}$
0	0	0	0	1
0	0	1	0	1
0	1	0	0	1
0	1	1	0	1
1	0	0	0	1
1	0	1	0	1
1	1	0	0	1
1	1	1	1	0

Table 10.4 Boolean identities

$A + 0 = A$	$A \cdot A = A$
$A + 1 = 1$	$A + \bar{A} = 1$
$A \cdot 0 = 0$	$A \cdot \bar{A} = 0$
$A \cdot 1 = A$	$\bar{\bar{A}} = A$
$A + A = A$	$\overline{A + B} = \bar{A} \cdot \bar{B}$
	$\overline{A \cdot B} = \bar{A} + \bar{B}$

The first example shows that brackets may be removed by multiplying out as in normal arithmetic. The second two examples have no arithmetic counterpart.

A useful manipulation technique is De Morgan's theorem which states that in any logical expression, AND can be replaced by OR, and vice versa, provided that each term is also replaced with its inverse complement. The resulting expression is then the inverse of the original.

Example 1 From ABC we negate to:

$$\overline{ABC} = \bar{A} + \bar{B} + \bar{C}$$

Hence

$$ABC = \overline{\bar{A} + \bar{B} + \bar{C}}$$

Example 2 From $F = AB + CD$ we negate to:

$$\bar{F} = (\bar{A} + \bar{B}) + (\bar{C} + \bar{D})$$

Applying De Morgan again

$$F = \overline{(\bar{A} + \bar{B}) \cdot (\bar{C} + \bar{D})}$$

The equivalence of the original and the final expressions in the above two examples may be checked using a truth table.

10.3.20 Digital electronic gates

The principles of Boolean algebra have been considered with respect to manually switched circuits. In modern digital systems the switches are formed with transistors for speed of operation and they are generally referred to as 'gates'. Over the years various technologies have been developed in the manufacture of logic gates. The earliest forms of electronic gate were based on the unidirectional conduction properties of diodes. Diode logic gates have now been superseded by transistor–transistor logic (TTL) gates or the more recent complementary metal oxide semiconductor (CMOS) family of logic gates.

The internal construction and operation of modern logic gates may be quite complex, but this is of little interest to the digital-systems designer. Generally, all that the designer needs to know is the power supply voltages, the transient switching times, the 'fan out' and the 'fan in'. 'Fan out' refers to the number of similar gates which can be driven from the output of one gate, whilst 'fan in' refers to the number of similar gate outputs which can be safely connected to the input of one gate.

10.3.20.1 Transistor–transistor logic gates

The TTL family is based on the bipolar junction transistor and was the first commonly available series of logic elements. TTL logic gates are rapid switching devices, the SN7400 for example takes just 15 ns to change state. The standard power supply is 5 V with a low tolerance band of ±0.25 V. This low tolerance necessitates a reliable power supply regulation which is reasonably facilitated through any one of the wide variety of supply regulators which are now available in integrated circuit form. For the SN74 series TTL integrated circuits, the fan out is about 10.

A TTL based system can draw quite large instananeous loads on a power supply and this can result in substantial interference 'spikes' in the power lines. Since the spikes can upset the normal operation of the system it is common practice to connect small capacitors directly across the power lines, as close to the TTL integrated circuits as possible. One capacitor (0.1–10 μF) per five integrated circuits is sufficient in most instances.

TTL circuits are continually being improved and a major recent advance has been the introduction of the low power 'Schottky' TTL circuits. These use the same generic code numbers as the standard series, but have 'LS' inserted before the type code, e.g. SN74LS00. The operating speed is about twice as fast and the power consumption is about 20% of the standard series. Schottky devices are, however, slightly more expensive.

10.3.20.2 Complementary metal oxide semiconductors

The problematic features of the power supply associated with the TTL family of logic devices has been largely responsible for the growth of its major competitor, CMOS. CMOS integrated circuits are based on the field effect transistor and can operate off a range of power supply voltages between ±3 V and ±18 V. CMOS devices dissipate very little power, are very cheap and are simple in operation. The fan out is about 50 and they have a far greater immunity to power-supply noise. The noise immunity of CMOS devices means that there is no requirement for smoothing capacitors to the extent that they are generally found in TTL circuitry.

There are also some disadvantages associated with CMOS devices. The main one being that CMOS is slower than TTL, roughly about one-tenth of the equivalent TTL circuit. CMOS integrated circuits are also very sensitive to electrostatic voltages. Manufacturers do build in some safety features to reduce the electrostatic sensitivity, but CMOS devices must still be handled with due care. A brief comparison of TTL and CMOS devices is given in *Table 10.5*.

10.3.21 Gate symbols

Having defined a system output in terms of a Boolean expression, the actual circuit can be constructed using the required gates selected from the logic family chosen. In general, the design will be centred round the more readily available NAND and NOR logic gates. In laying out a gate interconnection diagram, standard symbols are used to represent the individual gates. Unfortunately, no universal set of

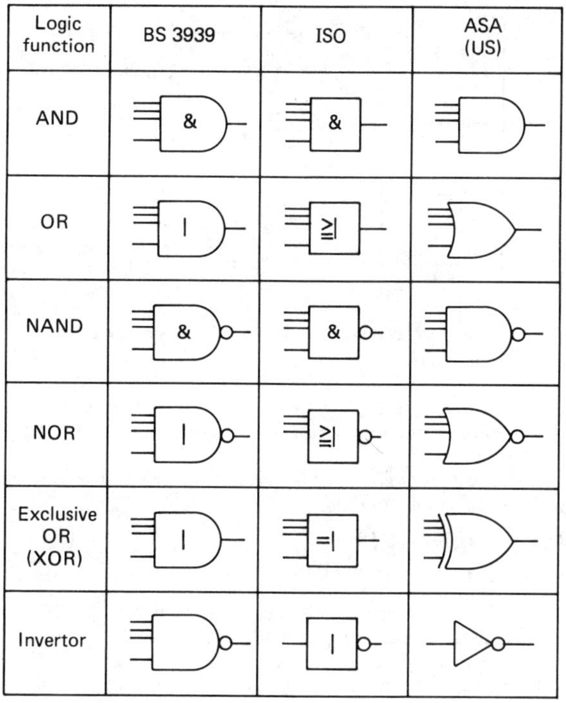

Logic function	BS 3939	ISO	ASA (US)
AND	&	&	
OR		≥	
NAND	&	&	
NOR		≥	
Exclusive OR (XOR)		=1	
Invertor		1	

Figure 10.58 Gate symbol systems in current usage

symbols has emerged and several systems are in current usage. *Figure 10.58* summarises the most common gate symbol systems.

10.3.22 Logic systems using simple gates

A vending machine which dispenses either tea or coffee serves as an illustrative example. The logic circuit may be realised using AND gates as shown in *Figure 10.59*. The money input is common to both gates. Although workable, the system has a minor fault in that if both buttons are pressed, after the money criterion is satisfied, then the output will be both tea and coffee. This fault can be designed out of the system by extending the logic circuit as shown in *Figure 10.60*. The extended system incorporates a NAND gate and an additional AND gate. If both buttons are now pressed then the output from G_3 will be 0. With the output 1 from G_1, the output from G_4 will be 0 and the machine will dispense tea. On pressing either button on its own and satisfying the money input

Table 10.5 Comparison of TTL and CMOS devices

Property	TTL	CMOS
Power supply	5 ± 0.25 V	3–18 V d.c.
Current required	Milliamperes	Microamperes
Input impedance	Low	Very high
Switching speed	Fast to 10 ns	Slow to 300 ns
Fan out	10	50

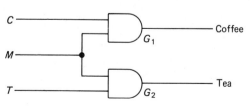

Figure 10.59 Logic circuit for a drinks vending machine

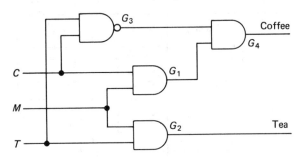

Figure 10.60   Extended logic circuit for a drinks vending machine

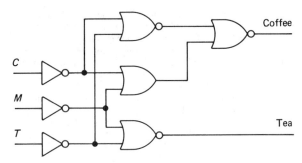

Figure 10.61   Logic circuit for a drinks vending machine using OR and NOR gates

criterion, the correct drink will be output. The operation of the extended system is verified in the truth table shown in *Table 10.6*.

By inspection of *Figure 10.60* the system can be represented in Boolean expressions as:

$$\text{Coffee} = (CM) \cdot \overline{(CT)} \tag{10.87}$$

$$\text{Tea} = TM \tag{10.88}$$

where C, T and M represent the coffee button, tea button and money input, respectively, and the overscore represents the inverse complement.

Using De Morgan's theorem, the system may alternatively be written as:

$$\text{Coffee} = \overline{(\bar{C} + \bar{M})} + \overline{(\bar{C} + \bar{T})} \tag{10.89}$$

$$\text{Tea} = \overline{\bar{T} + \bar{M}} \tag{10.90}$$

Thus the same logic system can be implemented using one OR and three NOR gates, as shown in *Figure 10.61*. The validity and equivalence of equations (10.87) to (10.90) may easily be checked using a truth table.

Four logic gates are again required but the circuit operates with inverted input signals. This means that three inverters are also required in the circuit as shown (*Figure 10.61*).

It is apparent that the logical function can be realised in several different ways, e.g.

$$\text{Coffee} = \overline{(\bar{C} + \bar{M})} \cdot (\bar{C} + \bar{T})$$

and

$$\text{Tea} = TM$$

Using the above realisation, the circuit takes the form shown in *Figure 10.62*.

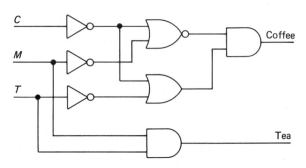

Figure 10.62   Alternative logic circuit for a drinks vending machine

10.3.23   Logic systems using NAND and NOR gates only

Logic gates are packaged as arrays of the same gate type in integrated circuit form. A typical example is SN7408, which is a 14-pin DIL package containing four separate two-input AND gates. Because the logic gates are marketed in this particular form, it is advantageous to design the logic circuit using only one type of gate. This normally minimises the number of integrated circuit packages required.

Figure 10.63 shows a two-input NAND gate driving into a single-input NAND gate. For the two-input NAND gate, the Boolean expression is:

$$F = \overline{A \cdot B}$$

Table 10.6   Truth table for a drinks vending machine

Inputs			Outputs	
C	M	T	Coffee	Tea
0	0	0	0	0
0	0	1	0	0
0	1	0	0	0
0	1	1	0	1
1	0	0	0	0
1	0	1	0	0
1	1	0	1	0
1	1	1	0	1

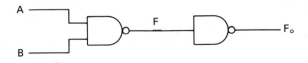

A	B	$A.B$	$\overline{A.B}$	$\overline{\overline{A.B}}$
0	0	0	1	0
0	1	0	1	0
1	0	0	1	0
1	1	1	0	1

Figure 10.63   AND realisation using NAND gates

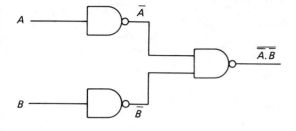

A	B	$\bar{A}$	$\bar{B}$	$\bar{A}.\bar{B}$	$\overline{\bar{A}.\bar{B}}$
0	0	1	1	1	0
0	1	1	0	0	1
1	0	0	1	0	1
1	1	0	0	0	1

Figure 10.64 OR realisation using NAND gates

As F is then fed into a single-input NAND gate, which operates as an inverter, the final output is:

$$F_o = \bar{\bar{F}} = \overline{\overline{A \cdot B}} = A \cdot B$$

It is apparent, therefore, that the circuit shown in *Figure 10.63*, using NAND gates, performs the same function as the logical AND operator.

Figure 10.64 shows two single-input NAND gates with their outputs driving into a two-input NAND gate.

Following through the truth table, it can be seen that the circuit performs the logical OR function. If the output F is then fed to another single-input NAND gate (not shown in the figure), then the function performed will be a logical NOR. Thus it can be seen that suitable combinations of NAND gates can be made to perform the logical functions AND, OR and NOR. In a similar manner, it can be shown that the AND and

OR functions can be realised using NOR gates only. This is illustrated in *Figure 10.65*. The conclusion which can be drawn is that any logic circuit can be realised using NAND gates, or NOR gates alone.

Considering again the drinks vending machine depicted in *Figure 10.61*, the single OR gate may be replaced with a two-input NOR gate which then feeds directly into a single-input NOR gate. This arrangement is shown in *Figure 10.66*. Note that NOR gates are also used in place of invertors in the input signal lines. By inspection of the circuit diagram, the governing Boolean expressions are:

$$\text{Coffee} = \overline{\overline{(\bar{C} + \bar{M})} + \overline{(\bar{C} + \bar{T})}}$$
$$= (\bar{C} + \bar{M}) + (\bar{C} + \bar{T}) \quad (10.91)$$
$$\text{Tea} = \bar{\bar{T}} + \bar{M} \quad (10.92)$$

Equations (10.91) and (10.92) are identical to equations (10.89) and (10.90), respectively. This, of course, must be true because the circuits from which the expressions were deduced perform identical logical functions.

Similarly, the circuit shown in *Figure 10.60*, involving one NAND and three AND gates, may be replaced by an equivalent circuit using only NAND gates. This equivalent circuit is shown in *Figure 10.67*. Inspection of the circuit gives the Boolean expressions:

$$\text{Coffee} = \overline{\overline{(C \cdot M)} \cdot \overline{(C \cdot T)}}$$
$$= (C \cdot M) \cdot \overline{(C \cdot T)} \quad (10.93)$$
$$\text{Tea} = \overline{(T \cdot M)} = TM \quad (10.94)$$

Perhaps as expected, the Boolean expressions are identical to equations (10.87) and (10.88), which were deduced from the logic circuit shown in *Figure 10.60*.

The realisation of Boolean expressions in either all NAND, or all NOR, gates can be achieved by following two simple rules.

(1) *NAND realisation*—First obtain the required Boolean expression in AND/OR form and construct the circuit required. The final output gate must be an OR gate.

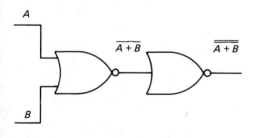

A	B	$\overline{A+B}$	$\overline{\overline{A+B}} = A+B$
0	0	1	0
0	1	0	1
1	0	0	1
1	1	0	1

A	B	$\bar{A}$	$\bar{B}$	$\overline{\bar{A}+\bar{B}} = A \cdot B$
0	0	1	1	0
0	1	1	0	0
1	0	0	1	0
1	1	0	0	1

Figure 10.65 OR and AND realisations using NOR gates only

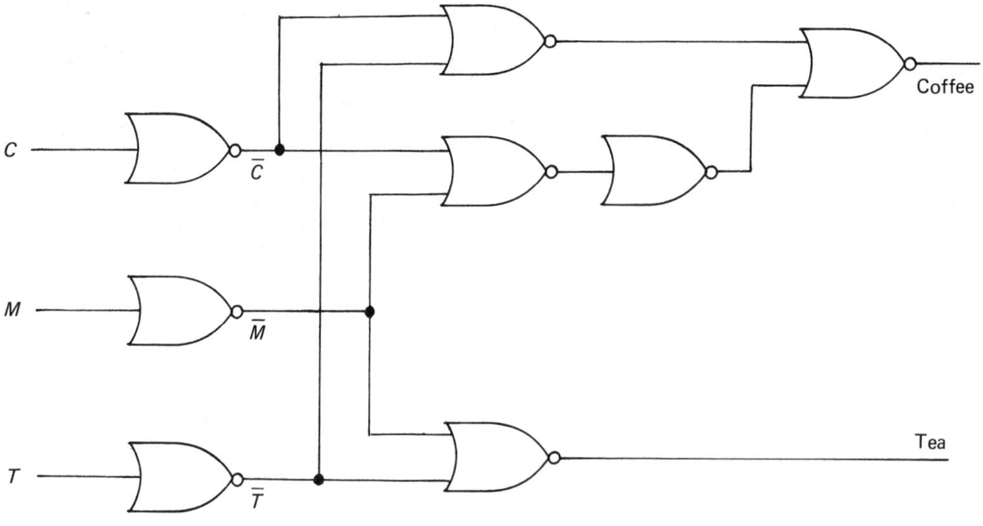

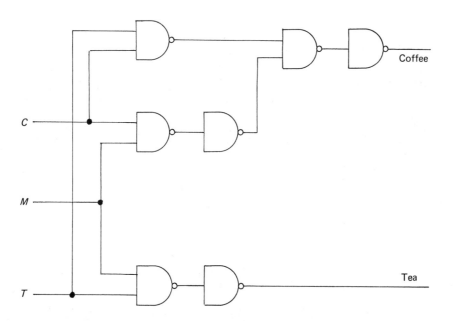

Figure 10.66 Logic circuit for a drinks vending machine using NOR gates only

Figure 10.67 Logic circuit for a drinks vending machine using NAND gates only

Replace all gates with NAND gates and, starting with the output gate, number each level of gates back through to the inputs. The logic level at the inputs to all 'odd' level gates must be inverted.

(2) *NOR realisation*—Obtain the required Boolean expression in OR/AND form. The final output gate must be an AND gate. Replace all gates with NOR gates and number each level of gates from the output back through to the input. The logic level at all inputs to 'odd' level gates must be inverted.

Application of these rules is best illustrated by an example.

Take the NAND realisation of $F = AB + C(D + E)$. *Figure 10.68* shows the realisation of the function in AND/OR form.

As inputs D and E appear at an odd level of gate input, they must be inverted. In terms of the actual circuit this will mean that inputs D and E are inverted, using NAND gates, prior to entering the NAND gate at level 4.

A similar procedure is adopted for a NOR realisation of a Boolean expression. The exclusive-OR function serves as an interesting example. Written as a Boolean expression, the exclusive-OR function is:

$$F = A \cdot \bar{B} + \bar{A} \cdot B \qquad (10.95)$$

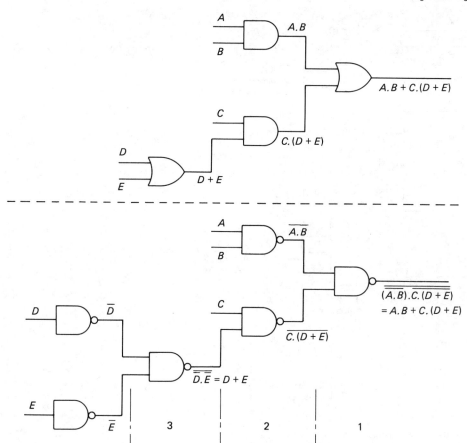

Figure 10.68 NAND realisation of a Boolean function

For the NOR realisation, however, it is necessary that the final output gate is an AND. The exclusive-OR function must therefore be manipulated such that the final logical function in the expression is an AND. Using De Morgan's theorem:

$$F = \overline{(A \cdot \bar{B}) \cdot (\bar{A} \cdot B)}$$

$$= (\bar{A} + B) \cdot (A + \bar{B})$$

Multiplying this expression out gives:

$$F = \bar{A} \cdot A + \bar{A} \cdot \bar{B} + A \cdot B + B \cdot \bar{B}$$

Since $A \cdot \bar{A} = B \cdot \bar{B} = 0$, the expression simplifies to:

$$F = \bar{A} \cdot \bar{B} + A \cdot B$$

$$= \overline{(\bar{A} \cdot \bar{B}) \cdot \overline{(A \cdot B)}}$$

Using De Morgan again gives:

$$F = (A + B) \cdot (\bar{A} + \bar{B}) \tag{10.96}$$

The realisation of equation (10.96) is shown in *Figure 10.69*.

10.3.24 Unused inputs

Multi-input gates are also commonly available. In practical circuits, however, it is important that any unused inputs are tied, i.e. they are either connected to the positive voltage supply, or to the zero voltage supply. The unused inputs are therefore set at either logic level 1, or at logic level 0, as required. In connecting an unused input to the positive supply, the connection should be made through a 1 kΩ resistor. Failure to connect any unused inputs can result in intermittent malfunction of the circuit, or in harmful oscillations with attendant overheating.

10.3.25 Latches

It is often useful to 'freeze' a particular binary sequence and devices called 'latches' are used for this purpose. A latch has four inputs and four outputs. Normally the outputs assume the same state as the inputs. However, when a control signal, known as a 'strobe' input, is taken to logic '1', the outputs are locked in whatever state they were in at the instant of the strobe input going high. This enables the binary sequence to be 'captured' without affecting the on-going processes, whatever they may be. The latch therefore serves as a temporary state recording device which may subsequently be referred to during various interrupt operations.

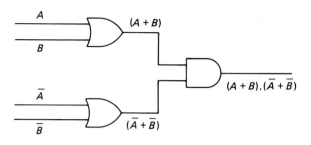

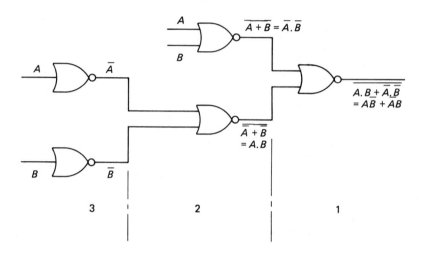

Figure 10.69 NOR realisation of the exlusive-OR function

10.3.26 The Karnaugh map

The Karnaugh map provides an alternative representation of a Boolean expression for all possible Boolean input combinations. In some respects the Karnaugh map is like a truth table in that identical logical expressions display an identical pattern on a Karnaugh map. The Karnaugh map, however, also has great utility in simplifying Boolean expressions in a systematic manner.

The Karnaugh map consists of a set of boxes in which each box represents one possible combination of the Boolean input variables. The boxes are assigned either a '1' or a '0' to indicate the value of the Boolean expression for the particular combination of input variables that the box represents. The number of boxes required is 2^n, where n is the total number of input variables. Although any number of input variables can be represented, a practical limitation is about seven. *Figure 10.70* shows the Karnaugh map for a four input system.

Within each box the unique Boolean input combination is represented by assigning each variable the logic values indicated along the horizontal and vertical axes. These values conform to the binary Gray code in which adjacent consecutive characters differ only in one variable. This imparts a property to the Karnaugh map in that adjacent squares, vertically or horizontally, differ only in one variable.

As an example, the Boolean expression

$$F = \bar{A}\bar{B}CD + A\bar{B}C\bar{D} + \bar{A}B\check{C}D$$

is represented by the Karnaugh map given in *Figure 10.71*.

The maps are drawn up by placing a '1' in each box for which the combination of input variables makes the logical expression have a value of 1. All the other boxes represent the combination of input variables which make the expression have a logical value of 0. The '0' is not usually entered in the box.

A second example for consideration is

$$F = \bar{A}\bar{B}CD + AC + \check{C}\bar{D}$$

CD \ AB	00	01	11	10
00	$\bar{A}\,\bar{B}\,\bar{C}\,\bar{D}$	$\bar{A}\,B\,\bar{C}\,\bar{D}$	$A\,B\,\bar{C}\,\bar{D}$	$A\,\bar{B}\,\bar{C}\,\bar{D}$
01	$\bar{A}\,\bar{B}\,\bar{C}\,D$	$\bar{A}\,B\,\bar{C}\,D$	$A\,B\,\bar{C}\,D$	$A\,\bar{B}\,\bar{C}\,D$
11	$\bar{A}\,\bar{B}\,C\,D$	$\bar{A}\,B\,C\,D$	$A\,B\,C\,D$	$A\,\bar{B}\,C\,D$
10	$\bar{A}\,\bar{B}\,C\,\bar{D}$	$\bar{A}\,B\,C\,\bar{D}$	$A\,B\,C\,\bar{D}$	$A\,\bar{B}\,C\,\bar{D}$

Figure 10.70 Karnaugh map for a four-input system

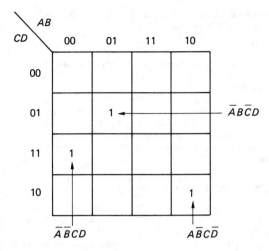

Figure 10.71 Karnaugh map for the Boolean expression
$F = \bar{A}\bar{B}CD + A\bar{B}C\bar{D} + \bar{A}\bar{B}C\bar{D}$

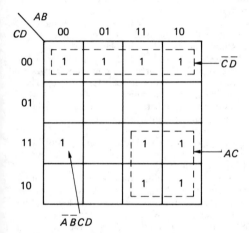

Figure 10.72 Karnaugh map for the Boolean expression
$F = \bar{A}\bar{B}CD + AC + \bar{C}\bar{D}$

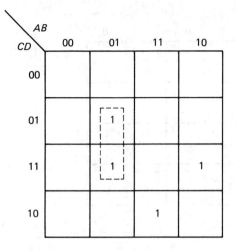

Figure 10.73 Karnaugh map for the arbitrary expression
$F = \bar{A}B\bar{C}D + \bar{A}BCD + ABC\bar{D} + A\bar{B}CD$

The Karnaugh map for this expression is shown in *Figure 10.72*. It can be seen that the term AC includes all four squares in which both A and C are included. Similarly, the term $\bar{C}\bar{D}$ also encompasses four squares on the map. It may be concluded that in a four-variable expression, any term which contains the four variables will occupy one square on the Karnaugh map. Any term which contains only three of the variables will occupy two squares, and any term which contains only two of the variables will occupy four squares. A term containing only one of the variables will occupy eight squares in the Karnaugh map.

The Karnaugh map may be used in a reverse mode to deduce the Boolean expression. This technique is applied to the map given in *Figure 10.73*. The expression may be read off as:

$$F = \bar{A}B\bar{C}D + \bar{A}BCD + ABC\bar{D} + A\bar{B}CD$$

Alternatively, there is an obvious grouping of '1's which can be taken together giving:

$$F = \bar{A}BD + ABC\bar{D} + A\bar{B}CD$$

In grouping '1's like this, the term (or terms) which is dropped out is always the one which is represented both as 0 and 1 within the grouping. The procedure results in two possible Boolean expressions, both of which are correct. The second expression, however, is simpler and this technique forms the basis of using the map to minimise Boolean expressions.

10.3.27 Minimisation of Boolean expressions

The principle of minimisation is based on the Boolean identity $A + \bar{A} = 1$. Thus

$$F = ABCD + ABC\bar{D} = ABC(D + \bar{D}) = ABC \qquad (10.97)$$

The grouping of squares along any axis therefore enables the minimisation which is typified by equation (10.97). An extension of this principle is shown in *Figure 10.74*. The Boolean expression depicted in the figure can be written as:

$$F = ABC\bar{D} + ABC\bar{D} + \bar{A}BCD + A\bar{B}CD$$
$$= AB\bar{D}(C + \bar{C}) + \bar{B}CD(\bar{A} + A)$$
$$= AB\bar{D} + \bar{B}CD$$

Minimisation in the above examples reduces the four terms in the expression to two terms, each involving three variables. The groupings in the example are akin to the idea of rolling the map into a cylinder about either axis to complete the two groupings as shown.

In extending the minimisation principle to five variables, the number of squares required is $2^5 = 32$. This is best handled as two sets of 16 squares in a top and bottom arrangement. The 16 square layers represent the first four input variables and each layer accommodates the two possible input combinations for the fifth variable. Higher numbers of input variables can be dealt with, but the map becomes increasingly more difficult to handle.

In certain situations involving a number of input variables, particular combinations of the variables never actually occur in practice. Under these circumstances the output which

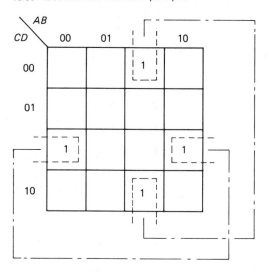

Figure 10.74 Extended minimisation principle

would occur with these combinations of variables is irrelevant. The output can therefore have any value since it is a never occurring situation. Such input combinations are called 'don't care' conditions and they can be incorporated in a system to allow a simpler circuit realisation. The principle can be illustrated by means of an example.

For the expression

$$F = \bar{A}B\bar{C}D + AB\bar{C}D + \bar{A}BCD$$

it is stated that the combination $ABCD$ will never occur. Including the don't care condition in the expression gives

$$F = \bar{A}B\bar{C}D + AB\bar{C}D + \bar{A}BCD + \{ABCD\}_x \qquad (10.98)$$

The don't care combination is usually enclosed within brackets with a subscript x or 0.

The Karnaugh representation for the expression is shown in *Figure 10.75*. The don't care condition is clearly indicated in the figure. By ignoring the don't care condition, the minimisation of the expression results in:

$$F = B\bar{C}D + \bar{A}BD \qquad (10.99)$$

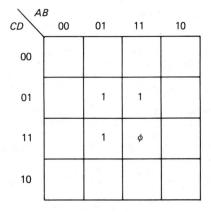

Figure 10.75 Karnaugh representation of
$F = \bar{A}B\bar{C}D + AB\bar{C}D + \bar{A}BCD + \{ABCD\}_x$

If the network output is allowed to be 1 for the don't care condition, the minimisation yields:

$$F = BD \qquad (10.100)$$

The example evidently shows that considerable savings in the realisation of an expression can be made by including a relevant don't care condition.

It is also worth bearing in mind that, although the Karnaugh map can yield a minimum gate solution to a given problem, it might not be an optimum solution. In the real world other considerations may well dictate in terms of parts, design, assembly costs and the number of integrated circuit packages required.

10.3.28 Positive and negative logic

In the above discussion of digital logic systems, no mention has been made of the significance of the logic levels in terms of the actual voltages applied. Two possibilities exist to differentiate between logic 1 and 0. In a positive logic system, logic level 1 is represented by a more positive voltage level than logic level 0. Both logic voltage levels could actually be negative, but many digital systems operate with a voltage of 0–0.8 V denoting logic level 0, and a voltage of 2.4–5 V denoting logic level 1. This standard is used in the TTL and CMOS series of logic devices.

In a negative logic system, logic level 1 is represented by a less positive voltage than logic level 0. This standard applies to data transmission interfaces where a voltage in the range −3 to −15 V denotes logic 1 and a voltage in the range +3 to +15 V denotes logic 0. The large differentiation between 0 and 1 ensures good immunity to electrical noise. These voltages, however, are not compatible with TTL and CMOS devices and interconversion integrated circuits are required within the data transmission interface.

As an alternative to using the terms logic 1 and logic 0, the terms 'high' and 'low' are often substituted. In a positive logic system a transition from logic 0 to logic 1 can be termed 'a transition from low to high'.

The logic level definitions also influence the function of the logic device. *Figure 10.76* shows two types of two input NOR gates. In *Figure 10.76(a)* the inputs are negative logic and the output is positive logic. The NOR gate therefore performs the logical AND function. In *Figure 10.76(b)* the inputs are positive logic while the output is negative logic. This NOR gate therefore performs the logical OR function.

10.3.29 Tri-state logic

Tri-state logic does not represent three logic levels but denotes three states which may be logic 1, logic 0, or 'unconnected'. A separate 'enable' input determines whether the output behaves as a normal output, or goes into the third, open-circuit

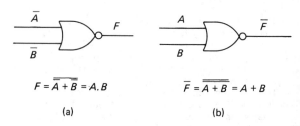

Figure 10.76 NOR gates using (a) negative and (b) positive logic input–output systems

state. Tri-state devices are used in applications where different logic devices are required to be connected into output lines which are common to other logic devices, for example computer data buses. While one set of logic devices is transmitting signals, the other set of devices is temporarily disconnected, or disabled.

10.3.30 Sequential logic circuits

The logic circuits so far considered are all examples of combinational logic systems where the output is determined by the combination of input variables present at that time. Sequential logic circuits are those in which the outputs depend on the sequence of prior inputs. The main difference between sequential and combinational logic systems is that the former circuits must possess some semblance of 'memory'. The basic memory element in sequential logic systems is provided by one of several 'bistable' gates, so called because of the two different, but stable, outputs which the gates produce.

10.3.30.1 The SR (bistable) flip flop

The term 'flip flop' is traditionally used with respect to basic memory elements, and in the term 'SR flip flop' the 'S' denotes 'set' and 'R' denotes 'reset'. The SR flip flop was an early development, commonly constructed using discrete transistors. The internal operation, in which two transistors alternate between the cut-off and saturated states, is of less importance than the external function which the device performs. Using the systems approach, the SR flip flop can be represented as shown in *Figure 10.77*.

The system shows the two inputs S and R and the two output lines traditionally denoted as Q and $\bar{Q}$. For sequential circuits the truth table is more usually called a state table. The state table for the SR flip flop is given in *Table 10.7*.

Each set of input variable values is considered for both possible states of the output. This is necessary because the output values do not depend uniquely on the input variable

values, but also on the current values of the outputs themselves.

The operation of the SR flip flop may be summarised as follows.

(1) With $S = 0$ and $R = 0$, the output is not affected and remains as it was.
(2) With $S = 1$ and $R = 0$, the output will change to $Q = 1$ if previously Q was 0. Q will remain at 1 if previously Q was 1.
(3) With $S = 0$ and $R = 1$, the output will change to $Q = 0$ if previously Q was 1. Q will remain at 0 if previously Q was 0.
(4) In all cases considered, the output $\bar{Q}$ will be the inverse complement of Q.

The SR flip flop may be constructed using cross-coupled NOR, or NAND gates as shown in *Figure 10.78*.

10.3.30.2 The T (trigger) flip flop

The T flip flop is another bistable circuit having two outputs (Q and $\bar{Q}$) but only one input (T). The T flip flop changes state on every T input signal and then remains in that state while the T input remains low.

10.3.30.3 The JK flip flop

The JK flip flop uses integrated circuit technology and, since it can perform both of the SR and T flip flop functions, it has become the most common flip flop in current use. *Figure 10.79* gives the state table and logic symbol for the JK flip flop.

The state table is identical to that for the SR flip flop, with the exception that the input condition $J = 1$, $K = 1$ is allowed. For these latter inputs the JK flip flop functions as a T flip flop using an input clock signal, in the form of a pulse train, as the trigger.

The JK flip flop operates in a clocked, or synchronous mode. In synchronous mode, the J and K inputs do not in themselves initiate a change in the logic outputs, but are used to control inputs to determine the change of state which is to

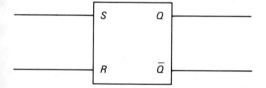

Figure 10.77 The SR flip flop

Table 10.7 State table for the SR flip flop

Input		Output change	
S	R	$Q_n \to Q_{n+1}$	$\bar{Q}_n \to \bar{Q}_{n+1}$
0	0	$0 \to 0$	$1 \to 1$
0	0	$1 \to 1$	$0 \to 0$
0	1	$0 \to 0$	$1 \to 1$
0	1	$1 \to 0$	$0 \to 1$
1	0	$0 \to 1$	$1 \to 0$
1	0	$1 \to 1$	$0 \to 0$
1	1	Not available	
1	1		

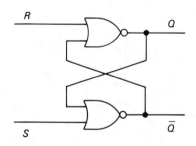
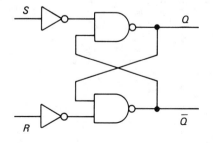

Figure 10.78 The SR flip flop using cross-coupled gates

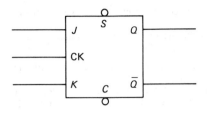

Inputs		Outputs
J	K	$Q_n - Q_{n+1}$
0	0	$0 \to 0$
0	0	$1 \to 1$
0	1	$0 \to 0$
0	1	$1 \to 0$
1	0	$0 \to 1$
1	0	$1 \to 1$
1	1	$0 \to 1$
1	1	$1 \to 0$

Figure 10.79   The JK flip flop and the corresponding state table

occur. A pulsed input to the clock terminal (CK) then determines the timing of the state changes. The clocked mode allows for precise timing of the state changes in a sequential circuit.

JK flip flops may also be provided with additional 'set' (S) and 'clear' (C) inputs which can be used to set output Q to 1, or clear output Q to 0 at any time.

Multiple J and K inputs are also commonly available to enable logical ANDing of multiple input signals.

A slightly more complicated flip flop arrangement is the JK master–slave flip flop. This consists of a pair of SR flip flops connected together by various logic gates as shown in *Figure 10.80*.

The JK master–slave flip flop differs from the simpler arrangement in that if the clock pulse is at logic 1, a logic 1 applied to either J or K will not set the outputs. The new data, however, are accepted by the master. When the clock pulse returns to 0, the master is isolated from the inputs but its data are transferred to the slave with the result that Q and $\bar{Q}$ can then change state. In a circuit involving many such flip flops, the advantage of the master–slave arrangement is that it can allow for synchronisation of all the output state changes.

10.3.31   Registers and counters

In the previous section it was shown that a logic level of 1 on a particular input line to a flip flop can set an output line to 1. In this way the flip flop can perform an elementary memory function. For a binary signal of length n bits, n flip flops are required to construct a memory device for the n-bit input signal. A group of flip flops used together in this manner constitutes a 'register'.

Data may be entered into the register in a serial, or a parallel manner. In the parallel method the n-bit binary 'word' is available on n input lines. Each line is connected to its own flip flop and the n data bits are entered simultaneously into the register. In the serial entry method, the data are available on only one input line in a time sequence. The data are entered consecutively and are timed into the register by a system clock. Serial entry registers are also called 'shift registers' since the data bits are entered into the first flip flop and moved consecutively along into the next flip flop as the next data bit arrives at the first flip flop and so on. The serial method of data entry requires as many shift and store operations as the number of bits in the binary word. This means that the serial entry method is much slower than the parallel method. Serial entry, however, is much less expensive than parallel entry.

Yet another type of register is the counting register. This consists of a number of flip flops arranged to store a binary

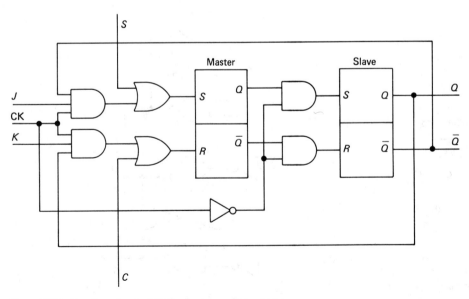

Figure 10.80   The JK master–slave flip flop

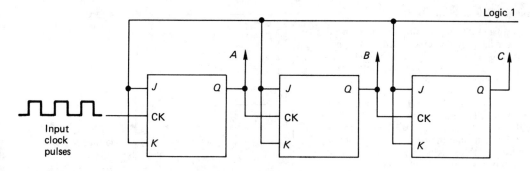

Figure 10.81 Asynchronous 3-bit binary counter

word which is representative of the number of input pulses applied at the input terminal. Using n flip flops, a total count of 2^n can be made.

Counting registers, or 'counters', may be synchronous in which the state changes in all the flip flops occur simultaneously, or asynchronous in which the state changes in various flip flops do not occur at the same time. *Figure 10.81* illustrates an asynchronous, 3-bit binary counter composed of JK flip flops. All J and K inputs are held at logic level 1 and the input signal consists of a pulse train fed to the clock input of the first flip flop. In this mode the JK flip flop is operating as a T flip flop. The output Q from the first flip flop provides an input for the clock of the second flip flop and so on through the network. The outputs Q also form the binary representation of the counter where A is the least significant bit, increasing through to C which represents the most significant bit. The state table and timing diagram for the counter are shown in *Figure 10.82*.

The state table shows that each flip flop changes state when the next less significant flip flop output changes from 1 to 0. The output signal from each flip flop, moving through the network, is at half the frequency of that of the previous flip flop. These output signals thus provide the correct binary count of the number of input pulses applied to the input. The 3-bit binary counter can count up to a maximum of eight decimal places. If a ninth pulse is applied at the input, the count reverts back to the initial zero setting and the count continues again as normal for further input pulses.

Asynchronous counters are also referred to as 'ripple' counters because of the way that the changes of state ripple through the network of flip flops.

A synchronous version of the counter can also be realised using a network of JK flip flops. The synchronous counter additionally uses the outputs Q and $\bar{Q}$ of each flip flop, in logic gate networks, to produce the necessary control signals for the J and K inputs. This ensures that all flip flops change state correctly to the desired state table for each clock pulse. The synchronous counter alleviates the problems associated with transient operation inherent in the asynchronous counter.

There are of course many other types of flip flop available, but the only other one of significant practical importance is the D flip flop. This is shown in *Figure 10.83*, where the 'D' refers to 'data'. In the D type flip flop, the D input is fed directly into the J input line and the inverse complement of D is fed to the K input line. This ensures that J and K are always the inverse complement of one another. A logic 0 or 1 on the data input will then flip, or flop, the outputs when the clock pulse is at logic 0.

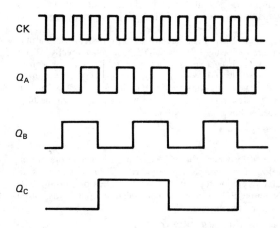

Clock pulses	Flip-flop		
	C	B	A
0	0	0	0
1	0	0	1
2	0	1	0
3	0	1	1
4	1	0	0
5	1	0	1
6	1	1	0
7	1	1	1
0	0	0	0

Figure 10.82 State table and timing diagram for a 3-bit binary counter

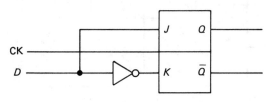

Figure 10.83 D type flip flop

10.3.32 Timers and pulse circuits

An essential feature of the flip flop circuits described in the previous two sections is the provision of a pulsed clock signal. Although timers can be designed using discrete components, it is normal to design round the commonly available timers which are already available in integrated circuit form. The most prevalent timer currently in use is the NE555, which is manufactured as an eight-pin dual-in-line (DIL) package. The CMOS equivalent to the TTL based NE555 is the ICM7555. These packages are essentially identical and ICM7555 can be used to replace NE555, although the converse is not always applicable. The so-called '555' timer is very versatile and can be used in either 'monostable', or 'astable' mode.

10.3.32.1 Monostable

Figure 10.84 shows the 555 wired up for monostable operation. When the 'trigger' is taken from +5 to 0 V (i.e. high to low), the output will go high for a period determined by the values selected for R and C. The length of the output pulse is given by $1.1RC$.

The timer can deliver currents of more than 100 mA and it can therefore be used to drive a DIL reed relay directly. When such a relay is switched off, however, the back e.m.f. generated by the relay coil could damage the timer. As a precaution a diode is normally connected in parallel with the relay coil, in the opposite direction to the current flow, to absorb the high induced voltage.

10.3.32.2 Astable

Figure 10.85 depicts the 555 wired up for astable operation. The 100 nF capacitor is only required for TTL based timers. In astable operation the output is a continuous pulse train. The 'ON' and 'OFF' times can be controlled independently within certain limitations with:

$$\text{ON time} = 0.693(R_1 + R_2)C \qquad (10.101)$$

$$\text{OFF time} = 0.693(R_2)C \qquad (10.102)$$

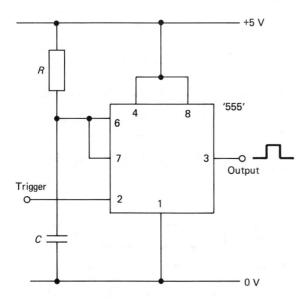

Figure 10.84 The 555 timer in monostable operation

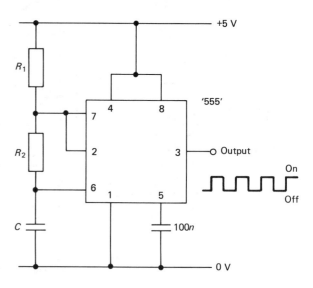

Figure 10.85 The 555 timer in astable operation

Obviously the ON time can only be equal to, or greater than the OFF time. The output signal, however, can always be inverted if in a particular application short duration positive pulses are required.

The maximum operating frequency for the 555 timer is about 500 kHz and the minimum frequency, limited by the leakage of the capacitor, is about one cycle per several hours.

Alternative pulsed output circuits can be constructed using TTL or CMOS gates (see Kaufman and Seidman[6] and Watson[7]).

10.3.33 Digital computers and microprocessors

No coverage of digital electronics, however brief, can fail to give some cognisance to the impact of the digital computer and its associated microprocessor. The modern digital computer, although a complex digital system, consists of no more than the basic logical subsystems previously discussed. This includes AND, OR, NAND and NOR gates, registers, counters and communication interfaces. Space limitations do not allow a detailed description of computer systems and microprocessors to be given in this chapter. However, Chapter 11 of this book contains specific details related to microprocessor technology, number systems and interfacing techniques for digital computers.

The main advantages of the microprocessor based system are that the logical functions for a particular application can be developed and implemented in software, as opposed to electronic hardware. In many instances the microprocessor based system may actually be the cheaper alternative to a hardwired logic gate circuit. The software is easy to alter in the event of incorrect system operation and the complete system can be tested as a simulation before being committed.

For relatively small logical switching applications, up to say 32 inputs, the single-card microcomputer, or the single-chip microcomputer, represents an ideal low-cost solution (see Milne and Fraser[8]). These micro-systems can be used as dedicated devices where all the system components reside on a single card or a single chip, respectively. The major applications for these devices are in the high volume production markets such as automotive electronics, washing machines, bus ticket machines and time attendance recorders.

10.3.34 Application specific integrated circuits

Application specific integrated circuits (ASICs) are programmble logic devices (PLDs) which have their internal logic configuration determined by the user, as opposed to the manfacturer. The systems design engineer therefore customises the actual silicon building blocks to meet the requirements of the system. Such customisation provides for performance, reliability, compactness, low cost and design security. PLDs are available in both TTL and CMOS technology. The latter are erasable and can be reprogrammed almost indefinitely. PLDs represent the fastest growing segment of the semiconductor industry in recent times and it can be expected that they will play an increasingly important role in the design of digital logic systems in the future.

Internally, PLDs consist of an array of AND gates connected to an array of OR gates, with input and output blocks containing registers, latches and feedback options. *Figure 10.86* shows the general architecture of a programmable logic device.

In customising the PLD, the user essentially determines which of the interconnections between the gate arrays will remain open and which will be closed. The customisation procedure, however, requires additional development 'tools' which consist of:

(1) a wordprocessor to generate the source code;
(2) development software to transform the high level language source code into a fuse pattern for the PLD—the code which is generated is referred to as a 'JEDEC' file; and
(3) a PLD programmer to implement the program within the device.

The PLD programmer, connected to the parallel printer port of an IBM-PC, or a true compatible, programs the PLD by 'burning' the fuse pattern in the memory array of the device. When returned to its normal operating mode, the PLD then performs the customised logic function. Horowitz and Hill[9] provide a reasonably detailed coverage on applications of programmable logic devices.

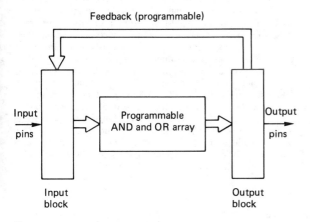

Figure 10.86 General architecture of a PLD

10.4 Electrical machines

The function of a rotating electrical machine is to convert mechanical power into electrical power, or vice versa. The conversion from mechanical to electrical power is done with a 'generator' and the conversion of electrical to mechanical power is done with a 'motor'. Electrical machines may be further subdivided into a.c. or d.c. machines. The major portion of all electrical energy generated in the world today is produced by a particular type of a.c. machine called an 'alternator'. The applications of electric motors are no less substantial and they are used in a vast variety of industrial drives. It is usually the mechanical features of a particular application which determines the type of electric motor to be employed and the torque–speed characteristics of the machine are therefore very important.

10.4.1 The d.c. generator

All conventional electrical machines consist of a stationary element and a rotating element which are separated by an air gap. In d.c. machines, generator or motor, the stationary element consists of salient 'poles' which are constructed as laminated assemblies with coils wound round them to produce a magnetic field. The function of the laminations is to reduce the losses incurred by eddy currents. The rotating element is traditionally called the 'armature' and this consists of a series of coils located between slots around the periphery of the armature. The armature is also fabricated in laminations which are usually keyed onto a locating shaft. A very simple form of d.c. generator is illustrated in *Figure 10.87(a)*.

The single coil is rotated at constant speed between the opposite poles, north and south, of a simple magnet. From Faraday's law (equation (10.25)) the voltage generated in the coil is equal to the rate of change of flux linkages. When the coil lies in the horizontal plane, there is maximum flux linking the coil but a minimum rate of change of flux linkages. On the other hand, when the coil lies in the vertical plane, there is zero flux linking the coil, but the rate of change of flux linkages is a maximum. The resultant variation in the voltage generated in the coil as it moves through one revolution is shown in *Figure 10.87(b)*. It is apparent that the generated voltage is alternating with positive and negative half-cycles. To change the a.c. output voltage into a d.c. voltage, a simple yet effective mechanical device called a 'commutator' is used. The commutator (*Figure 10.88*) incorporates brass segments separated by insulating mica strips. External connection to the armature coil is made by stationary carbon 'brushes' which make sliding contact with the commutator. Referring to *Figures 10.87(a)* and *10.88(a)*, as the coil rotates from the horizontal plane through 180° the right-hand side of the coil is under the north pole and is connected via the commutator to the upper brush. Meanwhile, the left-hand side of the coil is under the south pole and is connected to the lower brush. A further 180° of rotation effectively switches the coil sides to the opposite brushes. In this manner the coil side passing the north pole is always connected to the positive upper brush while the coil side passing the south pole is always connected to the negative lower brush. The resultant output voltage waveform is shown in *Figure 10.88(b)*.

If two coils physically displaced by 90° are used, the output brush voltage becomes virtually constant, as shown in *Figure 10.89*. With the introduction of a second coil, the commutator requires four separate segments. In a typical d.c. machine there may be as many as 36 coils which would require a 72-segment commutator.

The simple d.c. generator shown in *Figure 10.87* can be improved in perhaps three obvious ways: the number of coils

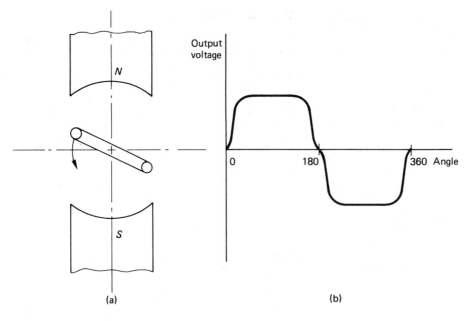

Figure 10.87 (a) A single-coil, two-pole d.c. generator, and (b) the coil voltage output

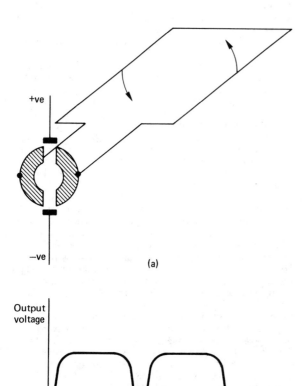

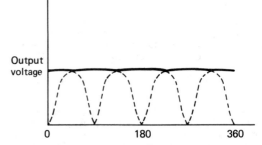

Figure 10.89 The output voltage of a two-coil, two-pole d.c. generator

Figure 10.88 (a) Commutator connections to an armature, and (b) the output voltage waveform

can be increased; the number of turns on each coil can be increased; and another pair of poles can be introduced. A typical d.c. machine would therefore normally incorporate four poles, wired in such a way that each consecutive pole has the opposite magnetic polarity to each of its neighbouring poles. If the e.m.f.s generated in the armature coils are to assist each other then, while one side of the coil is moving under a north pole, the other side of the coil must be moving under a south pole. With a two-pole machine the armature coils must be wound such that one side of the coil is diametrically opposite to the other. With a four-pole machine the armature coils can be wound with one side of the coil physically displaced 90° from the other. The size of the machine will generally dictate how many coils, and the number of turns on each coil which can be used.

10.4.1.1 Armature e.m.f.

If a conductor cuts flux then a voltage of 1 V will be induced in the conductor if the flux is cut at the rate of 1 W s^{-1}. Denoting

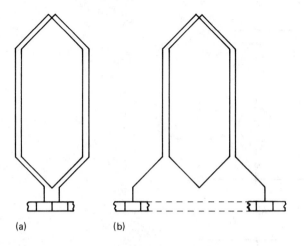

(a)   **(b)**

Figure 10.90 Lap (a) and wave (b) windings for a two-turn coil

the flux per pole as Φ and the speed as N (rev s^{-1}) for the single-turn coil and two-pole generator shown in *Figure 10.87(a)*, the e.m.f. induced in the coil is:

$$E_{\text{coil}} = \frac{\text{Flux per pole}}{\text{Time for half rev.}} = \frac{\Phi}{1/(2N)} = 2N\Phi$$

For a machine having Z_s armature conductors connected in series, i.e. ($Z_s/2$ turns), and $2p$ magnetic poles, the total induced e.m.f. is:

$$E = 2N\Phi\frac{Z_s}{2}2p = 2N\Phi Z_s p \text{ (volts)} \qquad (10.103)$$

Z_s depends on the type of armature winding, the two main types being 'lap-wound' and 'wave-wound'. These windings are shown in *Figure 10.90*.

The lap winding is characterised by the fact that the number of parallel paths through the winding is equal to the number of poles. In wave winding, the number of parallel paths through the winding is always equal to 2. If Z denotes the total number of armature conductors, then for the lap winding

$$Z_s = \frac{Z}{\text{No. of parallel paths}} = \frac{Z}{\text{No. of poles}} = \frac{Z}{2p} \qquad (10.104)$$

and for the wave winding

$$Z_s = \frac{Z}{\text{No. of parallel paths}} = \frac{Z}{2} \qquad (10.105)$$

Lap windings are generally used in low-voltage, heavy current machines and wave windings are used in all other cases.

10.4.1.2 Armature torque

The force on a current-carrying conductor is given by equation (10.27), i.e.

$$F = BlI$$

The torque on one armature conductor, therefore, is

$$T = Fr = B_{\text{av}}lI_a r \qquad (10.106)$$

where r is the radius of the armature conductor about the centre of rotation, I_a is the current flowing in the armature

conductor, l is the axial length of the conductor, and B_{av} is the average flux density under a pole. Note that

$$B_{\text{av}} = \frac{\Phi}{(2\pi rl)/2p}$$

The resultant torque per conductor is

$$T = \frac{\Phi 2plI_a r}{2\pi rl} = \frac{\Phi p I_a}{\pi}$$

For Z_s armature conductors connected in series, the total torque on the armature is

$$T = \frac{\Phi p I_a Z_s}{\pi} \text{ (N-m)} \qquad (10.107)$$

10.4.1.3 Terminal voltage

Denoting the terminal voltage by V, the induced e.m.f. by E, and the armature resistance by R_a, then

$$V = E - I_a R_a \text{ \{for a generator\}} \qquad (10.108)$$

$$V = E + I_a R_a \text{ \{for a motor\}} \qquad (10.109)$$

For the motor, the induced e.m.f. is often called the 'back e.m.f.'.

10.4.2 Methods of connection

The methods of connecting the field and armature windings may be grouped as follows.

(1) *Separately excited*, where the field winding is connected to a source of supply independently of the armature supply.
(2) *Self-excited*, which may be further subdivided into:
    (a) *shunt wound*, where the field winding is connected across the armature terminals;
    (b) *series wound*, where the field winding is connected in series with the armature winding.
(3) *Compound wound*, which is combination of shunt and series windings.

The four alternative methods of connection are illustrated in *Figure 10.91*.

10.4.3 The separately excited generator

In the separately excited generator (*Figure 10.91(a)*), running at a constant rated speed with no load across the output, it is assumed that initially the poles were completely demagnetised. If the field current, and hence the magnetic field, is gradually increased, then a plot of terminal voltage against field current takes the form shown in *Figure 10.92*.

As the field current increases, the iron poles begin to saturate and the proportionality between the flux and the field current no longer exists. If the field current is then reduced, the magnetic hysteresis causes the terminal voltage to have a slightly greater value than that obtained when the field current was being increased. When the field current is reduced to zero, a 'residual voltage' remains. On increasing the field current once more, the curve follows the broken line to merge with the original lower curve. These curves are termed the 'open-circuit characteristics' of the machine.

If the generator is now connected to a variable external load and driven at constant speed with a constant field current I_f, the terminal voltage variation with armature current is as shown in *Figure 10.93*.

The decrease in terminal voltage with increase in load is due mainly to the voltage drop across the armature resistance R_a.

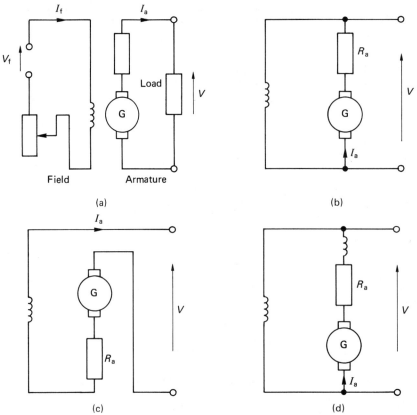

Figure 10.91 Methods of field connection: (a) separately excited; (b) shunt wound; (c) series wound; (d) compound wound

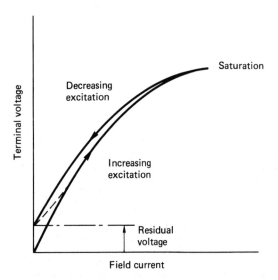

Figure 10.92 Open-circuit characteristic of a separately excited generator

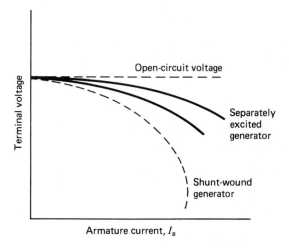

Figure 10.93 Load characteristic of a separately excited generator

In addition, the decrease in terminal voltage is attributed to a decrease in flux caused both by the demagnetising ampere-turns of the armature and also the magnetic saturation in the armature teeth. These effects are collectively known as 'arma-ture reaction'. A plot such as that shown in *Figure 10.93* is referred to as the 'load characteristic' of the generator.

The separately excited generator has the disadvantage inherent with a separate source of direct current required for the field coils. They are used, however, in cases where a wide range in terminal voltage is required.

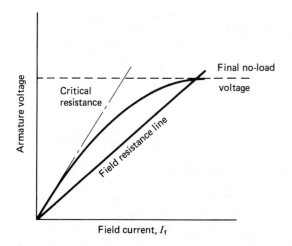

Figure 10.94 No-load characteristic for a shunt-wound generator

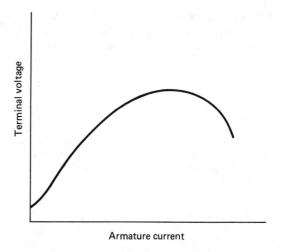

Figure 10.95 Costant-speed load characteristic for the series-wound generator

10.4.4 The shunt-wound generator

The field winding in the shunt-wound generator is connected across the armature terminals as shown in *Figure 10.91(b)* and is therefore in parallel, or 'shunt', with the load.

A shunt generator will excite only if the poles have some residual magnetism and the resistance of the shunt circuit is less than some critical value.

If when running at constant speed, the field is disconnected from the armature, the voltage generated across the armature brushes is very small and entirely due to residual magnetism in the iron. When the field is connected, the small residual voltage generates a flow of current in the field winding. The total flux in the field winding will gradually build up and the final terminal voltage will depend on the resistance of the field winding and the magnetisation curve of the machine. The general characteristic is shown in *Figure 10.94*.

When connected to an external load, the shunt-wound generator exhibits a drop in terminal voltage as the armature current is increased (see *Figure 10.93*). The drop in voltage in the shunt-wound generator is much greater than that in the separately excited generator. This stems from the fact that, as the terminal voltage drops, the field current also drops which causes a further drop in terminal voltage.

The shunt-wound machine is the most common type of d.c. generator employed. The load current, however, must be limited to a value well below the maximum value to avoid excessive variation in terminal voltage.

10.4.5 The series-wound generator

For the series-wound generator, the field winding is connected in series with the armature terminals as shown in *Figure 10.91(c)*. The armature current therefore determines the flux. The constant-speed load characteristic (*Figure 10.95*) exhibits an increase in terminal voltage as the armature, or load, current increases.

At large values of load current, the armature resistance and reactance effects cause the terminal voltage to decrease. It is apparent from the figure that the series-wound generator is totally unsuitable if the terminal voltage is required to be reasonably constant over a wide range of load current.

10.4.6 The compound-wound generator

The compound-wound generator (*Figure 10.91(d)*) is a hybrid between the shunt-wound and the series-wound generators. Normally a small series field is arranged to assist the main shunt field. This is termed 'cumulative compounding'. The shape of the load characteristic (*Figure 10.96*) depends upon the number of turns on the series winding. If the series field is arranged to oppose the main shunt field, 'differentially compounded', a rapidly falling load characteristic is obtained.

The number of turns on the series coil can be varied to give an over-compounded, level-compounded or an under-compounded characteristic, as shown in the figure.

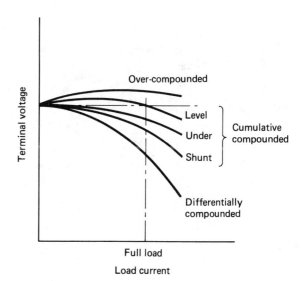

Figure 10.96 Load characteristic for a compound-wound generator

10.4.7 The d.c. motor

There is no difference in basic construction between a d.c. generator and a d.c. motor. The only significant distinction between the two machines is quantified by equations (10.108) and (10.109). These equations illustrate the fact that, for a d.c. generator, the e.m.f. generated is greater than the terminal voltage; for the d.c. motor, the generated e.m.f. is less than the terminal voltage.

Equation (10.103), which gives the relationship between the induced e.m.f. and the speed of a d.c. generator, applies equally as well to the d.c. motor. Since the number of poles and number of armature conductors are fixed, a proportionality relationship can be derived to relate speed as a function of induced e.m.f. and flux:

$$N \propto E/\Phi \qquad (10.110)$$

Or, using equation (10.109)

$$N = (V - I_a R_a)/\Phi \qquad (10.111)$$

The value of $I_a R_a$ is usually less than about 5% of the terminal voltage such that, to a reasonable approximation,

$$N \approx V/\Phi \qquad (10.112)$$

In a similar manner, equation (10.107), which gives the armature torque on a d.c. generator, also applies to the d.c. motor. A proportionality relationship for the d.c. motor torque is, therefore,

$$T \propto I_a \Phi \qquad (10.113)$$

Equation (10.112) shows that the speed of a d.c. motor is approximately proportional to the voltage applied to the armature and inversely proportional to the flux. All methods of controlling the speed of d.c. motors are based on these proportionality relationships.

Equation (10.113) indicates that the torque of a given d.c. motor is directly proportional to the product of the armature current and the flux per pole.

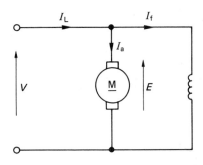

Figure 10.97 The shunt-wound motor

10.4.8 The shunt-wound motor

The shunt-wound motor is shown schematically in *Figure 10.97*. Under normal operating conditions, the field current will be constant. As the armature current increases, however, the armature reaction effect will weaken the field and the speed will tend to increase. However, the induced voltage will decrease due to the increasing armature voltage drop and this will tend to decrease the speed. The two effects are not self-cancelling and, overall, the motor speed will fall slightly as the armature current increases.

The motor torque increases approximately linearly with the armature current until the armature reaction starts to weaken the field. These general characteristics are shown in *Figure 10.98* along with the derived torque–speed characteristic.

Figure 10.98(a) shows that no torque is developed until the armature current is large enough to supply the constant losses in the machine. Since the torque increases dramatically for a slight decrease in speed, the shunt-wound motor is particularly suitable for driving equipment like pumps, compressors and machine tool elements where the speed must remain 'constant' over a wide range of load.

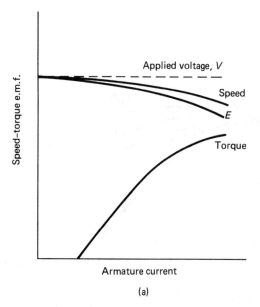

(a)

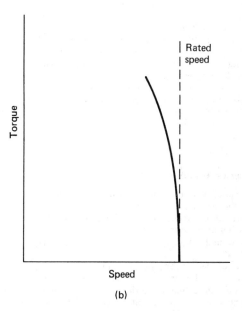

(b)

Figure 10.98 Load characteristics of the shunt-wound motor

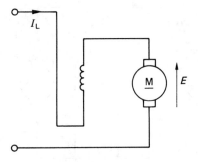

Figure 10.99 The series-wound motor

10.4.9 The series-wound motor

The series-wound motor is shown in *Figure 10.99*. As the load current increases, the induced voltage E will decrease due to the armature and field resistance voltage drops. Because the field winding is connected in series with the armature, then the flux is directly proportional to the armature current. Equation (10.112) therefore suggests that the speed–armature current characteristic will take the form of a rectangular hyperbola. Similarly, equation (10.113) indicates that the torque–armature current characteristic will be approximately parabolic. These general characteristics are illustrated in *Figure 10.100* along with the derived torque–speed characteristic.

The general characteristics indicate that, if the load falls to a particularly low value, the speed may become dangerously high. Therefore a series-wound motor should never be used in situations where the load is likely to be suddenly relaxed.

The main advantage of the series motor is that it provides a large torque at low speeds. Series motors are eminently suitable, therefore, for applications where a large starting torque is required. This includes, for example, lifts, hoists, cranes and electric trains.

10.4.10 The compound-wound motor

Compound-wound motors, like compound generators, are produced by including both series and shunt fields. The resulting characteristics of the compound-wound motor fall somewhere in between those of the series- and the shunt-wound machines.

10.4.11 Starting d.c. motors

With the armature stationary, the induced e.m.f. is zero. If, while at rest, the full voltage is applied across the armature winding, then the current drawn would be massive. This current would undoubtedly blow the fuses and thereby cut off the supply to the machine. To limit the starting current a variable external resistance is connected in series with the armature. On start-up the full resistance is connected in series. As the machine builds up speed and increases the back e.m.f., the external resistance can be reduced until at rated speed the series resistance is disconnected.

Variable-resistance 'starters' are also usually equipped with a return spring and an electromagnetic 'catch plate'. The catch plate keeps the starter in the zero resistance position while the machine is running at its rated speed. The electromagnet is powered by the field current and, in the event of a supply failure, the electromagnet is de-energised and the return spring pulls the starter back to the full resistance 'off' position. This ensures that the full starting resistance is always in series with the armature winding when the machine is restarted.

An overload cut-out switch is another normal feature incorporated in the starter mechanism. The overload cut-out is another electromagnetic switch; this is powered by the supply current. The overload switch is normally 'off' but, if the supply current becomes excessive, the switch is activated and it short circuits the supply to the electromagnetic catch plate. This in turn de-energises the catch plate and the return spring takes the starter back to the 'off' position. *Figure 10.101* illustrates the essential features of a starter device for a shunt-wound motor.

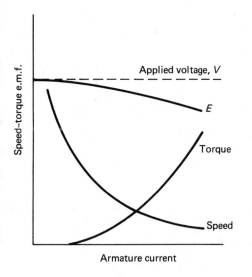
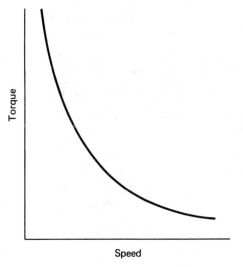

Figure 10.100 Load characteristics of the series-wound motor

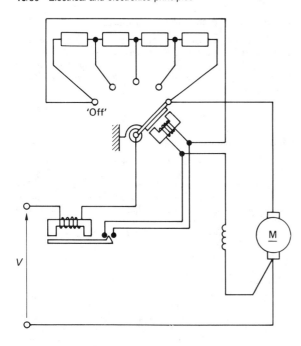

Figure 10.101 Starter device for d.c. machines

10.4.12 Speed control of d.c. motors

Equation (10.112) shows that the speed of a d.c. motor is influenced by both the applied voltage and the flux. A variation in either one of these parameters will therefore effect a variation in the motor speed.

10.4.12.1 Field regulator

For shunt and compound motors a variable resistor, called a 'field regulator', can be incorporated in series with the field winding to reduce the flux. For the series motor the variable resistor is connected in parallel with the field winding and is called a 'diverter'. *Figure 10.102* shows the various methods of weakening the field flux for shunt, compound and series-wound motors.

In all of the above methods of speed control, the flux can only be reduced and from equation (10.112) this implies that the speed can only be increased above the rated speed. The speed may in fact be increased to about three or four times the rated speed. The increased speed, however, is at the expense of reduced torque, since the torque is directly proportional to the flux which is reduced.

10.4.12.2 Variable armature voltage

Alternatively, the speed can be increased from standstill to rated speed by varying the armature voltage from zero to the rated value. *Figure 10.103* illustrates one method of achieving this. The potential divider, however, carries the same current as the motor and this limits this method of speed control to small machines. In addition, much of the input energy is dissipated in the controller which consequently renders the system inefficient.

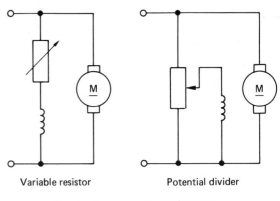

Variable resistor   Potential divider

Shunt and compound-wound motors

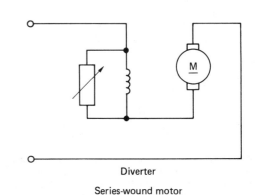

Diverter

Series-wound motor

Figure 10.102 Speed control by flux reduction

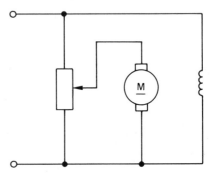

Figure 10.103 Speed control by varying armature voltage

10.4.12.3 Ward Leonard drive

In this case the variable d.c. voltage for the speed-controlled motor is obtained from a separate d.c. generator which is itself driven by an induction motor (see *Figure 10.104*). The field coil for the d.c. generator is supplied from a centre-tapped potential divider. When the wiper arm is moved from O to A, the armature voltage of the d.c. motor is increased from zero and the motor speed will increase. In moving the

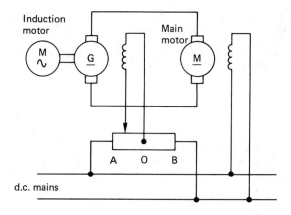

Figure 10.104 Ward Leonard drive

wiper from A to O and on through to B, the motor will decelerate to a standstill and then increase in speed again, but in the opposite direction of rotation. The Ward Leonard drive is smooth and accurate in either direction and also provides for very responsive braking. The complexity, however, makes it a very expensive system and it is only used in high-quality applications.

10.4.12.4 Chopper control

Figure 10.105 shows a thyristor circuit connected in series with the armature of a d.c. motor. The thyristor circuit is triggered such that it operates essentially as a high speed ON/OFF switch. The output waveform across the armature terminals is depicted in *Figure 10.106*.

The ratio of time on to time off, i.e. the 'mark/space ratio', can be varied with the result that the average voltage supplied to the armature is effectively varied between zero and fully on. The frequency of the signal may be up to about 3 kHz and the timing circuit is necessarily complex. However, speed control of d.c. motors using thyristors is effective and relatively inexpensive.

10.4.13 Efficiency of d.c. machines

The losses in d.c. machines can be generally classified as follows.

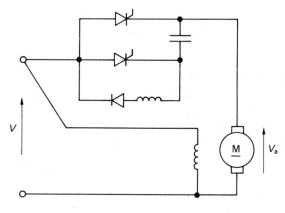

Figure 10.105 Speed control using thyristors

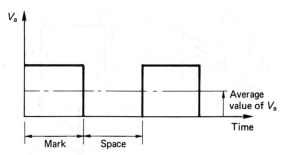

Figure 10.106 The voltage across armature terminals

(1) *Armature losses*—the I^2R loss in the armature winding, often referred to as the 'copper loss'.
(2) *Iron loss*—this is attributable to magnetic hysteresis and eddy currents in the armature and field cores.
(3) *Commutator losses*—this is related to the contact resistance between the commutator brushes and segments. The total commutator loss is due both to mechanical friction and a voltage loss across the brushes.
(4) *Excitation loss*—in shunt-wound machines, this power loss is to the product of the shunt current and the terminal voltage.
(5) *Bearing friction and windage*—bearing friction is approximately proportional to the speed, but windage loss varies with the cube of the speed. Both these losses are fairly minor unless the machine is fitted with a cooling fan, in which case the windage loss can be quite significant.

Despite the variety and nature of the losses associated with d.c. machines, they nonetheless have a very good performance, with overall efficiencies often in excess of 90%.

10.4.14 Three-phase circuits

Since a.c. machines are generally associated with three-phase systems, it is necessary to consider some aspect of three-phase circuits before a meaningful discussion of a.c. machines can be undertaken.

The limiting factor of a d.c. machine is related to the commutator, which restricts the maximum voltage which can be generated. Because of their efficiency and performance, three-phase machines have emerged as the dominant type of electrical generator and motor and, on a world-wide basis, three-phase electrical distribution networks are the norm.

10.4.15 Generation of three-phase e.m.f.s

Figure 10.107 shows three similar coils displaced at 120° relative to each other. Each loop terminates in a pair of 'slip-rings' and, if the coils are to be isolated from one another, six slip-rings are required in total.

If the three coils are rotated in the anti-clockwise direction at constant speed, each coil generates a sinusoidally varying e.m.f. with a phase shift of 120° between them.

10.4.16 Star and delta connections

The three coils shown in *Figure 10.107* can be connected together in one of two symmetrical patterns: the 'star' (or 'wye') connection and the 'delta' (or 'mesh') connection. The two types of connection are shown in *Figure 10.108*.

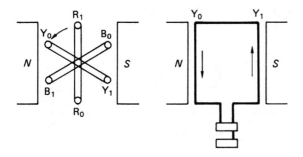

Figure 10.107 Generation of three-phase e.m.f.s

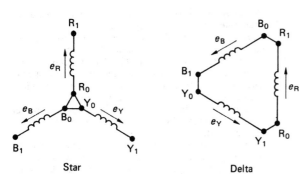

Star   Delta

Figure 10.108 Star and delta connections for three-phase systems

The star pattern is made by joining R_0, Y_0 and B_0 together. This connection point is referred to as the 'neutral point'. The delta pattern is formed by connecting R_0 to Y_1, Y_0 to B_1 and B_0 to R_1.

10.4.17 Three-phase voltage and current relationships

Figure 10.109 shows a three-phase star connected alternator supplying currents I_R, I_Y and I_B to a balanced, or equal, resistive–inductive load. This gives the usual 'four-wire' star

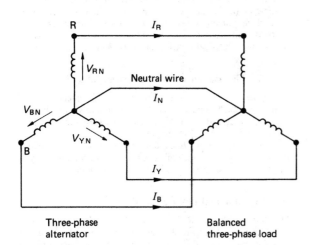

Three-phase
alternator

Balanced
three-phase load

Figure 10.109 Three-phase supply connections

connected system. Since there are only four transmission cables involved, the alternator connected in a star pattern will only require four slip-rings.

For a balanced system the phase voltages V_{RN}, V_{YN} and V_{BN} are all equal in magnitude and equally displaced by a phase angle of 120°. The currents I_R, I_Y and I_B are also equal in magnitude and equally displaced in phase angle but they all lag their respective phase voltages by some angle ϕ. Phasor addition of the currents shows that the neutral current I_N is zero.

The voltages between the transmission cables are called the 'line voltages'. If the phase voltages are all equal then the phasor addition shows that the line voltages are given by:

$$V_{LINE} = 2V_{PHASE} \cos(30)$$

or

$$V_L = \sqrt{3}V_p \qquad (10.114)$$

For the star connection, the line currents I_L are equal to the phase currents I_p.

Figure 10.110 shows the alternator windings connected up in the delta pattern. In the delta pattern the line voltages are equal to the phase voltages. Phasor addition of the currents shows that, if the phase currents are equal, the line currents are given by:

$$I_L = \sqrt{3}I_p \qquad (10.115)$$

10.4.18 Power in three-phase circuits

The power per phase is given by:

$$P_{PHASE} = V_p I_p \cos(\phi) \qquad (10.116)$$

where V_p is the phase voltage, I_p is the phase current, and ϕ is the phase angle between V_p and I_p.

The total power for a three-phase circuit is simply three times the power for one of the phases, i.e. three times equation (10.116).

For a star connection:

$$P = 3 \frac{V_L}{\sqrt{3}} I_L \cos(\phi) = \sqrt{3}V_L I_L \cos(\phi) \qquad (10.117)$$

For the delta connection:

$$P = 3 V_L \frac{I_L}{\sqrt{3}} \cos(\phi) = \sqrt{3}V_L I_L \cos(\phi)$$

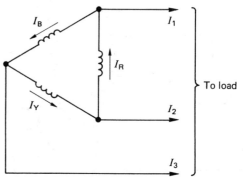

To load

Figure 10.110 Alternator windings in a delta connection

Thus exactly the same relationship is obtained for both types of connection. In terms of line voltages and currents, therefore, the power in a three-phase circuit is independent of the winding connection and is given by equation (10.117).

Equation (10.117) does not apply, however, if the system is unbalanced. In an unbalanced system the total power can only be obtained as the summation of the powers in each of the individual phases.

10.4.19 Three-phase alternators

Alternators are constructed with a stationary a.c. winding and a rotating field system. This reduces the number of slip-rings required to two and these have to carry only the field exciting current as opposed to the generated current. The construction is thereby simplified and the slip-ring losses are minimised. In addition, the simpler arrangement enables heavier insulation to be used and, consequently, much higher voltages can be generated. The robust mechanical construction of the rotor also means that higher speeds are possible and substantially higher power outputs can be generated with an alternator. A simple form of three-phase generator is depicted in *Figure 10.111*.

The three coils on the stator are displaced by 120°, and the rotor (which is a salient-pole type) is supplied via the two slip-rings with a d.c. current. As the rotor is driven by some form of prime mover, a rotating magnetic field is established and the e.m.f.s generated in the coils are displaced with a phase shift of 120°. The magnitudes of the generated voltages are dependent on the flux produced by the rotor, the number of turns on the stator coils and the speed of rotation of the rotor. The rotor speed will also dictate the frequency of the generated voltage.

The no-load and load characteristics of an alternator are very similar to those of the d.c. separately excited generator (*Figures 10.92* and *10.93*, respectively). In constant-speed operation, the terminal voltage exhibits a drooping character-

istic where the decrease in terminal voltage is due to 'armature' resistance and reactance effects. For an alternator, the term 'armature' is taken to imply the stator windings.

As the load on an alternator is increased, the speed of the prime mover drops. This is an unacceptable situation because the speed controls the frequency of the generated voltage. To maintain a constant frequency, the prime mover must be governed to run at constant speed over the entire range of expected loads. This is particularly important where many alternators are to be run in parallel to supply a distribution system such as the National Grid. In such cases the prime movers are always speed controlled and the output voltage is regulated to comply with the rated values. In the UK, alternators are usually two-pole machines driven at 3000 rev. $\min^{-1}$ to produce the rated frequency of 50 Hz. In the USA, a great deal of the electrical power consumed is generated from hydroelectric power stations. The water turbines used in these installations are fairly low speed machines and the alternators, which are directly driven, are equipped with multiple poles to produce the rated frequency of 60 Hz. An alternator running at 240 rev. $\min^{-1}$, for example, would require 30 poles to give the rated output frequency.

The production of the rotating magnetic field may also be actioned using three, 120° displaced, rotor coils supplied with three-phase current. The rotational speed of the field is related to the frequency of the currents:

$$N_s = \frac{f \times 60}{\text{No. of pole pairs}} \qquad (10.118)$$

where N_s is the speed of the field (in rev. $\min^{-1}$), and f is the frequency of the supply currents.

The speed of the rotating field is termed the 'synchronous speed' and for an equivalent single pair of poles, i.e. three coils, this is 3000 rev. $\min^{-1}$ when the frequency of the supply currents is at 50 Hz.

The use of a.c. excited rotor coils to produce the rotating magnetic field simplifies the mechanical construction of the rotor and greatly facilitates the dynamic balancing of the machine. An added advantage is that the waveform of the generated voltage is improved. The a.c. method of exciting the field is used extensively in large alternators. The use of salient-pole rotors is normally restricted to the smaller machines.

10.4.20 Synchronous motors

Synchronous motors are so called because they operate at only one speed, i.e. the speed of the rotating field. The mechanical construction is exactly the same as that of the alternator shown in *Figure 10.111*. The field is supplied from a d.c. source and the stator coils are supplied with a three-phase current. The rotating magnetic field is induced by the stator coils and the rotor, which may be likened to a permanent bar magnet, aligns itself to the rotating flux produced in the stator. When a mechanical load is driven by the shaft, the field produced by the rotor is pulled out of alignment with that produced by the stator. The angle of misalignment is called the 'load angle'. The characteristics of synchronous motors are normally presented in terms of torque against load angle, as shown in *Figure 10.112*.

The torque characteristic is basically sinusoidal with

$$T = T_{\max} \sin(\delta) \qquad (10.119)$$

where $T_{\max}$ is the maximum rated torque, and δ is the load angle.

It is evident from equation (10.119) that synchronous motors have no starting torque and the rotor must be run up to

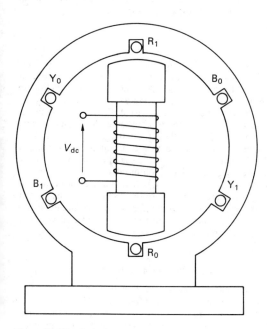

Figure 10.111 A simple three-phase generator

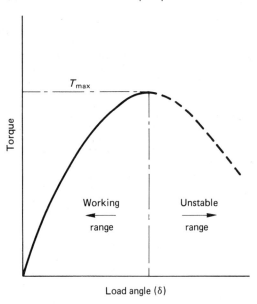

Figure 10.112 The torque characteristic for a synchronous motor

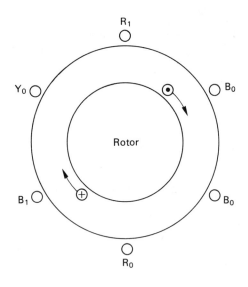

Figure 10.113 Schematic representation of an induction motor

synchronous speed by some alternative means. One method utilises a series of short-circuited copper bars inserted through the outer extremities of the salient poles. The rotating magnetic flux induces currents in these 'grids' and the machine accelerates as if it were a cage-type induction motor (see Section 10.4.21). A second method uses a wound rotor similar to a slip-ring induction motor. The machine is run up to speed as an induction motor and is then pulled into synchronism to operate as a synchronous motor.

The advantages of the synchronous motor are the ease with which the power factor can be controlled and the constant rotational speed of the machine, irrespective of the applied load. However, synchronous motors are generally more expensive, and a d.c. supply is a necessary feature of the rotor excitation. These disadvantages coupled with the requirement for an independent starting mode make the use of synchronous motors much less common than induction motors.

10.4.21 Induction motors

The stator of an induction motor is much like that of an alternator and in the case of a machine supplied with three-phase currents, a rotating magnetic flux is produced. The rotor may be one of two basic configurations which are the 'squirrel cage', or the slip-ring type. In the squirrel cage motor the rotor core is laminated and the conductors consist of uninsulated copper, or aluminium, bars driven through the rotor slots. The bars are brazed or welded at each end to rings or plates to produce a completely short-circuited set of conductors. The slip-ring machine has a laminated core and a conventional three-phase winding, similar to the stator and connected to three slip-rings on the locating shaft.

Figure 10.113 shows a schematic representation of an induction motor having three stator coils displaced by 120°. If the stator coils are supplied with three-phase currents, a rotating magnetic field is produced in the stator. Considering the single-rotor coil shown in the figure, at standstill the rotating field will induce a voltage in the rotor coil since there is a rate of change of flux linking the coil. If the coil forms a closed

circuit then the induced e.m.f. will circulate a current in the coil. The resultant force on the current carrying conductor is a consequence of equation (10.27) and this will produce a torque which will accelerate the rotor. The rotor speed will increase until the electromagnetic torque is balanced by the mechanical load torque. The induction motor will never attain synchronous speed because, if it did so, there would be no relative motion between the rotor coils and the rotating field. Under these circumstances there would be no e.m.f. induced in the rotor coils and, consequently, no electromagnetic torque. Induction motors therefore always run at something less than synchronous speed. The ratio of the difference between the synchronous speed and the rotor speed to the synchronous speed is called the 'slip' s:

$$s = \frac{N_s - N}{N_s} \qquad (10.120)$$

The torque–slip characteristic is shown in *Figure 10.114*. With the rotor speed equal to the synchronous speed, i.e. $s = 0$, the torque is zero. As the rotor falls below the synchronous speed the torque increases almost linearly to a maximum value dictated by the total of the load torque and that required to overcome the rotor losses. The value of slip at full load varies between 0.02 and 0.06. The induction motor may therefore be regarded as a constant-speed machine. In fact the difficulties of varying the speed constitute one of the main disadvantages of induction motors.

On start-up, the slip is 1 and the starting torque is sufficiently large to accelerate the rotor. As the rotor runs up to its full-load speed, the torque increases in essentially inverse proportion to the slip. The start-up and running curves merge at the full-load position.

10.4.22 Starting induction motors

As with d.c. motors, the current drawn during starting of a.c. motors is very large, up to about five times full-load current. A number of devices are therefore employed to limit the starting current, but they all involve the use of auxiliary equipment, which is usually quite expensive.

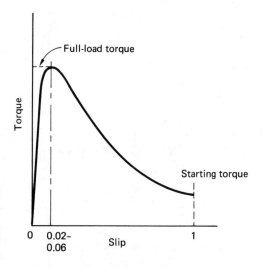

Figure 10.114   The torque–slip characteristic for an induction motor

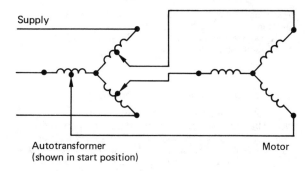

Figure 10.116   Autotransformer starter

10.4.22.1   Star–delta starter

The star–delta switch (*Figure 10.115*) is the cheapest and most common method employed. With the machine at standstill and the starter in the 'start' position, the stator coils are connected in the star pattern. As the machine accelerates up to running speed, the switch is quickly moved over to the 'run' position, which reconnects the stator windings in the delta pattern. By this simple expedient, the starting supply current is reduced to one-third of what it would have been had the stator windings been connected up in the delta pattern on start-up.

10.4.22.2   Autotransformer starter

The autotransformer represents an alternative method of reducing the starting current drawn by an induction motor. *Figure 10.116* shows a three-phase, star-connected autotransformer with a mid-point tapping on each phase. The voltage supplied to the stator is, therefore, one-half of the supply voltage. With such an arrangement the supply current and the starting torque are both only one quarter of the values which would be applied to the motor when the full voltage is supplied. After the motor has accelerated, the starter device is moved to the 'run' position, thereby connecting the motor directly across the supply and opening the star connection of the autotransformer.

Unfortunately, the starting torque is also reduced and the device is generally expensive because it must have the same rating as the motor.

10.4.22.3   Rotor resistance

With slip-ring induction motors it is possible to include additional resistance in series with the rotor circuit. The inclusion of extra resistance in the rotor provides for reduced starting current and improved starting torque.

10.4.23   Braking induction motors

Induction motors may be brought to a standstill by either 'plugging' or by 'dynamic braking'.

10.4.23.1   Plugging

This refers to the technique where the direction of the rotating magnetic field is reversed. This is brought about by reversing any two of the supply leads to the stator. The current drawn during plugging is very large, however, and machines which are regularly plugged must be specially rated.

10.4.23.2   Dynamic braking

In this braking technique the stator is disconnected from the a.c. supply and reconnected to a d.c. source. The direct current in the stator produces a stationary unidirectional field and, as the rotor will always tend to align itself with the field, it will therefore come to a standstill.

10.4.24   Speed control of induction motors

Under normal circumstances, the running speed of an induction motor will be about 94–98% of the synchronous speed,

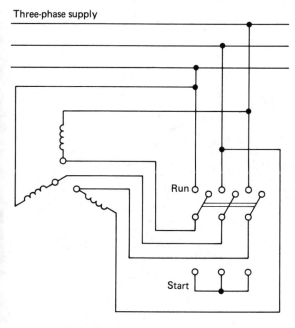

Figure 10.115   Star–delta starter

depending on the load. With the synchronous speed given by equation (10.118), it is clear that the speed may be varied either by changing the frequency of the supply current, or by changing the number of poles.

10.4.24.1 Change of supply current frequency

Solid-state variable-frequency drives first began to appear in 1968. They were originally applied to the control of synchronous a.c. motors in the synthetic fibre industry and rapidly gained acceptance in that particular market. More recently they have been used in pumping, synchronised press lines, conveyor lines and, to a lesser extent, in the machine-tool industry as spindle drives. Modern a.c. variable-frequency motors are available in power ratings of 1–750 kW and speeds of 10/1 to 100/1.

The synchronous and squirrel cage induction motors are the types most commonly used in conjunction with solid-state, ajdustable frequency inverter systems. In operation the motor runs at, or near, the synchronous speed determined by the input current frequency. The torque available at low speed, however, is decreased and the motor may have to be somewhat oversized to ensure adequate performance at the lower speeds. The most advanced systems incorporate a digital tachogenerator to supply a corrective feedback signal which is compared with a reference frequency. This gives a speed regulation of about ±3%. Consequently, the a.c. variable-frequency drive is generally used only for moderate to high power velocity control applications, where a wide range of speed is not required. The comparative simplicity of the a.c. induction motor is usually sacrificed to the complexity and cost of the control electronics.

10.4.24.2 Change of number of poles

By bringing out the ends of the stator coils to a specially designed switch it becomes possible to change an induction motor from one pole configuration to another. To obtain three different pole numbers, and hence three different speeds, a fairly complex switching device would be required.

Changing the number of poles gives a discrete change in motor speed with little variation in speed over the switched range. For many applications, however, two discrete speeds are all that is required, and changing the number of poles is a simple and effective method of achieving this.

10.4.24.3 Changing the rotor resistance

For slip-ring induction motors additional resistance can be coupled in series with the rotor circuit. It has already been stated that this is a common enough method used to limit the starting current of such machines. It can also be used as a method of marginal speed control. *Figure 10.117* shows the torque characteristics of a slip-ring induction motor for a range of different resistances connected in series with the rotor windings.

As the external resistance is increased from R_1 to R_3, a corresponding reduction in speed is achieved at any particular torque. The range of speeds is increased at the higher torques.

The method is simple and, therefore, inexpensive, but the reduction in speed is accompanied by a reduction in overall efficiency. Furthermore, with a large resistance in the rotor circuit, i.e. R_3, the speed changes considerably with variations in torque.

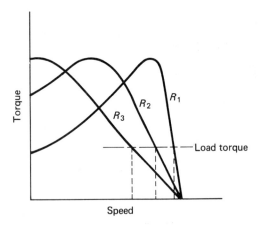

Figure 10.117 Torque–speed characteristics for various rotor resistances

10.4.24.4 Reduced stator voltage

By reducing the applied stator voltage a family of torque–speed characteristics are obtained as shown in *Figure 10.118*. It is evident that as the stator voltage is reduced from V_1 to V_3, a change in speed is effected at any particular value of torque, provided, of course, that the torque does not exceed the maximum load torque available at the reduced stator voltage. The latter point is obviously a limiting factor which places a constraint on this method of speed control. In general, only very small speed ranges can be obtained using variable stator supply voltage.

10.4.25 Single-phase induction motors

The operation of an induction motor depends on the creation of a rotating magnetic field. A single stator coil cannot achieve this and all of the so-called single-phase induction motors use some or other external means of generating an approximation

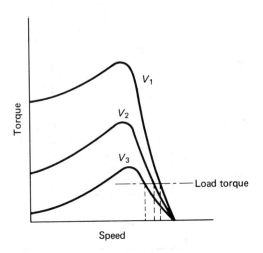

Figure 10.118 Torque–speed characteristics for various stator voltages

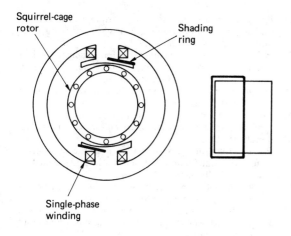

Squirrel-cage rotor

Shading ring

Single-phase winding

Figure 10.119 Shaded-pole motor

to a two-phase stator supply. Two stator coils are therefore used and these are displaced by 90°. Ideally, the currents which supply each coil should have a phase difference of 90°. This then gives the two-phase equivalent of the three-phase induction motor.

10.4.25.1 The shaded-pole motor

The stator of the shaded pole motor consists of a salient-pole single-phase winding and the rotor is of the squirrel cage type (see *Figure 10.119*). When the exciting coil is supplied with a.c. the flux produced induces a current in the 'shading ring'. The phase difference between the currents in the exciting coil and the shading ring is relatively small and the rotating field produced is far from ideal. In consequence, the shaded-pole motor has a poor performance and an equally poor efficiency due to the continuous losses in the shading rings.

Shaded-pole motors have a low starting torque and are used only in light-duty applications such as small fans and blowers or other easily started equipment. Their advantage lies in their simplicity and low cost of manufacture.

10.4.25.2 The capacitor motor

A schematic layout of a capacitor motor is given in *Figure 10.120*. The stator has two windings physically displaced by 90°. A capacitor is connected in series with the auxiliary winding such that the currents in the two windings have a large phase displacement. The current phase displacement can be

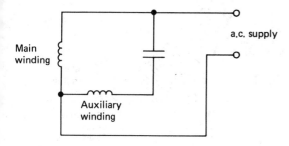

Main winding

a.c. supply

Auxiliary winding

Figure 10.120 Capacitor motor

made to approach the ideal 90° and the performance of the capacitor motor closely resembles that of the three-phase induction motor.

10.4.25.3 The universal motor

These are small d.c. series-wound motors which operate at about the same speed and power on d.c., or on single-phase current with approximately the same root-mean-square voltage. The universal, or plain-series motor, is used mainly in small domestic appliances such as hair dryers, electric drills, vacuum cleaners, and hedge trimmers.

10.4.26 The d.c. permanent magnet motor

The d.c. permanent magnet motor is a continuous rotation electromagnetic actuator which can be directly coupled to its load. *Figure 10.121* shows a schematic representation of a d.c. permanent magnet motor. The permanent magnet motor consists of an annular brush ring assembly, a permanent magnet stator ring and a laminated wound rotor. Such motors are particularly suitable for servo systems where size, weight, power and response times must be minimised and where high position and rate accuracies are required.

The response times for permanent magnet motors are very fast and the torque increases directly with the input current, independently of the speed or the angular position. Multiple-pole machines maximise the output torque per watt of rotor power. Commercial permanent magnet motors are available in many sizes from 35 mN-m at about 25 mm diameter, to 13.5 N-m at about 3 m diameter.

Direct drive rate and position systems using permanent magnet motors utilise d.c. tachogenerators and position sensors in various forms of closed-loop feedback paths for control purposes.

10.4.27 The stepper motor

A stepper motor is a device which converts a d.c. voltage pulse train into a proportional mechanical rotation of its shaft. The stepper motor thus functions both as an actuator and as a position transducer. The discrete motion of the stepper motor makes it ideally suited for use with a digitally based control system such as a microcomputer.

The speed of a stepper motor may be varied by altering the rate of the pulse train input. Thus if a stepper motor requires 48 pulses to rotate through one complete revolution, then an input signal of 96 pulses per second will cause the motor to rotate at 120 rev. min^{-1}. The rotation is actually carried out in finite increments of time, but this is visually indiscernible at all but the lowest speeds.

Stepper motors are capable of driving a 2.2 kW load with stepping rates from 1000 to 20 000 per second in angular increments from 45° down to 0.75°.

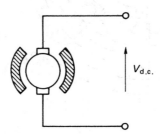

$V_{\text{d.c.}}$

Figure 10.121 A d.c. permanent magnet motor

There are three basic types of stepper motor.

(1) *Variable reluctance*—this has a soft iron multi-toothed rotor with a wound stator. The number of teeth on the rotor and stator, together with the winding configuration and excitation determines the step angle. This type of stepper motor provides small to medium sized step angles and is capable of operation at high stepping rates.

(2) *Permanent magnet*—the rotor used in this stepper motor consists of a circular permanent magnet mounted onto the shaft. Permanent magnet stepper motors give a large step angle ranging from 45° to 120°.

(3) *Hybrid*—this stepper motor is a combination of the previous two types. Typically the stator has eight salient poles which are energised by a two-phase winding. The rotor consists of a cylindrical magnet which is axially magnetised. The step angle depends on the method of construction and is generally in the range 0.9–5°. The most popular step angle is 1.8°.

The principle of operation of a stepper motor can be illustrated with reference to a variable reluctance, four-phase machine. This motor usually has eight stator teeth and six rotor teeth (see *Figure 10.122*).

If phase 1 of the stator is activated alone, then two diametrically opposite rotor teeth align themselves with the phase 1 teeth of the stator. The next adjacent set of rotor teeth in the clockwise direction are then 15° out of step with those of the stator. Activation of the phase 2 winding on its own, would cause the rotor to rotate a further 15° in the anti-clockwise direction to align the adjacent pair of diametrically opposite rotor teeth. If the stator windings are excited in the sequence 1, 2, 3, 4 then the rotor will move in consecutive 15° steps in the anti-clockwise direction. Reversing the excitation sequence will cause a clockwise rotation of the rotor.

The terminology used in association with stepper motors is listed below.

(1) *Pull-out torque*—the maximum torque which can be applied to a motor, running at a given stepping rate, without losing synchronism.

(2) *Pull-in torque*—the maximum torque against which a motor will start, at a given pulse rate, and reach synchronism without losing a step.

(3) *Dynamic torque*—the torque developed by the motor at very slow stepping speeds.

(4) *Holding torque*—the maximum torque which can be applied to an energised stationary motor without causing spindle rotation.

(5) *Pull-out rate*—the maximum switching rate at which a motor will remain in synchronism while the switching rate is gradually increased.

(6) *Pull-in rate*—the maximum switching rate at which a loaded motor can start without losing steps.

(7) *Slew range*—the range of switching rates between pull-in and pull-out in which a motor will run in synchronism but cannot start or reverse.

The general characteristics of a typical stepper motor are shown in *Figure 10.123*.

During the application of each sequential pulse, the rotor of a stepper motor accelerates rapidly towards the new step position. However, on reaching the new position there will be some overshoot and oscillation unless sufficient retarding torque is provided to prevent this happening. These oscillations can cause rotor resonance at certain pulse frequencies resulting in loss of torque, or perhaps even pull-out conditions. As variable reluctance motors have very little inherent damping, they are more suceptible to resonances than either the permanent magnet, or the hybrid types. Mechanical and electronic dampers are available which can be used to minimise the adverse effects of rotor resonance. If at all possible, however, the motor should be selected such that its resonant frequencies are not critical to the application under consideration.

Owing to their unique characteristics, stepper motors are widely used in applications involving positioning, speed control, timing and synchronised actuation. They are prevalent in x–y plotters, punched tape readers, floppy disc head drives, printer carriage drives, numerically controlled machine tool slide drives and camera iris control mechanisms.

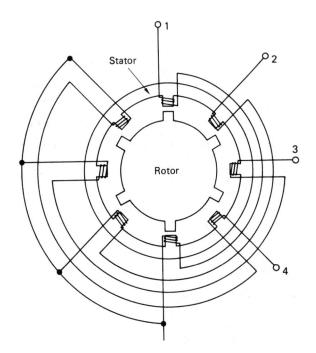

Figure 10.122 Variable reluctance stepper motor

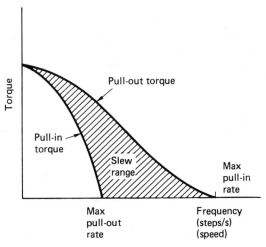

Figure 10.123 Stepper motor characteristics

By far the most severe limitation on the purely electric stepper motor is its power-handling capability. Currently this is restricted to about 2.25 kW.

10.4.28 Brushless d.c. motors

These motors have position feedback of some kind so that the input waveforms can be kept in the proper timing with respect to the rotor position. Solid state switching devices are used to control the input signals and the brushless d.c. motor can be operated at much higher speeds with full torque available at those speeds. The brushless motor can normally be rapidly accelerated from zero to operating speed as a permanent magnet d.c. motor. On reaching operating speed, the motor can then be switched over to synchronous operation.

The brushless motor system consists of a wound stator, a permanent magnet rotor, a rotor position sensor and a solid state switching assembly. The wound stator can be made with two, or more input phases. *Figure 10.124* gives the schematic representation of a two-phase brushless motor.

The torque output of phase A is:

$$T_A = I_A(Z\Phi/2\pi) \sin(p\theta/2) = I_A K_T \sin(p\theta/2) \qquad (10.121)$$

where I_A is the current in phase A, $K_T = (Z\Phi/2\pi)$ is the torque constant of the motor, p is the number of poles, and θ is the angular position of the rotor. In the expression for the torque constant, Z is the total number of conductors and Φ is the magnetic flux.

In a similar manner, the torque output of phase B is:

$$T_B = I_B K_T \cos(p\theta/2) \qquad (10.122)$$

If the motor currents are arranged to be supplied in the relationships

$$I_A = I \sin(p\theta/2)$$

and

$$I_B = I \cos(p\theta/2)$$

then the total torque for a two-pole motor becomes

$$T = T_A + T_B = I K_T[\sin^2(\theta) + \cos^2(\theta)]$$

$$= I K_T \qquad (10.123)$$

Equation (10.123) shows that, if all of the above conditions are satisfied, then the brushless d.c. motor operates in a similar manner to the conventional d.c. motor, i.e. the torque is directly proportional to the armature current. Note that the armature current in this context refers to the stator windings.

Excitation of the phases may be implemented with sinusoidal or square wave inputs. The sine wave drive is the most efficient but the output transistors in the drive electronics must be capable of dissipating more power than that dissipated in square wave operation. Square wave drive offers the added advantage that the drive electronics can be digitally based.

The brushless d.c. motor will duplicate the performance characteristics of a conventional d.c. motor only if it is properly commutated. Proper commutation involves exciting the stator windings in a sequence that keeps the magnetic field produced by the stator approximately 90 electrical degrees ahead of the rotor field. The brushless d.c. motor therefore relies heavily on the position feedback system for effective commutation. It might also be apparent that the brushless motor as described is not strictly a d.c. machine, but a form of a.c. machine with position feedback.

The further development of the brushless d.c. motor will depend to a large extent on future advances in semiconductor power transistor technology. It is likely, however, that within the next decade the true brushless d.c. motor, using solid state switching, will become commercially viable and will progressively dominate the d.c. servo-system market.

This brief discussion of rotating electrical machines is in no way comprehensive. A fuller discourse on a.c. and d.c. machines is given by both Gray[10] and Sen.[11] Orthwein[12] presents an interesting practical discussion on the mechanical applications of a.c. and d.c. motors and Kenjo and Nagamori[13] provide a detailed in-depth study of permanent magnet d.c. motors.

10.4.29 Transformers

One of the major advantages of a.c. transmission and distribution is the ease with which an alternating voltage can be increased or decreased. Common practice in the UK is to generate voltages at 11–22 kV and then transform up to 33 kV, or 132 kV for transmission on the National Grid to the consumer centres. At these centres, the voltages are transformed back down to 415 V, or 240 V, and then distributed for industrial and domestic use.

10.4.30 Basic transformer action

Figure 10.125 illustrates a simple single-phase transformer in which two separate coils are wound onto a ferrous core. The coil which is connected to the supply is called the 'primary winding' and that which is connected to the load is called the 'secondary winding'. The ferrous core is made in laminations, which are insulated from one another to reduce eddy current losses.

If a sinusoidal voltage V_1 is applied across the primary winding a current I_1 in the coil will induce a magnetic flux ϕ in

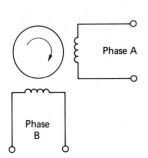

Figure 10.124 Two-phase brushless motor

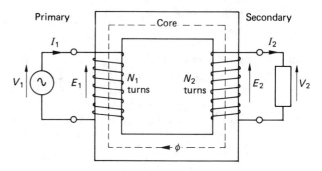

Figure 10.125 Single-phase transformer

the core. From Faraday's law, equation (10.25), the induced e.m.f. in the primary coil is

$$E_1 = N_1(d\phi/dt) \tag{10.124}$$

Since the magnetic flux is common to both coils then the e.m.f. induced in the secondary winding is

$$E_2 = N_2(d\phi/dt) \tag{10.125}$$

Hence

$$\frac{E_1}{E_2} = \frac{N_1}{N_2} \tag{10.126}$$

The ratio of primary coil turns to secondary coil turns (N_1/N_2) is called the 'transformation ratio'. The primary and secondary winding impedances (Z_1 and Z_2, respectively) are both very small such that when the secondary winding is on open circuit, $V_1 \doteqdot E_1$ and $V_2 \doteqdot E_2$. Therefore,

$$\frac{V_1}{V_2} = \frac{N_1}{N_2} \tag{10.127}$$

When a load is connected across the secondary winding, a current I_2 will flow in the secondary winding. From Lenz's law, this will set up a flux which will tend to oppose the main flux ϕ. If the main flux is reduced then E_1 would be correspondingly reduced and the primary current I_1 would then increase. The increased primary current would tend to produce a flux to oppose that induced by the secondary current. In this manner the main flux is generally maintained. In steady state, the ampere-turns in the primary and secondary windings are balanced, i.e.

$$I_1 N_1 = I_2 N_2$$

or

$$\frac{I_1}{I_2} = \frac{N_2}{N_1} \tag{10.128}$$

10.4.31 Transformer voltage equation

In normal operation, the flux may be considered to be a sinusoidally varying quantity, i.e.

$$\phi = \Phi \sin(\omega t) \tag{10.129}$$

From Faraday's law, the induced e.m.f. in the primary side is

$$e_1 = N_1(d\phi/dt) = N_1 \Phi \omega \cos(\omega t)$$

The r.m.s. value of the induced e.m.f. is

$$E_1 = \frac{2\pi f N_1 \Phi}{\sqrt{2}} = 4.44 \, f N_1 \Phi \tag{10.130}$$

Similarly, for the secondary side,

$$E_2 = 4.44 \, f N_2 \Phi$$

10.4.32 Transformer losses

Equations (10.127) and (10.128) define the ideal transformer in which there are no resistive, or inductive losses. An actual transformer, of course, does involve some losses.

(1) *Copper losses*—These are associated with the I^2R loss in both coils. They may be represented, therefore, as a resistance in series with each coil.
(2) *Iron loss*—These are associated with magnetic hysteresis effects and eddy current losses in the iron core. The iron losses are essentially constant for a particular value of

supply voltage. Iron losses can be represented as a resistor in parallel with the primary coil.
(3) *Flux leakage*—The useful, or main flux, is that which effectively links both coils. In practice, some of the flux will escape, or otherwise fail to link both coils. The e.m.f.s produced by the leakage fluxes are proportional to the fluxes and lead the fluxes by 90°. The effect of flux leakage may be likened, therefore, to having an additional inductive coil in series with the primary and secondary coils. In practice, the flux leakage loss is usually lumped together with the iron loss.

10.4.33 Determination of transformer losses

10.4.33.1 Open-circuit test

The secondary coil is on open circuit and the full rated voltage is applied to the primary winding. The transformer takes a small no-load current to supply the iron loss in the core and the copper losses are essentially zero. Since the normal voltage and frequency are applied, a wattmeter connected to the primary side will give a measure of the iron loss. The iron loss can then be taken as a constant, irrespective of the load.

10.4.33.2 Closed-circuit test

With the secondary winding short circuited, the transformer requires only a small input voltage to circulate the full load current. The wattmeter on the primary side then gives an indication of the full load copper losses. If the load is expressed as a fraction of the full load, then the copper losses at reduced loads are proportional to the load squared. At half load, for example, the copper losses are one-quarter of the full-load value.

10.4.34 Referred values

In dealing with transformers it is usual to base all calculations on one side of the transformer. Parameters on the neglected side are accounted for by 'referring' them over to the side on which the calculation is to be based. The transformation ratio is used to scale the equivalent values.

For example, the copper loss on the secondary side ($I_2^2 R_2$) can be referred to the primary side through the relationship:

$$I_2'^2 R_2' = I_2^2 R_2 \tag{10.131}$$

where the prime denotes the referred values.

Using equation (10.128), the referred resistance becomes:

$$R_2' = (N_1/N_2)^2 R_2 \tag{10.132}$$

Thus equation (10.132) gives an equivalent resistance R_2' in the primary side which accounts for the actual resistance R_2 of the secondary winding. Similarly, reactances may be referred to one or other side of the transformer for calculation purposes.

10.4.35 Transformer efficiency

The transformer efficiency, as with any machine, is the ratio of the output power to the input power. The difference between the output and the input power is the sum of the losses, which for the case of a transformer is the copper and the iron losses:

$$\eta = \frac{\text{Output}}{\text{Input}} = \frac{\text{Output}}{\text{Output} + \text{Copper loss} + \text{Iron loss}}$$

Therefore,

$$\eta = \frac{V_2 I_2 \cos(\theta_2)}{V_2 I_2 \cos(\theta_2) + I_2^2 R_e + F_e} \qquad (10.133)$$

Note that R_e represents an equivalent resistance which consists of the resistance of the secondary winding and the resistance of the primary winding referred over to the secondary side:

$$R_e = R_2 + (N_2/N_1)^2 R_1 \qquad (10.134)$$

The iron loss F_e is assumed to be constant and $\cos(\theta_2)$ is the load power factor, which is also assumed to be constant.

By dividing the numerator and the denominator of equation (10.133) by I_2 and then differentiating the denominator with respect to I_2, and equating the result to zero, it can be shown that for maximum efficiency $I_2^2 R_e = F_e$. Maximum transformer efficiency then occurs when the copper loss is equal to the iron loss. The general efficiency characteristics for a transformer are shown in *Figure 10.126*.

Equation (10.133) also shows that the output will be influenced by the load power factor. At unity power factor the output, and hence also the efficiency, is maximised. As the power factor decreases, the transformer efficiency also decreases proportionally.

10.4.36 Voltage regulation

As the load current drawn from a transformer is increased, the terminal voltage decreases. The difference between the no-load output voltage and the output voltage on load is called the 'regulation'. The percentage regulation is defined as:

$$\% \text{ Regulation} = \frac{\text{No-load voltage} - \text{Load voltage}}{\text{No-load voltage}} \times 100$$

$$(10.135)$$

Figure 10.127 shows the two voltages in terms of phasors referred to the primary side. In the figure, V_1 is the no-load primary voltage and V_2' is the secondary side voltage referred to the primary. R_e and X_e denote the equivalent resistance and reactance, respectively, including the referred secondary

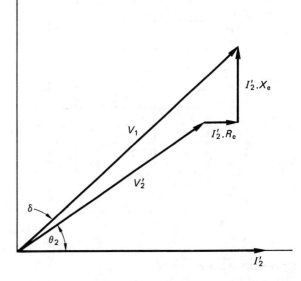

Figure 10.127 Phasor diagram for a transformer with a lagging power factor load current

values. Since δ is very small, then to a reasonable approximation

$$V_1 = V_2' + I_2' R_e \cos(\theta_2) + I_2' X_e \sin(\theta_2) \qquad (10.136)$$

The percentage regulation is, therefore,

$$\% \text{ Regulation} =$$
$$(100/V_1)[I_2' R_e \cos(\theta_2) + I_2' X_e \sin(\theta_2)] \qquad (10.137)$$

Equation (10.137) is based on the assumption that the load power factor is lagging. This is the normal situation. If, however, the load power factor is leading, then the plus operator within the square-bracketed term must be replaced with a minus operator.

10.4.37 Three-phase transformers

Modern large three-phase transformers are usually constructed with three limbs, as shown in *Figure 10.128*. In the figure the primary windings are star connected and the secondary windings are delta connected. In actual fact, the primary and secondary windings can be connected in any pattern depending upon the conditions under which the transformer is to operate. It is important, however, to know how the three-phase transformer is connected, particularly when two or more transformers are to be operated in parallel. It is essential, for instance, that parallel-operation transformers belong in the same main group and that their voltage ratios are perfectly compatible.

10.4.38 Autotransformers

The autotransformer is characterised by having part of its winding common to both the primary and secondary circuits (see *Figure 10.129*). The main application of the autotransformer is to provide a variable voltage, and they are used, for example, to limit the starting current drawn by an induction motor (see Section 10.4.22).

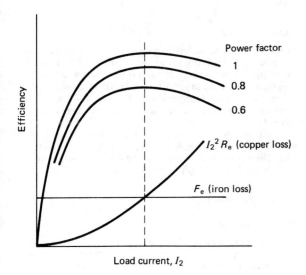

Figure 10.126 Transformer efficiency characteristic

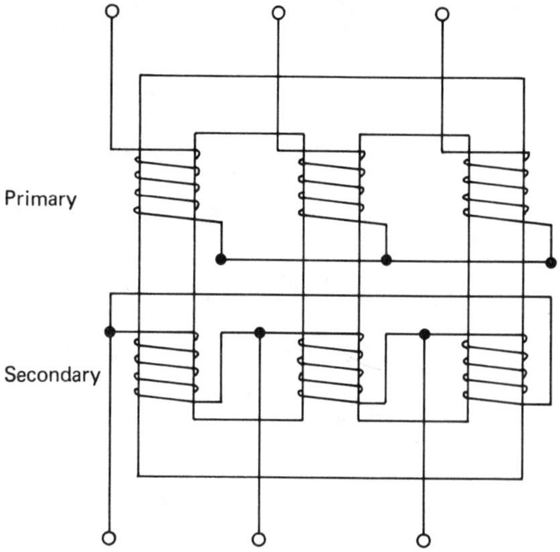

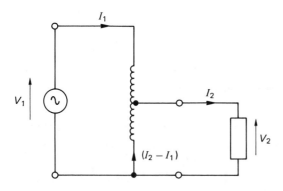

Figure 10.128 Three-phase transformer

Figure 10.129 Autotransformer

A major disadvantage of the autotransformer is that the primary and secondary windings are not electrically isolated from one another. This presents a serious risk of shock and, therefore, autotransformers cannot be used to interconnect high-voltage and low-voltage systems.

10.5 Electrical safety

The definitive document on electrical safety is the regulation cited as The Electricity at Work Regulations 1989, which came into force on 1 April 1990. The full text of the regulations is set out in Statutory Instrument 1989, No. 635, available from HMSO. The notes included here are based on the Health & Safety series booklet HS(R)25, which forms part of the Electricity at Work Regulations.[14] The regulations are made under the Health and Safety at Work Act 1974, and the Act imposes the responsibility for people at work on both the employer and the employee. It is partly the individual's responsibility, therefore, to ensure that safe working practices are observed in the workplace.

10.5.1 Electric shock

Most serious injuries and deaths from electric shock occur from contact with the mains electricity supply. In the UK, the mains supply is about 240 V a.c. and in the USA it is about 110 V a.c. The live wire (colour-coded brown) is at a higher potential with respect to the earth. If a person is in contact with the ground, then an electric shock can be sustained by touching the live wire only. When a shock is received the passage of the electric current through the body may cause muscular contractions, respiratory failure, fibrillation of the heart, cardiac arrest, or injury from internal burns. Any one of these can be fatal. The greatest danger occurs when the current flows across the chest. This can occur either when the current flows from one arm to the other, or when the current flows from one arm to the opposite leg.

The magnitude of an electric shock depends on the strength of the current, which in turn depends on the voltage and the ohmic resistance of the body. The resistance of the human body varies between people and is primarily dependent on the resistance of the skin. This variability in skin resistance means that a 'safe' voltage cannot be readily specified, and all voltages in excess of about 50 V must therefore be regarded as being potentially lethal. If the skin is damp, either from water or from perspiration, then skin resistance is dramatically reduced. Under such circumstances the chances of an electric shock proving fatal are greatly increased.

Injury can also be caused by a minor shock, not serious in itself, but which has the effect of contracting the muscles sufficiently to result in a fall.

10.5.2 Electric burn

Burns can be caused either by the passage of heavy current through the body, or by direct contact with an electrically heated surface. Burns may also be caused by the arcing across a short circuit, or as a result of an electrically originated fire or explosion.

Any circuit brought near an induction heater will receive energy and heat will be generated extremely rapidly. Therefore, no rings or other metal trinkets should be worn when in the vicinity of an induction heater. Neither should any metal be held in the hands. In general, no part of the body should come within close proximity (about 1 m) of an induction heater.

10.5.3 Rescue

To render assistance to a person undergoing an electric shock, the rescuer should first attempt to isolate the circuit by switching off the supply. If the rescuer cannot isolate the supply, he should try to break the victim's contact with the live apparatus using insulating material. The insulating material is essential to prevent the rescuer from becoming a second victim.

If the victim is unconscious, the rescuer should send someone for medical assistance and start artificial resuscitation. This assumes, of course, that the rescuer is trained in modern methods of artificial resuscitation.

10.5.4 Protection

10.5.4.1 Insulation

Electrical cables consist of one or more metal cores which may be single wires, but are more usually stranded wire and surrounded by insulation. The insulation serves to contain the flow of current and prevent a person from touching the live metal and, thereby, receiving a shock.

10.5.4.2 Fuses

A fuse is a device which will melt when the current exceeds a predetermined value. In operation, fuses serve as current-limiting devices and they are used for overload protection of electrical equipment. Two types of fuse are generally available: the rewireable type and the cartridge type.

Fuses are specified by the maximum current that they can transmit and the correctly rated fuse must be used at all times. A high rating fuse must never be substituted for a low rating one.

10.5.4.3 Circuit breakers

A circuit breaker is a mechanical device in the form of a switch which opens automatically if the circuit which it controls becomes overloaded. Circuit breaks may be operated magnetically or thermally and they can also be manually reset and adjusted. Plug-in circuit breakers are available and are recommended for use with small electric power tools.

10.5.4.4 Earth leakage protection

Normally there is no net flow of electricity to an electrical device. The flow in from the live wire is exactly balanced by the return flow in the neutral cable. If an earth fault develops, however, a leakage current will result and this can be detected by the earth leakage apparatus. Modern earth leakage devices are so sensitive that the supply is immediately disconnected before a lethal current can be drawn from the mains. It should be noted that earth leakage protection will operate only when the fault occurs between line and ground.

10.5.4.5 Isolation

Effective means of disconnecting cables or apparatus from the source of supply must be provided so that maintenance, repair or alteration may be carried out safely. This is achieved by isolating switches which have no automatic features.

Various circuits and motors should not 'share' an isolation switch, unless it is clear that under no circumstances will it be necessary, or convenient, to use one circuit while the other is being serviced.

Isolation switches should be capable of being locked in the 'off' position, but not in the 'on' position. If the isolation switch cannot be locked it should be possible to remove the fuse on the power line so that the line cannot be energised by inadvertent closing of the isolation switch.

10.5.5 Earthing

The external metal casing of electrical apparatus and cables must be earthed for three reasons:

(1) to prevent the casing rising to a dangerous voltage if there is a fault such as a short circuit between the conductor and the casing;
(2) to conduct any current away by a safe path; and
(3) to ensure that the faulty circuit is automatically disconnected from the supply by drawing sufficient current to blow the protective fuse, or operate the circuit breaker.

Earthing consists of connecting the metal casing by means of a conductor to an earth electrode. The earth electrode may be a buried pipe, or other such conductor which is known to be making an effective connection to the 'general mass of the earth'. Where the earth connection to a casing is made with a nut and bolt, a spring washer or other similar locking device must be used.

Earthing is a legal requirement and must be effective at all times.

10.5.6 Double insulation

Although the electricity regulations require all portable apparatus used at normal mains voltage to have an earthing conductor, these can introduce their own hazards. As a result 'double-insulated', or 'all-insulated' apparatus are made which do not require earthing. Double insulation means what the name suggests and all live conductors are separated from the outside world by two separate and distinctive layers of insulation. Each layer of insulation would adequately insulate the conductor on its own, but together they virtually negate the possibility of danger arising from insulation failure. Double insulation avoids the requirement for any external metalwork of the equipment to be protected by an earth conductor.

10.5.7 Low-voltage supplies

Portable tools, particularly hand-held inspection lamps, can be a source of danger because they are subject to severe wear and tear and they are likely to be used in confined spaces where skin resistance could easily be reduced by damp conditions.

In cases where work is carried out within confined metal enclosures, mains voltage equipment must not be used. A double wound transformer with a secondary centre tap to earth is allowable in these cases. This transformer gives 50 V for lighting and 110 V for portable tools.

These notes on general electricity safety are by no means extensive or authoritative. Reference should always be made to the full guide to the regulations.[14] Further recommended reading on electrical safety guidelines is given in references 15 and 16.

References

1 BRITISH STANDARDS INSTITUTION, *BS 3939: Part 1 Guide for graphical symbols for electrical power, telecommunications and electronics diagrams* (13 parts), Milton Keynes (1986)
2 BRITISH STANDARDS INSTITUTION, *PD 5686 The International System of Units*, Milton Keynes (1970)
3 HUGHES, W., *Electrical Technology*, 6th edn (revised by I. McKenzie Smith), Longman, London (1987)
4 BELL, E. C. and WHITEHEAD, R. W., *Basic Electrical Engineering and Instrumentation for Engineers*, 3rd edn, Granada, London (1987)
5 BELL, D. A., *Fundamentals of Electric Circuits*, 3rd edn, Reston Publishing, Prentice-Hall, Reston, Virginia (1984)
6 KAUFMAN, M. and SEIDMAN, A. H., *Handbook of Electronics Calculations for Engineers and Technicians*, 2nd edn, McGraw-Hill, New York (1988)
7 WATSON, J., *Mastering Electronics*, Macmillan, London (1983)
8 MILNE, J. S. and FRASER, C. J., 'Development of a mechatronics learning facility', *Mechatronic Systems Engineering*, Vol. 1, 31–40, Kluwer, Dordrecht (1990)
9 HOROWITZ, P. and HILL, W., *The Art of Electronics*, 2nd edn, Cambridge University Press, Cambridge (1989)
10 GRAY, C. B., *Electrical Machines and Drive Systems*, Longman Scientific & Technical, London (1989)
11 SEN, P. C., *Principles of Electric Machines and Power Electronics*, Wiley, New York (1989)
12 ORTHWEIN, W., *Machine Component Design*, West Publishing, St. Paul, Minnesota (1990)
13 KENJO, T. and NAGAMORI, S., *Permanent-Magnet and Brushless d.c. Motors (Monographs in Electrical and Electronic Engineering)*, Clarendon Press, Oxford (1985)

14 HEALTH & SAFETY EXECUTIVE, 'Memorandum of guidance on the electricity at work regulations', *Health & Safety Series Booklet HS(R)25*, HMSO, London (1989)
15 IMPERIAL COLLEGE OF SCIENCE AND TECHNOLOGY, *Safety Precautions in the use of Electrical Equipment*, 3rd edn, Imperial College, London (1976)
16 REEVES, E. A., *Handbook of Electrical Installation Practice*, Vol. 1, *Systems, Standards and Safety* (ed. E. A. Reeves), Granada, London (1984)

Bibliography

FITZGERALD, A. E., HIGGINBOTHAM, D. G. and GRABEL, A., *Basic Electrical Engineering*, 5th edn, McGraw-Hill, New York (1981)
NASAR, S. A., *Handbook of Electric Machines*, McGraw-Hill, New York (1987)
SAY, M. G., *Alternating Current Machines*, Pitman, London (1983)

11 Microprocessors, Instrumentation and Control

C J Fraser

J S Milne

Contents

11.1 Basic control systems

11.1.1 Introduction

Control engineering is based on the linear systems analysis associated with the development of feedback theory. A control system is constituted as an interconnection between the components which make up the system. These individual components may be electrical, mechanical, hydraulic, pneumatic, thermal or chemical in nature, and the well designed control system will provide the 'best' response of the complete system to external, time-dependent disturbances operating on the system. In the widest sense, the fundamentals of control engineering are also applicable to the dynamics of commercial enterprise, social and political systems and other non-rigorously defined concepts. In the engineering context, however, control principles are more generally applied to much more tangible and recognisable systems and subsystems.

Invariably, the system to be controlled can be represented as a block diagram, as in *Figure 11.1*. The system is a group of physical components combined to perform a specific function. The variable controlled may be temperature, pressure, flow rate, liquid level, speed, voltage, position, or perhaps some combination of these. Analogue (continuous) or digital (discrete) techniques may individually, or simultaneously, be employed to implement the desired control action. More recently, the advances made in microelectronics have resulted in a growing emphasis on digital techniques, and the majority of modern control systems are now microprocessor based.

11.1.1.1. *Classification of control systems*

Engineering control systems are classified according to their application and these include the following.

Servomechanisms Servomechanisms are control systems in which the controlled variable, or output, is a position or a speed. D.c. motors, stepper motor position control systems and some linear actuators are the most commonly encoun-tered examples of servomechanisms. These are especially prevalent in robotic arms and manipulators.

Sequential control A system operating with sequential control is one where a set of prescribed operations is performed in sequence. The control may be implemented as 'event based', where the next action cannot be performed until the previous action is completed. An alternative mode of sequential control is termed 'time based', where the series of operations is sequenced with respect to time. Event-based sequential control is intrinsically a more reliable 'fail-safe' mode than the time-based type. Consider, for example, an industrial process in which a tank is to be filled with a liquid and the liquid subsequently heated. The two control systems are depicted in *Figure 11.2*.

The time-based sequential control system (*Figure 11.2(a)*) is the simplest. The pump is switched on for an interval which would discharge enough liquid into the tank to fill it to approximately the correct level. Following this, the pump is switched off and the heater is switched on. Heating is similarly allowed to continue for a preset time, after which the liquid temperature would approximately have reached the desired value. Note that the control function is inexact and there are no fail-safe features. If the drive shaft between the motor and the pump becomes disengaged, or broken, then the heater will still come on at the prescribed time, irrespective of whether there is liquid in the tank or not. The event-based sequential control system has fail-safe features built-in and is much more exact. In operation the pump is switched on until the liquid-level sensor indicates that the tank is filled. Then, and only then, is the pump switched off and the heater switched on. The temperature of the liquid is also monitored with a sensor and heating is applied until such time that the temperature reaches the desired value. Obviously, with two additional sensors, the event-based system is the more expensive. The advantages it offers over the time-based system, however, far outweigh its disadvantages, and event-based sequentially controlled systems are by far the most common. Time-based systems do exist nonetheless, and they are found in applications where the results of malfunction would be far less potentially catastro-phic than those occurring in the example described. The essential difference between the two systems is that event-based sequential control incorporates a check that any opera-tion has been completed before the next is allowed to proceed. The modern automatic washing machine and automatic dish-washer are examples of sequentially controlled systems.

Figure 11.1 System to be controlled

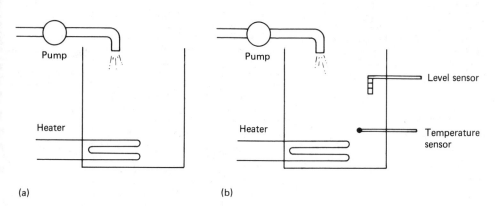

(a) (b)

Figure 11.2 Simple sequential control systems: (a) time based; (b) event based

Numerical control In a system using numerical control, the numerical information is stored in the form of digital codes on a control medium which may be a paper tape, a magnetic-sensitive tape, or a magnetic-sensitive disc. This information is used to operate the system in order to control such variables as position, direction, velocity and speed. There is a wide variety of manufacturing operations involving machine tools which utilise this versatile method of control.

Process control In this type of control the variables associated with any process are monitored and subsequent control actions are implemented to maintain the variables within the predetermined process constraints. The word 'process' is all-encompassing and might include, for example, electrical power generation. The generation of 'electricity' can be considered as a manufacturing process where the 'product' is kilowatt-hours. In the control of power generation, the variables which are measured include temperature, pressure, liquid level, speed, flow rate, voltage, current and a range of various gas concentrations. This is further complicated by the need to satisfy the power demand, and it is apparent that the control of such a system is necessarily complex. Similarly complex examples exist in the oil and paper making industries, in the automotive assembly industries and in any entity which aspires to the designation of a 'flexible manufacturing system'.

11.1.1.2 Open- and closed-loop control

The basic open-loop system is shown in *Figure 11.1* and extended in *Figure 11.3* to give a more complete picture. The input element supplies information regarding the desired value X of the controlled variable. This information is then acted on by the controller to alter the output Y through the load. External disturbances are fed in as shown and will cause the output to vary from the desired value. The open-loop system may be likened to the driving of a vehicle where the driver constitutes the input element. Essentially, two variables are controlled by the driver and they are the speed and the direction of motion of the vehicle. The controller, in the case of speed, is the engine throttle valve and in the case of direction, is the steering system.

In order for the system to become a closed loop, two further elements must be added:

(1) a monitoring element to measure the output Y; and
(2) a comparing element to measure the difference between the actual output and the desired value X.

The monitoring and comparing elements are connected through the 'feedback' link, as shown in *Figure 11.4*.

It is a moot point and can be argued that the driver in the above example also performs the functions of monitoring and comparing. The vehicle driver, therefore, if considered to be part of the complete system, constitutes a closed-loop feedback control. For the purpose of definition, however, any system which incorporates some form of feedback is termed a 'closed-loop system'. With no feedback mechanism, the

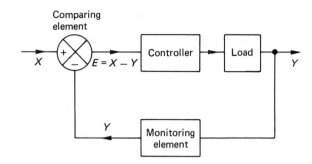

Figure 11.4 Closed-loop feedback control system

system is categorised as 'open loop'. For the most practical engineering purposes, control systems are of the closed-loop variety to take advantage of the benefits of feedback, which may be either 'positive' or 'negative'. A positive feedback signal aids the input signal. It is possible, therefore, to have output with no input when using a positive-feedback signal and, since this is detrimental to the control function, positive-feedback systems are very rare.

11.1.1.3 Linear and non-linear control systems

For a control system to be linear it must satisfy both the amplitude proportionality criteria and the principle of super-position. If a system output at a given time is $Y(t)$ for a given input $X(t)$, then an input of $kX(t)$ must produce an output of $kY(t)$ if amplitude proportionality is satisfied. In a similar manner, if an input of $X_1(t)$ produces an output of $Y_1(t)$, while an input of $X_2(t)$ produces an output of $Y_2(t)$ then, if an input of $[X_1(t) + X_2(t)]$ produces an output of $[Y_1(t) + Y_2(t)]$ the superposition principle is satisfied. Non-linear systems do not necessarily satisfy both these criteria and these systems are generally 'compensated' such that their behaviour approaches that of an equivalent linear system.

11.1.1.4 Characteristics of control systems

The characteristics of a control system are related to the output behaviour of the system in response to any given input. The parameters used to define the characteristics of a control system are: stability, accuracy, speed of response and sensitivity.

The system is said to be 'stable' if the output attains a certain value in a finite interval after the input has undergone a change. When the output reaches a constant value the system is said to be in the 'steady state'. The system is unstable if the output increases with time. In any practical control system, stability is absolutely essential. Systems involving a 'time delay', or a 'dead time' may tend to be unstable, and extra care must be taken in the design of such systems to ensure stability. The stability of control systems can be analysed using various analytical and graphical techniques. These include the Routh–Hurwitz criteria and the Bode, Nichols and Nyquist graphical methods.

The accuracy of a system is a measure of the deviation of the actual controlled value in relation to its desired value. Accuracy and stability are interactive and one can in fact be counter-productive to the other. The accuracy of a system might be improved but, in refining the limits of the desired output, the stability of the system might be adversely affected. The converse also applies.

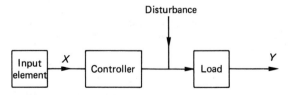

Figure 11.3 Open-loop control system

The speed of response is a measure of how quickly the output attains a steady-state value after the input has been altered.

Sensitivity is an important factor and is a measure of how the system output responds to external environmental conditions. Ideally, the output should be a function only of the input and should not be influenced by undesirable extraneous signals.

11.1.1.5 Dynamic performance of systems

The dynamic performance of a control system is assessed by mathematically modelling, or experimentally measuring, the output of the system in response to a particular set of test input conditions.

(1) *Step input*—This is perhaps the most important test input, since a system which is stable to a step input will also be stable under any of the other forms of input. The step input (*Figure 11.5*) is applied to gauge the transient response of the system and gives a measure of how the system can cope with a sudden change in the input.

(2) *Ramp input*—A ramp input (*Figure 11.6*) is used to indicate the steady-state error in a system attempting to follow a linearly increasing input.

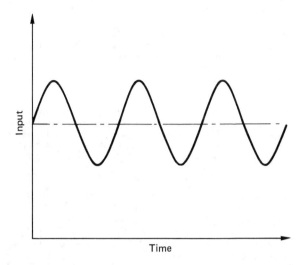

Figure 11.7 Sinusoidal input

(3) *Sinusoidal input*—The sinusoidal input (*Figure 11.7*) over a varying range of input frequencies, is the standard test input used to determine the frequency response characteristics of the system.

Although the three standard test inputs may not be strict representations of the actual inputs to which the system will be subject, they do cover a comprehensive range. A system which performs satisfactorily under these inputs will, in general, perform well under a more natural range of inputs. The system response to a parabolically varying test input can also be analysed, or measured, but this is a less commonly used test signal compared with the above three.

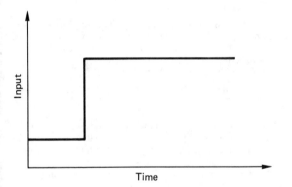

Figure 11.5 Step input

11.1.1.6 Time domain and frequency domain

The time domain model of a system results in an output $Y(t)$ with respect to time for an input $X(t)$. The time domain system model is expressed as a differential equation, the solution of which is displayed as a graph of output against time.

In contrast, a frequency domain model describes the system in terms of the effect that the system has on the amplitude and phase of sinusoidal inputs. Typically, the system performance is displayed in plots of amplitude ratio $[Y(t)/X(t)]$ or $20 \log_{10}[Y(t)/X(t)]$, and phase angle, against input signal frequency.

Neither system model has an overriding advantage over the other, and both are used to good effect in describing system performance and behaviour.

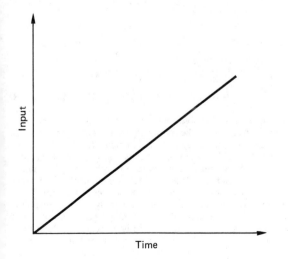

11.1.2 Mathematical models of systems—time domain analysis

Differential equations are used to model the relationship between the input and output of a system. The most widely used models in control engineering are based on first- or second-order linear differential equations.

Figure 11.6 Ramp input

11.1.2.1 First-order systems

Some simple control systems, including those for the control of temperature, level and speed, can be modelled as a first-order linear differential equation:

$$\tau \frac{dY}{dt} + Y = kX \tag{11.1}$$

where X and Y are the input and output, respectively, τ denotes the system time constant, and k is the system gain. When the input X is a step of amplitude A then the solution to equation (11.1) gives the result shown in *Figure 11.8*. The solution curve shown in *Figure 11.8* has the analytical form:

$$Y(t) = kA[1 - e^{-t/\tau}] \tag{11.2}$$

Equation (11.2), which is the time-domain solution, is an exponential function which approaches the value kA as t approaches infinity. Theoretically, the output never reaches kA and the response is termed an 'exponential lag'. The time constant τ represents the time which the output would take to reach the value kA if the initial rate of response were maintained. This is indicated in *Figure 11.8* by the broken line which is tangential to the solution curve at time $t = 0$. For practical purposes, the final steady-state output is taken to have been reached in a time of about 5τ.

If the input is a ramp function, then the response of a first-order system is as shown in *Figure 11.9*. The ramp input is simulated by making the right-hand side of equation (11.1) a linear function of time, i.e. kAt. With this input, the time domain solution becomes:

$$Y(t) = kA[t - \tau(1 - e^{-t/\tau})] \tag{11.3}$$

The solution equation shows that, as t becomes large, the output tends to $kA(t - \tau)$. Therefore the output response is asymptotic to a steady-state lag $kA\tau$.

The response of a first-order system to a sinusoidal input can be obtained by setting the right-hand side of equation (11.1) equal to $kA \sin(\omega t)$, where ω is a constant circular frequency (in rad s^{-1}). The time domain solution yields:

$$Y(t) = \frac{kA}{\sqrt{(1 + \tau^2\omega^2)}} [\sin \alpha e^{-t/\tau} + \sin(\omega t - \alpha)] \tag{11.4}$$

where $\alpha = \tan^{-1}(\tau\omega)$.

The response is shown in *Figure 11.10*. The output response exhibits a decaying transient amplitude in combination with a steady-state sinusoidal behaviour of amplitude, $kA[\sqrt{(1 + \tau^2\omega^2)}]$ and lagging the input by the angle α.

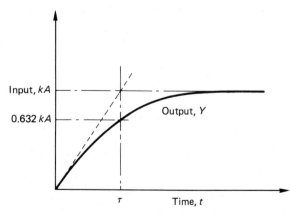

Figure 11.8 Response of a first-order system to a step input

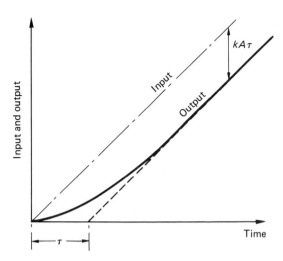

Figure 11.9 First-order system response to a ramp input

11.1.2.2 Second-order systems

While some control systems may be adequately modelled as a first-order linear differential equation, many more practical systems, including position control, are more usually represented by a differential equation of second order. The second-order differential equation has the general form:

$$\frac{d^2Y}{dt^2} + 2\zeta\omega_n \frac{dY}{dt} + \omega_n^2 Y = kX \tag{11.5}$$

where ζ is termed the 'damping ratio' and is defined as the ratio of the actual damping in the system to the damping which would produce critical damping. ω_n is the undamped natural frequency of the system, and k is the system gain.

The time-domain solution depends on the magnitude of ζ and three solutions for a step input are possible.

(1) *Light damping, $\zeta < 1$*:

$$Y(t) = \exp(-\zeta\omega_n t)[A \cos(\omega_n\sqrt{1 - \zeta^2}t)$$
$$+ B \sin(\omega_n\sqrt{1 - \zeta^2}t)] \tag{11.6}$$

(2) *Critical damping, $\zeta = 1$*:

$$Y(t) = \exp(-\zeta\omega_n t)[At + B] \tag{11.7}$$

(3) *Heavy damping, $\zeta > 1$*:

$$Y(t) = A\exp[-\zeta\omega_n t - (\omega_n\sqrt{\zeta^2 - 1})t]$$
$$+ B \exp[-\zeta\omega_n t + (\omega_n\sqrt{\zeta^2 - 1})t] \tag{11.8}$$

The three possible solutions, for a step input, are shown in *Figure 11.11*, where the output $Y(t)$ is plotted as a percentage of the step input X against the parameter $(\omega_n t)$.

For $\zeta = 1$, the system is critically damped and the steady-state value is attained in the shortest possible time without any oscillatory response. With $\zeta > 1$, the system is overdamped and the response curve is again exponential in form. Overdamped systems may have an undesirably sluggish response. Indeed, since the effect of $\zeta > 1$ simply delays the response to the steady-state value, then there is no real advantage to be gained in using high ζ values. For cases where $\zeta < 1$, the system is said to be 'underdamped' and the response curve is oscillatory with an exponential decay. A number of perfor-

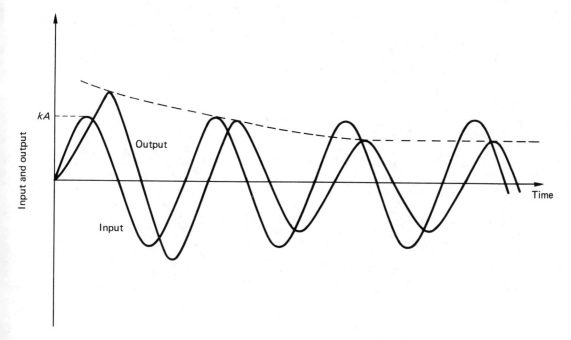

Figure 11.10 First-order system response to a sinusoidal input

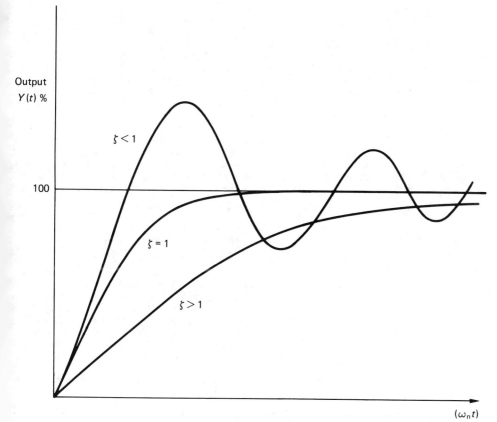

Figure 11.11 Response of a second-order system to a step input

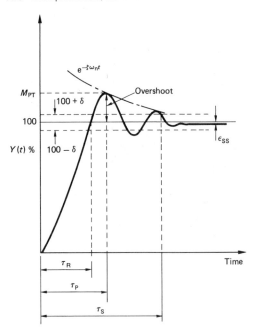

Figure 11.12 Response curve for an underdamped system to a step input

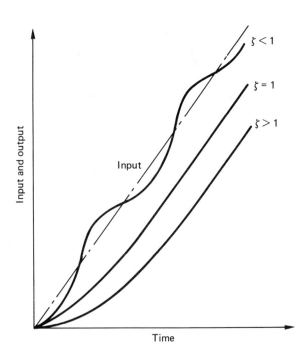

Figure 11.13 Response of a second-order system to a ramp input

mance measures are used to describe the response of an underdamped system to a step input. These are illustrated in *Figure 11.12*.

The speed of the response is reflected by the rise time τ_R and the peak time τ_p. For underdamped systems, the rise time is the time taken for the output to reach 100% of the step input. The peak time is the time taken to the first maximum in the output response. For critically damped and overdamped systems, the time taken for the output to change between 10% and 90% of the input is used as a measure of the speed of the response.

The degree in which the actual output response matches the input is measured by the percentage overshoot (PO) and the settling time. The percentage overshoot is defined as:

$$PO = \frac{M_{PT} - 100}{100} \qquad (11.9)$$

where M_{PT} is the peak value of the output.

It may further be shown that the percentage overshoot is given analytically as:

$$PO = 100\, e^{-\zeta\pi/\sqrt{1-\zeta^2}} \qquad (11.10)$$

Another useful relationship is derived from the ratio of successive peaks, i.e.

$$\ln\left(\frac{A_n}{A_{n+1}}\right) = \frac{2\zeta\pi}{\sqrt{1-\zeta^2}} \qquad (11.11)$$

where n is an integer denoting the peak number, i.e. first, second, etc. Equation (11.11) is referred to as the 'logarithmic decrement'.

The settling time τ_s is the time taken for the oscillatory response to decay below a percentage of the input amplitude δ, which is often taken as $\pm 2\%$.

Finally, we have the steady-state error ε_{ss} which is self-explanatory.

The response of the second-order system to a ramp input is shown in *Figure 11.13*. The form of the response curves again depends on the value of the damping ratio, but in each case the output asymptotes to a steady-state lag. The lag is not the same in each case, however, since it is also dependent on the damping ratio.

The response of a second-order system to a sinusoidal input may also be considered. In general, the output response will lag behind the input with a transient decaying amplitude depending on the nature of the damping ratio. It is more informative, however, to study the response of second-order systems to sinusoidal inputs using frequency domain methods.

11.1.3 Laplace notation for differential equations—frequency domain analysis

For analyses in the frequency domain it is customary to write the differential equation in terms of the Laplace operator s. This gives rise to the system 'transfer function' which is formed by replacing the input and output (X and Y, respectively) with their corresponding Laplace transforms ($X(s)$ and $Y(s)$). The method applies only to linear differential equations. In practice, many systems would contain some degree of non-linearity, and various assumptions would have to be made to simplify and approximately linearise the governing equation.

The advantage of using the Laplace transform method is that it allows the differential equation to be expressed as an equivalent algebraic relation in s. Differentiation is represented by multiplication by the Laplace variable s. Thus dY/dt becomes $sY(s)$ and d^2Y/dt^2 becomes $s^2Y(s)$.

11.1.3.1 First-order systems

The governing equation is rewritten with the appropriate Laplace transforms replacing the differential operators. Thus equation (11.1) becomes:

$$\tau s Y(s) + Y(s) = kX(s) \tag{11.12}$$

Hence,

$$Y(s)[1 + \tau s] = kX(s) \tag{11.13}$$

The system transfer function is defined as the ratio of the output to the input:

$$\frac{Y(s)}{X(s)} = \frac{k}{1 + \tau s} \tag{11.14}$$

Equation (11.14) enables the convenient facility of incorporating the transfer function within the usual block-structure representation of a control system. Thus a first-order, open-loop control system can be systematically depicted as shown in *Figure 11.14*.

For analyses in the frequency domain, we are predominantly concerned with the system response to sinusoidal inputs. Differentiation, or integration, of a sinusoidal function does not alter the shape or frequency. There is simply a change in amplitude and phase, e.g.

$$\text{Input} = A \sin(\omega t) \tag{11.15}$$

$$\frac{d}{dt}[A \sin(\omega t)] = \omega A \cos(\omega t)$$

$$= \omega A \sin\left(\omega t + \frac{\pi}{2}\right) \tag{11.16}$$

Comparison of equations (11.15) and (11.16) shows that differentiation has changed the amplitude from A to ωA and that there is a phase shift of 90° associated with the process.

Equation (11.16) thus describes the steady-state output from a first-order, open-loop control system. The transient part of the output, which the time-domain solution illustrates (*Figure 11.10*), is not apparent in the frequency-domain solution.

The Laplace operator s may be replaced with $j\omega$, where j is the complex operator $\sqrt{-1}$. Equation (11.14) then becomes

$$\frac{Y}{X} = \frac{k}{1 + j\omega\tau}$$

Using the complex conjugate, it may be shown that the modulus of the amplitude ratio is

$$\left|\frac{Y}{X}\right| = \frac{k}{\sqrt{1 + \omega^2\tau^2}} \tag{11.17}$$

Equation (11.17) shows how the output amplitude will be influenced by the input sinusoidal frequency. Note that this agrees with the time domain solution (equation (11.4)) for large values of t, after which the steady-state is achieved.

The technique described above is general and may be used to determine the amplitude ratio for any second, or higher, order system.

N.B. Common practice, especially in graphical representations, is to express the amplitude ratio in decibels, i.e.

$$dB = 20 \log_{10}(Y/X) \tag{11.18}$$

11.1.3.2 Second-order systems

Using the Laplace transfer operator, the governing equation (equation (11.5)) may be rewritten as

$$s^2 Y(s) + 2\zeta\omega_n s Y(s) + \omega_n^2 Y(s) = kX(s) \tag{11.19}$$

Therefore,

$$Y(s)[s^2 + 2\zeta\omega_n s + \omega_n^2] = kX(s) \tag{11.20}$$

The system transfer function is

$$\frac{Y(s)}{X(s)} = \frac{k}{s^2 + 2\zeta\omega_n s + \omega_n^2} \tag{11.21}$$

Thus a second-order, open-loop control system can be represented schematically as shown in *Figure 11.15*.

For a sinusoidal input of the form $X = X_0 \sin(\omega t)$, the frequency-domain analysis gives the following steady-state solutions for the amplitude ratio and the phase lag:

$$\left|\frac{Y}{X}\right| = \frac{1}{\sqrt{(1 - r^2)^2 + (2\zeta r)^2}} \tag{11.22}$$

and

$$\phi = \tan^{-1}[(2\zeta r)/(1 - r^2)] \tag{11.23}$$

where $r = (\omega/\omega_n)$.

The above frequency-response characteristics are shown in *Figure 11.16*.

When the input signal frequency is equal to the system's natural frequency, the amplitude ratio has the value of $(1/2\zeta)$ and the phase lag is −90°. Note that, if the damping ratio is zero, then the amplitude ratio theoretically approaches infinity under this resonance condition. In practice, if the damping ratio is moderately low, then very large output amplitudes can be expected if the input frequency is in the vicinity of the natural frequency of the system.

Thus far we have considered the open-loop system response for first- and second-order systems. Such systems are unconditionally stable. The addition of a feedback loop, however, increases the order of the system, and there is always the possibility that the second-order system with feedback may be unstable. Furthermore, if any system, either first or second order, incorporates a 'time delay' (also known as a 'deadtime' or a 'transportation lag'), then unstable operation is more likely to occur.

11.1.4 Stability criteria

Time delays are very difficult to handle mathematically when they occur in differential equations, and the inclusion of

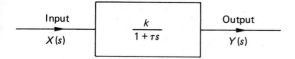

Figure 11.14 First-order, open-loop control system

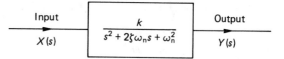

Figure 11.15 Second-order, open-loop control system

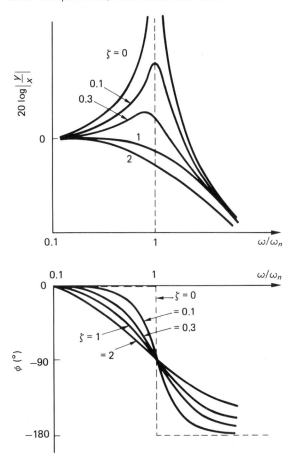

Figure 11.16 Second-order, open-loop control system frequency response

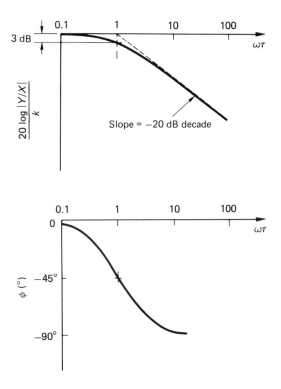

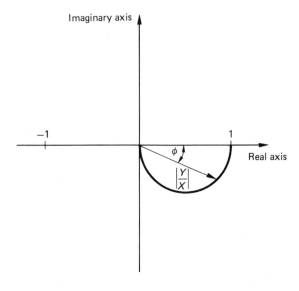

Figure 11.17 Bode plot for an open-loop, first-order system

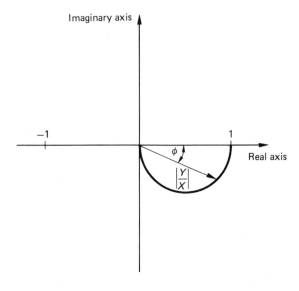

Figure 11.18 Nyquist plot for an open-loop, first-order system

multiple feedback loops can greatly increase the order of the governing equation. For these two reasons solutions in the time domain become extremely difficult, and frequency-domain methods are almost exclusively used to assess the behaviour of more complex control systems. The main consideration in frequency-domain analyses is the stability of the system and how the system can be adjusted if it happens to be unstable. Various graphical methods are used and these include the Bode and Nyquist plots.

The Bode plot is a graph of amplitude ratio and phase angle variation with input signal frequency. The resulting normalised plot for an open-loop, first-order system is shown in *Figure 11.17*. Note that, when the input frequency is equal to the inverse of the system time constant, the output amplitude has been decreased, or attenuated, by 3 dB. The phase lag at this point is $-45°$. This is characteristic of first-order systems.

The Nyquist plot represents the same information in an alternative form. The plot is in polar coordinates and combines the amplitude ratio and phase lag in a single diagram. *Figure 11.18* shows the Nyquist plot for the open-loop, first-order system.

11.1.4.1 Bode and Nyquist stability criteria

The Bode[1] criterion for stability is: *the system is stable if the amplitude ratio is less than 0 dB when the phase angle is $-180°$*

This is illustrated graphically in *Figure 11.19*, which represents a stable system because Bode's criterion is satisfied. The 'gain margin' (GM) and 'phase margin' (PM) are used as measures of how close the frequency-response curves are to the 0 dB and $-180°$ points, and are indicative of the relative stability of the system.

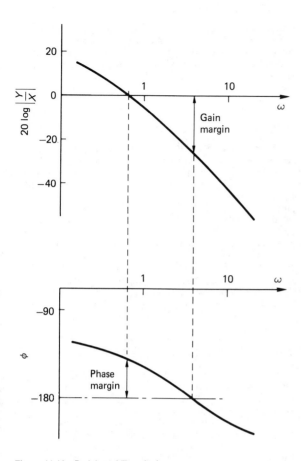

Figure 11.19 Bode's stability criterion

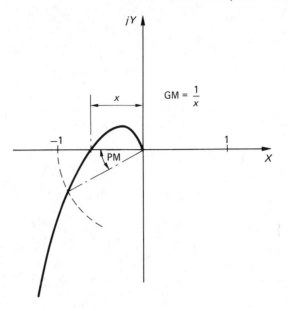

Figure 11.20 Nyquist's stability criterion

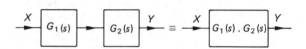

Figure 11.21 Transfer functions in series

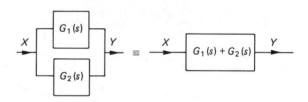

Figure 11.22 Transfer functions in parallel

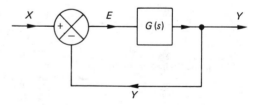

Figure 11.23 Control system with unity feedback

The Nyquist[2] criterion for stability is: *the system is stable if the amplitude ratio is greater than −1 at a phase angle of −180°.* In effect, this means that the locus of the plot of amplitude ratio and phase angle must not enclose the point −1 on the real axis. A stable response curve is shown plotted in *Figure 11.20*. Also indicated in *Figure 11.20* are the gain margin and phase margin in the context of the Nyquist plot.

11.1.4.2 System stability with feedback

In a closed-loop system, the transfer function becomes modified by the feedback loop. The first task, therefore, is to determine the overall transfer function for the complete system.

For simple open-loop systems the transfer functions are combined according to the following rules:

(1) for elements in series, the overall transfer function is given by the product of the individual transfer functions (see *Figure 11.21*); and

(2) for elements in parallel, the overall transfer function is given by the sum of the individual transfer functions (see *Figure 11.22*).

For a system with feedback, the overall transfer function can be evaluated using a consistent step-by-step procedure. Series and parallel control elements are combined in the manner shown above in order to reduce the system to a single block which then represents the overall transfer function.

Consider the simple control system depicted in *Figure 11.23*. Since the feedback line does not include any transfer function, it is termed a 'unity feedback' system, i.e. the output is compared directly with the input to produce the error signal. The closed-loop transfer function is obtained as follows:

$$Y = G(s)E = G(s)[X - Y]$$

Thus

$$\frac{Y}{X} = \frac{G(s)}{1 + G(s)} \tag{11.24}$$

If the element whose open-loop transfer function $G(s)$ is a first-order subsystem, then $G(s)$ may be replaced by the expression given in equation (11.14). The closed-loop transfer function may then be written as

$$\frac{Y}{X} = \left(\frac{k}{1 + \tau s}\right) \bigg/ \left(1 + \frac{k}{1 + \tau s}\right)$$

$$= \frac{k}{1 + k + \tau s} \tag{11.25}$$

Dividing top and bottom by $(1 + k)$ results in

$$\frac{Y}{X} = \left(\frac{k}{1 + k}\right) \bigg/ \left(1 + \frac{\tau s}{1 + k}\right) \tag{11.26}$$

Defining the following terms

$$k_c = \frac{k}{1 + k} \tag{11.27}$$

and

$$\tau_c s = \frac{\tau s}{1 + k} \tag{11.28}$$

where k_c is the closed-loop gain and τ_c is the closed-loop system time constant. The final closed-loop transfer function may be expressed as

$$\frac{Y}{X} = \frac{k_c}{1 + \tau_c s} \tag{11.29}$$

Equations (11.27) and (11.28) show, respectively, that both the closed-loop system gain and the time constant are less than those associated with the open-loop system. This means that the closed-loop response is faster than the open-loop response. At the same time, however, the closed-loop gain is reduced.

Using the procedures outlined in the example, any other complex control system may be similarly analysed to determine the closed-loop transfer function. Thus, knowing the gain constants and other characteristics of the elements which make up the system, the frequency response may be obtained. The stability of the system may then be assessed and any corrective measures taken as necessary. In practice, it is often found that the gain of some of the system elements must be altered in order to ensure stable operation. Another commonly applied corrective measure is to add a phase advance circuit into the system. The procedure might also be operated in reverse—by starting with a desired response, a suitable control system can be configured and adjusted to meet the response.

The practising control engineer will use many techniques to assess system stability. These might include the numerical Routh–Hurwitz criterion, which determines only whether a system is stable or not. Alternative graphical methods include the use of Hall charts, Nichols charts, inverse Nyquist plots and root locus plots. The graphical methods also indicate the relative stability of a system. Numerous commercial computer packages are available (see, for example, references 3 to 5) to assist the designer of control systems. These packages include the usual graphical representations and can be obtained from the suppliers whose addresses are given in the references. The reader is also referred to the Bibliography given at the end of this chapter for a more comprehensive coverage of these methods and techniques.

11.1.4.3 Effect of trasnport delay

The influence of a transport, or time, delay on the response of an underdamped second-order system to a step input is displayed in *Figure 11.24*. Although it is virtually impossible to account for a time delay in a differential equation, it is simply accommodated in the frequency-domain model as an additional element in the system block diagram. In the frequency-domain model, the time delay effects a phase shift of $-\omega T$ and can be expressed as

$$\text{Time delay} = \mathrm{e}^{-sT} = \mathrm{e}^{-j\omega T} \tag{11.30}$$

Consider the open-loop response of a first-order system incorporating a time delay as illustrated in *Figure 11.25*. The open-loop transfer function becomes

$$\frac{Y}{X} = \frac{Y(s)}{X(s)} = G(s) = \frac{k\mathrm{e}^{-sT}}{1 + \tau s} \tag{11.31}$$

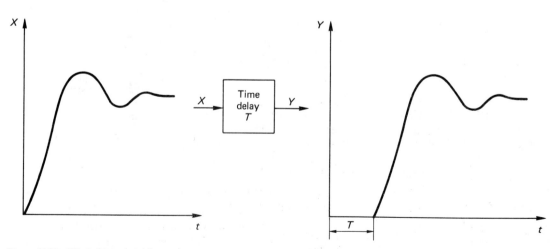

Figure 11.24 Effect of transport delay

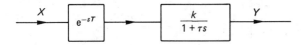

Figure 11.25 First-order system with a time delay

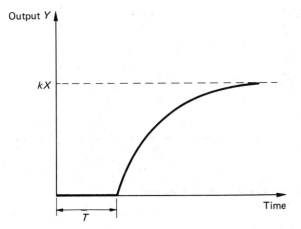

Figure 11.26 First-order, open-loop system response with a time delay

$$\text{Amplitude ratio} = \left| \frac{Y}{X} \right| = \frac{k}{\sqrt{1 + \omega^2 \tau^2}} \qquad (11.32)$$

$$\text{Phase lag} = \phi = -[\omega T + \tan^{-1}(\omega \tau)] \qquad (11.33)$$

The first-order, open-loop system response with a time delay is shown in *Figure 11.26*. Similarly, a time delay in a system with feedback does not alter the amplitude ratio but adds a phase shift to the frequency response. The system therefore becomes less stable and, in some cases, it may be necessary to reduce the closed-loop gain in order to obtain a stable response. The penalty to be paid for increasing the stability in this manner is an increase in the steady-state error.

11.2 Control strategies

The basic closed-loop system with common symbol representation is shown in *Figure 11.27*. The nomenclature used in the figure is defined as follows:

$SP(s)$ is the set point (the required value $r(t)$ is sometimes used);

$PV(s)$ is the process value (the corrected value $c(t)$ is sometimes used);

$E(s)$ is the error signal, which is the difference between SP and PV;

$U(s)$ is the control effort output from the controller to the process;

$C(s)$ is the controller transfer function; and

$G(s)$ is the process transfer function.

The transfer function for the closed-loop system is obtained as before:

$$PV(s) = C(s)G(s)E(s) \qquad (11.34)$$
$$= C(s)G(s)[SP(s) - PV(s)]$$

Hence,

$$\frac{PV(s)}{SP(s)} = \frac{C(s)G(s)}{1 + C(s)G(s)} \qquad (11.35)$$

11.2.1 ON/OFF control

In many applications, a simple ON/OFF strategy is perfectly adequate to control the output variable within preset limits. The ON/OFF control action results in either full or zero power being applied to the process under control. A mechanical type of thermostat proides a good example of an ON/OFF based controller. The ON/OFF control strategy results in an output which fluctuates about the set point, as illustrated in *Figure 11.28*.

ON/OFF controllers usually incorporate a 'deadband' over which no control action is applied. The deadband is necessary to limit the frequency of switching between the ON and OFF states. For example, in a temperature control system, the ON/OFF control strategy would be:

(1) if temperature $< T_{min}$, then heater is to be switched ON;
(2) if temperature $> T_{max}$, then heater is to be switched OFF.

The deadband in the above case is $(T_{max} - T_{min})$, and while the temperature remains within the deadband no switching will occur. A large deadband will result in a correspondingly large fluctuation in the process value about the set point. Reducing the deadband will reduce the level of fluctuation but will increase the frequency of switching. The simple ON/OFF control strategy is mostly applicable to processes and systems which have long time constants and, consequently, have relatively slow response times, e.g. temperature and level control.

While being simple in concept, ON/OFF control systems are, in fact, highly non-linear and they require some complex non-linear technique to investigate their stability characteristics.

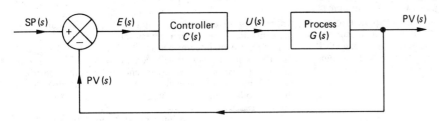

Figure 11.27 Basic closed-loop control system

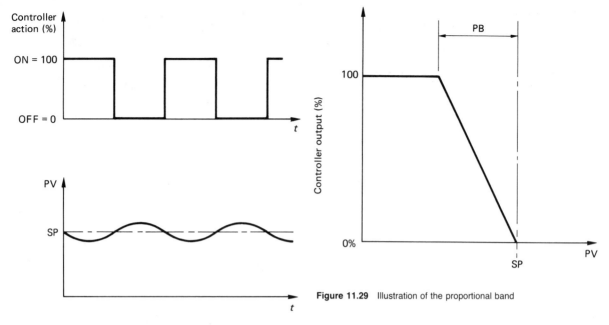

Figure 11.28 Output variation with ON/OFF control

Figure 11.29 Illustration of the proportional band

11.2.2 Three-term, or PID, control

Since complicated transfer functions can be very difficult to model, the most common strategy used to define the controller transfer function is the so-called 'three term', or PID, controller. PID is the popular acronym for 'proportional, integral and derivative'. The three elements of the controller action U based on the evaluated error E are described below.

11.2.2.1 Proportional action

Controller output = KE (11.36)

where K is the controller gain.

Manufacturers of three-term controllers tend to favour the parameter 'proportional band' (PB) in preference to gain K. The proportional band represents the range of the input over which the output is proportional to the input. The PB is usually expressed as a percentage of the input normalised between 0 and 100% (see *Figure 11.29*).

To illustrate the concept of proportional band, a temperature control application can be considered where the set point is, say, 80°C and the proportional band is set to, say, 5% over a measured temperature span of 0–100°C. The actual proportional band is, therefore, 5°C and proportional action will apply over the temperature range 75–80°C. If the temperature is below 75°C, then 100% of the available power will be supplied to the heating device. Between 75°C and 80°C, a proportion of the available power will be applied to the heating device, as shown in *Figure 11.29*. For temperatures in excess of 80°C, 0% of the available power is supplied. It should be apparent that 'proportional band' is a more meaningful term than 'gain'. The two parameters, however, are very simply related:

PB% = $100/K$ (11.37)

It is also apparent from *Figure 11.29* that, as the proportional band is decreased, then the control action tends towards an

ON/OFF strategy. A very large proportional band will result in a somewhat sluggish response.

It must also be noted that, for proportional control only, there must always be an error in order to produce a control action. From equation (11.35), proportional control only gives a transfer function of the form:

$$\frac{PV(s)}{SP(s)} = \frac{KG(s)}{1 + KG(s)} = \frac{1}{(1/KG(s)) + 1}$$ (11.38)

For steady-state conditions, s tends to 0 and $G(s)$ tends to a constant value. Equation (11.38) shows, therefore, that the gain must theoretically tend to infinity if PV = SP, and the steady-state error is to approach zero.

This is simply another manifestation of the classical control problem; i.e. stability at the expense of accuracy, and vice versa. With a very high gain, i.e. a low proportional band, the steady-state error can be very much reduced. A low proportional band, however, tends to ON/OFF control action, and in sensitive systems a violent unstable oscillation may result.

11.2.2.2 Integral action

The limitations of proportional control can be partly alleviated by adding a controller action which gives an output contribution that is related to the integral of the error value with respect to time:

Controller output = $K_i \int E dt$ (11.39)

where K_i is the controller integral gain. $K_i = K/T_i$ and T_i is the controller 'integral time', or 'reset'.

The nature of integral action (equation (11.39)) suggests that the controller output will increase monotonically as long as an error exists. As the error tends to zero the controller output tends towards a steady value. The general behaviour of the controller output with integral action is shown in *Figure 11.30*.

If T_i is very large, the integral action contribution will be low and the error may persist for a considerable time. If, on

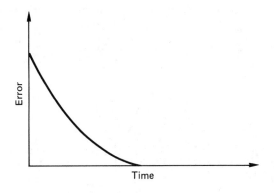

Figure 11.30 Controller output with integral action

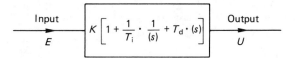

Figure 11.31 Three-term, or PID, control

In Laplace notation, the three-term controller transfer function is as shown in *Figure 11.31*.

11.2.3 Empirical rules for PID controller settings

A simple, and still popular, technique for obtaining the controller settings to produce a stable control condition is that proposed by Ziegler and Nichols.[6] The technique is purely empirical and is based on existing, or measurable operating records of the system to be controlled.

11.2.3.1 Open-loop 'reaction curve' method

The process to be controlled is subjected to a step input excitation and the system open-loop response is measured. A typical open-loop response curve is shown in *Figure 11.32*.

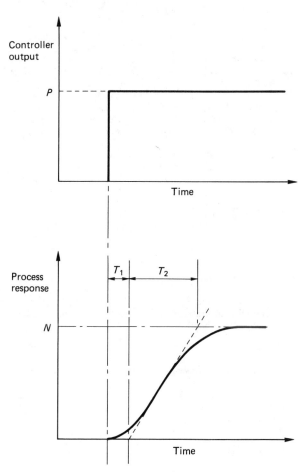

the other hand, T_i is too small, the magnitude of the integral term may cause excessive overshoot in the output response. Unstable operation is also possible when T_i is too small and the controller output value then increases continuously with time.

11.2.2.3 Derivative action

The stability of a system can be improved and any tendency to overshoot reduced by adding derivative action. Derivative action is based on the rate of change of the error, i.e.

$$\text{Controller output} = K_d \frac{dE}{dt} \quad (11.40)$$

where $K_d = KT_d$ is the controller derivative gain, and T_d is the controller 'derivative time', or 'rate'.

Equation (11.40) indicates that the derivative action is dependent on how quickly, or otherwise, the error changes. Derivative action tends, therefore, only to come into operation during the early transient part of the response of a system.

The full three-term control strategy may be written as:

$$K\left[E + \frac{1}{T_i} \int E dt + T_d \frac{dE}{dt} \right] \quad (11.41)$$

To summarise, the proportional action governs the speed of the response, the integral action improves the accuracy of the final steady state, and the derivative action improves the stability. Note that derivative action may result in poor performance of the system if the error signal is particularly noisy.

Figure 11.32 Open-loop system response to a step input

Any system which has a response similar to that shown in *Figure 11.32* has a transfer function which approximates to a first-order system with a time delay, i.e.

$$G(s) = \frac{ke^{-sT_1}}{1 + T_2s} \tag{11.42}$$

In general industrial applications, oscillatory open-loop responses are extremely rare and *Figure 11.32* is in fact representative of quite a large number of real practical processes.

If N is the process steady-state value for a controller step output of P, the system steady-state gain is

$$k = N/P \tag{11.43}$$

The 'apparent dead time' T_1 and the 'apparent time constant' T_2 can be measured directly from the process response curve. The three parameters k, T_1 and T_2 are then used in a set of empirical rules to estimate the optimum controller settings. The recommended controller settings are given in *Table 11.1*.

In fast-acting servomechanisms, where T_1 may be very small, the method is not very successful. For moderate response systems, however, the method will yield very reasonable first approximation controller settings.

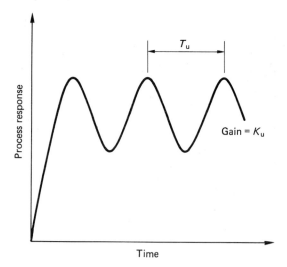

Figure 11.33 Continuous cycling method

11.2.3.2 Closed-loop 'continuous cycling' method

The process to be controlled is connected to the PID controller and the integral and derivative terms are eliminated by setting $T_d = 0$ and $T_i = \infty$. In some industrial controllers the integral term is eliminated, with $T_i = 0$. A step change is introduced and the system run with a small controller gain value K. The gain is gradually increased for a step input until constant-amplitude oscillations are obtained, as illustrated in *Figure 11.33*.

The gain K_u, which produces the constant amplitude condition, is noted and the period of the oscillation T_u is measured. These two values are then used to estimate the optimum controller settings according to the empirical rules listed in *Table 11.2*.

For a temperature control system, typical values of T_u are about 10 s for a tungsten filament lamp, 2 min for a 25 W soldering iron and 10–30 min for a 3 kW heat treatment furnace.

The PID settings obtained according to the Ziegler and Nichols methods are approximate only, and some 'fine tuning' would almost certainly be required in practice.

Table 11.2 Optimum controller settings according to Ziegler and Nichols[6]

Control action	K	T_i	T_d
P	0.5 K_u	—	—
P + I	0.45 K_u	$T_u/1.2$	—
P + I + D	0.6 K_u	$T_u/2$	$T_u/8$

11.2.4 Three-term controller with a first-order system

The block diagram of the system is depicted in *Figure 11.27* and equation (11.35) defines the closed-loop transfer function.

If a P + I controller is to be used, i.e. no derivative action, then the controller transfer function is

$$C(s) = K\left[1 + \frac{1}{T_i s}\right] \tag{11.44}$$

The process is modelled as a first-order system and its open-loop transfer function is given by equation (11.14).

Substituting equations (11.14) and (11.44) into equation (11.35) results, after some manipulation, in

$$\frac{PV(s)}{SP(s)} = \frac{\dfrac{kK}{\tau T_i}(1 + sT_i)}{\left(s^2 + \left(\dfrac{1}{\tau} + \dfrac{kK}{\tau}\right)s + \dfrac{kK}{\tau T_i}\right)} \tag{11.45}$$

Comparing the denominator with that for the generalised second-order system, i.e. equation (11.21), it can be shown that

$$2\zeta\omega_n = \left(\frac{1}{\tau} + \frac{kK}{\tau}\right) \tag{11.46}$$

and

$$\omega_n^2 = (kK/T_i\tau) \tag{11.47}$$

For the system being controlled, both k and τ are known either via a mathematical model, or an open-loop test. The controller settings K and T_i can then be calculated for a chosen damping ratio ζ and natural frequency ω_n. Alternatively, a

Table 11.1 Optimum controller settings according to Ziegler and Nichols[6]

Control action	K	T_i	T_d
P	$T_2/(T_1 k)$	—	—
P + I	$(0.9\ T_2)/(T_1\ k)$	$T_1/0.3$	—
P + I + D	$(1.2\ T_2)/(T_1\ k)$	$2T_1$	$0.5\ T_1$

controller gain can be imposed and the corresponding natural frequency evaluated.

For full PID control, an initial value of $T_d = T_i/4$ can be used. Other systems can be similarly handled to obtain the approximate PID controller settings. In all cases some fine adjustment would probably be necessary to obtain the optimum output response.

11.2.5 Disturbance sensitivity

The main problem with the classical single-loop control system is that it is not truly representative of the natural environment in which the system operates. In an ideal single-loop control system the controlled output is a function only of the input. In most practical systems, however, the control loop is merely a part of a larger system and is therefore subject to the constraints and vagaries of the larger system. This larger system, which includes the local ambient, can be a major source of disturbing influences on the controlled variable. The disturbance may be regarded as an additional input signal to the control system. Therefore, any technique which is designed to counter the effect of the disturbance must be based on a knowledge of the time-dependent nature of the disturbance and also its point of entry into the control system. Two methods commonly used to reduce the effect of external disturbances are 'feedforward' and 'cascade' control.

11.2.5.1 Feedforward control

The principle of a feedback loop is that the output is compared with the desired input and a resultant error signal acted upon by the controller to alter the output as required. This is a control action which is implemented 'after the fact'. In other words, the corrective measures are taken after the external disturbance has influenced the output. An alternative control strategy is to use a feedforward system where the disturbance is measured. If the effect of the disturbance on the output is known, then theoretically the corrective action can be taken before the disturbance can significantly influence the output. Feedforward can be a practical solution if the external disturbances are few and can be quantified and measured. A block diagram illustrating the feedforward concept is shown in *Figure 11.34*.

Feedforward control can be difficult to implement if there are too many, or perhaps unexpected, external disturbances. In *Figure 11.34*, the path which provides the corrective signal appears to go back. The strategy is still feedforward, however, because it is the disturbance which is measured and the corrective action which is taken is based on the disturbance and not the output signal. Some control systems can be optimised by using a combination of feedforward and feedback control.

11.2.5.2 Cascade control

Cascade control is implemented with the inclusion of a second feedback loop and a second controller embodied within a main feedback loop in a control system (see *Figure 11.35*). The second feedback loop is only possible in practice if there is an intermediate variable which can be measured within the overall process. Cascade control generally gives an improvement over single-loop control in coping with disturbance inputs. The time constant for the inner loop is less than that for the component it encloses and the undamped natural frequency of the system is increased. The overall effects of cascade control are an increase in the system bandwidth and a reduction in the sensitivity to disturbances entering the inner loop. Disturbances entering the outer loop are unaffected. Cascade control works best when the inner loop has a smaller time constant than the outer loop.

11.2.6 Direct digital control

Most of the standard texts on control engineering are centred on the mathematical modelling of systems and processes, and subsequently on the stability considerations of these entities. This approach requires detailed knowledge of the system constituent parts to enable the formulation of a suitable differential equation to describe the dynamic behaviour. It is often only in the idealised world of servomechanisms that adequate models can be derived. For many real processes an adequate system model can be difficult, if not impossible, to obtain.

The modern emphasis is therefore on the application of computer based control strategies which can be made to work with real systems. The recent developments in microelectronics, particularly microprocessors, has made microcomputer devices the natural choice as the controller for many systems. The microcomputer allows the implementation of such functions as arithmetic and logic manipulation, timing and counting. With many analogue input/output modules available to interface to the microcomputer, the overall 'intelligence' of the system is greatly enhanced.

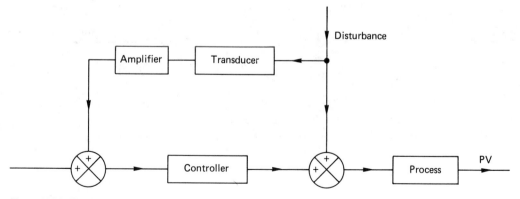

Figure 11.34 Feedforward control system

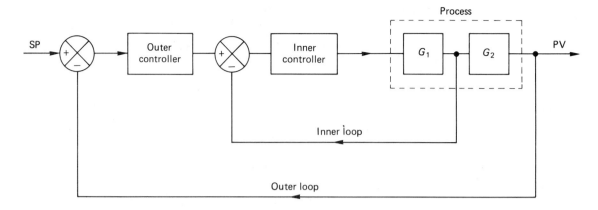

Figure 11.35 Cascade control system

The basic elements of the computer based control system include the microprocessor, memory, an input interface to measure the process variable and an output interface to supply power to the controlled process. The control effort output to the process is determined by the control strategy which takes the form of an algorithm incorporated within the computer software. The fundamental digital based control system is depicted in *Figure 11.36*.

In the generalised layout given in *Figure 11.36*, the microcomputer performs a number of tasks which would require separate elements in an equivalent analogue system. The two inputs to the microcomputer are the desired set point and a signal from the process via a feedback loop. The term 'process' is being used in a quite arbitrary sense in this context. The microcomputer firstly performs the function of comparing the process value with the set point to establish the error. The control strategy is then applied to the error value to determine the corrective action necessary. The microcomputer subsequently outputs the appropriate signal to the process via other additional elements in the system. These additional elements include the input–output interfaces between the digital-based

computer and the otherwise analogue-based control system. Digital to analogue converters (DACs) and analogue to digital converters (ADCs) are featured in Section 11.5. The transducer, which provides the link between the physical world and the electronically based control system, is covered in its various forms in Section 11.3. The essential fundamentals of microprocessor technology are outlined in Section 11.4, and the applications of microcomputer based control are described in Section 11.6.

11.2.7 Adaptive and self-tuning control

The concept of adaptive control is based on the ability to measure the system behaviour at any time and to alter the controller settings automatically to provide an optimum system response. Adaptive control has been a very active research topic over the last 10 years, but it is only recently that practical applications using adaptive controllers have appeared.

The simplest approach to adaptive control is the so-called 'gain scheduling' method (*Figure 11.37*). The principle of gain

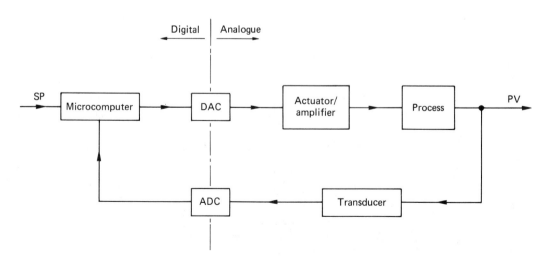

Figure 11.36 Fundamental digital-based control system

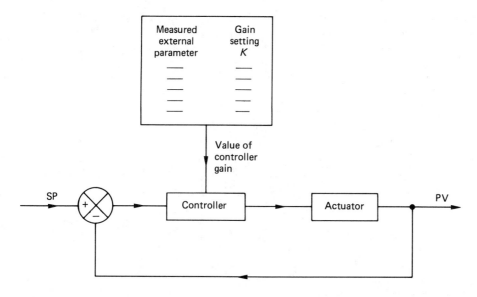

Figure 11.37 Adaptive control by gain scheduling

scheduling is that some relevant external parameter is measured and an appropriate value of gain is then selected for the controller. Gain scheduling was first developed for aileron control in high-altitude aircraft. The low air density at high altitude has a profound effect on the in-flight dynamics and the purpose of gain scheduling was to provide the pilot with a more consistent 'feel' of the aircraft's handling, independent of altitude. Gain scheduling has the advantage that the system stability margins can be well established for any value of gain and the technique is generally fast acting. The method is limited, however, since the gain adjustment is a function of only one measured parameter. In most systems the process may be subject to any number of external parameters, and the more modern adaptive controllers use some mathematical model as a basis of comparison with the actual control system (*Figure 11.38*).

The mathematical model shown in *Figure 11.38* receives the same input as the actual system and an error is created relating the difference between the actual and the model system output. The error may then be used as a basis for altering the controller settings. Obviously, the quality of the control will depend on how well the model reflects the actual system. The usual implementation of model reference adaptive control is illustrated in *Figure 11.39*.

It is worth noting that the original feedback loop is left intact such that failure of the adaptive loop will not render the system inoperative. External disturbances operating on the actual plant will change the actual/model error signal and provide the basis for retuning the controller settings via the adaptive loop. The adjustment of the controller settings implies that there must be some well-defined strategy to determine the level and nature of the adjustments made.

Self-tuning control takes the adaptive concept one stage further, in that the mathematical model of the system is also updated as more input and output data from the actual system are acquired. A schematic diagram of a self-tuning controller is shown in *Figure 11.40*. The computer based self-tuning controller estimates the system dynamics and then uses this estimate to implement the optimum controller settings. The continuous updating of the system parameters at each sampling interval is called 'recursive parameter estimation'. Previously estimated parameters are also available, and these can be used in perhaps a 'least-squares' method to provide some overall smoothing of the control function. With the latest system parameters available, the self-tuning controller then goes through some 'design' procedure to optimise the controller settings. This 'design' is usually based on the desired output response of the system. One particular design procedure is based on the root locus method for stability analysis. By adjustment of gains and time constants in the control algorithm, the method seeks to tune the transfer function and thereby govern the output response. Other procedures are often based on the rules of Ziegler and Nichols.[6] The final process in the self-tuning control cycle is the physical imposition of the optimised controller settings on the actual system.

Self-tuning control is generally applied to the more complex processes where transportation delays, non-linearities and multiple control loops greatly add to the complexity. The stability of such systems is in most cases non-deterministic since there is no generalised theory available. Traditionally then, most self-tuning controllers are based on well-

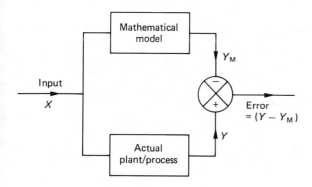

Figure 11.38 Model/actual error generation

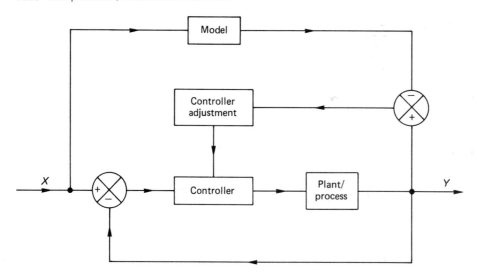

Figure 11.39 Model reference adaptive control

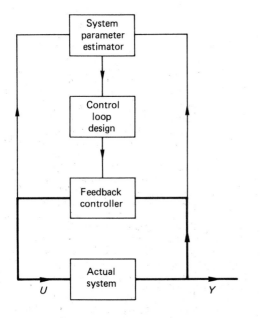

Figure 11.40 Self-tuning controller

established three-term control principles, but with the added enhancement of adaptability. A number of proprietary self-tuning controllers are available commercially (e.g. Kraus and Myron[7] describe the Foxboro Company's 'EXACT' controller). The EXACT controller is based on PID principles and uses the Ziegler and Nichols[6] rules in the self-tuning mode.

11.2.8 Sampled-data systems

The two previous subsections gave an overview of direct digital control and the natural progression to adaptive and

self-tuning controllers. The common factor which relates these concepts is the use of a computer, or microcomputer, as a central feature of the control system. The computer acts as the compensator in the control loop and the analogue-to-digital and digital-to-analogue interfaces provide the link between the digital-based computer and the otherwise analogue-based controlled system. Being digitally based, the computer operates in discrete time intervals and, indeed, the control strategy, which exists in the software, must also take a finite time for its evaluation and implementation. Time delays are also inevitable in the analogue-to-digital and the digital-to-analogue conversion processes and these cumulative time delays result in what is called a 'sampled-data system'. The difference between a sampled datum, or a discrete signal, and its continuous counterpart is displayed in *Figure 11.41*. The closure time q is the time taken to complete the digitisation of the instantaneous signal. Generally, $q \ll T$.

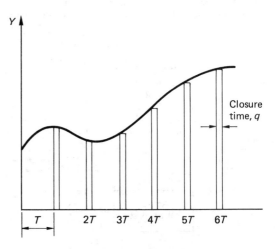

Figure 11.41 Digitisation of a continuous signal

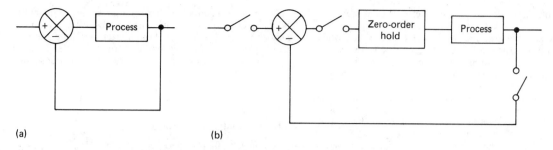

Figure 11.42 Continuous (a) and discrete (b) closed-loop control systems

It is apparent that much less information is available in the sampled-data signal as it exists only as a pulse train, interspaced with gaps in the information between the sampled points. If the sampling frequency is high enough, then this need not be troublesome (see also Section 11.6.4). The inevitable additional time delays in a sampled-data system, however, have implications regarding the overall stability of the system.

The Laplace transform method cannot be used to analyse a sampled-data system, but there is a related transform which is applicable to discrete time systems known as the 'z-transform'. The relation is:

$$z = e^{sT} \qquad (11.48)$$

The symbol z is associated with a time shift in a difference equation in the same way that s is associated with differentiation in a differential equation. Equation (11.48) then gives a conformal mapping from the s plane to the z plane and provides the means for analysing discrete time systems. The general method of solution involves the derivation of the closed-loop transfer function in terms of the Laplace variable. The equivalent discrete time system is then represented by introducing a 'zero-order hold' to account for the additional time delays in the discrete system (*Figure 11.42*).

The transfer function for a zero-order hold is:

$$\frac{1 - e^{-sT}}{s} \qquad (11.49)$$

The zero-order hold is simply included in the evaluation of the closed-loop transfer function for the discrete-time system. The next step is to replace the Laplace transforms with their equivalent z transforms. The resulting transfer function in terms of z transforms can then be analysed for stability in much the same manner as the root-locus method is used for continuous systems. A comprehensive coverage of z transform techniques and their application to the stability analysis of sampled-data systems is given by Leigh.[8]

11.2.9 Hierarchical control systems

The ultimate aim in industrial optimisation is the efficient control of complex interactive systems. Recent hardware developments and microprocessor-based controllers with extensive data-handling power and enhanced communications have opened up possibilities for the control of interlinked systems. What is required, but not yet realised, is a theoretical framework on which to base the analysis of such systems. Nonetheless, and in the absence of theory, hierarchical control systems do exist and are currently being used effectively in the control of various large-scale plant and processes.

The usual approach adopted is to subdivide the complex system into a number of more manageable parts. This is the

concept of hierarchical control which might be thought of as a subdivision in decreasing order of importance. Hierarchical control exists in two basic forms: multilayer and multilevel. Multilayer control is that in which the control tasks are subdivided in order of complexity. Multilevel control, on the other hand, is that where local control tasks are coordinated by an upper echelon of supervisory controllers. Multilayer control is illustrated concisely in an elaborate adaptive-type controller. The hierarchy is depicted in *Figure 11.43*.

The first level is that of regulation which is characterised by the classical single-closed-loop control system. Moving up the hierarchy we have optimisation of the controller parameters. Optimisation is representative of the basic adaptive controller, using simple gain scheduling or some model reference criterion. The next highest level is that of parameter adaptation. Parameter adaptation is embodied in the self-tuning controller which represents the beginnings of an 'expert system' approach. The highest level is that of model adaptation which is based on long-term comparisons between the model and the actual performance. If the system is modelled accurately to

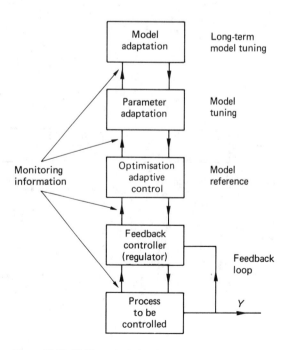

Figure 11.43 Multilayer control system

begin with, the model adaptation level might only rarely be entered.

Multilevel control is characterised as local controllers whose actions are governed by higher levels of supervisory controllers. The local controllers operate independently to achieve local targets. The function of the supervisory controller is to reconcile the interaction of the local controllers to achieve the 'best' overall performance. the multilevel concept has some similarity with cascade control but is not so amenable to analysis. Multilevel control gives rise to a pyramid-like structure, typified by that shown in *Figure 11.44*.

At the base of the pyramid are the local controllers, monitoring and adjusting individual parameters in the overall process. At the next highest level the supervisory controllers 'oversee' a more complete picture of the process. The intermediate supervisory controllers have more input data to contend with, and they might perhaps relax the control on one of the process variables while tightening up on another. This, of course, would only be done to benefit the process overall. The highest level of supervisory controller has the responsibility for the entire process. This controller may have access to additional input data which are not available to any of the lower level controllers. The main supervisory controller then is in overall 'command' and can influence any of the 'subordinate' controllers. The similarity between multilevel control and the organisation structure of an industrial company is not merely coincidental. The latter is the structural model upon which the former is based.

11.3 Instrumentation and measurement

The implementation of any control action is based on the error which exists between the process value and the required set point. Establishment of the error requires a measurement to be made of the process value. In some cases, the set point is also a measured parameter. For an electronically based control system, either analogue or digital, the connection between the real physical world and the control system is transmitted by some form or other of transducer.

11.3.1 Analogue transducers

The function of a transducer is, in the broadest sense, one of energy conversion. To this end an electric motor could be considered as a transducer which converts electrical power into mechanical power. In the context of instrumentation and measurement, the transducer provides the means of conversion between some physical variable, which is representative of the process, to a proportional voltage or current. The proportional voltage, or current, then forms the basic signal upon which the control of the process is based.

In manufacturing engineering, the most commonly measured parameters, for control purposes, are position and speed. These two parameters form the basic feedback variables associated with the control of machine tools, robots and automatic guided vehicles. Feedback control may also be based on a measurement of force, or on some visual criterion, particularly in the control of robotic arms and manipulators. In the general field of process control, the measured variables might also include temperature, flow rate, torque, power, displacement, angle, strain, humidity, light intensity, acidity, sound or perhaps simply time. The list as given probably represents about 90% of all the measurements made in the industrial environment. With the appropriate transducer and signal processing, all of the above parameters can be converted into a proportional output voltage.

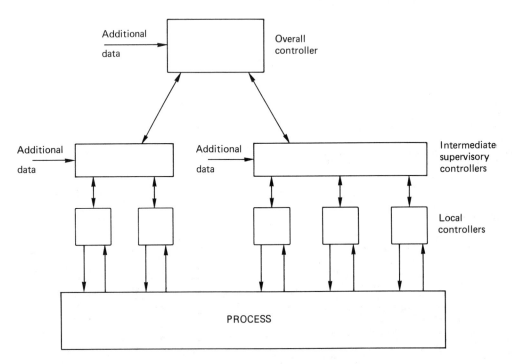

Figure 11.44 Multilevel control system

11.3.2 Digital transducers

Most conventional transducers have an analogue output voltage which could be measured using fairly traditional instrumentation, including for example an oscilloscope or a voltmeter. The output of many transducers, however, is in the millivolt range and, more often than not, the basic output must be further processed, or conditioned, to produce a useful signal. The processing required might only involve simple amplification but, if the controller is digitally based, then additional signal processing is needed to convert the analogue voltage into a digital representation. This latter processing is performed using an analogue-to-digital converter (see Section 11.5.3). An obvious advantage would be a transducer which can produce a direct digital output. Such a device could then dispense with the intricacies both of signal conditioning and analogue to digital conversion. There are, however, very few devices available which can easily produce a digital output in response to a physical variable. The absolute digital shaft encoder (*Figure 11.45*) is one of the few transducers which satisfy the digital-output specification in the strictest sense.

The device incorporates a sectioned disc which is mounted on a rotating element. The disc is divided into a number of tracks, three in this case, and a number of sectors. If the number of tracks is denoted by m, then the number of sectors is numerically given by 2^m. With $m = 3$, the number of sectors is 8 and this also denotes the number of angular resolvable sections on the disc. For each resolvable section a unique pattern is generated. These patterns may be detected by optical, magnetic or direct contact methods. Resolution may be improved with the addition of more tracks. Twelve tracks, for example, give an angular resolution to within $\pm 0.09°$ rotation. The absolute shaft encoder may be used to measure either rotational speed or angular displacement.

In *Figure 11.45* a simple binary code is used to illustrate the principle of the absolute shaft encoder. It can be seen that between sectors 3 and 4 and sectors 7 and 0, all the tracks undergo a change in pattern. When all the tracks change there must always be one which changes first before the others. Any misalignment of the track sensors, therefore, could generate a considerable error in rotational-position measurement. For this reason, practical shaft encoders use some 'unit distance code' in preference to a binary code. A unit distance code, such as the 'Gray code' described in Section 11.4.3, is one where only one track changes pattern for each consecutive resolvable position.

11.3.3 Measurement of angular position

The absolute shaft encoder, discussed above, is a suitable transducer for the measurement of angular position and has the added advantage of a digital output which can be interfaced directly to a computer. A close relative, but one which operates on a rather different principle, is the incremental shaft encoder. The incremental shaft encoder consists of a disc in which a series of holes is arranged around the periphery. A light source, usually a light emitting diode (LED), is positioned on one side of the disc and a photosensitive transistor on the other (*Figure 11.46*).

Rotation of the disc generates a series of light pulses which are registered as a square pulse train output from the photosensitive transistor. By counting the number of pulses relative to a known datum, the angular position of the disc can be determined. By using two light sources and two photosensitive transistors, with a 90° phase difference between the two output signals, the direction of rotation may also be ascertained. The phase relationship between the two signals for clockwise and anti-clockwise rotation is illustrated in *Figure 11.47*. A similar system may equally well be used to measure linear position and direction.

In clockwise, or right-hand movement, the signal from S_2 comes on while the signal from S_1 is still on. For anti-clockwise, or left-hand movement, S_2 comes on while the signal from S_1 is off. This relative switching of the two signals forms the basis of the direction sensing.

In manufacturing applications, high resolution incremental encoders are often used as position feedback sensors, either rotating with a lead screw, or in linear motion with a machine

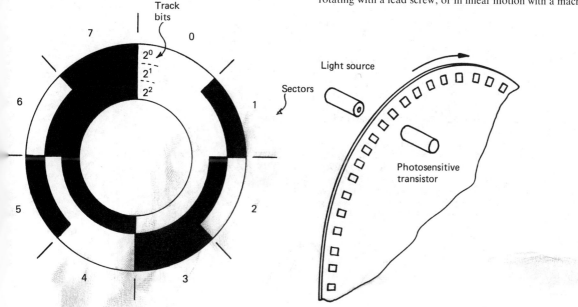

Figure 11.45 Absolute digital shaft encoder (3-bit binary code)

Figure 11.46 Incremental shaft encoder

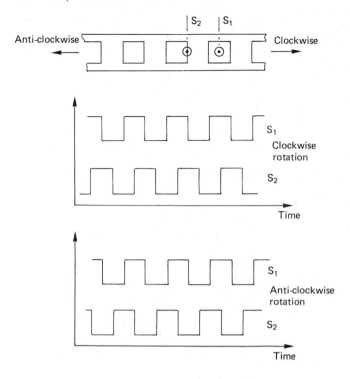

Figure 11.47 Direction sensing with an incremental shaft encoder

slide. They are also used to measure the joint positions in robotic manipulators. In these applications, where high accuracy is essential, the encoders consist of two glass gratings, one of which is fixed and the other which moves. Light which is transmitted through both gratings varies in intensity with the relative displacement of the gratings. The light intensity is picked up by two phototransistors and the signals are used to measure relative position and direction in much the same manner as described previously. The grating pitch is typically 100 per millimetre, but the resolution can be greatly increased by using the moiré fringe effect. Moiré fringes are a diffraction effect which results when the gratings are inclined at a slight angle to each other. The important feature is that the spacing of the interference fringes is many times the spacing of the grid lines. A grid inclination angle of just less than 0.5°, with 100 grid lines per millimetre, would produce a fringe pattern with a 1 mm spacing.

The main disadvantages of the incremental encoder are the possibility of missing a pulse during the count and the loss of the datum reference in the event of power failure. Electrical noise and dirt between the light source and the phototransistor are also problematical. Nonetheless, incremental encoders are cheaper and generally more accurate than absolute encoders.

When accuracy is less important, an angular position transducer based on the variable potential divider (*Figure 11.48*) can be used.

By measuring the voltage between points A and B, a linear relationship between angular displacement, θ and voltage v is obtained:

$$\theta = v(\Theta/V) \tag{11.50}$$

The maximum voltage occurs when the sliding contact at B reaches the maximum rotation at Θ.

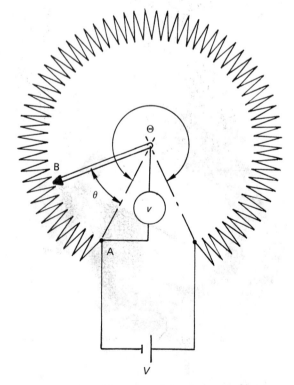

Figure 11.48 Variable potential divider position transducer

As an angular position transducer, however, the variable potential divider has a number of limitations. Equation (11.50) is only strictly applicable if the internal resistance of the voltmeter used to measure v can be considered to be infinite. Taking the internal resistance of the voltmeter R_i into account, Ohm's law for parallel resistors can be used to show that the total resistance between points A and B is:

$$R_{\text{total}} = (R_i R_\theta)/(R_i + R_\theta) = R_\theta/([1 + (R_\theta/R_i)] \qquad (11.51)$$

where R_θ is the resistance along the arc of resistor element enclosed by the angle θ.

Equation (11.51) shows that R_{total} will approach the value of R_θ only when R_i tends to infinity. This is referred to as a 'loading error' which causes the net resistance between A and B to be reduced due to the additional circuit through the voltmeter.

Further limitations are associated with the sliding contact at B which is a source of friction and inevitable wear.

11.3.4 Measurement of linear displacement

The incremental encoder and the variable potential divider can each be easily adapted for the measurement of linear displacement. The most common transducer used to measure linear displacement however is the linear variable differential transformer (LVDT). LVDTs incorporate an iron core which can move freely within a primary, or power coil and two secondary coils (*Figure 11.49*).

The secondary coils are connected in series opposition and are equally positioned with respect to the primary coil. When the core is centrally located the e.m.f.s generated in the secondary coils are equal and opposite and the net output voltage is zero. Displacement of the core relative to the central zero produces a proportional a.c. output which is converted to a d.c. output via rectification and smoothing circuits. LVDT transducers are essentially immune to friction and wear problems, they have infinite resolution and are highly linear and accurate. LVDTs are available with measurement ranges up to about 1 m.

There are, in addition, a wide range of other displacement transducers which rely on inductive or capacitive effects. These are mostly restricted to the measurement of relatively small displacements of the order of 0.1 mm.

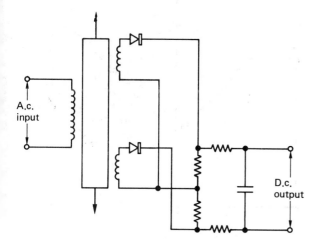

Figure 11.49 Linear variable differential transformer

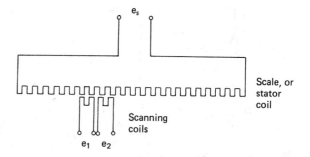

Figure 11.50 Linear scale, or 'inductosyn'

Linear scales, also known as 'inductosyns', are frequently used for the precision measurement of linear position (*Figure 11.50*). They are especially prevalent as a position sensor in machine tool slides. The scale coil is in the form of a linear 'stator' and the scanning coils operate as movable 'rotors'. The pitch of the stator and rotor coils, typically 2 mm, is accurately formed and the two rotor coils are displaced relative to each other by 0.25 of the pitch. The unit operates in much the same way as a transformer, with the rotor coils representing the primary windings and the stator coil the secondary windings. An a.c. voltage is supplied to each of the rotor coils with a phase shift of 90° between the two signals. The voltage generated in the stator coil is the sum of the two rotor coil voltages. The scale coil output can be decoded through measurement of the phase changes induced by movement of the rotor coils. Accuracies of 0.001 mm are possible, but it is necessary to count the number of cycles generated since the output signal is periodic. The inductosyn scale was developed from the synchroresolver which was the main shaft position transducer prior to the introduction of modern digital systems.

11.3.5 Measurement of rotational speed

Rotational speed is a commonly monitored parameter in general engineering measurement applications and there is considerable variety in the types of transducer used for the purpose. The absolute digital shaft encoder is suitable, but the incremental shaft encoder is more easily adapted for the measurement of rotational speed. The output waveform from any one of the photosensitive transistors on the incremental encoder is either a square pulse train, or a sinusoidal like signal. Measurement of the output signal frequency can then be simply converted to the rotational speed. The signal frequency can be determined in software using the timing facilities available on microcomputers. Alternatively, the signal can be processed through a frequency-to-voltage converter, to produce a d.c. output voltage which can be calibrated in direct proportion with the rotational speed. The latter method represents a more complete speed-measuring sensor which facilitates easier interfacing to a computer. Frequency-to-voltage converters are readily available in a range of integrated circuit forms.

The frequency-to-voltage converter also forms an integral part of many other types of speed-measuring transducer. One fairly common speed-measuring device is the magnetic pick-up. This passive transducer, requiring no external power supply, responds to the movement of ferrous parts past its pole piece. *Figure 11.51* shows the basic configuration which requires the transducer to be in close proximity to a toothed wheel. The output signal is approximately sinusoidal and a

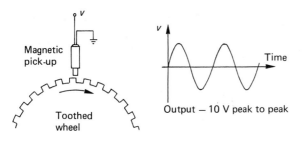

Figure 11.51 Magnetic pick-up speed sensor

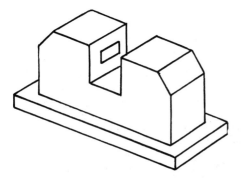

Figure 11.52 Slotted opto-switch

frequency-to-voltage converter completes the signal processing for adaptation as a speed sensor.

Optical sensing devices are also fairly common and these include slotted opto-switches (*Figure 11.52*). These usually incorporate a LED as a light source and a photosensitive transistor as a receiving device. A toothed wheel passing between the receiver and transmitter will generate the necessary pulsed output signal. Frequency-to-voltage conversion again completes the system for speed measurement.

Optical devices containing both the receiver and the transmitter on the one unit are also available (*Figure 11.53*). These units respond to the passage of a reflective surface as shown. This particular type of optical sensor is often used in hand-held tachometers.

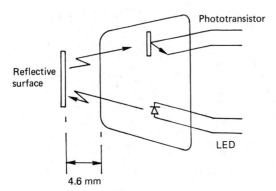

Figure 11.53 Opto-electrical sensor

Last, but not least, in the list is the well established and fairly traditional tachogenerator. These are often found in older systems where they were used as the standard transducer for speed measurement. The tachogenerator is a small permanent magnet alternator with an output voltage almost exactly proportional to speed. The voltage generated at full speed is typically of the order of 120 V. This voltage is much too high to be interfaced to a computer-based system and the output must be rectified and reduced through a voltage divider to be compatible with a digital system.

The frequency of the tachogenerator output is also proportional to speed, and alternative signal conditioning circuitry can be added to take advantage of the frequency characteristics. Conditioning circuits based on the frequency of the tachogenerator output either use a pulse counting system, or a frequency-to-voltage converter.

Due to the high cost and the high driving power required, tachogenerators are now only used on fairly large rotating machine elements.

11.3.6 Transducers based on strain measurement

In transducer technology, 'strain bridge' circuits are featured predominantly. The underlying reason for this is simply that the strain, which is essentially a displacement, can be measured as the resultant of a wide range of physical causes. The versatility of strain measurement is such that it finds application in devices adapted for the measurement of force or load, torque, pressure, flow rate, acceleration, displacement and seismic activity amongst others. The common factor involved is that the physical variable to be measured ultimately causes, by design, a deflection in an elastic member. Direct measurement of the strain associated with the deflection can then be related back to the physical variable through a calibration.

Strain-sensitive sensors are invariably incorporated into a Wheatstone bridge circuit (*Figure 11.54*). The Wheatstone

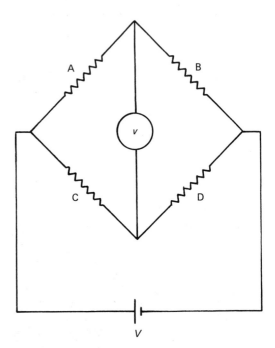

Figure 11.54 Resistive element Wheatstone bridge

bridge consists of four resistors, a voltage source and, usually, a high impedance instrument to record the bridge output voltage. When the bridge is 'balanced', no current is drawn through the voltmeter. From Ohm's law it can be shown that for balanced conditions the ratios A/C and B/D must be equal. Thus if C is a variable resistor, A may be determined if C, B and D are all known.

In most cases the sensing element constitutes one 'arm' of the bridge, while the other three arms have constant resistances, usually equal to the sensing-element resistance at particular reference conditions. These bridges normally operate in an unbalanced mode where the voltmeter reading is used to quantify the changing conditions at the sensing element, being the result of changes in the physical variable which is measured.

Assuming initially that the bridge is in balance with all elements having a resistance R, the sensing element is then subjected to a condition which causes its resistance to change to $(R + \Delta R)$. The bridge thus becomes unbalanced and a potential difference v is generated and registered on the voltmeter. Analysis of the circuit shows that provided $R \gg \Delta R$ an approximately linear relationship exists between v and ΔR:

$$v = (V\Delta R)/(4R) \tag{11.52}$$

A detailed proof of equation (11.52) is given by Fraser and Milne.[9]

Even if the voltage recording instrument has a low internal resistance, e.g. a centre-zero galvanometer, then Thevenin's theorem shows that a near-linear relationship still applies provided that ΔR remains small.

Wheatstone bridges may also be energised with a.c., in which case they become known as 'impedance bridges'. The same principles still apply, although for equal ratios of bridge impedances only the magnitude of the voltages will be equal. The potentials may still be out of phase and a phase-sensitive instrument would indicate imbalance.

The simple resistive element Wheatstone bridge provides the basic circuit to be used with a variable-resistance sensor to form a working transducer. The sensor could be a thermistor in a temperature-sensitive system, or most commonly a strain gauge in a displacement-sensitive system.

The principle embodied in a strain gauge is that a conductor, of initially uniform dimensions, will alter its proportions when subjected to an applied stress. Since the resistance of the conductor is also dependent on the physical proportions, then the stress will also influence the resistance of the conductor.

The resistance of a conductor is given by

$$R = \beta(L/A) \tag{11.53}$$

where β is the resistivity of the material, L is the length of the conductor, and A is the cross-sectional area of the conductor. For a cylindrical conductor, the cross-sectional area is $\pi D^2/4$, where D is the diameter of the cylinder.

Differentiating equation (11.53) gives

$$(dR/R) = (d\beta/\beta) + (dL/L) - (dA/A)$$

Now

$$(dA/A) = 2(dD/D)$$

Thus

$$(dR/R) = (d\beta/\beta) + (dL/L) - 2(dD/D)$$

Multiplying through by (L/dL) gives

$$(dR/R)(L/dL) = (d\beta/\beta)(L/dL) + 1 - 2(dD/D)(L/dL) \tag{11.54}$$

where (dL/L) is the ratio of the change in length to the original length which, by definition, is the axial strain ε_A.

Similarly, (dD/D) is, by definition, the transverse strain ε_T.

Also $(dD/D)(L/dL) = (\varepsilon_T/\varepsilon_A) = -\nu$, where ν is Poisson's ratio for the material.

Substitution of these parameters into equation (11.54) gives

$$(dR/R)(L/dL) = (d\beta/\beta)(1/\varepsilon_A) + 1 + 2\nu \tag{11.55}$$

The subject of equation (11.55), $(dR/R)(L/dL)$, is called the 'gauge factor' and is traditionally denoted by G. The constancy of the right-hand side of equation (11.55) determines the suitability of any material to function as a strain gauge.

Since

$$G = (dR/R)(L/dL) = (dR/R)(1/\varepsilon_A)$$

Then

$$dR = GR\varepsilon_A \tag{11.56}$$

Values of G and R are specified by the manufacturer of the strain gauge, and equation (11.56) shows that the change in resistance dR is directly proportional to the axial strain ε_A.

Evidently a high value of G would be advantageous in producing a high change in resistance for any input axial strain. Metallic strain-gauge materials have gauge factors of about 2, although some semiconductor materials have gauge factors in excess of 100. The commonest form of strain gauge is the bonded-foil type (*Figure 11.55*).

Bonded foil strain gauges are rigidly fixed to the strained member and are electrically insulated from it. In typical applications, the change in resistance is very small (of the order of 0.25 mΩ per microstrain) and, as a result, the measurement is very sensitive to temperature variations. Unless special temperature-compensating gauges are available, it becomes necessary to provide some other means of compensation for temperature drift. The usual method is to incorporate a 'dummy' gauge in the bridge circuit (*Figure 11.56*).

The dummy gauge is at all times subject to the same ambient temperature as the active gauge, but it is attached to a separate, unstrained piece of the same material. In this manner resistance changes due to temperature variation are cancelled out and the output voltage is a result only of applied strain on the active gauge. The extra 'compensation lead' shown in the diagram is used to negate the effect of temperature variations in the wiring to the active gauge. This is not an

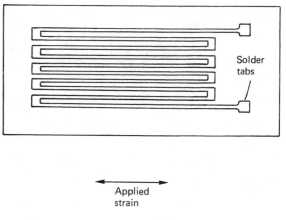

Solder tabs

Applied strain

Figure 11.55 Bonded-foil-type strain gauge

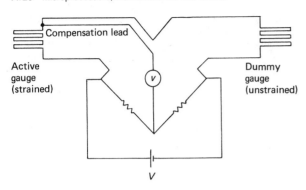

Figure 11.56 Strain bridge with temperature compensation

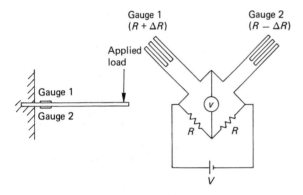

Figure 11.57 Cantilever-type load transducer

essential part of the circuit, but is advisable if the connecting cables to the active gauge are particularly long.

The sensitivity of the circuit can be increased if it is known that the flexural member is subject to equal and opposite strains on its opposite surfaces. *Figure 11.57* illustrates an application to a load transducer in the form of a simple cantilever.

The bridge output with two active gauges is:

$$v = (V/2)/(\Delta R/R) \qquad (11.57)$$

By using active gauges on all four arms of the bridge, the sensitivity can again be doubled. In either case, temperature compensation is automatic. The three forms of strain bridge described, using one, two and four active gauges are known as quarter-, half- and full-bridge circuits, respectively.

Even with four active elements in the bridge circuit the output is still measured in terms of millivolts and amplification of the output signal is essential. Furthermore, since it is almost impossible to ensure equality of the resistances within the tolerance demanded by the sensitivity of the circuit, a zero-adjustment potentiometer is also virtually essential. Other problems are associated with noise, zero drift and non-linearity, but custom-built precision amplifiers are readily available to condition output signals from resistive strain bridge circuits.

Strain bridges are often used as the sensing element in force transducers and commercial load cells. Proprietary units are available in a large variety of types and sizes for the measure-

ment of both tensile and compressive loads. Strain bridges are also attached to tubular elements to function in the capacity of a torque meter. Flexible diaphragms are also available with a strain bridge output and these are used extensively in pressure-sensing applications.

The basic cantilevered structure depicted in *Figure 11.57* can further be adapted for use as an accelerometer, or a seismic sensitive device. In these two cases it is the force due to an acceleration which is sensed through the measurement of strain. The acceleration is then determined via Newton's law relating force and acceleration. Impulsive and shock forces may also be monitored using a cantilevered element strain bridge system.

Volumetric and mass flow rate can equally well be measured. This is done using a strain bridge which is attached to a flexural member which deflects under the drag force exerted by the flowing fluid.

The examples cited are by no means exhaustive and, indeed, it may be the measurement of the strain itself which is the primary concern. This is certainly the case in experimental stress analysis.

The strain bridge circuit undoubtedly exhibits great flexibility in its applications as a commercial transducer. Furthermore, it is probably the most extensively used sensor and circuit in many custom-built measurement applications.

11.3.7 General force, pressure and acceleration transducers

In the previous section the utility of the resistive strain bridge was highlighted in its varied applications as a transducer. In all of the examples considered it is the displacement of a flexural member which is actually measured. The measured displacement is then related back to the physical variable of interest through a calibration. Since the measurement parameter is displacement, then there is no reason why any of the other methods featured in Section 11.3.4 cannot be used as an alternative to the resistive strain bridge. In practice, and with the exception of the linear-type encoders, this is actually found to be the case. *Figure 11.58* shows four common types of commercial pressure transducer.

In all types, a change in pressure causes a deflection in a diaphragm which can be sensed using the various displacement transducers. Transducer (*a*) generates a change in capacitance between the flexible diaphragm and a central electrode. The conditioning circuitry required is quite complex and includes a capacitive bridge. The high sensitivity of the capacitance transducer, however, makes it a popular choice for many applications. The common capacitive microphone operates on these principles.

Type (*b*), which is illustrated as a differential pressure transducer, relies on the induced eddy currents generated in the coils for the sensing of the diaphragm deflection. Types (*c*) and (*d*), respectively, are miniaturised versions of the linear variable differential transformer and the variable potential divider.

Many other mechanical configurations are possible and the basic sensing principles outlined above find general application in a wide range of force, pressure and acceleration transducers.

Force, pressure and acceleration transducers based on the piezoelectric effect belong in a totally separate class. A piezoelectric crystal, e.g. quartz, produces an electrostatic charge when subjected to a compressive strain. A difficulty exists, however, since the charge is generated as a consequence of the application of the mechanical load. If the charge is subsequently dissipated, then no new charge would be generated until the loading state again changes. This feature

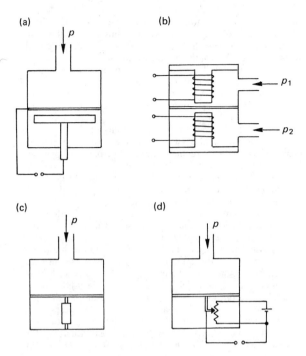

(a) p

(b) p_1 p_2

(c) p

(d) p

Figure 11.58 Commercial pressure transducers

makes piezoelectric devices unsuitable for static measurements. They are eminently suitable, however, for the measurement of dynamic loads. Therefore, piezoelectric transducers are used extensively to monitor the time variation of pressure within the cylinders of internal combustion engines and as an accelerometer in vibration monitoring.

Measurement of the charge, however, without dissipating it, is problematic and piezoelectric transducers require sophisticated and expensive signal conditioning.

11.3.8 Proximity sensing

Proximity sensing has many applications in manufacturing engineering ranging from simple component counting functions to the ON/OFF switching of actuators and other devices in some overall control context. In most cases the basic requirement is for a 'GO–NO GO' type measurement, and proximity switches are extensively used in these applications.

Spring-loaded microswitches are perhaps the simplest form of proximity sensor. These require mechanical contact to activate the switch and can be used effectively in the sequence control of pneumatic and hydraulic cylinders. In typical operations, the movement of a ram is initiated and allowed to continue until a microswitch is tripped by a collar attached to the ram. The microswitch then perhaps activates a solenoid operated control valve to reverse, or cut off the supply to the ram. Such systems can be extended as required to build up a fairly simple automated sequential process.

As the name might suggest, however, proximity switches are basically non-contacting devices. The magnetic pick-up and the slotted opto-switch, described for rotational speed measurement in Section 11.3.5, are both proximity devices and they can be used directly in this capacity. Many inductive and capacitive devices are also designed to function as a proximity switch. The most common type of proximity switch

sensors nonetheless are the devices which operate on the 'Hall effect'. The Hall effect, named after its discoverer, is the phenomenon whereby an electron passing through a semiconductor material, and in the presence of a magnetic field, will experience a force which will deflect the electron from its normal path. The effect is similar to that used to position the light spot on a cathode ray oscilloscope, but on a much smaller scale. Modern Hall-effect sensors are supplied as complete integrated circuits with a convenient voltage output. A small permanent magnet is also usually supplied for attachment to the component to be sensed. Hall-effect sensors have the advantages of high-speed operation, no moving parts and are free from switch 'bouncing' effects.

All the above devices essentially rely on close proximity (a few millimetres or so) for their actuation. For longer range proximity sensing, light-beam transmitting and receiving devices can be used. A typical application is in the batch counting of components moving along a conveyor belt. In this type of application, a light beam is transmitted across a space, (which may be up to about 15 m) and is continuously monitored by a receiving device. The passage of a non-transparent object interrupts the beam and triggers a voltage pulse which can be measured and recorded. Reflective type proximity sensors, with transmitter and receiver housed in the one unit, are also available.

11.3.9 Measurement of flow rate

The measurement of flow rate in process control covers a range of different transducers which can be subdivided as:

(1) pressure differential devices,
(2) rotary devices,
(3) vortex shedding devices, and
(4) non-intrusive devices.

Pressure differential devices include Venturi-meters, orifice plates, flow nozzles, or any similar device which presents a constriction in the flow path. For incompressible flows, the constriction causes the fluid to accelerate with an attendant drop in pressure. The operating principle of these devices is based on two fundamental laws of fluid dynamics. These are the equation of mass continuity and the steady flow energy equation, or Bernoulli equation. Application of these laws across any type of constriction results in an equation of the form:

$$\text{Flow rate} = B\sqrt{\Delta p} \qquad (11.58)$$

where B is, in general, an experimentally determined constant, and Δp is the measured pressure differential across the constriction.

The measurement of Δp is performed with a suitable differential pressure transducer and, provided B is a well behaved 'constant', the flow rate can be accurately calibrated against the square root of the output voltage from the pressure transducer.

If the fluid is a gas it is usually necessary to measure the local gas temperature and the absolute barometric pressure in addition to the meter differential pressure. The common differential pressure flow metering devices are shown in *Figure 11.59*.

Rotary flow meters may either be of the positive displacement type or the turbine type. In either case it is the speed of the rotational element which is measured and converted into an equivalent flow rate. In positive displacement meters, the total number of revolutions is also counted and used as a measure of the cumulative volume of flow. *Figure 11.60* illustrates three contemporary rotary-type flow meters.

(a)   (b)   (c)

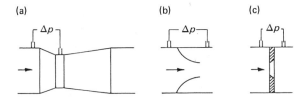

Figure 11.59   Differential pressure flow meters: (a) Venturi; (b) flow nozzle; (c) orifice

(a)

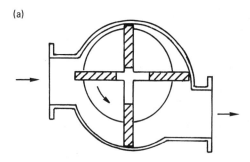

(b)

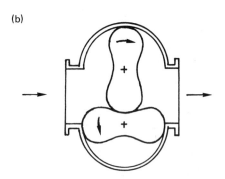

(c)

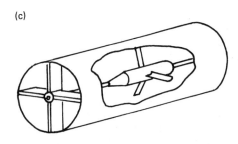

Figure 11.60   Rotary type volumetric flow meters: (a) rotary vane; (b) lobed rotor; (c) turbine

The rotary-vane and lobed rotor types are positive displacement meters in that, for each revolution of the rotor, a fixed quantity of fluid is passed through from inlet to outlet. The turbine type meter is not a positive displacement device.

In most rotary-type flowmeters, the rotational speed is detected with some form of proximity sensor which responds

to the passage of a vane, lobe or blade as the case may be. A frequency-to-voltage converter is normally used in the signal conditioning process to produce a steady output voltage in proportion to the volumetric flow rate.

Vortex shedding flowmeters are comparative newcomers in the field of flow measurement, but they have become well established as a general purpose flow monitoring device. The operating principle stems from the fact that when a 'bluff' body is exposed to an oncoming flow, the fluid particles cannot follow the severe surface contours of the body. The flow therefore breaks away from the body and periodic vortices, or eddies, are shed alternately from the upper and lower extremities of the body (*Figure 11.61*).

Over a reasonable range, the periodicity of the vortex shedding is in direct proportion to the fluid velocity. A measurement of the shedding frequency can, therefore, be related to the mean velocity and hence also to the flow rate. The vortex shedding frequency is sensed by means of a variety of methods. These include pressure variation with a piezoelectric pressure transducer, temperature variation with a heated thermistor and the interference of an ultrasonic beam transmitted across the wake behind the body. Vortex shedding flowmeters have no moving parts and can be used with any fluid. The output signal is quite noisy, however, and the linear operating range might be too restrictive in some cases.

The last group of flow metering devices are the non-instrusive types. These are by far the most expensive flowmeters, but they have the advantage of leaving the flow undisturbed with no attendant loss in pressure.

Laser–Doppler anemometry is a non-intrusive method which transmits a low power laser beam into the flow. The light scattered by minute particles flowing with the fluid transmits a signal back to a receiver and a measurement of the Doppler shift frequency can be used to determine the local fluid velocity. The Dopper shift frequency is also similarly used in some of the proprietary types of ultrasonic flow meters.

Many other ultrasonic flowmeters operate on a beat-frequency measurement. These meters incorporate two separate ultrasonic transmitter/receiver probes which pass signals along opposite directions across the flow path (*Figure 11.62*).

Due to the flow of fluid, the transit times of the two signals are unequal and, by multiplying the signals together, a beat frequency can be measured. The beat frequency is proportionally related to the fluid velocity, and its measurement can therefore be calibrated for volumetric flow rate.

The electromagnetic type of non-intrusive flowmeter is worthy of mention due to its wide usage in industrial applications. These flowmeters require the fluid to be conducting and their operating principle is based on Faraday's laws of electromagnetic induction. A magnetic flux is generated in a direc-

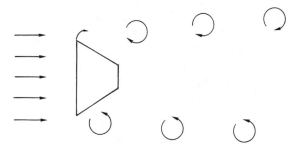

Figure 11.61   Vortex shedding from a bluff body

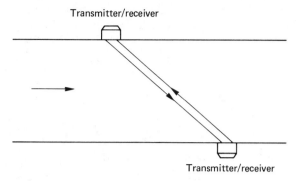

Figure 11.62 Ultrasonic 'beat frequency' flowmeter

tion perpendicular to the flow. As the flow constitutes a moving conductor, then a measurable voltage is generated between two electrodes positioned perpendicularly both to the magnetic field and the flow path. The magnetic field is energised with an alternating supply to prevent electrolysis and other problems associated with magneto-hydrodynamic action.

11.3.10 Temperature measurement

The measurement of temperature is often required in system monitoring and general process-control applications. For digital-based systems, thermoelectric devices are commonly utilised to generate the output signal in voltage form. The thermoelectric effect is a consequence of the fact that the number of free electrons which can exist in a metal is a function both of the temperature and the composition of the metal. For practical purposes it is sufficient to know that, when two dissimilar metals are joined together at their ends, a potential difference will exist if the two ends are held at different temperatures. This physical phenomenon provides the basis of a voltage-output temperature transducer (*Figure 11.63*).

In wiring up to the voltmeter a third metal (usually copper) has been introduced into the circuit. This has no adverse effect provided that the temperatures at the two copper connections are both equal.

The thermocouple, however, is a differential rather than an absolute temperature measuring device. To function as a temperature sensor, therefore, the temperature at one of the junctions (the reference junction) must be known. The most convenient reproducible reference temperature available is that of melting ice at, or near, standard atmospheric pressure. This is 0°C, and a range of specially selected materials has been rigorously calibrated against this reference to be used as standard thermocouple combinations in industrial applications. A popular pair is nickel–chromium (Chromel) and nickel–aluminium (Alumel). These two metals used in combination give the standardised 'K' calibration, approximately 4 mV per 100°C temperature difference.

The reference temperature need not be 0°C but can be any other known temperature, often room temperature (15–20°C). In these cases the reduced potential due to the higher reference temperature can be compensated for by adding the potential which would exist between 0°C and room temperature. This introduces a small error, however, since the calibrated relationship between temperature and voltage, although very near, is not quite linear. The output voltage, in addition, must be amplified to be compatible with a digital interface. Purpose-built amplifiers are readily available, however, particularly for K-type thermocouples, and these also usually incorporate a built-in ice-point compensation circuit.

If high accuracy and long-term stability are required, then a 'platinum-resistance thermometer' may be used as an alternative to the thermocouple. The temperature–resistance characteristics of the platinum-resistance thermometer are standardised in BS 1904: 1981.[10] In practice, the platinum sensor coil ultimately forms one arm of a Wheatstone bridge and matched compensating leads must be used in wiring up to the bridge (*Figure 11.64*).

If absolute accuracy is not so important, then thermally sensitive resistors, or 'thermistors' as they are generally called, can be used. Thermistors are metal oxide semiconductors with a large, usually negative, temperature coefficient of resistance. Thermistors are available in bead, rod, disc, washer and probe form. They are extremely sensitive and have a very fast transient response. However, the resistance–temperature characteristic is very non-linear and this has been a significant limitation on their general use in industry. The non-linearity is much less of a problem in a microprocessor-based system, since the non-linear effects can be easily accommodated in software, and thermistors are being increasingly used over much larger ranges than before. Thermistors with positive temperature coefficients of resistance are also available and these have particular application in over-temperature protection circuits.

All the temperature sensing devices so far considered have an upper limiting temperature of about 1000°C. The S-type thermocouple can operate up to 1700°C, but for higher temperatures, radiation-detection devices must be used. Radiation sensors incorporate a pyroelectric element which functions as a radiation-sensitive capacitance. The voltage developed across the element decreases as the frequency of the incident radiation increases. Commercial temperature-sensing devices are used in the steelmaking industries and in intruder alarm systems. The basic sensor principle can also be adapted to function as a gas detector, and it is used as such to measure the presence and density of carbon monoxide, carbon dioxide and methane in gas flows.

11.3.11 Level sensing

The simplest form of level sensor is the float-type switch. These switches are either ON or OFF depending on whether the liquid level is above or below the float. A series arrange-

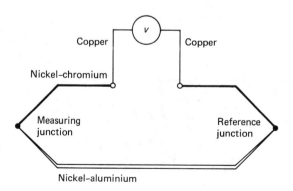

Figure 11.63 Basic thermocouple temperature transducer

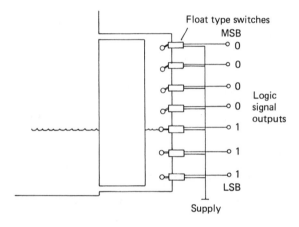

Figure 11.64 Platinum-resistance thermometer

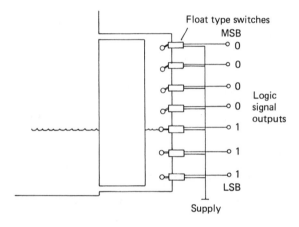

Figure 11.65 Digital output level transducer. MSB, most significant bit; LSB, least significant bit

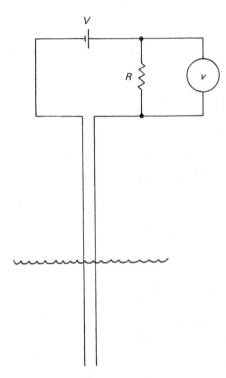

Figure 11.66 Variable output level sensor

ment of float switches (see *Figure 11.65*) could be adapted to operate as a digital level transducer.

Other digital output level transducers, similar in principle to that shown in *Figure 11.65*, use capacitive, inductive, or ultrasonic sensors to detect the liquid level. Ultrasonic and optical sensors are also used in level devices which measure the transit time of a beam reflected off the liquid free surface.

A variable output, but non-linear, level sensor is shown in *Figure 11.66*. Provided the liquid is conducting, the resistance between the two unconnected wires will depend on their depth of immersion, i.e. the liquid level. The voltage measured across the fixed resistor is inversely proportional to the depth of immersion of the two probe wires. Commercial sensors of this type, marketed as wave monitors, use an a.c. supply to prevent the adverse effects of electrolysis.

11.3.12 Bar-code readers

A form of optical data input which is finding increasing application is the bar-code reader. The code consists of a series of black and white vertical lines which are printed onto the object (*Figure 11.67*).

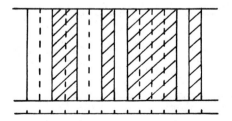

Figure 11.67 Section of a bar code

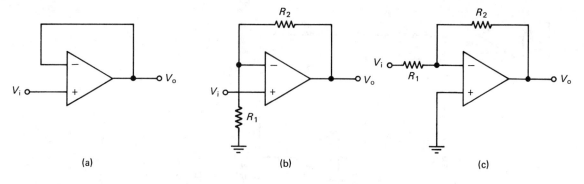

Figure 11.68 Operational amplifier circuits: (a) non-inverting unity gain; (b) non-inverting gain $V_0 = [(R_1 + R_2)/R_1]V_i$; (c) inverting gain $V_0 = -(R_2/R_1)V_i$

The code is read by an optical sensor which incorporates a lamp, a phototransistor and a number of optical focusing lenses. The decoding software is necessarily complex, however, since the speed at which the code is read can vary.

A typical bar code might consist of a start pattern (101), five 7-bit characters, a check sequence, a second group of five characters and an end pattern. Two consecutive black bands represent a bit value of 1, while two consecutive white bands represent a bit value of 0. The code is designed such that every character starts with a white band and ends with a black band. This ensures that every character starts and ends with a 1–0 transition. In addition, every character code includes at least one 1–0 transition within the code.

The decoding program must include a timing loop to determine the speed of reading. The timing is usually based on a count of the 1–0 transitions. With the reading speed established, the code can then easily be translated. Bar-code readers are very evident in large supermarkets and libraries, but they have applications in manufacturing stock control and automatic assembly lines for component counting and identification purposes.

11.3.13 Signal conditioning

Although the output signal from many sensors and transducers is already in the necessary voltage form, it is only very rarely that the voltage is high enough to be directly interfaced to a digital system. In most instances the transducer output will require some form of amplification prior to conversion from analogue to digital form.

The term 'amplifier' is a shortened form for the full description 'voltage amplifier'. Most amplifiers amplify voltage, but other types are encountered and these are normally given their full description, e.g. current amplifier, charge amplifier and power amplifier. Amplifiers are further classified as a.c. or d.c. The d.c. amplifier will accept a.c. inputs but the a.c. amplifier will block any d.c. input. The input signal may be either single ended with one active signal line, or differential with two active signal lines. Lastly, the amplifier may be 'inverting', with a reversal of sign at the output, or 'non-inverting'.

The commonest amplifier configuration embodies a differential input, single-ended output voltage amplifier with either inverted or non-inverted output. These amplifiers are referred to as 'operational amplifiers'. *Figure 11.68* illustrates some typical idealised operational amplifier circuits.

The important characteristics of an operational amplifier are its gain and its bandwidth. The gain, which can be as high as 10^6, is related to the combination of passive resistors which make up the external circuit. The gain is constant only over a restricted range of input signal frequencies. Outside this range the gain is attenuated as shown in *Figure 11.69*.

The 3 dB 'cut-off' frequencies are used to define the bandwidth of the amplifier. The product of gain and bandwidth (GBW) is quoted by some manufacturers as a quality parameter. Other descriptive parameters associated with amplifiers are:

(1) *common mode rejection ratio*—the ability of an amplifier to reject differential input gain variation;

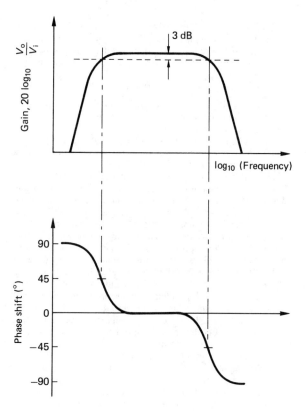

Figure 11.69 Gain and phase shift characteristics of an operational amplifier

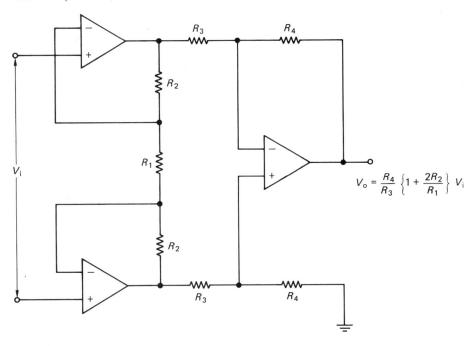

$$V_o = \frac{R_4}{R_3} \left\{ 1 + \frac{2R_2}{R_1} \right\} V_i$$

Figure 11.70 Instrumentation amplifier

(2) *offset voltage*—the voltage output attributed to input and output voltages generated by component variations;
(3) *offset drift*—temperature dependent voltage output;
(4) *non-linearity*—departures from linear input–output characteristics; and
(5) *distortion*—frequency dependent non-linearities.

Popular and inexpensive general purpose operational amplifiers are those in the so-called '741' family. For high accuracy measurements, where low drift and low noise are also essential, an instrumentation amplifier with a high input impedance and high common mode rejection ratio would normally be required (*Figure 11.70*).

The problem of impedance mismatch is frequently encountered in instrumentation systems. This occurs when a transducer having an internal resistance and functioning as a small voltage source, is coupled to a recording instrument which also has an internal resistance. If the internal resistance of the transducer is many times smaller than that of the recording instrument then the problem is insignificant. If, on the other hand, the internal resistances are of similar orders of magnitude then the error in the measurement and the loss of the signal can be quite large. The standard solution to the impedance mismatch problem is to interface an amplifier with a high input impedance and a low output impedance between the transducer and the recording equipment. Such amplifiers are usually configured to have unity gain since their primary function is the prevention of serious signal loss.

For many measurement applications, custom built special purpose amplifiers are already available in integrated circuit form. These greatly ease and facilitate the signal conditioning circuitry. Fraser and Milne[9] give a practical guide to many of the signal conditioning interfaces required for the amplification of signals from strain bridges, thermocouples and other low level voltage output sensors.

11.3.14 Analogue and digital filtering

Noise is inherently present in all physical systems where measurements are made. In sampled data systems the effect of noise, illustrated in *Figure 11.71*, can give rise to further misinterpretation in the form of 'aliases' (see also Section 11.6.4).

Discrete sampling (see above) results in an output signal which suggests that the measured variable is increasing linearly, but with a superimposed sinusoidal fluctuation. The apparent sinusoidal variation is entirely the effect of background noise and it is obviously good practice, therefore, to try and eliminate noise in the measurement system. It is perhaps fortuitous in mechanical systems that background noise is generally manifested at much higher frequencies than the frequency associated with the primary variable of interest.

The sources of noise are varied and may originate from thermoelectric effects, electrochemical action, electrostatic and electromagnetic pick-up, self-generated component noise, offset voltages and common earth loops. If the frequency content of the signal to be measured is known beforehand, then positive steps can be taken to eliminate most of the unwanted effects of noise by the inclusion of suitable filters. Filters exist in three broad categories: low pass, high pass and band pass. The gain characteristics for each type are shown in *Figure 11.72*.

A low-pass filter is one which allows the transmission of signals below a particular cut-off frequency. Signals at frequencies above the selected cut-off are progressively attenuated. The high-pass filter, in contrast, transmits only that part of the signal which is at frequencies above the cut-off value. The band-pass filter transmits, without attenuation, the signal contained within an upper and a lower cut-off value. The cut-off frequency is defined as that at which the signal attenuation is −3dB.

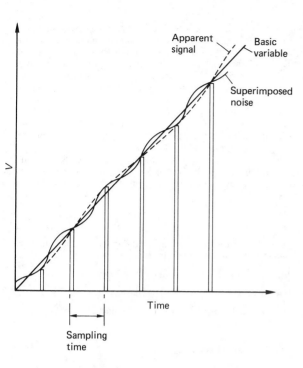

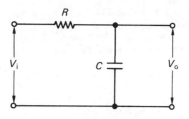

Figure 11.73 Low-pass filter

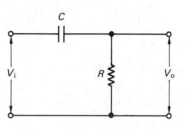

Figure 11.74 High-pass filter

Figure 11.71 Noise generated 'aliases' in a sampled data signal

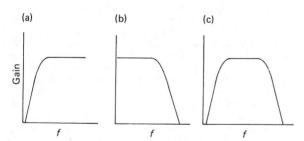

(a)   (b)   (c)

Figure 11.72 Filter performance curves: (a) high-pass filter; (b) low-pass filter; (c) band-pass filter

The simplest forms of analogue filter are those which incorporate only passive resistive, capacitive or inductive elements.

Low-pass filter The transfer function for a low-pass filter (*Figure 11.73*) is given as equation (11.14), with the gain $k = 1$. The time constant τ is equal to the product of the values of resistance and capacitance (RC). The -3 dB cut-off frequency is given by:

$$f = 1/(2\pi RC) \tag{11.59}$$

A suitable choice of resistor and capacitor, therefore, allows any desired cut-off frequency to be imposed in the signal conditioning train.

High-pass filter The transfer function, in terms of the Laplacian variable, for the high-pass filter (*Figure 11.74*) is:

$$(V_o/V_i) = (sRC)/(1 + sRC) \tag{11.60}$$

The cut-off frequency is similarly selected through a judicious choice of resistor and capacitor.

Band-pass filters might be thought of as a series arrangement of a low-pass and a high-pass filter. With the appropriate combinations of resistors and capacitors, the low- and high-frequency noise components in the signal can be suitably attenuated as required.

Filters which include an amplifier in the circuit are referred to as 'active filters' and the relationship between input and output is a much more complex function of time. Williams[11] gives a comprehensive discourse on filter design.

Figures 11.73 and *11.74* illustrate the simplest forms of passive analogue filter which are used to suppress background noise. The governing equations ((11.14) and (11.60)) may equally well be expressed in terms of finite differences. In finite-difference form, the equations can be used to action the filtering process on a discrete version of the input signal. This is the basis of a digital filter which can be implemented in software and requires no external hardwired components.

The setting of the cut-off frequencies in the digital filter are achieved through adjustment of the constants appearing in the finite difference approximating function. These numerical constants are simply related to the physical time constant in the equivalent analogue filter and also the digital sampling rate.

The advantages that the digital filter has over its analogue counterpart are the ease with which the cut-off frequencies can be adjusted. The -3 dB cut-off frequency can also be set exactly, since no hardwired components, which have physical tolerance bands, are used. The digital signal may also be filtered any number of times simply by processing the data repetitively through the filtering algorithm. The disadvantage incurred is that digital filtering takes longer to perform in real time. The end result of digital filtering on the signal is exactly the same as would be obtained using an analogue filter, i.e. any time varying signal of a frequency outside the cut-off value is subject to attenuation with a corresponding phase shift.

11.4 Microprocessor technology

11.4.1 Summary of number systems

In the manipulation of data within a computer, a two-state numbering system is used. This is termed the 'binary system' and is based on a simple ON/OFF principle. For the semiconductor integrated circuits which make up the computer system, 5 V denotes ON, or logic level '1', while 0 V denotes OFF, or logic level '0'. In practice, a tolerance band is adopted with 2.4–5 V representing logic '1' and 0–0.8 V representing logic '0'.

The microelectronic devices in the system handle the transfer of information in 1s and 0s which are referred to as 'bits', being a short form for 'binary digit'. A group of eight bits is termed a 'byte' and a number of computer systems are based on 8-bit technology with the handling of data codes as 8-bit 'words'. 16-bit and 32-bit machines are also available.

The computer operates with three numbering systems: decimal, binary and hexadecimal (often simply called 'hex'). Numerical data would normally be entered by a human operator in decimal form since this is the most familiar number system. The computer, however, must ultimately convert the decimal number into a binary code since this is the eventual form in which the number will be processed and stored. The hexadecimal system is an in-between state and represents a particularly compact method of handling binary numbers as groups of four bits.

In binary representation the only possible logic levels are 0 and 1. The base is chosen as 2, and integer numbers can be represented using 8-bit codes as shown below.

Bit number	7	6	5	4	3	2	1	0
	2^7	2^6	2^5	2^4	2^3	2^2	2^1	2^0
	128	64	32	16	8	4	2	1
	Most significant bit (MSB)							Least significant bit (LSB)

The conversion from binary to decimal is as follows:

Binary number	1	0	1	1	1	0	0	1	
Giving	$128 + 0 + 32 + 16 + 8 + 0 + 0 + 1 = 185$ decimal								

Conversion from decimal to binary is the reverse process to the above.

It is apparent that the highest number which can be accommodated in 8-bit binary notation is 1111 1111, which is equivalent to 255 decimal.

Generally, therefore, computer systems handle integer numbers in four consecutive bytes, i.e. as 32 bits. The most significant bit is used to denote the sign of the number and the resulting range of integer numbers is:

-2^{31} to $2^{31} -1$
or
$-2\ 147\ 483\ 648$ to $2\ 147\ 483\ 647$

Real numbers are handled in five bytes with the most significant byte representing an exponent and a sign bit and the other four bytes representing the mantissa and a sign bit. The resultant range of real numbers is 0.5×2^{127} to 2^{127}, with either a positive, or a negative sign for the mantissa.

The handling of numbers in binary notation is extremely cumbersome for a human and a short-hand notation is adopted for convenience. This is the hex system in which the binary number is arranged in groups of four bits. Four bits, which is half of a byte, is called a 'nibble'. A byte therefore consists of an upper and a lower nibble.

Since there are only 10 unique symbols in the decimal numbering system, the first six letters of the alphabet are used to denote the additional six symbols in the hexadecimal system.

Decimal	0	1	2	3	4	5	6	7	8	9	10	11	12	13	14	15
Hexadecimal	0	1	2	3	4	5	6	7	8	9	A	B	C	D	E	F

Thus, using hex notation, 8-bit binary numbers may be replaced by two hex symbols, e.g.

167 decimal $\equiv$ 1010 0111 binary $\equiv$ A7 hex

Higher numbers are handled similarly, e.g.

6836 decimal $\equiv$ 0001 1010 1011 0100 binary $\equiv$ 1AB4 hex

11.4.2 ASCII code

In the interchange of information between the constituent parts of a computer, or a peripheral device, a binary code is used to represent the alphanumeric characters. The most commonly used code for digital communication links is the American Standard Code for Information Interchange (ASCII pronounced Askey). ASCII is a 7-bit code which can accommodate 128 definable characters.

When communication takes place in a serial fashion, the ASCII code is extended to 8 bits, usually by inserting a zero in the most significant bit. In addition, one or two start bits, a parity bit and a stop bit are also included. The start bit(s) informs the receiving device that a character code follows. The parity bit provides a check that no bits have been corrupted during transmission, by ensuring that the sum of all the 1s in the ASCII group give either an even number for 'even parity', or an odd number for 'odd parity'. The stop bit, set to logic '1', terminates the transmission of the character.

The transmission rate in bits/second is termed 'baud'. Since there are 11 bits associated with the transmission of one character, a speed of 2400 baud corresponds to 2400/11 = 218 characters/second.

11.4.3 Gray code

The Gray code is just one of many binary codes in which only one of the digits changes between successive consecutive numbers. The main application is in the sensing of rotational and translational position in mechanical systems (see Section 11.3.2).

In converting from Gray to binary code, the most significant bit of the binary number B is equal to the most significant bit of the Gray code G. For all other bits, the relationship between binary and Gray is given by

$$B(n) = G(n) \oplus B(n + 1) \qquad (11.61)$$

where n denotes the bit reference number and $\oplus$ represents an exclusive-OR logic comparison.

The conversion of 1101 Gray to binary is shown in *Figure 11.75*; 1101 Gray $\equiv$ 1001 binary.

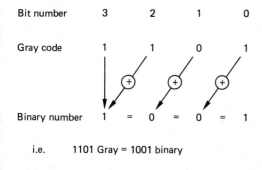

Figure 11.75 Conversion of 1101 Gray to binary

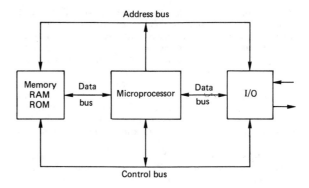

Figure 11.76 Basic components in a microcomputer

In practical position-sensing applications, the conversion process can be programmed in the software, or implemented in a hardwired logic circuit using exclusive-OR gates.

11.4.4 System architecture

Since its inception in the mid-1970s, a number of different microprocessor designs have been developed and these are available in several versions. The 'popular' designs are produced by a few manufacturers and these include such companies as Intel, Motorola, Rockwell, Texas Instruments and Zilog.

Microprocessor-based systems require additional family support chips and, in true digital form, all microelectronic components which constitute a microcomputer are designated numerically rather than by name. Some of the more popular microprocessors available are described below.

(1) *8-bit*—Binary data are handled in a word, 8-bits wide, defining the accuracy of the number handling representation, e.g. Intel 8080 and 8085, Motorola 6800 series, Rockwell 6502 series, and Zilog Z80.
(2) *16-bit*—Binary data are handled in a word of 16-bits width, e.g. Intel 8086, Motorola 68000, and Zilog Z8000.

The 8086 is one of the most powerful and versatile 16-bit microprocessors available, and it has been widely used in industry. Further enhancements include the 32-bit versions (80286, 80386 and 80486) which provide increased processing power.

Although it is unnecessary for the user of the technology to understand in detail how each individual chip actually functions, it becomes essential to have at least a working knowledge of the logical organisation of the system hardware and how each component relates to each other. The composition of this hardware structure is known as the 'system architecture'.

A digital computer system comprises three main constituent parts: the microprocessor, the memory and the input–output (I/O). Digital signals which have a common function are transmitted between the main components by a group of wires, or conduction tracks, termed a 'bus'. In a microcomputer, there are three buses: the data bus, the address bus and the control bus.

The interconnection between the basic hardware components in a microcomputer is illustrated in *Figure 11.76*.

The microprocessor is a very-large-scale integrated (VLSI) circuit, which is the brain of the microcomputer system and acts as the central processing unit (CPU). Integrated circuits are generally classified according to the number of components on the silicon chip—the VLSI chip has tens of thousands

of components. The Intel 8086 microprocessor, introduced in 1979, has 29 000 transistors packed on the 225 m^2 chip.

The main feature of the microprocessor is the arithmetic and logic unit (ALU). The ALU allows the arithmetic manipulation of data, addition and subtraction with 8-bit systems and multiplication with 16-bit systems. Logical operations (AND, OR, etc.) can also be performed. In addition to the ALU, the CPU contains a number of temporary data-storage registers to fetch, decode and execute instructions, and to time and control circuits to manage the internal and external operation of the complete microcomputer system.

The processing power of the CPU is influenced by such factors as word length, instruction set, addressing modes, number of available registers and information transfer rates. For word processors, or the manipulation of large quantities of data as in computer-aided design (CAD) packages, 16- or 32-bit microprocessors are essential. In the field of measurement and control, 8-bit systems are usually adequate.

The system clock, accurately controlled by a quartz crystal maintaining a constant frequency, acts as the heartbeat for the system and provides all the timing reference points.

All the basic components (CPU, memory, and I/O) and their interconnections may reside in a complete microcomputer system encompassing the traditional keyboard, monitor, etc. Alternatively, they may reside on a single card, or even on a single chip to give a single-chip microcomputer.

11.4.5 Bus structure

As outlined in Section 11.4.4, the connection between the system components is made by an arrangement of three buses.

(1) The *data bus* transmits the words in binary form representing the program instructions for the CPU to execute. It can also carry the information transmitted between the CPU and memory or I/O devices. Although the popular PCs use a 16-bit data bus, 8-bit data transfer operations remain the norm in data acquisition and control applications.
(2) The *address bus* transmits the memory address related to the data bus information. In 8-bit systems this bus commonly has 16 lines to give 64k of addresses. PCs usually have an effective 20 lines to give 1 Mbyte of available addresses, although software limitations often restrict this to 640k.
(3) The *control bus* transmits the collection of timing and control signals which supervise the overall control of the system operation.

The physical format of a bussing system is basically a circuit board with a number of connectors. Different types of microprocessor require different hardware interfaces and, to alleviate the problems, standard bus structures have been developed in order to facilitate the connection of hardware components. In industrial-type systems, cards for various microcomputer functions such as processor, memory, digital and analogue I/O, power switching, slot into a standard backplane or motherboard rack. This offers the advantage of being able to plug any specific card, designed to the bus standard, into a free slot in the rack to build up the system as required.

The physical form of the bus is represented by its mechanical and electrical characteristics. Such information as card dimensions, input and output pin-out connections, signal levels, loading capability and type of output gates must be known.

Standard buses are compatible with cards from different manufacturers. The most popular bus structures include Multibus, S-100 Bus, STD Bus, and the STE Bus.

11.4.6 Memory devices

Memory devices consist of those used to store binary data, which represents the user program instructions, and those which are necessary for the user to operate the system.

Memory takes the form of one or more integrated circuits. These basically hold locations capable of storing a binary word. Each location is assigned a unique address within the system and data can be selected through the address bus. As a binary code is deposited by the CPU on the address bus, defining a specific location in memory, the contents of that location are selected and placed on the data bus. The appropriate piece of memory hardware and specific location is selected by means of an address decoding circuit built up from logic gates within the microcomputer system. The end result is a highly flexible data-manipulation arrangement.

In an 8-bit microcomputer, i.e. 8-bit data bus, the address bus is 16-bits wide. This enables $2^{16} = 65\,536$ locations to be addressed, and the total memory capacity of the machine is 64k.

The memory is further subdivided into pages with the high-order byte of the address denoting the page number and the low-order byte indicating one of the 256 locations available on each page.

In machines with 8086/8088 microprocessors, an additional 4 bits are effectively made available on the address bus. This theoretically constitutes 1 Mbyte of addressable memory. Microcomputers such as the IBM-PS2, employing the 80286 and 80386 microprocessors, have an address bus which is 24 bits wide and can address up to 16 Mbyte of physical memory. This releases new levels of processing power to accelerate the processing speed in measurement and control applications.

The types of memory chips built into the system basically divide into two categories: random-access memory and read-only memory.

11.4.6.1 Random-access memory (RAM)

In the RAM, data can be read from or written to any specified location. RAM is more correctly defined as read/write memory and data retention is dependent on power being applied to the device. This type of memory is normally employed for the temporary storage of computer programs, at the editing or execution stage, or the storage of data from measuring transducers, prior to permanent storage as a disc file. In a number of the systems available, the RAM is made non-volatile by providing battery back-up.

11.4.6.2 Read-only memory (ROM)

In the ROM data are held in a secure manner and can be read in any specified sequence. Once the chip has been configured it cannot be overwritten and the programs which specify the system operation, termed the 'monitor program', are 'burnt' into ROM when they are known to operate in a satisfactory manner. Basic ROM is inflexible since the software contained therein is developed by the system manufacturer. It is often useful, however, to have all programs which are to be permanently stored in the microcomputer in a non-volatile form, held in an erasable and programmable read-only memory (EPROM). This is undoubtedly the most popular type of ROM used because the write process is reversible. These chips come in popular memory capacities of 2k, 4k, 8k, 16k and 32k and they are designated by name as 2716, 2732, 2764, 27128 and 27256, respectively. The numbers following the '27' indicate the number of kilobits of memory available within the device.

EPROMS are supplied in an uncommitted form with each location holding FF hex. They are configured using an EPROM programmer which 'burns' or 'blows' the required data, in machine code form, onto the chip. If an error in the data exists, or an alteration is to be made, then the complete EPROM can be returned to its uncommitted state by exposing the small 'window' in the device to intense ultraviolet light for about 20–30 min. EPROM erasers are available for this purpose. Once programmed as required, it is usual to cover the window with opaque material. If uncovered, it would normally take some months before program corruption was experienced through the effects of natural sunlight.

A similar type of memory device is an electrically erasable read only memory (EEPROM or E^{2}PROM). This is essentially similar to the EPROM, but enables the user to alter any particular byte of data rather than wiping the entire chip. E^{2}PROM is not so popular as EPROM for economic reasons.

11.4.7 Input/output structure

With the microprocessor acting as the brain of the microcomputer system and the memory chips storing the system operating software and application programs, the other essential hardware required is that associated with the input and output of data in essentially binary form. Interface support chips associated with the various microprocessor families are available to enable communication with such hardware essentials as keyboards, display monitors, disc drives and printers.

The same I/O interface circuits are used in measurement and control applications and the main functions required of the devices are:

(1) digital I/O logic lines which can be read or set by the microprocessor;
(2) data direction register to configure lines as either input or output;
(3) handshake lines to supervise data transfer via the I/O lines; and
(4) timing and counting facilities.

The software used for controlling the communication between the microcomputer and other external devices is dependent on the I/O interfacing technique employed. The two most common methods are 'memory mapped' and 'dedicated port addressed'.

11.4.7.1 Memory mapped I/O

In this method the I/O chip is connected into the system in the same way as the memory illustrated in *Figure 11.76*. The I/O lines are contained in groups of 8 bits termed a 'port' and this byte is addressed in the same manner as any other location in memory. The port is accessed using memory transfer instructions like PEEK and POKE in high level BASIC, or LDA and STA in low level 6502 assembly language.

Since the interface is connected into the bus structure in exactly the same way as the RAM and ROM, no additional decoding hardware is required. Memory addresses are, however, used up for I/O and, as a result, communication is slower than the port addressed alternative.

11.4.7.2 Dedicated port addressed I/O

This method involves a second dedicated I/O data bus as shown in *Figure 11.77*. When data are to be input or output, the necessary control signals are sent from the CPU to the I/O interface chip and the port data are transmitted via the dedicated I/O data bus. This does not affect the addressing of memory within the system and results in faster data transfer than with the memory mapped technique. The ports, or channels, are assigned unique addresses (numbers) on the dedicated bus and are accessed using the additional software instructions of IN (or INP) and OUT in both low- and high-level programming languages.

11.4.7.3 Types of I/O support chip

Although a number of I/O support chips are available, there are essentially two which figure prominently with the memory mapped and port addressed techniques. These are the 6522 versatile interface adapter (VIA), usually associated with the memory mapped 6502 microprocessor systems, and the 8255 programmable peripheral interface (PPI), which is associated with such processors as the 8080, Z80 and 8086 in port-addressed systems.

The 6522 VIA   This is a general interface chip which provides such interface functions as two 8-bit parallel bidirectional ports, each with a pair of handshake lines and two 16-bit counter timers.

The ports, often designated as 'data registers A and B' (DRA and DRB), each have an associated data direction register (DDRA and DDRB) which is used for setting a bit on a port as either an input or an output. The addresses follow the sequence PORTB, PORTA, DDRB and DDRA.

If a.c. or d.c. loads such as solenoids, motors or lamps are to be driven from the port logic signal levels, then a power scaling interface, such as a Darlington driver, compatible with the microprocessor VIA must be used (see Section 11.5.1).

The VIA control lines CA1, CA2, CB1 and CB2 can be set to operate in various read/write modes. This is achieved through the peripheral control register (PCR) in the VIA. CA1 has no output capability, but CA2, CB1 and CB2 can all be used as either input or output. However, these lines are incapable of switching a power scaling device.

The two programmable timers within the 6522 are generally referred to as 'T_1' and 'T_2'. These are 16 bits wide and implemented as two 8-bit registers with a low-byte/high-byte arrangement. The modes of operation, selected by writing the appropriate code to the auxiliary control register (ACR), are:

(1) generate a single time interval;
(2) generate continuous time intervals (T_1 only);
(3) produce a single or continuous pulse on bit 7 of DRB (T_1 only); and
(4) count high to low transitions on bit 6 of DRB (T_2 only).

The 8255 PPI   All microprocessor families have parallel I/O interfaces and these are designed for use with the particular type of CPU. The 8255 PPI is used basically with Intel 8- and 16-bit devices such as the 8080 and 8086/8088.

The 8255 PPI provides three 8-bit bidirectional ports which may be operated in three modes. No other functions such as timing or additional handshaking are available. The ports are designated as 'A', 'B' and 'C' and data direction is specified by writing to a write-only control register.

If hardware timing is required then a separate counter/timer device must be used. One commonly adopted for use with an 8255 PPI is the Intel 8253 chip, which provides three independent 16-bit counters each with a count rate of up to 2.6 MHz. The 8253 has various modes of operation but works basically on the same principle as the timers in the 6522 VIA.

11.4.7.4 Direct memory access

In data acquisition systems, involving analogue and digital signals suitably conditioned for inputting to a microcomputer, there is a limitation of about 100 kHz on the sampling rate when using direct program control to transfer data to memory. If it is necessary to acquire the maximum amount of data at the highest speed, using the maximum amount of the computer's resources, then the DMA technique might be employed.

This is a hardware technique which causes the microprocessor momentarily to abandon control of the system buses so that the direct memory access (DMA) device can access the memory directly. The DMA controller, connected to the I/O interface, needs to know how many bytes are to be transferred, and where in memory the input data are to be stored. The data transfer rate is much faster than in an interrupt servicing method and data-sampling rates of the order of 1 MHz are possible for most microcomputers.

11.4.8 Memory map

The memory locations in RAM and ROM, which the processor can address, must accommodate space for such requirements as system monitor and utilities, user software and I/O. The manufacturer of the microcomputer assigns an area of memory for each functional requirement and provides the necessary information in a system memory map.

In 8-bit systems, with 64k of addressable memory, the memory map is usually composed of 32k of RAM and 32k of ROM or EPROM. The ROM holds the operating system software and normally some space is available in EPROM form for user firmware. In addition to providing space for user programs, the RAM area contains the system stack and the

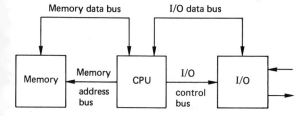

Figure 11.77 Port addressed I/O

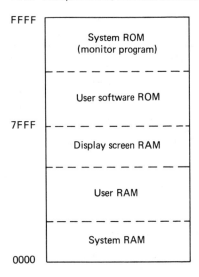

Figure 11.78 System memory map

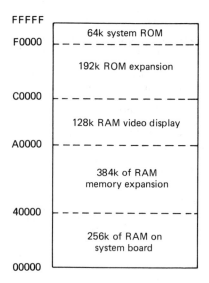

Figure 11.79 General memory map for a PC

visual monitor data storage. The I/O facilities are also assigned an area of memory in a memory mapped system.

In a 6502 or 6800 based system, the RAM is usually low down in memory and the ROM is high up. A typical memory map is shown in *Figure 11.78*. The I/O is accommodated anywhere within the structure and varies from one manufacturer to another.

The display-screen RAM is often fitted with a movable boundary in order to set variation resolution modes for graphics. Typically, the screen memory need only be 1k for a text mode to give a 40 × 25 character display. This would be insufficient for computer graphics and a high resolution would require 20k of memory to MAP the screen. It should be noted that this greatly reduces the amount of RAM available to the user.

8080 and Z80 based systems have a distribution of memory similar to that shown in *Figure 11.78*. The ROM, however, is usually low down and the RAM high up in the memory map.

The 16-bit PCs, with a 20-bit address bus, have 1 Mbyte of addressable memory, with the RAM comprising the first three-quarters and the ROM occupying the last quarter. A general memory map, showing the distribution of RAM/ROM, for a PC is shown in *Figure 11.79*.

A familiarity of the memory map of the system, to be used in any data-acquisition application is essential because it indicates the areas reserved for the operating system. The programmer can then knowledgeably determine the locations available for data storage and machine code programs.

11.4.9 Communication standards

Various standards have been drawn up to define the protocol for the transmission of binary data from within the microcomputer bus structure to external devices such as display monitors, printers and other peripheral equipment. Most microcomputers are equipped with this facility and manufacturers of data measurement and control instrumentation usually offer an external communication port as an extra.

The most commonly accepted standards are those defined by the American Electronic Industries Association (EIA) and the Institute of Electrical and Electronic Engineers (IEEE). The standards fall into the two categories of serial and parallel

data communication. The difference between the two relates to the number of bits of information transmitted simultaneously between the devices. The serial method is the slower of the two, with the bits denoting the characters of information travelling sequentially along a single path. In the parallel method, the data word is sent as a parallel code, invariably 8-bits wide, resulting in a 'bit parallel, byte serial' transmission of information.

11.4.9.1 Serial communication

Serial communication is the most common method used for the interconnection of a microcomputer to the relatively slow peripheral hardware, or between two computers, when transferring a low volume of information.

The (EIA) RS232C, or its successors the RS422 and RS423, is the most widely adopted standard employed and connection between devices is made via a standard 25-pin connector. This allows communication with one peripheral device only. Twenty-one of the signal lines are defined in the standard, although only five (or even three) are usually required.

The three main connections are 'transmitted data' (pin 2), 'received data' pin 3, and 'signal ground or common return' (pin 7). These would normally be connected as shown in *Figure 11.80*. For communication in both directions, i.e. full

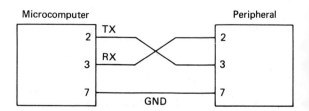

Figure 11.80 The three main connections in serial communication. TX, transmitted data; RX, received data; GND, signal ground or common return

duplex, the two handshaking control lines ('request to send' (pin 4) and 'clear to send' (pin 5)) are also required.

The standard applies to data transmission interchange usually at rates between 110 and 9600 baud. A logic '1' is represented by a voltage in the range of -3 to -15 V and a logic '0' by a range of $+3$ to $+15$ V. This large differentiation between '1' and '0' ensures good immunity against electrical noise. However, the voltages used are not compatible with the transistor/transistor logic (TTL) semiconductor family, and interconversion chips are required within the interface.

The RS232C is limited to short communication links of about 30 m, although the RS422 and RS423 standards (succeeding the RS232) have extended communication distances and increased transmission speeds. The RS423, which is compatible with the RS232, has superior driving and receiving interfaces, allowing communication over distances of up to 1500 m at 9600 baud, or 15 m at 100 kbaud.

It should be noted that, whilst the voltages and signal connections for the plug are defined in the standard, the data protocol is not identified. This must be known for the devices which are to be connected and can be set accordingly by software. The requirements are:

(1) baud rate;
(2) number of bits in the ASCII group defining the character being transmitted;
(3) odd, even or no parity; and
(4) number of stop bits.

11.4.9.2 Parallel communication

The RS232 serial standard for communication was developed essentially for the connection of microcomputers via a telephone link. The parallel standard emerged from the need to establish a means of interfacing a variety of instruments for data-logging applications. The most common standard for the integration of automated test systems, developed by Hewlett-Packard, is referred to as the IEEE-488 interface bus and has achieved wide recognition amongst instrument manufacturers since the start of the 1980s.

The bus consists of 24 lines, accommodated within standard stacked type connections. The eight birdirectional data lines carry information as 7-bit ASCII codes between the microcomputer (controller) and an instrument (listener) on the bus. The roles may be reversed when data are being logged. To process the information on the data bus, up to eight control and status signals are available.

The bus is designed to interface with up to 15 instruments, within a localised area, involving a total cable length of not more than 20 m. Each instrument is uniquely numbered within the range 0–30 and the overall activity is controlled by one of the devices, termed the 'controller'. This controller is usually the microcomputer with an appropriate interface. Each device number is switch selectable within the instrument. Other functional aspects of the devices on the bus are that they must be capable of acting as a 'listener' or a 'talker'. A listener is a device which can receive data over the bus and a talker is one capable of transmitting data. There may be several listeners active on the bus at any one time, but there can only be one talker. Most devices, including the microcomputer controller, can act as either listeners or talkers.

When setting up an instrument to measure some physical variable, codes devised by the instrument manufacturer are sent on the bus (in ASCII format) as a data string to the numbered device. In the case of a multichannel DVM this could take the form of the channel number to be monitored, voltage range to be selected and a terminating character. An example of the corresponding string to be put on the bus is

'C9R2T', which denotes channel 9, range number 2, say 0–10 V, and 'T' as the string terminating character recognised by the instrument.

Manufacturers of add-on cards, to give IEEE-488 facilities with microcomputers, usually supply software for initialising the bus, setting it up for transmitting data from controller to instrument, and returning data from instrument to controller.

The measured quantity is also sent to the computer in the form of an ASCII string from which the actual numerical value can be extracted.

One of the most important management control lines is the service request (SRQ). This is a type of interrupt line that is activated low by a device residing on the bus and needing service from the controller. It is used as a means of indicating that the instrument is ready to transmit the current reading onto the bus.

Thus, a typical software sequence for implementing the control of an instrument on the IEEE-488 bus for data acquisition is:

(1) initialise the bus and set the instrument as a listener;
(2) put a control string on the bus to set up the instrument as required;
(3) check for SRQ line to go low indicating that data can be read;
(4) set the instrument as a talker; and
(5) read returned string and convert it into a numerical value.

When operating in high-level BASIC, high data-collection rates are not possible. However, since most instrument manufacturers offer the standard as an option, it provides an intelligently controlled flexible arrangement for test and measuring instruments.

11.5 Interfacing of computers to systems

The serial and parallel communication standards are the basic interfacing links between computers and their associated peripheral devices, or between computers and a comprehensive range of measurement instrumentation. In general computer control applications, however, two other common interfaces are fundamental. These are the digital interface which implements the controller output and the analogue interface which is associated with the measured variable input.

11.5.1 Digital interfacing

The computer output port may be used to transmit control signals on any one of the available lines by writing the appropriate number to the port address. When a line, or bit, is set 'high', i.e. at a logic level of 1, the voltage on the line is approximately 5 V. The current available, however, is fairly minimal (of the order of 1 mA) and no load can be connected directly to the port. There is grave danger, in fact, of causing extensive damage to the computer by connecting a load directly to the port. An interface must therefore be provided to enable the computer to switch in power loads using the logic level control signals from the output port. The most common interface device used for this purpose is the power transistor. In typical applications the power transistor, operating on logic level control signals, switches in a mechanical relay which in turn switches in the load.

The 'Darlington driver' is a popular power transistor which is available as an integrated circuit and normally includes a number of separate stages. *Figure 11.81* shows the wiring diagram for a single stage in a Darlington driver. The Darling-

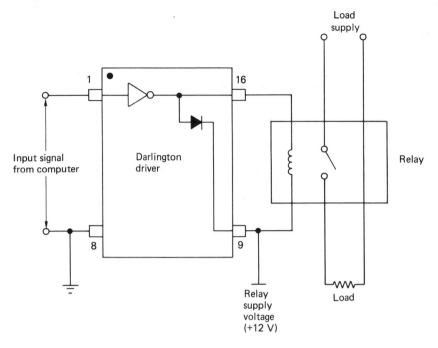

Figure 11.81 Power switching with a Darlington driver

ton driver can switch up to 500 mA at 50 V and each stage is diode protected for the switching of inductive loads.

To provide total isolation from high voltages, port output signals can be coupled through an 'opto-isolator'. The opto-isolator, interfaced between the computer output port and the power control device, is not an essential element in the digital interface. Opto-isolation, however, ensures that no hard-wired connections are made between the computer and the power device. An added advantage is that the opto-isolator acts as a buffer to spurious noise signals which can corrupt the digital logic values being transmitted on the buses.

The opto-isolator (*Figure 11.82*) transmits signals by means of infra-red radiation, emitting from a source and sensed at a phototransistor.

When a computer based on transistor/transistor logic (TTL) is powered up, the state of the lines of the output port 'float high'. That is to say each output line becomes set to a logic value of 1. Since a logic 1 is normally associated with the function of switching a device ON, then a port which floats high could inadvertently activate some power device. Obviously this is a dangerous precedent which requires an additional element in the digital interface to counteract the effect. The device commonly used is the 'inverter', or NOT gate, which has the simple function of inverting all logic signals from 1 to 0, and vice versa (*Figure 11.83*).

Following power up, a logic 0 must then be sent to the relevant line of the output port, to become a logic 1 after inversion and to operate the control function. The composite digital interface for a computer output port, suitable for power switching is depicted in *Figure 11.84*.

In high frequency switching applications, electromechanical relays are not suitable. The semiconductor devices such as silicon controller rectifiers (SRCs, alternatively called thyristors or triacs) may be more appropriate. Also particularly suitable are the various solid-state relays which can operate directly from logic level signals.

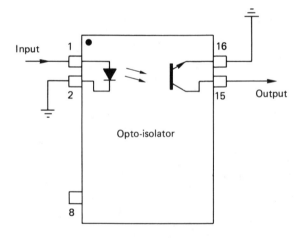

Figure 11.82 Opto-isolator

11.5.2 Controller output interface hardware

The digital interfaces discussed above are suitable for switching in power loads in an OF/OFF control system. For a digital control algorithm based on a PID strategy, some means is required of discretely varying the output power supplied to the controlled device (*Figure 11.85*).

A number of different methods are used to supply variable power to the system and some of these are described below.

11.5.2.1 The digital-to-analogue converter

The required control effort value U is calculated in the program according to the control strategy employed. This

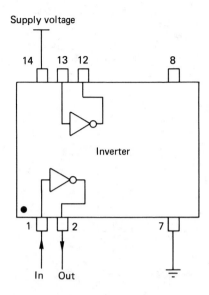

Supply voltage

14 13 12 8

Inverter

1 2 7

In Out

Figure 11.83 Inverter integrated circuit

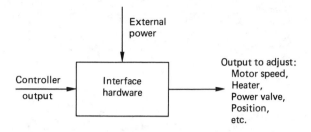

External power

Controller output

Interface hardware

Output to adjust:
Motor speed,
Heater,
Power valve,
Position,
etc.

Figure 11.85 Variable power output device

value is converted to an equivalent binary number and output to the computer port where it is then transmitted to a digital-to-analogue converter (DAC). The DAC converts the binary input into a proportional output voltage which may then be suitably amplified to drive the controlled device. The controlled device could be, for example, a d.c. motor the speed of which is directly related to the supply voltage. The interface is illustrated in *Figure 11.86*.

Two basic types of DAC are available: the adder converter and the ladder converter. The adder converter can be illus-

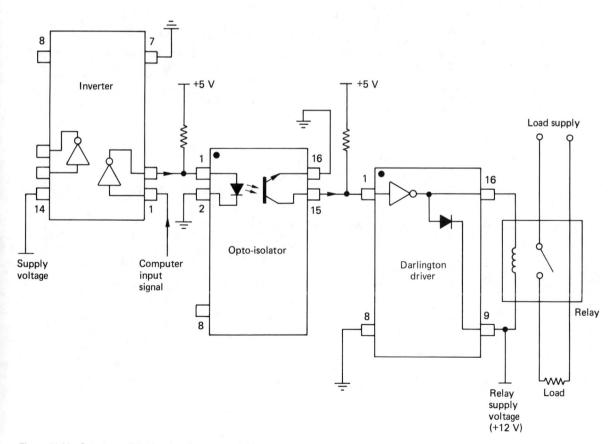

Figure 11.84 Output port digital interface for power switching

Computer
output
port

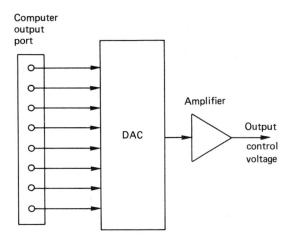

Figure 11.86 Variable power output using a DAC

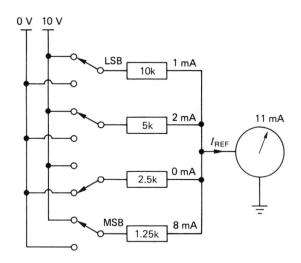

Figure 11.87 4-bit, adder-type DAC

trated as a simple example of Ohm's law. A 4-bit adder type DAC is shown in *Figure 11.87*. The resistance values of the line resistors are halved for each consecutive increasing 'bit' and the supply voltage is common. The current drawn through each line, if it is connected, is 1, 2, 4 and 8 mA corresponding to bits 0, 1, 2 and 3, respectively. The summation of the currents at output therefore equates on a decimal to binary basis with the input. *Figure 11.87* shows a digital, or switched input of 1011 binary. This gives a current summation of 11 mA, equivalent to the corresponding decimal number. The output, a proportional current, can be converted to a proportional voltage through an operational amplifier. Because of the range of resistor values required, the adder converter is less popular than the ladder converter which uses only two resistor values, i.e. R and $2R$.

The ladder converter DAC (*Figure 11.88*) must be analysed using the network theorems of Thevenin and Norton. The end result is similar, however, with a proportional voltage output corresponding to the digital switched input on a decimal to binary basis.

11.5.2.2 Pulse-width modulation

Output power from the interface hardware can be varied by sending ON/OFF pulses to the power device. The frequency range is normally 2–10 kHz. If the time ON and the time OFF periods are equal, then rapid switching of the power supply will transmit 50% of the total power available. Due to the

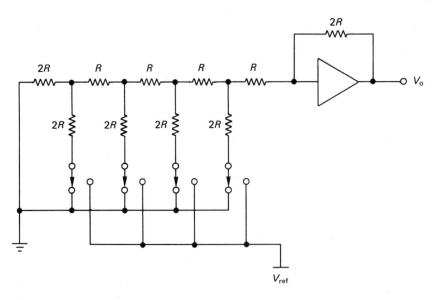

Figure 11.88 Ladder-type DAC

relatively high switching frequencies and the levels of power transmitted, solid-state relays are more suitable for this application.

The time ON is called the 'MARK' and the time OFF is the 'SPACE'. The MARK/SPACE ratio can be evaluated from the controller output value calculated in the software.

PWM can be achieved either by keeping the MARK fixed and varying the SPACE, i.e. varying the signal frequency, or by varying both the MARK and the SPACE within a constant period.

If the pulse-width-modulation (PWM) output is supplied by the controller, then it is not possible to allow the MARK+SPACE period to extend over the complete control loop cycle. There must be some time allowed to sample the process transducer and to calculate the required controller output value. The problem is overcome by using a synchronisation ('sync') pulse signal with a longer period than the MARK+SPACE time. The excess time during the sync pulse is used to read the transducer output and to process the information. The MARK/SPACE ratio is then implemented over the remainder of the sync pulse. Two separate timing loops are used to control the MARK and SPACE times, respectively. Due to the relatively high frequency of the output signal, the control software must be written in assembly language for speed. An alternative to this is to use hardware support chips to output the PWM signal under the control of the computer's CPU.

11.5.2.3 Controlling a.c. power by control of thyristor phase angle

Various applications, such as temperature control, require a.c. power adjustment and solid-state relays can be used effectively to vary output power between 0 and 100%. The power is controlled by varying the phase angle between the supply voltage and that which appears across the load when current conduction begins. A phase control device operating off a 0–5 V signal from a DAC can be used to alter the phase angle between voltage and current in the range 0–180°. The control of the power output to the load is non-linear, but linearisation between output and input can be accomplished in the software.

11.5.2.4 Controlling flow control valves

In level control systems, fluid flow rates are controlled by varying the degree of opening of a gate-type valve. Because of the forces involved, a pneumatic 'actuator' is normally employed, working off a controlled pressure in the range 3–15 lb in.$^{-2}$ (i.e. psi). (N.B. The industry still favours the Imperial unitary system, but the approximately equivalent pressures in the Metric system are 20–100 kN m^{-2}.) An applied pressure of 3 psi is equivalent to the valve being full open, 15 psi corresponds to the valve being fully closed.

Most flow control valves are fitted with a 'positioner', which operates off a current signal in the range 4–20 mA. The current range generally corresponds on a linearly proportional basis to the pressure range. Since the computer interface usually involves a DAC, then an additional element, a voltage-to-current converter, is required to interface between the DAC and the valve positioner. The complete interface is shown in *Figure 11.89*.

11.5.3 Analogue interfacing

The basic role of the analogue interfacing is one of conversion of the continuous analogue signals, from process-measuring transducers, to the digital representation that the computer requires to operate on. In all practical applications, the monitoring and acquisition of data is the necessary precursor to the subsequent control functions which might be actioned.

The process variables are ultimately represented as voltages. Using the appropriate signal conditioning circuits, these voltages would ideally be processed to range between zero and some reference value. The final task is the digitisation of the analogue signal, which is accomplished through an analogue-to-digital converter (ADC). The ADC samples the analogue signal, performs the conversion and outputs a digitally encoded binary number which is directly proportional to the magnitude of the input voltage. The essential elements in the signal train are shown in *Figure 11.90*.

Figure 11.90 indicates a sample and hold (S/H) element between the signal conditioner and the ADC. Since the analogue input may vary while the conversion is taking place, there is a degree of uncertainty in deciding the instant in time which the output code represents. The S/H element removes this uncertainty by capturing an instantaneous 'snapshot' of the input for the ADC to convert before moving on to the next sample. The S/H element is only essential if the input signal varies very rapidly. The ADC and S/H functions are often packaged within a composite integrated circuit.

ADCs are available in a number of different forms, some of which are described below.

11.5.3.1 Staircase and comparator

The staircase and comparator is the simplest form of ADC (see *Figure 11.91*). The device incorporates a DAC which generates a voltage increasing in small steps. At each step the staircase input is compared with the analogue input. When the generated staircase is approximately equal to the input, the process is halted and a binary count is made of the number of steps taken during the process. The binary count from zero represents the coded digital output.

The staircase and comparator ADCs have relatively slow conversion times, typically 20 ms. However, they are cheap and are essentially immune to electronic noise.

11.5.3.2 Integrating type ADC (or dual slope)

The major elements comprising a dual slope ADC are illustrated in *Figure 11.92*. At the start of conversion, a voltage-to-current converter is switched to the integrator causing it to ramp up a slope which is proportional to V_{in}. This occurs over a fixed period of time at the end of which the input is switched over to the reference current source. At the instant of switching the integrator output voltage is proportional to V_{in}, a counter is enabled and counting begins at a rate set by the internal clock. In the meantime, the reference current causes the integrator to ramp down at a slope which is proportional to V_{ref}, i.e. a constant slope. When the integrator output again reaches ground the comparator switches the counter off and the counter then contains a digitally encoded value proportional to V_{in}. *Figure 11.93* shows the voltage variation at the integrator output. It can be seen from this figure that there are two similar triangles such that:

$$V_{in} = V_{ref}(T_v - T_f)/(T_{max} - T_f) \qquad (11.62)$$

T_v is directly proportional to the counter output and with T_{max}, T_v and V_{ref} all known; the input voltage V_{in} is determined by proportion.

The integrating type of ADC has similar operating characteristics and conversion times to that of the staircase and comparator types. For faster analogue-to-digital conversion, the 'successive approximation' or 'counter' types are generally employed.

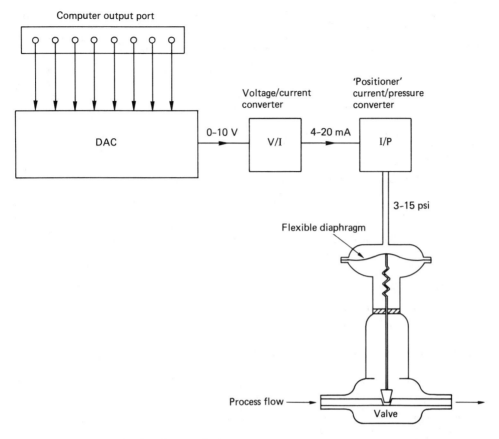

Figure 11.89   Digital interface for a flow control valve

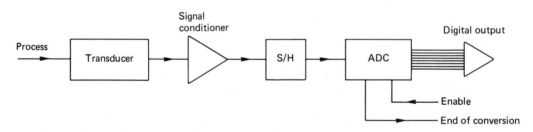

Figure 11.90   Analogue-to-digital conversion

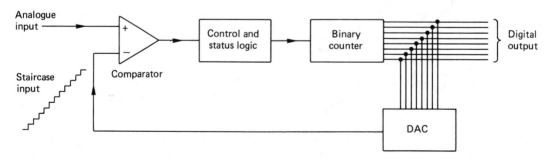

Figure 11.91   Staircase and comparator type ADC

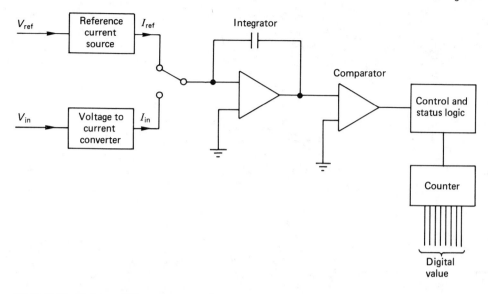

Figure 11.92   Dual-slope ADC

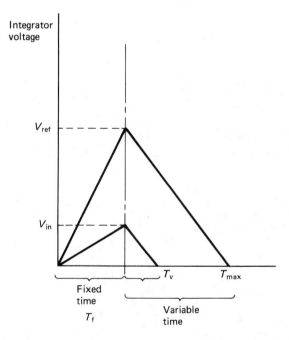

Figure 11.93   Integrator voltage variation

11.5.3.3   Successive-approximation type ADC

In this ADC, the input signal is compared with a number of standard reference voltages, generated from a DAC, until the combination of standard voltages required to make up the input value has been determined. The main components of the converter are a clock, a counter, a comparator and a DAC.

When an analogue signal is input to the converter, the counter starts a count and passes a digital value to the DAC.

The DAC generates a voltage to represent the most significant bit and the comparator assesses this against the analogue input. If the analogue signal is greater than the voltage from the DAC, then the logic 1 in the MSB is retained; if the analogue signal is smaller, then a logic 0 is assigned to the MSB. This process is then repeated on the next most significant bit and so on for all the other bits down to the least significant bit. The conversion time for this type of converter may be of the order of 10–25 μs, depending on the hardware design. *Figure 11.94* outlines the essential features of a successive approximation ADC.

11.5.3.4   Parallel-conversion type ADC

The parallel type ADC has by far the fastest conversion time, at about 1 μs, but it is also the most expensive. With parallel conversion, the analogue input is fed simultaneously to a number of comparator circuits, each having a different reference voltage. The resulting comparator outputs are fed to a logical coding network which generates the appropriate digital values to represent the state of the comparator outputs.

Regardless of the type of ADC used, the pin functions on the integrated circuit are basically similar and generally comprise the power supply, the data bits, the start conversion pin (SC or $\overline{\text{CONVERT}}$) and the end of conversion pin ($\overline{\text{EOC}}$ or $\overline{\text{STATUS}}$). The overscore signifies that the pin is active low.

The conversion is software initiated by sending a 'pulse' (logic 0, followed by logic 1) to the $\overline{\text{CONVERT}}$ pin. On the negative edge of this pulse the counter in the successive approximation ADC is set to zero and on the positive edge the counter starts incrementing. At the start of conversion the $\overline{\text{STATUS}}$ pin goes from low to high and when it again goes low, the conversion is complete (see *Figure 11.95*).

The end of conversion may be readily detected using suitable software. As an alternative it is possible to include a software-generated time delay following the start conversion pulse to allow conversion to complete before reading the value at the input port. The length of the delay can generally be found by trial and error.

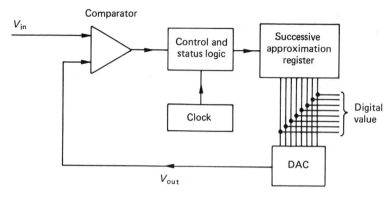

Figure 11.94 Successive-approximation ADC

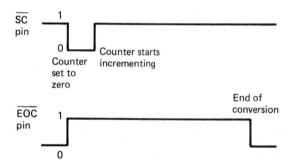

Figure 11.95 Start conversion and end of conversion pin signals

Table 11.3 Summary of the relationship between number of bits and resolution

No. of bits, n	2^n	Resolution (%)
8	256	0.4
10	1024	0.1
12	4096	0.025
16	65536	0.0015

In choosing the appropriate ADC for a particular application the four main features to be considered are: conversion time, resolution, accuracy, and cost.

Conversion time The conversion time is a measure of the operating speed of the converter and is the time taken for the complete translation of an analogue signal to digital form. The conversion time in many of the staircase and comparator and the intergrating types of ADC may depend on the level of the analogue input signal. Faster conversion times are obtained with low-level inputs due to the manner in which the conversion is completed. Successive approximation and parallel conversion types of ADC have a fixed conversion time. This is because the exact same conversion process is performed regardless of the analogue input level.

However, the conversion time of the ADC does not indicate the fastest rate at which data can be captured. If the data are to be stored in the computer's RAM, then this must be done in a sequential and ordered manner. This involves setting a base address and incrementing various registers to step the storage address of each byte of data placed in memory. Further time delays could be accrued in a sample and hold device. Therefore the minimum data capture period is often many times greater than the specified conversion time of the ADC.

Resolution The resolution of an ADC is the number employed to represent the digital output in binary form. For example, the resolution of an 8-bit ADC is limited to one part in 256 of the maximum voltage corresponding to the full-scale setting. An improvement in resolution can be obtained with a 12-bit converter, with one part in 4096. *Table 11.3* summarises the relationship between the number of bits and the resolution.

Accuracy The accuracy is related to linearity defects, zero error and calibration difficiencies in the electronics of the converter, and should not be confused with the resolution.

Cost Cost will depend on the quality required in the three factors described above, and on the means of conversion employed. The cost is closely associated with the speed of conversion and with the resolution and accuracy. Cost generally increases with increases in one or all three of these variables.

11.5.4 Multiplexing

In applications where a number of transducers are to be sampled, a multiplexer (MUX) can be used to switch in various channels as and when required to a single ADC. The switching is software controlled from the computer. The basic principles are illustrated in *Figure 11.96*.

The multiplexer and ADC often form an integral part of a complete system. In some cases, even the signal conditioning can be software controlled, with all the necessary hardware mounted on a single 'card' and plugged directly into the bus system of the computer. Multiplexers, or analogue switches, are available with various numbers of input channels.

Minimum-cost conditions usually dictate whether multiplexing will be implemented or not, but the reduced cost must be balanced against an inevitable reduction in sampling rate. *Figure 11.97* shows three possible arrangements of signal conditioning, multiplexing and conversion for analogue interfaces. System A is the most common, while B and C can

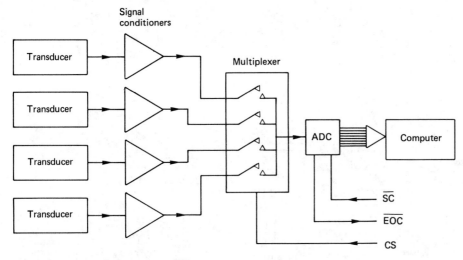

Figure 11.96   Multiplexer for multiple inputs

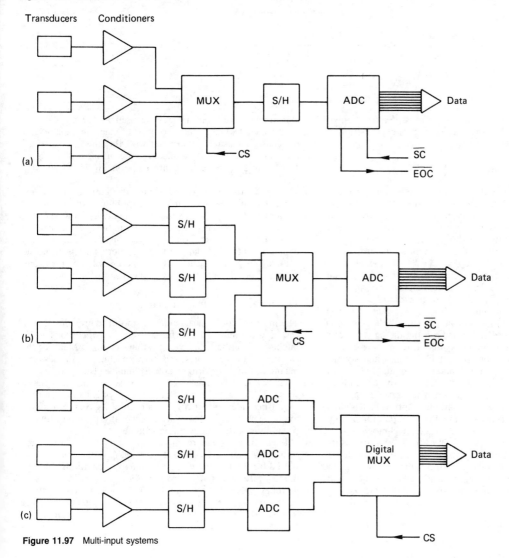

Figure 11.97   Multi-input systems

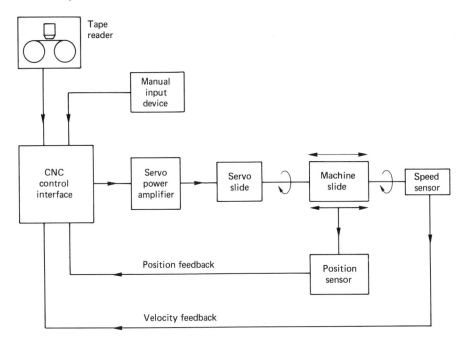

Figure 11.98 Machine–tool slide control system

provide for virtually simultaneous sampling. System C gives the most representative snapshot at a particular period in time, but it is also the most costly.

11.5.5   Machine–tool interfaces

The control system for a machine–tool slide is shown in *Figure 11.98*. Typically there are two negative-feedback loops, one for position and one for velocity, in a cascade arrangement as shown. The position sensor is usually an optical grating device, or an inductosyn and the speed sensor is a tachometer.

The CNC interface initially has to decode the manual, or control tape, input data. The data consist of a sequence of commands including feed and speed data, essential dimensional reference points and other constraints to be observed by the machine during its operation. In operation, the interface is required to monitor the slide position and speed and to check various limit switch settings for compliance with the sequential program instructions. The transducer input signals would normally be switched in through a multiplexer prior to digitisation with a fast-conversion-type ADC. Limit switches would also be checked, or set, through additional digital I/O lines. If any errors are detected, the interface must be able to indicate these and take the appropriate action. The interface includes a real-time clock which generates an interrupt every few milliseconds. The clock acts as a monitor of operator actions, enables the output of error signals to the machine servos and checks all current signals from each of the feedback sensors.

For a typical CNC milling machine, there are three independent axes and each would have the same monitoring and control functions applied to each axis. In addition, the spindle speed would be monitored and controlled and the machine might also incorporate a tool-changing facility based on a simplified robot arm.

Further refinements could include a load transducer in an additional feedback loop to measure the cutting forces during machining. Force sensing may be used as the basis for an adaptive control loop. In the context of machine tools, adaptive control is usually associated with the alteration of feed rates and cutting speeds to maximise the cutting power. *Figure 11.99* shows an adaptive control option on an NC turning machine.

The adaptive loop can optimise the cutting operations, prevent spindle overload, maximise tool life, reduce time loss in 'air cuts' and simplify the programming. However, the additional sensors and their protection in the harsh machining environment means that the adaptive loop is much more costly to implement. The adaptive control interface, which has no manual input data facility, is also necessarily complex, and requires considerable memory capacity.

11.5.6   Robot control interfaces

The machine–tool interface described in the previous section can be programmed to perform a series of operations which might be described as 'sequenced automation'. Many of the simpler robots, e.g. pick and place machines, use the same technology and perform essentially similar tasks.

However, these machines are not robots in the strictest sense. The essential feature of a true robot is its capability of exercising independent control in each of its axes, or rotating joints, such that its 'hand' can reach any position and any orientation within the working volume.

Each joint on the robot has an actuator, an associated position sensor and a velocity sensor. Six actuators are required for full flexibility in position and orientation, although in most cases only five or less are used. The computer must at all times be able to ascertain the current and desired locations of the hand. The position sensor data processing therefore involves the manipulation of various coordinate transformation matrices in the definition and control of the hand location.

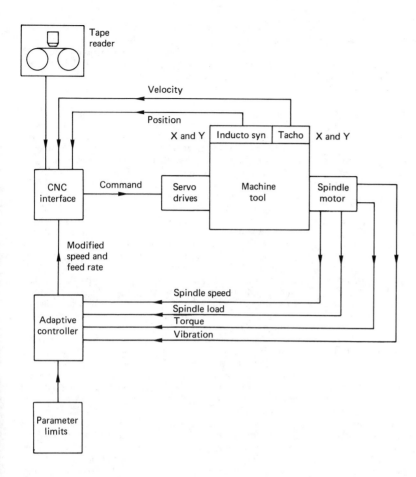

Figure 11.99 Adaptive control on a NC turning machine

Force sensing within the 'gripper' is commonly featured in many industrial robots. The obvious application is to prevent the proverbial 'vice-like' grip on some fragile object. Force sensing is also incorporated by inserting a sensor between the gripper and the wrist. Many wrist force sensors are strain-gauge based and can respond to applied forces and moments in each of the three Cartesian directions. The wrist sensor provides feedback signals for static and inertial loading on the robot arm. Robots which operate a control strategy based on wrist force sensor data are termed 'active compliant'. The active compliant robot system has the capability of modifying the joint positions and motion to minimise the effect of external forces. Many wrist force sensors have been successfully developed for robot manipulators engaged in pin-insertion-type operations. Whitney[12] reports on some detailed investigations on wrist-force-sensor applications.

The sensor producing signals related to position, velocity and force form the main feedback loops in many industrial robot assemblies. These signals are converted through ADCs and the digital control algorithm, often based on a PID strategy, is then solved to determine the required controller output values at that given time. Power is subsequently output to the actuators either through DACs, or using pulse-width-modulation techniques. In most respects the above process is exactly the same as that which would be used to control a CNC machine tool with an adaptive loop. The robot-arm motion, however, is much more complex and the computation involved is far more extensive than that required for the CNC machine tool. The computer used in robotic manipulators therefore requires a large memory capacity and a fast processing speed to accommodate the multiplicity of tasks associated with the control functions. The control algorithm is in itself complicated by the multiple feedback loops for position, velocity and force, and a 'hybrid' control strategy is normally adopted. The hybrid strategy allows force to be controlled along certain Cartesian axes, while position is controlled along the remaining degrees of freedom. Velocity is controlled along the same degrees of freedom as position. The computational effort is significant due to the many coordinate transformations required, and hybrid control can only be implemented when the robot is moving slowly.

Recent advances in robot technology are centred on the development of tactile sensors, range finders and machine vision. Tactile sensors ideally simulate the human sense of touch. They should, therefore, be able to detect presence, shape, location and orientation. In addition, the ideal tactile sensor should be able to respond to applied forces and moments in terms of magnitude, direction and location. Varying degrees of success have been achieved with tactile sensors (see Bejczy,[13] Harmon[14] and Rebman and Trull[15]).

Many range-finding devices are based on optical proximity sensors or on ultrasonic echo-sounding transducers in applications to short- and long-range finding, respectively. The function of the range finder is two-fold; either to locate the gripper in the correct position relative to the object to be picked up, or to avoid potential collisions with other objects. Like tactile sensors, however, many range-finding devices have not yet been developed to the stage where they are consistently reliable for routine use.

Perhaps the most useful additional sensory attribute that a robot can have is that of vision. Industrial machine vision is currently the subject of considerable world-wide research and development, and a number of effective systems are already available. The first problem with a machine vision system is the formation of a digital image. There are a number of different digital imaging devices in use, but they all have the common function of converting light energy into voltage (see Ballard and Brown[16]). The voltage output is inevitably processed through a fast ADC to produce the digital representation of the light intensity, in terms of 256 possible grey levels. The resolution of the digital image depends on the number of discrete lines and points per line which are used in the construction of the picture. Using 512 lines, with 512 discrete points on each line, results in 262 144 individual picture cells, or 'pixels', to define the image. Each pixel is represented as an 8-bit number and these can be stored in memory as a two-dimensional array. Note that this would use up 262 144 byte of computer memory for the storage. With such vast amounts of data to be handled, a separate 'preprocessor' is often used in the system. The preprocessor interprets the visual information and then transmits the result to the robot controller.

Since the image can be represented numerically and stored in memory, it can also be retrieved and processed according to the particular requirements. In many robot vision applications the vision task is that of part recognition and orientation of the part. Often a two-dimensional projection of the part shape is all that is required and a single solid-state camera can be used to generate the two-dimensional image. Part recognition is based on the part silhouette, and the image-processing technique requires only a brightness threshold level to distinguish between 'object' and 'background'. The level of vision is crude but can be sufficient for many automated assembly processes. Control of the lighting arrangements can greatly enhance the vision system but, in general, machine vision still leaves much to be desired at the current stage of development.

The three major development areas of tactile sensing, range finding and machine vision are all being actively researched at the present time. It is apparent that these sensory functions all involve considerable amounts of additional input data and this imposes further limitations on the system in general. The limitations are related to the data-processing speed and the communication interfaces. Recent new microprocessors, e.g. Intel's 80486, operate at much higher speeds than any of their predecessors, and future developments in communications are likely to be centred on applications using light signals transmitted along fibre-optic cables. These advances should then open up possibilities for further enhancements in robot sensory perception systems.

11.5.7 Mobility and automated guided vehicles

Mobility in robot systems ranges from simple slideways and tracks, as in gantry-type robots, to the autonomous free roving automated guided vehicles (AGVs). Full autonomy in AGVs is not yet realised, but partial autonomy has been developed for some unmanned space vehicles. The partial autonomy incorporated in a space vehicle becomes more necessary as the distance from the command station increases. Radio transmissions have a round-trip delay of 2.6 s from the Moon, but can

be 10–40 minutes from Mars. These transmission times are unacceptable, particularly if the vehicle is required to take some emergency action. For this reason, the unmanned vehicles sent to Mars were equipped with preprogrammed instruction sets. Following initialising radio signals from Earth, they then operated essentially in autonomous mode.

The AGVs used in manufacturing engineering are rather less sophisticated than the unmanned space vehicles, and full autonomy for manufacturing AGVs is some way in the future. Nonetheless, AGVs are increasingly being used in manufacturing functions and their further development is inevitable.

Typical AGV systems contain an on-board computer and they can follow a route which has been predetermined by laying a wire in a shallow groove in the floor, which is then filled flush. The track wire is energised with a.c. power, and two magnetic-sensitive detectors, placed either side of the track on the AGV, can be used to steer the vehicle on course. Branches of the main track can be energised using a different frequency to indicate the branch. Alternative route sensors are optically based and follow a reflective or fluorescent strip on the floor.

The AGV can be programmed to follow a particular route and to perform simple lifting and retrieving tasks. The program instructions are input either through a local terminal, temporarily or permanently attached to the AGV, or relayed through radio signals from a remote central control computer.

The AGV must carry its own power source and this is often battery based.

Hydraulically powered AGVs are used extensively in underwater inspection and maintenance operations. The control signals and the hydraulic power are supplied via an umbilical line from the surface vessel. Submersible AGVs are more commonly termed 'remotely operated vehicles' (ROVs).

The last essential function of the AGV is that it must have some means of avoiding obstacles, including humans. Proximity sensors are used extensively in this role. The sensing range is dependent on the speed of the vehicle and must obviously be long enough to allow the vehicle to stop before actual contact is made.

11.6 Microprocessor-based control

Technological developments in microcomputers with their associated I/O hardware and software tools have enabled the designers of automatic control systems to incorporate a higher degree of intelligence than was possible in the past. Digital computers are now used extensively to control machines and processes. The physical appearance of these controllers varies considerably according to the application, and may range from single-chip microcontrollers (SCMs), where all microcomputer components reside on one integrated circuit, to desktop personal computers (PCs).

The SCM provides very cheap computing power and is mainly associated with high-volume applications such as electronic washing machines, automotive electronics, taxi meters, ticket machines and time attendance recorders. However, they can just as easily be used in the control of manufacturing processes in the same way as PLCs, industrial rack-based controllers and PCs.

Since its first appearance in 1981, the IBM PC and its associated compatibles has been adopted as an industry standard. In addition to an increase in processing power, there are a number of advantages in using a PC-based control system.

This integration of the disciplines of microelectronics, computer science and mechanical engineering is the basis of the developing technology of mechatronics. This has been defined as the synergetic combination of mechanical engineering, microelectronics and systems thinking in the design of products and processes.

Typical examples of mechantronic products include robots, CNC machine tools, automatic guided vehicles, video recorders, automatic cameras and autoteller machines.

11.6.1 Direct digital control

Direct digital control (DDC), as outlined in Section 11.2.6, is employed in systems where such physical quantities as temperature, pressure, flow, position and speed are to be constantly monitored and regulated. The design and operation of a DDC system incorporates formal control theory with computer related hardware and software to achieve a chosen control strategy. The end product of a real-time control system is a computer-based system which runs the plant efficiently and safely under all operating conditions.

In addition to providing the necessary control functions related to the direct manipulation of valves, drives and other actuators, there are a number of non-control functions available from the system, such as:

(1) logging and storage of data;
(2) processing and display of data;
(3) communicating with operators through graphic displays of the plant; and
(4) informing of abnormal operating conditions through alarm facilities.

The benefits of DDC are directly identifiable with more precise control which leads to increased production, improved quality of product and the efficient use of energy and raw materials. There is also a reduction in maintenance requirements and a saving in capital investment through minimisation of the number of recorders, instruments and other control elements required.

The intangible benefits include an improvement in the handling of engineering and accounting data, increased knowledge of the process from the data collected and displayed, and improved safety.

11.6.2 Hardware requirements

Figure 11.36 displays the elements of a real-time direct digital control system which constitutes a microprocessor-based controller with the associated I/O interfaces. Digital and analogue interfacing techniques for connecting to external sensors and actuators are described in Section 11.5.

In industrial applications, the transducers used for measuring the loop variable normally transmit a current in the range 4–20 mA in proportion to the measured value. This is in preference to using a voltage due to the attenuation over long distances. Furthermore, current signalling can offer a better performance than voltage signalling methods in rejecting electrical noise. At the immediate controller input, the 4–20 mA current signal is converted into a voltage prior to the digital conversion. In control loops the actuators may also operate on a current standard. Since a variable voltage is usually generated through a DAC, then voltage to current conversion is also required.

The function of the controller is to monitor the operating conditions of the process and to evaluate, according to a specified strategy, the necessary output control action to ensure that the controlled system operates in a safe and efficient manner. A large variety of digital control devices are available to the system designer, ranging from single-chip microcontrollers, single-card computers and programmable

logic controllers (see Section 11.7), to complete PC systems (see Section 11.6.7).

11.6.2.1 Low level control devices

Following the inception of the microprocessor in the mid-1970s, the logical progression was to increase component density and incorporate memory and I/O interface facilities into one device. These so-called single-chip microcomputers (SCMs) are usually referred to as 'microcontrollers' and they provide the intelligence required for such applications as measurement and industrial control.

The architecture of a typical SCM is similar to that of a traditional 'single-card computer' and comprises:

(1) a CPU;
(2) memory (RAM and ROM);
(3) parallel and serial ports;
(4) timers;
(5) hardware interrupt lines; and
(6) ADCs.

A large number of SCMs is now available and some display all the above features. This greatly simplifies the designer's task, since many of the functions that previously necessitated separate chips in a microprocessor system are included on a single integrated circuit.

Examples of some popular devices currently available are described below.

The Mitsubishi series 740 of 8-bit CMOS microcontrollers These are ultra-low-cost devices which are available in a variety of different forms. For example, a Mitsubishi designer's kit based on the M50734 SCM including a user's manual and software manual, along with designer's notes, can be purchased for less than £40. This particular model comes complete with such facilities as four 8-bit I/O ports, a four-channel multiplexed 8-bit ADC, timers and counters, a pulse-width-modulation output and a single-channel full duplex universal asynchronous receiver transmitter (UART) with a built-in baud rate generator.

There is, unfortunately, no internal memory and external RAM and EPROM must be memory mapped through appropriate decoding chips into the system. With a 16-bit address bus, 64 kbyte of memory can be accessed but an area is specified for special function registers associated with the I/O availability, system stack and interrupt vectors. A typical arrangement for the memory map in a control application would be 2 kbyte of RAM at the bottom end of memory, i.e. 0000–07FF, and an 8k EPROM at the top end, i.e. C000–FFFF.

The additional hardware required to make the microcontroller functional is a 5 V ±10% power supply, a crystal clock oscillator at 8 MHz and a RESET control.

The architecture is based on six main registers similar to those used in the popular 6502 microprocessor and the instruction set is upwards compatible but includes additional mnemonic coding.

An alternative to the Mitsubishi M50734 is the M50747-PGYS which has an 8k EPROM, mounted piggyback style on the top of the SCM. Since 192 byte of internal RAM are available, no external memory with associated decoding is necessary. This model has a number of bidirectional 8-bit I/O ports and timers but has no analogue-to-digital conversion facilities.

The Motorola MC 68705R3 microcontroller[17] This is a powerful SCM which comprises an 8-bit CPU with 112 byte of RAM, about 4 kbyte of EPROM, four 8-bit I/O ports, a timer and a four-channel ADC. The integrated circuit which makes this possible requires more on-board transistors than the 16-bit 68000 microprocessor used in the Apple Macintosh microcomputer.

The EPROM facility for the user software is not of the piggyback variety, but is embedded into the package which has the disadvantage of requiring a specialised software development system. The inherent ADC is extremely useful in control systems which involve analogue measurements. It uses the successive approximation technique with an internal sample and hold circuit and the conversion time is 30 μs with a 4 MHz crystal clock. The external reference voltage which is to correspond to the full 8-bit resolution is connected directly onto the SCM package.

The Rockwell R6500/1 one-chip microcomputer This is an easy to use SCM which contains 64 byte of RAM and 2k (R6500/1EB1) or 3k (R6500/1EB3) of addressable EPROM fixed piggyback fashion onto the package. This allows a standard EPROM such as a 2716 or 2732 to be easily removed, programmed or reprogrammed, then reinserted as often as desired.

Thirty-two bidirectional TTL compatible I/O lines from four ports are available with a 16-bit timer/counter that can operate in four modes. This makes a useful and versatile sequential controller which can be used with pneumatic or hydraulic systems.

To make the integrated circuit package functional, a power supply, clock and reset switch must be connected.

The EPROM contains the reset vectors which direct the program counter to the start address of the code which is to be executed in sequence to carry out the specified control task. The reset vectors are also within the same EPROM. The RAM available must also be used for stack operation and the stack pointer must be set up at the start of the applications software.

11.6.2.2 Industrial controllers

These are usually modular industrial microcomputers which are built to internationally recognised bus standards such as STD or EURO and selected cards are held in a standard rack. The ability of these systems to acquire data, control equipment and analyse the logged data make them particularly suited to process-control applications where performance monitoring is required. The system also lends itself well to laboratory data acquisition where the function is to collect data from a range of different instruments.

The development of applications software, usually in some form of real-time high-level language, is now usually carried out by communication between the controller and an IBM PC or compatible. This makes life easier for the system builder by adding enhanced editing and data storage facilities. These rack mounted industrial controllers are generally more expensive than PLCs, although they offer more flexibility.

Since most practical industrial processes tend to be poorly defined and it is often difficult to derive an accurate mathematical model of the system, the control strategy usually adopted is the 'three term' (PID) strategy. This has led to the emergence of microprocessor based PID controllers, the control output of which is based on the error signal (evaluated from the declared set point and measured process variable) and the chosen settings of gain, integral and derivative time. Self-tuning PID controllers are now available.[7] Some of these employ a process 'pattern-recognition' technique with an 'expert system' approach based on the tuning rules usually employed by skilled control engineers.

11.6.3 Software considerations

The power in the digital control of a system is in the software and any controller requires a real-time language. This is one which can synchronise its operations with events occurring in the 'real world' and thereby respond to and control interfaced mechanisms and processes.

The facilities required of a real-time software language are:

(1) ease of switching external devices on and off using software;
(2) ease of deciding whether external devices are on or off at any time;
(3) ease of timing the duration of a process;
(4) ease of making a process run for a predetermined time; and
(5) ease of making the program respond to things that happen in the outside world.

Although high-level languages such as BASICA, QUICK-BASIC, TURBOBASIC, FORTRAN and PASCAL can to some extent meet the above requirements, there are a variety of real-time languages which have been specifically designed for control applications.

11.6.3.1 Real-time high-level languages

BASIC This is a programming language commonly used for engineering applications, and real-time interpreted dialects such as CONTROL BASIC and RTBASIC[18] figure prominently. These can be used with both memory mapped and port addressed I/O organisation (see Section 11.4.7) and provide keywords which enable the individual bits on a digital port to be read or written to for switching devices on or off. ADCs and DACs are also easily operated through keywords not usually available in ordinary dialects of BASICs. Timing operations can also be easily implemented.

In addition to the digital and analogue I/O keywords available to assist in the development of control software, floating point arithmetic is included for data-processing requirements.

FORTH[19] This is an interpreted threaded language developed specifically for control applications. The instructions which constitute an application are stored as a list of previously defined routines. This list is threaded together during the entry of source code from either the computer keyboard or the mass storage buffers. The process of producing the list is often termed 'compilation', but this is not strictly correct since the result of the true compiling of source code produces pure machine code.

FORTH is a most unusual language since arithmetic calculations do not follow traditional methodologies. Before calculations can be made it is first necessary to understand how the stack operates in FORTH. Most high-level languages use one or more stacks for their internal operations, but languages such as PASCAL and FORTRAN are designed so that the user does not have to understand how the stack functions. FORTH allows the user full control of the values stored in the stack and their manipulation.

One of the main features of FORTH, and one of the most powerful, is that once the program has been written, a single word which defines the complete program can be entered into the FORTH dictionary as a command. This dictionary contains words defining, say, routine control operations, and a single word could trigger a series of actions as it threads its way throughout the dictionary.

Programs can be typed in at the keyboard and executed directly but, to save programs for execution at a later date, the mass storage buffers and the FORTH editor must be used. Unlike other high-level languages, FORTH performs only a very limited number of error checks. The errors detected are those which are most likely to cause the system to crash if allowed to pass undetected. The main reason for the lack of error checking is that is would slow FORTH down, and since the main use of FORTH is in time critical control situations, this would be prohibitive.

C This is a general-purpose programming language which cannot be truly classed as either a high-level or a low-level language. It has all the features expected of a high-level language but allows the programmer to access the computer's hardware and has the high performance usually expected of low-level languages.

Although the language was developed during the 1970s, it is now gaining rapid popularity in a range of industrial applications which include real-time control. There are a number of versions of C available, but the accepted standard for the programming language can be found in the book by Kerningham and Ritchie.[20]

Accessing the input and output of data at external ports is easily accomplished, and another exceptionally important feature is the use of timing operations which have a resolution of the computer system clock period.

Using C it is possible to construct concise, well-structured and well-documented programs that can include a variety of useful library functions such as needed for DOS I/O calls for hardware, screen handling and real-time clock timing facilities.

11.6.3.2 Low-level languages

At the machine level the program is stored in the memory in a binary format. It is conceivable to write the program immediately in machine code but this would entail an indefensible amount of labour. It is easier to use mnemonic programming where each instruction is accorded a symbolic name close to colloquial language. The name is easier to remember and the resulting program is much easier to read than pure machine code. This so-called 'assembly language program' can be translated into the necessary machine code for a particular microprocessor by using an assembler program.

The main attraction of using machine code for a particular application is its speed of operation and, if a single-chip microcontroller is to be used, the problem of software development must nonetheless be faced. This requires the writing of a program into the system memory when no keyboard, display monitor or operating system software exists.

The inexpensive approach is to choose a SCM which supports an external EPROM in piggyback fashion. The software code for the specific task may then be developed using an appropriate assembler program which operates on an IBM PC, or a compatible machine. The resulting code is then used to blow the EPROM. This method can, however, be very time consuming since, in order to achieve the specified requirements, the debugging process will require the continual updating of the EPROM data as the program is edited.

An alternative is to use an EPROM emulator inserted into the target system to temporarily hold the program which is transmitted from the host computer memory. This enables the user to alter the program easily as required and then finally blow the EPROM for permanent use. EPROM emulation should not be confused with the technique of 'in-circuit emulation' which involves the removal of the microprocessor from the target system. The in-circuit emulator then takes complete control by emulating, in real time, all the functions of the removed microprocessor it replaces.

A software development system should therefore comprise the following.

Hardware

(1) A host microcomputer with keyboard and display monitor;
(2) a printer for a hardcopy of listings and disassemblies;
(3) disc drives for the permanent storage of data and programs;
(3) an EPROM emulator;
(5) an EPROM programmer.

Software

(1) An editor; and
(2) translation tools such as an assembler or cross-assembler, disassembler, debugger and linker.

Traditional cross-assemblers will only assemble for one microprocessor, and this can be very expensive if a number of upgrades or different types are to be accommodated. These very sophisticated and dedicated development systems are generally too expensive and complicated for producing the operational programs to control a fairly basic and ordered sequence of events as usually occurs in mechanical systems.

An alternative is to employ an IBM PC or compatible, with an assembler which can handle multiple instruction sets. The processor to be used is specified in the source code and the assembler adapts automatically to the correct format.

Tailoring a general purpose and readily available machine such as the IBM PC to a semidedicated role requires relatively low investment and low software production costs. For one-off applications this approach is the only real proposition economically.

11.6.4 Sampling frequency in digital control loops

A direct digital control (DDC) loop contains both hardware and software contributions. Transducer measurements of the process variable must be made regularly and the invariably analogue values converted into sequences of numbers that can be handled by the controller. The measured value is compared with some set condition and a control algorithm implemented in software evaluates the necessary control effort. This effort is calculated as a numerical value in the computer and a conversion of the output is necessary to obtain a form which is suitable to drive the required control elements or actuators. A sampling rule for the measured variable is used to determine the rate at which the sampling is performed.

Digital sampling gives a sequence of 'snapshots' of an analogue variable. The controller only holds representations of the variables at the times when the samples are taken (see *Figure 11.100*). The sampling rate must obviously be matched to the rapidity of the variations in the process variable. High performance control systems require the sampling interval Δt to be short, although very rapid sampling will increase the computational load. Δt is usually specified in terms of other system parameters.

The classic reference related to digital sampling is that due to Shannon and Weaver[21] which states that a signal which is bandwidth limited can be recovered without any distortion, provided samples are taken at a rate of twice that of the highest frequency. Twice this frequency may be regarded as an absolute minimum and a value of five to ten times the highest frequency produces a more realistic digital representation of a sinusoidally varying continuous signal.

Real industrial plant have limited bandwidth, and in applications with long time constants, a sampling frequency of

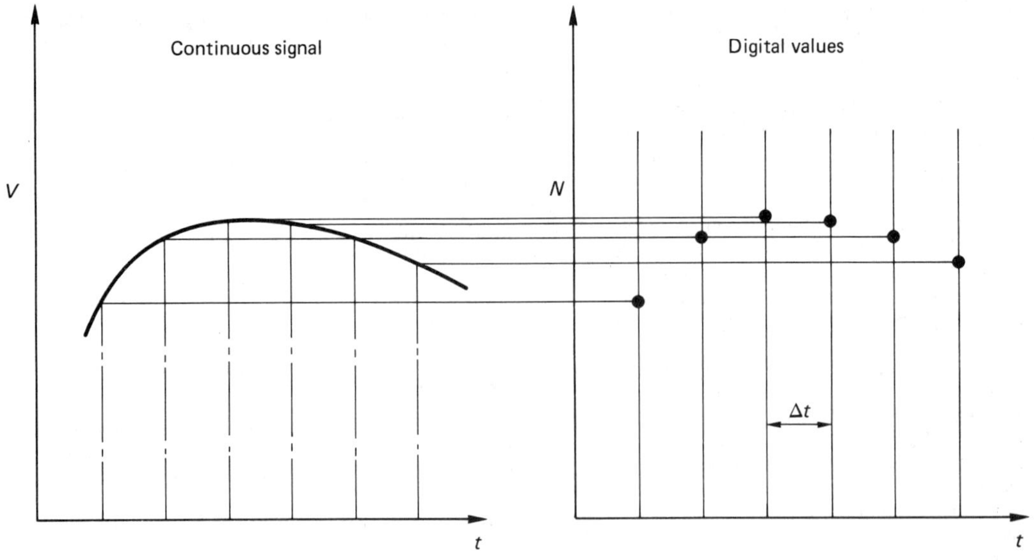

Figure 11.100 Digital sampling of a continuous signal

twice the highest plant frequency is adequate. In practice, for the vast majority of control loops involving the control of such variables as temperature, pressure, flow and level, sampling intervals of 0.2–1 s usually prove to be fast enough. These are the rates normally fixed in the commercially available industrial PID controllers.

Fast acting electromechanical servosystems require much shorter sampling intervals (down to a few milliseconds). Shannon's sampling theorem cannot strictly be applied if the system maximum frequency is unknown. However, practical experience and simulation have produced useful empirical rules for the specification of the minimum sampling rate. One such rule is based on the dominant time constant (see Section 11.1.2), which can be obtained from an open-loop test on the process. A suitable sampling interval for use in a closed-loop control digital algorithm is:

$$\Delta t < (\tau/10) \tag{11.63}$$

This may, however, prove to be unsatisfactory with systems which have a large time constant when a fast closed-loop response is forced by the controller settings.

11.6.5 PID digital control algorithm

A mathematical model which produces a transfer function (see Section 11.1.3) for the process to be controlled can usually be derived for servomechanisms, and specifications in terms of damping ratio, natural frequency and bandwidth can realistically be defined. This is not the case, however, for the majority of industrial processes. For this reason a three-term controller implementation (see Section 11.2.2) is preferred in practice.

The five essential steps required to achieve good process control are as follows.

(1) The variable (PV) that best represents the desired condition of the final product must be measured. Measurements such as temperature, pressure, flow, position and speed are commonly used.

(2) This measurement must then be compared with the desired set point (SP) value of the variable to yield an error signal E, where $E = \text{SP} - \text{PV}$.
(3) This error is then applied as an input to the controller.
(4) The controller output U must then be applied to a final control device such as a powerstat, a valve or a motor drive system.
(5) A source of energy or material flow is then adjusted in response to the controller output until the deviation between PV and SP will be as near zero as the sensitivity of the system permits.

This is the basic principle of negative feedback as used in most automatic control systems. The sequence of operations at a chosen sample rate is: measure, compare and correct.

The function of the digital controller is to apply a control algorithm based on the error value. An algorithm is a computer procedure that performs mathematical functions on a given set of variables. The computational procedure that converts (SP-PV) into a controller output is commonly called the 'PID algorithm'. This algorithm is robust and performs well in practice. Although it may appear inferior as a scientific method, it is difficult to improve on significantly.

The controller output from a PID control strategy is as given by equation (11.41):

$$U = K[E + (1/T_i) \int E \, dt + T_d(dE/dt)] \tag{11.41}*$$

For implementation on a digital controller this must be transformed into the appropriate software for the system being controlled. This transformation from a continuous to a discrete form can be achieved by either the use of Z transforms or difference equations. The latter is easier to understand and implement, and the steps in deriving any digital algorithm by the difference equation method are as follows.

(1) Express the output requirement in a differential equation form.

(2) Replace the equation in difference form using the discrete digital approximation:

$$dY/dt = (Y_i - Y_{i-1})/\Delta t \qquad (11.64)$$

where Δt is the sampling interval.

(3) Solve for the present value of the variable Y_i from the previous value Y_{i-1}.

Applying these steps to the three-term controller yields:

$$U_i = K[E_i + (\Delta t/T_i)\Sigma E_i + (T_d/\Delta t)(E_i - E_{i-1})] \qquad (11.65)$$

The error at any particular time i is best evaluated as a percentage of the set point and the above algorithm would give the controller output as a percentage value which must be contained within the range 0–100%.

This PID digital algorithm can be easily programmed directly in a high-level language and the framework for a program in BASIC is given below.

(1) The values for the chosen set point and the controller setting requirements K, T_i and T_d must first be input to the program.

(2) A sampling time Δt is chosen and specified in the program.

(3) Since a summation of the error value is required, a variable termed, say, 'SUM' must be set to zero. This also applies to the 'previous error' value E_{i-1} in order to start the control loop.

(4) Numerical constants can be calculated, e.g.

IC = $\Delta t/T_i$ and DC = $T_d/\Delta t$

(5) The program listing continues:

```
100   REM a routine is required to measure the process
      variable value
200   E=100*(1-(PV/SP))/SPAN
300   SUM=SUM+E
400   OUTPUTI=IC*SUM
500   OUTPUTD=DC*(E-EP)
600   U=K*(E+OUTPUTI+OUTPUTD)
700   IF U<0 THEN U=0
800   IF U>100 THEN U=100
900   REM the output U must then be output from the
      controller in a form
1000  REM which is appropriate to the power require-
      ment of the process being controlled
1010  REM eg- an A/D converter or PWM (see Section
      11.5.2)
1020  EP=E
1030  REM repeat control loop from line 100
```

N.B. It is essential that the sampling interval Δt is greater than the time taken to complete the control algorithm. The above routine would generally be called from a main program which contains a time adjustment such that the complete sequence is contained exactly within the specified sampling time Δt.

The program reveals the general method of programming PID control algorithms for use with a computer and forms the basis of the software incorporated in various commercially available controllers.

11.6.6 Speed control

Many industrial processes require variable-speed drives using electric motors. Electromechanical methods of motor speed control, such as Ward–Leonard motor/generator sets, have largely been superseded by using digital controllers interfaced to power semiconductor devices. The common types of motor are those which are operated either by d.c. or by a.c. A.c. machines are also referred to as 'synchronous motors'.

11.6.6.1 D.c. drives

D.c. motors can rotate in either a clockwise or an anti-clockwise direction, depending on the direction of current flow to the coils via the brushes. The speed is load dependent and can be varied by altering the supply voltage. Arcing at the brushes generates interference which may cause problems when using computer control through corruption of logic levels representing I/O data. It is essential that the spikes generated at the high interference frequencies are filtered out. This can be achieved by fitting a capacitor across the contacts close to the motor, and adding an inductance in series with the power supply.

A typical direct digital control loop is illustrated in *Figure 11.101*. Details of typical methods of measuring rotational

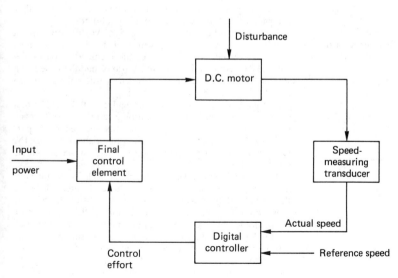

Figure 11.101 Closed-loop control of a d.c. motor

speed are given in Section 11.3.5. The digital controller invariably implements a three-term strategy as outlined in Section 11.6.5, and the numerical value calculated must be converted into a motor input power through a final control element. This may take the form of a DAC and a power amplifier. D.c. servomotor control modules are commercially available for a range of motors and with a variable input from the DAC the motor speed can be varied over its complete operating range. The speed and nature of response to an external disturbance such as a change in load are dependent on the controller settings.

11.6.6.2 A.c. drives

The a.c. motor does not have brushes and hence the problem with interference that arises when using digital control does not exist. A.c. drives are also less expensive, more robust and require less maintenance since they do not have a commutator. However, being a synchronous device, the speed is locked to the frequency of the supply. To control the speed of an a.c. induction motor over a wide range, both the frequency and the amplitude of the applied voltage must be varied. The classic method of achieving this is to use a d.c./a.c. inverter in which a positive and negative d.c. supply is alternately switched to the motor. This switching arrangement is shown in *Figure 11.102*.

In order to avoid short circuiting the d.c. supply, any one of the two switches must be off before the other is switched on. The switching may be achieved by using either transistors or thyristors which can operate with a turn-on/turn-off time of typically 5 μs. Thyristors have been developed for this purpose and have current- and voltage-handling capabilities of 1000 A and 4000 V, respectively; the corresponding figures for power transistors are 400 A and 800 V.

The control loop for the a.c. drive is identical in principle to that depicted in *Figure 11.101* where the final control element constitutes the high-frequency switching of an inverter circuit to produce a pulse-width modulation technique (see Section 11.5.2) for varying the input power to the machine. The transistor or thyristor based on a.c. drive at powers up to 500 kW is a viable proposition and has the advantages of close speed control in both directions, controlled acceleration or deceleration and high efficiency with a good power factor.

11.6.6.3 Stepper motors

An alternative method of obtaining a variable speed is to use a stepper motor. This is a power device which converts a d.c. voltage pulse train into a proportional mechanical movement of its shaft. There is a range of motors commercially available which are designed for a variety of operating conditions.

The motor is designed with a number of phases which must be supplied with current in a specified preset continuous sequence according to the number of phases. Reversing the sequence causes the motor to rotate in the opposite direction. The motor shaft rotates by a finite amount for each pulse received. Thus if a motor is designed for 48 pulses/revolution then a pulse train received at the rate of 96 pulses/second will cause the motor to rotate at a speed of 120 rev. min^{-1}. Stepper motors having outputs in the order of kilowatts are available and they are used for applications involving accurate position or speed control such as X–Y plotters, numerically controlled machine tool slide drives and in carburettors for internal combustion engines.

Interfacing a stepper motor to a digital computer is relatively straightforward, and a number of integrated circuits and power driver cards are available for a variety of popular digital controllers.

11.6.7 The PC as a controller

The IBM PC and its compatibles are firmly established as the standard for industrial computing throughout the UK and Europe. They are relatively cheap, plentiful and fairly easy to use. As they are integrated into the industrial environment, it is evident that they have a high potential as control devices on the factory floor.

In addition to an increase in processing power, there are many other advantages in using a PC-based control system.

(1) There is a large choice of software which is not available for dedicated controllers.
(2) There is a large choice of available tools to produce applications software efficiently.
(3) The PC is available in a variety of forms, e.g. a single card, a portable, a desk top and an industrial version for use on the factory floor.
(4) Expansion plug-in slots to the PC bus structure are available and a large range of cards for digital or analogue I/O have been produced by a number of manufacturers.
(5) The PC-based controller is more flexible than the dedicated, or mini-computer system and can be easily configured indefinitely to suit different applications.

Data acquisition and control add-ons for PCs are either external rack-mounted systems or plug-in boards.

The external-box approach usually involves attaching a separate rack-type enclosure with power supply to the host PC. The connection is via either the included serial or the parallel data communication link. Various modules based on a standard card format, such as the Eurocard, can be plugged into the enclosure housing as required.

There are two options for capturing data with PCs. The first option is to use an ADC card that plugs directly into the back plane of the host computer. The cards are generally 'port addressed' and may be driven by any language having IN/OUT commands. The base address is usually switch selectable on the card.

The second option uses instruments such as digital voltmeters and frequency meters which have an interface board that enables data transfer from or to the controlling PC. The most common standard is the IEEE-488 (GPIB), where data in bit parallel, byte serial manner are transmitted from the PC, as an ASCII string, to the instrument informing it of the settings required for the measurement to be made. Once read, a control signal is sensed and a string representing the measured value is returned to the PC. It is of course necessary to have an IEEE-488 card installed into an available slot in the

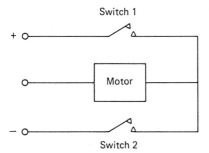

Switch 1

+

Motor

−

Switch 2

Figure 11.102 Principle of a d.c./a.c. inverter

PC. Up to 15 instruments can be accessed in this way from one card.

The quickest, easiest and least expensive way to get measured data into the PC, or control signals out, is to use I/O cards. These are available for many applications and include:

(1) multichannel digital I/O with opto-isolation and Darlington driver facilities;
(2) pulse counting and timing;
(3) multiplexed analogue-to-digital conversion with programmable gain;
(4) digital-to-analogue conversion; and
(5) thermocouple input.

Recently there have been rapid advancements made in the software available for data capture and control. Packages are now available which provide the user with a development system having an environment of windows and pull-down menus. The National Instruments[22] 'Lab Windows', for use on PCs, is a library of function modules for programming specific instruments with the IEEE-488 interface. These modules are accessed in the development program via function panels to interactively set up and acquire data from instruments. In addition, a suite of programs is included for data presentation, analysis and formatting in an interactive environment using either QUICKBASIC or C.

A number of manufacturers of rack-mounted microcomputer controllers and programmable logic controllers (PLCs) have united their product with a PC and provide user-friendly software to assist in the development of the required control programs. Such aids are invaluable for displaying, storing and printing PLC ladder relay diagrams as an alternative to the hand-held programmer and EPROM blower. Logged data can also be transferred from the controller to the PC where they can be displayed on a mimic diagram of the plant with animated symbols, updated values of controlled parameters, alarm messages and bar graphs.

Software development for PCs has at last begun to catch up with the hardware, and the data acquisition and control market has long awaited reliable development systems.

11.7 Programmable logic controllers

Automation systems generally involve the application of such microprocessor related equipment as CNC machine tools, pick-and-place machines and industrial robots. As shown in *Table 11.4* there has been a greater involvement in the application of programmable logic controllers (PLCs) in manufacturing industry than any of the other automation devices.[23] The figures in *Table 11.4* are taken from a survey of all UK manufacturing industries employing 20 or more people.

PLCs are mostly employed in the relatively straightforward control of a single process or piece of equipment. They are particularly common in the food processing, chemical and automotive industries. PLCs first appeared in the 1970s in place of relay circuits and they have been continually developed as a result of the rapid progress in microelectronics. Some of the current more powerful PLCs overlap with microcomputers or process computers and it is often impossible to distinguish between one and the other.

11.7.1 The PLC in automation systems

The PLC is particularly useful for controlling manufacturing processes which constitute a sequence of operations that are either event or time driven. PLCs have had a significant impact on industrial control because of their ruggedness, versatility and reliability. They have virtually replaced hard-wired relay, counter and timer logic systems, due to their cost, flexibility and relative size. In addition, the number of applications in which PLCs are being used has increased substantially in recent times.

The principal criterion for determining size is the I/O availability and this can broadly be divided into the following categories.

(1) *Small PLCs*—up to 128 I/O lines, with typically 12 inputs and 8 outputs in the basic form. These are designed as stand-alone items and can usually be expanded with respect to digital I/O requirements. A typical system with the associated hand-held controller is shown in *Figure 11.103*. Such systems basically perform logic, counting and timing operations, but have no arithmetical manipulation capabilities.
(2) *Medium-sized systems*—128–1000 I/O lines, which are generally contained in a rack arrangement. These offer more extensive I/O including analogue-to-digital and digital-to-analogue options with some enhanced programming features.
(3) *More powerful rack mounted systems, with over 1000 I/O lines*—These PLCs offer data communication for operation within a complete computer integrated manufacturing arrangement. They embody quite sophisticated man–machine interfaces which employ real time operating systems and advanced computer graphics. Three term PID control, digital filtering and vision capabilities can be implemented in many modern PLCs and a variety of operator interfaces are available to allow easy entry of desired settings and process variables.

A typical small PLC is shown in *Figure 11.103*.

Table 11.4 Usage of automation equipment[23]

Equipment used	1983	1985	1987	1989 (estimate)
CNC machine tools	17	19	27	42
PLCs	34	59	73	110
Pick and place machines	4	11	10	15
Robots	1	2	3	6

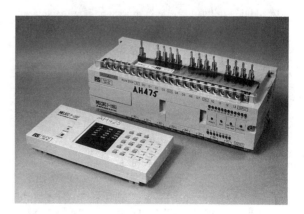

Figure 11.103 A typical small PLC

Recent developments have seen a move from the automation of single machines towards the automation of the whole manufacturing process. The PLC is a typical representative. The modern concept is the automated factory where computer integrated manufacturing (CIM) combines the basic functions of production into a single system. The system incorporates the highest level of automation economically feasible for each function involved. The PLC constitutes one of the building blocks in such a distributed hierarchical control system which is based on a 'pyramid' structure. The peak of such an arrangement contains the mainframe computer which is responsible for handling the databases for production scheduling, sales and other management needs. The PLC or microcomputer immediately precedes the machine and process applications located at the base of the pyramid.

However, although the small single-purpose basic PLC is relatively secure in its role in the market, the medium and large PLCs face considerable competition from microcomputers. This is particularly the case in applications where a lot of arithmetical manipulation has to be performed. It is now possible to get IBM PCs and other microcomputers in a 19 in. rack format which can be mounted within standard control panels. This represents a serious competitor to many PLC based systems.

11.7.2 The PLC vs. the microcomputer

The PLC and the microcomputer contain the same basic constituent components such as central processing unit, memory devices, input/output interfaces, decoding logic and a connecting bus structure. They are, however, entirely different in physical appearance and operation.

11.7.2.1 The case for the microcomputer

The dominance of the International Business Machines Company (IBM) in the microcomputer market has led to an accepted industry standard machine in the form of the IBM PC. Since its appearance, several manufacturers have produced so-called 'compatibles' which are based on the associated 16-bit and 32-bit technology. The latest processors now used allow the programmer to write programs that use more memory than is actually available in a given system. This is done by exchanging data between the main memory and secondary storage devices. It is this processing power, memory availability and peripheral hardware that has made a dramatic impact on engineers' acceptance of PCs for control applications.

In addition to this increase in processing power, there are many other advantages in using a PC-based control system.

(1) The large choice of available software ranging from high-level language interpreters and compilers to sophisticated man–machine software interfaces which include a range of selectable data processing and graphical display routines.
(2) The large choice of available tools to produce applications software efficiently.
(3) The variety of forms available which range from the small portable type for use in the field to the rugged industrial version for use on the factory floor.
(4) The large range of plug-in cards available for use with the PC bus structure to provide both digital and analogue I/O facilities.
(5) The ease with which the system can be reconfigured to enable, for example, the application of wordprocessing, spreadsheets or databases to be carried out. There is no need to purchase a whole new system for every new application.

However, in order to truly harness the flexibility that accompanies the personal computer, some knowledge of microprocessor technology is required and the ability to understand or develop applications software is essential.

11.7.2.2 The case for the PLC

The alternative to the microcomputer for control applications is the PLC which may be described as less intelligent but is ideally suited for carrying out logical sequential operations to produce outputs which are conditional of input states. The advantages of using a PLC for controlling machines and processes may be summarised as follows.

(1) The PLC is relatively inexpensive compared with a microcomputer.
(2) The construction is exceptionally rugged and it requires very little space in the control cabinet.
(3) Reliability is high, immunity to electrical noise is good and maintenance is low.
(4) Memory is used economically due to the method adopted for the processing of the data. The PLC processor is tailor-made to execute logical operations (AND, OR, etc.) as they occur in the control program.
(5) The operating speed is fast.
(6) An in-built interface provides for the easy connection to a variety of input–output devices connected directly to the machine or process.

The PLC is, however, less interactive than the microcomputer, partly due to the fact that relay ladder logic is generally used as the programming language because of its alleged wide acceptance by those who design, operate and maintain control systems. During program execution the ladder is continuously scanned and outputs are set in accordance with input conditions. It is reputed that ladder logic is straightforward and easily understood, but software development can be a problem for the non-specialist in systems which involve a number of timer and counter requirements.

In conclusion, there is probably equal support, at the present time, for PC and PLC systems. Both systems then, should continue to play an important part in manufacturing operations for the foreseeable future.

11.7.3 Ladder logic programming

The most commonly used programming language for use with a PLC is the ladder diagram. For the popular small PLCs it is usually possible to enter the input, output, timer and counter instructions directly into the controller using a hand-held programmer. The mnemonic codes are displayed on a small screen within the programmer prior to conversion into machine language and transfer into the PLC battery backup memory. Although the program can be edited the main disadvantage of using this method is that there is no way of saving and printing the developed program.

A preferable alternative is to use a PC with an appropriate software package available from the PLC manufacturer to develop and display the ladder diagram on the PC monitor. Use is made of the PC disc drives and printer as required. Once completed, the resulting file is transmitted from the PC to the PLC for execution. This is by far the best method of developing ladder logic software.

In a ladder diagram the power source is represented by the two vertical rails of the ladder and the various instructional requirements which represent the control circuits make up the

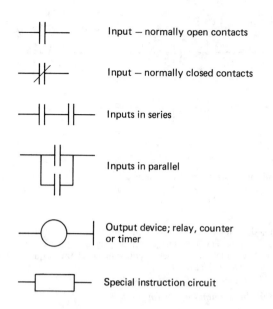

Figure 11.104 Instruction set ladder logic symbols

rungs. The symbolic ladder circuit layout is constructed using standard graphic logic symbols to represent input contacts, output loads, timers and counters, as shown in *Figure 11.104*. When using a PC to develop the program, comments can be added to each instruction rung. This makes the program easier to understand and facilitates fault finding.

Each rung on the ladder starts with an input condition and finishes with an output to a load, timer or counter. When used to automate a machine, the ladder is continuously scanned in a sequential manner and sets the outputs on or off according to the requirements based on the sensed input conditions.

11.7.3.1 Inputs

The most common symbol used in programming a PLC is the input contact which may be active in either the normally open (NO) or normally closed (NC) mode of operation. The NO contact symbol is a request to examine if the switch is ON in order to proceed along the rung. For the NC condition to be satisfied the contact must be OFF.

A popular real-world input device is the switch. This is basically used to open and close an electrical circuit. The PLC supplies the voltage across the switch which is usually either 24 V d.c. or 110 V a.c. A large number of switch designs are used to control systems. The limit switch is one such example and it is designed to open and close when a machine part reaches a specified position, or a physical variable reaches a required value. Once the limit switch condition has been sensed the PLC takes appropriate action as defined in the ladder program.

11.7.3.2 Outputs

The most common output devices are the electromagnetic relay and the solenoid. The relay provides isolation of the load voltage and current from the PLC and is ideal for the switching of high loads via the controller's in-built I/O interface. Alternatively, the solenoid, which consists of wire surrounding a movable plunger, allows electrical control of any mechanical device attached to the plunger. This arrangement is applied in

electropneumatic systems to control the position of a valve spool in order to direct the air path through a pneumatic system.

Due to the nature of the sensed input and the switched output load, the PLC I/O module normally provides a safe electrical separation between internal and external circuits. This is effectively carried out by an opto-isolator which ensures that there is no physcial hard-wired connection between the PLC and the external I/O device.

A typical rung in a ladder diagram including I/O with logic operations is shown in *Figure 11.105*. This rung may be written for a particular PLC as shown in *Table 11.5*.

Different PLCs use various ways of translating the ladder programs into mnemonics, but the principle is basically the same as that illustrated in the table. This translation is necessary if the program is to be entered into the controller via a hand-held programmer. However, if a PC-based software development system is used then the ladder diagram is entered directly by pressing the appropriately specified function keys.

11.7.3.3 Timers

Time delays are commonly used in control applications, and in a PLC program a timer is used to activate an output following an elapsed time as specified by a value stored in the memory. The timing base units are usually tenths or hundredths of a second.

There is no standard way of specifying timer routines. Manufacturers use different methods, but the same basic principles of operation apply. This entails an enabling input condition to start the timer. At the end of the timed period the timer must be reset by disenabling the input.

It must be noted that, during the program execution, the ladder is *continuously* scanned rung by rung and outputs set according to input conditions. There can be no hold on any particular rung. However, in implementing timing operations the timer output can be used as an input on any rung, thus

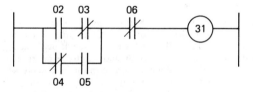

Figure 11.105 A typical ladder rung

Table 11.5 A rung for a particular PLC

Step	Instruction	I/O number
000	LD	02
001	AND NOT	03
002	LD NOT	04
003	AND	05
004	OR BLK	
005	AND NOT	06
006	OUT	31
007	END	

Note: the 'OR BLK' statement 'ORs' the blocks grouped under the preceding two load, i.e. LD, statements.

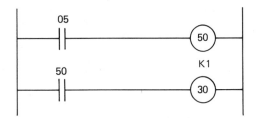

Figure 11.106   Example of a timing operation

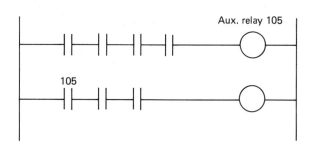

Figure 11.108   Use of an auxiliary relay

making outputs dependent on the specified time period to have expired. This is illustrated in *Figure 11.106*.

In the ladder diagram, the timer number 50 is set for a timing operation of 1 s and is enabled by switching input number 05 to an ON state. In the next rung the output number 30 will not be energised until the specified time has elapsed. The timer must then be reset by setting the input 05 to OFF.

11.7.3.4   Counters

A counter is used to activate an output after a predetermined number of counts as stored in the PLC memory. It operates in much the same manner as a timer but, rather than counting time, the counter counts the number of times that an event occurs. The event is some specified input condition occurring in the process. The counter is reset from another input switch and it can be used as an input to any other rung in the ladder. This is illustrated in *Figure 11.107*.

Each time that switch 02 is activated a count is produced up to the specified value of 10. Once the count is completed the output load number 30 is switched ON. The counter is then reset by switching input 01 OFF. An incrementing or decrementing mode for the counter operation is usually possible with most PLCs.

11.7.3.5   Auxiliary relays

Most PLCs are equipped with various special function I/O facilities that exist in the development of safe and efficient control programs. The auxiliary relay is one such device which operates internally and is assigned an address that is different from that of any other real output device in the system. It is basically a 'dummy relay' which is used as an aid in developing the program. For example, consider an output condition

which is dependent on, say, six inputs and a rung is restricted to four inputs. The auxiliary control relay can be used to connect two rungs into a logical equivalence of the requirement as shown in *Figure 11.108*.

11.7.3.6   Input/output numbering

Different PLC manufacturers use different numbering systems for the controller's available I/O. The addresses assigned for the small Mitsubishi type F-20M PLC are typical (note the octal numbering system adopted):

12 inputs:	00–07; 10–13
8 outputs:	30–37
8 timers with a range of 0.1–99 s:	50–57
8 counters with a range of 1–99 counts:	60–67
48 auxiliary relays:	100–157

11.7.4   Controlling pneumatic and hydraulic systems

Consider the sequencing of a double acting pneumatic cylinder with digital sensing devices of either mechanical, optical or magnetic construction to detect the end of stroke condition. The direction of motion is controlled by an electropneumatic five-port spool valve which is solenoid/pilot operated. The air supply to the system manifold is enabled by an ON/OFF switch connected to a solenoid operated three-port valve.

The control program is required to switch on the air, extend the cylinder rod to the end of the stroke, wait 5 s then retract the rod and repeat to complete 10 cycles. The main ON/OFF switch is to be used to reset the counter and it must also stop the cycling at any time. The input/outputs are assigned as follows:

Inputs
ON/OFF switch	00
Sensor to detect retracted position	01
Sensor to detect extended position	02

Outputs
Air supply valve	30
Direction control valve solenoid to extend cylinder rod	31
Direction control valve solenoid to retract cylinder rod	32

Timer   50

Counter   60

Figure 11.109 shows the ladder diagram to satisfy the specified sequence. Note that deactivating the ON/OFF switch will stop the system at any time and that the rod will not

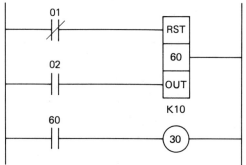

Figure 11.107   Example of a counter operation

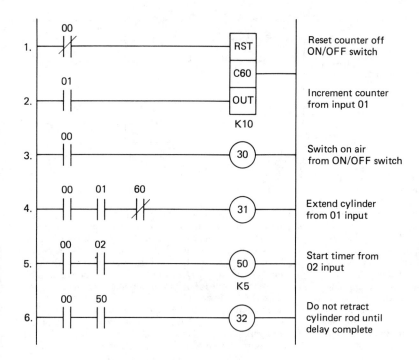

Figure 11.109 Ladder diagram for a specified example

extend once the counter has completed the required number of cycles.

With solenoid operated pilot valves, the solenoid need only be pulsed since the pilot line in the valve supplies air to move the spool and it is usually unnecessary to maintain the solenoid in an energised condition. This is evident in the ladder diagram with inputs 01 and 02 from the end of stroke sensors returning to the OFF state once the cylinder rod moves from the end condition.

With solenoid only operated valves however, it is necessary to latch the output load ON in order to maintain the rod motion. The ladder relay logic for rung 4 to latch on solenoid 31 in order to extend the cylinder rod is illustrated in *Figure 11.110*.

Once the rod has left the sensor 01, power is transmitted to the output via the switched output acting as an input. This latches the control solenoid ON and it is switched OFF once the end of stroke is reached and sensor 02 is switched ON.

11.7.5 Safety

Since the equipment generally associated with PLCs often involves high voltages and currents, electrical safety procedures must be followed at all times. In addition, when testing a developed program to carry out a specific task it is essential that a simulation is first carried out in order to avoid the incorrect movement of large pieces of machinery which would result in dangerous operating conditions. Most PLCs contain a test-mode facility where outputs can be simulated by LED indication for a switched input requirement. The process is simulated by the manual manipulation of switches at specific times. The effects of the control program on the process may then be observed by the status of the PLC LED indicators on the I/O modules. If necessary, the program can be edited in the program mode and then retested prior to connecting to the actual system which is to be controlled.

11.7.6 Networking of PLCs

Computer integrated manufacturing incorporates the highest level of automation economically feasible from each computer-based activity within a complete system. The overall control task is divided into a number of individual routines assigned to specific control elements such as CNC machine tools, robots and PLCs. The PLC thus constitutes one of the building blocks of the automated factory pyramid concept. This concept requires a suitable network to enable the various elements to exchange information and utilise a control database.

The manufacturers of PLCs have addressed the problem of networking, but the major disadvantage is that each manufacturer adopts their own standards. Communicating with a variety of other control devices is not a strength of PLC

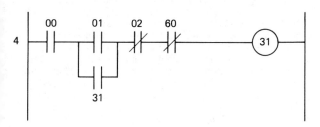

Figure 11.110 Latching of output for solenoid operated valves

Table 11.6 Typical PLC networks

Manufacturer	Network
Allen Bradley	Allen Bradley Data Highway
Festo	PC-IFS
General Electric	GE Net Factory LAN
Mitsubishi	Melsec-NET
Texas Instruments	TI-WAY

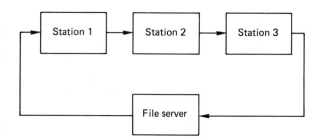

Figure 11.112 A ring network

networks due to the number of data-transmission protocols and data formats adopted. However, most manufacturers provide a dedicated local area network (LAN) for communication between controllers within their own product range. Examples of these are given in *Table 11.6*.

The basic concepts relating to the standards of communication in digital systems are covered in Section 11.4.9. In a decentralised network structure, the sensing, actuating and processing associated with each controlled element in a system is connected to a supervisory or master computer. This evaluates signals supplied by each component and returns the necessary control requirements. A PLC, PC, single-board or single-chip computer could be used as the controller for the controlled element and a process computer, PC, minicomputer or mainframe can be used in the supervisory role.

11.7.6.1 Network structures

Although there are various methods adopted for the networking of computer-based products they all possess certain common features. Each element must be equipped with a suitable interface associated with the LAN chosen and each element in the system must be connected into the network by means of wire or fibre-optic cabling to transmit the data from one station to another. Software is also necessary to handle all data transfer within the system correctly.

The physical arrangement of the elements or stations on the network is usually of a 'star' or 'ring' pattern.

In the star-shaped network (*Figure 11.111*), the stations are connected on a line parallel to one another and connected to a central computer referred to as the 'file server'. With this method, each user station must decide on whether data sent by the file server are for itself or another station. The cable length is limited in this system, but signal amplifiers can be used if required.

In the ring structure (*Figure 11.112*), the file server master computer transmits data to the first station on the network. These data are checked, evaluated and passed on to the second station if not required. The data are thus transmitted from one station to another until the user for whom the data are intended is found. Data can therefore be passed around great distances, but the failure of any one station causes the system to break down.

The integration of all intelligent elements in an automation system requires that all devices communicate with each other. Due to different suppliers having different communication specifications, international communication standards based on the open system interconnection (OSI) model have evolved. Widespread adoption of such standards will make it easier and cheaper to link devices together.

One such standard is the manufacturing automation protocol (MAP) which was initiated by General Motors to integrate all levels of control systems such as PLCs, robots, welding systems, vision systems, etc., irrespective of the manufacturer. Another very popular LAN standard adopted for industrial and commercial applications is ETHERNET which was created by Digital Equipment Corporation, Intel and Xerox.

Users of automation systems will make their purchasing decisions based on the amount of software packages and support services that the distributed control system vendors can provide.[24] Only those vendors who are capable of supplying such products and services to provide a complete solution to a control problem will be successful in the future.

11.8 Robot applications

Many of the first applications of robotic devices were used to carry out tasks in hazardous, or unpleasant environments. The best-known examples are the paint-spraying and spot-welding operations in the automotive industries. Other robotic applications were subsequently developed to replace manual workers performing monotonous activities and to carry out basic materials handling and transportation functions. These tasks were relatively easy to automate since the robots used were programmed to follow well-defined and fairly consistent paths. The functions, in general, were also repetitive with little requirement for additional sensory perception. The recent developments in sensor technology have provided the basis for enhanced sophistication in robotic applications. Modern robots, with additional sensing capabilities are increasingly finding application in the more difficult areas like arc welding, fettling and deburring, inspection and measurement and, most significantly, in automated assembly.

The enhanced capabilities of the more recent robotic devices have fostered the modern concept of the flexible

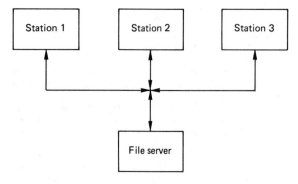

Figure 11.111 A star-shaped network

Figure 11.113   Flexible manufacturing system (Dundee Institute of Technology)

manufacturing system (FMS). The flexible manufacturing system is an integrated conglomeration of CNC machines and robots with the capability of producing a range of products with a minimum amount of human intervention. Materials handling and transportation is actioned through automated guided vehicles (AGVs), or motorised conveyors and the process starts from raw materials input through to quality control and finished product at output. Overall control of the system is governed by a central computer in a hierarchical type structure. Inventory control, procurement of materials, orders and computer aided design may also form an integral part of the overall process to be controlled.

Figure 11.113 illustrates a flexible manufacturing system, developed at the Dundee Institute of Technology. The system receives raw material direct from a continuous casting machine. In a typical application, the manufacturing route produces a range of pipe fittings which are assembled by a robot at the end of the production line.

The flexibility of the system makes it adaptable for a wide range of products. Flexibility would also normally incorporate the ability to re-route the process in the event of a failure at one of the machining workstations. *Figure 11.113* is a fairly representative example of typical robot applications in a flexible manufacturing system.

11.8.1   Robot geometry

In choosing a robot system for a range of specific tasks, some consideration should be given to the geometrical configuration of the robot system. The geometrical coordinate system relates to the relative positioning of the hand and has ramifications pertaining to the computational effort required in the position and velocity control functions.

The Cartesian geometry is typified in the gantry-type robot. These robots allow movement in three rectangular orthogonal directions and the hand motion corresponds in direct proportion to the motion along the three independent axes. The Cartesian system is by far the simplest in terms of computational effort. The gantry-type robot has much similarity with the typical overhead travelling crane.

Cylindrical polar geometries involve one rotation in combination with two linear-motion axes. The linear-motion axes are normally height h and reach r. This results in a cylindrical working envelope as shown in *Figure 11.114*.

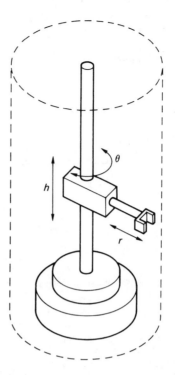

Figure 11.114   Cylindrical polar geometry

Linear hand motion across the r, θ plane involves simultaneous rotation with variable reach r. The relationship between the cylindrical polar coordinates and the Cartesian reference frame are computed from the expressions:

$$z = h \tag{11.66}$$

$$x = r \cos(\theta) \tag{11.67}$$

$$y = r \sin(\theta) \tag{11.68}$$

Spherical polar geometries involve two mutually perpendicular rotations in combination with one linear axis (*Figure 11.115*).

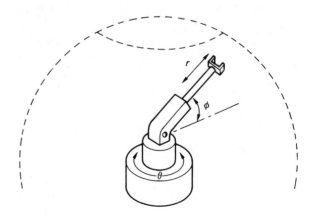

Figure 11.115   Spherical polar geometry

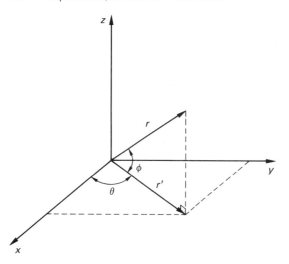

Figure 11.116 Transformation of spherical polar coordinates to Cartesian coordinates

For position and velocity control, the computer must be able to relate the spherical polar geometry back to the Cartesian reference frame (*Figure 11.116*).

It can be seen from *Figure 11.116* that the following relationships apply:

$$y = r' \sin(\theta); \; x = r' \cos(\theta) \tag{11.69}$$

$$z = r \sin(\phi); \; r' = r \cos(\phi) \tag{11.70}$$

$$r^2 = x^2 + y^2 + z^2 \tag{11.71}$$

Thus

$$(y/x) = \tan(\theta); \text{ or } \theta = \tan^{-1}(y/x) \tag{11.72}$$

and

$$(z/r') = \tan(\phi), \text{ or } \phi = \tan^{-1}(z/r') \tag{11.73}$$

Hence with the two rotations and the reach all measured, the hand motion can be transformed into an equivalent Cartesian reference frame.

The necessity of the coordinate transformations stems from the fact that the robot will, in normal circumstances, be programmed to move relative to locations specified in a Cartesian reference frame, e.g. 'point to point programming'.

Articulated geometries are commonly featured in robot-arm manipulators. These have the distinct advantage that they more closely emulate the motion of the human arm. There are no linear axes involved, but full flexibility is obtained by incorporating three rotations for the first three joints as shown in *Figure 11.117*.

The transformation from the $(\theta_1, \theta_2, \theta_3)$ to the (x, y, z) reference frame is obviously more complicated, but the articulated geometry enables the manipulator arm to reach over obstacles more easily. Typical industrial robots utilising the articulate geometry configuration include the Cincinnati Milacron and the Unimation PUMA series.

The SCARA-type robot represents a fundamentally different structure. The SCARA (selective compliance assembly robot arm) is depicted in *Figure 11.118*. There is a certain degree of similarity between the SCARA and the articulated geometry, but the rotary joints in the SCARA system enable rotation in the horizontal, as opposed to the vertical plane.

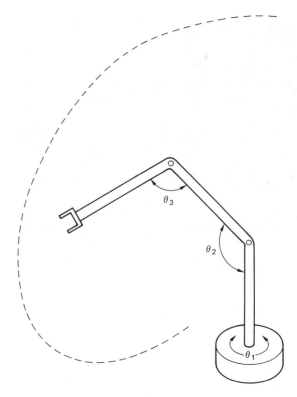

Figure 11.117 Articulated geometry

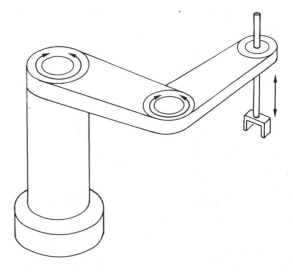

Figure 11.118 SCARA-type robot configuration

The SCARA system also incorporates a vertical-lift axis attached to the end effector. The SCARA configuration embodies the best features of the articulated and the cylindrical polar geometries. Since there is no applied torque due to the lifted load on the rotating actuators, the basic structure

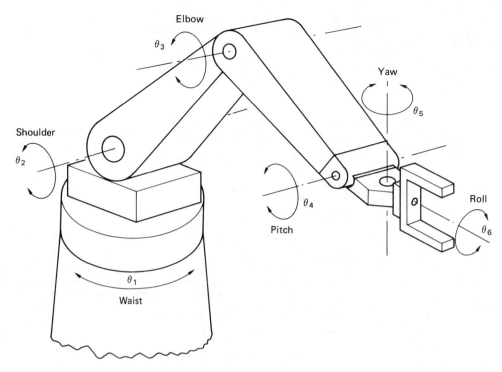

Figure 11.119 Articulated geometry robot with wrist pitch, roll and yaw

is more rigid. Therefore SCARA-type robots can generally lift heavier loads than equivalent articulated geometry robots. SCARA-type robots are employed extensively in precision assembly applications.

Further complications are introduced when the wrist joint incorporates, as is normal, additional degrees of freedom in the form of pitch, yaw and roll. A representative example is shown schematically in *Figure 11.119*. The additional degrees of freedom invoke more complex coordinate transformations, but the flexibility is greatly enhanced. The pitch, yaw and roll, together with the three translational degrees of freedom, enable the robot both to position and orientate the hand in any location and at any angle as desired. Subject, of course, to the constraints of the resolution of the sensors used to measure the rotations.

Similarly, the rate of change of the joint rotations with respect to time can be processed through the transformation matrices to generate the velocity-feedback signals. Velocity feedback is normally used to allow the hand to approach a target area rapidly at first and at a slower rate finally as the object is approached.

11.8.2 Robot programming

The primary concern of the manufacturing engineer relating to the control of a robot is the means by which a desired motion and sequence of events can be implemented on the machine. At shopfloor level, the finer details of sensor technology, analogue to digital conversion and feedback control are only secondary considerations. These are the major considerations in the design and perhaps the purchase of a new robot. However, for most practical purposes, the problem facing the manufacturing engineer is how best to utilise the limited

capabilities of an existing machine. The main considerations are therefore centred round the programming of the robot to perform specific tasks and the means of the path control to be adopted.

The simplest form of path control is based on a point-to-point strategy, e.g.

Go to point A, stop.
Go to point B, stop.
Open gripper.
Go to point C, stop.
Close gripper.
Go to point B, stop.
Go to point A, stop.

The required joint rotations are determined in the software and the robot arm travels along an unconstrained path between the stopping points. Point-to-point path control is adequate for many 'pick and place' type operations. The stop–start motion, however, is unsuitable for paint spraying, sanding, arc welding and other operations requiring a more rigorously controlled motion.

A common method of exercising rigorous control over the arm motion is on-line programming, or path recording. On-line programming is the facility whereby the robot system is 'taught', by example, how to perform a specific task. Considering, for instance a paint-spraying operation, the operator guides the robot hand through the actual process. Meanwhile the joint position sensors are monitored and recorded every 5 ms or so. On playback, the recorded sensor positions are simply fed to the joint servos and the robot reproduces the 'taught' sequence. The sequence also includes the switching of the spray gun as required. The simplicity of on-line programming is attractive since the operator does not need a knowledge of any programming language. There are, however, additional complications. Since the robot is a massive structure, the operator cannot physically move the arm against the gravitational and inertial loads. The solution adopted is to use a lightweight 'teaching arm' which has an identical kinematic structure to that of the actual robot. This form of on-line programming is referred to as 'walk-through'.

In contrast to walk-through programming, 'lead-through' systems, using a 'teach pendant', are still used extensively for point-to-point control. The teach pendant acts as a remote controller with which an operator can drive the robot arm to the required positions. At the lowest level, the teach pendant features a series of switches to drive each joint and a 'teach button' to record the end positions. More sophisticated teach pendants can allow for time delays, speed control, holds for other input data, a coordinate selection system and a variety of special library software routines. The lead-through program which is generated is much simpler than that in a walk-through based program since only the end-point locations are committed to memory. A lead-through generated program is therefore reasonably easy to edit and modify. In general, if a walk-through program requires alteration then the normal procedure adopted is to reprogram the entire operation from the start.

Off-line textual programming is only used for relatively simple robot tasks. The programming language syntax typically allows the programmer to rotate a given joint through a prescribed number of degrees in a specified direction. The biggest disadvantage of textual programming languages is the fact that there are so many of them in existence. A general lack of standardisation has somewhat hampered their development. Some of the many languages available include Unimation's VAL and IBM's AML (a manufacturing language). AML is a well structured and powerful language and with the world-wide influence of the IBM Corporation, it is the most likely candidate to be adopted as a standard. The advantages invested in off-line programming are that the robot system is free for other work during the software development stage. In addition, the new programs written build up a useful software library and the robot can eventually form an integral part of a totally computerised manufacturing system.

11.8.3 Path control

Path recording (described in the previous section) represents an effective method of controlling the trajectory of a robot arm. Cartesian control is an alternative in which the trajectory is evaluated prior to the motion. Cartesian control involves many coordinate transformations and off-line computation is normally completed well in advance of the actual motion. With modern high-speed microprocessors, the transformations can be computed in real time as the arm is moving. The computation consists of choosing a total time for the motion and dividing that time into suitable increments. For each increment, the hand configuration is computed and the transformation matrices evaluated to determine the joint rotations. The software accommodates acceleration and deceleration by imposing variable Cartesian displacements over equal time increments at the beginning and end of the traverse. Thus, for motion along a straight line, the arm accelerates uniformly to a constant velocity at the start of the manoeuvre and decelerates to zero velocity at the end. This is accomplished in the software by specifying the Cartesian trajectories in terms of fourth-order polynomials in time. The polynomial coefficients are determined by the position, velocity and acceleration boundary conditions which are imposed. The technique results in a series of coordinated set points and a PID control strategy is employed to regulate the motion between the set points.

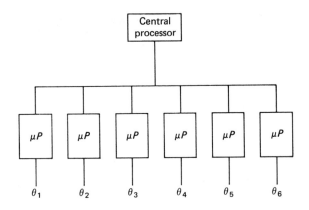

Figure 11.120 Unimate PUMA architecture

The intermediate computation is extensive, but some simplification can be obtained by dividing the path into more manageable segments and interpolating between the discrete segment points. This still involves a lot of computation and, in most cases, Cartesian control is only feasible when the robot arm is moving slowly.

11.8.4. Computational hardware

In any industrial robot system one of the main limiting factors is the time penalty associated with the computational effort. Distributed computing, in which several computing units interact in real time has immediate application to robotics. The Unimate PUMA exemplifies this concept by using a separate microprocessor to handle the input and output signals to each of six independent rotating joints (*Figure 11.120*). The central processor, in a hierarchical structure, reads the angular position via each of the joint processors, calculates the coordinate transformations and outputs the required set points to the joint processors. The joint processors then implement the control of the individual joints in parallel.

Further developments in parallel processing incorporate distributed calculation of the kinematics and distributed joint controllers. These systems share a central, multiport memory which is available to all the other processors. In this manner the computation is evenly distributed and thus accelerated.

Other time-saving expedients include fast matrix-inversion techniques and integer-based calculation procedures. Direct memory access (see Section 11.4.7) is another hardware technique used to speed up data-transfer operations within and between processors.

References

1 BODE, H. W., 'Relations between attenuation and phase in feedback amplifier design', *Bell Systems Tech. J.*, 421–454 (Jul. 1940); also in *Automatic Control: Classical Linear Theory* (ed. G. J. Thaler), 145–178, Dowden, Hutchinson & Ross, Stroudsberg, Pennsylvania (1974)

2 NYQUIST, H., 'Regeneration theory', *Bell Systems Tech. J.*, 126–147 (Jan. 1932); also in *Automatic Control: Classical Linear Theory* (ed. G. J. Thaler), 105–126, Dowden, Hutchinson & Ross, Stroudsberg, Pennysylvania (1974)

3 GOLTEN & VERWER PARTNERS, *Control System Design and Simulation for the PC (CODAS)*, Golten & Verwer Partners, Cheadle Hume, Cheshire (1988)

4 CAMBRIDGE CONTROL LTD, *SIMBOL 2- Control System Design and Simulation on IBM-PC or PS/2*, Cambridge Control Ltd, Cambridge (1988)

5 ARTHUR F. SAUNDERS, *Laplace Systems Analysis Program*, Laplace Systems–Arthur F Saunders, Woolavington, Bridgwater (1989)

6 ZEIGLER, J. G. and NICHOLS, N. B., 'Optimum settings for automatic controllers', *Trans. ASME*, **64**, 759 (1942)

7 KRAUS, T. W. and MYRON, T. J., 'Self-tuning PID controller uses pattern recognition approach', *Control Eng.* (Jun. 1984)

8 LEIGH, J. R., *Applied Digital Control*, Prentice-Hall, Englewood Cliffs, NJ (1985)

9 FRASER, C. J. and MILNE, J. S., *Microcomputer Applications in Measurement Systems*, Macmillan Education Ltd, Hong Kong (1990)

10 BRITISH STANDARDS INSTITUTION, *BS 1904 Industrial platinum resistance thermometer elements*, Milton Keynes (1961, 1981)

11 WILLIAMS, A. B., *Electronic Filter Design Handbook*, McGraw-Hill, New York (1981)

12 WHITNEY, D. E., 'Quasi-static assembly of compliantly supported rigid parts', *J. Dynam. Systems, Measurement Control*, 65–77 (Mar. 1982)

13 BEJCZY, A. K., 'Effect of hand-based sensors on manipulator control performance', in *Mechanism and Machine Theory*, Vol. 12, 547–567, Pergamon Press, New York (1977)

14 HARMON, L. D., 'Automated tactile sensing', *Int. J. Robotics Res.*, **1**(2), 3–32 (1982)

15 REBMAN, J. and TRULL, N., 'A robot tactile sensor for robot applications', *ASME Conf. Computers in Engineering*, Chicago, IL (1983)

16 BALLARD, D. H. and BROWN, C. M., *Computer Vision*, Prentice-Hall, New York (1983)

17 CAHILL, S. J., *The Single Chip Microcomputer*, Prentice-Hall, New York (1987)

18 CONTROL UNIVERSAL LTD, 'Real time BASIC', *Cube Technical Manual*, 138 Ditton Walk, Cambridge CB5 8QF

19 BRODIE, L., *Starting FORTH*, Prentice-Hall, New York (1981)

20 KERNINGHAM, B. W. and RITCHIE, D. M., *The C Programming Language*, Prentice-Hall, New York (1988)

21 SHANNON, C. E. and WEAVER, W., *The Mathematical Theory of Communication*, University of Illinois Press, Urbana, IL (1972)

22 NATIONAL INSTRUMENTS, 21 Kingfisher Court, Hanbridge Road, Newbury RG14 5SJ

23 NORTHCOTT, J. and WALLING, A., *The Impact of Microelectronics—Diffusion, Benefits and Problems in British Industry*, PSI Publications, London (1988)

24 BABB, M., 'Implementing distribution control into the 1990s', *Control Engineering*, 2–4 (Aug. 1989)

Bibliography

BANNISTER, B. R. and WHITEHEAD, D. G., *Transducers and Interfacing*, Van Nostrand Reinhold, London (1986)

BARNEY, G. C., *Intelligent Instrumentation*, Prentice-Hall, New York (1985)

BOLLINGER, J. G. and DUFFIE, N. A., *Computer Control of Machines and Processes*, Addison-Wesley, Reading, Massachusetts (1988)

BURR BROWN, *The Handbook of Personal Computer Instrumentation—for Data Acquisition, Test Measurement and Control*, Burr Brown Corporation, Tucson, Arizona (Jan. 1988)

CASSEL, D. A., *Microcomputers and Modern Control Engineering*, Reston Publishing, Prentice-Hall, Reston, Virginia (1983)

CLULEY, J. C., *Transducers for Microprocessor Systems*, Macmillan, Hong Kong (1985)

DOEBELIN, E. O., *Control System Principles and Design*, Wiley, New York (1985)

DORF, R. C., *Modern Control Systems*, 3rd edn, Addison-Wesley, Reading, Massachusetts (1980)

GAYAKWAD, R. and SOKOLOFF, L., *Analog and Digital Control Systems*, Prentice-Hall, New York (1988)

HUNT, V. D., *Mechatronics: Japan's newest threat*, Chapman & Hall, New York (1988)

ISMAIL, A. R. and ROONEY, V. M., *Microprocessor Hardware and Software Concepts*, Collier Macmillan, New York (1987)

KAFRISSEN, E. and STEPHANS, M., *Industrial Robots and Robotics*, Reston Publishing, Prentice-Hall, Reston, Virginia (1984)

KIEF, H. B., OLLING, G. and WATERS, T. F., *Flexible Automation—The International CNC Reference Book*, Becker Publishing, Bonn (1986)

MEADOWS, R. and PARSONS, A. J., *Microprocessors: Essentials, Components and Systems*, Pitman, Avon (1985)

SCOTT, P. B., *The Robotics Revolution*, Basil Blackwell, Oxford (1984)

SHARON, D., HARSTEIN, J. and YANTIAN, G., *Robotics and Automated Manufacturing*, Pitman, Avon (1987)

SNYDER, W. E., *Industrial Robots, Computer Interfacing and Control*, Prentice-Hall, New York (1985)

WARNOCK, I. G., *Programmable Controllers: Operation and Application*, Prentice-Hall, London (1988)

12

Machine Tool Control Elements

G M Mair

Contents

12.1 Machine tool control system—overview

This chapter provides an overview of machine tool control system concepts. It allows an appreciation of the functions needed to be performed by each part of the system. It also describes these elements and explains how they relate to each other.

Machine tools may be controlled manually or automatically. Automatic machines may employ 'hard' automation in the form of cams and fixed sequence mechanical or pneumatic systems, or 'soft' automation in the form of numerical control or other easily reprogrammable systems. Although most of this chapter is concerned with the elements used in numerical control machine tools, much of what is described is common to other types of manufacturing equipment.

Cam-controlled machines are employed in mass-production systems where frequent changes in machine-element movements and their sequences are unlikely. Cams were the earliest means of automatic control and are still to be found today on automatic lathes and dedicated assembly machines. *Figure 12.1* shows some examples.

Termed 'brain wheels' on Christopher Spencer's original automatic lathe cams are still found on machine tools used for mass production. They can be used to simply trigger ON/OFF signals by activating pneumatic or electric switches. They usually control the pattern of movement of machine tool elements such as slides or cutting tools, in this case the cam profile is cut to provide the desired movement to be imparted to the element by the cam follower. The most common type of cam is the flat-plate cam cut from steel plate. This type may have rotary or linear input as shown in the figure. Other types of cam are also used, e.g. the radial closed track and the drum cams. In each case the control signal is represented by the cam surface relative to a set datum. This is translated into movement via the cam follower, one end of which rests on the cam surface and is constrained to respond to the cam movement.

Numerically controlled machine tools may be *open loop* or *closed loop*, they will also have a *path control* system specified.

An additional designation used is *servo control* or *non-servo control*. Originally, 'servocontrol' implied a system that involved amplification of a control signal, e.g. power steering in a motor car, where the manual force at the steering wheel is amplified by the servosystem before transmission to the road wheels. In machine tools this type of servo control is always present, whether the control signal is a pulse of low-pressure air to a spool valve in a pneumatic circuit, or a voltage to an amplifier in an electrical system, the signal always results in an amplified response. Thus, in the context of industrial equipment, the term 'servo control' normally has a different connotation. It usually indicates the capacity of the system or element for closed-loop control. These terms are now described in more detail.

12.1.1 Open-loop control

This type of control is found, for example, on simple machines that are under direct manual control, or cam-operated machines, or machines under computer control that are only subjected to small loads. *Figure 12.2* illustrates the principle. A signal is sent from the operator, or computer, to the amplifier and actuator. It is assumed that the machine responds in an appropriate manner to the command. In practice, the loop is often closed in a rudimentary manner, e.g. a human operator may confirm the response visually, or a microswitch might be mechanically activated when a desired position is reached. In light-duty applications, electric stepper motor and pneumatic systems operate in this way.

12.1.2 Closed-loop control

For most computer-controlled medium to heavy duty industrial equipment this is the normal method of control. It involves the use of *feedback*. The principle is illustrated in *Figure 12.3*. A control signal is sent from the controller to the amplifier and actuator, the response of the system is monitored by feedback sensors. These sensors send signals, propor-

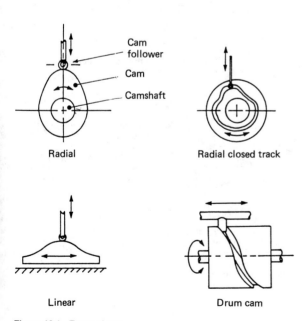

Radial

Radial closed track

Linear

Drum cam

Figure 12.1 Types of cam

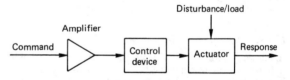

Figure 12.2 Open-loop system

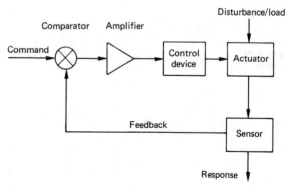

Figure 12.3 Closed-loop system

tional to the variable being measured, back to the controller. The controller compares the feedback signals with its original control signal and then modifies its output to compensate. Thus a 'loop' is formed which keeps the system under control. The British Standards Institution defines a closed-loop control system as, 'a control system possessing monitoring feedback, the deviation signal formed as a result of this feedback being used to control the action of a final control element in such a way as to tend to reduce the deviation to zero'.

12.1.3 Control-system stability

One of the factors to be considered when designing a control system is its *stability*. This is determined by the manner in which the system responds to a suddenly applied signal. As noted above, in a closed-loop system, self-correction occurs; however, if overcorrection is present then *oscillations* or 'hunting' may appear. This happens when the system overcorrects itself first in one direction and then in another. If this is too severe then the system does not settle down and in a severe case will run out of control, just like a car in a skid for which the driver overcorrects. As this condition is obviously unsatisfactory it is normal to slow the system response down by *damping*. Critical damping occurs when the system produces the desired result exactly, overdamping occurs when the system response is too sluggish, and underdamping occurs if the system oscillates a few times before settling. In an electrical system damping can be achieved by increasing the electrical resistance, this is analogous to a mechanical system, such as a shock absorber, in which oil is forced through a small orifice. Another method of preventing oscillations is the use of a large 'dead-band' region, or 'differential gap', on either side of the desired value. Thus the system stops correcting itself as soon as the feedback signal enters the dead-band region. As the dead band increases, so the settling time and precision of the system decreases.

12.1.4 Controller types

There are six common types of controller incorporating the four basic control actions of, ON/OFF, proportional, derivative, and integral control. These types are outlined below.

12.1.4.1 ON/OFF control

In this type of controller the control signal is either fully on or fully off. To prevent hunting here it is usually necessary to have a dead-band region for the error signal to pass through before switching takes place. Using a car analogy again, the engine coolant system thermostat is a good example of an ON/OFF controller.

12.1.4.2 Proportional control

This gives a smoother control action. The controller basically performs as an amplifier and provides a control signal proportional to the error.

12.1.4.3 Integral control

Here the rate of change of the control signal is proportional to the error signal. A large error signal produces a rapid increase in the control signal, a small error signal produces a slow control signal increase. A zero error produces a constant control signal. These controllers can be used when there is a constant load on the system. This is because a control signal will still be produced to counteract the load even when no error signal is present.

12.1.4.4 Proportional plus integral control

Some of the disadvantages of the previous two systems are complementary and can be overcome by combining them. For example, a proportional controller needs an error signal to be able to counteract a load, whereas the integral type can cope with zero error but often provides a slow response.

12.1.4.5 Proportional plus derivative control

In derivative control the control signal is proportional to the rate of change of the error signal. This is seldom used on its own as, without a changing error signal, the controller would produce zero output. The advantage of derivative control is that it anticipates changes in the error, thus providing a more rapid response to changes.

12.1.4.6 Proportional plus integral plus derivative (PID) control

This is the commonly used PID controller formed by a combination of the three main control actions. It provides a fast response and good stability and precision.

12.1.5 Path control

This defines how the components of the machine move. In the case of a machine tool the terminology is usually, 'point to point', 'straight line', and 'contouring'. In the case of an industrial robot, 'point to point', 'point to point with coordinated path', and 'continuous path' would probably be used. These terms are described in *Figure 12.4*.

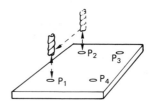

Point-to-point control on an NC drilling machine.
The path followed between points is unimportant.
The machine moves between points to obtain minimum cycle time

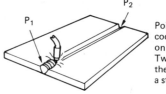

Point-to-point with coordinated-path control used on a welding robot.
Two points are specified and the control system interpolates a straight line between them

Complex form internal cavity being milled using continuous-path control of three axes.
More complex shapes may demand up to six-axis control

Figure 12.4 Point-to-point, straight-line and continuous-control paths

12.1.6 Control-system hardware

The control system in a machine tool is made up of a number of physical elements. Firstly there is the controller cabinet which, in a modern machine tool, will probably house the microprocessor control system or 'computer'. This contains the logic units, interfacing boards, and data-handling equipment, etc., i.e. the low-voltage electronics. In an electrically powered machine there will also be the motor control and amplifier circuits to provide the high power output to the electric motors. In fluid power machines, i.e. pneumatic and hydraulic, control of the actuators will be implemented by solenoid operated valves and hydraulic servovalves interfaced to the controller. The movement of the machine tool components is effected by electric motors of various types acting via transmission components such as gears, ballscrews, etc. In the case of industrial robots, pneumatic and hydraulic rams and rotary actuators are also used. Feedback to the controller is provided through the use of sensors such as optical shaft encoders, tachometers, etc.

12.2 Electric motors

There are essentially three types of electric motor normally used in machine tools: d.c. permanent magnet and 'brushless' servomotors for drives, and much less frequently, stepping motors for light-duty positioning. The term 'servomotor' is used to signify that the motor includes an integral feedback device, usually a tachogenerator or optical shaft encoder. A brief description of each type of motor is given below together with comments on their characteristics and advantages and disadvantages.

12.2.1 D.c. permanent-magnet servomotors

The basic construction of a d.c. permanent magnet servomotor is shown in *Figure 12.5*. Permanent magnets are used to provide a permanently energised magnetic field. These magnets are increasingly of the rare-earth type, e.g. samarium cobalt and neodymium iron boron, which provide a relatively high flux density in relation to the earlier Alnico or ferrite magnets. This reduces size and weight in relation to torque

produced. The magnet segments (there may be four, six, or eight poles) are fixed to the steel outer housing by adhesive or bolts. This type of construction, as opposed to separately excited types, facilitates control, improves reliability, increases efficiency, reduces heating, and simplifies construction. Since the field flux is constant the torque generated is directly proportional to the armature current, i.e. speed changes can be effected by altering the armature voltage. There is a current regulator incorporated in the system to compensate for speed drops that may occur due to static and dynamic loads on the motor.

The cylindrical type of motor is the most common in machine-tool construction. However, where fast response and high acceleration are required, and there is little danger of overloading, then motors with a disk rotor may be used. The disk-shaped armature is lightweight as it is not composed of iron. The conductors are stamped from sheet copper and lie in layers on the disk which rotates between the exciter and magnets. It is important that these motors are not overloaded as they as susceptible to burning out due to excess current.

In situations where continual high torques and speeds are required, the d.c. permanent magnet servomotor is likely to prove suitable. It provides smooth operation over a wide range of speeds and is especially efficient at high speeds, it is also relatively inexpensive. However, the brushes in a d.c. motor create problems. For example, they restrict its use in applications that require a high frequency of starts, stops, and direction changes; the possibility of sparking at the brushes prohibits its use in flammable or explosive atmospheres; as the brushes and commutator wear small particles are created, this makes its use in 'clean' environments difficult; and of course maintenance and replacement of the brushes is regularly necessary. For these reasons, and for heavier duty applications, brushless motors are increasing in popularity.

12.2.2 Brushless servomotors

Though remaining a more expensive solution than the d.c. servomotor, the brushless servomotor has a number of advantages. Since there are no brushes, maintenance costs due to brush and commutator wear are reduced and such motors can be used in some hazardous areas. They are quiet and reliable. Rotor inertia is reduced, they are capable of very high speed

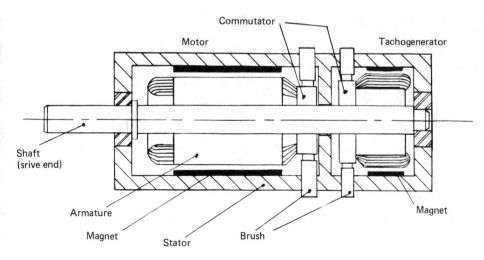

Figure 12.5 Basic construction of a d.c. permanent-magnet servomotor

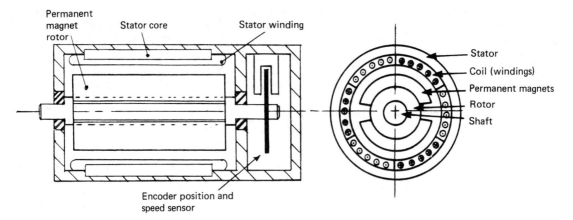

Figure 12.6 Basic construction of a 'brushless' servomotor

operation, and high peak torques are obtainable. They have smaller size and weight-to-power ratios than other types. Heat dissipation is better as heat is more easily lost from the stator than the rotor. However, despite the basic motor being of simple construction and, therefore, less expensive, the control system is relatively complex making a full system more expensive than a conventional d.c. alternative.

The basic construction of a brushless motor is shown in *Figure 12.6*. In this type of motor, permanent magnets are mounted on the rotor and the windings, or coils, are wound on the stator, i.e. the inverse of the d.c. permanent-magnet servomotor design. The motor operates in a synchronous manner, that is to say the rotor rotates in synchronism with the frequency and direction of the current supplied to the coils. The field coils in the stator are electronically commutated by switching the current from one coil to another. By controlling voltage and frequency, the motor speed under load is also controlled, i.e. the rotor speed is the same as the rotational frequency of the magnetic field. The synchronisation of the rotor speed and rotating magnetic field is essential. This is obtained by continuously monitoring the angular position of the rotor by the use of precision feedback elements such as shaft encoders or resolvers. These position data are used to control the electronic commutation. These 'brushless motors' may be a.c. or d.c. The physical construction of each type is similar, it is the form in which the current is supplied to the coils that differs. In the a.c. type a polyphase sine wave a.c. is used, in the d.c. type it is a d.c. pulse. After conditioning of the signal in both types and modification of the waveforms and pulses, there may in fact be very little difference. Thus the term 'brushless motor' is used to cover both types.

12.2.3 Stepping motors

On modern machine tools, high feed and traverse rates in conjunction with relatively high torques are required. Stepping motors are not particularly suited to these conditions. However, since there are often opportunities for using them on special-purpose and custom-made machines, they are discussed briefly here. The basic construction of a stepping motor is shown in *Figure 12.7*. The motor is composed of a multipole permanent-magnet rotor and a stator with a number of poles energised by windings. By passing current through these windings the poles on the rotor are caused to align with the stator field. The stator poles are energised sequentially so that

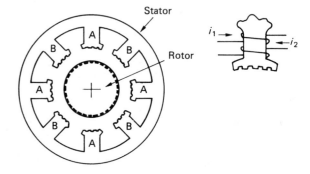

Figure 12.7 Basic construction of a stepper motor. The windings have been omitted for clarity; there is one set of windings for poles A and one set for poles B. i_1 and i_2 produce the north and south stator poles, respectively

a stepping action is produced as the rotor follows the rotating field. There are three types of stepping motor:

(1) *permanent magnet*—normally with large step angles of 20° to 90°;
(2) *hybrid*—these have smaller step angles between 1.8° and 2.5° and provide higher dynamic and holding torque. Both permanent-magnet and hybrid types exhibit a detent torque when power is removed, this retains rotor position against small loads; and
(3) *variable reluctance*—this type has step angles between those of the permanent-magnet and hybrid types, lower torque and higher speed than the permanent-magnet type, but does not exhibit a detent torque.

The advantages of stepping motors are as follows. They have a simple, low-cost construction. Under small load conditions they can directly convert digital instructions into incremental rotational displacement, with no necessity for feedback. They are very reliable and negligible maintenance is necessary. Torque is good at low speeds. The disadvantages of stepper motors prevent them being widely used on heavy duty machine tools. Some disadvantages are as follows. Their resolution and stepping frequencies are too low, this results in low feed and traverse rates on the machine tool axes. Although torque is good at low speeds, maximum torque is usually poor and this results in poor acceleration. Where medium loads are experienced then feedback sensors must be built into the system, otherwise rotational steps might be missed. Thus low-speed, high-torque conditions requiring short, fast, repetitive movements are suitable for stepping motor application.

12.3 The servomotor control amplifier

Suppliers of electric servomotors normally provide complete control systems including the control amplifiers. In the case of d.c. permanent-magnet and brushless motors these are now mostly transistor types using pulse width modulation. Amplifiers incorporating thyristors are not now very common due to poorer regulation capabilities. Stepping motors are driven via amplifiers containing purpose-built integrated-circuit chips. These units can take direct input from microprocessor-based digital controllers. There are at least three inputs to each amplifier unit: one carries the start/stop signal, one the rotational direction signal, and one the drive pulses. The output from the unit is the power pulses to the stator windings on the motor. Since, as mentioned above, stepping motors are not widely used on machine tools, and thyristor control for d.c. motors is not now common, the remainder of this subsection briefly considers the principle of the pulse-width-modulation method of control.

The control amplifier has essentially two functions, i.e. to supply power to the motor to cause it to run, and to control this power so as to maintain the required speed. The low-voltage command to the amplifier is usually in the region of -10 V to $+10$ V. Thus a 0 V command means standstill, -10 V means full speed reverse, and $+10$ V means full speed forward. If the input voltage is changed from -5 V to $+5$V, the motor must respond by changing from half speed reverse to half speed forward. The motor cannot do this instantaneously, it must overcome inertia and decelerate and stop before reversing direction and accelerating to the new speed. To control a motor efficiently in this way *four-quadrant control* is necessary. This allows control under the following conditions.

(1) *Motoring forward*—In this case the motor is accelerating the load and rotating in a clockwise direction, i.e. the motor is driving the load.
(2) *Braking forward*—In this case the motor is decelerating the load and rotating in a clockwise direction. The torque is reversed from (1), i.e. the load is driving the motor forward.
(3) *Motoring reverse*—In this case the motor is accelerating the load and rotating in an anti-clockwise direction. Torque is reversed from (1), i.e. the motor is driving the load in reverse.
(4) *Braking reverse*—In this case the motor is decelerating the load and rotating in an anti-clockwise direction. The torque is the same as in (1), i.e. the load is driving the motor in reverse.

To implement this control, the previously mentioned pulse-width-modulation amplifier is used. The detail of the construction of each amplifier depends on the type of power supply, type of power output, and type of motor being used. For this reason it is simply the principle of operation that is shown here.

Initially the three-phase mains supply is rectified to produce a constant d.c. voltage which will provide the high power supply to the motor. This d.c. voltage is broken up by the pulse-width-modulation unit into pulses of the desired width

and frequency. The pulses occur at frequencies of several kiloherz and are filtered by the inductance of the motor coils, thus preventing torque pulsation. The principle of operation is shown in *Figure 12.8(a)*, only two pairs of transistors are shown for simplicity. The transistor controller switches the power transistor pairs on or off, depending on the input signal it receives from the system controller. Current is allowed to flow in the forward direction by switching one pair of transistors, say AA', on and the other pair, BB', off. Current flow is reversed by switching the state of both pairs, there is usually a very small interval during which all transistors are off, this prevents them being damaged. In *Figure 12.8(b)*, a zero net

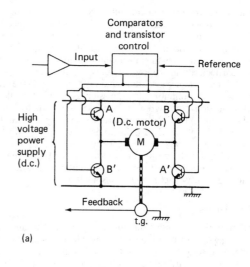

(a)

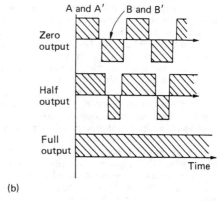

(b)

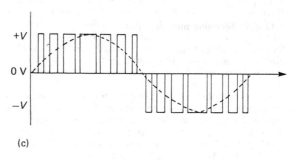

(c)

Figure 12.8 Pulse-width-modulation control

torque is produced by the motor when both pairs of transistors are switched on for an equal duration, a forward torque is produced when pair AA' is switched on for longer than pair BB', and when pair AA' is fully on and BB' fully off then the motor draws maximum current in the forward direction, but speed control is not possible. *Figure 12.8(c)* shows the pulse-width-modulation waveform for one type of brushless motor which requires pulses in a form that provides an approximation of a sine wave.

12.4 Transmission elements

In order to transmit movement and force from an actuator to the axis of a machine tool, industrial robot, or other piece of machinery, some sort of transmission system is required. These systems are mainly associated with electrically powered machines which are now the most popular in industry. In a few cases, particularly in robotics, direct electric drives are employed that require no transmission elements; however, these are still uncommon. Transmission systems also allow speed reductions or increases between the actuator and machine axis, and the transformation of rotary motion to linear, and vice versa. The most common transmission elements used are gears and screw–nut systems, these are described below. There are also special gear systems such as harmonic and cycloidal drives, and elements such as timing belts for light duty synchronous transmission.

12.4.1 Gears

Gears comprise the most ubiquitous type of transmission system; they are to be found in almost all machines. Gears provide good torque transmission, can be purchased 'off the shelf' if necessary, are reliable, and can be produced in many types to suit a variety of applications. For example, spur gears are the simplest type and have their teeth parallel to the gear shaft axis. They are relatively inexpensive but are not suited to high loading and they can be noisy. For these reasons, helical gears are popular as they have a greater area of tooth contact and the load is distributed over two or more teeth at any moment. This means that, although more expensive, they can transmit higher loads and are smoother and quieter in operation. However, the geometry of helical gears causes end thrust on their shafts, which in turn necessitates the use of thrust bearings. At some further cost this problem can be overcome by the use of herringbone or double helical gears. For non-parallel shafts, crossed helical gears can be used, although they have a lower load capacity. For shafts whose axes intersect at right angles, bevel gears are used. The straight bevel is analogous to the spur gear and the spiral bevel to the helical gear. For high reduction ratios on right-angled shafts, worm gears can be used; these also have a high load capacity. To transform rotary motion to linear, and vice versa, rack and pinion systems are used. The rack is effectively a gear of infinite diameter. The rack may be pushed by a linear actuator, so causing rotation of the pinion, or the pinion may be driven by a motor so causing either linear movement of the rack or the pinion to be driven along the rack. These gears are shown in *Figure 12.9*.

'Backlash' is the term used to describe the lost movement between the input and output shafts due to clearance between the gear teeth, it may also be referred to as 'play' or 'give'. It is usually unwanted but often some has to be built in to prevent binding, jamming, overloading, overheating, and noise such as high speed whistling. Thus backlash is evidenced by non-movement of the output shaft for an amount of time when the

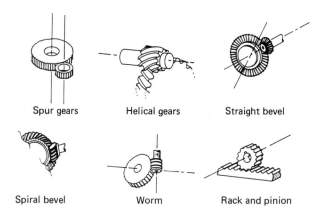

Figure 12.9 Gear types

input shaft changes direction. As well as built-in backlash, unintentional play can develop in a system due to wear of the contacting surfaces of the components. For high-precision work, such as is found in machine tools and industrial robots, backlash generally needs to be kept to a minimum. Various techniques can be applied to achieve this, each one suited to particular conditions; a few are noted here. The simplest method is to use spring-loaded split, or 'scissor' gears. Here the gear is in two halves, i.e. it is split along a plane parallel to its face and at right angles to its axis. One half is fixed to the gear shaft, the other is spring loaded and free to slide over the face of the fixed half. Thus, when mated with another gear, the springs force the two halves to rotate relative to each other until no backlash exists between the mating gears. This method is only suitable for low-torque applications. For higher torque situations split gears can be used. The gears are locked in position, after adjustment, by using screws or rivets. Another method is to use adjustable centres where the centre distance between gear pairs is manually adjusted before clamping the gear in position. Spring-loaded centres may also be used for lighter loads. Alternatively, spur or helical gears can be cut at a slight angle to produce tapered teeth. The gears can then be moved in an axial direction relative to each other and thus reduce or remove backlash.

12.4.2 Screw–nut systems

Also extremely common, screw–nut systems are found on almost all machine tools, many industrial robots, and other pieces of manufacturing machinery. While gearing is used to effect speed reduction and variance, screw–nut systems are used to translate rotational movement into linear movement of the machine elements. The screw is normally driven by an electric servomotor, either directly or through a gear system. In a machine tool the nut is fixed to one member, e.g. the machine table, and the screw to the other, e.g. the machine bed. As the screw rotates the nut is driven linearly along the axis of the screw, thus causing the table movement. This type of transmission system provides reliability, rigidity, repeatability, and resistance to vibration.

It is usual for ball or roller screws to be used because, in machine tools, it is important that smooth operation of the elements occurs. One factor that can prevent this is the phenomenon of stick–slip. This occurs at slow speeds due to the system experiencing fluctuations between static and sliding friction which in turn is caused by the lubricant film spora-

Figure 12.10 Ball nut–screw arrangement. (Courtesy of Transrol SKF Ltd)

dically breaking down. One means of overcoming this is to use rolling friction elements to reduce the coefficient of friction; this also has other beneficial effects such as reducing temperature rise at high speeds and reducing power requirements. A section through a ball screw–nut assembly is shown in *Figure 12.10*. Ball screws with internal and external ball recirculation are obtainable with efficiencies of over 90%. The external method is less expensive than the internal one, but a larger nut is necessary. To increase the load-carrying capacity the number and area of contact points between the nut and screw needs to be increased; this is done by using roller screws.

Backlash can be eliminated in screw–nut systems by adjusting preload through the use of shims. This is done by splitting the nut at right angles to the screw and inserting a shim, i.e. a packing piece, between the nut halves. The nut is then clamped together, e.g. by using cap screws or rivets. If the shim has been accurately ground, no significant increase in friction will occur.

Other transmission systems are available, but conventional gear train systems and screw–nut systems are the major ones used for machine tools. Industrial robots utilise a wider range of transmission elements, e.g. synchronous belts, chains, pretensioned metal bands, pulleys, and harmonic and cycloidal gear drives, which provide extremely high speed-reduction ratios in a compact package.

12.5 Fluid power actuators

The use of hydraulic drives in machine tools has decreased considerably. This has been due to various factors such as high noise levels from power packs, vibrations transmitted through the fluid from the pump, leaks, and analogue control systems using sensitive servovalves with which few maintenance engineers are familiar. However, although hydraulic drives are no longer common on machine tools, they do still have some advantages over electric drives, i.e. they can withstand high shock loads, hydraulic force can be applied directly where required without the use of mechanical transmission elements, and they can be used in fire-risk areas as the voltages required by the servovalves are very small. For these reasons there are still industrial applications for the simpler hydraulic actuators, e.g. some robots employ linear and rotary hydraulic actuators.

For simple, high-speed, inexpensive movement, pneumatic actuators are still extremely popular. Being intrinsically safe they can be used in high fire-risk areas and explosive atmospheres. They are seldom used in servocontrolled applications or for moving heavy loads. This is due to the compressibility of air which prevents reliable positioning and speed control. Pneumatic power systems and devices are ubiquitous in industry and are well known by almost every factory worker.

Since hydraulic and pneumatic actuators are both fluid power devices, their basic construction is similar. The main differences lie in the wall thicknesses and material, as hydraulic systems have to operate at internal pressures of around 20 MPa, whereas pneumatic devices operate at about 0.7 MPa.

Hydraulic and pneumatic motors are not considered here because they are seldom, if ever, used on modern machine tools. Pneumatic motors are used for driving small tools such as automatic screw drivers, and hand-held deburring tools. For heavy-duty work, hydraulic torque amplifiers are available; these are, in effect, servocontrolled hydraulic motors. Pneumatic linear actuators are available in the normal ram-type-cylinder form or as 'rodless' cylinders. Hydraulic linear actuators are available as simple cylinders or as 'linear amplifier' units which incorporate an electric pilot motor, a servovalve, and a hydraulic ram. *Figures 12.11* to *12.13* illustrate the configuration and method of operation of the fluid power cylinder, pneumatic rodless cylinder, and hydraulic linear amplifier. Fluid power rotary actuators are normally of the rack and pinion or rotary vane type; their method of operation is shown in *Figures 12.14* and *12.15*.

12.6 Fluid power actuator control valves

Located between the power source (i.e. a compressor or a pump for pneumatics or hydraulics, respectively) and the actuators, are the control valves. Their function is to direct the energy contained in the fluid to the appropriate parts of the system.

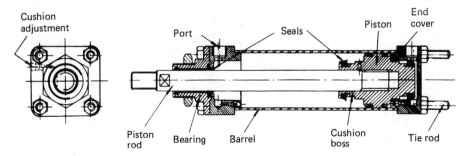

Figure 12.11 Double acting pneumatic cylinder. (Courtesy of Norgren Martonair)

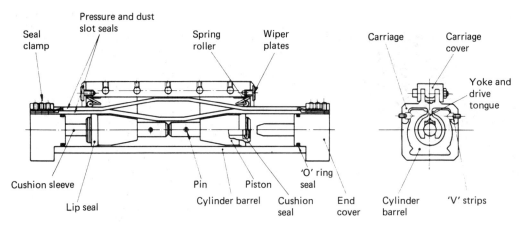

Figure 12.12   Double acting pneumatic rodless cylinder. (Courtesy of Norgren Martonair)

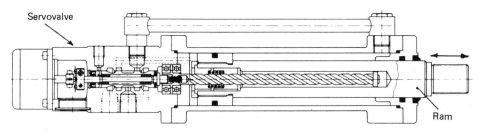

Figure 12.13   Hydraulic linear amplifier

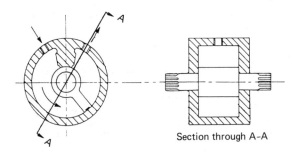

Figure 12.14   Fluid power rotary vane actuator

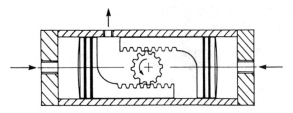

Figure 12.15   Fluid power rack and pinion rotary actuator

12.6.1   Control-valve specification

For fluid power control valves the International Standard ISO 1219 prescribes the symbols to be used to represent these valves. Examples of these are shown in *Figure 12.16*.

In order to fully define a fluid power control valve, a number of factors must be specified.

(1)   The number of ports on the valve. These are the main ports necessary for the control of the air to and from the actuators; the number does not include the pilot signal ports.

(2)   The number of positions of the internal switching element. This is either two or three, the former being more common in pneumatic circuits and the latter in hydraulics. The number of ports and the number of positions available for a valve are represented by stating the numbers with a solidus separating them, e.g. '5/2 spool valve' represents a five-port, two-position valve.

(3)   The control method used to cause the internal switching. Methods include pneumatic, electric, or some form of mechanical action such as a spring, a lever, or a manual push button. Their equivalent symbols are shown in *Figure 12.17*.

(4)   The rest position, or state, of the valve, i.e. whether normally closed (NC) or normally open (NO).

(5)   Whether the valve is monostable or bistable. A monostable valve has only one stable position. This is often the rest position in the absence of a control signal, it is

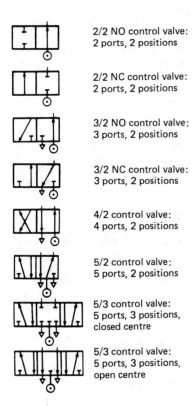

Figure 12.16 Fluid power control valve symbols

Type of control

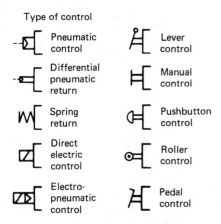

Figure 12.17 Control method symbols

usually achieved by the action of a spring or differential air pressure. The solenoid operated poppet valve described later is a monostable device. A bistable valve remains in the last position attained whether or not a command signal is present. A command signal is required for every change of state. Because of its nature, a monostable device has, in effect, a memory function, i.e. it 'remembers' the last command. The spool valve, also described later, is an example of a device that is normally supplied in a bistable form, although monostable versions are available.

12.6.2 Pneumatic control valves

In pneumatic systems the flow of air to the actuators is controlled by pilot and spool valves. These are non-servo systems, similar valves are also used in non-servo hydraulic systems. Pilot valves receive the control signal, either mechanical or electrical, and translate this into a low pressure air pulse to the spool valve. Pilot valves are usually of the poppet type. These have an internal switching element which opens or blocks the movement of air by moving a seal off of, or on to, a seat. The switching element may be operated mechanically, but in modern control systems solenoid valves are common. A solenoid valve takes an electrical signal and converts it into a pneumatic signal pulse to a spool valve. The solenoid coil is energised by a d.c. or an a.c. current which is switched to it by the system controller, e.g. a programmable logic controller (PLC). When this happens the core is caused to rise, thus opening the air passage to the spool valve. A spring is usually incorporated to return the valve to its rest position after the coil has been de-energised. Poppet valves are available in two rest position states, i.e. normally closed (NC), and normally open (NO).

Solenoid operated pilot valves are often mounted on the spool valve as illustrated in *Figure 12.18*. The type of spool valve shown is a five-port linear slide valve, i.e. it has a high pressure input port, two outlet ports to an actuator, and two exhaust ports.

The air pulse from the pilot valve alters the position of the switching element in the spool valve, thus directing high pressure air to the appropriate actuator chamber. The switching element is a shouldered shaft or 'spool', and movement of this within the valve chamber allows either high-pressure air to flow to the actuator, or exhaust air from the actuator to flow to atmosphere. A simple pneumatic system is shown schematically in *Figure 12.19*; this incorporates a double-acting pneumatic cylinder, a spool valve, and two mechanically operated pilot valves. This circuit provides the piston with a reciprocating action.

Other elements are also available. For example, the shuttle valve is used to direct flow when two alternative air signals are to be used in a system. It is a very simple bistable device allowing air to flow only in the direction determined by the last command signal. Flow regulators are another example; these are used to restrict air flow and can be used on the exhaust ports of actuators to control speed of movement. Conversely, the speed of actuator movement can be increased by use of

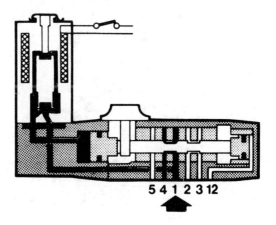

Figure 12.18 Solenoid operated pilot valve. (Courtesy of Joucomatic/Techno-Nathan International)

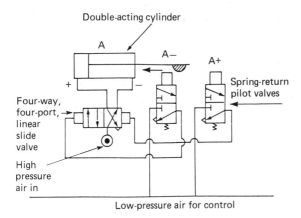

Figure 12.19 A pneumatic circuit producing a reciprocating action

quick exhaust valves fitted directly to the actuator exhaust ports. There are also electropneumatic and pneumatic–electric interface modules available. These convert electric signals to pneumatic and pneumatic signals to electric, respectively, thus allowing simple interfacing to electronic control systems.

12.6.3 The hydraulic servovalve

All the valves mentioned above operate in either the fully on or the fully off state. In hydraulic systems, where the fluid used is incompressible, it is possible closely to control the actuators by controlling the fluid flow. This closed-loop servo control of force, displacement, and speed is effected by the hydraulic servovalve. The type described here indicates the general principle of operation, although it is only one of the many types available.

The valve shown in *Figure 12.20* comprises three main elements: a torque motor, a pilot stage, and a spool-valve assembly. These elements are integrated into a single compact

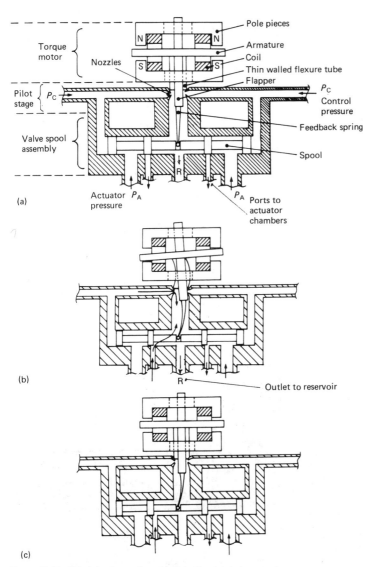

Figure 12.20 Principle of operation of a Moog flow control servovalve

unit in the actual product. The valve's function is analogous to that of the spool valve discussed in the previous section, with the exception that the present one allows fine control of the fluid flow to the actuator. Here, a flapper amplifier stage is used; other valve types have spool, deflector jet, and jet pipe amplifiers. In the figure, control fluid P_c enters the valve at the right- and left-hand sides and initially flows freely through the valve to the return port R. The high pressure fluid P_a required to operate the actuator can be seen entering the spool chamber through two ports at the base of the valve. Ports to the actuator chambers are shown adjacent.

The main elements of the first stage, i.e. the torque motor, are the upper and lower pole pieces, two permanent magnets, an armature and a coil. Attached to the armature is the flapper and feedback spring. These are surrounded by a thin-walled flexure tube which prevents hydraulic fluid from entering the motor chamber. The permanent magnets polarise the pole pieces, so creating a magnetic field across the gap in which the armature is supported. A d.c. current proportional to the command signal from the system controller is passed through the coil. The current flowing in the coil creates a torque on the armature which causes it to rotate through a small angle. This torque is controlled by the d.c. current.

On experiencing torque the armature causes the flapper to block the flow of fluid P_c through one of the nozzles in the second stage. For example, if the armature experiences an anti-clockwise rotation then the flapper will be pushed against the right-hand nozzle, this will increase the pressure on the right-hand side of the spool, pushing it to the left. A path is now opened in the valve's third stage for the high-pressure fluid to flow from the pump, through the servovalve, to the appropriate actuator chamber. Fluid in the low-pressure side of the actuator can also now flow back through the valve to the hydraulic reservoir.

As is evident from the figure, when the spool moves to the left it will also move the feedback spring to the left. This produces a clockwise torque on the armature, thus tending to pull the flapper away from the nozzle again. When this 'feedback torque' equals the torque created by the current flow, the flapper will return to its central position and the spool will remain in its new position as P_c will be equal on each side. Thus, assuming constant pressure, the flow of hydraulic fluid to the actuator will be proportional to spool position and the spool position will be proportional to the input current.

Servovalves are used in conjunction with feedback transducers to provide full closed-loop control, e.g. a potentiometer for position servoing, a tachometer for velocity servoing, and a load cell containing a strain gauge for force servoing. *Figure 12.21* shows a basic system.

12.7 Feedback transducers

The previous sections in this chapter have covered the principles of operation of a machine tool control system, the actuators necessary to effect movement, and the methods of control of these actuators. This section examines the elements necessary to provide feedback to the system controller on position, proximity, displacement, velocity, and force. There is a wide range of these feedback sensors and, therefore, only the most common ones are covered here.

12.7.1 Position and proximity sensors

12.7.1.1 Limit switches

These are widely used as simple position feedback devices. Their main component is usually a 'microswitch'. This is a small switch that requires only a very slight movement to cause a mechanical snap action which opens or closes the electrical connection. The casing and means of activating the microswitch are available in a number of forms, e.g. plunger, roller, lever, and roller lever (see *Figure 12.22*).

The switches are located at the desired limits of travel of the machine-tool elements. They often act as safety devices, being fixed at a point just beyond the expected software-set stopping point of a machine tool table or robot arm. A mechanical trip or 'dog' contacts the switch to complete, or open, an electric circuit and shut off the machine tool. These switches can be

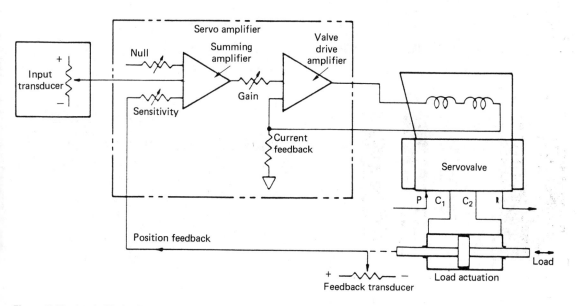

Figure 12.21 A typical hydraulic servovalve positioning system. (Courtesy of Moog Inc.)

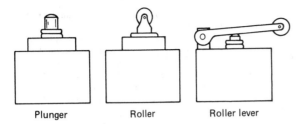

Figure 12.22 Some limit-switch configurations

used in a normally open or a normally closed mode. Use outside the limits of the machine tool is also common, e.g. they are incorporated in safety barriers or interlocked gates, and are used as component detection or pallet identification devices.

Microswitches have also been used for sequence control. For example, a linear array of microswitches can be linked to various fluid power solenoid valves. The switches can then be switched on or off in the required sequence by setting pegs on a rotating drum to contact the microswitches in the appropriate order, i.e. the pattern of pegs will determine the sequence, and the speed of drum rotation will determine the sequence cycle time.

12.7.1.2 Reed switches and magnets

These are usually found on pneumatic cylinders where their function is to detect the position of the piston within the cylinder chamber. The switch component is mounted on the outside cylinder wall at the position at which the piston is to be sensed. The casing contains a ferromagnetic flexible blade, or 'reed', type switch. The piston within the cylinder contains a permanent magnet. The cylinder itself is non-magnetic, being made of aluminium, non-magnetic stainless steel, or brass. As the piston comes within close proximity of the switch, the magnetic field from the magnet causes the contacts to close. An electric circuit is thus completed, allowing a control signal to pass to the controller (see *Figure 12.23*). These devices can enable detection of the piston to close tolerances, and some have a light-emitting diode (LED) fitted to signal closure of the switch. They have the advantage of being non-contacting devices, thus providing greater reliability. Being small they

provide a compact position detection system without the need for additional sensors such as limit switches, which occupy often much needed space.

12.7.1.3 Inductive sensors

These are used for non-contact proximity sensing of metallic machine elements or other objects. Often called 'magnetic pick-ups' they appear in the form of a coil wound round a permanent magnet and enclosed in a small cylindrical case (see *Figure 12.24*). The control system is connected to the coil in such a way as to enable the voltage across the coil to be measured. Initially the magnet produces a steady magnetic field around the case and no voltage is induced in the coil. However, as a metallic object approaches the pick-up, so the magnetic field around the coil is changed, thus causing a corresponding change in the induced voltage in the coil. A threshold value can be set in the controller to determine at what voltage level the object should be deemed to be in adequate proximity to the pick-up. It is possible, over very short distances, to calibrate the pick-up in such a manner as to determine the distance of the object very accurately by monitoring the induced voltage.

12.7.1.4 Capacitive sensors

These are used for non-contact proximity sensing of non-metallic objects. They can be used in a safety system around the machine tool to sense the presence of personnel, or in non-contact keypads where the proximity of a finger or hand would be sufficient to operate a switch. An oscillator-type circuit is shown in *Figure 12.25*. Here a 'capacitor' is formed comprising the metal plate, the atmosphere which acts as the dielectric, and the non-metallic object which acts as the other

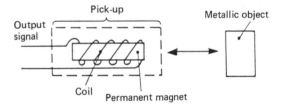

Figure 12.24 The inductive sensor

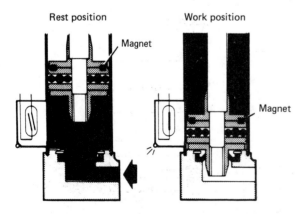

Figure 12.23 Reed switch and magnet used on a pneumatic cylinder. (Courtesy of Joucomatic/Techno-Nathan International)

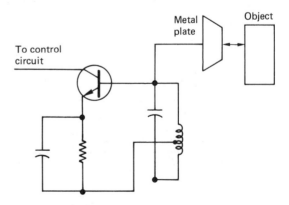

Figure 12.25 The capacitive sensor

'plate'. When the non-metallic object is not present, circuit oscillation is maintained by the feedback, compensating for losses occurring in the coil and capacitor circuit. The circuit is so adjusted that if the circuit load is increased the oscillations stop, so causing a current change which can be used to operate a switch. The presence of the non-metallic object causes the necessary additional load to stop the oscillations.

12.7.1.5 Photosensors

These are used for non-contact sensing in many situations, e.g. for detecting the presence of a part in an automatic machining or assembly system, for counting components passing along a conveyor belt, or for constructing light curtains in safety barriers. They are also widely used in optical displacement and velocity feedback elements. Photocells are available that are sensitive to light in the visible or infra-red spectrum. The most common type is the 'photoconductive' or 'photoresistive' sensor, this is composed of materials the electrical resistance of which is decreased when exposed to light.

In the cadmium sulphide type (*Figure 12.26*), a layer of cadmium sulphide is sandwiched between two metal electrodes (alternatively, a layer of lead selenide can be used for infra-red sensing). The top electrode can be made as a thin wire mesh of tin oxide in order to allow light to reach the photoconductive layer. If the cell is in total darkness and a voltage is applied across the cell, the cell provides a resistance of several megohms, and a negligible current will flow. If light then impinges on the cell the resistance drops to a few ohms and a relatively large current will flow. Since the resistance varies inversely with the light intensity, the light level can be determined by measuring the resistance.

Another type of photocell utilises the photosensitive properties of many semiconductors such as diodes and transistors (see *Figure 12.27*). Again when a voltage is applied while the photocell is exposed to light, a current proportional to the light intensity will flow.

Both the above-mentioned photocells require the application of a voltage from an external source. A different type is the photovoltaic sensor. Commonly known as a 'solar cell' this produces a voltage when exposed to light. Photovoltaic materials used are selenium, silicon and germanium. As in the inductive sensor mentioned earlier, the photocell can be used as a simple ON/OFF device by setting a threshold value, or it can be used as an analogue device. In its ON/OFF form, light is often supplied from a LED from which light passes in a beam to the cell. When the beam is broken by a person or component the change in voltage causes a signal to be passed to the controller. The cell may also be used to register the presence of a reflective or bright surface against a dark background. The main problem with the use of light sensors in the industrial environment is that the cell can be occluded by contaminants such as oil or dust.

12.7.1.6 Pneumatic sensors

Sometimes called 'air-jet sensors', these are used for non-contact sensing in areas where contaminants prohibit the use of photocells. They are suitable for use in high-temperature, corrosive, and explosive atmospheres and in situations where the object to be detected has an imprecise location. Two types are common: the back-pressure sensor and the interruptible-jet sensor.

Figure 12.28 shows the principle of operation of the back-pressure sensor. A tube-shaped probe is supplied by air at constant pressure P_c, this supplies air to the control chamber via a fixed orifice. The pressure of the air in the control chamber is determined by the distance between the end of the probe and the object to be detected. As the object approaches the sensor the escape of air decreases and the pressure P_c increases. In fact, when the distance x is less than $0.25D$ the pressure P_c varies proportionally with distance. Air loss and sensitivity to strong currents usually limit the diameter D to a maximum of about 4 mm, which limits the dynamic range to about 1 mm. Thus the device can be used as a simple proximity detector over distances of about 1–15 mm, but can be used as an analogue distance measuring device at distances under 1 mm.

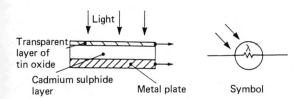

Figure 12.26 Cadmium sulphide photocell

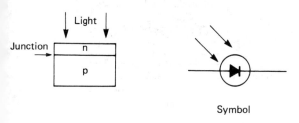

Figure 12.27 Photodiode photocell; n, electron rich material; p, electron deficient material

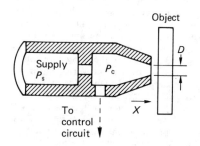

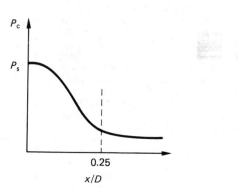

Figure 12.28 Pneumatic back-pressure proximity sensor

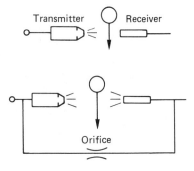

Figure 12.29 Pneumatic interruptible-jet proximity sensor

The interruptible-jet sensor can detect objects passing be-tween the emitter and the detector. It is analogous in opera-tion to the LED and photocell arrangement. *Figure 12.29* shows the principle of operation. A jet of air from a low-pressure air source is projected across a gap to a receiving orifice. The air pressure within the orifice is monitored. If an object now passes through the gap and breaks the jet, the pressure in the receiver drops and this is noted by the controller. For use in dusty environments, the receiver can be pressurised as shown to prevent the possibility of clogging. Supply pressures range from 0.07 to 0.2 bar. Typically the transmitter may have a diameter of around 3 mm and the receiver a diameter of about 6 mm. Combined transmitter and receiver units have detection distances of around 6 mm, and separate units can detect up to 100 mm.

12.7.2 Displacement transducers

These may be linear or rotary, analogue or digital, and incremental or absolute. They can all be used to provide direct or indirect feedback. Direct feedback occurs when the trans-ducer is mounted integrally with the elements being moni-tored. For example, if the movement of a machine-tool table is being measured relative to a fixed bed, then a feedback transducer mounted on the table and/or the bed will provide a direct measurement of displacement. However, if the trans-ducer is mounted on, say, the shaft of an electric drive motor, then the feedback will be indirect because the movement being measured must still be transmitted to the machine-tool table. The latter method is less precise as mechanical inaccura-cies and backlash may be present in the transmission system. Direct systems are, therefore, to be preferred, although there are many cases where they cannot be used, e.g. 'revolute' geometry industrial robots have a design that prevents prac-tical direct position feedback of their end effector.

Linear transducers are commonly used on machine-tool tables and slideways. They provide reliable and precise direct displacement feedback using optical gratings and encoders. Linear potentiometers can be used over short distances, but are not used on machine-tool axes. Rotary transducers in the form of optical shaft encoders, potentiometers and resolvers are popular for mounting integral with servomotor assemblies for indirect displacement feedback.

Analogue transducers produce a continuously changing signal analogous to the value being measured; potentiometers and resolvers are analogue transducers. Digital transducers produce digital signals, e.g. an optical grating will produce, after signal conditioning, a train of square-wave pulses for input to an electronic counter.

Incremental transducers provide signals that must be added as they arrive at the counter to determine the displacement from a known starting, or 'zero', point. However, with an absolute transducer each signal pattern specifies uniquely where the device is. Thus with an incremental transducer, when powering up the system, or restoring power after a failure, the machine axes must be returned to the 'zero' point, but with an absolute encoder this is not necessary. The following describes the operation of some position trans-ducers.

12.7.2.1 Potentiometers

These are available in both linear and rotary form, the latter probably being the most popular. They are analogue devices with a voltage output proportional to the displacement of a wiper blade in contact with a resistive element. Their principle of operation is shown in *Figure 12.30*. A d.c. voltage V is applied across the resistive element R. The resulting voltage v across the distance r from the positive terminal to the wiper blade is measured. In the case of a rotary potentiometer this voltage is proportional to the angle of the wiper blade relative to the datum; in the case of a linear potentiometer it is simply proportional to the distance travelled. The voltage v is equal to Vr/R. It is possible to derive a 'voltage constant' for the pot-entiometer. If we let the voltage constant be K, then $v = Kr$, where K is the voltage constant of the potentiometer in volts per radian for rotary potentiometers and volts per millimetre for linear potentiometers. The value of the measured voltage therefore indicates displacement, and direction is indicated by increasing or decreasing voltage values. In practice, the shaft of a rotary potentiometer may be connected to the element being monitored either directly, through a gear train, via a rack and pinion, or by a synchronous belt.

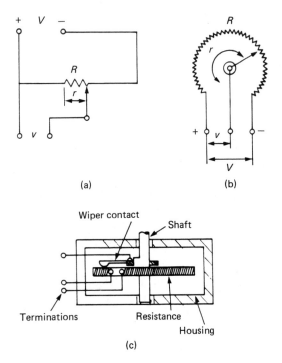

Figure 12.30 The potentiometer: (a) circuit; (b) layout; (c) construction

12.7.2.2 Resolvers

Again these are analogue devices, but in this case the input is an a.c. voltage. Resolvers are essentially rotating transformers with an output proportional to the angle of a rotating element (the rotor) with respect to a fixed element (the stator). Physically, they resemble small electric motors. Typically, there is a single winding on the rotor and two on the stator. The electrical connections to the rotor are made by slip rings and brushes. The principle of operation is shown in *Figure 12.31*. The two stator coils are 90° apart and, when the rotor is excited by a signal $V \sin(\omega t)$ the voltage induced across the stator coils is:

$$v_1 = V \sin(\omega t) \sin \theta$$

and

$$v_2 = V \sin(\omega t) \cos \theta$$

where θ is the angle of the rotor with respect to the stator.

In practice, the stator coils may be fed with an input voltage and an output voltage induced in the rotor coil. For example, the stator coils may be supplied with two square-wave voltages 'sin' and 'cos' at a frequency of about 2 kHz, their amplitudes will be the same but they will be 90° out of phase with each other. The voltage induced in the rotor coil can be filtered and conditioned to produce a square-wave voltage signal the phase position of which relative to, say, the 'sin' voltage is a measure of the angular position of the rotor relative to the stator.

It is also possible to get multipole resolvers. The difference in operation occurs in the relationship of the phase shifts to the physical angle of rotation. For example, if in a two-pole resolver an electrical rotational angle of 360° corresponds to a physical rotation of 360°, in a 10-pole resolver the same physical rotation will produce a electrical rotational angle of $5 \times 360°$.

The rotor shaft can be connected to the motor shaft or other elements in the same way as the potentiometer. The output signal is connected to an electronic counter which records both the number of full revolutions of the rotor and the relative angular displacement at any point in time.

Another device similar to the resolver is the *synchro*. This is similar in construction to the resolver except that it has three coils wound on the stator, each 120° apart. Again, when the rotor coil is energised, voltages are induced in the stator coils the amplitude and phase of which are indicators of the angular displacement of the rotor relative to the stator.

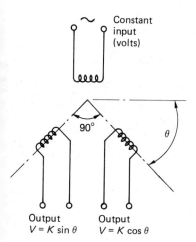

Figure 12.31 The resolver

12.7.2.3 Inductosyns

These are not normally found on new machines and are now mostly sold for replacement. They can be obtained in both linear and rotary forms. The rotary type can be considered as a flat, many-poled, synchro. For example, an inductosyn with 2000 poles would have an electrical rotational angle of 360° for a physical rotation of the rotor of 0.36°. In linear inductosyns a voltage is induced in the windings in the long linear scale analogous to a rotor, with the slider and its windings being analogous to the synchro stator.

12.7.2.4 Optical encoders

These are photoelectric devices obtainable in linear and rotary configurations. They operate by using a photosensor to monitor changing light intensity created by optical gratings. A grating comprises opaque lines and transparent spaces of equal width. These gratings are commonly produced by depositing very thin chrome layers onto a glass substrate using thin-film processes such as vacuum deposition or sputtering. Alternatively, the gratings may be etched onto stainless-steel strips so that reflected light can be used. These strips are bonded onto rigid steel bars which have a similar coefficient of expansion to that of the machine slide. It is also possible to etch the gratings directly onto a solid stainless-steel bar. Reflected light systems avoid the problems inherent in using glass scales, i.e. different coefficients of expansion to the machine elements and sensitivity to contamination. With linear devices, grating periods of 20, 10 and 8 μm, and with rotary devices up to 9000 lines per revolution, are typical.

Method of operation Two types of encoder exist, i.e. the incremental and the absolute. The following applies to the incremental type, the absolute type is described later. *Figure 12.32* shows the principle of operation. The encoder is basically composed of two parts; the graduated glass scale which may be either straight line or circular, and the scanning unit. This unit comprises the light source, a collimating lens, a scanning reticle with its own gratings, and the photocells which often are of the photovoltaic type. In linear encoders the scale is usually fixed to the machine bed and the scanning unit fixed to the machine table. In the rotary types the scanning unit is held relatively fixed in the encoder housing, while the optical disc is connected to a rotating shaft which is coupled to a motor drive shaft, leadscrew, etc. (see *Figure 12.33*).

There are three configurations of scanning unit. In one configuration the lines in the reticle and scale are parallel. In this case the light intensity falling on the cells increases or decreases, as the lines either occlude the light or coincide with each other, so forming a space for it to pass, as shown in *Figure 12.34*.

In the second configuration the graduations are magnified by the use of moiré fringes. This effect is produced when two lined patterns are superimposed with one pattern of lines slightly inclined to the other (see *Figure 12.35*). When illuminated by either transmitted or reflected light, fringes are created which lie approximately at right angles to the graduations. The 'wavelength' or spacing of these fringes is directly related to θ, the angle of inclination between the lines, i.e. the 'wavelength' increases as the angle decreases. In practice, the scanning unit contains the reticle with the lines inclined to those on the fixed grating. As the unit passes along the grating the fringes move vertically up or down, depending on the direction of horizontal travel. For each horizontal movement of the unit by one grating, the fringe moves vertically by one 'wavelength'. Thus, if a mask with a narrow horizontal slot is placed over the reticle, then the intensity of light passing

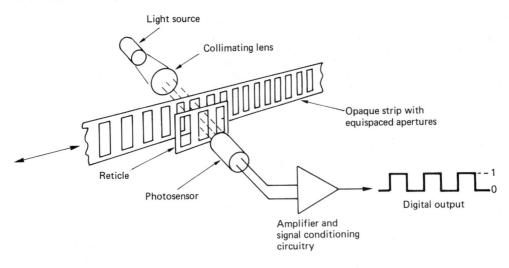

Figure 12.32 Principle of operation of an optical encoder

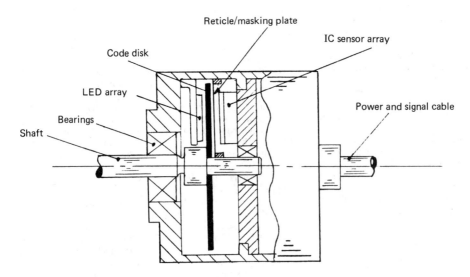

Figure 12.33 Section through an optical shaft encoder

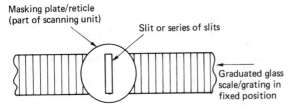

Figure 12.34 Optical encoder arrangement using parallel slits

through the slot will wax and wane as the fringes pass under it. The distance between the fringes is much greater than that between the gratings. This means that a number of slots can be used to monitor the movement of the fringe over one 'wavelength'; for example, four slots could be used with four photosensors to provide four signal pulses for one grating pitch movement.

The third method uses the interferential scanning principle for obtaining extremely fine resolutions and high accuracies. For example, the Heidenhain LIP 101R incremental linear encoder has a grating period of 8 μm, a signal period of 4 μm, and the interpolation is capable of giving measuring steps of 0.01 μm, although under practical conditions maximum accuracies of ± 0.2 μm could be expected. The encoder uses the reflected-light method. It consists of a solid steel rectangular bar onto which a grating has been etched, and a scanning unit containing the reticle and light source. Briefly, the system operates as follows. The grating period is too fine to allow the use of moiré fringes and, therefore, the interference principle is used to produce light and dark bands which can be observed by the photocells. When light waves of the same wavelength are in phase they combine to increase the resultant amplitude,

Both scales have equal groove spacing

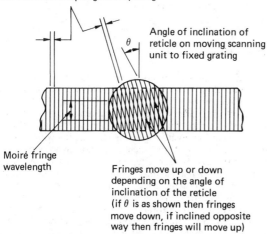

Angle of inclination of reticle on moving scanning unit to fixed grating

Moiré fringe wavelength

Fringes move up or down depending on the angle of inclination of the reticle (if θ is as shown then fringes move down, if inclined opposite way then fringes will move up)

Figure 12.35 Optical encoder arrangement using moiré fringes

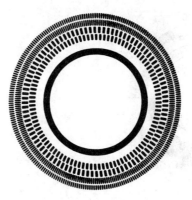

Figure 12.36 Incremental encoder disk with two pairs of tracks. (Courtesy of Gaebridge)

and hence light intensity, to a maximum. When they are out of phase by 180°, i.e. half a wavelength, the resultant amplitude which is the algebraic sum of both wave amplitudes, is at a minimum. If the waves are of the same amplitude and 180° out of phase then the algebraic sum of the amplitudes is zero, interference occurs and darkness is observed.

Light of the same wavelength can be produced from a monochromatic light source or by diffraction, i.e. splitting the light into its constituent elements by passing it through a narrow slot. In the interferential scanning principle, as collimated light passes through the gratings in the reticle most of its goes straight through with zero diffraction. However, some light is diffracted to the right and left of the grating in various degrees. Due to the nature of this diffraction and the fixed phase relationship that the waves initially have to each other, it is possible to observe interference patterns that are created as the phase relationships change due to the relative movement of the reticle. It is these interference patterns that are scanned by the photosensors to provide the displacement measurement.

Signal processing Irrespective of which of the three methods of operation is employed, a light signal fluctuating in intensity is produced. This light signal causes a photocell to generate a voltage in the form of a sine wave corresponding to the increase and decrease in light value. Interpolation and digitising circuitry takes this sinusoidal signal and transforms it into a square wave pulse train. The sinusoidal signal is first amplified, then interpolated to provide measuring steps of from 1/20 to 1/200 of the grating period. Finally, using Schmitt triggers, the signal is changed into a square wave. The resulting pulse train can then be used as direct input to the counting circuitry of the machine controller or display unit.

The system can therefore count the number of pulses from a datum to determine displacement; using an internal clock, the system can use the displacement to derive velocity and acceleration. In incremental encoders, an additional grating is usually employed to act as a local zero datum. In order to determine the direction of travel the signal phase relationships are usually examined. For example, in a rotary encoder two tracks and two sensors can be employed (see *Figure 12.36*). These tracks generate signals with a 90° phase shift, which is equivalent to a quarter-period time displacement. Either of

the tracks can be used to provide the displacement data, but it is the comparison of their signals to determine the relative phase which provides direction information. It is also possible simply to use two sensors over one track. These sensors are positioned exactly one-half segment width apart. Again the signals produced by the sensors will be phase shifted by 90° relative to each other and, by noting which sensor responds first, the control system can determine the direction of movement.

On incremental shaft encoder discs the gratings can be arranged to provide direct angular feedback, e.g. 360 or 3600 gratings per revolution, or chosen in combination with a mechanical gearing system such that metric or imperial readouts can be obtained. For example, if the gearing is arranged so that one revolution of the disc corresponds to a mechanical movement of, say, a machine table of 1 in., then a disc with two sets of tracks could be used. One set would contain 1000 gratings per revolution, and the other 2540, thus allowing signals corresponding to a resolution of 0.001 in. and 0.01 mm to be selected, as desired.

Absolute encoders Though available in linear and rotary types, absolute encoders are more often used in the latter form. Two problems with incremental encoders are:

(1) they only produce output signals while power is applied and the slideway is moving or shaft is rotating; and
(2) if power loss occurs the actual position reading is also lost, which means that on powering up again the encoders must be brought back to a datum or 'home' position from which to resume counting.

Absolute encoders overcome these problems by allowing the position of, say, a shaft to be uniquely defined immediately on power up, whether or not there is movement. The absolute encoder is a more complex device than the incremental type and may cost twice as much. It has a large number of concentric tracks with one sensor per track (see *Figure 12.37*). A twelve track encoder, for example, provides 4096 uniquely defined positions for each revolution. Resolution can be increased by increasing the number of tracks and by using multiturn encoders.

It is possible to use a pure binary coded disc, but this means that all digits on, for example, a four-track disc must change simultaneously from 15 to 16—this can produce ambiguity problems which may result in the recording of incorrect information. Methods of overcoming this include using antiambiguity tracks on the disc, or tightening up mechanical

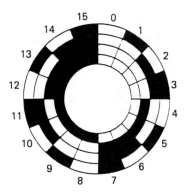

Figure 12.37 Scheme of an absolute disk using binary code

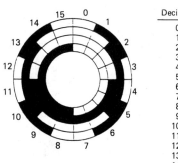

Decimal	Natural binary	Gray code
0	0000	0000
1	0001	0001
2	0010	0011
3	0011	0010
4	0100	0110
5	0101	0111
6	0110	0101
7	0111	0100
8	1000	1100
9	1001	1101
10	1010	1111
11	1011	1110
12	1100	1010
13	1101	1011
14	1110	1001
15	1111	1000

Figure 12.38 Scheme of an absolute disk using Gray code and a table showing a comparison of codes

tolerances. The most common method, however, is to use Gray coding (see *Figure 12.38*) in which only one digit changes at any moment. Conversion to true binary, or other code, if required, can be done using electronics within the encoder itself.

Encoders in operation Linear encoders are normally mounted directly onto the bed and table of a machine tool. Rotary encoders are indirect in that they are connected to the motor drive, leadscrew, rack and pinion, or other gearing, usually by using collet or sleeve and grub screw arrangements. All types of encoder are available in a wide range of physical sizes. They operate from 5 or 12 V power supplies with loadings of around 60 or 120 mA, depending on type. Interfacing capability is usually provided, e.g. RS232 and RS422,

with some encoders incorporating their own microprocessor systems for error checking, etc.

12.8 Conclusion

This chapter has described the individual control elements used in machine tools. For very high-volume production runs some machines use 'hard' automation. Examples of this are cam-controlled multispindle lathes where the cams control the sequence and movement of the cutting tools, and dedicated assembly machines where cams, in conjunction with pneumatic elements, control the sequence and movement of pick-and-place units, presses, inspection probes, etc. At the other end of the spectrum, manual machines used for single-unit production such as prototyping, employ only the basic elements of drive unit, gearing system and, probably, leadscrew. An electronic aid to control may involve the use of a linear encoder linked to a digital readout. These readouts give easily read feedback on tool position and can provide direct conversion from metric to imperial scales if required. Fully programmable numerically controlled machine tools take the output from the feedback devices and use this as input to the control unit. It is these numerically controlled machine tools that have been the main concern of the chapter.

Further reading

BARBER, A., *Pneumatic Handbook*, 7th edn, Trade and Technical Press (1989)

BRADLEY, D. A. *et al.*, *Mechatronics*, Chapman and Hall, London (1991)

BRINDLEY, K., *Sensors and Transducers*, Butterworth-Heinemann, Oxford (1988)

CRISPIN, A. J., *Programmable Logic Controllers and Their Engineering Applications*, McGraw-Hill, New York (1990)

DOTE. Y and KIROSHITA, S., *Brushless Servomotors*, Oxford University Press, Oxford (1989)

ELECTRA-CRAFT CORPORATION, *DC Motors, Speed Control, and Servo Systems*, Electra-Craft Corporation (1980)

GAYAKWAD and SOKOLOFF, *Analogue and Digital Control Systems*, Prentice-Hall, New York (1988)

HUGHES, A., *Electric Motors and Drives*, Butterworth-Heinemann, Oxford (1990)

JACOB, J. M., *Industrial Control Electronics*, Prentice-Hall, New York (1988)

KENJO, T., *Stepping Motors And Their Microprocessor Controls*, Oxford University Press, Oxford (1985)

MAIR, G. M., *Industrial Robotics*, Prentice-Hall, New York (1988)

MORRIS, N., *Control Engineering*, McGraw-Hill, New York (1991)

PARR, A., *Hydraulics and Pneumatics—A Technician and Engineer's Guide*, Butterworth-Heinemann, Oxford (1991)

PARR, E. A., *Industrial Control Handbook, Vol. 1 Transducers, and Vol. 2 Techniques*, Blackwell Scientific, Oxford (1987)

SINCLAIR, I. R., *Sensors and Transducers—A Guide for Technicians*, 2nd edn, Butterworth-Heinemann, Oxford (1992)

13 Communication and Integration Systems

G Miller

B Etter

Contents

This chapter deals with the hardware and software requirements of computer-aided and computer-integrated manufacturing systems. Such topics as computer architecture, operating systems, languages, peripheral devices, man–machine interfaces, networks, databases, and expert systems are described. The incorporation of computer technology into a manufacturing environment is a complicated and expensive proposition, and should be thoroughly planned before implementation. However, the use of computers in a manufacturing and industrial environment can improve productivity through increased work throughput, increased efficiency, and better quality control. Further information regarding each of these subjects can be found in the publications listed in the Further Reading section at the end of this chapter. In order to understand the information regarding computer systems and communications hardware and software, a glossary of commonly used computer terms is given in *Table 13.1*.

Table 13.1 Glossary of terms commonly used in computing

Term	Description
Address	Logical location of information in a memory system
ALU	Arithmetic logic unit: functional part of the CPU which performs arithmetic operations on data
Architecture	Physical structure of a computing system
Array	Multidimensional group of numbers; multiple vectors
Bit	Most basic form of information in a digital system with discrete values of 0 (OFF) and 1 (ON)
Bus	Series of parallel logic channels used to transmit information
Byte	Pattern of 8 bits
CCD	Charge coupled device: a secondary NDRO, volatile, storage device
CPU	Central processing unit: main computational/controlling device in a computer system
Data	Information used or modified by instructions
DRO	Destructive readout device: device whose contents are destroyed when read
Duplex	Method of data transmission where information is transmitted in two directions simultaneously
GPIB	General purpose interface bus: class of data links used for communication with instrumentation
I/O	Input/output: devices used for input or output
Instruction	Information controlling the movement or alteration of data
LAN	Local area network
MODEM	Modulator/demodulator

Table 13.1 Cont'd

Term	Description
MBM	Magnetic bubble memory: a secondary NDRO, non-volatile, storage device
Non-volatile	Storage device which loses information after power is lost
NDRO	Non-destructive readout: device whose contents are not destroyed when read
Program	Series of instructions ordered to achieve a certain result
RAM	Random access memory: memory which is accessed in the same time cycle, independent of location
ROM	Read only memory: memory which is accessed only during read operations
RWM	Read write memory: memory which is accessed for reading or writing
Register	High speed storage device normally located within the CPU
Simplex	Method of data transmission where information is transmitted in one direction only
Vector	One-dimensional group of numbers
Volatile	Storage device which loses information after power is lost
Word	Multiple groups of bits, the size of which depends on the computer system

13.1 Computer architecture

13.1.1 Basic computer architecture

Figure 13.1 illustrates the basic components of a modern computer system. The ALU is the component of the CPU which actually performs computation on the data presented to

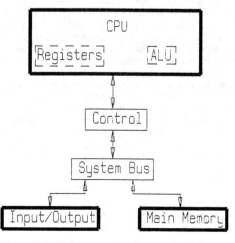

Figure 13.1 Basic components of a computer system

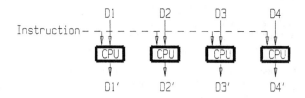

Figure 12.3 Computer system bus structure

Figure 13.4 SIMD instruction and data flow

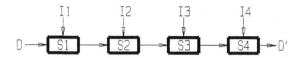

Figure 13.5 MISD instruction and data flow

the CPU via the system bus. The control logic decodes the incoming instructions and is responsible for synchronisation in the CPU and with the system bus. The register's function is to facilitate the movement of data and instructions within the CPU. The system bus is the physical connection between the CPU and memory and external I/O devices.

The basic structure of the system busses of a common computing architecture is illustrated in *Figure 13.2*. Here an additional I/O processor is used to control operations with slower mechanical devices. This additional processor allows the host CPU to perform computational tasks independent of the I/O processes. The control bus contains the instructions to be performed between the processors and memory. The address bus contains the address of the memory location being accessed by the controller. Data coming into or out of the processors are sent via the data bus.

13.1.2 Computer architecture classification

The flow of information to and from the CPU is commonly referred to as a 'stream'. The architecture of modern computers can be divided into four basic groups based on the instruction and data streams.

The first class of computers is the simplest, where the CPU executes one instruction with one segment of data. These machines are classified as single instruction stream, single data stream (SISD). The instruction and data flow are illustrated in *Figure 13.3*. Most microcomputers and many minicomputers fall into this category. Most of the DEC PDP-11 series minicomputers portray this type of architecture.

The next step in the hierarchy is a single instruction stream, multiple data stream (SIMD) computer. This type of machine normally includes one program control unit and multiple execution units. The data and instruction paths are shown in *Figure 13.4*. Most of the machines with this type of architecture are minicomputers. Among these are the ILLIAC IV and the Goodyear STARAN.

The complement of the above architecture is the multiple instruction stream, single data stream (MISD) machines. Pipeline processing is the predominant architecture in this class. This technique is illustrated in *Figure 13.5*. The data flow through the processor in a linear, pipeline fashion. This pipeline is divided into *m* segments. Each segment may be the operand of a separate instruction. This allows a single pro-

cessor to act as an *m*-unit processor in many applications. Most of these machines are classified as mainframes or supercomputers. Examples of this type of architecture include the CRAY-1, CDC STAR 100 and the TI ASC.

The last category is that of multiple instruction stream multiple data stream architectures. This category includes machines capable of executing independent programs simultaneously. Many experts refer to this architecture as the class of 'multiprocessors'. This structure is illustrated in *Figure 13.6*. These computers may also be classified as vector computers with the inclusion of vector operations in the machine instruction sets. Machines with this type of architecture are expensive and are normally reserved for applications where computational speed is critical. This type of architecture is normally achieved with multiple SIMD or MISD architecture computers linked together.

13.1.3 Computer size classification

The four classes of computer architecture can be implemented in almost any size of computer system. The classification of system size is normally based on a system's computational speed and intended usage. Most multiuser systems are minicomputers and mainframes; however, a multiuser operating system can be implemented on microcomputers. A workstation is normally distinguished as a high-power single-user system; however, many minicomputers are implemented in a workstation configuration. Whether implemented in a single or a multiuser environment, workstations are normally used for dedicated processes which are computationally extensive, e.g. graphics applications.

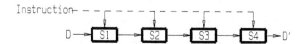

Figure 13.3 SISD instruction and data flow

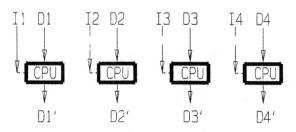

Figure 13.6 MIMD instruction and data flow

The smallest size modern digital computer is classified as a microcomputer. Most of these systems are miniaturised versions of the larger computer systems. Microcomputers have the following general architectural charcteristics:

(1) data and memory busses are normally 8 or 16 bits wide;
(2) ALU operations are normally register–register or register–memory operations;
(3) all data types are either integer or decimal; and
(4) interrupts are used to transfer control to peripheral devices.

One of the major limiting factors in microcomputers is the smaller bus width. Memory and I/O access are slowed down due to this physical limitation. Recent developments have increased the physical address bus to 32 bits, making the abilities of this class of computer comparable to many larger systems. Other physical-size constraints prevent the use of microcomputers in larger system applications.

Minicomputers are distinguished from microcomputers primarily by the physical address size capabilities. Most modern minicomputers have address widths of 32 bits or more. This permits the use of faster bus access without complex addressing schemes. The larger bus widths also provide for an increase in computational accuracy due to the increase in significant digits (bits) in each execution cycle. This class of computer also provides for an increase in data types. This permits single-cycle execution for many operations which require multiple executions in a smaller system. The resulting increased instruction-set size frequently exceeds 200 commands.

The class of mainframe computers, or supercomputers, is distinguished from the smaller systems primarily by computational speed and price. Several minicomputers have similar architectures to the larger mainframes, only the implementation in the large computer maximises all available hardware technologies to provide a more optimal system. Another common distinction is an increased width, which increases computational accuracy and speed. The increased bus width also provides for more addressing structures and data types, which increases the size of the basic instruction set.

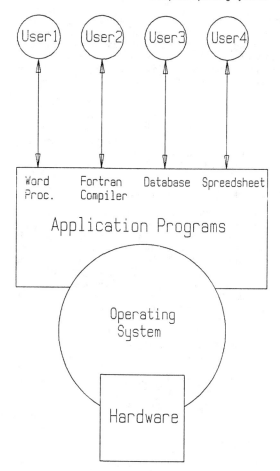

Figure 13.7 Operating system relationships

13.2 Computer operating systems

A computer operating system is the software (programs and data) that controls the hardware and applications software of a computer so that they can constitute a useful system for carrying out calculations. From a computer user's perspective, an operating system provides a means of accessing and controlling applications software. From a computer system manager's perspective, an operating system controls the central processor, various internal and external memories, various peripheral devices, and physical communications hardware in order to allow the computer as a whole to be optimally managed and operated. Operating systems are used to control the number of users of a computer system, the amount of computer time and memory allocated to each user, the interaction between the computer user and peripheral devices, and the access to applications software systems. *Figure 13.7* depicts the relationship between the computer operating system and the hardware, applications software, and computer user.

13.2.1 Operating system techniques

The most basic operating system is a single-user system which executes one program at a time. This type of system is most prevalent in dedicated systems such as microcomputers or workstations. Single-user systems can be used with larger computers; however, such an implementation would not use the computer to its maximum efficiency.

Batch systems are used in multiuser computer applications. Each individual's job is processed in sequential order of submittal: first in, first out (FIFO). This operating system can cause major bottlenecks in system performance when time-consuming jobs are being done.

A modification of the batch operating system is the multiprogramming system. This mode divides the CPU execution time among several different submitted programs, in order of priority and sequence. Programs that exceed a predetermined execution time are temporarily suspended to allow the next program to operate for the same allotted amount of CPU time. The overhead of switching between programs can become excessive, depending on the architecture of the system and the amount of CPU time allotted. This method also aids CPU operations when interacting with slower devices in I/O operations. It should be noted that batch systems and multiprogramming systems would perform faster on a system with a parallel architecture; however, multiprogramming does not imply true simultaneous processing.

The principle of multiprogramming can be developed to provide for an interactive multiuser system. This technique is classified as 'time sharing'. Most modern computer systems are capable of time-sharing functions. Each user interacts with the system as though it were a dedicated computer. This is the predominant operating system technique currently used in multiuser computer systems. Many systems use a combination of batch and time sharing to provide optimal flexibility and performance.

A variation of the single-user system is multitasking, where a single user uses several programs 'simultaneously'. The operating system actually uses a time-sharing technique for each of the enacted programs. The CPU time allotments are such that the user is unaware of the switching between the various programs (tasks).

13.2.2 Addressing and memory management schemes

The various operating systems schemes can rapidly deplete the available amount of main memory for most computer architectures. The operating system must provide for access to data in secondary memory storage. This secondary storage is commonly a magnetic-medium device, or other high-storage device. These devices have slower access times; therefore access by the primary CPU(s) should be minimised.

Most computer programs exhibit a high degree of structure. Current trends in computer science encourage the development of structured programming. Programs may be segmented into modules based on this structure for ease of transfer from secondary memory devices. Instructions and operands are transferred to and from the secondary and main memories for operation by the primary CPU(s). These segments vary in length based on the amount of main memory available and the length of the program module. Although segmentation corresponds to program module limits, it results in a great deal of fragmentation of main memory. A complex memory allocation operating system is required to avoid this fragmentation in segmentation methods.

A similar process known as 'paging' transfers fixed blocks of data to and from secondary memory. Paging avoids the problem of main memory fragmentation because fixed blocks of memory are moved; however, the page boundaries normally do not coincide with program boundaries.

A virtual memory addressing scheme comprises paging, segmentation, or both, to provide the user with an apparent infinite memory area. There are many different variations of this memory allocation scheme, but *Figure 13.8* illustrates the basic differences between segmentation, paging, and virtual memory. Certain programs may require such frequent swapping of memory that the system spends more time moving memory than executing the program. This is known as 'thrashing'. The detection and prevention of thrashing has resulted in rather complex virtual memory operating systems.

Computer architectures with multiple processors frequently access memory at the same time. The process of interleaving enables two or more processors to access memory by partitioning the memory addresses into separate modules, as indicated in *Figure 13.9*. If two or more processors attempt to access the same module at the same time, a contention occurs. The interleaving process aids in access of most memory address; however, branching instructions in the program may cause departures from the normal process.

The cache memory is a small section of memory located between the processor and the main memory. This technique is similar to paging; however, the cache is between the main memory and the processor rather than between the secondary and main memories. The primary difference is that cache memories are even faster than the main memory, similar to the dedicated CPU registers. There is a risk of thrashing due to the similarity to virtual memory systems. The control of a

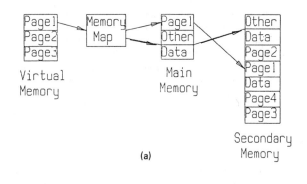

(a)

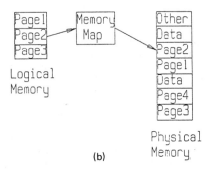

(b)

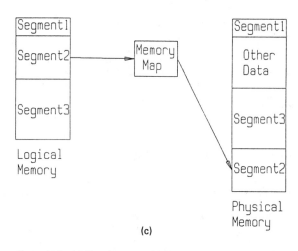

(c)

Figure 13.8 (a) Virtual memory. (b) Memory paging. (c) Memory segmentation

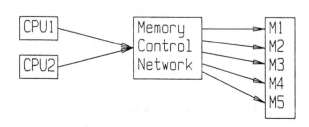

Figure 13.9 Memory interleaving

virtual memory system is handled with operating system software, whereas the control for cache is processed with dedicated logic circuits. The size of the pages is much smaller with the cache than with the virtual memory system. The primary purpose of the cache is to increase system speed by making the memory access time similar to the main processor speed.

13.2.3 Commonly used operating systems

Unix is a general-purpose time-sharing system which was originally developed at Bell Laboratories for the PDP 11/20 computer.[1] This system has blossomed into widespread use throughout the world on a variety of computers. However, this system is primarily used on the PDP/11, VAX 11/780, Amdahl 470, Univac 1100, Interdata 7/32 and 8/32, and various microcomputer systems.

This system is usually written in the C language and entails approximately 8000 lines of code with another 800 lines of assembly language code, making this system quite small compared with most other operating systems. The user interface is terse, clean and relatively easy to use. Installation on a new computer system is relatively easy and the system is portable to other compatible computer systems with ease.

Unix supports multiple processes. The CPU scheduling gives priority to I/O bound jobs. However, priorities are continually recomputed every second, and the system defaults to a round-robin time sharing control if no I/O commands are present. The key Unix features include:

(1) a programmable command language interpreter,
(2) the ability to control asynchronous processes,
(3) compatible file, device, and interprocess I/O,
(4) a hierarchical file system with demountable volumes, and
(5) a high degree of portability.

IBM has developed the *OS/360* operating system for its 360 and 370 class of computers.[2] The original intent was to develop a single operating system which could be used for a wide variety of computers (size and function) marketed by IBM. OS/360 is actually a series of operating systems which provide a vast array of functions and controls and support a wide range of applications programs. Thus, the OS/360 operating system is very large and very expensive. The system comprises over 200 000 lines of code.

OS/360 provides support for many higher level computing languages such as Cobol, Fortran, and PL/1, as well as support for assembly language code. The job control language is very sophisticated and detailed. The file control system is also very detailed and contains a large number of utility programs to control user-generated files. This system supports interactive programming and contains specialised subsystems to support various user-specific areas such as business databases. Due to the complexity of the OS/360 operating system, the interaction between software and hardware must be specified at the time of initial installation. The OS/360 system is not portable between computers and is not easy to use. However, due to its large number of features and the vast number of IBM systems in use today, the OS/360 system is widely used.

There are two basic versions of this system: OS/MFT uses fixed memory regions, and OS/MVT uses variable memory regions. The MVT version allows several user jobs to run concurrently, with each job assigned a section of computer memory. This system does not support paging. Thus, the memory initially assigned to a job remains with the job until its conclusion. With the advent of the IBM 370 architecture, virtual memory was added to the MFT version of OS/360. Two versions of MFT were subsequently developed to employ virtual memory: VS1 creates one large virtual address space, and VS2 (now called MVS) provides each user with his own virtual memory.

The *1100 Executive* system is employed by the Univac 1100 series of computers. This system supports various batch and interactive languages including Fortran and Basic. This system provides good support for data communications and time-sharing operations.

The *MCP* operating system supports the Burroughs B5000, B6000 and B7000 series of computers and is primarily intended for business applications including support of COBOL. The operating system is written in Algol and provides excellent support for large databases and high-level programming.

The *TOPS* operating system was written to support the DEC system 10 and system 20 computers. This system is primarily designed for scientific applications involving the Fortran language. Networking and time sharing are both supported. The *VMS* operating system has been written for the various VAX computer systems and contains features that are similar to that of TOPS.

The *DOS* (disk operating system) system is an industry standard for microcomputer operating control. This system is primarily intended for single-user, single-job applications, although the latest versions have attempted to address the expanding field of multifunction operation, which has been developed for the 386 series of microcomputers. The DOS system supports a wide variety of applications programs in several languages, including Fortran, Basic and Pascal.

13.3 Computer languages

C is a general-purpose programming language featuring economy of expression, modern control flow, data-structure capabilities, and a variety of operators and data types. It is best known as the primary language of the Unix operating system, although it is used in many other environments and on a vast array of computer models and sizes, from mainframe to microcomputer. The C language has been used with numerical, scientific, text-processing and database-management applications. It is relatively easy to use. The system is very portable and is not tied to any one particular application area.

Fortran was one of the first high-level languages developed for general use. The name is an abbreviation for 'formula translator'. Fortran is primarily intended for scientific programming and it supports various data types. Fortran was one of the first computer languages to provide a structured, yet user programmable, system for handling data input and output, complex and double precision operations, and control of data-storage allocation. It is a memory-intensive language that is not easily transportable between computer systems. The latest version, Fortran 77, is designed to support structured programming, which previous versions did not allow. The WATFIV version of Fortran contains excellent error diagnostics and rapid program compilation. Fortran is available for virtually all types and sizes of computer, from microcomputers to mainframes. Because it was one of the original scientific programming languages, Fortran is still widely used throughout the world.

Cobol is another of the older, high-level computer languages; it was developed during the late 1950s, near the time of Fortran development. Cobol is primarily designed for business applications and is an abbreviation for 'common business oriented language'. The original aim of Cobol was to be used on a variety of machines without difficulty. Thus, user training and program installation is relatively simple. A Cobol program is split into four divisions: identification, environment, data, and procedure. The 'identification' division merely states the title of the application. The 'environment' division links the data and procedure to a specific type of computer. The 'data' division describes the files and records which will be employed by the program. The 'procedure' division specifies and controls the processing and computation which will be accomplished. Cobol, like Fortran, has been so widely used, that it has now been developed as an ANSI standard.

Algol is a successor to Fortran and is an abbreviation for 'algorithmic language'. It was developed as a method of developing algorithms that is machine independent. Its major features include a flexible, yet powerful, method of handling loop calculations and the ability to combine a set of program statements into a single compound statement. Algol can handle large amounts of data efficiently by applying dynamic data-storage allocation, which assigns a location for data at the time that it is needed in a calculation. This storage is then released after the calculation is completed. Algol has been widely used in in Europe, although its use in the USA has been limited. Recently, Pascal has supplanted Algol as the most widely used algorithmic language in Europe.

Pascal is a successor to Algol which has been designed for structured programming. Pascal is an easy language to learn and use and is very portable. However, it does not contain many of the features of other algorithmic languages, including dynamic data storage. In addition, the number of predefined functions within Pascal is less than that of other languages. However, Pascal allows the programmer to define various types of data, which makes this a unique language in that regard.

PL/1 was developed for the IBM 360 computers and draws on many features found in Fortran, Cobol and Algol. It is widely used for IBM based mainframe systems and supports both scientific and business applications. However, because it contains the features of many other languages, it is a large, memory-intensive language that is not easy to use or learn.

APL is a procedures-oriented language that is usually used in a real-time interactive mode by a user. A key feature of APL is the ability to use single-statement, multiple-function operations that would require the use of numerous statements in other languages. Data are structured in arrays and calculating functions have no specific hierarchy. The use of APL can often generate very concise, yet powerful, programs. However, input and output control is limited, since the system is primarily designed for use at an interactive terminal. Use of APL has not been widespread, because it has not been used for a wide variety of applications.

Basic is an abbreviation for 'beginners all purpose symbolic instruction code'. It was designed as a simple, easy to use, easy to learn, interactive computer language. It has been widely used by beginners and is primarily employed on microcomputers. Basic has severe limitations in data storage and data manipulation and is primarily designed for use in small programs. Basic employs relatively simple user-defined functions, subroutines and matrix operations, and was originally conceived as a simpler, more rapid version of Fortran. Unlike Fortran, Basic lacks widespread control of input and output functions. However, Basic can be used as a rapid, interactive controller to access other programs written in various languages, and for various applications.

13.4 Peripheral devices

13.4.1 Classification of peripheral devices

Peripheral devices are normally perceived to be devices located outside (peripheral) the main computer, although they

are actually distinguished by function, not location. Peripheral devices may be located within the computer housing itself, e.g. disk devices and secondary memory. These devices are distinguished by a slower access time; however, they offer a much lower cost of storage of information.

13.4.2 Storage devices

Most of the attention regarding peripheral devices is focused on storage devices. Most of these devices may be used for both input and output; however, some are dedicated for input. The information stored in these devices is performed before the user accesses the media. There are two primary factors involved in the selection of storage media: cost and access time. *Figure 13.10* illustrates the trade-off in access speed and cost per bit of some typical storage devices.

Most main memory is classified as random access memory (RAM) where the processor can access any address, independent of physical location. Information accessed in a serial storage device must be accessed in a predetermined sequential manner. *Figure 13.11* portrays the distinction between these two addressing modes.

Another important characteristic of storage devices is the manner in which they are accessed during reading operations. The information in most devices is not altered during a read process, being classified as non-destructive readout (NDRO). The contents of some devices are altered or destroyed when reading, and are therefore classified as destructive readout (DRO) storage devices.

13.4.3 Semiconductor storage

The information in most RAM disappears when the power is turned off; therefore, it is characterised as 'volatile'. Storage devices which retain information after power down are 'non-volatile'. Dynamic RAM (DRAM) is RAM which is a DRO storage device. This is the fastest semiconductor memory device; however, it does require the additional overhead of refreshing memory due to its DRO nature. Static RAM (SRAM) is the class of semiconductor memory which does not

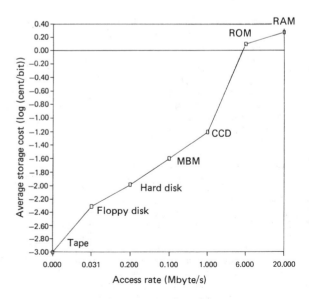

Figure 13.10 Storage category performance relationships

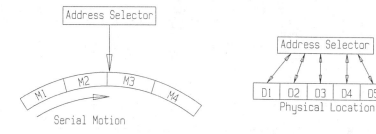

Figure 13.11 Conceptual address structures for SAM and RAM storage

require refreshing during reading operations (NDRO). This memory is still volatile, and has a slower access time than does the DRAM. It is frequently used as a temporary secondary storage device.

Recent technology has allowed for the development of non-volatile RAM (NVRAM); however, most RAM in this category is read-only memory (ROM). A ROM is a semi-conductor device used for storing programs or other information as read-only information. ROMs function as input devices for dedicated functions. ROM is preprogrammed with a special device known as a 'programmable ROM' (PROM) burner. This process permanently alters the ROM. It should be noted that the terms PROM and ROM are interchangeable.

Two alterable versions of ROM are also available: the erasable ROM (EPROM), and the electrically erasable ROM (EEPROM). The EPROM is erased by exposure to ultraviolet light while the EEPROM is erased by a special control signal. All ROM devices are normally programmed off-line in a special programming unit. ROM technology has been an extremely cost-effective means of non-volatile storage of programs or dedicated instructions. This is available at a low cost per bit storage.

Two other semiconductor technologies have been used for storage. Magnetic bubble memories (MBM) provide non-volatile storage at a medium cost per bit, as shown in *Figure 13.10*. The bits in a MBM are stored as magnetic bubbles in cylindrical regions. These bubbles are moved in a serial manner to access the appropriate address for read/write operations. Charge coupled devices (CCD) operate in a similar fashion, charges being stored in an array of capacitors. The access rate from CCDs is faster than from MBM; however, CCD are volatile devices, and are therefore only suitable for temporary secondary storage.

13.4.4 Magnetic storage devices

Magnetic media have been used as traditional storage media for computers for some time. These devices include floppy disks, magnetic tape, magnetic drums and magnetic disks. Magnetic-tape units were the first magnetic storage devices used in computers, and still provide a cost-effective method of information storage. Most magnetic tapes have slower access times than magnetic disks and drums; therefore, most tape units are used as a tertiary back-up device.

The floppy disk is another popular cost-effective storage magnetic device. Data are storage in a serial manner in concentric tracks which are divided into sectors, as illustrated in *Figure 13.12*. The disk spins to provide serial access to the different sectors in the circumferential direction, while the head is positioned to provide access to discrete tracks in the radial direction. This addressing technique provides for the

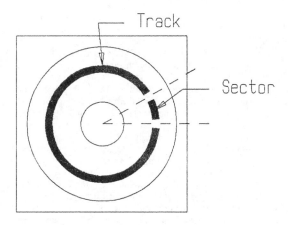

Figure 13.12 Floppy disk data address structure

addressing of data in a rectangular coordinate scheme, even though data are physically stored on a radial disk. Although the data are accessed in a serial manner, they are addressed directly and the floppy disk is characterised as direct access memory.

The predecessor of the floppy disk was the hard disk. Access is similar to that of the floppy disk, except the data are stored in much greater density and, therefore, less mechanical movement is needed to access data. Unlike the floppy disk, there can be several hard disks to a hard disk unit, or pack. Since there are multiple disks, there will be identical tracks in the same disk unit. The groups of identical tracks are termed 'cylinders' (see *Figure 13.13*).

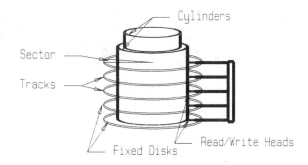

Figure 13.13 Moveable-head hard-disk cylinder with address structure

These disk packs are usually sealed, ensuring higher reliability; however, many of the larger systems provide for removable disk packs. Hard disks are not intended for frequent removal and are more expensive to use in terms of cost per bit than are floppy disks as previously indicated in *Figure 13.10*. This technology does provide a significant increase in access speed and is the most popular mode of secondary memory.

The removable disk packs and many of the permanent packs use a movable head, although some units use a fixed-head system with multiple read units (see *Figure 13.14*). The fixed-head system provides an increase in reliability, speed, and storage capacity with a corresponding increase in cost per bit.

13.4.5 Hard copy output devices

Most hard-copy devices are classified as printers or plotters, where the data are transferred to paper in some fashion. Other methods for obtaining hard copy include videotape records of graphics images. Other photographic processes may be included for permanent hard-copy storage.

Graphics images are frequently stored as hard copy due to the tremendous amount of storage required to represent the image. These images may be stored during the development of a project; however, system managers will eventually be forced to erase these images to make room for more. This housekeeping of a system's storage facilities requires the back up of old information. Hard copy results may also be used for ease of transportability of the qualitative information.

The resolution obtained with the various hard-copy devices varies, and is a heavily advertised subject. One of the most economical devices is the dot-matrix printer. The dots are actually miniaturised solenoids patterned in a matrix formation. These solenoids actuate small strikers which impact an ink ribbon, leaving an ink dot on the paper. The resolution of the printer is determined by the number of dots in the print head and the size of the print head. The print head may be shifted slightly to fill the gaps between dots while printing. This mode is commonly referred to as 'letter quality'. The letter-quality mode normally requires two passes of the print head to print each line of text. Graphics images are bit mapped onto the page, requiring an extensive number of passes per inch of the print head. The printing of graphics images is time consuming due to the extensive movement of the print head.

High-speed dot-matrix printers have a print head which does not move. These print heads are constructed of a series of dot-matrix heads in line, allowing the simultaneous printing of a line of text. The only mechanical movement required is the movement of the paper past the print head.

One of the most recent technologies implemented for hard-copy devices is the laser printer. Laser printers provide very high-quality output at moderate cost. The speed of these printers is normally higher than that of dot-matrix printers, and the operational cost is higher. These printers are frequently used for graphics output devices due to their high quality.

13.5 Human–computer interfaces

Many computer-controlled functions within a manufacturing environment operate without continuous interaction with a human operator. These systems use preprogrammed, self-opoerating functions which require initial programming input by a human. However, many other computer-controlled manufacturing functions use continuous human input to monitor and control various manufacturing operations. The interface hardware and software components between a computer and a human operator depend on several factors including:

(1) the manufacturing task to be monitored and/or controlled, including throughput timing and numbers of tasks to be monitored;
(2) the level and sophistication of the computer hardware and software;
(3) the number and type of tasks required of the human operator; and
(4) the degree of accuracy and quality control required of both the computer and the human operator.

The interface between a human operator and a computer requires communication hardware and software. These can be categorised as either (1) computer output devices which communicate the state of computer operation to the human operator, or (2) computer input devices which allow human input to control computer-driven manufacturing functions.

13.5.1 Computer output interface devices

13.5.1.1 Terminals, monitors and displays

The vast majority of peripheral devices which communicate the state of computer function to a human operator use visual output such as terminals, monitors and text and/or graphics displays. Many of these monitors are generic in nature and can be interconnected to various types and levels of computer. Other outputs are specialised in performance and can only be integrated into single-function computer systems.

Factors which should be considered when using visual displays include:

(1) whether the computer system is generic or specialised,
(2) the level of text information to be displayed,
(3) the level of graphics information to be displayed,
(4) the size of the display unit,
(5) the available space for a display unit,

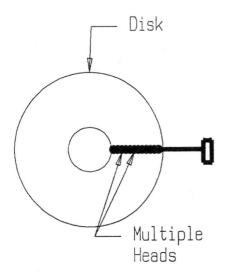

Figure 13.14 Fixed-head hard-disk cylinder

(6) the level of ambient light and glare, and
(7) the distance between the display and the human operator.

Due to the developments in computer technology over the last decade, it is usually possible to match the computer display to the intended use and for an intended environment. Display systems can be purchased that are text specific, graphics intensive, and have touch and/or light-pen input capabilities. Large display systems are available for distant viewing, and glare screens are available for areas where the ambient light may otherwise adversely affect viewing of the screen.

13.5.2 Computer input interface devices

13.5.2.1 Keyboards

The keyboard represents the most widely used device for transmitting text information from a human operator to a computer. The design and operation of many keyboards have been standardised over many years of usage as data-entry devices. A keyboard may be a standard model which is commonly used with all levels and types of computer, or may be specialised for use with a specific computer or specialised equipment. The use of a standard keyboard as a data-entry device offers several advantages, including:

(1) flexible text input,
(2) user-defined function keys,
(3) independent numerical keypad,
(4) independent screen cursor control,
(5) simplicity of operation, and
(6) standardised hardware connection to the computer.

A specialised keyboard offers such advantages as predefined multifunction keys and user-specific layouts.

The most common keyboard layout uses the QWERTY system (see *Figure 13.15*). This system is named after the left-most six keys on the top row, and has become an international standard for keyboard layouts in English-speaking countries.[3] Variations of this system exist in France (as AWERTY) and in German-speaking countries (as QWERTZ). This layout is similar to that of typewriter key-boards with commonly used letter keys placed in close proximity. Approximately 10 key postions out of a total of 100 available positions (a combination of 50 lower and upper case) can be user specific for specialised applications (see *Figure 13.16*).

A separate numerical keypad is located to one side, as is a separate screen-cursor-control keypad. The numerical keypad is arranged in a 3 number by 3 number array with the zero at the top or the bottom of the array (depending on whether the array is arranged in ascending or descending numerical order). In addition, a full number set is situated as lower-case keys on the top row of the main key area. The cursor-control layout usually appears in a T-shaped or cross-like arrangement. The latter configuration usually contains a 'home' key which, when pressed, sends the cursor to top left-hand corner of the computer display.[4]

A set of function keys may be placed at either the top or on the left-hand side of the keyboard. The use of top-mounted function keys is more appropriate for keyboards placed to the left of the user because this arrangement minimises the reach distance of the human operator. These function keys may be user defined or may have predetermined functions, depending on the specific application software in operation.

Figure 13.15 QWERTY keyboard

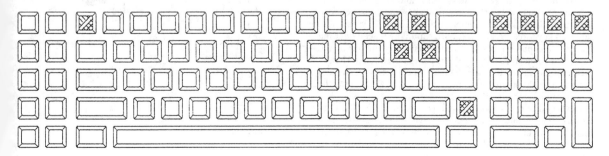

Figure 13.16 QWERTY key arrangement with user-specific keys

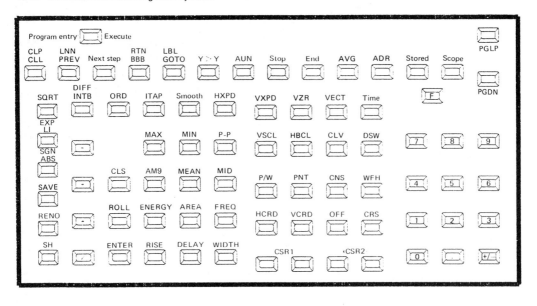

Figure 13.17 Specialised keyboard

The DVORAK keyboard system is a common variation on the QWERTY system, with a few letter and punctuation keys moved to different keyboard locations in order to allow both hands to use the keyboard simultaneously without excessive awkwardness.[4] However, for computer applications where the operator will be using only one hand to input data, the QWERTY system is more advantageous. This system overloads the left hand, and thus a QWERTY keyboard is more appropriately placed to the left of the operator.

Standard keyboard layouts can also vary in terms of keyboard dimensions, key sizes, and tactile sensation. A typical key size is 13 mm × 13 mm with 19 mm spacing (centre to centre) between keys. The key-activation force varies from 0.25 to 1.5 N with key displacement varying from 1 mm to 6 mm. The selection of a keyboard interface should incorporate the intended application, including anticipated key force and whether a bare or gloved hand will be used to input data.[5]

An example of a specialised keyboard is shown in *Figure 13.17*. Specialised keyboards are usually smaller than standard keyboards and are less flexible. There are fewer keys and the function keys are predefined for the intended application. Such keyboards are routinely used as an interface to computerised electronic testing equipment, as well as for robotic controllers.

Considerations for keyboard selection and usage include the placement of function keys, numerical keys and cursor-control keys as well as overall keyboard dimensions, key size and required key force.[6,7] The keyboard should be placed near the attached computer in order to minimise interference of the keyboard connecting wires with on-going operations. This, in turn, should affect the selection of function-key placement design. The dominant hand of the user will also affect the optimum placement of the keyboard.

Keyboard usage is most advantageous for input of text or specialised functions via function keys. Thus, its use as a manual computer interface device is intended for situations where variability and flexibility of alphanumerical input is most important, or where specific, single-key functions can be elicited. It is not advantageous in situations where the human operator does not have at least one hand free or where the computer function is menu driven or graphics intensive.

13.5.2.2 Mice and joysticks

A mouse (*Figure 13.18*) is a hand-operated input device which is attached by a wire to a computer through a serial communications port. Typically, it is a small, rectangular plastic box which is placed under the palm of the user's dominant hand and is slid along a flat surface to control the cursor position on a computer screen. A mouse contains one to three buttons on the top which can be clicked by the index or middle finger to activate functions such as selecting computer screen menus, drawing objects on screen, or specifying selected menu options. A mouse may be either mechanical or optical in its operation. Both types operate by converting sensor information from mechanical wheels or optical sensors into x–y coordinates of cursor position on the computer monitor/screen.[8–10]

A mouse is a rapid cursor-control device. Its use in the operation of menu-driven software is far superior to that of the

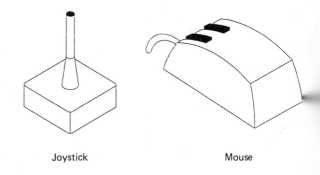

Joystick   Mouse

Figure 13.18 Mouse and joystick

cursor keypad controls on a standard computer keyboard.[11] However, a mouse is exclusively intended as a cursor-positioning device and as a limited input-selection device. For those applications which require more flexible inputs, a mouse cannot be used as the sole data-input system. In addition, the use of a mouse requires additional table space over and above that of a keyboard (usually a 20 cm × 20 cm square). The advantages of a mouse include its rapid control of cursor position, the ability of the user to look at the screen without the need to look down at the mouse during operation, and its use as a free-hand-drawing device. However, usage of a mouse for accurate drawing purposes is limited as the mouse is somewhat awkward as a drawing tool and its resolution is limited. It is an excellent input device for applications which are primarily menu driven, including CAD systems. *Table 13.2* lists various mouse manufacturers and the features and relevant supported software systems of commonly used mice.

A joystick (*Figure 13.18*) is a lever which is mounted vertically in a fixed base. It operates by tilting the lever forward, backward, left or right in order to control the cursor position on a computer screen in much the same fashion as the control elicited by a mouse. A button on the top of the lever is used as a single input-selection device, similar to that of a mouse button. A joystick is connected by wire to a serial communications port of a computer similar to that of a mouse. A joystick is advantageous where there is limited table space available, in that one can control the cursor position from a single spot, whereas a mouse requires hand movement over a small area.[12] However, unlike a mouse, a joystick cannot be used as a drawing device. It is primarily used for cursor position and menu selection. A typical mouse and joystick are shown in *Figure 13.18*.

Both mice and joysticks are low-cost devices with limited usefulness. Their resolution is limited. Thus, the number of menu selections and the size of each menu item displayed on screen may be a limiting factor in the use of either of these pointing devices.[13,14] The use of a mouse for long periods appears to be less tiring than the use of a joystick, which eventually affects the wrist muscles of the user.

13.5.2.3 Touch screens

A touch screen provides a direct method of cursor-position input to a computer monitor or screen in response to a touch of a finger on the screen at the intended location. It is primarily used as a menu-selection system where the menu is placed in a one- or two-dimensional linear array, as shown in *Figure 13.19*. This differs from the typical use of a mouse as a menu-selection device in that mouse-driven systems usually use pull-down menus from a highlighted band of selections at the top of a computer screen. Touch screens can operate with physical screen overlays, or by conductive, optical, ultrasonic or physically sensitive grids that are incorporated into the computer screen, as shown in *Figure 13.20*. Because a touch screen is a specialised computer monitor which is quite different from a standard computer monitor, its usage is

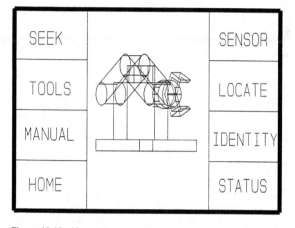

Figure 13.19 Menu-driven computer screen

Table 13.2 Mouse features

System	Sensor	Interface	Buttons	CAD Support	Windows support
Logitech Mouse, Logitech Inc.	Optomechanical	Serial, bus	3	Yes	Yes
Microsoft Mouse, Microsoft Corp.	Mechanical	Serial, bus	2	Yes	Yes
PC Mouse, Mouse Systems Corp.	Optical	Serial, bus	3	Yes	Yes
IBM PS/2 Mouse, IBM Corp.	Mechanical	Special-ised	2	Yes	Yes
SummaMouse, SummaGraphics Corp.	Optical	Serial	3	Yes	Yes
IMSI Mouse, IMSI Inc.	Optical, mechanical	Serial	3	Yes	Yes
American Mouse, American Computer and Peripheral Inc.	Mechanical	Serial, bus	3	Yes	Yes
MultiMouse, American West Engineering Corp.	Optomechanical	Serial keyboard	6	Yes	Yes
Manager Mouse, Torrington Co.	Mechanical	Serial	3	Yes	Yes

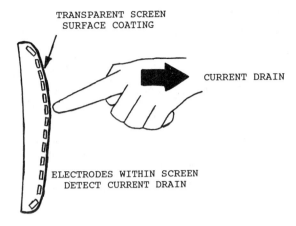

TRANSPARENT SCREEN
SURFACE COATING

CURRENT DRAIN

ELECTRODES WITHIN SCREEN
DETECT CURRENT DRAIN

Figure 13.20 Touch-screen operation

intended for specialised applications where a standard monitor is not required. Thus, the use of alternative input devices (which require a standard computer monitor) in concert with a touch screen is usually precluded. There are a few touch-screen systems which can be installed on standard computer monitors, as opposed to purchasing a dedicated computer/touch-screen system. However, their performance in terms of resolution and speed of operation is inferior to that of dedicated systems.[9]

A major advantage of the touch screen is that the hand–eye coordination is maximal because the input and output devices are one and the same, namely the computer monitor screen. Since the number of menu selections and functions is limited, there are usually far less errors in the use of a touch screen as an input device than occur with other input devices such as a keyboard or mouse. The software that is incorporated for use with touch screens is often designed to minimise user training. Thus, the use of touch screens is often much simpler than that of other input devices in that the user is often 'led through' a series of menus leading towards the selection of a particular computer-controlled function.[15]

However, there are also several disadvantages to the use of touch screens. Arm fatigue is common after even moderate use, since the operator must usually lift an arm in order to reach the computer screen. A touch screen is limited as a computer input device to menu-driven applications. This is typical of some robotic functions and many electronic testing applications. Data entry is slowed by arm movement and the

screen is temporarily blocked from view as a menu selection is made. The selection of small items on a screen may also be problematic. It is possible inadvertently to select the wrong menu item, especially after prolonged use. Due to constant contact with fingers, touch screens require frequent cleaning in order to remain effective.

Touch screens are not available for all computer systems. They are typically used with selected microprocessor-based systems and have seen only limited use as a computer input device, primarily as an adjunct to computerised electronic circuit-testing systems. Some keyboards may be purchased with small, specialised, user-definable touch pads incorporated in the lower right-hand-side of the keyboard, near the cursor-control keypad.

13.5.2.4 Light pens

A light pen is essentially a pencil-sized electronic version of one's finger which is used to point to a menu selection on a light-sensitive computer screen. Thus, it is quite similar in function to a touch screen. The use of a light pen requires a light-sensitive CRT display which is quite different from a standard computer monitor.[9] Thus, as with a touch screen, the use of a light pen often precludes the use of alternative input devices. A typical light pen is shown in *Figure 13.21*.

As with a touch screen, a light pen is most often used as a pointer to select items from a menu array displayed on screen. One merely points at a particular menu selection and enables the pen by a push button located at its tip. However, a light pen can also be used as a drawing device by dragging the pen along the screen while it is enabled. Since there is not direct contact of fingers with the computer screen, the screen does

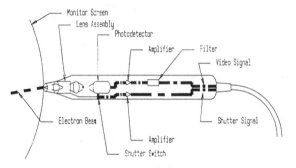

Monitor Screen
Lens Assembly
Photodetector
Amplifier
Filter
Video Signal
Electron Beam
Amplifier
Shutter Signal
Shutter Switch

Figure 13.21 Light-pen operation

Table 13.3 Light pens

System	Graphics adapter	CAD support	Windows support
L-PC Lite Pen, The Light-Pen Co.	CGA, EGA, mono.	No	Yes
FT-156 Light Pen, FTG Data Systems Inc.	CGA, EGA	Yes	Yes
Fast-point Light Pen, HEI Inc.	CGA	Yes	Yes
Warp Speed Light Pen	CGA, EGA, mono.	Yes	Yes

CGA, colour graphics adapter; EGA, enhanced graphics adapter; mono., monochrome.

not need frequent cleaning as does the touch screen. *Table 13.3* lists the major manufacturers and the features of commonly used light-pen systems that are additions to otherwise standard computer systems. It is also possible to purchase a dedicated computer system with a built-in light-pen system that is designed for specialised applications such as electronic-circuit testing and CAD systems. [16]

The disadvantages of the light pen are similar to those of the touch screen. Arm fatigue is common. The screen is blocked during the menu-selection process. Resolution is limited and incorrect selections are quite possible. In addition, light pens and light-sensitive computer screens are not common and are not available for all computers. As with the touch screen, light-pen systems are marketed for selected microprocessor-based systems. Typical applications are menu-driven functions such as those found with electronic-circuit-testing systems.

13.5.2.5 *Graphics tablets and digitisers*

A graphics tablet is a flat panel which is connected by a wire to a serial communications port in a computer. These are typically available only for microprocessor-based systems. A pencil-sized stylus is attached by a wire to the tablet and is dragged along the tablet surface. The tablet mimics the computer screen in that the path of the stylus is depicted directly on the screen. Thus, a graphics tablet is primarily used as a graphics-input device. However, unlike a mouse or light pen, the use of a graphics tablet allows drawing to be accomplished horizontally in a manner similar to one's natural drawing mode on a flat horizontal surface.

A graphics tablet is useful for drafting and drawing with the input appearing directly on the screen. The resulting graphic display can be stored in the computer memory for later retrieval and modification. One can trace an existing drawing by placing it onto the tablet and dragging the stylus over the drawing. [17] A drawback in its usage is that the tablet and the resulting display on the computer screen are physically separated. Thus, there is limited hand–eye coordination between what is being drawn and the resulting image on the screen. In addition, a graphics tablet requires considerable table space (typically a 50 cm × 50 cm square) and has limited resolution. [18] Data entry is slower than with many other input devices. *Table 13.4* lists the manufacturers and the features of commonly used graphics tablets.

An alternative method of direct graphic data entry is a graphic digitiser. This device scans an existing paper copy and digitises the image for storage in computer memory. Such a system offers greater resolution than a graphics tablet, albeit at considerable expense. However, a scanning digitiser allows direct entry of circuit diagrams and similar detailed graphic information into computer memory. Both the graphics tablet and the scanning digitiser are primarily graphic-image generators and their application in a computerised manufacturing environment is limited.

13.5.2.6 *Eye position and gaze input*

The use of eye position as an input to a computer is primarily intended for applications where both hands are occupied with other tasks and are not available for data entry by keyboard, mouse or other input device. Such a system requires the use of specialised glasses which are connected by wires to the computer through an electronic interface. The computer connection is similar to that of a mouse, which is accomplished through a serial communications port. An eye position input system is also similar to a mouse in that it acts as a cursor positioning device. However, as it is intended as a 'hands-free' system, there is no analogue to the mouse button. Instead, switching and selection is accomplished by fixing one's gaze on a single spot on the computer screen for a fixed time period, which is selectable by the operator.

An eye position input system operates via detection of infra-red light which is shone into and reflected from the eye. An alternative system tracks eye position through the us of light emitting diodes (LEDs). Such systems require calibration of the operator's gaze with computer screen cursor position. These systems are heavily dependent on head motion. They are primarily intended for use in the selection of menu items on screen, in a similar fashion to that of the touch screen, light pen, or mouse. As the time required to fix one's gaze in selecting a menu item is programmable, the input response time is variable. However, currently employed systems can produce menu selections in less than 0.2 s, which is much faster than the response time for touch screens or mice, which incorporate a relatively slower hand motion in the selection process. [19]

The use of eye position systems requires specialised equipment which is relatively expensive compared with other computer input devices. A major disadvantage is the need for specialised glasses which may interfere with other operator tasks. However, the use of eye position input devices to control computer functions allows 'hands-free' operation, unlike other input devices. The technology of these systems is still relatively young compared with that of other input

Table 13.4 Graphics tablets

System	Size (cm × cm)	Programmable	CAD support	Windows support
IS/ONE Series Tablet, Kurta Corp.	41 × 38	Yes	Yes	Yes
VersaWriter, Versa Computing Inc.	30 × 34	No	No	No
SummaSketch 1201, SummaGraphics Corp.	41 × 41	No	Yes	Yes
KoalaPad + Touch Tablet, Koala Technologies Inc.	15 × 20	No	No	No
Digi-Pad Type 5A, GTCO Corp.	40 × 40	Yes	Yes	Yes
Pencept Penpad 320 Pencept Inc.	38 × 38	Yes	Yes	Yes

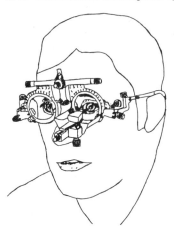

Figure 13.22 Specialised glasses for use with an eye-controlled computer system

devices. However, the use of eye tracking systems for such applications as aircraft controls has advanced the technology in this area to such a degree that it may become a premier 'hands-free' computer control system in the near future.[20] A typical eye position control system is shown in *Figure 13.22*.

13.5.2.7 Voice input

The use of human speech as an input to control computer function represents the optimum 'hands-free' methodology presently available. The use of speech as a controlling input allows the operator to use his/her hands for other functions and to gaze at the computer monitor while issuing commands. Speech-recognition systems are available from many commercial sources and can be interfaced to various types of computers, including mainframe, mini- and micro-computers. However, most of these systems are designed for application with microcomputer-based systems, where a solid-state expansion card fits into an existing personal computer (PC). Other systems operate as stand-alone systems with a built-in microprocessor which is connected to an existing computer through a serial communications port. The expansion card versions of speech-recognition systems are slower, but are also far less expensive than stand-alone models.

The voice signal is monitored by a microphone which is incorporated into either a light-weight headset worn by the operator or placed on a table-mounted free-standing base. Frequency modulated (FM) telemetry may be employed in conjunction with the headset and a battery powered FM amplifier may be used to allow the operator to move about without being physically tethered to the computer. Thus, robotic functions, assembly-line operations, materials (inventory) handling and hazardous materials handling may all be controlled by voice through computer controls.[21]

The voice signal is transduced by the microphone and preprocessed by an analogue filter/amplifier network in order to eliminate spurious ambient noise. This analogue signal is then digitised via an analogue-to-digital converter within the processing computer. The digital signal is then divided into its frequency components by either a digital filter network or through the use of a fast Fourier transform (FFT) or other time domain-to-frequency domain calculation. Each uttered word or brief phrase is thus transformed into a stored pattern of amplitude vs. frequency. Typically, the processed speech is limited to the telephone bandwidth (300–3000 Hz).[22–25]

A working vocabulary must be created which is relevant to the intended application and whose words are phonetically different so that their stored patterns are dissimilar. In general, short words with similar sounds, such as 'go' and 'no', are not dissimilar enough to be readily distinguishable by the speech recognition system, and should therefore not be included in a working vocabulary as a group. The frequency components of the entire working vocabulary are stored as a series of templates during a brief vocabulary training period. For many speech recognition systems, the training period may be as brief as 1 min, depending on the vocabulary size. Due to the variability between individual voices, all operators who would use a voice control system must train and record their respective vocabulary templates. Once this has been done, an operator simply keys in his identity so that the system calls up the proper template. During control operation, each spoken word is compared with the stored template to determine the proper word match. Once this has been accomplished, an appropriate process control function may be initiated by the computer. The processing speed of such systems is rapid, such that the human operator can initiate 'real time' control of computer driven manufacturing functions.

Voice processing systems are evaluated on the basis of:

(1) overall word recognition performance,
(2) training requirements,
(3) vocabulary capabilities,
(4) noise immunity, and
(5) computer interface requirements.

A typical voice processing system contains a microphone, a noise cancelling system, an analogue-to-digital converter, a time-to-frequency converter, a frequency template development system, a training module, a performance analayis module, and perhaps a telemetry system for remote operator utility (see *Figure 13.23*). *Table 13.5* lists the concerns which must be addressed for each component of the voice recognition system.

In particular, microphone directionality, noise immunity, and processed word error reduction are of considerable concern to the operation of a voice recognition system as an integral part of a computerised control system for a manufacturing environment. When using a table-based microphone, the speaker must take great care to speak towards the microphone in order to reduce the errors associated with speech recognition. Even when using headset-mounted microphones, it is important for the operator to speak towards the microphone, because many such systems employ swing away microphones. In addition, the tone of the operator's voice should remain relatively constant in order to ensure proper accuracy in word recognition. Thus, operations which might severely affect the tone of the human voice would be poor candidates for speech controlled processes.[26,27]

Noise cancellation in an industrial setting is a significant concern in the use of speech processors. Industrial noise may incorporate broadband noise resulting from blowers and fans, discrete frequency noise resulting from various periodic sources, and aperiodic impulse noise resulting from stamping and pressing machines and related sources. Centre clipping of the voice signal as well as other data-processing methodologies have significantly reduced the effects of broadband noise on processed speech. Periodic noise has been successfully treated with predictive filtering techniques. However, impulse noise of sufficient amplitude to mask the voice signal remains a serious problem. Noise cancellation may be accomplished through an analogue circuit placed between the microphone and computer interface, or by means of digital processing within the computer itself.[28,29]

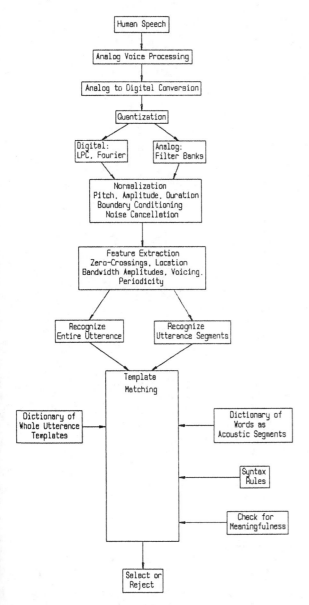

Figure 13.23 Voice processing system. LPC, linear predictive coding

Errors resulting from incorrect voice processing may be due to either omission or commission. Omission errors are those in which the spoken word is not recognised and no computer control command is elicited. Commission errors are those in which the wrong word in the template is matched and an incorrect computer control command is elicited. Correct noise cancellation techniques can significantly reduce the incidence of errors during speech processing. Virtually error-free operation has been demonstrated for broadband noise up to 110 dB.

Voice processing as an input to computer-controlled manufacturing processes is a reliable method of 'hands'-free' input and can serve as a useful adjunct to other input devices such as a keyboard or mouse. Voice based systems can serve as a

Table 13.5 Areas of concern for speech recognition systems

System component	Areas of concern
Voice processing board	Speaker dependency
	Speed of processing
	Bandwidth
	Noise cancellation method
	Word recognition method
	Training requirements
	Accuracy
	Vocabulary
Microphone	Type of element
	Directional sensitivity
	Amplitude sensitivity
	Noise cancellation
Computer interface	Analogue-to-digital conversion rate
	Analogue-to-digital conversion
Accuracy	Power requirements
	Connection slot or bus
	Bandwidth
Telemetry	Bandwidth
	Amplitude sensitivity
	Directional sensitivity
	Accuracy

keyboard emulator, thus producing any input of a standard keyboard entry, or as a mouse emulator to provide menu selection operation. Code words or short phrases can be used as an alternative to user-defined function keys on a keyboard. *Table 13.6* lists the most commonly used computer based speech recognition systems and their respective features.

Speaker independence, impulse noise cancellation, and limited vocabulary size are areas where future research is targeted in order to expand voice-controlled applications in a manufacturing environment.[30-32]

13.6 Networks

The term 'computer network' is used to describe situations in which: (1) physically remote terminals are connected to a central processing computer, (2) many remote small computers are interconnected with the aid of a central processor or storage unit, or (3) independent large computer systems are linked in order to share hardware peripherals or software systems.

A glossary of network system terms is given in *Table 13.7*.

13.6.1 Basic communications classification

The flow of information is critical in modern manufacturing environments. Many processes are now controlled or synchronised by a host computer or by several computers in an integrated system. Speed and reliability are critical to these information networks.

The flow of information between two devices may be categorised as one-directional (simplex), alternating informa-

Table 13.6 Voice recognition systems

System	Unit placement	Uses own microprocessor	Training passes required	Vocabulary size
Voice Scribe 1000, Dragon Systems Inc.	Internal	No	5	1000
Voice Card VPC2150, Votan Corp.	Internal	Yes	1	64
Kurzweil Voicesystem, Kurzweil Applied Intelligence Corp.	External	Yes	3	1000
IBM Voice Communications Option, IBM Corp.	Internal	Yes	5	160
TI-Speech System, Texas Instruments Inc.	Internal	Yes	3	50
KB5152V, Key Tronic Corp.	External	Yes	3	160
Pronounce, Microphonics Technology Corp.	Internal	No	1	128
Voice Developer System, Verbex Voice Industries Corp.	External	Yes	4	100
VocaLink SRB-LC Interstate Voice Products Inc.	Internal	No	3	400
VoiceKey, Roar Technology Inc.	Internal	No	2	512
VoiceCommand, Interpath Corp.	Internal	No	4	200
IntroVoice VI, The Voice Connection	Internal	No	3	400
Coretechs VET 3 Speech Terminal, Scott Instruments Corp.	External	Yes	3	200

Table 13.7 Glossary of network system terms

Term	Description
Access protocol	The traffic rules that LAN workstations abide by to avoid data collisions when sending signals over shared network media. Common examples are carrier sense multiple access (CSMA) and token passing
APPC	Advanced program-to-program communications: this IBM protocol is analogous to the OSI session layer, since it sets up the necessary conditions that enable application programs to send data to each other through the network
Application layer	The highest level of the OSI model. It describes the way that application programs interact with the network operating system
ARCnet	Attached resources computing: a networking architecture using a token-passing bus architecture, usually on coaxial cable
Bridge	A type of product that links different local area networks, enabling users on one network to use all the resources available on the other

Table 13.7 Cont'd

Term	Description
Buffered repeater	A device that amplifies and regenerates signals so that they can travel further along a cable. This type of repeater also controls the flow of messages to prevent collisions
Bus	A 'broadcast' arrangement in which all network stations receive the same message through the cable at the same time
Cache	An amount of RAM set aside to hold data that network stations are expected to access again. The second access, which comes from RAM, is very fast
Coaxial cable	A type of network medium. Coaxial cable contains a copper inner conductor surrounded by plastic insulation and then a woven copper or foil shield
CSMA	Carrier sense multiple access: a media-sharing scheme in which stations listen in to what is happening on the network media; if the cable is not in use, the station is permitted to transmit its message

Table 13.7 Cont'd

Term	Description
Data link layer	The second layer of the OSI model. Protocols functioning in this layer manage the flow of data leaving a network device and work with the receiving station to ensure that the data arrive safely
Ethernet	A network cable and access protocol scheme originally developed by Xerox, now marketed mainly by DEC and 3Com
Fibre optics	A data transmission method that uses light pulses sent over glass cables
File server	A type of server that holds files in private and shared subdirectories for LAN users
Frame	A packet on a token-ring network
Gateway	A portal though which networked stations can access resources on mainframe and minicomputer hosts as well as on nationwide data networks
ISO	International Standards Organization (ISO), Paris, which developed the open system interconnection (OSI) model
Locking	A method of protecting shared data. When an application program opens a file, file locking either prevents simultaneous access by a second program or limits such access to 'read only'
MAC	Media access control
MAP	Manufacturing automation protocol: a token-passing bus LAN designed by General Motors
Media	The cabling or wiring used to carry network signals. Typical examples are coaxial, fibre-optic, and twisted-pair wire
NetBIOS	Network basic input/output system: a layer of software originally developed by IBM and Sytek to link a network operating system with specific hardware. It can also open communications between workstations on a network at the session layer
Network layer	The third level of the OSI model, containing the logic and rules that determine the path to be taken by data flowing through a network
NFS	Network file system: one of many distributed file system protocols that allow computers on a network to use the files and peripherals of another networked computer as if they were local
OSI	Open system interconnection: a model developed by the ISO describing the network communications process
Packet	A block of data sent over the network transmitting the identities of the sending and receiving stations, error-control information, and a message
Peer-to-peer resource sharing	A software architecture that lets any station contribute resources to the network while still running local application programs
Physical layer	The first layer of the OSI model. It consists of network wiring and cable
Presentation layer	The sixth layer of the OSI model, which formats data for screen presentation and translates incompatible file formats

Table 13.7 Cont'd

Term	Description
Print server	A computer on the network that makes one or more attached printers available to other users. The server usually requires a hard disk to spool the print jobs while they wait in a queue for the printer
Protocol	A specification that describes the rules and procedures that products should follow to perform activities on a network, such as transmitting data. Protocols allow products from different vendors to communicate on the same network
Redirector	A software module loaded into every network workstation. It captures application programs' requests for file- and equipment-sharing services and routes them through the network
Repeater	A device that amplifies and regenerates signals so they can travel farther on a cable
RFS	Remote file service: one of the many distributed file system network protocols that allow one computer to use the files and peripherals of another as if they were local. Developed by AT&T and adopted by other vendors as a part of UNIX V
Router	A machine in a large network that reads the destination of a message and selects the best route
SAA	Systems application architecture: a set of specifications written by IBM describing how users, application programs, and communications programs interface
Server	(a) A computer with large power supply and cabinet. (b) Any computer on a network that makes file, print, or communications services available to other network stations
Session layer	The fifth layer of the OSI model, which sets up the conditions whereby individual nodes on the network can communicate or send data to each other. The functions of this layer are used for many purposes, including network gateway communications
SMB	Server message block: a distributed file system network protocol that allows one computer to use the files and peripherals of another as if they were local
SNA	System network architecture: IBM's model of a communications system
StarLAN	A networking system developed by AT&T that uses CSMA protocols on twisted-pair telephone wire
Token passing	An access protocol in which a special message (token) circulates among the network nodes giving them permission to transmit
Token ring	Refers to the wire and the access protocol scheme whereby stations relay packets around in a logical ring configuration
Topology	The map of the network. The physical topology describes how the wires or cables are laid out, and the logical or electrical topology describes how the messages flow

Table 13.7 Cont'd

Term	Description
Transport layer	The fourth layer of the OSI model. Software in this layer checks the integrity of and formats the data carried by the physical layer, managed by the data layer, and routed by the network layer
Twisted-pair wiring	Cable comprising two wires twisted together at 6 turns per inch to provide electrical self-shielding
X-25	An international standard describing how computers can access packet-switched networks. Typical X-25 networks include Tymmet, Telenet, and MCI Mail
XNS	Xerox network services: a multilayer protocol system developed by Xerox and adopted, at least in part, by Novell and other vendors. XNS is one of the many distributed file system protocols that allow network stations to use other computers' files and peripherals as if they were local

LAN, local area network.

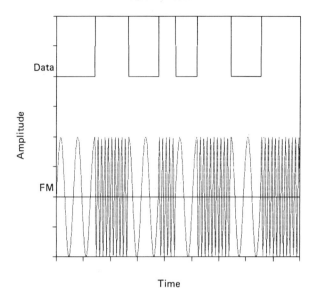

Frequency Modulation

Figure 13.24 Frequency modulation

tion flow (half duplex), or both directions simultaneously (full duplex). A permanent data path between two discrete devices is dedicated, whereas a path with multiple devices is shared. The ideal system would have dedicated links between all devices; however, this is not normally physically or economically feasible. Therefore, most systems comprise a network of shared data links.

This flow of data may be modulated and demodulated (MODEM) with a carrier frequency. The carrier wave is altered (modulated) in accordance with the value of the input data. The frequency, phase, or amplitude are the most common parameters which are altered. *Figure 13.24* illustrates the frequency modulation (FM) technique. This modulated signal is demodulated to extract the data at the receiving point. This method extends the transmission distance of many signals. Frequency modulation techniques usually enhance the noise rejection characteristics of the system as well.

Data transmitted over phone lines normally have a carrier frequency of 1000 Hz. MODEM systems with one frequency are termed 'baseband', whereas systems with multiple carrier frequencies are classified as 'broadband'. A single data link between several transmission devices can transmit several different logical transmissions simultaneously at different frequencies. This data link can therefore provide for multiple logical paths over the same physical link using MODEM transmission techniques.

The flow of data in any system can be categorised as synchronous or asynchronous. The information flow inside a computer system is passed from one point to another at discrete instants controlled by a system clock. This method synchronises (synchronous) the data flow within the computer. When connecting several on-going processing units, an asynchronous link is normally used. This method permits the flow of data without the need to synchronise the transmitting and receiving devices. The synchronous system must interrupt the transmitting and receiving points to synchronise them and transfer the data. An asynchronous system transmits a start bit and stop bit to delineate the beginning and end of data words, respectively.

13.6.2 Transmission media

There are several different physical means for the transmission of data. *Table 13.8* lists some of the more common media with their respective data bandwidths. The data bandwidth of a system is defined as the difference between the minimum and maximum data frequency transmission rates. Fibre optic technology has been used extensively in recent years in industrial settings due to its immunity to electromagnetic interference. A fibre optic line cannot be 'tapped' easily, thus increasing data security while decreasing the ease of splicing for repair or expansion.

The baseband coaxial or twisted pair are still used extensively due to their relative cheapness and simplicity. These systems are much easier to repair, and are easier to modify for additional installations. However, this ease of repair and modification reduces the system security.

The implementation of either coaxial or twisted pair can be connected as an unbalanced transmission link, as shown in

Table 13.8 Data transmission medium characteristics

	Twisted pair	Baseband coaxial	Broadband coaxial	Fibre optic
Baud rate (Mbyte/s)	0.1	10	400	1700
Cost	Lowest	Low	Higher	High
Noise immunity	Poor	Intermediate	Good	Immune
Security	Lowest	Fair	Improved	Very high
Maintenance cost	Lowest	Low	Higher	High
Equipment complexity	Lowest	Low	High	Intermediate
Maintenance skill	Lowest	Low	High	High
Other	High transmission loss		Conduit somtimes required. MODEMs required	Small size. Point–point links. Difficult to repair

Figure 13.25 Unbalanced transmission link

Figure 13.26 Balanced transmission link

Figure 13.25. The difference in ground potential (ground loop) between the transmitter and receiver may introduce noise into the system. The balanced link is depicted in *Figure 13.26*. The differential amplifier configuration will eliminate most ground loops as well as reduce electromagnetic noise in the signal.

The theoretical analysis of communications networks is closely related to systems theory analysis of manufacturing systems. Data networks transfer information at much greater rates than typical manufacturing systems, increasing the possibility for errors. Deliberate redundancy and inefficiency may be required in a data transmission system to ensure reliable transmission.

13.6.3 Network topologies

The topology of a network is functionally identical to the architecture of a computer. It represents the physical map of the data links in the network. A few of the common network topologies such as the star, distributed or ring configurations are illustrated in *Figure 13.27*. The star version of a network

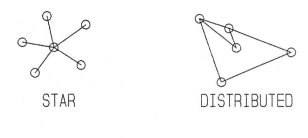

STAR   DISTRIBUTED

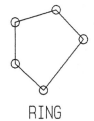

RING

Figure 13.27 Typical network configurations

requires all communications to travel through a central element. A distributed version allows for alternative communication paths between elements. The ring version of a network requires communication through a circle of elements, in effect through all elements in the path between the transmitting and receiving elements.

Networks are composed of *nodes* (the elements described above), with *circuits*, *channels*, or *links* connecting them. A node could vary from a small piece of computer hardware up to a large computer within the network. The channels could consist of hardwire connections, microwave link ups, satellite links, radio links or cable television links. However, the method of linking is usually transparent to the user. Messages between two modes are usually split into small *packets* of finite length. Packets may vary from 1000 to 2000 bits in length.

It is likely that a manufacturing system will incorporate several different styles of networking to maximise the transmission characteristics of each network for the particular application. The linear networks resemble data bus lines within the computer itself. The possibility for possible data collision is higher than that for loop networks. The mixing of these different systems is commonly referred to as an 'irregular tree network'.

13.6.4 Network protocols

Logic protocols are used with the physical network topology to prevent data collision on the shared data links. One of the primary factors limiting the efficient development of data networks is the disagreement between transmission standards. Some of the standards in current use are listed in *Table 13.9*.

Table 13.9 Data transmission standards

Standard	Description
IEEE-488 (GPIB)	Bus defining physical and data-link control layer; relates to digital communication between computers and various laboratory instrumentation and equipment
RS-232C	EIA standard with several configurations; serial interface between computers and peripheral equipment; uses bit parallel data transmission
RS-366	Standard governing computer-controlled phone dialling
RS-408	EIA standard for numerical control interface
RS-422	EIA standard for balanced voltage digital interface
RS-423	EIA standard for unbalanced voltage digital interface
RS-449	EIA standard for general purpose digital interface; intended improvement to RS-232C
802	Specifies physical and data link layers
802.3	Bus using CSMA/CD method; used in Ethernet applications
802.4	Bus using token passing technique
802.5	Ring (loop) bus using token passing technique
802.6	Bus using MAN access technique

EIA, Electronics Industry Association.

Most of the current network protocol development is intended for local area networks (LAN). A LAN is normally considered to be a network within a single building or close group of buildings. A local area network (LAN) is an intermediate speed data link for restricted distances. Networks with longer data links are classified as wide area networks (WAN). The transmission media required for these long-distance networks are different; however, much of the communication protocol may be the same.

Most current LAN development is based on the International Standards Organization (ISO) Open Systems Interconnection (OSI) model. This model is composed of seven layers which describe the physical and logical nature of the system.

(1) The physical layer describes the physical transmission medium: wires, cables, fibre optics, etc.
(2) The data layer describes the protocol involved with the flow of data in the LAN and ensures correct transmission.
(3) The network layer determines which physical link should be used in a transmission.
(4) The transport layer performs data transfer and error recovery; it ensures reliable transmission.
(5) The session layer provides for access to resources between links.
(6) The presentation layer regulates the appearance and format of the data.
(7) The application layer governs the manner in which programs interact with the network system.

One of the more common hardware connection standards in current use is the RS-232C. This standard was one of the first proposed by the Electronics Industry Association (EIA). It is normally configured as the serial interface between a computer and its peripheral equipment. A typical pin configuration for the RS-232C is shown in *Figure 13.28*. The RS-232C standard has shown several limitations, including distance limitations (normally 50 ft without line loss), transmission speed limitations (20 000 bit/s), configuration inconsistency, limited data bandwidth, cross-talk between data lines, and electromagnetic and radiofrequency (RF) interference.

The EIA RS-449, RS-422A and RS-423A are newer standards intended to alleviate these problems with the RS-232C.

These standards specify a 37-pin connector (D-37). There are separate provisions for balanced and unbalanced transmissions.

The IEEE-488 bus is also referred to as the general purpose interface bus (GPIB). This bus has been used extensively in recent years for interfacing to various types of instrumentation.

13.6.5 Typical networks

Typical networks in wide use today are described below.

Arpanet This is a widespread network based throughout the USA. It was originally developed by the US Defense Advanced Research Projects Agency (DARPA). This network consists of over 100 nodes which act as hosts for connection on a regional basis. Each host node is called an 'interface message processor' (IMP) and can serve several connections to the network. Interconnection between IMPs occurs at a variable rate between 50 and 1000 kbit/s. A specialised Network Control Center has been established to monitor failures in communication within this distributed network and can also measure traffic volume throughout the network.

Telenet This is designed similar to Arpanet as a distributed system. There are nine major nodes and hundreds of minor nodes. Many of the local connections use time-sharing operations. Access exists throughout the USA and in many other countries. Telenet is a public, packet switching network.

Prestel This is a network constructed by the British Post Office which uses domestic television sets for local displays and telephone lines for communication outwards to the network. Thus, the input rate to the network (75 bit/s) and the output rate from the network (1000 bit/s) are quite different. Various news agencies and advertisers access the network in order to send information into thousands of homes through a variety of databases.

Euronet This provides information retrieval services to European countries using databases located at several host

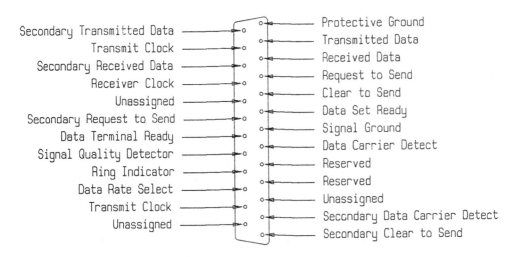

Figure 13.28 EIA RS-232C pin configuration

organisations. This packet-switched network uses 48 kbit/s lines. A French version of this network is named *Transpac*.

Ethernet This is a local area network developed by the Xerox Corporation. It links various nodes within a building or in relatively close proximity via coaxial cables.

In addition, a wide variety of PC based networks have been developed in recent years with the advent of high speed, large storage microcomputers in the 286 and 386 class. Such networks require a large memory unit to serve as a central processor, which is hardwire connected to other units, usually in a star-type network.

13.7 Databases

The term 'database' refers to a collection of stored data which contains some associations or relationships between all or any parts of the data. In general, this collection of data is very large, thus necessitating the use of computer access as opposed to manual data handling. The computer program which controls data input to the database, user access to the database, calculations conducted with data, and output from the database is called a *database management system* (DBMS).

The smallest set of related data which is stored in a database is called a *record*. A record may have very few or many data elements (words, numbers, or lines of information) which are often called *fields*; however, all of these elements or fields should be related in some fashion in order to exist within the same record. An example might be a listing of personnel information for an employee including name, address, job title, salary, social security number, etc. Each item of data occupies a single field within the overall record.

A set of records is called a *file*. In this example, a file would contain all employee personnel records, and might be entitled 'employee personnel data'. The database may contain other files which contain other relevant information regarding the operation of a company. A database may be accessed through a single user environment, such as a PC, or through a multi-user environment such as a network terminal. In the latter case, the database may be shared among several users simultaneously, each performing a different calculation or function with the same set of data.

Databases can be used to generate mailing lists of clients, store personnel records, store and update inventory information, maintain equipment maintenance schedules, handling job cost accounting, and storing look-up tables for various calculations and manufacturing functions. Databases can be operated on a variety of computer types and sizes from mainframes to microcomputers. However, due to the usually large amount of data which is present in a database, it is more appropriate to store the data in a computer with appropriate memory capacity. This would suggest that a minicomputer or mainframe system is most appropriate for database management. However, with proper data communication and/or networking capabilities, it is possible to access a mainframe-based database with a microcomputer or remote terminal.

The database management system is used to provide several key functions. These include

(1) the organisation of data into a variety of requirements as specified by each and every user of the database;
(2) the output of data to a variety of output peripheral devices and in a variety of configurations which can be user specified;

(3) the ability to perform mathematical computations and various sorting functions with all or part of the data; and
(4) the control of database access and sharing among many users.

Because database management systems must provide a variety of features to many users, such software systems are usually quite large and cumbersome. Many such systems require computer programming in special languages that are specific to a given database program. These languages are not usually similar to any standard programming language or operating system. Thus, it may be difficult to learn to develop a database. However, once the database program has been written, it is usually easy to *use* the database, because the user is not aware of the magnitude of the source code or the specific method of database operation. The speed of operation of a database is dependent on the type of database and the number of records which are being accessed.

Database programs may be one of three different categories: (1) hierarchical, (2) network, or (3) relational. A hierarchical database contains a tree-like structure, so that data queries are routed up and down the tree to the final point of selection. This kind of data structuring is very natural and easy to understand. Data queries are accomplished rapidly and the storage of data is managed efficiently in computer memory. However, in order to use a hierarchical database, it is usually necessary to know about the entire tree structure of the data, which may be a burden to the user. The mainframe database manager *Focus* uses a hierarchical database system, as does the micrcomputer version *PC Focus*.

The *network* version of a database management system uses mathematical set theory and is capable of managing very large sets of data. However, this type of database is very difficult to develop, because the relationship between the various data elements is extremely complex. Each set of data has individualised ownership and membership characteristics within the network. This type of database system is difficult to conceive and plan, and is extremely difficult to alter after the system has been developed. However, a network type of database management system is extremely powerful and is well suited for large-scale applications. The *MDBS* series of database programs is the best known of the network database managers.

The *relational* version of database management systems allows concurrent access to multiple data files which are related through common fields of data. Thus, such a system is designed to allow the user to collect and compare data from various records and files. A limitation to the relational database management system is that such comparisons usually require initial data field entry to be consistent among all records and files. The most powerful and well known of the relational database systems is *DBASE*, which has been developed for minicomputers and microcomputers. However, such a system can also be operated in a mainframe computer environment.

The development of database systems for manufacturing and business environments has been widespread. However, such development is an extensive task requiring significant programming experience and proper planning for the implementation and usage of such a system. Often, the selection of a particular database management system will impact the selection of computer hardware and associated interface devices. Thus, the implementation of a database in a manufacturing environment is manpower and hardware intensive, with the development of a database requiring up to 1 year to complete. The actual cost of the database management software is usually the least expensive component of the planning and development process, compared with the programming manpower and hardware purchase components.

13.8 Knowledge-based systems

A knowledge-based system (KBS) is a computer program which stores information relevant to a particular application area and can be used to solve problems in that area in an intelligent manner. Such a system is also called an *expert system* and is sometimes classified under the general heading of *artificial intelligence*. A knowledge-based system contains two major elements: a *knowledge base*, which contains the storehouse of knowledge in a particular area in the form of facts and rules, and an *inference procedure* or *inference engine*, which operates on the knowledge base and performs logical inferences and deductions leading towards a final solution to a posed problem. The use of the term 'expert system' refers to the capacity of a knowledge base to simulate the knowledge of a supposed human expert in the field relevant to the problem at hand. A knowledge-based system with no particular application is called an *expert system shell*.

A knowledge-based system should be oriented to a single application, as opposed to a standard computer language which has more generic functions. A knowledge-based system is 'rule based', meaning that the data are manipulated according to the rules established for a particular application solution (IF...THEN). These rules may be used in an ever-changing order, depending on the information received by the user and interpreted by the knowledge base. This differs from standard computer languages which operate on data in a sequential stepwise manner and is generally 'data driven'. A standard computer program uses easily definable algorithms, while an expert system uses less definable inference techniques.

A knowledge-based system should also contain the following features.

(1) It should be user friendly. The major purpose of a knowledge-based system is to impart expert knowledge towards the solution of a problem which can be used by a non-expert. In fact, its final use is likely intended for a computer novice who may have little understanding of the original design of the software. Thus, the dialogue between the program and the user should allow the user to be directed through the program operation towards a realistic solution of a particular problem. In a manner of speaking, the knowledge base actually gives advice to the user leading towards the solution of a problem.

(2) The method of arriving at a solution in terms of 'how the expert thinks' should be easily interpreted and understood by the user. It is important for the user to believe that the program will arrive at the most logical solution to a problem. The knowledge-based system should be able to describe to the user how the inferences are being made at any point during the program.

(3) The knowledge base should be easily updated, because the knowledge in a particular field will probably advance with time.

Knowledge-based systems usually involve on-line interaction with a user with the aid of a terminal, because the user is often queried continually by the knowledge base in an effort to arrive at the proper deductions. This is similar to a doctor asking a patient questions regarding medical history and symptoms before arriving at a diagnosis.

Specialised computer languages have been created for the development of expert system shells, which can then be used to create specific expert systems or knowledge bases for a particular application area. The most common of these languages are *LISP* and *PROLOG*. These languages involve the use of symbolic processing which uses manipulation of abstract quantities such as words, phrases and terms relevant to the application area in question.[33] Although the terms and expressions in expert system languages are relatively limited and simple, they are quite different from standard computer language expressions and terms. Thus, the development of an expert system usually requires a computer programmer with specific knowledge of expert system development.

In order to implement an expert system, it is necessary to obtain a software system which allows the development of expert system shells, from which particular expert systems can be developed for individual applications. It is important to note that each expert system is specific to a given application area or area of particular expertise. In contrast, the expert system shell is a somewhat generic program that can be used in the creation of many application-specific expert systems.

The use of expert systems usually requires specialised computers in order properly to interpret the rule-based methodology of a knowledge base. Only within the last decade have such hardware and software systems been significantly developed to the point that they have received widespread usage. During this period, expert systems have been used in manufacturing and industrial environments in ever-increasing numbers.[34] Expert systems have been used in the analysis and troubleshooting of electronic circuits, in the selection of appropriate machine parts in an assembly operation, in the analysis of graphic images, and in the planning of facilities design.[34] The costs associated with implementation of an expert system include the likely purchase of a specialised minicomputer or super-microcomputer and associated expert system shell software, and the acquisition of programming expertise that is specific to expert system development.

References

1 RITCHIE, D. M. and THOMPSON, K., The Unix time sharing system, *Commnications of the ACM*, **17**(7), 365–375 (1974)
2 MEALY, G. H., WITT, B. I. and CLARK, W. A., The functional structure of OS/360, *IBM Systems Journal*, **5**(1), 3–51 (1966)
3 NOYES, J., The QWERTY keyboard: a review, *International Journal of Man–Machine Studies*, **18**, 265–281 (1983)
4 SEIBEL, R., Data entry devices and procedures, in H. P.Van Cott and R. G. Kinkade (Eds), *Human Engineering Guide to Equipment Design*, US Government Printing Office, Washington, DC, 311–344 (1972)
5 TAYLOR, R. M. and BERMAN, J. U. F., Ergonomic aspects of aircraft keyboard design: the effects of gloves and sensory feedback on keying performance, *Ergonomics*, **25**, 1109–1123 (1982)
6 NORMAN, D. A. and FISHER, D., Why alphabetic keyboards are not easy to use; keyboard layout doesn't matter, *Human Factors*, **24**, 509–519 (1982)
7 POLARD, D. and COOPER, M. B., The effect of feedback on keying performance, *Applied Ergonomics*, **10**, 194–200 (1979)
8 OHLSON, M., System design considerations for graphics input devices, *Computer*, **11**, 9–18 (1978)
9 RITCHIE, G. J. and TURNER, J. A., Input devices for interactive graphics, *International Journal for Man–Machine Studies*, **7**, 639–660 (1975)
10 SOMERSON, P., The tale of the mouse, *PC Magazine*, **1**(10), 66–71 (1983)
11 WARFIELD, R. W., The new interface technology: an introduction to windows and mice, *Byte*, **8**(12), 218–230 (1983)
12 BURKE, D. and GIBBS, C. N., A comparison of free moving and pressure levers in a positional control system, *Ergonomics*, **8**, 23–29 (1965)
13 ALBERT, A. E., The effect of graphic input devices on performance in a cursor positioning task, *Proceedings of the 26th Annual Meeting of the Human Factors Society*, Seattle, Washington, 54–58 (1982)
14 CARD, S. K., ENGLISH, W. K. and BURR, B. J., Evaluation of mouse, rate controlled isometric joystick, step

keys and text keys for text selection on a CRT, *Ergonomics*, **21**, 601–613 (1978)

15 PFAUTH, M. and PRIEST, J., Person–computer interface using touch screen devices, *Proceedings of the Human Factors Society 25th Annual Meeting*, Santa Monica, California, 500–504 (1981)

16 MIMS, F. M., A few quick pointers, *Computers and Electronics*, 64–117 (May 1984)

17 WHITFIELD, D., BALL, R. G. and BIRD, J. M., Some comparisons of on-display and off-display touch input devices for interaction with computer generated displays, *Ergonomics*, **26**, 1033–1053 (1983)

18 FOLEY, J. D. and WALLACE, V. L., The art of natural graphic man–machine conversation, *Proceedings of IEEE*, **62**, 462–471 (1974)

19 SHEA, C. H. and MILLER, G. E., Eye tracking control for robotic systems, in *Advanced Topics in Manufacturing Technology*, ASME, New York, 39–48 (1987)

20 CALHOUN, G. C., ARBAK, C. L. and BOFF, K. R., Eye controlled switching for crew station design, *Proceedings of the Human Factors Society 28th Annual Meeting*, Santa Monica, California, 258–262 (1984)

21 MILLER, G. E., ETTER, B. D. and SHEA, C., Voice control of manufacturing systems, in *Advanced Topics in Manufacturing Technology*, ASME, New York, 35–38 (1987)

22 DAUTRICH, B. A., RABINER, L. R. and MARTIN, T. B., On the use of filter bank features for isolated word recognition, *Proceedings of the IEEE International Conference on Acoustics, Speech, and Signal Processing,* 1061–1064 (1983)

23 HATON, J. P., Speech recognition in Western Europe, in W. A. Lea (Ed.), *Trends in Speech Recognition*, Prentice-Hall, Englewood Cliffs, NJ, 512–526 (1979)

24 LEA, W. A., Towards versatile speech communication with computers, *International Journal of Man–Machine Studies*, **2**, 107–155 (1970)

25 DODDINGTON, G. R. and SCHALK, T. B., Speech recognition: turning theory into practice, *IEEE Spectrum*, 26–32 (Sep. 1981)

26 LEA, W. A., Evidence that stressed syllables are the most readily decoded portions of continuous speech, *Journal of the Acoustic Society of America*, **55**, 410A (1973)

27 PICKETT, J. M., Effects of vocal force on the intelligibility of speech sounds, *Journal of the Acoustic Society of America*, **28**, 902–905 (1956)

28 FLANAGAN, J. C., Automatic speech recognition in severe environments, *Final Report of the Committee on Computerized Speech Recognition*, National Research Council, Washington, DC (1984)

29 POLLOCK, I. and PICKETT, J. M., Masking of speech by noise at high sound levels, *Journal of the Acoustic Society of America*, **30**, 127–130 (1958)

30 MARTIN, T. B., Practical applications of voice input to machines, *Proceedings of IEEE*, **64**(4), 487–501 (1976)

31 MARTIN, S. B. and WELCH, J., Practical speech recognizers and some performance evaluation parameters, in W. A. Lea (Ed.), *Trends in Speech Recognition*, Prentice-Hall, Englewood Cliffs, NJ, 24–38 (1980)

32 WOODARD, J. P. and LEA, W. A., New measures of performance for speech recognition systems, *Proceedings of the IEEE Conference on Acoustics, Speech, and Signal Processing*, **9**(4), 1–4 (1984)

33 CHADWICK, M. and HANNAH, J., *Expert Systems for Microcomputers*, Tab books, Blue Ridge Summit, PA, 31–44 (1987)

34 KRIZ, J., *Knowledge Based Expert Systems in Industry*, Ellis Horwood/Wiley, Chichester/New York, 11–16 (1987)

Further reading

Computer architecture

GORSLINE, G. W., *Computer Organization Hardware/Software*, Prentice-Hall, Englewood Cliffs, NJ (1980)

HAYES, J. P., *Computer Architecture and Organization*, McGraw-Hill, New York (1978)

LOREN, H., *Introduction to Computer Architecture and Organization*, Wiley, New York (1982)

MYERS, G. J., *Advances in Computer Architecture*, 2nd edn, Wiley, New York (1982)

RALSTON, A. and REILLY, E.D., *Encyclopedia of Computer Science and Engineering*, Van Nostrand Reinhold, New York (1983)

WALKER, R. S., *Understanding Computer Science*, Texas Instruments Corp., Dallas, TX (1984)

Operating systems

PETERSON, J. and SILBERSCHATZ, A., *Operating System Concepts*, Addison Wesley, Amsterdam (1984)

RALSTON, A. and REILLY, E. D. (Eds), *Encyclopedia of Computer Science and Engineering*, Van Nostrand Reinhold, New York (1983)

Computer languages

HELMS, H. L., *Computer Language Reference Guide*, 2nd edn, Sams & Co., Indianapolis (1984)

HOROWITZ, E., *Programming Languages: A Grand Tour*, 3rd edn., Computer Science Press, Rockville, MA (1987)

RALSTON, A. and REILLY, E. D. (Eds), *Encyclopedia of Computer Science and Engineering*, Van Nostrand Reinhold, New York (1983)

Visual displays, keyboards, mice, touch screens, graphics tablets

ANON., 36 High Tech Input devices, *PC Magazine*, **6**(14), 95–202 (1987)

BOFF, K. R. and LINCOLN, J. E. (Eds), *Engineering Data Compendium: Human Performance and Perception*, USAF Armstrong Medical Research Laboratory, Wiley, New York (1988)

RUBINSTEIN, R. and HIRSH, H. M., *The Human Factor: Designing Computer Systems for People*, Digital Press, Burlington, MA (1984)

SALVENDY, G. (Ed.), *Handbook of Human Factors*, Wiley Interscience, New York (1987)

VAN COTT, H. P. and KINDALE, R. G. (Eds), *Human Engineering Guide to Equipment Design*, revised edition, US Government Printing Office, Washington, DC (1963)

Speech recognition systems

ANON., Understanding the Masters Voice, *PC Magazine*, **6**(18), 261–305 (1987)

BRISTOW, G. (Ed.), *Electronic Speech Recognition Techniques, Technology, and Applications*, McGraw-Hill, New York (1986)

LEA, W. A. (Ed.), *Trends in Speech Recognition*, Prentice-Hall, Englewood Cliffs, NJ (1980)

LEVINSON, S. E. and LIBERMAIN, M. Y., Speech recognition by computer, *Scientific American*, **4**, 64–76 (1981)

Networks and data communications

DERFLER, F. J., Software connectivity's new frontier, *PC Magazine*, **7**(11), 93–278 (1988)

FRIEND, G. E., FIKE, J. L., BAKER, H. C. and BELLAMY, J. C., *Understanding Data Communications*, Texas Instruments Corp., Dallas, TX (1984)

Local Networks: Distributed Office and Factory Systems, Proceedings of the Localnet 1983 Conference, Online Publications Ltd, Pinner

MAYNE, A. J., *Linked Local Area Networks*, 2nd edn, Wiley, New York (1986)

RORABAUGH, B., *Data Communications and Local Area Networking Handbook*, Tab Books, Blue Ridge Summit, PA (1985)

Databases

ANON., Programmable databases, *PC Magazine*, **7**(9), 93–306 (1988)

DATE, C. J., *An Introduction to Database Systems*, Addison Wesley, New York (1976)

DEAKIN, R., *Data Base Primer*, New American Library, New York (1984)

HSIAO, D. K., *Advanced Database Machine Architecture*, Prentice-Hall, Englewood Cliffs, NJ (1983)

RUMBLE, J. R. and HAMPEL, V. E. (Eds), *Database Management in Science and Technology*, North-Holland, Amsterdam (1984)

Expert and knowledge-based systems

BRYANT, N., *Managing Expert Systems*, Wiley, New York (1988)

CHADWICK, M. and HANNAH, J., *Expert Systems for Microcomputers: An Introduction to Artificial Intelligence*, Tab Books, Blue Ridge Summit, PA (1987)

KRIZ, J., *Knowledge-Based Expert Systems in Industry*, Wiley/Horwood, New York/Chichester (1987)

14 Computers in Manufacturing

J Hunt

G M Mair

Contents

14.1 Introduction

Computers are now used in practically all the activities within a manufacturing environment and this chapter covers the major areas of the application of computer systems in the design and production of manufactured goods. The majority of departments in manufacturing companies will have already implemented stand-alone computer systems in support of their own work. Whilst the separate application of computer power has brought about many advantages, it has created 'islands of automation' and has confirmed traditional functional demarcations because the different computer systems and software could not communicate with each other.

The increasing power of computer hardware and software has promoted the development of computer-integrated manufacture (CIM). The aim of CIM is the integration of separate functional systems into a single plant or enterprise-wide system. The greater integration of all manufacturing activities is the key to future competitiveness and even company survival. It is the route that leading companies are already pursuing, and the majority will have to follow to some degree in order to be successful.

In this chapter the term 'manufacturing' is used in relation to the wider scope of the complete manufacturing organisation including the indirect support activities. 'Production' is used in relation to the specific role of the shop-floor departments in making the product.

14.2 Computer-integrated manufacturing

14.2.1 The manufacturing organisation and CIM

'Computer-integrated manufacturing' (CIM) is a general term used to describe the bringing together and integration of the many computer-based activities used in manufacturing. It is not a single technique or application program. The objective of CIM is to control and coordinate the activities of a manufacturing organisation via the use of a single integrated computer system. To do this it must integrate or oversee the many separate computer systems used in manufacturing today.

The separate computer based systems already in major use are materials requirements planning (MRP) and computer-aided design (CAD), engineering (CAE), manufacturing (CAM), test (CAT) and production management (CAPM) applications. Some integration is already occurring as computers and software become more powerful and cheaper. The relationships and development path of separate systems toward full CIM is shown in *Figure 14.1*. Early separate programs are now being developed as modules in integrated suites of manufacturing software. Numerical engineering in the form of CAM is combining with CAD to give CADCAM. Computerised numerical control (CNC) and direct numerical control (DNC) have been combined with planning software to allow the development of the integrated control systems needed to

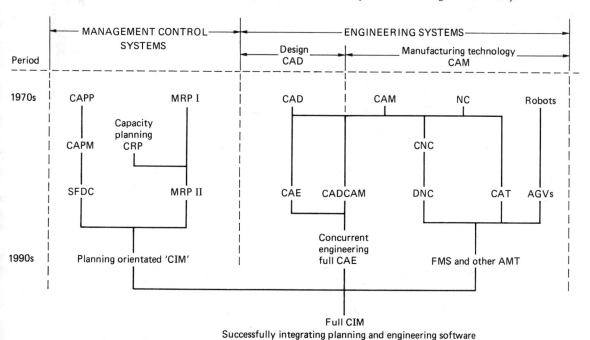

Figure 14.1 Progression of stand-alone computer-aided technologies to full CIM. The figure shows the general application of computer-aided technologies by function and over time. It is a simplified statement and does not show the gradual introduction and continuous development of the technologies over the time period covered.

AGVs	automated guided vehicles	CNC	computerised numerical control
AMT	advanced manufacturing technology	CRP	capacity requirements planning
CAD	computer-aided design	DNC	direct numerical control
CADCAM	computer-aided design and manufacture	FMS	flexible manufacturing systems
CAE	computer-aided engineering	MRP I	materials-requirements planning
CAM	computer-aided manufacturing	MRP II	manufacturing-resource planning
CAPM	computer-aided production management	NC	numerical control
CAPP	computer-aided process planning	SFDC	shop floor data collection
CAT	computer-aided test		

run flexible manufacturing installations. MRP, capacity requirements planning (CRP) and other planning programs have been integrated as MRP II.

14.2.2 CIM systems

Computer-aided systems in manufacturing fall into two main groups: manufacturing control systems and engineering systems.

Manufacturing control systems (also known as computer-aided production management (CAPM) or manufacturing information systems (MIS) support the management, planning, scheduling, inventory control and financial functions of a manufacturing unit, i.e.:

(1) manufacturing management controls,
(2) financial controls and budgets,
(3) sales-order control,
(4) production planning and control,
(5) inventory control and purchasing, and
(6) shop-floor management and scheduling.

Engineering systems (computer-aided engineering (CAE) in the widest sense) support the technical functions which process the engineering data relating to the product and its manufacture. Product designs and process control are major features of this software, i.e.:

(1) mechanical product design and engineering analysis (mechanical CAD (MCAD))
(2) electronic product design and engineering analysis (electronic CAD (ECAD)),
(3) manufacturing and process engineering (CAM),
(4) numerical control and technology robotics (CNC/DNC),
(5) quality systems and engineering—automatic testing (CAT), and
(6) software engineering (computer-aided software engineering (CASE)).

Many systems are currently described as being CIM systems, implying wide integration, when in fact they only cover a proportion of the total manufacturing environment—usually within either the management or engineering fields with some limited integration between the two. There are few examples of complete corporate or manufacturing unit-wide CIM systems; this is due to three major reasons.

(1) The necessary powerful and all-embracing integrating software is still being developed.
(2) The cost of such systems is high because the programs are large and complex. In addition, the software ideally needs to be tailored to the particular organisation, products and processes.
(3) The commercial risks involved in changing from a set of separate but functioning computer systems to one central system are very high. Companies are, therefore, cautious of wide-scale CIM implementation on both operational and financial grounds.

There are, however, a number of CIM installations that have proved successful in the integration of the activities within their scope. The subintegration of planning activities on the one hand and engineering activities on the other, has proceeded through the 1980s and produced many successful installations. The key and most difficult integration remaining is that of the management and engineering databases and operational programs.

14.2.3 Selecting a CIM system

A CIM project must be a matter of manufacturing strategy set by senior executives, because the cost and effect on the performance of the company in either success or failure will be significant. Company policies on matters such as attaining world-class manufacturing (WCM) standards, reorganisation into 'flatter' team/product organisation structures, product developments, just-in-time (JIT) working, sourcing policies and future manufacturing locations will have a bearing on the final CIM solution.

The introduction of CIM should be a catalyst for simplifying and restructuring the organisation and operations, as this is likely to bring greater benefits than just the integration, automating or updating of existing practices. When considering the introduction of a CIM system, it must be realised that it will affect every activity and piece of data produced by the operating areas within its scope. This means that every activity and operational requirement should be reviewed.

Check list for a CIM project

(1) Appoint a project team and leader responsible to a senior executive. Set clear objectives for the project scope and team's responsibilities.
(2) Collect a list of desired applications and their advantages from operational managers *and* staff.
(3) Use a series of review teams (having both producers and customers of the data and outputs) to challenge and rank the applications on the basis of:
  (a) data sources and form of input and lead times,
  (b) data destinations and form of output and lead times,
  (c) interrelation of data sets and criticality to each other,
  (d) effect of data loss, corruption or delays,
  (e) security of and access to data, and
  (f) other operational benefits and disadvantages.
(4) Produce an agreed applications list ranked in criticality to the total operation and the costs of success, failure and non-implementation.
(5) Review the labour and skills requirements for both implementation and ongoing support of the new system—operational staff, support staff needed for hardware, software applications and software systems support.
(6) Review additional hardware likely to be needed, and locations.
(7) Review the *practical* lead times for the project implementation and phase out of existing systems—this needs a great deal of care.
(8) Write a system specification based on the above considerations.
(9) Obtain proposals and tenders from a number of system vendors.
(10) Hold a review with the two or three most likely vendors to ensure that they:
  (a) fully understand the specification;
  (b) fully meet the specification for software performance;
  (c) have an adequate specification of the hardware needed;
  (d) have real experience in the specific area of application, i.e. can provide good training for operator, supervisors and support;
  (e) can meet the desired and/or proposed timetable—be particularly concerned with time needed to write/tailor and prove software;

(f) can provide good support during implementation;
(g) can supply sufficient and clear support doumentation;
(h) can provide good support during service at a reasonable cost (N.B. software support is more critical than hardware support).

(11) Visit other vendor installations before selecting the best vendor.
(12) Revise your implementation and running costs on the basis of the vendor reviews in which you will have learnt a great deal of what is really needed.
(13) Only now put forward the expenditure application for final approval.
(14) Agree a contract with non-performance penalties for the vendor in operation and support as well as implementation.
(15) Appoint a joint project team with the vendor and have corresponding members from both sides for your own security and continuity.
(16) Run a pilot scheme to test major assumptions and critical areas.
(17) Hold regular project meetings and reviews for senior management and users.

14.3 Manufacturing control systems

Computer based manufacturing control systems normally include a suite of integrated software modules for the management and planning of production and inventory and the processing of related financial and engineering data. The degree of integration between the modules varies and a prospective purchaser must ensure that the modules offered provide a smooth continuity of operation within the particular manufacturing environment.

The major software applications and modules in use are:

Materials-requirements planning (MRP)
Manufacturing-resource planning (MRP II)
Capacity requirements planning (CRP)
Optimum production technology (OPT)
Shop floor data collection (SFDC)
Simulation programs

14.3.1 Materials requirements planning

14.3.1.1 Background

MRP has been the major module used in computer based manufacturing control over the last 20 years. The data sets needed to control the material flows of a manufacturing unit are fundamental to the operation of the unit and the data are produced or used by all the core departments in the unit. Hence a MRP system will occupy a pivotal position in the control and planning of manufacturing. This role is illustrated in *Figure 14.2*.

Even a small company or manufacturing plant may have over 5000 part numbers to handle, and large units may have to deal with 50 000 part numbers. This represents a large, constant workload in deciding, purchasing and controlling the flow of production materials in a factory. A large amount of data must be processed for each part number on a regular basis, creating a significant data processing workload. This task, which is common to every manufacturing unit, was an obvious early and commercial candidate for automation on a computer.

The programming techniques developed to handle this task are called materials requirements planning (MRP). This is now sometimes called MRP I to distinguish it from a later related, but broader, development called 'manufacturing resource planning' (MRP II) (see Section 14.3.2).

Because of the large amount of data processing and memory needed for MRP, most early systems were installed on the company or factory mainframe or mini computer which had already been installed for financial and sales work. With the dramatic increase in the processing power of supermicros (the 80386 and 80486 machines) the reduction in memory costs and the introduction of cheap local area networks (LAN), it is now possible to run all but the largest MRP programs on an appropriate desk-top hardware system.

14.3.1.2 System operation

Types of data needed Two types of data are needed: static and dynamic data.

Static data are data that are standard for some time. Such data comprise:

(1) Details of the product structure, parts lists and material specifications, as produced by the design office. These are entered into the MRP database either manually or, increasingly, automatically from CAD.
(2) Order rules and production and purchasing lead times for each part number.
(3) Shop-floor department numbers and material routings.
(4) Factory standard costs and times; purchase costs.
(5) Stock location numbers in stores (if an appropriate location system is used).

Dynamic data are data that change with sales demand and production ouput.

(1) A master production schedule (MPS) is entered from production control. This specifies the gross output required and is derived from a sales forecast that normally looks at least 3 months forward and often up to 1 year. In some cases the sales forecast may be taken directly.
(2) Stock levels—opening, closing, issued, committed, free, safety, production, spares, repairs.
(3) Completions—MRP will be sent reports on stores issues against requisitions for line stock or JIT, production completions, stores receipts, purchases receipts, work in progress (WIP) and scrap.

Calculations A new MPS gross requirement is entered periodically—usually weekly, bi-weekly or monthly, depending on the nature of the business. A MRP run is a large and lengthy computer operation and produces a large number of printed reports. Thus the frequency of the run must be the optimum for the rate of sales-demand changes, the response flexibility of production, the capacity of people and the system to handle large amounts of changing detail and the costs of doing so. The system calculates the nett requirements by taking account of total stock, stock already committed to orders, safety stock levels, scrap allowances and the resultant free stock quantities. Production and material purchase lead times are applied to arrive at completion due-dates and order-release dates.

Outputs A materials-requirements report is issued, normally to inventory control, purchasing and production control departments. Purchasing recommendation reports are issued to the purchasing department who will review the situation and initiate the release of purchase orders—either producing them manually or sanctioning the release from a computer based purchasing module linked to MRP.

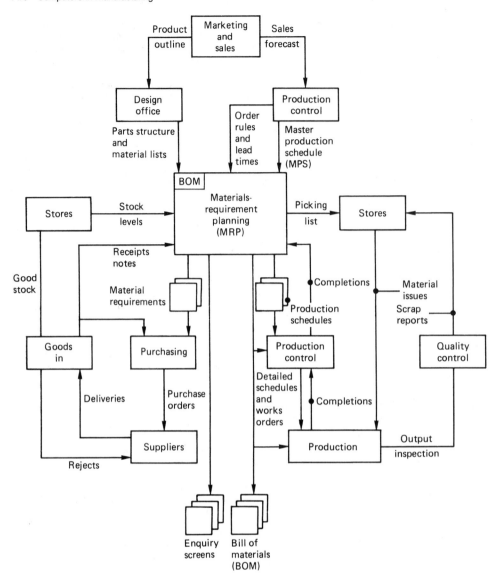

Figure 14.2 The pivotal role of MRP in the control and planning of manufacturing

Stock picking lists will be issued to the various stores. The actual quantity picked and shortfalls will be fed back.

A bill of materials (BOM) for each product and subassembly is produced. This reflects the order and method of production, which is usually different in structure from the parts list produced by the design office. This difference should decrease as the drive to closer integration and design for manufacture techniques takes effect.

Stock status, inventory value and many other reports are normally produced.

Data screens   There is a constant demand for instant access to data (particularly stock statuses and inventory costs) from a number of departments, i.e. inventory control, purchasing, production control, production, manufacturing engineering,

design and finance. Consequently, it is necessary to have data available at terminal monitors in these departments. The data will be set out on a number of 'screens' that detail specific areas of data. The screens are related in a formal structure and are usually accessed via a menu and 'next screen' facility. A typical stock status screen is shown in *Figure 14.3*.

Personal access to read most data is normally unrestricted or moderately restricted by password because a large number of people need access. Conversely, the ability to modify data will be strictly controlled and only possible at certain terminals—normally within the 'owning' departments of inventory control and/or production control. Terminals will be connected to the computer holding the processing and central memory by a network. Most read-only outlets will be 'dumb terminals' at least insofar as the MRP application is concerned.

Part No.	abc12345		Date 11/05/92
Revision	3		Time 15:03:00
Effective	10/4/92		

Used In	xyz5677	xyz5678	xyz5679
Qty per part	4	4	2

				p
Material Part No.	AL6789	Labour cost		24.123
Revision	2	Material		10.123
Effective	20/10/89	Overhead		5.123
Annual demand	8000	Standard cost		39.369

Stock status

Total		2000	
Committed – Production	1000		
– Spares	200		
– Repairs	100		
Total		1300	
Free		700	
On order		1000	due 12/6/92
Covered		1700	

Demand

Period—months	5	6	7	8	9	10
Qty	500	600	1000	900	800	800
Covered	500	600	600			
To cover			400	900	800	800

Issues

		Qty	Date	Order No.
This period	1	100	10/4/92	456789
	2	200	20/4/92	456789
	3			
	4			

Last period	600
Year to date	2400

Figure 14.3 Typical stock status screen

14.3.2 Manufacturing resources planning (MRP II)

Materials requirements planning (MRP) systems have a major limitation in that, having determined the demand in terms of part numbers, volumes and material required, they do not address the next major issue, i.e. the capacity of the factory to meet the demand. If there is a significant mismatch between the two factors, then MRP will produce an unobtainable target, which will result in wasted time and overstocking as the factory capacity fails to convert materials and parts into finished goods.

MRP II addresses the subject of capacity and financial planning and integrates the whole manufacturing control and planning environment from sales forecast to shop-floor control. In this sense it can be regarded as a major component of a CIM environment. The major elements of a MRP II system are shown in *Figure 14.4* and a hardware environment is shown in *Figure 14.5*. It can be seen that the MRP holds a central position, but the computer-based controls and planning are extended to cover:

(1) sales-order management,
(2) purchase-order management,
(3) works-order management,
(4) capacity requirements planning,
(5) shop-floor control, and
(6) cost accounting.

These are present in the form of software modules that are integrated with the necessary links to each other. The system has the capability of accepting a sales or business plan and developing a master production schedule from it. This would then drive a rough cut capacity plan to see if capacity matched demand. If the capacity is regarded as finite and is not sufficient, the original sales/business plan can be modified. Once a master production schedule has been accepted, the MRP and production control modules can proceed.

Surveys have shown that MRP II has met with some initial problems in a number of installations. There are a number of reasons for this, but one of the most common is the difficulties met in organisations having to absorb such a wide ranging

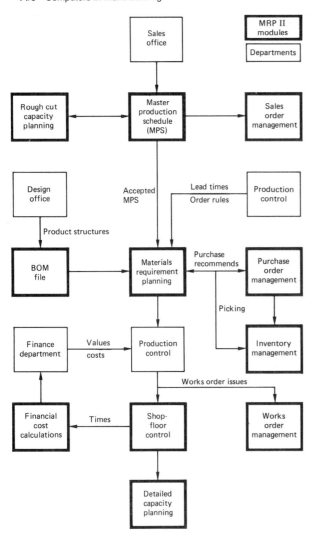

Figure 14.4 A MRP II structure. BOM, bill of materials

system and the many changes it enforces. Secondly, there has been (as with so many major projects) an underestimation of the time needed for successful implementation. Thirdly a reluctance by some customers to spend sufficient staff time on training in the new systems has decreased their success rate.

A major disadvantage with MRP II systems is that they are expensive because of their size and wide scope. The structure of these systems and their cost tends to retain centralisation of activities and decision-making in the traditional service departments. This runs contrary to current manufacturing philosophy which is to product-oriented working groups (i.e. group technology for a complete product and its support services) and 'flatter' structures. In this environment the decision-making and support services are being devolved to the semiautonomous product groups. Also the use of JIT is removing the need for in-depth capacity planning on a large-batch basis.

14.3.3 Capacity requirements planning

Some planning systems, including MRP, assume that infinite capacity is available. This is clearly not so, and at some point capacity requirements must be taken into account. This can be done by using capacity requirements planning (CRP) which is a module within MRP II and can also be used in conjunction with MRP.

CRP modules need data on the process routings, sequence, set-up times and process times in order to calculate the capacity situation. Advanced programs provide simulation and a 'what if' capability for capacity planning.

14.3.4 Optimised production technology

Optimised production technology (OPT), is both a measurement philosophy for managing a manufacturing unit and a proprietary scheduling technique.

14.3.4.1 Philosphy

The 'philosophy' is a method of defining the goals and measuring the performance of manufacturing units. It uses three parameters which, it is claimed, are more easily understood than traditional financial measurements. The OPT measurements are said to be more readily identifiable with the global aims of a manufacturing unit.

Traditional measure	OPT measure
Net profit	Throughput
Retain on investment (ROI)	Inventory
Cash flow	Operating expense

Throughput is the rate at which the manufacturing unit generates money from actual sales—it does not value production output (which would not highlight finished stock still held as inventory).

Inventory is a measure of raw material value only. Added value of manufacturing appears as operating expenses. Finished production unsold is also included.

Operating expense is the money spent to convert inventory (raw material) into throughput (sales revenue).

The goal of OPT is to increase throughput (sales) whilst decreasing inventory and operating expense. Supporters of the method believe that the measures give a clearer understanding of the overall goals of a manufacturing plant through the visibility of the measurement of conversion of raw materials to sales at minimum cost—something that can be lost in the use of traditional accounting methods. Ongoing improvements can thus be promoted and implemented more clearly when their performance is measured in this way.

OPT involves building a mathematical model of the factory from which simulation and scheduling can be created.

14.3.4.2 OPT scheduling

The OPT scheduling method is an alternative to the demand and due-date backward scheduling of CRP and MRP II. There are a number of rules to be followed, some of the most fundamental being as follows.

(1) Production flow not production capacity should be balanced.
(2) Capacities should be identified as 'constraining' or 'non-constraining' because the constrained resource will directly affect throughput (sales) performance. Con-

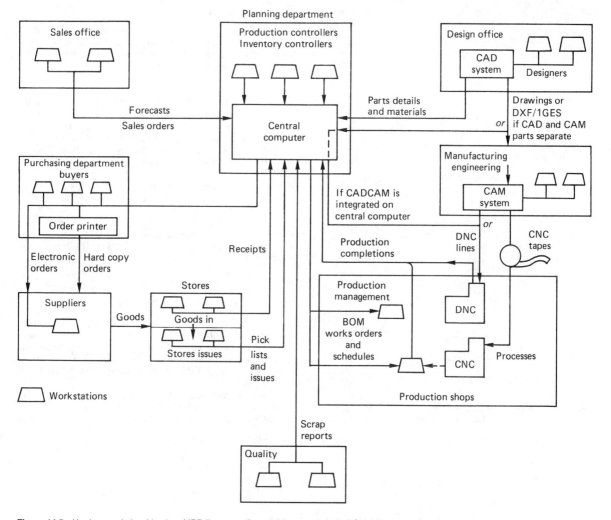

Figure 14.5 Hardware relationships in a MRP II system (financial lines excluded). BOM, bill of materials; CAD computer-aided design; CAM, computer-aided manufacturing; CNC, computerised numerical control; DNC, direct numerical control

strained resources are forward scheduled to their finite capacity, whilst non-constrained ones are backwards scheduled to infinite capacity. Any buffers required are expressed in terms of time, not inventory.

(3) There are two types of batch quantity—transfer batch sizes and process batch sizes. By balancing flow through balanced transfer batch sizes, lead times can be reduced. This is shown in *Figure 14.6*.

Because OPT balances flow it tends to work on smaller batch sizes. This can bring large reductions in inventory and lead times but, like JIT, it requires a new attitude and commitment to the philosophy from staff—in the face of the traditional seemingly more efficient production obtainable from large batch sizes. OPT can be used in conjunction with MRP II and JIT.

The functions of the OPT system are shown in *Figure 14.7*.

14.3.5   Shop floor data collection

Shop floor data collection (SFDC) provides an efficient method of collecting data from a number of locations around a

factory and feeding them directly to a central computer where they are processed. The types of information collected include:

(1) production batch completions—codes and quantities;
(2) job time booking—process, set up, waiting, breakdown;
(3) staff attendance and time booking;
(4) quality reports—SPC;
(5) work in progress and stock recording;
(6) tooling and engineering records;
(7) machine output (direct count from machine); and
(8) machine time logs.

Programs normally allow the user to design report and/or screen layouts.

Data are collected by using:

(1) real-time direct machine-to-computer communication;
(2) data-input terminals;
(3) a combination of (1) and (2);
(4) the input devices are connected to the central computer by a local area network (LAN).

(a)

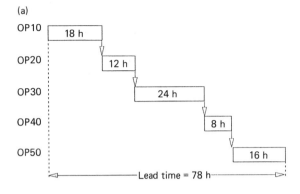

(b)

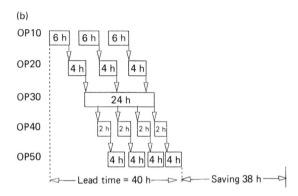

Figure 14.6 Predictions of lead time from OPT. (a) Transfer batch equal to process batch. (b) Transfer batch not equal to process batch

14.3.5.1 Input devices

Shop-floor terminals These are specially designed units which are simplified and ruggedised for shop-floor use. They have a small keyboard and display which allow staff to key in data. Some units have a card-reading facility which sends identification cards that are similar to credit cards.

Personal computers Personal computers (PCs) are used for input and also as monitoring devices to give the shop floor and supervision a view and feedback of the data being transmitted. In many cases the 'PC' is a remote terminal to the central computer. It will have a monitor screen and keyboard, but no processing or memory capabilities.

Bar-code readers These read bar codes put on control documentation and products. Their major advantage is the elimination of the keying errors and speed.

Machine data units These are fitted to machines to automatically capture data on output and other features of the machine utilisation.

14.3.5.2 Central computer system

In some applications, data are fed to the main planning computer (a mainframe, mini or supermicro). In other more independent data collection systems, a PC is used as a file server or the actual computer. A 286 processor and 40 Mbyte hard disk are usually sufficient for systems having less than 100 input points.

Data are transmitted from the input points to the computer via a local area network (LAN).

A SFDC system is shown in *Figure 14.8*.

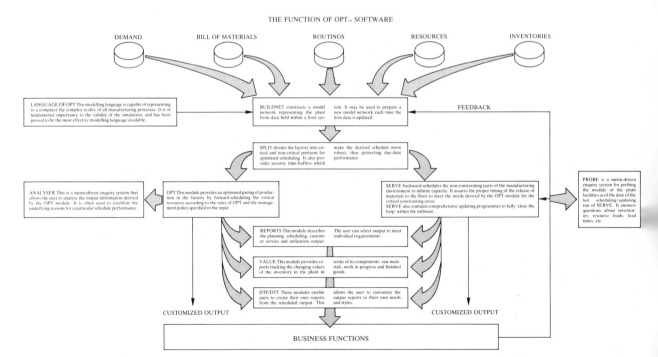

Figure 14.7 The function of OPT software. (Courtesy of Timegate Manufacturing Solutions)

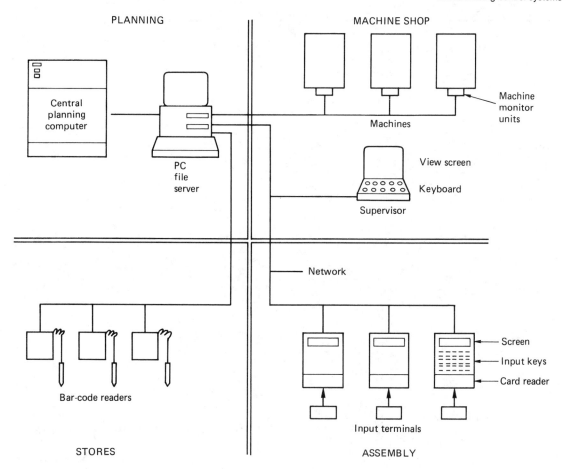

Figure 14.8 Elements of a SFDC system

14.3.6 Computer based simulation techniques

14.3.6.1 Introduction

Manufacturing systems are complex and dynamic. They combine a large number of activities, facilities and objectives in order to achieve the design and making of a product. Their complexity is compounded by the fact that the systems are dynamic—conditions within them are constantly changing as orders are received, plans initiated, materials flow and tasks completed.

The use of computer-based simulation is growing because it is a tool that allows detailed forecasting which was previously impossible to carry out by manual methods, due to the amount of data to be processed in even a simple system. The ability of a computer to hold a model of a system as a series of mathematical algorithms and its ability to store and process vast amounts of detailed data, now makes simulation of systems a practical possibility.

Simulation is normally carried out by a module within an application program. A fundamental requirement of simulation is the construction of a mathematical model of the system under consideration. The mathematical model can then compute likely outcomes from the input conditions specified by the user. Areas in which simulation is used include:

(1) production control and management,
(2) MRP II,
(3) OTP scheduling,
(4) factory planning and layout,
(5) flexible manufacturing systems (FMS) design,
(6) computer-aided engineering (CAE),
(7) computer-aided design (CAD), and
(8) computer-aided manufacture (CAM).

The output of simulation can be as text, but is increasingly graphical for ease of understanding. *Figure 14.9* shows a graphic simulation from a factory planning program.

14.3.6.2 Major elements of simulation

'Simulation' is defined in the *Oxford English Dictionary* as: 'the technique of imitating the behaviour of some situation or system (economic, military, mechanical, etc.) by means of an analogous situation, model or apparatus, either to gain information more conveniently or to train personnel'.

As far as FMS design is concerned this may be re-expressed as: 'a technique that assists design by testing a model of the proposed system with the range of likely operational conditions'.

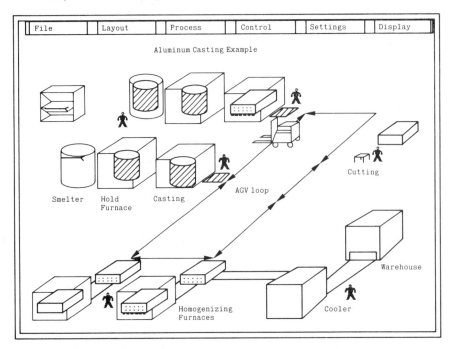

Figure 14.9 Factory planning simulation graphic of a factory. (Courtesy of CACI Products)

A 'model' is defined as: 'a simplified or idealised description of a system, situation or process, often in mathematical terms, devised to facilitate calculations and predictions'. Mathematical models will demonstrate the effect of the operational inputs, outputs and loads on the system, and this is the simulation technique that is discussed here.

Types of simulation There are three basic types of simulation, all of which use the technique of recording changes in the state of a model at points in time throughout a defined period. The method of establishing and using the points in time is one of the major differences between the three types. The three types are described below.

(1) *Discrete*—assumes that changes only occur at distinct points in time and that they occur instantaneously. No changes take place between points. Manufacturing simulation software packages are based on this method.
(2) *Continuous*—assumes that changes are continuous throughout the period. The simulation of continuous material flows, levels of materials and the application of mechanical forces are typical applications.
(3) *Combined*—includes both discrete and continuous simulations. This is the more common situation in real-life systems where many activities of either type impinge upon each other.

14.4 Computer-aided design and manufacture

14.4.1 Introduction

The development of CAD, CAM and CADCAM has made a great impact on design and manufacturing and now offers a wide choice of software applications and systems hardware. A number of terms are used to describe the fields of application and are generally taken to mean the following.

CAE (computer-aided engineering)—(1) in the widest sense the term covers all engineering activities that are run on computers, e.g. CAD, CAM, CAT and CASE. (2) In the narrower sense, it encompasses design and analytical activities such as finite element analysis (FEA), and dynamic simulation programs, but does not include computer-aided design or draughting.

CAD (computer-aided design)—this term is commonly used to describe most computer-aided design and draughting facilities. Two-dimensional CAD is predominantly a draughting medium with some limited design capabilities. Three-dimensional CAD provides enhanced draughting plus geometric modelling.

CADD (computer-aided design and draughting)—this is a fuller description of the normal design and draughting facilities, but as there is now a general trend to use 'CAD' when meaning 'CADD', the latter term is becoming redundant.

CAM (computer-aided manufacture)—(1) In the widest sense this encompasses all modern manufacturing technologies that use computers in a central role. Thus all CNC and DNC machines, robots and flexible manufacturing systems are elements of CAM. (2) In the narrower sense, CAM refers to the computer-aided part programming systems for CNC/DNC machines. An in-built or linked CAD facility is used to create the part geometry prior to programming and can also be used to design tooling and fixtures. Modern CAM software also has production planning and scheduling capabilities to assist in the control of production on the programmed machines.

CADCAM (computer-aided design and computer-aided manufacture)—these are systems in which the traditional CAD and

CAM elements are integrated. The CAD and CAM programs share a common database, which allows the part geometry needed for CAM can be taken directly from the drawing files. Parts can be machined under the direct control of the software (via DNC) once the design is established.

CAE has been widely applied to aeronautical, civil, chemical, electrical and electronic manufacturing and mechanical engineering. The applications that the manufacturing engineer is most likley to encounter are CAD, CAM and CADCAM systems used in the design and manufacture of mechanical and electrical/electronic products.

Figure 14.10 shows the relationships between the modules commonly used within CAE, CAD and CAM.

14.4.2 CAD/CAM systems

14.4.2.1 System configurations

Centralised systems (Figure 14.11(a)) These employ one central computer (a 'host computer') which carries out all the processing, and controls all the terminals and other hardware elements of the system. This has the advantage of there only being the need to maintain one computer but, if the computer

also serves other functions such as finance and production, there are likely to be operational delays because CADCAM is a heavy user of processing power and memory. The ensuing slower response time is likely to be unacceptable for the interactive nature of CADCAM work. A host computer dedicated to CADCAM is clearly a better operational proposition than is a share facility, always assuming that it is justified economically.

Multiple-host systems (Figure 14.11(b)) A number of host computers are used to service groups of terminals with most of the peripherals connected to one host. This gives a faster response because the processing capacity is increased and more localised. This provides additional security over having a number of host computers. The hosts are usually 32-bit minicomputers the individual cost of which will be lower than the required single mainframe host, but collectively the cost may be about the same. Multiple host systems may be the best solution to a functional or geographical split in the total CADCAM environment.

Distributed systems (Figure 14.11(c)) In this case the processing and random access memory (RAM) and read-only memory (ROM) capacities are distributed to the terminal

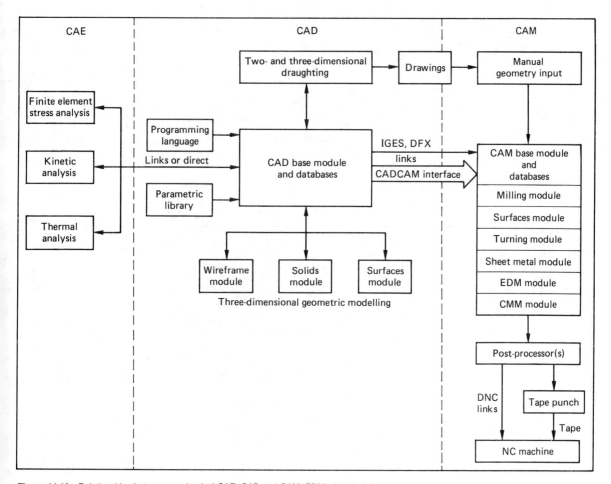

Figure 14.10 Relationships between mechanical CAE, CAD and CAM. EDM, electrical discharge machining; CMM, coordinate measuring machines

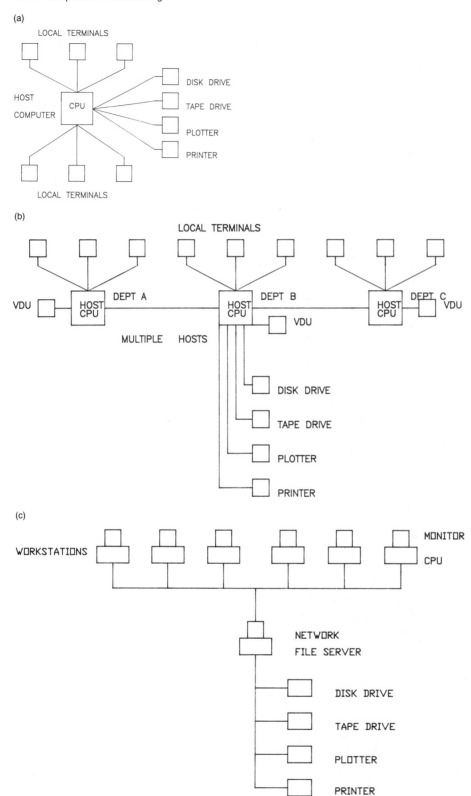

Figure 14.11   System configurations. (a) Centralised system. (b) Multiple-host system. (c) Distributed system via a local area network. CPU, central processing unit

positions by replacing the 'dumb' terminals with intelligent workstations that are connected by a local area network (LAN). In order to control the communication of the workstations with the shared peripherals, a computer known as a 'fileserver' is also needed. This carries out the communications and network control functions of central hosts, but leaves the applications processing at the individual workstations. Response times are kept to a minimum because each workstation has its own processor.

Stand-alone (single-workstation) systems This is the basic configuration in which a single computer is connected to its own plotter, printer and other peripherals. These may be mainframe or mini computers but a very large proportion are PCs. Stand-alone systems are widely used in industry and professional engineering because they have been bought to satisfy an immediate demand of individuals or small departments. However, with the general increase in the number of workstations being installed there is a growing trend to connect these into a local area network (LAN), thereby creating a distributed system for sharing plotters, printers, discs and other peripherals.

14.4.2.2 System hardware

A wide range of hardware has been developed in order to replicate the resources and capabilities of the traditional design, draughting and manufacturing engineering functions. The combinations of hardware used will depend on the system configuration. A typical CADCAM set-up is shown in *Figure 14.12* and the individual types of hardware are discussed in more detail in Section 14.4.6.

Host computers These are central computers that provide the processing, memory and system communications and controls with the terminals acting only as input/output devices. All printers, plotters, tape punches and other peripherals are controlled by the host.

Terminals These provide the point of interface between the engineer and a host computer. Usually comprising a monitor, keyboard and other input devices, they have no or limited processing and memory capacity.

Figure 14.12 Typical PC-based CADCAM system with two screens, a keyboard and a mouse. (Courtesy of Pathtrace Engineering Systems)

Workstations These are 'terminals' that are computers with processing and memory capabilities. The engineer or designer works at a graphics monitor using a variety of input methods such as keyboard, digitising tablet, mouse, tracker ball or light pen. The workstation may be a purpose-designed unit or a PC with sufficient processing power and memory. A workstation may have one monitor, but more usually has two—a full-colour high-resolution screen for graphics work and a second (often mono) screen for the display of commands, data listings and system management activities. Some systems provide two high-resolution graphics screens to allow parallel working in two levels or views.

Fileserver This is a computer in a networked (distributed) system that is used to supervise the communications between workstations and the shared peripherals. On some systems it will also hold the file mass storage capacity.

Mass storage CAD and CAM systems need a large amount of memory for long-term storage of designs and data. This usually takes the form of hard-disk drives on most machines. Tape storage was the first medium to be used (on mainframes) but is being superseded by disks which have much shorter access times and, being sealed units, are less susceptible to damage.

Plotters These produce drawings automatically by a variety of technologies. They are usually shared by workstations via a network. Standard sizes range from A4 to A0. Larger plotters are available for special applications.

Printers These produce hard copy of programs and data, normally on A4 or A3 size paper.

Tape streamer This is used to back up all files onto a tape at regular intervals (every few hours, or daily), in order to avoid the complete loss of files or the work of the current period due to an accident of some sort. Tape is acceptable in this case because real-time fast access is not a requirement.

Tape punch This is used for the production of paper tapes for numerical control machines.

14.4.2.3 Software

In common with normal computing practice, three types of software are used:

(1) operating systems software which controls the functions of the hardware;
(2) communications software which provides the communication between the hardware; and
(3) applications software which carries out the specific task required by the user.

Operating systems software Some of the most widely used operating systems are:

(1) *VMS*—the DEC (Digital Equipment Corporation) operating system for the VAX model of minicomputers which are a major hardware platform for CADCAM systems.
(2) *UNIX*—a hardware-independent, multitasking system developed by Bell Laboratories, which has been applied to many platforms. UNIX has many variants such as XENIX, ULTRIX, HP-UX and Sun OS which are supplied by other vendors.
(3) *DOS*—the 'disk operating system' developed by Microsoft for stand-alone PCs. It has the disadvantage of being

a single-task system which is unsuitable for multiuser or multitasking operatings.

The operating system that seems destined to have become the most universally used is UNIX and its variants, because of their machine independence, multitasking and multiuser applications plus the already established wide usage which makes it a *de facto* 'industry standard'. All the major CAD applications packages have been written to work on UNIX machines. With the general move towards more 'open' (e.g. hardware-independent) systems, modern PCs are being supplied with both DOS and UNIX type operating systems.

Communication software (protocols) With the increasing emphasis on CIM and open systems, it is important that different computer systems can communicate, i.e. are capable of securely transferring data and files between one another. In order to do this they must have a common software protocol to carry out the transfer. Communication protocols in wide use include DXF (Data Exchange File) from AutoCAD, and IGES (Initial Graphics Exchange Specification). Most system suppliers provide these links plus other protocols for communication with various widely used systems.

Languages used The majority of early CAD software was written in FORTRAN because this was the primary language used in engineering and scientific fields. Many modern packages are now being written in C and C++ because of the development of UNIX (which is written in C) versions of the software. Most modern packages also include an appropriate high-level-language programming module which can be used to customise areas of the application software to particular requirements.

Applications software CAD and CAM software applications in manufacturing fall into four broad groups.

Mechanical engineering
(1) Two- and three-dimensional mechanical computer-aided draughting (MCAD);
(2) three-dimensional computer-aided design for geometric modelling;
(3) CAE for the engineering analysis of designs; and
(4) CAM for mechanical components—mainly parts machining and fabrication.

There is now a strong movement to integrate mechanical CAD and CAM in the form of CADCAM in order to reduce errors, costs and the time taken to bring a new product to the market.

Electrical/electronic engineering
(1) electronic CAD (ECAD) for printed circuit board (PCB) and printed wiring assembly (PWA) design;
(2) ECAD for integrated and hybrid circuit design and engineering;
(3) CAE for circuit and product engineering analysis; and
(4) CAM for PCB and PWA manufacture.

Electrical CAD and CAM have been more integrated than mechanical CADCAM for some time, because their joint development was needed to enable the efficient design and production of integrated circuits and complex multilayer PCBs and PWAs.

14.4.3 Mechanical CAD and CAE

The usual application areas for mechanical CAD and CAE software are:

(1) two- and three-dimensional computer-aided draughting (CAD),

(2) geometric modelling of designs (CAD), and
(3) engineering analysis of designs (CAE).

14.4.3.1 Computer-aided draughting

CAD has the following advantages over manual draughting.

(1) Once proficient, a CAD user's overall productivity can be 3–5 times better than manual draughting. Improved advantage ratios can be obtained and are often claimed by vendors, but the lower figures give a truer average picture over all operations.
(2) Geometric shapes and features can be easily and rapidly created, copied, moved, amended and projected.
(3) Designs, shapes and features can be saved for future use on other drawings.
(4) Additional views can be created automatically.
(5) Dimensions are calculated automatically by computing the X–Y coordinates.
(6) Features such as hatching, chamfering, fillets, etc., are automatic once defined.
(7) A system of drawing levels allows segregation of assemblies, subassemblies and component details as on manual drawings, but the features can be transferred/copied between layers.
(8) The use of different colours for lines, features and dimensions improves the clarity of complicated views both on the screen and in the plotted drawing.
(9) The quality of the drawings from the plotters is usually superior to other methods of reproduction.
(10) Drawings can be produced at any scale from the same digital information.

CAD has the following disadvantages versus manual draughting.

(1) Capital investment is higher. A minimum of CAD software, a workstation and a plotter is needed to replace a drawing board.
(2) CAD uses computers, plotters, etc., that can malfunction or breakdown. Therefore, the additional costs of hardware and software maintenance agreements are necessary if lost time is to be kept to a minimum.
(3) If there is a fault with a shared piece of equipment, e.g. a fileserver or plotter, then several people will be inconvenienced.
(4) Training requirements vary with the complexity and user-friendliness of the software, but several days are usually needed to cover training on the most frequently used features. Even with this it will normally be several months (depending on the usage rate) before a designer becomes fast and efficient in using the program.
(5) It takes several minutes to bring a system on line and to produce a required drawing. This cannot compete on speed and convenience with manual drawing when the requirement is for a small and simple sketch. N.B. three-dimensional sketching packages that are very versatile are just coming on to the market, but these are additional to the basic two- and three-dimensional packages and cost approximately £800.

The operation of CAD systems In CAD the graphics display screen replaces the two-dimensional drawing board. All programs provide a basic two-dimensional draughting package which varies in the functionality offered. Some programs also offer three-dimensional draughting by adding depth and perspective to the two-dimensional base by various methods.

CAD programs incorporate a large number of features that enable the creation of the geometric shapes, symbols, dimensioning systems and informational elements normally used in design and drawing. In addition, they offer drawing operations and options that derive from computer graphics technology—these include moving drawing elements around the screen, choice of many colours for any line or element, choice of line widths and style, different text sizes and fonts, view zooming, and other image-handling operations.

CAD systems are operated via a system in which the available features are presented in listings and sublistings called 'menus'. The desired feature is selected by placing an arrow or cursor cross-wires over the name of the feature in the menu which is displayed on the screen or a graphics digitising tablet. The feature selection and drawing actions are usually input by the user via either a stylus or puck on a digitising tablet, a mouse, tracker, ball, joystick or light pen. The computer keyboard is, by comparison, used only occasionally for the input of dimensions, data or text. More details of the hardware mentioned are given in Section 14.4.6.

Some basic menu elements are shown in *Figure 14.13* and a typical menu screen is shown in *Figure 14.14*.

14.4.3.2 Geometric modelling

Geometric modelling is used to provide three-dimensional constructions which:

(1) produce a realistic three-dimensional image of the design to give better visualisation;
(2) provide data that can be used in calculating various engineering analyses of the design;
(3) provide data for the subsequent production of detail drawings; and
(4) provide output for DNC, CNC and other CAM processes.

Modelling can give tremendous advantages in the design process because it can save expensive and time-consuming experimentation and the reworking of unsuccessful designs. Three methods of producing three-dimensional models are used: wireframe, surface and solids.

Wireframe modelling This method creates a model by connecting points in space. The connecting lines so drawn are perceived by the viewer as being the edges of the viewed object. Initially all lines are shown. This gives an effect of transparency in which features such as back edges that would normally be hidden from the viewer can be seen. Such an image is acceptable with simple objects but can become confusing with complicated ones. However, a computer graphics technique of hidden-line removal is employed which clarifies the image and seems to present only the normally seen surfaces. This is achieved by the viewer's visual interpretation of the graphic, not a mathematical construction of the model. Most CAD packages can now provide automatic surfacing of wire frames which can be rendered in colour shades to produce images with surfaces, perspective and depth. A wireframe model is shown in *Figure 14.15*.

Surface modelling This involves the construction of models by the mathematical definition of surfaces. Surfaces are defined by applying various 'surface element' functions such as: surface of revolution, tabulated cylinder, fillet surface, or free-form sculptured surface. Surface modelling is particularly useful in modelling complex sculptured surfaces such as aircraft sections, car bodies and mould forms for plastics and castings. The surface data can be taken for use in CAM machining modules for the production of part program tool paths in either DNC or CNC operations. This is particularly

Main menu	Submenu	Purpose
FILE		CONTROLS GENERAL MANAGEMENT OF FILES
	help	Provides guidance on all features
	status	Indicates status of the system
	directory	Directory of files
	create	Open a new drawing file
	load	Load an existing drawing file
	return	Return to prior drawing
	save	Save the working file
	save as	Save the working file under another name
	copy	Copy the current drawing
	rename	Rename the current drawing
	end	End the working session
VIEW		CONTROLS VIEWING OPTIONS OF THE SCREEN
	2D	Two-dimensional perspective working
	3D	Three-dimensional perspective working
	zoom	Zoom view in and out
	pan	Move view around
	redraw	Redraw the current screen
GEOMETRY		PROVIDES THE DRAWING GEOMETRY FEATURES
	lines types	Various types of line can be drawn
	arcs methods	Various methods of production
	circles methods	Various methods of production
	ellipse methods	Various methods of production
	polygon methods	Various methods of production
	splines methods	Various methods of production
	tangent methods	Various methods of production
DRAW		PROVIDES THE ACTIONS USED IN DRAWING
	trim	
	snap	
	rotate	
	mirror	
	stretch	
	fillet	
	chamfer	
EDIT		EDITS DRAWING OR FEATURES
	erase	
	move	
	copy	
	undo	
	re-do	
ATTRIBUTES		PROVIDES PERIPHERAL FEATURES TO DRAWING
	levels	
	colours	
	text	
DIMENSIONS		CONTROLS TYPE AND STYLE OF DIMENSIONING
	inch	
	metric	
	+/-	
	geometric	
	arrows	
OUTPUT		CONTROLS METHODS OF OUTPUT
	plot	
	print	
	export	

Figure 14.13 Basic features included in most CAD draughting packages. Most CAD systems contain basic features which are expanded as the specification increases on more expensive packages. Features are accessed by the selection of the feature name from a menu listing displayed on the screen. The menus have a hierarchical structure with top-level menus splitting down into submenus and sub-submenus. The nomenclature and range of features varies between software packages

useful in the multi-axis machines that are needed to create complex surface contours.

Solids modelling Two methods are used: constructive solid geometry (CSG) and boundary representation (B-Rep).

CSG combines simple geometric elements to form more complex shapes. The elements are known as 'primitives' or 'entities' and are in the form of blocks, cones, cylinders,

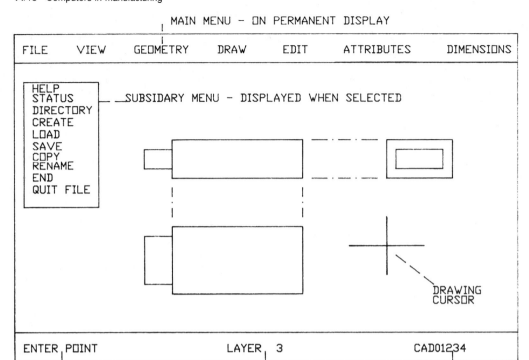

Figure 14.14   Layout of a CAD screen

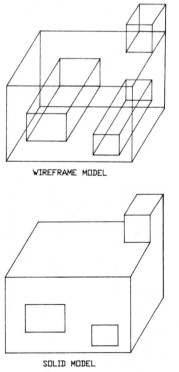

Figure 14.15   Types of model construction

spheres and tori. These can be combined in the Boolean relationships of union, difference and intersection. This method works better with regular shapes than with irregular shapes. The Boolean combinations are illustrated in *Figure 14.16*.

B-Rep uses boundary data, i.e. definitions of the constituent surfaces/planes, to define the solid model. The program is designed to define how the individual details are joined together. Complex non-uniform shapes need extensive

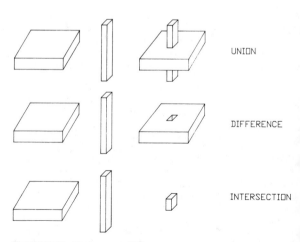

Figure 14.16   Boolean operations

description of the boundaries of the faces, but others with uniform cross-section can be more easily created by developing the shape from the appropriate axes (e.g. with solids of revolution such as cylinder, the radial cross-section can be swept for 360° around the centre line). Uniform bodies may be 'extruded' to any desired length.

Solid models can also be created from conversions of wireframe models, surface models and two-dimensional details.

14.4.3.3 Computer-aided engineering

Computer-aided engineering (CAE) encompasses many applications in the engineering analysis of designs. This work originally needed mini or mainframe computers because of the size of the programs which contain many calculations, but with the increased capacity of workstations and PCs many CAD suppliers are now including CAE modules in their program suites. This is possible because the data used to construct the three-dimensional models can also be used for a variety of engineering analyses, e.g.

(1) mass properties—volume, weight, centriods and moments of inertia;
(2) stresses;
(3) thermal properties;
(4) interferences; and
(5) movement paths.

The major tool for carrying out these analyses is finite element analysis (FEA). This is a technique that breaks the model down into small simple elements of a size appropriate to the local calculation of the condition being analysed. Each element is considered to be joined to its neighbour at points called 'nodes', and may be one, two or three dimensional. The network of elements over the whole model is known as the 'finite element mesh' (FEM) (see *Figure 14.17*).

FEA involves a considerable amount of work in the model definition and data preparation before the actual element analysis takes place. Interpretation of the results of the analysis also creates a large workload. Consequently, software houses have developed pre-processors and post-processors to handle the preparation and interpretation tasks. These may be used in conjunction with a resident FEA program from the

software supplier or be interfaced to other proprietary FEA packages.

14.4.3.4 Future trends in CAD and CAE

Software will continue to show the normal development trend of the lower specification programs gradually being upgraded to incorporate features currently available only in higher specification programs. Relatively cheap CAD software suites will incorporate more parametric design, geometric modelling, full-colour three-dimensional visualisation, engineering analysis and kinetic simulation modules.

Two-dimensional draughting has until now been the most common application, but three-dimensional draughting and modelling suites with engineering analysis options will become far more widespread. Three-dimensional sketching packages with links to databases and spreadsheets will come into use for initial design work.

Open systems will become widespread with all hardware and software products being tailored, where necessary, for open working.

The majority of applications programs will be written with graphical user interfaces (GUIs) of the MS-Windows (TM) type of cascading menus.

High-specification PCs will run on UNIX or a variant of it as well as on DOS, and UNIX will be consolidated as the most widely used operating system.

Low-cost workstations and high-specification PCs will compete on price and be the major platforms for CAD and CAM systems. These will provide reduced instruction set computers (RISC) or 80486 and later microprocessors, increased operating speeds, larger memories and enhanced graphics. 64-bit processors will be put into machines in this size range. The distinction between workstations and high-specification PCs will disappear.

14.4.4 Computer-aided manufacture

14.4.4.1 Introduction

In the widest sense CAM can be taken to encompass all CAM technology including numerical engineering, flexible manufac-

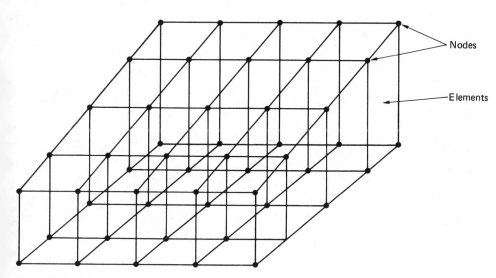

Figure 14.17 Finite element mesh

turing systems (FMS), flexible assembly and robotic welding and handling. The narrower and more usually understood scope of CAM within the context of CAD/CAM is in the computer-aided programming and related engineering activities for numerically controlled manufacturing processes. The narrower definition is used here because FMS and robotics are discussed in other sections.

CAM packages are offered to the market in three forms: CAM, CAMCAD and CADCAM packages.

CAM packages are entirely separate from any CAD packages. In this case a series of instructions is entered into the computer to define the geometry of the part. This early technique has been largley overtaken by the use of interactive CAD facilities embodied in CAMCAD and CADCAM packages.

CAMCAD packages contain an in-built CAD program which is used to define the part geometry. This is done by the engineer working with a graphics screen to draw the part in a similar fashion to CAD draughting. In more recent packages, these facilities have been greatly enhanced to the point where some three-dimensional work is possible. Modern packages also include the communications capabilities (IGES and DFX, etc.) to import part geometry from other CAD systems. These programs are known as 'CAMCAD' packages because they have been developed from early CAM programs by companies primarily serving the numerical control market, not the product design market. The structure of a CAMCAD package is shown in *Figure 14.18*.

CADCAM packages contain CAM facilities which comprise a series of modules that are integrated with the CAD within a complete CADCAM suite. In this case the part geometry is created in the CAD suite and held in a common database from which it can be downloaded to the CAM modules. This is a more efficient and potentially more error-free method of working than using separate CAD and CAM.

In each of the above cases the type of numerical control machine requires specifically tailored software. This is supplied in the form of modules that can be purchased separately, usually as additions to a mandatory base module. The most commonly used modules are:

(1) milling—($2\frac{1}{2}$-dimensional and three-dimensional, surfaces, and multiplane modules);
(2) turning (two axis, multi-axis and C axis);
(3) grinding;
(4) sheet metal punching;
(5) laser cutting; and
(6) electrical discharge machining.

Typical screens for milling, turning and sheet metal work are shown in *Figures 14.19* to *14.21*, respectively.

14.4.4.2 Software

Part programming has been computer aided since the introduction of the first automatic programming tool (APT) languages, but those early programs did not have the advantages of the more recently developed interactive graphical programming. This technique is now becoming the standard in industry with the development of powerful workstations and PCs that include good quality colour graphic screens.

Interactive graphical programming provides immediate visual interaction between the programmer and the computer. Instructions are input via the selection of commands from command menus displayed on the screen or on a digital tablet. The results are shown on the graphics screen in the form of drawings, geometric elements, dimensions and text displays, giving the programmer an immediate two- or three-dimensional indication of the action just taken.

The major elements of CAM software are:

(1) part geometry definition;
(2) tool definition, libraries and management;
(3) production of the cutter location file (CLF);
(4) post-processing of the CLF;
(5) tool-path verification displayed on screen;
(6) storage of parameters for use on other parts;
(7) DNC and other communications links; and
(8) production control facilities.

Part geometry definition If the part geometry cannot be imported from a CAD file it must be created by the programmer. To this end, the CAM software contains CAD features that allow the production of part shapes and automatically registers the commanded and resultant dimensions and coordinates.

Tool definition The cutting tools and their dimensions must be defined in order to allow the correct calculation of tool paths.

Cutter location file When the part geometry and tooling data are completed, the software writes the program which defines the cutter path. This is usually known as the 'cutter location file' (CLF).

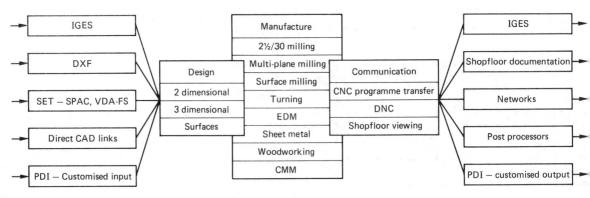

Figure 14.18 CAMCAD system. (Courtesy of Pathtrace Engineering Systems)

Post-processing The computer produces a CLF that must be converted to the input requirement of the specific machine to be used. The conversion operation is carried out by a separate software program known as a 'post-processor'. There are two ways in which this task is handled.

(1) Post-processors that are specific to a particular machine and must be purchased (normally from the machine supplier) for each machine type. These are the type used in smaller and cheaper CAM installations.

(2) A universal post-processor that can be customised to provide a post-processor for the machine to be used. These are normally available with advanced CAMCAD and CADCAM packages.

Tool verification Once the tool paths have been calculated, they can be illustrated on the screen, in sequence, so that the progress of the machining operation can be seen. Verification is available in both two- and three-dimensional solids presentations. Automatic verification will check for tool collisions with the work piece, clamping, fixtures and machine parts.

Storage of parameters for other parts It is possible and very useful to have the parameters of a part available for future use in the programming of a similar part or family of parts.

DNC and other communications links Complete programs may be down-loaded at one transfer if the machine tool buffer memory is large enough. If not (normally only the case with large programs), the blocks can be drip-fed to the machine. In earlier systems, links between the host computer and machines was created by the addition of a receiving unit on each machine and a manual switching box to route programs to the correct machine. The current trend is to add a communication card to the machine control and route the programs via a fileserver which controls a LAN running to all the machines.

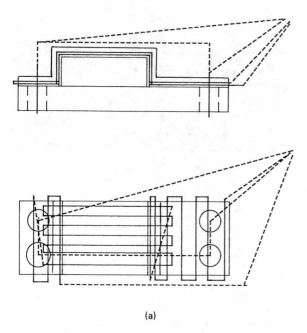

(a)

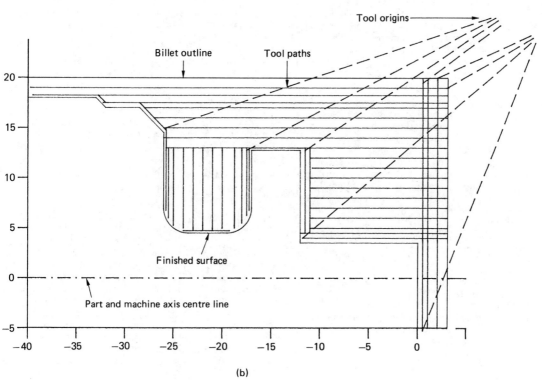

(b)

Figure 14.19 (a) Screen for 2½-dimensional milling and drilling operations. (b) CAM screen for turning

(a)

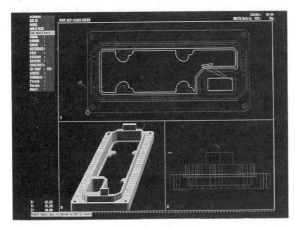

(b)

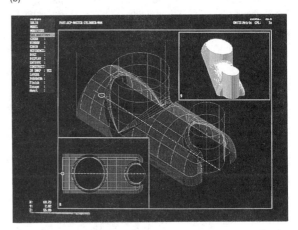

Figure 14.20 (a) Screen for 2½-dimensional milling. (b) Screen for surface milling. (Courtesy of Pathtrace Engineering Systems)

Production control facilities These provide the capability to control and gain data on production batches on the machines that are linked to the workstation or host computer by a DNC link or LAN. The trend to integrate production control activities with the programming work will increase as CIM becomes more widely implemented.

14.4.4.3 Hardware

CAM systems hardware is very similar to CAD hardware with the addition of tape punching. In an integrated CADCAM or CAMCAD system all the activities are usually performed on the same equipment. A typical system comprises the elements described below. For further details of CAD and CAM hardware see Section 14.4.6.

These should comprise the following elements.

Workstations (1) Central system—with a processor of at least 80286 but preferably 80386 level. Also a maths coprocessor should be included to increase the speed of the many calculations involved.

(2) Graphics screen—(needed for part and tool path representation in interactive working with the programmer). A video graphics adapter (VGA) grade should be used with a resolution of at least 1024 × 768 for continuous use.

(3) RAM—should be a minimum of 640kbyte with 1Mbyte preferred for heavy use. (1 or 2Mbyte are now becoming standard offerings).

(4) Keyboard—either standard (83/84) or enhanced (110/111) types.

(5) Input device—either a digital tablet, mouse, tracker ball or similar device for part geometry definition. Digital tablets or a mouse are becoming the preferred method due to ease of handling.

Tape punch This is needed to produce new and duplicate numerical control tapes.

Plotter This is used to plot product outlines and tool paths. It is a very useful visual aid. It can also be used for associated jig and tool design.

Printer This is used to print copies of the numerical control programs, tooling and other data listings.

Tape streamer This is used to take back-up copies of files. Back-up can be done on floppy or hard discs, but a large system will need a tape streamer.

File server This is needed in a network of several workstations and peripherals and for communication to the machines in a DNC environment.

Communications Serial communications ports for RS232C, RS422 or Ethernet or token ring LANs should be provided for linking to numerical control machines. A parallel port should be used for connection to a printer. Communication methods have limitations on the distance between elements of hardware and the total network length. These need to be checked against the total and likely future requirements and cost limitations.

14.4.5 Electronic CADCAM

14.4.5.1 Introduction

The rapid expansion of the electronics industry resulted in the early development of electronic CAD and CAM for the integrated design and manufacture of electronic products. This is because the combined effect of micro-sized and multiple-layered circuits has made the use of computer-aided techniques essential for efficient and high-quality design and production, as the tasks became too complex and miniature to be either possible or economical to do by manual methods.

The software is completely different from mechanical CADCAM software, but the hardware used is generally the same, requiring an equivalent quality of graphics screens, RAM capacities and mathematical calculation handling via maths coprocessors. For a description of CAD hardware see Section 14.4.3.

A large range of computer-aided applications is available, one of the most widely used being the design and production

(a)

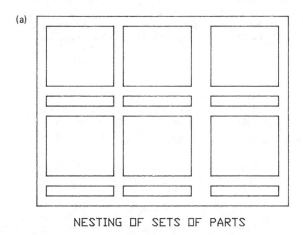

NESTING OF SETS OF PARTS

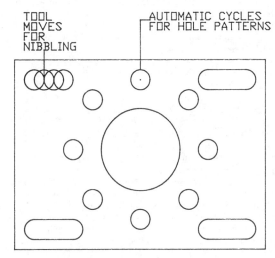

TOOL
MOVES
FOR
NIBBLING

AUTOMATIC CYCLES
FOR HOLE PATTERNS

(b)

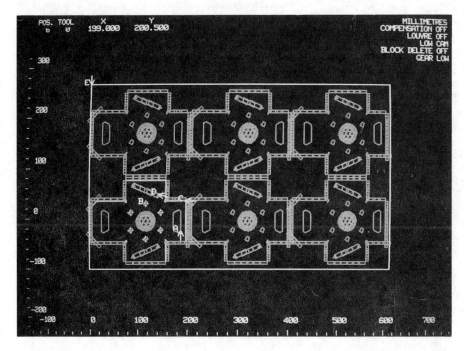

Figure 14.21   (a) Screens for sheet metal operations. (b) Screen showing sheet metal punching. (Courtesy of Pathtrace Engineering Systems)

of printed circuit boards (PCBs). As with mechanical CAD-CAM, the capabilities of electronic CADCAM programs vary with price, the more expensive packages (£3000 and above) offering a greater degree of automation, flexibility and analysis than basic programs costing a few hundred pounds.

14.4.5.2 CADCAM systems for PCB design and production

The main elements of a CADCAM system for PCB design and production are described below.

Schematic design capture This is an interactive technique in which the designer constructs the proposed circuit by placing component symbols and connections on the screen. This is a functional representation of the circuit design intent, not a physical layout. The great advantage of the technique is that it allows the use of predrawn symbols, thus saving considerable draughting time, and facilitates easy editing and modification of the design via normal CAD methods adopted for this special application. Schematic capture programs can also contain error-checking facilities which check for errors such as wires going nowhere, shorted outputs and bus contention.

Component/symbol database Large system databases can hold details of over 2000 components and 100 pad shapes which are reproduced on the schematic design when selected by the designer. The data held include component name, type, value, and pin number. Such a database is usually organised as a number of sublibraries, one for each of the common types of component used. A variety of coding systems is used but the best employ the industry standard numbering systems that define various component types and ranges, e.g. the 74000 series of components.

Net lists These provide hard copy of data about the circuit and are produced automatically from the calls on the component database made when producing the schematic. Commonly used lists include bills of materials, connection tracks and pins.

Circuit layout The schematic is converted into a physical layout of components on the board. The components can be placed either interactively by the designer or automatically.

Routing This defines the routes of the tracks between the components and around the board. This may be done interactively or automatically. Some routing programs use a grid as the basic template for defining routes, whilst others work without a grid by using special algorithms. Algorithms can give denser track layouts because they are not restricted to the pitch of the grid.

Outputs from the design process Only a computer can efficiently and accurately produce, modify, align and keep track of the coordinates and geometry of the successive layers of the board.

(1) *Output to photo-plotters*—Photographic/chemical processes are widely used in the manufacturing of a number of electronic products. In PCB production these are required for producing the complex layouts of the copper tracks, connection holes and solder resist layers. The most popular method of converting the design data to a format suitable for use by a photo-plotter is by converting the net lists to produce Gerber format files. these are used by the PCB manufacturer (often a specialist subcontractor) to produce photographic negatives which become the master artwork for the production of the various layers of the PCB fabrication. Gerber editors and

viewers allow the lines and holes that are to be photo-plotted to be seen prior to production.

(2) *Output to NC drilling machines*—The holes needed to connect tracks of copper from one board layer to another are normally produced by special NC drilling machines. The coordinate positions of the holes are produced from the design process and transmitted as a file or tape to the NC machine control.

(3) *Output of drawings and documentation*—Drawings of the various stages of the design process and for each layer are plotted from the CAD system. Bills of materials, and net and other listings are printed.

Engineering analysis Programs are available for the analysis of various conditions of the circuit and PCB designs. These include thermal stress, impedance and signal noise analysis.

Circuit simulation Simulation programs are available for analogue, digital and mixed circuits. They can save considerable time and cost in experimentation and bread-boarding by creating the circuit operation and answering 'what if' questions. Simulation programs use net lists from the schematic and device data from libraries as inputs and apply mathematical modelling by the use of algorithms from a library of algorithms.

14.4.6 Hardware review

The range of specifications, sizes and technologies employed in CAD and CAM hardware is enormous. Consequently this section can only give an indication of the more commonly used hardware and typical specifications.

14.4.6.1 Workstations/host computers

System workstations and host computers fall into three categories:

(1) PCs—general-purpose PC designs adapted for CAD and CAM work (60% of installations and increasing);
(2) workstations—purpose designed for CAD and CAE (20% of installations);
(3) mini computers for large and high specification systems (19% of installations).

PC hardware Early PCs were mainly used for stand-alone or small networked systems using low or medium specification CAD and CAM software, whilst higher specification programs were run on workstations, mini computers or mainframes. With the introduction of 16-bit processors and larger memories, high specification programs can now be run on PCs.

The major PC platforms for CAD and CAM work are now machines containing Intel 80386 and 80486 or equivalent processors. Earlier speeds for these were 12 and 16 MHz, but 20, 25, and 33 MHz are available and better suited to CAD work. An accompanying chip for maths coprocessing (e.g. Intel 80287 or 80387) is also needed because of the high number of calculations involved in graphics work.

Previously, systems were supplied as standard with up to 256 kbyte system RAM with extensions to 640 kbyte. This is adequate for simple applications but 1 Mbyte system RAM or more is advisable when handling sophisticated applications. Manufacturers now offer 1, 2 or 4 Mbyte as standard with optional expansion capability to 8, 16, 32 and 64 Mbyte of RAM.

The minimum amount of video RAM needed for the currently available editions of software is 256 kbyte and this should suit most two-dimensional applications. 512 kbyte and

1.2 Mbyte of RAM is becoming used as additional features are introduced. Three-dimensional working and engineering analysis modules need a large memory capacity. Video RAM is housed on graphics adaptor cards of which there are many models and suppliers. It is possible, therefore, to upgrade graphics capabilities without replacing the central system. Some graphics cards have expansion capabilities on board.

Hard disc storage is the most frequently used medium for mass storage in PC-based systems. The minimum storage for a simple system without a heavy load is now probably 40 Mbyte, and the price of disc storage is now so reduced that a purchaser will probably be offered this as standard with options for 60 and 80 Mbyte. Fileservers addressing several machines should have between 200 and 1000 Mbyte (1 Gbyte), depending on the size of the system and its load.

The performance of systems is greatly affected by the memory access time and bandwidth. This is particuarly so in 32-bit microprocessors with high clock speeds where the processor will continually be waiting for access to memory. To avoid this condition, additional high-speed (cache) memory is used as a buffer between the processor and memory. The cache memory may be built into the processor chip or on the main system board.

Due to the large number of IBM or IBM clones installed, the most commonly used operating systems are versions of disc operating system (DOS) which was originally written by Microsoft for the IBM PC—the IBM version is called 'PC-DOS'. This and other PC-oriented systems were designed to control single-computer installations running one task at a time and are not suitable for controlling multitask or multiuser systems. Modern PCs are, however, providing an in-built migration path from DOS systems to multitask operating systems such as OS/2 and miltuser operating systems such as Novell and UNIX.

The more powerful UNIX based PCs are challenging the use of entry level and bottom of the range workstations as the specifications of each become less distinguishable.

Workstations Technical workstations have been specifically designed to incorporate the enhanced processing power, memory, graphics monitors and input devices needed by medium- and high-specification CAD and CAM systems. Workstations normally use a multitask operating system such as UNIX, XENIX or VMS (DEC systems).

Workstations can be used as stand-alone units, connected to a plotter or printer. More usually they are incorporated into larger systems where they may be networked to common output and storage peripherals or act as a satellite terminal to a central system computer.

In the past, workstations were sufficiently expensive (approximately £60 000 to £100 000) to deter small and medium sized companies from investing in a higher level of CAD and CAM. However, hardware prices are reducing, particularly with the introduction of reduced instruction set computer (RISC) architectures. There is a very strong trend in this direction being pursued by many workstation manufacturers and the general availability of less expensive workstations (at less than £3000) that are indistinguishable from top-specification PCs will increase greatly.

Mini and mainframe computers In the past, large computers were required to provide the necessary processing, memory and mass storage capabilities needed for medium and large installations running sophisticated software. The need for a mainframe computer restricted applications to large or high-technology companies in industries such as automobiles and aerospace. The introduction of mini-computers widened the user base substantially, with systems such as the VAX series

from DEC. With the increasing use of workstations and PCs as system processors rather than satellite terminals, the need for mini and mainframe computers is now reduced to all but the largest systems.

14.4.6.2 *Graphics display monitors*

The graphics monitor is the major vehicle for interaction between the designer and the CAD program. In addition to being a substitute for the two-dimensional drawing board it is also a means of presenting three-dimensional images and the command menus used to operate the system. The sizes normally used are 14, 17 and 20 in. for low- and medium-specification systems. High-specification systems may use 26 in. displays.

Two types of monitor in common use are *raster* and *vector* displays. Both use the principle employed in television sets; a cathode ray tube projects a beam of electrons onto the internal surface of the screen, which has a coating of phosphor. The electron beam causes the phosphor to glow and it is this phosphorescence that the viewer sees. The effect lasts for only a fraction of a second and so, in order to maintain a seemingly constant image, the whole screen must be 'refreshed' with further passes of the beam at a rate of 30–80 kHz in the horizontal and 30–120 Hz in the vertical, depending on the model.

Raster displays These are the most commonly used type for commercial CAD and CAM displays. The majority of CAD-CAM displays are colour monitors in which separate phosphor dots emit red, blue or green light in response to electron excitation. The display contains lines of *pixels* which contain groups of three points of phosphor arranged in a triangle (triad). The pitch of the triads (known as the 'dot pitch') is typically 0.28 or 0.31 mm. By digitally directing three beams (one for each colour) at the pixels and by varying beam intensities, it is possible to create a wide range of colours based on mixtures of the primary red, blue and green. Early models could only display four colous, but 16, 64 and 256 colours are now common on PCs and 4096 colours on higher specification machines. The number of colours depends on the number and processing methods of the bits used per pixel, not on the screen specification.

The *resolution* of the display monitor is a measure of its clarity, which is dependent upon the number of pixels in each horizontal line and in the vertical. Resolution is expressed as the number of pixels and is a function of the number of bits used to program the pixel, not the quality of the screen. If 4 bits are used per pixel, then a display with a resolution of 640×480 pixels needs:

$$640 \times 480 \times 4 = 1\ 228\ 800 \text{ bits}$$

at 8 bit/byte

$$= 153\ 600 \text{ or } 150 \text{ kbyte of video RAM}$$

The development of improved resolution with successive models of monitor is shown in *Table 14.1*. Currently VGA is the normal standard supplied with super (SVGA) and extended super (ESVGA) gaining increasing usage. The earlier models are now practically unobtainable.

The overall performance of the graphics display is dependent on the processing power (now housed in one chip for a VGA display) and the amount of random access memory (RAM)—typically 256 kbyte to 1.0 Mbyte is used. These may be housed on the main system board or on a separate graphics adaptor card. There are many models of such cards on the market and these can be purchased to enhance the performance of the graphics.

Table 14.1 Resolution of graphics monitors (graphics and colour mode, not mono and text)

Monitor type	Maximum resolution	No. of colours
PC colour graphics adaptor (CGA)	320 × 200	4
PC enhanced graphics adaptor (EGA)	640 × 350	16
PC video graphics array (VGA)	640 × 480	16
PC super video graphics array (SVGA)	800 × 600	256
PC enhanced video graphics array (EVGA)	1024 × 768	256
Workstations	1280 × 1024	≥1000

Vector displays These were developed before raster displays and were first applied to the early monochrome screens. In this type of monitor, a single electron beam draws a continuous straight line between two coordinate points anywhere on the screen. With a typical resolution of 4096 × 4096 the line definition is better than in the raster type because vector displays are drawn with a single sweep of the electron gun, whereas the raster method must employ a group of pixels. The reduced resolution on raster displays shows up particularly on diagonal lines the edges of which are in the form of a series of steps.

14.4.6.3 Input devices

Hand held interactive devices The main method of communicating with the computer for CAD and CAM work is the use of various hand-held devices whose input produces an immediate response on the graphics screen. This allows the designer to work 'interactively' with the computer.

These devices control the position of a cursor arrow and/or coordinate cross-wires to produce the desired geometry 'drawing' and other images on the screen. They are also used to select the program menu options by pointing the cursor/cross-wires at the desired option description from a list (i.e. menu) displayed on the screen. There are several types of device in general use.

Digitising tablets are flat boards which represent the drawing and screen area. Any coordinate point within a defined area can be selected and transmitted to the computer. Digitising tablets are normally used where resolutions of better than 0.015 in. are needed for the transfer of existing drawings (i.e. digitising) into the computer. Several methods of tablet construction and point definition are available, but the most widely used has a network of fine wires beneath the surface of the board and a hand held stylus or puck. When these are placed over a point, its x–y coordinate position is defined by the electrical interaction of the wire mesh and the hand-held device. Pressing the end of the stylus on the tablet activates a switch. A puck has cross-wires which are placed over the desired point. Keys are used for control, as in a mouse.

Floor mounted A0 size digitising tables are available, but the usual size for CAD and CAM is a desk-top tablet of between approximately 12 in. × 12 in. to 15 in. × 18 in. Tablets normally have an area dedicated to the drawing activity and another area displaying command menus and possibly a full alpha-numeric key layout. This is indicated on a printed template and the stylus or puck is placed over the printed command or key in order to select it.

Tablets are often part of the larger work surface found in specially designed CAD workstations.

A *mouse* is a very widely used device, particularly on PC-based systems. A ball or set of wheels is housed in a small hand-held device. As the device is moved across any flat surface, the ball or wheels rotate through friction with the surface. This rotation produces signals that are converted to indicate the movement of the mouse relative to its starting position. Function keys incorporated in the mouse are then used to give 'enter' or other commands after the movement (which is displayed instantaneously on the screen) has been completed or menu selection made. Early models were analogue devices using variable-resistor systems. Most modern devices are digital with resolutions of up to 400 dots per square inch (dpi) (0.025 in.).

Tracker balls, *thumbwheels* and *joysticks* work in the opposite mode to a mouse, in that a base unit remains static and a large roller in the tracker ball, thumb wheels or joystick are rotated in the base by hand action. Tracker balls and joysticks produce a combined x and y movement whilst thumbwheels produce an x or y movement—the latter can be an advantage in work requiring accurate location.

Light pens are similar to a ball-point pen in shape. When the pen is held against the screen and an internal switch activated it detects the presence of light. A signal is transmitted to the processor which can then deduce the position of the pen. The accuracy of a pen is not as good as other methods which limits its use for drawing work. It is used, however, as a general pointing device. Pens are not suitable for all screens which is another factor that limits their wide use.

Keyboards Keyboards are needed to input text, numerical data and some commands but are not the major means of interaction between the user and the computer (this is normally a mouse, digitising tablet or other similar hand-controlled input device).

Two layouts of keyboard are in general use in the UK. Both use the QWERTY layout for the alphabetical keys but have different arrangements for the numeric/cursor keypad and function keys. The two layouts are:

(1) *83/84 type*—this was the first keyboard developed. It has 83 keys in the US version and 84 in the UK version (the additional key being required to accommodate the £ sign). There are 10 function keys. This board transmits 8 bits on each depression and another 8 bits upon release of a key.
(2) *110/111 type*—known as an 'enhanced keyboard' this has 110 and 111 keys in the US and UK versions, respectively. It has 12 function keys and additional control and cursor keys. It transmits 11 bits for each key depression and release.

Because of the different bit scan rates, the keyboards are not interchangeable. Either can be used with CAD and CAM systems, but the enhanced keyboard provides separate cursor and numeric key operations and more functionality.

Various suppliers assign different tasks to the function keys and operations obtained by combinations of the shift/alt/control and alpha-numeric keys. These are normally indicated by additional inscriptions on the keys and/or templates that can be laid around the edges of the keyboard.

14.4.6.4 Output devices

Plotters Plotters are used to produce line drawings from CAD or CAM system plot files onto a variety of media including drawing paper, glossy paper, polyester film, vellum and transparency film. Most plotters accept the plot file format Hewlett Packard Graphics Language (HPGL), but other formats used include Calcomp 960/970, Versatec V80 and VRF. Plotters have a memory known as a 'buffer' to hold the plot files downloaded from the computer. Standard buffer sizes range from a few kilobyte on small machines to 20 Mbyte on top-of-the-range models. RS232C serial and Centronics parallel interfaces with the computer are normally available.

'Intelligent' plotters also have a microprocessor incorporated. This takes on the management tasks, vector-to-raster conversion and the creation of shapes such as circles, arcs, etc. from a library of 'primitives'. These capabilities release the system computer from such tasks and increase the speed of throughput. Typical resolutions for plotters are 200, 300 and 400 dpi or 0.25 and 0.0125 mm.

Machines vary in size from standard A4 up to standard A0 with larger machine sizes up to 2 m × 3 m. A4 and A3 models are usually desk mounted and the larger machines floor mounted.

Several types of plotter are in use. Pen plotters are the cheapest device for producing coloured line drawings because they only have to carry different coloured pens to do so. Alternative designs such as electrostatic, thermal and laser plotters are better for producing areas of colour shading because they use a raster technique similar to the graphics monitor principle. This necessitates the inclusion of three colour systems providing the primary subtractive colours of yellow, magenta and cyan which are mixed to produce all the other colours. This together with the higher cost of the internal processing systems makes them 2–5 times more expensive than pen plotters of equivalent size.

Pen plotters are vector devices that draw lines between coordinates in any direction by using x and y axes movements simultaneously. The axes movements are obtained by moving either pen or paper, depending on the model. Lines and text are drawn with fibre-tip, wet-ink or ball-point pens which are automatically lifted where no line or text is needed. Pen acceleration and maximum speed should be considered when selecting a plotter as these will dictate the throughput rate and also influence line quality. Standard quality plotters have a resolution of 0.025 mm, whilst expensive models can give 0.00015 mm. A variety of pen colours and thicknesses is available. Both cut sheets and rolls of drawing material are used. A4 and A3 plotters are usually the desk-mounted flat-bed type in which the bed may be horizontal, vertical or inclined. Larger sizes are usually floor mounted and may be flat-bed or drum type.

Flat bed pen plotters hold the paper fixed to a flat surface. A number of pens (usually 4 to 12) is held in a magazine and selected by the head when it is programmed to do so. Pen movement is obtained by either moving the pen head in both x and y directions or, alternatively, by moving the pen in only one direction with movement of the bed providing the other axis.

Drum pen plotters have the paper fixed to a drum which is horizontal. The drum rotates back and forth to provide one axis of movement, allowing the paper that is not in contact with the drum to hang vertically. Pens move along the axis of the drum to provide the other direction of travel. Because one axis relies on the rotational accuracy of the drum, they are not as accurate as flat-bed plotters. Drum plotters have the advantage of taking less floor space than flat-bed plotters because the drawing medium hangs vertically.

Electrostatic plotters are raster devices that produce an electrostatic charge onto a dielectrically coated medium via rows of styluses set in a write head. A raster scan technique is used to deposit small electrostatic dots on the medium. This then passes through a toner section in which particles of the toner powder attach to the charged medium. The image is then fixed by heat. The medium is drawn off reels of various widths and fed through the plotter. Plotting is very fast with possible throughput rates 5–20 times as fast as that of pen plotters. However, they typically cost 2–4 times as much as pen plotters of comparable size.

Thermal plotters use a process of applying wax-based ink onto the print medium by a combination of heat and pressure. They use a full page raster technique and are typically faster than a pen plotter. Resolutions of 300 and 400 dpi are achieved. The majority of machines are monotone, colour being substantially more expensive. Thermal plotters have few moving parts and, consequently, are quieter than pen plotters.

Printers In *laser printers/plotters* a laser is used to project a light beam onto a mirror which deflects the beam onto the paper. Rotation of the mirror provides movement of the beam across the paper. The paper is supported by a drum which rotates to provide movement from one line to the next. The drum is photosensitive and the image is built up pixel by pixel. Laser printers have gained in popularity because, with resolutions of 200–400 dpi, the quality level of the images is superior to that of dot-matrix-printer output. They are faster than dot-matrix printers, particularly for multiple copies, and much quieter, but they are more expensive.

Dot matrix printers are one of the most widely used devices for producing hard copy from computer systems and are used for text and low-resolution line work. An array of mechanical pins is used, each pin hitting an ink ribbon to deposit a dot of ink on the paper. The most usual pin-array pattern is 9 vertical by 5 horizontal giving a rectangle of 45 pins from which to produce the desired character. Selection of pin usage by the software can create a wide range of characters and symbols in a variety of font styles. The printer prints a line from left to right and at the end of the line automatically returns the carriage and advances the paper. Better images can be created by overprinting on the carriage return right to left or by the inclusion of more pins in the print-head design. A second set of pins can be offset slightly to produce a greater density of print on the paper.

14.5 Numerical engineering and control

14.5.1 Introduction

14.5.1.1 Definition of terms

There are a number of acronyms and terms used in this field which are commonly understood to mean:

Numerical engineering (NE)—a general term used to cover all numerically based engineering activity.
Numerical control (NC)—control of a machine by the repetitive reading of a paper or magnetic tape.
Computer numerical control (CNC)—control of machine by its own resident computer.
Direct numerical control (DNC)—the control of a number of machines by a remote computer which communicates with each machine's resident computer.

14.5.1.2 The advantages of numerical engineering

The development of numerical engineering and control technology has been a major factor in the improvement of manufacturing efficiency over the last 30 years. NC, CNC and DNC machines have now largely replaced manually and electromechanically controlled equipment and have been applied to many machining, forming, assembly and inspection processes.

NE has greatly improved productivity, quality and flexibility in manufacturing processes by providing machines that have the following advantages.

(1) Improved productivity:
  (a) shorter machining cycle times through faster and deeper cuts and faster tool approach speeds;
  (b) automatic loading/unloading of parts; and
  (c) unmanned running.
(2) Improved quality:
  (a) greater accuracy;
  (b) greater repeatability of accuracy achieved;
  (c) better tool setting systems and monitoring;
  (d) easier modification of the process parameters; and
  (e) in-process probing and inspection.
(3) Greater flexibility:
  (a) faster changeover/setting times;
  (b) more economic small- and medium-batch runs (i.e. less inventory);
  (c) more operations on one machine (one-stop machining); and
  (d) better optimisation of use via improved control capabilities.

14.5.1.3 Brief history

Numerical control (NC) The first applications were seen in the early 1950s as the automatic programming tool (APT) language was developed at the Massachusetts Institute of Technology (MIT). Other versions of APT such as ADAPT, EXAPT and UNIAPT have been introduced subsequently. The early machines were either:

(1) hardwired, in which circuits used fixed logic components to produce a standard control capability; or
(2) used punched paper or magnetic tape for software program storage off the machine and input to it—the tape would be cycled through a tape reader on the machine every time a component was produced because the machine control does not have any memory to retain data or the allied computing power needed to process it.

Computer numerical control (CNC) A major development in the 1970s was the introduction of computing capabilities into the machine controllers. As microprocessors became available at an economical price, they were incorporated in the controllers. These were then able to store programs in RAM (random access memory) and read them when instructed which meant that the program need only be entered into the machine once and could then subsequently be modified and updated at the machine. This has greatly increased the speed, capability and versatility of the controller and, consequently, the machine. The majority of numerically controlled machines are now CNC.

Direct and distributed numerical control (DNC) Direct numerical control was first developed in the 1960s to allow the direct control in real time of a number of machines from a single central controller. It was intended to reduce the cost of numerical engineering by reducing the number of the then relatively expensive machine controllers and by using the greater memory capacity and computing power of a central mainframe or mini computer for the storage of large and/or a large number of programs. The introduction of controllers with larger memories and greater computing power on each machine (through the development of 8, 16 and 32 bit microprocessors), has overtaken this concept and restricted its wider application.

DNC is now more generally regarded as 'distributed numerical control' which stores and distributes programs to a number of machines from a central 'host' computer, but does not directly control the machines in real time at operational levels. It is concerned more with the communication of the machine(s) to the outside world, and is one of the enabling technologies for the introduction of flexible manufacturing.

Flexible manufacturing systems (FMS) CNC/DNC is the core technology of FMS which is basically the expansion of numerical control to encompass many processes and mechanical handling and its integration with production planning activities to form a unique, computer-controlled production facility.

Current situation NE is still developing and the current and future states of the technology are due to continuing, parallel development in four major areas:

(1) machine design and performance,
(2) computing power of the machine controllers,
(3) software and programming methods (CADCAM), and
(4) tooling design and performance.

14.5.2 Numerical control systems

14.5.2.1 NC systems

NC-only installations are becoming fewer, with replacement of older tape controlled machines by CNC versions. Paper tape (eight channel, 1 in. wide) is the most commonly used medium for general machines, whilst magnetic tape is used on some specialist machines. Magnetic tape can store much more data per inch (typically 500 characters/inch) than paper tape at 10 or 12 characters/inch. However, being a magnetic medium, it requires greater protection from damage and the corruption of data.

Comparing NC systems with CNC, the advantages of NC paper tape systems are:

(1) cheaper machine controls than CNC;
(2) paper tape can withstand more mishandling, oil and dust than can disks;
(3) magnetic tape can store many characters; and
(4) security of storage on the shop floor is less risky than with reuseable floppy disks.

The disadvantages of NC paper tape systems are:

(1) NC has no memory or ability to make decisions from machine data feedback;
(2) tape must be read for every part produced;
(3) tapes wear with use and must be replaced;
(4) tape readers in constant use break down more often than do CNC controllers;
(5) the control signal input is slower than CNC; and
(6) tape systems cannot be incorporated easily into the DNC of FMS installations.

Programming for NC and the use of tapes and tape codes are discussed in Section 14.5.7.4.

Updating NC controls It is possible to update some NC machine controls to CNC or DNC modes by the addition of a behind-the-tape reader (BTR) system. These are systems that provide additional control boards that are installed 'between' the tape reader and the machine controller. A number of companies offer such a product.

14.5.2.2 CNC systems

CNC is now the most widely used type of numerical engineering. The advantages of CNC systems are:

(1) programs are stored on the machine ready for use;
(2) programs and data can be modified on the machine;

(3) variable tape formats are acceptable, e.g. ISO and EIA;
(4) Imperial and metric working is switchable from within the program;
(5) commonly used machining patterns and instructions can be stored as subprograms;
(6) tool compensations and offsets can be stored;
(7) machining conditions can be optimised through feedback and adjustment;
(8) many control tasks have been automated, giving more speed and accuracy;
(9) more control options are selectable from within the program;
(10) machine diagnostic programs can be used; and
(11) communications with other computer systems are possible, e.g. CAM and DNC.

The major elements of a CNC system are shown in *Figure 14.22*.

Programming There is a number of methods of producing a program and these are described in detail in Section 14.5.8. The programmer defines the part geometry, movement required and other features. The computer program then produces a cutter location file (CLFile or CLF) which defines the necessary tool paths. This is in a generalised format and is not readable by a CNC machine.

Post-processing A post-processor is a program that converts CLFs into the word address format suitable for a particular machine model. Different models require different instruction formats. The G and M codes and other machine management data are added here.

Input to the machine This is only done once because the program is then stored in the controller RAM. This allows the program to be read and executed when required, and also to be viewed and edited to some extent at the machine.

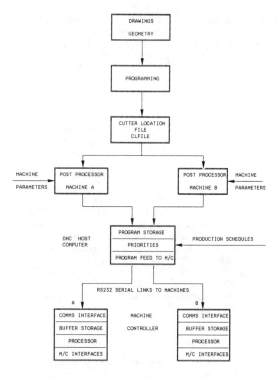

Figure 14.23 A DNC system

14.5.2.3 DNC systems

The main elements of a DNC system are shown in *Figure 14.23*. There are two types of DNC: direct and distributed.

Direct numerical control This was first developed in the 1960s to allow the direct control in real time of a number of NC machines from a single central computer. It was intended to provide the advantages of CNC to NC tape machines that did not have the computing and memory capabilities in their controllers. However, the introduction of CNC on individual machines overtook the advantages of early direct numerical control in most investment cases, and has restricted its wider application.

Distributed numerical control This was developed to provide central storage, engineering and management of part programs for a number of CNC machines from a central computer. A major benefit is the better scheduling and optimisation of the time available on machines across multiple shifts, whilst leaving the lower level internal control functions at the individual machines where local control is more appropriate and convenient. There are two types of distributed control system: maximum flexibility and minimum cost systems.

In *maximum flexibility systems* complete part programs are fed to the machine which then needs sufficient memory to accept it. This avoids any interruption of the part production which could be caused by conflicts at the control computer delaying the drip feed of the program blocks. It also reduces the chances of data-transmission errors.

In *minimum cost (drip feed) systems* part programs are held in the central computer, but are fed a block at a time into a buffer store in the machine control which commonly accommodates up to 4 kbyte or 100 blocks. This system is useful for

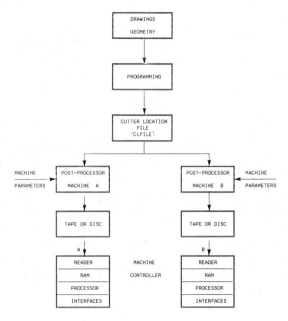

Figure 14.22 A CNC system

very large programs which can be run without the need for a large memory on the machines, which consequently can have cheaper control. However, with the reduction in memory cost and the corresponding increases in the standard memory size now offered (typically 1 Mbyte) this advantage is reduced.

DNC links from the central computer to the machines are made via an RS232 serial communications cable. This feeds data a bit at a time, but is fast enough for the transmission between the computer and machine buffer store. However, in order to achieve the data-transfer speeds needed during the running of the program on the machine, a parallel link which transmits a word (8 bits) at a time is needed.

The application of DNC In small environments where the operators play a large engineering and supervisory role, the independent control of machines is more appropriate and less expensive than either direct or distributed control. This is particularly so with the current trend to interactive graphical (conversational) programming at the machine.

In many large and small environments the current move is towards flexible manufacturing. This widens the scope of control from the purely technical to include production planning and management. In this situation DNC is being used as one of the enabling technologies in creating a self-contained computer-controlled flexible manufacturing system.

14.5.3 CNC machines—general features

CNC technology has been applied to many manufacturing processes. The most common applications are in:

(1) machining centres—horizontal and vertical;
(2) turning centres and lathes;
(3) sheet metal—punch presses, laser cutting, bending;
(4) grinding—surface, cylindrical;
(5) coordinate measuring machines (CMM); and
(6) electrical discharge machining (EDM)

This section covers features that are common to a number of these, whilst the subsequent sections deal in more detail with particular machines.

14.5.3.1 NC axis conventions

NC axis classification follows the three-dimensional cartesian coordinate system and is established in BS 3635: 1972: Part 1. *Figure 14.24* shows the three primary axes and the associated rotational axes.

The z axis is parallel to the main spindle of the machine. It will be horizontal on a lathe or horizontal machining centre and vertical on a vertical machining centre.

The x axis is always horizontal and at 90° to z.

The y axis is at right angles to both the x and z axes.

The axes A, B and C are the rotary movements about x, y and z, respectively.

The w axis is used for secondary spindles or slideways parallel to z.

The r axis is used for tertiary spindles or slideways parallel to z.

The u axis is the traverse movement related to the w axis.

14.5.3.2 Cutting spindles and workheads

Most machines for general use have one cutting spindle or workhead, the axis orientation of which defines it as being either a vertical or horizontal machine. The spindle axis is normally fixed in one plane but machines with tilting spindles

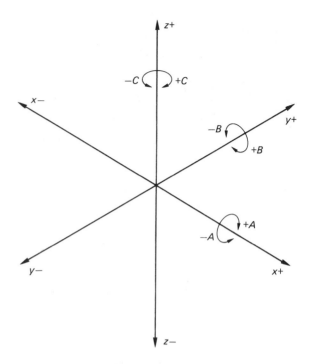

Figure 14.24 Coordinate axes convention

(typically up to 150° swing) are available. Multispindle machines are used for specialist tasks such as aerospace-component contouring or machining heavy engineering products.

Machines will be selected to provide the spindle speed ranges and horsepower to suit the materials to be cut. Aluminium work requires higher speeds and lower horsepower than ferrous metals and alloys.

Distortion and inaccuracies due to the thermal effects of continuous high speeds and heavy cutting forces are a major consideration in CNC machines. Main spindle assemblies are normally cooled by a recirculating oil system to overcome this problem.

Tables 14.2 and *14.3* show typical spindle and axis capabilities for a range of machining centres.

14.5.3.3 Drive systems

A.c and d.c. servodrives are the most widely employed types of drive, but stepping motors, hydraulic drives and pneumatic drives are also used.

Servodrives The majority of CNC machines employ a.c. or d.c. infinitely variable speed servodrives, in which the drive speeds are varied by electrical means—varying the voltage on d.c. motors and the frequency on a.c. motors. Many machines have a two-ratio gearing system for spindle drives in order to cover the wide speed and torque ranges offered. Generally, CNC drives do not employ mechanical gearboxes and clutches to give a stepped range of speeds, as described previously for conventional machine tools. The advantages that electrically variable stepless drives have over a geared system are:

(1) faster response of the drive system to the commanded changes;

Table 14.2 Typical axis and spindle characteristics of vertical machining centres without pallet shuttle*

| Feature | Machine size | | |
	Small	Medium	Large
Axis travel (mm)†			
x	600 (23.6)	1270 (50.0)	2600 (102.4)
y	410 (16.1)	650 (25.6)	814 (32.0)
z	510 (20.1)	650 (25.6)	750 (29.5)
Feed rate, xyz (mm/min)‡	0.001–5000 (17)	0.001–5000 (17)	0.001–3600 (12)
Rapid traverse (m/min)‡	20 (65)	15 (49)	10 (33)
Axis drive (kw)			
x	0.9	2.8	3.3
y	0.9	2.8	3.3
z	2.8	3.3	3.3
Table loading (lb/kg)	1540/700	4400/2000	6600/3000
Spindle speed (rpm)	80–8000	14–6000	10–3500
Spindle drive (kw)§	7.5 (5.5)	15 (11)	15 (11)
Drive type	A.c. direct servo	A.c. direct servo	A.c. direct servo
Tool-shank No.	40	40, 50	50
Tool storage			
Standard	20	30	40
Optimum	30	40	50

* Courtesy of Mori-Seiki (UK) Ltd.
† Numbers in parentheses are inches.
‡ Numbers in parentheses are feet per minute.
§ Numbers in parentheses are for 30 min.

(2) more rapid and smoother acceleration and deceleration;
(3) infinite number of speeds available—not limited by gearing ratios;
(4) full torque available over (typically) top 97% of the speed range;
(5) fast and efficient reversal of direction when needed;
(6) less noise than geared drives; and
(7) less maintenance than geared drives.

Where an indirect drive is appropriate, toothed belts are used to connect motor output shafts to the driven system. The belts are quieter and cheaper than gear trains.

Servodrive feedback Servodrive systems need to have closed-loop feedback devices to indicate actual position and speed. Such devices are necessary for position feedback because the rotational position of the servodrive motor is too variable to be used to calculate accurately (together with the known lead of the leadscrew) the actual new position of the driven structure. Positional feedback devices such as encoders or linear scales are needed and these are discussed in Section 14.5.3.5.

Feedback devices are necessary for velocity feedback (speeds and feed rates) because both a.c. and d.c. servodrives given an increased speed with an increased signal input, but the speed obtained cannot be forecast accurately enough for good control conditions. The resultant changed output speed

must be measured and fed back to modify the input signal, if necessary. This is a continuous velocity feedback closed-loop situation. The device used to exert the control is a tachogenerator which produces a voltage proportional to its speed. This voltage is fed back to the input signal controller and provides the data for further corrective action. The tachogenerator is normally housed within the servodrive unit.

Stepping motors These are driven by the receipt of digital pulses which cause the shaft to rotate by a precise amount. The frequency of pulses determines the speed and the total pulses gives the degree of movement. The digital nature of the stepper motor provides a sufficiently accurate measurement of position and speed for many applications, so a feedback system is not normally required, making the cost of the total system lower. A disadvantage is that the achievable power output is lower than with servodrives, so applications are limited to lighter duty machines.

Hydraulic drives These are used on a number of machines, e.g. the table longitudinal feed on CNC surface grinding machines. In this case hydraulic drives are more suitable for the very frequent directional reversals and rapid speeds needed for the table longitudinal feed, where high positional accuracy is not necessary. Servodrives are used, however, on the wheel-feed (y) and cross-feed (z) axes where high preci-

Table 14.3 Typical axis and spindle characteristics of horizontal machining centres with pallet shuttle

Feature	Machine type		
	Small FMT-100	Medium FMT-200	Large FMT-300
Axis travel (mm)			
x	750	1300	1750
y	650	1000	1000
z	500	1000	1000
Feed rate, xyz	1–2400 mm/min in 1 mm/min steps		
Rapid traverse	2400 m/min (80 ft/min)		
Axis drive			
Axis positioning	Encoder (standard); linear scale (optional)		
Max. pallet load (kg)	750	2000	2000
Spindle speed	5–4500 rpm in 1 rpm steps		
Spindle drive			
Spindle power	22 kW at S6–60%; 17 kW continuous		
Tool-shank No.	50	50	50
Tool storage capacity	48 standard; 60 or 120 optional		

* Courtesy of FMT Ltd.

sion is essential and the axis drive/positioning system needs to incorporate a linear scale positional feedback device.

Pneumatic drives Pneumatic drives can be used in a number of peripheral applications on machines, but are not commonly used for main slideway drives. Recently, however, pneumatic systems have been replacing some hydraulic drives on the grounds of lower cost, noise, cleanliness and heat generation.

14.5.3.4 Slide movement elements

Slide movements are normally driven by the a.c. or d.c. servodrives described in the previous section. These drive either a recirculating ball and leadscrew, or rack and pinion systems.

Recirculating ball and leadscrew This is the most commonly used method on CNC machines. It comprises an accurately ground or rolled leadscrew running through an enclosed set of continuously circulating steel balls. The mechanical efficiency of this system is very high—typically over 90% which is much more than a conventional open leadscrew and nut. The system is preloaded in both directions to eliminate backlash and torsional wind-up.

Rack and pinion These are used on large machines with very long (e.g. several metres) slide travels where the adequate support of a long leadscrew would be very difficult. The mechanical degree of accuracy is not as high as a ball/

leadscrew but with the use of positional feedback systems (described in the next section) this is overcome sufficiently to achieve the required machine accuracies.

Moving table and structures are usually supported by low friction recirculating multiroller bearings acting on three or four sides of two or three hardened and ground slideways. This design minimises stick–slip and aids smooth acceleration. The drive movement is normally mounted between the slides to minimise skewing.

14.5.3.5 Slide position measurement

Slide position can be measured by a number of methods, which fall into two categories: rotary (angular) and linear transducers. *Rotary transducers* are mounted on the drive shaft or leadscrew and measure the displacement of a rotating disc. This is then used to calculate the linear position—the method is, therefore, an indirect measurement of slide position. *Linear transducers* are mounted adjacent to the axis slideway and measure the linear position of the slide directly.

Note that both methods are 'indirect' in the sense that the real measurement requirement is to know where the cutting-tool edge is located in relation to the work piece. Reference to the position of a point on a machine structure is, therefore, an indirect measure of the tool–work piece relationship. There are also several measurement technologies employed, often for both rotary and linear devices. These are:

(1) binary encoders,
(2) engraved gratings,
(3) moiré fringe gratings, and
(4) synchro-resolvers (linear inductosyns).

The two most commonly used methods are rotary encoders fitted to the drive unit and linear transducer positional systems.

Rotary encoders fitted to the drive unit This is widely used as the standard offering for position measurement. The encoder has a rotating disc that has a binary code or grating marked on it. The position and rotations of the disc are sensed by photocells which initiate a pulse when the binary code or grating markings are detected. The angular movement produced by one pulse is known (it varies with design) and is used to calculate the total rotational and linear displacement of the structure.

Linear transducer positional systems (linear digital scales) These are used where the machining accuracies required demand a more direct measurement of linear axis position. They are mandatory on grinding machines and are offered as an option on machining centres and lathes.

Inductosyns These are used by some manufacturers. An inductosyn has a scale fixed along the length of the axis travel and carries a single winding wire laid out in a U-shaped pattern along it. The moving carriage has a short length of scale with two windings of the same size and pattern as the larger scale. The position of one of the windings is displaced by a quarter of the pitch, i.e. at a phase angle of 90°. A constant voltage supplied to the slider windings induces a voltage in the fixed scale. If the carriage is in the correct position no current flows and the drive system stops. If the position is incorrect, current flows and the drive system moves in the direction relative to the current flow.

Positional accuracies The accuracy depends on the design of the machine and the cumulative inaccuracy of the drive and

Table 14.4 Manufacturers' quoted accuracies for general machining centres

	Positional accuracy (full range)	Repeatability
Typical highest positional accuracy quoted		
in.	±0.0002	±0.00004
mm	±0.005	±0.001
Typical lowest positional accuracy quoted		
in.	±0.0004	±0.00008
mm	±0.01	±0.002

measuring systems. Typical figures currently quoted by leading manufacturers of machining centres are given in *Table 14.4*.

14.5.3.6 In-process probing

In-process inspection and sensing is a valuable aid to the efficient running of CNC machines. It is used for:

(1) true position sensing of work piece–job setting;
(2) true position sensing of tool–tool setting;
(3) measurement of machined work piece;
(4) datum (calibration) work;
(5) component identification;
(6) tool-wear measurement;
(7) tool-breakage detection; and
(8) tool-length measurement and offset adjustments.

The effect of this capability is to add intelligence to the machine, which will allow:

(1) longer periods of unmanned running;
(2) better quality through closed-loop machining control;
(3) less downtime due to poor quality or tooling problems; and
(4) reduced setting times.

The equipment is normally a surface sensor that will detect when its stylus touches a surface. The stylus position is related to a datum so it is known where the surface is in the *xyz* frame.

There are two types of sensor in common use: sensor fixed to the machine table, and touch trigger probes mounted on the table or in tool holders ready for transfer to the machine spindle.

Fixed sensor *Figure 14.25* shows a sensor supplied by Cincinatti Milacron on their machining centres. The sensor is orientated for vertical or horizontal spindles/tool presentation and controlled by software in the machine controller.

Touch trigger probes These were developed for coordinate measuring machines (CMM), but have now found wide use in CNC machines. *Figure 14.26* shows different types of probe produced by Renishaw Metrology Ltd. The probes have a touch-sensitive stylus and a signal is sent to the CNC controller when it touches a surface.

Probes may be table mounted or held in the machine tool magazine ready for transfer to the spindle. Three methods of signal transmission are used.

(1) *Inductive transmission*—when placed in the spindle an inductive transmission interface aligns with one mounted on the machine body.

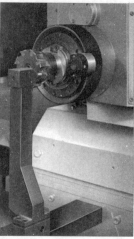

Figure 14.25 Sensors fixed to a table. (Courtesy of Cincinnati Milacron)

(2) *Hardwired transmission*—used for linking static tools in the standard manner.
(3) *Optical transmission*—optical transmitters and receivers are mounted within each other's view on the probe and anywhere outside the machine's working envelope. Light emitting diodes (LEDs) on the probe head send an infra-red signal that is detected by the receiver unit.

A range of software programs is available to determine:

(1) x, y, z datums;
(2) x, y, z position and error;
(3) bore and boss diameter and x, y centre coordinate;
(4) web or pocket width and x, y centre coordinate;
(5) internal and external x, y intersections;
(6) protection positioning for clamps, etc.;
(7) tool setting; and
(8) vector programs.

14.5.3.7 Swarf and coolant control

Automatic swarf and coolant clearance systems are needed because of the unmanned nature of the machines and the large amounts of swarf produced by the high productivity of the machines.

CNC machines are designed to allow the swarf and coolant to fall and drain away from the working area and machine surfaces. The swarf and coolant fall into a conveyor system which takes them to one end of the machine. Here, a unit separates the swarf from the coolant. The coolant is filtered and returned to the circulating system, whilst the swarf is either collected from that point or an elevator and discharge hopper are fitted. The hopper is set at a height which allows a barrow to be placed under its discharge chute. The swarf is then taken away manually.

Automatic swarf-clearance systems serving several machines by using overground or underground systems have been installed, but the large capital costs of providing and maintaining these has not proved to be an attractive economic proposition for their general implementation in typical machine shops.

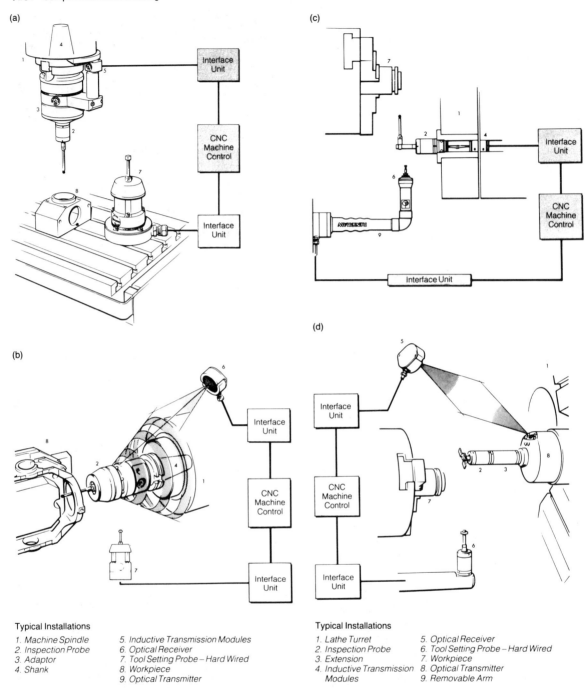

Typical Installations

1. Machine Spindle	5. Inductive Transmission Modules
2. Inspection Probe	6. Optical Receiver
3. Adaptor	7. Tool Setting Probe – Hard Wired
4. Shank	8. Workpiece
	9. Optical Transmitter

Typical Installations

1. Lathe Turret	5. Optical Receiver
2. Inspection Probe	6. Tool Setting Probe – Hard Wired
3. Extension	7. Workpiece
4. Inductive Transmission	8. Optical Transmitter
Modules	9. Removable Arm

Figure 14.26 (a, b) Probes on CNC machining centres. (c, d) Probes on lathes. (a, c) Touch probes. (b, d) Optical probes. (Courtesy of Renishaw Metrology Ltd)

14.5.4 Machining centres

14.5.4.1 Typical configurations

CNC machining centres are now the major machine tool used for the production of prismatic parts. They combine operations formerly carried out on mills, drills, small borers and shaping machines, and are in essence a development of conventional milling machines into a universal role. In fact the major classification of machining centres follows that of milling machines in being designed with vertical or horizontal cutting spindles.

Modern design of machining centres has concentrated on obtaining high rigidity, force/vibration absorption and thermal

Table 14.5 Typical types and sizes of machining centre (other configurations are available)

Category	Table size (mm)	Spindle				No. of axes	No. of tools	Pallet handling	Applications
		Horizontal/ vertical	Single/ multiple	Max. power (kW)	Max. speed (rpm)				
Drill and tap	600 × 300	V	S	5	10000	3	12	N	High-speed drill and tap on small parts
Small	900 × 450	V, H	S	10	12000	3	40	Y	Small parts, aluminium
Medium	2000 × 600	V, H	S	20	6000	4	80	Y	General casting, forgings
Large	3000 × 1000	V, H	S	40	4000	5	120	Y	General casting, forgings
Gantry	⩾6000 × ⩾2000	V (mainly)	M	80	10000	6	60	N	Aerospace parts and heavy industry

N, no; Y, yes.
Note the drill & tap and travelling gantry machines are included because they process prismatic parts and have many features that are common to the machining centres

balance in order to give repeatable accuracies and long life under the high cutting and machine movement forces that are generated by modern cutting technology. *Table 14.5* lists details of a range of machining centre sizes.

Vertical machining centres These range in size and application from small machines for light machining, drilling or tapping to large twin-column machines for heavy engineering work. *Table 14.2* gives details of the specification of three machines from the range offered by Mori-Seiki Ltd. This is typical of the range that leading suppliers produce in the general machining field, whilst other companies concentrate on a particular size range. *Figure 14.27* shows the features of a medium sized vertical machining centre for general machining.

Pallet change systems are available on most models, either as standard or as an optional extra.

Most machines for general machining have a single fixed column supporting the vertical z axis spindle movement, the y-axis movement being supplied by table horizontal motion. Travelling column machines are available where the y axis motion is provided by horizontal movement of the column.

A recent development has been the introduction of 'compact' machines for small general machining. These typically have a 10–20 tool carousel for 40 taper formats, and no pallet-handling facilities. They are designed as stand-alone machines for small batches of light work and the whole machine is usually enclosed within a guarding system similar in character to that of a CNC lathe enclosure. The floor area occupied by compact machines is typically 40–50 ft^2.

Horizontal machining centres These range in size from small to very large machines for heavy engineering work. They are a common component of flexible manufacturing systems installations where pallet changing by guided vehicle is used.

Other types of machine A variety of CNC borers, plano-mills and profiling machines are used for heavy engineering or specialist work. These include many configurations combining some of the following features:

(1) fixed or moving bed,
(2) single, two or three columns,
(3) fixed or moving columns,
(4) single or multiple spindles, and
(5) fixed or swivelling spindles.

Figure 14.28(a) shows a five-axis machining centre with a spindle that can swivel through 150° for contouring work. *Figure 14.28(b)* shows the usual axis conventions for such machines. These follow the conventions outlined in Section 14.5.3.1.

Figure 14.29 shows a high speed gantry machining centre with a fixed bed for holding the work and a moving gantry supported on parallel rails. The operator travels in a cab on the gantry. Components can be up to 5 m (16 ft) wide.

14.5.4.2 Workholding systems

Pallet systems Pallets carry the work-holding fixtures to and from the working head. The pallet is located accurately at the workhead (usually by tapered cones) and locked in position to ensure repeatability of accuracy and safety. It is common for fixtures to be dedicated to one pallet to save on set-up time and setting inaccuracies. Pallets are supplied with either tee-slots or a pattern of lapped holes for fixture bolts.

Pallet changing is a common feature of many stand-alone horizontal machining centres and is essential for machines in a flexible manufacturing system (FMS) or flexible manufacturing cell (FMC). Two types of system are used.

(1) Twin pallet:
  (a) rotary pallet changer, and
  (b) in-line pallet shuttle.
(2) Multi-pallet pool:
  (a) rotary pallet carousel, and
  (b) linear (usually chain conveyor) carousel.

Figure 14.30 shows alternative layouts for pallet systems.

Twin pallet systems may form part of two system configurations. (1) Where the pallets are on a single machine, the operator loads and unloads one pallet whilst the other is being machined. This system allows uninterrupted machining and lightly manned running (depending on the machining cycle times). (2) Twin pallet systems also form part of the input/output system of a FMC or FMS where pallets are fed to and from the CNC machine by rail guided vehicles (RGV) or automatic guided vehicles (AGV). In this system one pallet load of parts is machined while the other pallet is transferred from machine to vehicle, and vice versa.

In a *multi-pallet pool*, 4–16 pallets are permanently connected to the machine. Systems of four to eight pallets are often held on a rotary table, but larger numbers are usually mounted on a conveyor system running in front of the machine. This type of installation is used for unattended running (say through a second and third shift), where the total number of components multiplied by the number of machining cycles gives a long machine running period. The pallet pool is also a way of providing within the machine envelope all the fixture pallets required for the product range to be processed on that machine. This provides a stand-alone machine with some of the flexibility of a FMS or FMC. If the machining centre is part of a flexible-manufacturing installation, then the pallets may be held in a rack system and retrieved by a stacker

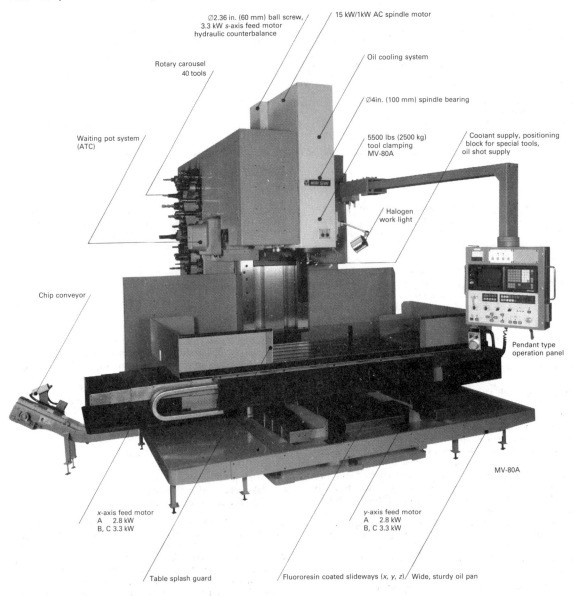

Figure Figure 14.27 A vertical machining centre. (Courtesy of Mori-Seiki Ltd)

crane and/or conveyor running in front of the machine. (see *Figure 14.59*).

Fixtures Fixtures are often held permanently on one pallet in order to reduce repetitive setting and maintain quality by having a consistent set-up. Multiple component fixtures are common and give greater efficiency because of:

(1) the reduction in the number of multiple pallet travels to and from the workhead for separately fixtured components, and

(2) the reduction in tool approach distances—these will be shorter for component-to-component travel around a multiple fixture.

A fixture may also be used to house sets of components that go to make up a subassembly, e.g. a body of some description and its mating cover. This ensures more accuracy in the fit of mating machined faces and hole centres on the different components and also reduces the risk of mismatched quantities at assembly.

Multiple fixtures are usually constructed in the form of a mass block with components mounted on two or four vertical faces. They can be used to carry out the same operations on all faces or a series of operations on successive faces of the fixture. Component turnover would be carried out manually at a station adjacent to the machine or pallet carrier. This approach is suitable for low-volume work, but for higher volumes it may be more economical to have a number o

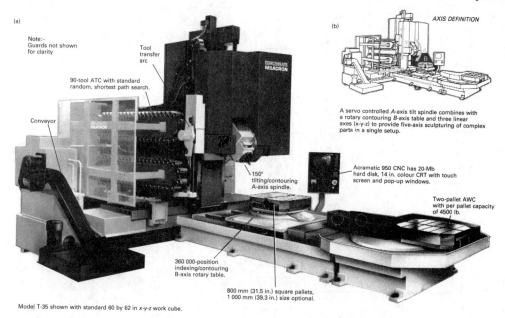

Figure 14.28 (a) A five-axis profiling machine (guards not shown for clarity). (b) Definition of axes. (Courtesy of Cincinnati Milacron)

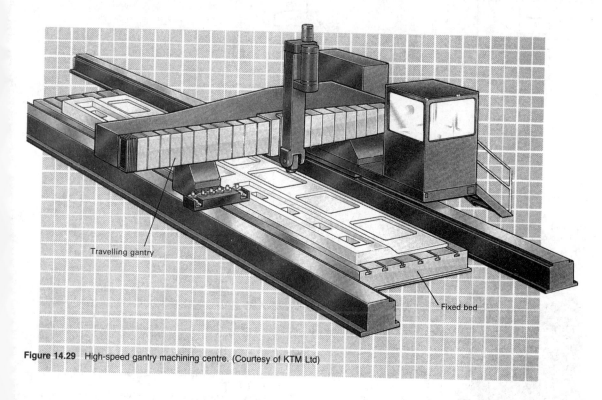

Figure 14.29 High-speed gantry machining centre. (Courtesy of KTM Ltd)

machines, each set for operations on the opposing faces of the component. A mass block with fixture bolt holes is shown in *Figure 14.31*.

Indexing tables These are used with components bolted directly to them or with fixtures mounted on them. They will

normally be linked to the machine control, their operation becoming a fourth or fifth axis of the machine. Indexing accuracies are of the order of ±3 in. of arc. Indexing increments are usually 360 000, i.e. are related to an increment in axis control of 1/1000.

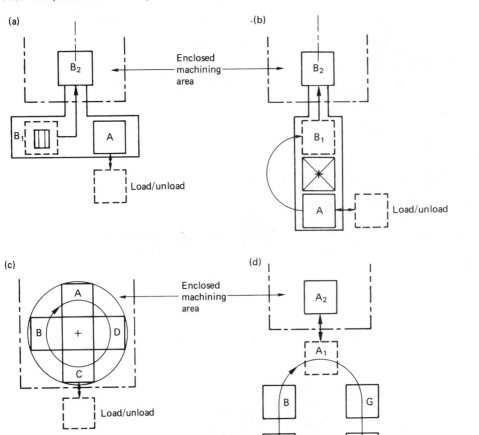

Figure 14.30 Pallet layouts. (a) Twin pallets, linear changer. (b) Twin pallets, rotary changer. (c) Multiple pallets, rotary carousel. (d) Multiple pallets, linear carousel. N.B. In (c), the complete table may be inside or outside the machining enclosure. If it is outside, a layout similar to (b) is needed

Figure 14.31 A mass block for holding work fixtures

14.5.4.3 Tooling systems

Automatic tool changers These are normally offered with tool capacities from 10 on small drill and tap machines to 400 on large machines. An automatic tool-changing arm transports tools from the magazine to the spindle. Whilst that tool is being used the tool changer will collect the next tool required and return to its spindle load/unload position ready to exchange tools at the end of the present cutting operation.

The four commonly used types of tool magazine are shown in *Figure 14.32*.

The *rotary carousel* type is used only on smaller machines due to the restricted number of tools that can be held at the periphery of a carousel that will fit the envelope of the machine. They may be vertically or horizontally mounted. The vertical type requires a tool-change arm, but some horizontal models align the tool pocket with the spindle to effect direct transfer.

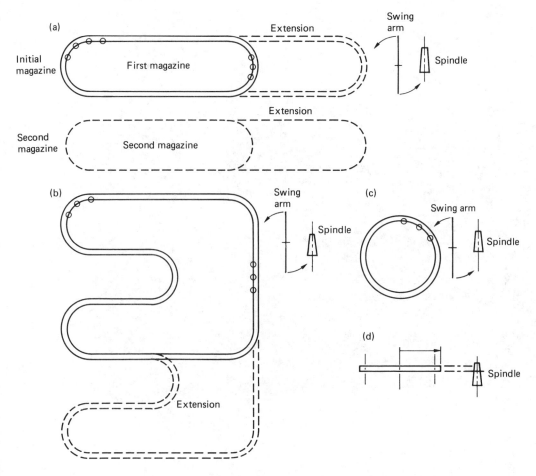

Figure 14.32 ATC magazine configurations: (a) linear; (b) linear looped; (c) rotary vertical; (d) rotary horizontal, direct change

In the *linear chain carousel* type the tool pockets are mounted on a chain conveyor system which can accommodate many more pockets per square foot of machine side area than a rotary carousel. An added advantage is that by doubling back on the chain route an even longer length of chain and, therefore, tool holders can be installed. Automatic tool changing machine tool magazine sizes are offered in steps, e.g. 40, 80, 120 and 160. Most machines can have two extended magazines, thus doubling the tool capacity to 200–400 tools.

Tool transfer from magazine to spindle In a *swing arm changer* a pocket at each end rotates to alternately accept and deposit tools at the spindle and tool magazine. Tool-to-tool change times of 3 s are claimed. All linear and some rotary carousel types of tooling system require a changer of this type.

In *direct carousel change*, if the carousel is designed to move the tool pocket onto the spindle centre line, direct transfer from pocket to spindle, and vice versa, can be made. This eliminates the need for a swing arm changer.

Magazines often have two or three sizes of tool pocket to cater for different tool sizes. Two systems are widely used for returning tools to the magazine:

(1) fixed address—the tool is returned to the same preprogrammed pocket that it came from; and

(2) random memory—the tool is returned to the nearest pocket of the correct size. This can save tool-change time.

Types of tooling CNC tooling has been developed from the traditional milling, drilling and boring tools with the mounting device to the spindle being one of the standard tapers used internationally. Most machining centres accommodate either ISO 40, 45, 50 or 60 taper sizes.

14.5.5 CNC turning centres and lathes

14.5.5.1 Introduction

CNC turning centres and lathes are the major machine tool used for the production of rotating parts. The distinction between 'turning centres' and 'lathes' is arbitrary and the following text uses 'lathe' to cover both. These CNC machines have the advantages of numerical control to conventional turning but, in addition, have incorporated dual spindle/turret work, second-operation work and automatic parts loading/unloading. They can also be included in FMS and FMC installations.

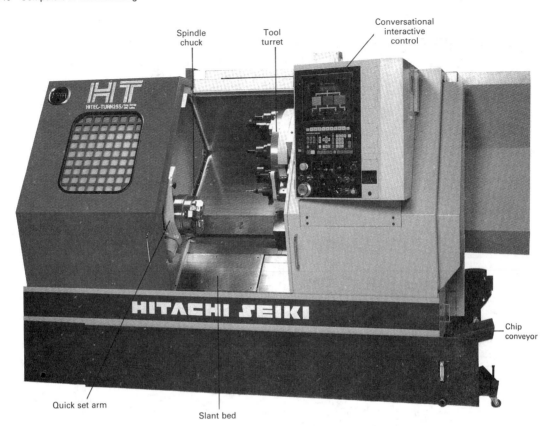

Figure 14.33 A CNC lathe. (Courtesy of Hitachi-Seiki)

Table 14.6 Range of typical features in CNC lathes

Size range	Max. turn diameter (mm)	Max. turn length (mm)	Bar cap or bore (mm)	Spindle drive motor (kW)	No. of axes	Max. No. of tools	Work handling system
Small	200	300	30	7	2	8	Small robot arm
Medium	600	700	90	25	3	12	Robot arm
Large	1200	10000	120	50	4	24 (two turrets)	Magazine, large robot gantry loader

CNC technology has been applied to the complete range of lathe sizes from small bench-top and training machines to the large long-bed machines used in heavy engineering.

The features of CNC lathes varies considerably in both scope and scale. The number of spindles and tool turrets as well as the work-piece envelope are combined to give machines designed to process a particular type of component, level of quality and rate of production output. *Figure 14.33* shows a medium sized slant-bed lathe and *Table 14.6* shows a typical range of common features.

CNC lathe construction has altered considerably due to the much heavier cutting forces and faster speeds and feeds now employed. The conventional flat-bed configuration has been superseded by slant-bed and vertical-bed designs. These give more rigidity and support to the machine and tool turrets, and also allow the large volumes of coolant swarf to fall below the working area onto a collection system. *Figure 14.34* shows the basic construction of a slant-bed machine and *Figure 14.35* shows a small flat-bed lathe.

Figure 14.34 CNC lathe construction. (Courtesy of Cincinnati Milacron)

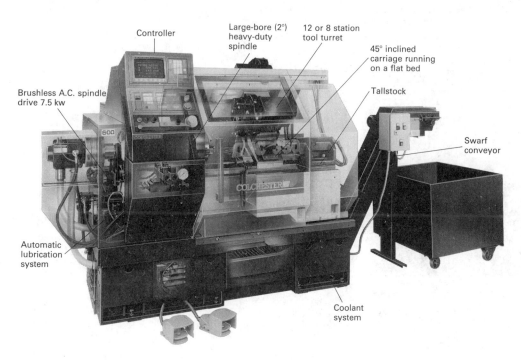

Figure 14.35 A small CNC lathe. (Courtesy of Colchester Lathe Co. Ltd)

14.5.5.2 Axes

The number of axes varies from 2 to 6. The convention normally used for lathes is:

z—the axis in line with the spindle and the longitudinal movement of the tooling carriages or turrets along the machine bed;

x—the axis at 90° to *z*, and normally the axis for cross-slide movement;

C—the rotational axis of *z*, used to specify the spindle radial position for second-operation work.

N.B. with dual-turret machines both turret movements are referred to as *z* or *x*. The program distinguishes which turret is being referred to.

14.5.5.3 Tooling

Tool holding and setting Tool holding and setting systems have been specifically developed for CNC lathes. These achieve two major objectives:

(1) accurate tool location in a uniform system of holders; and
(2) short tool-changeover times through the use of quick one-action location and locking devices.

The systems employ a hierarchy of tool and mounting block units that fit into the tool turrets. The quick location and clamping is achieved by one-key or twist-type mounting designs. A tapered polygon mounting system is shown in *Figures 14.36* and *14.37*. The reduction in set-up and tool-change times is significant. Complete set-up times can be

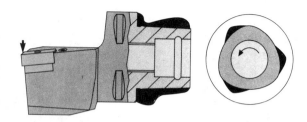

Figure 14.36 Collet activated taper polygon tool mounting. The length of the taper and the precision contact surfaces result in low surface pressures. This means long tool life and great repetitive accuracy. The stress curves lack peaks, avoiding the risk of vibration or deformation. (Courtesy of Sandvik Ltd)

reduced from 60 to 10 min and a single tool change from 10 to 1 min.

Other features that aid fast changeover times are:

(1) in-built probes used to aid tool offsets—feeds data directly to the machine controller;
(2) quick-change jaws in chucks;
(3) quick-change chucks; and
(4) presetting fixtures which are used to preset tooling off of the machine.

Machining envelope CNC lathes hold and employ a large number of tools in a small area which must also include the

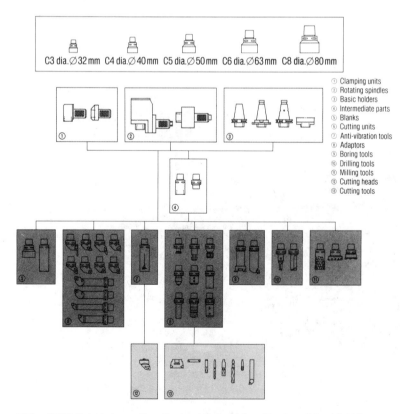

Figure 14.37 Twist lock mounting—Coromant Capto system. (Courtesy of Sandvik Ltd)

work piece and, where used, chucks, tailstocks, parts-handling equipment and setting/inspection probes. The position, shape and movement paths of all these elements vary considerably, so the possibility of collisions needs careful consideration. The two major aids in avoiding collisions are manufacturer's machining envelope diagrams and program verification.

Manufacturer's machining envelope diagrams—each machine model has the machining envelope defined. *Figure 14.38* shows an envelope diagram which specifies the positions and travel limitations of features.

Program verification—It is essential that the validity of a program and the set-up is verified before cutting metal. This can be done physically by the trial machining of plastic prototype materials which are hard enough to maintain the machined form but sufficiently soft to avoid serious damage should a collision occur.

Verification is now possible during the programming stage. Modern graphical and conversational programming software includes the facility to show the work piece and the path of each tool cut on the screen. *Figure 14.39* shows the tool paths for some G codes and the magnified scaling of geometrical features as shown on the Fanuc OTC control of a Colchester CNC lathe.

14.5.5.4 Work handling

With the increasing use of FMS and unmanned running, work-handling systems have been developed for CNC chucking lathes. The most usual forms are:

(1) the work piece magazine feeds directly into machine;
(2) a robot transfers the work piece into the machine from a magazine or conveyor; and
(3) overhead gantry supports a feed arm/robot arm transferring work pieces into the machine from a magazine or conveyor.

Direct-feed magazines are the cheapest method, but are usually limited to feeding small symmetrically shaped parts in order to prevent parts jamming in the feeder tracks.

Robots may be small purpose-built units that are a sub-system of the lathe, or larger free-standing units, particularly

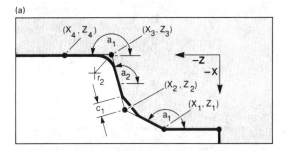

(a)

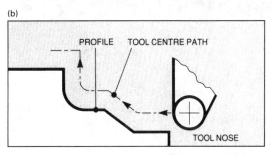

(b)

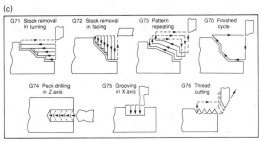

(c)

Figure 14.39 Tool path and geometry displays on a cathode-ray tube. (a) Direct drawing dimension programming. (b) Tool nose radius compensation. (c) Multirepetitive cycles. (Courtesy of Colchester Lathe Co. Ltd)

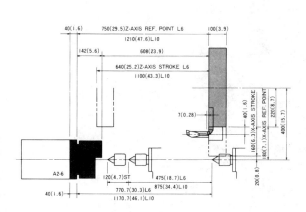

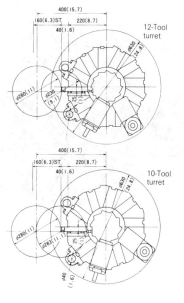

Figure 14.38 Turning capacity. (Courtesy of Rockwell Machine Tools Ltd)

Figure 14.40   Robot feeding CNC lathe. (Courtesy of Hitachi-Seiki)

for large parts. Robots can more easily handle unsymmetrical parts that will not feed in magazine tracks and also parts that need to be reversed for a second operation. *Figure 14.40* shows a robot that is a subsystem of the lathe.

Gantry systems are usually employed on machines turning large parts. They straddle the machine from a separate structure and support a moving arm on a horizontal beam. *Figure 14.41* shows a gantry system that loads parts and tools.

14.5.6   Other CNC machines

14.5.6.1   *Sheet metal punch presses*

These are used to punch components from material thicknesses of up to 0.250 in. (6.35 mm), although the major-ity of components produced by this method are in the 0.032–0.120 in. range. The sheets of material (most commonly plain or coated steel) are supplied in standard sizes of typically 2 m × 1 m or 6 ft × 4 ft.

The sheets are laid horizontally and clamped at the edges on an *xy* table which is moved under a punch head by the program. The head is then activated to punch holes with a series of standard shaped and sized tool-steel punches, or to nibble a contour with a single punch. By using this process on large sheets of material, a large number of components can be produced from the same sheet. The punch-hit rate is now up to 200 per minute, depending on the traverse distance between holes.

There are two types of machine:

(1) a rotating turret head which contains all the tools re-quired (similar to an ATC carousel on a machining centre)—a punch ram moves directly through the turret tool pocket to push the punch into the material; and
(2) a single punch head which is fed with tools from a separate magazine.

Both types may have either an overhung C frame or a bridge-type construction to suspend the punch head over the *xy* table. An example of a C frame machine and the tool holder and punching area are shown in *Figures 14.42* and *14.43*.

Programming is usually performed on punching modules of CAM suites that also contain milling and turning modules. Two prime objectives of the punching programs are to:

(1) minimise total sheet processing time by optimising the punching sequence and the *xy* table travel routings; and
(2) maximise material yield by optimising the number of components that can be produced from a sheet.

The latter involves laying out the component geometry and arranging patterns and shape *nesting* to give the best yield. Modern computer-aided programs do this automatically and present a plot of the components and tools on screen graphics and hard copy.

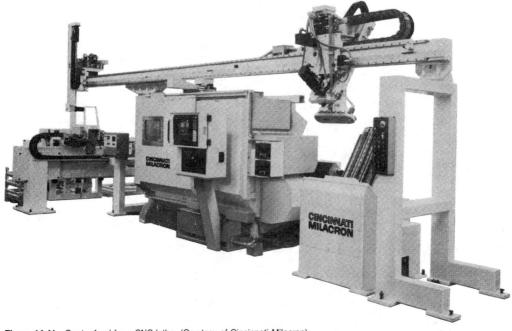

Figure 14.41   Gantry feed for a CNC lathe. (Courtesy of Cincinnati Milacron)

Figure 14.42   A C frame punch press. (Courtesy of Trumf Ltd)

14.5.6.2   Grinding machines

CNC technology has been introduced to grinding processes with many of the features of CNC machining centres and lathes successfully applied. This section describes the CNC developments in cylindrical grinding.

The Jones and Shipman CNC cylindrical grinder model Format 15 is taken as an example of a modern machine design and control system (*Figure 14.44*).

Machine design   The aspects of machine design are described below.

(1) Longitudinal table slide (z axis)—the drive systems are similar to other types of CNC machine. A d.c. servomotor drives a preloaded ballscrew for table positioning. The drive motor is mounted on the end of the leadscrew assembly which runs between the two slideways. Position is verified by a linear-scale feedback device.

(2) Wheel slide (x axis)—A d.c. servomotor is mounted directly on a hardened and ground ball screw running between the slideways. A linear-scale feedback device ensures extremely accurate wheel-slide positioning.

(3) Wheel-head swivelling (B axis)—where automatic (i.e. programmed) wheel-head swivelling is provided it is accomplished by means of a d.c. servomotor with a rotary encoder for positional feedback.

(4) Work-head swivelling (D axis)—Work-heads are driven by a d.c. servomotor via a vee belt. The work-head speeds are programmable and the head can be swivelled as the D axis with positional feedback from a rotary encoder.

(5) Work-head wheel speed—The wheel is driven by a d.c. servomotor. A program option is to have constant peripheral wheel speed which maintains the optimum cutting velocity, despite the reduction in wheel diameter.

(i) Punching area

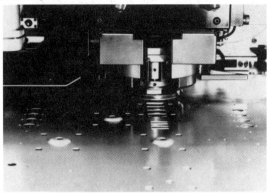

(ii) Tool holder

(iii) Tooling

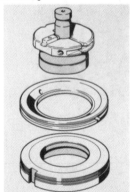

Figure 14.43 (a) Punching area, (b) tool holder, and (c) tooling for the C frame punch press shown in *Figure 14.42*. (Courtesy of Trumf Ltd)

Machine control and programming Grinding software is usually based on standard programming codes similar to those found in machining centres and lathes. Input can be manual data input (MDI), tape or direct numerical control (DNC) from off-line programming facilities. As an example, the main features of a Jones and Shipman Series 10 machine control and software are:

(1)   *x* and *z* axes are program controlled;

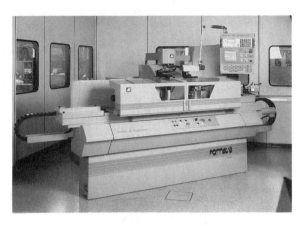

Figure 14.44 A CNC cylindrical grinding machine. (Courtesy of Jones and Shipman Ltd)

(2)   d.c. servodrives with linear scale feedback on both axes;
(3)   servo input resolutions (wheel slide 0.0002 mm; work slide 0.001 mm);
(4)   menu on the cathode ray tube takes operator through machine set-up;
(5)   dry run and prove out facility;
(6)   in-process diameter and shoulder length gauging;
(7)   programmable wheel dressing;
(8)   constant surface speed program;
(9)   circular and linear interpolation;
(10)   part program memory of 30 m (100 ft);
(11)   memory extension by RAM or bubble up to 1220 m (4000 ft);
(12)   canned cycles in battery-backed RAM; and
(13)   G & M codes for canned cycles—
   G61   contour generation grinding,
   G78   shoulder grinding
   G82   plunge grinding with gauging,
   G83   combined shoulder and plunge grinding with gauging,
   G84   multiplunge grinding,
   G86   traverse grinding,
   G87   traverse grinding with gauging,
   G88   plunge grinding,
   G89   combined shoulder and plunge grinding, and
   M91   contour generation dressing.

Some grinding macros are shown in *Figure 14.45*.

14.5.7   Machine controllers

14.5.7.1   Introduction

A CNC machine is controlled by a combination of the properties and actions of the controller, the physical control elements of the machine and the program software. This subsection deals with controllers, whilst Section 14.5.3 covers the machine elements and Section 14.5.8 deals with programming methods in detail. Like many other areas of computer application, the power and flexibility of machine controllers have progressively increased with the introduction of 8-, 16- and 32-bit microprocessors. In the immediate future, 64-bit reduced instruction set computers (RISC) will be introduced.

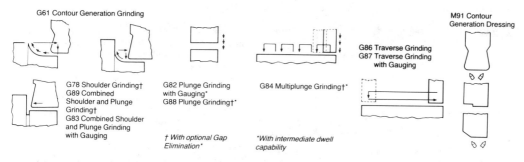

Figure 14.45 Grinding paramacros. (Courtesy of Jones and Shipman Ltd)

Controllers are available at various levels of capability in order to match the very wide range of CNC machines on the market. Many machine suppliers standardise on one supplier of controllers across their machine range. This greatly helps in the design of machine–control interfaces, the economics of supplier's inventory and the servicing and support of controllers in the field. A number of machine suppliers use their own controllers or models made by a sister company. In some cases the purchaser may have a choice between these and one other independent controller supplier.

Some of the major factors that influence the capabilities of a controller are:

(1) types of machine and processes to be controlled—machining centres and turning centres commonly have a different variant of a basic controller model;
(2) number and type of axes and how many are to be simultaneously controlled;
(3) spindle speeds and power;
(4) feed rates and torques;
(5) acceleration/deceleration profiles;
(6) accuracy and repeatability required;
(7) tool-management needed;
(8) special machining cycles (interpolations, etc.);
(9) standard canned cycles needed;
(10) type(s) of programming to be used (MDI conversational, PC graphical, APT);
(11) input/output devices (key types, CRT, tape readers, discs, printers);
(12) size and type of memory (ROM, RAM) needed; and
(13) communications required (DNC, RS232, etc.).

Before looking at controllers in more detail, the basic types of control are discussed in the next section.

14.5.7.2 Control theory

Control activities can be classified as working in either open-loop or closed-loop systems. CNC machines and controllers work in both modes.

Open-loop systems In open-loop systems (*Figure 14.46*) there is no feedback of the result of the command or energy input and, consequently, there is no automatic regulation of the input to achieve and maintain the desired output. An on–off switched process is the simplest form of open-loop system.

Closed-loop systems These can be divided into those that use feedback control and those that use feedforward control.

(a) *Feedback control*—in closed-loop systems the output condition is measured and fed back to the controller which

Figure 14.46 Flow diagram for an open-loop system

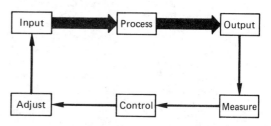

Figure 14.47 Flow diagram for a closed-loop system with feedback control

compares it with the command value or a preset standard (*Figure 14.47*). The controller normally has the intelligence to:

(1) compare the actual output against the desired standard;
(2) decide/calculate the degree of corrective action needed; and
(3) instruct an adjustment device or the origin of the input to act.

Many CNC control cycles are normally continuous as long as the 'process' (in this case the programmed machine activity) is running. They can usually be intermittent by setting the controller, or a portion of it, to the manual mode.

Applying closed-loop analysis to a typical CNC activity such as a table axis movement gives the loop diagram shown in *Figure 14.48*. The logic sequence is:

(1) the machine control program commands the drive control to drive the motor;
(2) the motor moves the table to a new position;
(3) the position transducer or encoder measures/signals the position of the table; and
(4) the drive control makes further drive action if necessary.

In practice, the drive controller, drive motor, position-measuring device and the axis drive element of the main controller can be integrated in one of a number of alternative configurations.

A characteristic of closed-loop systems is the time delay or lag between the measurement of the output and the effective response of the adjusted input. This delay is caused by the time taken for each element of the feedback loop to carry out

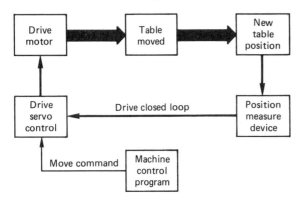

Figure 14.48 Flow diagram for a closed-loop system

its task and the time needed for the transmission of signals. These may be small individually but can accumulate into a discernable lag, which can cause problems in fast time-dependent operations. One area where this can occur is in the accurate positioning of slides where the loop delay is called 'servo lag'. A remedy is to use feed-forward control.

(b) *Feedforward control (predictive control)*—This is a control technique that looks ahead of the current situation and predicts conditions that could disturb the controlled variable, i.e. output. This prediction is converted (outside of any feedback loop) into corrective action to minimise the deviations of the output. This is useful in high-precision machining and is a feature of some drive servocontrol systems. It minimises the effect of the feedback system lag between the actual machine position and the command position.

14.5.7.3 CNC controllers

Current developments This section deals with the latest developments in CNC controllers, as these represent the main type of control used in numerical engineering. The latest controllers are substantially more powerful than earlier models because of the introduction of:

(1) faster microprocessors—16- and 32-bit reduced instruction set computers (RISCs) are now in common use and 64-bit RISCs are becoming available;
(2) improved memory—improved ability to store more technical data and part programs;
(3) programmable logic controller (PLC) applications—use of fast acting PLCs for the control of some axis independent of the CNC control;
(4) digital electronics—greater use of digital electronics giving faster operation of control functions;
(5) better graphics—use of 32-bit microprocessors allows faster and better graphics which also enhances the use of conversational programming; and
(6) surface-mount electronics—use of surface-mount components and printed circuit boards increases reliability and reduces the size of control units.

Controllers are usually designated for use with one of the major machine categories and are usually suffixed with a code, e.g.

Fanuc 0M — machining centres,
Fanuc 0T — turning centres,
Fanuc 0G — grinding machines.

The 0M and 0T models are shown in *Figure 14.49*. These are variants of the basic model 0 controller, the differences being

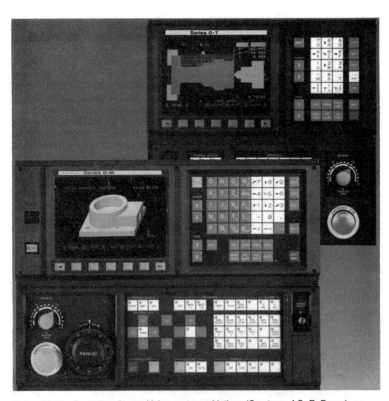

Figure 14.49 Controllers for machining centre and lathes. (Courtesy of G. E. Fanuc)

in those functions that deal with machine or programming requirements unique to a process.

There are four main areas of development in controllers:

(1) conversational programming at the controller;
(2) faster and more accurate control of functions involving speeds, feeds and position—particularly in the higher performance range;
(3) improved communications ability for DNC and networking (LANs and MAP 3.0); and
(4) use of a PLC for some functions and the introduction of ladder logic programming in PASCAL or C with the ladder display on the cathode ray tube.

Controller functions The major functions of top of the range controllers for machining centres and lathes are given in *Tables 14.7* and *14.8*. The majority of the functions listed are standard on most comparable models.

The major controller functions are:

(1) number of axes controlled—the primary parameter of the controller;
(2) axes controlled simultaneously—allows linear and circular interpolation;
(3) dimensional designation—all controllers allow working in absolute or incremental dimensional values;
(4) Imperial/metric designation—all allow working in decimal or metric;

Table 14.7 Typical control options for a CNC machining centre

Option	Scope
Conversational programming	Included
Controlled axes	3
Simultaneous controlled axes	3
Interpolation	Linear, circular
Designation system	Incremental, absolute
Least input increment	0.001 mm (0.0001 in.)
Tape code input	EIA/ISO, autorecognition
Spindle-speed command	S code, speed direct command
Function	G2/3 M2 T4
Spindle-speed override	50–200% (10% steps)
Feed-rate command	F code, speed direct command
Threading	F/E code direct command
Feed-rate override	0–200% (10% steps)
Override cancel	Included
Rapid traverse override	25%, 50%, 100%
Manual feed functions	Rapid, jog, manual pulse
Manual pulse generator	×1, ×10, ×100 (×50)
Dwell	Stop time command G04
Reference-point return	Manual/auto G27 to G29
Coordinate-system setting	G50
Automatic-coordinate setting	T code
Work coordinate system shift	Included
Incremental offset	Included
Decimal-point input	Included
Programme No. search	Included
Sequence No. search	Included
Label skip	Included
Optional block skip	Included
Fixed cycle	G90, G92, G94
Multiple repetitive cycle	G70 to G76
Radius programming on arc	Included
Chamfering corner R	Included
Memory lock key	Included
Number of tool offsets	50
Automatic tool-nose-radius compensation	Included
Backlash compensation	Included
Buffer storage	Included
Machine lock	Included
Custom macro	Common variables, 100
Second reference point return	Included
Program check function	Dry run and spindle stop
Dry run	Included
Single block	Included
Cathode-ray-tube display	14 in. colour
Part-program storage and editing	40 m (132 ft)
Background editing	Included
No. of registrable programs	99
Self-diagnosis	Included
Manual absolute	'On' fixed
Constant surface speed	Included
Imperial/metric conversion	Included
Cape punch interface	Included
Run-hour display	Included
Groove-width offset (ID, OD, end)	Included
Tape reader	Included

ID, internal diameter; OD, outer diameter; ISO, International Organisation for Standardisation; EIA, Electrical Industries Association.

Table 14.8 Typical control options for a CNC lathe

Option	Scope
Conversational programming	Included
Controlled axes	2
Simultaneous controlled axes	2
Interpolation	Linear, circular
Designation system	Incremental, absolute
Least input increment	0.001 mm (0.0001 in.)
Tape code	EIA/ISO, autorecognition
Spindle-speed command	S code, speed direct command
Function	G2/3 M2 T4
Spindle-speed override	50–200% (10% steps)
Feed-rate command	F code, speed direct command
Threading	F/E code direct command
Feed-rate override	0–200%
Override cancel	Included
Rapid traverse override	1%, 5%, 25%, 100%
Manual feed functions	Rapid, jog, manual pulse
Manual pulse generator	×1, ×10, ×100 (×50)
Dwell	Included
Reference-point return	Manual/auto G27 to G29
Coordinate-system setting	G50
Automatic-coordinate setting	T code
Work coordinate system shift	Included
Incremental offset	Included
Decimal-point input	Included
Programme No. search	Included
Sequence No. search	Included
Label skip	Included
Optional block skip	Included
Fixed cycle	G90, G92, G94
Multiple repetitive cycle	G70 to G76
Radius programming on arc	Included
Chamfering corner R	Included
Memory lock key	Included
Number of tool offsets	50
Automatic tool-nose-radius compensation	Included
Backlash compensation	Included
Buffer storage	Included
Machine lock	Included
Custom macro	Common variables, 100
Second reference point return	Included
Program check function	Dry run and spindle stop
Dry run	Included
Single block	Included
Cathode-ray-tube display	14 in. colour
Part-program storage and editing	40 m (132 ft)
Background editing	Included
No. of registrable programs	99
Self-diagnosis	Included
Manual absolute	'On' fixed
Constant surface speed	Included
Imperial/metric conversion	Included
Cape punch interface	Included
Run-hour display	Included
Groove-width offset (ID, OD, end)	Included
Tape reader	Included

ID, internal diameter; OD, outer diameter; ISO, International Organisation for Standardisation; EIA, Electrical Industries Association.

(5) least increment—most controllers work to 0.001 mm (0.0001 in.);

(6) tape codes—all controllers work to both ISO and EIA;

(7) program functions—all controllers incorporate G, M. F, S and T programming functions (see Section 14.5.8);

(8) program-storage capacity—normally quoted in metres or feet of tape length;

(9) manual feed—rapid, jog manual pulse by the operator;

(10) custom macros—allows customer creation of macro routines;

(11) background editing—allows editing of a program whilst another one is running on the machine; and

(12) constant cutting speed—automatically adjusts spindle speed to give a constant surface cutting speed as the part diameter reduces in machining.

Controller hardware Leading models usually incorporate the following hardware.

(1) Processors—16 bit or 32 bit.

(2) Memory—1 Mbyte.

(3) Programmable logic controllers (PLCs)—these are incorporated for axis control, as described above.

(4) Cathode ray tube—9, 10, 12 or 14 in. colour screens are normally supplied.

(5) Paper tape reader/punch.

(6) Floppy disc drive.

(7) Keyboard.

(8) Interfaces to the machine: slideway drives; spindle drives; and tooling systems.

14.5.7.4 Tape control

Tape formats The most usual form of program storage and transfer to a machine is by paper tape which is read by the machine controller. Paper, magnetic and plastic tape is used, the most common being a thin plastic tape (Mylar). Paper and plastic tapes are usually 1 in. wide and carry seven or eight tracks of holes punched along their length. An arrangement of holes in seven or eight tracks across the tape is used to represent one character in the program. *Figure 14.50* shows an eight-track tape. The tape has seven tracks for data characters, one track is used for a parity check, and a row of smaller holes for the tape sprocket feed engagement.

The arrangement of a complete set of all the characters used to program numerical control (NC) machines is embodied in the two most universally used tape codes: the International Organisation for Standisation (ISO) and the Electrical Industries Association (EIA) (American). All modern machine controllers read both codes.

Tape codes (a) *ISO code*—this is based on the American Standard Code for Information Exchange (ASCII) in which a numerical value is used to represent each character. It is widely used in computing and communications. The basis of the hole values in the tape is that of the binary coded decimal (BCD) code representing the ASCII code for the character. An example of such coding is shown in *Table 14.9* and the ISO code with the corresponding tape hole punching is shown in *Table 14.10*.

The ISO parity check uses the insertion (or not) of a hole in the eighth row which will give, in total, an even number of holes. This is known as an 'even parity' system.

(b) *EIA code*—this is similar in principle to the ISO code, but uses different decimal equivalent codes (except for numbers), and the 'sixth' track for parity check. A hole is added to establish an odd number of holes as the parity check. A table of EIA codes is shown in *Table 14.11*.

Paper-tape readers Tape readers feed a tape by means of a sprocket driving in the sprocket holes track of the tape. As the tape passes over the read position the presence or not of holes is detected by one of three means: pneumatic, electro-mechanical or optical.

(a) *Pneumatic*—a small jet projects a low pressure compressed air stream at each track position one side of the tape. If there is no hole then the air back-pressure is sensed and the stable '0' binary state is maintained. Where a hole exists the passage of air is unrestricted and the loss of back pressure casues a binary '1' signal to be initiated.

(b) *Electromechanical*—in this case a pin smaller than the hole size is presented to each tape track. The pins actuate electrical contacts. If a hole is present the pin rises into the hole and the closing of a contact causes a voltage which registers a binary '1'. No hole gives an open contact and binary '0'.

(c) *Optical*—a photocell is used to detect the holes in each track. When a hole is present a light beam passes through it and lands on a receiver cell which creates a voltage and, therefore, a binary signal '1'. No hole produces no signal and a binary '0'. Optical readers are the fastest and can reach speeds of over 1500 characters per second.

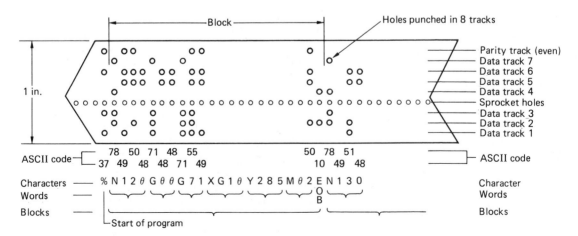

Figure 14.50 Paper tape punched according to the ISO code. EOB, end of block

Table 14.9 Example of binary coded decimal code representing the ASCII code for some characters

Character	Decimal code (ASCII code)	Tape code							
		Tape hole track: 8	7	6	5	4	3	2	1
		BCD code value:	64	32	16	8	4	2	1
Carriage return	13	P				•	•		•
%	37	a	•				•		•
1	49	r	•	•					•
9	57	i	•	•	•				•
A	65	t	•						•
B	66	y	•					•	
Z	90		•		•	•		•	

In NC machines the ends of smaller tapes are joined to form a continuous loop which is fed through the reader for every component produced. Longer tapes use spools for input and output to avoid jamming and damage. In CNC machines the tape is used only to load the program to the controller RAM. It is then retained as a record in a tape archive.

14.5.7.5 Magnetic data media

Magnetic tape These are plastic tapes coated with a magnetic medium on one side. The data are entered by magnetising spots on the surface in the same manner as for domestic music tapes. An advantage of magnetic tape is that the data capacity is in excess of 100 characters per inch compared with 10 characters per inch for paper tape. This makes magnetic tape a better option for storing multiple programs on one tape and gives more efficient archiving. Like other magnetic media the tape needs protection from external electromagnetic forces, grease, dust and heat. Another disadvantage is that the data are recorded serially and, therefore, longer times are needed to access program files than are needed on a disc. Files should be backed up on a tape streamer which is a separate tape drive that can be connected to any computer in order to take copies of files. This is often done at the end of a shift or working day for all files that have been altered or, in some cases, all files.

Discs The discs used are the standard 5.25 and 3.5 in. diameter computer discs in general use. They are used in the normal manner, being loaded with programs in a disc drive connected to the computer where the programming has taken place, and read in a disc drive in the controller. Access to the correct program file is faster on discs because they are a parallel storage device. A back-up copy should be taken either as another disc or on a tape streamer. In the absence of direct numerical control or networked programming terminals, discs are also used to transport and load programs to local workshop program file storage and distribution centres.

14.5.8 Programming

Machine controllers manage the machine by issuing electronic control signals for activities such as axis positioning (i.e. the part dimensional requirements), spindle speeds, feed rates, tool selection and the control of ancillary equipment. Prior to this, the part and machining requirements must be established and presented to the controller in the form of a part program. The program is a sequential series of instructions written in the form of numerical data and alpha-numeric codes. Part pro-grams are written and fed into the machine control by means of one of several methods.

(1) In NC and CNC machines, off-machine programming facilities are used to create and then transfer the program to the machine in the form of punch tape, magnetic tape or magnetic disc.
(2) In machines having a conversational programming capability, the program is automatically constructed by the machine controller software with data entered at the controller keyboard in response to a question/answer menu (see Section 14.5.8.3).
(3) In DNC systems, off-machine programming facilities are used to create and then store and transfer the program electronically from a central computer storing a number of programs.

The relationships between programming methods and CNC/DNC machining are shown in *Figure 14.51*.

There are two main types of programming method.

(1) Point to point—this can be used where the route and speed of movement from one machining point to the next is unimportant. Operations such as drilling/tapping and sheet metal punching can use this method.
(2) Continuous—this is used where the route and speed of tool movement are important. Thus milling and turning operations where a form is generated by the movement of the tool, are programmed in the continuous mode. Contouring and profiling machines also use this type of program.

The four methods of programming in general use are:

(1) manual programming directly in the word address mode;
(2) APT language programming;
(3) conversational programming on-machine with interactive graphics; and
(4) graphical programming off-machine (CAM and CAD-CAM).

14.5.8.1 Manual programming

Athough this is being overtaken by the use of the more automated graphical and conversational methods, an appreciation of manual programming will assist in understanding the methods now being employed.

NC programs are structured to order the machine actions in a predefined sequence and to give all the information needed for each action before the controller processes the next one. It is, in computer terms, an interpreted program, not a compiled

Table 14.10   ISO control codes*

Decimal equivalent	Binary representation							Name of character	Character symbol	Representation in punched tape								
	b_7	b_6	b_5	b_4	b_3	b_2	b_1			P	7 b_7	6 b_6	5 b_5	4 b_4	F	3 b_3	2 b_2	1 b_1
0	0	0	0	0	0	0	0	Null	NUL						•			
8	0	0	0	1	0	0	0	Backspace	BS	•				•	•			
9	0	0	0	1	0	0	1	Tabulate	TAB					•	•			•
10	0	0	0	1	0	1	0	End of block	LF					•	•		•	
13	0	0	0	1	1	0	1	Carriage return	CR	•				•	•	•		•
32	0	1	0	0	0	0	0	Space	SP	•		•			•			
37	0	1	0	0	1	0	1	Program start	%	•		•			•	•		•
40	0	1	0	1	0	0	0	Control out	(			•		•	•			
41	0	1	0	1	0	0	1	Control in	)	•		•		•	•			•
43	0	1	0	1	0	1	1	Plus sign	+			•		•	•		•	•
45	0	1	0	1	1	0	1	Minus sign	−			•		•	•	•		•
47	0	1	0	1	1	1	1	Optional block skip	/	•		•		•	•	•	•	•
48	0	1	1	0	0	0	0		Ø			•	•		•			
49	0	1	1	0	0	0	1		1	•		•	•		•			•
50	0	1	1	0	0	1	0		2	•		•	•		•		•	
51	0	1	1	0	0	1	1		3			•	•		•		•	•
52	0	1	1	0	1	0	0		4	•		•	•		•	•		
53	0	1	1	0	1	0	1		5			•	•		•	•		•
54	0	1	1	0	1	1	0		6			•	•		•	•	•	
55	0	1	1	0	1	1	1		7	•		•	•		•	•	•	•
56	0	1	1	1	0	0	0		8	•		•	•	•	•			
57	0	1	1	1	0	0	1		9			•	•	•	•			•
58	0	1	1	1	0	1	0	Alignment function	:			•	•	•	•		•	
65	1	0	0	0	0	0	1		A		•				•			•
66	1	0	0	0	0	1	0		B		•				•		•	
67	1	0	0	0	0	1	1		C	•	•				•		•	•
68	1	0	0	0	1	0	0		D		•				•	•		
69	1	0	0	0	1	0	1		E	•	•				•	•		•
70	1	0	0	0	1	1	0		F	•	•				•	•	•	
71	1	0	0	0	1	1	1		G		•				•	•	•	•
72	1	0	0	1	0	0	0		H		•			•	•			
73	1	0	0	1	0	0	1		I	•	•			•	•			•
74	1	0	0	1	0	1	0		J	•	•			•	•		•	
75	1	0	0	1	0	1	1		K		•			•	•		•	•
76	1	0	0	1	1	0	0		L	•	•			•	•	•		
77	1	0	0	1	1	0	1		M		•			•	•	•		•
78	1	0	0	1	1	1	0		N		•			•	•	•	•	
79	1	0	0	1	1	1	1		O	•	•			•	•	•	•	•
80	1	0	1	0	0	0	0		P		•	•			•			
81	1	0	1	0	0	0	1		Q	•	•	•			•			•
82	1	0	1	0	0	1	0		R	•	•	•			•		•	
83	1	0	1	0	0	1	1		S		•	•			•		•	•
84	1	0	1	0	1	0	0		T	•	•	•			•	•		
85	1	0	1	0	1	0	1		U		•	•			•	•		•
86	1	0	1	0	1	1	0		V		•	•			•	•	•	
87	1	0	1	0	1	1	1		W	•	•	•			•	•	•	•
88	1	0	1	1	0	0	0		X	•	•	•		•	•			
89	1	0	1	1	0	0	1		Y		•	•		•	•			•
90	1	0	1	1	0	1	0		Z		•	•		•	•		•	
127	1	1	1	1	1	1	1	Delete	DEL	•	•	•	•	•	•	•	•	•

* Courtesy of Pitman, London.

Table 14.11 EIA control codes*

Decimal equivalent	b_7	b_6	b_5	b_4	b_3	b_2	b_1	Name of character	Character symbol	8 (b_7)	7 (b_6)	6 (b_5)	P	4 (b_4)	F	3 (b_3)	2 (b_2)	1 (b_1)
0	0	0	0	0	0	0	0	Space	SP						•			
1	0	0	0	0	0	0	1		1				•		•			•
2	0	0	0	0	0	1	0		2				•		•		•	
3	0	0	0	0	0	1	1		3						•		•	•
4	0	0	0	0	1	0	0		4				•		•	•		
5	0	0	0	0	1	0	1		5						•	•		•
6	0	0	0	0	1	1	0		6						•	•	•	
7	0	0	0	0	1	1	1		7				•		•	•	•	•
8	0	0	0	1	0	0	0		8				•	•	•			
9	0	0	0	1	0	0	1		9					•	•			•
11	0	0	0	1	0	1	1	End of block	EOB				•	•	•		•	•
16	0	0	1	0	0	0	0		Ø			•	•		•			
17	0	0	1	0	0	0	1	Optional block skip	/			•			•			•
18	0	0	1	0	0	1	0		S			•			•		•	
19	0	0	1	0	0	1	1		T			•	•		•		•	•
20	0	0	1	0	1	0	0		U			•			•	•		
21	0	0	1	0	1	0	1		V			•	•		•	•		•
22	0	0	1	0	1	1	0		W			•	•		•	•	•	
23	0	0	1	0	1	1	1		X			•			•	•	•	•
24	0	0	1	1	0	0	0		Y			•		•	•			
25	0	0	1	1	0	0	1		Z			•	•	•	•			•
26	0	0	1	1	0	1	0	Backspace	BS			•	•	•	•		•	
27	0	0	1	1	0	1	1	Comma	,			•		•	•		•	•
29	0	0	1	1	1	0	1	Tabulate	TAB			•		•	•	•		•
32	0	1	0	0	0	0	0	Minus sign	−		•		•		•			
33	0	1	0	0	0	0	1		J		•				•			•
34	0	1	0	0	0	1	0		K		•				•		•	
35	0	1	0	0	0	1	1		L		•		•		•		•	•
36	0	1	0	0	1	0	0		M		•				•	•		
37	0	1	0	0	1	0	1		N		•		•		•	•		•
38	0	1	0	0	1	1	0		O		•		•		•	•	•	
39	0	1	0	0	1	1	1		P		•				•	•	•	•
40	0	1	0	1	0	0	0		Q		•			•	•			
41	0	1	0	1	0	0	1		R		•		•	•	•			•
43	0	1	0	1	0	1	1	Program start	%		•			•	•		•	•
48	0	1	1	0	0	0	0	Plus sign	+		•	•			•			
49	0	1	1	0	0	0	1		A		•	•	•		•			•
50	0	1	1	0	0	1	0		B		•	•	•		•		•	
51	0	1	1	0	0	1	1		C		•	•			•		•	•
52	0	1	1	0	1	0	0		D		•	•	•		•	•		
53	0	1	1	0	1	0	1		E		•	•			•	•		•
54	0	1	1	0	1	1	0		F		•	•			•	•	•	
55	0	1	1	0	1	1	1		G		•	•	•		•	•	•	•
56	0	1	1	1	0	0	0		H		•	•	•	•	•			
57	0	1	1	1	0	0	1		I		•	•		•	•			•
58	0	1	1	1	0	1	0	Lower case			•	•		•	•		•	
60	0	1	1	1	1	0	0	Upper case			•	•		•	•	•		
63	0	1	1	1	1	1	1	Delete	DEL		•	•		•	•	•	•	•
64	1	0	0	0	0	0	0	Carriage return	CR	•			•		•			

* Courtesy of Pitman, London.

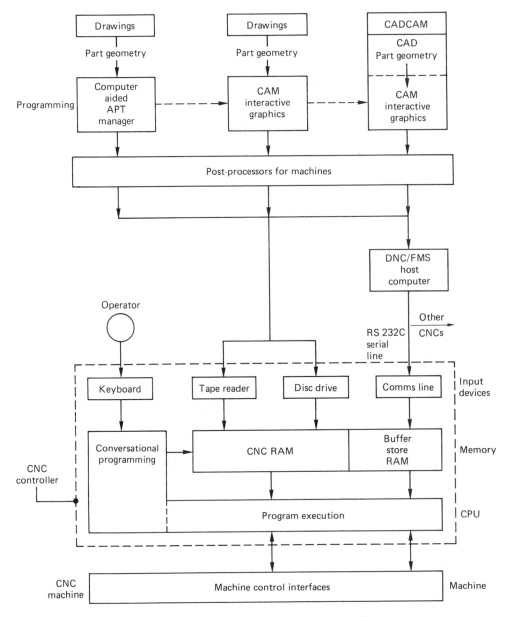

Figure 14.51 Relationship between computer-aided programming and control methods

one. This means that for each machine action the position, speeds, feeds, tools needed and ancillary machine actions are all defined in the same instruction *block*. When the action has been completed, the machine control then activates the next instruction block in the defined sequence.

Block format The *word address* is the most widely used structure of a program instruction block. Each block contains a number of *words*. A word is a complete piece of a datum relating to one required condition, e.g. a position, feed rate or spindle speed. The word address format uses a capital letter as an 'address' followed by a number of numerical data characters. In *Figure 14.52* a data character is represented by 'x'. The spaces between words are only included for clarity. The

number of data characters allowed in the words varies with controller and is defined by the machine control manufacturer. These must be adhered to rigidly or program errors and machine malfunctions will occur.

The word addresses shown in *Figure 14.52* are common to most controllers. Other addresses used by controllers include:

D   tool diameter offset,
H   tool length offset, and
—   various codes for fixture offset.

The advantage of the word address format in which the address of every word is included, is that it allows repeated instructions to be omitted from succeeding instruction blocks. This makes the programs shorter and programming faster than

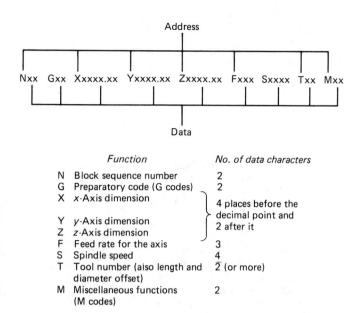

Function	No. of data characters
N Block sequence number	2
G Preparatory code (G codes)	2
X x-Axis dimension	4 places before the decimal point and 2 after it
Y y-Axis dimension	
Z z-Axis dimension	
F Feed rate for the axis	3
S Spindle speed	4
T Tool number (also length and diameter offset)	2 (or more)
M Miscellaneous functions (M codes)	2

Figure 14.52 Word address structure

in the fixed sequential format (in which all blocks must be repeated) and tab sequential format in which all address tabs must be repeated.

Other symbols used in the word address format are:

%	Start of program.
/	Skip block—used when one program produces similar parts.
[	Control in—the control ignores every character after the bracket.
]	Control out—the control regards every character after the bracket. This allows sections of a block to be jumped e.g. [xxxxx] tells the control to ignore xxxxx.
[ ]	Calculate values within the brackets.
( )	Display information between the brackets—on the screen.
EOB	End of block.

N (block address) N... defines the blocks with a sequential numbering system, e.g. N010, N015, N020, N..., N995. The numbers used are normally 5 or 10 digits apart to allow the subsequent insertion of blocks without having to renumber the following ones. Most blocks are short enough to be printed on one line of a print out, and in this way are numbered and displayed in the same manner as the program lines of other types of computer program.

G codes These are used to call up commonly used programmed routines, thereby saving programming time. A list of codes is shown in *Table 14.12*. G and M code functions are either modal or non-modal, which refer to the codes applied continuity (or not) from one block to succeeding blocks.

(1) *Modal codes*—Once called up, these will continue in operation in subsequent blocks without having to be written in every block. They are cancelled by another modal code of the same word address.
(2) *Non-modal codes*—these operate only in the specified block.

Absolute/incremental values Absolute (G90) is always a dimension taken from the datum zero. All points are thereby related to that zero point. Incremental (G91) gives values of *x*, *y* and *z* that are the distance from the existing position to the next one.

Zero suppression Older controls used zero suppression in order to avoid putting all the zeros into the fixed-word-length format. This reduces program writing time (see *Table 14.13*). However, most modern machines use *decimal-point* programming in which the position of the decimal point is sufficient to define the value of the number.

Other addresses Other addresses used are listed below.

F (feed rate)—this word is used to specify the feed rate of the tool.
S (spindle speed)—calls up the spindle speed required.
T (tool number)—calls up the tool required by using a preassigned tool number.
M (miscellaneous) codes—these initiate miscellaneous machine activities and program stop routines such as optional program stop, spindle on counterclockwise, flood coolant on. They also control external facilities such as indexing tables and clamps. M codes can be modal or non-modal (see definition under G codes) and are listed in *Table 14.12*.

Macros These are self-contained programs that can accept variables from the main program and process them. A macro might take a printed circuit board dimension and a number of holes and from these calculate the *x* and *y* coordinates of the hole centres. A macro may be defined at the beginning of a program, but will only be activated when it is 'called' by the program. Macros are particularly useful in drilling and milling operations where a sequence is repeated.

Subroutines Subroutines or subprograms are a series of instructions that can be activated and looped for repetitive

Table 14.12 Programming codes

A number of standard codes are used to reduce the amount of programming effort needed to command commonly used machining operations, instructions and conditions. These are commonly known as:

G codes—call up machining commands
M codes—call up machine control activities
F codes—call up feed rates
S codes—call up cutting speeds
T codes—call up tool selection

G codes (preparatory codes)

The majority of manufacturers follow the same practice in designation of codes, but their detailed implementation may differ.

Sample G codes

G00 Rapid movement for position
G01 Linear interpolation used for straight-line feed
G02 Circular interpolation, clockwise
G03 Circular interpolartion, counterclockwise
G04 Dwell, a programmed stop to the tool movement
G17 Circular interpolation *xy* plane
G18 Circular interpolation *xz* plane
G19 Circular interpolation *yz* plane
G20 Inch units
G21 Millimetre units
G28 Return to home position
G29 Return from home position
G31 Reverses programmed direction of *x* axis
G32 Reverses programmed direction of *y* axis
G41 Tool radius compensation left
G42 Tool radius compensation right
G43 Tool length compensation—positive direction
G44 Tool length compensation—negative direction
G70 Imperial unit
G71 Metric units
G80 Cancel canned cycle
G81 Drilling cycle
G82 Drilling cycle with dwell
G83 Deep hole drilling
G84 Tapping cycle
G85 89—boring cycles
G90 Absolute mode
G91 Incremental mode

There are two types of code:

(1) modal codes remain active after being entered, unless they are cancelled by another G code; and
(2) non-modal codes are only active in the program block in which they appear.

M codes

These control the auxiliary functions of the machine.
M00 Program stop
M02 End of program
M03 Spindle on, clockwise
M04 Spindle on, counterclockwise
M05 Spindle off
M06 Tool change
M07 Oil mist coolant on
M08 Flood coolant on
M09 Coolant off
M30 End of tape

Table 14.13 Zero suppression

Word format, decimal	Leading zero supresion Y±042, word	Trailing zero suppression Y±420, word
0.11	Y0.11	Y0000.11
11.01	Y11.01	Y0011.01
110.00	Y110.00	Y0110
1105.10	Y1105.10	Y1105.1

application. The major difference with a macro is that the subprogram can only be defined after it has been called in a block of the main program. After execution the program flow returns to the main program at a point immediately after the original call point.

Canned cycles (fixed cycles) These are common machining routines that are prewritten into the controller by the manufacturer. Some of the G-code functions are canned cycles, e.g.

G81 drilling,
G82 drilling with dwell,
G83 Peck drilling for deep holes,
G84 tapping,
G85 boring, and
G80 canned cycle cancel.

14.5.8.2 *APT programming*

The automatic programming tool (APT) was the first computer-aided programming language and was developed in the 1950s at the Massachusetts Institute of Technology. It was generally one of the most successful languages of this type and many versions were subsequently produced. APT is a high-level language using English-type statements in a manner similar to other high-level languages. This is then converted by further software into the particular low-level machine code needed by the machine being programmed. In the case of NC this conversion software is known as a 'post-processor'.

Structure of APT The language uses four types of statement:

(1) geometry statements—these define the geometry of the part, e.g. lines, circles, points, curves;
(2) motion statements—these describe the tool path required;
(3) post-processor statements—these provide specific information on machine speeds, feeds and other machine control parameters; and
(4) auxiliary statements—these give the part and tool tolerances.

The general form is:

Part No.
Machine
Geometry statements
Motion statements
Post-processor and auxiliary statements

Commands are entered as single-line statements, as opposed to the serial block format used in manual programming.

Geometry statements:

P = point
L = line
C = circle

e.g. P1 = POINT/100,-50
P2 = POINT/80.40,-10

C1 = CIRCLE/10,-5,12
C2 = CIRCLE/-15,-10,1.5

L1 = LINE/100,50.6,-10
L2 = LINE/50.6,50,-5

Motion and other statements:

GOTO/P1
GOFWD/
GOLFT/

Post-processor and auxiliary statements:

CUTTER/
SPINDL/
FEDRAT/

The information is compiled in a form known as the 'cutter location file' (CLFile or CLF). This is in a general form unsuitable for any particular machine and needs to be further processed in a post-processing operation.

Post-processor This is software specific to a particular machine and, in order to produce a program suitable to a specific machine, carries out the following functions:

(1) checks that the syntax of the part programme is correct;
(2) inserts G and M codes into the program; and
(3) coordinates are entered in the correct place.

Post-processor programs are purchased from the software supplier when the machine type is specified.

APT requires a large computing capacity and, since the development of desk-top microprocessor-based computers, has been superseded in many applications by graphical and conversational programming.

14.5.8.3 Conversational programming

This is a recent development that allows programming to be done at the machine by trained operators. The principle is that the 'programmer' works at the keyboard and CRT of the controller. The cathode-ray tube shows screens that guide the operator through the programming of the machine. The controller software contains functions for:

(1) conversational input,
(2) program writing and editing,
(3) part drawing, and
(4) calculations.

A large amount of machining data is held resident in the controller software and this avoids the need for the programmer to be a manufacturing engineer with training in programming at some level. Machining speeds and feeds for commonly used materials are in the control database, but these can be modified during programming.

Machining centres A conversational programming sequence on a machining centre for the component would appear on the menu screen as:

(1) Initial setting.
(2) Side cut prepare.
(3) Side cutting figures.
(4) Pocket prepare.
(5) Pocket figures.

(6) Hole cutting.
(7) Hole patterns.
(8) End of program.
(9) List of tools used.
(10) Graphic check.

Lathes Conversational programming has been developed for CNC lathes. The programmer is taken through a series of screens which present many features of the program in graphical form, e.g.

Graphical representation of the part shape.
Graphical representation of tool positions.
Graphical simulation of the tool cut paths.

An example is the Moric I system by Mori-Seiki. *Figure 14.53* shows the screens for such a program. The menu options to be selected are on the bottom of the screen and the resultant program feature is displayed above these. The stages are:

(1) Data input.
(2) Part drawing input.
(3) Program editing.
(4) Tool path check.
(5) Set-up assist.

The use of conversational programming has been hastened by the introduction of 32-bit processors which increase processing speeds between two- and five-fold. Many machine controllers now have the facility for conversational programming and this trend will continue. Conversational programming can usually be performed whilst the machine is running to another program. This is known as 'background programming'.

14.5.8.4 Graphical programming

A major development in the 1980s was that of graphical numerical programming (GNC). In this method, the programmer des not write lines of program as in manual or APT type methods. Instead, the software creates a program from data entered by the programmer via a keyboard, mouse and/or digitiser tablet. The method allows the programmer to construct the geometry of the part on the screen via a CAD facility and then see the tool paths around the part portrayed both in sequence and total. Alternatively, there are integrated CAD-CAM software suites in which the part geometry will already exist. This subject is discussed in greater detail in Section 14.4.2.

The basic stages in preparing a program are:

(1) create the part geometry (or import from CADCAM);
(2) enter machining parameters; and
(3) the computer then:
    (a) calculates dimensions,
    (b) writes the program, and
    (c) produces tool listings.

The program can be verified by being simulated graphically on the screen. This shows each tool cut path, the part shape and any potential collisions.

The graphical programming equipment that is normally required is:

(1) computer system (an 80386 or 80486 machine),
(2) graphics screen (a VGA or SVGA is preferable),
(3) data screen,
(4) keyboard,
(5) digitising tablet/mouse,
(6) printer,
(7) plotter, and
(8) tape punch.

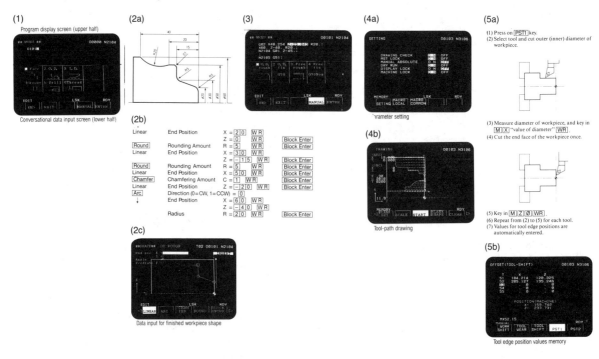

Figure 14.53   Conversational programming for a CNC lathe. (Courtesy of Mori-Seiki (UK) Ltd)

14.6   Flexible manufacturing systems

14.6.1   Definition of flexible manufacturing

'Flexible manufacturing' (FM) is a term used to describe a manufacturing activity and facility that is computer controlled so that it can automatically process a number of different products in any desired quantity and priority.

The flexibility of a FM facility comes from the fact that it is:

(1)   an automated system in which the majority of the resources employed are programmable;
(2)   fully tooled and programmed to process a variety of dissimilar products—as opposed to a dedicated automation line which normally produces only one or a restricted family of like products;
(3)   controlled by a supervising computer which interprets the overall production schedule and creates a system schedule of product type, priority and quantity. It will then control the processes, materials handling equipment and tooling to produce to the system schedule; and
(4)   capable of producing product in any priority or quantity.

FM facilities take three major forms: cells (FMC), systems (FMS), and flexible transfer lines. The main distinctions between these forms are the number of part varieties handled, the degree of integration with other external facilities, overall size, the extent of the transport systems needed and capabilities of the computer control system and software.

Flexible manufacturing cells (FMC)   A cell is a group of typically 2–6 machines that are controlled by a computer within or adjacent to the cell. The machines have automatic component feeding and handling in the form of fixtures on pallets for machining or assembly centres, plus robots or stacking magazines for other machines. The cell is a stand-alone unit with a minimum of interfacing and control from other sources. A cell should be designed so that it can be upgraded to or incorporated in a FMS if future demand and economics make this viable.

Flexible manufacturing systems (FMS)   A system is designed to fully schedule and control the automatic processing of a variety of products through a number of cells and other production facilities, the total operation being controlled by a central computer. A FMS may have a large amount of equipment providing extensive processing, materials handling, inspection and tooling management capabilities. The central (host) computer is networked to the facilities to provide either direct control or to communicate with the autonomous controls of the equipment.

Flexible transfer lines   These are an application of FMS to high-volume production and are used in the automotive industry as an alternative to the dedicated transfer lines previously used. There are installations in the UK that produce only one or two similar component designs, but the basic flexibility of the system and the ability to reassign the constituent computer numerical control (CNC) and programmable equipment makes FM a better long-term investment than a dedicated transfer line. This is an advantage in an industry which must frequently introduce new product designs.

Cell layout (Figure 14.54)   A robot loads and unloads lathes, cleaning and assembly modules. Components are fed into and out of the cell via chutes and magazines to avoid the robot arm swing envelope. The cell controller, which may be a microcomputer or a programmable controller, is outside the cell.

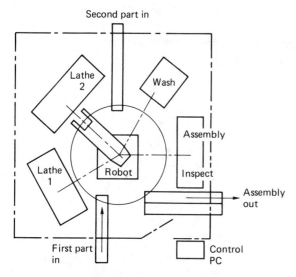

Figure 14.54 A typical layout of a flexible cell using a robot to feed machines with two parts from magazines.

For a stand-alone cell that is *not* part of a larger FMS, the control has no wide degree of integration with other facilities. Conversely, if part of a FMS it is supervised by the executive computer.

Linear cell or small system (Figure 14.55) The arrangement shown is widely used for small groups of CNC machining centres that may be independent or part of a larger system. The system layout and flow is linear and work is moved

around by a rail guided vehicle (RGV). Manual load/unload station(s) are used to put the work pieces onto fixtures that are mounted on pallets, which in turn are transported by the RGV to the machines. Pallets are stored alongside the RGV track and can act as product buffer storage. They are usually dedicated to one fixture to avoid frequent set-up of fixtures on a pallet. The machines have a two-pallet shuttle arrangement to allow the movement of one pallet whilst the machine is processing the other. Access to the machine working area, control and tool magazines is from the rear of the cell. The system control can be at a number of levels of capability, depending on the requirements, and can be self-contained or integrated into a larger system.

Full FMS (Figure 14.56) This layout features a number of cells (a modular approach is normally preferred on both processing and system security grounds), automatic guided vehicles (AGVs) to transport work, an executive computer with the capability of scheduling the system and supervising subsidiary computers, a tool service area and an automatic tool distribution system. The tool distribution system is usually the most optional of these elements. Many FMS employ a mixture of AGVs for inter-cell transport and RGVs within some cells. A large system designed to handle several hundred different components in a year poses considerable system scheduling and resource management problems. A high level of executive control software is needed and pallet transport control and tool supply management become key issues in achieving an efficient system. Computer simulation of models of the system is necessary to obtain the correct design and then continue to use it at maximum efficiency. A flexible 'transfer line' may well have a non-linear layout and the same elements as a multi-product FMS.

The system elements mentioned in the descriptions above are covered in greater detail in Sections 14.6.5 to 14.6.11.

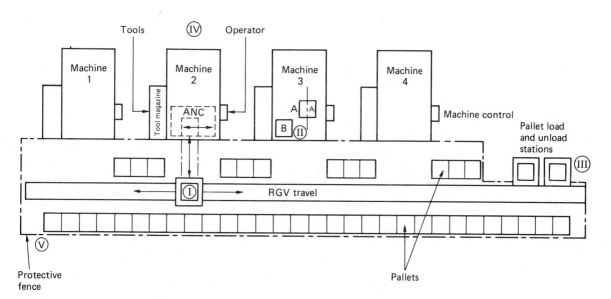

Figure 14.55 A typical layout of a linear cell or small system, with four work centres. I, Pallets are supplied by a rail-guided vehicle (RGV); II, pallets are exchanged on machines by an automatic work changer (AWC); III, the pallets are loaded and unloaded mainly at one end of the rail; IV, access to the tool magazines, machine control and spindle area is from the 'back'; V, a protective fence encloses the RGV and moving pallets

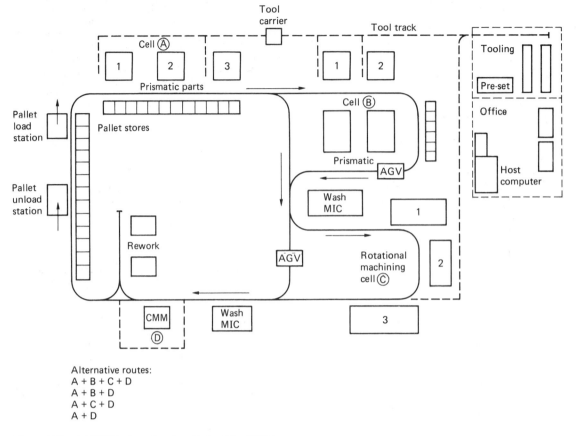

Alternative routes:
A + B + C + D
A + B + D
A + C + D
A + D

Figure 14.56 A full FMS using automatic guided vehicles (AGVs)

14.6.2 Advantages and disadvantages of flexible manufacturing

14.6.2.1 Advantages

(1) The ability to run small and varied batches of product, as required.
(2) Greater flexibility in changing products, priorities and quantities.
(3) Increased productivity over using a comparable number of stand-alone CNC machines and processes.
(4) Less lost time for setting and tool changing.
(5) The above results give better machine utilisation (>90% against 50%).
(6) Reduced labour costs through unmanned running or minimum manning.
(7) Reduced lead time (savings of greater than 50% are common).
(8) Reduced inventory levels (savings of greater than 50% are common).
(9) FM is very compatible with just-in-time (JIT) working.
(10) Can be reassigned in the future to other products (it is not dedicated).
(11) FM is a major step toward computer-integrated manufacturing (CIM).

These advantages give production management considerable process-cost savings plus the flexibility to introduce immediate changes in the products to be processed, priorities and quantities.

FM provides the ability to run smaller and more mixed batches of work, which in turn reduces the lead times from order acceptance to delivery. The reduction in lead time and batch quantities gives a reduction in the inventory levels and costs carried by the company. This is normally a major part of the financial justification of a FMS.

14.6.2.2 Disadvantages

(1) High capital cost—the larger and more complex the system, the greater the amount of investment needed for non-value-adding elements such as computing power and work handling equipment. On the other hand, small cells with a minimum of handling equipment are normally competitive with the benefits/costs performance of equivalent stand-alone machines.
(2) Investment in FM should be part of a long-term manufacturing strategy. Payback can take 4–5 years, which will concern management used to the short-term 2-year payback targets common in the UK and the USA.
(3) Introduction of a FMS will consume a large amount of managers' and specialists' time. Companies must be prepared to carry the apparent and hidden costs of this in one form or another.

(4) FM will introduce new working practices and organisation. Managers and staff must be capable of adapting, and be willing to do so.
(5) Skills that are not already in the organisation or are too few will be needed for planning, implementing and running a FM. This will entail retraining and possibly recruitment.

N.B. A very practical solution that overcomes the above points is offered by the major FMS suppliers. This is to start with one cell (of an almost standard design that they have developed) and build on it over several years.

14.6.3 Application areas and development of flexible manufacturing

14.6.3.1 Products and process technologies

There are many applications of FM and they all differ according to the type of product being processed, the volume/variety mix and the manufacturing technologies employed. Any combination of these is possible, but typical examples for a given cell/system are given in *Table 14.14*.

The majority of large systems have been installed in automotive, aerospace and high volume/value OEM producers who have the demand levels and financial strength to justify capital expenditure of millions of pounds. A greater proportion of cells and small systems are found in medium/small OEM and subcontract machining (see *Table 14.15*). In addition to primary processing machines, many installations include stations for deburring, cleaning, inspection, tool setting and buffer storage. As seen previously the detailed form of any of these applications may be as one cell, multiple cells or a fully integrated system.

14.6.3.2 The increasing use of FM

Four broad categories of production technology have been employed up to the present time: conventional (non-numerical control (NC)) stand-alone machines; NC and CNC machines; dedicated transfer lines; and early FMC/FMS installations. With the increasing demand for manufacturing effi-ciency, flexibility and inventory reduction, the emphasis in manufacturing investment and technology is moving away from the first three categories and towards FM. This trend is illustrated in *Figure 14.57*.

(1) In the recent past, old conventional machines were most likely to be replaced by stand-alone CNC equivalents. There is now the opportunity to make a step change to FM, which will normally involve CNC or direct numerical control (DNC) machines as the basic work centres.
(2) Early NC or CNC machines may be replaced by later CNC and DNC versions, but greater advantages can be gained if these are now incorporated in a FMC/FMS.
(3) Early FMC/FMS installations will be replaced or expanded into versions with the latest technology and more powerful software.
(4,5) In some cases dedicated transfer lines or special purpose machines are being superseded by FM. Where higher product variety and lower volumes are now needed, a FMS is appropriate (4). Where high volumes of a few products are still the requirement, a flexible transfer line with the potential for reassignment is a better (5) investment than a traditional dedicated line that can only be converted at great cost.
(6) A FMS incorporates the integration of planning, engineering and production activities and, as such, is a large component in the move towards CIM. A good CIM strategy includes the parallel development and eventual integration of FM facilities with the other manufacturing activities.

Table 14.14 Typical combinations of volume and mix for different products produced by the same cell/system

Product	Volume	Mix
Automotive components	High	Low
Subcontract machining	Variable	High
Industrial OEM	Low	High
Aerospace components	Low	High
Heavy engineering	Low	Low

Table 14.15 The type of technology/equipment used in producing different products

Product	Predominant technology/equipment
Machined components	CNC machining centres, lathes and grinders
Sheet metal/fabrication	CNC presses, laser cutting and bending
Assemblies	Flexible assembly systems, robotic assembly
Printed circuit boards	Automatic printed circuit board production, assembly and test

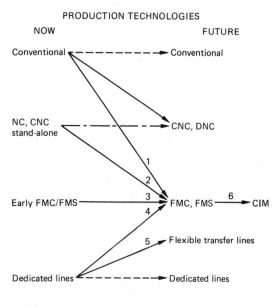

Figure 14.57 The trends in the utilisation of FM. For explanation of 1 to 6 see text

14.6.3.3 Technical developments

FM is the result of many years of work in the development of new machine designs, materials handling systems, tooling, control technology, computing power and applications software. Some of the major features of improvement have occurred in:

(1) machine design—making structures more rigid and increased power;
(2) materials handling—the development of automatic guided vehicles;
(3) tooling—improved life and quick change methods;
(4) controls and computers—the introduction of cheap 16- and 32-bit processors has enabled the rapid developments seen in CNC, DNC and computer-aided manufacturing (CAM) applications; and
(5) software—more powerful controls and computers plus growing experience has led to the production of better software.

The developments have been led by machine tool manufacturers who anticipated the market need for greater flexibility and who, in many cases, first developed and used flexible systems for the production of their own CNC products. This in turn provided the impetus for parallel development in different companies of the other elements in a flexible installation.

14.6.4 Elements of flexible manufacturing systems

A number of common elements may be found in FM. Size and complexity varies with each system, but many contain the elements illustrated in the typical layouts shown in *Figures 14.54* to *14.56* and listed in *Table 14.16*.

A very common application of FM is in the machining of components. It is appropriate, therefore, to describe such a system in order to illustrate the fundamental aspects of FM.

The following sections discuss the role and main features of the major elements, but the range of equipment on offer is too large for a detailed review to be given here. Engineers will need to contact potential suppliers during both the feasibility study and the detailed specification stages of a FMS project.

14.6.5 Work centres

The majority of these are CNC machines programmed using CAM software. They should incorporate as much tooling and inspection/probing equipment as is economically justified, in order to:

(1) make them as versatile as the likely range of applications need;
(2) guarantee high quality (thereby reducing scrap and stoppage losses); and
(3) alert the system of any problems.

Machining centres Machining centres for prismatic parts form the core machining capability of many systems. They are either of the horizontal or vertical type and normally have a minimum of two pallet positions in order to allow simultaneous processing and feed-in/feed-out of pallets loaded with the product. Tool capacities range from 20 on small machines to 300 on larger machines using multiple magazines, and often carry more than one of a number of tools to avoid mid-batch tool replacement. The total tool capacity should include some spare pockets in order to cater for any extra tooling needed for additional or changed products. The machining axis varies from 3 to 6 depending on the number of indexing tables or spindle tilt axes that can be handled by the machine control. Automatic probing, coolant supply and swarf clearance systems are needed.

Table 14.16 Common elements of a FMS

Element	Purpose
Work centres	The primary work units of the system
Pallets	To carry the work-holding fixtures into the work centres and locate/hold them whilst processing
Fixtures	To hold and accurately locate products whilst they are worked on
Work changers	To load and unload fixtures and pallets
Tooling	Held by each work centre or a central point to cover machining of all the products programmed
Transport vehicles	To move pallets, fixtures and materials and work in progress around the system
Buffer storage	Temporary storage for balancing the system. This may be the storage of pallets
Inspection stations	Gauging and test of product
Cleaning	Pre- and post-operative cleaning of product
Supervising computer	Control of the complete system

Turning machines Turning machines for rotational parts will usually be CNC single or twin spindle with up to five axes and a variety of options for cross-slide and end turrets with driven tooling to given second operation work such as drilling and milling. Chucking and bar-feed machines are employed, and automatic coolant and swarf clearance systems are needed. Automatic probing and quick-change tooling systems are advisable. Part loading/unloading by robot arm, gantry systems and magazines is used.

Other metal cutting machines are also normally CNC and may include grinding wire erosion equipment, laser cutting, etc.

14.6.6 Pallets

Pallets are used in two major roles: as carrier pallets and as fixture pallets.

Carrier pallets hold batches of work and act only as a work transport medium. These are mainly used in support of robot and conveying systems.

Fixture pallets are used in machining centres. They carry the work-holding fixtures to and from the working head. The pallet is located accurately and locked in position at the work head to ensure repeatability of quality and safety. It is common for pallets to be dedicated to one fixture to save on set-up times, process delays and setting inaccuracies.

14.6.6.1 Pallet handling methods

Adjacent locations and RGV Pallets are held on both sides of the RGV track and transported by the vehicle (see *Figure 14.55*).

Remote pallet locations and AGV Pallets may be remote from a cell or adjacent to it and moved to the machines by an AGV (see *Figure 14.56*).

Pallet conveyors Pallets are held on rotary tables or conveyors. These may be adjacent to the machines, i.e. feeding directly onto them or remote and used as automatic work-change stations fed by AGV. *Figure 14.58* shows a pallet rotary table and conveyors from Cincinnati Milacron.

Automatic storage and retrieval (AS/AR) systems These are racking systems which can hold pallets in multiple levels. A stacking crane or robot stores and retrieves the pallets and either feeds them directly to the machines and work-change stations or to a conveyor that does so. *Figure 14.59* shows two forms of AS/AR systems offered by Mori-Seiki UK.

14.6.6.2 Work-change stations

These are where the product is loaded and unloaded from the fixture/pallet. These can be automatic (automatic work changer (AWC)), but are more often manually operated because of the high capital cost of providing automatic changing for a wide variety of components. Robots can be used if the product is supplied in the correct orientation in racks or magazines. Loading and unloading is often done at the same station, particularly on small linear track installations, but if the system is large with multiple cells and feed tracks it may be done at separate locations.

14.6.7 Fixtures

Fixtures are often held permanently on one pallet to reduce repetitive setting and maintain quality by having a consistent set-up. Multiple component fixtures are common and give greater efficiency due to:

(1) reduction in multiple pallet travels to and from the work head for separately fixtured components; and
(2) reduction in tool approach distances—these will be shorter for component-to-component travel around a multiple fixture.

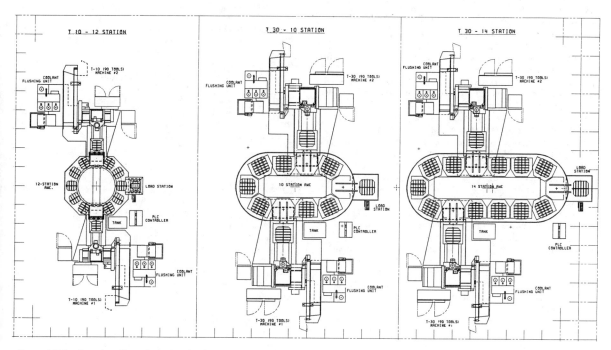

Figure 14.58 An automatic work changer feeding two machining centres. (Courtesy of Cincinnati Milacron)

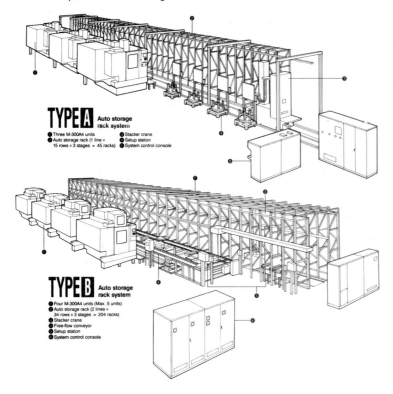

Figure 14.59 The Partner series line form pallet AS/AR systems. (Courtesy of Mori-Seiki UK)

A fixture may also be used to house sets of components that go to make up a subassembly, e.g. a body of some description and its mating cover. This ensures more accuracy in the fit of mating machined faces and hole centres on the different components and also reduces the risk of mismatched quantities at assembly.

Multiple fixtures are usually constructed in the form of a cube with components mounted on two or four vertical faces. They can be used to carry out the same operations on all faces or a series of operations on successive faces of the fixture. Component turnover would be carried out manually at a station adjacent to the machine or pallet carrier. This approach is suitable for low volume work, but for higher volumes it may be more economic to have a number of machines, each set for operations the opposing faces of the component.

Indexing tables are also used with components bolted directly to them or with fixtures mounted on them. These will normally be linked to the machine control, their operation becoming a fourth or fifth axis of the machine.

Fixture pallets may be stored and issued to individual machines, cells or central points for the system. These alternatives are illustrated in the various system layouts shown in the figures.

The best design, combination and allocation of fixtures, pallets, tooling and machines varies with each installation, but the main objective must be (within acceptable cost) to avoid conflict for these resources and the resultant delays, in order to minimise the total in-system time and to maximise throughput.

14.6.8 Tooling

Tooling is normally taken to cover consumable cutting bits, tool holders and complete tooling systems. In FM installations, tooling needs careful planning if a good balance between operational requirements and tooling costs is to be achieved. A detailed study of the machining requirements of all the components to be machined by a system or on a particular work centre should lead to some rationalisation of tooling. This is desirable not only from the point of initial cost but also the continuing cost of carrying and maintaining the necessary spare tooling.

14.6.8.1 Management and supply systems

The availability, tracking and servicing of tooling is a crucial element of FM. The best tooling system solution will vary considerably with the type and total number of components and processes. There are a number of options for tool supply and management: machine, cell and system centred.

Machine centred A work centre doing many short cycle operations to different levels of accuracy and surface finish may need its own full complement of tooling permanently in position.

Cell centred Tooling can also be organised in sets against a particular group of work centres or cells. The tool set can then be issued to any of the designated cells as required, and tools replaced individually.

System centred If a system has a number of similar machines carrying out like operations, or if there is a logistical and economic justification, then all work centres can be served from a central point.

Tool management A tool management system is required by the central control computer so that it can program tooling supply to meet the production schedule. FM suppliers have integrated tooling management software modules in their system controls, and proprietary software packages exist with varying levels of capability to match the FM requirements.

14.6.8.2 Tool changing

Automatic tool changers (ATCs) change tools at the work spindle and are a standard feature of machining centres. There are several designs, but the most widely used is the swing arm which rotates through 180° from the tool magazine to the spindle, picking up the used tool at one end of the arm and depositing the new tool with the other end.

Tool magazines are normally of the recirculating chain type which give a greater tool capacity and are more easily extendable than the rotary designs seen on smaller machines.

Tool changing in the magazine within the production cycle at a work centre is usually automatic for safety reasons. Changing 'out of cycle', i.e. at the end of a batch run, can be either manual or automatic.

With a fully tooled work centre where the tool magazine capacity is large enough to hold all the tools required, large scale tool changes should not be necessary, and manual changing or replacement for tool wear should be suitable.

Where there is a number of similar machines, their operations can share commonly used tools, or if a work centre is used for a large number of products that need different tools then a central tool store and distribution system may be appropriate.

Central tool storage Tools can be stored in a central area with the location known by the computer. When a component is scheduled onto a machine, the tools required are selected and fed to the machine. The tool store is stocked and maintained with tools sharpened and pre-set by tool engineers working in an area, ideally an area next to the store.

Automatic tool distribution systems As with other features of a FMS, many alternative methods and layouts have been used. Each unique combination of requirements dictates which option to select. Some major alternatives are:

(1) tools are stored adjacent to the work centres being supplied (small systems and common operations);
(2) tools are stored remote from most work centres (in large systems);
(3) tools are transported by adjacent linear track or robot; and
(4) tools are transported by AGVs.

Where tools are stored adjacent to work centres, machining centre tool magazines are usually accessed from the side or rear of the machine. This avoids conflict with space for product and operator control access at the machine. It also allows in many cases the provision of an automatic tool distribution system running along the back of the machines. The distribution method from this type of magazine may be by linear track tooling pallet or robot.

Where tools are stored remote from work centres, if a store is feeding a well-dispersed system it will be located some distance from a proportion of the work centres. The transport

medium in this case is most probably an AGV, because AGVs will already be in use for product pallet distribution.

14.6.9 Transportation methods

There are two common methods of linking work centres over distances that are longer than the reach of a robot arm: RGVs and AGVs.

RGVs are mounted on two parallel rails that run in a straight line above the floor surface—hence the description 'linear track installations'. The rail system is usually engineered very close to the work centres, and normally only one or two vehicles are used because their range of movement and flexibility is restricted by running on one track. RGVs can serve in-line layouts by moving up and down the straight track but are rather limited in complex floor layouts involving many work centres or cells.

AGVs run directly on the floor and are guided by a control that follows a guide wire buried in the floor. They are used in linear installations but are more usually applied to larger systems which cover a wide floor area. They are able to negotiate the directional changes that a large system layout would require, and are entirely separate from the work centres.

A comparison of the advantages and disadvantages of RGVs and AGVs is given in *Table 14.17*. Currently, major suppliers are seeing a move away from AGVs to RGVs, particularly in small systems and machining centre cells (up to six machines) in which one vehicle is sufficient for all the transportation needs. Two contributory reasons for this are the lower initial cost of and the smaller total floor area needed for a RGV.

14.6.10 Control and planning systems

14.6.10.1 Introduction

A FMS control system normally resides in a central executive computer which is often known as the 'host computer'. This

Table 14.17 Comparison of transporter methods*

AGVs	+/−†	RGVs
Little constraint on layout	+ −	Usually linear arrangement only
Leaves floor free of major obstruction	+ −	Restricts access to equipment
Travel distance not a major cost factor	+ −	Cost of rail directly related to distance
Requires wider aisles for turning	− +	Allows close separation of machines
Expensive to use individual pallet stands	− +	Lends easily to pallet stands
Vehicle is self-contained	+ −	Vehicle requires trailing leads, or contactor rail
Little guarding required	+ −	More difficult to guard
Higher initial cost for small system	− +	Lower initial cost for small system
Longer journey time (including docking)	− +	Shorter journey time over shorter route
Requires more complex control	− +	Less complex control
Easy to have two or more vehicles	+ −	Harder to manage two vehicles on the same track
Requires alignment mechanism for pallet discharge	− +	Rail gives a natural alignment
Vehicle requires recharging	− +	Can work from direct supply
Easy to configure track	+ −	Difficult to reconfigure

* Courtesy of Cincinnati Milacron.
† +, Advantage; −, disadvantage.

brings together segments of other systems already being used in manufacturing and applies or supervises them to give an integrated control for the FM system.

The existing types of manufacturing system that are incorporated to a greater or lesser degree, depending on the scope of the FMS are described below.

Planning systems Production planning software packages or modules of larger software suites are used by production management and production control for the planning and scheduling of production operations. Such a program will reside in the host computer to carry out these functions for the FMS and will be the top-level element for driving the rest of the system.

Engineering systems Computer-aided manufacturing (CAM) systems are used to produce the part programs for CNC machines and the supporting tooling management. In addition, CAD facilities are used for the design of tooling and fixtures. Increasingly, CADCAM is available in which designs can be drawn from CAD software into CAM for the manufacturing engineering work. Part programs can be held in the host computer for downloading, or in the machine controls for initiation. Tooling and other technical data are held in the host computer for reference in its planning decisions.

Machine controls The controls will have been developed by the machine suppliers and controls companies and reside in the machines. CNC machines and robots are examples. The host computer supervises, not supersedes, the machine controls. In addition, DNC systems are used to control a number of machines from a central point. Programmable controllers (PCs) are also supervised by the host computer.

Communications systems Communications networks, software and hardware are used to connect the control system to the operating equipment. UNIX and DOS based computers can be connected to remote equipment by the addition of a communications card at each equipment and network cabling/hardware running between them.

Business controls The FMS planning control needs to receive master production requirements from the main manufacturing system computer. These can be downloaded to the FMS via a network or direct link. Depending upon the degree of computer integrated manufacturing (CIM) that the company is operating, links with modules of the main computer can be made. As CIM becomes established in a company, a FMS can become a subsystem of the CIM system.

14.6.10.2 Functions of a FMS control system

The system control software resides in the system executive (often called the 'host') computer. Its functions are to:

(1) accept production plans from the main manufacturing business software suites, e.g. MRP;
(2) schedule the system and issue production orders accordingly;
(3) capture and record performance and status monitoring and issue reports;
(4) supervise subsidiary systems for:
    (a) components routing and pallet control,
    (b) part program control,
    (c) machine control and status monitor,
    (d) transporter control (AGC and RGV),
    (e) tooling management and life status, and
    (f) quality control and reporting;

(5) run simulation models of the proposed workload prior to issue (see Section 14.6.11); and
(6) oversee the communications network and local terminals.

Figure 14.60 shows a simplified structure of a control system. Details of the control systems used by two leading FM suppliers that take advantage of the new generation of 32-bit processor capabilities are shown in *Figures 14.61* to *14.64*.

The SATE system from FMT Ltd, is illustrated in *Figures 14.61* and *14.62*. *Figure 14.61* depicts part of a FMS with one item of each of the major elements shown. The control system modules and communication links are superimposed on this. *Figure 14.62* shows the data structure of the system:

(1) constant data are mainly of an engineering nature, relating to the machining requirements of the products loaded to the system;

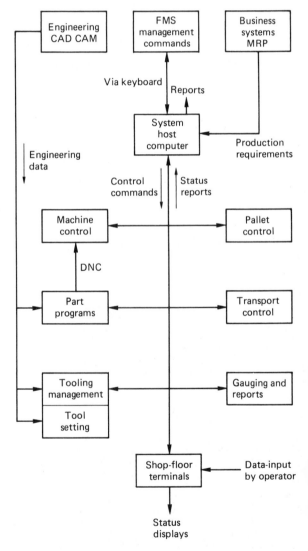

Figure 14.60 A FMS control system

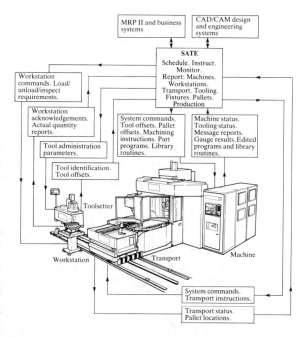

Figure 14.61 Part of the SATE FMS system showing one item of each of the major elements. (Courtesy of FMT Ltd)

14.6.10.3 Systems hardware

Executive or host computers The computing power needed is dependent on the complexity and size of the system. Minicomputers have been and are used (the DEC PDP11 series is a widely used machine), but the increased power now available from 16- and 32-bit desk-top PCs is sufficient to run many systems. Advances in multitasking operating systems, networking and communications architectures (MAP, OSI, etc.) will increase the capabilities of computers and communications to control complex FMS and factory systems. The host computer is normally situated in an office adjacent to or within the FMS, where local management, supervision and manufacturing or systems engineers can access the computer. A typical specification may include the following.

(1) Executive computer:
   (a) 32-bit processor with a 16-bit maths coprocessor,
   (b) 1.2 Mbyte of system memory,
   (c) $3\frac{1}{2}$ in. and $5\frac{1}{4}$ in. disk drives,
   (d) 40–100 Mbyte hard disk,
   (e) power supplied from an uninterruptable power supply (UPS),
   (f) keyboard for manual data entry,
   (g) mouse and tablet for menu-driven work, and
   (h) back-up disks or tape streamer.
(2) A 14 in. colour graphics screen for data, system status and simulations.
(3) An A4 or A3 printer for report output.
(4) An A4 or A3 plotter.

(2) dynamic data are the actual production-demand data that are subject to change, as is pallet allocation to the parts needed; and
(3) status data are the monitored status data of the working units of the system.

The computer hardware currently being installed is the SUN SPARC station IPC. This is used as both the central host computer and satellite terminals, which may be sited in the tool setting area, engineering and management offices.

The FMS 20-32 control system from Cincinnati Milacron, is shown in *Figures 14.63* and *14.64*. *Figure 14.63* shows a system network, the FMS20-32 master computer and satellite computers for cell management, tool setting, simulation and communications interfaces. *Figure 14.64* shows the data structure, software and hardware configurations.

Local computers These carry out specialist tasks (at the request of the host), that require processing power to make local decisions. Two common uses are as transport controllers, which control the activities of pallet transporters, and tool management controllers, which supervise all the tooling servicing, supply and distribution activities. The shop-floor environment is a factor to be considered and 'hardened' computers, terminals and PCs may be needed to withstand harsh conditions or operator treatment.

Local terminals These may be data acquisition terminals (DAT) or personal computers (PC). They are used by operators and other staff for data input and output at key points around the system layout.

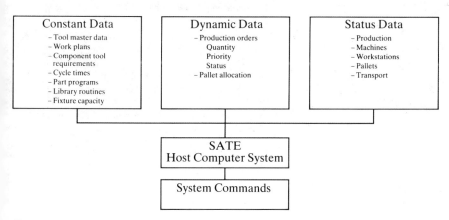

Figure 14.62 The data structure of the SATE FMS system. (Courtesy of FMT Ltd)

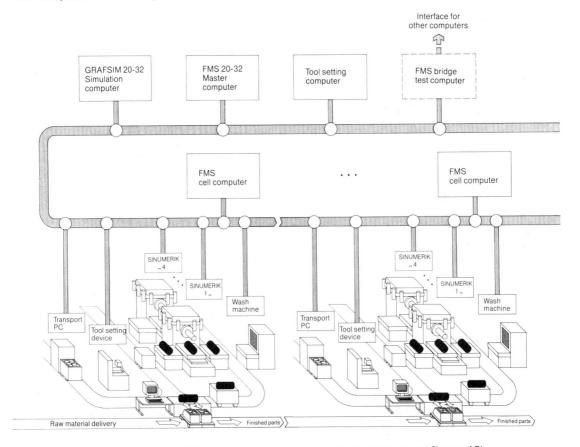

Figure 14.63 Functional structure of the FMS20-32 control system. (Courtesy of Cincinnati Milacron and Siemens AG)

14.6.11 Simulation of flexible manufacturing systems

14.6.11.1 Introduction

A FMS is very complex. It combines a large number of activities and facilities in order to achieve the making of a number of relatively small quantities of product. Its complexity is increased by the fact that the system is dynamic—conditions within it are constantly changing as orders are received, plans initiated, materials flow and operations are completed.

Given the high cost of introducing a FMS in terms of both capital investment and human effort, it is vital that the system operates at maximum efficiency. In order to achieve this aim, considerable work needs to be done in the planning and design stage and simulation is a tool to be used in this respect. Simulation techniques have been used in industry for many years and there is a large number of simulation software packages on the market. Major FMS suppliers use one of these or their own simulation software in configuring a FMS, and also offer a simulation facility as part of their system software.

14.6.11.2 FMS simulation

The simulation programs used for FMS can be either general purpose programs with a FMS module or the FMS suppliers own program packages. Two examples of the former are SEE-WHY from ATT/Istel which is used by FMT and

HOCUS from P.E. Consulting Limited which is one of two packages used by Cincinnati Milacron; the other is a GRAF-SIM, a Siemens program.

Many simulation packages are written in Fortran or other high-level languages, and early editions needed the user to be able to program in the language in order to set up the model of the system to be simulated. With recent advances this is not necessary because the user works through a series of menu-driven screen displays with interactive graphics that provide easy data-input and model-building activities.

A simulation package is used to:

(1) design the system—to run simulations of various configurations and operating conditions during the design process; and
(2) use the system—to simulate the effect that various production demand patterns will have, i.e. a 'what if' capability.

Modern packages use interactive graphics that allow the user to see the effect of inputs immediately, and not have to wait for the completion of a simulation run.

14.6.12 Planning a flexible manufacturing system

14.6.12.1 FMS strategy

As has already been shown in many ways, a FM project is an operational and technical challenge and, probably, a large

Software: indivdual function modules for your own FMC solution

- Master, Control, Status data administration
- Production order Planning
- Tool Requirements
- Tool Balance
- NC Programme Administration
- Tool Flow Control
- Load/Unload Dialog
- Material Flow Control
- SINUMERIK 3/8/850 Interface
- LAN Connection
- CIM Interface
- Tool Setter Connection
- Mesuring Machine Interface
- Wash Machine Interface
- System Image
- Back-up Data Administration
- Testprogramme Interfaces
  (using point-to-point and SINEC H1)

Hardware: SICOMP WS 20-32 as a cell computer

CPU

32 Bit Processor NEC 32032, 16 Bit Prozessor Intel 80286, 6 MHz, no Wait States, buffered clock with date, buffered CMOS RAM, 8 interfaces (6 available), 7 DMA channels.

Main Memory

2/4M Byte + 640K Byte

Peripheral Memory

1.2 Mega Byte Floppy Disk, 20 or 70M Byte Winchester Disk, further units connectable

Operating System

C-DOS 4.11, multitasking, MSDOS-Program running capability

LAN-Connection

SINEC H1 (based on Ethernet)

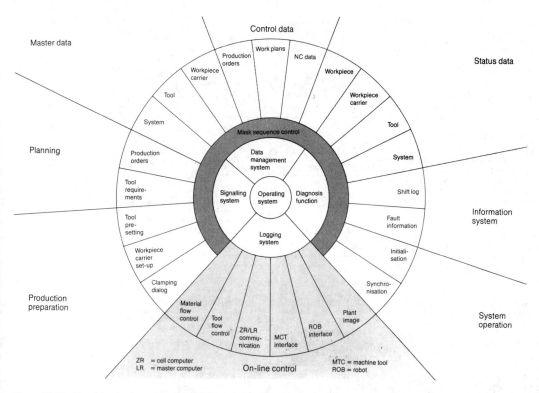

Figure 14.64   The data structure of the FMS20-32 control system. (Courtesy of Siemens AG)

financial risk. Such an undertaking will require top management approval and strong commitment from all those involved. Any significant industrial project is phased to reach certain points in the route to the eventual goal. For most companies, this is the approach that should be taken with a FM project.

A step-by-step approach is currently recommended by several leading UK suppliers who now have considerable experience of the challenges created by a FM project.

Broadly, a FM program should be started by introducing one self-contained cell and then adding to it as experience is gained. This will give:

(1) a better guarantee of early success than will a 'one-hit' large scheme;
(2) a lower rate of cash flow out of the company;
(3) a faster return on investment if the correct segment is selected for the initial work;

(a)

Stage 1 Software options ◈ □

The stand-alone machine, with its minimum manned capability allows the user to build-up experience in palletising work and presenting mixed batches of work to the machine. The FM 100 can have two or four pallets. The latter is a useful configuration if there is likely to be an extended time span between stage 1 and 2.

Benefits
A CNC machining centre with all the essential elements for extended periods of unmanned operation. Shop floor personnel can rapidly assimilate the principles of the new technologies.

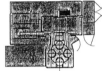

Stage 2 Software options ◈ □

The addition of a rail-guided vehicle, and up to 15 pallet stands enhances performance in two ways. The range of components that can be handled — without operator involvement — increases. And, the periods of unmanned operation can be greatly extended.

Benefits
Extended unmanned capability and a wider range of component can be available for immediate manufacture, so all components for a complete assembly can be machined sequentially.

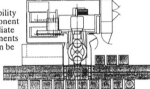

Stage 3 Software options ◈ □

Additional machines up to four, each with up to 15 pallet stands can be added. At this level, pallets are dedicated to individual machines. Load/unload is carried out at a load/unload station allocated to each machine. The rail guided vehicle responds to calls from the individual machines' control system.

Benefits
Increased manufacturing capacity from additional machines with the ability to further reduce inventory.

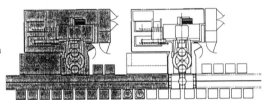

Stage 4 Software options ◈ □

The same system hardware parameters as stage 3 plus an independent transport control. Any pallet can now be loaded to any machine. Dedicated load/unload stations improve operator efficiency. The transporter works to priorities established by the operator instead of answering machine calls for the next component on a queuing basis.

Benefits
Improved productivity from balanced scheduling leads to increased cell efficiency.

Intelligent car control, VDU and keyboard

Stage 5 Software options ◈ □ ○ △ ⬡

With the addition of a host computer the cell can be expanded to seven machines, inclusive of support operations including inspection and wash facilities. A host computer up-grades the system to exploit all the facets of Automated Manufacturing Technology.

Benefits
The final step to AMT. Linked into CAD, MRP and other business packages, the cell offers the user the fastest product development combined with rapid re-direction of resources to reflect changing market patterns.

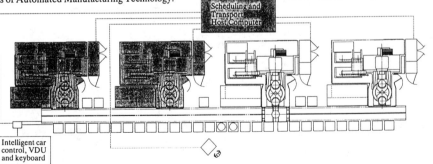

Scheduling and Transport Host Computer

Intelligent car control, VDU and keyboard

Figure 14.65 The processing development of a cell. (Courtesy of FMT Ltd)

(b)

 ## Direct Numerical Control

This facility, allows part programs to be stored away from the machines in a central system then transmitted to the machines' control as dictated by production schedules. Such a system offers considerable advantages when compared to program transfer by conventional paper tape. The facility to transfer programs back to the central storage system, after prove out on the shop floor, provides a secure method of updating records.

 ## CAD/CAM

It is now increasingly popular for companies to adopt computer aided design (CAD) to generate engineering designs. CAD data can be processed to produce part programs for CNC machines. This coupled with a DNC facility, provides a paperless environment from design to finished part.

 ## Host Computer — Manufacturing Cell

The incorporation of a host computer improves significantly management efficiency. The computer software will co-ordinate instructions within the cell and allow automatic scheduling of work to be planned in advance. Other benefits that the cell computer brings include: component routing control; tool management; part program control; status monitoring; and quality control. The cell host is the key to integration with other factory systems.

 ## Link to Business Systems

Of key interest to executives with overall responsibility for a plant is the linking of the company business systems including MRP II. This allows orders to be downloaded to the cell host and order status to be transmitted in real time — again in a paperless environment.

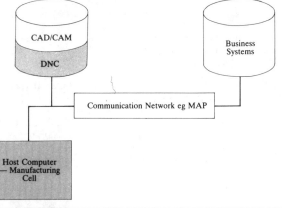

Computer Integrated Manufacturing

The goal of the fully integrated company is increasingly to utilise electronic communications as its medium. CIM is the integration of manufacturing and business processes, wherein financial, commercial and manufacturing departments make use of a common data base which has been designed to ensure security and consistency of information throughout the organisation. Management can perform advanced planning, taking account of the many variables in the manufacturing environment, including the control of material replenishment, inventory, production schedules and the many aspects of business finance.

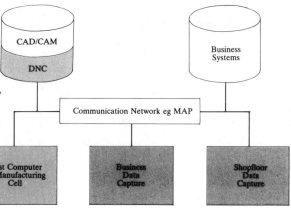

(4) more accurate assessment of the operational and financial benefits; and

(5) avoidance of large peaks in the demand on managers' and staff time.

The level of system control can be upgraded to handle the increasing complexity as the initial installation is expanded. *Figure 14.65* shows how this can be achieved for a linear machining centre cell, but the principles are equally applicable to most potentially extensive FM installations.

14.6.12.2 Factors to be considered in designing a FMS

The design of a FMS needs a detailed study of all the many and varied operating elements involved. A large system is really a 'factory within a factory' and, as such, needs support services. The system design should be the best balance of the often conflicting requirements of the individual elements, and also the normally attendant financial, labour and space constraints.

The elements and factors to be considered in designing a FMS are described below.

Products and processes
(1) The range of products to be processed—types, volumes, frequencies.
(2) The processes involved—types, inputs and accuracies needed.
(3) Operation times and other time periods needed.
(4) The operation flows and scheduling priorities.
(5) Material flows and buffer storage policies.

Equipment
(1) The machinery and equipment best suited to the process tasks.
(2) The grouping of machines and equipment to achieve the desired flows.
(3) The pallets and fixtures needed.
(4) The location, storage, distribution and control of pallets and fixtures.
(5) The pallet and fixture loading and unloading units to be used.
(6) The tooling needed.
(7) The location, storage, distribution and control of tooling.
(8) The transportation equipment and routes to be used.
(9) The control computer, local terminals and the network required.
(10) The location, control, back up and individual access to the computer system.
(11) The software used and its control, up-date, back-up and access by individuals.

Human resources
(1) The skill levels and numbers of operators, manufacturing engineers, CNC programmers, maintenance engineers, tooling engineers, planners and system software support, supervision and management.

Space
(1) The space needed for the flows and equipment listed above.
(2) The working areas and access for all staff —both within and outside the FMS.
(3) Rest areas and access.
(4) Buffer storage areas.
(5) Service runs.

Services
(1) Power, lighting, compressed air, water, drainage, ventilation and fume extraction, battery vehicle charging, toilets, washing facilities, vending machines, telephones, fax, and computer networks.

Project team
(1) The feasibility study, detailed design, installation and successful commissioning of such an undertaking is obviously a multi-disciplinary task that will normally be carried out by a project team. The team should include specialists from the suppliers of the equipment, software and other services, as well as the appropriate internal departments.

14.6.12.3 A FM project

Because a FM project is a capital investment in manufacturing capacity and technology it should involve a considerable amount of manufacturing engineering effort, but must also have a large input from other areas such as production control, manufacturing management, site engineering and the FM vendors. It is advisable, therefore, to create a project team with representatives from all the major contributors, in order to successfully manage and carry out the project.

The project team and inputs required A FM project will require a very wide range of inputs because its scope will involve all the normal activities of a manufacturing organisation. The functions and areas of interest are listed in *Table 14.18*.

Table 14.18 The project teams and inputs required in a FM project

Function	Input and area of interest
Project manager	A senior manager responsible to the executive body
Manufacturing management	Overall operation and manning
Manufacturing engineering	Processes, tooling, CNC, CAM, layouts
Production control	Shop scheduling, priority rules
Inventory control	Inventory levels and flow
Purchasing	Project and production materials
Systems/software	Host computer systems and software
Quality assurance	Quality systems and control
Quality engineering	Inspection and gauging equipment
Product design	Design changes for manufacture; new products to go on FM
Site engineering	Plant services and buildings
Maintenance	System maintenance routines
Cost accountant	Project and operational costings
The turnkey suppliers team	Supplier's project manager and specialists
Other suppliers	Suppliers reporting to the company or to the turnkey manager

The number of interested parties is very large and they must be included in the planning and commissioning stages as appropriate. A project team having this many permanent members would, however, be unwieldy. There should be a smaller core team to manage the project under the chairmanship of the project manager, and it should include representatives from:

(1) production management,
(2) manufacturing engineering,
(2) production control, and
(4) inventory control.

If the intention is to set up the FM as a self-contained unit with permanent staff from those areas, then they should be on the core teams.

In a small company or FM scheme, several of the functions may be combined or represented by one individual.

Vendors turnkey project responsibility   There will normally be a number of equipment and systems suppliers with whom good communications must be established. These can be managed individually by the company or by the appointment of one major supplier who would act as the turnkey supplier/project manager and manage the other suppliers on a subcontract basis. This arrangement is advisable for any but the smallest projects because the turnkey supplier:

(1) has far more experience of this type of system than does the customer company and its employees;
(2) will submit proposals for the design and implementation of the complete scheme—these can be based on computer simulations of the operation of alternative schemes or sizing;
(3) will contribute a core of design experience based on other applications; and

(4) will carry legal and financial responsibility for the successful integration and performance of the individual suppliers elements within the complete system.

14.7 Industrial robotics and automation

This section examines the 'robotic' equipment used in manufacturing industry and includes reprogrammable automation equipment. Numerical control (NC) machine tools and automated assembly equipment are excluded except where industrial robots are involved. Similarly, 'hard' automation (e.g. cam-controlled mass-production equipment) is omitted. Thus the main subjects of interest are industrial robotics, artificial vision and automatic identification systems.

14.7.1 Industrial robotics

14.7.1.1 Definition

It is important, in an industrial context, to have an unambiguous definition of a robot. All the major industrialised countries have adopted definitions with some more complex than others. The British Robot Association definition is succinct and is therefore selected for use here, i.e. 'An industrial robot is a reprogrammable device designed to both manipulate and transport parts, tools or specialised manufacturing implements through variable programmed motions for the performance of specific manufacturing tasks'. Industrial robots are therefore *reprogrammable manipulators* of tools, materials or components. A typical robot is shown in *Figure 14.66*. They should be easily reprogrammed to carry on new tasks and they should have a degree of dexterity. Thus a NC machine tool is not a robot since, although it is easily

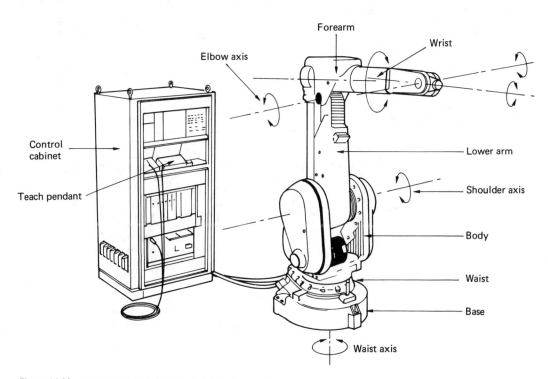

Figure 14.66   ASEA IRB 2000 industrial robot. (Courtesy of ASEA)

reprogrammable, it is not designed to do anything other than cut material. Neither is the type of arm used to manipulate radioactive material in the nuclear industry; these devices are constantly under the control of a human operator and are therefore remotely controlled, i.e. they are not programmed to operate autonomously.

14.7.1.2 Reasons for using industrial robots

In modern manufacturing there are many advantages in using robots rather than human labour or hard automation. In relation to human labour:

(1) robots work to a constant level of quality;
(2) waste, scrap and rework are minimised;
(3) they can work in areas that are hazardous or unpleasant to humans;
(4) no jobs are boring, tiring or stressful to robots;
(5) continuous 24-h production is possible for many days;
(6) they are a single investment—salaries do not have to paid each year at increasing rates, and there are no burden costs such as pension and insurance schemes, holidays, sick pay, etc.;
(7) investment in a robot involves a once-only capital expenditure, whereas human labour requires an ongoing salary cost that increases annually; and
(8) robots are advantageous where strength is required, and in many applications they are also faster than humans.

Also, in relation to special-purpose dedicated equipment, robots are more easily reprogrammed to cope with new products or changes in the design of existing ones. Dedicated equipment usually requires expensive strip-down and rebuild in these situations and often has to be discarded as obsolete.

As well as these obvious reasons there are other, indirect, advantages to be gained from robotisation.

(1) When changing from manual to robotic methods, the product components will often have to be redesigned to provide simplicity of presentation, positive gripping points, unambiguous orientation and location, adoption of the stacking principle for assembly, and ease of location for screws, etc. This usually results in a simplified, better and cheaper design for the product.
(2) Quality will be improved in many areas as automatic inspection techniques are adopted.
(3) Design changes can be implemented more quickly and new products introduced efficiently.
(4) Lead times can be reduced.
(5) Work in progress can be reduced.
(6) In comparison to dedicated equipment, smaller batch sizes can be handled and downtime between product changeovers is reduced.

It should be remembered that robots are simply one alternative and that human labour and dedicated special-purpose equipment also have their place in the manufacturing environment. Although much simplified, *Figure 14.67* shows the cost-to-volume region in which robots are most attractively employed. Generally, unless there are severe environmental or hazardous conditions, human labour is suitable for low-volume high-variety work. Conversely, for very high-volume low-variety work, dedicated equipment, or 'hard' automation, is probably the most cost-effective.

14.7.1.3 The construction of industrial robots

Essentially, an industrial robot consists of two elements—the manipulator (or 'arm') and the robot controller. The con-

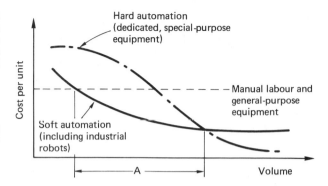

Figure 14.67 Product-volume region A is most suited to robots

troller contains the microprocessor system and the power control units. Hydraulic and pneumatic robots also have pump and compressor units, respectively. The particular geometry of the arm will provide an associated work envelope; the arm will be powered by electrical, hydraulic or pneumatic means; it will be non-servo or servo controlled; it will be programmed on-line, off-line or both; and will be capable of point-to-point, point-to-point with coordinated path, or continuous path movement. A brief explanation of each of these terms is given below.

The control unit This unit interfaces with the robot's internal and external sensors, drive units, peripheral equipment, and the programmer and operator. It is therefore usually capable of handling serial and parallel data transmission at various rates, and can carry out digital-to-analogue and analogue-to-digital conversion as necessary. Communication with the programmer is via a visual display unit and keyboard, or a teach pendant. There will also be floppy disk drives for loading and saving programs and a printer for hard copy.

Program interpretation is carried out within the controller. Some robots are capable of using more than one language. In this case the language and operating system are usually loaded from disk at the beginning of the programming session. Within the unit there is ample memory space to store all necessary data for coordinate transformation, trajectory computation, monitoring and decision-making.

The unit also performs the functions necessary for full servo control. In some robots each axis has its own microprocessor system supervised by a 'master' system. There may also be a system dedicated to handling the sensory data input. Thus in a six-axis robot there may be eight integrated microprocessor systems within the control unit.

Control of the power units is effected by the control unit. In electric-drive robots, low-voltage control signals are sent out to the motor drive amplifiers, one for each axis, to produce power for the motors. Servo control is maintained by monitoring feedback from internal sensors. In fluid power systems, solenoids and servo-valves are also controlled in this way.

Robot geometry The robot arm is composed of links and joints. The joints, also referred to as 'articulations' or 'kinematic pairs', normally each have only one degree of freedom. In robots this means that a joint is probably (1) revolute (i.e. rotation about one axis), (2) prismatic (i.e. linear movement along one axis), or (3) screw (i.e. a combination of linear and rotational movement along one axis with translation defined by the screw pitch and the rotational displacement). Other types of joint with more than one degree of freedom (e.g. a ball joint) are more difficult to control.

The number and relationship of these link and joint arrangements defines the dexterity of the robot arm. There is a maximum number of six degrees of freedom available to any free body in space. For many industrial applications (e.g. arc welding, fettling and spray painting) six degrees of freedom are desirable. With a minimum of six joints it is possible to achieve this. However, the design of the robot must be carefully considered as there will be 'no-go' regions into which the robot arm will not be able to reach due to its physical limitations. Also, even though a robot may have a large number of joints, it may not have the equivalent number of degrees of freedom. For example, an arm with six revolute joints whose axes are all parallel has only three degrees of freedom.

Usually a robot arm has three major axes providing three degrees of freedom. These axes allow the robot to position its end effector or gripper at any point in space within its work envelope. In addition to the major axes there are also one, two or three additional axes, normally in the form of a 'wrist' at the extremity of the arm. These allow the robot to orientate its end effector, which is fixed to the wrist, about any point in space. As mentioned previously, the number of degrees of freedom required depends on the task to be performed. For example, population of a printed circuit board will only require four degrees of freedom. This operation is essentially of the pick-and-place type, with one vertical, two horizontal and one rotational movement for component orientation. However, for arc welding a complex three-dimensional seam (say, at the intersection of two cylinders) a robot with the full six degrees of freedom are necessary. This is to ensure that the welding gun is constantly maintained at the correct orientation to the work as welding proceeds. Indeed, this particular task may demand additional axes to be employed by clamping the work on a multi-axis servo-motor controlled work table whose movements are integrated with those of the robot and controlled by the same control unit. Some typical robot configurations and work envelopes are shown in *Figure 14.68*.

In the *cartesian* configuration a rectangular work envelope is obtained. This configuration appears in two forms. One is the gantry type in which the robot arm is suspended from a gantry held within a rectangular frame. This provides a rigid, simple construction but has the disadvantage of having a small work volume in relation to the floor area occupied, this is sometimes termed the robot's 'footprint'. The other type is similar to that shown in the figure and this may be floor or bench mounted. Larger cartesian robots are used for palletising and loading conveyors, etc., and the simpler types can be found servicing injection moulding machines. Smaller types can be used for assembly, although in most cases they have now been superseded by the SCARA types (see below).

Cylindrical coordinate robots basically consist of a rotatable pillar to which a horizontal arm is attached. This arm can move vertically and horizontally with respect to the pillar, thus allowing a cylindrical work volume to be created. Their long reach and small footprint make them useful for machine tool servicing, palletising, and other material handling tasks.

The *polar* coordinate robot was one of the first types developed and was hydraulically powered. This construction and drive system made them suitable for heavy duty handling and spot welding. Recent robot designs seldom adopt this configuration since path control is relatively complex but dexterity is limited.

Maximum flexibility is achieved with the *revolute* configuration. These robots have one vertical and two horizontal revolute joints for their main axes. They have a small footprint/work volume ratio and have a long reach which can extend below their base if necessary. Applications include spot welding, arc welding, spray painting, adhesive and sealant application and material handling. They have also been used for assembly tasks but with proper product design the less complex SCARA robots are less expensive and simpler to program.

Specially designed for assembly the *SCARA* configuration has become extremely popular for assembly work where the components are either all on one plane as in a printed circuit board population, or the product has been designed for assembly by vertical straight line stacking. The acronym SCARA stands for 'selective compliance arm for robotic assembly'. The selective compliance exists in the horizontal plane, this allows slight movement for 'peg in the hole' type insertions where a chamfer is provided on the mating components. The vertical revolute axes ensure vertical rigidity. Whereas the previously considered configurations had three major axes with the ability to affix a 'wrist' with a further one, two, or three axes, the SCARA robot is designed specifically for four-axis operation. The two major rotational axes provide movement within a partially circular area, a linear axis which may be pneumatic non-servo or electric servo provides vertical movement, and rotation of the gripper produces component orientation.

Other configurations are available. For example, the pendulum arm type was originally designed for assembly but is now also found in other applications, and the multiple joint, or 'elephant's trunk', type is designed for maximum flexibility in spray painting tasks.

Robot drive and control systems   Many of the first industrial robots were hydraulically driven. However, most robots now produced are electrically powered. The previous definition of a robot implies relatively sophisticated control, and this is verified by the fact that nearly all industrial-quality robots are fully servo controlled. Non-servo controlled robots, just on the borderline of the definition between robots and simple pick-and-place units, are usually driven pneumatically. Some very light-duty devices, and often the orientation axes on SCARA robots, use non-servoed stepping motors.

Electric robots are usually driven by d.c. permanent magnet servo-motors, brushless motors or, occasionally, stepping motors. Electric-drive systems are relatively clean and quiet when compared with fluid power machines. They are easily maintained and repaired and are well suited to electronic control. Recent developments in the use of rare-earth materials for permanent magnets mean that power-to-size ratios are increasing, and the use of brushless motors reduces maintenance costs. Brushless motor drives can also be used in areas such as clean rooms, since contamination particles are reduced, and in situations where there would previously have been a fire risk due to the possibility of brush arcing. Unless incorporating direct-drive motors, electrically driven robots do have the disadvantage of requiring transmission systems. These add cost and weight, and also reduce precision due to gear backlash and other unwanted movement.

Hydraulically powered robots still have some advantages. For example, they have very good power-to-size ratios and hydraulic force can be applied directly at the desired point without the need for a transmission system. Hydraulic fluid is incompressible and, therefore, there are no backlash problems. Assuming the power pack, which contains the electrically driven hydraulic pump, is located remotely, then the robot can be used in high fire-risk areas. This is because only very low voltages for control and feedback purposes are present on the actual robot arm. Because of their necessarily sturdy construction due to the high hydraulic pressures experienced, they can withstand higher shock loads than can other

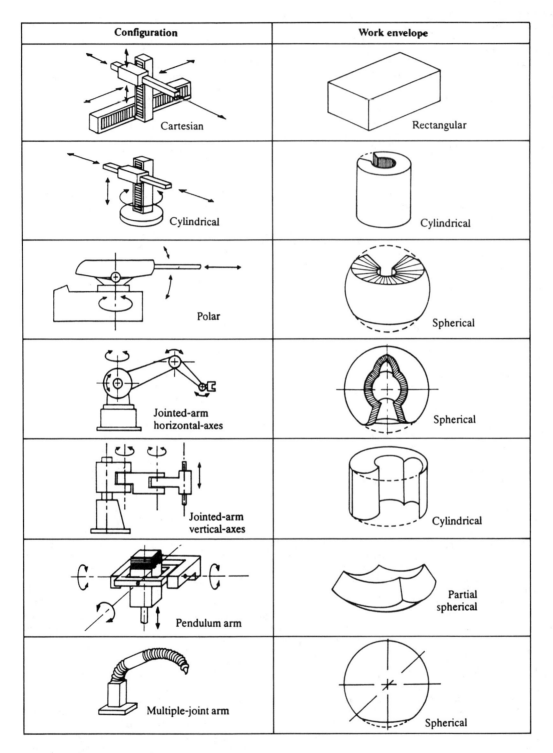

Figure 14.68 Robot classification by geometric configuration and work envelope

robot types. However, it is their disadvantages that have led to their reduction in popularity. A noisy power pack, even when protected by an acoustic muffler, makes them environmentally unattractive. Historically, they have tended to be less reliable than electric robots with leakages occurring which contaminate work areas and cause performance loss. Servo control of hydraulics is not as simple as that for electrics and the availability of skilled personnel is lower. The viscosity of the hydraulic fluid can be affected by temperature and this can cause variations in performance. Finally, cost is not directly proportional to size. Smaller hydraulic robots tend to be much more expensive than their electric counterparts.

Pneumatic powered robots are the cheapest and least sophisticated type. They are usually not servo controlled but can carry out complex movement sequences if necessary. They are fast, simple, reliable and easily understood by most factory technicians and maintenance personnel. They also have the advantage of being intrinsically safe and can therefore be used in explosive atmospheres. The major disadvantage of pneumatic robots is that precise servo control is not practical. This is due to the compressibility of air, particularly when moving heavier loads. Thus pneumatic robots are usually found in limited-sequence, light-load fixed-speed applications.

14.7.1.4 Path control

The application will influence the choice of the robot path control system. Robots with simple point-to-point control are suitable for assembly, palletising and other materials-handling tasks. Point-to-point with coordinated path control is suitable for tasks such as arc welding, sealant application and spot welding of moving components (e.g. car bodies on a conveyor system). Continuous path control is used where the dexterous movement of a human operator has to be mimicked (e.g. in spray painting, or where complex contouring movements are necessary).

In point-to-point control the robot will move between defined points without regard to the path taken between them. In some robots an additional software facility allows the choice of movement between points in either the shortest travel time or in a straight line. In revolute robots movement in an 'elbow up' or 'elbow down' mode can also be selected. When using robots with simple point-to-point control particular care has to be exercised to ensure that collisions with obstacles are avoided when the robot is running.

In point-to-point with coordinated path the control software allows the path the end effector will follow between points to be determined. Straight lines, circles, arcs and other curves can be defined. Two points are all that is necessary to define a line; circular movement can be programmed by specifying three points on a circumference or a centre point and a radius.

Full continuous path control is most often obtained by playing back the information recorded from physically leading the robot through a desired task. Every movement of the arm is recorded in real time by sampling joint positions at a high frequency, and this is used in on-line programming. Continuous path control is also employed in off-line programming in some systems where there exists the facility to insert the mathematical equations for the desired curves. These curves are then followed by the end effector.

14.7.1.5 Robot programming methods

Industrial robots may be programmed using a number of techniques. The most basic methods, employed in some very early hydraulic robots, used rotating drums on which pegs could be set to close microswitches. These switches operated solenoid valves, so controlling the flow of fluid to the ac-

tuators. Adjustable mechanical stops were fixed on the moving members to contact limit switches at approapriate points. These mechanical programming methods provided control of sequence and distance moved. More recently, the use of programmable logic controllers has been applied to sequence the movement of pneumatic robots and modular units. These are all non-servo control systems employing simple feedback from limit switches or proximity sensors.

Full servo controlled robots employ dedicated microprocessor-based controller units as described earlier. This allows sophisticated on- and off-line programming techniques to be employed.

On-line programming Here the robot arm itself is used during the direct programming operation. This method has the following advantages.

(1) The robot can be observed as programming progresses; this increases confidence in the finished program since possible collisions and other robot 'no-go' areas are easily identified.
(2) In applications such as spray painting and welding where human experience is important, the direct programming of the arm by an expert effectively produces a transfer of skill from the human to the robot.
(3) This type of programming is easy to learn and can often be accomplished by the personnel that the robot is replacing.
(4) Less expensive computing hardware and software are involved than in on-line programming.

On-line programming also has one main disadvantage. Where the task being programmed is complex the programming time may be prohibitively long. For example, if a deburring operation is to be carried out and the robot has to be taught the shape of every hole and curve, the programmer will find the exercise extremely tedious and the cost benefits may be trivial. It is therefore a technique suited to tasks which are highly repetitive in nature, i.e. one sequence of movement can be repeated automatically a number of times.

There are basically two methods of on-line programming: teach by lead-through and teach by pendant. It is also possible to program directly from a computer terminal attached to the controller, which is similar in principle to the teach by pendant method.

(1) *Teach by lead through.* This is used for programming continuous path operations such as are found in spray painting. The programmer grasps a pistol attachment on the end of the robot wrist and proceeds to lead the robot 'by the nose' through the required program. A button or trigger on the pistol is used by the programmer to record the start and end points of each sequence of movements. Where a heavy duty robot is involved a slave arm may be used instead of the actual robot arm. This will be an arm dimensionally identical to the main arm but will be of a much lighter construction; in fact a lightweight frame or tubular structure will suffice. In any case the feedback sensors within the arm will monitor the movements made by the operator and send the appropriate signals back to the memory in the controller. The system monitors the arm joint positions many times per second during the programming exercise and, therefore, the controller memory capacity needs to be large.
(2) *Teach by pendant.* For work that requires point to point, and point to point with coordinated path, movements this is the normal method of programming. It involves the programmer using a hand-held pendant which transmits commands through a cable to the robot controller, the

robot then responds to these commands. In this way the programmer can lead the robot through a task. Teach pendants have as many different configurations as there are robot models. Essentially, however, they all contain sufficient controls to send the necessary instruction to the controller. For example, one type is composed of a small keyboard which allows around 80 different program instructions to be transmitted. These commands can be observed on an LED display just above the keyboard capable of showing current and previous program lines. Below the keyboard is a joystick and speed control device. The robot arm is programmed by moving the end effector to a desired position using the joystick, and once the arm is at the required position and orientation a key is pressed to record the point in memory. On a six-axis robot this will be recorded as a six-coordinate location x, y, z, α, β, γ (see *Figure 14.69*). In this way all the points to which the robot is desired to go will be recorded in memory within a 'point file'. An 'instruction file' is then created using the keyboard command. This file contains the instructions as to what the robot should do between each point, e.g. the robot may be instructed to move from point 1 to point 2 at a speed of x mm/s. At point 2 it may be instructed to open its gripper, operate a spot welding gun, or open a valve to allow adhesive to be dispensed. Thus two files will be constructed, one with the desired end-effector locations, and one with the instructions concerning the robot operation and sequence and speed of movements.

In many cases, as well as a teach pendant, there will be the facility of programming the robot on line using a computer terminal. This makes more complex programs easier to create, especially if a good programming language is available. These programming languages are discussed below.

Off-line programming Off-line programming involves creating the program for a robot task, without the need to be connected physically to the robot or even to be anywhere near its physical presence. In fact, when coupled with simulation techniques, off-line programming can be carried out before deciding on which robot to purchase for a specific application.

Some of the advantages of off-line programming are as follows.

(1) The robot for which the program is being made can continue working on its old task until ready for the new program. This obviously reduces robot down-time and increases productivity.
(2) If the control system and language allows, it is possible to build into the program collision avoidance, error recovery and other contingency routines.
(3) Compared to on-line programming, it is easier to make alterations to cope with variations in products and design changes.
(4) Off-line programming is suited to full computer integration of a facility. For example, if a robotised computer-controlled machining cell is in operation, then with off-line programming the problems of downloading programmes to the robot at the appropriate times are greatly reduced.

There are also some disadvantages:

(1) real-world contact is lost;
(2) more expensive hardware and software are required;
(3) more programing skill is needed; and
(4) fine adjustments under production conditions on the shopfloor are usually necessary.

Programming languages When using on-line programming methods, instructions can be given to the robot by using control switches, knobs and buttons in conjunction with simple coded commands. This method allows simple programs to be constructed. For more complex programs a robot language must be used and for off-line working a programming language is obviously essential.

Commercial robot languages are all, at present, termed 'explicit' languages. This signifies that all commands must be specified explicitly, i.e. points are defined in space using the appropriate number of coordinates, and the arm is given explicit instructions regarding what to do between and at those points.

Still at the research stage are languages termed as 'implicit' or 'world model' languages. These languages will allow the

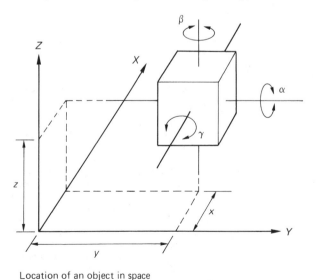

Location of an object in space

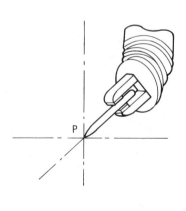

Point P defined
by six coordinates
ie P = x, y, z, α, β, γ

Figure 14.69 Definition of robot tool centre-point position

robot to be programmed using very simple commands which contain within them the implication of a large number of required actions. For example, a simple statement such as 'Build 500 microswitches of type X by 09.00 hours tomorrow morning' implies that the robot has a tremendous amount of inherent knowledge and skill. It will require a 'world model' of its local environment and full knowledge of the shape, size and weight of the switch components. It will need to know such things as how switches are assembled, the sequence of operations necessary, and how long it should take to complete one switch. The robot will also require a wide range of sensors, e.g. vision, touch and force. It will definitely need considerable decision-making ability and embody a number of artificial intelligence techniques. Thus so much expense would be involved in creating a robot of this type that they are at present commercially impractical.

Returning to the explicit languages, we find that these vary considerably in sophistication. 'Primitive' motion languages contain simple coded commands or words that connote the desired action. However, they are not as elegant or powerful as the 'structured' programming languages. These have capabilities similar to those used in computer programming. They can incorporate subroutines, branching, logical statements and loops. They also have additional commands, not needed in a conventional computing language, necessary for programming the physical actions of the robot. The ability to insert commands to allow external sensors to be interrogated and data to be transmitted and received to and from peripheral equipment will also be present. Each robot manufacturer provides their own language with the control system. Although some language software is designed so that it can be purchased separately, there is as yet no industry standard. Some powerful languages available at present are Karel from FANUC, AML from IBM, and VAL originally from Unimation but which is now used on other robots.

Simulation Graphical simulation of the robot and its environment offers many benefits to the industrial engineer and programmer. *Figure 14.70* shows an example of a low-cost simulation system display. This particular software, called 'Workspace', runs on a PC. Simulation packages provide the ability to model the robot kinematically and often dynamically to given an animated, real-time, visual representation of how the robot will work under programmed conditions. Three-dimensional wire-frame—sometimes with colour shading or solid modelling—techniques are used.

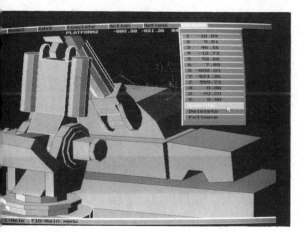

Figure 14.70 Robot simulation as an aid to programming. (Courtesy of TQ International)

Firstly, a model of the robot is created from design data supplied by the robot manufacturer. For the system user this may simply involve selecting the required robot from a 'library' of robot models contained within the software supplied by the simulation system vendor. The surrounding environment, e.g. a machining cell, is also modelled by the user. The robot can then be manipulated on screen to observe how best to locate it to optimise performance. Should the first robot selected prove unsuitable, other models can be tried until the most suitable configuration and size is found. Once certain that the desired robot is being used, a detailed program can be constructed. Assuming the simulation system is designed to cope with robots from any manufacturer, a post-processor will be needed to handle the transformation of the simulation data into the particular language used by the selected robot. Using this method the industrial engineer or programmer does not require an intimate knowledge of the particular language of the robot selected.

There are a number of advantages associated with simulation.

(1) Simulation allows a prospective robot purchaser to try out various models at relatively little cost before making a decision. Parameters such as work envelope, cycle times and joint configuration limitations can be compared for all the robots held in the library.
(2) The immediate robot work area can be simulated and various permutations of machines and operation sequences experimented with before finalising the layout.
(3) Potential collisions can be detected at an early stage and programs and layouts modified to suit.
(4) Simulation is very suitable for teaching and training purposes. Mistakes in programming can be observed and learned from without the hazards and potential costly damage that would be experienced in the real world.
(5) At the robot design stage, simulation delays the need for physical prototypes to be built, thus reducing research and development costs.

These advantages, coupled with improving computing power-to-cost ratios, mean that simulation is becoming an increasingly popular robot-programming tool in industry, education and research.

14.7.1.6 *Industrial robot deployment and applications*

This section considers the manner in which industrial robots have been adopted from both a population distribution viewpoint and an examination of the tasks to which they have been applied.

Although mechanical and hard-wired handling devices had been in use in industry for some time, it was not until 1973 that the first computer controlled industrial robots emerged. These were produced by ASEA in Sweden and Cincinnati Milacron in the USA. Interest and development in industrial robots increased steadily with a proliferation of manufacturers and agents, but this peaked in the early to mid-1980s. By this time there was much publicity and a brouhaha arose around the subject, the media aroused public awareness, and expectations of the technology began to overtake the capabilities of commercial robots and even the research teams. Eventually, many robots were being applied to tasks unsuited to their abilities and special advantages, in tasks where humans or special-purpose machines would have been more appropriate. By the time one survey of manufacturing organisations in the UK was carried out in the late 1980s, over 75% of respondents claimed poor returns from their robots. Industrial robots are no more

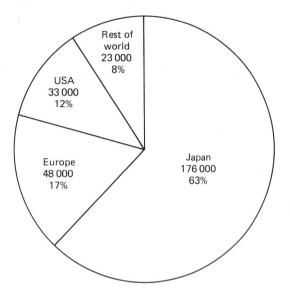

Figure 14.71 World robot distribution (total in 1988, 280 000). (Courtesy of International Federation of Robotics)

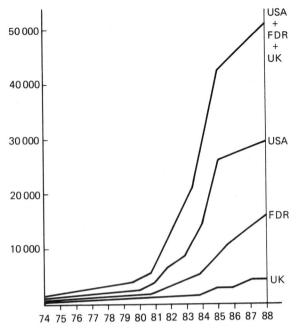

Figure 14.72 Approximate robot population growth in the USA, West Germany and the UK from 1974 to 1988

special than other industrial machines that must be appraised and selected carefully for specified tasks. The converse is also true, i.e. the tasks chosen for robotising must be appropriate for robot capabilities.

The steady increase in robot installations world-wide has continued, although the rate varies from country to country. The International Federation of Robotics (IFR) produces world figures based on information supplied by individual national robot associations. It should be noted, however, that the actual statistics used are of varying reliability depending on the source and they should be considered as 'ball park' figures only. *Figure 14.71* shows the world robot population at the end of 1988. As can be seen, Japan has by far the biggest share, although a broad definition used by the Japanese probably accounts for such a difference between them and the USA which has approximately twice the human population. The rate of growth of robot installations is shown in *Figure 14.72* for the USA, West Germany and the UK. This indicates a slow growth until about 1980 followed by a rapid increase to around 1986. Although the growth rate has now reduced, it is apparent from the graph that industrial robots are still regarded as desirable elements in the manufacturing system, provided they are installed wisely.

An analysis of the UK robot applications for 1989 is shown in *Figure 14.73*. It is interesting to note that over recent years the relative percentages have remained roughly similar. Although constituting the largest percentage, injection moulding robots are mostly of the least expensive non-servo type. Next to these in usage are the spot welding robots; conversely these are often of the more expensive, heavy duty, fully servoed type. With improving software, increasing sensor sophistication, and accumulating experience, arc welding has steadily gained in popularity as a robot application. Also, increasing speeds and more efficient designs have led in recent years to increasing numbers of robots in assembly. A few notes are now provided on these and other applications.

Injection moulding The robots used in this application are normally pneumatic, non-servoed, rectangular coordinate devices. They are essentially sequence controlled pick-and-place units clamped rigidly onto the frame of the injection

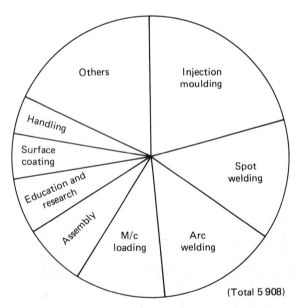

Figure 14.73 UK robot application distribution 1989. (Based on figures supplied by B.R.A.)

moulding machine. The sequence of movements is usually as follows: as the injection moulding machine opens, ejector pins push the moulded component and attached sprue and runner system out of the die; the robot arm reaches vertically downwards and grasps the moulding; the arm retracts allowing a new moulding cycle to begin; the arm moves horizontally to one side of the machine; and the arm either deposits the moulding onto a conveyor or into a bin. The arm may also place the components into a degating fixture where the sprue-and-runner system will be removed or, if fitted with pincers, the arm will remove the gate-and-runner system itself, thus allowing the components to be separated before deposition. Mechanical or vacuum end effectors may be used.

The injection moulding process is a particularly suitable application for a number of reasons. By using robotics, cycle times can be reduced and made more consistent when compared with manual mould removal. This increases production rates and improves product quality. Since hot plastic is being injected under pressure, the work area is hazardous and often unpleasant for human operators. The moving tool platens are also a hazard due to the high clamping forces involved, and the heat and fumes produced by the process can produce an unpleasant atmosphere.

Spot welding Traditionally the major application for the larger servo controlled industrial robots, spot welding is widely used in motor-car assembly. Early installations used polar coordinate hydraulic robots; however, most are now electrically driven revolute types. On car assembly lines it is possible to mount the robots on rails to travel with the car, but the most common technique is to keep the robot in a fixed position and program it to spot weld as the car moves through the robot's operating 'window'. This is termed 'line tracking', since the robot joint coordinates are constantly modified using velocity feedback information from the line. The flexibility demanded for this operation usually necessitates a robot with a full six degrees of freedom and a relatively long reach.

The noisy and generally unpleasant environment in which spot welding normally takes place makes this an ideal application for robotics. The welding guns are large, heavy and unwieldy for manual workers, and the work pace set by the assembly line is fatiguing. Furthermore, the wide variety of car models and options desired by customers means that the reprogrammable nature of robots can be fully exploited. The weld precision and quality is, over a period of time, better than that of human welders, due to the fatigue factor. A result of this more consistent welding is the ability to specify fewer welds, reduce cycle time, and obtain a stronger fabrication.

Arc welding Although available for some time, robotic arc welding took longer to become popular than did spot welding. This was due to the need to gain experience in design of the fabrication and work fixturing method suitable for robotic welding. Also the development of seam tracking methods had to improve considerably before they could be applied reliably in shop floor situations. The robots most commonly used are again of the six-axis, electrically powered, revolute type. To ensure that the welding gun is kept at the proper orientation to the seam it is sometimes necessary to add further axes to the system. This is done by providing a multiaxis servo-controlled worktable capable of having its control integrated with that of the robot. Thus for complex work a nine-axis robot and table system might be used. Seam tracking methods often use a laser system attached to the robot wrist, or even integrated in the welding head, to follow the path of the weld preparation.

Arc welding is an unpleasant job for human workers. The flash, heat and fumes that are produced make the work unhealthy, and for this reason it is often done in isolation from other workers. This can make the work unsociable. The type of highly skilled worker necessary to obtain good quality welds does not always want to work under these unpleasant, unhealthy, and unsociable conditions. This has led to a skilled welder shortage in some areas; thus the attractiveness of using one welder to 'train' a large number of arc welding robots. The unpleasant conditions also lead to the need for frequent rest breaks for welders; with robot welding this is unnecessary and highly increased 'arc on' times are obtained.

Machine loading With the increasing use of unmanned machining cells and in some cases full flexible manufacturing systems, the use of robots to load and unload machine tools is an essential ingredient. In this case they are secondary material handling devices removing the material to be machined from automated guided vehicles, pallets, or conveyors, inserting it into the machine, then unloading it after the work cycle has been completed and placing it back in a carrier or the primary material handling device. Typically the machines being loaded/unloaded will be NC machining centres, lathes, or milling machines. Some robots have been specifically designed to fit directly onto the NC machine tool, whilst others used are of the revolute and cylindrical coordinate type. Using point-to-point control these electrically powered robots are interfaced to the machine control and safety systems. In some cases double grippers are used. This allows unloading of a machined component and loading of a blank to be carried out by simply indexing the gripper through 180°, thus maximising machining time. By using machines that can be accessed by the robot from the rear, space can be better utilised, safety is improved, and the machine can be operated from the front by a human worker should the need arise.

Assembly Use of robots in assembly has proliferated in recent years, mainly due to the advent of robots specially designed for the task and increasing recognition of the need for 'design for assembly' at the product design stage. It has long been known that the simplest, quickest and cheapest way to assemble a product is to put it together by 'stacking' one component on top of another using simple vertical movements, horizontal movements also being necessary for picking and placing. When human beings were used, their dexterity and speed meant that the necessity for designing products so that they could be assembled by the 'stacking' method was not always obvious. However, with the advent of automated assembly methods this principle became more widely recognised. Recently, as more attempts have been made to robotise jobs, improvements in assembly design have progressed.

The knowledge that most assembly operations comprise relatively short horizontal and vertical movements for picking and placing components, coupled with a single rotational movement for component orientation, led to the development of the SCARA type robots described earlier. Other types of robot used in assembly are cartesian and cylindrical robots. Originally, revolute robots were used, but these are now regarded as expensive with control systems unnecessarily complex for the straight-line movements required for assembly. Assembly robots require high speed and precision, they are normally electrically driven, some with direct drives, and point-to-point control is used.

Parts assembled are usually of small size, and they are likely to be used in assemblies that are part of large batch orders. Where mass production volumes are involved, dedicated equipment is used. This is most evident in the population of printed circuit boards with electronic components. Where only a few boards are to be built, human labour will be used; for very high volume production, dedicated printed circuit board component insertion machines with extremely high insertion rates are adopted; for batch production the assembly robots are used. They are easily reprogrammed to cope with different board designs, but they are not as fast as the dedicated equipment.

Surface coating Surface coating of products, particularly spray painting, is a popular application due to the highly unpleasant environment usually created. The paint being sprayed is often toxic, the operation itself uses compressed air

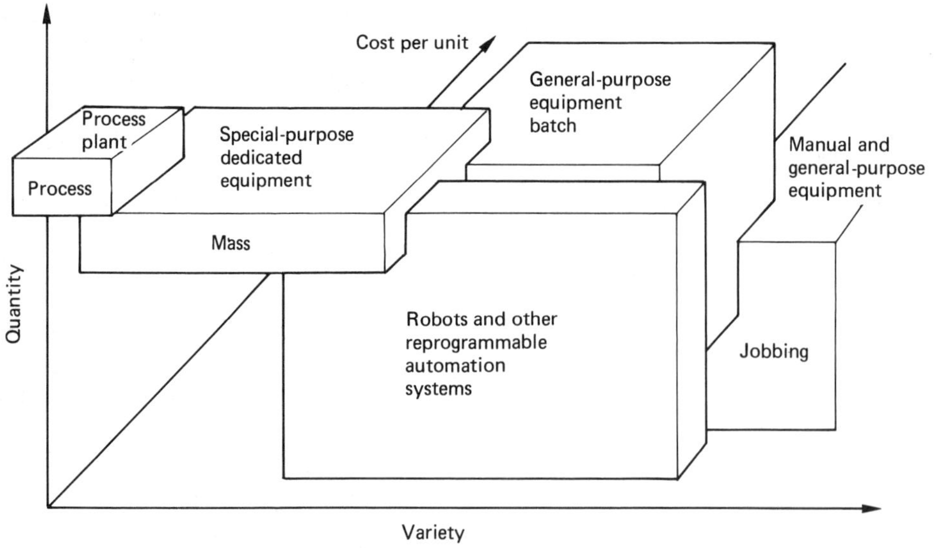

Figure 14.74 The effect of robots and other reprogrammable automation systems on the cost per unit

and is therefore noisy, and the protective clothing that needs to be worn is often hot and uncomfortable. Use of robots in spray painting also improves quality and ensures a consistent spray throughout the working day and week. Hydraulic, revolute robots, with continuous path control, have been popular in these applications.

The other applications noted earlier in the statistics continue to grow, and as robot capabilities improve so new applications will be added. The addition of increasingly sophisticated sensors and artificial intelligence will see more applications appear outside the factory environment. Conversely, there will also be more robots of simple design, e.g. the SCARA, created for specific tasks.

14.7.1.7 Using industrial robots

Suitable applications The potential advantages of robotisation can be maximised by making wise application selections. Industrial robots realise their full economic potential in applications where the product volume is large enough to recoup the expenditure on hardware, programming and engineering costs, yet is sufficiently low to prevent justification of dedicated special-purpose equipment. However, high-volume work with frequent model or option changes, such as is found in automobile assembly, is suitable for robotisation. The effect of robots on the cost per unit, in relation to volume, is shown in *Figure 14.74*. The following are some further indicators as to applications that should provide suitable opportunities.

(1) Tasks which are carried out in (or create) an unpleasant or hazardous environment. For example, toxic or flammable atmospheres are created by processes such as arc welding and spray painting, and removing human operators from these jobs can improve quality and increase production rates.
(2) Jobs that are tiring or boring. Robotisation of these eliminates absenteeism and labour turnover problems and usually improves quality.
(3) Repetitive and simple operations requiring simple movements allow the least expensive robots to be used and minimise installation and programming problems.

(4) Desired cycle times should not be too short. For example, if the cycle time is greater than, say, 3 s then the choice of robot is relatively wide. However, if very short cycle times are required, as in printed circuit board component placement, then more specialised and expensive high-speed robots are necessary.
(5) The tolerances on the components and tools should allow robots of average precision to tackle the work.
(6) The variety of products expected to be handled by the robot should not be large nor changes from one product to another too frequent. This keeps engineering and reprogramming costs to a minimum.

The following points are also relevant. If the task being considered has an integral inspection element, then additional costs are incurred when vision or other sensing methods are added to carry out that inspection. In materials-handling applications very heavy loads demand the use of larger and more expensive robots. If an ordered environment exists (or can be made to exist) around the robot then robotisation is simplified. If possible, work should be oriented and positioned at the previous operation before presentation to the robot. Most robots available commercially have limited reasoning ability, therefore, tasks should demand little in the way of intelligence or judgement.

Once the task to be robotised has been selected, the next stage is the selection of an appropriate robot.

Selecting the robot Robot selection should be carried out after listing task demands such as cycle time, payload required, necessary precision, and cost. These demands are then compared against the specifications provided by the robot supplier or manufacturer, some of which are listed below.

(1) *Speed* Having decided the speed required of the robot from the work analysis, the detailed specification should now be examined. Some manufacturers may give maximum speeds for each axis of the robot, some the maximum speed of the end effector. These should be given for maximum load and at maximum reach as well as for under optimal condtions. It should be remembered, however, that maximum speed is not necessarily a very

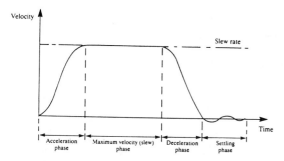

Figure 14.75 Robot velocity curve

useful piece of information, because a robot arm must accelerate to and decelerate from this speed. *Figure 14.75* shows a typical robot velocity curve. For some applications, particularly assembly, a 'goalpost' time is a more useful specification. This is the time, supplied by the robot manufacturer, that it should take the robot to complete a standard series of movements carrying a standard load. For example, the movement may be: close gripper, move up 30 mm, move across 300 mm, move down 30 mm, open gripper. This will prove more useful for estimating than will a maximum speed figure.

(2) *Payload* The maximum load expected to be encountered will have been determined, and a robot with sufficient strength to handle a considerably greater load should be selected. The specification should show whether the maximum load capacity is given with the arm close to the body or at full extension where the capacity will be much less due to leverage. Robots are available with capacities ranging from a few grams to 2 t.

(3) *Precision* The overall precision of a robot is composed of three elements, i.e. resolution, repeatability, and accuracy. The resolution of a robot is normally a feature that is transparent to the user and therefore not included in standard specification sheets. It refers to the smallest controlled movement the end effector is capable of making. This is determined by (a) the resolution of the computer controller (i.e. the number of bits used to define a position over a given range), (b) the resolution of the drive system (e.g. the number of steps per revolution provided by a stepper motor and associated gearing), and (c) the resolution of the feedback elements such as shaft encoders. The repeatability of a robot is determined by its resolution plus clearances and wear on moving parts plus any other inaccuracies and errors in the total system. It is a statistical term describing how well the robot can return consistently to a taught point. This is the most common figure relating to precision to be included in robot specification sheets. Repeatabilities of ± 1 or 2 mm for medium-duty work, ± 0.05 to ± 0.01 mm for medium assembly, and ± 0.02 mm for precision assembly are typical.

(4) *Accuracy* Assume that a computer-controlled robot is to move to a point in space. This point is defined by entering the coordinates into the control system. The difference between the taught target point and the actual position achieved by the robot, in the real world, is the 'accuracy'. This is determined by the resolution, inaccuracies in the 'model' of the robot held in memory and other factors such as bending or thermal expansion of the robot arm. The relationship between accuracy and repeatability is shown in *Figure 14.76*.

(5) *Configuration* The supplier will provide information on the geometric configuration and dimensions for the effective work envelope of the robot. These can then be used to construct templates either on card or on computer to enable an appropriate work layout to be designed.

(6) *Control system and programming method* The specification will provide information on whether point-to-point, point-to-point with coordinated path or continuous path control is provided. It will also state the programming methods used. For CP programming by lead-through a slave arm may be available and for point-to-point or point-to-point CP teach by pendant methods may be used. For many robots programming using a computer terminal and a high-level language will also be available.

(7) *Cost* The cost of a complete robot installation can vary considerably from that of the basic robot. The robot chosen can influence this total cost. Ease of programming, and interfacing capabilities will influence the engineering costs. Cost of fixturing, parts presentation and orientation devices, and end-of-arm tooling must be included in the total. Also, if working to a fixed budget for the robot, there will probably have to be a trade-off between precision, speed, strength and reach.

Other specifications that should be considered include drive system, number of degrees of freedom, type and number of input and output ports, and memory size.

Robot safety As well as presenting the normal safety problems associated with moving equipment that is electrically, hydraulically or pneumatically powered, and machines that are under microprocessor control, industrial robots present some additional problems that are unique:

(1) While executing a program the robot appears to the inexperienced observer to be moving spontaneously and unpredictably, each movement being difficult to anticipate. This applies particularly when the robot is at a 'dwell' point in its work cycle. It may appear to be deactivated but in fact it will spring into action as soon as it receives an appropriate command from the system controller.

(2) Most robot arms sweep out a work volume much larger than that occupied by their base. With a six degrees of freedom robot the arm movements and positions are therefore difficult to visualise.

(3) The integrity of the control system hardware and software is particularly crucial since faults will produce erratic and unpredictable behaviour.

(4) Heavy-duty robots are built to be rugged and inelastic, and fast-moving arms are therefore extremely dangerous. Size is not necessarily important as was proved when one person was killed when struck on the back of the neck by a small teaching robot.

Accidents can be caused by human carelessness, insufficient training, poor robot or installation design, poor quality components used in the system, and software errors. Most accidents occur to those familiar with the robot such as programmer, maintenance engineers and operators. Those unfamiliar with robots tend to be more wary—it is the complacency caused by familiarity that is dangerous.

Industrial robot safety should be considered at the stages of robot design, supply, installation, programming and everyday usage. The designer should ensure all controls conform to good ergonomic practice. Controls and displays should obey standard conventions, mushroom-shaped stop buttons should

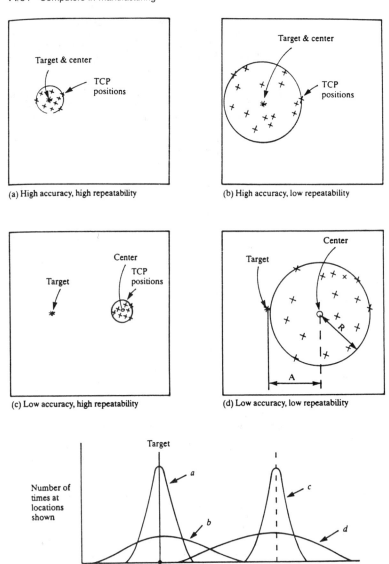

Figure 14.76   Schematic representation of accuracy and repeatability

protrude from surfaces and there should only be one start button, which should be recessed. All emergency stops should be hardwired into the power supply and not rely on software execution. Conventional good design practice should be observed, moving parts should not be exposed and there should be no trapping points for limbs or fingers, with no unnecessary protrusions capable of inflicting injury.

The robot supplier should ensure that proper instruction and training is given to appropriate personnel designated by the purchaser. The supplier should also make the user fully aware of the robot's limitations and any possible hazards that may be encountered.

When the user is planning and implementing the installation full consideration should be given to the robot's position within the factory, e.g. it should not be located near any trapping points such as roof pillars or stanchions, and it should not be possible for it to reach into passageways or manual

work areas. Preparation of safety manuals, or safe working procedure documentation may also be carried out at this stage. Reference should always be made to appropriate rules, regulations and guidelines. In the UK there is the Health and Safety Executive (HSE) guidance booklet *Industrial Robot Safety*, and the MTTA booklets *Safeguarding Industrial Robots*, Parts 1 and 2. There is also the British Standard BS 5304 *Code of Practice—Safeguarding of Machinery* which contains the basic principle of safeguarding, i.e. unless a danger point or area is safe by virtue of its position, the machinery should be provided with an appropriate safeguard which eliminates or reduces danger. Light curtains and pressure-sensitive mats are commonly used around the immediate vicinity of the robot. For maximum safety a 2 m high cage around the robot is recommended. This should have doors electrically interlocked to the power supply to ensure that unauthorised entry deactivates the robot.

During everyday operation management must ensure that only fully trained personnel operate the robot. Established safety procedures must be adhered to, appropriate warning signs given high visibility and, generally, a state of continual safety 'awareness' cultivated.

14.7.2 Industrial vision systems

14.7.2.1 Vision system components

The basic elements of an industrial vision system are shown in *Figure 14.77*. A camera is first necessary to acquire an image. This camera may be a vacuum-tube or a solid-state type, the latter now being the most popular. The signal from the camera is then processed in the vision system computer. The image observed by the camera and the digitised image used for computer processing are observed on a monitor which is switchable between them, or two monitors may be used, one for each image. A means of communicating with the system is necessary and this would take the form of a computer terminal and visual display unit. A means of allowing an automated physical reaction in response to the vision analysis is required. This demands an interfacing unit connected to a robot or other device (say, a simple pneumatic cylinder for rejecting bad parts). Finally, to ensure optimum viewing conditions special lighting arrangements may be necessary to avoid distracting shadows or glare.

14.7.2.2 Vision system types and operation

Vacuum-tube cameras provide an analogue voltage proportional to the light intensity falling on a photoconductive target electrode. This electrode is scanned by an electron beam and the resulting signal is sampled periodically to obtain a series of discrete time analogue signals. These signals are then used to obtain digital approximations suitable for further processing. For example, if an analogue-to-digital converter has a sampling capability of 100 ns, the image is scanned at 25 frames per second and each scan is composed of 625 lines, then there will be 640 picture elements, or 'pixels', per line. Since some time is lost as the electron beam switches off when moving from one line to the next, the number of pixels in a frame will be 625×625, which gives a total of almost 400 000 pixels. This is difficult and expensive to handle computationally in real time, especially if mathematical analysis of the image is to be carried out. For this reason, the number of pixels can be reduced, depending on the application, to provide a more manageable image.

Solid-state cameras use arrays of photosensitive elements mounted on integrated circuits. The light from the scene is focused by the camera lens onto the IC chip. Charge-coupled devices (CCDs) or photodiode arrays are scanned to provide a voltage signal from each light-sensitive element. These solid-state arrays are available in many densities, often from 32×32 to 1000×1000 pixel arrays. The larger pixel densities are too high for real-time vision analysis but they do provide high-quality video pictures. The voltage signals from the photosites are again digitised before further processing. In a 'line scan' camera linear arrays of photosites are used rather than the area type. Line scan cameras can be used where the object is moving steadily across the field of view. For example, an object passing under the camera on a conveyor belt can be scanned repeatedly and an image built up in the vision system memory.

Recently, solid-state cameras have become widely available and relatively inexpensive due to their large-volume production. They have a number of advantages over the vacuum-tube type, i.e. they are much smaller and lighter, more robust, more reliable, use less power, have a broader temperature operating range at lower temperatures, and are less likely to be damaged by high light intensities.

The pixel voltage signals from the camera are now assigned to a finite number of defined amplitude levels. The number of these 'quantisation' levels is the number of 'grey levels' used by the system. An 8-bit converter will allow 256 grey levels to be defined. In practice, this number of levels is often unnecessary and processing time can be reduced by using only 16 grey levels. In some cases only two grey levels, i.e. black and white, are necessary, this is termed 'binary' vision.

Each grey level is next 'encoded', i.e. it is given a digital code, and the data stored in memory. In a computer vision system this is done for one picture 'frame' and the data stored in memory in what is termed a 'frame buffer' or 'picture or frame store'. Various algorithms are then used to minimise the data for analysis and organise them in such a way as to allow feature extraction and object recognition. Objects are usually recognised by the system by first showing it a sample of the object. The system 'remembers' the object by storing information on features such as object area, perimeter length, number of holes, and minimum and maximum radii from the centre of gravity. The sequence of all these processes is shown in *Figure 14.78*.

14.7.2.3 Vision system applications

As the capabilities of vision systems increase so also does their popularity. Custom-designed hardware and developments in algorithms mean that the systems are becoming faster and more reliable. They are now found in a wide variety of industrial applications and are sometimes supplied as integral components or programmable electronic component placement machines and other robotic systems. Three main areas of application are listed below.

Identification Here the system is used to identify a product or individual component. For example, it may involve character recognition, as in reading alpha-numeric data on a product label or recognising a component on a workbench prior to assembly.

Inspection This is one of the major applications of vision systems as it is estimated that visual inspection accounts for around 10% of total manufacturing labour costs. This percentage can be very much higher in some industries (e.g. electronic product manufacturing such as printed circuit boards, computers and other consumer goods). Sensible application choices can prove very cost-effective. Inspection is generally further divided into 'qualitive' and 'quantitive' inspection. In qualitive inspection it is attributes that are examined (e.g.

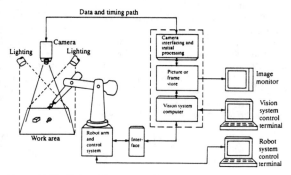

Figure 14.77 Robot vision system elements

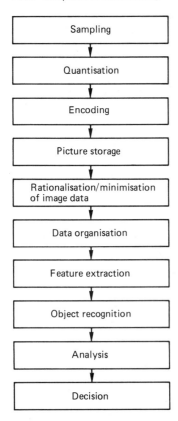

Figure 14.78 Typical operation sequence implemented in a vision system

glass bottles may be checked for flaws or castings checked for cracks, or the number of pins on an integrated circuit chip verified). In quantitive inspection dimensional or geometric features of a product are measured and checked (e.g. the diameter of a component turned on a lathe or the width of a steel strip coming from a rolling mill).

Decision-making This is a general term which implies a number of applications. For example, a vision system could be used to guide the welding head of a robotic welder along the seam of a fabrication, or it could assist an automatic guided vehicle find its way around a factory shopfloor. In conjunction with artifical intelligence techniques vision can be used to provide the information input necessary to provide autonomous working of robotic devices in unstructured environments.

In conclusion, vision system technology can be said to be rapidly improving and, in conjunction with advances in related technologies, it will continue to make a significant impact on factory automation for many years to come.

14.7.3 Automatic identification systems

Automatic identification systems are being increasingly applied throughout manufacturing industry as well as in non-industrial environments. Within the manufacturing context the term 'automatic factory data collection' (AFDC) is sometimes used. If full factory automation, or computer integrated manufacturing, is desired then automatic identifica-

tion is an essential ingredient to ensure throughput times are minimised and high-integrity information is readily available. It allows real-time updating of records and tracking of materials, components and products, thoughout the factory from goods inward to final dispatch. Material held in store, work in progress, quality records and component histories can all be monitored and recorded automatically. With some systems it is possible to leave the quality and other product history information with the product so that the automatic identification information can be interrogated at a later date when the product is in use in the marketplace. Some of these systems are briefly described below.

14.7.3.1 Bar coding

This is the most widely used and familiar method. It is very versatile as the same bar-code symbol can be read by a number of different scanners each suited to its point of use. For example, the bar codes on the cartons of a mass-produced product, say a video recorder, may be read at the packing stage by a fixed-position high-speed scanner as they pass along a conveyor line, a hand-held scanner may be used to identify the cartons at a distribution centre, and in the local retail store a shop assistant may use a wand scanner to identify and record the arrival and sale of the product.

In a bar-code numeric, or alphanumeric, information is represented as a series of bars of varying thickness and separations, examples can be seen on almost any prepackaged product from chocolate bars to washing machines. Various systems of symbols have been developed, each suited to particular applications such as the retail trade, shipping, manufacturing or libraries. Irrespective of the particular code adopted, the principle of operation is similar. Each digit is represented by one or more black or white bars of varying thickness. These are read by a scanner which shines a light, often a laser, at the bar code and monitors the amount of reflected light falling on a sensor. This produces a series of binary voltage levels which are decoded to provide the identification data. These data are transmitted to an appropriate device, e.g. a material control computer.

14.7.3.2 Optical character recognition

Instead of the series of bars used in bar coding, this system uses stylised alphanumeric characters that can easily be read by people as well as machines. In optical character recognition the characters are scanned by a light source, the scanner then decoding the information before transmission to the control computer. Vision systems may also be used. In this case a camera observes the characters in its field of view and the vision system computer analyses the image.

14.7.3.3 Magnetic stripes

These are strips of material containing information encoded in the form of electromagnetic charges. A common example of this is the brown band found on the back of credit, banking and employee identification cards. The information contained within the band is read by a decoder for transmission to a computer. The decoder may be fixed, e.g. wall mounted, or it may be incorporated in a hand-held wand. In the factory context they can be used for such things as product identification or operator time attendance cards.

14.7.3.4 Radiofrequency identification

Radiofrequency data communication (RFDC or RFID) is used in applications when a read-and-write facility is required, or there is no direct line of sight between the scanner and the

RFID 'tag'. This tag contains a custom-built programmable integrated circuit and a long-life battery. The system comprises a transponder mounted on the object to be identified, or tracked, and a transmitter/reader. When the transponder, or 'tag', comes within the normal operating range of the transmitter/reader, it is interrogated and sends out a unique train of pulses, or 'signature', at a specific radiofrequency. The reader decodes this information and then transmits it to the system control computer. RFID is also useful in dirty or dusty environments, or in applications where other identification means would be unsuitable, e.g. surface coating of car bodies.

14.7.3.5 *Voice data entry*

Just as the previous systems recognise optical or electro-magnetic patterns, so voice data-entry systems recognise the sound patterns of the operator's speech. Instructions are input vocally through a microphone and the words or phrases recognised by the system. The operator is therefore free to use his hands which would otherwise have been employed on a keyboard. These systems are being continuously improved, but at present it is still necessary to 'teach' the system the patterns of each operator's voice. Every person has a unique pattern, in fact 'voiceprints' can be used for identification purposes in the same way as fingerprints. Although this means that these systems can only be used by the individuals who have taught the system and the vocabulary must be limited, it also means that unauthorised users do not have access and that the problems of external noise are minimised.

15

Manufacturing and Operations Management

J L Burbidge

Contents

15.1 Introduction to manufacturing management

In the early 1900s, the dictionary defined 'manufacturing' as 'the making of wares by hand or machine', and defined 'production' as 'the manufacture and distribution of goods'. Manufacturing was then, and even with the changes in its meaning with time, is still, that part of production which is concerned with making the product.

It is impossible, however, to make an intelligent study of the making of products without some consideration of the markets for which they are made. The distinction between making products (in the factory) and distributing them (in the market) is not therefore a useful one, for those who wish to study either production or manufacturing in depth.

Today, without any attempt to redefine the difference between the terms 'production' and 'manufacturing', the latter has become the fashionable word. It is proposed here to treat the two words as synonyms. The distinction between them was made by the early economists; it may still have some value in the science of economics, but it has little or nothing to offer to the science which we are studying. Manufacturing management is, therefore, a large part, or perhaps all, of the science of 'production management'. It is an applied science which draws on most of the other sciences. In this chapter we will look mainly at those functions concerned with the making of products and will deal only briefly with the remaining functions.

15.1.1 Management processes and functions

Management can be defined as: the art and science of planning, directing and controlling human effort, so that the object-

ives of an enterprise may be attained. It can be seen as a set of thousands of different tasks devoted to this objective. The processes of management are listed below.

(1) *Planning*, or deciding what to do in the future.
(2) *Direction*, or 'the process of management by which plans are caused to be implemented'. Direction covers both the issue of orders and the means used to persuade people to carry them out.
(3) *Control*, or 'the process of management which constrains events to follow plans'. Control includes monitoring which compares plans with actual output, and feedback which transmits information about significant variances between plans and output to management, so that action can be taken to correct them.

All management tasks require the application of all three of these processes for their successful completion.

Management tasks can also be sorted into sets of closely related tasks, which require similar skills for their efficient performance. These sets of tasks are known as 'management functions' and form the basic classification of the science of production management.

Table 15.1 lists the eight main management functions, describes the types of tasks which they include, notes the types of skill needed for each of them, and lists some of their principal inputs and outputs, and the special controls used with each of them.

Although the functions are partly autonomous, including many tasks which are of no interest outside their own function, the functions are nevertheless closely related: first because, for the efficiency of the enterprise as a whole, they need to share common goals and objectives; and second, because a part of the data input needed by each function comes from

Table 15.1 The management functions

Function*	Type of task	Controls	Inputs	Outputs
Product design (design)	Plans final form of product	Quality control	Ideas; market research; R&D	Parts list; drawings
Production planning (production engineer)	Plans how product is to be made	Process control; maintenance	Parts list; drawings; sales forecast	Plant list; layout; operating times; tooling and routes
Production control (work scheduling)	Plans material supply and processing activities	Progressing; loading; inventory control	Sales progress; parts list; plant list; routes; operating times	Programmes; orders; purchase delivery schedules
Purchasing (commercial)	Finds sources; makes supply contracts	Purchase progressing	Purchase delivery schedules	Purchase orders
Marketing (commercial)	Finds/develops markets; sales and distribution	Sales control	Sales orders	Sales records; sales programmes
Finance (financial)	Plans investment, profit and cash flow	Budgetary control; standard costing	Annual programmes; invoices in/out; bank payments in/out	Budgets; accounts; balance sheet; profit and loss account
Personnel (people)	Plans employment; conditions; welfare; training; promotion	Merit rating; attendance	Hired; fired; promoted; retired	Employee list; conditions of employment
Secretarial (data processing)	Plans communications data process and store	Data control	Production system design	Software

* The main skills required in each function are given in parentheses.

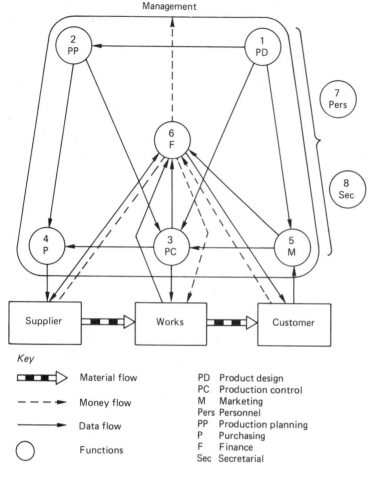

Management

Key

⬛▶ Material flow

- - - ▶ Money flow

———▶ Data flow

◯ Functions

PD Product design
PC Production control
M   Marketing
Pers Personnel
PP Production planning
P   Purchasing
F   Finance
Sec Secretarial

Figure 15.1   The arterial flow system

some other function or functions, and a part of their output will again provide the input for some other function or functions, or provide instructions to the works, the company suppliers or its customers.

Figure 15.1 shows the main information flow between the functions, by means of a network diagram. This primary flow of information is called the 'arterial flow system'. Note that this chart also shows the flow of information between the management functions, the works, and the company's customers and suppliers, and shows the principal money flow routes.

15.2   Types of production

There are many different ways in which production enterprises can be classified by type. They include, for example:

(1)   classification by type of product produced, e.g. boats, machine tools, aero engines;
(2)   classification by volume of output, e.g. mass production and job lot; and
(3)   classification by the amount of capital employed, e.g. capital intensive, and manual or craft.

For manufacturing management, however, we need a classification that brings together factories which are faced with the same types of management task. For this purpose, a two-part classification has been adopted which combines a material conversion type classification, based on the ratio of material varieties to product varieties, and a classification by the type of market served (i.e. large contract, jobbing, batch or continuous flow).

15.2.1   Material conversion classification

Figure 15.2 shows the material conversion classification of industries into four main types and gives examples of each of them. Process industries are industries in which a small number of material varieties are converted into an equally small number of product varieties. Material conversion is generally concerned with a small number of processes always used in the same sequence.

Implosive industries manufacture parts and are those in which a small number of material varieties are converted into a large number of part varieties. Like process industries, they usually use a small number of processes in the same sequence, but one or two of these processes divides the material flow into many different components. The foundry industry is a typical

	Process $m \mathrel{\square} p$	Implosive $m \mathrel{\triangleleft} p$	Square $m \mathrel{\square} p$	Explosive $m \mathrel{\triangleright} p$
Material input Product output Material flow type	Bulk material Bulk material Line flow	Bulk, or general material Components Batch/line flow	Components Components Batch flow	General or special materials Assemblies Batch flow

Examples	B Cement
	E Ore treat (mine)
	F Milk
	F Sugar
	F Distilleries
	C Gases (oxygen, nitrogen)
	B Bricks
	B Timber
	F Breweries
	    Tanneries
	    Paper
		E Foundries
		D Potteries
		D Glass
		C Decorative laminates
		T Spinning, fibres
		E Brake linings
		    Printing
		T Knitting
		F Bakeries
		E Rolling mills
		E Wire drawing
			E Jobbing machine
			T Dyeing textiles
			T Finishing textiles
			E Heat treatment
			    X-ray
			E Painting
			E Electroplating
			E Metal spraying
			    Polishing
				E Automobile
				E Electronics
				D Consumer durables
				E Machine tools
				E Internal combustion engines
				E Electrical
				T Weaving
				D Clothing and shoes
				C Chem. dyestuffs
				D Furniture
				E Welding

Figure 15.2 Types of industry according to the material-conversion classification. m, Number of material varieties; p, number of product varieties. Industry types: F, food; C, chemicals; T, textiles; E, engineering; B, building; D, domestic

example: the main materials are sand, coke and pig iron; the processes are metal melting, moulding, pouring, knocking-out and fettling; diversification to produce a large number of different castings takes place at the moulding stage.

Square industries also manufacture components and are those in which a large number of material varieties are converted into an equally large number of component varieties. Square industries are often service industries which provide a special service to other types of industry.

Explosive industries are those which convert a large number of material varieties into a relatively small number of product varieties. These are assembly industries which combine components to produce assembled products.

15.2.2 Classification by type of market

The second type of classification used in this chapter is based on the type of market served. Again there are four main classes:

(1) large contract (non-repetitive),
(2) jobbing (non-repetitive),
(3) batch (repetitive), and
(4) continuous (repetitive).

The large-contract type covers non-repetitive products such as ships, special machines, power stations and new factories. The regulation and control of the production of such products is a type of project management. It uses such techniques as critical path analysis (CPA), or the programme evaluation and review technique (PERT).

The term 'jobbing' is generally reserved for non-repetitive orders for small parts or assembled products which are made in factory workshops. Because the processes are non-repetitive they are usually 'made-to-order' products. It is possible that some of these items may be ordered again later, but as there is no certainty that this will happen, they can only be treated as non-repetitive.

Batch-production products are repetitive products which are made intermittently in batches. Such products generally share the capacity (machine hours) of the machine tools or other production facilities used to make them with other products in the product range.

Continuous-production products are repetitive products which are made on a continuous flow line. The machines and other processing facilities are laid out in a line in the sequence in which they are used, and are reserved for the manufacture of one particular part, or of a 'family' of similar parts.

15.2.3 Combined classification

The above two classifications, by material conversion type and by type of market, can be combined to give 16 classes, as shown in *Figure 15.3*. Four of these classes (CP, CI, CS and JP) are trivial, being very unlikely to occur in practice. Three others (LI, LS and LE) are unlikely to be found in the pure form, but may be found in partial form combined with some other class. The LE class, which in the form of the automated transfer line, for some years appeared to be the ultimate objective for explosive products, has lost ground to the BE (batch/explosive) class, and its automated form of the flexible manufacturing system (FMS), due to the growing demand of markets for a wider choice of product varieties.

Figure 15.3 also shows the form of material and the form of product produced by each type of industry. In particular, only the explosive industries produce assembled products, whilst the remainder produce either bulk products (measured in general units of weight, volume, area or length) or components.

15.3 Systems theory

Systems theory provides a means for studying complex systems. It has been particularly useful for the study of manufacturing systems.

A system can be defined as: a set of variables which together form an easily identified and definable entity, and are so related that a change in any one of them will induce, or be induced by, a change in at least one of the others. A manufacturing system is then a set of all those variables involved in production, which are so related that changes in their values cause the values of some of the other variables in the set to change.

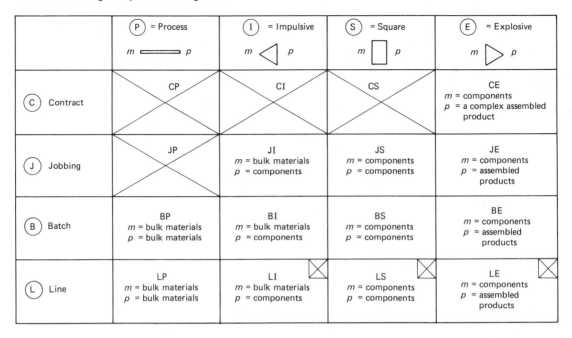

Figure 15.3 Combined classification of production. *m*, Number of material varieties; *p*, number of product varieties

15.3.1 Types of variable

The variables in any system can be divided into four main types: parameters, output, input and system design.

Parameters Also known as *regulatory parameters*, these are variables to which a manager can assign arbitrary values at will. Examples from manufacturing are order quantity, run quantity and selling price. Within certain limits, a manager can assign any value to such variables, and this will be their real value. The manager may be influenced in his choice of parameter values by his knowledge of the value of other variables, but changes in the values of these other variables have no direct effect on the value of the parameters.

Output variables These are variables to which a manager cannot assign arbitrary values. He can only change their values indirectly by changing the value of appropriate parameters. Examples of output variables are cost, stock, profit and scrap. In the case of stock, for example, the manager cannot change the stock value directly. He can, however, reduce the stock volume by increasing the value of the run frequency parameter (runs per year) which at constant output rate will also reduce the value of the 'run quantity', or number of parts made together as a batch before changing to make some other part.

Input variables Also known as *uncontrollable input variables*, these are variables imposed by the environment in which the enterprise operates. The manager has little chance to influence the values of these variables, but they do have an important effect on the values of the output variables. Examples are the weather, the exchange value of the pound (£), and the rate of interest charged on loans.

System design parameters These are parameters in that a manager can assign arbitrary values to them at will. In this case, however, the choice for each variable is between two alternative values, unlike the regulatory parameters where the manager chooses a value from a range of values. Examples are make or buy, multi- or single-cycle ordering, and process or product organisation. The design of a system can be seen as the choice between large numbers of alternatives such as these.

15.3.2 Feedback control

Figure 15.4 is a diagrammatic representation of a system. The rectangle represents the system, and the arrows leading into and out of it represent the different types of variable which influence its performance.

The same figure also shows the feedback control loop, with which the manager tries to control the output variable values. It is assumed that the manager has planned a series of periodic values for the output variables which he wants to achieve. He

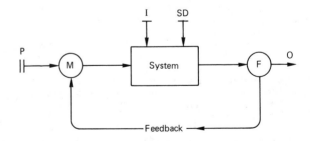

Figure 15.4 Diagrammatic representation of a system. P, parameter; O, output variable; I, input variable; M, manager; F, monitoring; SD, system design variable

arranges for the actual values of these output variables to be measured at the end of each period, and for these values to be compared with the planned values. This process is known as 'monitoring'. Details of any significant variances between the plans and actual output values are fed back to the appropriate manager, who either changes parameter values to bring output values back into line with the plans, or changes the plans for the same reason.

In simple terms, the manager regulates the output of the system by applying the management processes of planning, direction and control. For example, in the case of planning the output of finished products from the factory, the manager first 'plans' a 'production programme' showing the number of products to be made in a series of future time periods. He then 'directs' that these plans should be implemented by issuing the 'production programme' as an order to the factory. Finally, he 'controls' to ensure that the plans are followed by 'monitoring' and by arranging that he is informed (feedback) if there are significant 'variances' between the planned and the actual output value.

The difficulties with this simple concept are: there are hundreds of output variables; the output variables interact in that the values achieved for one variable will affect the values of some of the others; and these relationships are extremely complex and difficult to forecast. It can be said that the production system is very complex, and that it comes into the class of cybernetic systems, or very complex systems, which we can never hope to understand in all their infinite detail.

15.3.3 The design of the system

Even though we may not understand the minute detail of how a production system works, we can learn to predict how it will behave in given circumstances, and how to steer the output variable values in a required direction.

The illustration in *Figure 15.4* was first drawn by a man called Mr Black and the rectangle in this figure is, therefore, widely known as the 'Black box'. Black showed that if we measure at regular intervals the values of the parameters, uncontrollable input variables and output variables, we can learn to anticipate how the system will react to different parameter changes, even though we do not understand in detail the way in which the system works.

We do not, however, have to accept the production system as we find it. We know that we can improve its performance by improving its design. We know, for example, from experience and from experiment, that some types of system are more 'sensitive' than others. In other words, a change in the parameter values in some systems induces a change in related output variable values much more quickly than in others. We also know that certain types of production system are more 'predictable' than others. In other words, there is a high probability that a given change in parameter values will induce a predictable change in the values of some output variables.

Thus we know that we can learn to design production systems which are very much more efficient than those being used at present. As was mentioned earlier in this section, we can see that system design requires a choice between a very large number of alternatives.

15.3.4 Connectance

The main outstanding problem in systems theory is: How can one learn to understand the relationship between the different variables in the system?

The author has postulated that it is impossible to make general statements about the quantitative relationship between production system variables, because these depend on the nature of the enterprise, on the design of its system, and on the values of the uncontrollable input variables. We can, however, make general statements about the direction of the change in value of one variable which is induced by a given direction of change in another related variable. We can hope, therefore, to learn how to steer system output variable values in required directions, even though we may be unable to predict the quantitative effects of such changes.

The author has catalogued several hundreds of these direction-of-change relationships in order to form a relational data bank. *Figure 15.5* for example, shows the variable 'run quantity', i.e. the number of parts of the same design which are run off at a particular work centre before changing to make some other part. The figure shows that reducing the run quantity at constant output rate, reduces

(1) cycle stocks,
(2) thoughput time, and
(3) stock holding costs.

It will be realised that changes in these output variables themselves induce further changes in other variables, which in turn induce even more changes. *Figure 15.6* shows the results achieved using a computer program called 'Cascade', which explores these sequential types of change.

Figure 15.7 illustrates the results achieved with a computer program called 'Cure', which reverses the Cascade solution, and says: I have this problem; what changes must I make to solve it?

15.4 Management and the functions

Manufacturing is concerned with the conversion of materials into products for sale to customers. This simple idea is illustrated in the diagram in *Figure 15.8*.

Production management is, directly or indirectly, concerned with the regulation and control of the flow of materials into the works from suppliers; between conversion processes inside the works, and finally of finished products to the company's customers. The hundreds of different management tasks involved in this work in any factory can be classified according to:

(1) function, and
(2) hierarchical level.

15.4.1 Management of functions

The division into management functions is listed in *Table 15.1*. These eight functions provide the basic classification of the science of production management.

For the study of management tasks and of the way in which they are related, the division into functions provides an ideal starting point. The majority of management tasks are specific to only one function. This makes it possible to divide management into largely independent specialised 'functions', or 'subjects for study'.

Difficulties with the functional approach only arise when we try to organise the work of management, or in other words to allocate the responsibility for different sets of tasks to people in factories. Traditionally, in most factories the management was organised functionally, with different departments for e.g. product design, production planning, production control, purchasing, marketing, finance, personnel and secretarial work. Although the work of these functional departments was loosely coordinated by the general management, in practice they operated largely as independent kingdoms.

Regulatory parameter			P	3	0	3

1. Name:
Run qty — Mfgr. (R.Q)

2. Definition:
The quantity of a part run off at a work centre before changing to produce some other part

3. Function:

1.	Product design		6.	Finance	
2.	Production planning		7.	Personnel	
3.	Production control	✓	8.	Secretarial	
4.	Purchasing		9.	General Manager	
5.	Marketing		0.		

(box: 3)

4 Limitation: L

1.	Run frequency		fv	P	3	0	4	3	2
2.	Transfer quantity		max.	P	3	0	5	5	1
3.									

5. Induction: C

1.	Stocks	0	3	4	0	5	1	0
2.	Cycle throughput time (Mfgr)	0	3	3	0	1	1	0
3.	Cost — stock holding	0	6	8	2	0	1	0
4.								
5.								

6. Notes

Stock charts demonstrate induced changes

(1) 4 runs pa of 42 = 168	(2) 12 runs pa of 14 = 168
Average stock = 36	Average stock = 12

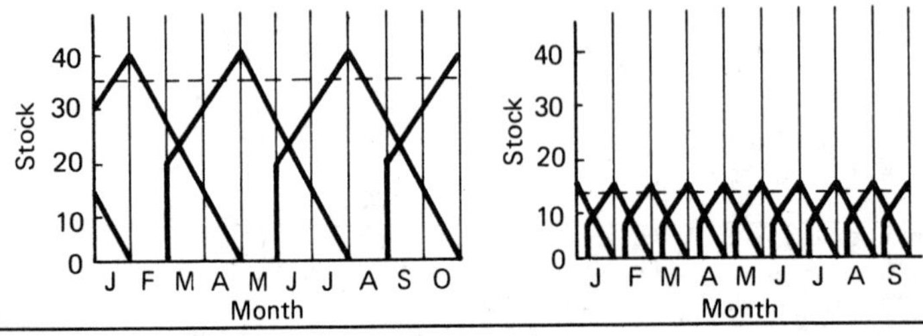

Figure 15.5 A regulatory parameter. Mfgr, manufacturing; fv, a direct function of the variable under consideration; pa, per annum

This type of organisation is strong in the coordination of tasks specific to particular functions, but weak in coordinating tasks which are common to several functions, and in coordinating functional plans to achieve the best overall company results.

15.4.2 Hierarchical levels

There is a general need in management for some form of hierarchical control. In other words, there is a need for levels of management, starting at the top with direction or general management which is responsible for strategy and the regulation and control of overall performance. One or more further levels are needed to elaborate the plans of general management in greater detail, leading to the supervisory level which is responsible for the regulation and control of the direct workers.

Figure 15.9 shows a company with three levels of management—general management, functional or middle management, and supervisory management—in the form of a pyra-

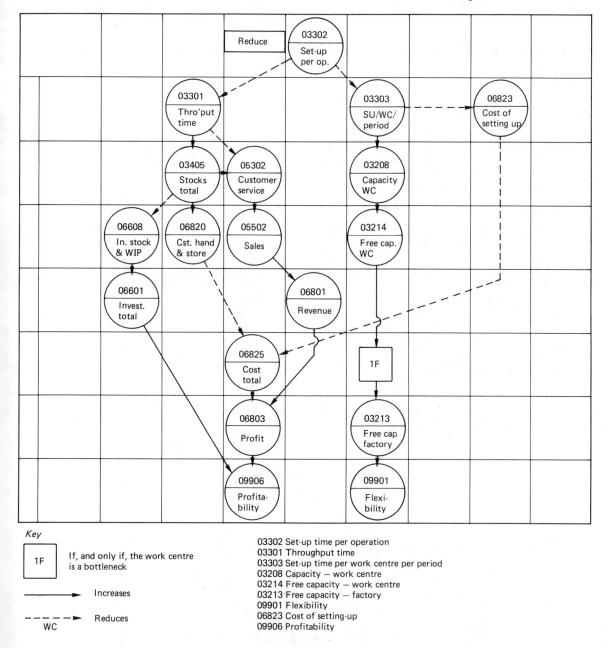

Key

1F	If, and only if, the work centre is a bottleneck

————————▶ Increases

— — — —▶ Reduces
WC

03302 Set-up time per operation
03301 Throughput time
03303 Set-up time per work centre per period
03208 Capacity — work centre
03214 Free capacity — work centre
03213 Free capacity — factory
09901 Flexibility
06823 Cost of setting-up
09906 Profitability

Figure 15.6 A cascade network showing the effect of reducing the set-up time per operation

mid. The base of the pyramid is the material flow system of the company. It shows machine tools and other work centres and the way in which materials flow between them during production. At each of the levels in *Figure 15.9*, a number of decision-making tasks must be carried out. These are connected with other tasks, both at the same level and also at other levels. The output from each task provides either an output from the system or an input to some other task inside the system. This representation of management as a pyramid

is borrowed from the GRAI method, developed by Professor Pun and Dr Guy Doumeingts of Bordeaux University. They have developed methods for planning management systems, based on the hierarchical decision-making flow illustrated in *Figure 15.9*.

The GRAI method, and other less formal methods which plan management systems hierarchically, tend to weaken the functional divisions in management, and to strengthen the central control of the enterprise as a whole and the connection

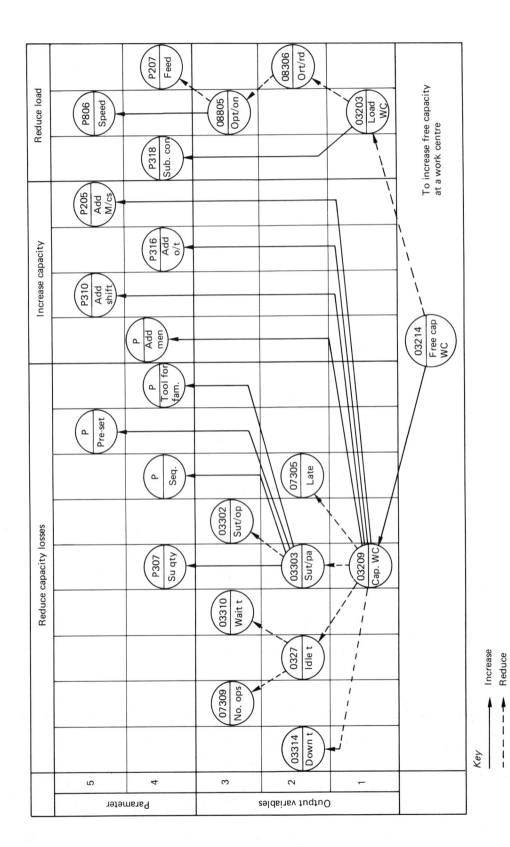

Figure 15.7 A cure network

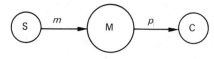

Key

S = Suppliers
M = Manufacture
C = Customers
m = materials
p = products

Figure 15.8 Manufacturing

between decisions made at different levels. Production control, for example, is the scheduling function in management. It operates at the levels of:

(1) *programming*—schedules the completion of finished products for sale;
(2) *ordering*—schedules the completion of made parts and the delivery of purchases; and
(3) *dispatching*—schedules the completion of operations in the manufacture of parts.

With the traditional functional organisation, all of this work would be done in a production-control department. With hierarchical planning, there is a tendency for top-level decisions to be made at the top or general management level, and for lower level decisions to be delegated to the lowest, or supervisory level of management.

15.4.3 Integrated and computer-integrated manufacturing

The functional organisation of management was simple to organise and to understand, but its failure to consider the enterprise as a single whole was a fundamental weakness, which cannot be tolerated in the future. Industry cannot be run efficiently by a collection of independent warring kingdoms.

The problems of the need for change in management organisation have been reduced by the development of computerised data processing. Once the strategies and policies for management have been fixed, and the primary operating plans have been agreed, what is left is systematic data processing, which can be automated using a computer.

It is still true that most management tasks are specific to one function only, even if we do not accept the functions as the basis for management organisation. The core of the computerised data-processing system is, therefore, a number of subsystems covering functional needs. The output of these subsystems forms the input to some other subsystem, or provides data for suppliers, the 'works', or customers. These major data flows form the arterial flow system (see *Figure 15.1*) which can also be automated.

Every manufacturing enterprise is unique, and thus the management system of each enterprise must also be unique. What industry really needs is tailor-made computer software, not standard operating packages. In the early days, computer operating system software was sold for universal or wide application. It was generally sold on the basis that the software program could be amended to suit the special needs of the customer. In practice, because these programs were written in first-generation languages such as Fortran, they were very

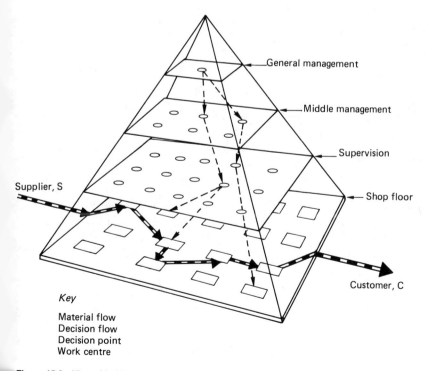

Key

Material flow
Decision flow
Decision point
Work centre

Figure 15.9 Hierarchical levels of management

difficult to change. Companies ended up using only a small part of a very expensive package, or trying to change their management system to suit the package. Present-day packages are not much better, but there is some hope for the future in the latest 'distributed databases' which, if used with fourth-generation computer languages, make it much simpler for companies to write their own programs.

The first need is to design an integrated manufacturing (IM) system in which: each task is done only once (data-processing tasks are not repeated in several functional subsystems); data are stored in one place only (but not necessarily all together in the same place); and the subsystems have been simplified as far as possible by integration to eliminate unnecessary variety.

It is only when the IM system has been designed that computerisation to produce a computer-integrated manufacturing system (CIM) can be profitably considered.

15.4.4 Recording a management system

Because management systems are generally very complex, a major problem is how to record such systems so that they can be checked efficiently, and can be communicated to those (e.g. software designers) who need to understand how they work, without risk of ambiguity.

One widely known method for recording system design is the IDEF system developed by the USA Space Agency. This is too complex for general use in management, and there have been several attempts to simplify it for this purpose. One of these, known as MODUS (developed by BIS Applied Systems) is illustrated in *Figure 15.10*. This figure shows the first level division of a production-control system into the progressive subsystems of programming, ordering and dis-

patching, together with the three main feedback control systems of progressing, loading and inventory control.

As an expansion of *Figure 15.10*, *Figure 15.11* shows the subsystem of 'ordering' in detail, to show how it works.

The MODUS method uses progressive charts such as these to illustrate a management system. The method of drawing the charts is obvious, but it might be mentioned that they are easier to read if the main data flows are from left to right, as shown in *Figures 15.10* and *15.11*, and only 'feedback' is in the reverse direction.

15.5 General management

General management is the management of the enterprise as a whole, and is the work done by the general manager or chief executive of the enterprise. He exercises direct authority over the senior managers—those of marketing, design, manufacturing, finance, personnel, etc.—and is directly responsible for strategy, the main lines of policy and policy coordination, organisation, personnel, and the general control of operations.

15.5.1 Strategy and policy

'Strategy' is a word borrowed from military science, which means 'generalship'. It covers the first level of planning in an enterprise, including the fixing of objectives, the organisation or the allocation of responsibility for different types of task to individuals, and the specification of general lines of action to be followed in order to achieve the selected objectives.

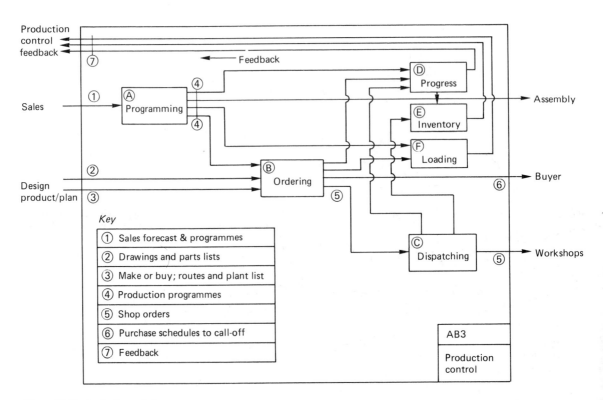

Figure 15.10 Production control

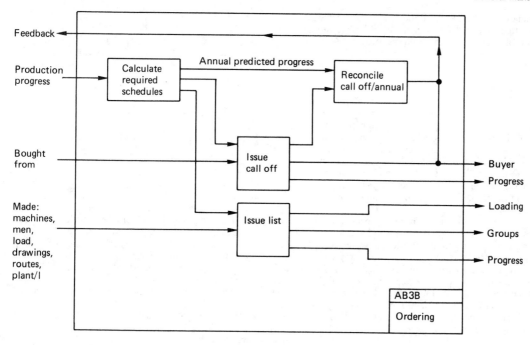

Figure 15.11 Ordering with period batch control (see Section 15.11.9.2). N.B. with period batch control, each product in the programme requires: (1) a standard number of parts to be made in each group; (2) a standard load in machine hours on each machine in each group; and (3) a standard number of parts to call off from each supplier

'Policy' can be defined as a plan of action. In military science, strategy is seen as management directed towards the winning of a war, while policies are concerned at a lower level with the winning of battles. For example, in manufacturing a strategy might be adopted: 'to minimise the stock investment' and policies which could be adopted to this end might include the introduction of group technology and just-in-time (JIT) production control.

A general manager may delegate the responsibility for the lower levels of policy, but he must coordinate its introduction to ensure that his managers do not introduce policies which conflict with the strategies he has adopted, or with each other. The strategies will generally be fixed at an overall company level; the policies are generally concerned with different functions.

15.5.2 Organisation

Organisation can be seen as the bringing together of tasks into sets, and the allocation of responsibility for those sets of tasks to people. Such allocations must generally be hierarchical. In other words, the general manager cannot directly control all the people in an enterprise. He will be responsible for the senior managers in the enterprise, and they in turn will have others reporting to them at the next level.

The type of organisation adopted by an enterprise will depend on the strategies and objectives which it adopts. Management organisation in recent years has been strongly functional, with largely independent departments concerned with each of the eight functions. This type of organisation is now changing. Strategies which emphasise the importance of flexibility—or the ability rapidly to follow changes in market demand, without losses due to materials obsolescence—have

tended to change organisational structures towards cross-functional, multidisciplinary forms, giving better coordination.

Again, in the past most workshops in manufacturing enterprises used process organisation. New strategies such as low stock manufacturing, total quality control, increased worker participation in decision-making, and flexibility, have forced the change from process to product organisation, and in particular towards group technology where organisational units complete 'families' of products or parts.

15.5.3 Personnel and general control

It has been said that the most important asset of any enterprise is its people. It is the major responsibility of any general manager to demonstrate his belief in this idea. Successful management styles differ greatly from one general manager to another. There is no standard approach which should be adopted by all managers. The main need is to show that he is interested in the people in the enterprise, and that he notices what they are doing and has some understanding of how they feel.

The main contact of the general manager of an enterprise will be with the senior first-line managers who report to him. To maintain control of the enterprise, he needs to monitor their performance, but he must try to do so in a way which will not kill their initiative and motivation.

To maintain full control and to coordinate decision-making in the different functions, the general manager will normally need to have direct control of the first level of decision-making in each function. In the product design function, for example, he should have the final word in the composition of the product range. In finance, the annual budget should be issued in his name, and in production control he should approve the annual programmes before they are adopted. If he is to build an efficient and effective team, however, he needs to delegate the further levels of decision-making to his subordinates.

15.6 Product design

The product design function includes all management tasks which are concerned with planning, directing and controlling the final form of the products produced by a manufacturing enterprise. This work requires an element of creativity, coupled with experimental verification and improvement through research and development (R&D), and with quality control (QC), based on comparison between the actual results achieved in manufacture and the required measures shown in product specifications.

Product design is simple to control in most process industries. It is closely associated with production planning and production control in jobbing and contract industries (see *Figure 15.3*), where product design is often a part of any sales contract. In the remaining types of industry which normally make standard products, the main problems in the product design function are: how to specify the product design; how to regulate modifications to the design, and how to 'control' the quality of the finished product. This section of this chapter deals only with these three problems.

15.6.1 The product specification

A product specification is a record of the decisions made when designing a product. According to the type of industry, it may take the form of a model, a pattern, a sample, a chemical formula, a series of tests or, in its most complex form as the specification for an assembled mechanical product, it may comprise dimensioned drawings, a parts list ('bill of materials' in the USA), and other supporting specifications.

Product specifications are needed in manufacturing as:

(1) the basis for sales contracts;
(2) the basis for purchase contracts;
(3) to provide the information needed by production planning to plan the methods of productions;
(4) the basis for quality control;
(5) the basis for production control; and
(6) the basis for computerised data processing.

For all these purposes, an accurate, unambiguous record which shows the required final form of the materials in the product is absolutely essential.

Consider the case of the specification for assembled mechanical engineering products. First, the drawings. Drawings are needed of the assembly as a whole, and also of each separate component included in the assembly. At least three views (or projections) are needed for each drawing—from the side, from the side at 90° to the left, and a view from overhead—to make them unambiguous.

The principal dimensions must be shown on the drawings in the form of maximum and minimum, or limiting, values. The difference between these 'limits' is known as the 'tolerance'. Limits are necessary because absolute perfection is impossible to obtain. The limits are fixed, (a) in order to achieve the required level of performance from the machine (operational dimensions), and (b) to ensure the interchangeability of components (see *Figure 15.12*).

The second requirement is a parts list. This must list every component included in the product. Each of these separate items must have its own name and also a code number which is

Limiting values: 1.201 and 1.2000
Tolerance: 0.001

Figure 15.12 Limits on drawings

specific to that item. Most engineering products can be divided into assemblies, subassemblies and so on. For convenience, in 'kitting out' or picking sets of parts for assembly, it is desirable that assemblies and subassemblies are shown separately in parts lists. This has the disadvantage that some items may appear more than once in the same parts list, but it is not difficult to produce consolidated lists of these items for use in component ordering.

The third specification requirement mentioned above, of other supporting specifications, may include material specifications, surface-finish specifications, and test specifications, describing the tests to be used to ensure that the machine is functioning satisfactorily.

15.6.2 Design modifications

During the life cycle of most products, there are occasions when the design of the product must be changed. It is desirable that the introduction of such 'modifications' should be carefully controlled, and that one should be able to tell which customers received products to the old design and which had the new modified design.

Modifications have different degrees of urgency. For example, if the change is needed to improve the safety of a product it may have to be introduced immediately in all products before dispatch, and even retrospectively in products which have already been sold. If, on the other hand, the modification is designed to reduce costs, then it is possible to wait for the best economic moment before it is introduced.

The retrospective change of products already sold is normally the most expensive form of modification. In other cases, material obsolescence is the most expensive cost induced by modification. When a modification is introduced, any parts or materials to the old design may only have scrap value. Where possible, the introduction of modifications should be arranged to allow time to use up old stocks. With traditional types of production control, this may require a wait of many months. It is one of the advantages of just-in-time low-stock production control systems that modifications can be introduced quickly without losses due to obsolescence.

A record is needed which shows the modification state of all products sold, so that the correct spare parts can be delivered if they are needed in the future.

15.6.3 Quality control

The feedback control system which constrains events to follow the plans contained in product design specifications, is called 'quality control'. The monitoring component of quality control which measures the actual dimensions achieved during manufacture compares them with the values shown on the drawings, and feeds back information about parts which are unacceptable, is called 'inspection'.

Systematic inspection is relatively new in industry. Before its introduction, quality was seen as the responsibility of the craftsmen who did the work, and they were expected to inspect their own work. Today, while it is accepted that there are parts of inspection requiring special skills (metrology, for example), there is a tendency to return to the old order. More and more companies are delegating inspection to the floor of the shop where the work is done, and using small quality-control units to check that the product is being inspected, and to check the work of the inspectors to ensure that the best methods are being used for inspection.

In the sense of manufacturing reliability or compliance with the product specification—measured in units of, say, inspection rejects per million parts produced—there is little doubt that for manual operations quality depends on the skill and motivation of the people who do the work. Inspection provides policing that eliminates items which are outside limits, but it does not generate an improvement in quality.

With the development of automation, however, the emphasis in manufacturing reliability is changing from control of the product to control of the process. If the process is controlled, then the product quality is also controlled.

In the present climate of high competition, quality control is more important than ever, but its scope is much wider than it used to be. Today, quality is measured in terms of customer satisfaction. Therefore, quality control is concerned not only with manufacture to specification, but also with design, the methods of manufacture, marketing, after-sales service and any other factor which affects customer satisfaction.

15.7 Production planning

Production planning is the function of management which plans, directs and controls the physical means used to manufacture products in the enterprise. It can conveniently be studied in two parts: (a) the design of the detailed methods to be used to make components and to assemble them to form products; and (b) the methods used to organise the work and to plan the material flow system and plant layout for the enterprise. These two aspects are described in this section and Section 15.8, respectively.

Before production planning can commence, the planners need two types of information. First, they need copies of the product specifications for all the different products to be produced. Second, they need a planning specification which fixes the criteria to be used in making the choice between alternative planning possibilities.

In the most complex form of production planning in explosive industries (e.g. mechanical engineering assembly), seven main types of decision have to be made one after the other for each component:

(1) make or buy,
(2) the choice of material form,
(3) the processing route,
(4) tooling requirements,
(5) time standards (work study),
(6) plant layout (see Section 15.8), and
(7) assembly layout (see Section 15.8).

In other types of industry not all of these planning stages are needed. It will be noted that in explosive industries, production planning is an on-going task, requiring a wide range of new decisions every time a product is modified, or a new product is introduced. In the process industries, however, there is a low rate of design change, and production planning is mainly concerned with pre-planning, or with the initial design of the plant and the plant layout for the factory. The others fit somewhere between these two extremes.

The main controls associated with production planning are process control and maintenance.

15.7.1 The planning specification

The decisions made in production planning depend, in part, on the first level strategic plans made for production. One must know, for example, what quantities of the product are likely to be required per year and the expected life of the product. This information takes the form of a long-term (several years ahead) sales programme.

Next, one needs to know the schedule for product introduction. In other words, when must the first product be produced, and how much time is available for planning and for obtaining any necessary materials and tools?

One also needs a general understanding of production control policy in the company. If, for example, the company is satisfied with an existing multicycle ordering system (stock control or MRP), then order frequencies will be low, and the main emphasis in production planning will be on the minimisation of operation times. If, however, the company needs to reduce stocks and is moving towards just-in-time (JIT) production control, then methods should be chosen which reduce setting-up and throughput times, and are more suitable for use with high order frequencies.

One also needs to know the 'sanction quantity' for the product. For how many products, in other words, do the directors authorise their managers to commit the company, by placing contracts with suppliers for future deliveries of materials and bought components?

Again, the planners must know what is the tooling 'write off quantity' for the product. In other words, what is the number of products over which the cost of any new tooling must be recovered in the selling price? The write-off quantity will affect the choice of selling price for the product and/or the possible investment in tooling. Finally, one needs to know how much money is available for investment in new machines and other processing facilities.

It is best if these decisions are written together in a 'planning specification', which clearly tells the production planners what criteria they are to use when choosing between alternatives during production planning. An example is given in *Figure 15.13*.

15.7.2 Make or buy

The first task in production planning is to go through the product parts lists and decide for each part and subassembly whether provision should be made to produce it in the factory or if it should be bought.

In its wider context, this decision is concerned with choosing whether to provide processing facilities for making castings, forgings, or gears for example. Today, with group technology, these questions can be rephrased to ask: Are we going to provide groups for gear manufacture, for heavy beds and cases, for small rotational parts from bar, for sheet metal pressings, for welded fabrications and others?

The main criteria for choice in this make-or-buy decision-making area are:

(1) economic,
(2) security, and
(3) flexibility.

The problem of the economic choice between make or buy is discussed in Section 15.12. It is much more complex than a simple comparison of manufacturing costs and purchase price.

Planning specification	
PRODUCT Blenker	SPECIFICATION NO. 207.30117

A SALES PROGRAMME

1969	1970	1971	1972	1973
50	1200	1500	1750	2000

B INTRODUCTION PROGRAMME

1. Prototype. *Already tested*
2. Pre-production batch. *150 sets already produced*
3. Production tooling. *Complete by 1.10.69*
4. Start prod. assembly. *1.12.69*

C CYCLE

1. Standard. *2 weeks. (Cat. A and B made. Cat. A bought)*
2. Cat 'B' bought. *8 weeks*
3. Cat 'C'. *10 weeks*

D 'SANCTION QUANTITY'

1. Up to 1250 products. *1250 sets*
2. After 1250 sets. *3250 sets (confirm Jan 1 1970)*

E TOOLING BUDGET

1. Special tooling. *£45 000*
2. Write off quantity. *4 500 (end of 1972)*

F TARGET PRODUCTION COST

1. Direct materials. £15.50
2. Direct labour. £ 7.02½
3. Direct expense. £ 6.08

     Total direct cost   £28.60½

Issued by. *J Smith*	Date. *1.1.69*

Figure 15.13 A planning specification

The fear that others may learn how to make the product and become competitors may persuade some companies that they need to make the more complex parts of their products themselves, for reasons of security.

Finally, the trend in most product markets today is towards an increase in the rate of design change. This makes flexibility—or the ability to change product designs rapidly with minimum losses due to obsolescence, making it possible rapidly to follow changes in market demand—a subject of growing importance. Companies which can make most of the parts they need are generally more flexible in this sense than are those who buy a large proportion of the parts they use.

Make-or-buy policies found in practice include: make all the parts needed; buy all the parts needed; and provide facilities so that most of the different parts needed can be made, but only provide full capacity for the security items, subcontracting some of the other items in order to achieve the necessary capacity.

15.7.3 The choice of material form

There is no problem in choosing the form of material in the process and implosive industries, because in both cases production starts with a small number of material varieties. There is little or no choice between alternative material forms. The square industries are mainly composed of service factories which subcontract for other industries. In this case the choice of material form is generally made by the customer.

It is only in the explosive industries that there is a choice. In essence, we need to decide on the point at which we wish to start processing. A steel part, for example, may start as a cast-steel billet, be rolled into bars, be machined, be electroplated, and then be assembled with other parts to make a finished mechanical product. In this case we need to decide:

(1) Do we want to cast billets?, or
(2) Do we want to buy billets and roll bars?, or
(3) Do we want to buy bars and machine parts?, or
(4) Do we want to buy machined parts, and then electroplate them?, or
(5) Do we want to buy electroplated finished parts, and then assemble them?

If we start at the beginning, we can hope for maximum flexibility because a small variety of materials can be used to produce a very large variety of parts. Because in this case materials are bought in their lowest form, material costs will be low, but the investment in plant and tooling will be high, as will be the expenditure on direct labour. If, on the other hand, we consider buying components in a form which is ready for assembly, we need to know a great deal more about the relative investments and expenditures before we can make a reasonable choice.

At each stage of processing a part, there will also be a choice. For example, if the first stage is to produce a casting ready for machining, one needs to choose between using sand castings, the Cosworth method, die castings, and the lost wax process, among others. Again, to make a gear one might start with a blank cut from bar, or pay more for a hot pressing, which would improve grain flow and quality, reduce the material waste per part, and also, if made in the factory, would help to reduce bought material costs.

15.7.4 The choice of processing route

The next problem is to choose the processing route for each component. A typical 'route card' for an engineering product is illustrated in *Figure 15.14*. The figure shows the 'process' of machining a particular part, divided into 'operations' which are listed and numbered according to their sequence of application. Details are given of the machine to be used for each operation (in *Figure 15.14* by means of a two-letter code) and the standard operation time.

Some companies also give details of machine set-up times. This information is needed for scheduling, but is of doubtful use for loading, because the set-up time varies greatly with the sequence in which different parts are processed. For loading it is more accurate to compare the net load (or the sum of the operation times) with the net capacity, or capacity available for useful work after correcting for set-up time and other losses.

The division into operations is made according to rules, such as the following.

(1) If following operations are carried out on different machines with different operators, they should have different operation names and numbers.

KL11-7		DERRICK CONTROL FAB.			PRN
1	9	BRACKET		M34099	4546

FABRICATION

31/ 982	41/ 503	40/ 434	40/ 434	31/ 982	71/ 987	VR
W	MV	DM	DM	W	SA	STO

10 W MV 31/982	Assemble and weld complete, except items B. *Note*: Tack F to B before loading to fixture P6607.	10 00 54
20 MV W DM 41/503	Load to table and mill feet to 3/8 in. diameter.	30 00 17
30 DM MV DM 40/434	Load to table pit jig and drill four 17/32 in. diameter holes in feet J4749.	20 00 06
40 DM DM W	Load to jig drill and ream 1.5005 in./1.4995 in. diameter holes J4748.	60 00 43 P.P.
50 W DM SA	Stud weld three studs. Part 'H' TM6606	— 00 $09\frac{1}{2}$
60 SA W VR 71/987 STO	Deburr including all holes. Final inspection.	00 00 04

Figure 15.14 A route card

(2) If one man operates two or three machines as a work centre, only one operation number is needed for the work done at the work centre.

(3) If a series of operations is carried out on one machine, but the first of these operations must be completed for the whole batch and the machine must then be re-set for the next operation, then each operation should have a different operation name and number.

There is often a wide choice possible in the number of operations per part. Reducing the number of operations, reduces waiting (queuing) time, and also reduces throughput times and stocks. The present trend is, therefore, towards a reduction in the number of operations.

The number of operations per part can be reduced by such methods as:

(1) Planning one complex operation on a machine, to take the place of several simple operations.

(2) Moving a succession of operations on simple machines onto one more complex machine, e.g. changing several drilling operations on single-spindle drills into one operation on a coordinate drilling machine, or changing several operations on milling, boring and drilling machines into one operation on a numerical control machining centre.

(3) Adding short manual operations, such as deburring and inspections, to machine operations with long operation times, so that they can be done by the machine operator between loading and unloading in the machining cycle.

15.7.5 Planning tooling requirements

Having decided on the route for a part, the next job is to plan the tooling, which must be provided at the workplace in order to carry out the operations done there. Tooling can be classified into three main types:

(1) fixtures,
(2) consumable tools, and
(3) jigs.

Fixtures hold materials or assemblies in a convenient position for work to be done on them. Consumable tools are items such as drills, reamers, turning tools, milling cutters, grinding wheels and files, which wear due to direct contact with materials, and have to be replaced at intervals. Jigs are like fixtures in that they hold materials, but they also guide tools so that they operate in the correct positions on the held materials. A typical example is a drilling jig which holds the materials to be drilled, and also guides drills by means of drilling bushes, so that holes are drilled in required positions.

In designing tooling, there are four main objectives:

(1) *quality*—the tooling must be sufficiently rigid and wear resistant that it is possible to produce consistent results;

(2) *speed*—the tooling must be sufficiently rigid and vibration free to work consistently at optimum speeds;

(3) *set-up time*—it must be possible to change the tooling set-up very quickly for one part after another; and

(4) *cost*—inside the limits set by the first three objectives, the cost of the tooling should be minimised.

A simple example of an assembly or manual processing fixture which follows these objectives is a vice with interchangeable soft jaws. Most fixtures for these purposes can follow the same principles of a basic fixture, with adaptive attachments to convert it for work on other similar components. A similar approach can also be used for machining fixtures. Examples are a collet chuck with interchangeable collets and an adjustable jaw chuck with interchangeable jaws.

In consumable tooling, the trend in the case of cutting tools has been towards small interchangeable tool 'bits'—often of hard tungsten carbide or ceramic materials—in accurately machined tooling holders. Tool bits designed in this way can be changed rapidly when they wear, and the tools can be pre-set in their holders for rapid set-up changes.

In the case of jigs, there is a trend towards a reduced demand for this type of tooling. The introduction of coordinate drilling machines and of numerical control machining centres, has for example, reduced the need for drill jigs. Numerical control machines also simplify, or eliminate the need for jigs, with other processes.

15.7.6 Work study

Work study is a technique designed for studying the methods used in industry to do work. It uses two subtechniques: method study and time study. Method study analyses existing or planned methods and, by eliminating unnecessary or wasted movements, attempts to find the simplest and most efficient way to complete a task. Time study measures the time required to complete tasks in order to provide the necessary data for scheduling and/or loading. Work study was originally developed to study work done in manufacturing workshops. A very similar technique called 'organisation and methods' (O&M) is used to study the work done in offices.

Method study generally starts by breaking down the work required to complete an operation, into steps. The steps are listed in sequence and classified to show for each step if it is: work, transportation, delay, storage or inspection. Transportation distances and work step times, are estimated and added to the chart. An attempt is then made to redesign the method to eliminate the transportation, storage and delay steps. An example is given in *Figure 15.15*.

A more advanced form of method study is called 'motion study'. This was developed in the USA by the late Frank Gilbreth and his wife. They broke down tasks into a series of fundamental elements which they called Therbligs (Gilbreth spelt backwards) see (*Figure 15.16*), and then rebuilt an ideal cycle from which all unnecessary elements were eliminated.

'Time study' was originally based on the time taken to do a task, as measured with a stop watch. The method was used as follows:

(1) *Preliminary study*—make sure that the correct method is being used.
(2) *Divide into elements*—the elements must be easily identified and measured.
(3) *Time the element*—take a number of readings of the time required per element.
(4) *Find the 'recorded element time'*—strike out abnormal times and use the average, minimum, modal or good time method, to find the recorded element time.
(5) *Levelling or performance rating*—adjust the recorded element times to the level that can be expected from an average man or woman.
(6) *Allowances*—add time allowances to cover delays, rest and personal needs.
(7) *Standard time per operation*——specify.

Time study with stop watches is not usually liked by the workers in factories. Synthetic times can be used to avoid the need to use a stop watch. With these methods, standard operation times are found by adding recorded standard time values for standard 'basic work elements' or motions.

Another method used for measuring operation times in small batch and jobbing production is known as 'slotting'. A list is made of the operations done on a machine. These are listed in their estimated operation time sequence. Time study is used to find the operation times for selected operations in the range (bench marks). New work is slotted into the series between bench marks (i.e. more time consuming than this task; less time consuming than the next one). An operation time is allocated to the new job which is half way between those for the adjacent bench marks.

15.8 Design of the material flow system

The material flow system in a manufacturing company is the system of routes along which materials flow between the places where work is done on them in order to change their shape, form or nature. Its nearest equivalent in everyday life is a detailed road map. There is evidence that the efficiency of a manufacturing system depends on the complexity of its material flow system. Simple flow systems are generally much more efficient than are complex flow systems. It is also readily apparent that the nature of the material flow system depends on the way in which the factory is organised at the level of the workshop. This can be taken as our point of departure for an examination of material flow systems.

15.8.1 Organisation at workshop level

Organisation has been defined as: 'the process of determining the necessary activities and positions within an enterprise, department or group, arranging them into the best functional relationships, clearly defining the authority, responsibilities and duties of each, and assigning them to individuals so that the available effort can be effectively and systematically applied and coordinated.

There are two main types of organisation used at workshop level:

(1) process organisation, and
(2) product organisation.

With process organisation, organisational units specialise in particular processes, e.g. turning, milling, grinding, drilling, assembly, heat treatment, painting, and press work. The people in such units specialise in the skills required to do each process.

With product organisation, organisational units complete particular products, assemblies, major stages in assembly, parts, or 'families' of similar parts. They are equipped with the variety of machines and other manufacturing facilities needed to complete such entities, and the people in these units have between them all the skills needed to do so.

There are two main types of product organisation found in industry:

(1) continuous line flow, and
(2) group technology.

With continuous line flow, the machines and/or other processing facilities used to make parts, are laid out in a line one after the other in the sequence in which they are used. The materials being processed flow between these workstations in a continuous stream.

With group technology, products or components are completed in their own special organisational units (or groups). In the case of components, different items may use different combinations of facilities in different sequences inside the groups. In the case of assembly groups, products may be completed by single groups, or by sets of groups working together. In both cases the groups produce continuously, and

Product: *B. B. Haymaker*	Date: *2.1.78*
Operation: *Riveting flier arm assembly (AA200)*	Study by: *J. Smith*
Department: *Sub-assembly*	Sheet No.: *1 of 1*

Symbols:
　　　　○ Work　　⇨ Transport　　D Delay　　▽ Storage　　□ Inspect

Time	Travel in m	Symbol	Step No.	Description
				Present Method
25 sec		○	1	Press pin (AA213) into bracket (AA215) on Hyton press (2073). Place in box pallette
	15	⇨	2	Transport to work in progress store
		▽	3	Store
	15	⇨	4	Transport to Hyton press (2073)
15 sec		○	5	Squeeze rivet end of pin. Place in pallette
	15	⇨	6	Transport to work in progress store
		▽	7	Store
	15	⇨	8	Transport to Hyton press (2073)
30 sec		○	9	Rivet arm (AA219) to bracket and pin assembly (AA210). Place in pallette
	35	⇨	10	Transport to welding department
				Proposed Method
50 sec		○	1	Press pin (AA213) into bracket (AA215) and rivet, then rivet on arm (AA219). New fixture for Hyton press. Place in pallette
	35	⇨	2	Transport to welding department

Note. In addition to savings in transportation and storage, estimate operation time saving of 20 seconds each.

Summary:	○ Steps	⇨ Steps	⇨ Metres	D Steps	▽ Steps	□ Steps
Present Method	3	5	95		2	
Proposed Method	1	1	35		0	*Q.C.*
Saving	2	4	60		2	

Figure 15.15　An operation analysis sheet

make most of their own subassemblies. With group assembly, a main objective is minimum throughput time and sub-assemblies are not usually made in batches for stock.

15.8.2　Continuous line flow

Continuous line flow is used in practice in: process industries (see *Table 15.1*); in the mass production of products in very large numbers; and for continuous manual operations, such as assembly and packaging, for segments of material flow systems where the production facilities are normally always used in the same sequence, and the operation times are similar in length.

Continuous line flow infers a need to balance the operation times at each station on the line. If this is not done, either stock will build up between stations where a long operation follows a short one, or there will be idle times when some operations must wait for materials at each cycle, from the previous operation. The automated version of continuous line flow is the automatic transfer line.

The ideal situation for continuous line flow is found in the process industries. These convert a small number of material varieties into an equally small number of product varieties (see *Figure 15.2*). Conversion is achieved using a small number of processes, generally in the same sequence. The only limitation to the use of continuous line flow in process industries is found

Therblig	Abbreviation	Symbol	Suggested by:
Search	Sh		Eye searching
Find	F		Eye straight
Select	St		Reaching for object
Grasp	G		Hand open to grasp
Transport loaded	TL		Hand with item in it
Position	P		Hand placing object
Assemble	A		Four parts together
Use	U		Word 'use'
Disassemble	DA		Assembly one part removed
Inspect	I		Magnifying lens
Pre-position	PP		A nine-pin set up for game
Release load	RL		Hand dropping item
Transport empty	TE		Empty hand
Rest	R		Man seated
Unavoidable delay	UD		Man falling down
Avoidable delay	AD		Man lying down
Plan	Pn		Finger to brow (thinking)
Hold	H		Magnet and iron bar

Figure 15.16 Therbligs

in those cases where there are some batch processes. The long-term expectation is that most of these batch processes can be converted to continuous flow. For the time being, they require intermediate batch processing stages in otherwise line-flow factories.

The use of continuous line flow for the mass production of components for sale or to feed assembly is diminishing due to the demands of markets for a greater variety of product designs. This trend is likely to continue, leading to a gradual reduction in the number of automatic transfer lines for component processing, and an increase in the use of flexible manufacturing systems (FMSs).

Continuous line flow in assembly can be found in the forms of machine-paced lines (mechanical conveyors) and operator-paced lines. The future use of machine-paced continuous line flow for manual assembly is likely to diminish, (a) because such lines are generally unpopular with the people who work on them; (b) due to the present explosion in the number of product variants, which reduces the efficiency of such lines; and (c) due to the gradual development of mechanised and automated assembly.

Simple assembly lines comprising a few stations making simple assemblies do have a future. In this case, balance is achieved by a team of workers who move between stations as necessary to maintain the flow of work. This type of layout is often used within assembly groups.

Where lines must be used, they have the advantage of short throughput times, leading potentially to low stocks and low stock-holding costs.

15.8.3 Group technology—component processing

The term 'group technology' (GT) was first used by Professor S. P. Mitrovanof of Leningrad University, as the name for his research into the relationship between the shape of components and the processing methods used to make them. One of his findings was that it was possible to set up lathes so that they could make a set of similar parts, one after the other at the same set up. Later, a company in Alsace added milling and drilling machines to a section containing lathes set up in the Mitrovanof manner, to form a group which completed all the parts it made. Later again, in Britain, it was shown that this method of organisation into groups which completed all the parts they make, could also be used with other processes, and in other types of industry.

At this stage group technology had evolved into a method of organisation which could be used in any factory where continuous line flow was not possible. Group technology makes it possible to gain the advantages of continuous line flow, with jobbing and batch production, in implosive, square, and explosive industries.

The difference between process organisation and group technology is illustrated diagrammatically in *Figure 15.17*. With process organisation, materials move from one processing unit and foreman to another during manufacture, until all the manufacturing processes have been completed. In each unit they join the queue of work waiting to be processed and must then wait for their turn to be processed. This type of organisation is typified by very long throughput times. It is not

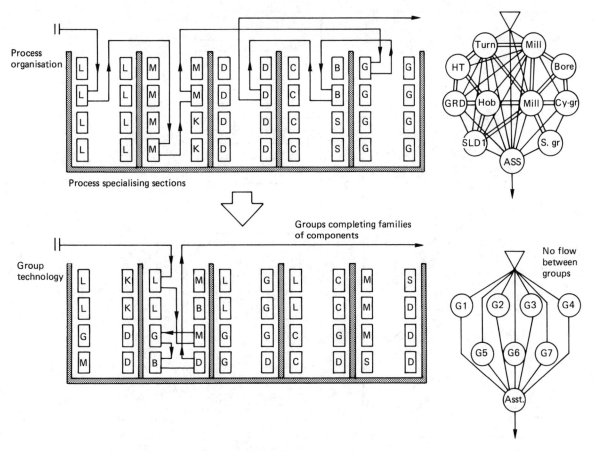

Figure 15.17   The difference between process organisation and group technology. Group technology is a form of product organisation

uncommon to find that the throughput time in a processing department with process organisation is over 100 times greater than the sum of the operation times. With group technology, each part is completed in its own special group under the supervision of its own foreman. In these circumstances it is possible to schedule the work to achieve very short throughput times.

Figure 15.18 shows a group installed about 25 years ago. It was designed to produce simple rotational parts made on bar lathes. This group completed all the parts it made, and was responsible for its own inspection, operation scheduling, setting-up, tool storage, minor tool maintenance, preventive maintenance inspections up to the minor inspection of machine tools, and for its own housekeeping. The group was, in effect, a semi-independent mini-factory.

The group shown in *Figure 15.18* was equipped with manually operated bar capstan lathes, mills and drilling machines, and with barrel deburring machines and cleaning tanks. Today, the capstan lathes have been replaced by CNC turning machines, one of which is capable of both first and second operation turning. Much of the milling and turning is done by power tooling on the turning centres, but a small machining centre and a drilling machine have also been installed.

Many of the product designs have changed, but there is still a need for small rotational parts. The present group produces a greater volume of work than the old one, has half as many

machines, and employs half as many people. It provides an example of the gradual evolution of groups towards automation.

Experience has shown that, providing one starts with a department which completes all the parts it makes, and which employs, say, 20 or more people, one can, with very few exceptions, divide it totally into groups which complete all the parts they make. The exceptions are generally processes needed by more than one group, of which X-ray and heat treatment are examples from practice. Where such processes cannot be divided for assignment to different groups, they are organised as a service group, which can provide a service to any group that needs it. Service groups should always be treated as undesirable exceptions which need to be eliminated as soon as possible.

The number of people in a group depends partly on the complexity of the product. If some complete parts are made with operations on 40 different machine tools, then there must be one group with at least 40 machines installed, with enough people to man them. Experience has shown that with group sizes in the range 10–15 people per group, with one or two groups in some factories employing a few more people for complex parts, then it is generally possible to divide into groups without buying any new machines.

Table 15.2 lists the main advantages of group technology. Most of these advantages arise because groups complete parts and their machines are close together under one foreman. The

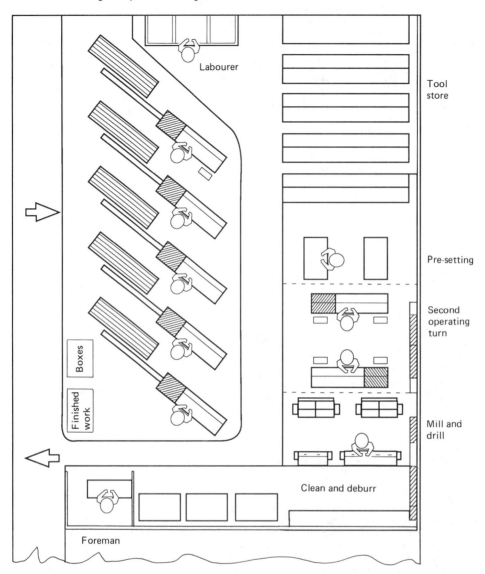

Figure 15.18 A group technology group. The group carries out its own operation scheduling, inspection, materials handling, setting up, tool storage and minor repairs, minor plant inspections and maintenance, and housekeeping

maximum advantages are only obtained, however, if the change to group technology is total and company wide. Group technology is not a technique which can only be used for some parts and not for others.

Figure 15.19 shows the effect of group technology on the material flow system of an enterprise. *Figure 15.19(a)* shows a network diagram which illustrates the flow system with process organisation. The circles represent processes and the arrows between them show the routes followed by materials between processes. This illustration is not some monstrosity chosen for effect, it is typical of most companies in the mechanical engineering industry that use process organisation.

Figure 15.19(b) shows the same company but with group technology. There is flow between groups at different processing stages, but because with very few exceptions the groups complete all the parts they make, there is no flow between groups at the same level or processing stage. It can be seen that group technology greatly simplifies the material flow system of an enterprise.

15.8.4 Classification and coding

The main difficulty in introducing group technology into industry has been in finding a way to plan the change from process organisation to group technology. Most of the early group technology applications were planned, using classification and coding (CC). This method classifies the information contained in part drawings, and is based on the idea that parts which are similar in shape can normally be machined by the same group of machines.

Table 15.2 Advantages of group technology

1. Short throughput times
Because machines are *close together* under one foreman, giving:
(a) low stocks,
(b) low stock holding costs, and
(c) better customer service

2. Better quality (fewer rejects)
Because groups *complete parts*, and machines are *close together under one foreman*

3. Lower materials handling costs
Because machines are *close together under one foreman*

4. Better delegation
Because groups *complete parts*. The foreman can be made responsible for cost, quality and completion by due-date giving:
(a) reduced indirect labour cost, and
(b) more reliable production

5. Training for promotion
Process organisation only produces specialists. Group technology trains generalists

6. Automation
Group technology is the first evolutionary step in automation. A group is a flexible manufacturing system with some manual operations

7. Morale & job satisfaction
Most workers prefer to work in groups than with process organisation

Although this idea is true for the majority of parts, it is not a reliable method for finding groups, because there are too many exceptional cases when it is untrue. For example, parts may be similar in shape but vary in size, manufacturing tolerances, requirement quantities, or materials, making it necessary to manufacture them on different sets of machines.

A second disadvantage of CC is that it fails to find the large numbers of parts in most factories which differ in shape or function from others, but ought to be made in the same groups because they are made on the same machines. For this reason, CC never finds a complete division of the parts made in a factory into families.

A third disadvantage of CC is that, because it is based on the analysis of component drawings, it does not find groups (of machines), but only families of parts. After CC has produced a list of parts arranged in 'families' together with a long list of exceptions, there is still a great deal of expensive work to be done in production planning for the groups.

Finally, CC ignores the effect of its decisions on the material flow system and, in particular, it fails to start with a division into departments which complete all the parts they make. Many of the early problems in planning group technology could be traced to the fact that it is impossible to divide a department which does not complete all the parts it makes into groups which do complete all the parts they make.

If one starts, for example, with a 'machine shop' (department) for which 15% of the parts have intermediate operations in some other department, and another 15% have intermediate subcontracted operations, there is no possibility of forming groups until these anomalies have been corrected. There is no mechanism in CC for making this necessary correction.

In an attempt to develop CC so that it can be used to plan group technology groups, some manufacturing and computer software companies have developed data with 60 or more digits per component. In these companies CC has become an esoteric and very expensive computer game. As long as these banks are restricted to the information contained in component drawings and parts lists, such systems are still unable to find an efficient division into group technology groups. Even if production data are added to the component data, they only provide the information needed for later production flow analysis, at unnecessarily high cost.

It should be stated that, although CC is an inefficient method for planning group technology, it can be used efficiently to reduce drawing-office costs (by finding existing designs which can be used again) and to reduce tooling costs (by finding tools for similar parts which can be adapted at a low cost for a new part, instead of making new tools). It is also possible that CC may be needed for the future development of computer-aided design and manufacturing.

15.8.5 Production flow analysis

An alternative method to CC for planning group technology groups and families is production flow analysis (PFA). Where CC finds 'families' of parts by analysing the information in component drawings, PFA finds both 'groups' of machines and 'families' of parts, by analysing the information in component route cards.

Where CC is based on the idea that parts which are similar in shape or function can be made by the same set of machines, PFA is more direct. PFA considers that parts that are made on the same set of machines should be made together in a group which contains this set of machines. Because most parts which are similar in shape are made on the same machines, PFA (like CC) tends to bring such parts together into the same groups, unless they vary greatly in materials, size, tolerance or requirement quantities, and are made for these reasons on different machines. PFA groups also include parts which are not similar in shape or function if they are made on the same machines, and for this reason is able to find a complete division of all made parts into families.

PFA uses five subtechniques one after the other to plan group technology. Four of these techniques are illustrated in *Figure 15.20*. Company flow analysis (CFA) is used to simplify the flow of materials between the different factories, or major product divisions, in a large company. Factory flow analysis (FFA) plans a simple material flow system between the departments in a factory. It does this by designing departments which, with very few exceptions, complete all the parts they make. Group analysis (GA) divides each department into groups which complete all the parts they make. Line analysis (LA) analyses the flow of materials between the machines in each group, to help in finding the best arrangement for group layout. Tooling analysis (TA) analyses the tooling used on each machine in each group, in order to rationalise the tooling and to find 'tooling families' of parts which can be made at the same set-up.

15.8.6 Factory flow analysis

In most manufacturing enterprises PFA starts with factory flow analysis (FFA). The first need is to record the existing material flow system as illustrated in the simple example given in *Figure 15.21*. The processes used in the factory are given a

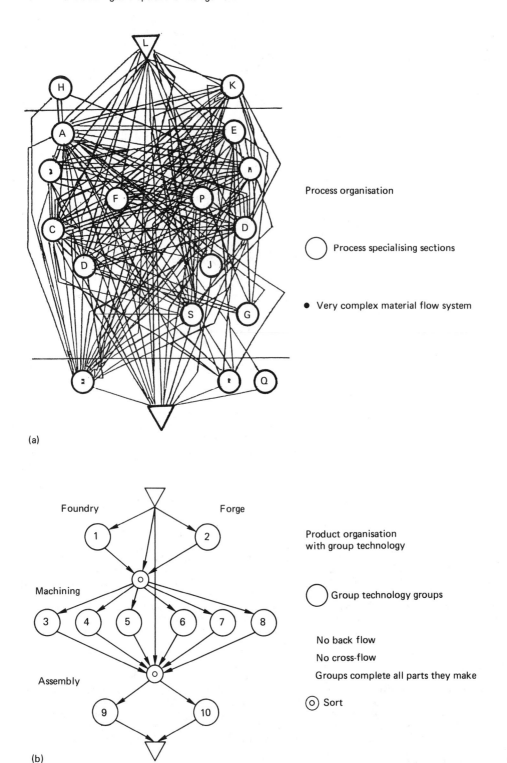

Process organisation

◯ Process specialising sections

● Very complex material flow system

(a)

Foundry   Forge

Product organisation
with group technology

◯ Group technology groups

Machining

No back flow
No cross-flow
Groups complete all parts they make

Assembly

⊙ Sort

(b)

Figure 15.19   The change engendered by changing from process to product organisation. It can be seen that the use of group technology greatly simplifies the material flow system

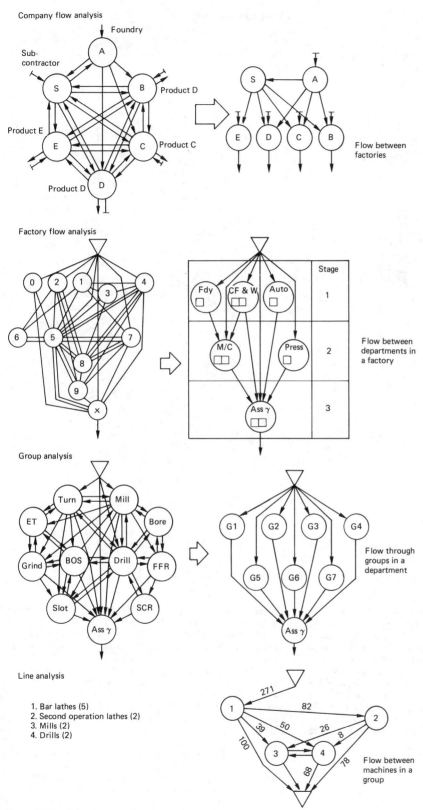

Figure 15.20   Subtechniques of production flow analysis

Process code

Code	Process
1	Blank production
2	Sheet metal
3	Forging
4	Welding
5	Machining
6	Assembly
9	Subcontracting

From/to chart

	To							From	Ends	Sum
From	1	2	3	4	5	6	9	From	Ends	Sum
1	—	10	27	133	57	1	—	228	—	228
2	—	—	1	1	—	43	—	45	—	45
3	1	3	—	5	5	32	—	46	—	46
4	3	1	—	—	8	50	1	63	145	208
5	1	3	3	10	—	80	3	200	—	200
6	—	—	—	1	1	—	—	2	365	367
9	—	—	—	—	3	2	—	5	—	5
Into	5	17	31	50	74	308	4	—	510	1099
Start	223	28	15	58	126	59	1	510	—	—
Sum	228	45	46	208	200	367	5	1099	—	—

PRN frequency chart

PRN	No.	PRN	No.	PRN	No.
1 2 3 4	1	1 5 4	3	4 1 4 6	2
1 2 6	8	1 5 4 6	1	4 1 5 6	1
1 3 2 4	1	1 5 4 9 9 5 6	1	4 2 6	1
1 3 2 6	1	1 5 6	142	4 5 4 6	1
1 3 4	3	1 5 6 4	1	4 5 6	7
1 3 5 6	3	1 5 9 5 6	1	4 6	46
1 3 6	18	1 6	1	5 4	4
1 4	131	2 6	28	5 6	120
1 5 1 3 2 6	1	3 1 2 6	1	5 9 5 6	1
1 5 2 6	3	3 4	1	5 9 6	1
1 5 3 5 6	1	3 5 6	1	6	59
1 5 3 6	2	3 6	12	9 6 3 6	1
Total					510

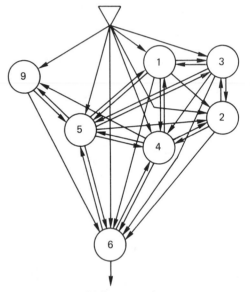

Material flow network

Figure 15.21 Factory flow analysis record of an existing flow system. PRN, process route number

code number. Each part is then given a process route number (PRN) showing the code numbers of the processes used to make it in their sequence of use. A PRN frequency chart and a from/to chart are produced and this information is used to draw a network diagram of the existing material flow system.

Simplification of this system starts by drawing the primary-flow system, which is based on the PRNs which include the largest number of parts. As shown in *Figure 15.22*, this primary-flow system includes a large proportion of the made parts, but uses a very small proportion of the different PRNs. It is much simpler than the original network based on all the PRNs.

The primary network is now simplified by:

(1) combination, and
(2) redeployment of plant.

Combination brings together closely related processes to form departments. As shown in *Figure 15.23*, this combination

makes it possible for many other PRNs and their parts to join the simplified system.

Redeployment of plant then makes it possible for these new departments to complete all the parts they make. As an example, a cold forging and welding department (see *Table 15.3*) sent 24% of the components it made to a machining department for intermediate drilling operations. Moving one radial drill to department 34 made it possible for them to complete all the components they produced. In other factories, boring machines and milling machines for plate edge milling have also been transferred to cold forging and welding departments.

It is not difficult with FFA to convert a traditional department based on process organisation to a group technology department which completes all the parts it makes. Nearly all the machines in each department will be the same after the change as they were before, and the number of exceptional parts requiring replanning seldom exceeds 1%.

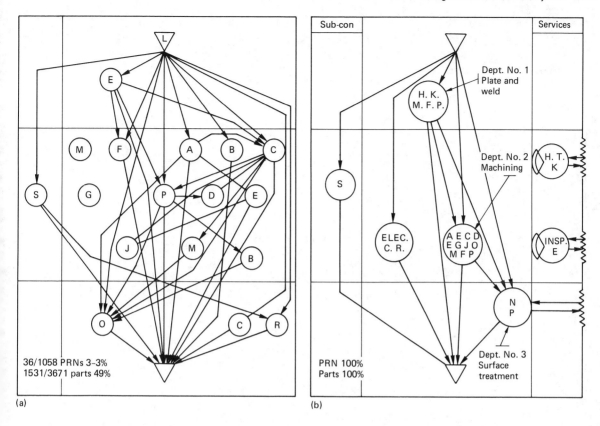

Figure 15.22 Factory flow analysis: (a) the primary-flow system; (b) a simplified flow system

15.8.7 Group analysis

Group analysis is the primary subtechnique of PFA. Providing one starts with a processing department which completes all the parts it makes, and providing this department employs, say, 20 people or more, it is always possible to divide it into smaller groups which complete parts.

As illustrated in *Figure 15.23*, group analysis can be seen as a problem of the resolution of a matrix. *Figure 15.23(a)* is a matrix with parts listed across the top in part number sequence, and machines listed vertically in machine number sequence. Ticks show which machines are used to make each part. It is always possible in such a case to rearrange the sequence in which the parts are listed, and to reorder the listing sequence for the machines in such a way that groups and families are formed. There will usually be some machines of which there are more than one machine of the type, which are needed in more than one group, and there will usually be some exceptional operations which do not fit. In general, however, the division into groups is a natural division which is not difficult to find.

The main problem with the matrix is usually its size. In a company with, say, 4000 current parts in production and 300 machine tools, the matrix is much too big for manual resolution. Even with the computer, it helps if one can divide the machines into sets (types) which can all do any operation on any machine in the set, and can divide the parts into sets (modules), for which there is a high probability that the parts

in any set will all fit happily into one group family. These modules are, in effect, mini-groups.

In the case of the machine tools, the ideal division into types would be such that any operation allocated to one machine of a type, could equally well be allocated to any other machine of the same type. In practice it is seldom possible to change the exsting plant code numbers to follow such an ideal, due to the thousands of changes which would be needed in the plant list, route cards and plant layout records. The analyst should know what would be the ideal arrangement, but cannot generally hope to use it.

15.8.8 Group analysis—the division into modules

In the case of the parts, the total matrix (see *Figure 15.23*) can be divided without great difficulty into a much smaller number of submatrices, or modules. Each module is based on one machine—the *key* machine—from the plant list, and contains all the remaining parts which have not been included in previous modules which have operations on the key machine.

Figure 15.24 illustrates a typical module. It consists essentially of a matrix showing the parts in the module, and the key machine together will all other machine types used to make them. For ease of presentation on the computer, additional details about the machines and the parts are generally shown separately, in addition to the matrix.

Table 15.3 A home-and-away chart

PRN	No.	Home dept.	Away Dept	Away Operation	Away Transfer	Revised PRN
1234	1	34	2	Press	Press to 34 from 2	134
1324	1	34	2	Press	Press to 34 from 2	134
1326	1	34	2	Press	Press to 34 from 2	136
1356	3	34	5	Drill	Drill to 34 from 5	136
151326	1	1	5	Drill	Drill to 1 from 5	136
		34	2	Press	Press to 34 from 2	—
1526	3	5	2	Press	Exception	—
15356	1	34	5	Drill	Drill to 34 from 5	136
1536	2	34	5	Drill	Drill to 34 from 5	136
154	3	34	5	Drill	Drill to 34 from 5	14
1546	1	34	5	Drill	Drill to 34 from 5	146
1549956	1	34	5	Drill	Exception	—
		34	9	SpHT	—	—
		34	9	X-ray	—	—
		34	5	Drill	—	—
1564	1	34	5	Drill	Drill to 34 from 5	14
		34	6	Assembly	Bench operation to 34	—
15956	1	5	9	Grind	Exception	—
3126	1	34	1	Saw	Saw to 34 from 1	36
			2	Press	Press to 34 from 2	—
356	1	34	5	Drill	Drill to 34 from 5	36
4146	2	34	1	Saw	Saw to 34 from 1	46
4156	1	34	1	Saw	Saw to 34 from 1	46
			5	Drill	Drill to 34 from 1	—
426	1	34	2	Press	Press to 34 from 2	46
4546	1	34	5	Drill	Drill to 34 from 5	46
456	7	34	5	Drill	Drill to 34 from 5	46
54	4	34	5	Drill	Drill to 34 from 5	4
5956	1	5	9	Mill	Exception	—
596	1	5	9	Drill	Transfer operation 9 to 5	56
9656	1	5	9	SpHT	Exception	—
			6	Fettle	Bench	—
24	41	—	—	—	—	—

PRN, Process route number. SpHT, Special heat treatment

It will be obvious that some machines—such as drilling machines in machine shops—are going to be needed in most of the groups, and others such as gear hobbing machines will probably be special to one group only. In order to form modules likely to be each in one group only, one must start forming modules based on the special machines. This is done by classifying the machines as follows.

S *Special machines*—there is only one machine of the type. It would be very difficult to transfer an operation on a machine of this type to any other type of machine.

I *Intermediate machines*—these are the same as the S-type machine, but there is more than one.

C *Common machines*—machines likely to be in more than one group, e.g. lathes, mills and drills in machine shops.

G *General machines*—these are widely used machines, of which there are too few to divide between groups, e.g. saw for cutting blanks.

E *Equipment*—this is used to assist manual operations, e.g. vices, surface plates, hand power tools etc.

A special plant list (SPL) is produced, showing the sequence in which the machines in the plant list are to be selected as the key machine. This lists the machines in the sequence SICGE, and in ascending *F*-value sequence within each category, where *F* is the number of parts with operations on each machine.

Figure 15.25 shows a module summary in the form of a matrix containing modules and machines. It is obvious that each module removes at least one machine (the key machine) from the data bank. These items are not considered again in forming later modules. The last 22 modules in *Figure 15.25* contain over 50% of the made parts, reflecting the fact that in most mechanical products most of the parts are simple items made on common machines such as lathes, mills and drills.

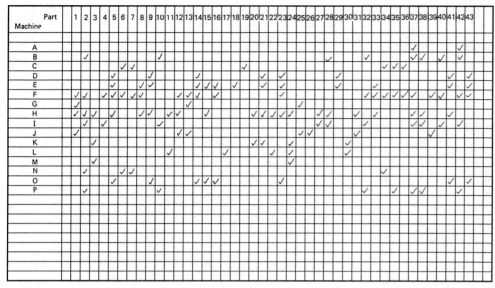

(a)

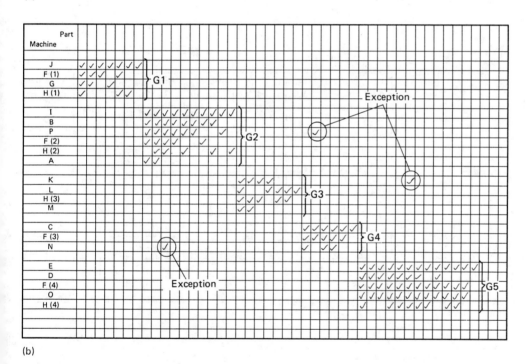

(b)

Figure 15.23 Group analysis matrices: (a) component machine chart; (b) division into groups and families

15.8.9 Finding groups and families

Groups are formed by combining modules. The method involves:

(1) choosing likely groups,
(2) checking the modules and module summary for group feasibility,

(3) selecting modules as nuclei for groups,
(4) using the computer to find best fit groups for remaining modules,
(5) allocating machines to groups,
(6) re-allocating exceptional operations,
(7) eliminating remaining exceptions, and
(8) checking the loads on machines allocated to two or more groups.

Machines

Module No.	Machine No.	Machine Name	N	F	SIC GE	f
08	231202	Beoach	1	9	S	9
24	060210	Cyl. grind	1	52	S	1
25	060901	Hone	1	68	S	1
47	040409	Mill vert.	1	46	S	1
26	040607	Engraver	1	92	S	1
30	130104	Geap cut	3	42	I	1
29	060301	Bore grind	2	41	I	1
63	050101	Jig borer	2	112	I	1
31	040606	Pantograph	2	57	I	1
45	010125	Lathe ctr.	1	35	C	1
56	060202	Cyl. grind	4	72	C	1
57	010105	Lathe ctr.	4	80	C	1
58	010123	Lathe ctr.	1	87	C	1
48	040512	Mill univ.	2	47	C	1
68	010119	Lathe ctr.	2	205	C	2
67	050902	Drill coor.	2	146	C	1
65	040510	Mill univ.	2	125	C	1
70	040504	Mill univ.	6	347	C	2
69	010126	Lathe ctr.	7	260	C	2

Matrix

Item	Part No.	231 202	060 210	060 901	040 409	040 607	060 301	130 104	040 606	050 101	010 125	040 512	060 202	010 105	010 123	040 510	050 902	010 119	010 126	040 504
1	28947273	⊗					⊗							⊗						
2	28948636	⊗															⊗			⊗
3	28949279	⊗																	⊗	⊗
4	28949303	⊗																	⊗	
5	28949428	⊗	⊗			⊗								⊗					⊗	
6	38444311	⊗		⊗	⊗			⊗		⊗	⊗	⊗				⊗			⊗	
7	39484005	⊗																		
8	39454624	⊗																		
9	39454731	⊗					⊗							⊗						

Parts

Item	Part No.	Part name	Mtl type	Mtl form
1	28947273	End cover	MS	Bar blank
2	28948636	Spacer	MS	Bar blank
3	28949279	End cover	MS	Bar blank
4	28949303	Spacer	MS	Bar blank
5	28949428	Bush	MS	Bar blank
6	38494311	Gear	MS	Forging
7	39454095	Bush	MS	Bar blank
8	39454624	Bearing	MS	Bar blank
9	39454731	Spacer	MS	Bar blank

Figure 15.24 A module

	Machine					Module number and (f)																																																																													
Key	No	Name	N	F	SICGE	1	5	6	7	8	9	10	11	12	13	14	15	16	17	18	19	20	21	22	23	24	25	26	27	28	29	30	31	32	33	34	35	36	37	38	39	40	41	42	43	44	45	46	47	48	49	50	51	52	53	54	55	56	57	58	59	60	61	62	63	64	65	66	67	68	69	70	71	72	73	74	75	76	77	78	79	80	81	82	83
5	080301	Slotter big	1	1	S		1																																																																											
6	060104	Surf. grind	1	2	S			2																																																																										
7	040906	Borer pre set	1	2	S				2																																																																									
8	040605	BMW brake	1	2	S					2																																																																								
9	010101	Ct lathe long	1	3	S						2																																																																							
10	040603	Copy mill	1	4	S							4																																																																						
11	060102	Surf grind	1	5	S								5																																																																			
12	040604	Copy mill	1	5	S							1		4																																																																
13	130401	Gear hob	1	6	S										6																																																																
14	010103	Ct lathe long	1	6	S											6																																																														
15	060205	Cyl. gr v. big	1	6	S										2		4																																																														
16	010106	Ct lathe long	1	7	S				1		2							4																																																													
17	060302	Bore gr shaft	1	7	S													3	4																																																												
18	060502	Bore gr	1	8	S															8																																																											
19	080303	Slotter small	1	8	S														1	1	6																																																										
20	108102	Broach	1	9	S										4							5																																																											
21	010303	V. Borer big	1	9	S		1				1												8																																																									
22	231202	Broach	1	9	S																			9																																																								
23	060902	Hone	1	14	S										1										13																																																							
24	010304	V. Borer v. big	1	14	S				1	1																12																																																						
25	060401	Thread grind	1	15	S										1					2							12																																																					
26	040908	Borer v. big	1	16	S		1																			3		11																																																			
27	010702	N.C. Turn ctr	1	17	S						1								1	2	1						1		11																																																						
28	231212	Broach	1	22	S								1						1		1									19																																																			
29	010305	Borer big	1	25	S																					2		2			21																																																	
30	040905	Borer v. big	1	26	S		1			1								2							1				5		2	12																																															
31	040511	V. Mill small	1	30	S																												30																																													
32	060206	cvl gr. big	1	30	S					1			1	1	4					2		1					1							19																																														
33	010705	Turn ctr big	1	35	S									1	4			2	1	1	1	3							2		1			4	15																																													
34	060103	Surf grinder	1	38	S																		1					1			3			2		31																																												
35	040514	Mill 3 spindle	1	41	S																																41																																											
36	040413	Mill CNC	1	41	S															2					1											4		34																																										
37	040904	Borer v. big	1	68	S		1		2							5				1			3			1		1	3		1	1							49																																								
38	060210	Cyl grinder	1	68	S										1									1			2			1			8	1	1		1			57																																								
39	060901	Hone	1	78	S																1			1			2			1			3	1	1	1	1			17	43																																						
40	040409	Mill vert NC	1	106	S																			1	1					1					3	1	1	1		5	5	87																																					
41	040607	Engraving M/C	1	135	S							1					1							1						1			3							16	16	8	88																																				
42	040909	M/C centre	2	25	I																1				2										2			1		2		1		16																																			
43	010128	Lathe Micron	2	27	I							2					2																2								11		1		9																																		
44	060101	Surf grinder	2	32	I					1										4					2									3	1	2	1			2				1		15																																	
45	130104	Gear cut M/C	3	37	I																			1						1			2							8	84		1																																				
46	040907	Borer CNC	2	46	I		1										3		1	2			1			4		1	4		3	1			1	1			3			6						14																																			
47	060301	Bore grinder	2	52	I					1			1		1					4	1	2		1	1		3		1	1	1	1		9	1						1					3			19																																	
48	050101	Jig borer	2	53	I																2			1						1				1	1	1	1	1		5	1	17	3							18																																
49	040606	Pantograph	2	84	I							1								1	1	1		1			3			1				1	1	4	2	2		1	1	1							1	1	60																																
50	060202	Cyl grinder	3	86	I									2	6		2	1	1	2	2	1		1	1		8		1	1				8	2									1		2			14		1	29																															
51	040501	Mill univ	1	15	C					1					2		1					1						1						2										2		2		1					2																														
52	010129	Lathe capstan	1	18	C																						2										2			2		1								1				10																														
53	020111	Lathe revolver	1	36	C																																			1			1											1	33																													
54	040406	Mill vert NC	1	38	C																														1	1			1			7		1		1		2	1							83																														
55	040101	Mill horiz Sm	3	41	C																						1								1	1		1								1			3			3			1		30																													
56	010125	Lathe centre	1	43	C																			1			1			1			1				1			8	3	15		1								1		1				7																											
57	040401	Mill vert	1	47	C			2							3		2													1	1				1	1		2			1		4		2						1				2				27																										
58	050308	Drill 4 spind	1	50	C							1	2												1										1	1		3	2										1	3	2	1					3			29																									
59	040408	Mill vert big	1	55	C													1		1			2			1		1			1	1						2				1				3		5			3					1			3		29																								
60	050201	Drill rad big	2	55	C														1		1					2		2			1			1	1	1			1										1		1		1			1			1		4	34																							
61	040407	Mill vert med	1	65	C															1	1	1												7	1	2							1							1	4					1				4		2	39																						
62	020107	Lathe capstan	1	73	C				1																3		4																1								2	6					7			3				47																					
63	050301	Drill 6 spind	1	73	C										2										2		2							2	2	3		3			1			1		3			2		3	1			2		4		1	1			1	10	24																				
64	051003	Drill capstan	1	74	C																																1	1	2					1		1			2	1	2									2		1	1		6	53																			
65	050801	Drill coor big	1	87	C								1				1			4	1		1			1			2		4			6	1		2		4							2					1	4	2			3			4		1	2	2	1			37																		
66	010707	Turn ctr	1	88	C							2		2											1		2										1	3		4	4	2	1			1			1	2	2	6		1			5			4			1	1	1	1		40																	
67	010105	Lathe ctr long	2	92	C			2	1	1			1		4		3	3	2	3	1	5	2	1		4			7		2			6	2	3			5									3			2	2	2			1						2				1	3		18																
68	010127	Lathe bar/chuck	1	93	C																												5		3		4			16	7	9	2	1						1				1											1					43															
69	010123	Lathe ctr	1	109	C																1			1						1			4	1						14	18	13	21							1	3	1																		1	29														
70	051003	Drill & tap	1	117	C							1				1				2									1		2			2	2	2	2					1		1		1		1	4		3					1			1	2			3	1	2	2		3		1		75													
71	040512	Mill univ NC	2	133	C																2						1			1			2				2			6	7	18	2	1						5				4		2		3		2										16	1	2	56												
72	010704	Turning ctr	2	147	C							1	1		1					1					1		3		1			1		10	5		5	2	2	4		19		3		2		1	13	4	1	3					2			5	1	1	3			4	2	3	1			8		33											
73	051101	Tapper v. small	2	161	C							1		1	1			1		1	1				3		2								1	9	1	7			1			2		3			3	3	6	3			2	1	4			5	2		5	6	2	7		3				4		1	69											
74	051001	Drill coor NC	1	174	C																													3	1	4		2								1				1	6	1					3			4	1		9		5	1	2		1					7	9	112									
75	010119	Lahte ctr sheet	2	217	C							2		1	1	1	3	1	1	4	1	1		2	6		4		2	2		2		8		3		4				3		1		2		1	16		5	12				1	4	1		6			2	2	2	2	2	12	1			6		14	1	4	68								
76	050902	Drill coor	2	230	C													1			2			1	4		2		1	1						8	1	1					2	1		6			4	1	11	6					5						2	5	3	3		11	1			3		4	79		9	82							
77	040510	Mill univ	2	247	C																			1	1					1			1	1		1	8			17	19	30				1				13	7	1		3	2			3		2			1		1			1		9	15		2						76						
78	040504	Mill univ sm	7	272	C							2			2			1		4	2	1		2	2		5			2						8	5	9			1			4		6			4	3	12	8				4	4			12		6	5	5	9	12		8	3			17		8	34	10	17	28		7					
79	010126	Lathe ctr	7	387	C									1						1	1			2	3		2			2		2	14	1		1				49	39	30	23	1						5	2	2		4	7			4		2			1		1		11	1		19	28	1	3						11		113				
80	170502	Saw band	1	2	G							1																																																	1				1																					
81	070601	Hand grinder	3	8	G							1		2						1	1										1		1				3	1																											1					1						1							4		
82	070502	Linisher	6	243	G							1		1											2			1				1	2				20	11				1				3				2	10	6				1	11		1	9	8	4	10	12	3	7	7	6	2			9	3	3	11	12	5	13		6				67	
83	170104	Saw circ.	2	526	G																		1		1																	31	13	5					2	5	20					10	6	2	8	7	4	2	16		9	25	11					37	28	5	47	7	1	46	26	2	3			15	94

Note: Modules 1 to 4 inclusive and their key machines were eliminated early in the analysis.
Figures in matrix show (f) = No. of parts with ops. on each machine

50% parts
26% modules

Figure 15.25 Module summary

In choosing the groups, one starts with a scale check. In practice it has been found that groups employing 10–15 people, with perhaps one or two larger groups for complex parts, can usually be formed without having to buy any new plant. If then an existing machine shop employs 90 people, one would expect to form between six and nine groups.

Knowing the number of groups required, one finds likely groups in two main ways. First by direct examination of the modules and the module summary, and second on the basis of past experience.

In the first case, direct examination will usually find small sets of modules which use the same S and I machines, providing one or two nuclei for likely groups. Direct examination can also be used to find one or two simple groups, using only common machines to make simple parts. In the second case, experience has shown that some types of part have formed groups in many different factories. If such parts are required in reasonable quantities, it is likely that they will again form groups. Examples of this second case are: small rotational parts made on bar lathes; gears; heavy machine beds, gear cases, etc.; and simple turned milled and drilled parts. In this case it is not difficult to find one or two modules to form the nucleus of each group.

The computer is used to find the best-fit group for each of the remaining unallocated modules, and the machines are allocated to the groups in proportion to the number of parts they machine. This step finds large numbers of exceptions, most of which can be eliminated as shown in *Figure 15.26*, by the re-allocation of operations to other similar machine types which are available in the groups. There may then be one or two difficult exceptions which have to be replanned, and a final load check is advisable, to ensure that machines required in more than one group are not overloaded in one of them.

Line analysis (LA) is then used to provide the information needed for plant layout, and tooling analysis (TA) is then used to plan the group tooling and to find 'tooling families'.

15.9 Assembly groups

Assembly groups are similar to group technology component processing groups, in that both types complete all the components they make, and both are dedicated to simple material flow systems and to the minimisation of throughput times.

Component processing and assembly groups have different historical backgrounds. The group technology component processing groups were developed by engineers to improve production efficiency. Assembly groups owe their origin to behavioural scientists and were introduced originally to improve workers' morale and job satisfaction.

Research carried out by the Tavistock Institute of Human Relations in the late 1940s had shown that workers preferred to work together as a team to complete products, rather than work alone as specialists in particular processes. In the early 1950s, the Norwegian government was rebuilding their industry after the war. They decided that they wanted to introduce group working and engaged two scientists (Trist and Emery) from the Tavistock Institute and one from Norway (Thorsrud) to introduce groups into Norwegian factories. Together they formed groups in over 100 factories. They had some successes, but failed to gain the full potential advantages of the change. The experiment ran out of steam and was dropped, but it generated a wide interest in group working throughout the industrial world.

In Sweden about this time, Volvo were having difficulties in recruiting Swedes to work on their assembly lines. They brought in foreigners, mainly from Finland, to do this work.

Gullenhamer, the president of Volvo, believed that there could be no long-term future for a company which had to bring in foreigners to do work which its own nationals refused. He instructed his production engineeers to study the work done in Norway, and to design an assembly factory for Volvo cars where Swedes would be happy to work. The result was the Volvo assembly plant at Kalmar, based on group assembly.

Other companies which were experiencing problems with long machine-paced assembly lines and who turned to groups at this time, included Philips in Holland, making television sets, and Olivetti making calculators in Italy. It was expected at the time that the change to groups would increase costs, but it was seen as necessary in order to eliminate the alienation of labour. In the event, the change in nearly every case reduced costs rather than increased them.

15.9.1 Planning assembly groups

There are five main ways of dividing assembly work into groups. These are illustrated in *Figure 15.27*. The first three types of group complete all the products they assemble. In the last two types—branched groups and groups in series—each group completes a major part of the product, and a number of groups work together to complete the finished product. It should be noted that even in the case of branched groups and groups in series, all assembly groups produce continuously. They do not assemble intermittently in batches for stock at any stage in assembly.

Inside the groups illustrated in *Figure 15.27* there are again five different ways in which the work can be organised. These are illustrated in *Figure 15.28*. With mono working, each worker completes products. With parallel working, several workers each complete the same type of product. With team, branched and line working, workers collaborate to complete products.

A number of groups from practice are illustrated in *Figure 15.29*. The first shows groups making television sets at one of the Philips factories at Eindhoven, Holland. Groups in parallel each made the same set. Inside the groups the work was organised for line flow. The second illustration shows groups making electronic calculators at an Olivetti factory at Ivrea, Italy. The groups were designed as branched groups, one assembling the printer unit, one the keyboard, one finishing the product, and a final group for testing. Inside the groups, some used mono working, and others used parallel and line working.

Finally, *Figure 15.29(c)* shows three of the 21 groups in the Volvo, Kalmar car assembly plant. Each group completed a logical part of the assembly, e.g. the suspension, the brakes, engine installation, electrics, upholstery, or doors, etc. The groups are organised internally for branched work, some workers making subassemblies, and others installing them on the cars. A buffer stock of up to three cars is allowed between groups, to enable groups to vary their pace of work and to take occasional breaks. To improve group team spirit, each group has its own rest room (coffee bar), showers, toilets, cloakroom and a sauna bath.

The design of assembly groups is not as complicated as the design of component processing groups. A special technique like production flow analysis is not necessary for planning. All one needs in designing assembly groups is to remember the objectives of completing products, minimum throughput time, minimum stocks and continuous flow.

15.9.2 Traditional assembly organisation

If one goes back far enough in time, factories were small and all assembly was based on mono groups. In the last 60 years,

(a) After machine allocation   (b) After reallocation of operations

MOD	NO	N	F	Cd	NAME	1	2	3	4	5	6	7	8	9	1	2	3	4	5	6	7	8	9
14	LC301	1	16	S	Lathe ctr.		1/15			(1)						1/21							
57	LC217	4	80	C	Lathe ctr.	(6)	(6)	(3)	1/9	1/13	(2)	2/41						1/11	1/18		2/67		
68	LC593	2	205	C	Lathe ctr.	(14)	(6)	(1)	(2)	(4)	(13)	1/132	(4)	1/29							1/132	(4)	1/49
58	LC092	1	87	C	Lathe ctr.	(2)	1/79					(6)				1/79							
45	LC067	1	35	C	Lathe ctr.	(1)	1/34									1/34							
69	LC253	7	260	C	Lathe ctr.	1/3	5/232					(2) 1/20		(3)	1/26	5/232					1/29		
41	LC201	1	27	C	Lathe ctr.		1/27									1/27							
46	LC202	2	39	C	Lathe ctr.		2/39									2/51							
19	BV013	1	26	S	Bore Vert		1/21			(5)						1/21							
17	BV096	1	24	S	Bore Vert			1/22		(2)							1/22						
12	BV111	1	13	S	Bore Vert				(1)	1/12								(1)	1/19				
35	TC019	1	14	C	Turn centre		(1)		(3)	1/10								1/9 →o					
64	TC020	2	125	C	Turn centre	(5)	1/23		(1)	(7)	(5)	1/79		(5)		1/41					1/79		
44	TC021	1	33	C	Turn centre	(1)	(5)	(1)	(4)	1/15	(2)	(4)		(1)				1/33					
59	TC083	1	91	C	Turn centre	(8)	(10)	(1)	(1)		(5)	1/56		(10)							1/66		
32	TC064	1	8	C	Turn centre		(1)					1/6		(1)	1/16 ←——o								
61	LP772	1	98	C	Lathe Cap	(2)	(1)	(1)		(1)	(7)	(10)		1/76		(1)				(7)	(10)		1/76
43	MV053	1	30	C	Mill Vert	(3)	(1)			(1)	(1)	(5)		1/19								1/19	
42	MV058	1	27	C	Mill Vert	(1)	1/22			(4)					1/10 →o								
49	MV171	1	49	C	Mill Vert		(2)	(6)		(10)		1/31							1/16 ——o				
47	MV234	1	46	C	Mill Vert	(1)	1/42					(3)				1/68							
52	MV169	1	66	C	Mill Vert	(5)	(1)			(1)	(4)	1/55									1/94		
34	MU069	1	13	C	Mill Univ.			(2)	(1)	1/10								(1)	1/13				
70	MU157	6	347	C	Mill Univ.	1/17	(○)	(1)	(2)	(3)	(11)	2/123	2/121	1/57	1/19		(2)				2/123	2/121	1/57
65	MU231	2	125	C	Mill Univ.	(2)	1/87					1/36				1/87					1/59		
36	MU232	1	15	C	Mill Univ.		1/15									1/51							
48	MU309	2	47	C	Mill Univ.		1/24					1/23				o→ 1/9			1/16 →o				
02	MS007	1	2	S	Mill 3sp.		1/2									1/2							
40	MC015	1	19	C	Ml. copy	(2)				(1)	(2)	1/14							(1)		1/26		
38	MC037	1	16	C	Ml. copy	(1)					(3)	1/12								1/8 ←o			
31	MP006	2	57	1	Pantograph	(4)	(2)	(1)	(2)		(3)	2/45			1/4 ←——	(1)	(2)			o /46			
26	XX010	1	92	S	Engraver	(1)	1/91								(1)	1/91							

Key ◯ = Exception (no/machine):

No. of machines per group — (top)
No. of parts — (bottom)
(box: 1 / 15)

Figure 15.26 Allocation of machines to groups and re-allocation of operations to machines

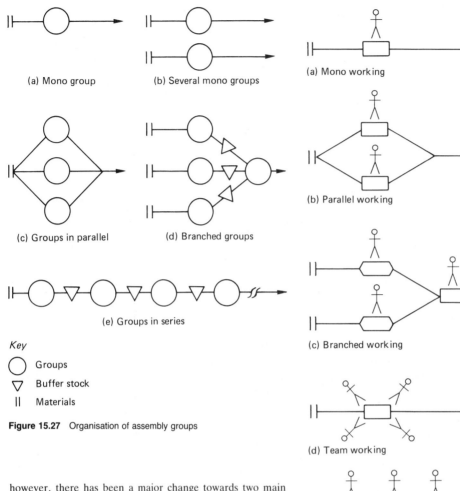

(a) Mono group

(b) Several mono groups

(c) Groups in parallel

(d) Branched groups

(e) Groups in series

Key

◯ Groups

▽ Buffer stock

‖ Materials

Figure 15.27 Organisation of assembly groups

(a) Mono working

(b) Parallel working

(c) Branched working

(d) Team working

(e) Line working

Key

▢ Assembly

⬡ Subassembly

Assembler

Figure 15.28 Work organisation within groups

however, there has been a major change towards two main types of assembly:

(1) machine paced assembly lines, and
(2) progressive assembly in batches.

Machine paced assembly lines are now being replaced by groups, mainly because machine paced lines are unpopular with workers. It has been found from experience, however, that such lines are very inflexible and that balancing losses are much larger than seems likely at first sight. Groups are generally more efficient than mechanised lines, and this is likely to accelerate the change.

Progressive assembly in batches is a more recent development than machine paced lines. With the development of flow control methods of production control, in which ordering is based on the calculation of parts requirements from a production programme, a need arose for parts lists (bills of material) to be arranged hierarchically to show assemblies, subassemblies, sub-subassemblies and so on.

Having this information readily available, many engineers decided that the most economical method of assembly would be to make assemblies, subassemblies and so on, progressively in batches for stock. *Figure 15.30* shows an example with five stages of assembly. After fixing a batch quantity and estimating the duration times for each stage, it was estimated that the throughput time to complete a product was 4 months. It was found that a skilled fitter with two assistants could complete one product in 2 days. This indicated that a change to group

assembly could reduce throughput times per product to 2 days, and decimate the investment in stocks.

It is true that manufacturing in large batches for stock reduced direct labour costs, but these savings were more than offset by increases in the stock investment and in indirect costs. Flexibility is very low in the case of progressive assembly in batches, leading to high obsolescence losses. Major savings are being achieved in practice in the explosive (assembly) industries, by changing from progressive assembly in batches to group assembly.

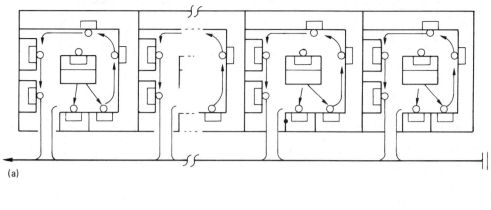

(a)

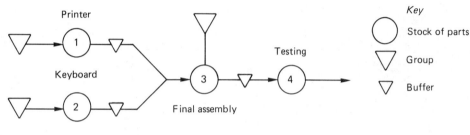

Key

○ Stock of parts

▽ Group

▽ Buffer

(b)

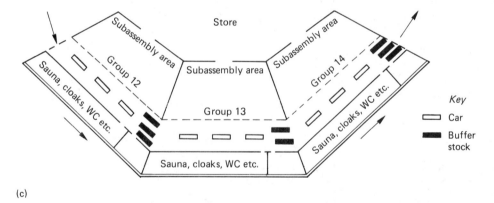

(c)

Figure 15.29 Examples of group assembly. (a) Groups in parallel, Philips, television sets; (b) branched groups, Olivetti, calculators (c) groups in series, Volvo, cars

15.10 Marketing

Marketing and sales are not an integral part of manufacturing management, but are so closely related to it that one cannot consider one without the other. There is no point in manufacturing products if they cannot be sold, and before they can start planning for manufacture, the manufacturing functions must receive from marketing a forecast of future sales, and details of sales orders as they are received.

In jobbing and contract industries (see *Figure 15.3*) the products are special to each order, and the design, production planning and production control functions are all involved with marketing. When companies make standard products,

however, production control on its own is the principal manufacturing function associated with marketing.

Two main sales programmes are required. First, an annual programme used in financial planning, the planning of purchase contracts, etc.; and second, a series of short-term period programmes produced at regular intervals through the year, used ideally to regulate manufacture and the delivery of purchases.

15.10.1 The long-term sales programme

The first requirement of manufacturing from the marketing function is an annual or other long-term sales forecast or sales

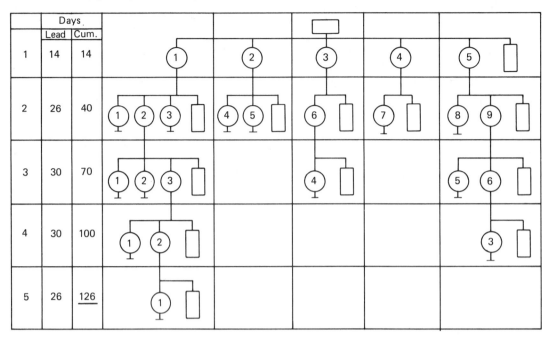

Key

☐ Final product

◯ Assemblies, subassemblies, etc.

▯ Parts

Figure 15.30 Batch assembly in progressive stages

Table 15.4 Annual sales programme

Product	J	F	M	A	M	J	J	A	S	O	N	D	Total
A	100	100	100	110	120	130	130	110	100	100	100	100	1300
B	350	350	370	370	390	400	390	375	360	350	350	350	4405
C	210	200	185	180	175	175	175	180	190	200	210	210	2290
D	195	175	160	160	160	160	160	160	170	180	195	195	1875
E	220	210	200	180	180	180	180	180	200	220	230	230	2410

programme, for all products sold by the company. This needs to show the quantities of each product which the company hopes to deliver to customers in a series of future planning periods (see *Table 15.4*).

As it is not given to human beings to be able to foretell the future, there is always an element of guesswork in such programmes. They are best seen as targets for future action, rather than as forecasts of a pre-ordained future. There are three main techniques which can be used to help in the selection of viable targets:

(1) statistical forecasting,
(2) market research, and
(3) business forecasting.

Statistical forecasting is based on the idea that the affairs of men continue to follow the same course unless they are subjected to strong external influences. It is believed, therefore, that one can project from a record of past results to determine probable future results. This idea would have been more useful to our grandfathers than it is to us, because the rate of change of the markets of the world is very much greater today than it was 50 years ago.

Figure 15.31 shows how statistical forecasting is used to predict an annual sales total for a product. Moving annual totals (MAT) of sales (the totals at the end of each period for the previous 12 months), are plotted for each period in previous years. The MAT curve smoothes random and seasonal variations. The total for the next 12 months is found by

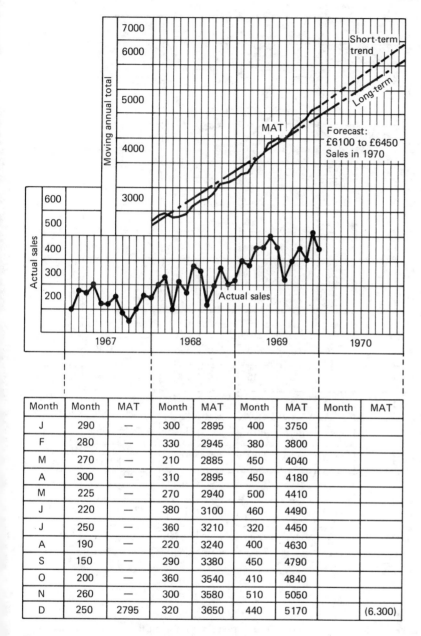

Figure 15.31 Use of moving annual total (MAT) to forecast sales

Month	Month	MAT	Month	MAT	Month	MAT	Month	MAT
J	290	—	300	2895	400	3750		
F	280	—	330	2945	380	3800		
M	270	—	210	2885	450	4040		
A	300	—	310	2895	450	4180		
M	225	—	270	2940	500	4410		
J	220	—	380	3100	460	4490		
J	250	—	360	3210	320	4450		
A	190	—	220	3240	400	4630		
S	150	—	290	3380	450	4790		
O	200	—	360	3540	410	4840		
N	260	—	300	3580	510	5050		
D	250	2795	320	3650	440	5170		(6.300)

projection, using the least-squares method. By basing projection on both annual and short-term past figures, a range of possible targets can be found.

Once the annual sales target has been fixed, the next task is to distribute this total between future time periods. *Figure 15.32* shows how to find the average percentage of annual sales which is likely to be sold each period.

Market research, the second of the three sales programming techniques listed above, uses questionnaires addressed to salesmen and to potential customers, to determine future sales trends and to fix sales targets.

Finally, business forecasting bases estimates of future sales on related published statistics. For example, records of the number of ships hulls laid down in shipyards, have been used to indicate future sales for ships fittings, and Ministry of Agriculture statistics of the number of acres of farmland 'put down to grass', have been used to predict future sales of grass machinery. In practice there are only very limited possibilities of using this method.

These three techniques do not provide a full scientific justification for the choice of any particular sales programme. There is still a need for flair and experience in this work, and

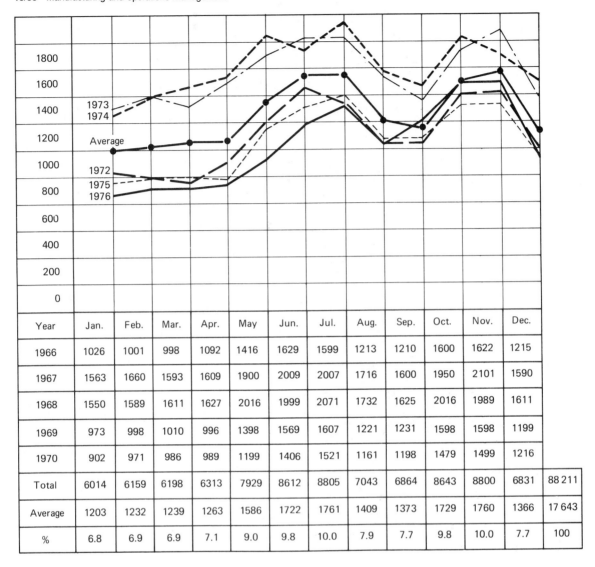

Year	Jan.	Feb.	Mar.	Apr.	May	Jun.	Jul.	Aug.	Sep.	Oct.	Nov.	Dec.	
1966	1026	1001	998	1092	1416	1629	1599	1213	1210	1600	1622	1215	
1967	1563	1660	1593	1609	1900	2009	2007	1716	1600	1950	2101	1590	
1968	1550	1589	1611	1627	2016	1999	2071	1732	1625	2016	1989	1611	
1969	973	998	1010	996	1398	1569	1607	1221	1231	1598	1598	1199	
1970	902	971	986	989	1199	1406	1521	1161	1198	1479	1499	1216	
Total	6014	6159	6198	6313	7929	8612	8805	7043	6864	8643	8800	6831	88 211
Average	1203	1232	1239	1263	1586	1722	1761	1409	1373	1729	1760	1366	17 643
%	6.8	6.9	6.9	7.1	9.0	9.8	10.0	7.9	7.7	9.8	10.0	7.7	100

Figure 15.32 Seasonal distribution of sales

all that can be said for the techniques is that they can help us to avoid making silly mistakes.

15.10.2 The series of short-term programmes

Programming based on a series of short-term programmes, planned progressively at short-term intervals, is known as 'flexible programming'. There are two main ways of planning these programmes. The first uses 'order accumulation' followed by smoothing (see *Figure 15.36*). The second method, used when throughput times for manufacture are long and there are wide variations in delivery promise times, plans a new sales programme each period, which attempts to deliver all orders by due-date, and at the same time to smooth production by bringing forward the manufacture of orders for future delivery into periods when the requirements for deliv-

ery in that period do not use all the available capacity (see *Figure 15.37*).

15.11 Production control

Production control is the function of management which plans, directs and controls the material supply and processing activities of an enterprise. It can also be described as the function of management which regulates the flow of materials through the material flow system of an enterprise. The material flow system can be seen as a static system, or arrangement of routes, and production control now regulates the dynamic flow of materials through the system.

As shown in *Figure 15.33*, production control takes place at a series of three planning levels, known as programming,

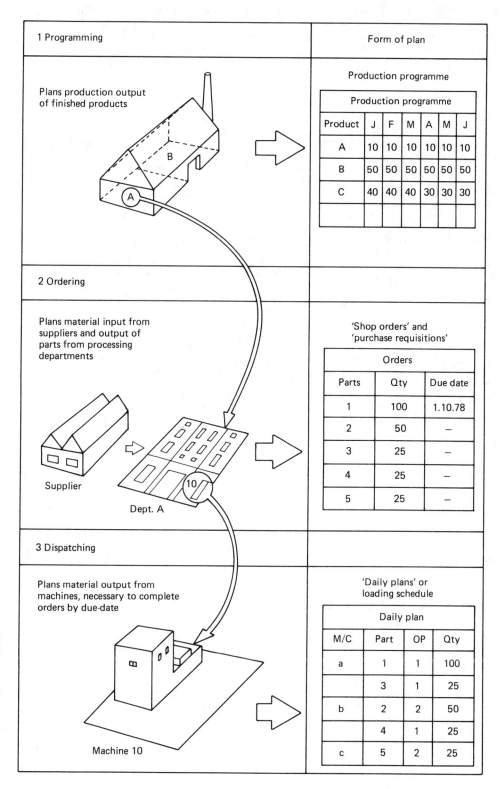

1 Programming	Form of plan

Plans production output of finished products

Production programme

Production programme						
Product	J	F	M	A	M	J
A	10	10	10	10	10	10
B	50	50	50	50	50	50
C	40	40	40	30	30	30

2 Ordering

Plans material input from suppliers and output of parts from processing departments

'Shop orders' and 'purchase requisitions'

Orders		
Parts	Qty	Due date
1	100	1.10.78
2	50	–
3	25	–
4	25	–
5	25	–

Supplier

Dept. A

3 Dispatching

Plans material output from machines, necessary to complete orders by due-date

'Daily plans' or loading schedule

Daily plan			
M/C	Part	OP	Qty
a	1	1	100
	3	1	25
b	2	2	50
	4	1	25
c	5	2	25

Machine 10

Figure 15.33 Levels of production control. (Courtesy of William Heinemann Ltd)

ordering and dispatching. Programming plans the output of finished products from the enterprise. Ordering next, plans the output of finished parts from company workshops and the input of materials and bought parts from suppliers, necessary to complete the production programme. Dispatching plans the completion of operations in workshops, leading to the completion of orders for parts and for assembly.

It will be realised that all three of these levels are concerned with scheduling, or with planning the start and/or finishing times for work tasks. Scheduling can be seen as the principal skill required for production control.

Although all three of these levels are required for all types of enterprise, their complexity varies with the type of industry. *Figure 15.34* repeats the classification of production types given in *Figure 15.2* and shows the effect of each type on the work at each production control level.

It should be noted that only the explosive industries need to order both components and material items. In process, implosive and square industries, the products in the programme are bulk materials, or components. Only bought materials are ordered at the second, or ordering level. Process industries are the simplest for production control. Programming and the ordering of materials, deal in this case with very few items, and the mechanised processing system generally covers most of the needs of dispatching automatically. Looking now at the next level of industrial classification, illustrated in *Figure 15.3*, contract explosive industries are the most difficult. They make small numbers of very complex products. They require the coordination of many different processes, generally including design and process planning, using project management techniques such as critical path analysis or the programme evaluation and review technique, and usually include the control of some jobbing workshops (JI, JS or JE). In implosive industries, programming and dispatching are complex, but the ordering of the small number of material varieties is generally

simple. Outside the contract explosive industries, the most difficult problems for production control occur in the square and implosive industries of the jobbing, batch and line types (JS, JE, BS, BE, LS and LE).

Production control also encompasses three main types of feedback control. First, *progressing* monitors material conversion at each level and feeds back information about late completion or excess production against programmes, orders and operation schedules. *Inventory control* monitors stock levels and feeds back information about shortages and excess stock. *Loading* measures and compares load and capacity, and feeds back information about overload and spare capacity. The three levels and these three feedback controls provide the main subjects to be studied in production control.

15.11.1 Programming—general

Production control programming is mainly concerned with the planning of production programmes, showing what products are to be made in future periods of time. However, as there is no point in making products if they cannot be sold, production control generally starts with a sales programme showing products which the sales department wish to send to customers in the future. Due to seasonal variation in sales, and the economic advantages of 'smoothing' to achieve an even rate of production, the sales and production programmes are generally different. Knowing the sales and production programmes, it is simple to calculate the probable levels of finished product stock at the end of each period, and to produce the related stock programme. *Table 15.5* shows such a set of sales, production and stock programmes, covering a 'term' ('programme horizon' in the USA) of 1 year for a product.

The programme term is the period of time into the future covered by a set of programmes. Most companies need programmes covering at least three main terms:

Key@ m = No. mtl. varieties p = No. prod. varieties		Process m ▭ p	Implosive m ◁ p	Square m ▢ p	Explosive m ▷ p
Material input Product output Material flow type		Bulk mt. Bulk mtl. Line flow	Bulk, or gen. mtl. Gen. mtl., or comp. Line flow	Components Components Batch flow	Gen. or sp. mtl. Assemblies Batch flow
Examples		B Cement E Ore treat (mine) F Milk F Sugar F Distilleries C Gases (oxygen, nitrogen) B Bricks B Timber F Breweries Tanneries Paper	E Foundries D Potteries D Glass C Decorative laminates T Spinning, fibres E Brake linings Printing T Knitting F Bakeries E Rolling mills E Wire drawing	E Jobbing machine D Diamond cutting E Machine overhaul T Dyeing textiles T Finishing textiles E Heat treatment X-ray E Painting E Electroplating E Metal spraying Polishing	E Automobile Electronics D Consumer durables E Machine tools E Internal combustion engines E Electrical T Weaving Clothing and shoes C Chemical dyestuffs D Furniture E Welding
1. Programming – annual		✓	✓	✓	✓
Programming – short term		✓	✓	✓	✓
2. Ordering – material		✓	✓	✓	✓
Ordering – parts		✕	✕	✕	✓
3. Dispatching		✕	✓	✓	✓

Figure 15.34 Types of industry and production control. F, food; C, chemicals; T, textiles; E, engineering; B, building; D, domestic

Table 15.5 Sales, production and stock programmes

	Week number												
	1	*2*	*3*	*4*	*5*	*6*	*7*	*8*	*9*	*10*	*11*	*12*	*13*
Sales	22	22	22	22	22	22	22	22	22	22	22	22	22
Production	23	23	23	23	23	23	23	23	23	23	23	23	23
Stock (13)	14	15	16	17	18	19	20	21	22	23	24	25	26

	Week number												
	14	*15*	*16*	*17*	*18*	*19*	*20*	*21*	*22*	*23*	*24*	*25*	*26*
Sales	22	23	24	25	25	25	26	30	28	32	32	32	32
Production	23	23	26	26	26	26	26	26	28	28	28	29	29
Stock	27	27	29	30	31	32	32	28	28	24	20	17	14

	Week number												
	27	*28*	*29*	*30*	*31*	*32*	*33*	*34*	*35*	*36*	*37*	*38*	*39*
Sales	32	32	32	30	30	30	28	26	26	24	24	23	23
Production	29	29	29	29	29	29	26	26	26	24	24	23	23
Stock	11	8	5	4	3	2	0	0	0	0	0	0	0

	Week number												
	40	*41*	*42*	*43*	*44*	*45*	*46*	*47*	*48*	*49*	*50*	*51*	*52*
Sales	22	22	22	22	22	22	22	22	22	22	22	22	22
Production	23	23	23	23	23	23	23	23	23	23	23	23	23
Stock	1	2	3	4	4	6	7	8	9	10	11	12	13

(1) *long-term programme*, say 6 years ahead, for corporate planning;

(2) *annual programme*, 1 year ahead, for financial and purchase planning; and

(3) *short-term programmes*, say 4 weeks, for ordering and for the call-off of deliveries from suppliers.

Because it is not given to human beings to foretell the future with any accuracy, the long-term and annual programmes are 'forecasts' which will need changing during their term, to keep them in line with the realities of the market. The short-term programme, on the other hand, will usually be based on orders already received and, in any case, because they only cover short periods of time and there are frequent opportunities for replanning, they can generally be 'fixed programmes', which should never need to be altered once issued.

Programming based on short-term programmes is known as 'flexible programming'. *Figure 15.35* shows diagrammatically why such programmes are flexible, or in other words are able quickly to follow changes in market demand. If material supply and processing are to be based on the actual, or forecast, needs given in a production programme, this programme must cover a term starting sufficiently far ahead to allow for the *throughput time* or time taken by materials to pass through the factory, plus the *lead time* or time required to obtain delivery of materials from suppliers. This total is known as the 'programme lead'. The only way to achieve a truly

flexible system is: first, to reduce throughput times by simplifying the material flow system (group technology); second, to reduce supplier lead times by using the call-off method of purchasing, in which deliveries are called-off at short intervals

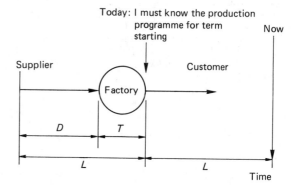

Figure 15.35 Programme lead. *L*, Programme lead, or the time ahead of today's date at which the programme must be issued if it is to be used to plan manufacture and material delivery. *T*, Manufacturing throughput time. *D*, Delivery lead time. N.B. Both *T* and *D* are partly dependent on the length of the programme term. $L = D + T$

as required, against a long-term contract; and third, by replanning the programme at regular short intervals so that it is mainly composed of firm orders.

15.11.2 The annual programmes

A set of annual programmes has already been illustrated in *Table 15.5*. In one sense these programmes are forecasts of what is likely to happen in the future. In another sense, however, they are targets fixed by management for future achievement. In the first sense, annual programmes are generally unreliable. It is not given to human beings to foretell the future, and the idea that there is a fixed future to forecast, is itself an extreme example of fatalism. In the second sense, if we treat programming as the task of setting reasonable targets for future output, and if all our other actions are directed towards the achievement of these targets, then there is a much greater chance that our programmes will be achieved.

In the final analysis, however, the achievement of an enterprise is greatly influenced by the state of the economy, by the action of its competitors, by fashion, and by other influences outside management control. Most companies need to change their annual programmes several times a year, to bring them into line with the realities of market demand.

All companies need annual programmes for a number of important reasons. First, they need annual programmes so that they can estimate future revenues, expenditure and profit, and forecast future cash flows. Second, they need annual programmes so that they can negotiate with material suppliers and make acceptable purchase contracts. Third, by showing when new products are to be introduced and old ones are to become obsolete, they show when new people must be employed, when training is needed, when the plant layout must be modified, the date by which stocks of an old product must be used up, and the achievement dates for other similar tasks.

A major problem with annual programmes is 'smoothing'. If the sales programme is very seasonal (as illustrated in *Table 15.6* in the case of an agricultural implement) manufacturing by the just-in-time method when the product can be sold, will waste capacity. This capacity can be saved by 'smoothing' or making for stock out-of-season, but in this case there will be very heavy stocks to be carried for most of the year, and the company will need much more capital to operate. In the agricultural industry this problem can be solved by making a seasonally balanced range of products, which can be manufactured and sold, each in its own particular season of the year. In other industries it may be necessary to vary working hours—with seasonal overtime in the peak demand period, and seasonal short time when there are no sales—and to offer discounts or long-term credit terms to customers who order out of season.

15.11.3 Flexible programming

In addition to the annual programmes, most manufacturing enterprises need to plan a series of short-term programmes at regular intervals throughout the year. Normally these are based on terms of 1 or 2 weeks. This gives companies the necessary flexibility to follow changes in market demand without accumulating unsaleable stocks. For this reason short-term programming is also called 'flexible programming'.

There are two main cases to be considered. First, in the case of process, implosive and square industries (see *Figure 15.3*) the products are either bulk materials or components. In this situation the sales programme is formed by accumulating orders over a period. In the second case, which is special to explosive or assembly industries, particularly where there is a wide variation in the delivery times for different orders, a new sales programme is planned each period. The aim is to deliver all products to customers by the promised delivery date, and at the same time to maintain an even workload during each period on the assembly department.

The first of the two cases is illustrated in *Figure 15.36*, by the example of a factory making decorative laminates, with a

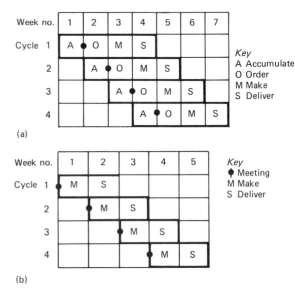

Figure 15.36 PBC schedules. Accumulation of sales orders to form the programme (a) Period batch control for decorative laminates. (b) Period batch control for a jobbing machine shop—orders are held in a 'waiting file' until ready for production and are then moved to a 'ready file' where they are stored by delivery week number, and are allocated to a programme at the weekly meetings

Table 15.6 Smoothing the production programme

	Month												
	J	*F*	*M*	*A*	*M*	*J*	*J*	*A*	*S*	*O*	*N*	*D*	*Total*
Sales	100	50	50	50	50	80	100	150	200	200	200	150	1380
Production	115	115	115	115	115	115	115	115	115	115	115	115	1380
Stocks	15	80	145	210	275	310	325	290	205	120	35	0	—

standard 3-week delivery time. Orders for different colours, patterns and sizes of laminate are collected in the first week of the cycle. In the second week, orders are sorted by product type; existing stocks are subtracted to find the 'must make' quantity; the capacity required is calculated and if there is spare capacity, additional quantities of the more popular lines are ordered for stock to smooth production. The products are made in the third week of the cycle and order sets are 'picked' and sent to wholesalers in the fourth week. The same figure also shows the standard schedule used in each cycle in a jobbing machine shop.

The use of flexible programming in the case of explosive industries is illustrated in *Figure 15.37*. With some simple products it may again be possible to use the accumulation method, but for many products a more flexible method is needed. *Figure 15.37* shows flexible programming, as used to control the assembly line in a tractor factory. A programme meeting was held on Friday afternoon each week. At this meeting the sales manager presented a sales programme showing the tractors he wished to deliver to customers in the week, starting 2 weeks ahead. At the same meeting, the production manager presented a production programme for the tractors he wanted to assemble during the week starting 1 week ahead. This programme attempted to meet the sales manager's requirements, but planned additional tractors for stock if orders were less than the capacity, or drew on existing stocks if orders were more than the assembly capacity. The week immediately following the meeting was used to chase shortages and to arrange deliveries, or call-off from suppliers. *Figure 15.37* also shows an alternative flexible programming schedule used in a company where a 2-week throughput time was needed to cover assembly, test and painting. In both these cases parts ordering was done by other independent systems.

Flexible programming provides the foundation for all just-in-time ordering systems. These are described in Section 15.11.7.

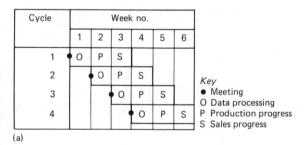

(a)

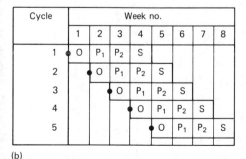

(b)

Figure 15.37 PBC flexible programming in explosive industries. (a) Flexible programme for tractor assembly. (b) Flexible programme for a product with a longer assembly throughput time

15.11.4 Programme reconciliation

It is necessary to compare the planned cumulative output from the annual programme with that from the series of short-term programmes, at regular intervals. If they show widely different cumulative values for output, the annual programme must be corrected to bring it into line with the actual market conditions, which are followed by the series of short-term programmes.

Table 15.7 shows such a reconciliation. If the cumulative output from the series of short-term programmes is less than that in the annual programme, the latter will eventually have to be changed, to make it possible to correct the financial budgets and forecasts. If some, or all, of the purchase deliveries are based on the annual programme, an additional reason for correcting the annual programme will be to reduce stocks. If the reverse is true and the output in the series of short-term programmes is more than that in the annual programme, the annual programme must again be amended to correct the financial plans, and in order to increase material supplies by correcting purchase delivery schedules.

15.11.5 Ordering

Ordering is the second level of progressive production control. It plans, directs and controls the output of components from company workshops, and the input of materials and bought parts from suppliers.

Because ordering (outside process and implosive industries) involves hundreds of different items, it is generally advisable to systematise the methods used, in order to increase their reliability and to reduce their cost.

Any method of ordering can be classified in three main ways:

(1) as either stock-base or flow-control methods; and
(2) as either single- or multi-cycle methods; and
(3) as just-in-time methods, or not.

Stock-base systems are those in which the release of orders is based on the level of stocks in a store, and flow-control methods are those in which ordering is based directly on calculations from a production programme. Single-cycle methods are those in which orders are released on a periodic series of ordering dates for completion by a related series of due-dates, and multi-cycle systems are those in which each part has its own special order date and due date. Just-in-time methods are those with which products are only made when they can be delivered to customers, and parts are only made and materials and bought parts are only accepted from suppliers when they are needed immediately for further processing or for assembly.

15.11.6 Single- and multi-cycle ordering

The difference between single- and multi-cycle ordering is illustrated in *Figure 15.38*. With single-cycle ordering, orders are issued at the beginning of each period for completion by the end of that period. All orders therefore have the same maximum throughput time. It is not required that all parts should be ordered every period. With this method parts are generally only made when they are needed for the production programme and parts are not normally made for stock.

Because the allowed throughput times with single-cycle ordering are short, its use is generally restricted to the simple material flow systems associated with continuous line flow, or group technology. The use of the single-cycle ordering, however, itself contributes to a reduction in throughput times.

Table 15.7 Programme reconciliation

Programme	Cumulative output of product X by month											
	J	*F*	*M*	*A*	*M*	*J*	*J*	*A*	*S*	*O*	*N*	*D*
Annual	100	200	300	400	500	600	700	800	900	1000	1100	1200
Short term	100	203	321	416	490							
+	—	3	21	16	—							
−	—	—	—	—	10							
Material stock sets	40	40	37	16	0	10						

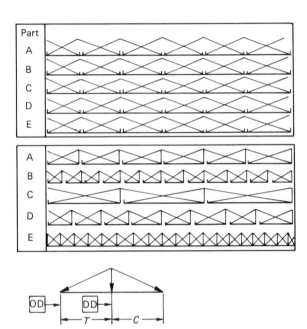

Figure 15.38 Single- and multi-cycle ordering. OD, Order date; DD, due date; T, throughput time; C, consumption time

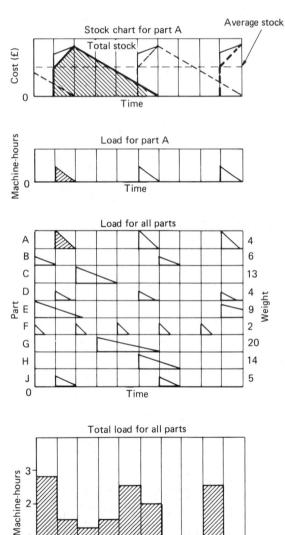

Figure 15.39 Surge effect multi-cycle systems and load surge. There are unpredictable changes in the total load as the peaks and troughs of the part cycles drift in and out of phase

Because all parts are ordered together on the same day, 'sequencing' can be used efficiently, making all parts in the same tooling families, one after the other at the same set-up.

With multi-cycle ordering, each part has its own order quantity and its own allowed throughput time (lead time). In production, therefore, each part has its own special order and due date. Multi-cycle ordering systems are more complicated than single-cycle systems, because each part must be treated as an individual item for control. They also have the major disadvantage that they are unstable. Because the peaks and troughs of the individual order cycles for different components drift into and out of phase in an unpredictable manner, multi-cycle systems are subject to unpredictable surges in stocks and load. This phenomenon is known as the 'surge effect'. It is illustrated for the case of load in *Figure 15.39*.

The use of multi-cycle ordering has been justified in the past by the theorem of the *economic batch quantity*. This theorem, published in Chicago in 1913 by the late E. N. Harris, says that for any part there is one economic batch quantity (EBQ) which gives minimum cost, and that for minimum total

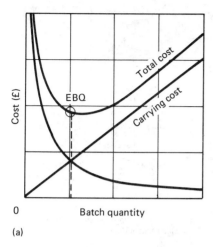

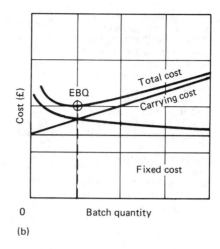

Parameter		Investment stocks	Costs	Profitability
OQ	Order quantity	▭	⬆ (Y)	⬇
RQ	Run quantity	⬇	⬇	⬆
SQ	Set-up quantity	▭	⬆ (W)	⬇
TQ	Transfer quantity	▭ (X)	⬆ (Z)	⬇

(c)

Figure 15.40 Economic effects of changes in batch quantity. (a) Traditional economic batch quantity (EBQ) curve. (b) Curve shown in (a) adjusted for 'fixed' costs. (c) The effect of reducing parameter values. N.B. In (a) and (b) OQ, RQ, SQ and TQ are all given the same value and are treated as a single parameter, the 'batch quantity', but (W) reducing SQ increases the number of set-ups per year and thus the cost of setting. (X) Reducing TQ makes possible close scheduling which can indirectly reduce throughput times and thus stocks. (Y) Reducing OQ increases the number of orders per year. (Z) Reducing TQ increases the number of handling moves and tends to increase costs

production cost all parts should be made in their own EBQ sized batches.

The EBQ theory is illustrated in *Figure 15.40(a)*. It assumes that costs can be divided into *preparation costs*—a fixed sum per batch, so that cost per piece falls exponentially with increases in batch quantity—and *carrying costs*, or the costs of holding stocks which rise in direct proportion when batch quantities are increased. The EBQ theory is false for four main reasons. First, it is the supreme example of the fallacy of suboptimisation. A large number of independently calculated optima for hundreds of different parts of a problem cannot possibly find the optimum solution for the problem as a whole. Second, as shown in *Figure 15.40(b)*, most industrial costs do not vary with changes over large ranges of batch quantity values. The variation of cost with batch quantity change is much less significant than that suggested by *Figure 15.40(a)*. Third, there are many different ways in which parts are batched for convenience in production. The most important in the present case are:

(1) *order quantity*—the quantity authorised for manufacture by an order;

(2) *run quantity*—the number of parts of a particular design, run off on a machine before changing to make some other part;

(3) *set-up quantity*—the number of parts, not necessarily all the same, which is made on a machine before changing the set up. Increasing the set-up quantity by making many parts at the same set-up, reduces both set-up costs and throughput times and increases capacity; and

(4) *transfer quantity*—the number of parts transferred together as a batch between machines for successive operations.

These four quantities are independent variables. One can vary the value of any of them within certain limits, without changing the value of the other three. As shown in *Figure 15.40(c)* each of these variables has a different effect on cost and investment. There are major economic advantages to be gained by keeping them as independent variables, for example by increasing set-up quantities and reducing run quantities at the same time. In EBQ theory all these four batch-quantity types are given the same values and are treated as a single variable, known as the 'batch quantity' (or 'lot size').

It is interesting to note that none of the above four batch-quantity types finds a minimum value in the curve linking cost and batch quantity. There is only a minimum EBQ point on the curve, in the case where all four are treated as a single batch quantity. Therefore, the EBQ theorem is again false, because it uses a type of batch quantity which would never be used in practice in industry.

Finally, the EBQ theorem is false because it assumes that the only way to overcome high set-up times is to invest in stocks. Experience, notably in Japan, shows that it is generally more profitable to invest in reducing set-up time than it is to invest in stocks, by making in large batches in order to spread set-up costs over more parts.

15.11.7 Just-in-time production control systems

The term 'just-in-time' (JIT) originated recently in Japan, but the idea is older. It refers to systems which—with minor variations for smoothing or for similar reasons—only make products when they can be delivered to customers, only make parts when they are needed immediately for assembly, and only accept deliveries from suppliers when they are needed immediately for further processing, or for assembly. It can be seen that the main objective of JIT systems is to minimise stock.

All JIT systems start with flexible programming. Only in this way is it possible to follow the needs of the market without accumulating stocks.

All JIT systems are single-cycle systems. Multi-cycle systems are never JIT because they generate heavy stocks, and because they are unstable due to the Surge effect.

There are three main JIT methods in use: period batch control (PBC), the kanban method, and optimum production technology (OPT). These methods are described in the following paragraphs.

15.11.8 Stock-based ordering methods

Ordering methods are either *stock-based* or *flow-control* methods. Stock-based methods are those where the release of orders is triggered by changes in the stock level. These methods can be further divided into *replacement*, *re-order level* or *requirement cover* methods.

15.11.8.1 Replacement methods

These are methods which fix the stock levels for all items, and take steps to maintain these levels by issuing orders immediately to replace any issues. Two such methods that have been used in practice are known as:

(1) base stock control, and
(2) the kanban method.

Base stock control was never widely used due to problems in batching for processing. Kanban solved this problem by adopting standard containers. Every time a container load is issued, a new container load is ordered to replace it. The container load is, in effect, the run quantity.

The kanban method is illustrated in *Figure 15.41*. The method was developed by the Toyota company in Japan, who have a physical rate of stock turnover of 50 times per year. In other words, they hold an average of 7.3 days stock of

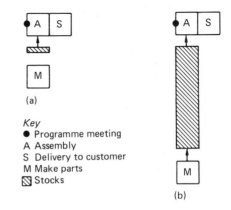

Key
● Programme meeting
A Assembly
S Delivery to customer
M Make parts
▧ Stocks

Figure 15.41 The kanban system: (a) single product, continuous demand; (b) many products, intermittent demand

materials (365/50) at different stages of processing. The method is best suited to mass-produced products such as cars, and is much less efficient in multiproduct, low-volume, intermittent production. Assembly with the kanban method is based on flexible programming using a series of fixed short-period programmes. For this reason ordering is effectively single cycle. The kanban method is, therefore, a JIT method.

15.11.8.2 Re-order level methods

These methods are those in which new orders are issued when the stock of any item drops to a pre-planned re-order level (ROL). In their simplest form these methods are known as 'stock control methods'. Such a method is illustrated in *Figure 15.42*.

As shown in *Figure 15.43*, stock-control methods have serious deficiencies for the general regulation of material flow in factories. First, because batch consumption times must be greater than lead times, they require a great amount of stock to make them work. Second, they are multi-cycle systems and are therefore subject to the surge effects with stocks and with load. Third, they are unsatisfactory in seasonal industries because they leave large sums of capital tied up in stocks during the dead season. Fourth, they make sequencing difficult, because few parts in the same tooling families are ever on order at the same time. Fifth, they are a major cause of materials obsolescence, because parts are never available in balanced product sets.

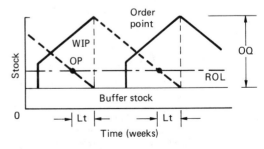

Figure 15.42 The stock-control method. A new order for the order quantity OQ is issued when the stock of parts drops to the re-order level (ROL). Lt, lead time; OP, order point; WIP, work in progress

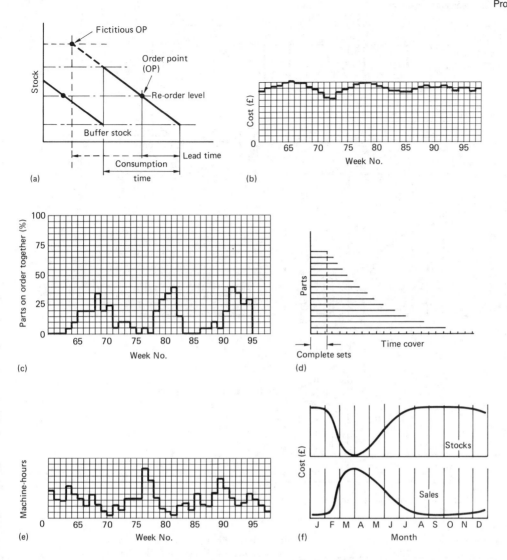

Figure 15.43 Deficiencies of stock-control ordering. (a) A large stock investment is required because the consumption time must be longer than the lead time. (b) Random order release produces highly variable and unpredictable stock levels. (c) Random order release does not permit family processing—parts in the same families are never all on order together. (d) There are obsolescence losses because parts are never in balanced product sets. If a design is changed, many parts must be scrapped. (e) Random order release gives highly variable and unpredictable load on the machines. (f) The method is costly because seasonal production stocks are at a maximum during the dead season when they are not needed

One further major defect of stock control is that, if it is used to regulate orders on a series of inventories, the demand variation is magnified at each step. This effect is well known in industry and has been explained by Forester in *Industrial Dynamics*.[1] *Figure 15.44* shows the transfer of demand from a final customer (varying by ±5%) through retailers and wholesalers to factory stock (demand varying by ±40%).

The most important variants of stock control are *stock and provision control* and *stock and allocation control*. With these methods, the re-order level is applied to either the stock plus outstanding orders, or to the stock less allocations. There are a few cases where such methods have advantages but, as general production control methods, they tend to be no more efficient than stock control, and to suffer the additional disadvantage that they are very complicated to administer.

15.11.8.3 Requirement cover methods

There are some types of product and some types of market for which it must be possible to deliver ex-stock on demand. The traditional method for ordering such items was stock control but, as already explained, stock control has serious deficiencies. An alternative method known as 'stock ratio optimisation' (STROP), is used to measure the demand rate and the stock of all items at regular period intervals, and by dividing the stock by the demand rate, to find the time cover (days supply) provided for each item. Ordering then attempts to raise the stocks of those items with the lowest time cover, until either the stock reaches the maximum permissible level, or all the capacity has been committed.

The STROP shown in *Table 15.8* is based on a knitting factory manufacturing a wide range of socks in different sizes,

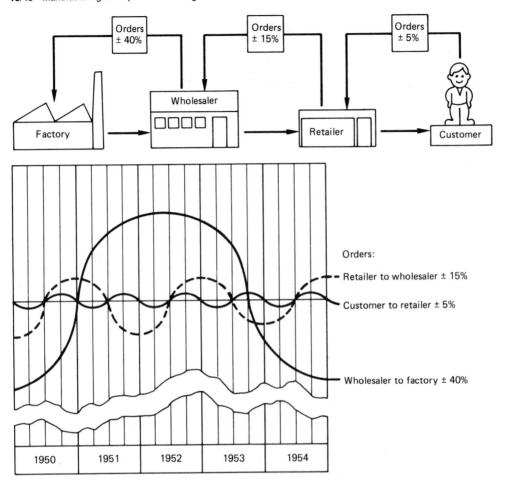

Figure 15.44   Industrial dynamics—the magnification effect. (Courtesy of J. W. Forester[1])

Table 15.8   Stock ratio optimisation

Item	Type code	Weeks cover	Order
1	1763	Nil	2.5
2	1411	0.01	2.5
3	1412	0.20	2.3
4	3605	0.25	2.25
5	4320	1.9	0.6
6	4007	1.9	0.6
7	3185	2.2	0.3
8	3192	2.5	Nil
9	1602	2.7	
10	4193	2.8	
11	1121	3.6	
12	1702	3.7	

Time cover = Stock/Demand rate
Demand rate/week = Moving annual total sales/52
   To equalise time cover:
(1)   Order enough of item 1 to increase time cover to the same as that for item 2;
(2)   order enough of items 1 and 2 to increase time cover to the same as that for item 3;
(3)   order 1, 2 and 3 to bring into line with 4; and
(4)   continue until all capacity is used, *or* until stock limit is reached.

patterns and colours. The demand in this case is based on the moving annual total sales for each item, adjusted for seasonal differences. In other cases, shorter term moving averages might be more appropriate.
   STROP has a future, not only in manufacturing, but also for the regulation of stocks in distribution warehouses.

15.11.9   Flow control ordering methods

Flow control ordering methods are those in which the quantities of items to be ordered are calculated from the production programme for the products in which they are used.
   The common factor in all flow-control systems is that they all start by calculating the 'requirement schedule' (the number of each item needed to complete products) using the parts list ('bill of materials' in the USA) and a production programme showing the number of each product to be manufactured.
   The next step is to decide on the production programme to be used for ordering (annual or short term) and to fix the batch quantity, order date and due date for each item. The main methods for doing this are illustrated in *Figure 15.45* and are listed below:

(1)   component scheduling,
(2)   period batch control (PBC),

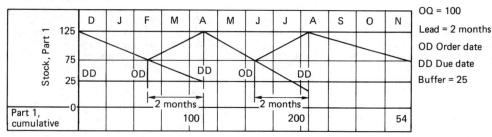

		J	F	M	A	M	J	J	A	S	O	N	D
Programme product A		10	11	12	12	12	12	14	14	9	7	6	8
Required schedule	Part 1	20	22	24	24	24	24	28	28	18	14	12	16
	Part 2												

(a) Requirement schedule (all flow control systems)

	D	J	F	M	A	M	J	J	A	S	O	N	
Part 1	20	22	24	24	24	24	28	28	18	14	12	16	254

(b) Component scheduling (one part only)

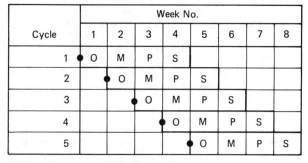

OQ = 100
Lead = 2 months
OD Order date
DD Due date
Buffer = 25

(c) MRP method (one part only)

Cycle	Week No.							
	1	2	3	4	5	6	7	8
1	● O	M	P	S				
2		● O	M	P	S			
3			● O	M	P	S		
4				● O	M	P	S	
5					● O	M	P	S

Key
● Programme meeting
O Order from groups
P Assemble
S Deliver to customer
M Make parts

(d) Period batch control (1–week periods)

Figure 15.45 Flow control ordering systems

(3) materials requirement scheduling (MRP), and
(4) optimum production technology (OPT).

15.11.9.1 Component scheduling

Component scheduling advances the requirement times shown in the requirement schedule by enough time to allow reliable accumulation of all items before the requirement date. This method has been mainly used for regulating purchase deliveries in mass production. The delivery schedules have necessarily been based on annual or other relatively long-term programmes. Such programmes, and also therefore such purchase delivery schedules, need frequent revision, and the method is gradually being replaced by the call-off method.

15.11.9.2 Period batch control

Period batch control (PBC) was developed by the late R. J. Gigli in the UK in the 1930s. It was installed before 1939 in 30 different factories in the UK, and was used to regulate the manufacture of Spitfire fighter aicraft during World War II. PBC is a JIT production control system. A typical schedule for PBC in an engineering factory is illustrated in *Figure 15.45(d)*. A programme meeting is held on each Friday and the sales programme (deliveries to customers) for week 4 and the production programme (assembly) for week 3 are fixed. The parts to be made in workshops and the items to be delivered on call-off by suppliers are calculated in week 1 for manufacture or delivery in week 2. In other industries and for other products, other schedules may be required. There are two limiting conditions for the use of PBC:

(1) the throughput time for any item at any major processing stage must not exceed one period; and
(2) set-up times must be reduced so that any increase in the number of set-ups per period does not reduce capacity.

Experience, notably in Japan, has shown that these are never impossible conditions. The basic solution in Japan has been simple material flow systems coupled with JIT production control or, in other words, group technology and PBC. It may be necessary in addition to make some technical improvements.

15.11.9.3 Materials requirement planning

Figure 14.45(c) shows a typical example of materials requirement planning (MRP). As with all flow-control systems, it starts by calculating the requirement for parts. It then selects order quantities for each item and estimates the lead times necessary to manufacture such batches. Batches of all parts are scheduled to provide the parts needed for assembly. Scheduling in this case must necesssarily be based on long-term programmes. A study carried out in the USA showed that 80% of companies using MRP, based ordering on a programme with a term of 8 months or more.

Although the above description is typical of the majority of MRP methods, other types of method are also covered by the same term. The problem is that, in the USA, the term 'MRP' is used in the same sense as the term 'flow-control system'. In the USA, therefore, all flow-control systems are MRP systems. In this sense, PBC is an MRP system, with a standard period ('bucket' in the USA), a fixed lead time ('lot for lot' in the USA), and a single cycle (same lead time for all parts). These ideas are being introduced in the USA under such terms as 'MRP with JIT'. Another complication with MRP is the recent introduction of MRPII, where MRP now means 'manufacturing resource planning'. MRPII has changed the meaning of the term 'MRP' to cover many of the ideas previously covered by the term 'computer integrated manufacturing' (CIM).

15.11.9.4 Optimum production technology

MRP made a major contribution by introducing the computer to production control. The method has become discredited in recent years because companies which used it were unable to turn their stocks over at the high rates achieved in Japan, and largely for this reason were unable to compete with Japanese companies. Optimum production technology (OPT) is a JIT ordering system developed in the USA by E. Goldratt. It was designed to achieve high rates of stock turnover, and has replaced MRP in a number of companies.

OPT is similar in principle to PBC. It also finds the quantities of parts to order, by explosion from a series of short-term programmes. The main difference between them is that PBC issues orders to groups and delegates operation scheduling to the group foreman, whereas OPT uses the computer to prepare operation schedules and issues these schedules to the workshops in lieu of orders.

15.11.10 Dispatching

Dispatching is the third level of production control. It plans, directs, and controls the completion of processing operations in workshops, needed to complete orders. In component processing departments or groups, dispatching covers such tasks as:

(1) storage and control of orders,
(2) storage and control of drawings,
(3) the scheduling of operations on work centres,
(4) the storage and control of tools and setting up,
(5) the handling of tools to and from work centres,
(6) the handling of materials to and from work centres,
(7) the removal of swarf from machines,
(8) the recovery of oil or coolant from swarf,
(9) the sweeping of floors and removal of rubbish,
(10) the inspection of work done, and
(11) preventive maintenance of machines and equipment.

In assembly departments, dispatching does all the same tasks except (7) and (8) above, plus *kitting-out* or sorting parts into assembly sets.

In traditional workshops, the inspection of work done and maintenance were normally controlled by independent departments. Operation scheduling was done by a dispatching clerk in the department, and he and the department manager supervised the execution of the remaining tasks. With group technology, all these tasks are normally delegated to the group foreman, with the exception of the recovery of oil, or coolant from swarf in machine shops, which is usually centralised. Although inspection is delegated to the groups, there is usually a small independent quality-control unit, which rechecks a proportion of the inspected parts, in order to maintain quality standards. Again, although the groups do their own preventative maintenance inspections, and minor repairs, there is normally a central maintenance department on whom they call for major repairs.

15.11.10.1 Operation scheduling in machine shops

The most difficult task in dispatching in machine shops is 'operation scheduling'. In traditional workshops two main methods were used: *due-date filling* and *Gantt chart scheduling*. These are illustrated in *Figures 15.46* and *15.47*. Due-date filling was simple and reliable in use, but throughput times were long, as each operation had to be completed on the whole batch, before starting the next one. Gantt charts, or planning boards, solved this problem, but were much less flexible. Any machine breakdown, absenteeism, or material shortages often made it necessary to redraw the whole Gantt chart.

With group technology, operation scheduling is generally delegated to the group foreman. With traditional methods of production control such as stock control and MRP, which are multi-cycle systems, there are difficulties in balancing the load between the groups. The most reliable production control method for use with groups is period batch control (PBC).

With PBC, the group foreman receives each period a *list order* showing the number of parts to be made, and a *load summary* showing the load (in machine-hours) imposed by this order on each machine in his group. The load summary tells him which of his machines are heavily loaded or bottlenecks. In addition to this information, he needs to know:

(1) the sequence of loading first operations, which will find work for all his operators in the first 15 min of the week, and get work started quickly on *critical parts* with many operations and long throughput times;
(2) the sequence in which parts should be loaded on the bottleneck machines, in order to minimise setting-up time and to maximise capacity; and
(3) the *critical parts* with many operations and long throughput times, which will have to be close scheduled.

If the foreman closely follows the first two of these sequences, if batches of critical parts are labelled for easy recognition and are given priority, and if parts are moved on completion of each operation to the machine for the next

A. On receipt of order

(1) Order filed in section for machine for first operation
(2) Due date copy of order filed by due date
(3) Job cards filed by part number and operation number

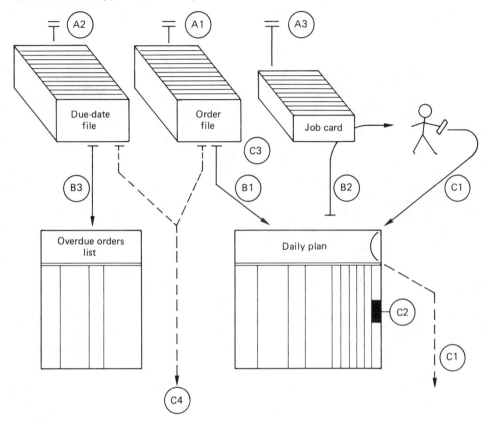

B. Daily

(1) Order file shows which jobs go on daily plan
(2) Daily plan shows which job to issue to worker
(3) Due date file shows which jobs are overdue

C. On completion of operation

(1) Worker returns job card, which goes to wages department
(2) Dispatcher brings daily plan up to date
(3) Order moved to section for machine for next operation
(4) After last operation, the order and copy are sent to the
    production control office

Figure 15.46 Due-date filing dispatching system

operation, the rest of the work can be scheduled, using the following simple rule: 'load next on each machine, a critical part—if any are waiting or have already started on the previous operation—or load the next part waiting in the queue'.

The only other operation scheduling task for the foreman, is to move operators between the lightly loaded machines in order to give priority to critical parts, and to eliminate queues.

15.11.10.2 Kitting out for assembly

In assembly groups, the most difficult task is kitting out or collecting parts into sets for assembly. In traditional assembly

shops this involved visiting the stores with a parts list, taking the required parts from the bins, and producing a 'shortage list' of missing parts. A shortage chaser, or expediter, then took special action to expedite completion of the shortages.

With groups, if the products are simple, parts produced in component processing groups can be moved direct to the assembly group on completion. If the products are complex, or there are many common parts, parts will be accumulated in a transit store before issue in sets to the assembly groups. In the transit store, batches of parts received will be counted, any surplus to the assembly requirements will be moved to the spares store (which will also hold any buffer stocks), and the remainder will be stored on racks for each assembly, which

A. On receipt of order

(1) Order scheduled on Gantt chart or planning board
(2) Order then filed by due date in order file
(3) Job cards filed by part number and operation number

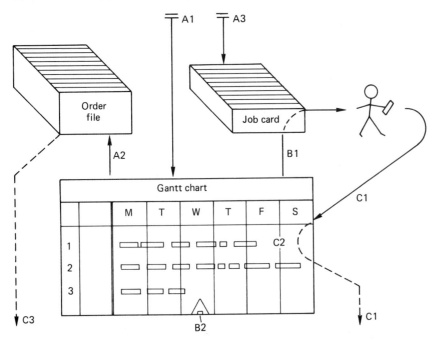

B. At all times

(1) Gantt chart shows which job to issue to worker
(2) Gantt chart shows which jobs are overdue

C. On completion of operation

(1) Worker returns job card which goes to wages department
(2) Dispatcher brings Gantt chart up to date
(3) After last operation, the order is sent to the
     production control office

Figure 15.47 Scheduling system for dispatching

will be moved to the correct assembly group at the end of each period.

15.12 Purchasing

Purchasing is the function of production management which finds sources for supplies, negotiates contracts with suppliers, and progresses deliveries of bought materials and components. Like marketing, purchasing is a commercial function, but it is more closely related to production control than to marketing. The specification of the delivery schedules for purchases, in order to regulate and control production and the investment in stocks, is a part of the production control function.

There are two main classes of purchases: bought items, which a company buys because it cannot make them; and subcontracted items, which a company could make (it has the necessary facilities, skills and know-how) but does not make

because it is short of capacity, or because it can buy them more cheaply than it can make them.

There are three main methods of buying: batch buying, schedule buying, and the call-off method. With batch buying, purchasing starts with a requisition for a specified quantity of an item, all for delivery by the same due date. Schedule buying starts in the same way with a requisition for a specified quantity, but the order is accompanied by a delivery schedule, calling for the delivery of specified quantities at a series of future delivery dates. The call-off method may again specify a total order quantity; it will usually be accompanied by a delivery forecast or schedule, showing the probable requirements for a series of future periods. However, the supplier may only deliver in each period the exact quantity specified in a call-off note, issued for that period.

15.12.1 Batch buying

Batch buying is mainly used in the purchasing of materials and parts in jobbing and large contract industries (see *Figure 15.3*).

In this instance, which includes the case where the batch quantity is 1, only the exact quantity needed to complete the sales order is ordered from suppliers. Batch buying is also the traditional form of purchasing used in continuous manufacture. Items were purchased in batches for stock. This method was generally associated with the cycle of:

(1) enquiry,
(2) quotations, and
(3) orders.

Enquiries were sent to a number of possible suppliers. Their quotations were compared, and the order was usually given to the supplier who quoted the lowest price. The problems with this approach were that it did not usually find the most reliable supplier and, because the supplier frequently changed, long lead times were necessary to give the supplier time to find materials, plan methods, and obtain tools for the work. Orders were normally delivered as a single delivery batch and quantity; stocks were therefore very high.

Some companies changed their methods to adopt one (or perhaps two or three) accepted suppliers. They used the stock-control ordering method to regulate the release of purchase orders, using a fixed order quantity (based on the economic batch quantity theorem), and a preset re-order level for each item. These changes made it possible to consider reliability when choosing suppliers, and also to reduce lead times. However, stocks were still high and, because this was a multi-cycle system, it was subject to the surge effect, making it very difficult to control the level of stocks.

15.12.2 Schedule buying

With schedule buying, there was again usually a fixed order quantity per purchase order, but deliveries were spread over a series of future delivery dates. This method tended to reduce the stock level, and also to reduce stock holding costs.

The main problem with this schedule buying method was that, as it attempted to fix periodic deliveries for periods a long time into the future, the schedule had to be found by calculation (explosion) based on long-term, often annual, programmes. As it is not given to human beings to foretell the future, such programmes require frequent revision, and any schedules based on these programmes also have to be revised at the same time.

In practice, schedule buying was mainly used to regulate the supply of special materials (castings, forgings, etc.) and bought parts, in mass production. This method was reasonably reliable if one allowed a high level of buffer stocks. Any attempt to reduce buffer stocks, however, accelerated the need for programme and schedule changes, and made the schedule buying method much less reliable.

15.12.3 The call-off method

The call-off method differs from the schedule-buying method in one major respect. There is still normally a fixed quantity ordered, and a forecast of future period requirements based on an annual or other long-term programme, but deliveries can only be made by suppliers against periodic delivery authorisations, known as 'call-off' notes, which are based on short-term programmes.

Purchase contracts made with this method are normally framed to protect the buyer—by allowing necessary modifications to component design, for example—and the supplier—by guaranteeing tooling costs if designs are changed quickly, and by guaranteeing replacement orders when old orders are cancelled early.

Table 15.9 shows the case of a purchased part for which the forecast demand was 100 per week. The call-off quantities given to the supplier on each Friday varied between 125 and zero. It can be seen that, in this case, the supplier was able to manufacture at the forecast rate of 100 per week, using a small quantity of buffer stock which never exceeded 100, or the average requirement for 1 week. For safety, the supplier should maintain a record of the moving average of the call-off quantities, so that he can adjust the manufacturing rate if the average demand rises or falls. The customer should also undertake to adjust his long-term delivery forecast if actual call-off varies from that forecast by more than a specified quantity.

In the case illustrated in *Table 15.9*, call-off is weekly, and each call-off covers the requirements for the week starting 1 week ahead. A call-off lead time of 1 week, as in this case, gives ample time for the supplier to pack and dispatch the required quantity each week.

15.12.4 The choice of supplier

In the past it was customary to choose the supplier who quoted the lowest price for an item, or to choose a number of suppliers whose prices were the lowest. This approach is largely discredited today.

The first problem is to decide between a single supplier (single sourcing) or several different suppliers for each item. The advantage of having several suppliers per item is that if one of them fails, or has a fire, or suffers a strike, the supply can be maintained by the others. The single source, on the other hand, has the advantages that:

Table 15.9 An example of the call-off method

1. Part No. ×100
2. Forecast: 100/week
3. Call-off

	Week No.									
	1	*2*	*3*	*4*	*5*	*6*	*7*	*8*	*9*	*10*
Call-off	100	86	79	125	—	**128**	127	120	97	101
Do. cumulative	100	186	265	390	390	518	645	765	862	963
Forecast cumulative	100	200	300	400	500	600	700	800	900	1000
Variance +	—	14	35	10	**110**	82	55	35	38	37
−	—	—	—	—	—	—	—	—	—	—

(1) the purchase price is often lower, due to larger order quantities;
(2) the purchasing costs are lower as there are fewer suppliers;
(3) the quality tends to be more consistent; and
(4) it is easier to maintain contact with a small number of suppliers.

Today suppliers are normally chosen in more advanced companies by a process of supplier assessment and periodic review. Potential suppliers are visited and assessed by specialists from the company. Selling price is still important, but other factors are also considered, including quality of work, delivery reliability, financial probity, labour relations and willingness to deliver on call-off.

The present trend is towards single sourcing for each item and for more items per source, or in other words, fewer sources.

With bought items, the choice of supplier is between external companies. With subcontracted items, the cost of making the item in-house must also be considered. *Figure 15.48* shows the case of a part which cost £6 to make, and could be purchased for £4 with an apparent saving of £2 each. In fact, the only saving was the cost of the materials. The overhead expenses still had to be paid, and there was no saving in labour because the reduction in work was not enough to release one employee.

In general, the choice between making and subcontracting is very difficult to evaluate. The easiest cause to justify is that where subcontracting is used to increase the supply of items made on bottleneck machines, thus increasing saleable output. The main advantage of this approach is the high increase in profits when output increases and passes the break-even point (see *Figure 15.49*).

15.13 Finance

The financial function is the function of management which plans, directs and controls the investment, cash flow, revenue,

TO MAKE

Material cost	£1 each
Direct labour	£1 each
Overheads	£4 each
Total	£6 each

TO BUY

Delivered price	£4 each
● To buy: apparent saving	£2 each

● But actual cost of buying instead of making:

Purchase cost	£4 each
Overheads (still have to be recovered)	£4 each
Direct labour (still to pay as reduction in work not enough to release one employee)	£1 each
	£9 each

● To buy: actual loss £5 each

Figure 15.48 A 'make or buy' cost comparison

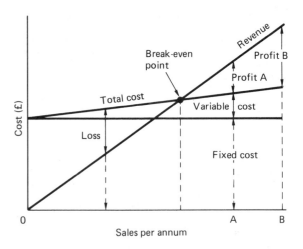

Figure 15.49 A break-even chart. Note that the chart is only approximate. Most cost changes vary in steps and few costs are directly proportional to output. Note also that a 15% increase in sales (A to B) doubles the profit

expenditure and profit for an enterprise. It records the amount of money in an enterprise in the form of classified 'financial accounts' and a 'balance sheet', and produces its plans for future revenues and expenditure in the form of budgets. A major objective of any manufacturing enterprise must be to make a profit. If it makes losses, it will eventually become bankrupt and cease trading. In a very real sense, financial accounting is there to warn the company when there is a danger that it may not survive.

15.13.1 Cost accounting

In addition to the financial accounts, most manufacturing companies also keep cost accounts which attempt to show how much it costs to make each product and component produced. The 'direct cost' of making a component comprises those items of expenditure which can be allocated directly to the component, i.e. materials incorporated, direct labour expended, and any easily measured direct expenditure such as metered power consumption. The 'indirect costs' of production include such items as rent, rates, taxes, management salaries, maintenance, etc., which cannot be allocated simply to components.

The methods of cost accounting were designed in the first half of the twentieth century. They were based on the idea of finding total costs, by sharing the overheads between the components, generally in proportion to their direct labour costs. This was not unreasonable when most machines were manually operated and total direct labour costs were about the same as total indirect costs. Today, with the growth of automation, direct labour cost has in many factories ceased to have any real meaning, and even where there are still manual labour costs to be measured the overheads are now 10 or more times greater. Cost accounting has become misleading rather than useful.

15.13.2 Break-even analysis

A second method of classifying costs divides expenditure into fixed and variable categories. Fixed costs are those which do not change with changes in product sales, and variable costs are those which change in direct proportion to sales output.

Figure 15.49 shows the relationship between total cost (fixed plus variable cost) and sales output, and between sales revenue and sales volume. It shows a point where the two are equal, i.e. the 'break-even point'. When output is below this point, the company makes a loss. When output is above the break-even point, there is a profit. Small increases in sales above the break-even point lead to major increases in profit, which largely explains the economic advantages of 'dumping'.

The problem with *Figure 15.49* is that most expenditure is fixed cost, for most of the time. Expenditure on direct labour, for example, only changes for hourly paid workers when the labour force is increased or reduced. Even with 'piece work', labour costs are only partly variable with sales output. The cost of purchases is probably the most variable type of expenditure, but in this case also the cost of running the purchasing department and of receiving and storing purchased items is also largely a fixed expense. Paradoxically, although most expenditure is a fixed cost, it tends in the long run to be variable as well. The fixed costs tend to rise in a series of steps as sales increase.

15.13.3 Conclusion

For most of the period since World War II, accountancy has been the master function, regulating investment and operations in manufacturing with little reference to the needs and aspirations of the market or of the other management functions. We still need strong accounting to monitor investment, revenues, expenditure and profit to ensure survival. There is a need today, however, for a more cooperative and holistic approach to management, and there is also a need for a new approach to cost accounting which is more in line with present-day needs.

15.14 Personnel

The personnel function plans, directs and controls employment, working conditions, welfare, training and promotion in an enterprise. This is one of the latest of the management functions to achieve specialist professional ranking.

In the early days of the twentieth century, companies were small and the leadership and control of labour were accepted as a part of the responsibility of general management. One cannot avoid the idea that the rise of the personnel function was partly due to the desire of general management to rid itself of an onerous duty. It is also possible that this change contributed to the rise of the trade unions, on the grounds that the wish of management to get rid of its leadership problems left a vacuum which the unions were happy to fill.

In the event, the enormous expansions in employment laws, coupled with the labour problems associated with the rapid changes in technology and management, gave the personnel function growing importance in the modern manufacturing enterprise.

One of the problems inherited by the personnel function is employee job satisfaction and morale. A great deal of research has been done by behavioural scientists in this field, but its findings are difficult to validate. The technological basis of manufacturing is found in the natural sciences of physics, chemistry and biology. Experiments in these sciences can be validated by using controlled experiments, in which one finally returns to repeat the original conditions to ensure that they have not changed.

Controlled experiments are impossible when studying human behaviour, because studying people changes their behaviour. Other forms of validation are needed for experiments and research in the field of the behavioural sciences.

The present position is that industry recognises that manufacturing can be seen as both a sociological system and a technological system. We recognise this fact by talking about socio-technological systems. This term, however, does not describe an integrated system, but rather two different ways of looking at the same system.

This leaves major gaps in the way in which we manage production today. We know, for example, that changes in employee attitudes can have a major effect on the economy of an enterprise. In estimating, in investment planning and in economic modelling, however, we ignore this fact. We do not consider the effects of changes in employee attitudes on our models of the enterprise, because we do not know how to express such relationships in mathematical or monetary terms.

We have already achieved a measure of integration between the technological and economic subsystems in production. Perhaps the greatest outstanding problem in production today is how to integrate the sociological subsystem with the technological/economic system.

15.15 Secretarial function

The secretarial function plans, directs and controls the communications, data-processing and data-storage activities in an enterprise. This function has changed out of all recognition since the introduction of the computer. Most of the data-processing requirements for manufacturing today can be handled by the personal computer, and the main problems at present are with the software.

Most data-processing tasks in manufacturing are specific to only one function. The functions do therefore form semi-independent data-processing units. The output from these units, however, either provides a necessary input to one or more other functions, or regulates the input from suppliers, the output from the factory, or the flow of finished products to customers. This flow of data between the functions is called the 'arterial flow system' (see *Figure 15.1*).

In the early days of the computer, the functions had their own computer systems and there was no communication between these so-called 'islands' of automation. Under these conditions the same data-processing tasks were often repeated in two or more functional units, and the same data were stored in several different places. Today, a main objective in system design is to integrate data processing and storage, so that each task is only done in one place, and any data that must be stored also have one unique location which can be accessed by any data-processing operation which needs it.

An important part of system design is the design of the database. The need is for rapid access to the data in any desired combination and sequence. Modern relational databases have this property and point the way to the future. One of the most difficult problems is how to keep the database up to date. Any change in product design, in processing methods, in plant, in suppliers, or in most other respects, can require major changes in the database. Routine corrections of this type are usually unpopular with those who must make them, who usually prefer more creative types of work. Unless the database is accurate, however, the computer output can only be rubbish. There is a need for both exacting routines for the rapid updating of the database when changes are necessary, and for periodic checks to test the accuracy of the database.

The final problem is how to design the software to carry out the range of data-processing tasks required, using the database provided. In recent years, the main effort has been towards the design of system packages for general use. Unfortunately, all factories are unique and there is no hope that a standard package will be equally efficient in factories of even approximately the same type. Until recently, packages of this type were written in first-generation computer languages such as Fortran. It has been found to be very difficult to amend such packages to suit the special needs of a particular company. Companies which have bought such packages have ended by either using only a small part of them, or attempting to change their own systems to suit the package.

The future need in software design is for tailor-made systems. There are certain highly standardised techniques which can be programmed for universal use, e.g. Pareto analysis, critical-path analysis, production-flow analysis, and statistical sales forecasting. Software for these techniques may be available in the future for purchase, for 'plugging in' to a company system. The total software system for any company, however, needs to be designed to meet the special requirements of that enterprise. The main hope for the future lies in the new fourth-generation computer languages which are beginning to make this a practical possibility.

15.16 Conclusion

This chapter has provided a general outline of manufacturing management. It started by looking at the different types of production found in factories. It then provided a brief introduction to systems theory, followed by the classification of management tasks into the eight 'functions', which provide the basic classification of the science of production management.

To give a total picture of the subject of production management, general management and each of the functions have been described in turn. However, because this is a book on manufacturing management, most of the text deals with those functions which are most closely associated with manufacturing.

References

1 FORESTER, J. W., *Industrial Dynamics*, MIT Press, Cambridge, Massachusetts (1965)

16

Manufacturing Strategy

J Hunt

Contents

16.1 Manufacturing strategy and organisation

16.1.1 The changing environment in world manufacturing

World trade has become increasingly competitive with the development of Japanese and other Asian manufacturing capabilities, the introduction of new technologies and vastly improved communications and transportation methods. In this situation, the implementation of a modern manufacturing strategy and organisation is central to the survival of a manufacturing company. Strategies and organisation structures for the 1990s and beyond will have to be fundamentally different from those of the past because of these developments.

The 1980s heralded a vast change in Western manufacturing philosophy, strategy and organisation. The continued increasing success of Japanese manufacturing in world markets plus generally fiercer international competition intruding into previously secure home markets has led Western manufacturing industry to adopt many Japanese manufacturing approaches. At the same time, improvements in UK labour relations and attitudes have smoothed the way for radical changes in manufacturing organisation and the greater involvement of all staff in the decision-making processes.

In addition, advances in technology and materials have created new industries (e.g. plastics, electronics and computers) and eliminated or severely reduced others. Existing industries that have survived the technological changes are facing an increased pace of competition. These factors, together with greater demands for product variety and higher quality, have led many leading companies to adopt new corporate and manufacturing strategies and organisation structures.

Table 16.1 shows a comparison between the general market situation for manufactured goods in the 1990s and that prior to the 1980s and the oil crisis of the 1970s. *Table 16.2* shows the

Table 16.1 The market environment

Aspect	Pre-1980s	1990s
Markets	Mainly domestic, some international	EEC per 1992, and world-wide
Source of competition	Other domestic companies, some European and North American	EEC, North America, Japan and Far East
Product range	Basic ranges of new products	More models and options needed to make sales in more competitive world markets
Product life	Varies with product, but typically 5–10 years	Varies with product, but typically 3–5 years
Consumer demand	Pre-oil-crisis boom as disposable incomes rose continuously	Prices slackening demand, plus some saturation
Industrial demand	Pre-oil-crisis; high demand on back of consumer demand	World overcapacity in many industries leads to low demand for capital goods
Quality expectation	Variable only in that demanded and achieved	Greater expectation of top quality by customer
Price demand	Booming Western economies could absorb rising costs and inflation	Stagnant economies sensitive to rising costs and inflation

Table 16.2 The manufacturing environment

Aspect	Pre-1980s	1990s
Product range	Increasing and high volumes	More model options with lower volumes
Product design	Manual drafting and early CAD	Wider use of CAD, CAE and CADCAM. Networking of design systems
Levels of volume	Steadily increasing as world demand expanded	Static, fluctuating or reducing with demand and technology changes
Volume production methods	Dedicated flow lines and machines	Flexible manufacturing systems
Batch production methods	Stand-alone machines	Flexible cells and JIT processing
Production technology	Single-purpose machines. Electromechanical controls	Multipurpose machines, CNC, computer controls
Manpower levels	Labour intensive	Reduction of both direct and overhead labour
Manpower skills needed	Single skill needed for one job only	Several skills needed for multiple tasks
Specialist skills	Organised in central service departments	Devolvement to the user departments
Quality assurance	Variable—only military and aerospace subject to quality-assurance systems	Universal demand by customers for implementation of quality assurance systems to ISO 9000; quality improvement programmes
Quality control	Inspection by roving inspectors and a central quality-control department	SPC and operator inspection/responsibility
Inventory policy	Large batch, make for stock, central stores	No stock. Make when needed (JIT)
Lead times	Variable with industry, but 12–16 weeks typical	Variable with industry, but 2–4 weeks demanded
Data processing, information technology	Centralised in data-processing department	Distributed to users as local processing power increases
Organisation structure	Hierarchical management with many levels	Team working with fewer management levels

CAD, computer-aided design; CAE, computer-aided engineering; CADCAM, computer-aided design and manufacture; JIT, just-in-time; CNC, computerised numerical control; SPC, statistical process control.

same comparison for the manufacturing environment. The comparisons are in the form of generalised statements that will apply to most industries and companies, with perhaps (but not necessarily) the exception of those in a strong monopoly or protected position.

The current situation is summarised below.

(1) Product competition is forcing the faster introduction of new product designs which will have a shorter life cycle than their predecessors.
(2) Price competition is forcing companies to be ever more cost efficient.
(3) Quality competition is demanding better working practices, methods and controls.
(4) Markets for manufactured goods have moved away from relatively insular domestic products and markets with prospects for continuous growth, to a world market of fluctuating demand, shorter product lives and fierce international competition invading domestic markets. The effect of this on the UK is illustrated in *Figure 16.1*. This figure shows the decline in the balance of UK exports over imports of manufactured goods from 1978 to 1989. In 1989 the UK had gone to a deficit of nearly £17 billion. The negative balance is now three times the positive balance at the start of the 1980s.
(5) The opening up of Eastern Europe and the continued development of the EEC will consolidate the total European market and effectively eliminate the UK as an isolated and separately supportable market for most products.

16.1.2 The 1980s environment and strategies

The overall effect of the factors described in *Table 16.1* led in many cases to sales being limited by productive capacity. The resulting supply-driven (i.e. production-driven) market gave manufacturing management the justification for employing more labour as the short-term solution for increasing production capacity. This in turn created extensive hierarchical

management structures to control a large labour force which was organised to do single repetitive tasks.

As technological advances came into place (plastics, micro-electronics, computers, and computerised numerical control (CNC) machining), additional specialist staff were required to apply the technologies. The specialists were usually organised into central service departments that could provide a service to all the 'user' departments.

The combination of a multilevel hierarchal management and central support services took nearly all responsibility and motivation away from local supervision and workforce. Despite this, the system seemed to work well because the continuous general high level of demand provided enough revenue to pay for extensive managerial and specialist resources. However, companies started to see severe problems when:

(1) demand dropped through (a) the general market situation, and (b) the oil crisis of the 1970s;
(2) the competitiveness and new technologies of Japanese products took markets away from European and North American companies; and
(3) the consumer (both private and industrial) started to demand more variety and better quality in addition to price competitiveness, i.e. more value for money.

A partial solution to these factors was seen in the 1970s with the introduction of group technology. This is primarily an organisational change, creating a multitasked group or cell to carry out all operations on a component or assembly instead of feeding them through a series of separate departments of like machines (lathes, mills, drills, etc.). This served to reduce lead times dramatically but, of equal if not greater importance, also gave some responsibility back to workers in their role as members of the cell team.

Group technology, although successful in itself, was still operated within the old structures of the typical manufacturing organisation and, as a consequence, failed to reach its full potential in many cases. However, its basic principles of cell

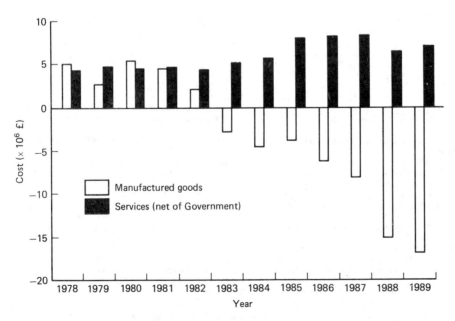

Figure 16.1 Balance of trade in manufactured goods and services. (Courtesy of CSO Pink Book)

manufacture, team working and wider responsibility for the individual worker are fundamental components of the new manufacturing strategies that are now emerging.

16.1.3 Future strategies

The general trend now seen as we move towards 2000 is being expressed in a number of strategies and techniques (see *Table 16.3*). It is arguable whether some of these are strategies in the sense of being high-level policies or operational techniques applicable at departmental level. Such definition is of minor interest, the more important point is that they are a group of activities that can be applied across a manufacturing organisation, and bring about a major improvement in competitiveness.

Many of the strategies listed in *Table 16.3* are already being implemented in companies—some at departmental and local levels and others such as world-class manufacturing (WCM) or computer-integrated manufacture (CIM) at entity level.

16.1.4 Key objectives of a manufacturing strategy

When considering a manufacturing strategy, management should keep in mind several key objectives that are at the centre of most modern strategies. These are:

(1) *Integration*—structure the total manufacturing system to work as one unit and to one well-understood strategy.

Table 16.3 Modern manufacturing strategies and techniques. There are a very large number of strategies, philosophies and techniques being used as methods of increasing competitiveness in cost, quality and delivery to the customer. Some of the more widely applied are given in the table

	Acronym
Enterprise organisation	
World-class manufacturing	WCM
Computer-integrated manufacture	CIM
Electronic data interchange	EDI
Networking—wide and local areas	WAN or LAN
Engineering	
Computer-aided engineering	CAE
Computer-aided design and manufacture	CADCAM
Concurrent or simultaneous engineering	CE/SE
Design for manufacture and assembly	DFMA
Engineering data management	EDM
Production	
Computer-aided production management	CAPM
Materials-requirements planning	MRP
Manufacturing-resource planning	MRPII
Optimised production technology	OPT
Just-in-time	JIT
Flexible manufacturing systems	FMS
Quality	
Company-wide quality improvement programs	QIP/CQP/CQI
Total quality management	TQM
Statistical process control	SPC
International quality system standards	ISO 9000/BS 5750

(2) *Flexibility*—create a situation where the minimum of hardware and human resources are capable of handling multiple products and operational demands.
(3) *Involvement*—involve staff more fully in consultation and decision-making in aspects that effect their work. This is a predominant feature of quality-improvement programmes, but should be generally applied as a company philosophy and manufacturing strategy.
(4) *Quality*—endow the complete organisation with a 'quality culture' that involves every individual in a continual quest for the improved quality of their performance and output.
(5) *Inventory*—create a situation where varying demand can be met from the lowest possible inventory.

It will be noted that measured cost and efficiency do not appear in these objectives. The philosophy behind this is that if the organisation can achieve a culture that meets the above objectives then lower costs and greater efficiency should come.

16.1.5 Key elements of a manufacturing strategy

When setting a manufacturing strategy the following elements need to be considered. Some are common to most manufacturing organisations, but the emphasis will vary from company to company.

The elements to consider include the following.

(1) Basic nature of operation—assembly only, full manufacturing of piece parts, subcontract content.
(2) Production volumes and mix.
(3) Inventory—inventory carrying policy and value targets; materials sources and costs.
(4) Manufacturing resources—what is needed after consideration of the above elements?
(5) Labour availability—requirements and availability of staff and their skills.
(6) Location—alternative sites for manufacture in relation to overall profitability, markets, available resources and investment needed.
(7) Existing product range—profitability, market life remaining, cut-off point and continued spares/service support.
(8) New products—development costs, potential profitability, lead time to market, product life cycle, distribution, service and spares.

16.1.6 The manufacturing organisation structure

16.1.6.1 Past structures

As noted before, most companies used to operate in a hierarchal way through many layers of management. *Figure 16.2(a)* shows a typical functional structure with the flow of product at 90° to the flow of responsibility up and down the hierarchy. Decisions, data and responsibility for all activities go up and down the organisational structure. This is not to say that interdepartmental communications do not take place at every level, but the need for say a manufacturing engineer, production supervisor and quality engineer to refer up the line and wait for separate responses in addition to trying to solve the problems at their level, does cause uncertainty and, possibly, a delayed solution.

The product flow is shown going across the page as materials flow and value is added between the goods-in and despatch doors. Systems are put in place to allow the product progress to proceed without constant reference up the management line, but there is not sufficient focus of individuals to

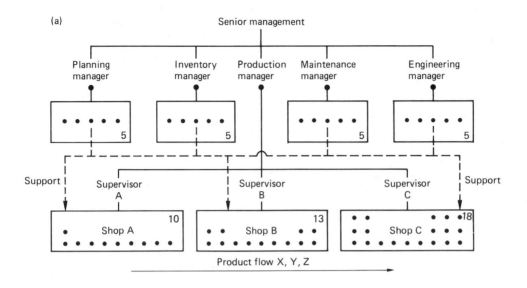

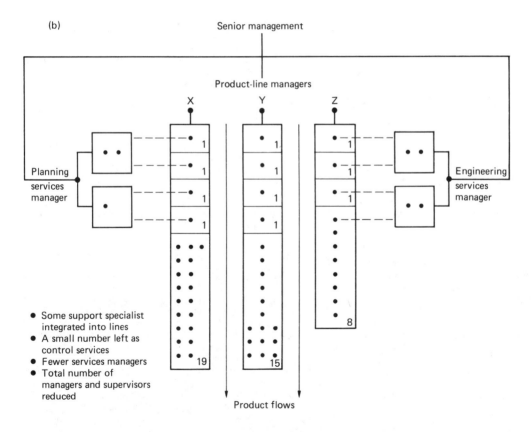

Figure 16.2 Manufacturing organisation structures. (a) Traditional organisation. (b) Product-line structure

products, except in dedicated and rather inflexible production lines. In any case the 'production line' is not the only contributor to the product flow. The focus of the centralised support and service functions is just as important and a lack of product focus in these areas is often a major block to the organisation's total efficiency door-to-door.

16.1.6.2 *Future structures*

Manufacturing organisation structures will be market and product oriented with the skills of the production and support staff aligned more to product groups than process groups. They will also be 'flatter' with fewer levels of management, as responsibilities are devolved back to the operating staff. *Figure 16.2(b)* shows such a structure which encourages:

(1) better communications between the individuals contributing to the design and production of the group product—giving a better and faster response to events and problems;
(2) a greater sense of 'ownership' of the product by the group members and, therefore, greater individual motivation;
(3) a greater input by individuals of possible solutions to problems; and
(4) multitasking and multiskilling becomes more acceptable to individuals in the interests of the group and their own development.

These factors will result in more highly motivated working and better quality, which in turn will create shorter lead times and lower costs.

16.2 Strategies for increasing manufacturing competitiveness

16.2.1 World-class manufacturing

World-class manufacturing (WCM) is the name given to the attainment of a level of customer service and manufacturing efficiency that ranks with the best in the world. With the increase in international competition, leading companies have realised that they will only survive if they are among the best in a world-wide market. They have also realised that in order to achieve such standing there must be a company-wide strategy and commitment to that level of attainment.

The adoption of a WCM strategy by top management leads to a commitment to vigorously pursue every avenue of improvement in the company. In particular, this is expressed as improvements in the *quality* of each person's contribution and all operational activities.

The *key objectives* of a WCM programme must be:

(1) 100% customer service/satisfaction,
(2) 100% quality in the product,
(3) competitive pricing,
(4) shortest delivery and manufacturing lead times possible, and
(5) lowest costs consistent with the above objectives.

Item (1), customer satisfaction, will come from the achievement of other objectives.

The *key philosophies* to be adopted in a WCM company relate to people and must be:

(1) every member of the company *must* adopt an attitude of personal responsibility for the highest possible quality of his/her work;

(2) the company and its departments are *teams* not warring groups or individuals;
(3) every member and working department of the company must think in terms of being a supplier to the next person or department in the chain, who must be regarded as customers; and
(4) the internal customers have as much right to the highest level of service possible as do the ultimate customers of the company.

The *operating techniques and methods* will vary from company to company, but some of the modern practices one would expect to see in use include:

(1) company-wide quality-improvement programmes;
(2) just-in-time (JIT) working;
(3) concurrent engineering;
(4) design for manufacture and assembly (DFMA); and
(5) computer-integrated manufacture (CIM)—
    (a) computer-aided production management (CAPM), and
    (b) computer-aided engineering (CAE and CAD-CAM).

For more details on item (5) the reader should consult Chapter 14 in this volume.

16.2.2 Quality improvement programmes

Quality-improvement programmes (QIPs) go under various titles. Some are proprietary products sold by consultants, whilst others are internal company schemes. They are vehicles for the application of a company's total commitment to attaining the highest level of quality throughout the organisation. QIPs are not a solution in themselves.

QIPs are designed to achieve the following.

(1) Increase each person's awareness of the need for high quality in their work.
(2) Apply QIP to all departments and activities—it is not a traditional quality inspection technique restricted to the shop floor only.
(3) Install a system that has formal procedures of working to follow, but which at the same time encourages people to use it in a personal and relaxed manner.
(4) Allow groups of people to examine their working situation and discuss and plan improvements both very minor and major.
(5) Provide a means of self-measurement of results plus local visual presentation of the plans and results.
(6) Train all people in their participation in the programme as individuals and as members of their working group. Shop-floor and office staff are trained in simple statistical and presentation techniques so that they have a greater interest through doing the analysis themselves.

16.2.3 Just-in-time (JIT) processing

Just-in-time (JIT) processing has been applied extensively in the last few years. It has proven to be an effective way of reducing lead times and inventory by working to the principles of a 'pull system', in which parts are made only as they are required by the customer process or department. In many companies, JIT has produced (at low capital cost) some of the inventory and cost reductions that high technology planning and manufacturing projects have failed to deliver. Reductions of over 50% in inventory and lead times have been achieved in many implementations.

A successful JIT implementation can be achieved by addressing the subjects of JIT philosophy, organisation, product manufacturing structures, batch sizes, and lead times.

16.2.3.1 JIT philosophy

As with most other activities, success can only be achieved by having the full commitment of the people involved. JIT is a physically visible system that is driven at shop-floor level rather being a top–down paper planning technique. In a well designed and understood JIT environment the shop-floor supervisors and staff become the owners as well as operators of the physical system. This provides a much better working atmosphere and interest because they can see the results of their own and others' work and, most importantly, make a positive contribution to the resolution of problems.

It is vital, therefore, that senior management engender the correct spirit of team work and involvement on the part of the shop-floor staff and their support services. JIT work teams should hold regular meetings (weekly if not daily) at their workplace to sort out current problems and for further education and the distribution of management information.

16.2.3.2 Organisation

The fundamental element in JIT systems is the *kanban* which is the repository of a JIT batch and may take the form of a tote bin(s), stillage(s) or other container(s) or may just be an area designated on the shop floor. The number of parts in the kanban is preset (but variable in response to situations), and becomes the process and transfer batch quantity. The number of kanbans will depend on the number of separate operations in the revised 'flatter' manufacturing structure.

JIT teams should be kept relatively small (say, less than 10 people) in order to maintain a close team spirit and sense of identity. The style of the team should be democratic with a supervisor or team leader advising, informing and guiding rather than ordering. Many JIT teams run with only a nominal hierarchical supervisor—in any well-mixed group natural leaders will emerge in different situations.

Communications should be kept simple, with paperwork at a minimum. Whiteboards, blackboards or flip charts should be used to display product and performance information in large characters. They should be sited in a prominent position within the JIT area and updated on a very frequent basis. Staff should take turns in this role to increase their involvement.

16.2.3.3 Product structures and bills of materials

A major technique for improving throughput and reducing lead times is to reduce the number of subassemblies and separate operations. This reduces the number of batches and their transfer operations, decreases work in progress (WIP) stocks and increases staff interest and their work. When introducing JIT the product structure and bill of materials (BOM) progress through manufacturing should be examined and modified to make them 'flatter', by removing unnecessary intermediate stages.

16.2.3.4 Batch sizes

Batch sizes are reduced to those that can be progressed in an appropriately small quantity. A batch of 1 may be ideal from the standpoint of having the lowest inventory value, particularly for high-value, long-process-time parts or assemblies. It is unlikely to be ideal for low-value, short-cycle processes because it creates too many demand pulls on the system,

thereby increasing the time and cost of progressing relative to the actual processing time.

Nevertheless, batches should be as small as is practicable. There are no generally applicable rules for this—each product and process must be carefully examined to determine the best balance of batch quantity and process time needed in relation to the whole production process of the product and, in particular, to the preceding and succeeding operations. The key objective is to balance the *flow* of the overall product in small efficient lots. The balancing of the start-out quantities of all the WIP is to be avoided in order to reduce the excessive inventory costs caused by varying rates of WIP consumption.

16.2.3.5 Lead times

The overall lead times of most products and assemblies are determined by the longest lead time within the constituent parts. Typically, one part with a lead time of, say, 10 weeks constrains the assembly of the rest of the parts which have lead times of 1 or 2 weeks. Long lead times can often be reduced by modifying the supply chain, BOM, process order and batch size of the parts and/or of the parent assembly.

The response time of the 'on-demand' pull system greatly reduces the lead time of a part. In other words, the direct action of the supplying department to the pull-system demand can avoid the delays often found when supposedly premanufactured parts are not in stock—the task of then getting these made quickly only further upsets a factory-wide scheduling system and capacity.

16.2.4 Concurrent engineering

Concurrent engineering (also called simultaneous or parallel engineering) is the technique of simultaneously considering the design, quality, manufacturing engineering, production and sourcing implications of a new product design *during the design process* and not as a series of afterthoughts. This methodology has long been the desire of manufacturing personnel, but has often been less welcome, and therefore unworkable, with design staff. The method has only become fashionable with the late realisation by non-manufacturing managers that it can significantly reduce the lead time to bring a new product from conception to the market place. Short lead times to market are becoming increasingly important for manufactured goods as fierce world-wide competition is demanding new or improved products at an ever-increasing rate. It is vital that this demand is met if a company is to maintain, let alone increase, its market share and profitability.

Concurrent engineering facilities are now being offered by computer systems as facilities on CADCAM and CAE software where the sharing of databases allows simultaneous or parallel working rather than serial working (where manufacturing engineering, quality engineering and production planning work started only after the substantial completion of the design work).

However, the main factor in concurrent engineering is multidisciplinary team working. The early participation of manufacturing personnel in the actual design process can make significant savings in development and production costs and lead time. Lead time is reduced by avoiding time spent on unmanufacturable or too costly designs and also in time savings through the overlapping and parallel working of the departments concerned.

A concurrent engineering programme is shown in *Figure 16.3*.

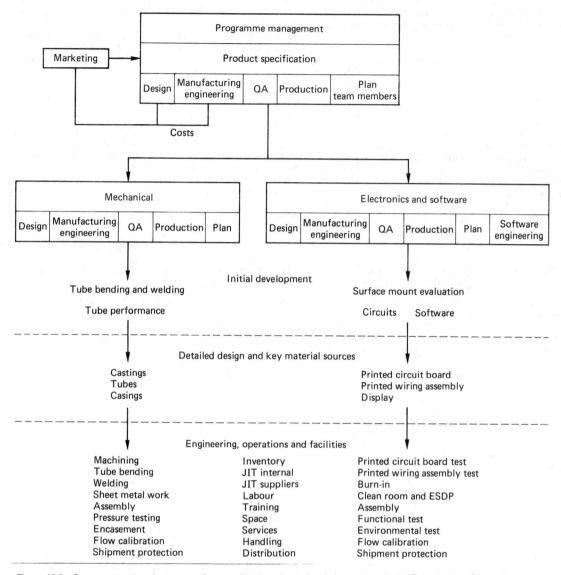

Figure 16.3 Concurrent engineering program for a combined mechanical and electronic product. JIT, just-in-time; QA, quality assurance; ESDP, electrostatic discharge protection. (Courtesy of Custom Networks Ltd)

16.2.4.1 A concurrent engineering project

Within a factory environment there are usually six functions that play a major role in the manufacture of products: production departments; manufacturing engineering; quality control; production control; inventory control; and purchasing. For successful project completion, all of these must be involved as early as is advisable.

Phase 1—Initial development and design As an absolute minimum for concurrency, an engineering team should be established comprising development engineers, product designers, manufacturing engineers and quality engineers. Being an engineering team, this group should be able to deal with all the technical considerations of the emerging design. In addition the manufacturing and quality engineers should also be

able to represent the general interests of the production floor because of their close daily working with it.

Phase 2—First design review In addition to the engineers, this should include staff from production, production control and inventory control in order to gain their first formal inputs to the preliminary design.

Phase 3—Full design and preproduction planning Two teams should now be set up. Whilst the engineering team(s) continue with the more detailed design of the product, the processes and their quality requirements, a manufacturing team will start preliminary planning of sourcing and production. It is recommended that a key designer, manufacturing engineer and quality engineer should sit on the manufacturing teams, and vice versa. This stops the project and its progress relying on one very large team meeting, but maintains the communications between the parallel engineering and manufacturing teams.

Phase 4—Final design review This may be preceded by intermediate reviews at appropriate programme or project further-authorisation points. This official authorisation step is essential if, in the rush to market, the company is to avoid using large amounts of resources in introducing a product that does not meet its performance, cost or market objectives. The review should be attended by key personnel from all the departments involved and by senior management. If the design is accepted this initiates full tooling for production and production materials sourcing, building on the preliminary work already undertaken. Because of the long lead times involved, some tooling, capital equipment and sourcing commitments may have to be made prior to full design acceptance, but this is a matter for management judgement.

16.2.5 Design for manufacture and assembly

Design for manufacture and assembly (DFMA) is a technique for analysing the cost efficiency of a proposed design during the design stage. It is a technique that has recently attracted attention in manufacturing and design circles with the realisation that unnecessary cost is best eliminated in the design work, not once a finalised design has reached the manufacturing domain. Commonly, about 90% of product cost is committed at the design stage (as logically it must be).

Manufacturing engineers have long wished for a greater consideration of manufacturability in the design but, as with concurrent engineering, it has taken fiercer world-wide competition to bring the message home to senior management and product designers. Many companies now operate a system of reviewing designs for manufacturability and cost efficiency. This involves a measure of concurrent engineering practice, but may well be performed under varying titles and engineering organisation structures.

Interest has also increased with the development of computer program versions of the basic design analysis methods. These methods address design and manufacturing factors and use a points or ranking system to evaluate them. The factors considered include:

(1) product and parts—output volumes, mix, and families;
(2) what production processes exist for part manufacture—cost of parts per process;
(3) part symmetry and other geometric features—ease of grasp and alignment; and
(4) type of assembly joining—snap fit, screw, bolt, adhesives.

The systems have libraries of data on which to draw as the user is taken sequentially through a question/answer or statement menu. The results can give guidance not only on the most efficient product cost, but also on the process and capital costs implied.

17

Control of Quality

H L Cather

Contents

17.1 The concept of quality

17.1.1 Defining quality

The word 'quality' is used in several different contexts which can lead to the meaning becoming slightly blurred, the last thing that is desired in quality. The mainstay of all matters relating to quality are the international and national standards, so where better to look for a definition of quality than in a British Standard?

BS 4778: 1971 is a British Standard devoted to a glossary of terms used in quality assurance. Within this standard, the following definitions are recommended to be used to differentiate between the three most common usages of the word 'quality':

Quality The totality of features and characteristics of a product, or service, which bears upon its ability to satisfy a given need, i.e. a fitness of purpose, or value for money, sense.

Grade When applied to a material or product, the quality grade is an indication of the degree of refinement; when applied to a service, the diversity of functions or facilities offered, i.e. this is a comparative or degree-of-excellence sense.

Quality level This is a general indication of the extent of departure from the ideal, usually a numerical value indicating either the degree of conformity or non-conformity of a product to its specification, i.e. a qualitative sense.

In order to control or improve quality, it is first necessary to identify, measure and evaluate it. The above definition of 'quality' calls for the identification of the characteristics and features bearing upon the fitness of purpose of a product or service. Furthermore, the ability to satisfy a given need must include economics as well as design, availability, maintainability, and all other characteristics that the need for the product or service involves.

This demonstrates that the quality is determined by all the parts of a business acting together. Quality can then be said to be the sum of

(1) recognising the customers' needs;
(2) designing products, or services, to meet them;
(3) producing reliable products to promised delivery dates; and
(4) maintaining effective after-sales service and feedback.

These elements add up to fitness for purpose, a perceived value for money and hence satisfied customers.

Figure 17.1 shows some of the determinants and measures of the quality of a product that bear upon its ability to satisfy a given need. Note that even in this figure, which is an extract from BS 4778: 1971 the term 'reliability' is used in two different senses—firstly in a general sense and then to denote particular characteristics which can be stated in quantitive terms.

It is important that quality is not only applicable to tangible artefacts and services, but also to verbal and written information.

17.1.2 Quality—a competitive weapon

In every market, an organisation has to operate against keen competition in order to gain its share of the available orders. It is, therefore, important that each organisation clearly identifies the criteria which will enable it to enter a market and then win orders against its competitors. The common order-winning criteria are:

(1) price,
(2) design,
(3) quality,
(4) product range,
(5) delivery speed,
(6) technical liaison, and
(7) delivery reliability.

Price by itself is seldom sufficient, therefore non-price factors such as quality are key determinants of an organisation's current order book and its survival into the future. It is very important that an organisation becomes able to assure prospective customers that their needs will be met by products

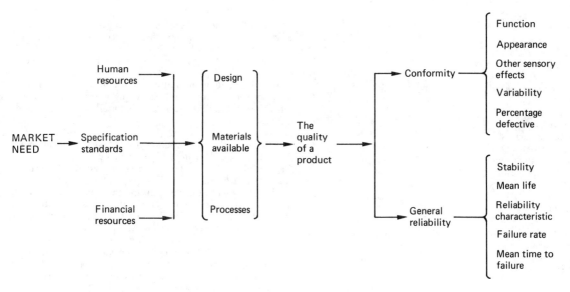

Figure 17.1 Some of the determinants and measures of the quality of a product

which are fit for their purpose. This assurance should not just be in the narrow sense of provision of proof of attainment of a particular quality level, but should be readily perceived by the customer.

Quality is of prime importance to a customer and few suppliers will not claim to be striving for good quality. Most, in fact, spend considerable time and resources on activities such as testing and inspection to detect and correct defective products. In addition, they have a large section which can speedily deal with any problems found during operational service. However, this effort, by itself, cannot ensure that quality is guaranteed. It is a form of quality control—a technique which sustains product quality to the specified requirements. Quality control is a subset of quality assurance, the latter being the key to efficient attainment of quality.

Quality assurance involves the whole of an organisation and covers all the activities concerned with the attainment of the product's quality. BS 4891: 1971 *A Guide to Quality Assurance*, gives the basic elements involved in quality assurance which are set out in *Table 17.1* and briefly described below.

Management objectives These should be closely linked to each market where an identifiable need exists. This requires an examination to determine for each product range the market entry and selling factors relating to quality, price, reliability, delivery, etc. Top management must give a clear lead to the importance attached to quality assurance by the

Table 17.1 The basic elements involved in quality assurance

	Market demand
Design specification control	Management
Specification	responsibility
Design engineering	for
Value analysis	
Reliability engineering	QUALITY
Development engineering	ASSURANCE
Documentation	usually
Design review	involves
	defining
Purchasing control	MANAGEMENT
Purchasing	OBJECTIVES
Vendor appraisal	
Vendor surveillance	establishing a
Receiving inspection	
	PROGRAMME &
Manufacturing control	PLANNING
Work study	
Process planning	operating a
Resource allocation	QUALITY
Tooling and gauges	CONTROL
Material processing	SYSTEM
Production and assembly	
Inspection and testing	taking
Packaging	advantage of
Storage	inherent
Shipping	knowledge
	and/or
	research and
Marketing and servicing	development,
Maintenance	subcontracting
Product support	subject to
After-sales service	
Market reaction	REVIEW &
Complaints	EVALUATION

organisation. Key personnel should be identified to resolve all quality-related matters.

Programme This sets out the steps to be followed in applying quality assurance throughout an organisation from the initial product conception to after-sales service.

Planning This seeks out any special requirements which may arise and lays down the procedures for dealing with any that do so.

Quality control This establishes the standards that will be met, the method of measuring comformity with these standards and the action to be taken when the standards are not met.

Design control Quality must be designed into the product to meet the customers' requirements through a design that allows practicable and economic manufacture and servicing. The design must encompass:

(1) minimum functional/operational requirements,
(2) maximum environmental expectations,
(3) technical and manufacturing limitations,
(4) safety requirements, and
(5) cost limitations.

Purchasing control All organisations are responsible for the outgoing quality of their products and they must, therefore, ensure that all bought-in material contributes towards meeting the customers' needs.

Manufacturing control Manufacturing is where the specification is finally translated into a finished article. All processing should therefore be carried out in controlled conditions to detailed instructions.

Material control This establishes a system to identify, segregate and protect all material from initial entry into the plant until reaching the end-user, including during transportation.

Marketing This is a direct link to the customers which should continually seek out and feed back market reactions. In particular marketing should:

(1) identify market requirements,
(2) consult manufacturing before accepting non-standard work or quality level, and
(3) ensure that all costs are considered in pricing products.

Servicing In addition to providing prompt efficient servicing and attention to complaints, servicing is responsible for the provision of technical documentation, user training and spares provision.

Effective, continuous market research is the starting point for achieving quality in the eye of the customer. Unless the customers' needs are known, and understood, a designer cannot meet them.

The designer must also ensure that the manufacturing processes are considered to ensure that the specification can be reached and held. The designer should also work closely with material and component suppliers to ensure that these also meet the specification requirement.

The entire manufacturing function must be committed to meeting the design specification each and every time. Quality must be built in, it cannot be inserted later by inspection.

Many customers cannot protect themselves by inspecting fully the finished product. It such cases, the supplier must provide some objective evidence that the specification has been achieved. This can be by test reports, or by guaranteeing adherence to an established national, or international, standard.

Marketing reinforces the customers' confidence in the product by back-up in technical liasion, servicing and spares availability.

17.1.3 Quality costs

The total of the quality-related costs within a company is very large. It has been estimated that in the UK some 10% of the Gross National Product can be classified as being quality related. Within any organisation they will range from 5% to 25% of the sales turnover value.

These costs are mainly due to failures and appraisal which add nothing to the real value of a product. They are reflected in additional costs and hence lower profits to every organisation which has them.

Many of these costs are preventable to some degree but, because they are not correctly identified, they often do not attract sufficient management input into this prevention.

There are considerable problems in clearly identifying all the costs which are quality related. BS 6143: 1992 *Guide to the Determination and Use of Quality Related Costs* does, however, identify the following four simple classifications.

External failure costs   These costs arise from the failure of the product outside the organisation. They are identified by items such as complaints, warranty work, returns, etc. In many cases they result in the loss of future customers, either through their direct experience or from passed on experiences of others. It is estimated to cost 10 times as much to gain a new customer as to hold onto an existing one, and hence this can be fatal.

Internal failure costs   These are costs arising inside the organisation by failure to meet the specified quality. They are identified by costs such as scrap, lost time, reworking, etc.

Appraisal costs   These are the basic costs of assessing the quality being attained, e.g. inspection, testing, equipment, etc. Some of these could in fact be reclassified as internal failure costs as their main function is to screen out defective products.

Prevention costs   These are the costs involved in any action taken to identify and eliminate the causes of quality failures. They include product design modifications, equipment maintenance and modification, vendor assessment, training, etc.

These costs are often not identified separately as quality costs by normal accounting procedures, but are submerged with other data as direct and indirect costs. Whereas the collection and reporting of these costs should not be a separate exercise from the remainder of the management accountancy information, they do require accurate separation in order to highlight them.

The quality costs can be established but are sometimes obscured. They must be separated out from the others. It is important to ensure that they carry an equitable share of overheads such as rates, heating, canteen, etc.

Figure 17.2 shows how these quality costs interrelate to one another. As more resources are committed to the prevention area, the occurrences of failure should reduce and hence their related cost should reduce. Note, however, that the reductions are often of a diminishing-returns nature. An economic balance between the prevention and failure costs can be

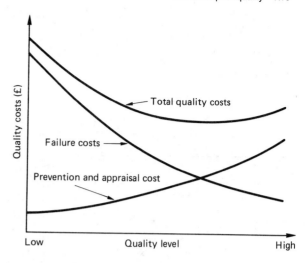

Figure 17.2 Build up of quality costs

breached by excess preventive costs as readily as by failure costs.

The basis to be used when justifying expenditure on preventive action must be the same as for any other revenue expenditure or capital investment. The cost of carrying out the preventive action should result in increased profit, be it through increased sales, a higher price differential or reduced costs. Quality prevention is only worthwhile when it pays for itself.

The other main reason for identifying quality costs is control. Ideally, there should be clear standards for comparing the quality costs against, but even without standards budget control can be carried out. Variations and high costs can be readily identified and tackled.

Measuring quality costs firstly gives an indication of how large a portion of the organisation's costs arise here. They can be used as an indicator of continued performance, or improvement. They focus attention and hence aid decisions on where resources could be directed to give the best chance of good returns. More importantly, measuring is a prerequisite for control.

Reports and ratios should be set up which are relevant to the organisation's objectives and sensitive to any change in cost or business trends. The following are commonly used:

(1) Labour base: $\dfrac{\text{Internal failure costs}}{\text{Direct labour cost}}$

(2) Cost base: $\dfrac{\text{Total failure costs}}{\text{Manufacturing costs}}$

(3) Sales base: $\dfrac{\text{External failure costs}}{\text{Sales revenue}}$

(4) Product base: $\dfrac{\text{Test and inspection costs}}{\text{Product units made}}$

(5) Value added base: $\dfrac{\text{Total quality costs}}{\text{Value added}}$

(6) Purchases base: $\dfrac{\text{Incoming inspection costs}}{\text{Value of purchased parts}}$

(7) Quality split: $\dfrac{\text{Prevention costs}}{\text{Total quality costs}}$

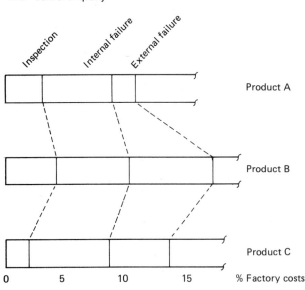

Figure 17.3 Comparative product quality costs

These should be periodically reviewed to reflect any changes which may have occurred in the base used, e.g. increased automation will reduce the direct labour cost, and a change in the product mix may affect the overall blend of quality and/or other costs.

As the organisation becomes skilled in identifying and reporting on quality costs, a more detailed analysis can be made of the breakdown of these quality costs by applying similar ratios to the above but on a departmental and/or product split, as demonstrated in *Figure 17.3*.

Charting techniques, amongst others, can aid in analysis and indicate when action is required.

17.1.4 Quality and the law

There are several areas of civil and criminal law in which quality plays an important part. It can be the ultimate adverse publicity to lose an action under the impartial, and public, gaze of the courts.

17.1.4.1 Civil law

There are three main areas where quality plays an important role.

Vicarious liability This states that an organisation is responsible for the actions of its employees if they occur in the course of that employment.

Contracts In civil law a contract is based on offer and acceptance. The customer may lay down certain specifications, inspections, etc. Terms related to conditions can be expressed in the contract, implied by custom and trade practice or by law, e.g. the Sale of Goods Act (1971) or the Supply of Goods and Services (1982). They include:

(1) goods/services must fit their description,
(2) goods must be fit for the purpose intended, and
(3) goods must be of merchantable quality.

Note that these are *absolute* duties of a seller and the seller is liable even though they have taken all possible care to ensure that these conditions are met.

Quality in all its aspects ensures that the specification can be met before the contract stage and is maintained afterwards. Penalties for failing to fulfil a contract include the customer rescinding the contract and leaving the seller with costs which it cannot recover. In some cases it can lead to penalties being applied in addition.

Consumer Protection Act (1987) This introduces *strict liability* regarding injuries received because of a defective product. A claimant need only prove that:

(1) he, or she, was injured,
(2) the product was faulty, and
(3) the defendant was responsible for the product's circulation.

Note that it is the producer, or importer, who is liable under this act. Negligence need not be proven, nor does *privity of contract*; the plaintiff only needs to show that, on the balance of probabilities, a defect in the product caused the damage.

17.1.4.2 Criminal law

There are several acts which stipulate criminal offences, where again quality plays a part in ensuring no liability occurs.

Trades Description Act (1968) This makes it an offence to make false statements about services or facilities.

Consumer Safety Act (1978) This allows the Government to set regulations about the standards which have to be achieved by specific products, e.g. motorcycle helmets, fabrics for children's nightwear, foam stuffing for furniture.

Health & Safety at Work Act (1974) This lays down an obligation on the part of suppliers to ensure that articles, or substances, be designed and constructed to be safe and without risk to health. Further duties are also laid down to carry out any necessary testing and examination, to carry out necessary research and to provide adequate information for the article's use, storage and transportation.

Consumer Protection Act (1987) This also lays down the General Safety Requirement which states it will be a criminal offence to supply unsafe consumer goods in the UK. In addition, this act consolidates and improves the regulation making power available under previous laws. Many British or international standards will be used as the basis for compliance under these regulations.

Note that in all court action, the court will take into account all relevant circumstances including:

(1) the manner in which the product is marketed;
(2) any instruction, or warning, given with the product;
(3) what reasonably might be expected to be done with the product; and
(4) the time the producer supplied the product.

In all cases, the available state of scientific and technical knowledge will also be taken into account. If it was possible for a defect to be discovered, then ignorance of this fact is seldom acceptable as a defence. It is seldom possible to contract out of these obligations, and an attempt to do so with consumers is in fact unlawful under the Unfair Contracts Act (1977).

In legal matters, the best way to avoid adverse actions is not to commit the offence. Quality is the key to prevention.

17.1.5 Organisation and the duties of a quality-assurance section

Quality assurance must have access to the highest levels within an organisation. Although personnel must have clear responsibility for certain quality tasks, the responsibility also resides in every position within the organisation. Tasks which may require separate personnel are listed below.

Quality engineering
(1) Establishment of quality requirements from the customer's needs.
(2) Analyse performances and recommend improvements.
(3) Assist in setting standards and review criteria.
(4) Determine and produce quality procedures.
(5) Conducts quality audits.
(6) Analysis of customers' complaints.

Process control engineering
(1) Appraise, interpret and implement quality plans.
(2) Conducts drives to reduce quality costs.
(3) Conducts specific tests on equipment for process quality studies.
(4) Calibrates test and measuring equipment.
(5) Records and analyses quality data.
(6) Performs testing, sampling and quality audits.

Information and equipment
(1) Designs, constructs and proves out test and measuring equipment.
(2) Provides calibration schedules for in-process control equipment and test/measuring equipment.

17.2 Quality through integrated design

17.2.1 The role of design with regard to quality

Design is an iterative process which proceeds from establishing the initial market need to selling to the customer. It involves going through a series of phases such as specification, conceptual design, detail design and manufacture.

It is during the specification and conceptual design phases that a product's ability to be fit for its intended purpose is initially realised. *Table 17.1* shows that this ability to meet a given need is determined by a combination of design, material and the processes involved in its manufacture.

It is estimated that, on average, failures during design are responsible for 80% of a product's quality problems. Design failures are also more liable to be passed on to the customer than are manufacturing failures.

Design must be based around the following criteria:

(1) minimum functional requirements,
(2) maximum environmental expectations,
(3) cost limitations, and
(4) safety requirements.

It is design's function to create exact and clear communications in the nature of:

(1) functional specifications,
(2) detail design, and
(3) manufacturing instructions.

Each of these should show clearly the acceptance and rejection criteria and how they have to be tested.

17.2.2 Specifications

The specification is the official means of communicating the needs or intentions of one party to another. It may be written from a using, designing, manufacturing or selling aspect. It should be written in clear unambiguous terms of the optimum rather than maximum or minimum quality grade required to fulfil the customers' need.

Where an appropriate British, or other, standard is already in existence which wholly covers the specification, this should be referred to. Where a standard only partially covers the need, it should be quoted with details of the variation.

Where no standard exists and a specification needs to be drawn up, then BS 7373: 1991 *Guide to the Preparation of Specifications*, gives a comprehensive list of items which may be required in the specification (see *Table 17.2*).

17.2.3 Modifications to design

Over time all designs will be modified. Strict control is essential to ensure that any necessary modifications are made in a cost-effective manner and that all parties concerned are notified and in agreement. It is very important from the quality angle that the correct modification is being referred to in manufacture and parts ordering.

Sometimes, temporary departures from the specification may be requested and these must be especially well controlled. In particular, the following must be fully documented.

(1) *Concession*—the authorisation to use or release a limited quantity of material, components or stores already manufactured but not complying with the specification.
(2) *Production permit*—the authorisation to proceed to manufacture a limited batch using a material or process which does not comply with the specification.

The use of the above should be strictly limited as they can often reflect on the quality image of the organisation.

Table 17.2 Items covered by BS 7373: 1991 *BSI Guide to the Preparation of Specifications*

1. Title
2. List of contents
3. Foreword
4. Scope of the specification
5. Role of equipment or material
6. Definitions
7. Relevant authorities to be consulted
8. Related documents and references
9. Conditions in which the item is to be installed, used, manufactured or stored
10. Characteristics
11. Performance
12. Life
13. Reliability
14. Control of quality
15. Packaging and protection
16. Information from supplier to user
17. After-sales service

17.2.4 Process selection

Each process has an inherent range of surface texture and dimensional tolerance which it can produce. The actual range attained being a function of the equipment design, age and the maintenance which it has received. It is imperative, therefore, that the process selected can attain the specified design tolerance required.

Within the possible process range, the cost of production increases substantially as the design tolerance specified tightens. It is therefore important that the widest possible tolerance which can still meet the functional requirements is specified.

Figure 17.4 demonstrates a nominal range of surface roughness values produced by different processes, and *Figure 17.5* demonstrates the relationship of texture to production time. A similar keen relationship exists between processing costs and achieved tolerance (see *Figure 17.6*).

17.2.5 Design techniques to aid quality

17.2.5.1 Product design specification check-list

This is used to produce the specification of what is needed, i.e. What needs must the product be designed to meet? *Table 17.3* lists some of the elements that must be queried to ensure that all aspects of the need are covered.

17.2.5.2 Conceptual design evaluation

Once the specification has been completed, many different ways of evaluating it may present themselves. A weighted evaluation of each concept against the design criteria can be made, as shown in *Table 17.4*.

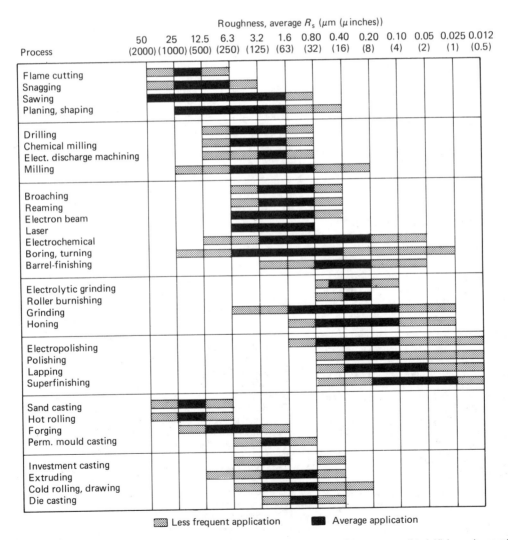

Figure 17.4 The texture capability of processes. The ranges shown are typical of the processes listed. Higher or lower values may be obtained under special conditions

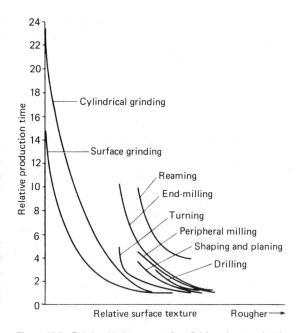

Figure 17.5 Relationship between surface finish and processing time

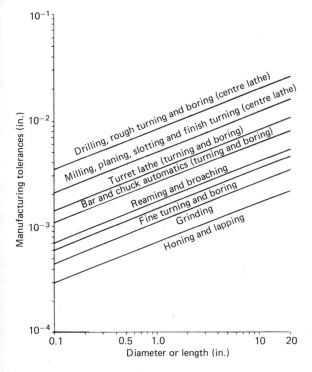

Figure 17.6 Typical tolerances for selected processes

Table 17.3 Design specification check-list

Factor	Points to be answered
Cost	Is it attainable? Market viability
Producibility	Existing processes. Experience in process. Desirability of new/old process. Supply of parts, material or equipment. Patents
Performance	Speed, frequency, continuity, etc. Noise and vibration level
Strength	Maximum loading. Type of loading (tensile, compression, torsion). Fatigue factors
Size and weight	Restrictions in processing, use or transport
Environment	Ambient conditions of use and storage. Resistance to corrosion
Standards	British or other standard. Testing requirements
Operation	Training, experience and attitude of operating and maintenance personnel
Maintenance	Serviceability. Replaceability. Specialised training and tooling
Life	Overall life. Time between services and overhauls
Reliability	Market need/expectation
Product life	Production span. Time till launch
Competition	Match with existing products
Storage	Method of packaging. Method of shipping. Method of bulk storage. Shelf-life
Safety	Interaction with people. Legal requirements. Materials to be avoided. Special features required. Effect on environment. Recyclability

grammes, and gives many examples of techniques and formulae which can be applied.

Reliability can be improved by minimising the parts in use, sticking to known reliable suppliers and components and de-rating an item's operating range, e.g. ensuring a unit rated for 15 A is used in a circuit normally operating at 5 A.

17.2.5.4 Failure mode, effect and criticality analysis

This is one of the techniques referred to in BS 5760: 1986. The aim of this technique is to apply a disciplined, documented approach to evaluating the design of a system, machine, component or process. It involves working through the design

17.2.5.3 Reliability assessment

BS 5760: 1986 *Reliability of Systems, Equipment and Components*, and its three constituent parts gives details of how to go about assessing reliability and managing reliability pro-

Table 17.4 Design analysis table for a mug washer

Attribute	Maximum points	Design A	B	C	D	E	F
Dries	10	10	10	7	9	2	10
Cleans insides	10	8	0	6	9	9	0
Cleans outsides	10	8	8	10	10	10	8
Range of mug size	10	10	10	2	8	8	10
Needs services	5	4	3	5	3	4	4
Design complexity	5	4	2	4	2	5	4
No. of components	10	7	3	8	1	9	9
Standard components	10	8	7	8	4	8	9
Ease of maintenance	10	8	8	7	9	5	9
Ease of manufacture	10	8	5	8	6	9	7
Ergonomic soundness	10	10	8	10	10	9	10
Reliability	10	9	7	9	8	8	8
Life expectancy	5	4	2	3	3	4	4
Size	5	5	3	5	4	5	4
Noise	10	10	9	8	6	5	9
Automatic cycle	10	10	10	10	10	7	10
Cycle time	5	4	4	4	5	5	4
Safety in use	10	8	8	7	8	9	8
Approximate cost	5	4	4	3	3	5	3
Total	160	139	111	124	118	126	130

considering in turn the effect and consequence of each failure or malfunction which *could* occur.

On complex systems, it is often necessary to break the main system down into various subsystems. Each subsystem can then be analysed with the result of these being recombined to give the analysis of the main system. The technique used is to complete the analysis form (see *Figure 17.7*) filling in the appropriate data:

Column 1 Description of component.
Column 2 Brief indication of component's function.
Column 3 List *all* possible failure modes of the component.
Column 4 Describe the effect that *each* failure will have on the overall function of the system.
Column 5 List *all* the possible causes of each failure.
Column 6 Indicate the *probability* of each failure cause arising: 1, low probability; 10, certainty.
Column 7 Rank the *severity* of each failure in terms of its effect on the function of the system: 1, only minor; 10, total failure or safety hazard.
Column 8 Rank the difficulty of *detecting* the failure: 1, easily detected; 10, unlikely to be detected.
Column 9 Calculate risk number (=product of columns 6, 7 and 8). This indicates the relative priority of each failure.
Column 10 Give a brief description of corrective action needed, who is responsible, and the expected completion date.

Data on probability can readily be obtained from records of past failures of components and systems coupled with the knowledge of users.

17.2.5.5 *Assembly analysis*

There are several proprietary systems which analyse an assembly with a view to reducing parts count and simplifying the assembly operations. These have a distinct effect on both quality and costs.

Figure 17.8 shows the loops involved in the design for manufacture procedure developed by Lucas Engineering & Systems.

17.3 Standards

17.3.1 British Standards Institution

Throughout the world each nation has standard institutions to ensure that products are marketable and interchangeable. In Britain this task is overseen by the British Standards Institution (BSI). This is an independent, non-profit-making organisation which receives most of its income directly as a result of supplying its services to Government, industry and commerce. Some of these services are described below.

17.3.1.1 *Preparation of British Standards*

British Standards are documents which spell out the essential technical requirements for a product, material or process to be fit for its intended purpose. They are prepared, published and up-dated by the BSI in response to the needs of the consumer, industry and the Government.

The use of standards assists in the simplification of manufacturing, stocking, cataloguing and specifying. In many cases the standard itself clarifies basic safety requirements.

Standards influence every stage of the industrial process—materials, dimensions, performance, requirements, codes of practice, test methods and even terminology. They are an essential ingredient for a safe, competitive and cost-effective business.

As one of the world's leading standards organisations, the BSI is active in the promulgation of international standards. This work is carried out through the International Organisation for Standardisation (ISO) and the International Electrotechnical Commission (IEC). Whenever possible, international standards are adopted as British Standards and are given dual ISO/BS or IEC/BS numbers.

17.3.1.2 *Information services*

The BSI has an extensive enquiry section which can identify any British, other national or international standard, place an order, and arrange translation and delivery. An extensive database is maintained which can be accessed directly by members' computing facilities.

17.3.1.3 *Technical help to exporters*

When selling goods abroad, knowledge of and compliance with foreign technical requirements is often critical to the success of any venture. This service can identify the appropriate standard for almost any product in most countries and assist in a detailed interpretation of the standard.

17.3.1.4 *Assessment and certification*

The use of third party certification systems allows manufacturers to demonstrate publicly their commitment to quality. The appearance of an independently awarded certification mark assures the purchasers of the quality of a product. Schemes which the BSI administers or assists in include the following.

Kitemark (Figure 17.9) If the Kitemark appears on a product, it means that the product has been independently tested

Product – I.C. engine	Subsystem – in line F.I. pump								F M E C A No. PR001
Process –	Location –								Date
System –	User –								Engineer

				P	Probability of occurrence				
				C	Severity of failure				
				D	Difficulty of detection				
				R.N.	Risk Number = $P \times C \times D$				

(1) Part No. Component	(2) Function	(3) Failure mode	(4) Effect of failure	(5) Cause of failure	(6) P	(7) C	(8) D	(9) R.N.	(10) How can failure be eliminated or reduced
Cams and follower rollers	Plunger drive	Fatigue pitting	Loss of power irregularity (or complete functional failure in extreme cases)	Overload (e.g. nozzle blockage, dirty or no oil, poor surface finish)	2	3	10	60	Supplier to determine failure definition. Limit operating pressure
Centre bearing bridge	Support camshaft centre bearing. Ties pump sides at centre	Cracked across centre of bridge	Mainly cosmetic. Slight oil seepage		1	2	10	20	Problem exists on present die cast pumps. Supplier likely to shell mould on our B cylinder pump therefore more material on bridge. To be assessed when samples are tested
Delivery valve holder		Erosion in bore	Nozzle blockage, loss of performance	High flow rates. Cavitation	5	3	10	150	To be assessed when samples are tested
Delivery valve	Unloading	Consistency U/L wear	Engine performance pump phase and delivery balance	Manufacture, adjustment poor filter/water	3	9	4	108	M/C test. L.P. systems detail
Delivery valve spring	Controls unloading	Fatigue/erosion/wear (omission)	Loss of engine performance, smoke	Hydraulic duty	4	4	4	64	Specification detail
Delivery stop peg	Control delivery lift	Wear—crushing (omission)	As above	As above	2	4	3	24	Injection rates
Plunger element		'F' slot polishing	Seizure or sticking leading to functional failure	Particulates produced by erosion of pump body in transfer gallery or erosion shield	1	10	10	100	Slots now roller into element prior to hardeing and grinding. To be assessed when samples are tested

Figure 17.7 Failure mode, effect and criticality analysis. (Courtesy of British Standards Institution)

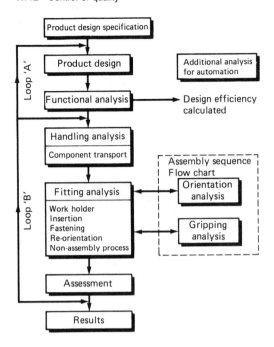

Figure 17.8   Design for assembly procedures. (Courtesy of Lucas Engineering & Systems Ltd)

Figure 17.9   The BSI Kitemark

Figure 17.10   The BSI safety mark

Figure 17.11   The BSI Registered Firm of Assessed Capability mark

against the relevant British Standard and that the company's quality system has been assessed against BS 5750: 1987. When a manufacturer has been licensed by the BSI, unannounced visits are made to the factory by the BSI Inspectorate to ensure that the agreed quality levels are being maintained.

Safety mark (Figure 17.10)   Similar procedures to the Kitemark apply, but with specific concern for safety in the product's normal use.

Registered firm of assessed capability   If a company conforms to the relevant part of BS 5750: 1987/ISO 9000, then they may be licensed to use the Registered Firm symbol (see *Figure 17.11*). For electrical components, the certification scheme is BS 9000: 1989.

17.3.1.5   BSI inspectorate

In addition to the assessment and routine surveillance for the BSI certification schemes, the BSI Inspectorate can also cater for the needs of an individual company or government body. This could entail assessment of a potential supplier, continually auditing quality levels, or ensuring that imported/exported goods are of the correct quality before despatch. This could be on a one-off basis or continuously.

17.3.1.6   Testing services

The BSI Testing Service carries out a portion of the testing required as part of the certification schemes, as part of its duties. Tests can also be carried out on a wide range of products against a variety of national and international standards, government regulations or company specifications.

17.3.2   Quality systems (BS 5750/ISO 9000)

Standards are the base point for quality levels and many standards are produced for quality techniques, and even for an organisation's quality system itself. To become registered under BS 5750: 1987/ISO 9000 has become very important, as this states that a company has the basic procedures and discipline to ensure that their output meets the customers' requirements.

In 1987, BSI revised the 1979 version of BS 5750 to become dual-numbered with ISO 9000, so that registered firms could deal on an international scale. The revised structure is:

Part 0: Section 0.1 *Guide to Selection and Use*
Part 0: Section 0.2 *Guide to Quality Management and Quality Systems Elements*
Part 1: *Specification for Design/Development, Production, Installation and Servicing*
Part 2: *Specification for Production and Installation*
Part 3: *Specification for Final Inspection and Test*

As can be seen in *Figure 17.12*, the coverage of the various parts is extensive and permeates all aspects of a company's organisation. It is worthwhile examining each clause in part 1 in a little more detail:

4.1 The management responsibility for maintaining quality must be clearly identified. This includes allocation of suitable personnel, the organising of quality audits and periodic review of the effectiveness of the quality system itself.

4.2 The quality system itself must be fully planned, documented and maintained. It should cover all aspects of the organisation including staff, equipment and records.

4.3 The contract with the customer sets out the specification that the organisation has to supply against. It must be carefully checked to ensure it is clear, unambiguous and capable of being met by the available resources.

4.4 Design control is needed to ensure that the design meets the customers' needs. This includes organising necessary interfaces and keeping records of design assumptions and calculations.

4.5 All procedures for changes/modifications must be especially tightly coordinated and authorised. Especially important is the procedure for ensuring the correct document is in place when required.

4.6 Correct quality depends on correct procurement procedures so that the items bought in meet the requirements. This involves vendor assessment, clear purchase specifications, incoming inspection and, of course, records.

4.7 Free issue material from the customer must be protected in use and storage when within the organisation's control.

4.8 Material control and traceability is necessary to ensure that all non-conforming material can be traced from suppliers, through the processing stages to customers. This is especially true in safety matters as it may give an opportunity to recover claims.

4.9 All manufacturing should take place to clear, proven instructions. The instructions should emphasise the control of any specialised processes and detail the necessary recording media.

4.10 All testing and inspection prior to, during and after manufacturing must be recorded to confirm compliance with specification. The techniques and equipment used must be clearly specified and their accuracy established. This should include certificates of compliance received.

4.11 Inspection, measuring and testing equipment must be capable of meeting the required precision consistently. Procedures should exist to control and maintain equipment, including scheduled calibration and traceability to recognised measurement standards.

4.12 The inspection status of all material must be clearly identified during all stages. The responsibility for releasing conforming material should be clearly established.

4.13 Non-conforming material must be clearly segregated and identified, including the reason. The subsequent disposition of this must be to clear procedures and the deciding authority recorded.

4.14 Corrective action should follow lapses in conformity as it is only the identification of causes that ensures subsequent conformity economically. This extends to vendors and subcontractors.

4.15 Handling, storage and transportation are all part of the supply train. Handling procedures should detail product protection to prevent damage and/or contamination which affects the products' usefulness.

4.16 Records are evidence of what occurs during any manufacturing and must contain all instructions, actions and results in an accessible system.

4.17 Auditing the quality system ensures it operates efficiently. Where improvements are identified, the responsibility for carrying them out must also be stated.

4.18 Proper training is the foundation of quality. Needs must be correctly identified and training must be effective, preferably by demonstrated records of competence reached.

4.19 Servicing quality must be handled in a similar manner to internal quality matters, including procedures for verification and records.

4.20 Statistical techniques, where used, must be specified, including any associated risk levels.

An organisation wishing to install a quality-control system such as BS 5750: 1987 will find it a useful marketing tool. In some cases, such as when supplying government agencies, it may be a prerequisite for tendering.

17.4 Material and process control

17.4.1 Incoming material

In addition to design and the processes involved, the other determinant of quality referred to in *Figure 17.1* is material. Section 17.6 details the sort of acceptance sampling schemes that can be applied to batches of incoming material. However, the most important method of ensuring one has suitable material available is to carry out a complete assessment of all suppliers' abilities to supply the correct material.

17.4.1.1 Vendor assessment

This assessment will involve visits to suppliers to examine their processing facilities and audit their quality procedures. Moreover, when examining suppliers, one should take into account all the factors which are of importance to the company:

(1) quality, i.e. the level of rejected batches;
(2) price per item, relative to other suppliers;
(3) delivery, i.e. deliveries made on time;
(4) service, i.e. the amount of back-up and flexibility demonstrated.

Vendor assessment is a simple rating of different suppliers against each other under headings so that their overall value to the company can be assessed. It is a matter of selecting the particular factors, such as those above, which are considered important, and applying a weighting to them in relation to their agreed impact on the company:

Clause (or subclause) No. in BS 5750: Part 0.2/ISO 9004	Title	Corresponding clause (or subclause) Nos in		
		BS 5750: Part 1/ISO 9001	BS 5750: Part 2/ISO 9002	BS 5750: Part 3/ISO 9003
4	Management responsibility	● 4.1	◐ 4.1	○ 4.1
5	Quality system principles	● 4.2	● 4.2	◐ 4.2
5.4	Auditing the quality system (internal)	● 4.17	◐ 4.16	—
6	Economics—quality-related cost considerations	—	—	—
7	Quality in marketing (contract review)	● 4.3	● 4.3	—
8	Quality in specification and design (design control)	● 4.4	—	—
9	Quality in procurement (purchasing)	● 4.6	● 4.5	—
10	Quality in production (process control)	● 4.9	● 4.8	—
11	Control of production	● 4.9	● 4.8	—
11.2	Material control and traceability (product identification and traceability)	● 4.8	● 4.7	● 4.4
11.7	Control of verification status (inspection and test status)	● 4.12	● 4.11	◐ 4.7
12	Product verification (inspection and testing)	● 4.10	● 4.9	● 4.5
13	Control of measuring and test equipment (inspection, measuring and test equipment)	● 4.11	● 4.10	● 4.6
14	Non-conformity (control of nonconforming product)	● 4.13	● 4.12	● 4.8
15	Corrective action	● 4.14	● 4.13	—
16	Handling and post-production functions (handling, storage, packaging and delivery)	● 4.15	● 4.14	● 4.9
16.2	After-sales servicing	● 4.19	—	—
17	Quality documentation and records (document control)	● 4.5	● 4.4	● 4.3
17.3	Quality records	● 4.16	● 4.15	● 4.10
18	Personnel (training)	● 4.18	◐ 4.17	○ 4.11
19	Product safety and liability	—	—	—
20	Use of statistical methods (statistical techniques)	● 4.20	● 4.18	● 4.12
—	Purchaser supplied product	● 4.7	● 4.6	—

● Full requirement.
◐ Less stringent than BS 5750: Part 1/ISO 9001.
○ Less stringent than BS 5750: Part 2/ISO 9002.
— Element not present.

Notes
1. The clause (or subclause) titles quoted in the table above have been taken from BS 5750: Part 0. Section 0.2/ISO 9004, the titles given in parentheses have been taken from the corresponding clauses and subclauses in BS 5750: Part 1/ISO 9001, BS 5750: Part 2/ISO 9002 and BS 5750: Part 3/ISO 9003.
2. Attention is drawn to the fact that the quality system element requirements in BS 5750: Part 1/ISO 9001, BS 5750: Part 2/ISO 9002 and BS 5750: Part 3/ISO 9003 are in many cases, but not in every case, identical.

Figure 17.12 List of quality system elements per part of BS 5750: 1987. (Courtesy of British Standards Institution)

Maximum points for quality   40
Maximum points for price   25
Maximum points for delivery   20
Maximum points for service   15

Each supplier can then be compared against these factors.

Example A supplier has the following profile:

90% of his deliveries are made on time.
The price is 105 p per item against the cheapest quote of 90 p.
95% of delivered batches are accepted on inspection.
His reactions meet our requirements on 80% of occasions.

This supplier would then be rated as:

Delivery rating = 90% of 20   = 18
Price rating = (90 p ÷ 105 p) × 25 = 21
Quality rating = 95% of 40   = 38
Service rating = 80% of 15   = 12

Total rating   89

This supplier's total rating can then be compared with those of all other suppliers for the same material or components, and it can be decided which supplier best serves the company's needs. The total rating can also be used in conjunction with selected suppliers to show where their total service is behind their competitors in order to drive up the general rating of all suppliers combined.

17.4.1.2 Vetted suppliers

Once a supplier has been selected to be the main, or sole, supplier, and we are convinced that the quality of their supply consistently meets our requirements, it should be possible to reduce the need to inspect their incoming goods.

It will still be necessary to monitor their quality, partially by means of occasional quality audits and partially from records of their material/components service within the plant and in service with our customers. To ensure that this can take place it is important to ensure that full traceability of all materials used is possible and maintained.

17.4.2 In-process control

In order to achieve control in any situation, we must know what we wish to produce and how the process cycle goes about achieving this. Ideally, we wish to set up a closed-loop control cycle as shown in *Figure 17.13*. In order to do so we need to be able to specify and quantify all the various aspects. Other sections deal with the process of measuring and comparing against standards and the adjustments that need to be made. This section discusses the inspection points at which measurement can take place.

In any feedback situation, two matters are important:

(1) the *time* that has elapsed since the process was carried out, and
(2) the *quantity* of produce that has been produced meanwhile.

With process control we wish to minimise both these factors.

17.4.2.1 Preprocess inspection

If we inspect the incoming material, as in *Figure 17.14*, we can decide on what process steps are required in order to reach specification. In the extreme, we can decide not to advance material for processing.

A common method is to incorporate a go/no go check on features and sizes into tooling fixtures so that material which is outside the incoming specification cannot be mounted prior to processing.

17.4.2.2 First-off inspection

When a process is first set up, it is important that the initial output is closely monitored to ensure it meets the specification on an ongoing basis.

17.4.2.3 Last-off inspection

When a batch has been completed, special care should be taken with the measurement of the final few items. This is to ensure that the fixtures and tooling are of sufficient standard to be used again. Where the output shows signs of approach-

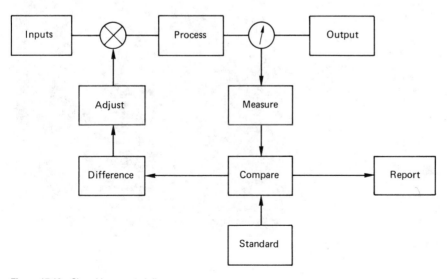

Figure 17.13 Closed-loop control diagram

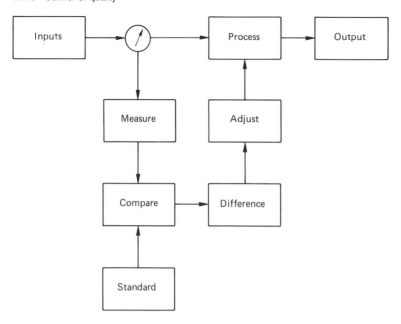

Figure 17.14 Preprocess control

ing the specification limits, then regenerative action on the equipment can be put in hand so that it is ready for a later run.

17.4.2.4 Post-process inspection

This is the normal stage at which the quality of the process is monitored. The location of this should be as close as possible to the process to minimise the opportunity for out-of-specification material to build up.

The first factor to consider is the frequency of post-processing inspection. How often need we carry out the inspection—after every item, once every 100 items, every 15 min, or when? This should be decided after examining the probability of the process going out of control, the integral drift that occurs, e.g. tool wear, and the costs involved both in inspection and in having out-of-specification items.

The second factor is the physical location of the inspection point. Ideally, the operative himself should be able to carry out, or oversee, the inspection process. Therefore, the inspection should take place close to the processing point to allow the operative access to the inspection equipment. If the inspection takes place at a remote station, e.g. a coordinate measuring machine, then the results need to be fed back to the process and/or the operative.

17.4.2.5 Adaptive control

Ideally any defect should be discovered as soon as it has occurred and corrective action initiated automatically. For this to happen adaptive control systems are needed that measure the process variable as it is being produced, or immediately afterwards.

The former can be achieved with in-process gauging, with further processing being determined by measuring the variation remaining until the specification is reached. *Figure 17.15* shows a typical electronic grinding gauge used in such a system.

Figure 17.15 An electronic grinding gauge. (Courtesy of Vernon Gauging Systems Ltd)

The latter can be carried out by probes mounted in the processing equipment itself. *Figure 17.61* shows such a probe mounted on a computer numerical control (CNC) machining centre.

The advantage of measuring in the processing location is in the timeliness and ease of feedback. There are disadvantages, however, in that specialised measuring equipment has to be duplicated, processing may have to be interrupted and the environment is less controlled, e.g. residual heat and stresses may be in the product.

17.5 Process capability

The most important determinant of the quality output of a process is the inherent capability of that process to produce in a consistent manner. Determining the range produced by a process and the movement, or drift, of that range is therefore important. These factors will determine if the process will consistently produce within specification or with a degree of substandard output. It also lays the basis for process improvement by identifying exactly where improvements may be possible, or even necessary.

17.5.1 Process variability

It is a relatively easy matter to take continuous measurements of the output of any process. We can then examine what happens to this value over a period of time. This can be examined in two main ways.

(1) *Dispersion* How does the total of the individual measurements vary over the whole of a production run? Does it vary between machines, between shifts or even at different times of the day? We can examine this by plotting a frequency diagram of the particular sizes being produced.
(2) *Drift* How do the individual measurements change as the process continues to operate? We can examine this by plotting individual measurements over time.

17.5.1.1 Dispersion patterns

The easiest method of quickly drawing a rough of the frequency distribution is to produce a tally check (see *Table 17.5*). This is only a matter of noting, by a cross, against each size when that size is produced, and gradually a histogram will be formed. The use of tally checks, however, is limited in the amount of data it can deal with.

Table 17.5 Tally chart of weights from a process

Weight (g)	Occurrences
220	XX
221	X
222	XXXX
223	XXXXXX
224	XXXX
225	XXXXXXXXX
226	XXXXX
227	XX
228	XXX
229	X

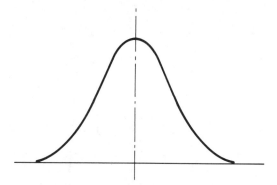

Figure 17.16 The normal distribution curve

It is more meaningful to collect a large amount of data over a significant period of time before forming a histogram or a frequency diagram for analysis. If the measurements vary in a purely random manner, the diagram produced will be the normal, bell-shaped distribution curve (*Figure 17.16*). This dispersion determines the capability of the process consistently to produce items within specification.

It is when the dispersion is set against the specification tolerances that the true capability is seen. When the dispersion is too great (see *Figure 17.17(a)*) then some of the produce from the process *must* be out of specification. This process is basically unsuitable for the present tolerances, unless something is done to bring them into a better match, e.g. widening the tolerance or improving the process. When the dispersion is much smaller (see *Figure 17.17(b)*) then the process can easily produce within the specification. If, however, the dispersion is substantially smaller, the process could be said to be 'too good', and is probably more expensive than necessary. If the dispersion is exactly the same as the tolerances (see *Figure 17.17(c)*) this is ideal. However, great care is required to ensure that the setting remains in line with the tolerances.

Ideally, we need a dispersion between *Figures 17.17(b)* and *17.17(c)*. It needs to be less than the tolerance limits to allow for some drift for unavoidable events such as tool wear, but not substantially so.

When the dispersion does not form a true normal curve, there is normally something wrong with the process itself. A departure from a random distribution is usually due to a cause. This shows up in the dispersion diagram as a bias, or as items occurring outside the normal curve. Investigation, if able to identify the causes, can then eliminate them.

17.5.1.2 Drift patterns

In addition to the range produced by a process, the way the output varies over time is also important. This can be seen by plotting the characteristic against a time base (see *Figure 17.18*). Again, it is when this plot includes tolerance zones that the true effect of any change is revealed.

Where the change is random, the range is clearly seen over a short period. A broad scatter (see *Figure 17.19*) means that care is required to ensure that the outermost items remain within tolerance. A tight band shows that less care is required in setting and less frequent adjustments are needed.

Where a gradual but distinct change occurs, i.e. drift, this can be used to determine the time interval between inspections. It also shows that initial settings can be made towards one of the tolerances so that the drift allows the maximum

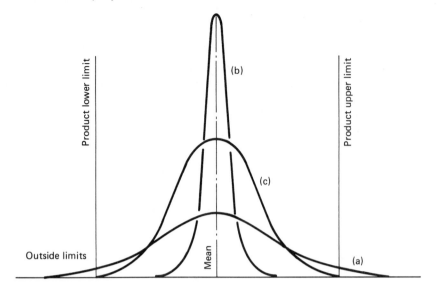

Figure 17.17 Comparison between process variability and product specification limits

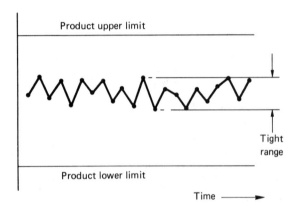

Figure 17.18 Component measurement over time—low variability

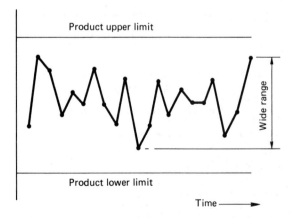

Figure 17.19 Component measurement over time—high variability

time before items exceed the other tolerance line (see *Figure 17.20*).

17.5.2 Problem identification and highlighting techniques

17.5.2.1 Pareto curves

It is important that effort in quality improvement be targeted to achieve the most beneficial results. Pareto charts effectively show up the relative frequency of particular types of faults occurring. If faults are collated by cause, the results can be charted on a histogram, with the heaviest occurrences at one end (see *Figure 17.21*). These can be converted into percentages of all faults for a fuller effect. Normally, from this analysis it can be seen that the majority of the faults occur in relatively few areas. It is in these areas that effort should be directed to show the largest quality improvements.

17.5.2.2 Fishbone diagrams

As there are many possible causes for variations in quality, it is often useful to carry out a brainstorming session to identify possible causes. During these sessions, it is helpful to produce a fishbone, or Ishikawa, diagram to show all possible causes. The effect being investigated is shown at the end of a horizontal arrow (see *Figure 17.22*). Potential causes are then shown as labelled arrows entering this horizontal arrow. Each main arrow may have other arrows entering it as possible subcauses. The subcauses may, in turn, be due to a number of possible sub-subcauses.

The process is continued, perhaps using a series of fishbone diagrams at different levels, until all possible causes, however unlikely, are listed. Once this has been done, each cause can be considered for further detailed investigation.

17.5.2.3 Cusum charts

The cusum (cumulative sum) chart can detect trends in any measured value very quickly. It is particularly useful for

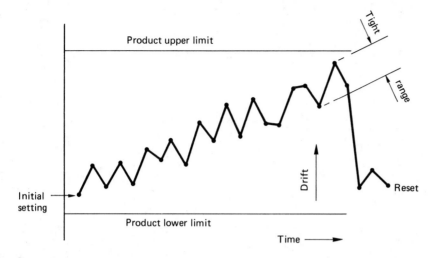

Figure 17.20 Initial setting when drift occurs

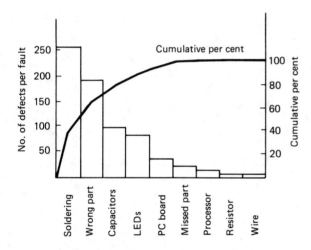

Figure 17.21 Pareto curve for electronic faults

plotting the evolution of processes because it presents data in a way that enables the eye to separate true trends from random variations. Cusum charts will detect small changes in mean values quickly and can be used for variables or attributes. In essence, a reference or target value is subtracted from each successive sample observation and this result accumulated with the previous ones. The cumulative value is plotted and trend lines can readily be seen on the resultant graph.

If the trend line is approximately horizontal, the average value of the sample is the same as the reference value. A downward slope shows that the average being achieved is less than the reference value; an upward slope shows that the average is greater. Where the slope is constant, the difference is also constant. This means that, although the average is different, it remains at a constant value. Hence only resetting is required to align the average with the reference value. When the slope is changing it shows the trend towards, or away from the reference value.

An example of some data, the resultant cusum and the actual-size control chart are shown in *Figure 17.23*.

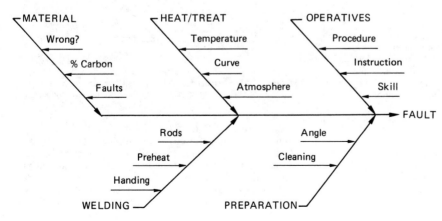

Figure 17.22 Fishbone diagram

Actual defects: 3, 4,   2, 6, 4,   5,   3,   3,   2, 4,   3, 5,   2, 4,   1,   5.
Deviation from 4: −1 0 −2 +2 0 +1 −1 −1 −2 0 −1 +1 −2 0 −3 +1

(a)

(b)

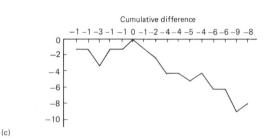

(c)

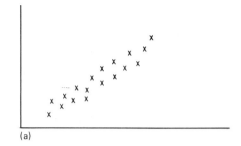

(a)

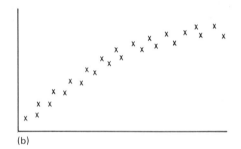

(b)

Figure 17.23   Example of a cusum chart. (a) The expected number of defects in each batch of 2000 items was 4. The actual number of defects in successive batches is shown. The cusum technique enables us quickly to determine if the present level is better or worse than expected. This gives rise to a number-of-defects chart (b) based on an expected value of 4 defects. It is difficult to see from (b) if the process is working at the expected average rate of defects or not. However, when the data are plotted as cumulative differences (c) it can readily be seen that the process is performing slightly better than expected, as in this particular case a negative change in the slope of the curve is good. The cumulative average is 3.5 defects as compared with the 4 allowed. The slope of the line in (c) shows the change in the deviation on a constant basis

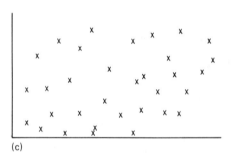

(c)

17.5.2.4   Correlation chart

Where there is thought to be a connection with a particular input variable and the quality level, it is beneficial to plot them together to determine if there is any correlation between them. *Figure 17.24* show various correlation patterns which could be apparent.

17.5.2.5   Measles charts

In complex assemblies, it is often useful to mark a sketch of the assembly with the location of any fault which arises. A clustering of marks (see *Figure 17.25*) readily identifies the location of recurring problems. This technique is particularly useful for electronic assemblies, but can also be used in pressure circuits and casting and moulding operations where porosity and cracking can occur.

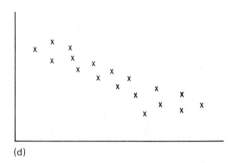

(d)

Figure 17.24   Correlation patterns: (a) linear, positive; (b) varying ratio of correlation; (c) negligible correlation; (d) negative correlation

17.6   Product acceptance sampling schemes

17.6.1   Product inspection frequency

It might be thought that the only way to ensure a batch of products entering, or leaving, a facility is made to specification

is manually to inspect every attribute of each item, i.e. to carry out 100% inspection. However, it has been proven that, especially in the case of repetitive production, 100% manual inspection often does *not* in fact give this assurance.

The reasons for the failure of 100% manual inspection to achieve its aim centre around normal human error in measuring and interpreting results, coupled with the attendant monotony associated with maintaining alertness when carrying out

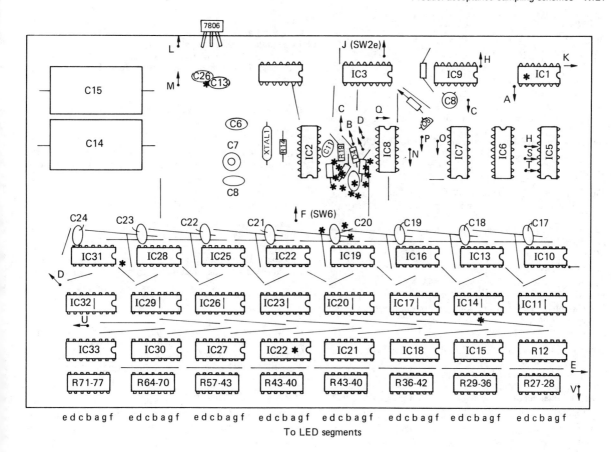

Figure 17.25 A measles chart of an assembly. * Denotes a fault

repetitive simple measuring tasks. In addition, measuring all the attributes of each item is a long and costly process.

In some cases it is impossible to test every item for a particular characteristic, e.g. ultimate strength, because the only available method for testing that characteristic involves testing to destruction.

In many industries it is common to determine the characteristics of the whole population by examining a relatively small sample, e.g. opinion polls are used to gauge the public's reaction to political events. It is therefore equally valid that taking a random sample from a batch of items and measuring pertinent attributes will give an impression of what the remaining, unmeasured items are like. This technique involves selecting a suitable sample size dependent on the batch size. Using probability theory and the normal quality level, the expected number of defectives in the sample can be calculated.

If the inspection of the sample produces less defectives than expected, then the whole batch is probably within the normal quality level and therefore can be accepted; conversely, if the number of defectives found is greater, then the whole batch is rejected, downgraded or sent for 100% screening.

17.6.2 Acceptance sampling schemes

BS 6001: 1991 *Sampling Procedures and Tables for Inspection by Attributes* details the following sampling schemes. The examples shown against each sampling scheme are all based on a batch of 1000 items with a normal quality level of 1% defectives being subjected to normal inspection.

17.6.2.1 Single sampling

This involves taking from a batch a single, representative sample (size n) and testing it for defects. The lot is accepted if the sample contains Ac or less defective items. The lot is rejected if it contains Re or more rejects. Normally, $Re = Ac + 1$.

Example

Batch lot size = 1000.
Sample size $n = 80$.
Accept lot if defectives in sample $Ac \leqslant 2$.
Reject lot if defectives in sample $Re \geqslant 3$.

17.6.2.2 Double sampling

Under this plan, a smaller sampler is taken initially than in single sampling. If the quality is such that the batch is either significantly good, or bad, then a decision to accept or reject the lot can be taken on the results from this initial sample. If the evidence is inconclusive, then a further sample (normally the same size as the first) is drawn and tested.

Example

Batch lot size = 1000.
First sample size $n_1 = 50$.
Accept lot if defectives in sample $Ac_1 = 0$.
Reject lot if defectives in sample $Re_1 \geqslant 3$.

If defectives in first sample is 1 or 2, then take second sample
of size $n_2 = 50$.
Accept lot if cumulative defectives from both samples
$Ac_2 \leqslant 3$.
Reject lot if cumulative defectives from both samples $Re_2 \geqslant 4$.

17.6.2.3 Multiple sampling

The principle here is similar to double sampling, except that
more than two successive samples are allowed for before a
decision is made. The samples are small and BS 6001: 1991
allows for up to seven samples to be drawn if the earlier ones
are inconclusive.

Example

Batch lot size = 1000.
First sample size $n_1 = 20$.
Lot cannot be accepted on basis of first sample, even if no
defectives found.
Reject lot if defectives in sample $Re_1 \geqslant 2$.

If defectives in first sample are less than 2, then take a second
sample of size $n_2 = 20$.
Accept lot if cumulative defectives from both samples $Ac_2 = 0$
(zero).
Reject lot if cumulative defectives from both samples $Re_2 \geqslant 3$.

If cumulative defectives from first two samples is 1 or 2, then a
third sample ($n_3 = 20$) is taken and a similar choice is made
whether to accept the lot, reject it or continue sampling and
testing.

17.6.2.4 Sequential sampling

Under this plan there are no fixed sample sizes. The result of
testing each sequential individual item is accumulated until
sufficient evidence exists to accept or reject the batch. The
scheme does allow for a maximum cumulative number of tests
at which the acceptance and rejection numbers under a
multiple-sampling plan would apply.

The procedure up to this point involves the use of two
factors—the handicap H, and the penalty p—both of which
are extracted from the relevant tables in BS 6001: 1991. A
continuous score is kept using the formula:

Score $= (H + \text{Number passed}) - (p \times \text{Number failed})$

(17.1)

The testing continues until either:

(1) the score reaches $2 \times H$, in which case the lot is ac-
    cepted; or
(2) the score falls to zero, in which case the lot is rejected; or
(3) the maximum sample size is reached, in which case the
    multiple-sampling-plan criteria are applied.

This procedure is too complex to be carried out in a normal
working environment. Tables showing break points are therefore issued where these schemes are in operation.

Example

Batch lot size = 1000.
Handicap $H = 40$.

Penalty $p = 32$.
Maximum sample size $n = 140$.
If maximum sample size is reached then:
    accept lot if accumulative defectives $Ac \leqslant 4$;
    reject lot if accumulative defectives $Re \geqslant 5$.

17.6.3 Factors affecting the choice of sampling plan

17.6.3.1 Simplicity

Single sampling is obviously the easiest to understand and
administer. As the number of samples possibly required
increases, so does the complexity.

17.6.3.2 Ease of drawing sample units

Where drawing samples is relatively easy, there is little
difference between using the different plans. If drawing
samples is difficult then it may be better to either use the single
sampling plan, or to draw out the maximum size sample
possibly required and select out the intermediate samples from
this.

17.6.3.3 Duration of test

If a test is lengthy yet can be applied to a number of sample
units simultaneously, then it is often better to do so. If, on the
other hand, only a small number of items can be tested at a
time, then multiple or sequential sampling may be best.

17.6.3.4 Multiple attributes

The more complex the item in terms of number and classification of possible defects, the more simple the scheme desirable.

17.6.3.5 Cost of test

Sequential sampling can normally give rise to a decision on a
smaller number of tests than can other plans. This may be
particularly important if the test is to destruction.

17.6.4 Use of BS 6001 sampling procedures and tables

Having selected what type of sample plan to use, e.g. single
sample, it is a simple matter to extract the pertinent details
from the tables within BS 6001: 1991.

17.6.4.1 Inspection level

BS 6001: 1991 allows for seven different inspection levels,
which are listed in *Table A17.1* in the Appendix to this
chapter. Unless otherwise specified, inspection level II should
be used. Level I may be specified where less discrimination is
required, or level III if greater discrimination is required. The
other four levels (S-1 to S-4) which are given may only be used
in special circumstances where very small sample sizes are
required, but these levels give a greater risk of producing
misleading results.

17.6.4.2 Code letter

The sample size is designated by a code letter (see Table
A17.1 in the Appendix to this chapter). The code letter is
selected from the combination of batch size and inspection
level. Therefore, for the previous example, with batch size of
1000 items the code letter to be used is J for inspection level II.

17.6.4.3   Acceptable quality level

The acceptable quality level is that level of defectives which can be considered satisfactory as a process average. At this level the sampling plan should select the majority of the lots from a process, assuming that this average is being maintained. The acceptable quality level is used with the code letter to determine the characteristics of the desired sampling plan.

17.6.4.4   Detail of sample plan

The code letter and the acceptable quality level are used in BS 6001 : 1991 to obtain the details of the sampling plan. BS 6001: 1991 contains three classes of inspection (normal, tightened and reduced) for each of the sampling schemes mentioned above.

The procedure is to turn to the table for the selected scheme and class and cross match the code letter against the acceptable quality levels listed. *Tables A17.2* and *A17.3* in the Appendix to this chapter are used to select a single and double normal sample plan for the example batch of 1000 items, respectively.

Initially one reads down the first column until the code letter is reached, in this case J. The second column then gives the sample size, i.e. for a single normal plan 80 on the same row as the J. If one continues along the row, the remainder of the sampling plan can be found by looking under the column with the specified acceptable quality level. Thus, for an acceptable quality level of 1%, we find an Ac (acceptance) number of 2 and a Re (rejection) number of 3. Similarly, for a double normal plan, the initial sample size is 50 with $Ac = 0$ and $Re = 3$. The second sample is also 50 with the revised values of $Ac = 3$ and $Re = 4$.

17.6.4.5   Selection of sample

It is important that the units selected for the sample are representative of the batch as a whole. Care must therefore be taken to select within the sample items taken from different areas of the batch.

17.6.4.6   Class of inspection

Normal inspection should initially be used, unless otherwise specified. This should continue to be used until:

(1) two out of five consecutive lots are rejected—then move to the tightened table until five consecutive lots are accepted, and then return to the normal table; *or*
(2) 10 consecutive lots have been accepted at the normal criteria—then move to the reduced table until one lot is rejected, and then return to the normal table.

17.6.4.7   Curtailment of sample testing

There will often be occasions before testing of the sample is complete on which the results being obtained make a final result *seem* probable. The desire to curtail testing at this point should be strongly resisted. Often full results are required to form a historical record of ongoing quality levels. Even where these are not being kept, the full result will tell clearly how well the normal quality level is being maintained.

It may appear permissible to stop testing once an accept/reject point is certain:

(1) if the reject number is reached before the full sample is tested; *or*
(2) if the number remaining is less than that required to reach a decision point.

If any such curtailment is to be allowed, it must take place under tight controls and be documented as such. It is often easier to insist on full testing than to administer this.

17.6.5   The operating-characteristic curve

Any sampling plan selected has a risk of producing a misleading result because it is based on only a portion of items. For each condition indicated by the sampling plan, there is a probability of either of two conditions being true, i.e. that indicated or the reverse.

These risks are important in two particular ways (see *Table 17.6*):

(1) a type I error can occur, i.e. the lot which is acceptable (at the acceptable quality level) is rejected by the producer;
(2) a type II error can occur, i.e. the customer receives a lot which contains more than the maximum defective level they are prepared to accept, i.e. the lot tolerant per cent defective (LTPD).

The risk over the complete range of possible defectives can be demonstrated using the operating-characteristic curve graph. One axis of the graph signifies the probability of any event occurring, the other the quality level, i.e. the percentage defective.

The curve represents the sampling plan. Any point on the graph shows the probability of the acceptance number (Ac) being the exact number of defectives found in the sample at that defective level.

If a sampling plan has no risk involved it will produce an operating-characteristic curve like the one shown in *Figure 17.26*. However, because we are dealing with samples, then

Table 17.6 Inspection transformation matrix

	True quality	
Condition indicated	Good	Bad
Good, batch accepted	OK	Error II
Bad, batch rejected	Error I	OK

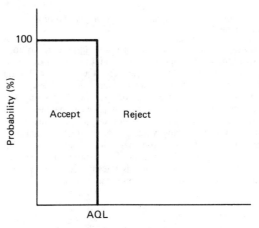

Figure 17.26 The ideal operating-characteristic curve. AQL, acceptable quality level

this ideal cannot be achieved, although we do wish the sampling plan to approach it as closely as possible.

17.6.5.1 Construction of the operating-characteristic curve

The starting point for drawing the operating-characteristic (OC) curve is the details of the sampling plan: the sample size n, and the acceptance number Ac. We then consult Poisson tables such as those given in Table A17.4 of the Appendix or Poisson charts such as the Thorndyke chart.

Before we can consult the Poisson table, we select the range of defectives that we wish to examine under the sampling plan. We then calculate the value of nd, which is simply the multiple of the sample size n and a set of defective levels d within the defective range. We then look up the Poisson table using the value of nd in the left-hand column against the sample value Ac, which heads each column. The resultant value is the cumulative probability P of finding Ac defectives or less in the sample at d defectives. It is this value that we plot to form the OC curve.

Example
Using the sampling plan for the batch of 1000, gave a sampling plan with $n = 80$ and $Ac = 2$.

$d\%$	0.1	0.2	0.5	1.0	2.0	5.0	10.0
nd	0.08	0.16	0.4	0.8	1.6	4.0	8.0
P	1.00	0.99	0.99	0.95	0.78	0.24	0.01

It is obvious that the selected range of values for d could do with some extra values between 2 and 10% to be included to enable a smoother OC curve to be drawn, i.e.

$d\%$	3.0	4.0	6.0	7.0	8.0	9.0
nd	2.4	3.2	4.8	5.6	6.4	7.2
P	0.57	0.38	0.14	0.08	0.05	0.02

The resultant value of the cumulative probability is plotted against the corresponding value of the defective to show the unique OC curve for this sampling plan (see *Figure 17.27*).

If we take the producer's risk as 5% and the customer's risk as 10%, then we can see that on this particular OC curve they have associated with them an acceptable quality level (AQL) of 1% and a lot tolerant per cent defective (LTPD) of 6.6%.

17.6.5.2 Effect of varying sampling-plan criteria on the operating-characteristic curve

Increasing the sample size n, even if keeping the acceptance number in the same proportion, will result in greater discrimination, but at an increased cost. The OC curve will tend towards the ideal rectangular shape (see *Figure 17.28*).

Reducing the acceptance number Ac, while holding the sample size at the same value, will mean more batches being rejected (see *Figure 17.29*). Not only will suspect batches be rejected, but so will many which are of satisfactory quality.

The effect of changing the inspection level is shown in *Figure 17.30*. Tightened inspection is more discriminatory, but costs more. Reduced inspection is the cheapest, but carries higher risks of passing lots with high defective rates.

17.6.5.3 Outgoing quality level

If all the rejected batches are subjected to 100% inspection and the defects made good, then it is possible to calculate what the average outgoing quality level (AOQL) is over the range of the sampling plan. The AOQL is obtained by using the

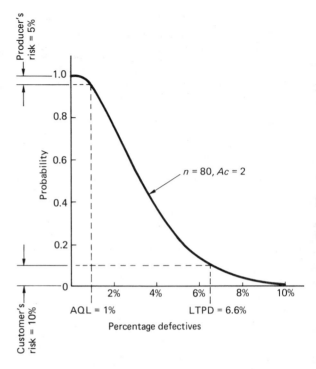

Figure 17.27 An operating characteristic curve. AQL, Acceptable quality level; LTPD, lot tolerant per cent defective

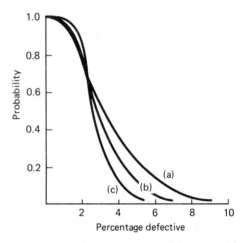

Figure 17.28 The effect on the operating-characteristic curve of increasing the sample size: (a) $n = 80$, $Ac = 2$; (b) $n = 160$, $Ac = 4$; (c) $n = 320$, $Ac = 8$

following formula, substituting in each case the appropriate values of the percentage defective d:

$$AOQL = \frac{P \times d}{P + [(1 - P) \times (1 - (d \div 100))]} \quad (17.2)$$

Using the information from the sampling plan used above, we can substitute values in this formula. When there is a

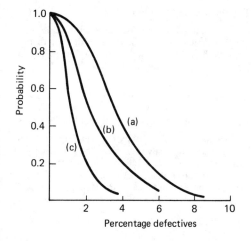

Figure 17.29 The effect on the operating-characteristic curve of varying the acceptance number: (a) $n = 80$, $Ac = 2$; (b) $n = 80$, $Ac = 1$; (c) $n = 80$, $Ac = 0$

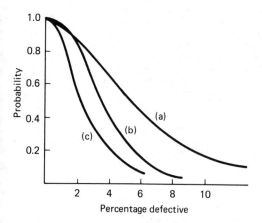

Figure 17.30 The effect on the operating-characteristic curve of varying the inspection level: (a) reduced inspection; (b) normal inspection; (c) tightened inspection

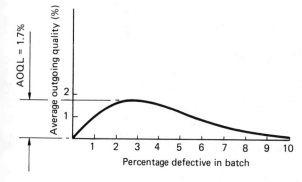

Figure 17.31 The outgoing quality level. AOQL, Average outgoing quality level

defective value of 1%, and an accumulative probability of 0.95, then:

$$AOQL = \frac{0.95 \times 1}{0.95 + [(1 - 0.95) \times (1 - (1 \div 100))]}$$

$$= 0.95$$

Covering the range of the sampling plan we can calculate the following relationships:

Lot defective	AOQL
0	0
0.5	0.50
1	0.95
2	1.56
3	1.73
4	1.56
5	1.25
6	0.89
7	0.60
8	0.43
9	0.20
10	0.11

These values are plotted in *Figure 17.31*. It can be seen that, if all rejected batches are subjected to 100% inspection, the maximum outgoing quality level occurs around the 3% incoming defective level, but is limited to a value much less than this, i.e. around 1.75% outgoing.

17.6.6 Calculating a non-standard quality plan

Whereas BS 6001: 1991 covers the normal requirements of acceptance sampling, there can be occasions when the customer requires a lot tolerant per cent defective (LTPD) different from that offered by the BS 6001: 1991 sampling plans. In these cases it is possible to work out a suitable sampling plan if the following factors are known:

(1) the desired producer's risk, α;
(2) the corresponding acceptable quality level, AQL;
(3) the desired customer's risk, β; and
(4) the corresponding LTPD.

We have to consult the Poisson probability charts (*Table A17.4* in the Appendix to this chapter) and tabulate the values of nd which lie under each of the acceptance numbers Ac for the desired risks α and β. We also produce a further column containing the ratio between these columns. When this ratio agrees with that between the AQL and the LTPD this gives us one part of the sampling plan—the acceptance number Ac. Knowing the value for nd we can then work back to the sample size as:

$$\frac{nd}{100} = \frac{AQL \times n}{100} = P \tag{17.3}$$

Example

Using data from the operating-characteristic curve shown in *Figure 17.27*: $\alpha = 5\%$, AQL $= 1\%$, $\beta = 10\%$, LTPD $= 6\%$. Values of nd at 95% and 10% and the ratio of these two nd values are given in *Table 17.7* against different Ac values.

The ratio LTPD/AQL $= 6/1 = 6$. The nearest value for this ratio in *Table 17.7* is that associated with an acceptance

Table 17.7 Data taken from *Figure 17.31* for use in the worked example of calculating a non-standard quality plan

Acceptance No., Ac	Value of nd at P = 95%	Value of nd at P = 10%	Ratio of columns
1	0.36	3.9	10.83
2	0.80	5.3	6.63
3	1.35	6.7	4.96
4	1.97	8.0	4.06
5	2.6	9.3	3.58

number of 2. Substituting the corresponding value of P, i.e. 0.8, in equation (17.3)

$$n = (100 \times 0.8)/1$$

$$= 80$$

Therefore, this method gives us a sampling plan with a sample size of 80 and an acceptance number of 2.

In this case, the calculated value is exactly the same as that suggested in BS 6001: 1991. This should not be surprising, as the data were derived from the operating-characteristic curve produced from the single sampling plan from BS 6001: 1991.

17.7 Statistical process control

In any production process, the important quality aspect is to ensure that the process continues within control. The use of statistical procedures and control charts enables this to be done in a cost-effective manner. The basis of the control lies in the continual calculation and charting of the actual results and comparing them with the probability of these results being within a normal frequency distribution.

17.7.1 The normal distribution

The normal distribution of an unbiased process should form a symmetrical, bell-shaped curve with tails extending, in theory, to infinity in both directions (see *Figure 17.16*). The important characteristics of the normal distribution are:

(1) the *mean*—this is simply the arithmetic average value of the variations; and
(2) the *distribution* of the values around the mean—this is denoted by the *standard deviation* of the distribution.

If we know the mean of a population μ, and its standard deviation σ, then we can easily determine the probability of any single value, or a range of values, being part of the same population by consulting statistical tables. *Figure 17.32* shows the relationship between multiples of the standard deviation and the proportion of its corresponding population being contained by it.

17.7.2 Variable control charts

Many processes produce items which are measurable in discrete quantities. In quality these measurements are termed 'variables' and they are particularly useful in controlling the variation of a process over time. It is always better to take more than one value when examining a process as the probability of an average value of a sample being different from the real population is less than that for one individual reading. This is due to the fact that the standard deviation σ_n of a set of samples is reduced from the standard deviation σ of a set of individual measurements in inverse proportion to the square root of the sample size n. This is shown in *Figure 17.33*, and is represented by the formula:

$$\sigma_n = \sigma/\sqrt{n} \tag{17.4}$$

The technique for controlling variables involves taking small samples from the produce of a machine or process and plotting from these the following two control charts.

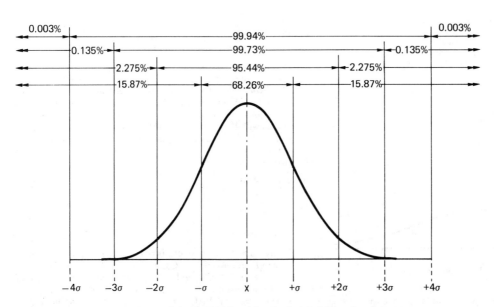

Figure 17.32 The coverage of a population by multiples of the standard deviation σ from the mean

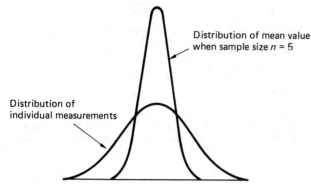

Distribution of mean value when sample size $n = 5$

Distribution of individual measurements

Figure 17.33 The change in the normal distribution curve with varying sample size

Table 17.8 Constants for limits on mean and range control charts

Sample size, n	Normal mean chart		Range chart		Modified limits	
	Warning limit, $A'_{0.025}$	Action limit, $A'_{0.001}$	Warning limit, $D_{0.975}$	Action limit, $D_{0.999}$	Warning limit, $A''_{0.025}$	Action limit, $A''_{0.001}$
2	1.128	1.937	2.81	4.12	1.51	0.80
3	0.668	1.054	2.17	2.98	1.16	0.77
4	0.476	0.750	1.93	2.57	1.02	0.75
5	0.377	0.594	1.81	2.34	0.95	0.73
6	0.316	0.498	1.72	2.21	0.90	0.71
7	0.274	0.432	1.66	2.11		
8	0.244	0.384	1.62	2.04		
9	0.220	0.347	1.58	1.99		
10	0.202	0.317	1.56	1.93		
11	0.186	0.294	1.53	1.91		
12	0.174	0.274	1.51	1.87		

(1) *Control chart for sample mean* ($\bar{X}$)—this allows us to check the movement of the process average. A drift in this value can indicate items such as tool wear, changes in setting, etc.
(2) *Control chart for sample range* ($\bar{w}$)—this enables us to keep a check on any changes in the variability of the process. A change here may occur due to loosening of setting mechanisms, machine wear, etc.

17.7.2.1 Collecting data to set up control charts

The procedure for collecting data comprises the following steps.

(1) Visit the machine at set intervals and take a small consecutive sample of 2–12 items. Measure each item.
(2) Calculate out each sample average x, and plot it on a blank mean control chart.
(3) Calculate each sample range w, i.e. the difference between the highest and lowest values, and plot it on a blank range control chart.

17.7.2.2 Setting control limits—without specification tolerances

When we have at least 10 (preferably more) settled points on the charts, we are ready to set up the control limits.

(1) Take the average of all the sample averages $\bar{x}$ to give a grand average $\bar{X}$. This is used as an estimated value of the true average of the process. Draw a line on the mean control chart to represent this grand average.
(2) Work out the mean range $\bar{w}$ by taking the average of all the sample ranges w. Draw a line on the range control chart to represent this mean range.
(3) Multiply the mean range by the constant $A'_{0.025}$ for the sample size n in *Table 17.8*. This gives the warning limit variable. Add this product to the grand average to give the upper warning limit (UWL); subtract it from the grand average to give the lower warning limit (LWL). Draw the UWL and LWL on the mean control chart.
(4) Multiply the mean range by the corresponding constant $A'_{0.001}$ to give the action limit variable. Add and subtract the product to/from the grand average to give the upper action limit (UAL) and lower action limit (LAL), respectively. Draw the UAL and LAL on the mean control chart.

(5) Multiply the mean range by the constants $D_{0.975}$ and $D_{0.999}$ to determine the UWL and UAL, respectively, for the sample ranges. Draw the UAL and UWL on the range control chart.

Example

The weights of samples taken from a prepackaging line are listed in *Table 17.9*.

Grand average = 2308.6/10 = 230.86
Mean range   = 64/10   =   6.4

Mean chart warning limits
    = Grand average ± (Mean range × $A'_{0.025}$)   (17.5)
    = 230.86 ± (6.4 × 0.377)
    = 230.86 ± 2.41
    = 233.27 (UWL) and
      228.45 (LWL)

Mean chart action limits
    = Grand average ± (Mean range × $A'_{0.001}$)   (17.6)
    = 230.86 ± (6.4 × 0.594)
    = 230.86 ± 3.80
    = 234.66 (UAL) and
      227.06 (LAL)

Table 17.9 Weights of samples taken from a prepackaging line

Sample No.	Individual weights (g)					Total weight (g)	Average (g)	Range (g)
1	230	230	233	226	233	1152	230.4	7
2	233	227	227	232	231	1151	230.2	6
3	232	232	230	234	227	1157	231.4	7
4	228	233	230	234	232	1160	232.0	5
5	230	230	228	230	234	1156	231.4	6
6	233	230	228	230	230	1156	231.4	5
7	230	228	234	226	227	1151	230.2	8
8	225	230	227	233	228	1150	230.0	8
9	230	232	228	226	226	1150	230.0	6
10	234	231	229	229	235	1158	231.6	6
Subtotal							2308.6	64

Range chart warning limit
$$= \text{Mean range} \times D_{0.975} \qquad (17.7)$$
$$= 6.4 \times 1.81$$
$$= 11.58$$
$$\simeq 12$$

Range chart action limit
$$= \text{Mean range} \times D_{0.999} \qquad (17.8)$$
$$= 6.4 \times 2.34$$
$$= 14.98$$
$$\simeq 15$$

The mean and range of the sample and the calculated control limits are shown on the control charts in *Figure 17.34*.

17.7.2.3   Setting control limits—with specification tolerances

We can classify the process variability relative to a specification tolerance by calculating a relative precision index. The latter can be used to determine the procedure to be followed by using the control charts.

The relative precision index (RPI) is calculated from:

$$\text{RPI} = \text{Specification tolerance/Mean range} \qquad (17.9)$$

This can give three possible classes of RPI, as shown in *Table 17.10*. The RPI value gives a clear indication of the possibility of producing within the specification limits using a particular process.

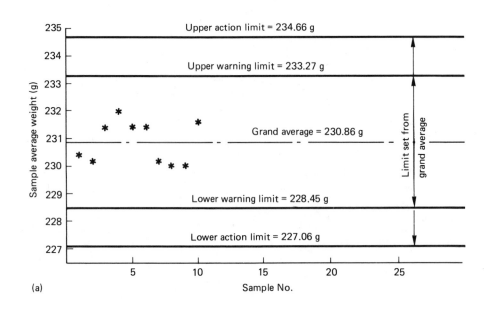

(a)

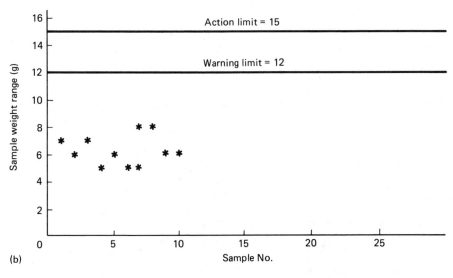

(b)

Figure 17.34   Mean (a) and range (b) control charts

Table 17.10 The three classes of the relative precision index (RPI)

Sample size/class	Low RPI	Medium RPI	High RPI
2	<6	6–7	>7
3	<4	4–5	>5
4	<3	3–4	>4
5, 6	<2.5	2.5–3.5	>3.5

Low RPI Some rejections are inevitable unless the index is increased by:

(1) increasing the specification tolerance, or
(2) reducing the process variability, i.e. the mean range.

Where neither of these are possible, then one must decide whether it may be more economical to offset the process mean (see *Figure 17.35*) in order to ensure that all out-of-tolerance items fall into an area where they are recoverable, and none into an irrecoverable set. Note that the total number of out-of-specification items will rise, therefore one has to consider the extra recovery costs against the savings through having no items discarded.

Medium RPI The process is just satisfactory as long as the mean is maintained.

High RPI This process is more than capable of producing to the required tolerance. In order to reduce resetting if normal control limits are used, *modified* control limits can be applied to take advantage of this relatively high precision.

17.7.2.4 Modified mean control limits

If the RPI is high, the warning and action limits on the mean control chart can be determined by using the $A''_{0.025}$ and $A''_{0.001}$ constants (*Table 17.8*) to replace the A' constants in the

formulae given in Section 17.7.2.2 (equations (17.5) to (17.8)). This time, however, the control limits are set *inwards* from the specification limits rather than outwards from the grand average.

Example

Using the same sample data as for normal control limits, but this time applying the data to a job with a specification of 230 ± 15 g. The Mean range remains at 6.4 and the range control chart is unchanged.

RPI = Specification tolerance/Mean range
  = 30/6.4
  = 4.69

thus the RPI is 'high', i.e. we can use modified control limits:

$$\text{UAL} = \text{Upper tolerance limit} - (\text{Mean range} \times A''_{0.001}) \tag{17.10}$$

  $= 245 - (6.4 \times 0.73)$
  $= 245 - 4.67$
  $= 240.33$

$$\text{UWL} = \text{Upper tolerance limit} - (\text{Mean range} \times A''_{0.025}) \tag{17.11}$$

  $= 245 - (6.4 \times 0.95)$
  $= 245 - 6.08$
  $= 238.92$

$$\text{LWL} = \text{Lower tolerance limit} + (\text{Mean range} \times A''_{0.025}) \tag{17.12}$$

  $= 215 + 6.08$
  $= 221.08$

$$\text{LAL} = \text{Lower tolerance limit} + (\text{Mean range} \times A''_{0.001}) \tag{17.13}$$

  $= 215 + 4.67$
  $= 219.67$

The modified mean control chart is shown in *Figure 17.36*; the range chart is the same as that constructed in Section 17.7.2.2 previously (see *Figure 17.36*).

Therefore, applying the modified limit in this case opens up the distance between the UAL and the LAL from 7.60 g, with normal control limits, to 20.66 g, with the modified control limits. In effect, 2.7 times the scope before resetting is necessary and hence a longer running time between settings. Where no information is available immediately on the process range, a suitable setting for action limits is approximately one-eighth of the specification tolerance inwards. The warning limits should be set to one-quarter of the specification tolerance inwards.

17.7.2.5 Operating variable control charts

The following rules are used with ongoing information on the charts to determine when action is required.
  The process can be said to be in control until:

(1) one point appears on or outside the action limit; *or*
(2) two points in 10 consecutive points are on, or outside, the warning limits; *or*
(3) three points in 20 consecutive points are on, or outside, the warning limits.

The process then should be considered to be out of control and stopped for corrective action to be taken.

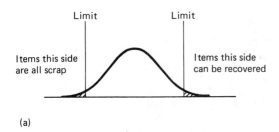

(a)

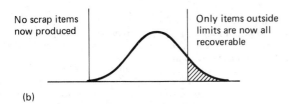

(b)

Figure 17.35 Offsetting the mean under low relative precision index (RPI) conditions. (a) Low RPI where the process mean is set midway between the limits. (b) The process mean set towards the limit outside of which recovery is possible

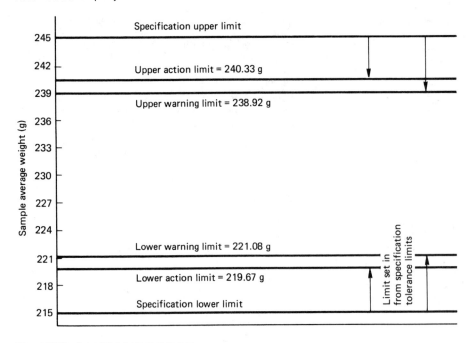

Figure 17.36 A modified mean control chart

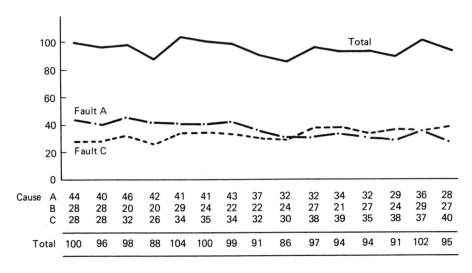

Cause	A	44	40	46	42	41	41	43	37	32	32	34	32	29	36	28
	B	28	28	20	20	29	24	22	22	24	27	21	27	24	29	27
	C	28	28	32	26	34	35	34	32	30	38	39	35	38	37	40
Total		100	96	98	88	104	100	99	91	86	97	94	94	91	102	95

Figure 17.37 An attribute chart showing the effect of combination

17.7.3 Attribute control charts

Athough variables can tell us more about a particular process than can attributes, the latter still play an important role in quality control because:

(1) many characteristics are difficult to measure as a variable, e.g. scratch-free surface;
(2) attributes are often quicker to measure and require less skill, e.g. go/no go gauges as opposed to a micrometer; and
(3) 100% inspection/testing may still be required, e.g. in electronic assembly.

An attribute may combine several characteristics, or operatives output (see *Figure 17.37*). In this case we should always be careful that improvements in one characteristic does not conceal worsening in another.

There are two main types of attribute control chart:

(1) *Number defective*—these charts are used when a relatively constant output is being achieved.
(2) *Percentage defective*—these are used when considering an output with substantial variation in size. This is much more complex, and some operatives and, truth to say, management, find it difficult to comprehend fully.

'Defective' in this case refers to the product failing, and such a failure may be due to more than one major defect.

Control limits for attribute control are set as for variables. There are, however, certain differences:

(1) no range chart is needed; and
(2) control is only over the number, or percentage, defective.

Although both upper and lower control limits can be set, the lower ones are often not shown. This is because a breach of a lower limit is thought of as being positive and hence should not even appear to be wrong, as breaching a limit may be thought of as being. However, if a lower attribute limit is breached, it may be due to a certain circumstance having occurred. If this circumstance can be identified and made permanent, then the overall defective level will drop and new upper limits can be established. When such action is required the same rules apply as for the variable control charts.

17.7.3.1   Number defective control chart

The procedure for developing the number defective control chart is similar to that described for the variable control chart.

(1) Take stipulated sample n and inspect/test for defectives, or decide on the basis of the breakpoint, i.e. daily production. In the latter case, the average output is calculated to take the value of n.
(2) Plot the number of defectives per sample on a chart.
(3) When the process appears to be in a steady state, work out the arithmetic average number of defectives per sample c. Draw this on the chart.
(4) Calculate the estimated standard deviation L using:

$$L = \sqrt{[c \times (1 - (c/n))]} \qquad (17.14)$$

(5) Using the average number of defectives c as a datum, plot a line at $2L$ upwards and downwards to set the *warning limits*.
(6) Set the *action limits* at $3L$ upwards and downwards from the average.

Example

If, for an average daily production of 1000 items, the average number of defective products is 20, then:

$$
\begin{aligned}
L &= \sqrt{[c \times (1 - (c/n))]} \\
&= \sqrt{[20 \times (1 - (20/1000))]} \\
&= \sqrt{[20 \times (1 - 0.02)]} \\
&= \sqrt{(20 \times 0.98)} \\
&= \sqrt{(19.6)} \\
&= 4.43
\end{aligned}
$$

Therefore, the warning limits are:

Upper warning limit $= c + 2L = 20 + 8.8 \approx 29$

Lower warning limit $= c - 2L = 20 - 8.8 \approx 11$

and the action limits are:

Upper action limit $= c + 3L = 20 + 13.2 \approx 33$

Lower action limit $= c - 3L = 20 - 13.2 \approx 7$

These are shown in *Figure 17.38*.

17.7.3.2   Percentage defective control chart

The procedure for constructing this chart is similar to that used for the number defective control chart.

(1) Work out the percentage defective of each batch under examination and plot these on the control chart.
(2) Calculate the *true* average percentage defective p by summing the *total number* of defectives found and dividing this by the *total number* of items produced, and then multiply by 100 (to convert to a percentage). Draw this value as the datum on the control chart.
(3) Calculate the average size of the batches under examination n by dividing the total produced by the number of batches.

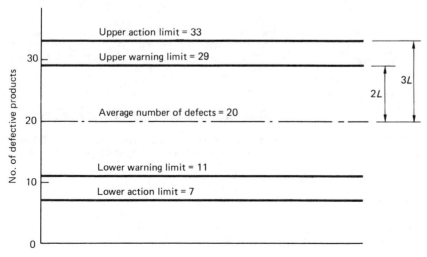

Figure 17.38   A number defective control chart

(4) Calculate the precontrol estimated standard deviation M using:

$$M = \sqrt{[(p \times (100 - p))/n]} \qquad (17.15)$$

(5) Set the *warning limits* at $2M$ inwards and outwards from the average p.

(6) Set the *action limits* at $3M$ inwards and outwards from the average p.

N.B. Because, in theory, the control limits are dependent on the size of the batch, the control limits should be adjusted depending on the lot size. This is often ignored because of the calculation involved and the limit remains constant. This is not ideal and is normally limited to differences in batch sizes of less than 25% and altering any result where the warning limit is breached to check the *true* variation. Alternatively, tables can be drawn up to show the variations of the limits in relation to batch size, or a computer can be used to analyse the results and decide on limit breaches.

Example

Consider the results from a daily production run shown in *Table 17.11*. The total number defective is 156 and the total produced is 9610. Thus:

Average percentage defective, $p = 100 \times$ (Total defectives/Total produced)

$= 100 \times (156/9610)$

$= 1.62$

Average batch size $=$ Total produced/Number of occasions

$= 9610/10$

$= 961$

Estimated standard deviation $M = \sqrt{[(p \times (100 - p))/n]}$

$= \sqrt{[(1.62 \times (100 - 1.62))/961]}$

$= \sqrt{[(1.62 \times 98.38)/961]}$

$= \sqrt{(159.38/961)}$

$= \sqrt{0.1659}$

$= 0.41$

Therefore, the limits are:

Upper action limit $= p + 3M = 1.62 + (3 \times 0.41)$

$= 1.62 + 1.23 = 2.85$

Upper warning limit $= p + 2M = 1.62 + (2 \times 0.41)$

$= 1.62 + 0.82 = 2.44$

Lower warning limit $= p - 2M = 1.62 - (2 \times 0.41)$

$= 1.62 - 0.82 = 0.80$

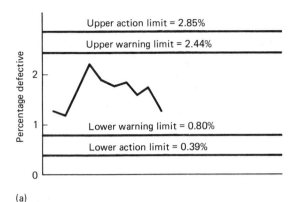

(a)

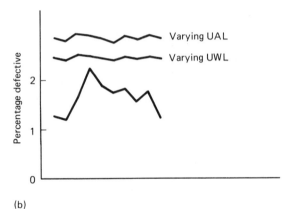

(b)

Figure 17.39 Percentage defective control charts: (a) with control limits kept constant; (b) with control limits varying depending on the batch size examined

Lower action limit $= p - 3M = 1.62 - (3 \times 0.41)$

$= 1.62 - 1.23 = 0.39$

These limits are shown in *Figure 17.39* which also shows the initial defective levels. The data are shown both with the control limits unadjusted for batch size and with the limits adjusted.

17.8 Measurement of form and surface

To maintain quality, it is necessary to measure with known accuracy each product or service against recognised standards.

Table 17.11 The results from a daily production run

No. defective	12	13	14	19	18	19	17	16	16	12
No. produced	950	1090	840	860	950	1080	940	1020	920	960
Percentage defective	1.26	1.19	1.67	2.21	1.89	1.76	1.81	1.57	1.74	1.25

The important factors in any measurement are:

(1) there must be an agreed standard between customer and supplier;
(2) all measurements are relative; and
(3) all measurements involve a degree of inaccuracy.

Therefore, before stating the degree of compliance of a product to its specification, we must be aware of the accuracy of measurement used.

17.8.1 National measurement system

Accuracy of measurement relies on the calibration of all the measuring equipment used against a known standard of a higher (known) accuracy. This requirement has led all industrialised countries to set up a national measurement system to ensure that all measurements take place against recognised standards.

The aims for a national measurement system are:

(1) to maintain national standards of sufficient accuracy to serve industry and commerce; and
(2) to provide a practical system for ensuring that the measurement of all standards used in industry and trade are traceable to a national standard.

In Britain, the National Physical Laboratory (NPL) holds the primary standards and is responsible for the management of the traceability network. The NPL also ensures compatibility with international standards.

The measurement traceability chain is shown in *Figure 17.40*. Each item should be checked against the item above it on the chain at set time intervals or when conditions call for it, e.g. if a gauge has fallen during use.

17.8.2 Characteristics of measuring instruments

Kinematic principles are the starting point in the selection of measuring instruments. The degrees of freedom should be limited to one direction only. In addition, instruments should be designed so that they will perform with the following characteristics:

(1) suitable sensitivity,
(2) known accuracy,
(3) consistency and repeatability, and
(4) suitable reactivity.

Sensitivity This is the rate of displacement of the output reading of the device relative to changes in the measured quantity. It is described in two ways:

(1) sensitivity over the full range of the instrument; and
(2) discrimination—the smallest change in the measured quantity that can be discriminated by the instrument.

Accuracy This is the degree of correction required to be made to the instrument reading to give the correct value of the measured quantity. This can also be looked at in two ways:

(1) the accuracy inherent in the device itself, and/or
(2) the accuracy due to the manner of the operation or operator.

Repeatability This is the range of variation obtained from the instrument when repeatedly measuring the same physical quantity. This should be carried out with the instrument advancing in one direction and then repeated using the reverse direction over a set range of known quantities.

Reactivity All instruments contain a degree of inertia which can show up in two ways:

(1) a time delay between a change in the measured quantity and that change being shown as an output from the instrument; and
(2) stepped response—slow minute changes in the measured quantity may not show up until sufficient forces are built up to overcome the inertia in the system.

The degree of each of the above characteristics required for the function that the instrument is intended to carry out needs to be incorporated in the instrument specification. As in most quality matters, fitness for purpose determines the quality level and hence the cost.

Principle of alignment This is a further important principle which states that: the line of measurement should be coincident with the dimension being measured. *Figures 17.41* and *17.42* show two common measuring instruments. The micrometer fits this principle but the vernier does not.

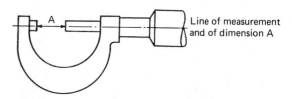

Figure 17.41 A micrometer. This instrument conforms with the principle of alignment

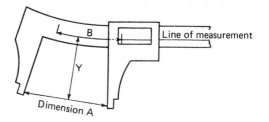

Figure 17.42 A vernier calliper. This instrument does not conform with the principle of alignment

Primary standard of length

↓

Secondary standards

↓

Reference/calibration grade slip gauges

↓

Inspection grade slip gauges

↓

Workshop grade slip gauges

↓

Workpiece gauges

Figure 17.40 The chain of traceability of measurement

Other errors In addition to errors introduced by the physical design of instruments, there is often a greater occurrence of errors arising whilst they are being used. One cause of these errors is environmental, e.g. dust, temperature, magnetic fields, etc. Another main one is human error, either in misapplication or in misreading of the instrument itself.

17.8.3 Linear measurement

Linear measurement uses a variety of equipment from simple steel rules to comparators. These are covered briefly in the following paragraphs in terms of use and accuracy. For more information the reader should consult the British Standards covering the accuracy and method of testing using each type of instrument.

17.8.3.1 Steel rules

Recommended lengths are 150, 300, 500 and 1000 mm. The edges should be straight and parallel to within 0.1 mm up to 300 mm in length. The tolerance between graduations is 0.1 mm for 150- and 300-mm rules, 0.2 mm for 500-mm rules and 0.25 mm for 1-m rules.

17.8.3.2 Feeler gauges

Recommended lengths are 75, 100, 150 and 300 mm with a width of 12 mm. Individual feelers are 0.03–1 mm thick. The tolerance on the thickness ranges from ±0.004 mm at 0.03/0.04 mm, up to ±0.01 mm above 0.65 mm. Because of the possibility of a build up in error on combining thicknesses, feeler gauges are supplied in four sets, each covering a representative set of sizes in varying combinations.

17.8.3.3 Dial gauges

Dial gauges (*Figure 17.43*) have a plunger movement parallel to the dial plane with scale divisions of 0.001 and 0.0001 in. or 0.01 mm. The inherent error in any one scale division is approximately 10%, and in any one revolution is approximately one scale division.

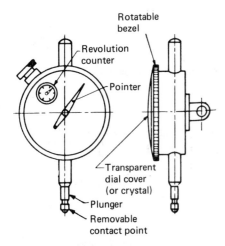

Figure 17.43 A dial gauge

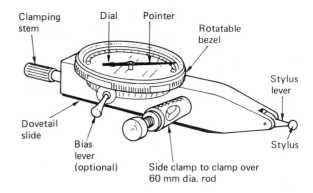

Figure 17.44 Dial test indicator

17.8.3.4 Dial test indicators

Dial test indicators (*Figure 17.44*) have dials normal to the direct of the stylus movement. Accuracy is about half of a similar dial gauge.

17.8.3.5 Vernier calipers and height gauges

Recommended caliper sizes are 150, 300, 600 and 1000 mm. The scale accuracy is ±0.02 mm for caliper and gauge sizes of up to 300 mm, ±0.04 mm for 300 and 600 mm, and ±0.06 mm above 600 mm. In addition, the faces may be up to 0.01 mm out of parallel.

Internal measurement requires an addition to be made to the scale reading equivalent to the combined width of the jaws. This is normally 5 mm for the 150 mm size and 10 mm for the others.

17.8.3.6 Micrometer heads

These are often used in the construction of measuring equipment and machine tools. They come in 13 mm and 25 mm travels with thimble diameters of <25 mm, 25–50 mm, and >50 mm. The maximum allowed scale error of all the types is 0.003 mm.

17.8.3.7 External micrometers

These come in a variety of sizes from 25 mm up to 600 mm at intervals of 25 mm plus a 0–13 mm size. Each micrometer above 25 mm in size should have a suitable gauge bar to set the zero reading. Tolerance in the alignment varies from 0.05 mm for the 13 mm size up to 0.75 mm for the 600 mm size on fixed anvils and is up to 50% greater with interchangeable anvils. In addition, there can be errors in the setting gauge, these ranging from ±0.002 mm at 50 mm to ±0.006 mm at 575 mm. Plain checking rings may be used to check built-up sizes.

17.8.3.8 Internal and stick micrometers

Internal micrometers normally come in sets with extension bars covering ranges of 25–150 mm up to 600–900 mm. Tolerance increases from 0.005 mm at 150 mm to 0.025 mm at 900 mm. Again, plain checking rings may be used to check built-up sizes.

17.8.3.9   Gauge (slip) blocks

These come in a variety of sizes up to 100 mm made up into working sets; typically six sets with a 1 mm base, three sets with a 2 mm base and three auxiliary sets. They also come in three general-use grades (0, I and II in relative coarseness), a calibration grade and a special high-accuracy grade (00). Tolerances in flatness increase from 0.05 μm for 00 and calibration grades to 0.25 μm for grade II. Parallelism tolerance is generally 0–0.10 μm greater than that for flatness. Tolerances on lengths are from 0.05 μm for grade 00 up to 20 mm, up to 2.4 μm for 100 mm grade II.

Blocks may be assembled together to make up a given dimension by *wringing* them together to eliminate all but a thin layer of oil between them. Because of the high adhesion of wrung blocks they must be *unwrung* to part them after use.

17.8.3.10   Length bars

These are similar to gauge blocks but are used for measuring larger sizes. They come in four grades: reference, calibration, grade I, and grade II. They are best used in a vertical plane or with support horizontally. Grade I and II bars have internal screwed ends so that they can be used in combination. Sizes vary from 25 mm up to 1200 mm for reference and calibration bars, but only up to 775 mm for grades I and II. Tolerances on flatness are 0.08–0.25 μm, on parallelism 0.08–0.8 μm and on length from 0.08 μm for the 25 mm reference grade up to 10.1 μm at 775 mm grade II.

17.8.3.11   Surface plates

Where blocks are used they require an accurate reference plane to be used against. These should be made to a similar tolerance as the blocks themselves. Surface plates are made out of aged cast iron or granite, carefully ground, lapped and polished. It is important that they be kept free of any residual stresses.

17.8.3.12   Comparators

A comparator is used to compare a known and an unknown dimension. The known dimension can be set by a special standard, a component of known size or a *stack* of gauge blocks. The difference between the standard and the component is then determined by a displacement sensor such as the dial gauge shown in *Figure 17.45*. Other comparators such as pneumatic and strain gauges can be used successfully.

When in use there is a combination of uncertainties produced by the device and the setting standard tolerances.

17.8.3.13   Microscopes

Microscopes are used in two main modes in measuring: in direct reading, and against a travelling scale.

When used as a fixed-scale microscope, a recticle such as that shown in *Figure 17.46(a)* is used. The scale is compared directly with the work dimensions. These scales are sometimes precalibrated, but even these may require recalibration by focusing on a calibration scale. A typical scale is 100 divisions of 0.1 mm or 0.004 in.

Travelling micrometers have a reticle like that shown in *Figure 17.46(b)*. The micrometer is used as an index with the work being moved using micrometer heads. Total travel is limited, normally to about 100–125 mm, with scales being of the order of 0.0025 mm.

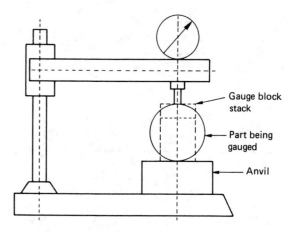

Figure 17.45   A comparator in use

(labels) Gauge block stack · Part being gauged · Anvil

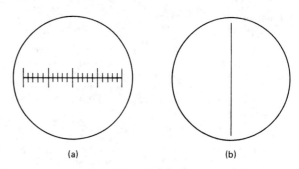

Figure 17.46   Measuring micrometer reticles: (a) fixed scale; (b) single filar

Proper sharp focusing is necessary with both types of instrument to ensure that the reticle image does not appear to move with respect to the work piece.

17.8.4   Angular measurement

It is possible to use a combination of linear measuring instruments to calculate an angular value, but the following equipment has been specifically developed for making angular measurements.

17.8.4.1   Bevel protractors

These are graduated in degrees and minutes. The graduations may be read directly from the scale (vernier) or through a magnifying system (optical).

17.8.4.2   Sine bars and tables

Sine bars come in three basic sizes: 100, 200 and 300 mm. They are used in conjunction with gauge blocks to calculate angles through set ups similar to that shown in *Figure 17.47*. In the figure, if l is the distance between the axes of the rollers and h is the height of the gauge block, then

$$\sin \theta = h/l \qquad (17.16)$$

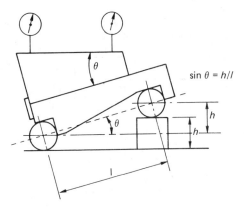

Figure 17.47 A sine bar set-up

Sine tables are similar to a wide, hinged sine bar. They have a table which is inclinable about a single axis. The angle is imparted by inserting gauge blocks under the setting roller.

17.8.4.3 Spirit levels

Spirit levels contain a curved glass tube, or vial, containing a bubble of air within a liquid. The bubble settles at the highest point of the vial. There are three types: graduations can be 5 or 10 seconds of arc in type 1 instruments, and 10 or 20 seconds in type 2 and 3 instruments. Maximum errors are approximately 20% of the scale division.

17.8.5 Limit gauges

Adoption of a system of limits and fits with unskilled or semi-skilled labour often leads to the use of limit gauges to determine whether or not a component is within the specified limits. They are quick to use and require less skill than direct measurement. However, it should be remembered that, as in the production of any item, it is impossible to make gauges to an exact dimension.

Limit gauges themselves therefore require tolerances for their manufacture. These tolerances should be applied in such a manner as to ensure, as near as possible, that the gauges fulfil their design function of accepting all suitable components whilst rejecting all the faulty items.

Although product tolerances are often dependent on the ease of achieving them, generally gauge tolerance is taken as 10% of the work tolerance. This, however, would mean that up to 20% of the allowed work tolerance may be lost. BS 969: 1982 suggests one possible policy to minimise this: the tolerance on a 'go' gauge should be *within* the work tolerance zone, and the tolerance on a 'not go' gauge should be *outside* the work tolerance zone. Additional tolerances may be added to allow for wear (up to 20%, depending on material and usage). These principles are demonstrated in *Figure 17.48*.

17.8.5.1 Taylor's theorem of gauging

This theorem is the key to the design of limit gauges and states:

(1) the 'go' gauge checks the maximum metal condition and should check as many dimensions as possible; and

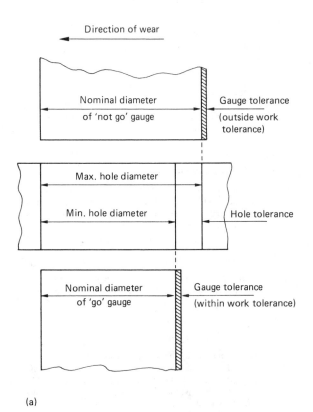

(a)

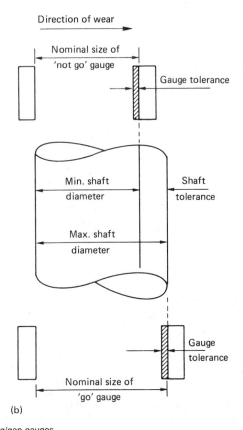

(b)

Figure 17.48 Disposition of gauge tolerances: (a) hole plug gauges; (b) shaft ring/gap gauges

(2) the 'not go' gauge checks the minimum metal condition and should check one dimension only, thus ensuring that the causes for the faulty dimension only are investigated.

This theory means that a hole 'go' gauge should, where possible, be a plug gauge about three to four diameters long. Similarly, a full form ring 'go' gauge should be used for shafts.

In practice, gap gauges are often used, but these should be accompanied by full form gauges, especially when eccentricity can occur in production.

17.8.5.2   External screw thread gauges

'Go' gauge   Ideally a full form ring gauge should be used, but this is often limited to small, standard sizes. A full form gap gauge is often used instead.

'Not go' gauge   A gap gauge checking the effective diameter only on a short length is used.

17.8.5.3   Internal screw thread gauges

'Go' gauge   This should be a full form plug gauge.

'Not go' gauge   This should have two gauges; one checking the pitch, the other the minor diameter. In practice, often only a simple plug gauge is used to check the minor diameter.

BS 919: 1960 lays down tolerance zones for the production of a limit gauge for screw thread checking.

17.8.5.4   Taper-limit gauges

Taper-limit gauges (*Figure 17.49*) do not in fact check the angle of the taper on a work piece. They only check that the diameter at a particular point on the taper is correct. This is achieved by grinding a step on one end of the gauge so that the diameters at the top and bottom of the step are the limits on

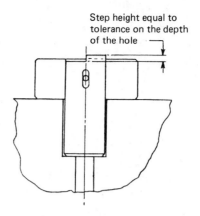

Step height equal to tolerance on the depth of the hole

Figure 17.50   Hole-depth gauge

the diameter. The limits are checked by ensuring that the component edge lies within the step.

17.8.5.5   Hole-depth gauges

Limit gauges for inspecting the depth of holes (*Figure 17.50*) are similar to those used to check tapers, but the former are in two parts: the sliding gauge and its sleeve. The edge of the sliding gauge should lie within the step. Note that tolerances are now required not only on the length of the gauge and the step but also on the thickness of the sleeve.

17.8.5.6   Large external and internal diameter gauges

The measurement of external diameters requires the use of a micrometer inserted into an angle piece, the included angle of which is accurately known. Using the configuration shown in *Figure 17.51* the radius R can be obtained using the formula:

$$R = (h + K_1) \times \cos(90 - \theta)/[1 - \cos(90 - \theta)] \qquad (17.17)$$

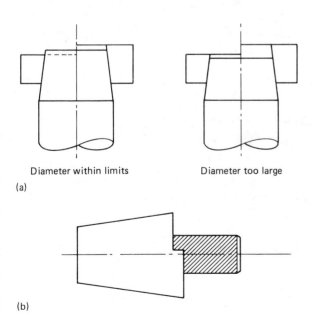

Diameter within limits   Diameter too large

(a)

(b)

Figure 17.49   Taper gauges: (a) taper ring gauge; (b) taper plug gauge

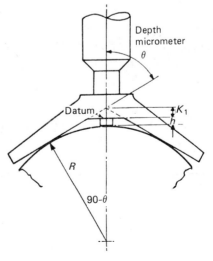

Depth micrometer

θ

Datum

K_1

h

R

$90 - \theta$

Figure 17.51   Measurement of a large external radius

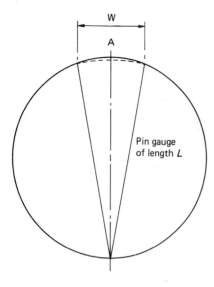

Figure 17.52 Measurement of large internal diameter

where R is the radius of the component, h is the height from the component to the datum face, K_1 is the distance from the intersection of the gauging faces to the datum face, and θ is the half included angle of the gauging faces.

The values of K_1 and $\cos(90 - \theta)/[1 - \cos(90 - \theta)]$ are constants for each gauge, and should be clearly marked on it.

An internal diameter is gauged by a pin gauge the length of which is accurately known. The pin is inserted into the bore to be gauged and the amount of rock about point A is measured, as shown in *Figure 17.52*.

The internal diameter D can be calculated using:

$$D = L + W^2/8L \tag{17.18}$$

or

$$W = \sqrt{8L(D - L)} \tag{17.19}$$

If the limiting values of D and the known length of the pin L are inserted into this formula, the limits on the amount of rock may be derived from equation (17.19).

Even with a steel rule, an accuracy of 1 mm can be achieved for the rock W, and this translates into an accuracy on the diameter D of some 0.05 mm when D is 400 mm.

17.8.5.7 Materials for gauges

The material selected for a gauge requires:

(1) *hardness*—to resist wear;
(2) *stability*—to maintain size and shape in use;
(3) *corrosion resistance*;
(4) *machinability*—to obtain desired shape, accuracy and finish; and
(5) *low coefficient of expansion*—especially where held by hand.

Normally a high carbon steel suitably heat treated will suffice, although other materials such as tool steel, Invar, and even glass can be used.

Note that a high quality surface finish substantially reduces the initial wear rate of a gauge. A maximum coefficient of linear expansion of 0.1 μm is ideal. Chromium plating of gauges is recommended where the gauges are subject to constant use.

17.8.6 Surface texture

17.8.6.1 Geometry of a surface

Any manufactured surface is made up of three main imperfections or errors (see *Figure 17.53*):

(1) *roughness*, caused by the finishing tool movement and shape;
(2) *waviness*, caused by variation in the process machinery; and
(3) *form*, gross errors in setting.

In addition, there is often a predominant pattern (*lay*) caused by the tool movement over the surface. Normally only the roughness and waviness are determined by the methods

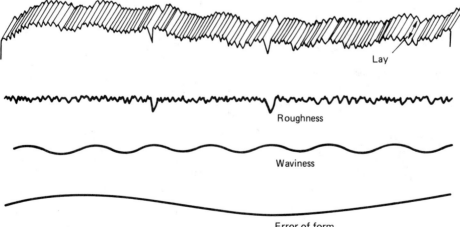

Lay

Roughness

Waviness

Error of form

Figure 17.53 Surface irregularities

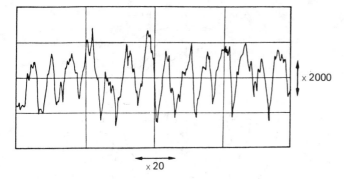

x 2000

x 20

Figure 17.54 A typical trace of a surface finish

discussed below as they can only measure a small portion of the total surface area at any one time.

17.8.6.2 Measurement

There are many methods for measuring surface texture, the most common being by drawing a stylus over the surface and amplifying the vertical and horizontal movements of the stylus to produce a trace (see *Figure 17.54*).

The stylus measurements are taken over a very short length, the actual size depending on the finishing process used (see *Table 17.12*). When a large surface area has to be checked, it is necessary to take a series of measurements and to use the mean of the values obtained.

In addition to the stylus method, the following have been found to be useful in determining surface texture.

(1) Visual comparison with a standard surface.
(2) A transparent plastic replica is made of the surface. After stripping, light is passed through this and measured. The surface roughness reduces the intensity of light by refraction.
(3) Reflection of light from the surface, measured by a photocell.
(4) Magnified (microscopic) inspection of the surface.

17.8.6.3 Expressing the surface finish

There are several different ways of expressing surface finish. The most common is the arithmetic mean deviation R_a, also called the centre-line average. This value is found by first measuring the area P along a set length L and finding the average height of the area H_m (see *Figure 17.55*):

$$H_m = P/L \tag{17.20}$$

Then, taking the areas above $(r_1, \ldots, r_n)$ and below $(s_1, \ldots, s_n)$ the centre line, we can calculate R_a (in micrometres):

$$R_a = \frac{\Sigma r + \Sigma s}{L} \times \frac{1000}{V_v} \tag{17.21}$$

where V_v is the vertical magnification, L is in millimetres and areas r and s are in square millimetres.

An alternative measurement is the R_x value, or 10-point height method. This comprises taking the five highest peaks and the five deepest valleys along the sample length L, as shown in *Figure 17.56* and calculating R_x using:

$$R_x = \frac{\Sigma \text{ Five peaks} - \Sigma \text{ Five valleys}}{5} \tag{17.22}$$

The actual use of the stylus affects the accuracy of the readings, as this determines how closely the profile recorded follows the real profile—if the surface is exceptionally rugged the stylus may not fit into the valleys, or it may round off the peaks as it is dragged over them. The stylus is normally made from a sharpened diamond for long wearing life.

17.8.7 Circular (roundness) measurement

17.8.7.1 Stylus instrument

Because of increasing technical advancement, the true roundness of a component often has to be determined. This has led to the development of a unit which measures the profile of a

Table 17.12 Suitable cut-off values for finishing processes

Typical finishing process	Designation	Meter cut-off (mm)				
		0.25	0.8	2.5	8.0	25.0
Milling	Mill		×	×	×	
Boring	Bore		×	×	×	
Turning	Turn		×	×		
Grinding	Grind	×	×	×		
Planing	Plane			×	×	×
Reaming	Ream		×	×		
Broaching	Broach		×	×		
Diamond boring	D. Bore	×	×			
Diamond turning	D. Turn	×	×			
Honing	Hone	×	×			
Lapping	Lap	×	×			
Superfinishing	S. Fin.	×	×			
Buffing	Buff	×	×			
Polishing	Pol.	×	×			
Shaping	Shape			×	×	×
Electrodischarge machining	EDM	×	×			
Burnishing	Burnish		×	×		
Drawing	Drawn		×	×		
Extruding	Extrude		×	×		
Moulding	Mould		×	×		
Electropolishing	El-Pol.		×	×		

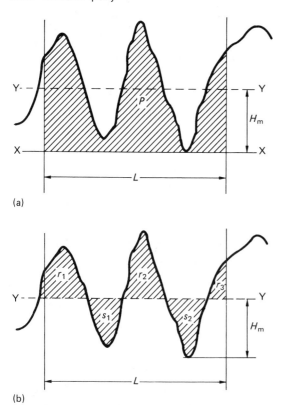

(a)

(b)

Figure 17.55   Calculation of R_a. (a) Determine the average height H_m. (b) The areas of r and s used in calculating R_a

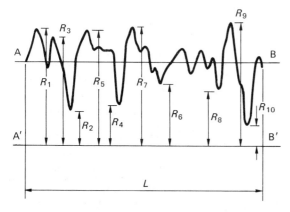

Figure 17.56   Calculation of R_x

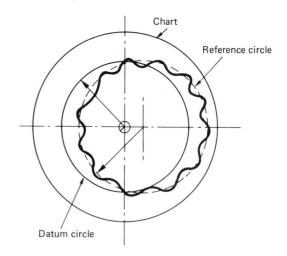

Figure 17.57   Evaluation of a surface roundness trace. The eccentricity of and variation in the surface radius are magnified, but the datum and reference circles are the normal size

The eccentricity and variation in radius from the circle of rotation are both shown on the chart to the same degree of magnification. However, a reference circle is drawn of the profile of the surface based on the centre of the real circle and variations in radii are measured from this reference.

17.8.7.2   Linear instrument methods

Where the stylus method is unavailable and/or a specification accuracy is not too high, roundness can be measured using linear instruments. Two-point measurement is seldom sufficient and three-point measurement is preferable (see *Figure 17.58*).

BS 3730: 1987 contains tables which give values of the deviation from the true value for each of these linear methods.

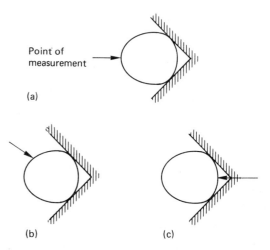

Figure 17.58   Three-point methods for measuring roundness: (a) symmetrical summit method; (b) asymmetrical summit method; (c) symmetrical rider method

component in a similar manner to the surface-texture-measurement instrument. The main difference is that the stylus is not drawn in a straight line but rotates in a true circle. The resultant trace is therefore circular, with magnification on radial measurement only (see *Figure 17.57*).

The assessment can be made on the basis of the departure from either the minimum inscribed circle or the maximum circumscribed circle.

Figure 17.59   The principle of air gauges

17.8.8   Air gauges

17.8.8.1   Principles

If a jet of air is escaping from an orifice as shown in *Figure 17.59*, then the pressure in the supply circuit is dependent on the distance from the orifice to an opposing face. By measuring this pressure, we can determine that distance.

The system is a comparative one and requires calibration against a standard unit. Tight tolerances can be measured readily with magnifications in the range 5000–10 000.

This method is dependent on the surface itself. With good surfaces the air flow can be directly onto the product and hence no contact, which reduces wear and other damage, need be made. When rough surfaces have to be measured, a contact probe may be used in which the orifice is internal (see *Figure 17.60*).

17.8.9   Electronic instruments

Any of the manual instruments, such as micrometer heads and air gauges, can be procured with digital read-out. Operations of these remove some of the operator-associated errors because the force used to set the anvils, etc., is normally applied and is limited by the device itself. This does not mean, however, that such instruments are completely free from operator error when used manually.

The mechanisms used to take the linear movement or pressure and translate it into the digital read-out need not concern the operator, except where special ancillary power is required. The operation is basically similar to the non-electronic instrument, with the added advantage of an electrical signal being capable of transmission to a remote point for automatic recording and/or decision-making.

17.8.10   Inspection probes

These probes allow direct gauging operations on products or tooling whilst in the processing machine itself. They can be permanently positioned on a machine bed, and hardwired, as shown in *Figure 17.61*. Alternatively, they can be mounted on machining spindles from tool-storage devices as and when required.

The probe shown in *Figure 17.62* consists of a stylus which gives a signal once it is deflected, including the direction

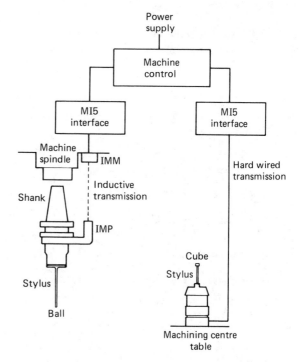

Figure 17.61   Probe system connections for computer numerical control (CNC) machining

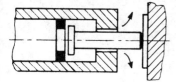

Figure 17.60   A contact gauging element

Figure 17.62   An inspection probe

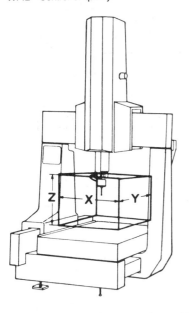

Figure 17.63 A coordinate measuring machine

taken. This is normally given in three-dimensional coordinates (x, y and z). The industry standard has inductive transmission, and the probe is activated as it is transferred into the machine spindle by the tool-changing device. Optical transmission can also be used.

17.8.11 Coordinate measuring machines

To measure a complex component quickly and accurately often requires sophisticated, programmable equipment dedicated to inspection. This is made possible by mounting probe devices on a computer-controlled multi-axis machine frame to produce a coordinated measuring machine such as that shown in *Figure 17.63*. These units normally come complete with measurement routines that enable an operative to move the probe quickly to a product feature and carry out a variety of measurements. This can be in any plane desired and can be made in relation to a fixed reference point set in three-dimensional space.

The programs themselves can carry instructions to the operatives which enables the probe to be used as the system requires. *Figure 17.64* gives some examples of the screens produced. Coordinated measuring machines can stand alone or can be integrated into a computer network to become a direct numerical control (DNC) unit. This enables instructions and programs to be passed down from the network and

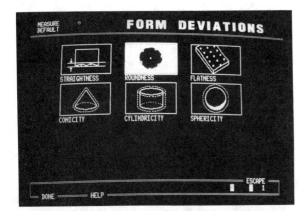

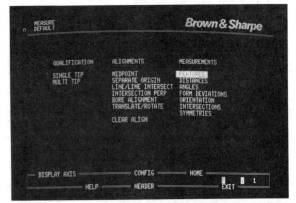

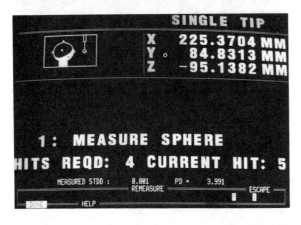

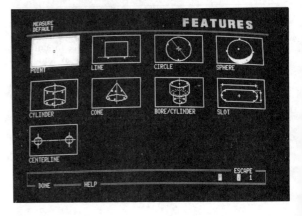

Figure 17.64 Some examples of the screens produced by a coordinate measuring machine

information regarding the results passed back up. The results may then be passed in turn to the processing unit for tooling adjustment.

17.8.12 Optical systems

17.8.12.1 Lasers

Although lasers are not normally used in direct measuring devices they can be used with triangulation techniques for range finding (see *Figure 17.65*). This enables them to be used to determine accurately small variations in distance from a fixed point, in effect similar to an air gauge. Commercial versions can measure distances to within 10 μm within a 3 mm range at 10 mm distance.

17.8.12.2 Machine vision

Computer vision analysis is basically a pattern-recognition system. At present it is, in metrology terms, a coarse system because of the large computer memory and associated analysis required to carry out a detailed test at high resolutions and speeds.

The system comprises three subsystems (*Figure 17.66*). The prime objective of the system is to recognise a profile of an object, irrespective of its position and orientation. It then may have to fix the position and orientation. It is highly advantageous, therefore, if at least the orientation is predetermined.

The basic minimum parameters contained in any pattern which require resolution are: area, perimeter, centre of area, and number of holes (including the area, perimeter and centre of the holes).

The camera basically produces a picture in a series of lines or small dots, or pixels, and it is the difference in grey scale (light level) of each of these pixels that is measured and then analysed. In a high resolution system the camera produces up to 256 pixels per line, 256 lines per frame (picture) and 25 frames per second. This requires 1.6 Mbit of information to be processed each second just at a single binary grey scale. In

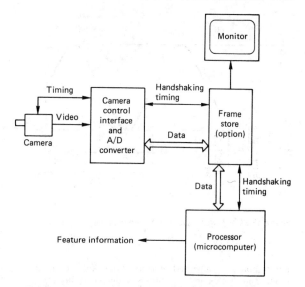

Figure 17.66 A machine vision processing system

higher grey-scale resolution systems up to 64 grey levels can be used, which requires six bits to give a total requirement of 9.83 Mbit/s.

Lower resolutions can be used if the field of vision of the camera is restricted to the need to discriminate. This reduces the processing load considerably. Resolution at low levels is often sufficient to carry out a check that features are in place, to differentiate between different products, or to determine the orientation of non-asymmetrical components.

The important factors in vision systems are the placement of the camera and the illumination because, if set correctly, these minimise parallax error. High contrasts reduce the problem of grey-scale differentiation and, therefore, back-lighting is preferable to reflected lighting.

17.8.13 Measurement of thread pitch

BS 84: 1956 shows that the tolerances on normal commercial threads are sufficiently large to be dealt with using the limit gauges. Thread gauges themselves and more precise threads, however, still require direct measurement to ensure that they are within their close tolerances.

Measurement of screw threads is complex, involving a number of interrelated characteristics, such as (*Figure 17.67*):

(1) major (crest) diameter;
(2) minor (root) diameter;
(3) form, particularly flank angles;
(4) pitch; and
(5) simple effective diameter.

Characteristics (3) to (5) comprise the virtual effective diameter. Measurement of the above dimensions requires specialised equipment, which is described below to show the range of techniques available. The equipment used is normally applied to external threads.

17.8.13.1 Measurement of the major diameter

The major diameter is the diameter of an imaginary cylinder which contains all the points on the crests of a thread. It is most conveniently measured by using a micrometer, or a comparator directly on the crest of the thread. The measure-

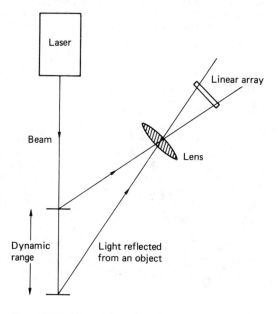

Figure 17.65 Triangulation using a linear array sensor

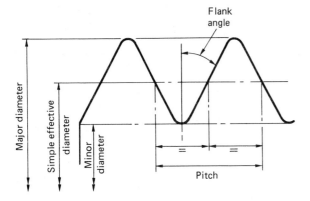

Figure 17.67 Characteristics of a vee-form thread

ment should be repeated at three axial positions to detect taper, and three angular positions to detect ovality.

17.8.13.2 Measurement of minor diameter

The minor diameter is the diameter of an imaginary cylinder containing all the points on the root of a thread. It is measured by a comparative process, similar to the method used for major diameter, except using hardened steel prisms to probe the root of the thread, as in *Figure 17.68*. The prisms must be offset by half the screw pitch. However, this could introduce an error in readings as the measuring pressure could set up a couple tending to rotate the component. To prevent this, the measurements should be carried out in a floating-carriage diameter-measuring machine. This allows the anvils to float at right angles to the thread axis. Again measurements should be taken at three axial and three angular positions to check taper and ovality.

17.8.13.3 Measurement of thread form (flank angle)

The flank angle is the angle made between the straight portion of the thread flank and a line normal to the thread axis. The effect of an error on a flank angle is to foul the mating thread

above, or below, the pitch line. In both cases the result is an increase in the effective diameter of an external thread, or a decrease in the case of an internal one. This results in a failure to mate with a correctly formed corresponding thread.

The actual effect is dependent on the thread type, but is approximately equivalent to a 0.25 mm difference at the effective diameter for every degree of error. If both flanks have an error then both errors contribute to the change in the effective diameter.

Normally, direct contact methods are impracticable, which leaves optical equipment as the only choice. There are two main methods used:

(1) projection, and
(2) microscope with a goniometric head.

17.8.13.4 Screw thread projection

There are many proprietary projectors used in this class such as the NPL projector developed by the National Physical Laboratory. They all work on similar principles.

The equipment consists of a parallel light projector, a set of centres and a screen onto which the enlarged image is projected in shadow form (see *Figure 17.69*). In order to avoid interference from the helix angle, the light is projected at an angle equivalent to the helix angle, with the thread axis parallel to the screen. The actual angle measurement is carried out using a protractor mounted on the screen. The horizontal axis of the protractor is adjusted until it is parallel to the thread crest or root as a datum.

17.8.13.5 Microscopic measurement of flank angle

This measurement is done using a microscope fitted with a goniometric head. The head has a clear screen in the lens carrying datum lines which can be rotated through 360°. The angle of rotation can be measured directly to 1 minute and estimated to fractions of a minute.

For measurement the microscope is focused using a bar set at the plane of the centre line of the thread. The thread is then set up in place of the bar, and the microscope is swung through the helix angle to avoid interference. The centres are mounted on slideways which can be accurately moved in two planes and rotated. The datum lines on the microscope head are aligned

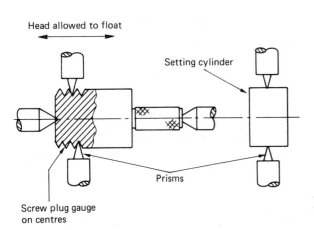

Figure 17.68 Measuring the minor diameter of the thread

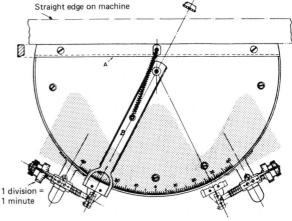

Figure 17.69 Screw thread projection

with the crest or root of the thread using the rotary table and the slideways, the table is then locked. The head itself can now be swivelled to measure the flank angles.

Note that the optical axis is not normal to the thread axis and some distortion of the image does occur. However, the accuracy achieved is sufficient for most purposes, and distortion can be allowed for, if necessary, by geometrical calculations.

17.8.13.6 Measurement of thread pitch

The pitch of a thread is the distance parallel to the thread axis between a point on a thread profile and the identical point on an adjacent thread. Whether the pitch error is positive or negative, the net effect is to increase the effective diameter of an external thread, or to decrease it in the case of an internal thread. The effect on the effective diameter is approximately twice the error in a single pitch. The actual effect depends on the thread type.

The pitch error can be of three types: progressive pitch error, periodic pitch error, or drunken thread error.

Progressive pitch error   This is caused when the linear tool movement/work rotation ratio is constant, but incorrect. Common causes include no suitable gear train combination, and faults in lead screws or generating machine. A graph of the cumulative pitch error is linear.

Periodic pitch error   This is caused when the tool/work ratio is not constant. Causes can be cyclic pitch errors in generating mechanism, or wear in radial or thrust bearings. Normally this type of error is cyclic and a graph of a cumulative pitch error is similar to a sine wave.

Drunken thread   Such errors are difficult to discover because, when measured parallel to the thread axis, the pitch appears correct. The thread, however, has not been cut to a true helix.

17.8.13.7 Measurement of pitch errors

To measure pitch error, a pitch-measuring machine, similar to that developed by the National Physical Laboratory, or one of the optical units described above for flank measurement can be used.

The NPL design relies on a stylus mounted on a block supported by a thin strip. The sideways pressure on the stylus is only constant at identical points on the thread profile, i.e. a pitch, or multiples of the pitch.

In the optical projection method the image line is set against one of the flanks. Therefore, when the same flank on the adjacent thread profile is set against this image line, one pitch has been traversed.

Both methods rely on accurately measuring the distance the thread moves axially between the same points on adjacent thread profiles. This can be for a single pitch, or a multiple of pitches.

17.8.13.8 Measurement of simple effective diameter

The simple effective diameter of a thread is the diameter of an imaginary cylinder parallel to the thread axis cutting the thread so that the distance between any adjacent interception points of cylinder and thread profile is exactly half a pitch (see *Figure 17.70*).

Measurement of the simple effective diameter is best carried out using the same basic unit as used to measure the minor diameter (see Section 17.8.13.2). The prisms are replaced by steel wires of such a size that they contact the thread profile at

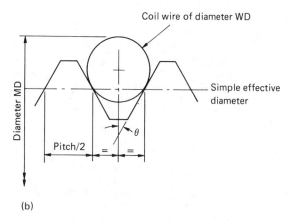

Figure 17.70 Measurement of the simple effective diameter: (a) using a measuring machine; (b) using a spring coil gauge

the effective diameter. These wires are termed 'best-size wires'.

It is normally possible to procure a standard best-size wire for a particular thread. If the thread is an unusual one, the diameter for the best-size wire can be found using:

$$d = p/2 \cos \theta \qquad (17.23)$$

where d is the diameter of the best-size wire, p is the thread pitch, and θ is the thread flank angle.

17.8.13.9 Determining the simple effective diameter

Measuring machine   An initial reading is taken with the wires over a known setting cylinder. The cylinder is replaced by the thread and the reading over the wires repeated. The difference, maintaining the sign, is then added to the setting cylinder diameter to produce the adjusted setting diameter AD. The effective diameter ED is then calculated using:

$$ED = AD + (0.5p \cot \theta) - WD \times (\text{cosec } \theta - 1) \qquad (17.24)$$

where p is the thread pitch, θ is the thread flank angle, and WD is the wire diameter.

Micrometer and spring coil gauge   Where the measuring machine is not accessible, it is possible to use a micrometer along with a spring coil gauge, of the correct 'best size', to determine the simple effective diameter. This involves springing the coil gauge over the thread and taking the size over the coil wires. The simple effective diameter ED is then found using:

$$ED = MD + (0.5p \cot \theta) - WD \times (\operatorname{cosec} \theta - 1) + 2WD \tag{17.25}$$

where MD is the measured distance over the coil gauge, p is the thread pitch, θ is the thread flank angle, and WD is the coil wire diameter.

17.8.13.10 *Measurement of virtual effective diameter*

The virtual effective diameter is defined as: the effective diameter of a thread of true form which will exactly engage with the produced thread form. Any pitch or flank angle errors will affect the actual effective diameter. In the case of an external thread, the simple effective diameter is enlarged by these errors. In internal threads the errors reduce the simple effective diameter.

The formula used is

$$VED = SED \pm Kp(\delta\theta_1 + \delta\theta_2) \pm (\delta p \cot \theta) \tag{17.26}$$

where VED is the virtual effective diameter, SED is the simple effective diameter, K is a constant depending on the thread type, θ is the thread flank angle, $\delta\theta_n$ are the thread flank errors, and δp is the pitch error over the length engaged.

The $\pm$ sign is $+$ for external threads and $-$ for internal threads.

17.8.13.11 *Measurement of internal or large threads*

Measuring the thread profiles of internal and large external threads is made by taking an accurate wax cast of several pitches of the thread and examining it using the projection method. The main problem is ensuring that the cast is taken and then examined in the correct plane. The root diameter of internal threads can then be measured and, using the data from the cast examination, it is possible to calculate both the simple and virtual effective diameters.

17.9 Non-destructive testing

17.9.1 Radiography

In the seventeenth century, it was known that the propagation of light can be represented by means of rays. If light from a very small, ideally a point, source is interrupted by an opaque obstacle, a very sharp shadow is formed. If the source is not small, the edge of the shadow becomes less than sharp. There is a dark shadow known as the 'umbra', and a diffuse edge known as the 'penumbra' (see *Figure 17.71*). This phenomenon occurs with all forms of radiated energy and can be applied to industrial radiography.

Radiation intensity The rate at which radiation energy crosses a surface is called the 'radiation flux'. The flux across any surface is the same as the rate of emission from the source. The flux per unit area is called the 'intensity of radiation'.

Inverse square law The intensity of radiation from a point source varies inversely as the square of the distance from the source. According to the inverse square law:

$$\frac{I_1}{I_2} = \frac{d_1^2}{d_2^2} \tag{17.27}$$

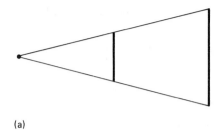

(a)

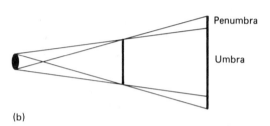

Penumbra

Umbra

(b)

Figure 17.71 The size of light source and the resulting shadows: (a) a point source gives a sharp shadow; (b) a non-point source gives a diffuse shadow

Strictly speaking, this law is applicable only to point sources. However, the error factor involved when using real sources is small enough to be ignored.

17.9.1.1 *X radiation*

Discovered by C. Rontgen, Professor of Physics at the University of Würzburg in 1885, X-rays are produced whenever high-speed electrons are stopped abruptly by allowing them to impinge upon a target of some material. X-rays are produced no matter what material is used for the target, but those made of materials of high atomic number give more intense radiation.

The modern X-ray tube comprises:

(1) a source of electrons—a heated element in an evacuated space;
(2) a target of heavy metal, usually tungsten; and
(3) an accelerating device—a high potential between the filament (cathode) and the target (anode).

Electrons produced at the filament are accelerated towards the target. They strike the surface with high energy and some of this energy is converted into X-rays. Most of the energy of the electrons is converted at the anode into heat. It is important, therefore, that the target material should be of high melting point. As the X-ray energy increases, the wavelength decreases, and short wavelength radiation therefore has the most penetrating power.

X-ray equipment (Figure 17.72) The electrons are produced by passing a current through the coil. They are accelerated to high speed by applying a high potential difference across the system, i.e. from target to coil. This forces the free electrons to rush towards the anodic target. The higher the potential difference, the greater the speed and, therefore, the greater

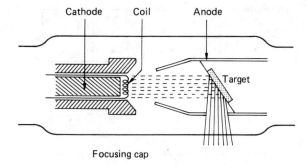

Focusing cap

Figure 17.72 An X-ray tube

the penetrating power of the X-rays. The tube is evacuated to prevent the propelled electrons being diverted or slowed down through colliding with electrons in the air.

Sometimes a conical target giving 360° delivery is used internally for X-raying butt welds in tubes. Efficiency is low, however, because of beam spread.

Factors affecting X-ray output The factors affecting the *intensity* of radiation emerging from the tube window are:

(1) the current (milliamperes, mA);
(2) the voltage (kilovolts, kV);
(3) the focal spot size; and
(4) the inherent filtration of the tube.

The intensity of radiation emerging from the tube window is proportional to the coil current. This current determines the number of electrons striking the target in a given time, and hence the density of the shot. Increasing density reduces the time of exposure. The density is usually 1.5–2.5 greyness factor.

The *velocity* of the radiation (*v*) is proportional to the potential difference (in kilovolts) between the anode and the cathode. The 'rule of thumb' is:

$$v = (\text{Thickness of metal} \times 4) + 50 \qquad (17.28)$$

where the metal thickness is in millimetres.

The larger the focal spot, the greater the intensity of radiation produced. However, it is also desirable, on account of definition (see Section 17.9.1), to maintain as small a focal spot as possible.

When a beam of X-rays passes through any material only a part of it emerges, due to scattering. The amount of scatter radiation reaching the film under practical conditions is not easy to determine, and depends on such factors as radiation quality, specimen shape, material and thickness. Scattered rays travelling at any angle tend to spoil definition and contrast. Scatter may be reduced by:

(1) limiting the size of the radiation field;
(2) backing the film with lead to reduce back-scatter;
(3) careful use of radiation energy;
(4) the use of filters; and
(5) blocking with a suitable medium, e.g. barium putty or lead shot.

17.9.1.2 Natural γ radiation

The discovery of radioactivity followed just 2 months after that of X-rays. Henri Becquerel discovered that uranium compounds fog a wrapped photographic plate, and concluded that some form of penetrating radiation was being emitted.

It was later discovered by Lord Rutherford that there are essentially three types of radioactive emission. These he called alpha (α), beta (β), and gamma (γ). These radiations came from the nuclei of the radioactive atoms.

(1) α particles are heavy and have low penetrating power.
(2) β particles are very fast with a range of up to about 2 m in air (impracticable for industrial use but used medically).
(3) γ rays have extremely short wavelengths and are, in consequence, of high velocity and penetrating power. They are used industrially.

The most common elements used in the production of γ radiation are: iridium-192, caesium-137 and cobalt-60.

Radioactive decay As the atoms of a radioactive source disintegrate to emit radiation, the source is said to be 'decaying'. A radioactive source decays exponentially, i.e. the rate of decay is always proportional to the amount of radioactivity remaining. Therefore, for a strong source, the initial decay is high, but the decay decreases progressively as the activity diminishes.

In an exponential decay there will be a certain period of time *T* after which the source will have decayed to half its original strength. After another interval of time *T* the strength will have decayed to one-half of one-half, and so on. Therefore, a source will decay to zero activity only after an infinite period of time. The 'half-life' should not be confused with the useful life of a source. Among other factors, the half-life of a source is a very important aspect from the point of view of practical industrial usage.

Strength of a source over time To calculate the strength of a γ-ray source after any number of half-lives is a matter of simple arithmetic. In order to determine the strength after any time, it is convenient to compile a decay chart. This can be done simply by plotting a graph of source strength against time. The strength at each successive half-life is known and the points can be joined by a smooth curve (see *Figure 17.73*). As can be seen from the figure, at any time the strength of the source can be found from this curve. A graph of this nature is supplied with all sources obtained for industrial use.

Requirements for a γ-ray source For an isotope to be of value in industrial radiography it must have:

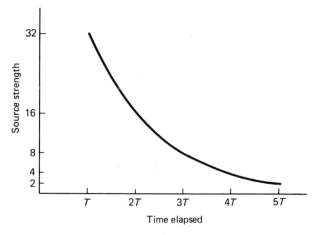

Figure 17.73 The decrease in source strength over time

(1) a reasonably long half-life;
(2) a reasonably short activation time, i.e. be manufactured by bombarding stable isotope in atomic pipe where nuclear reactions take place changing the stable material into an unstable radioactive source; and
(3) a useful γ-ray spectrum, i.e. it must generate wavelengths which will penetrate a reasonable range of material thicknesses.

Iridium-192, caesium-137 and cobalt-60 fulfil these criteria to varying degrees.

17.9.1.3 Production of radiographs

As previously mentioned, photographic plate is affected by radiation in a similar fashion to light. On exposure to radiation the plate is darkened to a degree dependent on the amount of radiation received. Accordingly, if a body is to be radiographed, it is placed at a distance D from the radiating source (X or γ), as shown in *Figure 17.74*.

The amount of radiated energy passing through the body will cause the film to darken. Should a defect such as a cavity be present in the material, a greater amount of energy will pass through to the film under that point (BC in *Figure 17.74*) because the air in the cavity absorbs less energy than does steel, or another metal, of a similar volume.

It can be seen from *Figure 17.71* that, to obtain the maximum sensitivity of the film, the distance from the source to the film should be at a maximum to avoid penumbral unsharpness. It should also be kept near enough to ensure that the exposure time is kept within reasonable limits. The inverse square law (equation (17.27)) can be used to calculate the appropriate distance.

For optimum film sensitivity, the source–film distance (see *Figure 17.74*) should be the maximum practicable, and the object–film distance should be the minimum practicable. It follows that, in the radiographic inspection of thick materials, a long source–film distance is necessary. On thin materials a short source–film distance can be used.

Sensitivity of flaw detection This is defined as:

$$\frac{\text{Size of the smallest visible defect} \times 100}{\text{Specimen thickness}}$$

Without actually sectioning samples radiographed, the size of the smallest defect is difficult to establish. An aid to obtaining a standard in this respect is the penetrameter or image-quality indicator.

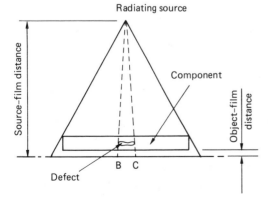

Figure 17.74 The set-up used for radiography

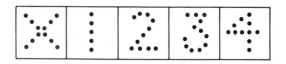

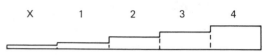

Figure 17.75 A step wedge penetrameter

Penetrameter sensitivity A penetrameter is an artificial defect of known thickness which is radiographed with the specimen. Various types are used, a common one being the step wedge (see *Figure 17.75*). The steps can have the following thicknesses:

Step	X	1	2	3	4
Thickness (mm)	0.1	0.2	0.3	0.4	0.5

The penetrameter sensitivity is defined as:

$$\frac{\text{Smallest visible penetrameter indication} \times 100}{\text{Specimen thickness}}$$

Thus with a No. 3 on a step wedge evident on a specimen thickness of 25 mm, the sensitivity is:

$$(0.4 \times 100)/25 = 1.6\%$$

Comparison between X and γ radiation The advantages of X radiation are that:

(1) it produces a film of higher contrast and sensitivity because of its longer wavelength and lower diffusion;
(2) an X-ray machine is switched off when not in use and does not constitute a safety hazard except when operational; and
(3) in general, shorter exposure times can be used.

The advantages of γ radiation are:

(1) the unit can be carried by one man;
(2) the cost of the source and container is low;
(3) no power supply is required;
(4) no maintenance is required;
(5) the source can be inserted where an X-ray tube cannot;
(6) the source gives uniform radiation in all directions, and thus many components can be radiographed simultaneously; and
(7) because of its constant output the decay time of the source can be calculated and compensated for.

17.9.1.4 Interpretation of weld radiographs

Some typical weld flaws are shown in *Figure 17.76*.

Slag inclusions During welding, a flux compound is used to clean the parent metal and shield the arc and molten pool from the air. The slag so produced floats on the surface of the molten metal pool. If the welder does not properly clear the corners at the root run to the parent metal intersection, slag can be readily trapped by subsequent filler runs. Slag inclusions are readily detected by radiography.

EXAMPLES OF WELDING DEFECTS

The radiographs of welding faults reproduced below are from the "Collection of Reference Radiographs of Welds" published by the International Institute of Welding.

(D) Traverse cracks

(A) Gas porosity (gas pockets)

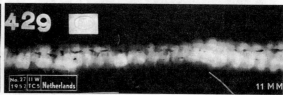

(E) Gas bubbles (piping)

(B) Lack of penetration

(F) Slag lines

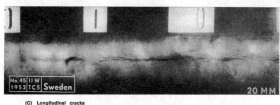

(C) Longitudinal cracks

(G) Undercutting

Figure 17.76 Examples of welding defects: (a) gas porosity (gas pockets); (b) lack of penetration; (c) longitudinal cracks; (d) traverse cracks; (e) gas bubbles (piping); (f) slag lines; (g) undercutting. (Courtesy of the International Institute of Welding)

Incomplete penetration This arises due to the weld gap not being filled during root run. It severely weakens the resultant weld.

Incomplete side wall fusion This defect is caused by:

(1) damp welding electrodes,
(2) oil on the electrodes or weld preparation, or
(3) use of a welding current that is too low to melt the parent metal.

The defect is difficult to determine by radiography because of its proximity to the slag zone and its small dimensions, lying at right angles to the ray's path. It looks similar to slag inclusions, but is lighter and feathery.

Porosity Porosity is caused by damp rods, or incorrect choice of welding electrodes. With damp rods porosity occurs at the beginning of each run of a rod for about 50 mm. As the rod heats up, the dampness vanishes and the weld becomes sound. Overlapping at the start of each rod, followed by grinding off the overlap helps to remove the region affected. Porosity is not hazardous in small amounts because it is spherical. Rods should be kept warm and dry, with preheating of the rods if necessary.

Piping The cause of piping is similar to that of porosity. Gases inside the molten pool endeavour to escape but are trapped by rapid cooling of the pool. It is frequently caused by welding thick material which gives rise to a large heat sink.

On film it appears as a tadpole-like shape. Unlike porosity, piping is a serious defect as the shape encompasses a point and hence a stress source. It must be removed.

Cracks There are many causes of cracks, most of which can be traced back to poor design. Shrinkage can occur as can other distortion; this causes build up of stress in the weld. Cracks normally show up on film as heavy black lines.

Undercutting This is caused by entering the weld pool at an angle or through using too high a welding current. It appears on the film as dark lines at the edge of the weld zone.

17.9.1.5 Safety aspects

It is important to remember that radiography is a hazardous operation involving risks through exposure to the sources. Both the degree (dose) of exposure and the time of exposure are important. This means that the operation must be carefully planned and controlled, i.e. access limited. This also applies to the storage and movement of any radioactive source. It is important that the dosage received by operators over time is recorded by means of film badges and medical examinations.

The layout of the operational area is important so that non-operators are protected by distance and/or suitable barriers. Where possible, the number of locations of use should be severely limited and equipment designed so that the direction of radiation is restricted.

17.9.2 Ultrasonics

17.9.2.1 Propagation of ultrasonic energy

Sound is the energy produced by a body in vibration. It is transferred from one part of a sound-conducting medium to another by means of a wave-like motion of particles. The wave motion gives rise to areas of increased pressure (compressions) and decreased pressure (rarefactions) which travel through the medium in the direction of propagation.

An obstacle (like the ear drum) in the path of the sound waves is therefore subjected to rapidly changing pressure variations which it records as 'sound'. The upper limit to the hearing of humans is 16–20 kilocycles per second. If the sound is propagated at a frequency above this, it is known as 'ultrasonic'. In all other aspects, ultrasound obeys the same laws of propagation, reflection, refraction, etc., as audible sound. It follows that special equipment is required to produce and receive high frequency ultrasonics.

Wave motion The term 'wave motion' denotes a condition that can be transmitted so that it can be experienced at a distance from its source. In ultrasonics the condition being propagated is a displacement of the particles of the medium in which the wave travels.

Any medium that has elasticity can propagate ultrasonic waves. Propagation takes the form of a displacement of successive elements of the material. As the substance is elastic, there is a restoring force that attempts to bring each element of material back to its original position. However, due to inertia, the particle continues to move past its starting position to take up a new one. From this position, it again passes through its starting position. It continues to vibrate around its starting position with constantly reducing amplitude. Elements of the material therefore execute different movements as the wave passes through them. As the wave moves through the material, successive elements experience displacements. Each successive element moves a little later than its preceding neighbour.

Ultrasonics obey the following formula:

$$\lambda = V/f \tag{17.29}$$

where λ is the wavelength, V is the velocity of the radiation, and f is the frequency.

Typical longitudinal wave velocities are:

Steel	5900 m/s
Aluminium	6400 m/s
Concrete	4200–5200 m/s
Iron	5930 m/s
Human bone	4000 m/s

Wave train (pulse) The term 'wave train' refers to a short group of waves before and after which there are no waves. Such a group is usually called a 'pulse' and may take several shapes (waveforms). *Figure 17.77* shows the waveforms used for different tasks.

Waveforms *Longitudinal waves* exist when the motion of the particles in a medium is parallel to the direction of propagation of the wave. This is the most commonly used waveform.

Shear waves exist when the motion of the particles of the medium is at right angles to the direction of propagation. Shear waves have a velocity approximately half that of longitu-

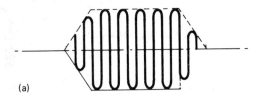

(a)

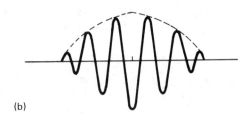

(b)

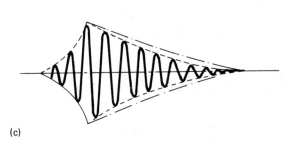

(c)

Figure 17.77 Wave train pulse forms. (a) Rapid rise and decay, square envelope; this form is used in autoamtic testing because of its well-defined shape. (b) Gradual rise and decay, triangular envelope. (c) Rapid rise and natural exponential decay; this is the most common form and it is often referred to as a 'damped train'

dinal waves. Shear waves cannot be generated in liquids or gases because there is little elasticity to shear in such materials. Shear waves are fairly commonly used.

Surface waves can only be generated on good surfaces and penetrate only the top two or three molecular layers. The velocity of surface waves is 90% of that of shear waves.

Lamb waves are a special type of waveform only possible in thin plates. The waveform depends on the material thickness, the angle of the incident beam, and the frequency of the incident beam.

The different types of wave are shown in *Figure 17.78*.

Refraction It is often necessary to induce ultrasonic waves in a material at an angle. When this is done, the beam introduced at one angle into the first material enters the second material at a different angle (see *Figure 17.79*). The difference is caused by the refraction which takes place at the boundary, or interface, between the two materials.

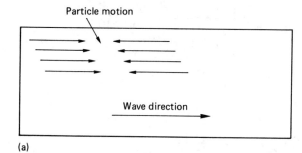

(a)

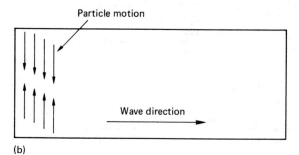

(b)

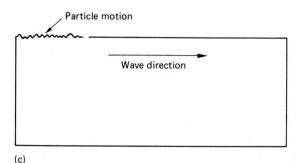

(c)

Figure 17.78 Types of wave: (a) longitudinal; (b) shear; (c) surface

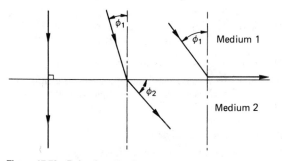

Figure 17.79 Refraction at surfaces

The angular relationships for plain wave propagation are the same as for light and are given by Snell's law:

$$\frac{\sin \phi_1}{\sin \phi_2} = \frac{V_1}{V_2} \tag{17.30}$$

where ϕ_1 is the angle of incidence, ϕ_2 is the angle of refraction, V_1 is the velocity in the first medium, and V_2 is the velocity in the second medium.

Total reflection at the interface occurs for $\phi_2 \geq 90°$.

Where shapes are complex the refraction pattern may be difficult to interpret. For example, in a part which is flat on one side and curved on the other, the wave can enter the flat side with little reflection but emerges from the curved side broken up, with various parts of the beam travelling at different angles.

Lenses can be designed to affect ultrasonic waves in the same manner as for light. They can be used to focus waves in materials and may be formed of plastic, metal or liquids.

Whether longitudinal, shear or surface waves are generated depends on the angle of incidence. As can be seen in *Figure 17.80*, there are angles of incidence where more than one type of wave is created. There are also angles at which the wave amplitude is at a maximum.

Beam spread A beam of ultrasonic energy is propagated through a material with very little divergence. However, some spreading does occur and is a function of the ratio λ/D, where λ is the wavelength of the ultrasonic wave and D is the diameter of the transducer. The relationship

$$\sin \alpha = (0.5\lambda)/D$$

generally holds, α being equal to half the angle of the beam spread.

The value of λ can be calculated if the velocity V through the material and the probe frequency is known:

$$\lambda = V/f \tag{17.31}$$

Prior to the divergence, a part of the ultrasonic beam is parallel. This part is called the 'near zone', or 'Fresnel zone', and beyond this is the 'far zone' (see *Figure 17.81*). The signal amplitude for a given probe system is at a maximum at the beginning of the far zone. Thus, this factor should be borne in mind when choosing the probe. The near zone (NZ) can be calculated from the relationship:

$$NZ = 0.25 \times D^2/\lambda \tag{17.32}$$

For ideal testing, from point of beam spread, the frequency and the diameter of the probe should be as high as possible; however, this increases the near zone. The sensitivity pattern achieved in the near zone fluctuates and thus all testing should be carried out in the far zone, as close as possible to the interface with the near zone.

17.9.2.2 Generation of ultrasonics

Ultrasonic sources There are a number of ways in which sound waves can be generated. The method chosen depends

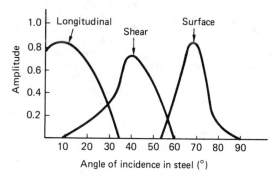

Figure 17.80 Types of wave generated according to the angle of incidence

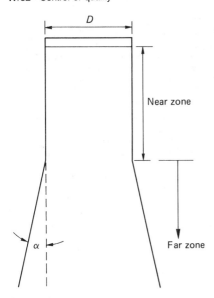

Figure 17.81 The divergence of an ultrasonic beam

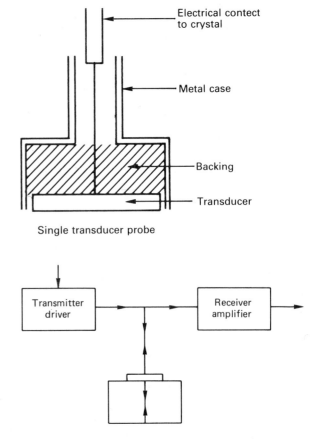

Figure 17.82 A combined transmitter and receiver probe

on the required power output and frequency range. Generators of the mechanical type, such as tuning forks can only be used up to 10^3 cycles per second and are, therefore, unsuitable for ultrasonic generation.

The most common and simplest method of producing high-frequency ultrasonics is the crystal transducer. This can generate frequencies up to 15×10^6 cycles per second.

Piezoelectric effect   Piezoelectric transducers make use of the expansion and contraction of various crystalline and ceramic materials under the influence of varying electrical fields. When an electric charge is applied to the face of a crystal, a change in the dimensions of the crystal takes place. The reverse is also true in that a mechanical strain applied to a crystal will create an electric signal. This makes a crystal capable of being a transmitter and/or a receiver of vibrations. Some of the materials used in piezoelectric transducers are natural substances, although synthetic crystalline and ceramic materials are also used.

Types of probe   Probes may have either a combined transmitter and receiver or a separate transmitter and receiver.

In combined transmitter/receiver probes (*Figure 17.82*) the same crystal generates and detects the ultrasonic energy. It has the advantages that damping is easier, giving sharp peaks, and that the direction of beam is easily targeted, thus making it easier to read signals. It has the disadvantage that it takes time to stop vibrating, which produces a dead zone in the metal.

In the separate transmitter/receiver probes (*Figure 17.83*) different crystals are used, with cork insulation preventing acoustic or surface reflected vibrations. The main advantage of such probes is that there is no dead zone.

17.9.2.3   Detection of a flaw

Reflection technique   This uses a common transmitter/ receiver. On good material, the ultrasonic energy generated is reflected from the back wall of the material only. If a flaw be present, part or all of the beam will be reflected back before it reaches the back wall (see *Figure 17.84*).

Transmission technique   In this technique a separate transmitter and receiver is used, one at each side of the material (see *Figure 17.85*). The presence of a flaw will reduce the signal reaching the receiver probe. The disadvantages of this technique are:

(1)   the specimen must be accessible from both sides;
(2)   two accurately positioned coupling points are required; and
(3)   no indication is given of the depth of the flaw.

Care should be taken when using ultrasonics to identify the presence of spurious echoes due to multiple wave modes being set up.

17.9.3   Magnetic particle testing

The requirements for magnetic particle testing are:

(1)   the material under examination must be capable of being magnetised;
(2)   there must be a means of producing the magnetic flux lines; and
(3)   there must be a medium to show up defects.

17.9.3.1   Basic principles

In the normal state, the individual particles of a piece of iron lie in random directions. When magnetised, the particles

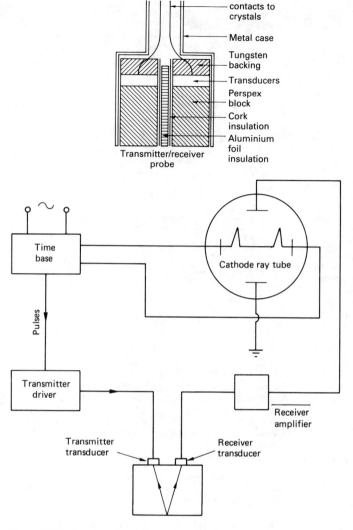

Electrical contacts to crystals
Metal case
Tungsten backing
Transducers
Perspex block
Cork insulation
Aluminium foil insulation

Transmitter/receiver probe

Time base

Pulses

Transmitter driver

Cathode ray tube

Receiver amplifier

Transmitter transducer

Receiver transducer

Figure 17.83 A separate transmitter and receiver probe

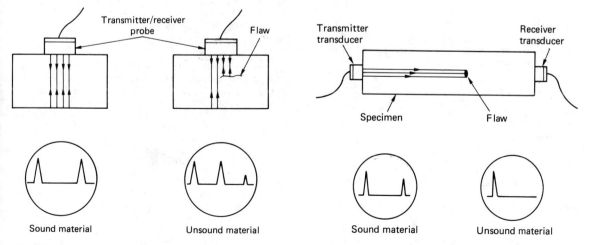

Transmitter/receiver probe

Flaw

Transmitter transducer

Receiver transducer

Specimen

Flaw

Sound material

Unsound material

Sound material

Unsound material

Figure 17.84 The reflection technique

Figure 17.85 The transmission technique

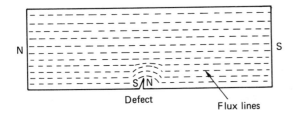

Figure 17.86   The formation of magnetic poles at a crack

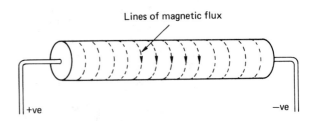

Figure 17.87   Flux flow in a bar

reorient themselves to form continuous lines of magnetic force (flux) along the direction of magnetism. When the magnetising force is removed, the particles are free to return to their random arrangement. The degree to which they do this depends on their characteristics.

When a bar is magnetised, the lines of the magnetic force will follow a regular path, as long as they pass through a consistent material structure. A break in the structure close to the surface, if correctly orientated to the flux, will disturb this regular path. In the vicinity of such a break the flux will pass round the break, instead of attempting to jump over it. This then sets up a pair of magnetic poles across the break. The poles will have sufficient strength to hold some magnetic detecting particles and hence show the position and length of the break (see *Figure 17.86*). The direction of the break must lie across the direction of the flux lines. If the break runs in the same direction, the flux will simply pass either side and produce minimal poles. Therefore, to examine a complex shape completely, several directions of magnetism must be sampled.

17.9.3.2   Magnetic detectors

Fine iron particles are used to highlight the presence of the poles. These can be applied wet or dry. In the wet process, the particles are suspended in a paraffin or water based medium. This is normally produced by diluting a purchased concentrate to the required strength. Alternatively, a dry powder may be applied by a puffing or dusting method. Whichever process is used, it is important that the concentration of the detector is maintained to ensure consistent results.

17.9.3.3   Viewing

The defect must be clearly and quickly seen. The particles must therefore be bright and contrast with the background of the material. Fluorescent magnetic ink enables defects to be spotted quickly under ultraviolet light.

17.9.3.4   Magnetisation

Magnetisation can be induced in a number of ways, but these are normally limited to:

(1)   current flow,
(2)   flux flow, and
(3)   coil shots

In each case, alternating or direct flow can be established.

Current flow technique   When an electric current is passed through a metal bar, a magnetic field is set up at right angles to the current flow. The magnet flux thus flows in a circular path round the bar (*Figure 17.87*). Similarly, a current may be

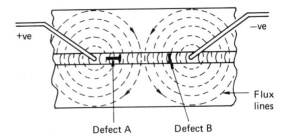

Figure 17.88   Flux flow in a plate

introduced into a plate through probes, with circular flux paths being created (*Figure 17.88*).

Defects that lie at right angles to the flux flow will be readily detected. Defects that lie along the flux paths, however, will escape detection, and the operation must therefore be repeated at right angles to the first. This is impracticable on bar which must be tested further using the flux flow or coil shot methods.

Flux flow technique   A permanent magnet or an electromagnet may be used to pass a flux through the specimen (*Figure 17.89*). The area of examination is limited to the area between the poles of the magnet, and these must be moved systematically to cover a whole surface. An irregular shape may have little flux at its extremities.

Coil shot technique   The specimen is placed inside a coil of wire through which a current is passed (*Figure 17.90*). An encircling coil enables flux to pass more uniformly around an irregular shaped specimen.

Type of current   For general work, a.c. is acceptable. This, however, restricts the test to the surface layers only. The maximum depth that can be tested using a.c. is 1.25 mm.

Where defects deeper than this are sought, d.c. must be used. Half-wave d.c. is even better and can detect defects up to 12 mm below the surface.

17.9.4   Penetrant testing

This method is only useful for locating defects which come to the surface. It can be used on metals, plastics, glassware, ceramics, etc., which are not normally porous in themselves. The basis of the method is that a liquid will enter small cracks on a surface. If the surface is wiped clean the liquid will seep out of the crack onto the surface, where it can be detected.

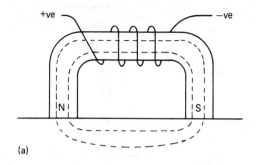

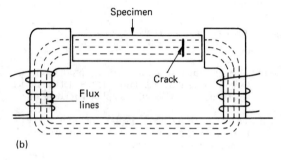

Figure 17.89 The flux flow technique: (a) creating flux flow in a plate; (b) creating flux flow in a bar

17.9.4.1 Precleaning

Prior to applying the penetrant, the specimen must be perfectly clean, including the inside of all surface cracks.

17.9.4.2 Penetrant application

The penetrant can be applied by dipping, spraying or brushing. Penetrants must have good penetrating properties and form an intimate contact with the surface under test.

Testing for porosity normally requires a different penetrant than that used for crack detection.

17.9.4.3 Post cleaning

In order to see defects, it is necessary to remove all the penetrant from the surface of the specimen. This can be done by one or more of the following means:

(1) light chemical cleaning,
(2) water washing,
(3) emulsification then water washing, and
(4) dust or powder brushing.

17.9.4.4 Development

With time, any penetrant held in a defect will tend to bleed out and thus indicate the position and nature of the flaw. In order to hasten this, a receptive layer of talc is normally applied to the surface. In the case of visible penetrants this further assists spotting through forming a contrasting background. Fluorescent penetrants may also be used. These draw the eye to defects and hence are useful where time is important. Very small cracks of only micrometre thickness can be detected.

17.9.5 Eddy-current testing

These are, in effect, electromagnetic tests. They rely on the differences in resistivity and/or magnetic properties between a reference component of known properties and a specimen under test.

17.9.5.1 Fixed head technique

A pair of matched coils is connected to a bridge circuit. To this is connected a signal supply voltage and a detector such as an instrument or cathode ray tube (see *Figure 17.91*). In the figure, B_1 and B_2 are static arms of the bridge which is completed by the balance coils. The power supply is stabilised and the detector device D is connected to an amplifier. Facilities R are provided for balancing the bridge.

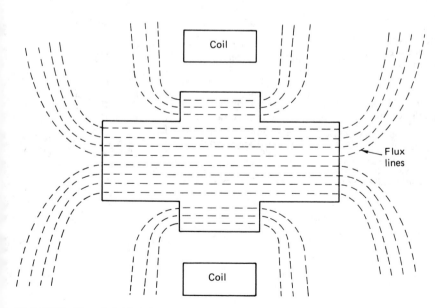

Figure 17.90 The coil shot technique

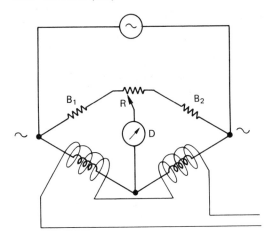

Figure 17.91 Scheme of a circuit for eddy-current testing

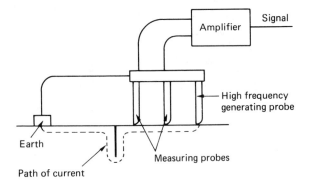

Figure 17.92 Measurement of electropotential

Built into each coil is an identical excitation winding fed from the same supply. The selection of the frequency and value of the exciting current is most important.

The reference component is placed in one coil and the components to be tested, one at a time, in the other. Differences in composition, hardness, chemical purity, strength, grain orientation, heat treatment, etc., are often accompanied by changes in resistivity and/or magnetic properties. It is by measuring and comparing the resultant imbalance in the bridge circuit that the difference between the components can be determined.

One method is to measure the resistive and reactive axes by using the display curves and loops on the cathode ray tube screen, but this requires a skilled operator. An alternative method is to measure these axes on a standard instrument, but this usually limits the detail obtained. This method is frequently linked to automatic sorting gates or spray marking equipment.

17.9.5.2 Search head technique

When the detection of a simple defect is required, a simple probe coil may be used to search a surface area. This relies on the spurious fields which exist in the proximity of a defect. This method can be automated.

17.9.6 Electropotential crack depth detection

This method of crack-depth detection is not normally used as a search method, but rather as a means of measuring the depth of a known crack. In *Figure 17.92* the two measuring probes measure the resistance of a unit length of surface. The probe is then moved so that the measuring probes stand one on each side of the crack. The increase in resistance produced by the electric wave following the crack down into the metal and back up the other side is read off the amplified reading of the potential difference.

The accuracy of this method is 5% at 1 mm, but reduces as the crack depth increases. Few accurate results can be obtained at crack depths of 100 mm.

Appendix

Table A17.1 Sample size code letters*

Lot or batch size	Special inspection levels				General inspection levels		
	S-1	*S-2*	*S-3*	*S-4*	*I*	*II*	*III*
2–8	A	A	A	A	A	A	B
9–15	A	A	A	A	A	B	C
16–25	A	A	B	B	B	C	D
26–50	A	B	B	C	C	D	E
51–90	B	B	C	C	C	E	F
91–150	B	B	C	D	D	F	G
151–280	B	C	D	E	E	G	H
281–500	B	C	D	E	F	H	J
501–1 200	C	C	E	F	G	J	K
1 201–3 200	C	D	E	G	H	K	L
3 201–10 000	C	D	F	G	J	L	M
10 001–35 000	C	D	F	H	K	M	N
35 001–150 000	D	E	G	J	L	N	P
150 001–500 000	D	E	G	J	M	P	Q
≥500 001	D	E	H	K	N	Q	R

* From ISO 2859 1: 1989 (E).

Table A17.2 Single sampling plans for normal inspection[*]

Acceptable quality levels (normal inspection)

Each AQL cell contains the acceptance number (Ac) and rejection number (Re), shown as "Ac Re".

Sample size code letter	Sample size	0.010	0.015	0.025	0.040	0.065	0.10	0.15	0.25	0.40	0.65	1.0	1.5	2.5	4.0	6.5	10	15	25	40	65	100	150	250	400	650	1000
A	2	↓	↓	↓	↓	↓	↓	↓	↓	↓	↓	↓	↓	↓	↓	↓	↓	0 1	1 2	2 3	3 4	5 6	7 8	10 11	14 15	21 22	30 31
B	3	↓	↓	↓	↓	↓	↓	↓	↓	↓	↓	↓	↓	↓	↓	↓	0 1	1 2	2 3	3 4	5 6	7 8	10 11	14 15	21 22	30 31	44 45
C	5	↓	↓	↓	↓	↓	↓	↓	↓	↓	↓	↓	↓	↓	↓	0 1	1 2	2 3	3 4	5 6	7 8	10 11	14 15	21 22	30 31	44 45	↑
D	8	↓	↓	↓	↓	↓	↓	↓	↓	↓	↓	↓	↓	↓	0 1	1 2	2 3	3 4	5 6	7 8	10 11	14 15	21 22	30 31	44 45	↑	↑
E	13	↓	↓	↓	↓	↓	↓	↓	↓	↓	↓	↓	↓	0 1	1 2	2 3	3 4	5 6	7 8	10 11	14 15	21 22	30 31	44 45	↑	↑	↑
F	20	↓	↓	↓	↓	↓	↓	↓	↓	↓	↓	↓	0 1	1 2	2 3	3 4	5 6	7 8	10 11	14 15	21 22	30 31	44 45	↑	↑	↑	↑
G	32	↓	↓	↓	↓	↓	↓	↓	↓	↓	↓	0 1	1 2	2 3	3 4	5 6	7 8	10 11	14 15	21 22	30 31	44 45	↑	↑	↑	↑	↑
H	50	↓	↓	↓	↓	↓	↓	↓	↓	↓	0 1	1 2	2 3	3 4	5 6	7 8	10 11	14 15	21 22	30 31	44 45	↑	↑	↑	↑	↑	↑
J	80	↓	↓	↓	↓	↓	↓	↓	↓	0 1	1 2	2 3	3 4	5 6	7 8	10 11	14 15	21 22	30 31	44 45	↑	↑	↑	↑	↑	↑	↑
K	125	↓	↓	↓	↓	↓	↓	↓	0 1	1 2	2 3	3 4	5 6	7 8	10 11	14 15	21 22	30 31	44 45	↑	↑	↑	↑	↑	↑	↑	↑
L	200	↓	↓	↓	↓	↓	↓	0 1	1 2	2 3	3 4	5 6	7 8	10 11	14 15	21 22	30 31	44 45	↑	↑	↑	↑	↑	↑	↑	↑	↑
M	315	↓	↓	↓	↓	↓	0 1	1 2	2 3	3 4	5 6	7 8	10 11	14 15	21 22	30 31	44 45	↑	↑	↑	↑	↑	↑	↑	↑	↑	↑
N	500	↓	↓	↓	↓	0 1	1 2	2 3	3 4	5 6	7 8	10 11	14 15	21 22	30 31	44 45	↑	↑	↑	↑	↑	↑	↑	↑	↑	↑	↑
P	800	↓	↓	↓	0 1	1 2	2 3	3 4	5 6	7 8	10 11	14 15	21 22	30 31	44 45	↑	↑	↑	↑	↑	↑	↑	↑	↑	↑	↑	↑
Q	1 250	↓	↓	0 1	1 2	2 3	3 4	5 6	7 8	10 11	14 15	21 22	30 31	44 45	↑	↑	↑	↑	↑	↑	↑	↑	↑	↑	↑	↑	↑
R	2 000	↓	0 1	1 2	2 3	3 4	5 6	7 8	10 11	14 15	21 22	30 31	44 45	↑	↑	↑	↑	↑	↑	↑	↑	↑	↑	↑	↑	↑	↑

[*] From ISO 2859 1: 1989 (E).

↓. Use first sampling plan below arrow. If sample size equals, or exceeds, lot or batch size, carry out 100% inspection.

↑. Use first sampling plan above arrow.

Ac. Acceptance number.

Re. Rejection number.

Table A17.3 Double sampling plans for normal inspection[†]

Acceptable quality levels (normal inspection)

Each AQL cell below gives **Ac Re** (acceptance / rejection numbers); the two stacked figures in the original correspond to the *First* and *Second* sample rows. Arrow symbols: ↓ = use first sampling plan below arrow; ↑ = use first sampling plan above arrow; * = use corresponding single sampling plan.

Code letter	Sample	Sample size	Cumulative sample size	0.010	0.015	0.025	0.040	0.065	0.10	0.15	0.25	0.40	0.65	1.0	1.5	2.5	4.0	6.5	10	15	25	40	65	100	150	250	400	650	1000
A				↓	↓	↓	↓	↓	↓	↓	↓	↓	↓	↓	↓	↓	↓	↓	↓	*	*	*	*	*	*	*	*	*	*
B	First	2	2	↓	↓	↓	↓	↓	↓	↓	↓	↓	↓	↓	↓	↓	↓	↓	*	0 2	0 3	1 4	2 5	3 7	5 9	7 11	11 16	17 22	25 31
	Second	2	4																	1 2	3 4	4 5	6 7	8 9	12 13	18 19	26 27	37 38	56 57
C	First	3	3	↓	↓	↓	↓	↓	↓	↓	↓	↓	↓	↓	↓	↓	↓	*	0 2	0 3	1 4	2 5	3 7	5 9	7 11	11 16	17 22	25 31	↑
	Second	3	6																1 2	3 4	4 5	6 7	8 9	12 13	18 19	26 27	37 38	56 57	
D	First	5	5	↓	↓	↓	↓	↓	↓	↓	↓	↓	↓	↓	↓	↓	*	0 2	0 3	1 4	2 5	3 7	5 9	7 11	11 16	17 22	25 31	↑	↑
	Second	5	10															1 2	3 4	4 5	6 7	8 9	12 13	18 19	26 27	37 38	56 57		
E	First	8	8	↓	↓	↓	↓	↓	↓	↓	↓	↓	↓	↓	↓	*	0 2	0 3	1 4	2 5	3 7	5 9	7 11	11 16	17 22	25 31	↑	↑	↑
	Second	8	16														1 2	3 4	4 5	6 7	8 9	12 13	18 19	26 27	37 38	56 57			
F	First	13	13	↓	↓	↓	↓	↓	↓	↓	↓	↓	↓	↓	*	0 2	0 3	1 4	2 5	3 7	5 9	7 11	11 16	17 22	25 31	↑	↑	↑	↑
	Second	13	26													1 2	3 4	4 5	6 7	8 9	12 13	18 19	26 27	37 38	56 57				
G	First	20	20	↓	↓	↓	↓	↓	↓	↓	↓	↓	↓	*	0 2	0 3	1 4	2 5	3 7	5 9	7 11	11 16	17 22	25 31	↑	↑	↑	↑	↑
	Second	20	40												1 2	3 4	4 5	6 7	8 9	12 13	18 19	26 27	37 38	56 57					
H	First	32	32	↓	↓	↓	↓	↓	↓	↓	↓	↓	*	0 2	0 3	1 4	2 5	3 7	5 9	7 11	11 16	17 22	25 31	↑	↑	↑	↑	↑	↑
	Second	32	64											1 2	3 4	4 5	6 7	8 9	12 13	18 19	26 27	37 38	56 57						
J	First	50	50	↓	↓	↓	↓	↓	↓	↓	↓	*	0 2	0 3	1 4	2 5	3 7	5 9	7 11	11 16	17 22	25 31	↑	↑	↑	↑	↑	↑	↑
	Second	50	100										1 2	3 4	4 5	6 7	8 9	12 13	18 19	26 27	37 38	56 57							
K	First	80	80	↓	↓	↓	↓	↓	↓	↓	*	0 2	0 3	1 4	2 5	3 7	5 9	7 11	11 16	17 22	25 31	↑	↑	↑	↑	↑	↑	↑	↑
	Second	80	160									1 2	3 4	4 5	6 7	8 9	12 13	18 19	26 27	37 38	56 57								
L	First	125	125	↓	↓	↓	↓	↓	↓	*	0 2	0 3	1 4	2 5	3 7	5 9	7 11	11 16	17 22	25 31	↑	↑	↑	↑	↑	↑	↑	↑	↑
	Second	125	250								1 2	3 4	4 5	6 7	8 9	12 13	18 19	26 27	37 38	56 57									
M	First	200	200	↓	↓	↓	↓	↓	*	0 2	0 3	1 4	2 5	3 7	5 9	7 11	11 16	17 22	25 31	↑	↑	↑	↑	↑	↑	↑	↑	↑	↑
	Second	200	400							1 2	3 4	4 5	6 7	8 9	12 13	18 19	26 27	37 38	56 57										
N	First	315	315	↓	↓	↓	↓	*	0 2	0 3	1 4	2 5	3 7	5 9	7 11	11 16	17 22	25 31	↑	↑	↑	↑	↑	↑	↑	↑	↑	↑	↑
	Second	315	630						1 2	3 4	4 5	6 7	8 9	12 13	18 19	26 27	37 38	56 57											
P	First	500	500	↓	↓	↓	*	0 2	0 3	1 4	2 5	3 7	5 9	7 11	11 16	17 22	25 31	↑	↑	↑	↑	↑	↑	↑	↑	↑	↑	↑	↑
	Second	500	1 000					1 2	3 4	4 5	6 7	8 9	12 13	18 19	26 27	37 38	56 57												
Q	First	800	800	↓	↓	*	0 2	0 3	1 4	2 5	3 7	5 9	7 11	11 16	17 22	25 31	↑	↑	↑	↑	↑	↑	↑	↑	↑	↑	↑	↑	↑
	Second	800	1 600				1 2	3 4	4 5	6 7	8 9	12 13	18 19	26 27	37 38	56 57													
R	First	1 250	1 250	↓	*	0 2	0 3	1 4	2 5	3 7	5 9	7 11	11 16	17 22	25 31	↑	↑	↑	↑	↑	↑	↑	↑	↑	↑	↑	↑	↑	↑
	Second	1 250	2 500			1 2	3 4	4 5	6 7	8 9	12 13	18 19	26 27	37 38	56 57														

↓ Use first sampling plan below arrow. If sample size equals, or exceeds, lot or batch size, carry out 100% inspection.

↑ Use first sampling plan above arrow.

Ac. Acceptance number.

Re. Rejection number.

* Use corresponding single sampling plan (or, alternatively, use double sampling plan below, where available).

[†] From ISO 2859 1: 1989 (E).

Table A17.4 Poisson cumulative probability tables (×1000) the probability of Ac or less defectives, where np is the average number of expected defectives

np	0	1	2	3	4	5	6	7	8	9	10	11	12	13	14	15	16	17	18	19	20	21	22	23	24	25	26	27	28	29	30
0.02	980	1000	1000	1000	1000	1000	1000	1000	1000	1000	1000	1000	1000	1000	1000	1000	1000	1000	1000	1000	1000	1000	1000	1000	1000	1000	1000	1000	1000	1000	1000
0.04	961	999	1000	1000	1000	1000	1000	1000	1000	1000	1000	1000	1000	1000	1000	1000	1000	1000	1000	1000	1000	1000	1000	1000	1000	1000	1000	1000	1000	1000	1000
0.06	942	998	1000	1000	1000	1000	1000	1000	1000	1000	1000	1000	1000	1000	1000	1000	1000	1000	1000	1000	1000	1000	1000	1000	1000	1000	1000	1000	1000	1000	1000
0.08	923	997	1000	1000	1000	1000	1000	1000	1000	1000	1000	1000	1000	1000	1000	1000	1000	1000	1000	1000	1000	1000	1000	1000	1000	1000	1000	1000	1000	1000	1000
0.10	905	995	1000	1000	1000	1000	1000	1000	1000	1000	1000	1000	1000	1000	1000	1000	1000	1000	1000	1000	1000	1000	1000	1000	1000	1000	1000	1000	1000	1000	1000
0.15	861	990	999	1000	1000	1000	1000	1000	1000	1000	1000	1000	1000	1000	1000	1000	1000	1000	1000	1000	1000	1000	1000	1000	1000	1000	1000	1000	1000	1000	1000
0.20	819	982	999	1000	1000	1000	1000	1000	1000	1000	1000	1000	1000	1000	1000	1000	1000	1000	1000	1000	1000	1000	1000	1000	1000	1000	1000	1000	1000	1000	1000
0.25	779	974	998	1000	1000	1000	1000	1000	1000	1000	1000	1000	1000	1000	1000	1000	1000	1000	1000	1000	1000	1000	1000	1000	1000	1000	1000	1000	1000	1000	1000
0.30	741	963	996	1000	1000	1000	1000	1000	1000	1000	1000	1000	1000	1000	1000	1000	1000	1000	1000	1000	1000	1000	1000	1000	1000	1000	1000	1000	1000	1000	1000
0.35	705	951	994	1000	1000	1000	1000	1000	1000	1000	1000	1000	1000	1000	1000	1000	1000	1000	1000	1000	1000	1000	1000	1000	1000	1000	1000	1000	1000	1000	1000
0.40	670	938	992	999	1000	1000	1000	1000	1000	1000	1000	1000	1000	1000	1000	1000	1000	1000	1000	1000	1000	1000	1000	1000	1000	1000	1000	1000	1000	1000	1000
0.45	638	925	989	999	1000	1000	1000	1000	1000	1000	1000	1000	1000	1000	1000	1000	1000	1000	1000	1000	1000	1000	1000	1000	1000	1000	1000	1000	1000	1000	1000
0.50	607	910	986	998	1000	1000	1000	1000	1000	1000	1000	1000	1000	1000	1000	1000	1000	1000	1000	1000	1000	1000	1000	1000	1000	1000	1000	1000	1000	1000	1000
0.55	577	894	982	998	1000	1000	1000	1000	1000	1000	1000	1000	1000	1000	1000	1000	1000	1000	1000	1000	1000	1000	1000	1000	1000	1000	1000	1000	1000	1000	1000
0.60	549	878	977	997	1000	1000	1000	1000	1000	1000	1000	1000	1000	1000	1000	1000	1000	1000	1000	1000	1000	1000	1000	1000	1000	1000	1000	1000	1000	1000	1000
0.65	522	861	972	996	999	1000	1000	1000	1000	1000	1000	1000	1000	1000	1000	1000	1000	1000	1000	1000	1000	1000	1000	1000	1000	1000	1000	1000	1000	1000	1000
0.70	497	844	966	994	999	1000	1000	1000	1000	1000	1000	1000	1000	1000	1000	1000	1000	1000	1000	1000	1000	1000	1000	1000	1000	1000	1000	1000	1000	1000	1000
0.75	472	827	959	993	999	1000	1000	1000	1000	1000	1000	1000	1000	1000	1000	1000	1000	1000	1000	1000	1000	1000	1000	1000	1000	1000	1000	1000	1000	1000	1000
0.80	449	809	953	991	999	1000	1000	1000	1000	1000	1000	1000	1000	1000	1000	1000	1000	1000	1000	1000	1000	1000	1000	1000	1000	1000	1000	1000	1000	1000	1000
0.85	427	791	945	989	998	1000	1000	1000	1000	1000	1000	1000	1000	1000	1000	1000	1000	1000	1000	1000	1000	1000	1000	1000	1000	1000	1000	1000	1000	1000	1000
0.90	407	772	937	987	998	1000	1000	1000	1000	1000	1000	1000	1000	1000	1000	1000	1000	1000	1000	1000	1000	1000	1000	1000	1000	1000	1000	1000	1000	1000	1000
0.95	387	754	929	984	997	1000	1000	1000	1000	1000	1000	1000	1000	1000	1000	1000	1000	1000	1000	1000	1000	1000	1000	1000	1000	1000	1000	1000	1000	1000	1000
1.00	368	736	920	981	996	999	1000	1000	1000	1000	1000	1000	1000	1000	1000	1000	1000	1000	1000	1000	1000	1000	1000	1000	1000	1000	1000	1000	1000	1000	1000
1.10	333	699	900	974	995	999	1000	1000	1000	1000	1000	1000	1000	1000	1000	1000	1000	1000	1000	1000	1000	1000	1000	1000	1000	1000	1000	1000	1000	1000	1000
1.20	301	663	879	966	992	998	1000	1000	1000	1000	1000	1000	1000	1000	1000	1000	1000	1000	1000	1000	1000	1000	1000	1000	1000	1000	1000	1000	1000	1000	1000
1.30	273	627	857	957	989	998	1000	1000	1000	1000	1000	1000	1000	1000	1000	1000	1000	1000	1000	1000	1000	1000	1000	1000	1000	1000	1000	1000	1000	1000	1000
1.40	247	592	833	946	986	997	999	1000	1000	1000	1000	1000	1000	1000	1000	1000	1000	1000	1000	1000	1000	1000	1000	1000	1000	1000	1000	1000	1000	1000	1000
1.50	223	558	809	934	981	996	999	1000	1000	1000	1000	1000	1000	1000	1000	1000	1000	1000	1000	1000	1000	1000	1000	1000	1000	1000	1000	1000	1000	1000	1000
1.60	202	525	783	921	976	994	999	1000	1000	1000	1000	1000	1000	1000	1000	1000	1000	1000	1000	1000	1000	1000	1000	1000	1000	1000	1000	1000	1000	1000	1000
1.70	183	493	757	907	970	992	998	1000	1000	1000	1000	1000	1000	1000	1000	1000	1000	1000	1000	1000	1000	1000	1000	1000	1000	1000	1000	1000	1000	1000	1000
1.80	165	463	731	891	964	990	997	999	1000	1000	1000	1000	1000	1000	1000	1000	1000	1000	1000	1000	1000	1000	1000	1000	1000	1000	1000	1000	1000	1000	1000
1.90	150	434	704	875	956	987	997	999	1000	1000	1000	1000	1000	1000	1000	1000	1000	1000	1000	1000	1000	1000	1000	1000	1000	1000	1000	1000	1000	1000	1000
2.00	135	406	677	857	947	983	995	999	1000	1000	1000	1000	1000	1000	1000	1000	1000	1000	1000	1000	1000	1000	1000	1000	1000	1000	1000	1000	1000	1000	1000
2.20	111	355	623	819	928	975	993	998	1000	1000	1000	1000	1000	1000	1000	1000	1000	1000	1000	1000	1000	1000	1000	1000	1000	1000	1000	1000	1000	1000	1000
2.40	91	308	570	779	904	964	988	997	999	1000	1000	1000	1000	1000	1000	1000	1000	1000	1000	1000	1000	1000	1000	1000	1000	1000	1000	1000	1000	1000	1000
2.60	74	267	518	736	877	951	983	995	999	1000	1000	1000	1000	1000	1000	1000	1000	1000	1000	1000	1000	1000	1000	1000	1000	1000	1000	1000	1000	1000	1000
2.80	61	231	469	692	848	935	976	992	998	999	1000	1000	1000	1000	1000	1000	1000	1000	1000	1000	1000	1000	1000	1000	1000	1000	1000	1000	1000	1000	1000
3.00	50	199	423	647	815	916	966	988	996	999	1000	1000	1000	1000	1000	1000	1000	1000	1000	1000	1000	1000	1000	1000	1000	1000	1000	1000	1000	1000	1000
3.20	41	171	380	603	781	895	955	983	994	998	1000	1000	1000	1000	1000	1000	1000	1000	1000	1000	1000	1000	1000	1000	1000	1000	1000	1000	1000	1000	1000
3.40	33	147	340	558	744	871	942	977	992	997	999	1000	1000	1000	1000	1000	1000	1000	1000	1000	1000	1000	1000	1000	1000	1000	1000	1000	1000	1000	1000
3.60	27	126	303	515	706	844	927	969	988	996	999	1000	1000	1000	1000	1000	1000	1000	1000	1000	1000	1000	1000	1000	1000	1000	1000	1000	1000	1000	1000
3.80	22	107	269	473	668	816	909	960	984	994	998	999	1000	1000	1000	1000	1000	1000	1000	1000	1000	1000	1000	1000	1000	1000	1000	1000	1000	1000	1000
4.00	18	92	238	433	629	785	889	949	979	992	997	999	1000	1000	1000	1000	1000	1000	1000	1000	1000	1000	1000	1000	1000	1000	1000	1000	1000	1000	1000
4.20	15	78	210	395	590	753	867	936	972	989	996	999	1000	1000	1000	1000	1000	1000	1000	1000	1000	1000	1000	1000	1000	1000	1000	1000	1000	1000	1000
4.40	12	66	185	359	551	720	844	921	964	985	994	998	999	1000	1000	1000	1000	1000	1000	1000	1000	1000	1000	1000	1000	1000	1000	1000	1000	1000	1000
4.60	10	56	163	326	513	686	818	905	955	980	992	997	999	1000	1000	1000	1000	1000	1000	1000	1000	1000	1000	1000	1000	1000	1000	1000	1000	1000	1000
4.80	8	48	143	294	476	651	791	887	944	975	990	996	999	1000	1000	1000	1000	1000	1000	1000	1000	1000	1000	1000	1000	1000	1000	1000	1000	1000	1000
5.00	7	40	125	265	440	616	762	867	932	968	986	995	998	999	1000	1000	1000	1000	1000	1000	1000	1000	1000	1000	1000	1000	1000	1000	1000	1000	1000
5.20	6	34	109	238	406	581	732	845	918	960	982	993	997	999	1000	1000	1000	1000	1000	1000	1000	1000	1000	1000	1000	1000	1000	1000	1000	1000	1000
5.40	5	29	95	213	373	546	702	822	903	951	977	990	996	999	1000	1000	1000	1000	1000	1000	1000	1000	1000	1000	1000	1000	1000	1000	1000	1000	1000
5.60	4	24	82	191	342	512	670	797	886	941	972	988	995	998	999	1000	1000	1000	1000	1000	1000	1000	1000	1000	1000	1000	1000	1000	1000	1000	1000
5.80	3	21	72	170	313	478	638	771	867	929	965	984	993	997	999	1000	1000	1000	1000	1000	1000	1000	1000	1000	1000	1000	1000	1000	1000	1000	1000
6.00	2	17	62	151	285	446	606	744	847	916	957	980	991	996	999	1000	1000	1000	1000	1000	1000	1000	1000	1000	1000	1000	1000	1000	1000	1000	1000
6.20	2	15	54	134	259	414	574	716	826	902	949	975	989	995	998	999	1000	1000	1000	1000	1000	1000	1000	1000	1000	1000	1000	1000	1000	1000	1000
6.40	2	12	46	119	235	384	542	687	803	886	939	969	986	994	997	999	1000	1000	1000	1000	1000	1000	1000	1000	1000	1000	1000	1000	1000	1000	1000
6.60	1	10	40	105	213	355	511	658	780	869	927	963	982	992	997	999	1000	1000	1000	1000	1000	1000	1000	1000	1000	1000	1000	1000	1000	1000	1000
6.80	1	9	34	93	192	327	480	628	755	850	915	955	978	990	996	998	999	1000	1000	1000	1000	1000	1000	1000	1000	1000	1000	1000	1000	1000	1000
7.00	1	7	30	82	173	301	450	599	729	830	901	947	973	987	994	998	999	1000	1000	1000	1000	1000	1000	1000	1000	1000	1000	1000	1000	1000	1000
7.20	1	6	25	72	156	276	420	569	703	810	887	937	967	984	993	997	999	1000	1000	1000	1000	1000	1000	1000	1000	1000	1000	1000	1000	1000	1000
7.40	1	5	22	63	140	253	392	539	676	788	871	926	961	980	991	996	998	999	1000	1000	1000	1000	1000	1000	1000	1000	1000	1000	1000	1000	1000
7.60	1	4	19	55	125	231	365	510	648	765	854	915	954	976	989	995	998	999	1000	1000	1000	1000	1000	1000	1000	1000	1000	1000	1000	1000	1000
7.80	0	4	16	48	112	210	338	481	620	741	835	902	945	971	986	993	997	999	1000	1000	1000	1000	1000	1000	1000	1000	1000	1000	1000	1000	1000
8.00	0	3	14	42	100	191	313	453	593	717	816	888	936	966	983	992	996	998	999	1000	1000	1000	1000	1000	1000	1000	1000	1000	1000	1000	1000
8.50	0	2	9	30	74	150	256	386	523	653	763	849	909	949	973	986	993	997	999	1000	1000	1000	1000	1000	1000	1000	1000	1000	1000	1000	1000
9.00	0	1	6	21	55	116	207	324	456	587	706	803	876	926	959	978	989	995	998	999	1000	1000	1000	1000	1000	1000	1000	1000	1000	1000	1000
9.50	0	1	4	15	40	89	165	269	392	522	645	752	836	898	940	967	982	991	996	998	999	1000	1000	1000	1000	1000	1000	1000	1000	1000	1000
10.00	0	0	3	10	29	67	130	220	333	458	583	697	792	864	917	951	973	986	993	997	998	999	1000	1000	1000	1000	1000	1000	1000	1000	1000
10.50	0	0	2	7	21	50	102	179	279	397	521	639	742	825	888	932	960	978	988	994	997	999	999	1000	1000	1000	1000	1000	1000	1000	1000
11.00	0	0	1	5	15	38	79	143	232	341	460	579	689	781	854	907	944	968	982	991	995	998	999	1000	1000	1000	1000	1000	1000	1000	1000
11.50	0	0	1	3	11	28	60	114	191	289	402	520	633	733	815	878	924	954	974	986	992	996	998	999	1000	1000	1000	1000	1000	1000	1000
12.00	0	0	0	2	8	20	46	90	155	242	347	462	576	682	772	844	899	937	963	979	988	994	997	999	999	1000	1000	1000	1000	1000	1000
12.50	0	0	0	2	5	15	35	70	125	201	297	406	519	628	725	806	869	916	948	969	983	991	995	998	999	999	1000	1000	1000	1000	1000
13.00	0	0	0	1	4	11	26	54	100	166	252	353	463	573	675	764	835	890	930	957	975	986	992	996	998	999	1000	1000	1000	1000	1000
13.50	0	0	0	1	3	8	19	41	79	135	211	304	409	518	623	718	798	861	908	942	965	980	989	994	997	998	999	1000	1000	1000	1000
14.00	0	0	0	0	2	6	14	32	62	109	176	260	358	464	570	669	756	827	883	923	952	971	983	991	995	997	999	999	1000	1000	1000
14.50	0	0	0	0	1	4	10	24	48	88	145	220	311	413	518	619	711	790	853	901	936	960	976	986	992	996	998	999	999	1000	1000
15.00	0	0	0	0	1	3	8	18	37	70	118	185	268	363	466	568	664	749	819	875	917	947	967	981	989	994	997	998	999	1000	1000
16.00	0	0	0	0	0	1	4	10	22	43	77	127	193	275	368	467	566	659	742	812	868	911	942	963	978	987	993	996	998	999	999
17.00	0	0	0	0	0	1	2	5	13	26	49	85	135	201	281	371	468	564	655	736	805	861	905	937	959	975	985	991	995	997	999
18.00	0	0	0	0	0	0	1	3	7	15	30	55	92	143	208	287	375	469	562	651	731	799	855	899	932	955	972	983	990	994	997
19.00	0	0	0	0	0	0	1	2	4	9	18	35	61	98	150	215	292	378	469	561	647	725	793	849	893	927	951	969	980	988	993
20.00	0	0	0	0	0	0	0	1	2	5	11	21	39	66	105	157	221	297	381	470	559	644	721	787	843	888	922	948	966	978	987
22.00	0	0	0	0	0	0	0	0	1	2	4	8	15	28	48	77	117	169	232	306	387	472	556	637	712	777	832	877	913	940	959
24.00	0	0	0	0	0	0	0	0	0	0	1	3	5	11	20	34	56	87	128	180	243	314	392	473	554	632	704	768	823	868	904
26.00	0	0	0	0	0	0	0	0	0	0	0	1	2	4	8	14	25	41	65	97	139	190	252	321	396	474	552	627	697	759	813
28.00	0	0	0	0	0	0	0	0	0	0	0	0	1	1	3	5	10	18	30	48	73	106	148	200	260	327	400	475	550	623	690
30.00	0	0	0	0	0	0	0	0	0	0	0	0	0	0	1	2	4	7	13	22	35	54	81	115	157	208	267	333	403	476	548

18 Terotechnology and Maintenance

William T File

Contents

18.1 Management of assets

18.1.1 Assets in manufacturing

In manufacturing terms, 'assets' may be hand tools, machine tools, process plant, manufacturing premises, warehousing, raw materials, work in progress and finished goods awaiting despatch. The definition may also be extended to human resources, the effective utilisation of which are usually manufacturing's principal asset.

Effective utilisation of human resources depends on the availability of the 'tools' of manufacturing, and the ultimate objective must be to have all the manufacturing equipment available for use 100% of the time. In reality, this figure is never likely to be achieved, but figures approaching 100% should be the target. Obviously, in common with all other hardware, 'you get what you pay for!' and availability is, in reality, a trade off with cost. This is discussed further later in this chapter.

The hardware elements of the list above (as opposed to the human resources) represent the capital employed in the production of goods. Before being put into use, these same hardware elements must be:

(1) specified,
(2) selected,
(3) designed,
(4) built,
(5) purchased,
(6) installed, and
(7) commissioned.

At some time in the future these hardware elements will have outlived their economic life and will either be sold or scrapped. This is the life cycle of manufacturing equipment; it begins when its acquisition is first considered and ends when disposal is finally completed. Typical relative costs associatd with each phase are shown diagrammatically in *Figure 18.1*.

It should be noted that the digram assumes some residual value in the equipment at the end of its useful life, and its sale potential will lower the cumulative life cost on disposal. One thing that is very clear from the diagram is that the costs in the early phases are relatively low compared with building/purchasing and operational/maintenance costs, so time spent ensuring that equipment is correctly specified will clearly be advantageous in the longer term.

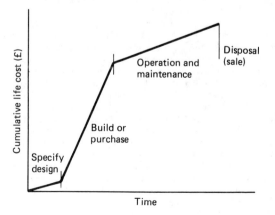

Figure 18.1 Cumulative life-cycle cost of an asset

18.1.2 Terotechnology practice

The term 'terotechnology' was initiated in the 1970s as a result of a project commissioned by the British Government to study and report on the state of maintenance in industry. As a result of the initial report of this study group, its terms of reference were widened. It was considered that maintenance could not be considered in isolation from other elements of engineering and the costs involved in supporting a manufacturing process. More recently, a British Standard, BS 3843: 1991 *Guide to Terotechnology (The Economic Management of Assets)* has been published. Part 1 introduces the concept, Part 2 describes the techniques involved and Part 3 continues with short notes on the use and application of the various techniques.

Terotechnology is a combination of management, financial, engineering, building and other practices applied to physical assets in pursuit of economic life costs. In practice, it is concerned with the specification and design for reliability and maintainability of plant, machinery, equipment, buildings and structures, with their installation, commissioning, maintenance, modification and replacement, and with feedback of information on design, performance and costs.

The key words in the above definition are 'economic life cycle costs'. 'Economic' implies that it may not necessarily be the cheapest, as in all good designs there is a specification to be met and inevitably a trade-off situation arises where money must be paid for quality or reliability.

The successful application of terotechnology principles to the management of assets involves all relevant functions within a company and depends on the systems in place to implement it. Terotechnology also requires the commitment of senior management to the process, as well as the commitment to feed back cost performance to enable comparison with the original targets set at the time of the initial specification and appraisal. In turn, this information will be utilised as a reference database for the purchasing or manufacture of future assets.

18.1.3 Terotechnology and asset management

Organisations within companies differ for many reasons, e.g. the size of the company, the diversity of products or services, and the geographical location of premises. Systems generally tend to be linked to organisation structures, and terotechnology, being a management system, must take due account of the organisation imposed by management. Having said that, such a system must be designed for those who will use it.

Companies which are organised to control their output by some form of product management system, whether through line or matrix management, have the simplest task, but for others there is no need to consider any reorganisation, providing line management and others accept the principles of terotechnology. Small companies will find that an informal system linked through their normal management processes will be sufficient. Larger companies, particularly if several different sites are involved with the same product or service, may perhaps require a nucleus of dedicated staff. However, it is essential that the costs of any asset-management system be kept to a minimum and that data presented to management are adequate to make the necessary decisions, but not excessive.

The majority of data required for the planning associated with terotechnology already exists in most companies. It will simply require collating and formatting to present a coherent life-cycle analysis.

The activities necessary for the implementation of terotechnology principles include the following.

(1) Decide which of the economically viable projects being considered should go forward.
(2) Carry out a detailed investment appraisal for these projects and then set targets for production costs and income from each.
(3) Decide what physical assets are needed and how to use them to achieve the targets over a specified period, taking into account their forecast costs of acquisition, production, maintenance and disposal.
(4) Specify the asset to be acquired and its performance.
(5) Assess tenders for the asset in terms of its life-cycle costs and economic performance. Select the tender which is preferable from a terotechnological point of view.
(6) Acquire, install and commission these assets.
(7) Operate and maintain the assets in a way which gives optimum economical performance.
(8) Monitor the use of the assets in terms of performance and cost.
(9) Modify to improve the economical performance of the assets using information from monitoring.
(10) Use the data collected to contribute to a decision regarding any replacements needed and the optimum time for these replacements to be carried out.

18.1.4 Maintenance and terotechnology

The application of terotechnology principles requires expertise in a variety of different job functions, and it can be seen from *Figure 18.2* that maintenance is the core of this process, in as much as the total maintenance function is included in terotechnology. By contrast, for example, the design function would be responsible for the technical decisions on alternative ways of meeting a product or process specification, independent of terotechnology, except where the cost implications of the decision may impact total life costs.

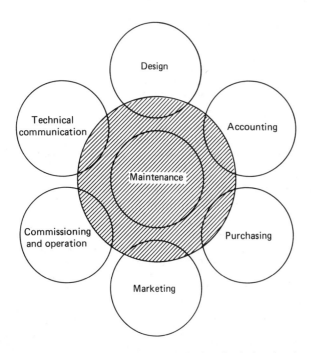

Figure 18.2 Relationship between terotechnology (hatched area) and different job functions

18.1.4.1 The role of the maintenance manager

The maintenance manager is, by definition, responsible for the care and satisfactory performance of the company's assets. He/she therefore plays a key role in controlling asset life costs and his/her involvement must begin at the tender and acquisition phase for new plant or process machinery. The manager has a valuable store of knowledge of the company's existing resources and capabilities, which, if tapped early enough, can be used to its best advantage in minimising life costs.

The following guidelines are among those which are key to the successful application of terotechnology as it affects the maintenance manager.

(1) The contribution from maintenance begins before plant is purchased.
(2) Information from the maintenance function should be used at the investment appraisal and tender assessment stages in the acquisition cycle.
(3) Adequate technical data and maintenance recommendations in an acceptable form should always be provided by the supplier of a new asset.
(4) Modern management and maintenance techniques should be applied to the company's assets to provide the highest availability consistent with overall economic performance over the whole life cycle.
(5) Comprehensive maintenance performance and cost information should be gathered and used to seek means of improvement.
(6) The maintenance function should make information concerning performance and costs associated with assets available to other functions within the company.
(7) Maintenance personnel should be encouraged to be aware of the cost implications of their actions.
(8) The maintenance manager should learn the 'language' of accountants and general management.

18.1.5 The asset register

The majority of organisations keep some records of the equipment that they use, either in hard-copy form or on a computer file. These can take various forms, but often only contain the information available at the time of purchase. Much valuable data is lost if these records are not updated regularly. This is usually due to the fact that the department holding the records has no direct interest in performance during the life of the equipment.

The maintenance manager, by virtue of his role in keeping his company's assets in a good state of repair, is ideally placed to be custodian of the company equipment register. On this register can be entered a summary of the maintenance cost and performance data relevant to the equipment concerned. This information will provide the basic input for life-cycle costing when the time comes to consider replacing the equipment. The information normally kept on the equipment register will not be in sufficient detail for day-to-day maintenance planning. However, there is an obvious advantage in being able periodically (and automatically) to transfer the total maintenance cost data from the maintenance reporting system to the equipment register, thus ensuring that it is always up to date and compatible with the annual operating plan. The register will be in the economics of each individual year, but with care it can provide an ideal base for long-term planning. *Figure 18.3* shows a typical format for an asset register.

		Equipment description						
Year	Planned costs	Actual costs			Performance			Comments
		Labour	Spares	Total Planned	Planned	Actual	Actual Planned	
Initial cost → 1 2 3		—	—	—	—	—	—	

Figure 18.3 Example of an asset register

18.1.6 Availability and related parameters

A recent Department of Trade and Industry survey showed that £8 billion is spent every year on maintaining direct production assets in the UK. This figure must be a fraction of the costs lost whilst production is interrupted for the maintenance process. In almost any situation, therefore, this represents a serious drain on the financial resources of the company and its ability to meet production targets. Few of the companies in the study attempted to quantify the indirect costs of poor and inefficient maintenance. Much of the indirect cost is incurred by the downtime of the equipment waiting for the repair to be completed, parts to be supplied, or perhaps just waiting for the maintenance engineer himself. Consequently, inadequate maintenance in one form or another results in poor manufacturing performance and, ultimately, in dissatisfied customers and lost markets.

It has been said that the competitive advantage gained by a company from expensive and complex manufacturing processes is only as good as the company's ability to ensure optimum availability. Whilst 100% availability is often desirable, it is seldom achievable. Perhaps the best example of 100% availability is to be seen in modern aircraft, where complete back-up systems or stand-by systems are available to ensure the safe arrival of passengers at their destination. From the manufacturing point of view, such back-up and stand-by systems would probably not make economic sense and, therefore, we are looking at something less than 100% availability. How much less is a question of economic and financial trade-offs.

Availability, therefore, is perhaps the most user oriented of all the indices, being in effect the user's perception of the reliability of the equipment.

$$\text{Availability} = \frac{\text{Possible operating time} - \text{Down time}}{\text{Possible operating time}}$$

In order to understand the concept of availability, it is important to understand the allocation of time (see *Figure 18.4*). One significant point which can be seen from *Figure 18.4* is the fact that maintenance can be performed either during the working day or during non-operative time. In the latter case there is the advantage that maintenance does not affect availability and, therefore, provides an interesting opportunity for restricting downtime by preventive maintenance done outside working hours. When purchasing new equipment, care should be taken to ensure that the derivation of the figures for the 'possible operating time' are those which correspond to the user demand on the equipment.

The costs associated with availability are shown in *Figure 18.5*. As can be seen from the figure, there is an optimum economic point for availability. As the availability approaches 100% the downtime costs will decrease and, inevitably, the maintenance costs will increase. The overall picture is, however, to some extent dependent on the capital cost of the equipment in the first place and, as can be appreciated, paying additional costs up front for easier maintenance and improved reliability would significantly alter the percentage points on the availability scale in *Figure 18.5*.

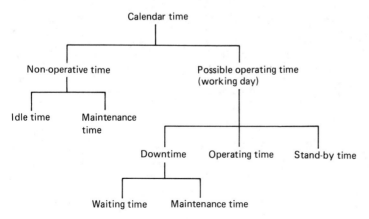

Figure 18.4 Time elements relating to maintenance

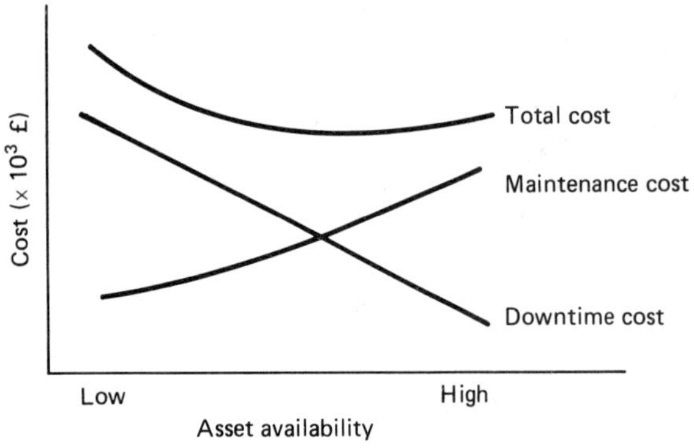

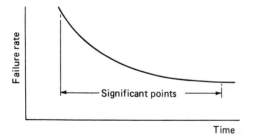

Figure 18.5 Costs associated with availability

Let us therefore look more closely at downtime.

Downtime = No. of failures × Time taken to restore equipment to its operational state

Obviously, downtime can be reduced by improving one or other of the factors on the right-hand side of the equation. The number of failures is directly dependent on the reliability specification of the equipment, whilst the time taken to restore the equipment will depend on:

(1) waiting time, and
(2) maintenance time.

In turn, the maintenance time will depend on:

(1) the skill of the maintenance engineer,
(2) adequate tools and spares, and
(3) the ease with which the equipment can be maintained.

This last factor is called *maintainability* and is defined as the property of the design which determines the effectiveness by which an equipment can be retained or restored to a state of readiness for operational use.

18.1.6.1 Reliability

Reliability can be expressed as the number of failures in a time period, which may be the expected life of the equipment. More often it is expressed as a failure rate, i.e. the *mean time between failures* (MTBF). This has the advantage that if, as usually happens, the failure rate varies over the life of the equipment, then different rates can be included in the specification related to the age of the equipment. These are usually designated as the *initial failure rate* (immediately after installation) and the *mature failure rate* (at some later defined time in the equipment life) (see *Figure 18.6*).

Figure 18.6 Specification of failure rates

18.1.6.2 The maximum and mean time to repair

These are two other important maintenance parameters. These will significantly influence the maintainability features of the equipment and, if coupled with the waiting time, can determine the level of spares inventory to be held, as well as geographical location relative to the equipment (see Section 18.8).

18.1.6.3 Operating cycle

Availability, reliability (and hence maintenance costs) and also the life of the equipment can show significant change in values with different operating and stand-by relationships. It is therefore essential that a clear understanding of the intended use of the equipment is established at an early stage in specification development. This can be demonstrated using a vending machine as an example.

Suppose the vending machine manufacturer quotes an availability of 98% based on reliability data of one equipment failure per 10 000 cups of beverages dispensed, how many cups per day does this relate to and is the day 8 or 24 h long? If it is a normal 8 h working day, then this availability figure would give a downtime of 6.4 h in 40 days, say one breakdown in 40 days allowing 6.4 h to cover both the response time and work time of the engineer.

Let us also assume that the vending machine is installed in a factory where there is shift working and a further 500 cups of beverage are dispensed during the two night shifts. The quoted availability will be the same at 98%, but a breakdown will occur every 13.3 days. Unfortunately (for the night shift), the maintenance engineer for the vending machine is unlikely to be on stand-by overnight. The probability of a failure during this time is 0.67, with a resulting penalty on response time of an average of at least 8 h. This has a considerable effect on the availability:

$$\text{Availability} = \frac{(13.3 \times 24) - [0.67(6.4 + 8) + (0.33 \times 6.4)]}{13.3 \times 24}$$

$$= 0.963 \text{ or } 96.3\%$$

Note that if the vending machine was operational over the weekends and the engineer was not available until Monday, then the availability would have been even lower.

We have already seen that the failure rate (or MTBF) makes a significant contribution to the availability of a piece of equipment, and in the examples above it was assumed that it was constant at 1 per 10 000 cups. In practice, this would have been correct in this instance because the average number of cups of beverage dispensed during each shift was also constant. However, had the machine been left in the 'ready' state all night, but principally only used during the 8 h day shift, its usage characteristics would change. In all probability the failure rate would have worsened marginally against the previous figure of 1 per 40 days due to the prolonged period over which the water was being kept hot in readiness for instant dispensing.

18.1.6.4 Usage rate

As can be seen from the above, usage rate is an important factor related to failure rates and *Figure 18.7* demonstrates the typical variation of MTBF with usage rate. The relationship shown in *Figure 18.7* is applicable to most situations, given that average (not peak or total) usage rate data are used. If the design is optimised for the intended usage, then the normal operating range would be expected to be found in the flat part of the curve.

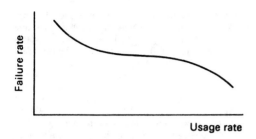

Figure 18.7 Typical variation in failure rate with usage rate

18.2 Life-cycle costing

18.2.1 Introduction to life-cycle costs

It is often assumed that the cost of maintaining equipment is constant over its expected life. This assumption has the drawback that it presupposes that the cost factors associated with maintenance remain unchanged over the years, and does not take infant mortality or wear out into account. Whilst performance over life may also vary according to a predefined pattern, various other factors relating to the cost of money do not vary in the same way, such as wage rates, inflation, etc. When a detailed justification for purchase of a new equipment is being prepared, then these factors must be considered. The process of life-cycle costing provides this opportunity, taking into account not only maintenance costs, but also all the other costs associated with the purchase and operation of an equipment that may be necessary.

If the equipment is to be purchased, for example, there will probably be capital costs in the first year and quite possibly a significant loan on which to pay interest and capital in subsequent years. Then there are inflation effects to consider, not only on the load, but also on the costs of maintenance itself. So, what does the equipment really cost, and what impact will its operational costs have on the company profits over a number of years?

Life-cycle costs The life-cycle costs associated with a piece of equipment are the costs of acquiring, using, keeping it in operational condition and final disposal (see *Figure 18.8*). These costs include the costs of feasibility studies, research, development, design, manufacture, installation, commissioning, operation, maintenance and replacement, and any credit for disposal. The costs of any supporting actions generated by these activities would also be included, such as training, documentation, tools and test equipment, inventory holdings, and any other additional facilities required as a direct result of the use of a piece of equipment. It can be stated that:

Life-cycle cost = Cost of operation + Cost of ownership

Cost of operation This comprises the direct operational costs such as energy, materials and labour used in the production of goods or services.

Cost of ownership This includes all the items defined above for life-cycle costs, with the exception of those listed under cost of operation. Cost of ownership can be subdivided into three parts, which are self-explicit: *cost of acquisition, cost of maintenance* and *cost of disposal*. The cost of ownership is, therefore, most relevant to the maintenance function.

Conventional company accounting is structured in such a way that direct costs are divided between those related to a product or service and indirect costs are shared as an overhead over the entire range of products or services offered. Life-cycle costing, which is a management accounting process, is intended to present the costs so that they are of direct assistance in the making of management decisions related to a particular product or service. Whilst the two approaches are entirely separate, they must be consistent in their treatment of cost data in order to be acceptable to all functions. In general terms this includes:

(1) labour overheads and utilisation factors;
(2) differing labour rates for different skill levels;
(3) parts overheads for inventory and distribution costs; and
(4) consistency with procedures for compiling operating budgets and long-term plans.

The principal use of the life-cycle-cost process is for comparative studies for alternative procedures, equipment purchases, trade-off decisions or similar exercises where a decision must be made on future courses of action. Situations may arise where trade-offs may be necessary between cost of ownership and cost of operation, or perhaps cost elements within either of these categories. A simple example of this would be the relatively high cost of installing thermal insulation, which would be offset by lower energy costs during subsequent years.

18.2.2 Life-cycle-cost process

Life-cycle costing requires clear terms of reference before embarking on any gathering of data towards decision-making. The results will only be as good as the input, and it is essential to obtain the commitment of all relevant functions involved to ensure the validity of the data obtained. Life-cycle costing is essentially a multifunctional task but, because of the central role of the maintenance manager in his/her responsibility for the upkeep of assets, he/she can well be the coordinating function in such a decision-making process.

Life-cycle costing requires:

(1) clear objectives;
(2) recognition of all alternative courses of action;
(3) realistic assumptions and forecasting of all data elements;
(4) consistent approach to the use and application of all data;
(5) analysis of difference in cost of each course of action, and all possible outcomes; and
(6) comparison of the final cost data with the objectives.

These requirements cannot be overemphasised, as life-cycle costing can be a time-consuming exercise and, if not approached correctly and comprehensively, will produce mis-

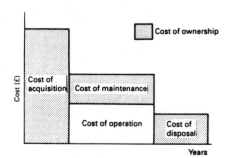

Figure 18.8 Histogram of life-cycle costs

Figure 18.9 Optimum life-cycle costs

leading results. When reviewing alternative courses of action, care should be taken to ensure that all assumptions made are varied according to the appropriate course of action and, where these are difficult to forecast, best and worst situations should be evaluated. Forecasting techniques used should be those which are already the standard within the company, and accepted by all other functions affected by the outcome of the investment decision-making process of which life-cycle costing will form a part.

A typical relationship showing the effect of asset availability with different acquisition costs and costs of maintenance and operation is shown in *Figure 18.9* (note the similarity with *Figure 18.5*). This demonstrates the point that within reason it is often worth incurring higher initial costs with a view to minimising the ongoing maintenance and operating costs.

Speculation phase When first considering the purchase of new equipment a graph such as that shown in *Figure 18.9* is a good way to demonstrate the cost impact of various alternative options. As these are reviewed, they may be plotted on this basis to determine the optimum cost performance of the equipment. If a number of alternatives is available which have projected life costs falling within the central low area of the total cost curve, then a considerable improvement in availability can be achieved by selecting one with the higher acquisition cost, without significantly affecting the total life cost.

Appraisal phase Having narrowed the field to perhaps two or three alternatives, detailed cash-flow projections should be prepared on a year-by-year basis, introducing alternative maintenance policies, and discounted to net present value (NPV). The discount factor used would normally be provided by the accounts department and would depend on a combination of inflation and market interest rate projections.

To show the effect of a discount factor, consider investing £100 today and assume that the best interest rate obtained will be 10%. This will provide £110 in 1 years time and £121 in 2 years time. It is said then, that the NPV of £121 in 2 years time is £100, given an interest rate of 10%. In other words, £100 is needed today to achieve £121 in 2 years time, the discount rate for the 2 years is then 100/121 = 0.83.

This can best be demonstrated by an example. Suppose a vehicle is due for replacement and a number of alternative makes with different specifications is available. The vehicle chosen after reviewing the market is a type costing £30 000, and this is entered in the first column in *Table 18.1* at 'year 0'. Had there been a loan involved then this would have been extended to perhaps 3 years together with appropriate interest payments. The operating costs are constant for the period under review at a total of £4000 per annum. Maintenance costs are shown to be increasing gradually from 'year 3' onwards, and coincident with this there are revenue losses shown due to the vehicle being off the road whilst under repair. The first row in the table is at constant economics at 'year 0', which is effectively today. For each successive year completed a 10% discount is estimated. This provides a cumulative total at NPV of £77 300, as the life cost taken over a 6-year span.

As any such project proceeds, the original factors assumed should be monitored and updated. They can then be used as a basis for monitoring and controlling performance and costs. Annual budgets, operating and long-term plans will often disguise fundamental project-related problems, including those of maintenance costs, which can only be highlighted by a life-cycle-cost review.

18.2.3 Marginal analysis

Supposing that there will be an option to overhaul the vehicle (used in the example in *Table 18.1*) in year 4 at a cost of £8000, with maintenance costs and therefore revenue losses being reduced thereafter, then a new life cost could be established. This time, however, any costs which are common to both situations, such as the original capital costs and the operating costs, will be ignored and the calculation based on the variable costs. This is known as 'marginal costing' and is usually used on a comparative basis as shown in *Table 18.2*. If the cumulative total shows a gain, then it should be a worthwhile proposition, assuming that the underlying assumptions are sound.

It is often worth varying the key assumptions to ensure that they are not critical to achieving the calculated result. For example, supposing that the maintenance and revenue cost savings projected above are not realised, and they are 30% higher than expected. Then substituting the new data in *Table 18.2* would show an expected cumulative loss at the end of year 6 of £100 instead of the expected £3600 gain. If there is a

Table 18.1 Example of life-cycle costing

Year	Capital ($\times 10^3$ £)	Operating costs ($\times 10^3$ £)	Maintenance costs ($\times 10^3$ £)	Revenue losses ($\times 10^3$ £)	Total (constant) ($\times 10^3$ £)	Discount	Total (NPV) ($\times 10^3$ £)
0	30				30.0		30.0
1		4.0	3.0		7.0	0.91	6.4
2		4.0	3.0		7.0	0.83	5.8
3		4.0	3.6	2.0	9.6	0.75	7.2
4		4.0	4.3	4.0	12.3	0.68	8.4
5		4.0	5.2	6.0	15.2	0.62	9.4
6		4.0	6.2	8.0	18.2	0.56	10.2
Cumulative total							77.3

Table 18.2 Marginal cost—overhaul option

| Year | Base case | | Option | | Gain | Discount | Gain at NPV |
	Maintenance cost ($\times 10^3$ £)	Revenue losses ($\times 10^3$ £)	Maintenance cost ($\times 10^3$ £)	Revenue losses ($\times 10^3$ £)	($\times 10^3$ £)		($\times 10^3$ £)
0					0		0
1	3.0		3.0		0	0.91	0
2	3.0		3.0		0	0.83	0
3	3.6	2.0	3.6	2.0	0	0.75	0
4	4.3	4.0	11.0	2.0	4.7*	0.68	3.2*
5	5.2	6.0	3.6	2.0	5.6	0.62	3.5
6	6.2	8.0	4.3	4.0	5.9	0.56	3.3
Cumulative total							3.6

* Loss.

high probability of this occurring, then obviously it would be unwise to make the investment in overhauling the vehicle in year 4.

18.2.4 Trade-off models

In *Table 18.2* and the accompanying narrative a trade-off situation was studied using the marginal costing technique, which compared two situations, one the 'base case' and the second introducing a major overhaul in year 4 of operation. If a number of such alternative strategies were to be considered, it would be worthwhile considering creating a trade-off model which could be used by flexing the base case. In the example quoted in *Table 18.3* the costings used in *Table 18.2* are reproduced individually at NPV, rather than adding them at yearly values and adjusting the totals by the discount factor.

The effect is to put all the costs at todays value of money and, therefore, direct comparisons can be made between the various cost factors, irrespective of when they occur. Hence the maintenance costs of £17 400 over the 6 year vehicle life is now directly comparable with the cost of the vehicle at £30 000 and it can be said, for example, that an additional £5000 can be spent on the vehicle if a saving of more than the same amount can be made on operating costs, maintenance costs or reduced revenue losses. This is, however, an investment decision, and cash-flow considerations at the time of purchase might prohibit this action because the return may only be realised later during the 6-year life of the vehicle.

As a further example, suppose that the opportunity exists to reduce the maintenance costs for a range of pumps by

Table 18.3 Costing of individual items at NPV

Year	Capital ($\times 10^3$ £)	Operating costs ($\times 10^3$ £)	Maintenance costs ($\times 10^3$ £)	Revenue losses ($\times 10^3$ £)	Total (NPV) ($\times 10^3$ £)
0	30				30.0
1		3.7	2.7		6.4
2		3.3	2.5		5.8
3		3.0	2.7	1.5	7.2
4		2.7	2.9	2.7	8.3
5		2.5	3.2	3.7	9.4
6		2.3	3.4	4.5	10.2
Total	30	17.5	17.4	12.4	77.3

additional capital investment to improve maintainability. It is assumed that the business under consideration is a major user of these pumps and, therefore, both the capital investment for the equipment and the maintenance costs are incurred by the same company.

Capital cost of each pump	£400
Installation rate of pump	600 per year for 3 years
Maintenance cost per pump year	£70
Total pump project cost (see *Table 18.4*)	= 597 000 + 314 000 = £911 000

In this particular instance, total life-cost figures, whilst understandable, are not of any immediate practical use. It is only a straightforward mathematical step to determine the direct relationship between, say, maintenance cost per pump year and capital cost per pump.

£1000 change in capital expenditure
$$= \frac{10^3}{1800}$$
$$= £0.55 \text{ per pump}$$

£1000 change in maintenance costs
$$= \frac{10^3}{6300}$$
$$= £0.159 \text{ per pump year}$$

At NPV,

Average capital cost per pump
$$= \frac{597 \times 10^3}{1800}$$
$$= £331.7$$

Average maintenance cost per pump year
$$= \frac{314 \times 10^3}{6300}$$
$$= £49.8$$

At todays value of money, therefore,

£1000 change in capital expenditure
$$= 0.55 \times \frac{400}{331.7}$$
$$= £0.66 \text{ per pump}$$

£1000 change in maintenance costs
$$= 0.159 \times \frac{70}{49.8}$$
$$= £0.22 \text{ per pump year}$$

Table 18.4 Pump project costing

Year	Population	Pump-years	Capital cost ($\times 10^3$ £)	Maintenance cost ($\times 10^3$ £)	Discount factor	Capital NPV ($\times 10^3$ £)	Maintenance NPV ($\times 10^3$ £)
0	0						
1	600	300	240	21	0.91	218	19.5
2	1200	900	240	63	0.83	199	52.3
3	1800	1500	240	105	0.75	180	78.8
4	1800	1800	—	126	0.68	—	85.7
5	1800	1800	—	126	0.62	—	78.1
Total		6300				597	314

Therefore, the total expenditure over 5 years will be reduced if less than £3 (i.e. £0.66/0.22) is added to the capital cost of each pump to save £1 per pump-year in maintenance cost, at todays value of money.

18.2.5 Minimising life-cycle costs

When considering the alternative options open to management regarding any proposed project, the principal objective will be to minimise life-cycle costs without degrading the required performance. It may well be that a lower performance is acceptable if the life-cost savings are significantly high, but this is a trade-off that can only be made after reviewing the individual circumstances.

Whether it is for this reason, or any other, the maintenance manager will probably be faced with the problem of optimising his/her approach rather than minimising it. Modification or overhaul of existing equipment is, therefore, an optimum measure which economically may make good sense when capital expenditure on new equipment either cannot be justified or cannot be afforded. In *Figure 18.10* the annual maintenance costs of a typical piece of equipment are plotted cumulatively in constant economics, showing an increasingly steep curve as time progresses.

The introduction of a modification, shown at point A in *Figure 18.10*, will improve reliability and, therefore, tend to reduce the gradient of the ongoing cumulative maintenance cost. It is important that such a modification is introduced sufficiently early in the life of the equipment to ensure that the cost of it (represented by the vertical rise in cumulative cost) is

recovered by the subsequent maintenance savings. If there is any doubt, then a marginal analysis should be undertaken to satisfy management of the wisdom of their actions.

The steep rise as wear out of the equipment begins to show on the graph can be checked (or at best, delayed) by over-hauling the equipment, and the effect of this is shown at point B in *Figure 18.10*.

18.2.6 Practical difficulties and limitations of life-cycle costing

As has been explained earlier, life-cycle costing is a project-related process which crosses all departmental boundaries within an organisation. One of the major difficulties which this creates, therefore, is to ensure that all data collected are complete and valid. It is very easy to overlook some of the less obvious implications of the project under investigation, particularly when considering a situation some years ahead.

When reviewing alternative courses of action, it is essential that all those providing data understand the assumptions being made and that their input is compatible with the course of action proposed. If these assumptions are difficult to forecast accurately, then a range should be evaluated.

The accuracy of the data projected for future years can also significantly affect the outcome of a life-cycle-cost exercise, and in this respect it is essential to adhere to existing forecasting techniques which are already a company standard and accepted by other functions.

Any decision which may be taken based on a marginal result must also be tested by making a small percentage variation to key input factors, to determine their sensitivity to the assumptions being made.

The company accountant's view may also constrain the introduction of a life-cycle-cost study, because his/her main concern is the annual budget and its impact on the profit-and-loss account. It is typical in many maintenance related projects to invest heavily in early years to gain the financial benefits later on. However, a significant investment can often give rise to a cash-flow problem.

For completeness, it should be mentioned here that there are three other methods of evaluating investments in projects which are in common use.

Return on investment (ROI) This is a simple calculation, expressing as a percentage, the savings or incremental earnings of an investment, divided by its cost. It has the disadvantage that it ignores both the effect of time and cash flows.

Payback This is an improvement over ROI in that it allows for the time element by calculating the length of time it takes

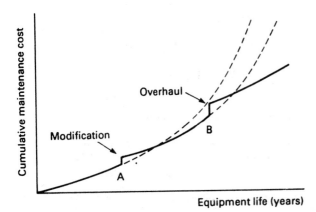

Figure 18.10 Plot of maintenance costs versus time

Table 18.5   Table of comparative accounting methods (all values in $\times 10^3$ £)

	Year			Total
	1	2	3	
Gain/loss, current accounting	(4.7)	5.6	5.9	6.8
Discount (10%)	0.91	0.83	0.75	—
LCC	(4.3)	4.6	4.4	4.7
ROI of 8.0	3.3	5.6	5.9	85% (14.8)
Payback on 8.0	3.3	8.9	—	>2 years
IRR	(2.5)	1.6	0.9	85% (0)

LCC, life-cycle-cost; ROI, return on investment; IRR, internal rate of return.

to recover the initial investment cost of a project. It is normally expressed in constant economics.

Internal rate of return (IRR)   The IRR factor is often used as a standard within a company to assess the financial benefits of new investments, but the establishment of this factor together with the calculations needed to establish it are best left to the accountants! It is a development of the life-cycle-costing process, and the IRR is defined as the percentage discount rate which must be achieved to give a zero net present value over a specified time. (See the life-cycle-costing example associated with *Table 18.1*, in order to understand this terminology.)

To understand the differences in presentation by using the above accounting methods, calculations for each method over a 3-year period are given in *Table 18.5*. The figures used are those of the overhaul option considered in *Table 18.2*, starting at the commencement of year 4 as 'today'.

Finally, if a decision is accepted, based on a life-cycle-cost analysis, then it is essential to follow it through with periodic reviews. As the project proceeds, all the relevant input and output factors should be monitored and updated and used as a basis for controlling costs. Annual budgets, operating and long-range plans will often disguise fundamental project-related problems, including maintenance costs which will only be highlighted by a life-cycle-cost review.

18.3   Plant selection and replacement

18.3.1   Plant-selection criteria

Perhaps the golden rule with plant selection is one of time—it is important not to be rushed into purchases of plant and equipment which subsequently prove to be inadequate to meet the needs of the manufacturing department. Careful selection will reap benefits for the company for many years to come.

The key to adequate plant selection is one of defining the requirements, in fact, drawing up the specification. Specification preparation is not a simple task to be completed in an odd afternoon. It is eventually to provide an absolute commitment between two parties to provide some equipment and/or service. It needs to be both detailed and yet clear and concise in its content in order that there can be no doubt about its meaning, and no room for misunderstanding. Inadequacies in design, performance, maintenance or operation can often be traced to inaccuracies or omissions in the original specification. It is essential that a dialogue should take place during the preparation of a specification between all interested functions, rather than leaving the task solely to one person to complete.

For plant selection the following general headings give a guide to the content of such a specification.

(1)   *Purpose of the equipment*—this is the technical specification and the function which the plant is expected to serve.

(2)   *The environment in which the plant is expected to operate*—is it to be in a normal workshop environment or is part of some external plant, possibly to be operated in tropical or sub-zero weather conditions?

(3)   *Production capacity*—the throughput of the plant in terms of whatever its intended function is to be, e.g. the production output expected from it such as units per day, gallons per minute.

(4)   *Availability expected*—this will provide some indication of the 'overcapacity' which will have to be provided in (3) to meet production demands. (See Section 18.1.6. for details of availability and related parameters.) At this point in time it is worth remembering that there are three shifts available in a day and it might possibly be a more economic proposition to operate the plant over two or more shifts, rather than confine it to a single day shift. This is often desirable if the capital cost involved is high compared with other parts of the plant in the production process.

(5)   *Life of plant*—expressed in appropriate units, e.g. years, millions of litres.

(6)   *Communications*—relevant authorities to be consulted, e.g. purchasing manager, manufacturing manager, maintenance manager, safety officer.

(7)   *Capital spend available*—indicate the preferred and maximum levels. (Note that when the specification is forwarded to the supplier this section should be amended to read the agreed price.)

(8)   *Acceptable on-going costs of operation and maintenance.*

(9)   *Comparison with all aspects of similar equipment currently in use*—this is particularly important where plant is being upgraded.

(10)   *Related equipment*—compatibility with other plant, especially important for process equipment.

(11)   *Maintenance requirements*—supporting documentation, tools and test equipment, spare parts, etc.

(12)   *Applicable standards*—full list of British or International Standards to which the new plant must comply.

(13)   *Certification (acceptance conditions)*—details to be given of any reports or test certificates required prior to acceptance of the plant.

(14)   *Life-cycle-cost summary*—the procedures outlined in Section 18.2 should be followed to produce a complete costing anticipated over the whole life of the plant.

18.3.2 Plant procurement

Having completed the specification of requirements, the next activity is to investigate the market place and select a suitable vendor. There are two distinct avenues of approach here, one of which will probably already have become apparent during the preparation of the plant specification. Is a standard range of equipment available to meet the need, possibly with some modifications? Alternatively, is a special unique piece of equipment required to be designed and built?

Whilst the user is always looking for improved performance and availability of his new equipment, he is unlikely to consider the full implications of maintenance and whether an investment in maintenance features would show benefits in cost of ownership. Unfortunately, this factor is rarely considered at the time of purchase of new equipment. Very often the maintenance manager is brought in late in the procurement cycle and sometimes only when the equipment arrives on the premises. The maintenance manager and his team have years of experience, which if correctly channelled could provide an investment in the future reduction of maintenance costs. Close cooperation with the equipment supplier, particularly in the early stages, provides a sound foundation for developing the best 'state of the art' technical approaches for maintenance.

In selecting the vendor, the company's previous experience with suppliers should be taken into account. It may be that the company has a vendor rating system in operation. If not an assessment may have to be made of an unknown supplier. In fact, it may well be necessary if a supplier who would normally supply a standard range of equipment is to be asked to make significant modifications or build a completely new piece of equipment to meet the specification.

Particular areas to bear in mind when selecting a vendor are:

(1) past performance or vendor rating;
(2) ability to meet the technical requirements of the specification;
(3) quality of equipment previously supplied and/or expected;
(4) delivery dates and the supplier's ability to meet them;
(5) price (value for money);
(6) on-going operational and maintenance costs of similar equipment produced by the vendor; and
(7) after-sales service available—this would include technical information available in the form of user and maintenance manuals, additional direct technical support available to resolve any problems which might subsequently arise, unique tools and spares.

Typical questions to be asked bfore finalising the purchase order for plant are as follows.

(1) Is the statement of requirements (the specification) adequate?
(2) Is the supplier fully conversant with these requirements and does he/she have the technical capability and production capacity to meet them?
(3) Will the plant be acceptable to operators in terms of safety and ergonomics?
(4) Will the plant meet the availability criteria defined—in terms of reliability and maintainability? Can the vendor provide test results and other data to support this requirement?
(5) Can the vendor provide the after sales service required, in particular spares support to meet the availability specified over the whole of the anticipated life of the plant? What are the inventory cost implications of this?
(6) Do the life-cycle costs meet the criteria originally established?

Before delivery of the plant, the purchaser should:

(1) establish effective lines of communication with the supplier, particularly if a long procurement cycle is involved;
(2) ensure that all the necessary services are available for installation and that the maintenance resources will be available;
(3) assess the requirement for trained personnel and ensure that adequate training of all staff concerned with the plant is undertaken, including operators and maintenance engineers; and
(4) ensure that all consumable items, on-site spares, tools and test equipment are available to meet expected demand.

18.3.3 After procurement of plant

There are a number of actions required by a purchaser of plant to complete the purchasing cycle. These are as follows.

(1) Check that all the acceptance conditions listed in the specification have been met.
(2) Ensure correct installation and commissioning. Many later problems can be traced back to hurried or inadequate activities at this stage.
(3) Ensure that the configuration of the plant is the same as that in the manuals and spare parts listing and that any modifications incorporated, or deviations allowed during manufacture, have been taken into account.
(4) Collect information on technical, maintenance or operational problems and maintenance and operational costs for future assessment of the next generation of plant. Feed this information back to the supplier as appropriate in order for him to improve the performance of subsequent plant for similar applications.

18.3.4 Plant replacement

In the life-cycle-cost examples described in Section 18.2, no account was taken of a replacement strategy, or the value which plant might have had in the second-hand market. Very often plant replacement is one of necessity for some reason or other, but using the life-cycle-cost process, it is possible to provide a justification for plant replacement on economic grounds. Strictly speaking, life-cycle costing can only be complete if a strategy for disposal and replacement is taken into consideration. It is often difficult to see a precise timing for disposal of equipment at the time of purchase, but there is usually some accounting practice or past history which will suggest a nominal life figure. Factors that will affect the disposal decision are:

(1) technological advances,
(2) change of requirement,
(3) reduced operating efficiency,
(4) increased maintenance costs,
(5) non-availability of spare parts, and
(6) prestige associated with new equipment.

Returning to the example of the vehicle (*Table 18.1*), and moving forward in time to the end of year 3, a new vehicle is now available at £40 000, with improved operating and maintenance costs. *Table 18.6* compares the options of purchasing a new vehicle and disposing of the older one for £10 000, against the alternative of continuing to run the older one to the end of year 6. As can be seen in this analysis, the total cost over the period under consideration is in favour of early replacement.

Table 18.6   Alternative replacement option

Year	Capital (×10³ £)	Operating costs (×10³ £)	Maintenance costs (×10³ £)	Revenue losses (×10³ £)	Total (constant) (×10³ £)	Discount	Total (NPV) (×10³ £)
3	40–(10)				30.0		30.0
4		3.5	2.0		5.5	0.91	5.0
5		3.5	2.0		5.5	0.83	4.6
6	(14)	3.5	2.0	1.5	(6.5)	0.75	(4.9)
Total							34.7
4		4	4.3	4	12.3	0.91	11.2
5		4	5.2	6	15.2	0.83	12.6
6	(2)	4	6.2	8	16.2	0.75	12.2
Total							36.0

18.4   Measurement of the effectiveness of maintenance

18.4.1   The need for measuring maintenance

Most companies know the direct costs of maintenance, but few companies really understand the opportunities associated with them. All too often the emphasis is either towards minimising the maintenance budget or repairing 'when it goes wrong', rather than taking the overall view of maintenance as a business opportunity.

The Asset Management Information Service (AMIS) is a Department of Trade and Industry supported scheme whereby a company can obtain assistance to conduct a self-audit of its maintenance activity. Some 450 of these audits have been carried out by MCP Management Consultants (the organisation funded to undertake these audits). They show that maintenance costs are typically between 1 and 8% of the annual sales turnover, the weighted average being 3.6%. These same surveys also point to relatively inefficient maintenance organisations. Through the scoring systems used in their audit they conclude that scores of around 50% are commonplace in areas described as 'general maintenance level', 'organisation and administration', 'work planning and control', 'cost control', 'productivity and maintenance effectiveness', 'materials management', 'failure analysis and betterment', 'training' and 'safety'. All these are key areas of the maintenance activity.

Repeat audits which were carried out 12–18 months after the original audit show that an improvement of 11–20% is quite achievable. This in turn demonstrates that significant improvements can be achieved in a relatively short time, providing key areas in which shortfalls are occurring can be identified, measured and monitored in such a way that corrective action can be taken to improve their performance.

The figures also indicate that 75% of the companies audited reported instances of unplanned work in excess of 20% of the time available, with over 35% having backlogs of work of between 1 and 5 weeks. Such figures do not augur well for the highest achievable measure of availability for their plant!

18.4.2   Maintenance costs

This is the traditional method of measuring maintenance, but whilst a cost figure without some qualification may be suitable as an overall budget or planning tool, it is extremely vulnerable to arbitrary manipulation by the accountants! The needs of the company accountant and the maintenance manager are very different. The former may not be concerned much beyond the total maintenance cost, whilst the latter needs a detailed breakdown of costs and their components to manage his day-to-day business. However, they do have one thing in common—that the totals must be compatible, and to achieve this it is important to ensure that no costs are omitted or double counted.

18.4.2.1   Maintenance hours

The maintenance engineers' time can be allocated to a number of different activities, and the individual maintenance manager will have to make the decision which to include in 'maintenance hours', but usually the term is applied to the actual work or 'hands-on' time spent by an individual engineer maintaining a piece of equipment.

18.4.2.2   Hourly rate

The engineers' time can be broken down into a minimum of the following categories:

(1)   work time,
(2)   travel time,
(3)   administration time,
(4)   waiting time, and
(5)   non-available time (e.g. training, meetings, holidays, sickness).

All of the above contribute to the hourly rate, amounting to a utilisation factor as a direct overhead on the work time. Additional payments made for overtime or bonuses also contribute to this rate. Depending on the accounting convention used in an individual company, it may also be necessary to add an appropriate proportion of any of the following to the hourly rate as the indirect overhead:

(1)   social and other benefits,
(2)   training and documentation costs,
(3)   capital costs, including tools and test equipment,
(4)   vehicle running and other travel costs,
(5)   clothing allowance,
(6)   occupany costs,
(7)   communication and on-going computer costs,
(8)   other internal recharges, and
(9)   outside services.

18.4.2.3   Parts cost

Parts can be internally manufactured items, repaired parts or bought in goods. Each type is likely to be recorded in the company accounts on a different basis.

If we start by considering the internally manufactured part, this has a value based on the materials and manpower required to produce it. It will also have had a proportion of overhead added to it, in exactly the same way as the overheads were added to the maintenance engineers' time. It may be held in inventory at this price, or perhaps a mark-up will also have been included in the transfer price between the manufacturing division and the stores department. This will depend on internal company accounting policies and organisation, principally whether the manufacturing division operates as a profit centre. Again the opportunity exists for a mark-up to be added when the stores issue a part. This time it would be to cover the overheads associated with holding the part in inventory together with any profit mark-up, if appropriate. Consequently, a part could easily be valued at double its production cost by the time it reaches the maintenance engineer.

The repaired part is more difficult to cost, depending on where it was repaired. At the time of removal from an equipment as a failed part it has zero value, being completely useless, and if there was no policy to repair it, it would be thrown away. If the maintenance engineer can repair it to its full working specification (and this implies that its life expectancy would be the same as that of a new part), then the part is available for use by the engineer at the sum of time and materials costs which were used in the repair. Care must be taken in accounting for this in recording maintenance costs.

If the repaired part is sent away from the maintenance function for repair, then it is usual for the inventory department to allow a credit to maintenance, and when this part is reissued to them as fully serviceable, it is accepted at the new part price.

The bought-in part has a price tag on arrival and it often depends on how far back down the chain of maintenance, stores and manufacturing that the purchasing function is located, as to the final cost to maintenance.

18.4.3 Maintenance indices

A number of different indices are used to measure maintainability and maintenance and, before discussing these in detail, a few general comments are appropriate.

The first is the data source. Any figures which are used in the compilation of an index must be compatible and measured on the same basis if any meaningful comparison is to be made. It follows, therefore, that the most effective use of an index is for the purpose of tracking from one point to another, using the same database, to determine the direction of trends. However, having said that, further notes are given below on each index. It is also essential that an appropriate index is chosen to suit a particular purpose. This is demonstrated in *Table 18.7* and the accompanying narrative.

Availability has already been defined in Section 18.1.6 as:

$$\text{Availability} = \frac{\text{Possible operating time} - \text{Down time}}{\text{Possible operating time}}$$

Maintenance effectiveness can be used as a measure of maintainability, and is a useful basis for comparing different equipment both from design projections and from historical data:

$$\text{Maintenance effectiveness} = \frac{\text{Actual equipment use}}{\text{Maintenance effort required}}$$

Units should be chosen to suit the working environment of the equipment and a typical parameter would be 'operating hours per maintenance hour'. The maintenance effort would be dependent on reliability and mean time to repair. The larger the resulting maintenance-effectiveness number, the better the maintainability.

In *Table 18.7* a comparison is made of the performance of four different equipment options which would perform a similar function, and which a user might consider purchasing. The various indices are calculated and a comparison can be made between them from a maintenance point of view. It will be seen that, despite the quite different failure rates expected, there is only one unit difference in availability over the range of the four options listed. It will also be seen that the maintenance effectiveness and maintenance cost figures are almost in the reverse order. Perhaps the figures used in the example are a little more extreme than would be found in practice, but it does demonstrate the variations found by using different indices.

The *prime maintenance index* is another factor often used for comparing one piece of equipment with another prior to purchase, and is favoured by accountants, but may be misleading from the maintenance manager's viewpoint.

$$\text{Prime maintenance index} = \frac{\text{Current annual maintenance costs}}{\text{Capital replacement cost}}$$

This index should be compared with the formula for maintenance effectiveness which performs the same function and is much more relevant to the maintenance activity. The significant difference between the two is that the maintenance effectiveness is related to usage of the equipment, whereas the prime maintenance index is simply a cost comparison. In the latter case the index can be influenced by outside factors such as purchasing discounts, equipment price changes and taxation considerations.

18.5 Operational aspects of maintenance

18.5.1 Maintenance planning and control

Many aspects of maintenance planning and control will be predetermined by company organisation and practices. There is no 'right' way which is applicable to every situation. However, of prime importance is that maintenance as a function is represented at a sufficiently high level within the

Table 18.7 Example showing a comparison of indices

Option	Failures per 1000 h	Average repair cost (£)	Average repair time (h)	Availability (%)	Maintenance effectiveness	Maintenance cost per 1000 h (£)
A	7	150	0.7	99.5	204	1050
B	10	100	1.0	99.0	100	1000
C	2	250	5.0	99.0	100	500
D	15	50	1.0	98.5	67	750

company to contribute to the overall profitability of the company.

The key to efficient use of maintenance resources is the work-control function. Ideally, this should be a centralised function across all maintenance resources, unless the maintenance function itself is extremely large, very diverse in trade skills or geographically widespread. In other words, its strength lies in the ability of the work controller to be able quickly to switch manpower from one activity to another according to the priorities of the moment. The work controller's function is two-fold:

(1) matching of the workload with resources, and
(2) allocate priorities within this activity.

It is important that the work controller has the authority to undertake these activities without recourse to higher management, other than in exceptional situations.

The Work Controller's tasks are as follows:

(1) receive requests for maintenance;
(2) check priority of request (immediate, within 2 hours, etc.);
(3) record relevant information regarding the request, including time received;
(4) allocate the job to the maintenance engineer or team (having the appropriate skills) and prioritise their activities;
(5) record details of work done, spares used, etc., as advised by the maintenance engineer and note the time that the work was completed; and
(6) update records as appropriate.

In its simplest form this may be done manually on record cards, or, as is more usual these days, details will be entered on to a computer terminal as they are received.

Inevitably, from time to time there will be insufficient manpower to meet all the immediate demands and the earlier a pending *overload situation* can be identified, the more chance there will be of taking remedial action sufficiently early to avoid any loss of availability of the equipment under the care of the maintenance department. Whether manual or computerised, the general approach to identifying and resolving such a situation will be the same.

When a request for maintenance is received, the work controller identifies the equipment and the maintenance engineer of the appropriate skill (e.g. electrician, mechanical fitter). The time is noted together with the maximum acceptable time for the engineer to arrive on site and the anticipated work time to resolve the problem. At any time, therefore, a manual or computerised record is available for each pending job in terms of the number of hours of work anticipated and the time frame in which the job should be carried out. If this is then matched to the availability of the maintenance engineers, any overload situation can be quickly identified. The task is then to eliminate the overload either by rescheduling non-urgent work or by bringing in extra manpower from another part of the plant which, hopefully, will not be overloaded at the same time.

18.5.2 Computers and maintenance

There are a number of suppliers of commercial software packages who can provide very comprehensive maintenance management systems which will incorporate this work queuing facility as well as many other features from plant records to stock control.

Any computer system must satisfy two fundamental requirements:

(1) its cost of implementation and operation must be offset by savings in direct resources or increased efficiency; and
(2) the data output must be as complete, timely and accurate as the manual system it is replacing.

An investigation into existing systems will identify a number of key factors such as:

(1) data sources, their completeness and accuracy;
(2) reports produced;
(3) capital and personnel resources currently in use;
(4) effectiveness of the system; and
(5) personnel involved in receiving, assessing and making decisions on the data.

This should reveal the main characteristics of the current system, its shortfalls and redundancies, enabling a detailed specification for a new system to be produced.

At this stage, in selecting a computer system, 100% reporting of everything directly by the maintenance engineer is possible—at a cost! Depending on the quantity and variation of equipment being maintained, together with the frequency of maintenance activities, can 5 min terminal input by the maintenance engineer be justified after every call-out? If it can (and the extra capital cost can also be justified), then it may be worth considering installing terminals for use of the maintenance engineer. There may be other advantages in doing this, such as the calling up of job routines on the computer screen and the ordering of parts required direct to and from the shop floor.

It is always preferable to identify the software first before considering the purchase of any hardware. One particular area which should be given close attention is the ability of the software to cope adequately with numbers from the existing part numbering system used by the maintenance function. It is not just the numbers themselves which may cause a problem, but the ability of the software system to sort them in a meaningful manner, e.g. differentiating between piece parts, purchased parts (possibly from different suppliers with their own unique numbering systems) and major assemblies.

If technical reporting coverage is to be included in the system, it will also be necessary for the system to be able to produce special reports such as reliability growth curves, both for specific parts (or groups of parts) and problems. The system also needs to be capable of batching reporting codes together to produce an historical database in order to progress identified problems and their solutions.

18.5.3 The profit-centre concept

Traditionally, maintenance has been regarded as an expense. If equipment is broken down then it will inevitably cost money to return it to its fully operational condition again. That is the generally accepted view of maintenance. Are there situations where it is not just an expense? Are there ways of offsetting the costs of maintenance? Can it be shown to make a profit? The answer to all three questions is 'yes', but it can only be achieved by a company-wide commitment to make profit an objective of the maintenance department.

Part of this commitment must include the ability of the accounting system to accept recharges between departments, and the introduction of transfer pricing policies for goods and services within the organisation. The latter will be necessary to determine the separate performances of both the 'selling' and 'buying' departments. This procedure may not be practicable for the smaller company, and care should be taken to ensure that any additional accounting practices are not more expensive to operate than the estimated overall profits that are likely to ensue from introducing the concept. It must be remembered that profits on internal transfers do not in themselves contribute to those of the company. In addition, if a manager decides to make a purchase outside the company in preference to purchasing from an internal department, he may unintentionally create underutilisation in the 'selling' department, thereby increasing the departmental overhead and reducing the overall profitability of the company.

There are, however, a number of intangible benefits to such a system including improved motivation, increased efficiency, better decision-making and the decentralisation of many management decisions.

The simple formula on which all businesses exist is:

Income = Costs + Profit

It follows, therefore, that to increase the profit, either the costs must be decreased or the income increased. There are clearly two opportunities for the maintenance function.

18.5.4 Direct and indirect costs of maintenance

Consider first the costs of maintenance. Part of the maintenance manager's job specification will inevitably be to minimise maintenance costs. So why not recognise this as an internal profit? If the role of the maintenance manager is compared with that of the production or sales manager, his contribution is equally as valuable to the overall profit of the company.

Improvements in sales means more production and some of that could have been brought about by improved availability of manufacturing equipment. Some of the reduction in production costs could also have been attributed to the improved availability. So, providing these gains can be isolated, they can be credited to the correct function, namely maintenance.

In Section 18.4.2 the constituent elements of maintenance cost were discussed and direct costs and indirect overheads were identified. The latter should not be confused with indirect costs, which are charges incurred (or in income terms, perhaps income derived) by other departments as a direct consequence of maintenance activity.

18.5.4.1 Direct costs of maintenance

These can usually be determined at the time of purchase of the equipment using the life-cycle-costing process, and the maintenance department would not be expected to deviate from these estimated costs without some reason being identified. It may be that the equipment does not perform as intended, perhaps it is not so easy to maintain as the manufacturer originally advised, or maybe its utilisation has increased over that which was planned at the time of purchase. Any of these reasons could justify additional direct cost being incurred by the maintenance department. On the other hand, less resources might be needed for the opposite reasons. These are changes to plan and should be treated separately from departmental profit and loss.

However, if genuine efficiencies are introduced, either in manpower or parts (either usage or stock holding, if the latter is part of the maintenance department budget), then these will lead to a reduction in direct maintenance costs. This in turn will contribute directly to the overall profit of the company, providing that the departmental targets for maintenance are being met and that undetected costs are not being incurred in other departments.

18.5.4.2 Indirect costs of maintenance

These arise in other departments of the business. In Section 18.2, for example, when considering life-cycle costs for the

purchase of a new vehicle, there was the opportunity to trade-off operating costs, maintenance costs and revenue losses to achieve the lowest overall cost. This was essentially a planning exercise done before any decision was made to purchase. Now, supposing that the overhaul option detailed in *Table 18.2* was not considered as part of the original purchase planning process, but was raised as an option by the maintenance department in year 4, then this overall profit to the company of £3600 would be achieved largely by eliminating expected revenue losses due to increased availability of the vehicle.

Other areas in which similar profits may be identified include:

(1) lowering of reject rates by improving maintenance;
(2) elimination of overcapacity planning by departments allowing for extra maintenance downtime; and
(3) reduction in support to maintenance in other departments due to more efficient maintenance.

Figure 18.11 shows a summary of the principal categories attributable to the cost of maintenance, which would provide opportunities for profit identification. Note that this is a positive approach to profit attributable to maintenance, rather than a cost-cutting exercise.

18.5.5 Maintenance revenue

Looking now at the 'income' side of the equation, maintenance can only contribute additional revenue if it can widen the range of services which it can provide to the user, and be able to charge for these services.

Depending on the environment in which the maintenance department operates, maintenance revenue may be derived from both within the company and from other quite separate organisations (see *Figure 18.12*). However, in a manufacturing environment it will probably be only appropriate to consider those opportunities which may exist within the company.

The first source of revenue is, therefore, to charge for maintenance. Internally, within a company, this may be unnecessarily cumbersome, particularly on a time and materials basis. However, there are merits in building the departmental budget in the form of a number of maintenance contracts to be agreed with the user departments (see Section 18.5.7). One of the principal advantages of this approach is that the maintenance manager has a clear plan of the number and variation of equipment that he/she will be expected to

Figure 18.11 The costs of maintenance

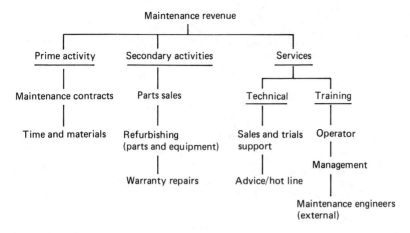

Figure 18.12 Sources of maintenance revenue

maintain during the period under review. Any additional requirements can then be added as a result of negotiation rather than surprise!

The second source is to increase the services offered by the maintenance department. It should be borne in mind that the prime objective of the maintenance department is to keep the equipment with which it is charged in fully operational order. It would be unwise to extend the services of the department to the extent that its activities detracted too much from this prime objective, but a review could be undertaken to identify complementary activities which would provide a more complete service to the user.

Typical complementary activities would include the following.

(1) *Training*—user or operator training in the correct use of the equipment, possibly extending to some elementary maintenance activities. There may also be a requirement to familiarise management and technical staff with the equipment.
(2) *Support for internal and external exhibitions*—any special equipment trials which might be undertaken, together with technical advice could also be considered.
(3) *Resiting and recommissioning*—this activity is often absorbed in the day-to-day operation of the maintenance department, but really should be recharged to the user department who require the equipment to be moved to its new location. When required, site surveys could all be undertaken as part of this activity.
(4) *Refurbishing of spares and equipment*—some refurbishing of spares may be undertaken by the maintenance department as part of its own economic approach to maintenance, in which case there would be no other department that could be recharged. However, where this activity can be performed for other equipment not normally maintained within the company, then the opportunity for recharging exists.

18.5.6 Pricing of maintenance

Recharges for services have been discussed above, but without any indication as to how these should be priced. A strategy on which recharges can be based should meet the following objectives.

(1) The price for maintenance activity should be consistent with the overall objectives of the company. For internal

maintenance they should be at such a level that does not encourage the use of external sources.
(2) The price level should be such that an accurate assessment can be made of the performance of the maintenance department, in particular with respect to its profit contribution to the company.
(3) The price level should be such as to ensure that the benefits of the profit centre concept are achieved, if possible independent of the action of other functions within the company.

There are three ways in which the price for maintenance can be set:

(1) by basing the level at a competitive market price,
(2) by cost-based pricing, and
(3) by an agreed price.

The preferred method must always be the first in the list above, but at times this may be difficult to establish. Cost-based pricing, using accepted labour rates and parts costs is a good alternative, and is often the preferred method for establishing internal recharges. The 'agreed' price usually forms the basis for the maintenance contract.

Maintenance costs are closely defined within an organisation, but maintenance pricing also depends on the level of service provided to the user. Manufacturing departments may be prepared to accept charges at a higher rate for shorter response times to their requests for maintenance on their key items of equipment, or the provision of maintenance cover outside the normal working day.

The allocation of overheads on maintenance costs should be examined. It is traditional to relate company overheads to man-hours within a departmental budget, but this may not be the most effective method from the point of view of pricing and establishing a satisfactory profit. It may well be possible to absorb a higher level of overhead on the maintenance of some equipment than others, to obtain a more competitive rate. Some alternative approaches to this are shown in *Table 18.8*, where the maintenance of three different equipment ranges is considered. As can be seen, if the pricing of maintenance for equipment C is not competitive when the overhead is based on labour-hours, this improves dramatically when the allocation is based on the number of pieces of equipment maintained, but severely penalises A. Two other examples show the effect of basing the allocation on call-outs and a random weighting by product range, the latter benefiting both A and C with a relatively small percentage impact on B.

Table 18.8 Example showing options for overhead allocation

	Equipment range		
	A	B	C
No. of equipments maintained	500	400	200
Call-outs/equipment/year	1	4	15
Average hours/call-out	1.0	1.5	2.0
Labour revenue/equipment (excluding overheads)	£20	£120	£600
Total overhead allocation based on labour-hours	£5000	£24 000	£60 000
Annual revenue options/equipment:			
Based on labour-hours	£30	£180	£900
Based on No. of equipments maintained	£101	£201	£681
Based on No. of call-outs	£38	£190	£862
Based on weighting by equipment range, 1:20:25	£24	£217	£842

18.5.7 Maintenance contracts

There are advantages to both the user and the maintenance function in obtaining agreement for maintenance on a fixed price contract basis. Both organisations have an agreed commitment (to pay and provide maintenance), and know exactly the conditions of this agreement and can plan for it with confidence. The user is assured of regular preventive maintenance (if such a scheme exists for the particular equipment), speedy response to any call for assistance, and a minimum of paperwork. From the point of view of the maintenance manager he has a known range of equipment to service at a fixed revenue, with only the need to raise regular user invoices, rather than one for each maintenance request.

However the basic price of a maintenance contract is established, it should be structured in the contract in such a way that it can be varied according to the equipment utilisation. This can either be by pricing it in relation to a meter reading (for example, £1.00 for every hour of operation), or by including an option to amend the charge should the utilisation vary beyond certain agreed limits. Above all, maintenance contracts must be competitive in the market place and provide an adequate rate of return. If a satisfactory rate of return at a reasonable market price is not achievable, then the maintenance manager will need to review his cost base. Some areas for examination are:

(1) Is the level of service (response time, labour skills, parts back-up and any other commitments to the user) higher than it need be?
(2) Is the overhead structure unfairly loading the cost base? Can it be divided any other way or spread more widely to produce a lower rate?
(3) Could the balance of the level of service offered to other departments or equipment users be varied to justify a differential charging rate (that is a higher rate for some special service)?

Failing that, there is little option other than to either accept a lower rate of return (perhaps only from some users) or decline to offer the maintenance contract. The latter course of action suggests an underfunded maintenance department!

18.5.8 Modifications

One method of reducing maintenance costs, and hence of improving profitability, which has not yet been discussed, is to improve the performance of the equipment. As has been said earlier, reliability and maintainability can only be designed into a piece of equipment, but designs can be changed and modifications introduced to improve either of these features.

The timing of such modifications is, however, very important. There is often a considerable lead time involved between identifying the need, designing the change, producing the hardware and instructions for fitting, and scheduling the work of modifying the equipment. Consequently, however attractive a modification looks at the time of its inception, an economic study needs to be undertaken, taking into account the lead time involved in fitting and the expected life of the equipment, to ensure that there is adequate payback to make it all worthwhile. If the parts for the modification must be purchased by the user, and particularly if freight and duty costs are involved, then this is another negative aspect to be taken into account.

Consider again the vehicle which was used as an example in Section 18.2.2 (*Table 18.1*). Supposing a gain of £300 per annum in operating costs is to be had for an outlay of parts at £600 and labour for fitting of £100. If this decision was taken in year 3, then at first glance an outlay of a total of £700 to gain £300 per year looks reasonably attractive. Supposing it will be early in year 4 before the modification is actually fitted, then this still leaves nearly 3 years to achieve the benefits. However, the marginal analysis shown in *Table 18.9* demonstrates that, if everything goes according to plan, the best that can be achieved is to break even. It is therefore probably not a worthwhile investment.

18.5.9 Refurbishing of parts and equipment

Refurbishing can be a useful source of profit to the maintenance department and, at the same time, can improve the overall utilisation of the staff. This can be achieved by utilising the 'waiting' time normally experienced by maintenance engineers. It is a useful buffer activity in terms of labour hours and, if successful, can often justify additional staff, either dedicated to refurbishing or simply to increase the overall resource level available for both maintenance and refurbishing activities. At the same time, having the additional staff may improve the response time to the calls for assistance from the user. However, it is important that the quality of the refurbished parts is of a high standard, as for all practical purposes a refurbished part should have an equivalent performance to that of a new part. If it does not have the same quality, then

Table 18.9 Cost and benefits of a modification

Year	Original case				Modified case			
	Operating cost ($\times 10^3$ £)	Maintenance cost ($\times 10^3$ £)	Discount	Total (NPV) ($\times 10^3$ £)	Operating cost ($\times 10^3$ £)	Maintenance cost ($\times 10^3$ £)	Parts cost ($\times 10^3$ £)	Total (NPV) ($\times 10^3$ £)
3	2.8	3.6	0.91	6.4	2.8	3.6		6.4
4	2.8	4.3	0.83	6.5	2.6	4.4	0.6	6.9
5	2.8	5.2	0.75	6.6	2.5	5.2		6.4
6	2.8	6.2		6.7	2.5	6.2		6.5
Total				26.2				26.2

this must be taken into account when assessing the success or otherwise of the refurbishing activity. For this reason, all refurbished parts should carry some identification to indicate their past history.

The income from refurbishing a part is equivalent to the price of a new part, less the cost of refurbishing, less any credit which might normally be available from the original supplier or for scrap. The credit possibility should not be overlooked because it might make a refurbishing opportunity non-viable.

In the example shown below, nominal credit has been allowed and added to an assumed revenue derived from other sources, the profit and loss account prior to establishing a refurbishing operation being as follows:

	$\times 10^3$£
Maintenance revenue	8500
Credit for parts x	100
Total income	8600
Maintenance costs Labour	4800
Parts x	400
Other parts	900
Fixed overhead	2400
Total costs	8500
Profit	8600 − 8500
	= 100

After establishing the refurbishing operation, the account will appear as follows. Note the division of the fixed overhead at 50% of labour costs, assuming no additional staff are needed to undertake the new activity.

	$\times 10^3$ £
Maintenance revenue	8600
Refurbishing costs Labour	300
Parts	50
Fixed overhead	150
Total refurbishing costs (parts x)	500
Maintenance costs Labour	4500
Parts x	500
Other parts	900
Fixed overhead	2250
Total costs	8150
Profit	8600 − 8150
	= 450

It can be seen that the refurbishing activity by itself is a loss-making activity, since parts x are refurbished at £500 000, whereas they could be purchased for £300 000 allowing for the credit. In spite of that the overall profit situation of the maintenance activity increases from £100 000 to £450 000! This increase would be sufficient to absorb some additional staff to assist with the maintenance activity, the numbers and resulting profit depending on how the labour-related fixed overhead increases.

This type of costing exercise demonstrates that significant savings are achievable by better utilisation of staff, which might not have been apparent unless a profit-centre concept for maintenance had been introduced.

18.6 Preventive maintenance

18.6.1 Classifying types of maintenance

There are two basic types of maintenance: the scheduled and the unscheduled. For maximum availability of the equipment or plant the unscheduled maintenance should be kept to a minimum. However, to some extent this minimising of maintenance needs to be tempered with the economic consequences of 'overmaintaining'. In a real situation, therefore, unscheduled maintenance will inevitably occur.

Where failures can be predicted, then maintenance can be scheduled. The origins of data for scheduled maintenance can be derived from a number of different sources, such as:

(1) safety requirements,
(2) manufacturer's recommended maintenance routines,
(3) operating experience,
(4) condition monitoring,
(5) non-essential performance deterioration, and
(6) non-essential equipment failures.

The first of these categories, safety requirements, is typified by a crane which would require weekly inspections and periodic testing of the wire rope. The next category is similar, but is a manufacturers's recommendation, rather than a mandatory requirement. It is usually related to recommended time intervals and includes such activities as lubrication, inspection, testing and replacing of some parts which have a short life. With experience these schedules are often added to or amended by the maintenance department themselves, producing a listing of activities to be performed at regular intervals. The last two categories in the list above are those items which will not affect the availability of the equipment in question, or perhaps allow it to continue to operate with some degraded aspect of performance which can be tolerated for a short time, until the next opportunity to carry out preventive maintenance.

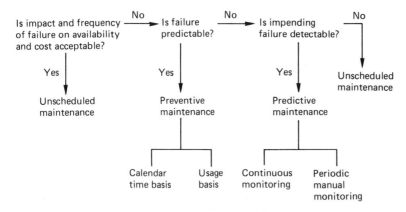

Figure 18.13 Decision-tree approach to determining the optimum maintenance period

Condition monitoring can be very effective where a failure can be anticipated by some form of deteriorating performance over a period of time to allow the maintenance engineer to take remedial action. This process is known as predictive maintenance and is dealt with in Section 18.7.

Therefore, the maintenance categories may be summarised as:

Maintenance = Unscheduled + Preventive + Predictive

The guiding principles are:

(1) 'as little as possible, as often as necessary', and
(2) 'maximise availability economically'.

Figure 18.13 shows a decision-tree approach to determining the optimum form of maintenance. This approach can be used for any particular application.

18.6.2 Failure patterns

Failures happen in many different ways and for many different reasons, not the least of which are during transport and storage. These are completely random and each should be treated as an individual problem, although obviously similar problems point to a need for investigation and correction. For operating failures it has been generally accepted for many years that the familiar bath-tub curve (*Figure 18.14*) represents a typical failure pattern.

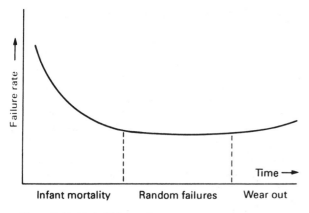

Figure 18.14 Typical failure pattern

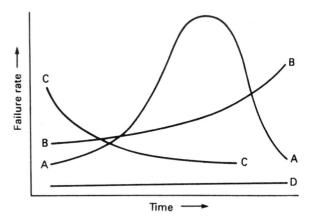

Figure 18.15 Different forms of failure pattern

In a well designed and manufactured equipment, the majority of failures are completely random. *Random failures* would occur in the bath-tub curve shown above during the period when the failure rate is at its lowest and, for most purposes, can be regarded as constant.

Other common forms of failure pattern are shown in *Figure 18.15*. Line A represents a wear-out failure for an item that has a finite life, such as a light bulb which is guaranteed for 1000 h, but the majority of which will fail before double that life is reached. Line B shows an item which has a failure pattern which is essentially the second two sections of the bath-tub curve, differing in as much as it does not exhibit any significant infant mortality. This type of curve is generally true for many mechanical moving parts which are subject to gradual wear, such as a bearing. Line C demonstrates the infant mortality pattern and is effectively the first section of the bath-tub curve. The failure of electronic components generally is typified by this curve. Line D can usually be applied to represent parts which seldom fail, or are subject to damage or loss during use or maintenance.

18.6.3 Establishing preventive-maintenance routines

The ideal contenders for inclusion in a preventive maintenance programme are those items which have a performance characteristic which falls below an acceptable level in a

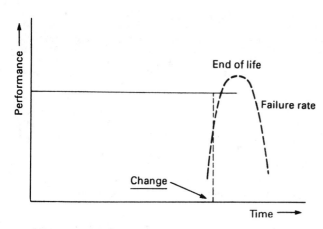

Figure 18.16 Preventive maintenance for a part showing instantaneous failure

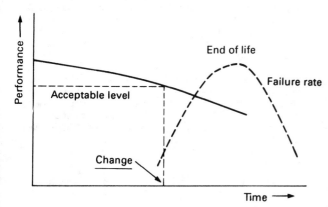

Figure 18.17 Preventive maintenance for a part showing degraded performance

Table 18.10 Impact of alternative preventive maintenance policies

Proposed PM interval (h)	Cumulative % parts failed	Costs per 1000 h (£)		
		PM	UM	Total
900	0.5	28	1	29
1000	3	24	4	28
1100	14	20	16	36
1200	36	14	38	52
1700	99	—	73	73

PM, preventive maintenance; UM, unscheduled maintenance.

preventive maintenance policies is shown in *Table 18.10*. It is assumed that preventive maintenance can be undertaken at a convenient time when there will not be any production losses. As can be seen, an interval of around 1000 h would be the most economic approach.

The costs given in the table are simplified in as much as all failures and replacements were calculated from time zero, without taking into account the effect of the clock resetting every time a part is replaced. This more complicated situation may be analysed by using computer based simulation programmes. This involves simulating failures, statistically distributed with the expected probability distribution function, following the logic of the proposed preventive maintenance strategy, and calculating the resulting costs. By carrying this process through a large number of failures and replacements, typically 100 or more, average costs can be obtained using the different strategies, and the optimum choice selected. This method of selection is known as a 'Monte Carlo simulation'.

That a part fails at a predicted time is no certainty that it is a suitable candidate to be incorporated in a preventative-maintenance programme. The effect of the failure is the significant factor. Does it cause a complete or partial shutdown of the equipment? Is the failure rate pattern sufficiently consistent to be able to predict an economic change point? When it fails, does it cause a safety hazard? The answer to questions such as these will determine the need and the appropriate interval for preventive maintenance.

Assuming that the replacement of individual critical parts determines an interval for preventive maintenance, then it will be economic to schedule other activities at the same time, such as lubrication. It may also be an economic proposition to bring forward other activities to align them with the primary preventive maintenance timing, so that the bulk of preventive maintenance work can be done at the same time.

18.7 Condition monitoring

18.7.1 Introduction to condition monitoring

Condition monitoring is a method of providing maintenance management with a means of determining the technical state of their machines or equipment on an on-going basis. This basic concept is known as 'predictive maintenance' (see Section 18.6.1), and its purpose is to anticipate when a breakdown is likely to occur. This approach involves some form of monitoring which could vary from the simple visual observation of dirt or smoke, the feeling of heat in a critical bearing housing, for example, to the highly sophisticated computer monitoring system.

There are three techniques generally used in condition based maintenance.

consistent interval of time. In this instance the word 'time' may be used to indicate elapsed time, operating time or an output quantity or usage. In fact, any way that can be measured relative to the equipment. Two types of performance degradation are common, and these are illustrated in *Figures 18.16* and *18.17*.

Figure 18.16 shows a type of failure that was discussed earlier in the previous section, such as the lamp bulb which either lights or does not, with failure which is instantaneous after perhaps 1000 h. *Figure 18.17* demonstrates another type of failure where the performance gradually degenerates over a period of time until it becomes unacceptable. Both graphs also show the distribution of failures with time and an economic change point can be determined statistically from the failure pattern (or design specification if no historical data exist).

The method of actually determining the most economic preventive maintenance interval is best demonstrated by an example. Suppose when an unscheduled maintenance activity is made, a part costing £5 is replaced at a total labour cost of £20 and production to the value of £100 is lost. The failure pattern of the part follows a normal distribution curve, similar to that shown in *Figure 18.16*. The impact of alternative

(1) *Inspection*—periodic checks on any measurable parameter where a degrading condition may be detected by the human senses.
(2) *Comparison*—routine checks of a parameter which must fall between certain specified limits.
(3) *Trend*—measurement of parameters recorded at set intervals and plotted to detect a changing situation.

The first technique involves traditional methods adopted by experienced maintenance engineers and relies entirely on the human senses such as feeling, hearing and seeing. None of these checks is entirely reliable or repeatable in modern industry. Perhaps the nearest check to this is the coin standing on edge on top of a bearing housing, or a similar piece of equipment.

In the previous section on preventive maintenance, *Figures 18.16* and *18.17* show two examples of different failure characteristics. In the first figure there is a sudden end to the working life of a component, and in the other there is a gradual degradation of performance. If this degradation is fully predictable over the life of the component, then replacement as part of a preventive-maintenance schedule will be the most economic. If, however, the failure might occur at an unpredictable time, through wear or a change in the usage pattern, for example, then it is an ideal candidate for a modern condition-monitoring installation.

It is important to realise that condition monitoring can be approached using one or both of two alternative strategies. Firstly, as a continuous process on line with the equipment; or secondly, by monitoring at set intervals either by using hand held portable equipment or by periodic scanning of fixed sensors from a central monitoring facility.

A good example of condition monitoring and its wide opportunities is a coolant system, where there are four parameters which may be very easily monitored. The first is the pressure at the pump outlet, the second is the vibration levels of the pump bearings, the third is the coolant level in the reservoir, and the fourth is the coolant temperature. Obviously, in practice it would not be necessary to monitor all these parameters, and in all probability two would suffice.

Today, manufacturing is becoming more and more capital intensive and with the introduction of such techniques as just-in-time (JIT), it is vitally important that equipment is kept operational. Hence reliable, detailed and readily available information on machine condition enables the maintenance manager to ensure that downtime during operational hours is kept to the lowest possible level.

18.7.2 Alternative condition-monitoring strategies

The first and most important task when considering introducing condition monitoring is to undertake a thorough survey of the maintenance activities and the likely breakdown characteristics of the equipment. With this knowledge the benefits of the introduction of a condition-monitoring programme may be assessed. There is an up-front cost in introducing these techniques, in planning time and monitoring equipment (sensors and data analysers) which should not be underestimated.

The first decision to be made in the selection of a condition-monitoring system is the choice of parameters to be measured and the frequency of monitoring. It is probably self-evident that the more frequently the equipment is monitored, the more expensive is the monitoring installation. However, a careful cost analysis may well show considerable longer term benefits in a high initial cost of fully automated monitoring equipment.

The alternatives available are reviewed below.

18.7.2.1 Automatic monitoring

This is the process of continuous on-line surveillance and interpretation of machine performance. It inevitably involves the setting up of a computerised monitoring facility with appropriate software to interpret the data being received from machine-mounted monitoring sensors. The characteristics of the equipment being monitored will be instantly transferred to a database, providing the maintenance engineer with immediate trend data from which a pending malfunction may be quickly and automatically identified.

18.7.2.2 Periodic monitoring

The alternative approach to continuous surveillance is periodic monitoring. As has been mentioned above, this may take two forms.

Automatic periodic monitoring This is essentially a simpler form of continuous surveillance, with the difference that the fixed monitoring points on the machine are interrogated at predetermined intervals from a single scanning installation. The intervals will depend on a combination of the likelihood of failure and the effect of the failure, that is to say how critical a failure would be to the overall performance of the plant. The advantage of such a system is that it does not involve the same amount of centralised computer hardware as the continuous-surveillance method.

Manual periodic monitoring This is a less capital-intensive approach, but does utilise a higher amount of labour. The procedure requires the use of a portable data collector, hand held by the maintenance technician, which is taken around from unit to unit to analyse the characteristics of individual parts of the machinery. Particular care is necessary in selecting monitoring points to ensure good repeatability and accuracy of the readings taken.

(1) The location of the monitoring point must be such that the pick-up module or probe can be quickly and simply located in its measuring position.
(2) The output from the monitored signals must be such that they are independent of the manner in which the pick-up module is attached in its measuring position.

One of the most commonly used methods of attachment of pick-up modules is to use a magnetic base (although this is unlikely to be suitable for any measurement involving eddy currents).

The results from such monitoring are continually plotted, either manually or through a computerised system, and any deviation from one reading to the next can be quickly detected and any appropriate corrective action taken.

The frequency of measurement must be determined in the light of experience with similar types of equipment, but a general guide is given in *Table 18.11*.

In order to maximise the efficiency of the measuring procedure, machines and measuring points should be grouped into 'routes' with regard to their measuring cycles, measurement equipment and geographical location. If a portable data collector is used, then this may be preprogrammed to provide the most cost-effective measuring sequence.

18.7.3 Condition-monitoring applications and sensing equipment

Depending on the design of the machine to be monitored, one or more parameters can be measured. Different types of

Table 18.11 Typical periodic monitoring routines

Frequency	Equipment category
Daily	Critical to production. Operating in a harsh environment. Failure would trigger faults in other areas. Failure would involve high repair costs
Weekly	Secondary equipment (not critical to continuous production). Difficult access for measurement
Monthly	Standard motors and similar equipment

sensor, known as 'pick-ups', can be used to suit the different parameters being monitored; these are reviewed below.

18.7.3.1 Absolute bearing vibration

This is generally measured on the outer loaded surface bearing housings, providing reliable criteria for establishing the condition of a machine fitted with ball or roller bearings. For the measurement of mechanical vibration and its conversion into an electrical signal, three different technologies are in common use:

(1) non-contacting pick-ups using the eddy current principle for the measurement of vibration displacement;
(2) electrodynamic pick-ups for the measurement of vibration velocity; and
(3) piezo-electric accelerometers for the measurement of vibration acceleration.

Damaged or worn rolling bearings will emit shock pulses which excite resonance in acceleration- or velocity-type pick-ups mounted on the loaded part of the bearing housing. The strength of individual pulses and the difference between strong and weak pulses provides the data for analysing the condition of the bearing. The choice of pick-up will be determined by the parameters to be monitored and the frequency ranges involved. The most common parameter in use is that of vibration velocity (or severity) which is typically measured in millimetres per second (V_{RMS}). At this point it must be stressed that adequate information regarding the 'acceptable' performance of a rolling bearing, or any other device being monitored, must be established and logged into the measurement database prior to any monitoring taking place.

Two standards have been published defining acceptable levels of vibration velocity for different types of rotating machinery commonly found in industry. These are:

BS 4675: Part 1: 1986 A Basis of Specifying Evaluation Standards for Rotating Machines with Operating Speeds of 10 to 200 Revolutions per Second
ISO 2372   Mechanical Vibration of Machines with Operating Speeds from 10 to 200 Revolutions/Second

Figure 18.18 is taken from ISO 2372 and shows the range of acceptable and unacceptable levels of vibration.

18.7.3.2 Relative shaft vibration

This is normally measured by using two non-contacting displacement transducers as pick-ups located at right angles to each other, in order to measure the vibration in two axes. Their principal application is in the measurement of the

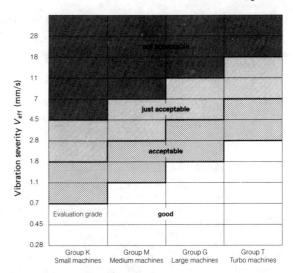

Figure 18.18 Evaluation limits for the vibrational behaviour of machines. (Courtesy of Schenk Ltd)

vibratory behaviour of a machine fitted with plain sleeve bearings.

18.7.3.3 Axial shaft vibration

This term is used to describe the shift of a shaft in the axial direction. As with relative shaft vibration, measurements are normally made with non-contacting displacement transducers, sensing the position of the shaft end, step, collar or flange.

18.7.3.4 Eccentricity

Turbine rotors and similar shafts can bend due to gravity or the effects of temperature whilst at rest. This can be detected and measured during run-up by using non-contacting displacement pick-ups. Any deviation will be identified and initiate the shut down of the machine before any significant damage can occur.

18.7.3.5 Relative shaft expansion

Again this is an application for the non-contacting pick-up to measure any difference in the length of a shaft compared with its casing. Such differences frequently occur where, say, a turbine rotor will heat up more quickly than its housing.

18.7.3.6 Rotational speed

Several alternative technologies are available for measuring rotational speed and their individual use depends on the accuracy required, as well as the particular application. Pick-ups in common use are those of the non-contacting variety, both displacement and inductive proximity tranducers. Equally, a wide range of tachometers using different technologies is commercially available.

18.7.3.7 Temperature measurement

This may be used in a variety of situations where friction occurs, or where changes may not necessarily indicate a pending failure condition, for example, temperature measurements of lubricants, motor or generator windings can provide valuable informative data on machine loadings at the time of measurement. In general, temperature measurements are made using thermocouples or resistance thermometers.

18.7.3.8 Pressure measurements

These are suitable for hydraulic, liquid or gas installations. The pick-up used for such applications usually comprises one or more pressure transducers to provide absolute or relative measurements.

18.7.3.9 Oil debris

This is another application for any machinery which has oil lubrication, hydraulic controls or drives. Particles suspended in oil may be isolated and identified as to their source. Debris may be collected using magnetic plugs or filters, from which it will be possible to identify debris in the range 100–1000 μm. For smaller particle sizes, ferrography enables microscopic examination and spectrographic analysis will determine the presence of particles of less than 10 μm in size.

One final note on condition monitoring concerns false alarms. Throughout the condition-monitoring process it is important that, where possible, duplicate sensors and circuits are used in order to highlight erroneous signals. If the application is suitable, then two different parameters should be measured before a failure signal is initiated. This obviously would not preclude investigation of signals from one pick-up which may be monitoring an apparently degrading situation. Such situations may already have been predicted as a result of trend data indicating an abnormal situation which may have been developing over a period of time.

18.8 Inventory and maintenance

18.8.1 Inventory objectives versus maintenance ideals

In some organisations, inventory may be the responsibility of the maintenance department, whilst in others it may be that of an independent function. More often than not the responsibility is a shared one, perhaps with maintenance being responsible for its own day-to-day requirements, drawing these from a centralised inventory and/or procurement department.

Whatever the organisation, the same fundamental conflict of interests will exist, whereby the maintenance department would like to have a quantity of every part instantly available and the inventory management (with an eye to the costs of holding the stock in readiness), will have other ideas. In reality, a balance has to be struck between availability of spare parts and stock holding costs. It is possible to do a complete cost analysis on this situation and, assuming the spares call-off rates are known, calculate the costs of holding each part of the equipment at each stocking echelon against the cost of delivery of the part from a higher echelon, together with the associated lost waiting time by the maintenance function. This, of course, does not take into account the lost production of the user, or his perception of the service that he is getting from the maintenance function. In practice, therefore, inventory must be optimised to provide the best possible service within whatever downtime is permissible and cost level is considered appropriate. This section is devoted to methods which can be adopted to achieve this objective.

Inventory is usually stocked at a number of different echelons, typically:

(1) with the equipment, perhaps in a locker only accessible to the maintenance engineer (unless there is an operator maintenance policy in being, which requires the operator to fit new parts or consumable supplies);
(2) with the maintenance engineer, in his/her vehicle if he/she is mobile, or again in a locker if he/she is stationed close to the equipment;
(3) in a local store, geographically central to a number of engineers; and
(4) in a central store, supplying parts direct to the base location of the engineer, or via a local store.

At each echelon, a goal should be set by which the efficiency of the supply can be measured. One method used comprises a logistics level of service (LOLOS) factor. A LOLOS of 98%, for example, would mean that a demand from stock can be met 98 times out of every 100 requests, the other two requests having to be referred for supplying from the next highest stocking echelon. It is important to understand that this factor applies to the overall operation of the store and not just the individual parts. For critical or high usage parts, sufficient stock must be held to meet almost every demand, otherwise a LOLOS of 98% overall would not be obtainable. Parts are normally classified according to their criticality and usage rates, providing a basis on which all inventory echelon and quantity decisions should be made. Remembering that the highly critical parts are those which affect safety and availability, these must therefore be readily available to the maintenance engineer, unless they have an extremely low replacement-rate prediction.

The factors which must be taken into account when determining inventory levels are:

(1) siting of stock at different locations;
(2) elapsed time betwen demand for replenishment or emergency order from one echelon to the next higher one;
(3) call-off by part number (replacement rate);
(4) stock holding cost;
(5) elapsed time between sending part for repair and its return as a fully serviceable part; and
(6) the LOLOS goal for each stock holding level.

Ideally, therefore, the overall inventory objective must be to minimise the direct and indirect costs of stock holding, whilst providing the highest LOLOS to the maintenance engineer. This can only be achieved through an awareness of all the cost implications, careful selection of parts, stocking echelons and order quantities.

18.8.2 Cost of stock holding

Stock is transferred in and out of the inventory system either at a standard cost or, if a profit-centre operation is in existence, a marked-up transfer price will apply to parts movements between some or all of the stocking echelons. The cost of holding stock depends on the following factors:

(1) interest which could be gained on the open market on capital which is invested in the stock;
(2) cost of storage facilities (rents, rates, insurance, etc.);
(3) cost of operation (staffing, transport and control systems); and
(4) deterioration and obsolescence.

An overhead figure of 25% of the cost of each part per annum would typically cover the costs of stock holding.

The 80/20 (Pareto) rule when applied to spare parts states that 20% of the parts account for 80% of the cost of parts usage (see *Figure 18.19*). Where these parts can be identified, this immediately indicates the priorities for close monitoring.

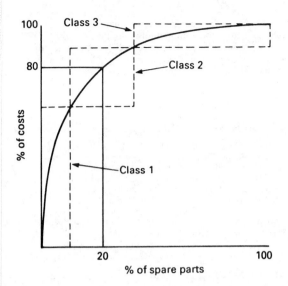

Figure 18.19 The 80/20 rule applied to spares

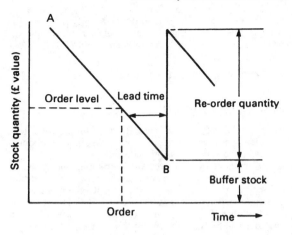

Figure 18.20 Stock movement and reorder trigger points

It is often convenient to classify these parts into three groups, as shown in *Figure 18.19*, allocating a usage value to each part number.

Usage value = Item cost × Usage per annum

From the point of view of controlling costs, priority should therefore be given to those items falling within class I and these should be subjected to strict inventory control.

18.8.2.1 Cost of stock-out

This brings into focus the implications of not having a part available when it is required. Specifically, these include:

(1) loss of equipment availability with consequential cost incurred in loss of output;
(2) additional costs involved in rushing small quantities of parts from the central warehouse to the maintenance engineer; and
(3) loss of goodwill, confidence in the equipment and the maintenance support, leading to loss of future sales when the time comes for equipment replacement.

The first two costs can obviously be quantified for any particular situation, but the third should not be overlooked, even if it is somewhat subjective, as this represents the future of the company (and the maintenance department!).

18.8.3 Safety stock and reorder quantities

However well planned the methods of supplying parts to the maintenance engineer, the system will only be successful if it is supported by effective stock-control procedures. In particular, keeping realistic levels of safety stock and ordering optimum quantities of new parts within the supplier's lead time. Normally, reorder systems depend on reaching a 'trigger' stock level at the central warehouse, or on the basis of a periodic review. The former is obviously the more efficient, and ideally lends itself to a computerised stock-control system. Where reordering is dependent on a periodic review, then the minimum stock levels must be sufficient to cover the call-off of parts between reviews, as well as during the reordering and replenishment cycle. Quite often a combination of both systems is used—the 'trigger' stock level for the more expensive parts and the periodic review for the cheaper parts.

Figure 18.20 shows diagrammatically the movement of stock against time. The slope of AB is based on historical usage, and

Table 18.12 Table to show the economic order quantity

Order quantity	Orders per year	Ordering cost (£)	Average stock	Holding cost (£)	Total cost (£)
500	20	1000	250	312	1250
1000	10	500	500	625	1125
1500	6.7	333	750	937	1270
2000	5	250	1000	1250	1500

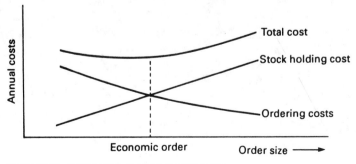

Figure 18.21 Determination of economic order quantity

the starting point in the parts-ordering activity is to establish the anticipated lead time involved between identifying the need to reorder and the receipt of the new stock. Having done this,

Buffer stock = Average usage during ordering lead time/3

or

Order level = 3 × Buffer stock

Note that the factor of one-third (or three), should be regarded as a good starting point, and experience will determine whether it is practical to reduce it.

The methodology involved in determining the economic order quantity (as shown in *Figure 18.21*) is best demonstrated by an example. Suppose that the annual spares usage of a particular part is 10 000, and its cost is £5. The inventory overheads are 25% of the parts in stock and the cost of raising an order and handling its delivery is £50. In *Table 18.12*,

$$\text{Orders per year} = \frac{10\ 000}{\text{Order quantity}}$$

Ordering cost = Orders per year × 50

Average stock variation = Order quantity/2

Note that, as the level of the buffer stock depends on usage and ordering lead times, it remains constant in this calculation and may be ignored.

Holding cost = Average stock variation × (0.25 × 5)

Total cost = Stock holding cost + Ordering cost

As can be seen from *Table 18.12*, the economic ordering quantity would be around 1000. It should be added that, if a supplier offers a discount for a larger quantity, then this effect would have to be incorporated into the figures in *Table 18.12*.

There are several good software packages that embody the inventory principles described above, both as stand-alone systems and those integrated with maintenance management systems.

Applicable standards

BRITISH STANDARDS INSTITUTION, *BS 3843 Guide to terotechnology (the economic management of assets)*, Milton Keynes (1991)

BRITISH STANDARDS INSTITUTION, *BS 4675 Mechanical vibration in rotating machinery*, Milton Keynes (1986)

INTERNATIONAL ORGANISATION FOR STANDARDISATION, *ISO 2372 Mechanical vibration of machines with operating speeds of 10 to 200 revolutions per second*, Geneva

Further reading

FILE, W. T., *Cost Effective Maintenance—Design and Implementation*, Butterworth-Heinemann, Oxford (1991)

KELLY, A., *Maintenance Planning and Control*, Butterworth-Heinemann, Oxford (1984)

NEALE, M., HENRY, T. *et al.*, Condition based maintenance. *Maintenance*, **1**, No. 4. Conference Communication, Farnham, Surrey (1986)

NEIBEL, B. W., *Engineering Maintenance Management*, Marcel Dekker, New York (1985)

THABIT, S. S., Life cycle costing—a decade of progress. *Chartered Mechanical Engineer*, Bury St Edmunds, Suffolk (May 1983)

19

Ergonomics

E N Corlett

Contents

19.1 Introduction

To argue for the importance of manufacturing in the nation's economy is probably unnecessary for readers of this book. It is an activity which we must pursue efficiently for the benefit of society at large; it is the vehicle by which the conditions of society are improved and the contributions of many aspects of intellectual activity are made of general utility. However, it can have its negative aspects, as we all know. The conditions of working people, and the ruination of nature, as a result of industrial working practices are the stuff of industrial history, of many social studies and, unfortunately, the results of much industrial activity today.

The enormous scale of environmental destruction and the widening of the realisation of the exploitation of people in some manufacturing situations world-wide, have brought a recognition that we must introduce new concepts into the assessment of industrial efficiency. Manufacturing is not the key factor in the economy before which all others must be subordinated, but is one of those which contributes to an acceptable form of society. Society comprises both the people who buy the manufactured products and the people who make them. To put it at its lowest, it would be foolish to disable (in any way) those working in manufacturing and to reduce their likelihood of buying the products. On a slightly higher level, the costs of making the product, in either economic, social or ecological terms, must not outweigh the benefits.

This preliminary is necessary because when we talk of occupational ergonomics we are dealing with the interactions between people and their work. As is shown in more detail later, this covers the physical, environmental, cognitive, organisational and social interactions to which we are all susceptible. It is not enough, for example, to consider just the reach distances and movement times in the design of a workplace, because optimum human performance is not achieved just by minimising these two factors. For a machine they may be appropriate, although the good designer would want to consider wear and expected machine life, maintenance and repairability, cost, manufacturability and several other factors. Therefore, even for a machine this pair of criteria is not enough, yet how many people believe that these criteria are adequate for judging the fitness of a person's workplace?

Underlying such a view is the perspective on a person as a simple machine, whose operation and motivation are independent of the effects of work activities. This implies a relatively simple system influenced by only a few variables of the most obvious kind. To reduce the argument to these terms is to reveal its absurdity. It also suggests that our personal knowledge of the complexity of human relationships and the effects of knowledge, difficulties and experiences on the performance of work as we all know of them, somehow does not apply to people doing industrial work. *They* are different, and can be categorised and valued broadly in terms of work study variables alone.

As a minimum we suggest that, when choosing the criteria on which to assess whether a work station is adequate, we must look beyond output in the short term; the interest is in quality performance in the long term. This requires that the work can be done day after day without detriment to the person's physical or mental health, with acceptable levels of accuracy and correctness, and with motivation. If any of these factors are absent, the firm will lose money and the workers are wasting their time. But this, i.e. that it shall not be harmful, is a negative level of acceptability. Why should acceptability not be set in positive terms?

Many professional people use skills in their work which they value, and take pleasure in their jobs. They put more into their work than is formally expected and enjoy the respect and good name which their performance engenders. Success, in such cases, breeds further success. Why should this situation not prevail at the shop floor or general office level? Why not set job-quality criteria beyond the minimal ('animal') level, in order to recognise and cater for the human dimensions of people?

Some readers may, by now, be thinking that such discussions go beyond ergonomics. A commonly held view, when one sees such phrases as 'ergonomically designed seating', is that the subject is primarily concerned with human body dimensions and their relationships with devices. But a more reliable short definition is 'matching work and its environment to people', and the environment can be illustrated as shown in *Figure 19.1*.

In the centre of the figure there is a 'man–machine interface'. Some process is influenced by a person, who senses the changes to the process, giving feedback of the effects on the process. The person is presented as a physiological system enclosing the sensing and effecting organs (the latter, of course, including speech), which in turn are linked by the cognitive system. Within this is an emotional aspect.

A feature of this diagram is that, wherever a boundary exists between any two areas, there are interactions across it. That is to say, taking as an example the sensory organs and the physical environment, the ambient lighting conditions will influence the ability to see, just as the ambient noise levels will affect what the person can hear. So the diagram warns us to explore some possible interaction effects which will influence a person's ability to achieve their objectives.

It is not possible in a short chapter to discuss all the areas for interaction, but in the light of the above discussion, it can be seen that the work organisation (e.g. piecework, shiftwork or group working) will affect people's emotional responses, as will their experiences and expectations arising from their social life (e.g. what their friends do, or what they see on television). These emotional responses will, in turn, influence the decisions they make concerning their work activities, and can directly affect their immediate responses to work demands. An example of the latter is the less well controlled driving activity that follows from seeing a serious accident or being angered by the behaviour of a fellow motorist.

Before describing some particular areas where ergonomics and advanced manufacturing technologies meet, some comment on the links between ergonomics and work study are appropriate. These two contributions to productivity are not totally opposed to one another, indeed there is no substitute, for example, to the use of predetermined motion–time systems in the evaluation of output and manning levels for new production systems involving people. There are also many useful measuring and recording procedures for dealing with work activities. The new contributions which ergonomics is making to work study are in the better understanding of the relationships between the arrangements which work study proposes, and in the performance, health and fatigue of people implementing these proposals. As mentioned earlier, the somewhat mechanistic view of how people do their work, presented by the pioneers at the turn of the century, is giving way to more human-oriented concepts and requirements. The principles of workplace design[1] given in *Table 19.1* are an example, replacing the limited concepts of early work study 'principles of easy movement', or similar titles. They offer an ordered sequence of requirements which match more closely the physiological needs of the human body. Increased knowledge of fatigue and recovery, as well as a better understanding of what people expect from work, enable jobs to be more productive with less of the aggravations which have been seen as a traditional part of, at least, much of British working relationships.

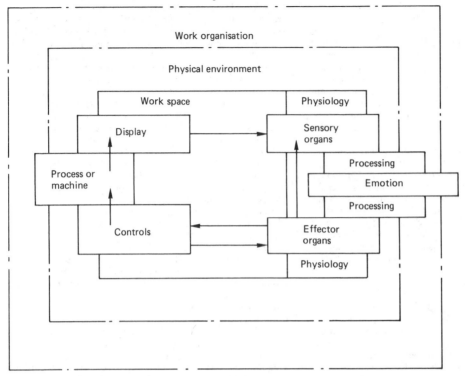

Figure 19.1 The man–machine system within its interactive environments

To close this section, one point which is important to ergonomics must be emphasised. The point of view of the ergonomist is that of the person doing the work and not as an observer watching the work being done. The sort of questions asked are 'How does this job match the person's capacities?' and 'What factors in the work situation give rise to these errors?' The direction of effort is to understand how the situation affects a person's performance, and hence how to change it so that people can achieve their objectives. It is *not* to assume that the workplace is 'optimum' and the person must be changed. Changing the circumstances enables people to modify their activities to give good performance. There are, of course, situations where we want people to change, hence the need for training, for example, but removing the obstructions to good performance is the initially preferred direction of the ergonomist.

19.2 Robotics

In the UK it is required that people and robots are separated, and access to the latter requires their isolation from the power supply or a reduction in their speed of movement. Until further developments in hardware and software this probably remains the safest procedure, for people and robots do not mix well. People cannot repeat the same actions consistently throughout a working day, nor can they continually bear in mind the actions of a neighbouring robot, to always be sure that they are not in its way. Although multibackup avoidance systems for robots are possible, they are expensive and rarely used; to assume the total reliability of a single level of control system for the robot and its work area is not acceptable.

Many ergonomics investigations have concentrated on the control box/teach control. The control of a robot from a separate control box has something in common with controlling the remote handling systems used in nuclear fuel processing plants. The robot, as it moves, changes its orientation so that what was at the bottom might now be at the top, or moving it to the left when its gripper is away from you, might require moving the control to the right when it is towards you. Even very experienced operators make frequent errors using these devices, and these are increased if they have two or three different robots, each with its individual design of teach control.

People's perceptions of movement do not always match reality. Nagamachi[2] has demonstrated that if people are asked to estimate whether they have enough time to pick up an object underneath a moving robot arm, they always misjudge its movement speed if it is moving towards them, believing they have more time than there really is. Judgement of clearances is also relatively poor by people, i.e. they will try to get through spaces which are too small. The use of computer graphics to assess pinch points in a computer's work space has been described by Yong et al.[3] and clearance problems are readily assessed by means of this technique.

Table 19.1 Ergonomically appropriate principles for use in the design of work and work spaces

1. The worker should be able to maintain an upright and forward facing posture during work

2. Where vision is a requirement of the task, the necessary work points must be adequately visible with the head and trunk upright or with just the head inclined slightly forward

3. All work activities should permit the worker to adopt several different, but equally healthy and safe, postures without reducing capability to do the work

4. Work should be arranged so that it may be done, at the worker's choice, in either a seated or standing position. When seated, the worker should be able to use the backrest of the chair at will, without necessitating a change of movements

5. The weight of the body, when standing, should be carried equally on both feet, and foot pedals designed accordingly

6. Work activities should be performed with the joints at about the midpoint of their range of movement. This applies particularly to the head, trunk and upper limbs

7. Where muscular force has to be exerted it should be by the largest appropriate muscle groups available and in a direction colinear with the limbs concerned

8. Work should not be performed consistently at or above the level of the heart; even the occasional performance where force is exerted above heart level should be avoided. Where light hand work must be performed above heart level, rests for the upper arms are a requirement

9. Where a force has to be exerted repeatedly, it should be possible to exert it with either of the arms, or either of the legs, without adjustment to the equipment

10. Rest pauses should allow for all loads experienced at work, including environmental and information loads, and the length of the work period between successive rest periods

19.3 Computer-aided manufacture

When machines were operated by hand, operators understood the whole machine system, how their actions affected surface finish, shape and output and how the machine's responses influenced the component. Whilst machines have become more integrated and their processing more complex, their operators have become more remote, in the sense that their control over what happens is through software rather than via direct influence. To achieve control now requires a 'mental model' of what the system does. An analogy is our knowledge of our local town. If our most direct way is blocked, during a journey, by a road being closed, we can 'reprogramme' our trip using side alleys and back streets to minimise the distance we have to travel. In the case of the operators of complex systems, they must understand the system sufficiently to know how it responds to their attempts to control it, under any

eventuality. It is not correct to assume that this is always best achieved by requiring the operator to follow the technical structure precisely. The form of presentation of information about the state of the process can assist or hinder the operator, whilst the way in which the program is structured can be crucial.

The difference can be exemplified by comparing a geographical representation of a railway system with the formalised display seen in a city underground map or a modern railway system control centre. The displays in these two situations are formalised to present the general directions and relationships between stations and other features relevant to the traveller or controller. To underground travellers the fact that their westbound train may at some time be travelling north is of no concern, because their mental map of the city tells them that their destination is broadly to the west of their starting point.

This is a broad subject, much studied by systems analysts and those who design process plant and nuclear generating plant control rooms. Not only does it have a major impact on the ease of learning and understanding the operation of complex manufacturing systems, but good design in this area can increase safety and reduce errors. It is obvious that, in emergencies, the controls, information presentation, etc., must be highly compatible with the operators' perceptions of the behaviour of the process. This compatibility is also important when the operator is tired, is on shift work or is interrupted or carries additional load due to some unforeseen circumstance. To put very expensive plant at risk by skimping on this key interface area is to create the sort of situation which gives rise to many dramatic tragedies, including the Three Mile Island atomic power station failure.

19.4 The automated office

Computation in the office has produced a revolution in operating practices. In many firms, the office and factory floor are tending to merge to bring information and control closer to the point of manufacture. Extensive studies of the appropriate arrangements of keyboards, screen layouts, lighting, seat and table designs have been made (see, for example, Cakir et al.[4]). A major effect arising from introducing computers is the increased amount of work which is done at the keyboard. Typescripts, filing and message despatch to other users' computers reduce the need for people to move from their chairs, increase the time they spend in a limited range of postures, and permit an increased number of keystrokes per day.

An extension of Tayloristic practices has led some companies to monitor keystrokes per day as a performance indicator, whilst others try pacing the keyboard operator by presenting information at a rate not under the operator's control. Several studies may be marshalled against these practices; perhaps the first point to be made is that the important criterion is most likely to be accuracy rather than speed, the old adage 'garbage in, garbage out' still holds, with the added comment that the garbage must be paid for. McAtamney[5] has demonstrated that pacing skilled keyboard operators even at a speed equivalent to their freely chosen pace, increases their error rate enormously, and that this does not appear to change with practice. What she also demonstrated was a recognisable increase in activity in the trapezius muscles, implying an increase in tension during the paced work periods.

This is interpreted as a contributory factor to the widespread effects (sometimes called 'repetitive strain injury') which arise in a considerable number of people due to too long periods at the keyboard. The constant tension in shoulder, neck and back muscles to maintain a suitable orientation at the keyboard gives rise to aches in these body parts which, if continued, can become chronic. In addition, the repeated use of the fingers on the keys, where 100 000 strokes a day are possible, and 50 000 not unusual, gives rise to, *inter alia*, inflammation of the tendon sheath in the wrist, causing pain and loss of grip strength. This damage, if the activity is persisted in, may result in permanent injury; certainly the initial symptoms should be heeded and the job changed to reduce exposure.[6]

As with so much in ergonomics, the key factor is variety of activity and posture (see *Table 19.1*). The temptation to long periods of sitting due to the flexibility that the computer provides can lead to neck, shoulder and back problems, and to eye strain because of the limited range of focusing of the eyes (and sometimes to the poor lighting arrangements provided). Another problem of long periods of sitting is the pooling of blood in the legs. Walking to perform various office tasks reduces this tendency, but 2 or 3 h of continuous sitting can result in significant swelling.[7] These changes are not such as the body normally experiences, which leads researchers to anticipate adverse consequences if measures are not taken to introduce more leg activity into the working day.

19.5 Manual assembly

The development of modern electronics has introduced new problems in their manufacture. Electrical products, in common with many others, are relatively large and their assembly requires many different movements and changes in posture. If cycle times are short the variety of postures is reduced and their frequency of use increased, giving rise to musculoskeletal symptoms due to over-use of a few muscles and joints. With electronic products, where they are assembled by hand, it is not unusual to have a 20-s cycle time during which the operator assembles eight or so components.

All of the assembly techniques require a pinch grip to hold, followed by an arching of the wrist to place the component perpendicularly into the printed circuit board. Simple arithmetic shows that this will involve over 10 000 cycles a day of the same activity, and wrist problems do occur in assembly-line workers. Strictly speaking, this is not a job for a person at all, because the joints, muscles, etc., are not designed for such extensive use of a limited number of actions (to say nothing of the limited mental activity required). As an interim measure, reducing wrist flexure, increasing the variety of movement by careful board design, and job rotation or enlargement to include some inspection or other tasks, would probably eliminate the problems completely. Whether or not the company sees this as economical is the concern discussed in Section 19.6.

The mass production of advanced electronic products is not always as modern as the products being produced. Short cycle times are still seen as the most productive method to adopt, with paced lines in widespread use. Surprisingly, the physical damage which a proportion of operators experience due to these highly repetitive conditions is not fully recognised by either management or some medical personnel, whilst the great majority of those who create these workplaces are entirely ignorant of the consequences of their designs.

The wider consequences in terms of work attitudes, motivation and quality performance are also well documented, but even less well known to many whose job it is to be up to date in the creation of efficient manufacturing systems. Even much of the education, in the UK at least, provides very little modern knowledge in this area, being still locked into pre-War work study, wage incentive schemes and personnel relations laws

and customs. Health-and-safety legislation, stimulated by Common Market pressure, is bringing changes, but there is still a long way to go if we are to match some of our major competitors' working performance and conditions.

19.6 The economics of ergonomics

It is not unusual to hear a manager discuss ergonomic proposals for workplace changes using a remark like, 'we can't afford to pay just to make the workers comfortable'. The implications are that the investment would not contribute (sufficiently) to productivity. Yet a moments thought would cause the manager to recognise that his/her secretary, the availability of tea and coffee, good desk and chair and easy chairs in the corner contributed much to comfort, and also much to the efficient and profitable use of the manager's time. Doing filing, searching for addresses or fending off unnecessary salesmen is not a good use of the time of a well-paid manager; equally, the appearance of an office can give confidence to a prospective customer, who might be unlikely to bring business to a manager working on a packing case in a corner of the stores. Exaggerated, perhaps, but the manager's comfort is a commercial advantage to the company—within limits of course!

What about our shop-floor workers? They might have worked for the company for several years, know its ways and its products. For them to spend part of their time and energies in overcoming needless obstacles of their equipment or needless bodily or mental fatigue is to be blind to the economic good of the company, which is paying them to produce its products, not to get tired and frustrated.

Many of our major competitors in Japan, Sweden or Germany, for example, do not make this mistake. In Japan, in particular, the people are a resource, not a cost, to be trained and retrained, encouraged and developed as part of the operational success of the enterprise. The costs of this activity are more than repaid by large quantities of high quality on-time products, a flexible workforce and people who, over the years, grow into a well-trained workforce with a considerable knowledge of the business.

This looks a costly operation to the typical Western manager. But how many companies relate their production policies to the costs of absence, lateness, hiring and training costs, and scrap, rework and returned deliveries?

Research in this field is rare, but one study[8] has examined the costs and returns from changes in the workplaces of women assembling telephone equipment. The data dealt with the situation before the ergonomic changes to the workplace (1967–1974) and the effects after the changes (1975–1987). Clearly such extensive data relate to no 'Hawthorne effect'. These authors assessed the costs arising from the purchase of ergonomic workstations, changes in ventilation and lighting and increased lighting costs. The benefits assessed were reductions in recruitment, training and instructors' salary costs and reductions in sick payments. All these factors were assessed for each of the years of the study. The costs to the company were NKr 338 992, whereas the savings were NKr 3 226 194; both figures reduced to a common value.

In this study no value was put on improvements in working relationships, which are reported in the paper, nor were changes in productivity evaluated and included. The documented reductions in sick leave and musculoskeletal ailments was a benefit to the workers themselves. The costs of such ailments in terms of a degraded social life are never borne by companies, but by society as a whole. The sharpening of health-and-safety legislation around the world may make a change in this respect, it could be an unwise company which ignored the possibility of a penalty for having work conditions which made some of its workers ill or unemployable.

19.7 Conclusion

This chapter has brushed the surface of the complex relationships between people and their work. Although there are many publications and much research to draw on, it will, we hope, also be clear that, by adopting a changed perspective on people at work, there are many possibilities for change which do not need much science, but the application of the logical consequences of that change. If people have to reach and bend unduly, repeat simple mindless activities endlessly, or strain their eyes or ears to get the information needed for their job, then at the very least you can assume you are paying a salary for someone to overcome the equipment's problems, and not for making the product.

But this is not to say that it is all common sense. Compared with a person, an internal combustion engine is a relatively simple device, yet deep knowledge and experience are required to design a better one. Our very familiarity is liable to cause us to overlook the complexity: 'I am a person, surely what I feel should be done must be right'. Everyone, however, is different, and dealing with these variations is one of the skills of the ergonomically trained person. Understanding the variability of people also helps in understanding what the relevant factors are which can be altered, and what the short- and long-term changes in the result are likely to be.

In manufacturing engineering terms, recognising the contributions of people and designing to enable their full benefits to be obtained makes sense. Assuming all is technology and that people will fit in around it not only does not make sense, but has been demonstrated decade after decade not to work by companies who have fallen further and further behind in productivity terms. Modern industry needs a modern approach to its aim of achieving a competitive market position, and one of the major components of that approach must be arranging for its workforce, at all levels, to be in the best position to put all its efforts into achieving the company objectives.

References

1 CORLETT, E. N., The human body at work: new principles for designing workspaces and methods. *Management Services*, **22**(5), 20–52 (1978)
2 NAGAMACHI, M., Human factors of industrial robots and robot safety management in Japan. *Applied Ergonomics*, **17**(1), 9–18 (1986)
3 YONG, Y. F., TAYLOR, N. K. and BONNEY, M. C., CAD—an aid to robot safety. In *Robot Safety* (eds M. C. Bonney and Y. F. Yong), IFS Publications, Bedford (1985)
4 CAKIR, A., HART, D. J. and STEWART, T. F. M., *Visual Display Terminals*, Wiley, Chichester (1979)
5 MCATAMNEY, L., Some effects of task loading in a VDU-based task. In *Contemporary Ergonomics 1989* (ed. E. D. Megaw), Taylor & Francis, London (1989)
6 KNAVE, B. and WIDEBACK, P. G., *Work with Display Units*. Elsevier, Amsterdam (1987)
7 WINKEL, J., On foot swelling during prolonged sedentary work, and the significance of leg activity. *Arbete och Halsa 1985–35*. National Board of Occupational Safety and Health, Solna, Sweden (1985)
8 SPILLING, S., EITRHEIM, J. and AARAS, A., Cost benefit analysis of work environment investment at STK's telephone plant at Kongsvinger. In: *The Ergonomics of Working Postures* (eds. N. Corlett, J. Wilson and I. Manenica). Taylor & Francis, London (1986)

Index

Compiled by D.C. Tyler